Secondary Structures of Polypeptides

The α-helix

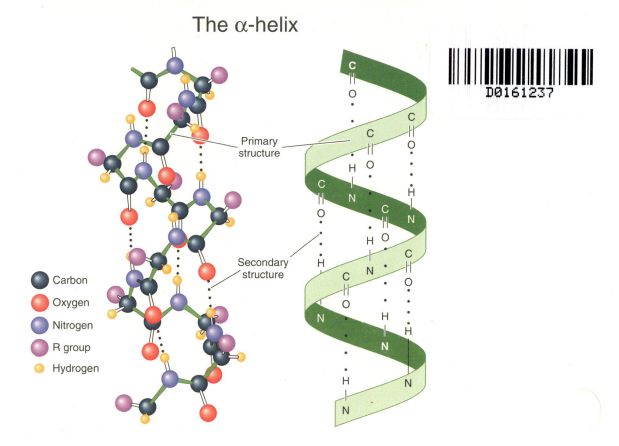

The α-helical secondary structure of a polypeptide chain.

The β-pleated Sheet

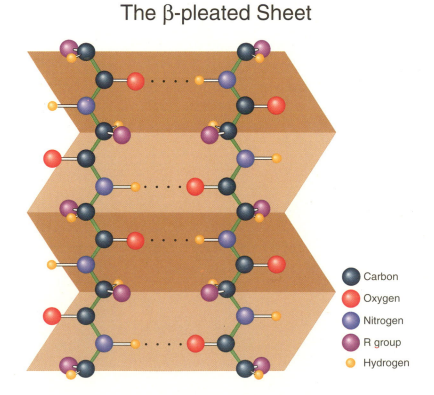

The β-pleated sheet secondary structure of polypeptides.

Principles of
MODERN GENETICS

Principles of
MODERN GENETICS

GERALD D. ELSETH

Bradley University

KANDY D. BAUMGARDNER

Utah State University

WEST PUBLISHING COMPANY

Minneapolis/St. Paul   New York   Los Angeles   San Francisco

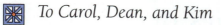 *To Carol, Dean, and Kim*

WEST'S COMMITMENT TO THE ENVIRONMENT

In 1906, West Publishing Company began recycling material left over from the production of books. This began a tradition of efficient and responsible use of resources. Today, up to 95 percent of our legal books and 70 percent of our college and school texts are printed on recycled, acid-free stock. West also recycles nearly 22 million pounds of scrap paper annually—the equivalent of 181,717 trees. Since the 1960s, West has devised ways to capture and recycle waste inks, solvents, oils, and vapors created in the printing process. We also recycle plastics of all kinds, wood, glass, corrugated cardboard, and batteries, and have eliminated the use of Styrofoam book packaging. We at West are proud of the longevity and the scope of our commitment to the environment.

Production, Prepress, Printing and Binding by West Publishing Company.

 TEXT IS PRINTED ON 10% POST CONSUMER RECYCLED PAPER    PRINTED WITH SOY INK™

PRODUCTION CREDITS

Cover Image: Carlyn Iverson
Copyediting: Pam McMurry
Interior and Cover Design: Diane Beasley
Composition: G&S Typesetters, Inc.
Photo and Illustration Credits follow the Index.

British Library Cataloguing-in-Publication Data. A catalogue record for this book is available from the British Library.

COPYRIGHT © 1995 BY WEST PUBLISHING COMPANY
                610 Opperman Drive
                P. O. Box 64526
                St. Paul, MN 55164–0526

02 01 00 99 98 97 96   8 7 6 5 4 3 2

LIBRARY OF CONGRESS CATALOGING-IN-PUBLICATION DATA
Elseth, G. D. (Gerald D.), 1936–
    Principles of modern genetics / Gerald D. Elseth, Kandy D. Baumgardner.
        p.   cm.
    Includes bibliographical references and index.
        ISBN 0-314-04207-5
    1. Genetics.   I. Baumgardner, Kandy D., 1946–   . II. Title.
QH430.E383   1994
    575.1—dc20
                                                        94–23062
                                                        CIP

GERALD D. ELSETH

Gerald D. Elseth is a professor of biology at Bradley University, where he teaches biochemistry, cell and molecular biology, general biology, and genetics. He received his B.A. degree in biology and chemistry from Moorhead State College and his M.S. and Ph.D. degrees in genetics from Utah State University. His research interests include genetics, theoretical biology, and molecular toxicology. He has received the Putnam award for teaching excellence. He also enjoys hiking, mathematical modeling, and listening to classical music.

KANDY D. BAUMGARDNER

Kandy D. Baumgardner is a professor of zoology at Eastern Illinois University and also serves as department chair. She teaches general and advanced genetics for students majoring in the biological sciences and a course in heredity and society for nonscience majors. She received her B.S. degree in biology from Bradley University in 1968 and her Ph.D. degree in zoology from Utah State University in 1973. Her nonprofessional interests include raising show pigeons, restoring native Illinois prairie and forest, and collecting old books.

CONTENTS IN BRIEF

I

MENDELISM: INTRODUCTION TO GENETIC ANALYSIS

II

THE CHEMICAL BASIS OF INHERITANCE

III

STRUCTURE AND REPLICATION OF CHROMOSOMES

IV

GENETIC VARIATION: MUTATIONS AND MUTANT STRAINS

V

LINKED GENES AND CHROMOSOME MAPPING

CONTENTS

I

MENDELISM: INTRODUCTION TO GENETIC ANALYSIS

Microtubules in chick embryo fibroblast mitosis.

II

THE CHEMICAL BASIS OF INHERITANCE

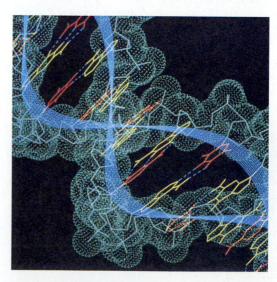

Computer model of the DNA double helix.

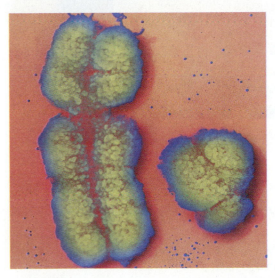

The human X and Y chromosomes.

IV

GENETIC VARIATION: MUTATIONS AND MUTANT STRAINS

CHAPTER 12
CHROMOSOME MUTATIONS 246

Child affected with Down syndrome.

LINKED GENES AND CHROMOSOME MAPPING

CHAPTER 13
MAPPING GENES IN EUKARYOTES 274

CHAPTER 14
MAPPING GENES IN PROKARYOTIC SYSTEMS 306

VI

GENE EXPRESSION AND CLONING

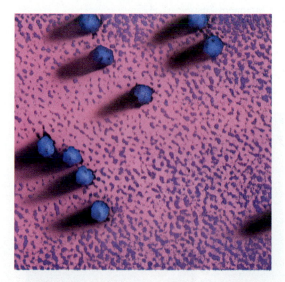

Scanning electron micrograph of ribosomes.

VII

REGULATION OF GENE EXPRESSION

CHAPTER 19
REGULATION IN PROKARYOTES   454

CHAPTER 20
REGULATION IN EUKARYOTES I: TRANSCRIPTIONAL ACTIVATION   483

CHAPTER 21
REGULATION IN EUKARYOTES II: DEVELOPMENT AND CANCER   505

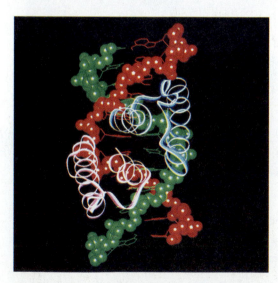

A basic helix-loop-helix protein bound to DNA.

VIII

MECHANISMS OF MUTATION, RECOMBINATION, AND REPAIR

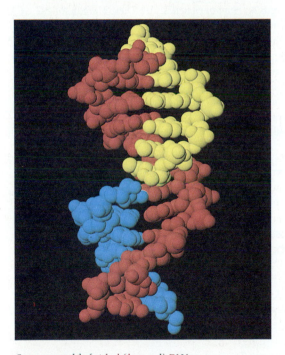

Computer model of nicked (damaged) DNA.

IX

THE GENETIC BASIS OF EVOLUTION

Color variants of the blood star, Henricia levinscula.

CHAPTER 25
GENETIC PROCESSES OF EVOLUTION   613

CHAPTER 26
MOLECULAR EVOLUTION   639

PREFACE

This book grew out of the need for a problems-oriented text based on a unified treatment of classical and molecular genetics. It is broad enough in scope to serve as the textbook for a one- or two-semester general genetics course, and it represents a distillation of our experiences in teaching genetics to undergraduates. This text provides an up-to-date, readable, and challenging introduction to genetics to students who have taken background courses in introductory biology and chemistry and who are majoring in biology or related areas. No previous courses in biochemistry, cell biology, or microbiology are necessary, since essential concepts in these areas are introduced where they are needed in the text.

Integrated Approach

Historically, genetics has developed along two different paths: classical genetics and molecular genetics. This dichotomy is reflected in the organization of many current genetics textbooks, which tend to treat classical and molecular genetics as distinct and somewhat isolated subjects. In contrast, this text presents a more integrated treatment of the major genetic concepts by discussing topics in a logical sequence along conceptual rather than disciplinary lines. To achieve a greater unification of principles (while retaining the flexibility of coverage needed to satisfy a variety of users), the text has been divided into nine parts. Each part consists of two to four chapters that approach the same general subject from different perspectives. For example, Part III (Chapters 8, 9, and 10) deals with chromosome structure and replication in terms of prokaryotes, eukaryotes, and subcellular genetic systems, while Part V (Chapters 13, 14, and 15) covers chromosome mapping and considers both classical and molecular approaches. Each part begins with a brief introduction to establish the conceptual framework that ties its component chapters together.

To facilitate integration of these two approaches, major chemical concepts are presented near the beginning of the text (Part II), so that the student can view genetics early on from the molecular as well as the cellular and organismic levels. In contrast to their treatment in most other general genetics texts, proteins are discussed in as much depth as the nucleic acids. The added emphasis on proteins reflects the greater background that today's students need in order to comprehend current work in genetics. The discussion of proteins and nucleic acids is preceded by an introduction to Mendelian analysis. Like most other texts in this area, this book begins with Mendel and ends with evolutionary genetics. Beginning the narrative with Mendel is a time-honored arrangement that gives the student a proper historical perspective and also introduces many of the mathematical concepts and tools that are applied throughout the text. The relationship between genes and chromosomes is also presented early in this text (Chapter 1), following a discussion of chromosomes, mitosis, and meiosis.

Flexible, Modular Organization

Although the topics are arranged in what we regard as a logical sequence, other teachers may prefer other arrangements. For this reason, most parts and chapters have been written as units that can be used interchangeably. For example, with a few modifications, Part VIII ("Mechanisms of Mutation, Recombination, and Repair") could immediately follow Part IV ("Genetic Variation: Mutations and Mutant Strains") or Part V ("Linked Genes and Chromosome Mapping"). Similarly, Chapter 6 ("Proteins") could be covered in Part VI ("Gene Expression and Cloning") without seriously affecting the flow of material. This flexibility also extends to the arrangement of topics in individual chapters. In most chapters, the less frequently covered topics are placed toward the end, so that they are easy to delete from reading assignments. Moreover, each major section is written as a unit, with its own in-chapter summary ("To Sum Up") and end-of-chapter questions and problems. Instructors can thus select a mix of topics from individual chapters to assemble the most suitable material for their courses.

Balanced, Comprehensive Coverage

Despite our individual preferences, we have tried to provide a balanced treatment of the classical and molecular approaches to give a broad, comprehensive coverage that reflects the scope of modern genetics. We have chosen to include a wide range of principles and techniques so that the book can be used for courses emphasizing slightly different topics and a variety of instructional approaches. For example, various syllabi can be designed around selected chapters to form a one-semester course in which only part of the textbook is covered. A one-semester general genetics approach could include parts of Chapters 1–4, 7–9, 11–13, 15–19, 21, 22, and 24. Because many of the chapters have a strong molecular orientation, a one-semester course in molecular genetics could be based on Chapters 6–11 and 13–23. The book also includes a thorough treatment of both the classical and molecular aspects of population and evolutionary genetics, so a one-semester course with a population and evolutionary emphasis could be based primarily on Chapters 1–5, 7, 11–13, 15–20, and 22–26.

Emphasis on Analysis

Genetics is an analytical branch of science in which principles are often expressed in mathematical terms. Thus, there is an emphasis on problem-solving and the analysis of research data throughout the book. To help the student develop the needed analytical skills, the book includes an entire chapter (Chapter 3) devoted to probability and genetic analysis.

Since students who enter a beginning genetics course usually have little experience with problem solving, each chapter includes several examples in the body of the text, along with one or more solved Examples. The Examples directly follow the locations in the text where pertinent concepts are discussed. After each solved Example, a Follow-Up Problem reinforces basic skills and encourages the student to extend the concepts being discussed. Solutions to all Follow-Up Problems are given in Appendix A at the back of the book. Numerous questions and problems are also included at the end of each chapter (a blue number indicates that the answer to the question or problem is given at the end of the book in Appendix B). These problems vary in character and degree of difficulty and range from exercises that provide necessary repetition of basic skills to questions that challenge the student to reason rather than just memorize facts.

The "Point to Ponder" feature deals with the impact of genetics on society and encourages the reader to reason through the ethical ramifications of recent genetic discoveries. This feature appears at the end of major sections in the chapter.

The boxed essays also help to sharpen analytical skills. The "Extensions and Techniques" boxes describe specific concepts and procedures that are used in the study and research of biological problems. The end-of-chapter questions dealing with the material discussed in these boxed essays are marked with an asterisk.

Clear and Inviting Presentation

In writing this text, our goal has been to present the principles of modern genetics in a clear, concise, and interesting manner without oversimplifying concepts and issues. Examples based on human genetics are provided throughout the text, and a series of boxed essays called "Focus on Human Genetics" discusses interesting and timely subjects such as the AIDS virus and human gene therapy. The emphasis in the text is on explanation rather than the mere presentation of facts. In keeping with this emphasis, the book is fully illustrated with figures that help to explain virtually every major point. The book also has an attractive, full-color art program enhanced with many color photographs to facilitate learning.

Other Pedagogical Features

The textbook contains several additional features to enhance student learning. Part openers introduce each major topic and help the student focus on the concepts discussed in each major section of the book. Each chapter has introductory text to ease the student into the chapter material. Frequent in-chapter summaries ("To Sum Up" sections) provide immediate reinforcement of new concepts, and chapter summaries give an overall review of the major concepts discussed in each chapter. Appendix A contains solutions to all Follow-Up Problems. Appendix B contains answers to selected end-of-chapter questions and problems. A list of selected readings for each chapter in Appendix C provides students with useful additional readings. There is also an extensive glossary and index.

ANCILLARIES

Instructor's Manual

An Instructor's Manual is available to adopters. This teaching aid includes over 900 test questions plus chapter commentary with lecture tips, discussion guidance for the Points to Ponder, suggested film and software lists, and a complete list of available acetates.

Solutions Manuals

Available upon request to all adopters of this text is a Solutions Manual prepared by the authors, which includes the answers and worked-out solutions to all end-of-chapter questions and problems. For the instructor's convenience, the Solutions Manual is available in two versions. Solutions Manual A contains solutions for all end-of-chapter problems in the text. Solutions Manual B contains solutions to selected problems that contain answers in Appendix B. Manual B is prepared for instructors who prefer that students do not have all of the answers available to them. In addition to detailed solutions, both manuals contain a section providing guidance and helpful hints to solving genetic problems. Both manuals are available for sale to students at the instructor's option.

Acetates

A set of over 200 full-color acetates of key art pieces, figures, and tables is available to adopters. A complete list is included in the Instructor's Manual.

WESTEST™ 3.1

The test bank containing questions from the Instructor's Manual is available on WESTEST™ 3.1 for Macintosh, DOS, and Windows. This easy-to-use package allows adopters to create tests and edit, add, modify, or delete questions. West's Classroom Management Software is also available to enable professors to record and store student data and generate various reports. Contact your West sales representative for details.

Videotapes

Selections from the West Life Science Video Library, which includes the "Secrets of Life" series, are available to qualified adopters for class use. Contact your West sales representative for further information.

ACKNOWLEDGMENTS

This project could not have been completed without the assistance of many people. We were fortunate to have the advice of many reviewers, whose valuable comments and suggestions helped to shape the text into its final form. The reviewers include:

Matthew T. Andrews
North Carolina State University

John Armstrong
University of Ottawa

L. Herbert Bruneau
Oklahoma State University

David P. Campbell
California State Polytechnic University—Pomona

Bruce A. Chase
University of Nebraska at Omaha

James Clark
University of Kentucky

Bruce J. Cochrane
University of South Florida

Diane M. B. Dodd
University of North Carolina—Wilmington

Max P. Dunford
New Mexico State University

David Evans
University of Guelph

David Foltz
Louisiana State University

David Francis
University of Delaware

David Fromson
California State University—Fullerton

Ranjan Ganguly
University of Tennessee—Knoxville

Elliott S. Goldstein
Arizona State University

Paul Goldstein
University of Texas at El Paso

Jeffrey C. Hall
Brandeis University

Wade Hazel
De Pauw University

Kaius Helenurm
San Diego State University

Margaret Hollingsworth
SUNY—Buffalo

George Hudock
Indiana University—Bloomington

R. B. Imberski
University of Maryland at College Park

James W. Jacobson
University of Houston

Mitrick A. Johns
Northern Illinois University

Kenneth C. Jones
California State University—Northridge

Mark Levinthal
Purdue University

Clint Magill
Texas A & M University

Kenneth A. Mason
Kansas University

Kenton S. Miller
University of Tulsa

Benjamin Pierce
Baylor University

H. James Price
Texas A & M University

Frank J. Rice
Fairfield University

R. H. Richardson
University of Texas—Austin

Mark F. Sanders
University of California—Davis

Katherine T. Schmeidler Sapiro
California State University—Long Beach

Randy Scholl
Ohio State University

Nancy N. Shontz
Grand Valley State University

Thomas Sneider
Colorado State University

Thomas P. Snyder
Michigan Technological University

Gilbert Starks
Central Michigan University

Michael Stock
Grant MacEwan Community College

Reginald Storms
Concordia University

Irv Tallan
University of Toronto

David R. Weisbrot
William Patterson College of New Jersey

Peter J. Wejksnora
University of Wisconsin—Milwaukee

Harrington Wells
University of Tulsa

Miriam Zolan
Indiana University

We also wish to thank Susan Mitchem for her assistance in preparing the manuscript and Connie Huber for her help in obtaining art and photo permissions and artwork. Many thanks also to the people who provided photographs and other illustrative material. Special thanks to copyeditor Pam McMurry, who lent her expertise and attention to detail to the final manuscript.

We are also grateful to the talented staff of West Educational Publishing and especially wish to thank acquiring editor Jerry Westby, developmental editor Betsy Friedman, promotions manager Mary Steiner, and the production team of Beth Olson, Deanna Quinn, and Mary Verrill for their help and support in bringing this book to reality.

Gerald D. Elseth

Kandy D. Baumgardner

CHAPTER 1

Genetics: Early History and Cytological Foundations

Early History of Genetics
The Cellular Basis of Heredity
   FOCUS ON HUMAN GENETICS: Human
     Genetics versus Eugenics

Genetics, the science of heredity, deals with the factors that are responsible for the similarities and differences between generations. These factors affect form and function at every level, from the molecules that compose each living cell through the organismal and population levels of biological organization. The concepts of genetics are therefore fundamental to all biological disciplines and serve as the unifying core in the study of modern biology.

There are several major areas of genetics. **Cytogenetics** combines cytology (the study of cells) and genetics in the study of chromosomes. **Molecular genetics** concerns itself with the molecular structure and expression of genes, how that expression is regulated, and the nature of the products of gene expression. **Population genetics** is the study of genetic variation in populations and the responses of gene and genotype frequencies to mating patterns and evolutionary forces. **Quantitative genetics** examines

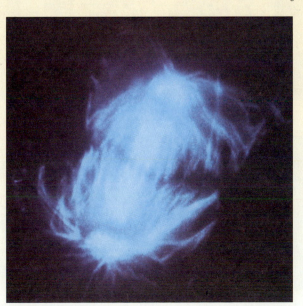

Microtubules in chick embryo fibroblast mitosis. Late anaphase.

traits that are determined by the interaction of many genes whose expression is often highly sensitive to environmental influences. **Transmission genetics**, frequently referred to as traditional or classical Mendelian genetics, deals with the patterns of inheritance of genes between generations. Each of these subject areas can be further subdivided into specialty disciplines. Furthermore, these areas overlap extensively, especially with the development of new molecular techniques to study gene structure and function. For example, population genetics is focusing increasingly on molecular variation in populations and its role in evolutionary processes.

Our understanding of genetics is based mostly on work with just a handful of organisms that are especially well suited for genetic research. These organisms include viruses (especially the bacterial viruses), bacteria (especially *Escherichia coli*), and several eukaryotes. Studies that began in the early 1940s on the genetics of bacteria and their viruses formed the root of molecular genetics; the discovery that DNA is the genetic material was part of this early work. Certain eukaryotic organisms, including microbes such as yeast and *Neurospora* (orange bread mold), plants such as the garden pea and corn, and animals such as *Drosophila* (a fruit fly) and mice have all proven to be excellent subjects for genetic analysis (▶ Figure 1–1).

There are three main reasons why these kinds of organisms make good genetic subjects: (1) they have relatively short generation times, (2) they produce large numbers of offspring, and (3) they can be experimentally manipulated in controlled crossing sequences. Although humans have also been popular genetic subjects, knowledge of human genetics has, until recently, lagged behind our understanding of the genetics of many other organisms. Humans do not meet the three criteria just listed, and thus are difficult to study using traditional genetic approaches. However, the advent of somatic cell genetics as well as newer molecular methods have allowed significant advances in the understanding of human genetics. Recently, the so-called "Human Genome Project" has been initiated; its goals are to identify all of the estimated 50,000 to 100,000 human genes, to locate their chromosomal positions, and to sequence the entire 3 billion nucleotide pairs of DNA that are present in human chromosomes. This formidable task marks the latest major development in genetics (■ Table 1–1).

EARLY HISTORY OF GENETICS

Genetics has developed as an experimental science during the twentieth century, following the discovery in 1900 of the work of the "father of genetics," Gregor Mendel. The roots of genetics can, however, be traced into antiquity. Deliberate breeding methods to produce improved plant and animal varieties useful to humans date back to the Babylonians some 6000 years ago. Speculations about how traits are inherited and why similarities and differences among

▶ **Figure 1–1** Eukaryotic organisms that are well suited for genetic research include (a) *Neurospora crassa* (common bread mold), (b) *Saccharomyces cerevisiae* (a yeast), (c) *Drosophila melanogaster* (a fruit fly), (d) *Mus musculus* (a mouse), (e) *Pisum sativum* (a garden pea), and (f) *Zea mays* (corn).

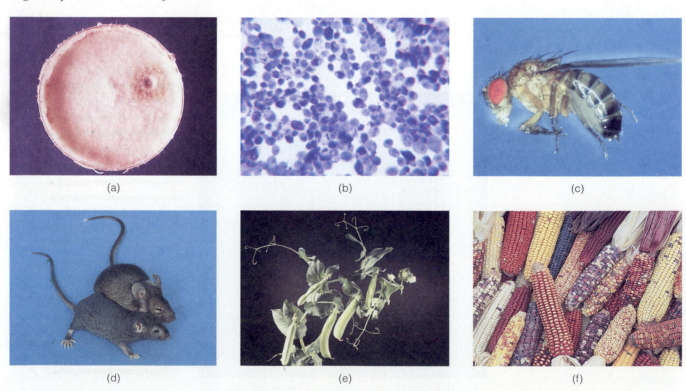

(a)   (b)   (c)

(d)   (e)   (f)

■ TABLE 1–1 Chronology of major events in genetics.

1865–66	Presentation and publication of Gregor Mendel's work in the *Proceedings of the Brunn Society for Natural History*.
1900	Discovery of Mendel's work by de Vries, Correns, and von Tschermak.
1902	Statement of the chromosome theory of inheritance by Sutton and Boveri.
1908	The Hardy-Weinberg law, describing the relationship of genotype frequencies to gene frequencies in randomly mating populations.
1909	Publication of Garrod's *Inborn Errors of Metabolism*.
1910	Morgan's description of sex-linked inheritance in *Drosophila*.
1913	Discovery of linkage and development of the recombination-distance concept of chromosome mapping.
1927	Müller's demonstration that X rays induce mutations.
1940	Demonstration of intragenic recombination.
1941	Beadle and Tatum's statement of the one gene–one enzyme hypothesis.
1944	Demonstration by Avery, MacCleod, and McCarty that DNA is the genetic material in bacteria.
1946	Lederberg and Tatum demonstrate conjugation and genetic recombination in bacteria.
1950	McClintock proposes the idea of transposable genetic elements.
1952	Demonstration by Hershey and Chase that DNA is the genetic material of phage.
1953	Watson and Crick's model for the structure of DNA.
1955	Fine structure mapping of the phage T4 *rII* locus by Benzer.
1956	Tjio and Levan demonstrate that the human chromosome number is 46.
1957	Fraenkel-Conrat and Singer show that RNA is the genetic material of tobacco mosaic virus.
1958	Meselson and Stahl demonstrate the semiconservative nature of DNA replication.
	Kornberg's isolation of DNA polymerase I from *E. coli*.
1959	Ochoa's discovery of the first RNA polymerase.
1961	Jacob and Monod propose the operon concept of gene regulation.
1964	Statement of colinearity principle by Yanofsky.
1965	First nucleic acid molecule, a tRNA, sequenced by Holley.
1966	Khorana, Matthaei, Nirenberg, and Ochoa crack the genetic code.
1970	Isolation of the first restriction endonuclease by Nathans and Smith.
	Discovery of reverse transcriptase by Baltimore and Temin.
1972	Berg produces the first recombinant DNA molecule.
1976	Demonstration of the relationship between proto-oncogenes and oncogenes by Bishop and Varmus.
1977	Discovery of introns in eukaryotic genes.
	Development of the Maxam-Gilbert and Sanger techniques for DNA sequencing.
1982	Report of the complete nucleotide sequence of phage lambda.
1983	Cech and Altman's discovery of catalytic RNA.
1985	Introduction of the polymerase chain reaction (PCR) technique for amplification of DNA.
1988	Initiation of the Human Genome Project.
1989	Cloning of the cystic fibrosis gene.
1990	Discovery of parental imprinting of genes.
	Anderson's first attempt at human gene therapy.
1991	Discovery of triplet-repeat mutations.
1993	Development of a high-resolution genetic linkage map of the mouse.
1994	Achievement of one of the first goals of the Human Genome Project: a comprehensive, high-density genetic map.

parents and offspring exist go back more than 2000 years to Aristotle, Hippocrates, and other ancient Greek philosophers. Some of these speculations centered around the belief that the environment could cause organisms to change and these acquired changes could then be inherited by the next generation. This idea was refined 2000 years later in 1809 by Jean Baptiste Lamarck and is known as the theory of inheritance of acquired characteristics. Lamarck proposed that an organism responds to its environment by gradually acquiring traits that it needs to be more adapted to that environment and that these acquired changes are then passed on to its offspring. This theory therefore describes a nonrandom, directed (by the environment) mechanism of acquiring variations within a population.

The development of several major biological theories in the middle-to-late nineteenth century marked a turning point in the history of biology. In 1859, Charles Darwin published his theory of evolution through the mechanism of natural selection. The major weaknesses of this theory were its lack of suitable explanations for how heritable variation arises in populations and how it is passed on to the offspring. (Darwin reluctantly accepted a Lamarckian view of inheritance of acquired variations.) Only after 1901, the year when Hugo de Vries stated the mutation theory, was an explanation for the origin of variation independent of environmental factors developed. Interestingly enough, de Vries originally presented his theory as an alternative to Darwinian evolution through natural selection, stating that evolution occurs by sudden and profound changes (mutations). However, the idea of mutation was gradually re-

Human Genetics versus Eugenics

The late nineteenth and early twentieth centuries saw the development of two approaches to human heredity—**human genetics** and **eugenics**. Human genetics is the study of heredity and variation in human populations through standard scientific techniques such as pedigree analysis, twin studies, chromosome analysis, biochemical testing, and DNA testing. In contrast, eugenics aims to "improve" the human species by selective breeding and/or the elimination of deleterious genes. Eugenics includes both **positive eugenics** (encouraging people with "desirable" traits to have children) and **negative eugenics** (discouraging people with "undesirable" traits from reproducing). It is very important to understand the difference between human genetics and eugenics, especially in light of the development of genetic technologies that purport to identify genes for a host of complex behavioral disorders.

Eugenics was sparked by Darwin's theory of evolution through natural selection. Although Darwin's books and papers contained almost no references to human evolution, Sir Francis Galton (a cousin of Darwin) and others looked upon selection as a framework for "bettering" the human population. Galton reasoned that if plant and animal populations enrich themselves by gaining adaptive traits and weeding out unadaptive traits by natural selection, then humans should exercise social control over their reproduction—they should practice selective breeding as a form of natural selection. He first used the term *eugenics* (from the Greek for "good birth") in 1883 to describe the use of controlled reproduction.

The idea that selective breeding can "improve" the human population by enriching the "fit" and eliminating the "unfit" has an important underlying premise—that all human traits are genetically based. In other words, eugenics assumes that nature is much more important than nurture in determining human characteristics and that human society can be improved "through better breeding." Galton came to this conclusion after studying the pedigrees of talented and successful people, such as artists, musicians, and military officers. He determined that these people very often had similarly successful ancestors and children and concluded that "genius is heredi-

The "genius" of violinist and composer Nicolo Paganini (1782–1840) was later theoretically attributed to his suffering from Marfan syndrome, a dominant-gene disorder. One symptom elasticized his joints and thus enhanced his playing ability.

tary," completely disregarding any possible influence of environmental factors on the development of the talents and "genius" of these individuals.

Galton's ideas had broad appeal, and the eugenics movement spread rapidly. The National Eugenics Labo-

---

fined to include the small, subtle changes in genes that provide the source of variation upon which natural selection acts, thereby actually strengthening Darwin's theory.

In 1868, Darwin brought back the old idea of **pangenesis**, first suggested in the 1600s, to explain heredity. According to the theory of pangenesis, each part of the body of an organism contains minute corpuscles or *gemmules* that represent the nature and identity of that part. These gemmules were thought to travel via the bloodstream to the gonads, where they collect, to be passed on to the next generation. Darwin proposed that modifications in these gemmules produce the heritable variations upon which natural selection acts.

The pangenesis theory implies that not all cells contain all hereditary factors; particular gemmules are found only in the cells of the body part they represent. Thus, the gemmules are analogous to the building blocks of these body parts, rather than to the blueprints for making these parts. Once fertilization occurs, the building blocks from the egg and sperm were assumed to mix and be in place as the building blocks of the body parts of the offspring. Of course, we now know that what is inherited is the blueprint, not the building blocks themselves, and we know that (nearly) all cells of an organism contain the exact same blueprint—the exact same set of genes.

Pangenesis was discredited by relatively simple experiments, such as the transfusion of blood from black rabbits into white ones and the subsequent observation that the transfused white rabbits produced no black offspring. It is obvious that hereditary gemmules do not travel through the

ratory was established in London, and the Eugenics Education Society was formed. After Gregor Mendel's basic principles of genetics were discovered in 1900, the movement gained even more momentum. A eugenics committee was established by a livestock breeders' association in the United States specifically to investigate heredity in the human race. The New York–based Eugenics Records Office was founded by a leading American geneticist; it collected pedigree data on inmates of prisons, asylums, and reformatories. In an attempt to apply Mendel's principles, the "inferior" traits of these inmates (as well as the "desirable" traits possessed by "gifted" individuals) were equated with single genes that followed Mendel's rules. The eugenics movement concluded that humans, through positive and negative eugenics measures, would be able to produce an improved human species in the foreseeable future.

The idea of using negative eugenics to eliminate "inferior" genes from the human population became especially popular. By the 1930s most states had sterilization laws that applied to sexual perverts, drug fiends, diseased and degenerate persons, drunkards, paupers, and those classified as feeble-minded. All these "traits" were deemed genetic—predetermined at birth—with no thought given to the role of the environment. The strict immigration laws passed by the federal government to restrict the immigration of "undesirable" races were also part of the eugenics movement.

By the 1920s, some scientists were attempting to restore a scientific basis to the application of genetics to humans. They pointed out several reasons why eugenics is not scientifically valid: First, the fundamental premise of eugenics—that human traits are solely determined by genes—was an assumption, not a proven fact. Second, Galton and others who followed him were biased in their analysis; they selected and studied families that fit their preformed ideas of "hereditary genius" and ignored families that did not fit the theory; such bias is not compatible with the scientific method of study. Third, eugenics assumed that a complex human trait such as artistic talent is the result of a single gene and therefore would be inherited through a family line according to Mendelian principles. There was, and still is, no evidence of such a simple genetic basis for complex traits. As we will see in Chapter 5, these traits result from the interaction of many genes and environmental factors, and their inheritance is impossible to predict—it is governed by unpredictability and variation among family members, not by simple single-gene inheritance schemes.

Despite its obvious lack of scientific rigor, the eugenics movement remained popular until the 1940s, when the atrocities of Nazi Germany were revealed. It is ironic that an extreme example of negative eugenics—the "race hygiene" practiced by the Nazis—was required before popular support for eugenics programs waned.

As students of genetics, you should evaluate ideas on the basis of their scientific merit and not become caught up in popular ideas of the time. Although eugenics fell into disfavor after World War II, it has regained some of its popularity with the development of new genetic testing methods and techniques for identifying genes associated with human disorders. We must all be careful in how we interpret and utilize such information, and we must not lose sight of the fact that linking a gene to a certain trait does not exclude a role for environmental factors (and for other genes) in the determination of that trait. As you proceed through this text, you will be asked to "ponder" various points of interest related to human genetics. As you consider these questions, try to keep a scientific perspective, or at least realize which aspects of your thoughts are scientifically valid and which are not.

bloodstream, yet, even today, we commonly speak of "bloodlines" when referring to inherited traits.

The **blending idea** of heredity that accompanied the pangenesis theory was also easily discredited. For example, according to the blending notion, a plant that inherits a red-flower gemmule from one parent and a white-flower gemmule from the other parent will be an intermediate (pink) color and will contain pink gemmules that are physically a blend of the red and white ones. Thus, if that pink-flowered plant is crossed to another pink-flowered plant, their progeny should be pink since no intact red or white gemmules remain in the pink-flowered plants. This prediction is not verified, however. Intermatings of pink-flowered plants produce red-flowered progeny and pink-flowered progeny, as well as some offspring with white flowers, showing that red and white hereditary factors are present in the pink-flowered plants. When we discuss Mendel's work in the next chapter, you will see that he described heredity in terms of particles that retain their identity from one generation to the next rather than physically blending together. Mendel thus observed a **particulate** rather than a blending mechanism of inheritance.

Lamarck's theory was directly challenged in 1892 by August Weisman, who developed the **theory of the germ plasm**. Weisman made a clear distinction between the part of the body important to heredity (the germ plasm) and the rest of the body (the somatoplasm). He emphasized the stability of the germ plasm, saying that the environment has little, if any, effect on the hereditary material, even though it may greatly modify external characteristics of an organism. According to Weisman, the hereditary material is thus transmitted essentially unchanged from one generation to the next. Of course, this theory has since been modified to incorporate the possibility of mutation, but its fundamental premise is well established and is not compatible with Lamarck's theory. Despite Weisman's theory, the idea that organisms acquire hereditary variations as a result of environmental conditions was not experimentally proven wrong until the middle part of this century, a point we will discuss later when we consider gene mutation.

The major biological discoveries of the nineteenth century concluded with the realization of the significance of Mendel's work as the outline of the basic principles of heredity. Mendel's principles not only provided an explanation for the transmission of genetic traits from one generation to the next; they also, when coupled with mutation theory, explained the source of hereditary variation in populations. Although Mendel published his work in 1866, it was not "discovered" until 1900. Biological historians have made various speculations as to why the significance of this work was not appreciated earlier. One important consideration is that researchers did not develop the techniques for accurate microscopic observations of chromosomes during mitosis and meiosis until the late 1880s. Knowledge of chromosome behavior during cell division was perhaps the most important factor in the acceptance of Mendel's laws of inheritance in 1900 because it provided known cellular structures (the chromosomes) that could be correlated with Mendel's hereditary factors. Before proceeding into Mendelian genetics, therefore, it is important to review cells, mitosis, and meiosis.

Point to Ponder 1.1

We have described eugenics as a late nineteenth and early twentieth century movement to encourage "desirable" individuals to have children and to discourage "undesirable" persons from reproducing. The laws of the early 1900s ordering sterilization of mentally disabled individuals and of criminals, the 1920s immigration laws prohibiting entry into the United States of "idiots, imbeciles, feebleminded, epileptics, and insane persons," and the atrocities committed by the Nazis in the 1930s and 1940s are examples of eugenic measures. The U.S. sterilization and immigration laws were reformed in the 1950s and 1960s, seemingly putting an end to eugenics in the United States. However, in the 1980s, California's Center for Germinal Choice was established as a place where Nobel Prize winners can deposit sperm for artificial insemination of "suitable" women. In addition, prenatal testing is being increasingly used to detect disorders in developing fetuses, and genetic testing is being used by health and life insurance companies and even employers to deny insurance and employment to some individuals because the testing reveals disease-associated genotypes. Do you view such activities as eugenics measures? Are modern technologies providing us with "acceptable" ways of using genetic knowledge for the improvement of human society?

THE CELLULAR BASIS OF HEREDITY

The cell is the smallest self-sufficient unit of living material, since it is the smallest structural unit that can carry out all the activities of life. Thus, the smallest living organisms are unicellular, that is, each consists of a single cell that functions as an independent unit. More complex organisms are multicellular and consist of many different types of cells that function together in the form of tissues or organs.

Cells come in a variety of shapes and forms, as shown in ▶ Figure 1–2. But despite the large diversity in outward appearance, all cells have certain features in common. One common feature is that they all are relatively small in size, some bordering on the limit of resolution of the light microscope. **Prokaryotes** (bacteria) are usually the smallest cells, generally ranging from 1 to 5 micrometers (μm) in diameter. **Eukaryotic cells** (all cells other than bacteria) tend to be at least an order of magnitude larger; most plant and animal cells, for example, have dimensions in the range of 10 to 50 μm. Because of their minute size, much of our knowledge of the structure of cells has been obtained from observations using the electron microscope. With this instrument, it is possible to investigate structures as small as 0.4 nanometer (nm) in size, providing a 500-fold increase in resolving power over that obtained with an ordinary light microscope.

Cells also share certain essential structural features (▶ Figure 1–3). All cells have a differentially permeable outer membrane called the **plasma membrane**, which separates the cell from the external environment. This membrane consists of a lipid bilayer (that is, two layers of phospholipids) with proteins integrated into or associated with the basic membrane structure. Contained within the membrane is the **cytoplasm**, in which the various chemical processes involved in cell metabolism occur. Dispersed within the cytoplasm of all cells are structures called **ribosomes**, which function in the process of protein synthesis.

In addition to a cytoplasm with ribosomes and a surrounding plasma membrane, all living cells also possess a structure in which the genetic information of the cell (deoxyribonucleic acid or DNA) is stored and replicated. In prokaryotes, this structure is a region within the cell called a **nucleoid** or **nuclear body**, which has no surrounding membrane. In eukaryotic cells, in contrast, the DNA is contained within a complex **nucleus** that is separated from the cytoplasm by a **nuclear envelope** (▶ Figure 1–4). The nuclear envelope is composed of two lipoprotein membranes that are joined around openings called **nuclear pores**. Dispersed throughout the nucleus of a nondividing cell is a network of fibers known as **chromatin**, which is composed of DNA and protein. Just before cell division, the chromatin fibers become organized into discrete bodies to form the chromosomes of the eukaryotic cell. Also suspended in the nucleus are one or more **nucleoli** (singular, **nucleolus**), which function in the assembly of ribosomes. In most eukaryotic cells, both the nucleolus and the nuclear envelope

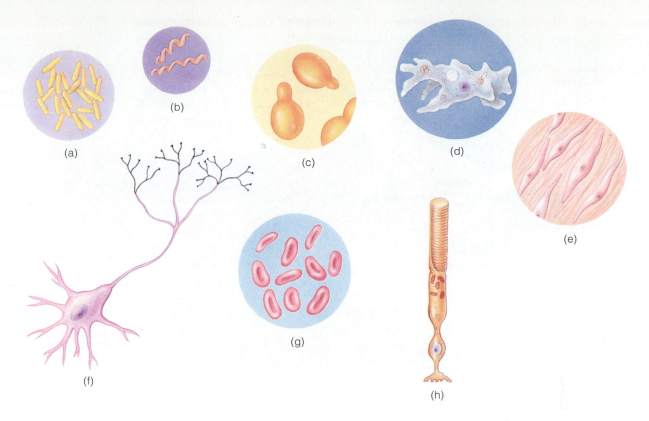

▶ **Figure 1–2** Some examples of different types of cells. (a) The bacterium *Escherichia coli,* (b) the mycoplasma *Spiroplasma citrii,* (c) budding yeast cells, (d) an amoeba, (e) human fibroblasts in loose connective tissue, (f) a human nerve cell, (g) human erythrocytes, and (h) a rod cell in the retina of the human eye.

are broken down into their component parts and later resynthesized during the course of cell division.

In addition to having a well-developed nucleus, eukaryotic cells also contain some cytoplasmic organelles that are not found in prokaryotes (see Figure 1–4). These structures include the endoplasmic reticulum (for biosynthesis and transport), lysosomes (for digestion), Golgi bodies (for secretion), mitochondria (for cellular respiration), and, in the case of plants, chloroplasts (for photosynthesis).

Chromosomes

The carriers of genetic information within cells are the structures called **chromosomes**, which are made up of DNA and protein. In eukaryotic cells, chromosomes are visible with the light microscope during cell division, appearing as rod-shaped, darkly staining structures (*chromosome* means "colored body"). The number of chromosomes in each cell nucleus varies considerably from species to species, as can be seen in ■ Table 1–2. The numbers listed in the table are those from somatic (non-sex) cell nuclei. In most higher animals and in many higher plants, somatic cells are **diploid,** that is, they contain two complete sets of chromosomes. One set of chromosomes in a sexually reproducing diploid

organism is derived from the maternal parent (the *maternal* set), and the other set from the paternal parent (the *paternal* set). Thus, chromosomes in the diploid state exist in

▶ **Figure 1–3** Essential structural features shared by all cells include the plasma membrane, the cytoplasm, ribosomes, and the nucleus (eukaryotes) or nuclear body (prokaryotes). The nucleus or nuclear body contains the genetic information of the cell. The cytoplasm consists of the cytosol and any cytoplasmic organelles suspended in it. Ribosomes function in the synthesis of proteins.

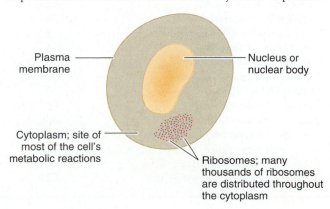

Plasma membrane

Nucleus or nuclear body

Cytoplasm; site of most of the cell's metabolic reactions

Ribosomes; many thousands of ribosomes are distributed throughout the cytoplasm

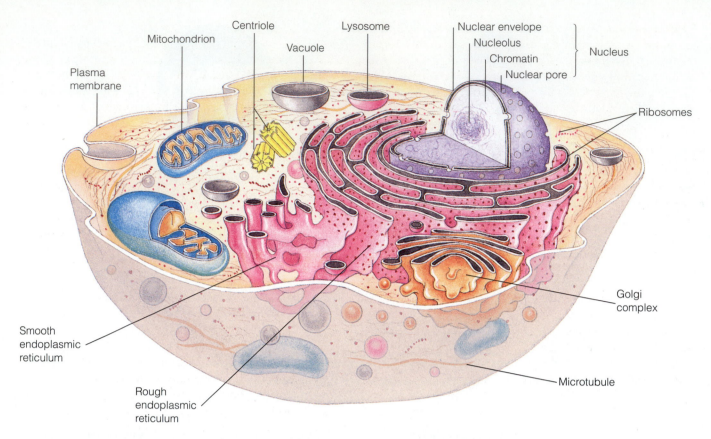

Plasma membrane
Mitochondrion
Centriole
Vacuole
Lysosome
Nuclear envelope
Nucleolus
Chromatin
Nuclear pore
Nucleus
Ribosomes
Golgi complex
Microtubule
Smooth endoplasmic reticulum
Rough endoplasmic reticulum

▶ **Figure 1–4** Generalized structure of a eukaryotic cell, showing major organelles.

pairs, with each pair consisting of one maternally- and one paternally-derived chromosome. We call the matched members of each pair **homologous chromosomes** or simply **homologs.**

Unlike the somatic cells of a diploid organism, the gametes (or sex cells) are **haploid,** that is, they contain only one complete set of chromosomes. Each gamete therefore contains only half the chromosomes present in the somatic

■ **TABLE 1–2** Diploid chromosome numbers in the somatic cells of selected animals and plants.

Organism	Chromosome number	Organism	Chromosome number
Boa constrictor (*Constrictor constrictor*)	36	Barley (*Hordeum vulgare*)	14
Cat (*Felis domesticus*)	38	Bean (*Phaseolus vulgaris*)	22
Cattle (*Bos taurus*)	60	Bread wheat (*Triticum aestivum*)	42
Chicken (*Gallus domesticus*)	78	Cabbage (*Brassica oleracea*)	18
Chimpanzee (*Pan troglodytes*)	48	Corn (*Zea mays*)	20
Donkey (*Equus asinus*)	62	Cucumber (*Cucumis sativus*)	14
Dog (*Canis familiaris*)	78	Cultivated oat (*Avena sativa*)	42
Fruit fly (*Drosophila melanogaster*)	8	Emmer wheat (*Triticum turgidum*)	28
Horse (*Equus caballus*)	64	Field horsetail (*Equisetum gigantia*)	216
Housefly (*Musca domestica*)	12	Green alga (*Acetabularia mediterranea*)	20
House mouse (*Mus musculus*)	40	Garden pea (*Pisum sativum*)	14
Human (*Homo sapiens*)	46	Indian fern (*Ophioglossum reticulatum*)	1260
Mosquito (*Culex pipiens*)	6	Potato (*Solanum tuberosum*)	48
Pigeon (*Columbia livia*)	80	Radish (*Raphanus sativus*)	18
Platyfish (*Platypoecilus maculatus*)	48	Rice (*Oryza sativa*)	24
Rat (*Rattus norvegicus*)	42	Rye (*Secale cereale*)	14
Starfish (*Asterias forbesi*)	36	Tobacco (*Nicotiana tabacum*)	48
Turkey (*Meleagris gallopavo*)	82	Tomato (*Solanum lycopersicum*)	24

cells of the same individual. For example, there are 23 chromosomes in a human egg or sperm, compared with 46 (23 pairs) in a human somatic cell. We symbolize the number of chromosomes in the gamete of an individual as n, defined as the gametic number; in humans, therefore, $n = 23$ and $2n$ (the somatic number) = 46.

The Genetic Material

As we shall see shortly, the study of chromosomes using the light microscope reveals their behavior during cell division, but it does not tell us anything about the nature of the genes found in these chromosomes or about the complexity of the genetic material. In subsequent chapters, we will discuss in detail the molecular basis of chromosome structure, chromosome replication, gene structure, and gene expression into a phenotypic trait. However, it will be helpful if we begin now to relate microscopic observations of chromosomes to the genetic material they contain, so that you can begin to develop a physical sense of how genetics works.

Each chromosome in a cell nucleus carries its genes in a single molecule of DNA. Thus, the DNA makes up the genetic material—the carrier of the genetic information—of chromosomes. In a nondividing cell, when the chromosomes are not visible using the light microscope, the chromatin threads are the DNA molecules of which we speak.

We will see in future chapters that the nature of DNA molecules is such that they explain the molecular basis of four requirements that must be met by the genetic material: (1) the storage of coded information (the genes) that dictates the various traits of organisms, (2) the expression of the coded information into these phenotypic traits, (3) the inheritance of the coded information from generation to generation, and (4) the production of variations in that information. These requirements are summarized in ▶ Figure 1−5. DNA, as the genetic material, is indeed complex enough to contain a sufficient number of different genes to account for the tremendous array of traits the genes control in all living organisms (Figure 1−5a). Expression of this information occurs through the processes of **transcription**, the production of messenger molecules that carry the genetic information from the nucleus to the cytoplasm, and **translation**, the decoding of these messages (Figure 1−5b). This decoding process results in the cell-based manufacture of proteins that, in complex and often poorly understood ways, give rise to the observed phenotypes for various traits. Through precise replication of the DNA molecules (Figure 1−5c), the genetic information is accurately copied and passed from one generation to the next. Finally, through occasional errors in the perpetuation of DNA, altered genes that code for mutant forms of proteins can be produced (Figure 1−5d). While it would be premature to further discuss the molecular details of these four processes now, keep in mind that a chromosome carries genes in the

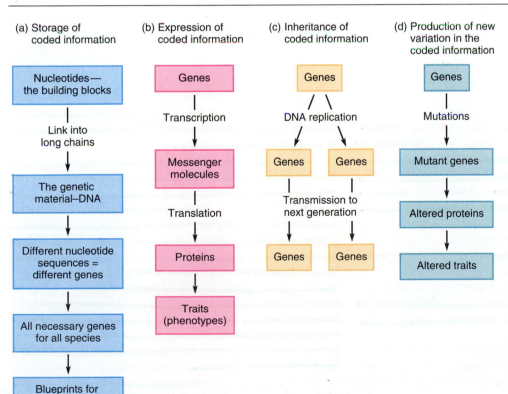

(a) Storage of coded information

Nucleotides— the building blocks
↓
Link into long chains
↓
The genetic material–DNA
↓
Different nucleotide sequences = different genes
↓
All necessary genes for all species
↓
Blueprints for all life forms

(b) Expression of coded information

Genes
↓ Transcription
Messenger molecules
↓ Translation
Proteins
↓
Traits (phenotypes)

(c) Inheritance of coded information

Genes
↙ ↘ DNA replication
Genes   Genes
↓   ↓
Transmission to next generation
↓   ↓
Genes   Genes

(d) Production of new variation in the coded information

Genes
↓ Mutations
Mutant genes
↓
Altered proteins
↓
Altered traits

▶ **Figure 1−5** Summary of the four requirements that must be met by the genetic material and how DNA meets those requirements. (a) Genetic information (genes) is stored as various sequences of nucleotides that make up different regions of DNA molecules. (b) The genetic information is expressed when genes are transcribed into messenger molecules that are then decoded (translated) to yield proteins. These proteins give rise to the phenotypic traits of an organism. (c) The genetic information is duplicated by the process of DNA replication and then passed on to the next generation. (d) The genetic information occasionally changes through mutation, resulting in altered genes that might then code for altered forms of proteins.

form of DNA as we proceed through a description of mitosis and meiosis.

Mitosis and Cell Reproduction

In addition to serving as the basic structural and functional units of living organisms, cells can also duplicate and pass on their genetic information through the process of cell reproduction. Free-living cells reproduce asexually by cell division. Most somatic cells of multicellular organisms can also divide, either for the growth and development of the organism or to replace and repair damaged tissues. The division of cells is a doubling process in which one cell becomes two identical daughter cells, both having the same amount and kind of genetic information as the original (parent) cell. The two cells then give rise to four, the four give rise to eight, and so on, during each successive division cycle. Each cell generation therefore doubles the number of cells, N, in a population in the geometric progression $N_0 \rightarrow 2N_0 \rightarrow 4N_0 \rightarrow 8N_0$, and so on, where N_0 is the initial number of cells. Each term in this progression can be represented by the general equation

$$N = N_0 2^t$$

where t is the number of division cycles or population doublings that have occurred. Progressing at this rate, it would take only 20 division cycles for a single cell to produce a population of 2^{20}, or about one million cells. With the exception of occasional mutations, all these cells will be genetically identical to the original cell. The significance of this process is illustrated by the production of a multicellular organism from a single fertilized egg (the zygote). An adult human, for example, contains approximately 10^{14} cells, all containing the same gene complement—that of the original zygote.

The division of eukaryotic cells is a complex activity that delicately balances two interrelated processes: (1) the division of the nucleus (**mitosis** or **karyokinesis**) to form two nuclei that are genetically and cytologically identical to each other and to the nucleus from which they came and (2) the division of the cytoplasm (**cytokinesis**), which results in the two nuclei being separated into different cells. In order for daughter nuclei to receive identical sets of chromosomes, cell division must be preceded by chromosome duplication. Chromosome duplication occurs during **interphase**, which is the interval between two successive mitoses and the time during which most growth and preparation for division take place. We can therefore divide the life cycle of somatic cells into three main stages: mitosis, cytokinesis, and interphase.

Figure 1–6 illustrates the replication of the DNA molecules that make up the chromatin threads. The chromatin fibers are not microscopically visible at this point as the structures we refer to as chromosomes, but we know that each duplicated chromatin fiber is genetically equivalent to

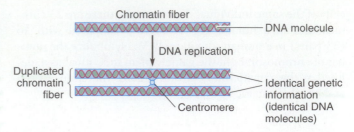

▶ **Figure 1–6** The replication of a chromatin fiber during interphase. A DNA molecule makes up the genetic material of each fiber; replication of the DNA produces identical DNA duplicates held together at a common centromere.

a duplicated DNA molecule. The duplicates carry identical genetic information. The identical chromatids are held together at their **centromere** region, which will become the point of attachment of the spindle fibers to the duplicated chromosomes during cell division.

The cell division process has long intrigued biologists. The events of mitosis are detectable cytologically (with a light microscope) and therefore were the first parts of the cell cycle to be studied. Though the events of mitosis actually proceed smoothly without extensive interruptions, mitosis is conventionally divided into four stages: **prophase, metaphase, anaphase, and telophase**. Certain landmark cytological events that serve to identify the four stages are shown in ▶ Figure 1–7 along with a brief description of each stage.

Prophase. Prophase, the first stage of mitosis, is characterized by the tight coiling (**condensation**) of the duplicated chromatin threads into shorter, thicker, cytologically-visible chromosomes. The beginning of prophase is therefore marked by the appearance of the threadlike chromosomes, each of which consists of two identical longitudinal halves held together at the narrowed region along their length (the centromere). The two longitudinal halves of each chromosome are called **sister chromatids** as long as they remain connected at the centromere.

As prophase progresses, the chromosomes continue to coil, becoming shorter and thicker and finally taking on the appearance of doubled rods; they are now referred to as **dyads** (▶ Figure 1–8). While the chromosomes are undergoing their characteristic structural changes, there is a gradual breakdown of both the nucleolus and the nuclear envelope. The region formerly occupied by the nucleus is then replaced by a cagelike structure called the **spindle**. The spindle is an organized system of hollow cylindrical fibers or microtubules that are assembled in the cytoplasm during prophase; these fibers are polymers of protein subunits called *tubulins* and can grow and shrink in length by polymerization and depolymerization at their ends. Located at the poles of the spindle in most animal cells and the cells of many lower plants are pairs of cylindrical organelles known as **centrioles**, which function in these cells in the development of the spindle microtubules.

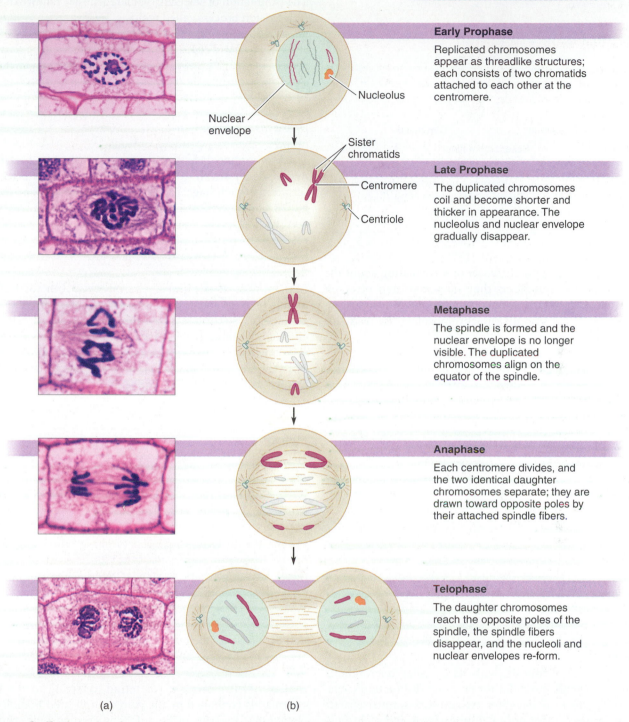

Early Prophase

Replicated chromosomes appear as threadlike structures; each consists of two chromatids attached to each other at the centromere.

Nucleolus

Nuclear envelope

Sister chromatids

Centromere

Centriole

Late Prophase

The duplicated chromosomes coil and become shorter and thicker in appearance. The nucleolus and nuclear envelope gradually disappear.

Metaphase

The spindle is formed and the nuclear envelope is no longer visible. The duplicated chromosomes align on the equator of the spindle.

Anaphase

Each centromere divides, and the two identical daughter chromosomes separate; they are drawn toward opposite poles by their attached spindle fibers.

Telophase

The daughter chromosomes reach the opposite poles of the spindle, the spindle fibers disappear, and the nucleoli and nuclear envelopes re-form.

(a)   (b)

▶ **Figure 1–7** The order of events during mitosis. (a) Photographs of mitosis occurring in a plant cell. (b) Diagrams of the nuclear events involved in mitosis in a hypothetical animal having a diploid chromosome number of 4.

Metaphase. Metaphase, the second stage of mitosis, begins when the nuclear envelope is no longer visible and the spindle is in place. The spindle is now composed of two types of microtubules: interpolar microtubules, which extend between the poles of the spindle without making connections to the chromosomes, and chromosomal mi-crotubules, which extend from the opposite poles to the centromeric regions of the chromosomes. During metaphase, the duplicated chromosomes (dyads) move to align themselves on the equatorial plane of the spindle, each attached at its centromere to chromosomal microtubules. Time-lapse photography reveals that this movement is

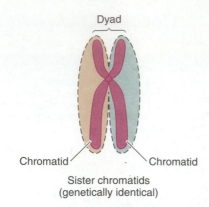

Dyad

Chromatid | Chromatid

Sister chromatids
(genetically identical)

▶ **Figure 1–8** A single duplicated chromosome, also referred to as a dyad. The dyad is composed of the two sister (identical) chromatids.

---

somewhat erratic, with chromosomes oscillating about the midplane of the spindle as they maneuver into position. When the centromeres of the chromosomes are aligned on the equator of the spindle, the chromosomes are ready for the next stage.

Anaphase. Anaphase, the third stage of mitosis, occurs once the centromeres divide and the sister chromatids begin to separate. The daughter chromosomes (formerly sister chromatids) then move apart, a process referred to as **disjunction.** The daughter chromosomes are drawn toward opposite poles by their attached microtubules. These microtubule-based movements occur through shortening of the chromosomal microtubules, which pulls the chromosomes toward the poles.

Telophase. Telophase, the fourth and final stage of mitosis, commences when the daughter chromosomes have completed their migrations to opposite poles. A nuclear envelope reassembles around the cluster of chromosomes at each pole, the spindle microtubules disappear, and the chromosomes gradually uncoil. With the reappearance of nucleoli, the daughter nuclei begin to assume their interphase morphology, as the chromosomes uncoil and become chromatin threads.

The process of mitosis, with its complex microtubule-based movements, provides for an exact disjunction (separation) of identical daughter chromosomes into separate nuclei. By the separation of duplicates of every chromosome, mitosis thus ensures that each daughter nucleus receives the same complement of chromosomes and therefore the same genetic information as the parent nucleus.

Meiosis and Sexual Reproduction

Almost all plants and animals and many of the eukaryotic microorganisms are capable of reproducing sexually

through the formation and subsequent union of gametes. The production of sex cells (gametogenesis) requires a special division process called **meiosis.** Unlike mitosis, meiosis results in the reduction of chromosome number and is restricted to certain specialized cells called **meiocytes** that are present within the reproductive tissues of an organism. A meiocyte is ordinarily diploid, having two copies of its genetic material, whereas a gamete is haploid, with only one copy. To achieve the haploid state, each gamete produced by meiosis receives only one chromosome of each pair found in the corresponding meiocyte.

Chromosome behavior during the different stages of meiosis is shown in ▶ Figure 1–9, along with a brief description of each stage. Observe that meiosis consists of two nuclear and cytoplasmic divisions but is preceded by only one duplication of chromosomes. As in mitosis, the chromosomes are duplicated during interphase, prior to the first meiotic stage. The first meiotic division (meiosis I) is a **reduction** division, which converts a meiocyte with a diploid ($2n$) number of duplicated chromosomes ($4n$ total chromatids or DNA molecules) into two daughter cells with a haploid (n) number of duplicated chromosomes ($2n$ total chromatids) in each. In contrast, the second meiotic division (meiosis II) is an **equational** division, producing cells that have the same number of chromosomes (but not chromatids) as their immediate progenitors.

Meiosis I

Meiosis I begins with condensation of the duplicated chromatin threads, the gradual breakdown of the nucleolus and nuclear envelope, and the formation of the spindle apparatus. The phases of meiosis I are:

Prophase I. In early prophase I (a phase called **leptonema** or the **leptotene stage**), the chromosomes begin to condense, becoming visible as slender, threadlike structures. As prophase I proceeds to **zygonema** (the **zygotene stage**), homologous chromosomes begin to align lengthwise through a pairing process known as **synapsis.** By **pachynema** (the mid-prophase I **pachytene stage**), homologous chromosomes have become closely paired and are aligned side by side. Each closely associated or synapsed pair of homologs is then called a **bivalent** (or **tetrad**, referring to the four chromatids contained by the paired duplicated homologs). Notice that synapsis of homologous chromosomes during prophase I of meiosis is a pattern of chromosome behavior that differs completely from what occurs during prophase of mitosis (where chromosomes never pair off).

Synapsis is accompanied by the appearance of a structure known as the **synaptonemal complex,** which is thought to be important in holding the paired homologs in close contact with each other. When viewed with an electron microscope, this structure appears to comprise three parallel bands of material: two outer bands called *lateral elements,* and an inner band or *central element* (▶ Figure 1–10). In the pachytene stage, the paired homologs of a bivalent exchange structural parts by means of a breakage and reunion process called **crossing-over**. Although it is not visible at this stage, the crossover process is known to result in the physical exchange of homologous segments between non-sister chromatids within a bivalent, as illustrated by ▶ Figure 1–11. Thus, if the homologs differ genetically, crossing-over can produce new combinations of genes in the nonsister chromatids. These new gene combinations are referred to as **recombinant types**, and their formation accounts for part of the genetic variability that is produced by meiosis.

In **diplonema** (the mid-to-late prophase I **diplotene** stage), the sister chromatids become clearly visible as the pairing forces between the homologs relax. Paired chromatids of the two homologs remain together at points of contact called **chiasmata** (singular, **chiasma**) (▶ Figure 1–12). Closer inspection reveals that the nonsister chromatids have undergone crossing-over at these points. Finally, in **diakinesis**, the last stage of prophase I, the homologs finish coiling into shorter and thicker structures. The chiasmata then undergo a process of **terminalization**, in which the points of contact move toward the ends of the chromosome arms. By the end of this stage, the spindle has formed and the bivalents are getting ready to align on the spindle equator.

Metaphase I. Later, in metaphase I, bivalents align at the center of the spindle. The alignment occurs in such a way that homologous centromeres are arranged on either side of the equatorial plane. The alignment of bivalents, rather than of individual duplicated chromosomes, is another important contrast between meiosis I and mitosis.

Anaphase I. Still later, in anaphase I, the homologous centromeres of each bivalent start to move toward opposite poles of the spindle, carrying along both chromatids of each chromosome. Thus, we see another contrast with mitosis: homologous dyads, rather than sister chromatids, disjoin (separate) in anaphase I, since centromeres do not divide at this time.

Telophase I. As a result of the disjunction of homologous dyads at anaphase I, each nucleus formed during telophase I receives only one member of each of the different chromosome pairs that existed within the parental cell. Therefore, meiosis I is a reduction division; its end result is two daughter cells, each containing a haploid complement of duplicated chromosomes. The haploid complement is composed of one representative of each original chromosome pair. In addition to generating haploid cells, meiosis I also differs from mitosis in that the products of meiotic cell division are not genetically identical.

Meiosis II

Meiosis II is an equational division process that is basically similar to mitosis in its pattern of chromosome behavior. The daughter cells produced by meiosis I proceed to the second meiotic division without further DNA replication. Meiosis II then converts these haploid products of the first meiotic division into four haploid cells with n unduplicated chromosomes in each. As in mitotic metaphase, individual chromosomes align on the spindle equator at **metaphase II**. Moreover, centromeres divide at **anaphase II**, so that only one chromatid of each duplicated chromosome enters a gamete produced at the end of **telophase II**. Thus we see that each gamete obtains a complete haploid set of nonhomologous chromosomes from the meiocyte.

The Genetic Significance of Meiosis

Two major genetic accomplishments of meiosis have been noted: (1) The products of meiosis are haploid, containing one member of each chromosome pair in the original meiocyte. Thus, sexual reproduction of diploid organisms involves the production of haploid gametes and their subsequent fertilization to regenerate the diploid state in the offspring. (2) Meiosis produces genetically variable gametes. The production of genetic variation is a crucial aspect of sexual reproduction. Genetic variation is the result of both a randomness of the alignment of the bivalents on the metaphase I spindle and the crossing-over process that takes place during prophase I. The random alignment of bivalents forms the basis of Mendelian genetics (Chapter 2) and will be discussed first.

Earlier we noted that one set of chromosomes in a sexually reproducing diploid organism is inherited from the maternal parent and the other set from the paternal parent. ▶ Figure 1–13 illustrates the two sets of chromosomes by color-coding them according to their origin. The color-coding indicates that the maternal member of each chromosome pair is different genetically from the paternal member of the pair, that is, although the members of a pair of homologous chromosomes contain the same kinds of genes on their DNA molecules, they are not genetically identical. For example, the paternal homolog may contain a mutant version of a gene while the maternal homolog contains the unmutated counterpart. Since each chromosome pair contains hundreds if not thousands of genes, the genetic differences between homologs can be quite substantial.

It is because of the genetic differences between the maternal and paternal homologs within each chromosome pair that the random alignment of maternally and paternally derived centromeres on either side of the spindle equator produces genetically different products in meiosis I. ▶ Figure 1–14 illustrates the situation for an organism having two pairs of chromosomes. The number of different chromosome combinations that can occur in gametes can be found using the formula 2^n, where n is the number of chromosome pairs. For an organism with just two pairs of chro-

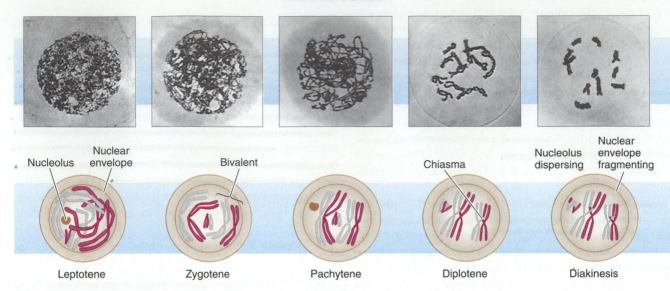

Nucleolus   Nuclear envelope   Bivalent   Chiasma   Nucleolus dispersing   Nuclear envelope fragmenting

Leptotene   Zygotene   Pachytene   Diplotene   Diakinesis

Prophase I

Replicated chromosomes appear as threadlike structures. Homologs pair to form bivalents.
An exchange of parts (crossing-over) occurs between homologs. Chromosomes shorten and thicken.

		Meiosis I	Meiosis II
Chromosomes:	2n ⟶	n ⟶	n
Chromatids:	4n ⟶	2n ⟶	n

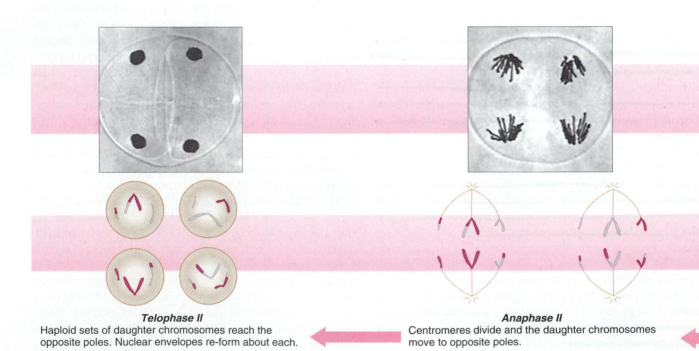

Telophase II
Haploid sets of daughter chromosomes reach the opposite poles. Nuclear envelopes re-form about each.

Anaphase II
Centromeres divide and the daughter chromosomes move to opposite poles.

▶ **Figure 1−9** A summary of the nuclear events involved in meiosis in an organism having a diploid chromosome number of 6.

Metaphase I
Bivalents align on the spindle equator.

Anaphase I
Homologs separate and move to opposite poles.
Centromeres do not divide at this time.

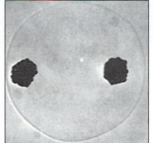

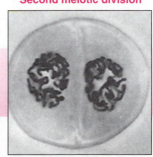

Telophase I
Haploid sets of duplicated chromosomes reach the
opposite poles. Nuclear envelopes re-form about each.

Second meiotic division

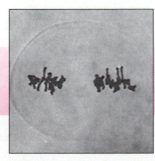

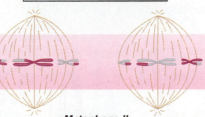

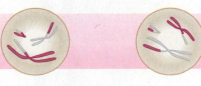

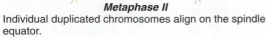

Metaphase II
Individual duplicated chromosomes align on the spindle
equator.

Prophase II
No further DNA synthesis occurs; haploid number of
duplicated chromosomes in each cell.

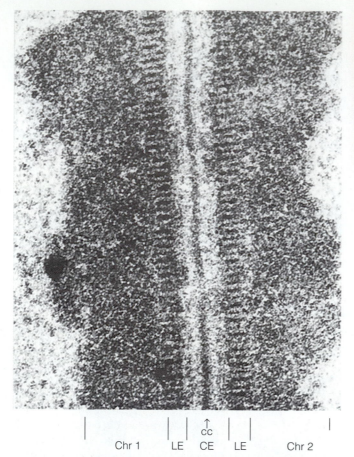

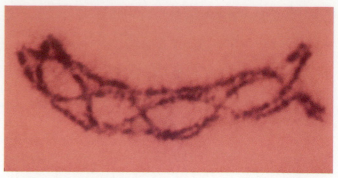

▶ **Figure 1–12** A photomicrograph showing chiasmata in a bivalent during meiosis.

mosomes, as in Figure 1–14, the number of genetically different gametes is $2^2 = 4$. In organisms with higher n values (see Table 1–2), the number of gamete types can become extremely large. For example, in humans where $n = 23$, 2^{23} (more than 8 million) gamete types are formed by the random alignment of maternal and paternal chromosomes on the metaphase I spindle. Thus, meiosis ensures that significant amounts of new genetic variability are produced each generation, simply through the random shuffling of homologs into gametes.

Crossing-over of course adds to this genetic variability. Homologous chromatids that have exchanged parts with each other have in the process exchanged segments of their DNA. The resulting crossover chromosomes that are incorporated into gametes are therefore recombinants, consisting of part maternal DNA and part paternal DNA. Although crossing-over at a particular place on a particular chromosome pair typically is rare, it can occur anywhere along a chromosome pair. This means that crossing-over will occur somewhere along each chromosome pair at every meiosis, thereby considerably increasing the number of different kinds of gametes that can be produced by an organism.

| | Chr 1 | LE | ↑ cc CE | LE | Chr 2 |

▶ **Figure 1–10** The synaptonemal complex of the fungus *Neotiella.* This is a longitudinal section of a bivalent, showing the dense central element (CE) containing the central component (cc); it lies between the banded lateral elements (LE) that are associated with the homologous chromatid pair (Chr 1 and Chr 2).

▶ **Figure 1–11** Crossing-over. If nonsister chromatids exchange homologous segments, the result is the formation of "hybrid" chromatids with recombinant gene arrangements.

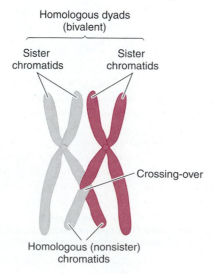

Homologous dyads
(bivalent)

Sister chromatids   Sister chromatids

Crossing-over

Homologous (nonsister) chromatids

▶ **Figure 1–13** The two sets of chromosomes in a diploid organism. One set (gray) has been inherited from the organism's maternal parent, the other set has been inherited from its paternal parent. In this case, the diploid chromosome number is 6, since each set consists of three chromosomes. Each member of a pair of homologous chromosomes has the same kinds of genes, but the members of a pair are not genetically identical.

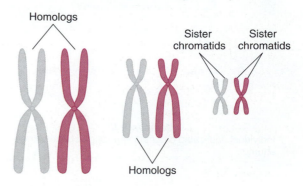

Homologs

Sister chromatids   Sister chromatids

Homologs

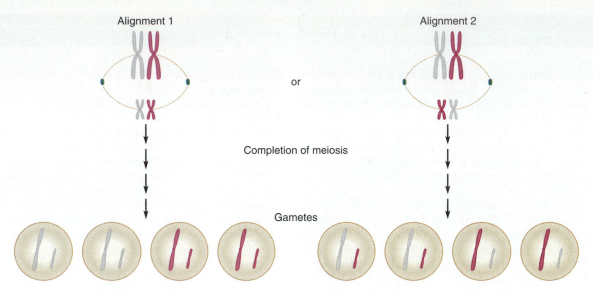

Alignment 1   or   Alignment 2

Completion of meiosis

Gametes

▶ **Figure 1–14** The random alignment of maternal and paternal chromosomes on the meiotic metaphase I spindle for two pairs of chromosomes ($2n = 4$). The homologous centromeres in each bivalent align randomly on the left versus right side of the spindle equator; in this case there are two alternative alignments. The result is that four different types of gametes ($2^n = 2^2 = 4$) can be formed by meiosis.

To Sum Up

1. Nuclear chromosomes occur in pairs in the somatic cells of diploid organisms. The matched members of each pair are called homologs. The gametes of a diploid organism contain one member of each pair of chromosomes; thus gametes are haploid— they have only half as many chromosomes as somatic cells.

2. The genetic information carried by each chromosome is contained in its DNA. The properties of DNA allow it to perform the four functions required of genetic material: (1) storage of genetic information, (2) expression of that information, (3) replication and inheritance of that information, and (4) production of variations in the information.

3. A cell cycle consists of interphase plus cell division. During interphase each chromatin fiber duplicates itself, resulting in duplication of the genetic information contained in each fiber.

4. Mitosis is the process of nuclear division that occurs in somatic cells. It produces identical daughter cells that have the same amount and kind of genetic information as the original cell. At the beginning of mitosis, chromosomes appear as threadlike structures that consist of two identical chromatids joined by a centromere. The chromatids are formed by the duplication of each chromosome during interphase.

5. We divide mitosis into four stages. During prophase the chromosomes shorten and thicken while the nuclear envelope and nucleolus disappear. By the end of prophase chromosomes can be microscopically observed as duplicated structures called dyads. The duplicated chromosomes align on the equator of the spindle during metaphase. Microtubules connect the centromeres of the duplicated chromosomes to the poles of the spindle. The sister chromatids separate during anaphase, and identical daughter chromosomes move to opposite poles of the spindle. The chromosomes uncoil during telophase as nucleoli reappear and a nuclear envelope re-forms around the chromosomes at each pole.

6. During gamete formation, specialized reproductive cells called meiocytes undergo meiosis. Meiosis consists of two successive nuclear divisions, each of which is composed of prophase, metaphase, anaphase, and telophase stages. Meiosis I, a reduction division, results in the separation of homologous chromosomes, while meiosis II, an equational division, is responsible for the separation of sister chromatids.

7. During prophase I, the duplicated chromosomes shorten and thicken, the nucleolus and nuclear envelope disappear, and homologous dyads pair or synapse to form bivalents (tetrads). Homologous chromosomes can also exchange segments during pairing through a process known as crossing-over. During metaphase I, the bivalents align on the equator of the spindle with microtubules connecting the homologous centromeres to the poles of the spindle. The homologous dyads separate during anaphase I, as the homologs move toward opposite poles of the spindle. At telophase I, a nuclear membrane re-forms around the haploid number of duplicated chromosomes at each pole.

8. Meiosis II is a mitotic-type division. The haploid number of duplicated chromosomes align along the spindle equator at metaphase II with each chromosome connected at its centromere to the poles of the spindle. Sister chromatids separate at anaphase II, and a haploid number of (unduplicated) chromosomes enters each gamete at telophase II.

9. In addition to reducing the number of chromosomes to the haploid condition in the gametes, meiosis is also responsible for the production of genetic variability. The random alignment of the maternal and paternal homologs at metaphase I produces 2^n genetically different kinds of gametes, where n is the haploid chromosome number. Additional variability is produced by crossing-over, which results in genetically recombinant chromosomes.

The field of genetics consists of several major areas of study in which the focus of research ranges from the structure and expression of genes at the molecular level to the origin and nature of genetic variation in populations.

The history of genetics dates back to ancient speculations about how traits are inherited. Genetics is nevertheless a comparatively young science that developed as a modern discipline only during the twentieth century, following the discovery of Gregor Mendel's work.

DNA is the genetic material of all cellular organisms. It is carried in chromosomes, which replicate during interphase and are transmitted to daughter cells during cell division. In eukaryotic cells, chromosomes are transmitted during cell division by the process of mitosis, which produces two daughter cells that are identical to each other and to the parent cell in chromosomal and genetic makeup.

Chromosomes occur in pairs in the diploid cells of eukaryotes. During meiosis, the members of each pair separate and are transmitted to different haploid nuclei. Unlike mitosis, meiosis consists of two division cycles following one duplication of DNA and thus results in a reduction of chromosome number. In further contrast to mitosis, meiosis produces genetic variability by forming random combinations of different maternal and paternal chromosomes in the gametes; these random combinations result from independent alignment of different chromosome pairs and from the exchange of segments of homologous chromosomes through crossing-over.

Questions and Problems

Early History of Genetics

1. Describe the blending idea of heredity and explain how it relates to the theory of pangenesis.

The Cellular Basis of Heredity

2. Define the following terms and distinguish between members of paired terms.
   bivalent and dyad
   centromere
   chromatids
   cytokinesis
   diploid and haploid
   homologous chromosomes and sister chromatids
   interphase
   meiocyte
   microtubules
   spindle
   synapsis
   synaptonemal complex
   tetrad

3. The following are onion root-tip cells in the process of mitosis. In which stage of mitosis is each of these cells? Describe what features you see that allow you to determine the mitotic stage.

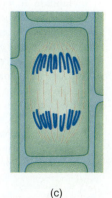

(a)   (b)   (c)

4. Give the specific stage in the mitotic cell cycle when each of the following events takes place: (a) Sister chromatids are first produced. (b) Sister chromatids are first visible with a light microscope. (c) Sister chromatids disjoin (separate).

5. Consider cells from an organism having a diploid chromosome number of 8 ($2n = 8$). (a) Draw the chromosomes as a cell of this organism proceeds through mitosis, and list or describe the key events at each mitotic stage. (b) Draw the chromosomes as a cell from this organism proceeds through meiosis I, listing or describing the key events at each stage. (c) Draw the chromosomes as the daughter cells go through meiosis II, listing or describing the key events.

6. Looking at the drawings and descriptions you made in answering question 5, list all similarities and differences as you compare and contrast (a) mitosis with meiosis I (b) mitosis with meiosis II (c) meiosis I with meiosis II.

7. A certain plant has 20 chromosomes in each somatic cell at interphase. How many chromosomes, centromeres, and chromatids are present in a single cell of this plant at (a) prophase (b) metaphase (c) anaphase (d) the close of telophase?

8. Suppose that a fertilized human egg undergoes cleavage (doubling of the number of cells) once every 48 hours throughout intrauterine development. Assume that at birth a baby has 10^{14} cells. (a) How long would it take to produce this number of cells under the assumed conditions? (b) Compare this time period with the actual gestation time. Why is there such a large difference?

9. Determine whether each of the following events occurs in mitosis, in meiosis, or in both kinds of nuclear division.
   (a) Chromosomes duplicate prior to the initial stage.
   (b) Homologs undergo pairing (synapsis).
   (c) Individual chromosomes align independently on the spindle equator.
   (d) Homologs separate and move toward opposite poles.
   (e) Diploid cells are the end result of the process.

10. The cell of a diploid organism observed under a microscope looked like this:

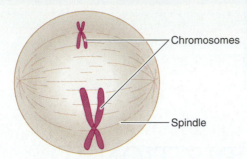

Assuming no other chromosomes are present within the cell, identify the stage of meiosis and/or mitosis in this cell.

11. A certain plant has 20 chromosomes in each meiocyte about to undergo meiosis. How many chromosomes, centromeres, and chromatids are present in each cell at (a) prophase I (b) metaphase I (c) anaphase I (d) the end of telophase I (e) metaphase II (f) anaphase II (g) the end of telophase II? (Assume that cytokinesis occurs at mid-telophase in both division cycles.)

12. How many different kinds of gametes can be produced by an organism having (a) 4 pairs of chromosomes, or (b) 8 pairs of chromosomes? (Just consider the random alignment of bivalents at metaphase I.)

13. Growth of normal somatic cells in culture is limited to a fixed number of division cycles. The number depends on the species and the age of the individual from which the cells were obtained. For example, human embryonic cells can undergo only around 50 cell cycles before the cells begin to deteriorate and division stops. How many progeny cells can a human embryonic cell theoretically produce during its normal life span? Compare this figure with the 10^{14} to 10^{15} cells that are present in a human adult.

14. The limited capacity of cells to divide in culture (described in question 13) is not altered by artificially halting their growth. For example, when human embryonic cells are placed in deep freeze to arrest cell growth, the only effect is to delay the inevitable deterioration of the culture. Freezing does not change the limit of 50 cell cycles that is characteristic of this cell type in culture. Suppose that a sample of 100 human embryonic cells from the twentieth cell cycle is frozen and later thawed. How many progeny cells can this sample theoretically produce by the end of its normal life span?

I

MENDELISM: INTRODUCTION TO GENETIC ANALYSIS

Many of the basic principles of genetics have their roots in the experiments performed by a nineteenth-century Austrian monk, Gregor Mendel. Mendel derived these principles from genetic crosses with the garden pea, *Pisum sativum,* and he published the results of his studies in 1866 in a paper entitled "Experiments in Plant Hybridization." The principles stated in the paper, collectively referred to as Mendelism, still stand as the cornerstone of modern genetics.

Mendel's work is considered a classic in experimental design and interpretation. He had the ingenuity to interpret his results using a mathematical model, even though he knew nothing about the physical basis of inheritance. In fact, at that time very little was known even about the chromosomes and the events of cell division discussed in Chapter 1. Mendel's principles therefore described the transmission of the hereditary units, which we call **genes,** in abstract mathematical terms. The importance of Mendel's work went completely unrecognized until 1900, several years after his death, because few scien-

Gregor Mendel (1822–1884).

tists understood the biological significance of his principles at the time they were proposed. By 1900 scientists knew much more about the biological meaning of Mendel's results. With the publication of the chromosome theory of heredity in 1902, biologists were able to give the first modern interpretation of the relationship between genes and chromosomes and provide a physical basis for Mendel's principles.

In the four chapters that comprise Part I, we will consider basic Mendelian genetics. Chapter 2 examines the principles themselves—the basic Mendelian model—and ties these principles to the behavior of chromosomes in meiosis. When we examine the link between genes and chromosomes, we will also see that not all genes behave in a Mendelian manner. The role of probability and statistics in genetic analysis will be explained in Chapter 3 to show how basic rules of probability provide a theoretical basis for Mendel's principles. Chapters 4 and 5 follow with a discussion of how Mendelian analysis can be extended to explain the genetic basis of more complicated traits.

Mendelian Principles

Mendelian genetics is concerned with the transmission of genes from one generation to the next and with the relationship between genes and hereditary traits. It is commonly said that the genes determine the structure, function, and development of an organism, but one should be careful to recognize that the expression of any given gene is quite likely to be influenced by the actions of other genes and/or by various environmental factors. In other words, even for a single trait, the relationship between the genetic constitution of an organism (its **genotype**) and its outward appearance (its **phenotype**) is often a very complex one. It would be more realistic to say that a given genotype determines the blueprint for the development of a particular trait while noting the important role that other genetic factors and the environment may play in expression of that genotype into a phenotype. We will discuss these complexities later in this text, as our picture of

Pea flower.

▶ **FIGURE 2–1** Monastery garden at Brno in Moravia (now part of the Czech Republic), where Mendel's experiments were conducted.

the gene and its expression gradually evolves. For now we will assume a predictable correspondence between genotype and phenotype, as we concentrate on elucidating the basic mechanisms of inheritance.

The basic rules governing the transmission of genes were established by Gregor Mendel by means of hybridization experiments conducted in the limited space of his monastery garden (▶ Figure 2–1). Mendel was not the first to investigate the mechanisms of heredity by hybridization, but he was the first to provide evidence for simple hereditary patterns and to interpret these patterns by relating the phenotype of an organism to its underlying genotype.

Mendel's success in establishing the basic rules of inheritance occurred for several reasons. One of the most important was his clear understanding of the kind of experimental design needed to answer the questions he was asking about the hereditary process. This clarity of understanding was reflected in his choice of experimental organism. Mendel chose to work with the garden pea plant because (1) it possesses traits that are easily recognizable and appear in widely contrasting forms and (2) it lends itself well to controlled crosses. Mendel realized that he had to be able to identify unambiguously the hereditary characteristics of each individual. He therefore limited his analysis to seven different traits, each expressed in a pair of contrasting forms (see ■ Table 2–1). The alternative characters of each trait were clearly distinguishable, leaving little room for error in identification, and they retained their contrasting appearance from one generation to the next.

Mendel also knew that he had to have complete control over the kinds of plants being mated. The pea plant is naturally self-fertilizing, but it can be used in genetic crosses if the anthers (male part) of the flower are removed and the pistil is then dusted with pollen from another source (▶ Figure 2–2). By using this technique, Mendel was able to dictate the precise sequence of matings in his experiments.

Because Mendel's experiments were designed to study the outcome of plant hybridization, it was crucial that pure-breeding varieties be used to initiate the experimental sequence of matings. **Pure-breeding** means that a plant, when self-fertilized, produces only offspring like itself. Mendel obtained pure-breeding plants from seedsmen and subjected them to two years of trial self-fertilizations to ensure that they were indeed pure types prior to beginning his hy-

■ **TABLE 2–1** Summary of the results of Mendel's experiments with seven traits of the garden pea.

Parental traits	F_1	F_2 (number) Dominant	F_2 (number) Recessive	F_2 (ratio)
Seed coat texture (round vs. wrinkled)	all round	5474 round	1850 wrinkled	2.96:1
Cotyledon color (yellow vs. green)	all yellow*	6022 yellow	2001 green	3.01:1
Flower color (purple vs. white)	all purple	705 purple	224 white	3.15:1
Pod shape (inflated vs. constricted)	all inflated	882 inflated	299 constricted	2.95:1
Pod color (green vs. yellow)	all green*	428 green	152 yellow	2.82:1
Flower location (axial vs. terminal)	all axial	651 axial	207 terminal	3.14:1
Stem length (long vs. short)	all long	787 long	277 short	2.84:1

*Even though two entirely different traits (cotyledon color and pod color) may exhibit the same contrasting characters (yellow vs. green), the character that is dominant in one case is not necessarily the dominant one in the other. A dominance relation holds constant for any one trait but is not necessarily the same for a different trait.

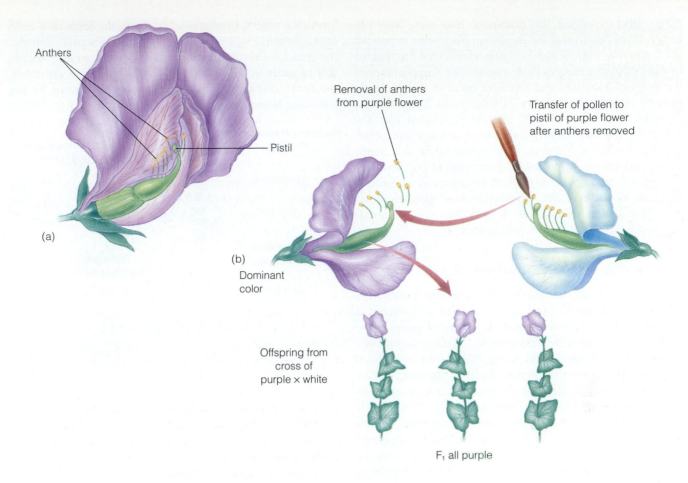

Anthers

Removal of anthers
from purple flower

Transfer of pollen to
pistil of purple flower
after anthers removed

Pistil

(a)

(b)
Dominant
color

Offspring from
cross of
purple × white

F_1 all purple

▶ **FIGURE 2–2** (a) The reproductive parts of the pea flower. (b) Controlled mating between purple-flowered and white-flowered plants. The anthers of the purple-flowered parent have previously been removed to prevent self-fertilization. Then the pollen from the white-flowered parent is dusted onto the pistil of the purple-flowered parent.

bridization experiments. As controls, he continued to self-fertilize plants from these pure lines during the entire course of his hybridization experiments, noting that they remained constant without any exceptions.

 ## THE PRINCIPLE OF GENETIC SEGREGATION

Mendel began his experiments by crossing pure-breeding plants that differed in only one pair of contrasting characters such as seed coat texture or pod color. By limiting his analysis to single character differences, Mendel avoided the overwhelming complexity that plagued the results of earlier studies, in which several characters were permitted to vary. He then analyzed the appearance of the hybrid offspring of this first cross between pure-breeding lines; these offspring are termed the F_1 (for first filial generation). Since these F_1 progeny were hybrid for a single pair of alternative charac-

ters, they are also called **monohybrids.** The F_1 monohybrids were then permitted to self-pollinate, as they normally would, to produce a second generation of offspring, the F_2 (or second filial generation), and this generation was also analyzed.

The results of Mendel's experiments on the seven pairs of contrasting characters are summarized in Table 2–1. In every case, all of the F_1 exhibited the characteristic of just one of the parental strains. For example, when pure-breeding plants with round seeds were crossed with pure-breeding plants with wrinkled seeds, all the F_1 seeds were round. Similarly, pure-breeding long-stemmed plants crossed with short-stemmed plants produced all long-stemmed progeny, and so on. Mendel termed the character that appeared in the F_1 the **dominant** trait and the character that failed to appear in the F_1 the **recessive** trait.

Unlike the F_1, the F_2 was composed of both dominant and recessive types. But in each of the seven experiments, plants that expressed the dominant trait were approximately three times more frequent than plants that expressed the recessive trait. That is, both characters reappeared in the F₂ but in a ratio of three to one. The reappearance of the original dominant and recessive traits in the F₂ clearly showed the blending theory of inheritance (described in Chapter 1) to be incorrect. Mendel concluded that the heredity factors do not physically blend into any type of hybrid factors in the offspring; they instead retain their identity as they are passed from one generation to the next. The inheritance of traits through the passing on of unblended particles that direct the development of phenotypes is called **particulate inheritance**; it is the Mendelian alternative to the old pangenesis/blending idea.

The Basic Model: Interpretation of Results

Mendel interpreted the results of his crossing sequence in the following manner: Each pair of alternative characters is determined by a pair of hereditary factors or **genes**. (The term *gene* was originally coined by W. L. Johannsen in 1903 to mean a Mendelian factor, so we will use this term from now on.) The genes that control a pair of alternative characters are known as **allelic genes** or simply **alleles**. In modern terminology, we say that the basic unit of inheritance is the gene, which exists in alternative allelic forms. The allelic forms of each gene can be symbolized by the uppercase and lowercase of the same letter (such as *A* and *a*), with the uppercase symbol designating the allele for the dominant trait (the dominant allele) and the lowercase symbol designating the recessive allele. (Other allelic symbols will be introduced later in the text.) In a pure-breeding line, we would write the genotype or gene constitution of the parents as consisting of either two *A* genes (*AA*), if they express the dominant trait, or two *a* genes (*aa*), if they express the recessive trait. Such pure-breeding individuals, in which both genes of a pair are alike (either *AA* or *aa*), are termed **homozygous** for the genes in question.

Another important feature of Mendel's model is that during reproduction, only one member of each pair of genes is included in a single gamete. Each F₁ monohybrid is therefore *Aa* in genotype, having been formed by the union of an *A*-carrying gamete from the dominant *AA* parent and an *a*-carrying gamete from the recessive *aa* parent. Since the hybrids carry contrasting alleles, they are termed **heterozygous** for these genes. The *Aa* heterozygote expresses the phenotype of the dominant parent, thus indicating that the presence of only one dominant *A* allele is sufficient to completely mask the presence of the recessive allele. Thus, both the *AA* homozygote and the *Aa* heterozygote will produce the dominant phenotype, even though they have a different genotype. The dominant phenotype is therefore only an outward manifestation of the presence of at least one dominant allele. It is not a very reliable indication of an individual's genotype.

Mendel's model has three important features: (1) each

form of a trait is determined by a specific form of a gene called an allele, (2) every individual carries two copies of the gene for a given trait, and (3) only one member of each pair of genes in an individual is transferred in a gamete to the next generation. These features are illustrated by the following sequence:

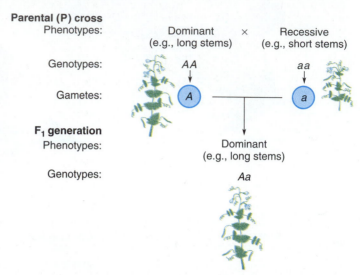

Parental (P) cross
Phenotypes: Dominant (e.g., long stems) × Recessive (e.g., short stems)

Genotypes: *AA*   *aa*

Gametes: *A*   *a*

F₁ generation
Phenotypes: Dominant (e.g., long stems)

Genotypes: *Aa*

If the F₁ reproduce, each gamete will contain just one of the alleles—either *A* or *a*—chosen at random from the pair present in the heterozygote. This third feature of Mendel's model is known as Mendel's **principle of genetic segregation**, which is formally stated as follows:

Rule THE PRINCIPLE OF GENETIC SEGREGATION

In the formation of gametes, the members of a pair of alleles separate (or segregate) cleanly from each other, so that only one member is included in each gamete.

The two types of gametes of an F₁ heterozygote are thus equally likely to occur, that is, each makes up one-half of the total gametes produced by the plant:

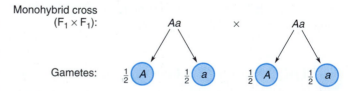

Monohybrid cross
(F₁ × F₁):   *Aa*   ×   *Aa*

Gametes: $\frac{1}{2}$ *A* $\frac{1}{2}$ *a* $\frac{1}{2}$ *A* $\frac{1}{2}$ *a*

Upon fertilization, the random combination of these gamete types will produce a genotype distribution among the progeny (the F₂) that is given by the product of the gamete proportions ($\frac{1}{2}A + \frac{1}{2}a$):

$$(\tfrac{1}{2}A + \tfrac{1}{2}a) \times (\tfrac{1}{2}A + \tfrac{1}{2}a) = \tfrac{1}{4}AA + \tfrac{1}{2}Aa + \tfrac{1}{4}aa$$

The F₂ genotypic ratio of 1 *AA* : 2 *Aa* : 1 *aa* expresses itself as a phenotypic ratio of three-fourths dominant ($\frac{1}{4}$ *AA* + $\frac{1}{2}$ *Aa*) to one-fourth recessive ($\frac{1}{4}$ *aa*). The phenotypic ratio can be further abbreviated to 3 *A*− : 1 *aa*, where *A*− = *AA* +

Monohybrid cross (F₁ × F₁):

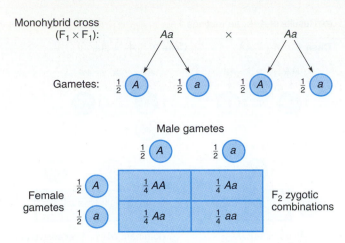

▶ **FIGURE 2–3** The segregation of alleles in a monohybrid cross. The ways that gametes can combine during fertilization are shown in the Punnett square.

Aa and is a shorthand way of designating the dominant phenotype.

When more detailed calculations are involved, it is often convenient to analyze the results of a genetic cross using a **Punnett square.** ▶ Figure 2–3 shows this approach for a monohybrid cross.

Progeny Testing: Detecting Heterozygotes

At this point, we can test the genetic interpretations of Mendel's results by predicting what would happen if the F_2 plants are subjected to further mating experiments. For example, if the F_2 plants with the recessive phenotype are allowed to self-pollinate, their F_3 progeny should all show the recessive phenotype, since the cross *aa* × *aa* yields only *aa* offspring. On the other hand, selfing the F_2 plants with the dominant phenotype should yield an F_3 of all dominant offspring or part dominant and part recessive offspring, depending on which of the dominant F_2 plants are being selfed. These outcomes are shown as follows:

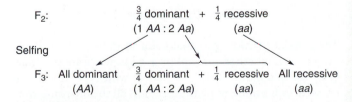

One-third of the dominant F_2 plants (one-fourth of the total F_2) are of the pure-breeding genotype *AA*. When selfed, these plants should produce only dominant progeny in the F_3. The other two-thirds of the dominant F_2 individuals have the *Aa* genotype and should yield a progeny of three-fourths dominant to one-fourth recessive when selfed. Mendel's data on F_3 generations confirmed these predictions.

In the preceding example, a procedure of **progeny test-**ing was used to verify Mendel's predictions. This procedure uses genetic crosses to evaluate the genotype of an individual based on the phenotypes of its offspring. Progeny tests are used extensively in breeding programs to determine whether a dominant individual is homozygous for the dominant allele or is a heterozygous carrier of the recessive allele.

Some examples of the results of progeny tests in maize (corn) are shown in ▶ Figure 2–4. An ear of corn is in many respects an ideal subject for analysis of genetic ratios. Each kernel represents a separate individual and is the result of an independent fertilization event. Therefore, as many as one thousand progeny from a single cross are conveniently located on one ear. The kernels must be planted to express the characters of the mature plant, but many character differences such as those affecting seed shape and seed color can be identified by direct examination of the kernel phenotypes. Progeny ratios can then be obtained by

▶ **FIGURE 2–4** *Zea mays* (maize, corn) showing (a) a 3:1 ratio of purple to yellow and (b) a 1:1 ratio of purple to yellow. The 3:1 ratio arose from a selfing of an F_1 hybrid. The 1:1 ratio was the result of a testcross of the F_1 hybrid.

(a)

(b)

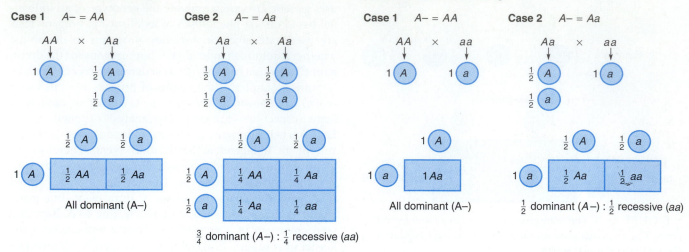

(a) Results of A– × Aa matings

Case 1   A– = AA

AA × Aa

$1\ A$ $\frac{1}{2}\ A$
$\frac{1}{2}\ a$

$\frac{1}{2}\ A$ $\frac{1}{2}\ a$

$1\ A$ | $\frac{1}{2}\ AA$ | $\frac{1}{2}\ Aa$

All dominant (A–)

Case 2   A– = Aa

Aa × Aa

$\frac{1}{2}\ A$ $\frac{1}{2}\ A$
$\frac{1}{2}\ a$ $\frac{1}{2}\ a$

$\frac{1}{2}\ A$ $\frac{1}{2}\ a$

$\frac{1}{2}\ A$ | $\frac{1}{4}\ AA$ | $\frac{1}{4}\ Aa$
$\frac{1}{2}\ a$ | $\frac{1}{4}\ Aa$ | $\frac{1}{4}\ aa$

$\frac{3}{4}$ dominant (A–) : $\frac{1}{4}$ recessive (aa)

(b) Results of A– × aa matings

Case 1   A– = AA

AA × aa

$1\ A$ $1\ a$

$1\ A$

$1\ a$ | $1\ Aa$

All dominant (A–)

Case 2   A– = Aa

Aa × aa

$\frac{1}{2}\ A$ $1\ a$
$\frac{1}{2}\ a$

$\frac{1}{2}\ A$ $\frac{1}{2}\ a$

$1\ a$ | $\frac{1}{2}\ Aa$ | $\frac{1}{2}\ aa$

$\frac{1}{2}$ dominant (A–) : $\frac{1}{2}$ recessive (aa)

▶ **FIGURE 2–5** Progeny tests that provide the criteria used to distinguish heterozygotes from dominant homozygotes. In (a) a dominant individual of unknown genotype is mated to a known heterozygote. In (b) the dominant individual is mated to one that is recessive. In both crosses, the unknown genotype is judged to be heterozygous if recessive offspring are produced.

merely counting the number of each type of kernel on individual ears of corn. For example, Figure 2–4 shows two kernel phenotypes, purple and yellow, with purple being dominant over yellow. If a dominant plant is heterozygous, self-pollination ($Aa \times Aa$) will yield a phenotypic ratio among the kernels of 3 purple : 1 yellow. The dominant plant can be further tested by crossing it to one that normally produces yellow kernels. (This cross is equivalent to $Aa \times aa$). A progeny ratio of 1 purple : 1 yellow is then expected, since half of the kernels will receive the recessive allele from both parents. If the dominant plant had been of a pure-breeding line (AA), neither mating would have produced yellow kernels.

As we have seen in this example, two types of matings can be used for progeny testing. In one type, a dominant individual suspected of being heterozygous is mated to a known carrier of the recessive allele ($A– \times Aa$) or to itself ($A– \times A–$) if it is normally a self-fertilizing plant. This approach is generally taken when individuals homozygous for the recessive allele are inviable, infertile, or otherwise uneconomical to produce. In the second type of mating, the dominant individual is crossed with a recessive homozygote ($A– \times aa$). Any dominant × recessive mating such as this is called a **testcross**; a testcross is often used to test whether dominant individuals in the cross are heterozygous or homozygous. The criteria for distinguishing heterozygotes from homozygotes in the two types of matings are shown in ▶ Figure 2–5.

Example 2.1

Example 2.1

In cattle, the polled (hornless) condition ($H–$) is dominant over the presence of horns (hh). A breeder who wishes to establish a pure-breeding herd of hornless cattle purchases a polled bull. To test whether the bull is a carrier of the recessive gene, he is mated to horned cows. The first mating produces an offspring that develops horns. (a) What is the genotype of the bull? (b) What percent of offspring from such matings is expected to be polled? (c) If the bull is mated only to cows that are known carriers of the recessive gene, what percent of such matings would be expected to give rise to horned calves?

Solution: (a) The bull must be heterozygous (Hh), since the recessive offspring (hh) must have received an h allele from both parents. None of the offspring could be horned if the bull had the HH genotype. (b) The mating is $Hh \times hh$. One-half (or 50 percent) of the offspring of such matings will be heterozygous (Hh) or polled. (c) The mating would then be $Hh \times Hh$. One-fourth (or 25%) of the offspring of such matings, on the average, will be horned (hh).

Follow-Up Problem 2.1

The standard (wild-type) coat color in mink is dark black-brown. Several other coat colors are known. One is a blue-gray (or platinum) color. When a wild-type mink was crossed with a mink of platinum color, all the offspring were wild type. Intercrossing the F_1 mink resulted in an F_2 consisting of 33 wild-type and 10 platinum animals. (a) Explain these results in terms of a single pair of genes. (b) How many wild-type mink in the F_2 are expected to be heterozygous?

Pedigree Analysis

Although many organisms such as peas and corn are well suited to genetic analysis by means of controlled crosses,

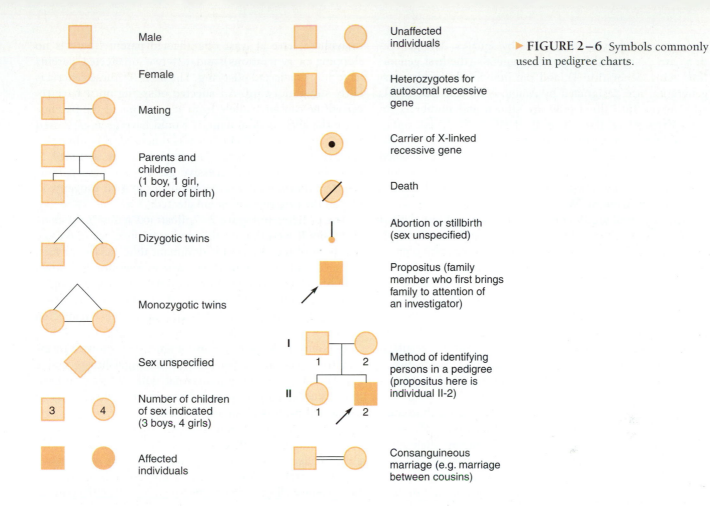

Male

Female

Mating

Parents and children (1 boy, 1 girl, in order of birth)

Dizygotic twins

Monozygotic twins

Sex unspecified

Number of children of sex indicated (3 boys, 4 girls)

Affected individuals

Unaffected individuals

Heterozygotes for autosomal recessive gene

Carrier of X-linked recessive gene

Death

Abortion or stillbirth (sex unspecified)

Propositus (family member who first brings family to attention of an investigator)

Method of identifying persons in a pedigree (propositus here is individual II-2)

Consanguineous marriage (e.g. marriage between cousins)

many other species are not. In these cases, performing controlled crosses may be improper (as in humans) or impractical because of long generations and small numbers of offspring. When information is available on the inheritance of a trait in two or more generations of a family or line of descent, one of the best practical alternatives to the study of controlled crosses is **pedigree analysis**. A pedigree is a diagram that represents two or more generations of related individuals and shows the transmission pattern for a particular genetic characteristic. To analyze a pedigree, one simply traces the history of a character back through a series of established matings in the hope of determining the mechanism of inheritance.

The standard symbols used in a pedigree or "family tree" are given in ► Figure 2−6. Many of these symbols also appear in the hypothetical four-generation human pedigree in ► Figure 2−7. This pedigree includes eight **affected** individuals, who are represented by the darker symbols. Females

► FIGURE 2−7 Hypothetical pedigree illustrating the symbols used in pedigree construction. Affected individuals in this pedigree express a dominant trait.

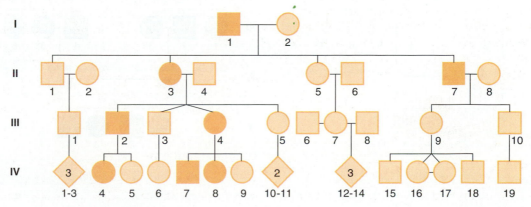

are represented by circles, males by squares. The generations are designated by Roman numerals (the first generation being generation I), and the individuals within each generation are designated by Arabic numbers. Individuals who marry into the family are shown and numbered in some cases (II–2, II–4, II–6, II–8, III–6, III–8) but not in others (mates of all members of generation III, except for III–7). Individuals who marry into the family are assumed to be unaffected unless there is definite evidence to the contrary, and for this reason they are often omitted to save space in the diagram. Within each group of siblings, individuals are numbered in order of birth from oldest (lowest number) to youngest (higher number). Individuals IV–1, IV–2, and IV–3 in Figure 2–7 have the same phenotype, as do individuals IV–10 and IV–11 and individuals IV–12, IV–13, and IV–14; they are grouped together under one symbol to conserve space. Siblings who are attached to the parental line at the same point are twins. Twins without a connecting line between them, such as III–3 and III–4, are fraternal (dizygotic) twins, which result from the fertilization of two eggs; twins with a connecting line between them, such as IV–16 and IV–17, are identical (monozygotic) twins, which arise from a single egg.

The pedigrees of some simple traits show inheritance patterns that are consistent with the transmission of a dominant gene. If the trait is caused by a dominant allele, then inheritance of that gene from one or both parents will produce the characteristic. Thus, an individual affected with a dominant characteristic can be heterozygous or homozygous for the causative allele. Two indications that a characteristic is caused by a dominant gene are (1) all affected

individuals have at least one affected parent (there is no skipping of generations) and (2) two unaffected parents have only unaffected offspring. The basis for these criteria is fairly straightforward. An affected offspring must have received the causative allele from at least one of the parents. Since the allele is dominant, it would have been expressed in the parent(s) from which it was inherited. If neither parent exhibits the characteristic, then both are assumed to have the homozygous recessive genotype. These parents can then transmit only recessive alleles to their progeny, so that all the progeny will be unaffected.

The pedigree in Figure 2–7 illustrates a dominant characteristic. If we examine this pedigree once again, we see that both of the criteria for dominant inheritance are met. All seven affected individuals whose ancestry is known have an affected parent. Furthermore, unaffected couples (which include family members II–1, II–5, III–1, III–3, III–5, III–7, III–9, and III–10) have only unaffected offspring.

In contrast, the pedigree in ▶ Figure 2–8 shows a recessive trait. The major feature that distinguishes recessive from dominant inheritance is that with recessive inheritance, unaffected parents can have affected offspring. The unaffected parents of an affected child would both be heterozygous. We would expect one-fourth of the offspring of such a mating to be affected. As we can see from the pedigree, unaffected parents IV–1 and IV–2 have affected offspring. Therefore, IV–1 and IV–2 must both be carriers of the recessive allele. The double line connecting the parents in generation IV means that they are related (in this case, second cousins).

▶ FIGURE 2–8 (a) A pedigree showing the inheritance of phenylketonuria (PKU), a human biochemical disorder caused by a recessive gene, *a*. (b) The pedigree with genotypes of the individuals given in the pedigree symbols.

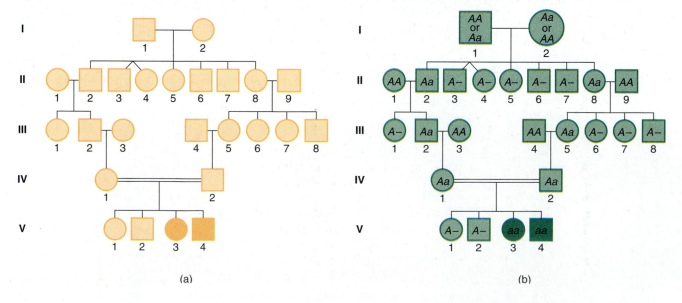

(a)

(b)

Example 2.2

A notch in the tip of the ear is the expression of a gene in Ayrshire cattle. In the following pedigree, the darker symbols represent individuals with notched ears and the lighter symbols represent individuals with normal ears. (a) Is the gene for notched ears dominant or recessive to the allele for normal ears? (b) Assume that III–2 and III–6 are mated. What is the chance that an offspring from this mating will have notched ears?

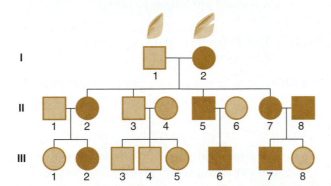

Solution: (a) Dominant. Parents II–7 and II–8, both with notched ears, produce an offspring with normal ears. This is possible only in the case of a monohybrid cross ($Aa \times Aa$) in which both parents are heterozygous for the dominant trait. (b) Since both III–2 and III–6 have a recessive parent, both are heterozygous, and their chance of having a dominant offspring is $\frac{3}{4}$.

The following is a pedigree of a fairly common human hereditary trait, which is expressed by individuals represented by the shaded symbols:

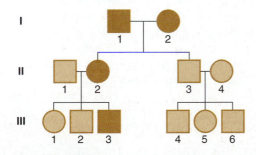

(a) Indicate whether the trait is dominant or recessive. (b) Using A to represent the dominant allele and a to represent the recessive allele, give the probable genotypes of the individuals in the pedigree.

Most pedigrees, particularly those involving humans, include a relatively small number of individuals. The number of offspring produced in a generation is therefore usually too small for the observed phenotypic ratio to be a reliable indicator of expected results. Substantial departures can arise simply owing to chance. However, if several different pedigrees for the same trait are examined and the data are pooled, all but one inheritance pattern can often be excluded. Pedigree analysis can therefore provide an important means of elucidating inheritance mechanisms.

To Sum Up

1. According to the basic Mendelian model of inheritance, genes are assumed to occur in pairs within the cells of higher organisms. The alternative forms of a gene are known as alleles.
2. The principle of genetic segregation holds for all sexually reproducing organisms. It states that only one member of a pair of alleles can enter a particular gamete. Fertilization then restores the genes to pairs.
3. Six different kinds of crosses are possible for a single pair of alleles. These crosses and the offspring they produce are as follows:

$$AA \times AA \rightarrow \text{all } AA$$
$$AA \times aa \rightarrow \text{all } Aa$$
$$AA \times Aa \rightarrow \tfrac{1}{2}AA : \tfrac{1}{2}Aa$$
$$Aa \times aa \rightarrow \tfrac{1}{2}Aa : \tfrac{1}{2}aa$$
$$Aa \times Aa \rightarrow \tfrac{1}{4}AA : \tfrac{1}{2}Aa : \tfrac{1}{4}aa$$
$$aa \times aa \rightarrow \text{all } aa$$

4. Progeny testing is used to determine whether a dominant individual is homozygous for the dominant allele or is a heterozygous carrier of the recessive allele. The dominant individual of unknown genotype is usually mated to either a known heterozygote or a recessive homozygote. In either case, recessive offspring can be produced only if the dominant individual is heterozygous. Progeny testing that involves a dominant × recessive mating is referred to as a testcross.
5. A pedigree illustrates data on a family to show the inheritance pattern of a genetic characteristic. In many cases, this pattern has features that suggest a dominant or recessive basis for the characteristic.

Points to Ponder 2.1

How much do you know about the genetic history of your family? Do you want to know more about that genetic history? Why or why not?

If a person is found to have a recessive genetic disorder, should all members of his or her family be required to have their genotype tested so that a family pedigree can be constructed? Should such data become part of the medical history of each family member? Who should be allowed to have access to such data? For example, should you have the right to inspect the genetic history of your future spouse?

Do you think that the "average" person understands enough about basic genetic principles to know how to interpret and apply information about their genes?

THE PRINCIPLE OF INDEPENDENT ASSORTMENT

Mendel did not restrict his analysis to the inheritance of single alternative traits. He also determined the ratios produced when two separate character differences are considered simultaneously in the same cross. For example, in one of his crosses he studied the inheritance of both seed coat texture (round vs. wrinkled) and seed leaf cover (yellow vs. green). The purpose of these experiments was to determine the influence that the inheritance of one trait might have on the transmission of another.

Different traits are specified by different genes and we indicate the genes using different letters. For example, if we let a be the recessive allele for wrinkled seeds and A the dominant allele for round seeds, we can similarly let b represent the recessive allele for green seeds and B represent the dominant allele for yellow seeds. The cross of pure-breeding round, yellow-seeded plants with wrinkled, green-seeded plants can then be represented as $AABB \times aabb$. According to Mendel's principle of genetic segregation, each gamete will contain only one member of each pair of alleles. A gamete must always contain one representative of each pair of genes that the parent possesses, so that the union of gametes through fertilization restores the proper number of genes in the next generation. The following sequence illustrates this point for two pairs of alleles:

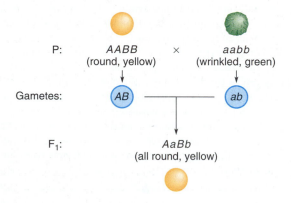

P: AABB   ×   aabb
(round, yellow)   (wrinkled, green)

Gametes: AB ————— ab

F₁: AaBb
(all round, yellow)

Note that the F₁ produced from this cross are **dihybrids**, because they are hybrid for two different pairs of contrasting characters.

Mendel continued the crossing sequence by allowing the dihybrid F₁ plants to self-pollinate, resulting in matings of the form $AaBb \times AaBb$. Since a gamete receives only one of each pair of alleles, four kinds of gametes will be formed by the F₁: AB, Ab, aB, and ab. What we do not know at this point is the expected proportion of each type. Are all gamete types equally likely to occur ($\frac{1}{4}$ or 25% each), or are certain gene combinations formed more frequently than others? For example, is there perhaps a tendency for gametes to contain all dominant or all recessive genes, so that AB and ab gametes are more likely than Ab and aB types? After all, the original parents were totally dominant and totally recessive.

The simplest hypothesis, which turns out to be the correct one in this case, is that nonallelic genes segregate independently of each other. The four gene combinations are equally likely to occur in gametes. Each gamete is formed in the F₁ as though we had randomly chosen an A or a allele and then randomly picked a B or b allele to go into the gamete with it. Whether the A or a allele is chosen has no bearing on whether the B or b allele is picked. Each possible combination of nonallelic genes is equally likely. This example illustrates Mendel's **principle of independent assortment**, which is stated formally as follows:

Rule   THE PRINCIPLE OF INDEPENDENT ASSORTMENT

Nonallelic genes segregate into gametes independently of each other, producing equal proportions of all possible gamete types.

Upon fertilization, the random combination of the gamete types in our example will produce a genotype distribution among the F₂ that is given by the product of the gamete proportions ($\frac{1}{4} AB + \frac{1}{4} Ab + \frac{1}{4} aB + \frac{1}{4} ab$):

$$(\tfrac{1}{4} AB + \tfrac{1}{4} Ab + \tfrac{1}{4} aB + \tfrac{1}{4} ab) \times (\tfrac{1}{4} AB + \tfrac{1}{4} Ab + \tfrac{1}{4} aB + \tfrac{1}{4} ab)$$
$$= \tfrac{1}{16} AABB + \tfrac{2}{16} AABb + \tfrac{1}{16} AAbb + \tfrac{2}{16} AaBB + \tfrac{4}{16} AaBb$$
$$+ \tfrac{2}{16} Aabb + \tfrac{1}{16} aaBB + \tfrac{2}{16} aaBb + \tfrac{1}{16} aabb$$

yielding a genotypic ratio of 1:2:1:2:4:2:1:2:1. As is shown in ▶ Figure 2−9, this product yields 16 (4×4) possible zygote combinations but only 9 different genotypes, since some genotypes occur more than once.

Because of dominance, the phenotypic ratio in the F₂ is 9 round, yellow (1 $AABB$ + 2 $AABb$ + 2 $AaBB$ + 4 $AaBb$):3 round, green (1 $AAbb$ + 2 $Aabb$):3 wrinkled, yellow (1 $aaBB$ + 2 $aaBb$):1 wrinkled, green (1 $aabb$). If we use a dash (−) to indicate that the second gene can be either dominant or recessive and still allow the first gene to produce a dominant phenotype for that pair, the phenotypic ratio can be expressed as 9 $A-B-$:3 $A-bb$:3 $aaB-$:1 $aabb$. The F₂ ratios of genotypes and phenotypes that result from the cross $AaBb \times AaBb$ are summarized in ■ Table 2−2.

Example 2.3

Coat color and spotting pattern in cocker spaniels depend on two gene pairs that segregate independently. A black coat color ($B-$) is dominant to red (bb) and solid color ($S-$) is dominant to white spotting (ss). A red-and-white spotted female ($bbss$) produces a litter of five pups: two red-and-white spotted, one black-and-white spotted, one solid red, and one solid black. Deduce the genotype and phenotype of the male parent in this cross.

Solution: The five pups have the following genotypes: red-and-white spotted = $bbss$, black-and-white spotted =

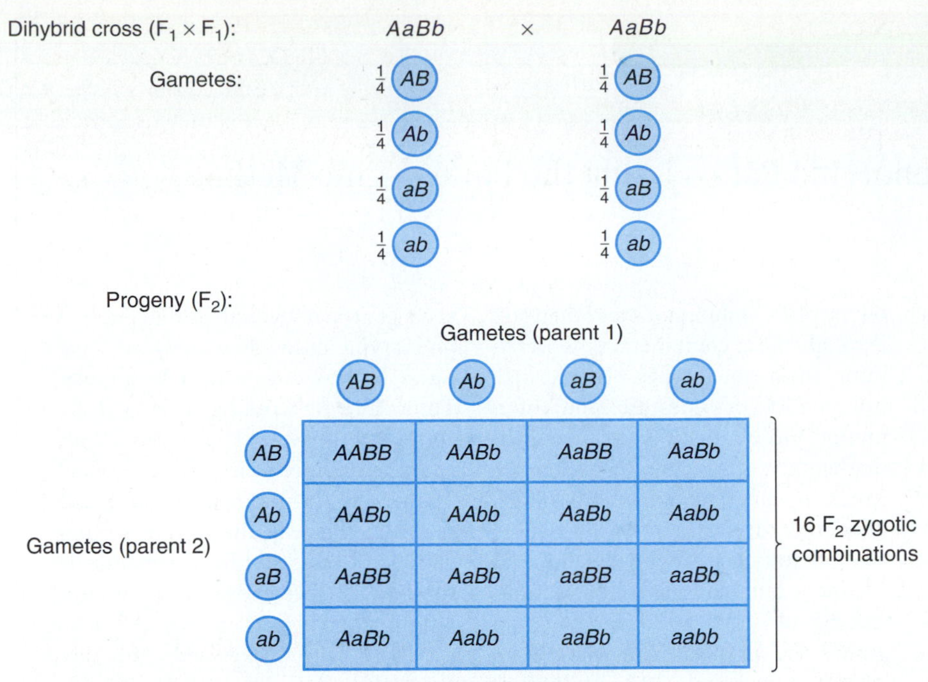

Dihybrid cross ($F_1 \times F_1$): $AaBb$ × $AaBb$

$B{-}ss$, solid red = $bbS{-}$, and solid black = $B{-}S{-}$. Observe that the solid black pup has at least one B and at least one S allele. Since the mother is $bbss$, the B and S alleles had to come from the male parent. Based on this information, we deduce that the genotype of the father is $B{-}S{-}$. The red-and-white spotted pups must have received both a b and an s allele from each parent, so the father must also carry a copy of both the b and s alleles. We now know that the male parent must be $BbSs$ in genotype and solid black in color.

Follow-Up Problem 2.3

In poultry, comb shape (rose or single) and feather color (black or red) are governed by different pairs of genes that segregate independently of each other. Matings between a rose-comb, black-feathered hen and a rose-comb, black-feathered rooster produced the following offspring: 45 with rose comb and black feathers, 16 with rose comb and red feathers, 15 with single comb and black feathers, and 4 with single comb and red feathers. (a) Which traits are dominant? (b) Using the first two letters of the alphabet for comb shape and feather color, respectively, give the genotypes of the rose-comb, black-feathered parents of these offspring.

Confirming Independence

The assumption that nonallelic genes associate independently during gamete formation was verified by Mendel's analysis of the F_2 results. His actual dihybrid cross ($AaBb \times AaBb$) for these traits produced 556 offspring as follows:

315 round, yellow	101 wrinkled, yellow
108 round, green	32 wrinkled, green

The ratios 315 to 101 and 315 to 108 are about 3 to 1, and the ratio 315 to 32 is about 9 to 1. The overall phenotypic ratio in Mendel's experiment is therefore 9:3:3:1, in agreement with predicted results. If we inspect the data more closely, however, we also find that each trait, when considered separately of the other, produces a 3:1 phenotypic ratio. The total number of round-seeded plants produced is $315 + 108 = 423$, and the total number of wrinkled-seeded plants produced is $101 + 32 = 133$, for a ratio of 423:133, or about 3:1. Similarly, the total number of yellow-seeded plants produced is $315 + 101 = 416$, and the number of green-seeded plants is $108 + 32 = 140$, again a ratio of about

■ **TABLE 2−2** Genotypic and phenotypic ratios resulting from the cross $AaBb \times AaBb$.

Genotype	Genotypic Proportions	Phenotype	Phenotypic Proportions
$AABB$	$\frac{1}{16}$		
$AABb$	$\frac{2}{16}$	round, yellow	$\frac{9}{16}$
$AaBB$	$\frac{2}{16}$		
$AaBb$	$\frac{4}{16}$		
$AAbb$	$\frac{1}{16}$	round, green	$\frac{3}{16}$
$Aabb$	$\frac{2}{16}$		
$aaBB$	$\frac{1}{16}$	wrinkled, yellow	$\frac{3}{16}$
$aaBb$	$\frac{2}{16}$		
$aabb$	$\frac{1}{16}$	wrinkled, green	$\frac{1}{16}$

Evaluating Phenotypic Ratios Using the Forked-Line Method

Calculating genotypic and phenotypic ratios using the Punnett square becomes extremely cumbersome when more than two gene pairs are involved. For this reason, other methods that permit more rapid solution of genetic problems have been devised. One useful approach is the forked-line method, a branch-diagram method for bringing together different combinations of independently assorting gene pairs. For example, suppose that we wanted to determine the phenotypic ratio produced by a trihybrid cross, $AaBbCc \times AaBbCc$. If we were to approach this problem using the Punnett square, we would have to construct a square with 64 zygotic combinations and then sort through these combinations to determine their phenotypes. The forked-line method provides the same information more quickly by applying the mathematical principle that when the genes for different traits behave independently during gamete formation, the phenotypic proportions for the various combinations of these traits can be obtained by multiplying together the proportions expected for each trait separately. (The multiplication rule of probability, which applies in this situation, is discussed in Chapter 3.) An application of this principle is shown in the table.

It is assumed in this method that each gene pair would produce a $3:1$ phenotypic ratio when analyzed separately by way of a monohybrid cross. Thus, $Aa \times Aa$ would give $3 A-:1 aa$, and $Bb \times Bb$ would give $3B-:1 bb$; when combined these would yield $9 A-B-:3 A-bb:3 aaB-:1 aabb$, and so on. In this way, the various combinations of traits can be systematically combined to give an overall phenotypic ratio for a trihybrid cross of $27:9:9:9:3:3:3:1$. As expected, this approach can be extended to any number of independently assorting gene pairs and can be used to obtain genotypic ratios as well as phenotypic ratios.

■ An application of the multiplication rule of probability.

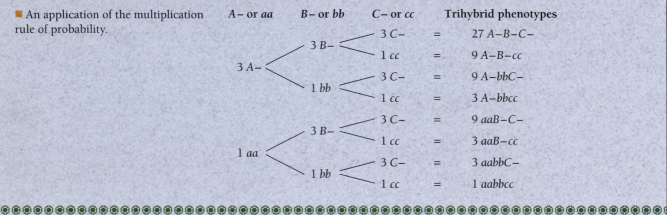

3:1. The 9:3:3:1 phenotypic ratio is thus seen to be the product of the two separate 3:1 ratios:

(3 round + 1 wrinkled) × (3 yellow + 1 green)
  = 9 round, yellow + 3 round, green + 3 wrinkled, yellow + 1 wrinkled, green

Recognizing that the 9:3:3:1 ratio is the mathematical product of the ratios for the two separate traits led Mendel to state the principle of independent assortment. He recognized that *multiplication* of the separate proportions for the individual traits to obtain the overall proportions for both traits considered together means that the different traits are behaving independently of each other. We will formally state this multiplication rule in the next chapter; however, it is helpful to point out its usefulness now so that genetic crosses involving independently assorting genes can be more quickly and easily worked through than they would be with Punnett squares. The Extensions and Techniques entitled "Evaluating Phenotypic Ratios Using the Forked-Line Method" outlines the reasoning followed in working crosses using the multiplication rule.

Mendel also used a testcross between a dihybrid ($AaBb$) and a double recessive homozygote to confirm independent assortment. The recessive parent (called the *testcross parent*) produces only the doubly recessive type of gamete, so the phenotypic ratio among the offspring will therefore depend

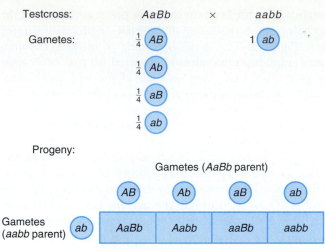

Testcross: AaBb   ×   aabb

Gametes: $\frac{1}{4}$ AB   1 ab

                $\frac{1}{4}$ Ab

                $\frac{1}{4}$ aB

                $\frac{1}{4}$ ab

Progeny:

Gametes (*AaBb* parent)

	AB	Ab	aB	ab
Gametes (*aabb* parent) ab	AaBb	Aabb	aaBb	aabb

Progeny genotypic ratio = progeny phenotypic ratio = 1:1:1:1

▶ **FIGURE 2–10** The independent assortment of nonallelic genes in a testcross in which the doubly dominant parent is dihybrid. The ways that gametes can combine during fertilization are shown in the Punnett square. Each possible zygotic combination in this cross constitutes $\frac{1}{4}$ of the total progeny.

---

solely on the ratio of gametes produced by the dihybrid parent. If nonallelic gene pairs segregate into gametes independently, then the four gamete types will be equally frequent, as will the four offspring classes produced upon fertilization. The predicted results are shown in ▶ Figure 2–10. When Mendel performed the testcross, his results were 55 round, yellow:51 round, green:49 wrinkled, yellow:52 wrinkled, green. These results are sufficiently close to the theoretical 1:1:1:1 ratio to confirm independent assortment. If the nonallelic gene pairs had not assorted independently, then some ratio other than 1:1:1:1 would have resulted.

To Sum Up

1. The principle of genetic segregation describes the distribution of allelic genes into gametes, and the principle of independent assortment describes the distribution of nonallelic genes. The latter principle states that members of different gene pairs behave independently of each other during gamete formation. In other words, nonallelic genes do not influence each other during their segregation into gametes. As a result, all of the different gamete genotypes are equally frequent.
2. If the nonallelic genes assort independently, a dihybrid cross typically gives a 9:3:3:1 phenotypic ratio in the progeny, and a dihybrid testcross gives a progeny ratio of 1:1:1:1. Thus, in addition to testing for the presence of a recessive gene in the heterozygous condition, a testcross can also be used to test for independent assortment of nonalleles.

▦ CHROMOSOMES AND MENDEL'S PRINCIPLES

Mendel's principles of inheritance were ignored for many years after their publication. Then in 1900 three botanists, Hugo de Vries of Holland, Carl Correns of Germany, and Erich von Tschermak-Seysenegg of Austria, simultaneously discovered Mendel's paper on heredity, recognized its importance, and cited it in their own publications. By this time much more was known about the structure of cells and the nature of chromosomes. The cytological work carried out in the nineteenth century had examined cell division in detail and had accumulated enough information to explain Mendel's observations in terms of the behavior of chromosomes. A clear parallel began to emerge between the numerical properties of chromosomes and those of genes, a parallel which you can understand by relating the description of chromosomes and meiosis discussed in Chapter 1 to the Mendelian principles discussed in this chapter. The three main properties linking chromosomes with genes are:

1. Chromosomes occur in pairs; so do genes.
2. Chromosomes of each pair segregate into different gametes; so do allelic genes.
3. Chromosomes of different pairs assort (align) independently; so do different pairs of nonallelic genes.

In brief, chromosomes obey Mendel's rules of inheritance. Thus, by studying the patterns of gene transmission, Mendel had in effect discovered the rules governing chromosome transmission, without knowing anything about chromosomes. This parallel behavior of genes and chromosomes led Walter Sutton and Theodor Boveri to propose in 1902 that genes are carried on chromosomes, a proposal that later became known as the chromosome theory of inheritance.

Meiotic Basis of Mendel's Rules

The pairing and disjunction of homologs during meiosis are responsible for the Mendelian behavior of chromosomes. Since genes are contained along the DNA of chromosomes, these pairing and disjunction events are also responsible for the Mendelian behavior of genes. For example, suppose that gene *A* is located on some chromosome of maternal

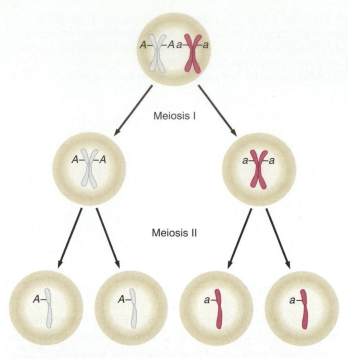

Meiosis I

Meiosis II

▶ **FIGURE 2–11** The chromosomal basis for the segregation of alleles. Maternally derived chromosomes are shown in white, paternally derived chromosomes in color. Because of chromosomal disjunction, only one member of each pair of alleles will enter a single gamete.

origin and its allele *a* is at the same position (**locus**) on the homologous chromosome of paternal origin, as illustrated in ▶ Figure 2–11. At the beginning of meiosis, chromosome duplication has already occurred, so that two copies of each allele exist, one on each of the two sister chromatids. Because only one of the four chromosome strands of the bivalent can enter a single gamete, we can readily see that the normal disjunction of homologs results in the segregation of alleles.

Normal pairing and disjunction of chromosomes also account for the independent assortment of nonallelic genes. To illustrate, let us consider the dihybrid *A/a B/b*, where the slashes (/) are used to indicate that the two pairs of genes are located on different pairs of homologous chromosomes. ▶ Figure 2–12 shows the alignment of both pairs of chromosomes at metaphase I. At this stage of meiosis, bivalents align with the maternally and paternally derived centromeres arranged at random on either side of the spindle equator. Two arrangements of nonallelic genes are therefore possible: one in which *A* and *B* go to one pole and *a* and *b* to the other, and another in which *A* and *b* go to one pole and *a* and *B* to the other. Since both arrangements occur with equal likelihood, the haploid cells produced by meiosis will have a genotype distribution of $\frac{1}{4}$ *AB*, $\frac{1}{4}$ *Ab*, $\frac{1}{4}$ *aB*, and $\frac{1}{4}$ *ab*. Thus, the independent alignment of each pair of chromosomes on the metaphase spindle results in the independent assortment of nonallelic genes.

▶ **FIGURE 2–12** The chromosomal basis for independent assortment of nonallelic gene pairs. The figure shows two alternative alignments of nonhomologous chromosomes at metaphase I. Since the alignments are strictly random, all four gamete types (*AB*, *Ab*, *aB*, and *ab*) are expected to occur in equal frequency.

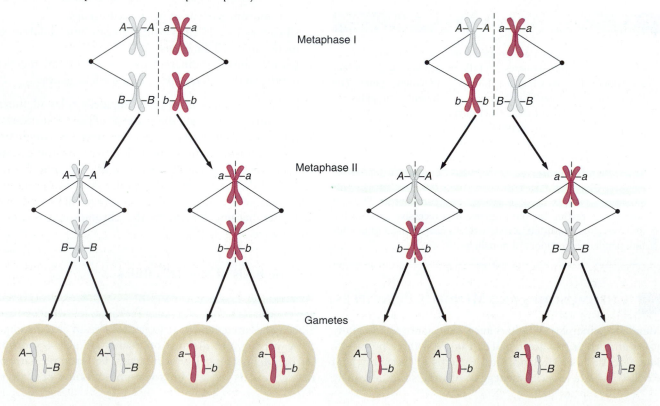

Metaphase I

Metaphase II

Gametes

Genomic Imprinting

Until recently geneticists assumed that once chromosomes from a sperm and egg combined to form the diploid genotype of the developing embryo, the two sets of genes work equally in determining the phenotype of the resulting offspring. Recently, however, human geneticists have realized that some genes function differently depending on whether they have been inherited from the mother or from the father. This phenomenon is called **genomic imprinting**.

Several human genetic diseases, including fragile X syndome (see the Focus on Human Genetics in Chapter 9), certain forms of diabetes, Angelman syndrome, and Prader-Willi syndrome, develop according to which parent provides the mutant genes. Angelman and Prader-Willi syndromes, for example, are forms of mental retar-

dation that are caused by the deletion of the same region on chromosome 15. Their behavioral symptoms, however, are almost exact opposites. An individual affected by Prader-Willi syndrome has inherited both chromosome 15s from the mother as a result of an error in meiotic segregation and exhibits mental retardation, small hands and feet, obesity, and slow movement. By contrast, inheritance of both paternal chromosome 15s results in Angelman syndrome, which is characterized by more severe mental retardation, a large mouth, red cheeks, and hyperactivity.

Research with insects in the 1960s first revealed genetic imprinting, but there was little interest in the subject until it was discovered in mammals in the 1980s. Perhaps the clearest evidence that there is more to diploidy

than just having two sets of chromosomes comes from mammalian embryos that have been manipulated so that both sets of chromosomes (both genomes) are either maternal or paternal. These diploid embryos do not develop to full term. In humans, pregnancies occasionally result from eggs that incorporate two paternal genomes but lose the maternal genome. These embryos, known as hydatidiform moles, die early in gestation and are characterized by overdeveloped placentas and underdeveloped embryos.

Researchers are now investigating the molecular modifications that occur in DNA in female versus male parents and govern the expression of that DNA in the next generation. We will have more to say about the possible mechanism of genomic imprinting in Chapter 20.

To Sum Up

The meiotic basis of Mendel's principles is seen in the parallel behavior of genes and chromosomes in meiosis I. The separation of members of each homologous pair of chromosomes during meiotic anaphase I results in the segregation of alleles. The independent alignment of each pair of homologs on the spindle at meiotic metaphase I followed by disjunction accounts for the independent assortment of nonallelic genes present on different chromosome pairs.

 EXCEPTIONS TO MENDEL'S PRINCIPLES

We have seen that the chromosome theory of inheritance provides a cytological basis for Mendel's rules. It also provides an explanation for certain important exceptions to Mendel's principles. Two major exceptions that were discovered shortly after the development of the chromosome theory concern the inheritance of nonallelic genes that are located on the same chromosome (**linked inheritance**) and the inheritance of genes that are located on chromosomes outside of the nucleus (**extranuclear inheritance**). In both

of these situations the patterns of gene transmission depart in important respects from the inheritance patterns observed by Mendel.

Linked Genes

In advocating the chromosome theory of inheritance, Sutton pointed out the possibility that there could be many more genes in an organism than there are chromosomes. Of course, we now know that the DNA of a single chromosome contains many different genes, each present at a different position (locus) on the chromosome. Genes that are found on the same chromosome but at different loci are said to be **linked**. Linked genes, being on the same chromosome, cannot assort independently of each other. Instead, these nonallelic genes tend to remain together during meiosis and to be transmitted in the same combination in which they were inherited. For example, consider a dihybrid *AaBb* individual formed from the fusion of *AB* and *ab* gametes, where the nonallelic genes are linked. The gametes that this dihybrid produces are then also likely to be *AB* and *ab* in genotype. The results are similar for a dihybrid *AaBb* formed from the fusion of *Ab* and *aB* gametes, but in this case the nonallelic genes tend to remain together in the *Ab* and *aB* combinations.

The first reported experimental results demonstrating linkage were published in 1905 by William Bateson and Reginald C. Punnett. Bateson and Punnett discovered this exception to Mendel's principle of independent assortment while studying the joint transmission of two gene pairs in the sweet pea: one pair affecting flower color (P for purple and its allele p for red) and the other pair affecting the shape of the pollen grains (L for long and its allele l for round). Upon self-fertilizing the F_1 $PpLl$ dihybrids obtained from crosses of $PPLL$ and $ppll$ parents, they obtained a ratio of approximately 226 $P-L-$:17 $P-ll$:17 $ppL-$:64 $ppll$ in the F_2. These results were significantly different from the 9:3:3:1 ratio that would be expected for independent assortment of nonallelic genes because there were too many offspring in the $P-L-$ and $ppll$ classes and too few in the $P-ll$ and $ppL-$ classes. In this case, the dominant P and L genes appear to be linked in some way, as do the recessive p and l genes, these linked genes showing a tendency to remain together in the formation of gametes.

At the time these results were published, Bateson and Punnett did not understand the physical basis for linkage. The first modern interpretation of linkage was not given until around 1911, when the geneticist Thomas Hunt Morgan, who had also observed similar departures from Mendel's principles, proposed that linked genes tend to be inherited together because they are located on the same chromosome. Moreover, since the tendency for linked genes to remain together is not complete (that is, only partial linkage is shown), Morgan also proposed that occasional new gene combinations are formed through crossing over. An example of the effects of crossing over is illustrated in ▶ Figure 2–13. You might recall that crossing over is a breakage and reunion process that results in the physical exchange of homologous segments between nonsister chromatids of a bivalent. As we can see in Figure 2–13, crossing over is also a **recombination process**, since one of its important effects is to produce new combinations of nonallelic genes on each of the chromatids involved in the exchange. We conventionally refer to the new combinations of non-allelic genes that are produced by crossing over as **recombinant types**, and to the original combinations of nonallelic genes that were inherited from the parents and are located on the chromatids not involved in crossing over as **parental types**.

▶ Figure 2–14 provides an explanation for the experimental results of Bateson and Punnett in terms of chromosomal linkage and crossing over. If we hypothesize that the selfed dihybrids produced gametes in a ratio of 8 PL:8 pl: 1 Pl:1 pL, with the first two gamete types being the parental types and the last two being the recombinant types, then combining the gametes through random fertilizations produces the observed ratio of 226 $P-L-$:17 $P-ll$:17 $ppL-$: 64 $ppll$. An important observation about these results is that when each trait is considered separately, the ratio reduces to the familiar 3 dominant:1 recessive. For example, looking at the purple vs. red trait, we get 226 + 17 = 243 purple

and 64 + 17 = 81 red, for a ratio of 243:81 or 3:1. A comparison of long versus round also gives a 243:81 = 3:1 ratio. These ratios of 3:1 for each trait alone show that Mendel's principle of segregation of alleles still applies to the members of each allele pair, regardless of whether the nonallelic genes are linked or are on different chromosomes. However, the fact that we are dealing with linked genes does make Mendel's principle of independent assortment invalid. In practical terms, this means that we cannot simply multiply $(3 + 1) \times (3 + 1)$ to predict the results of a dihybrid cross when the genes are linked, since linked genes do not behave independently.

Linkage of genes is most easily analyzed by studying the offspring from a testcross of the dihybrid, rather than through the more complicated cross between two dihy-

▶ **FIGURE 2–13** The recombination of linked genes through crossing over. The linked genes are assumed to be present on the same chromosome but at separate loci. Crossing over is the mechanism responsible for the formation of new combinations of linked genes.

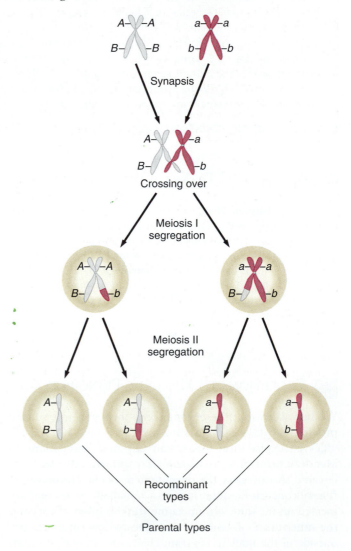

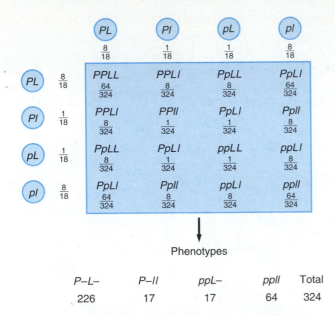

		PL $\frac{8}{18}$	Pl $\frac{1}{18}$	pL $\frac{1}{18}$	pl $\frac{8}{18}$
PL	$\frac{8}{18}$	PPLL $\frac{64}{324}$	PPLl $\frac{8}{324}$	PpLL $\frac{8}{324}$	PpLl $\frac{64}{324}$
Pl	$\frac{1}{18}$	PPLl $\frac{8}{324}$	PPll $\frac{1}{324}$	PpLl $\frac{1}{324}$	Ppll $\frac{8}{324}$
pL	$\frac{1}{18}$	PpLL $\frac{8}{324}$	PpLl $\frac{1}{324}$	ppLL $\frac{1}{324}$	ppLl $\frac{8}{324}$
pl	$\frac{8}{18}$	PpLl $\frac{64}{324}$	Ppll $\frac{8}{324}$	ppLl $\frac{8}{324}$	ppll $\frac{64}{324}$

↓

Phenotypes

P–L–	P–ll	ppL–	ppll	Total
226	17	17	64	324

▶ FIGURE 2–14 Results of Bateson and Punnett shown in terms of a dihybrid cross involving linked genes. The gamete ratio is hypothesized to be $8:1:1:8$. The expected proportion of each zygote class is then calculated by multiplication of appropriate gametic frequencies, and the phenotypic ratio is determined by summation of the relevant genotypic classes. The result is the ratio of phenotypes observed by Bateson and Punnett.

brids. For example, if our hypothesized gametic ratio of 8 PL : 8 pl : 1 Pl : 1 pL is correct, then a testcross PpLl × ppll will yield offspring in the ratio of 8 PpLl : 8 ppll : 1 Ppll : 1 ppLl, since offspring testcross ratios directly reflect the ratio of gametes produced by the dihybrid. In fact, we can reverse the analysis and use the observed frequencies of offspring from a testcross to estimate the frequency of genetic recombination between the nonallelic genes. We will return to a discussion of linked genes in Chapter 13, where we will consider the topic of genetic mapping—using the frequency of recombination as a measure of genetic distance between linked genes. For now, the important points are (1) genes linked on the same chromosome do not assort independently of each other, (2) linkage is a common situation since the DNA of a single chromosome can contain many genes that determine a variety of traits, and (3) in explaining linkage we more clearly see the chromosomal and meiotic basis of inheritance.

Example 2.4

In tomatoes, round fruit (O) is dominant to elongate fruit (o) and smooth fruit skin (P) is dominant to fuzzy skin (p). A homozygous round, fuzzy plant is crossed to a homozygous elongate, smooth plant, and their offspring (the F$_1$) are then testcrossed. The results among the progeny from the

testcross were 420 round, fuzzy : 57 round, smooth : 63 elongate, fuzzy : 460 elongate, smooth. Interpret these results.

Solution: The dihybrid testcross does not give the $1:1:1:1$ ratio expected if the nonallelic genes had assorted independently. Instead we see that the parental combinations of nonallelic genes (O with p and o with P) tend to be inherited together, giving many more Op and oP gametes than OP and op types. These latter gamete classes represent recombinant types produced by crossing over. Thus, we conclude that the nonallelic genes are linked.

Follow-Up Problem 2.4

A dihybrid testcross of heterozygous round, smooth plants to elongate, fuzzy plants gives 53 round, fuzzy : 445 round, smooth : 435 elongate, fuzzy : 67 elongate, smooth. Interpret these results.

Extranuclear Genes

Not all genes are found in the nucleus of a cell. For example, some of the genes controlling mitochondria and chloroplasts are contained in the DNA within these cytoplasmic organelles rather than within nuclear DNA. The genes within mitochondria and chloroplasts are carried on chromosomes whose size and structure differ from those of chromosomes within the nucleus (a topic discussed in detail in Chapter 9). The mechanism for transmission of organellar chromosomes into daughter cells at cell division also differs from that of nuclear chromosomes. As a consequence, the genes of organellar chromosomes show an inheritance pattern that is different from the simple Mendelian pattern of nuclear genes. This inheritance pattern is known as cytoplasmic (or extranuclear) inheritance in order to distinguish it from the inheritance patterns that we associate with nuclear genes. Cytoplasmic inheritance is characteristic not just of organelles but also of certain intracellular parasites and symbionts that have their own genetic material.

The existence of extranuclear genes clearly complicates the analysis of genetic characteristics. Not only must we establish that a trait is genetically caused, but we must also determine whether the trait, if inherited, is the expression of nuclear or cytoplasmic genes.

■ Table 2–3 lists some of the criteria that are useful in distinguishing between nuclear and cytoplasmic inheritance. One criterion is a difference in the relative genetic contribution of the parents. Since intact organelles are generally transmitted to the progeny in the egg, not in the male gamete, organellar genes often exhibit a phenomenon known as maternal inheritance. In maternal inheritance, the female parent has the deciding influence on the phenotype of the offspring. Thus for two alternative traits, say A^+ and A^-, we might expect to see cross results of the following

TABLE 2-3 Differences between extranuclear and nuclear inheritance patterns.

	Nuclear Genes	Extranuclear Genes
Relative genetic contribution of the parents	Male and female parents make equal genetic contributions to their offspring.	The female makes a greater genetic contribution to the offspring than the male parent.
Reciprocal cross results	Except for sex linkage (see Chapter 4), reciprocal crosses give the same progeny.	Reciprocal crosses yield different progenies.
Segregation ratios	Mendelian segregation ratios based on behavior of chromosomes during meiosis are observed.	Non-Mendelian segregation ratios are observed.
Linkage results	Genes are linked to other nuclear genes.	Genes show no linkage to known nuclear genes but are often carried on organellar chromosomes.

All white branch

Main shoot is variegated, having two different colors

All green branch

▶ **FIGURE 2-15** Variegation in the four-o'clock plant.

type: A^+ female × A^- male → all A^+ progeny, and A^- female × A^+ male → all A^- progeny. Maternal inheritance is therefore characterized by a difference in reciprocal cross results, since the trait is inherited exclusively through the female parent with no contribution by the male.

One example of maternal inheritance is demonstrated by variegated forms of the ornamental four-o'clock, *Mirabilis jalapa* (▶ Figure 2-15). Variegated plants have yellow or white patches or streaks on an otherwise green leaf or stem. The nongreen regions of the plant develop from cells that contain only colorless proplastids without chlorophyll. The cells in these regions would die were it not for the transfer of nutrients from the green plant regions that do contain chlorophyll. If a plant embryo contains a mixture of normal chloroplasts and colorless proplastids, some of the cells that are produced by division as the plant grows will receive normal chloroplasts and will produce green tissue, while others will receive only proplastids and will produce white tissue, and still others will inherit both types of plastids, giving rise to variegated tissue.

One of the pioneering studies on the inheritance of variegation was conducted on the four-o'clock by C. Correns in the early 1900s. The results of his studies are summarized in ■ Table 2-4. Notice that the phenotype of the progeny depends only on the phenotype of the female (or seed-

bearing) parent and not on the male (or pollen-bearing) plant. Thus, ovules derived from green portions of the plant produce only green progeny, ovules derived from white branches yield only white offspring, and ovules derived from variegated branches give rise to all three (green, white, and variegated) progeny types, regardless of the source of the pollen.

However, an inheritance pattern that shows a strong maternal influence does not always mean that the causative gene is located in the cytoplasm. Since the female contributes much more cytoplasm to the zygote than does the male, maternal substances of nuclear origin that are already present in the cytoplasm of the egg at fertilization can influence the phenotype of the offspring. In such instances, the maternal effect would depend on the nuclear genotype of the mother and not on the organellar genes. An excellent example of such a situation involves maternal effects on early embryonic development, which will be discussed in Chapter 21.

Furthermore, traits that show cytoplasmic inheritance in one case may demonstrate nuclear inheritance in another. Variegation provides an example of this difference. The recessive gene *j* on a nuclear chromosome in corn produces a green-and-white striped appearance in homozygous *jj* plants. Unlike variegation in the four-o'clock, maternal in-

■ TABLE 2−4 Progeny resulting from crosses between different phenotypes of the four-o'clock plant.

Pollen Source (Type of Branch)	Origin of Pollinated Flowers (Type of Branch)	Progeny Grown from Seed
White	white	white
	green	green
	variegated	white, green, variegated
Green	white	white
	green	green
	variegated	white, green, variegated
Variegated	white	white
	green	green
	variegated	white, green, variegated

heritance is not the pattern in this instance, since the cross $JJ \times jj$ yields only green offspring regardless of whether the variegated jj parent is the egg or pollen donor. Moreover, the cross $Jj \times Jj$ results in the classical 3 green (J–):1 striped (jj) ratio that is expected for the segregation of nuclear genes.

To Sum Up

1. Linked genes are nonallelic genes that are carried on the same chromosome; they are characterized by their tendency to remain together in the same combination as they were inherited so that they cannot assort independently. Thus linked genes do not obey Mendel's principle of independent assortment.
2. Linkage is demonstrated in the results of a dihybrid cross by a significant departure of the phenotypic ratio from the 9:3:3:1 ratio that is expected when genes assort independently. This departure from the 9:3:3:1 ratio occurs because the different kinds of gametes are not equally likely when genes are linked—some occur more frequently than others. Thus although each trait considered separately still obeys Mendel's first principle, the separate 3:1 monohybrid ratios cannot simply be multiplied to get the overall dihybrid ratio in the case of linked genes.
3. Gametes that contain the linked genes in the same combination as they were inherited are referred to as parental type gametes. Although the frequency of parental type gametes is relatively high, new gene combinations, referred to as recombinant type gametes, can be formed through the recombination process of crossing over.
4. Extranuclear genes are another exception to Mendel's second principle. The genes of mitochondria and chloroplasts exhibit characteristics of extranuclear inheritance, such as differences in the results of reciprocal crosses and non-Mendelian ratios. Both of these phenomena are due to the inheritance of the genes through the cytoplasm rather than through the nucleus.
5. The genes of cytoplasmic organelles sometimes exhibit inheritance patterns in which the phenotypic effects are influenced by genes in the nucleus. This interaction between cytoplasmic and nuclear genes can be complex, and traits that show cytoplasmic inheritance in one instance may demonstrate nuclear inheritance in another.

Point to Ponder 2.2

Mendel chose the seven traits listed in Table 2−1 for his work. As we have seen, all of these traits show independent assortment, meaning that all seven pairs of alleles that determine these traits are carried on different pairs of chromosomes. The garden pea has just seven pairs of chromosomes. How likely do you think it is that Mendel just happened to be so "lucky" in his choice of traits that all seven would show independent assortment? Do your thoughts in this regard lead you in any way to question Mendel's research ethics?

Chapter Summary

Gregor Mendel established the fundamental principles of gene transmission through his experiments with the garden pea in the late 1800s. Mendel accounted for his results on the inheritance of single pairs of contrasting characters by postulating that every individual carries two genes for a given trait and that each form of a trait is determined by a specific form of a gene (later called an allele). The alleles in Mendel's experiments exhibited dominant/recessive inheritance; heterozygous individuals therefore exhibited only the phenotype expressed by the dominant allele. Mendel also postulated that only one member of each pair of genes is transferred in a gamete to the next generation, and thus members of a pair of alleles must segregate during gamete formation. To account for his results on the inheritance of more than one pair of contrasting characters, Mendel postulated that different gene pairs segregate and assort into gametes independently of each other. All possible combinations of nonallelic genes are thus produced with equal probability in the gametes of an individual.

Mendel's principles of heredity are the basis for determining modes of inheritance through pedigree analysis and progeny testing. His principles also demonstrate a clear parallel between the behavior of chromosomes and genes and were thus instrumental in providing the first experimental support for the chromosome theory of inheritance.

Not all genes obey Mendel's principles of inheritance. Nonallelic genes that are carried on the same chromosome (linked genes) often fail to assort independently of each other, while genes that are carried on organellar chromosomes (extranuclear genes) follow rules of transmission that differ substantially from those of nuclear genes.

The Principle of Genetic Segregation

1. Define the following terms, and distinguish between the members of paired terms:

   alleles and homologs
   dominant and recessive
   genotype and phenotype
   homozygous and heterozygous
   pure-breeding
   testcross

2. Sheep can be either black or white in color. These colors are determined by a single pair of alleles. A white ram (male) and white ewe (female) produce a black lamb. Which color is dominant? Explain your answer.

3. Wire-haired versus smooth hair texture are alternative characteristics in dogs. A mating between two wire-haired dogs results in a litter of three wire-haired and two smooth-haired pups. (a) Which trait is dominant? (b) If the same parents produce several other litters, what ratio of wire-haired to smooth-haired pups is expected among the collective offspring?

4. In guinea pigs, coat color can be black or white. Matings between black guinea pigs sometimes yield offspring in a ratio of 3 black:1 white and other times yield all black offspring. Crosses of black × white sometimes produce all black progeny and other times produce a ratio of 1 black:1 white. White × white crosses produce only white offspring. Explain these results, and indicate which characteristic is dominant. Write the genotypes of the parents and offspring in each cross mentioned.

5. In tomatoes, stems can be purple or green. Several crosses are made involving parents of known phenotypes but unknown genotypes. The results of these crosses are given below. Using *A* and *a* to represent the alleles, designate the genotypes of the parents in each cross.

Parental Crosses	Purple Progeny	Green Progeny
purple × green	121	0
purple × purple	97	31
green × green	0	101
purple × green	63	67
purple × purple	76	0

6. Members of the Holstein breed of dairy cattle normally have a black-and-white spotted coat. On occasion, calves with a recessive red-and-white spotted coat are born. A dairy farmer purchases a prized black-and-white spotted bull. To the farmer's dismay, the bull produces a calf with the recessive coloration when bred to one of his black-and-white cows. (a) What is the genotype of the farmer's bull? (Use *R* and *r* for the color alleles.) (b) What phenotypic ratio is expected in the offspring if the bull is mated to red-and-white spotted cows?

7. A normally pigmented couple has several children. Both the husband and wife come from families in which one of the parents is albino (a Mendelian recessive trait). What fraction of the offspring of this couple is expected to have normal pigmentation?

8. Indicate whether the trait in each of the following pedigrees is most likely due to a dominant or a recessive gene:

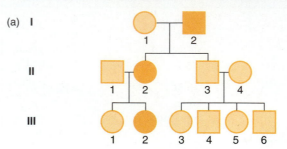

(a)

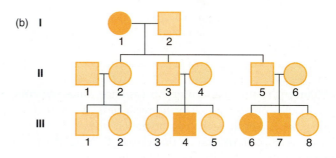

(b)

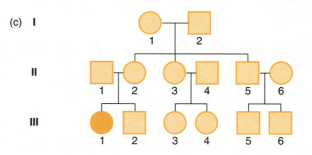

(c)

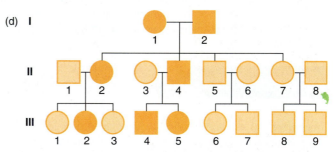

(d)

9. In the pedigrees given in problem 8, indicate which individuals are (a) definitely heterozygous, and (b) possibly heterozygous. (c) If individual (b)III−4 marries individual (b)III−6, what phenotypic ratio would be expected among their offspring?

The Principle of Independent Assortment

10. Determine the genotypic ratio expected for each of the following crosses:
    (a) *AAbb × aaBB*
    (b) *AaBB × AaBb*
    (c) *Aabb × aaBb*

11. Suppose that a geneticist crosses two true-breeding pea plants: one with purple flowers and long stems and the other with white flowers and short stems. (a) Describe the appearance of the F_1 (see Table 2–1 for information concerning dominance and recessiveness in these traits). (b) If the F_1 individuals are selfed, what phenotypes will appear in the F_2 and in what proportions? (c) If the F_1 are testcrossed to white, short plants, what phenotypic ratio will appear among their progeny?

12. In pigeons, checkered feather pattern (B) is dominant over barred (b), and grizzled (spotted) color pattern (G) is dominant over plain color (g). What phenotypic ratio would you expect among the progeny from each of the following crosses?
(a) $BBgg \times bbGG$
(b) $BbGg \times bbgg$
(c) $Bbgg \times bbGg$
(d) $BbGG \times bbgg$
(e) $BbGG \times BbGg$
(f) $BbGg \times Bbgg$

13. The fruit of the watermelon can be either short ($A–$) or elongate (aa), and its skin can be either green ($G–$) or striped (gg). A cross is made between a short, striped variety and a long, green variety. Four different phenotypes appear among the offspring, with approximately 25% of the F_1 plants producing fruits that are long and striped. What are the genotypes of the parental varieties in this cross?

14. Coat color ($C–$) in rabbits is dominant to albino (cc), and short hair ($L–$) is dominant to long (ll). Suppose that in a series of crosses, parents of known phenotypes but unknown genotypes produce the results shown in the table below. What are the most likely genotypes of the parents in each cross?

	Progeny			
Parental Crosses	colored, short	colored, long	albino, short	albino, long
color, short × color, short	74	26	24	8
color, long × albino, short	34	31	33	32
color, short × color, long	49	47	15	17
color, long × albino, short	37	32	0	0
color, short × albino, short	26	0	26	0

*15. Consider the cross $AaBbCcDd \times aaBbccDd$, where all gene pairs assort independently. What phenotypic ratio will be produced among the progeny? (Hint: Use the forked-line method rather than a Punnett square.)

Chromosomes and Mendel's Principles

16. Mendel discovered the rules of gene transmission by studying the diploid garden pea. Since many flowering plants are tetraploid (they have four copies of each chromosome), let us consider what might have happened had Mendel been unlucky enough to choose a plant with 4 genes of a kind rather than 2. Suppose that Mendel had started with a parental cross $AAAA \times aaaa$. Predict the numerical consequences of this cross in terms of the phenotypic ratios that Mendel would have observed in (a) the F_1 and (b) the F_2, assuming that the A allele is dominant to a and that the gametes of each plant receive two of the genes in random assortment. Compare your predictions with the results that Mendel actually observed.

Exceptions to Mendel's Principles

17. In corn, the aleurone can be colored (C) or colorless (c) and the endosperm can be starchy (Wx) or smooth (wx). A cross of a heterozygous colored, starchy plant to a colorless, smooth variety gives the following results: 335 colored, starchy: 167 colored, smooth: 166 colorless, starchy: 332 colorless, smooth. (a) Explain the origin of each of the offspring classes. (b) What ratio of gametes was produced by the colored, starchy plant?

18. If the colored, starchy plant in problem 17 is selfed, rather than testcrossed, what phenotypic ratio is expected among the offspring?

19. In barley, virescent leaves are the result of either a cytoplasmic factor (L_1 = normal leaves, L_2 = virescent leaves) or the recessive nuclear gene v (vv = virescent leaves). What genotypic and phenotypic results would you expect to obtain from each of the following crosses? (a) Pure-breeding normal female × L_1vv male. (b) L_1vv female × pure-breeding normal male. (c) pure-breeding normal female × L_2vv male. (d) L_2vv female × pure-breeding normal male. (e) Female F_1 from cross (a) × male F_1 from cross (d). (f) Male F_1 from cross (a) × female F_1 from cross (d).

20. In a certain strain of barley, foliage can be green or yellow, and ears can be long-awned or short-awned. When a true-breeding green, long-awned plant is fertilized by pollen derived from a true-breeding yellow, short-awned plant, all the offspring are green and long-awned. Intercrosses among these F_1 yield F_2 consisting of 624 green, long-awned and 290 green, short-awned plants. In contrast, when yellow, short-awned plants are pollinated by green, long-awned plants, the F_1 are all yellow and long-awned, and the F_2 consist of 648 yellow, long-awned and 310 yellow, short-awned plants. Explain the genetic bases for these color and ear traits.

*An asterisk indicates that the question or problem is based on information in the Extensions and Techniques text.

CHAPTER 3

Chance and Mendelian Inheritance

Even though very little was known about chromosomes and the nature of meiosis at the time of Mendel's experiments, Mendel was able to describe the basic principles of heredity by using a mathematical model. The principles of genetic segregation of allelic genes and independent assortment of nonallelic genes treat segregation and assortment of genes into gametes and their subsequent combination into zygotes as random events. Gamete formation and fertilization can thus be likened to a game of chance, such as coin tossing. One of the alleles in a monohybrid (*Aa*) is included in each gamete at random, resulting in two equally likely outcomes, *A* and *a*. The toss of a coin also results in two equally likely outcomes, heads (*H*) and tails (*T*). When two hybrids are crossed (*Aa* × *Aa*), the alleles in the gametes combine at random during fertilization to produce three genotypes in a ratio of 1 *AA* : 2 *Aa* : 1 *aa*. Similarly, when two coins labeled 1 and 2 are tossed together, heads and tails

Genetic variation in spotting of offspring.

combine at random so that three results—heads on both; heads on one, tails on the other; and tails on both—occur in a ratio of 1 (H_1H_2) : 2 $(H_1T_2 + T_1H_2)$: 1 (T_1T_2). Since the basic laws of probability can be used to predict the outcome of any chance event, such as the toss of a coin, it follows that these laws must also apply to the random processes of gamete formation and fertilization, thereby providing a theoretical basis for Mendel's principles.

✳ PROBABILITY AND GENETICS

In the general case involving a finite number of outcomes, we can define the probability of a particular event happening as the ratio of the number of ways the event can occur to the total number of possible outcomes. The probability of an event, symbolized $P(E)$, can then be calculated as

$$P(E) = \frac{\text{number of ways that event } E \text{ can occur}}{\text{total number of possible outcomes}}$$

For example, in tossing two coins, 1 and 2, the desired event might be getting one with heads and one with tails, HT. This event can occur in two ways: heads on coin 1, tails on coin 2 (H_1T_2); or tails on coin 1, heads on coin 2 (T_1H_2). Since there are four equally likely outcomes in all $(H_1H_2, H_1T_2, T_1H_2, T_1T_2)$ when two coins are tossed, the probability of the event HT is $\frac{2}{4} = \frac{1}{2} = 50\%$. Note that by defining the probability as the ratio of desired outcomes to total possible outcomes, it becomes a fractional number falling between the limits of 0 and 1. If the probability is 0, the event cannot occur. If the probability is 1, the event is a certainty.

One must be careful to distinguish between the *theoretical* relative frequency (or probability) and the *empirically* determined proportion of successes. When the coin-tossing experiment is actually performed, the relative frequency observed may or may not be 0.5. Suppose the two coins are tossed 10 times, with three tosses giving H_1T_2 and three giving T_1H_2. The observed frequency of HT is then $\frac{3}{10} + \frac{3}{10} = \frac{6}{10}$, or 0.6. This value is close to the theoretical value but not equal to it. If the experiment (tossing the coins 10 times) is repeated several times and the observed relative frequency of HT is calculated each time, we will probably obtain an array of different values. Some of these values may be close to 0.5; some may not. The observed variability may be particularly great in this case since the number of tosses is not very large.

Now suppose the experiment is redesigned and the pair of coins is tossed 100 times. If this 100-toss procedure is repeated several times, we will probably find that more of the observed HT frequencies will be closer to 0.5, the theoretical value. In mathematical terms, this result demonstrates the **law of large numbers.** This law states that as the sample size becomes very large, the observed relative frequency of an event approaches its theoretical value (probability). The expected or theoretical value should therefore be a good

predictor of what is actually observed if the sample size is large. The converse is also true. If the sample size is small, the observed relative frequency will often not be the same as the expected frequency. This tendency results solely from the small number of trials; it is not an indication that anything is wrong with the theory itself.

When evaluating a probability, all pertinent information must be taken into account if the calculated value is to be an accurate reflection of the level of uncertainty. A case in point involves determining the probability that an individual is a carrier of a recessive allele on the basis of family data. For example, suppose that you wish to determine the probability that a man whose sister is affected with cystic fibrosis (a disease caused by a recessive gene) is a carrier of the allele for this disorder. The fact that the man's sister is affected indicates that both his parents (who are not affected) are heterozygous (Aa). Since the man is known to be unaffected by the disease, he cannot be aa. He is therefore either AA or Aa, but the chances of being AA and Aa are not equal. The two possible kinds of $A-$ offspring are produced from the cross $Aa \times Aa$ in a ratio of 2 Aa : 1 AA. The probability that the man is a carrier, given that he has the dominant phenotype, symbolized $P(Aa|A-)$, is then

$$P(Aa|A-) = \frac{2(Aa)}{2(Aa) + 1(AA)} = \frac{2}{3}$$

There is thus a $\frac{2}{3}$ chance that the man is a carrier, given the condition that he is normal (dominant) in phenotype.

Probability determinations like the one in the preceding example are often performed in human genetics. Genetic counselors frequently meet with prospective parents who come from families with a history of a particular recessive disease and want to know if they carry the causative allele in the heterozygous condition. Breeding experiments, such as testcrosses, are obviously not applicable to humans. For some disorders there are biochemical tests that can directly determine whether an individual with a normal phenotype is heterozygous or homozygous, but for many inherited diseases, there is still no way such a determination can be made with certainty. The best that can be done is to calculate the probability that an individual is a carrier. The calculated probability, $P(Aa|A-)$, is a **conditional probability** because it is contingent on the condition that the individual in question is unaffected. If the parents of this individual can be established as carriers (for example, through an affected sibling of the individual in question), then the chance that the individual is a carrier is $\frac{2}{3}$.

Example 3.1

In the following pedigree, the darker symbols represent individuals with galactosemia, a disease that results from the lack of an enzyme that converts the sugar galactose to glucose.

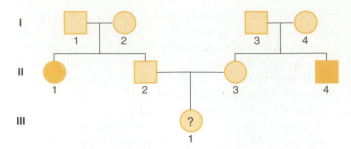

(a) Is the gene for galactosemia dominant or recessive?
(b) What is the chance that newborn female III–1 will have the disease?

Solution: (a) Galactosemia is a recessive trait, since both II–1 and II–4, who are affected, were born to normal (*Aa*) parents. (b) $\frac{1}{9}$; the chance of a recessive offspring is equal to the product of three conditional probabilities: (the probability that II–2 is *Aa*, given that he is *A*–) × (the probability that II–3 is *Aa*, given that she is *A*–) × (the probability that III–1 is *aa*, assuming that her parents are *Aa* × *Aa*), which is equal to $(\frac{2}{3}) \times (\frac{2}{3}) \times (\frac{1}{4}) = \frac{1}{9}$.

Follow-Up Problem 3.1

A man whose sister is affected with galactosemia marries a woman whose maternal aunt (mother's sister) is affected with the same disease. What is the chance that their first child will have galactosemia?

◆

Fundamental Rules of Probability

Two basic rules are helpful in calculating probability values, the addition rule and the multiplication rule.

Rule ADDITION RULE

If E_1 and E_2 are mutually exclusive events (the occurrence of one precludes the occurrence of the other), the probability that either of them will happen is the sum of their individual probabilities. Symbolically,

$$P(E_1 \text{ or } E_2) = P(E_1) + P(E_2)$$

The word *or* signals the use of the addition rule. In our coin-tossing experiment, the event *HT* can occur in one of two mutually exclusive ways: H_1T_2 or T_1H_2. The total probability of this event, $P(HT)$, can therefore be calculated from the sum of the individual probabilities: $P(H_1T_2) + P(T_1H_2)$. In genetics, we often apply the addition rule without even realizing that we are using it. For example, when we stated that three-fourths of the progeny from a monohybrid cross have the dominant phenotype, we have added the $\frac{2}{4}$ *Aa* and $\frac{1}{4}$ *AA* classes to obtain $\frac{3}{4}$. The event *A*– can be satisfied by either *Aa* or *AA*. Since they are mutually exclusive possibilities (only one can happen in a single individual), their probabilities have been summed.

A somewhat less obvious genetic example involves the dihybrid cross *AaBb* × *AaBb*. This cross yields 16 zygotic combinations, consisting of $\frac{9}{16}$ *A–B–*, $\frac{3}{16}$ *A–bb*, $\frac{3}{16}$ *aaB–*, and $\frac{1}{16}$ *aabb*. This example can be applied to a form of deafness that occurs in humans when an individual lacks one or both dominant nonallelic genes (i.e., normal hearing requires the *A–B–* condition). If both parents are *AaBb*, we then might be interested in asking what fraction of their offspring is expected to be deaf? The outcomes that satisfy this event are *A–bb*, *aaB–*, and *aabb*. Therefore, the chance that a single offspring from such a cross will be deaf is $P(A–bb \text{ or } aaB– \text{ or } aabb) = P(A–bb) + P(aaB–) + P(aabb) = \frac{3}{16} + \frac{3}{16} + \frac{1}{16} = \frac{7}{16}$.

Rule MULTIPLICATION RULE

If E_1 and E_2 are independent events (the occurrence of one does not affect the chance of occurrence of the other), the probability that they will both happen is the product of their individual probabilities. Symbolically,

$$P(E_1 \text{ and } E_2) = P(E_1)P(E_2)$$

The word *and* signals the use of the multiplication rule. Thus for two coin tosses, the probability of getting heads on coin 1 and tails on coin 2, $P(H_1T_2)$, is the product of the separate probabilities: $P(H_1)P(T_2)$.

One genetic application of the multiplication rule involves calculating the expected proportion of a gamete type produced by an individual in a cross. To illustrate, let us consider the genotype *AaBbCcDd*. What proportion of all gametes from such individuals will consist of the genotype *ABCd*? As long as the gene pairs are known to segregate independently, we can work with each pair individually and then combine the results through the multiplication rule. When each gene pair is considered separately, the chance of a gamete's receiving the designated gene is $\frac{1}{2}$. The probability of getting a gamete of genotype *ABCd* is then $P(A \text{ and } B \text{ and } C \text{ and } d) = P(A)P(B)P(C)P(d) = (\frac{1}{2})^4 = \frac{1}{16}$.

Calculating Mendelian Ratios

These rules of probability can be used to mathematically predict the expected outcome of a genetic cross, thus eliminating the need to work through the mechanics of a Punnett square. The multiplication rule is particularly useful when dealing with several gene pairs. The multiplication rule for independent events applies whenever nonallelic genes are randomly associated in the gametes produced by the parents. Since the gametes combine at random during fertilization, all gene pairs that assort independently into gametes are also transmitted independently to the offspring. In practical terms, this means that when we analyze a genetic cross, we can work with each gene pair individually and then combine the results through the multiplication rule. This concept was introduced in the last chapter, when we pointed out that Mendel's recognition of the 9 : 3 : 3 : 1 ratio as the mathematical product of the separate 3:1 ratios led him to state the principle of the independent assortment of nonalleles into gametes.

Use of the multiplication rule saves considerable time in working through crosses involving several pairs of independently assorting genes. For example, consider the dihybrid cross $AaBb \times AaBb$. Recall that we used the Punnett square method to determine that this cross yields a genotypic ratio of $1:2:1:2:4:2:1:2:1$ and a phenotypic ratio of $9:3:3:1$. These ratios can be calculated directly if we analyze the cross one gene pair at a time by first obtaining the results from each of the two unit monohybrid crosses, $Aa \times Aa$ and $Bb \times Bb$, and then combining the results through multiplication. In this case, both unit crosses yield a genotypic ratio of $1:2:1$ and a phenotypic ratio of $3:1$. Therefore, the genotypic ratio of the dihybrid cross can be directly calculated from the product

$(\frac{1}{4}AA + \frac{1}{2}Aa + \frac{1}{4}aa) \times (\frac{1}{4}BB + \frac{1}{2}Bb + \frac{1}{4}bb)$
$= 1\ AABB:2\ AABb:1\ AAbb:2\ AaBB:4\ AaBb:2\ Aabb:$
$1\ aaBB:2\ aaBb:1\ aabb$

and the phenotypic ratio from the product

$(\frac{3}{4}A- + \frac{1}{4}aa) \times (\frac{3}{4}B- + \frac{1}{4}bb)$
$= 9\ A-B-:3\ A-bb:3\ aaB-:1\ aabb$

The results thus yield $3 \times 3 = 9$ genotypic classes, $2 \times 2 = 4$ phenotypic classes, and a phenotypic ratio of $9:3:3:1$ derived from $(3+1)^2$.

Use of the multiplication rule is not restricted to two gene pairs. Any cross, regardless of the number of allelic pairs, can be treated in a manner similar to a dihybrid cross as long as the nonallelic genes segregate independently. For example, the results of the trihybrid cross $AaBbCc \times AaBbCc$ are the products of three monohybrid crosses $(Aa \times Aa)$ $(Bb \times Bb)(Cc \times Cc)$. In this case, there are $3^3 = 27$ genotypic classes, $2^3 = 8$ phenotypic classes, and a phenotypic ratio of $27:9:9:9:3:3:3:1$ from $(3+1)^3$.

The multiplication rule is especially useful when we are interested in the proportion of only one class, not the entire ratio. For example, suppose the cross is $AaBbCcDd \times AabbCcDD$. What proportion of the progeny of this cross is expected to be $A-B-C-D-$ in phenotype? To answer this question, we analyze the cross in terms of its separate unit crosses $(Aa \times Aa)(Bb \times bb)(Cc \times Cc)(Dd \times DD)$. By knowing the cross results in each case, we can then calculate the answer by multiplying $\frac{3}{4}$ $(A-$ from $Aa \times Aa) \times \frac{1}{2}$ $(B-$ from $Bb \times bb) \times \frac{3}{4}$ $(C-$ from $Cc \times Cc) \times 1$ $(D-$ from $Dd \times DD)$ to get $\frac{9}{32}$. This approach obviously saves time. To write out the cross in a Punnett square, we would have to write down and sort through a total of $4 \times 2 \times 4 \times 2 = 64$ zygotic combinations to obtain the desired result.

To Sum Up

1. Two rules, the addition rule and the multiplication rule, can be applied to the calculation of the probability that an event occurs. Genotypic and phenotypic proportions can be calculated directly by using these rules, and thus the time-consuming Punnett square method can be avoided.

2. The addition rule is used in "either/or" situations. If there are two or more alternative outcomes that can satisfy the desired event, then the probability that the event occurs is equal to the sum of the probabilities of the separate alternative outcomes.

3. The multiplication rule is used when the desired outcome includes the simultaneous occurrence of two or more events. The probability that both events occur together is the product of their separate probabilities.

Point to Ponder 3.1

A good fit to predicted genetic ratios is expected when the sample size is large; in contrast, human families are so small that expected ratios are seldom achieved. For example, consider a husband and wife who are both heterozygous for the recessive gene that causes cystic fibrosis. Each child that they might have will thus have a one in four chance of being affected with this disease. Suppose they already have two children and both are affected. How could you explain to them that the odds per child of being affected are only one in four and that if they have a third child, its chance of not being affected is great—three out of four? After the experience they have had so far, do you think they would believe you?

PROBABILITY DISTRIBUTIONS

So far, our main concern has been calculating the probabilities associated with a single trial of an experiment or a specific cross. In many problems in genetics, however, we must deal with the probabilities associated with repeated trials, such as those corresponding to the outcomes when the same experiment is performed several times in succession. For example, suppose that we are interested in the number of hybrid offspring produced by repeated matings of two monohybrids $(Aa \times Aa)$. If we designate this number as x, the value of x is a random variable that can vary from zero to the total number of offspring produced. Thus when two offspring are produced by repeated matings, there are three possible values for x (0, 1, and 2), simply as a result of chance. When the possible values of a random variable x are represented in tabular, graphical, or equation form, along with their corresponding probabilities $P(x)$, the resulting relationship is called a **probability distribution**.

Several different probability distributions are useful in genetics, but in the sections that follow, we will limit our discussion to three main types: the binomial, multinomial, and Poisson distributions.

Binomial and Multinomial Distributions

Suppose only two mutually exclusive outcomes are possible for a given trial; we will label these outcomes success and failure. Let $x = 1$ for a success and $x = 0$ for a failure, and let us represent the probabilities of success and failure by the symbols p and q, respectively. The random variable x is then said to have a **binomial distribution** described by the following probability model:

$$
\begin{array}{ccc}
x: & 1 & 0 \\
P(x): & p & q
\end{array}
$$

where $p + q = 1$. For example, the birth of a single child can be treated as a trial in which the sex of the child is a binomial random variable. Thus if we let $x = 0$ if the child is a male and $x = 1$ if the child is a female, we can represent the chance that the child is a girl by p and the chance that the child is a boy by $q = 1 - p$. In this case, the sex of the child is either male or female with equal likelihood, so $p = q = 0.5$.

As long as successive trials are independent, the binomial distribution can be extended to predict the number of successes in any number of trials. If we again express the probability of an event occurring in a single trial as p and the probability of it not occurring as q (which equals $1 - p$), the probability that the event occurs exactly x times out of n independent trials is then given by the formula

$$
P(x) = \frac{n!}{x!(n - x)!} p^x q^{n-x} \tag{3.1}
$$

where the symbol ! is read "factorial" and means the product of all integers down to one (e.g., $4! = 4 \times 3 \times 2 \times 1 = 24$). It should be noted that $0!$ is defined as 1, as is any number to the zero power.

Equation 3.1 is known as the **binomial equation**, since it represents a general term of the binomial expansion $(p + q)^n$. In this equation,

$$
\frac{n!}{x!(n - x)!}
$$

gives the number of arrangements of x successes and $n - x$ failures in n total trials, and $p^x q^{n-x}$ gives the probability of any one particular arrangement. For example, what is the probability in a family of four children that two are girls and two are boys? The solution can be obtained by letting p equal the chance that any child is a girl ($p = 0.5$) and q be the chance that any child is a boy ($q = 0.5$). In this case, there are

$$
\frac{4!}{2!2!} = 6
$$

possible arrangements of two girls and two boys in order of birth (GGBB, BBGG, GBGB, BGBG, GBBG, and BGGB). Since the sex of each child is independent of the sex of each of the preceding children, the probability of every arrangement is the same and is given by the product $p^2 q^2 = (\frac{1}{2})^2 (\frac{1}{2})^2 = \frac{1}{16}$. Therefore, by multiplying the number of arrangements of two girls and two boys by the probability of each arrangement, the probability of $x = 2$ girls and $n - x = 2$ boys in a family of 4 children is calculated as

$$
\frac{4!}{2!2!} \left(\frac{1}{2}\right)^2 \left(\frac{1}{2}\right)^2 = \frac{6}{16} = 0.375
$$

When the probability $P(x)$ is calculated for all possible values of x, we can obtain the probability distribution for a particular set of values of n and p. The probability distribution for sex phenotype in families of four children ($n = 4$, $p = \frac{1}{2}$) is given below as an example. The distribution is plotted in ▶ Figure 3–1a in the form of a histogram or bar graph.

$$
\begin{array}{cccccc}
x: & 4 & 3 & 2 & 1 & 0 & \text{girls} \\
n - x: & 0 & 1 & 2 & 3 & 4 & \text{boys} \\
P(x): & \frac{1}{16} & \frac{4}{16} & \frac{6}{16} & \frac{4}{16} & \frac{1}{16} &
\end{array}
$$

Five outcomes are possible in a family with four children, and their probabilities correspond to the terms of the binomial expansion $(p + q)^4 = \frac{1}{16} + \frac{4}{16} + \frac{6}{16} + \frac{4}{16} + \frac{1}{16}$. The sum of these probabilities is 1, since one of these outcomes must occur in each family with four children. By examining a complete probability distribution such as this one, we can determine which events are fairly common and which are rare. In this case, only $\frac{1}{16}$ (or 6.25%) of the families are expected to be made up of all girls, and the same proportion is expected to be made up of all boys. In contrast, $\frac{6}{16}$ (or 37.5%) of the families are expected to be made up of two girls and two boys. We might expect most families to have two girls and two boys, since each sex has an equal probability of occurring. The chance of this event is indeed

---

▶ **FIGURE 3–1** Histograms showing binomial distributions for (a) $n = 4$, $p = \frac{1}{2}$ and (b) $n = 5$, $p = \frac{3}{4}$. The abscissa gives the number of successes (x) and the ordinate shows the proportion of the population expected to have that number of successes.

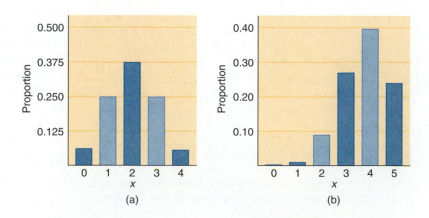

Human Sex Ratios

Population data on sex distribution in human families of different sizes fit the binomial expectations for $p = \frac{1}{2}$, $q = \frac{1}{2}$ reasonably well. It is worth noting, however, that the actual sex ratio in human live births is not exactly 1 boy : 1 girl. Although direct measurements at the time of conception are not possible, the ratio of male to female zygotes (the primary sex ratio) is estimated to be as high as 1.6 : 1 (160 males for every 100 females). The sex ratio at birth (the secondary sex ratio), however, drops considerably below the 1.6 : 1 value. Still, in all parts of the world there is an excess of male newborns; the secondary sex ratio in North America is 1.05 : 1 (105 boys for every 100 girls). As you know, the sex ratio continues to drop throughout life. It is 1 male : 1 female at about 20–25 years of age (the prime reproductive years), and it falls below 1 : 1, with an ever-increasing proportion of females, from the third decade of life onward.

Parental preference for one sex or the other also has some effect on the sex distribution in families. For example, suppose some parents decide to have children only until a girl is born. In such cases, all one-child families will consist of one girl, all two-child families will consist of a boy and then a girl, and so on. The distribution of the sexes in these families will be quite different from the binomial distribution described in the text. However, the overall distribution in the population will still approximate the binomial ($p = \frac{1}{2}$ and $q = \frac{1}{2}$) because the overall sex ratio in the population at birth is so close to 1 : 1.

The sex ratio of offspring can vary widely from family to family.

---

higher than that of any other, but you may be surprised to notice that although $\frac{6}{16}$ is the highest proportion, it is less than 50%.

Example 3.2

A husband and wife are both carriers of the recessive gene for galactosemia. What is the chance that among five of their children, four will be normal and one will be affected with galactosemia?

Solution: The situation clearly conforms to the binomial formula. The values of the symbols in the formula are $N = 5$, $x = 4$ normal children, $n - x = 1$ galactosemic child, p (the chance of a child's being normal) $= \frac{3}{4}$, and q (the chance of a child's having galactosemia) $= \frac{1}{4}$. The overall probability of the event (having four normal children and one galactosemic child) is then

$$\frac{5!}{4!1!}\left(\frac{3}{4}\right)^4\left(\frac{1}{4}\right)^1 = \frac{405}{1024} = 0.40$$

Recognizing that this situation can be handled using the binomial formula greatly reduces the calculations needed to determine the probability of the event in question. The alternative would be to determine the number of possible arrangements giving four normal and one affected child by listing them ($A-A-A-A-aa$, $A-A-A-aaA-$, $A-A-aaA-A-$, $A-aaA-A-A-$, and $aaA-A-A-A-$, for a total of 5, where $A- =$ normal and $aa =$ affected) and then adding their individual probabilities. The list is not that extensive in this case, but it would be if the n and x values were larger (try listing the alternatives for $n = 10$ and $x = 5$—

there are 252 of them!). Using the factorial term in the binomial formula as a substitute for listing the alternatives saves a great deal of time in working through probability situations of this type.

Follow-Up Problem 3.2

A husband and wife are both carriers of the recessive gene for galactosemia. What is the chance that among four of their children, two will be normal boys and two will be affected girls?

In the binomial distribution, there are only two alternatives (which we have labeled success and failure) for each event. However, many problems in genetics involve more than two alternatives. For example, the probability of getting 1 AA, 2 Aa, and 1 aa offspring from a monohybrid cross involves three alternatives, while the probability of getting 9 A–B–, 3 A–bb, 3 aaB–, and 1 aabb offspring from a dihybrid cross involves four alternatives. We can extend our model to include more than two alternatives for each event by expressing the probability as a general term of the multinomial expansion $(p + q + r + \cdots)^n$ where $p + q + r + \cdots = 1$. The probability equation for the **multinomial distribution** is

$$P(x, y, z, \cdots) = \frac{n!}{x!y!z!\cdots} p^x q^y r^z \cdots \qquad (3.2)$$

where $x + y + z + \cdots = n$. Thus, the probability that the cross $Aa \times Aa$ will produce 1 AA, 2 Aa, and 1 aa in a sibship of 4 is

$$\frac{4!}{1!2!1!}\left(\frac{1}{4}\right)^1\left(\frac{1}{2}\right)^2\left(\frac{1}{4}\right)^1 = 0.1875$$

Similarly, the probability that the cross $AaBb \times AaBb$ will produce 9 A–B–, 3 A–bb, 3 aaB–, and 1 aabb in a sibship of 16 is

$$\frac{16!}{9!3!3!1!}\left(\frac{9}{16}\right)^9\left(\frac{3}{16}\right)^3\left(\frac{3}{16}\right)^3\left(\frac{1}{16}\right)^1 = 0.0244$$

Observe that the probability of getting exactly the theoretical ratio in both of these crosses is substantially less than one.

Poisson Distribution

The binomial distribution is applicable to cases where successes and failures are fairly common, so that neither p nor q is close to 0 or to 1. The resulting distribution thus has a mean, or expected, number of successes that is near the center of the distribution. In cases where the probability of success (p) is very small but the number of trials (n) is large so that the product np is not negligible, the binomial distribution is closely approximated by another distribution that is important in genetics, the **Poisson distribution**. The terms of the Poisson distribution are given by the general equation

$$P(x) = \frac{m^x}{x!} e^{-m} \qquad (3.3)$$

where $P(x)$ is the probability of x successes in n independent trials, m is the mean number of successes (equal to np), and e is the base of natural logarithms (2.718 . . .).

The Poisson distribution is often used in genetics when dealing with rare events. For example, about 1 child in 700 is born with Down syndrome (a rare chromosomal abnormality). Births with this disorder are seemingly random in occurrence. If there are 1400 births per year in a large metropolitan hospital, what is the probability that one child with Down syndrome is born in this hospital during a year? To solve this problem, let $p = \frac{1}{700}$ and $n = 1400$. The mean number of children born with Down syndrome per year in this facility is then $(\frac{1}{700})(1400) = 2.0$. Since the rare occurrence of this disorder ($\frac{1}{700}$) and large number of independent trials (1400) correspond to conditions covered by the Poisson distribution, the probability of having a single child with Down syndrome ($x = 1$) can be calculated using equation 3.3:

$$\frac{2^1}{1!} e^{-2} = (2)(0.135) = 0.27$$

Unlike the binomial distribution, the Poisson distribution tends to show a high degree of "skewness" or lopsidedness, as can be seen in the first seven terms of the Poisson distribution for $m = 2$:

x:	0	1	2	3	4	5	6	...
$\frac{2^x}{x!} e^{-2}$:	0.135	0.270	0.270	0.180	0.090	0.036	0.012	...

A plot of this distribution is shown in ▶ Figure 3–2. It is a highly skewed distribution in which the maximum value of $P(x)$ is found at lower values of x.

▶ FIGURE 3–2 Histogram showing the Poisson distribution with a mean (or expected value) of $m = 2$.

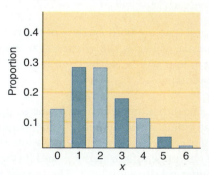

1. We have discussed three probability distributions: the binomial, the multinomial, and the Poisson distributions. The binomial distribution involves a total number of trials (n) partitioned into two groups called successes and failures. The probability of x successes in n total trials is calculated using the binomial probability term:

$$P(x) = \frac{n!}{x!(n-x)!} \, p^x q^{n-x}$$

where p and q stand for the chance of a success and chance of a failure, respectively, in each individual trial. The factorial term gives the number of ways that the event can occur, while the product $p^x q^{n-x}$ gives the probability of each possible way. By multiplying the two terms together, we get the probability of the binomial event occurring in any possible sequence of x successes and $n - x$ failures.

2. The binomial probability distribution is a special case of the more general multinomial distribution. In the multinomial distribution a total number of trials is partitioned into three or more categories. The desired event in such a case is the occurrence of a particular number of each of the three or more alternative outcomes, calculated as

$$P(x, y, z, \cdots) = \frac{n!}{x!y!z!\ldots} \, p^x q^y r^z \ldots$$

3. The Poisson probability distribution is used as an approximation to the binomial distribution when the probability of success is very small (success is a rare event) and the number of trials is large. In such cases, the average number of successes in n trials, m, is not negligible—it is equal to np. The probability of x successes in n trials is then calculated as

$$P(x) = \frac{m^x}{x!} \, e^{-m}$$

DETERMINING GOODNESS-OF-FIT: THE CHI-SQUARE TEST

In genetics, as in other branches of science, the observed numerical results of an experiment are often compared with those expected on the basis of some hypothesis. For example, suppose that in a crossing experiment with the fruit fly, *Drosophila*, a dihybrid cross ($AaBb \times AaBb$) yields 240 offspring in a ratio of 143 $A–B–$:36 $A–bb$:49 $aaB–$:12 *aabb*. We hypothesize that the two gene pairs are assorting independently. This hypothesis predicts a $9:3:3:1$ ratio for these four phenotypic classes, so the expected values in this case are 135 ($\frac{9}{16} \times 240$):45 ($\frac{3}{16} \times 240$):45:15 ($\frac{1}{16} \times 240$). In each of the four classes, however, the observed number deviates from the value predicted by our hypothesis. We must therefore decide whether these deviations are **significant** (reflecting a real difference between observation and the-

ory) or **insignificant** (reflecting random sampling error). If the deviations are significant, we conclude that they are too large to be accounted for merely on the basis of chance alone, and we must then reject the hypothesis of independent assortment.

In determining whether the deviations are significant or insignificant, we base our judgment on the likelihood that random sampling error alone would result in departures from the expected values that are as large as or larger than those observed. In statistical terms we ask, if the proposed genetic mechanism is correct (if the expected values calculated on the basis of the hypothesis are correct), what is the probability of obtaining results that differ by this much or more from their expected values? One approach to the problem would be to calculate the probability from the appropriate terms of the multinomial distribution (equation 3.2). Obviously, this would be an extremely tedious procedure if the numbers are at all large. Fortunately, a good approximation to this probability can be obtained by employing the statistical procedure known as the **chi-square (χ^2) test**. The chi-square test is a statistical test designed to evaluate the significance of deviations between observed and expected values in two or more categories. The calculations involved in the test transform the deviations into a single χ^2 value by means of the following formula:

$$\chi^2 = \sum \frac{(O - E)^2}{E} \qquad (3.4)$$

where O and E are the observed and expected values in a given class and Σ indicates summing over all classes. For the preceding dihybrid cross results,

$$\chi^2 = \frac{(143 - 135)^2}{135} + \frac{(36 - 45)^2}{45} + \frac{(49 - 45)^2}{45} + \frac{(12 - 15)^2}{15}$$

$$= 3.23$$

Observe that squaring each term eliminates negative values in the sum. Squared units are avoided by dividing each squared deviation by the expected value, so that each deviation is in proportion to the expected size for that class.

Once the deviations have been transformed into a χ^2 value, the probability of obtaining a χ^2 this large or larger is found by consulting ■ Table 3–1. To use the table, the number of degrees of freedom (df) must first be determined. In general, df in a chi-square test is one less than the number of classes, minus an additional unit for each parameter that must be estimated from the data in order to calculate expected values. For the preceding cross results, no parameter had to be estimated from the data, so df = 4 − 1 = 3. In a statistical analysis, df represents the number of classes that can be freely or independently filled. If we place 135 of the 240 offspring in the $A–B–$ class, 45 into the $A–bb$, and another 45 into the $aaB–$ class, we have no choice but to place the remaining 15 into the *aabb* class. There is freedom in filling all classes except for the last one filled, which must be composed of the remaining offspring.

df	\multicolumn{10}{c}{Probabilities}
	0.99	*0.90*	*0.80*	*0.70*	*0.50*	*0.30*	*0.20*	*0.10*	*0.05*	*0.01*
1	0.000	0.016	0.064	0.15	0.46	1.07	1.64	2.71	3.84	6.64
2	0.02	0.21	0.45	0.71	1.39	2.41	3.22	4.61	5.99	9.21
3	0.12	0.58	1.00	1.42	2.37	3.67	4.64	6.25	7.82	11.35
4	0.30	1.06	1.65	2.20	3.36	4.88	5.99	7.78	9.49	13.28
5	0.55	1.61	2.34	3.00	4.35	6.06	7.29	9.24	11.07	15.09
6	0.87	2.20	3.07	3.83	5.35	7.23	8.56	10.65	12.59	16.81
7	1.24	2.83	3.82	4.67	6.35	8.38	9.80	12.02	14.07	18.48
8	1.65	3.49	4.59	5.53	7.34	9.52	11.03	13.36	15.51	20.09
9	2.09	4.17	5.38	6.39	8.34	10.66	12.24	14.68	16.92	21.67
10	2.56	4.87	6.18	7.27	9.34	11.78	13.44	15.99	18.31	23.21
15	5.23	8.55	10.31	11.72	14.34	17.32	19.31	22.31	25.00	30.58
20	8.26	12.44	14.58	16.27	19.34	22.78	25.04	28.41	31.41	37.57
25	11.52	16.47	18.94	20.87	23.34	28.17	30.68	34.38	37.65	44.31
30	14.95	20.60	23.36	25.51	29.34	33.53	36.25	40.26	43.77	50.89

Using 3 df, our calculated χ^2 value (3.23) is found to lie between 2.37 and 3.67 in Table 3–1, which corresponds to a probability between 0.5 and 0.3. Thus the probability that random sampling error alone would produce a χ^2 value as large as or larger than 3.23, if the hypothesis being tested is true, is between 30% and 50%.

To make the final decision as to whether the observed results can be explained by the hypothesis, we must decide on the **level of significance** to use. The most widely used level is 0.05; using this level helps to minimize the chance of accepting a wrong hypothesis without overly increasing the chance of rejecting a correct one. Thus if the probability value associated with χ^2 is less than 0.05, the deviations between the observed and expected values are significant, and the hypothesis proposed to account for the data must be rejected. In our example, the probability is greater than 0.05, so we conclude that the genes are assorting independently. There is no significant difference between the observed results and those expected on the basis of independent assortment.

A few words of caution about using the chi-square test are necessary. The test is valid only when whole numbers are used; it will not work with values that are expressed in fractional or percentage form. The test is also very sensitive to small sample sizes. As a rule of thumb, the expected number in each class should be greater than or equal to 5 for the test to be accurate. Finally, the decision to accept or reject the hypothesis must not be taken as proof that the proposed genetic ratio is or is not the true one. A statistical test can only support or fail to support a proposed hypothesis; it can never actually prove that the hypothesis is true or false. Thus, by failing to reject a proposed hypothesis, we do not necessarily mean that there is nothing wrong with it, only that we have failed to detect anything wrong with it.

Example 3.3

A dihybrid cross ($dp^+dp\ st^+st \times dp^+dp\ st^+st$) that involves differences in wings and eye color in *Drosophila* yields the following results among the offspring (wild-type vs. dumpy wings; wild-type vs. scarlet eyes):

Offspring phenotypes	Number of offspring
wild, wild	380
wild, scarlet	122
dumpy, wild	96
dumpy, scarlet	42
	Total 640

If we assume a 9 wild, wild : 3 wild, scarlet : 3 dumpy, wild : 1 dumpy, scarlet phenotypic ratio (indicating a hypothesis of independent assortment), check the goodness-of-fit by means of the chi-square test. Use a 0.05 level of significance for acceptance or rejection of the hypothesis.

Solution: The expected numbers of offspring under the stated hypothesis are $(\frac{9}{16})(640) = 360$ wild, wild; $(\frac{3}{16})(640) = 120$ wild, scarlet; 120 dumpy, wild; and $(\frac{1}{16})(640) = 40$ dumpy, scarlet. The chi-square value is then calculated as follows:

$$\chi^2 = \frac{(380-360)^2}{360} + \frac{(122-120)^2}{120} + \frac{(96-120)^2}{120} + \frac{(42-40)^2}{40}$$

$$= \frac{(20)^2}{360} + \frac{(2)^2}{120} + \frac{(-24)^2}{120} + \frac{(2)^2}{40}$$

$$= 1.11 + 0.03 + 4.80 + 0.10$$

$$= 6.04$$

In this case, df = 4 − 1 = 3. The calculated χ^2 value is between 4.64 and 6.25 in the chi-square table, which corresponds to a P value between 0.20 and 0.10. Since this probability is greater than 0.05, we do not regard the deviations of the observed values from the expected values as significant, and we accept the hypothesis of a 9:3:3:1 ratio.

Follow-Up Problem 3.3

When a purple-flowered plant is selfed, it produces an F_1 consisting of 60 purple-flowered and 40 white-flowered offspring. Is this result consistent with the 3:1 ratio expected from a monohybrid cross? Check this hypothesis by means of a chi-square test.

Point to Ponder 3.2

A crucial step in the use of the chi-square test is the formulation of a hypothesis. The investigator makes an *educated guess* of the genetic ratio that the observed data are following and then tests that guess statistically, accepting or rejecting as the chi-square test dictates. Keep in mind that statistics can never "prove" that a particular hypothesis is correct or incorrect; a statistical test only accepts or rejects the educated guess as it applies to one particular set of data.

Students often have difficulty "guessing" the genetic ratio, i.e., deriving the hypothesis. For each of the following sets of data, see if you can hypothesize a genetic ratio that seems to describe the observed values: (a) 295:102, (b) 157:46, (c) 235:278, (d) 93:36:29:12, (e) 621:217:199:69, (f) 83:74:96:88, (g) 37:65:34.

What kind of reasoning did you use to come up with the hypotheses? Think about what you did and then calculate the expected numbers in each class.

Chapter Summary

Because of the random nature of gene transmission, the frequencies of genotypes and phenotypes produced by a genetic cross can be expressed as probabilities. Several different concepts of probability are useful in a genetic analysis: the addition rule for calculating the probability of occurrence of either of two (or more) mutually exclusive events, the multiplication rule for calculating the probability of the joint occurrence of two (or more) independent events, and the theory of conditional probabilities, in which the chance of an event is contingent on given circumstances.

When multiple trials of an event occur, the probability of each specific outcome can be expressed in the form of a probability distribution. Three distributions are particularly useful for predicting

discrete probabilities in genetics: the binomial distribution, in which there are two alternative outcomes for each trial, the multinomial distribution, in which more than two outcomes are possible for each trial, and the Poisson distribution, in which the outcome of interest is a rare event.

The chi-square test is a useful statistical test for evaluating the significance of deviations between observed and expected values. A chi-square value greater than some critical value (typically a value that occurs as a result of chance with a probability ≤ 0.05) is accepted as evidence that the hypothesis used to predict the expected values is incorrect.

Questions and Problems

Probability and Genetics

1. Calculate the probability of obtaining each of the following: (a) an ace *or* a king upon drawing a single card from a deck, (b) an ace *and* a king upon drawing two cards, each from a different deck, (c) four aces from a single deck when four cards are drawn without replacement, (d) an even number on the roll of an honest die, (e) numbers adding to eight when two dice are rolled.

2. A husband and wife are both carriers of the recessive gene for the metabolic disorder galactosemia. They plan to have two children. Calculate the probability of each of the following events: (a) the first child will be galactosemic, (b) both children will be galactosemic, (c) the first child will be galactosemic and the second will not, (d) only one child will be galactosemic.

3. How many different gamete classes can be formed by an individual who is heterozygous for (a) three pairs of genes, (b) four pairs of genes, (c) five pairs of genes, (d) n pairs of genes?

4. How many different genotypic classes in the progeny can be produced by selfing a plant that is heterozygous for (a) three pairs of genes, (b) four pairs of genes, (c) five pairs of genes, (d) n pairs of genes?

5. How many different genotypic classes in the progeny can be produced by a testcross in which one parent is heterozygous for (a) three, (b) four, (c) five, (d) n pairs of genes?

6. Assume that five pairs of genes assort independently. Calculate the probability of each of the following events: (a) an *AbcDE* gamete from an individual of genotype *AaBbccDdEe*, (b) an *AaBbCCDdee* offspring from the cross *AaBbCCddEe* ×

AaBbCcDdee, (c) an *A–bbC–D–ee* offspring from the cross *AaBbCCddEe × AabbCcDdee,* (d) an *A–B–C–D–E–* offspring from the cross *AabbCcddEe × aaBbccDdee.*

7. Albinism and attached ear lobes are both recessive traits in humans. Parents with normal pigmentation and free ear lobes have an albino child with attached lobes. What is the probability that their next child will be a normally pigmented, free-lobed son?

8. A woman whose grandfather is albino (a recessive condition), marries a man whose mother is albino. What is the probability that their first child will be albino?

9. Phenylketonuria (PKU) is a serious metabolic defect occurring in individuals who are homozygous for a recessive gene. Two unaffected parents have a daughter with the disease and an unaffected son. What is the probability that the son is a carrier (heterozygous) of the PKU allele?

10. Suppose the son in the preceding question marries an unaffected woman whose father has PKU. What is the probability that their first-born child is affected with the disease?

11. Suppose the son in question 9 marries an unaffected woman whose parents are also unaffected but whose sister suffers from PKU. What is the probability that their first-born child is affected with the disease?

Probability Distributions

12. Determine the probability that in a family of six children, (a) four are girls and two are boys, (b) the four eldest are girls and the two youngest are boys, (c) at least four are girls, (d) all are of the same sex.

13. Albinism is a recessive trait. A couple with normal pigmentation has an albino daughter. Calculate the probability that (a) the next child is albino, (b) of the next four children, the first is albino and the next three are normal, (c) of the next four children, one is albino, (d) the next child is an albino girl, (e) of the next two children, the first is an albino girl and the second is a normal boy, (f) of the next four children, two are albino and two are normal, (g) of the next four children, all are normal girls, (h) of the next four children, two are albino boys. (i) Among four children of this couple, what is the probability that one is a normal girl, two are normal boys, and one is an affected girl?

14. In dogs, black color (*B–*) is dominant to red (*bb*), and short hair (*L–*) is dominant to long (*ll*). The two gene pairs assort independently. A mating between dihybrid black, short-haired dogs yields a litter of five pups. (a) What is the probability that two of the offspring will have black, short hair? (b) What is the probability that the five pups consist of two with black, short hair, one with black, long hair, one with red, long hair, and one with red, short hair?

15. A rare genetic disorder occurs in 0.02% of a population. A sample of 20,000 people chosen at random is tested for the disease. (a) What is the probability that there will be no cases of the disease in this sample? (b) Calculate the probabilities of one, two, three, and four cases occurring in this sample.

16. About 1 in 10,000 newborns suffer from Edwards syndrome, a rare chromosomal abnormality that is fatal within the first months of life. Births with this disorder occur at random in populations. Suppose there are 10,000 births per year in a certain large city. (a) What is the mean number of children born with Edwards syndrome per year in this city? (b) What is the probability that there is one child born with this syndrome in this city during a year?

17. Fruit color in summer squash is determined by two interacting gene pairs, *A,a* and *B,b*. Color can be white (*A–B–* or *A–bb*), yellow (*aaB–*), or green (*aabb*). Fruit shape in this plant is controlled by two other gene pairs, *C,c* and *D,d*. Shape may be disk (*C–D–*), sphere (*C–dd* or *ccD–*), or elongate (*ccdd*). Assume independent assortment for all gene pairs. (a) How many different genotypes are responsible for the white, sphere-shaped phenotype? (b) If a tetrahybrid white, disk-shaped plant is selfed, what fraction of the offspring will be yellow and sphere-shaped? (c) Among 16 offspring produced by the tetrahybrid cross, what is the probability that 12 are white, 3 are yellow, and 1 is green?

Determining Goodness-of-Fit: The Chi-Square Test

18. Determine the number of degrees of freedom involved in testing deviations from the following ratios: (a) 3:1, (b) 1:2:1, (c) 9:3:3:1, (d) 3:6:3:1:2:1, (e) 1:2:1:2:4:2:1:2:1.

19. In tomatoes, red fruit (*A–*) is dominant to yellow (*aa*), and a tall stem (*B–*) is dominant to dwarf (*bb*). A series of matings between dihybrid tomato plants yields the following offspring: 557 tall, red-fruited plants, 187 tall, yellow-fruited plants, 192 dwarf, red-fruited plants, and 64 dwarf, yellow-fruited plants. Use the chi-square test to check these data for the hypothesis that independent assortment has occurred (i.e., a 9:3:3:1 ratio). What conclusion do you reach?

20. Two of the cross results obtained by Mendel are shown below, along with the corresponding theoretical ratios:

Cross	Results	Hypothesis
green pod × yellow pod	428:152	3:1
round, yellow seeds ×		
wrinkled, green seeds	31:26:27:26	1:1:1:1

Check the results of each cross for goodness-of-fit by means of a chi-square test. Use a 0.05 level of significance for acceptance or rejection of the hypothesis.

Extensions of Mendelian Analysis

After the discovery of Mendel's work, interest in genetics grew and many researchers entered the field. It quickly became apparent that Mendel's principles of segregation and independent assortment explained a wide variety of inheritance patterns, in some cases even when the typical $3:1$ and $9:3:3:1$ F_2 ratios were not observed. In such instances, segregation of alleles still occurs and nonallelic genes still assort independently (i.e., the genes are not linked), but the results are modified by the way in which genotypes are expressed and organized into phenotypic classes. In this chapter, we will examine the main extensions to Mendel's laws, showing how the complexities of gene expression provide major additional insight into the nature of heredity. We will also see that these extensions help provide further evidence of the validity and usefulness of Mendel's laws.

Antirrhinum (snapdragons) illustrate incomplete dominance.

MULTIPLE ALLELES

All the inheritance patterns we have discussed so far involve just two alternative forms of each gene, such as A and a or B and b. After the discovery of Mendel's work, it soon became clear that the number of allelic forms of a gene is not limited to two. Several allelic forms of a gene can exist in a population, even though a single individual can possess at most two alleles (if heterozygous) and may possess only one (if homozygous). When three or more allelic forms of a gene are present in a population, they are referred to as **multiple alleles** and are said to constitute a **multiple allelic series**.

Combining Alleles into Genotypes

The presence of more than two alleles in a population can significantly increase the number of possible genotypes. For instance, consider the case of three alleles: A, A', and A''. These alleles yield three kinds of homozygotes—AA, $A'A'$, and $A''A''$—and three kinds of heterozygotes—AA', AA'', and $A'A''$—for a total of six different genotypes. By the addition of just one allele beyond the conventional Mendelian number of two, we have doubled the number of genotypes from three to six.

The increase in number of genotypes becomes even greater with a larger number of alleles, as is illustrated by the inheritance of coat color in rabbits (▶ Figure 4–1). This trait is determined by a series of four alleles c^+, c^{ch}, c^h, and c. The c^+ allele determines the wild-type coat color (agouti), while the c^{ch}, c^h and c alleles are responsible for the variations in coat pigmentation called chinchilla, Himalayan, and albino, respectively. The four alleles can be combined into ten genotypes, with four of the genotypes expressing the wild-type color (c^+c^+, c^+c^{ch}, c^+c^h, and c^+c), three expressing chinchilla ($c^{ch}c^{ch}$, $c^{ch}c^h$, and $c^{ch}c$), two Himalayan (c^hc^h and c^hc), and one albino (cc). The alleles thus show a simple dominance hierarchy in which c^+ is dominant to all the other alleles, c^{ch} is dominant to c^h and c, and c^h is dominant to c (this relationship is summarized as $c^+ > c^{ch} > c^h > c$).

We can easily arrive at a relationship between the number of alleles and number of genotypes by considering a general case of N alleles. When N alleles are combined into different genotypes, they yield N kinds of homozygotes (one for each kind of allele) and $N(N-1)/2$ kinds of heterozygotes. The expression $N(N-1)/2$ comes from the combinations formula that gives the number of ways of combining N different alleles two at a time. You can show yourself that

$$\frac{N!}{2!(N-2)!} = \frac{N(N-1)}{2}$$

Upon adding the numbers of homozygotes and heterozygotes, the total number of possible genotypes then becomes

$$N + \frac{N(N-1)}{2} = \frac{N(N+1)}{2}$$

Thus if $N = 10$, there would be 10 kinds of homozygotes and 45 kinds of heterozygotes, for a total of 55 genotypes. Note that the number of different genotypes increases sharply with an increase in the number of alleles. For example, doubling the number of alleles from 10 to 20 results in almost quadrupling the number of genotypes from 55 to 210.

To Sum Up

When more than two allelic forms for a gene exist, that gene is said to have multiple alleles. The existence of multiple alleles increases the number of possible genotypes. If N represents the number of alleles in the series, then the total number of possible genotypes is calculated as

$$\underset{\substack{\text{number of} \\ \text{homozygotes}}}{N} + \underset{\substack{\text{number of} \\ \text{heterozygotes}}}{\frac{N(N-1)}{2}} = \underset{\substack{\text{number of} \\ \text{genotypes}}}{\frac{N(N+1)}{2}}$$

SEX LINKAGE

Until now we have assumed that the pattern of gene transmission is the same in males and females. In Mendel's ex-

---

▶ **FIGURE 4–1** Coat color in rabbits. (a) Wild-type or agouti color is the result of a black or brown tip on each hair, succeeded by a yellow band, with the portion of the shaft nearest the skin being gray. (b) Chinchilla rabbits lack the yellow band and so appear silver-gray. (c) Himalayan color is white with black extremities. (d) Albino rabbits totally lack pigment.

(a)   (b)   (c)   (d)

periments, for example, identical cross results were obtained regardless of which parent (pistillate or staminate) exhibited the recessive trait; that is, reciprocal crosses gave the same results among the offspring. The first definitive evidence to the contrary came in 1910 with the discovery of genes that are carried on chromosomes involved in sex determination. These so-called **sex-linked genes** exhibit an inheritance pattern that is related to the sex of the individual.

Sex Chromosomes and Sex-Linked Inheritance

In the majority of species with separate sexes, such as most animals, males and females differ with regard to a pair of **sex chromosomes** that are involved in sex determination. One sex has a matched pair of sex chromosomes, while the other sex (the male in mammals and in many insects, but the female in birds) has an unmatched pair or an unpaired chromosome. For example, female mammals have a pair of X chromosomes and produce only X-carrying eggs. Male mammals, in contrast, have one X and one Y chromosome, so that half the sperm cells produced by a male will carry an X chromosome and half will carry a Y chromosome. To distinguish between the sex chromosomes and those chromosomes not associated with the sex of the bearer, we refer to all chromosomes other than the sex chromosomes as **autosomes**. Unlike the sex chromosomes, autosomes occur in matching pairs in both sexes.

Although the sex chromosomes segregate as a pair during gamete formation, they differ significantly in genetic makeup. Some of the genes carried on sex chromosomes are found exclusively on the X chromosome (are **X-linked**) without a corresponding locus on the Y chromosome. Very few genes have been identified as being on the Y chromosome, while the X chromosome bears as many genes as would an autosome of similar size. Because nearly all sex-linked genes are restricted to the X chromosome, the term sex-linkage almost always indicates X-linked rather than Y-linked inheritance. (Y linkage is referred to as **holandric inheritance**.)

The phenomenon of sex linkage was discovered in 1910 by Thomas Hunt Morgan while he was studying the inheritance of eye color in the fruit fly, *Drosophila melanogaster*. The standard (wild-type) eye color in this fly is red, but many variant eye colors have been discovered, one of which is white (▶ Figure 4–2). When red-eyed females are crossed to white-eyed males, all the F_1 have red eyes, indicating that the allele for white eyes is recessive. Intercrosses among the F_1 give what at first glance appears to be a typical 3 dominant:1 recessive ratio. However, closer examination reveals that all the white-eyed F_2 flies are males; no white-eyed females are found. The F_2 ratio of eye color is thus 2 red-eyed females:1 red-eyed male:1 white-eyed male.

An explanation for these results is shown in ▶ Figure 4–3, where white eyes in *Drosophila* are attributed to an X-linked gene w and red eyes to the wild-type allele w^+. Like mammals, female fruit flies have XX and males have XY sex chromosomes. Males can therefore possess only a single

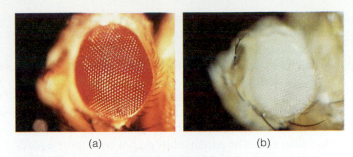

(a)   (b)

▶ **FIGURE 4–2** (a) The standard wild-type *Drosophila* eye color is red. (b) A sex-linked recessive gene causes white eyes.

dose of an X-linked gene and will be limited to two possible genotypes, in this case w^+Y and wY. On the other hand, females have three possible genotypes: w^+w^+, w^+w, and ww. As we can see from this example, terms such as homozygous and heterozygous do not apply to males in the case of

▶ **FIGURE 4–3** Chromosomal basis for eye color inheritance in *Drosophila*. The inheritance pattern is typical of an X-linked pair of alleles when the crossing sequence is started with a homozygous dominant female and a hemizygous recessive male.

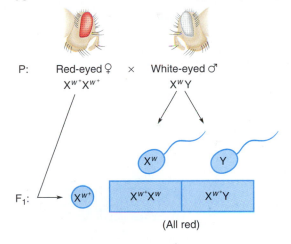

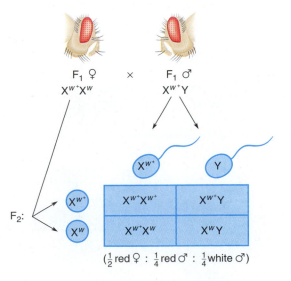

X-linked genes. The term **hemizygous** is used instead to refer to the genotype that is found in the male and consists of a single allele.

The crossing sequence in Figure 4–3 clearly shows that the mechanism for gamete formation and gene expression for X-linked genes follows Mendelian principles. The only difference between X-linked and autosomal inheritance is that with X linkage, one sex has only one copy of the gene rather than two. The parents thus do not contribute equally to the genetic constitution of their offspring. X linkage has three main diagnostic features:

1. *A difference in phenotypic ratios between the sexes.* The proportions of males and females that express an X-linked character will usually differ. This difference reflects the fact that females carry two doses of an X-linked gene, whereas males have only one.
2. *Reciprocal crosses give different results.* When the progeny from reciprocal crosses are compared, different phenotypic ratios will be observed if the trait is X-linked. For example, in ▶ Figure 4–4 the results of the cross $w^+w^+ \times wY$ are compared with the results of the cross $ww \times w^+Y$. Reciprocal crosses give the same result if the genes are on autosomes.
3. *Crisscross pattern of inheritance.* Since a male must receive his single X chromosome from his mother, X-linked genes are transmitted from mother to son. The resulting inheritance pattern can be similar to that in Figure 4–3, where the trait itself appears to pass from mother to son. An X-linked character cannot be passed from father to son, since the son inherits only a Y chromosome from his father.

Morgan recognized that these genetic results of sex linkage were consistent with the behavior of the X and Y chromosomes during gamete formation, where one sex donates only an X chromosome to its gametes and the other sex donates either an X or a Y. This discovery of parallel behavior in genes and chromosomes provided the first supporting evidence for the chromosome theory of inheritance.

Since the discovery of sex linkage, many different X-linked genes have been identified in various organisms, including humans. One example of an X-linked recessive trait in humans is the blood clotting disorder hemophilia A. Despite the rare nature of this disorder (it occurs at a frequency of 1 in 10,000 males), hemophilia A has received considerable attention because it was spread through the royal families of Europe by the descendants of Queen Victoria of England (▶ Figure 4–5); a pedigree of this disor-

▶ **FIGURE 4–4** Failure of X-linked genes to give the same results in reciprocal crosses.

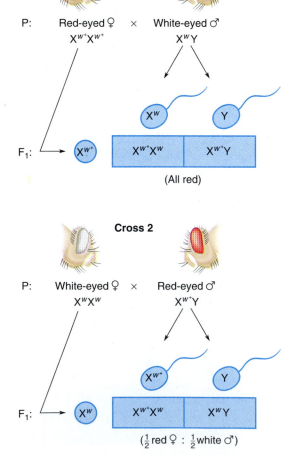

Cross 1

P: Red-eyed ♀ × White-eyed ♂
    $X^{w^+}X^{w^+}$ X^wY

F₁: X^{w^+} $X^{w^+}X^w$ | $X^{w^+}Y$

(All red)

Cross 2

P: White-eyed ♀ × Red-eyed ♂
    X^wX^w $X^{w^+}Y$

F₁: X^w $X^{w^+}X^w$ | X^wY

($\frac{1}{2}$ red ♀ : $\frac{1}{2}$ white ♂)

▶ **FIGURE 4–5** Queen Victoria and some of her family, as they posed for a photograph in the late 1800s.

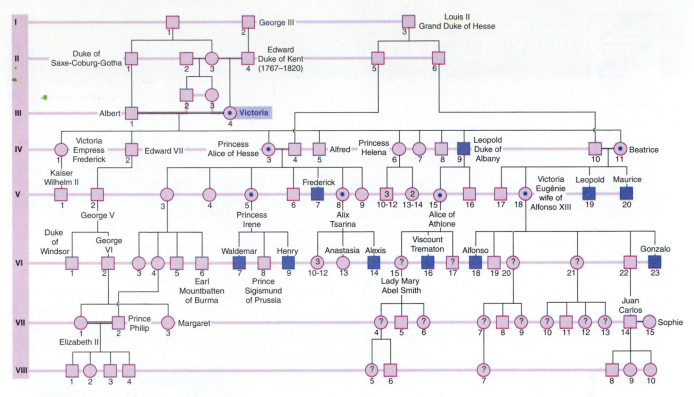

► **FIGURE 4–6** A pedigree of the family of Queen Victoria (III–4), showing the inheritance of hemophilia A in the royal families of Europe. Some individuals of interest are Prince Albert (III–1), King Edward VII of England (IV–2), Princess Alice of Hesse Darmstadt (IV–3), Prince Leopold, Duke of Albany (IV–9), Princess Beatrice of Battenberg (IV–11), Princess Irene of Hessen (V–5), Princess Alix, wife of Czar Nicholas of Russia (V–8), Princess Alice, wife of Alexander, Prince of Teck (V–15), Princess Victoria, wife of King Alfonso of Spain (V–18), Czarevitch Alexis of Russia (VI–14), Lord Trematon (VI–16) Prince Alfonso (VI–18), and Prince Gonzalo (VI–23). The last three died after automobile accidents. Queen Victoria passed the hemophilia gene to her daughters Princesses Alice and Beatrice, and to her son Prince Leopold. *Adapted from:* Michael R. Cummings, *Human Heredity: Principles and Issues,* 3d ed. (West Publishing Company, 1994), p. 117.

---

der in her descendants is shown in ► Figure 4–6. Males are affected more often than females, because a single copy of the recessive gene is all that is needed for expression in males. This is a common feature of rare, recessive X-linked traits.

■ Table 4–1 lists several human traits that are caused by X-linked genes. Notice the wide variety of characteristics listed. The traits that are controlled by the vast majority of X-linked genes have nothing whatsoever to do with sexual function. The type of trait itself therefore gives us no clue as to whether it is X-linked or autosomally inherited. We must base our conclusions about the chromosomal location of a gene solely on the pattern of inheritance, not on the type of trait that it produces.

While sex determination in humans and many other species follows the XX-female, XY-male system, there are notable examples of organisms that employ a different sex chromosome mechanism for determination of sex. For example, in birds the female has an unmatched pair of sex chromosomes and the male, a matched pair, as mentioned earlier. Birds follow a ZW-female, ZZ-male system, where

■ **TABLE 4–1** Examples of X-linked characteristics in humans.

Agammaglobulinemia (unusual proneness to bacterial infection)
*Color blindness
*Congenital deafness
*Diabetes
Glucose 6-phosphate dehydrogenase deficiency
Hemophilia
Lesch-Nyhan syndrome (mental retardation, "self-destructive" behavior)
*Ichthyosis (scaling of the skin)
*Immunodeficiency diseases
Juvenile (Duchenne) muscular dystrophy
*Night blindness
Ocular albinism
*Pituitary dwarfism
*Retinitis pigmentosa (progressive degeneration of the retina)
*Testicular feminization syndrome
Vitamin D-resistant rickets

*Disease has two or more different genetic causes, X-linkage being but one of the possibilities.

Color Blindness

Three genes carry the information for members of a class of proteins that allow humans to see the three primary colors in light—red, green, and blue (all other colors are mixtures of these three primary ones). These genes direct the production of visual pigment proteins called *opsins* that are located on photoreceptor cells (the cone cells) in the retinas of our eyes. The opsins function by binding to visual pigments in the red-, green-, or blue-cone cells, creating opsin/pigment complexes that absorb light of a particular wavelength. The blue-sensitive pigment complex absorbs maximally at 420 nm, the green-sensitive pigment complex at 530 nm, and the red-sensitive pigment at 560 nm. If one of the opsin proteins is defective or absent, the function of its corresponding cone cells is impaired, and inability to distinguish a particular color results.

The opsins that register red and green are determined by X-linked genes, whereas the blue-registering opsin is determined by a gene located on an autosome. Both red color blindness and green color blindness are X-linked recessive traits, and blue color blindness, which is very rare, is an autosomal dominant condition. The most common form of color blindness is known as red/green color blindness; it affects about 8% of the males of northern European descent in the United States. Males who have a mutation in the gene that determines the red-registering opsin are unable to distinguish red as a distinct color, a condition referred to as *protanopia* (they see both red and green as green). Similarly, males with a mutation in the green-registering opsin have green color blindness or *deuteranopia* (they see both red and green as red). Of the 8% of males affected with red/green color blindness, about 25% have protanopia and 75% deuteranopia.

The red and green color-blind conditions are combined into an overall category termed red/green color blindness in part because the two opsin genes are located next to one another on the X chromosome. In reality, a man's red/green color vision genotype should be symbolized using both gene pairs. The mutant gene for protan color blindness is symbolized by *p*, and its normal allele is *P*; the mutant allele for deutan color blindness is symbolized by *d*, and its normal vision allele is *D*. Thus, completely normal red/green color vision requires the presence of both dominant nonallelic genes. Similar genes are also present in other primates and in a wide variety of other vertebrates, including birds.

The three primary colors in light—red, green, and blue.

---

the sex chromosomes are represented by Z and W rather than X and Y. Because either sex can possess the matched (or the unmatched) sex chromosomes, depending on the organism, we use the term **homogametic sex** to refer to the sex that has matched sex chromosomes and **heterogametic sex** to refer to the sex having unmatched sex chromosomes.

To Sum Up

1. Dimorphism of the sex chromosomes is common among species with separate sexes. For example, in mammals females are XX and males are XY.

2. In X-linked inheritance, sexes usually exhibit different phenotypic ratios. If females are XX and males are XY, for example, then more males than females express the recessive phenotype, and more females than males express the dominant trait. The reverse is true in birds, where the male has the matched pair of sex chromosomes.

3. X-linked inheritance also causes different results in the offspring of reciprocal crosses. In contrast, reciprocal crosses give the same offspring results if inheritance is autosomal.

4. A crisscross inheritance pattern is another indication of X linkage. Grandfather-to-grandson and mother-to-son patterns are frequently observed for X-linked recessive characteristics. A father-to-daughter pattern is typical of an X-linked dominant gene.

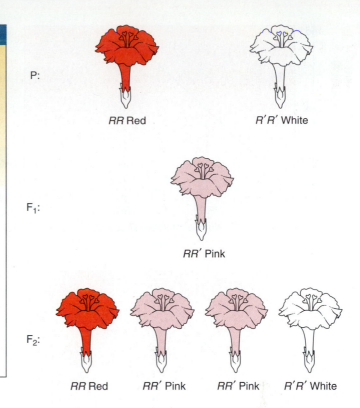

P: RR Red   R'R' White

F₁: RR' Pink

F₂: RR Red   RR' Pink   RR' Pink   R'R' White

▶ **FIGURE 4–7** Incomplete dominance of alleles for flower color in four-o'clocks. A cross between red-flowered and white-flowered plants gives an F_1 with pink flowers. When intercrossed, these F_1 hybrids yield a 1 red:2 pink:1 white phenotypic ratio in the F_2.

$$\text{chemical precursors} \xrightarrow{} \xrightarrow{} \xrightarrow{} \xrightarrow{} \xrightarrow{} \text{colorless substrate} \xrightarrow[\text{enzyme}]{R\text{ gene}} \text{red pigment}$$

Points to Ponder 4.1

Fragile X syndrome is a sex-linked form of mental retardation that occurs most often in males. Its exact genetic basis and mode of inheritance are complex and will be described later in Chapter 22. Fragile X syndrome is the second most common genetic cause of mental impairment, after Down syndrome. Its severity varies a great deal, ranging from severe retardation to learning disabilities, poor speech, hyperactivity, social anxiety, and short attention span.

Lesch-Nyhan syndrome is another sex-linked recessive disorder. Affected infants seem normal at first but by six months of age begin to experience neurological difficulties that worsen over time, leading to severe mental retardation, possible seizures, and aggressive and self-mutilative behavior. Death usually occurs by age 30.

Suppose that three babies are born with heart defects. Genetic testing reveals that one baby has fragile X syndrome, another Lesch-Nyhan syndrome, and the third has no additional abnormalities. Without surgery the babies will die. Does each baby have the right to receive the surgery? Should each baby receive the surgery? Who should be involved in making this decision?

DOMINANCE RELATIONSHIPS

In addition to assuming that two genes of each kind occur in both sexes, the basic Mendelian model also assumed that one allele is always completely dominant over the other allele. However, not all alleles show complete dominance. For example, flower color in carnations, snapdragons, and four-o'clocks can be red (RR), pink (RR'), or white ($R'R'$) (▶ Figure 4–7). Thus the flowers of the heterozygotes in these species have a diluted, intermediate color (in contrast, heterozygotes of the garden pea have the same flower color as dominant homozygotes), and neither allele appears to mask the other completely. The R and R' alleles are thus said to exhibit **incomplete dominance**. Note that the intercross between heterozygotes ($RR' \times RR'$) yields a ratio of 1 red (RR):2 pink (RR'):1 white ($R'R'$), instead of the familiar 3:1 ratio; thus the phenotypic ratio is identical to the genotypic ratio.

We can readily explain how incomplete dominance operates in flower color if we assume that the dilution effect is due to a complete or partial loss of gene activity. Suppose, for example, that a red flower pigment is produced from colorless precursors through a complex series of chemical reactions, each catalyzed by a different enzyme. Furthermore, assume that the amount of enzyme catalyzing the last reaction is under the control of gene R, so that a one-to-one relationship exists between the activity of the enzyme and the number of R alleles per genotype. (Possible mechanisms for this gene-enzyme relationship will be discussed in later chapters.) We can represent the stepwise conversion of precursors to pigment in the following manner:

Note that if the R' allele lacks function, so that a functional enzyme can be formed only in the presence of the R allele, $R'R'$ homozygotes will be unable to produce red pigment. The flowers of $R'R'$ homozygotes will therefore be white.

Flower pigments are complex substances. Ordinarily, several different chemical steps are required to synthesize them from simple precursor molecules. Despite the complexity of the process, the overall rate of pigment synthesis tends to be limited by the rate of the slowest reaction (the rate-limiting step). If the reaction in question is the rate-limiting step, then any increase in the activity of the enzyme under the control of gene R should result in a corresponding increase in the amount of red pigment and hence in the intensity of flower color. This allelic effect is shown in ▶ Figure 4–8, where flower color is plotted against the genotype. At low levels of gene (and enzyme) activity, heterozygotes show phenotypic effects that are intermediate to those of the RR and $R'R'$ homozygotes (Figure 4–8a). The R allele in this case exhibits incomplete dominance.

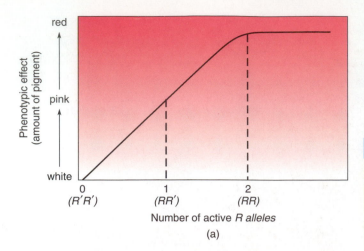

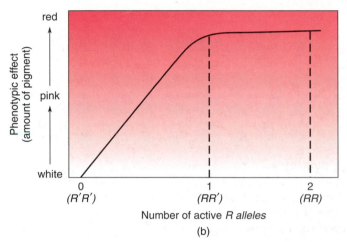

▶ **FIGURE 4–8** The distinction between complete and incomplete dominance. The graphs show a postulated increase in the degree of flower pigmentation corresponding to an increase in the number of active *R* alleles. It is assumed that the *R* allele controls a rate-limiting chemical step involved in the synthesis of red pigment. The amount of red pigment will rise up to a maximum level at which point another chemical reaction becomes the limiting step in pigment production. (a) Incomplete dominance, in which the effect of *RR'* falls below the maximum and is intermediate between that of *R'R'* and *RR*. (b) Complete dominance, in which both *RR* and *RR'* have effects at the plateau level and are indistinguishable from one another in phenotype.

Increases in pigment intensity cannot continue indefinitely. Ultimately, a plateau is reached at which some other chemical reaction involved in pigment formation cannot keep pace and becomes the rate-limiting step. If the presence of a single *R* allele is sufficient to raise the rate of the reaction under its control to the plateau level, additional *R* alleles would have no effect, and in this case heterozygotes would express a flower color indistinguishable from that of the *RR* homozygotes (Figure 4–8b). The *R* allele would thus show complete dominance over the *R'* allele.

In this model of flower color, dominance does not result from a masking effect at all; instead it reflects a difference in allele activity. We see now that the degree of dominance can vary along a continuous scale ranging from no dominance when heterozygotes are exactly intermediate in character, through various forms of partial dominance, to complete dominance at the other extreme when heterozygotes and dominant homozygotes are indistinguishable in appearance.

Codominance and Blood Group Inheritance

In addition to quantitative differences in the degree of phenotypic effect, alleles may also show qualitative differences in the kinds of effects they produce. Such qualitative differences in allele expression form the basis of another dominance relation called **codominance.** Two alleles are codominant if both are fully functional and express themselves individually when they occur in the heterozygous state. In codominance, as in incomplete dominance, heterozygotes are distinguishable on the basis of phenotype; but in contrast to incomplete dominance, a dilution effect does not occur in codominance. Heterozygotes show instead an unblended mixture of the separate effects of both alleles.

A well-known example of codominance in humans occurs in the ABO blood group system. These different blood groups are based on the kind of **antigen** present on the surface of the red blood cells. (Any molecule that stimulates the production of specific antibodies or that binds specifically to an antibody is called an antigen.) In the ABO system, the A and B antigens are determined, respectively, by the I^A and I^B alleles. Both of these alleles code for functional enzymes, and both produce their corresponding antigens through specific enzymatic modifications of a short polymer of sugars—an oligosaccharide (called H substance)—protruding from the membrane of human erythrocytes (▶ Figure 4–9). The I^A allele is responsible for an enzyme that at-

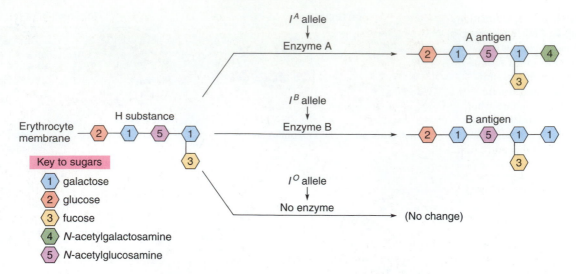

Key to sugars
1. galactose
2. glucose
3. fucose
4. N-acetylgalactosamine
5. N-acetylglucosamine

▶ **FIGURE 4–9** Biosynthesis of the A and B antigens on the red blood cell. By the action of specific enzymes, different sugars (N-acetylgalactosamine for antigen A and galactose for antigen B) are added to the end of an oligosaccharide called H substance. *Source:* A. P. Mange and E. J. Mange, *Genetics: Human Aspects* (Sunderland, MA: Sinauer Associates, Inc., 1989), p. 343.

taches an N-acetylgalactosamine residue to the end of the oligosaccharide, whereas the I^B allele is responsible for an enzyme that attaches a galactose residue to the basic structure. When both these alleles are present in the same individual, both antigens are formed. Alleles I^A and I^B are therefore codominant, with the $I^A I^B$ heterozygotes having a mixture of A and B antigens on their red blood cells and expressing both alleles in full.

In addition to the I^A and I^B alleles, there is also a third major allele in the ABO system, designated I^O. The I^O allele does not produce a specific antigen; it is therefore fully recessive to both I^A and I^B. Thus, individuals with the $I^A I^A$ or $I^A I^O$ genotype have only type A antigen, those with the $I^B I^B$ or $I^B I^O$ genotype have only type B antigen, $I^A I^B$ heterozygotes have both antigens, and $I^O I^O$ individuals have neither antigen. The phenotype and genotype relationships for this blood group system are given in ■ Table 4–2.

The classical test for ABO blood type is based on an antigen-antibody reaction that occurs when red blood cells that contain one or both antigens are mixed with a serum containing a specific antibody. The reaction causes the erythrocytes to adhere to one another so that they precipitate out

of the suspension of blood, forming a clumped mass of cells (▶ Figure 4–10). Because of the specificity of the antigen-antibody recognition, the A antigen reacts only with anti-A antibody and the B antigen reacts only with anti-B antibody.

▶ **FIGURE 4–10** The reaction between red blood cell antigens and antibodies against them. (a) The appearance of erythrocytes under the microscope (about 500× magnification). (b) The erythrocytes have been clumped by the addition of antibody against the cell surface antigens. (c) A molecular view of this reaction, showing that clumping arises from the ability of each antibody to combine with two antigens on different cells.

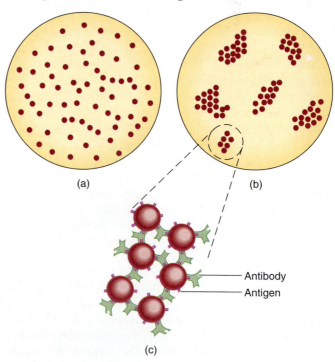

■ **TABLE 4–2** The ABO blood groups.

Genotype	Blood Type	Antigen on Cells	Antibody in Serum
$I^A I^A$, $I^A I^O$	A	type A only	anti-B only
$I^B I^B$, $I^B I^O$	B	type B only	anti-A only
$I^A I^B$	AB	types A and B	neither anti-A nor anti-B
$I^O I^O$	O	neither A nor B	both anti-A and anti-B

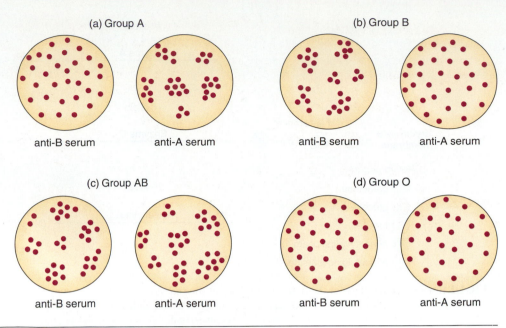

► **FIGURE 4–11** Agglutination reactions of the ABO blood groups. (a) Type A blood clumps only when exposed to anti-A serum. (b) Type B blood clumps only when exposed to anti-B serum. (c) Type AB blood clumps when exposed to either antiserum. (d) Type O blood will not clump when exposed to either antiserum.

(a) Group A

anti-B serum · anti-A serum

(b) Group B

anti-B serum · anti-A serum

(c) Group AB

anti-B serum · anti-A serum

(d) Group O

anti-B serum · anti-A serum

The diagnostic agglutination reactions of the different blood types are shown in ► Figure 4–11.

A remarkable feature of the ABO system is that each person's blood automatically contains antibodies against whichever antigen is not present on his or her red blood cells. In other words, prior contact with a foreign A or B antigen is not a prerequisite for antibody production. In most other immune systems of the body, actual contact with a foreign antigen is required before the specific antibodies are produced. Since clumping in the bloodstream could be fatal, knowledge of the ABO system is essential for determining which combinations are incompatible in blood transfusions. One type of incompatible combination occurs when the erythrocytes of the donor carry an antigen that corresponds to the antibodies in the serum of the recipient. For example, a type A donor and a type B recipient would be an incompatible combination, since the serum of a type B person contains anti-A antibodies.

The presence of the antibody in the serum of the donor is also a factor in compatibility for transfusions. Type O individuals, who have neither antigen, are referred to as "universal donors." Type O blood does contain both antibodies, however, so that in practice it is transfused to persons with other blood types only in true emergencies. In such situations, the blood is transfused very slowly, to allow the antibodies present in the donor serum to dilute rapidly in the blood of the recipient. ■ Table 4–3 lists the ABO combinations that are permitted and prohibited in transfusions.

Blood tests involving the ABO series (as well as other blood groups) are also used in medical and legal situations. In cases of disputed parentage, the results of these tests can be of considerable value in determining whether a person is the parent of a particular child. Evidence from blood tests is conclusive only in excluding the possibility of parentage, however. These tests cannot prove that a person is the parent of a child. Thus, in cases of illegitimacy or abandon-

ment, most courts will admit this genetic evidence only when it provides information that rules out parentage by the person in question.

Example 4.1

A family has four children, one with blood type A, one with type B, one with type AB, and one with type O. What are the genotypes of the parents?

Solution: The type O offspring is $I^O I^O$ in genotype and must have received an I^O allele from each of the parents. The AB offspring is $I^A I^B$ in genotype and therefore must have received an I^A allele from one parent and an I^B allele from the other. Thus, one parent must be heterozygous A ($I^A I^O$), and the other must be heterozygous B ($I^B I^O$).

■ **TABLE 4–3** The compatibility of ABO blood types for transfusions.

		Donor Blood Type			
		A	B	AB	O
Recipient Blood Type	A	C	I	I	E
	B	I	C	I	E
	AB	E	E	C	E
	O	I	I	I	C

C = Compatible; transfusion permitted as long as all other blood group systems are also compatible.

I = Incompatible; transfusion not permitted because of reaction of antigen on donor cells with antibody in recipient's plasma.

E = Transfusion permitted under emergency circumstances; antibodies in donor serum react with antigens of recipient.

Lectins (plant extracts that agglutinate red blood cells) can be used to distinguish different subgroups of type A blood. The two major divisions are A_1 and A_2; approximately 80% of type A individuals are A_1. The alleles that govern the two major subgroups of type A blood show the following dominance relationships to each other and to the other alleles in the ABO series: $I^A = I^B > I^O$; $I^{A_1} > I^{A_2}$. (a) If we include the two major subgroups of type A, how many different genotypes are possible in the ABO system? (b) How many different blood types do these genotypes produce?

To Sum Up

1. Different degrees of dominance can be explained by differences in the activity of alleles. Quantitative differences between alleles (differences in the degree of the phenotypic effect) can lead to either complete or incomplete dominance. In complete dominance, one allele has enough activity to give the maximum phenotypic effect, so that genotypes homozygous and heterozygous for this allele have the same phenotype. In partial and incomplete dominance, one allele has only enough activity to give a partial phenotypic effect, and thus homozygotes and heterozygotes have different phenotypes. Homozygotes and heterozygotes also differ in phenotype when codominance occurs; in codominance, qualitative differences (differences in the kind of phenotypic effect) are produced by two different alleles, both of which are functional. A key feature of both partial/incomplete and codominance is that heterozygotes have a different phenotype than either homozygote.

2. The ABO blood group is an example of a multiple allelic series in which the alleles exhibit both codominance and complete dominance with regard to other alleles in the series. Alleles I^A and I^B are codominant with each other, and both are completely dominant over I^O. The blood types in this system are characterized by the presence (or absence) of the A and B antigens on the red blood cells and by the ability (or inability) to produce certain antibodies against these antigens.

✳ NONALLELIC GENE INTERACTIONS AND ENVIRONMENTAL EFFECTS

Allelic interactions, such as dominance, are not the only mechanism that can alter gene expression. The expression of a gene can also be influenced by the actions of other genes and by the nongenetic environment.

Interactions between Nonallelic Genes

So far, we have dealt with genetic patterns in which each trait is determined by a single gene. We will now look at some situations in which a trait is determined by the interaction of two or more different genes. In these cases the products of the nonallelic genes act together to produce the phenotype. Effects produced by multiple genes are quite common in organisms, because even seemingly simple traits tend to be formed through a complex sequence of developmental steps, with different genes controlling different parts of the overall process.

One consequence of multiple genes is that more phenotypic classes may occur than can be accounted for by the action of only one pair of alleles. A classic example is the inheritance of comb shape in domestic breeds of chickens (▶ Figure 4–12). The Leghorn breed has a "single" comb. Wyandottes have a structurally different type of comb, a "rose" comb, whereas Brahmas have "pea" combs. When Wyandottes and Brahmas are crossed, the F_1 hybrids all have a "walnut" type of comb, whose shape is different from that of either parent. Intercrosses among the F_1 give four kinds of F_2 progeny—too many to be accounted for by a single pair of genes. Moreover, the progeny occur in a ratio of 9 walnut : 3 rose : 3 pea : 1 single, which is the classical

▶ FIGURE 4–12 Comb shapes in domestic breeds of chickens: (a) single, (b) pea, (c) rose, and (d) walnut.

(a)

(b)

(c)

(d)

EXTENSIONS AND TECHNIQUES

Tests for Allelic versus Nonallelic Genes

Alternative phenotypes for a single trait can result from two or more alleles at a single gene locus (allelic genes) or from the expression of genes at two or more separate loci (nonallelic genes). When a new phenotype appears in a population, therefore, geneticists usually begin their investigation by attempting to determine whether the gene for the variant phenotype is allelic or nonallelic to any other gene known to affect the trait in question.

For example, suppose that in a certain species of plant that normally produces purple flowers, we discover two pure-breeding variants: one with red flowers and one with yellow flowers. When each variant is mated back to the standard type, only purple-flowered offspring are produced. If we hypothesize that the three colors are due to allelic genes, we could designate the true-breeding types as AA (purple), a_1a_1 (red), and a_2a_2 (yellow), where A is dominant to the other two alleles. In contrast, if we hy-

pothesize nonallelic genes, we might consider the interaction of two gene pairs, A,a and B,b, and we might then designate the pure-breeding varieties as $aaBB$ (red), $AAbb$ (yellow), and $AABB$ (purple). In either case, crossing the variant types back to the purple variety would produce only purple-flowered offspring, as was observed:

For alleles: $a_1a_1 \times AA \rightarrow Aa_1$
and $a_2a_2 \times AA \rightarrow Aa_2$

For nonalleles: $aaBB \times AABB \rightarrow AaBB$
and $AAbb \times AABB \rightarrow AABb$

In order to determine whether the genes are allelic or nonallelic, we can perform a **complementation test**, in which we cross the newly discovered true-breeding variants with each other and observe the phenotype of the F_1. If the genes for the contrasting phenotypes are allelic, then all the offspring will exhibit a mutant phenotype (either red or yellow or an intermediate color, depending on the dominance re-

lationship between the variant alleles):

$a_1a_1 \times a_2a_2 \rightarrow a_1a_2$
(mutant phenotype)

On the other hand, if the genes for the alternate phenotypes are nonallelic, then the F_1 will all exhibit the standard wild-type color, because the wild-type forms of both gene pairs will complement each other in the heterozygous state:

$aaBB \times AAbb \rightarrow AaBb$
(wild-type phenotype)

Another test for allelic versus nonallelic genes is the **analysis of segregation ratios** produced upon intercrossing the F_1. If the genes are allelic, intercrossing the F_1 would produce a monohybrid ratio of $3:1$ or $1:2:1$ depending on the dominance relationship. Nonallelic genes, in contrast, would give an F_2 ratio of $9:3:3:1$ or a modification of that ratio if an epistatic relationship exists.

Mendelian dihybrid ratio with complete dominance in each allelic pair. Using the symbols A,a and B,b to denote the two gene pairs involved in comb shape, the genotypes and their corresponding phenotypes are $A-B-$ = walnut, $A-bb$ = rose, $aaB-$ = pea, and $aabb$ = single. The Wyandotte breed is $AAbb$, the Brahmas are $aaBB$, Leghorns are $aabb$, and the F_1 walnuts are $AaBb$. This example illustrates how one character can exhibit several novel phenotypes because two gene pairs, rather than one, specify the trait.

In the preceding example, we were able to recognize the effects of two gene pairs from the expected ratio of a dihybrid cross, but it is not always possible to do so. When two or more gene pairs affect the development of the same character, a form of nonallelic gene interaction called **epistatic interaction** can modify the expected phenotypic ratios. The term **epistasis** was originally used to describe the masking

of the expression of one or both members of a pair of alleles by a nonallelic gene. Epistasis is similar to dominance but differs in an important way. Dominance refers to the relationship between alleles of the same gene, whereas epistasis involves an interaction between entirely different genes. As with dominance, the conventional definition of epistasis, which implies an inhibiting effect, should not be taken literally. Only in certain limited cases does a member of one gene pair actually act to suppress the action of another gene.

A classic example of epistatic interaction is provided by common white clover. Some strains of this species are high in cyanide (HCN) content, while others test negatively for the substance. Cyanide is associated with richer vegetative growth in white clover but does not seem to harm cattle that eat these plants, despite its usual toxicity. High cyanide production has been shown to be the result of one dominant allele at each pair of genes (A–B–). If there is not at least one dominant allele at each pair of genes, the plants fail to produce HCN in detectable amounts. Thus a dihybrid cross ($AaBb \times AaBb$) of cyanide-producing plants gives rise to a 9:7 phenotypic ratio in terms of cyanide production:

Dominant
allele at
each gene
pair yields   *aa* and/or *bb* results in no HCN
high HCN   production

9 A–B–:(3 A–bb + 3 aaB– + 1 $aabb$)

= 9 HCN positive:7 HCN negative

At this point it is useful to look at a simple model of the underlying biochemical basis of cyanide production in clover. Such a model helps to relate a Mendelian phenotypic ratio for a trait to the actual functioning of the underlying genotype. Cyanide is produced in clover cells from the substance cyanogenic glucoside through a biochemical reaction that is catalyzed by a specific enzyme (enzyme β). Another enzyme (enzyme α) is responsible for the formation of cyanogenic glucoside from precursor substances. The fact that two different enzymes as well as two different genes are involved in cyanide production suggests that the A and B genes may function by controlling the formation of enzymes α and β. One possible sequence of biochemical reactions might be:

gene A gene B
↓   ↓
enzyme α   enzyme β
precursor ⟶ cyanogenic glucoside ⟶ cyanide

We can now see why genes A and B must both be present in their functional form (that is, as A–B–) in order for both reactions to occur and for the precursor compound to be transformed sequentially into HCN. Plants lacking a functional A gene, as in the case of aaB– and $aabb$ genotypes, also lack enzyme α and are unable to produce cyanogenic glucoside. Without the substrate needed for the second reaction, such plants cannot form HCN. Similarly, plants lacking a functional B gene are HCN negative because they lack enzyme β and are unable to convert cyanogenic glucoside into cyanide.

Experimental tests of the above model have been performed by measuring cyanide production in extracts from the leaves of plants following the administration of either enzyme β or cyanogenic glucoside. The results of these tests are summarized in ■ Table 4–4. Recall that A–bb and $aabb$ plants lack enzyme β and that neither aaB– nor $aabb$ genotypes can produce cyanogenic glucoside. The observed effects of glucoside and enzyme β are in agreement with the model.

Several other modifications of the 9:3:3:1 dihybrid ratio can also occur. These modified ratios depend on the genes involved and the nature of the epistatic interaction and include 15:1, 13:3, 12:3:1, and 9:3:4 (■ Table 4–5). (Later in the text we will see that these ratios, like the 9:7

■ **TABLE 4–4** Tests of leaf extracts for cyanide production.

Phenotype	Leaf Extract Alone	Leaf Extract and Glucoside	Leaf Extract and Enzyme β
A–B–	+	+	+
aaB–	–	+	–
A–bb	–	–	+
$aabb$	–	–	–

+ = Cyanide production.
– = No cyanide production.

ratio, can be explained by specific kinds of underlying biochemical pathways.) Although these modified ratios have often been used to identify different types of epistasis, caution is necessary, since the phenotypic ratio produced by a particular cross depends to a certain extent on how the phenotype is defined and measured. For example, more recent tests reveal that A–bb plants can produce HCN, but they produce it much more slowly than A–B– plants. If we were to define the phenotype in white clover on the basis of whether plants can produce cyanide rapidly, slowly, or not at all, a dihybrid cross ($AaBb \times AaBb$) would then yield a ratio of 9 fast (A–B–):3 slow (A–bb):4 negative (aaB– and $aabb$), rather than the ratio of 9:7 we discussed earlier. Thus the ratio alone does not necessarily imply that a particular epistatic interaction is operating.

Penetrance and Expressivity

As we have seen, the expression of a gene can be suppressed or otherwise modified by the action (or inaction) of other genes. The external environment can also alter gene expression. For example, in *Drosophila*, the normal wings are full and extend beyond the end of the abdomen (▶ Figure 4–13). A recessive mutant gene, *vg*, causes the wings of homozygotes for this gene to develop improperly at ordinary temperatures, so that only vestigial, or rudimentary, wings are formed. However, if these mutant flies are raised at abnormally high temperatures, such as about 30°C, many of these flies fail to express the expected vestigial-wing trait and instead develop normal wings. In this case, the environmental stress causes an extreme modification of gene expression. We can readily establish that the condition is environmentally induced, because the long-winged trait is not passed on to the next generation when environmental conditions return to normal: if the seemingly normal but genetically mutant flies are interbred at ordinary temperatures, their offspring will develop vestigial wings.

A genotype may thus fail to express its expected phenotype for a variety of different reasons. When this occurs, the gene is said to lack **penetrance**. The penetrance of a gene is measured quantitatively as the percent of individuals of a certain genotype that show the expected phenotype. For

Interaction	Nature of Interaction	Example	Ratio among Dihybrid Cross Offspring	
Complementary gene interaction	Expression of one gene is needed for expression of the other.	Coat color and pattern in mice	9 agouti 3 plain 4 white	9 A–B– 3 aaB– { 3 A–bb 1 aabb
		Flower color in sweet peas	9 purple 7 white	9 A–B– { 3 A–bb { 3 aaB– { 1 aabb
Modifier gene interaction	Expression of one gene is suppressed by action of the other.	Fruit color in summer squash	12 white 3 yellow 1 green	{ 9 A–B– { 3 A–bb 3 aaB– 1 aabb
		Feather color in chickens	13 white 3 color	{ 9 A–B– { 3 A–bb { 1 aabb 3 aaB–
Duplicate gene interaction	Different genes have identical functions.	Color in swine	9 red 6 sandy 1 white	9 A–B– { 3 A–bb { 3 aaB– 1 aabb
		Seed capsule shape in shepherd's purse	15 triangle 1 ovoid	{ 9 A–B– { 3 A–bb { 3 aaB– 1 aabb

▶ **FIGURE 4–13** Wing length in *Drosophila*. (a) An individual with wild-type wings. (b) One with vestigial wings.

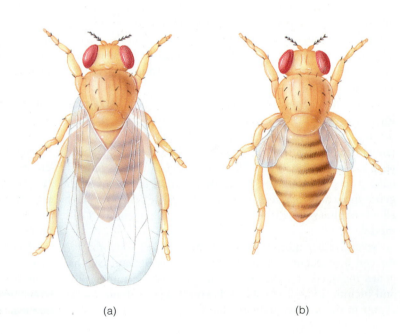

(a)   (b)

example, if 60 out of 100 flies homozygous for the *vg* gene show vestigial wings when raised at high temperatures and the other 40 have the normal phenotype, the penetrance of the recessive allele in the population is 60%. Penetrance depends greatly on both environmental conditions and the influence of other genes. In another group of homozygous flies, the *vg* gene might show 20% or even 90% penetrance. Because the penetrance of a gene can vary so much from one population to the next, we use the catchall term **incomplete penetrance** to describe any gene that expresses the expected phenotype less than 100% of the time.

Even if a gene is 100% penetrant, its **expressivity** (degree of expression) may vary among individuals. For example, flies that are homozygous for the *vg* gene and are raised at 21°C all have vestigial wings, but the precise wing size varies. In such a case, the gene is said to exhibit **variable expressivity**. As in incomplete penetrance, the reasons for such variation in phenotypic expression can be genetic or environmental or a combination of the two and are usually unknown.

Example 4.2

The failure of II−2 in the pedigree below to express the trait designated by the filled-in symbols violates the patterns expected for simple dominant/recessive inheritance. Show that this apparent case of incomplete penetrance can be explained on the basis of two complementary gene pairs that interact to produce a single trait.

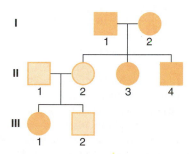

Solution: Assume that individuals with at least one dominant allele for each of two gene pairs (A–B–) show the trait in question. The other genotypes (A–bb, aaB–, and $aabb$) do not. One of several possible combinations of genotypes that could explain the pedigree is:

```
         AaBb ——— AaBb
           |
  aaBb — Aabb   A–B–   A–B–
    |
  A–B–   aaB–
         (or A–bb)
```

Follow-Up Problem 4.2

While performing some of their early experimental work in genetics, Bateson and Punnett made the surprising discovery that crossing two pure-breeding white-flowered varieties of the sweet pea gave rise to an F_1 with all purple flowers. Intercrossing the F_1 yielded an F_2 ratio of 9 purple : 7 white. Explain these results and suggest how two white-flowered plants could produce all purple-flowered offspring.

◆

Sex-Limited and Sex-Influenced Traits

In addition to elements in the external environment, the sex hormones provide an important internal environmental influence on the action of certain genes. In the extreme case, a given genotype is so dependent on the presence of these hormones that its expression is limited exclusively to one sex. The result is a **sex-limited trait**.

Sex-limited traits primarily concern the secondary sex characteristics. Beard development in males and milk production in females are examples. Both males and females possess the genes for these characteristics, but the sex hormones limit the expression of these genes to one sex. Thus, both parents contribute equally to the genetic basis of these traits in their offspring. Breeders of dairy cattle are therefore just as concerned with the quality of the milk production genes carried by the bull as with those of the cow. The bull does not express these genes himself, but he transmits them to his daughters, who do express them.

Sex limitation is an extreme example of how the expression of a gene can be controlled by hormones. In other less extreme cases of sex-controlled characteristics, only the dominance relationship of the two alleles is affected. Characteristics of this type are known as **sex-influenced traits**.

Pattern baldness in humans is an example of a sex-influenced trait. This trait is characterized by the premature loss of hair from the front and top of the head. It is more common in males than in females. Women who have the genotype for pattern baldness typically show only thinning of the hair rather than complete loss. A pair of alleles, B_1 and B_2, produce the trait. The presence of at least one B_1 allele results in baldness in males. The trait is sex-influenced in that the B_1 allele behaves as the dominant allele in males but as the recessive allele in females:

Genotype	Males	Females
B_1B_1	baldness	baldness
B_1B_2	baldness	no baldness
B_2B_2	no baldness	no baldness

Examples of sex-influenced traits in other organisms include horns in sheep (dominant in males, recessive in females), beards in goats (dominant in males, recessive in females), and spotting in cattle (mahogany-and-white is dominant in males, red-and-white is dominant in females).

We must emphasize that sex-limited or sex-influenced expression of a gene is not the same as sex linkage. Because of their association with the sex of the individual, sex-limited and sex-influenced traits are sometimes mistakenly thought to be sex-linked. Although a few genes of this sort are located on the X chromosome, most are autosomal. In the majority of cases, therefore, the inheritance pattern of a sex-limited or sex-influenced character is distinct from that of a sex-linked trait.

Example 4.3

In the pedigree presented below, the trait represented by the filled-in symbols is a fairly common human hereditary characteristic. Which of the following patterns of transmission are consistent with the pedigree: (a) autosomal dominant gene, (b) autosomal recessive gene, (c) X-linked dominant gene, (d) X-linked recessive gene, (e) Y-linked gene, (f) sex-limited autosomal gene, (g) sex-influenced autosomal gene?

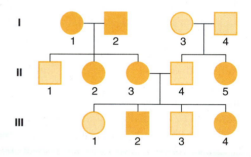

Solution: (g) A sex-influenced autosomal gene that is dominant in females, recessive in males. The cross involving I–1 and I–2 rules out simple autosomal recessive or X-linked recessive inheritance. Similarly, the cross involving I–3 and I–4 rules out simple autosomal dominant or X-linked dominant inheritance. The fact that both males and females show the trait rules out the possibility that it is a Y-linked or sex-limited characteristic, but if the gene (call it A_1) is dominant in females and recessive in males, the hereditary pattern could then be accounted for as follows:

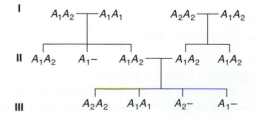

Follow-Up Problem 4.3

Which of the following patterns of inheritance are consistent with the pedigree shown below: (a) X-linked dominant gene, (b) X-linked recessive gene, (c) sex-influenced autosomal gene dominant in males, (d) sex-influenced autosomal gene dominant in females?

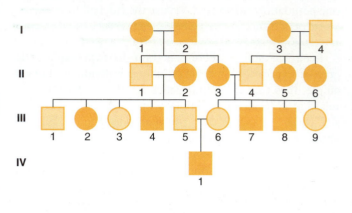

To Sum Up

1. Most traits are formed through a complex series of developmental steps, with different genes controlling different steps in the overall process. Therefore, not only can many characteristics be affected by a single gene, but many genes acting together can affect a single trait. Because of this multiple gene effect, genetic crosses that involve character differences for a single trait can sometimes give rise to phenotypic ratios that would be expected for the independent segregation of two or more gene pairs.

2. The expression of a genotype containing two or more pairs of genes depends on a number of factors, including dominance and interactions between nonallelic genes. Certain modifications of the basic dihybrid ratio of 9:3:3:1 should be immediately recognized as indicating epistatic gene interaction; these modified ratios include 9:7, 9:3:4, 13:3, 12:3:1, and 15:1.

3. Inheritance patterns are sometimes complicated by environmental and genetic factors that influence the expression of genes. Genes that are incompletely penetrant are not always expressed by the individuals who carry them. Even if a genotype is completely penetrant, it can exhibit variable expressivity, meaning that individuals of the same genotype differ from one another in their degree of phenotypic expression.

4. Sex-limited characters are expressed by only one sex. These characteristics are usually controlled by autosomal genes whose action depends on the presence or absence of certain sex hormones.

5. Sex-influenced traits are determined by genes whose dominance relationship is controlled by the sex of the individual. The allele that is dominant in one sex is recessive in the other.

The principles of segregation and independent assortment can be extended to inheritance patterns that differ from those observed by Mendel. These patterns include the inheritance of multiple alleles, in which there are more than two allelic forms of a gene and thus many different genotypes in a population, and sex-linkage, in which alleles display a sex-related pattern of transmission because they are carried on a particular sex chromosome. The inheritance pattern for X-linked genes differs from that of autosomal genes because the heterogametic sex (which has only one X chromosome and a single dose of each X-linked gene) expresses all X-linked alleles in full.

Mendel's principles also apply when genes interact to produce certain modified phenotypic ratios. The most common types of gene interaction include incomplete or partial dominance, which is characterized by the expression of an intermediate phenotype in the heterozygote, codominance, in which different alleles are expressed individually in the heterozygote, and epistasis, a form of gene interaction that can occur when two or more (nonallelic) genes affect a single trait. Variations due to dominance and epistasis can often be distinguished on the basis of modified Mendelian ratios in the F_2.

Mendel's principles can also be extended to cases in which a gene lacks complete penetrance (it fails to be expressed in all individuals of a specific genotype) and/or varies in expressivity (degree of expression) because of environmental effects. Sex-limited and sex-influenced traits, in which the penetrance and expressivity of a gene are affected by the sex of the individual, are two examples of incomplete penetrance and variable expressivity.

Multiple Alleles

1. Three of the many coat colors in mink (*Mustela vison*) are the standard wild-type (black-brown), a blue-gray color known as silverblu (or platinum), and a darker blue-gray color known as steelblu. Matings between wild-type mink can produce all wild-type, 3 wild-type:1 silverblu, or 3 wild-type:1 steelblu. Crosses of steelblu × steelblu sometimes produce all steelblu and other times produce 3 steelblu:1 silverblu. Crosses of silverblu × silverblu produce only silverblu. (a) Arrange the colors in order of dominance. (b) How many alleles are responsible for the inheritance of these coat colors? (c) How many different genotypes are responsible for steelblu color? for silverblu? (d) A particular mating yields progeny in a ratio of 1 steelblu:2 wild-type:1 silverblu. What are the coat colors of the parents in this cross?

2. The pattern of white spotting that characterizes various breeds of cattle is a multiple allelic trait that is of interest to animal breeders. Four spotting patterns with the following order of dominance are known: Dutch belt > white-faced or Hereford-type spotting > solid color > Holstein- and Guernsey-type spotting. How many different genotypes (in total) are involved in the determination of spotting pattern?

3. In rabbits wild-type or agouti coat color (c^+) is dominant over chinchilla (c^{ch}), which is dominant over Himalayan (c^h), which is dominant over albino (c). Determine the phenotypic ratio expected from each of the following crosses:
   (a) $c^+c^{ch} \times c^hc$ (b) $c^{ch}c^h \times c^{ch}c^h$ (c) $c^+c \times c^{ch}c$

4. In a certain species of plant, three alleles (A_1, A_2, and A_3) of a certain gene have their locus on chromosome 1 and five alleles of another gene (B_1, B_2, B_3, B_4, and B_5) have their locus on chromosome 2. (a) How many different genotypes are theoretically possible for each allelic series considered separately? (b) How many genotypes are possible for the two allelic series considered together? (c) Studies conducted on the three allelic forms of gene A show the following dominance relationships: $A_1 > A_2 > A_3$. How many different phenotypes can be produced by this allelic series alone? (d) How many different phenotypes can be produced by the cross $A_1A_2B_1B_2 \times A_1A_3B_3B_4$, if the B alleles show a simple hierarchy of dominance ($B_1 > B_2 > B_3 > B_4 > B_5$)?

Sex Linkage

5. In the house cat (*Felis domesticus*), coat color among females can be black, yellow, or tortoiseshell (a mixture of black and yellow); males are either black or yellow. These colors are determined by an X-linked pair of alleles. (a) Using the first letter of the alphabet, give the genotypes that produce these different phenotypes for coat color in female and male cats. (b) Suppose that a mating produces a litter of three kittens—two black males and one yellow female. What are the colors of the female and male parents?

6. Red-green color blindness is a sex-linked recessive trait. A couple has a daughter who is color blind and a son with normal color vision. What are the genotypes of the parents?

7. A woman with normal color vision whose father is color-blind has four sons. We know nothing about the color vision of her husband. What is the expected ratio of normal vision to color blindness among these sons?

8. Consider a rare X-linked dominant gene that causes a form of defective tooth enamel and dental discoloration in humans. (a) What percent of the sons and what percent of the daughters of an affected man are expected to be affected? (b) What percent of the sons and what percent of the daughters of an affected woman are expected to be affected?

9. The barred pigmentation pattern of the Barred Plymouth Rock breed of chickens consists of white wing bars on black feathers. This pattern is determined by a dominant Z-linked gene (*B*). The recessive allele of this gene (*b*) produces a nonbarred (plain) wing coloration. (a) What phenotypic ratio is expected among the progeny of a cross between a barred female and a nonbarred male? (b) Another mating, in which both the parents

are barred, produces a nonbarred female offspring. What are the genotypes of the parents in this cross?

10. For which of the following pedigrees is X-linked recessive inheritance a possibility? For which is X-linked dominant inheritance a possibility?

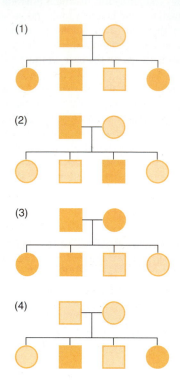

(1)

(2)

(3)

(4)

11. Which of the following modes of inheritance are possible for each of the pedigrees shown below: (a) autosomal dominant gene, (b) autosomal recessive gene, (c) X-linked dominant gene, (d) X-linked recessive gene?

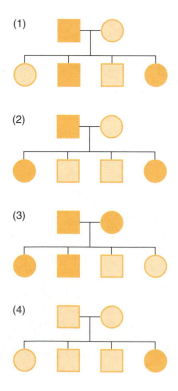

(1)

(2)

(3)

(4)

12. Consider the pedigrees in problem 11. If the traits designated by the dark symbols in pedigrees 1 and 2 are rare in the population, which is the most likely mode of inheritance for each pedigree?

Dominance Relationships

13. Matings between palomino horses yield an overall phenotypic ratio among their offspring of 1 chestnut : 2 palomino : 1 cremello. Predict the phenotypic ratios produced by the following crosses: (a) palomino × chestnut, (b) palomino × cremello, (c) cremello × cremello. (d) Would it be possible to develop a pure-breeding variety of palomino horses? Explain your answer.

14. Many traits that appear to be completely dominant when viewed at a superficial level actually show incomplete dominance on closer examination. One such case is that of round vs. wrinkled seeds in pea plants. The wrinkled seeds convert less of their sugar content into starch than the round seeds do. As a consequence, less water is retained when the seeds mature, and they take on a wrinkled or shrunken appearance. Microscopic examination of the seed contents reveals that despite their round external shape, heterozygous seeds contain starch granules that are intermediate in size and amount. The granules in heterozygotes are distinctly smaller and less numerous than those in dominant homozygotes. Suppose that a heterozygous plant self-pollinates. Give the phenotypic ratios expected among the seeds of this plant in terms of (a) external appearance and (b) size and quantity of starch granules.

15. The Blue Andalusian variety of chickens has slate blue feathers edged with black. This variety is produced when birds with splashed-white feathers (white with scattered slate blue flecks) are mated to birds with solid black feathers. Matings between Blue Andalusians produce a ratio in the progeny of approximately 1 black : 2 Blue Andalusian : 1 splashed-white. (a) What appears to be the genetic basis for feather color in the Blue Andalusian fowl? (b) Would it be possible to develop a pure-breeding variety of Blue Andalusians? Why or why not?

16. Another character difference in chickens is normal vs. frizzled feathers. Homozygous (FF) birds have normal straight feathers, while F'F' homozygotes show an extreme frizzled condition in which abnormally brittle feathers curve upward and forward. Heterozygotes (FF') are mildly frizzled, with only slightly abnormal feathers. Suppose that mildly frizzled Blue Andalusian fowl are mated. What phenotypic ratio will be expected among their progeny?

17. In tomatoes, the degree of pubescence (hairy covering) for fruit varies independently of that for stems and leaves. The texture of the fruit covering can be smooth (P–) or peach (fuzzy, pp). Stems and leaves can either be hairy (HH), have scattered hairs (Hh), or be hairless (hh). Suppose that seeds from a particular cross are planted and yield 50 tomato plants with the following characteristics: 9 smooth, hairy : 19 smooth, scattered : 8 smooth, hairless : 3 peach, hairy : 7 peach, scattered : 4 peach, hairless. What are the most likely genotypes of the parents?

18. Considering only the three ABO blood alleles I^A, I^B, and I^O, give the genotypes of the parents in a cross that produces offspring with (a) 4 blood types, (b) 3 blood types, of which type B is most common, (c) 3 blood types, of which type AB is most common, (d) 2 blood types, of which type A is most common.

19. Blood serum is taken from each of five students in a genetics laboratory. Each sample is then tested with red blood cells obtained from each of the other four students. Agglutination of red blood cells occurs in some sera (designated by +) but not in others (designated by −). The following table gives the results observed:

	Red blood cells from student				
	1	2	3	4	5
Blood serum from student 1	−	−	+	−	+
2	+	−	+	+	+
3	−	−	−	−	−
4	−	−	+	−	+
5	+	−	+	+	−

If the blood type of student 1 is known to be type A, what are the blood types of the four remaining students?

20. The ABO blood group system is valuable in helping to settle cases of disputed parentage. The following table lists the blood types of various mother-child combinations. In each case, list the blood types that can be *excluded* as possibilities for the father.

Blood type of child	Blood type of mother	Blood types that father cannot have
O	O	
O	B	
A	B	
B	O	
AB	A	
AB	B	

21. Three babies are born in a hospital on the same night. The blood types of the babies are A, B, and AB. If we know that the blood types of the three pairs of parents are (a) A and B, (b) AB and O, and (c) B and B, assign each baby to its proper parents.

22. At least thirty other blood groups, in addition to the ABO system, have been identified in humans. These other blood group systems stem from a minimum of thirty different genes involved in antigen determination and include MN, Duffy, Diego, Kell, Lewis, Lutheran, and Rh, to name just a few. The MN blood groups are the result of two alleles, L^M and L^N. L^M produces the M antigen that is found on red blood cells, while L^N produces the N antigen. Heterozygous individuals possess both proteins on their red blood cells and are said to be of blood type MN. (a) What type of hereditary pattern is being exhibited by these alleles? (b) What phenotypic ratio would be expected from the mating of two individuals with type MN blood?

23. Consider the rare instance of an accidental baby switch in a hospital. Mr. and Mrs. X believe that the baby they brought home from the hospital may not really be theirs. Blood tests reveal that Mr. X is type A M, Mrs. X is AB MN, and the baby is O MN. Does the baby belong to Mr. and Mrs. X? Explain your answer.

Nonallelic Gene Interactions and Environmental Effects

24. Susceptibility of corn to a particular plant disease depends on two independently assorting gene pairs. Plants with an $A-B-$ genotype are highly resistant to infection, $A-bb$ plants are moderately resistant, $aaB-$ plants are slightly resistant, and $aabb$ plants are not resistant (are highly susceptible) to infection. Suppose that you have access to moderately resistant and slightly resistant plants. Describe how you would proceed to develop a pure-breeding strain that is highly resistant to the disease.

25. Mice with at least one A gene and one B gene ($A-B-$) are agouti black, $aaB-$ mice are plain black (lack a yellow band near the tip of each colored hair), $A-bb$ are agouti brown (or cinnamon), and $aabb$ mice are plain brown. (a) A cinnamon male and an agouti black female produce approximately $\frac{3}{8}$ agouti black, $\frac{3}{8}$ cinnamon, $\frac{1}{8}$ black, and $\frac{1}{8}$ brown offspring. What are the genotypes of the parents? (b) In repeated matings, a black male and a cinnamon female produce a total of 34 agouti black and 36 cinnamon offspring. What are the genotypes of the parents?

26. Two independently assorting genes, D and E, interact in a complementary fashion to produce normal hearing in humans. Individuals who are homozygous for either or both of the recessive d or e genes ($ddE-$, $D-ee$, or $ddee$) are born deaf. (a) If a couple who are both $DdEe$ in genotype have a number of children, what proportion of their children is expected to have normal hearing? (b) Show how a deaf couple could have children with normal hearing.

27. In corn, different varieties produce kernels that turn either red, orange, or pink when exposed to sunlight. In contrast, normal kernels remain yellow. Crosses are made with the following results:

P	F_1	F_2
red × orange	all red	752 red : 259 orange
red × pink	all red	620 red : 202 pink
orange × pink	all orange	741 orange : 236 pink

Judging from these results, are the genes that determine the different colors allelic (a multiple allelic series) or are they nonallelic? Explain.

28. When pure-breeding *Drosophila* with scarlet-colored eyes are crossed to pure-breeding flies having brown eyes, the F_1 all have red (wild-type) eyes. Intercrosses of the F_1 give F_2 in the following ratio: 9 red : 3 scarlet : 3 brown : 1 white. Are the genes for scarlet and brown allelic or nonallelic? Explain your answer.

29. Fruit color in summer squash can be white, yellow, or green. A parental cross of pure-breeding white- and yellow-fruited varieties produces an F_1 of all white-fruited plants. Selfing the F_1 gives an F_2 of 12 white : 3 yellow : 1 green. (a) How many gene pairs appear to be responsible for fruit color in these crosses? (b) Give the genotypes of the white- and yellow-fruited parental varieties and of the F_1, and give the general genotypes for the phenotypic classes in the F_2.

30. The Angus breed of cattle is black, while the Jersey breed is a color called black-and-red. In crosses between the two breeds, the F_1 are always black. Intercrossing these F_1 gives three colors among the F_2: black, black-and-red, and red. One such group of F_2 consisted of 122 black : 29 black-and-red : 9 red. Use this phenotypic ratio to explain the genetic basis of this trait, assigning genotypes to all parents and offspring.

31. The color of the Duroc Jersey breed of swine can be red, sandy, or white. A certain cross of red × white yields all red in the F_1, as does a certain cross of sandy × sandy. Several mat-

ings of the F$_1$ offspring from the first cross with the F$_1$ from the second cross give the following totals in the F$_2$: 178 red: 123 sandy: 20 white. Explain the genetic basis of this trait, and assign genotypes to all parents and offspring.

32. Several genes affect coat color in horses. The recessive *e* gene, for example, reduces the intensity of the darker pigments without affecting the lighter red and yellow pigments. Thus, while *E−* horses are chestnut, *ee* horses are sorrel, with a light mane and tail. Another gene, *G*, acts in a dominant manner to replace the original hair color with gray. This graying affect often occurs early in life, so that the phenotype of mature *G−* horses is gray, regardless of the presence of the other color genes. Horses with the *gg* genotype retain their original color. (a) Suppose that crosses are made between gray horses that are *EeGg* in genotype. What colors are expected among their offspring and in what proportions? (b) A gray mare is bred to a sorrel stallion. The first mating results in a chestnut foal that later becomes gray. The second mating produces a sorrel colt that retains its color to maturity. What are the genotypes of the mare and stallion?

33. Three independently assorting genes (*A*, *C*, and *R*) work together in a complementary fashion to produce kernel color in corn. Kernels with at least one dominant allele for each of these three genes (*A−C−R−*) show color; all others are white. Give the ratio of colored to white kernels expected for each of the following crosses:

    (a) *AACCRr × AACCRr*   (c) *AaCcRr × aaccrr*

    (b) *AaCcrr × AaCcRR*   (d) *AaCcRr × AaCcRr*

34. What phenotypic ratio is expected from the testcross *AaBb × aabb* if the corresponding ratio from the dihybrid cross *AaBb × AaBb* is (a) 1:2:1:2:4:2:1:2:1, (b) 3:6:3:1:2:1, (c) 9:3:4, (d) 12:3:1, (e) 9:7, (f) 13:3, (g) 15:1, (h) 9:6:1?

35. In humans the sex-influenced gene for pattern baldness is dominant in males and recessive in females. A nonbald man marries a nonbald woman whose mother is bald. What is the chance that their first child is a male with the genetic predisposition for baldness?

36. Mahogany-and-white spotting in Ayrshire cattle is dominant in males and recessive in females. The reverse is true of red-and-white spotting. A farmer who has a herd of mahogany-and-white cattle purchases a red-and-white spotted bull. Describe the breeding procedure the farmer must use to establish a herd that is pure breeding for red-and-white spotting.

37. Cock-feathering is a sex-limited trait that is recessive in males. (a) A mating between a cock-feathered male and a hen-feathered female produces both cock-feathered and hen-feathered male offspring. What are the genotypes of the parents in this cross? (b) Suppose that the female in part (a), whose feathers are also barred (a dominant Z-linked trait), is mated to a hen-feathered, nonbarred male. They produce several offspring; the first chick to hatch becomes a cock-feathered, barred male. What phenotypic proportions are expected among their other offspring?

38. Each of the following pedigrees involves a different hereditary trait in humans. Pedigree 1 shows the transmission of a fairly common characteristic, whereas pedigree 2 shows the transmission of a trait that is rare in the general population. Which of the following patterns of transmission are consistent with each pedigree: (a) dominant X-linked gene, (b) recessive X-

linked gene, (c) sex-influenced gene, (d) sex-limited gene? Give reasons for your answers.

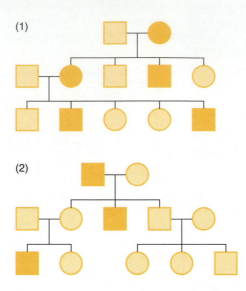

39. Quite often information is needed from more than one pedigree to establish correctly the mode of inheritance for a particular trait. Two pedigrees for the same hereditary trait are shown below. Determine which of the following patterns of transmission are consistent when each pedigree is considered separately and then when both pedigrees are considered jointly: (a) autosomal dominant gene, (b) autosomal recessive gene, (c) X-linked dominant gene, (d) X-linked recessive gene, (e) sex-influenced gene.

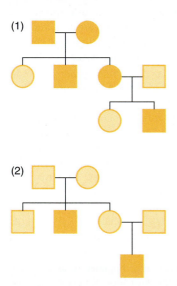

40. Which of the following statements are true of a sex-linked trait, of a sex-influenced trait, or of a sex-limited trait? (Assume this is a human trait.)

    (a) Gene can be transmitted from either parent to both sons and daughters.

    (b) Gene can be transmitted from father to both sons and daughters.

    (c) Gene can be transmitted from mother to both sons and daughters.

(d) Gene can be transmitted only from father to his sons.

(e) Trait is expressed more often in one sex than in the other.

41. The dark, light, and crosshatched symbols in the following pedigree designate three different coat colors in a certain species of mammal. Determine whether the inheritance pattern is consistent with (a) a single pair of incompletely dominant or codominant alleles, (b) three alleles at a single locus that show a simple hierarchy of dominance, (c) two different pairs of alleles with a complementary form of interaction (as in problem 26).

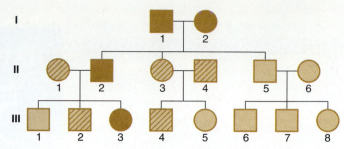

Quantitative Inheritance

So far we have been concerned with the inheritance of sharply defined, contrasting characteristics, such as round and wrinkled pea seeds and A, B, AB, and O blood types, in which there is a distinct qualitative difference between genetic types. These traits are generally characterized by discrete phenotypic classes with no blending or gradation of one type into another. In contrast, traits such as height, weight, and fertility vary in degree (rather than in kind) along a continuous scale of measurement. Such characteristics that can generally be measured on a quantitative scale are known as **quantitative traits**, and their pattern of inheritance is referred to as quantitative inheritance. There is perhaps no better illustration of the usefulness of probability and statistics in genetics than in the analysis of quantitative inheritance.

Variation in tomatoes from the former Soviet republics.

POLYGENES AND QUANTITATIVE TRAITS

Quantitative traits are **polygenic**: their expression is influenced by several or many genes, and it is often highly susceptible to modification by the environment. Many quantitative traits, including height and weight, exhibit a pattern of continuous variation that approximates a normal (bell-shaped) curve when the phenotypic value is plotted in the form of a frequency distribution (▶ Figure 5–1). A normal curve implies that regardless of how narrowly we subdivide the scale of measurement, we can always expect to find some individuals in every class if the population is large enough.

Although continuous variation is the rule, certain quantitative traits, such as number of offspring (litter size) and number of bristles on a fruit fly, are *discontinuous* because they have discrete phenotypic classes. Discontinuous quantitative traits are also called *countable* or *meristic traits,* since the phenotypic value is a discrete variable that is represented by a limited and thus countable series of integers. At their extreme, discontinuous quantitative traits may show only two discrete classes, such as presence or absence of a disease, as in the case of so-called **threshold traits** (see ▶ Figure 5–2). Threshold traits, like other quantitative traits, have an underlying polygenic basis, but only those individuals with a genetic susceptibility (liability) that exceeds some threshold value will differ from the others and express the alternative characteristic.

The analysis of quantitative traits is extremely important in the study of human genetics and in agriculture. Most traits of economic importance, such as milk production in cattle, monthly weight gain in swine, and grain yield in cereal crops, are quantitative in nature. Most of these traits show a continuous variation pattern that lacks distinct phenotypic classes. The effects of individual genes are therefore obscured and cannot be measured directly. For this reason, the study of these traits must rely on methods that analyze the combined actions of several genes, rather than on techniques that analyze genes individually on the basis of their phenotypic ratios. As you will discover, however, the underlying genetic basis of quantitative traits is what we focused on in previous chapters—particulate Mendelian inheritance.

Nature and Sources of Variation

Two main factors are responsible for the variation in phenotype for a quantitative trait: genotype and environment. We can summarize these effects quantitatively if we express the measured or observed value of a quantitative character as the sum

$$P = G + E$$

where P is the phenotypic value, G is the genotypic value, and E is the environmental deviation. In this model, G is the theoretical value of P in the absence of any environmental modification (i.e., when $E = 0$). Moreover, when the environment produces departures in phenotype from the value of G, they are assumed to occur about as often in one direction as in the other. Thus the average value of E for a given genotype, when measured under all environmental conditions to which the population is normally exposed, is equal to zero, and the average value of P for a given genotype will equal G. ▶ Figure 5–3 illustrates the effect of the environment on a quantitative trait expressed by two alleles. Observe that when the environmental departures are

▶ **FIGURE 5–1** Continuous variation pattern observed for height of human males. *Source:* Daniel L. Hartl, *Principles of Population Genetics* (Sinauer, 1980).

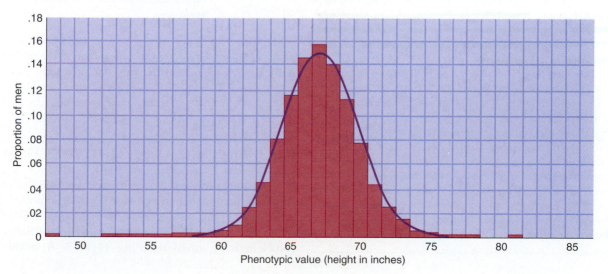

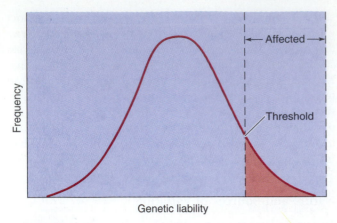

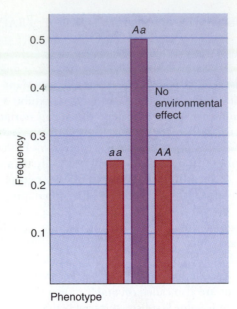

▶ **FIGURE 5–2** Genetic determination of a threshold trait. This model explains a threshold effect in terms of a potentially continuous background of genetic susceptibility. Only those individuals with a genetic liability that exceeds a threshold value are affected.

large enough, variability at even a single gene pair can obscure the boundaries between the phenotypic classes and produce a single continuous distribution.

In the model, the environmental deviation (E) is treated as a catchall term that includes any departure from the theoretical value of P that is not directly attributable to genetic effects. Among the more common causes of environmental deviations are nutritional and climatic factors such as availability and quality of food for animals and suitability of light, temperature, humidity, and soil characteristics for plants. Such environmental factors can induce variability at all stages in life, but they are particularly critical during periods of active growth and development.

To Sum Up

1. Quantitative traits, such as height, weight, and fertility, often show continuous variation over their entire range of values. These traits are controlled by the action of many genes (called polygenes), whose expression is often highly sensitive to environmental factors. Since it is usually impossible to separate individuals into discrete phenotypic classes, the individual contributions of polygenes cannot be analyzed on the basis of ratios among the phenotypic classes.

2. Some quantitative traits show a threshold effect. These traits have a polygenic basis but show a discontinuous form of variation in which individuals whose values fall below a certain threshold level have one phenotype and those whose values exceed the threshold have another.

3. The phenotypic variability of a quantitative trait results from variations in individual genotypes and from differences in the environmental conditions under which these genotypes are expressed. These effects are summarized mathematically by expressing the phenotypic value (P) of an individual as the sum of the genotypic value (G) and the deviation in phenotype due to the environment (E), so that $P = G + E$.

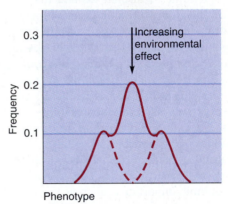

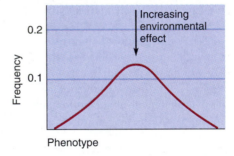

▶ **FIGURE 5–3** Effects of increasing environmental variation on the phenotypes expressed by a single pair of alleles. The genotypes *AA*, *Aa*, and *aa* are assumed to exist in the population at a ratio of 1:2:1.

▦ INHERITANCE OF QUANTITATIVE TRAITS

Genetic studies conducted during the early 1900s did much to establish a Mendelian basis for the inheritance of quantitative traits. The results of one series of crosses between strains of tobacco (*Nicotiana longiflora*) that differed in corolla length are summarized in ▶ Figure 5–4. In this cross-

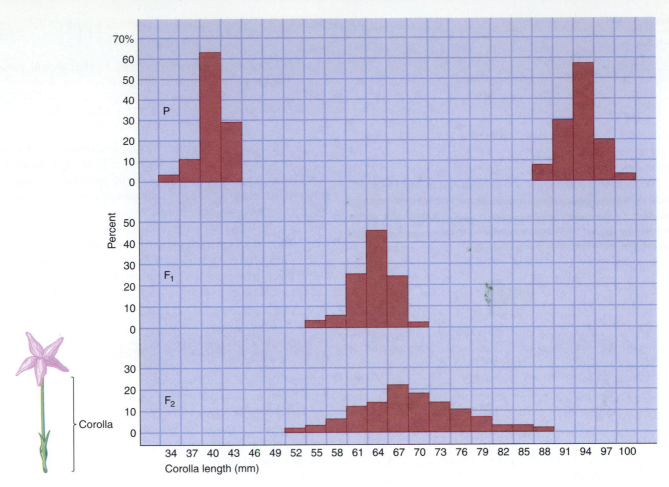

▶ **FIGURE 5–4** The results of crosses between strains of tobacco (*Nicotiana longiflora*) that differ in flower (corolla) length.

ing sequence, the parental strains are highly inbred and presumably homozygous for the genetic differences between them. The variation in each parental line can therefore be attributed to environmental rather than genetic causes. The F_1 hybrids, like their parents, are genetically uniform and thus show only environmental variation, but they are intermediate in average corolla length between the two parental lines. The F_2 offspring also show an intermediate average corolla length that is similar in magnitude to that of the F_1. However, the variability of the F_2 is considerably greater, indicating that the phenotypic variation observed in the F_2 incorporates variability that is both genetic and environmental in origin.

The Additive Polygene Model

Results such as those observed in the inheritance of corolla length have led geneticists to derive a simple *additive* Mendelian model in order to describe the behavior of polygenes. In this model, each gene is represented by two alleles: a contributing (or active) allele designated by an uppercase letter and a noncontributing (null) allele symbolized by a lowercase letter. The model also assumes that all gene pairs

assort independently and that all active alleles contribute an equal amount to the overall trait in an additive fashion.

If the parents differ in only a single gene pair (for example, $AA \times aa$), the model would predict a monohybrid (Aa) F_1 and the following genotypic distribution in the F_2:

Active Alleles	Genotype	Frequency	Value
2	*AA*	$\frac{1}{4}$	2α
1	*Aa*	$\frac{1}{2}$	α
0	*aa*	$\frac{1}{4}$	0

where α is the contribution of each active allele. The frequencies in the F_2 correspond to the terms of the binomial expansion $(p + q)^2 = p^2 + 2pq + q^2$, where p and q are the expected proportions of the active allele and null allele, respectively. Since both p and q equal $\frac{1}{2}$ in this series of crosses, we get a ratio of $1:2:1$ in the F_2.

We can make the model more realistic by assuming more gene pairs. For example, if the parents differ in two gene pairs ($AABB \times aabb$), the model would then predict a dihybrid ($AaBb$) F_1 and the following genotypic distribution in the F_2:

Active Genes	Genotypes	Frequency	Value
4	$\frac{1}{16}$ AABB	$\frac{1}{16}$	4α
3	$\frac{2}{16}$ AABb + $\frac{2}{16}$ AaBB	$\frac{4}{16}$	3α
2	$\frac{4}{16}$ AaBb + $\frac{1}{16}$ AAbb + $\frac{1}{16}$ aaBB	$\frac{6}{16}$	2α
1	$\frac{2}{16}$ Aabb + $\frac{2}{16}$ aaBb	$\frac{4}{16}$	α
0	$\frac{1}{16}$ aabb	$\frac{1}{16}$	0

The class frequencies now follow the binomial expansion $(p + q)^4 = p^4 + 4p^3q + 6p^2q^2 + 4pq^3 + q^4$, where p and q are again equal to $\frac{1}{2}$.

We can generalize from the preceding pattern to predict that if n is the number of independently segregating gene pairs and R is the difference between the phenotypes of the parental strains (R is equal to 2α in the case of one gene pair and 4α in the case of two gene pairs), the contribution of each active gene in the model will be

$$\alpha = \frac{R}{2n}$$

In the absence of environmental variation, the F_2 should segregate $2n + 1$ phenotypic classes that correspond in frequency to the $2n + 1$ terms of the binomial expansion $(p + q)^{2n}$. Thus, if the parents differ in three gene pairs ($AABBCC \times aabbcc$), the F_1 will be trihybrid ($AaBbCc$) and the F_2 should segregate $2n + 1 = 7$ classes in a ratio of $1:6:15:20:15:6:1$.

Since both p and q are $\frac{1}{2}$, it is possible to predict the expected proportion in each F_2 class using the binomial formula (see Chapter 3):

$$P(x \text{ active genes}) = \frac{(2n)!}{x!(2n - x)!}\left(\frac{1}{2}\right)^x\left(\frac{1}{2}\right)^{2n-x} \quad (5.1)$$

Thus, the expected proportion of the F_2 that would have four active genes out of a total of $2n = 6$ is

$$P(4 \text{ active genes}) = \frac{6!}{4!2!}\left(\frac{1}{2}\right)^4\left(\frac{1}{2}\right)^2 = \frac{15}{64}$$

The basic pattern that we have just described is shown in ▶ Figure 5–5 for increasing numbers of gene pairs. In addition to the expected distribution of the parental, F_1, and F_2 generations under uniform environmental conditions, the F_2 distribution is also modified to show the effect of significant environmental contribution to the phenotypic variation. When modified by the environment, the F_2 are not separable into discrete phenotypic classes, and thus the class frequencies cannot be compared directly with the terms of the binomial distribution.

Example 5.1

Suppose that the number of leaves on a particular plant is determined by four pairs of polygenes. Plants that are heterozygous for the four gene pairs are crossed and produce many offspring. How many phenotypic classes are expected among the progeny? Assuming independent assortment, what fraction of the offspring is expected to have the smallest number of leaves?

▶ FIGURE 5–5 Summary of the additive polygene model for different numbers of gene pairs. The highly variable F_2 distributions are shown as they would appear both in the absence and in the presence of environmental variation.

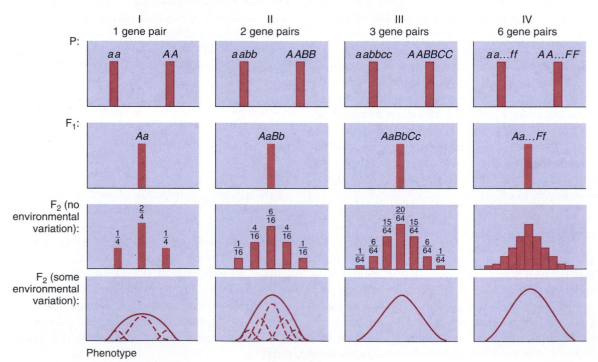

Solution: When $n = 4$ gene pairs, there will be $2n + 1 = 9$ phenotypic classes, assuming there are no environmental effects. The fraction of the offspring with the smallest number of leaves corresponds to either the first or last term of the binomial expansion $(\frac{1}{2} + \frac{1}{2})^8$, since the distribution is symmetrical. The expected frequency of this class is then $(\frac{1}{2})^8 = \frac{1}{256}$.

Follow-Up Problem 5.1

The number of leaves per stem on a plant (not the plant in Example 5.1) is known to be a quantitative trait. Suppose that a cross between plants that each have 14 leaves per stem yields offspring in the following ratio: 1 with 10 leaves:8 with 11 leaves:28 with 12 leaves:56 with 13 leaves:70 with 14 leaves:56 with 15 leaves:28 with 16 leaves:8 with 17 leaves:1 with 18 leaves. Using the first letter of the alphabet and as many more letters as necessary, give the genotypes of the parents of this offspring distribution.

Although highly oversimplified, the model that we have considered does provide insight into the behavior of polygenes and seems to fit the pattern of inheritance of some traits very well. For example, several quantitative traits in domestic plants and animals, such as plant height and egg production in chickens, yield results that are consistent with the model.

While the model is best suited for studies of experimental plants and animals, the principles involved have been applied with reasonable success to the determination of the number of genes contributing to human skin color differences in blacks and Caucasians (▶ Figure 5–6). The genetic basis of skin color differences was examined by the English anthropologists G. A. Harrison and J. J. T. Owens, who used a reflectance spectrophotometer to estimate the amount of melanin pigment from the percentage of light reflected by the skin at different wavelengths. Since light reflectance and skin pigmentation are inversely related, darker skin will reflect less light. These investigators examined blacks and Caucasians and F_1, F_2, and backcross progeny from interracial matings (children of interracial marriages who marry individuals largely of one race or the other). Their results clearly indicate a polygenic basis of inheritance, in that pigmentation ranged from black to white through various intermediate shades. Moreover, the F_1 mean is essentially between the two parental values (see Figure 5–6), providing support for an additive gene effect. A statistical analysis of their results suggests the likelihood that either three or four gene pairs are responsible for human skin color.

Point to Ponder 5.1

Some recent research reports have indicated the possibility of a partial genetic basis for homosexuality. (At this point the data are ambiguous; there are arguments for and against a genetic basis.) Do you think that society's attitudes toward homosexuality would change if it were indeed shown that it is a quantitative trait with both genetic and environmental components?

▶ **FIGURE 5–6** Distribution of human skin color as measured by the skin reflectance of red light at a wavelength of 685 nm. The F_2 distributions are theoretical curves based on the additive polygene model, assuming an environmental effect and different numbers of gene pairs. *Source:* W. Bodmer and L. Cavalli-Sforza, *Genetics, Evolution and Man.* Copyright © 1976 by W. H. Freeman & Company. Reprinted with permission.

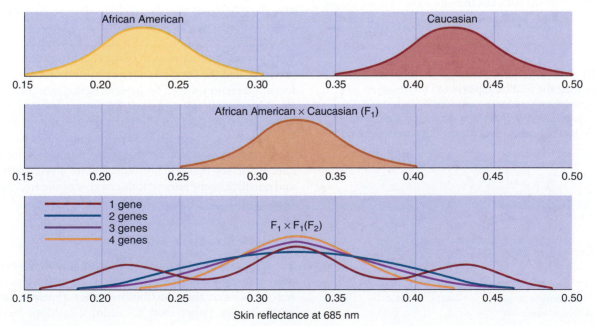

Dominance and Nonallelic Gene Interactions

Up to this point, we have assumed that all contributing genes have equal and additive effects on the expression of a quantitative trait. However, polygenes, like other genes, may interact in a nonadditive fashion and contribute unequally to the phenotype of an individual. There are two basic types of interactions involving polygenes: **dominance interactions**, which are interactions between alleles, and **interlocus interactions**, which are interactions between nonallelic genes (e.g., epistasis). The two types of interactions can be expressed as components of the genotypic value of an individual as follows:

$$G = A + D + I$$

where A is the additive value of a genotype and D and I are the dominance and interlocus deviations, respectively. In this expression, D and I represent departures from the theoretical value of a genotype that is assumed in the additive model. Thus, when D and I are zero, dominance and interlocus interactions do not occur and the genotypic value is strictly additive.

We can clarify the concept that dominance produces a departure from an additive gene effect if we consider a single gene pair in which one allele is completely dominant over the other. An example of this type of interaction is shown in ▶ Figure 5–7, where phenotypic value is plotted against the number of contributing alleles. According to the additive model, the phenotypic value of a genotype should increase linearly with the number of active alleles; the three points on the straight line in Figure 5–7 correspond to the theoretical additive values of the different genotypes. The differences between these points and the real or observed values constitute the dominance deviations.

The effect of complete dominance on a quantitative character is illustrated in ▶ Figure 5–8, which shows the phenotypic distribution patterns that are expected for different cross results. Observe that with dominance, the phenotypic

distribution of the F_1 will always be shifted in the direction of the dominant parent. The effect of dominance on the appearance of the F_2 distribution is not quite as obvious; the effect depends on the number of gene pairs. With one gene pair and complete dominance, a $\frac{3}{4}(A-) + \frac{1}{4}(aa)$ distribution is expected in the F_2, assuming no environmental variation. When two gene pairs are assorting independently in an $F_1 \times F_1$ cross, three phenotypes are expected in the relationship $(\frac{3}{4} + \frac{1}{4})^2 = \frac{9}{16}(A-B-) + \frac{6}{16}(A-bb + aaB-) + \frac{1}{16}(aabb)$. In general, therefore, n independently segregating gene pairs with dominance will produce an F_2 distribution of $n + 1$ classes, which correspond in frequency to the terms of the binomial expansion $(\frac{3}{4} + \frac{1}{4})^n$. Note that when n is small, we can easily recognize the presence of dominance by the asymmetrical appearance of the F_2 distribution. As n increases, however, the degree of asymmetry declines, ultimately producing an F_2 distribution that is difficult to distinguish from the distribution for a purely additive gene effect.

In our illustrations of dominance so far, we have assumed that nonallelic genes not only segregate and assort independently but also express themselves in an independent and additive fashion. Our assumptions correspond to the gene contributions in the following system:

	$\frac{3}{4}A-$ (α)	$\frac{1}{4}aa$ (0)
$\frac{3}{4}B- (\beta)$	$\frac{9}{16}A-B-$ $(\alpha + \beta)$	$\frac{3}{16}aaB-$ (β)
$\frac{1}{4}bb (0)$	$\frac{3}{16}A-bb$ (α)	$\frac{1}{16}aabb$ (0)

where α and β are the phenotypic contributions of $A-$ and $B-$, respectively. Although dominance interactions are assumed to occur at each of the loci, interlocus interactions are absent, since the different gene pairs contribute to the overall trait in an additive fashion. In the absence of interlocus interactions, $G = A + D$ for each genotype, because $I = 0$; but if, for example, the b allele in the bb homozygote were to mask the expression of the A gene, so that the $A-bb$ genotype had the same phenotypic value as $aabb$, then interlocus interaction would be present in the form of epistasis, in addition to the dominance effects. A numerical example of an epistatic effect on the phenotypic value is shown in ▶ Figure 5–9, where phenotypic values with epistasis are compared to those expected in the absence of interlocus effects. We see that when epistasis is involved, both allelic and nonallelic gene interactions can contribute to departures from additivity.

Multiplicative Gene Effects

The basic theory of polygenes was initially developed with the concept of additive gene action. In many quantitative traits, however, the contributing genes appear to act in a multiplicative rather than an additive fashion, increasing

▶ **FIGURE 5–7** A graph of phenotype vs. number of active alleles , showing the relationship between the observed phenotypic values when dominance occurs (dark circles) and the theoretical additive values (light circles). The departures, represented by the broken lines, are the dominance deviations.

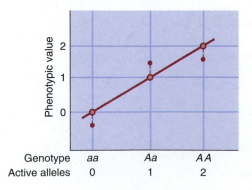

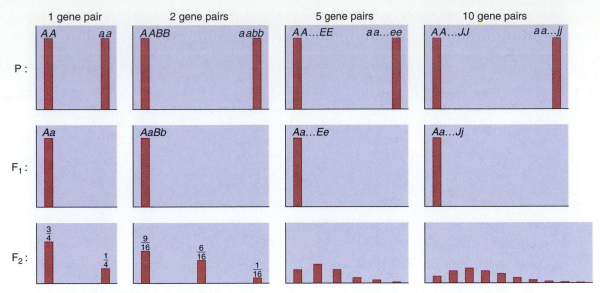

▶ **FIGURE 5–8** The effect of dominance on a quantitative trait for different numbers of gene pairs. The phenotypic distributions of the F_1 and F_2 are shifted in the direction of the dominant parental line.

the phenotypic value by a constant percentage rather than by the addition of a constant absolute amount. Suppose, for example, that the corolla length in a particular plant is increased by a factor of 10% for each contributing gene. The incremental increase per added gene will then be multiplicative in overall effect, and the size of the increase will depend on the existing genotype. For instance, the corolla length of a plant that already has the genotype for a potential length of 10 mm will increase to only $10 + (0.1)(10) = 11$ mm upon the addition of another active gene, since the phenotype would change by 10% of 10 mm—a 1 mm difference. However, the corolla length of a plant that has the genotype for a potential length of 100 mm would increase to $100 + (0.1)(100) = 110$ mm, for a change of 10 mm. In contrast, an active gene for an additive effect might add 1 mm to the flower length regardless of the existing genotype; in this case it does not matter whether the plant already has the potential for a length of 10 mm or 100 mm.

A multiplying gene effect is often associated with traits that are the net result of a biological growth process (e.g., fruit weight in tomatoes). Quantitative characters of this type often vary exponentially (rather than arithmetically) with the number of contributing genes. This exponential effect gives rise to phenotypic values that vary systematically along a geometric (or logarithmic) scale rather than an arithmetic scale. For example, suppose that fruit weight in tomatoes is doubled by the addition of each active gene. The phenotypic value, in this case, will be proportional to 2^x, where x is the number of active genes per genotype. The net effect is a geometric series of relative weights (i.e., 1, 2, 4, 8, 16, . . .). If these data are plotted on an arithmetic scale (1, 2, 3, 4, 5, . . .), the resulting distribution will be asymmetrical (skewed) in appearance, with a peak that is dis-

placed toward the lower end of the curve and an upper tail that approaches zero very slowly (▶ Figure 5–10a). Asymmetrical distributions of this type give the impression that the strains with the largest average phenotypes are also the most variable. However, this is merely a scale effect that disappears when the data are plotted along a logarithmic horizontal axis, which gives a distribution that is symmetrical

▶ **FIGURE 5–9** Effect of epistasis on phenotypic values. (a) Dominance occurs at each locus ($\alpha = 3$, $\beta = 2$), but there is no interaction between loci. (b) Recessive epistasis in which bb masks the expression of $A-$.

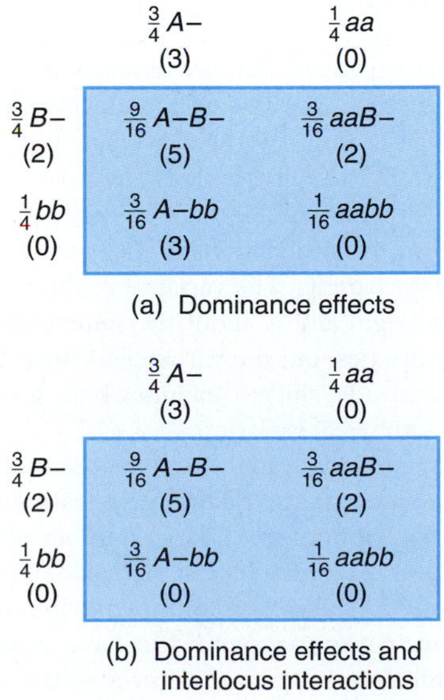

	$\frac{3}{4} A-$ (3)	$\frac{1}{4} aa$ (0)
$\frac{3}{4} B-$ (2)	$\frac{9}{16} A-B-$ (5)	$\frac{3}{16} aaB-$ (2)
$\frac{1}{4} bb$ (0)	$\frac{3}{16} A-bb$ (3)	$\frac{1}{16} aabb$ (0)

(a) Dominance effects

	$\frac{3}{4} A-$ (3)	$\frac{1}{4} aa$ (0)
$\frac{3}{4} B-$ (2)	$\frac{9}{16} A-B-$ (5)	$\frac{3}{16} aaB-$ (2)
$\frac{1}{4} bb$ (0)	$\frac{3}{16} A-bb$ (0)	$\frac{1}{16} aabb$ (0)

(b) Dominance effects and
interlocus interactions

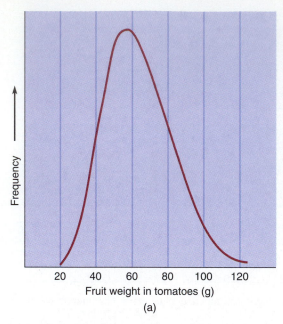

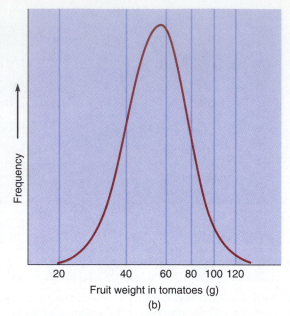

FIGURE 5–10 Distribution of a character that is determined by multiplicative gene effects. The distribution is asymmetrical in appearance when the phenotypic values are plotted along an arithmetic scale (a) but is symmetrical in appearance when the values are plotted along a logarithmic scale (b).

in appearance (Figure 5–10b). In mathematical terms, the skewing effect of the arithmetic scale is overcome by changing the increasing distances between the values in a geometric series (such as $2^0, 2^1, 2^2, 2^3, 2^4, \ldots$) to distances of equal length in the corresponding arithmetic series (0, 1, 2, 3, 4, . . .) by plotting the data on a logarithmic scale. (Recall that the logarithm of 2^x is directly proportional to the exponent x.) Thus if the gene effects are truly multiplicative, we would expect the transformed data to fit a symmetrical bell-shaped curve.

To Sum Up

1. The inheritance patterns that are associated with quantitative traits can be studied by crossing homozygous parental strains that differ significantly in their average phenotypic values. When the F_1 from these crosses are intercrossed to produce the F_2, the average phenotypes of the F_1 and F_2 populations tend to be similar to one another in value and intermediate between the values of the parents. The variance of the F_1 is strictly environmental in origin and is about the same as the variances of the parental varieties, but the variance of the F_2 is significantly greater in magnitude, since it includes both genotypic and environmental sources of variation.

2. The simplest model that can be used to account for quantitative inheritance assumes that a quantitative trait is determined by the additive contributions of a large number of independently segregating gene pairs. Each gene pair is composed of a contributing (or active) allele and a noncontributing (or null) allele. The active genes are assumed to have actions that are indistinguishable from one another in a statistical sense. Each

active gene contributes such a small amount to the overall trait that its effect is obscured by the genotype as a whole and by the environmental variation.

3. Several factors, including dominance and nonallelic gene interactions (epistasis), are known to be responsible for departures from the additive polygene model. The effects of dominance and epistasis can be statistically described in terms of differences between the actual genotypic values and the values that would be expected on the basis of additivity.

4. Polygenes for many quantitative traits, especially those that result from the biological process of growth, appear to act in a multiplicative rather than additive fashion; they increase the phenotypic value by a constant factor instead of by the addition of a constant amount.

⊞ STATISTICAL ANALYSIS OF QUANTITATIVE TRAITS

The analysis and interpretation of phenotypic variation is a major concern of genetics. Since much of this variability is quantitative in nature, involving the dispersion of measured attributes of a population, genetic analyses frequently require the use of special statistical techniques. In this section, we will consider certain mathematical approaches that are useful in analyzing quantitative data, with special emphasis on the analysis of quantitative traits.

Because quantitative traits are continuously varying characters expressing the cumulative actions of polygenes, they are among the most difficult to analyze. Since the individual contributions of polygenes are usually quite small, the ef-

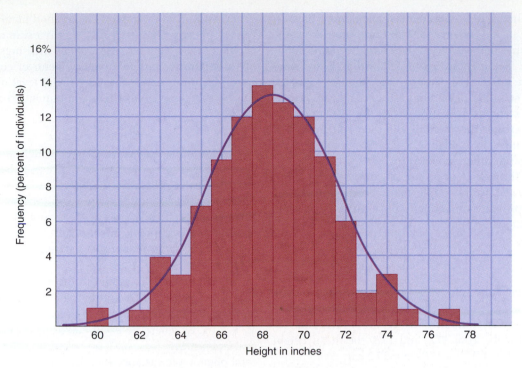

▶ FIGURE 5–11 Frequency distribution of mature body height of human males. The histogram gives the heights of 100 men by grouping the measurements into 1-inch class intervals. Superimposed on the histogram is the normal distribution of body height in a very large population.

fect each one has on the overall phenotype tends to be obscured by the genotype as a whole and by the effects of the environment. Thus, unlike simple qualitative characters, the phenotypic variation that is normally expressed by a quantitative trait does not reflect a direct correspondence between genotype and phenotype.

Mean and Variance

When quantitative traits are measured, the data are usually analyzed by grouping the measured phenotypic values into designated classes and computing the frequency of individuals in each class. The phenotypic values and their corresponding frequencies can then be graphed to give a frequency distribution. An example of a frequency distribution is shown in ▶ Figure 5–11, which also shows the graph of the **normal curve**. The normal curve is a theoretical distribution that is continuous and is symmetrical around a centrally occurring value. Two descriptive measures, the **mean** (m) and the **variance** (σ^2 or V), are used to characterize the general location of a normal curve on the horizontal axis and the breadth of its distribution. The relationships of the mean and the square root of the variance (called the **standard deviation**) to the normal curve are shown in ▶ Figure 5–12. In general, the mean is the average value; it is one of several measures that can be used to describe the location of centrally occurring values in a distribution. For the normal curve, the mean is the central value (or **median**),

and it is also the most frequently occurring value (or **mode**). Although the mean equals the median and the mode in a symmetrical normal curve, this property does not necessarily hold true for other distributions, many of which are asymmetrical or skewed in appearance.

The variance and its square root (the standard deviation) measure the degree of dispersion or breadth of the distribution. The standard deviation measures the horizontal distance from the mean to each inflection point on the normal

▶ FIGURE 5–12 A normal frequency distribution showing the mean and standard deviation.

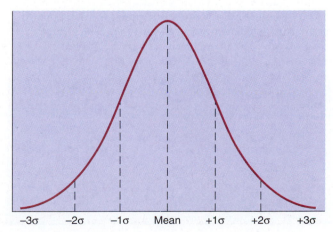

curve (see Figure 5–12). Mathematical calculations have established that in any normal distribution, approximately 68.3% of all values fall within one standard deviation above and below the mean, about 95.4% fall within ± two standard deviation units of the mean, and about 99.7% fall within ± three standard deviation units of the mean. Hence, the larger the standard deviation (and thus the larger the variance), the greater the variability in the population. The overlapping curves in ▶ Figure 5–13, for example, have the same mean but different variances.

Ordinarily, the values of m and σ^2 are not known but must be estimated from sample data. The sample mean, designated $\bar{x}$, provides an estimate of the true population mean and is calculated as

$$\bar{x} = \frac{1}{N} \sum x_i \qquad (5.2)$$

where Σ designates the sum, x_i is the value of one particular measurement, and N is the number of individuals in the sample. For example, if the heights (in cm) of three individuals are found to be 166, 172, and 178, the mean height would then be $(166 + 172 + 178)/3 = 172$ cm.

Similarly, the sample variance (s^2) provides an estimate of the true population variance; it is computed as the average of the squared deviations from the mean:

$$s^2 = \frac{1}{N - 1} \sum (x_i - \bar{x})^2 \qquad (5.3)$$

For example, for the three heights and computed mean given in the preceding example, the variance can be calculated as $[(166 - 172)^2 + (172 - 172)^2 + (178 - 172)^2]/2 = 36$ cm^2.

Analysis of Variation and Resemblance

The underlying contributions of genes and the environment to the overall variation of a quantitative trait can be studied statistically using an **analysis of variance**. This procedure was originally applied to genetic studies in the 1920s by the geneticist Ronald A. Fisher. The method involves partitioning the phenotypic variance (V_P) into a sum of component variances, so that each component measures the amount of variation in a particular causal factor or combination of causal factors. At the most superficial level of analysis, V_P can be partitioned into three component parts as follows:

$$V_P = V_G + V_E + V_{GE}$$

where V_G is the **genetic variance**, which arises from the presence of different genotypes in the population, V_E is the **environmental variance**, which is due to environmental differences, and V_{GE} is the **genotype-environment interaction variance**. If each specific departure in environmental conditions has the same effect on all genotypes, then V_{GE} is zero and the total phenotypic variance is simply the sum $V_G + V_E$.

We can extend the model to incorporate the variability that is associated with additive gene effects, dominance, and nonallelic gene interactions by expressing V_G as the sum $V_A + V_D + V_I$, where V_A is the **additive variance**, V_D is the **dominance variance**, and V_I is the **interaction variance**. The total phenotypic variance then becomes

$$V_P = V_A + V_D + V_I + V_E + V_{GE}$$

This equation is a general expression that incorporates many of the sources of variation in a quantitative trait. In practice, V_{GE} is usually assumed to be zero and V_D and V_I are frequently combined into a single interaction variance, so that evaluation of V_A and V_E is the major concern.

The value of V_E is usually estimated by measuring the variation in a trait expressed by individuals of the same genotype. With experimental organisms, such measurements are typically performed on homozygous (inbred) strains, while with humans, measurements of this sort are made on identical twins who have been raised in separate (independent) environments. In either case, $V_G = 0$, so that all variation is environmentally induced.

The value of V_A can often be estimated by comparing the strength of correlation in the trait in pairs of individuals with different degrees of genetic relationship between them (e.g., unrelated persons, parent and child, siblings, dizygotic twins, and monozygotic twins). The strength of correlation is expressed in terms of the **product-moment correlation coefficient** (r). The correlation coefficient gives the degree of correspondence in the phenotypic values between the persons concerned. For a group of paired individuals, the correlation coefficient is equal to 1 if the phenotypic values are always equal or if the phenotypic value of one individual varies in exact proportion to the phenotypic value of the other individual; it is 0 if the phenotypic values vary independently in all pairs. The correlation coefficient can be calculated from sample data as follows:

$$r = \frac{\Sigma(x_i - \bar{x})(y_i - \bar{y})}{\sqrt{\Sigma(x_i - \bar{x})^2 \Sigma(y_i - \bar{y})^2}} \qquad (5.4)$$

▶ **FIGURE 5–13** Frequency distributions with the same mean but with different variances. Curve (b) has a greater variance than curve (a) and includes a greater amount of variation.

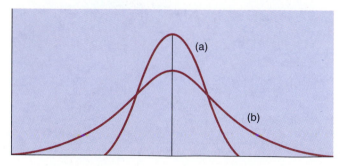

In this formula, the phenotypic value of one individual is represented by x_i, the phenotypic value of the other individual of a pair by y_i. $\bar{x}$ and $\bar{y}$ are the means of the two sets of variables, and $(x_i - \bar{x})$ and $(y_i - \bar{y})$ are deviations from the means. The term $(x_i - \bar{x})(y_i - \bar{y})$ shows that for each pair, one measures the deviation of the value of x from the mean and the deviation of the value of y from the mean and multiplies these two deviations together.

From ▶ Figure 5–14, we see that the correlation coefficient can range between −1 and +1. It is equal to +1 when there is perfect positive correlation; it equals −1 for perfect negative correlation; and it is zero when the variables are uncorrelated (that is, when the variables are independent). Intermediate values of r indicate varying degrees of correlation, from slight, when r is close to zero, to very strong, when r is close to −1 or +1.

The primary importance of the correlation coefficient in evaluating V_A lies in its relationship to a genetic parameter known as the **heritability (h^2)**. The heritability is defined by the ratio

$$h^2 = \frac{V_A}{V_P}$$

and therefore measures the proportion of the total variance that is caused by additive gene effects. The value of h^2 can vary from 0 to 1. It is equal to 1 when variation in the trait is strictly additive ($V_P = V_A$); it equals 0 when all variation is environmentally induced ($V_A = 0$). The heritability by this definition is commonly referred to as the heritability in the narrow sense, in order to distinguish it from a similar but more broadly defined statistic called the **broad-sense heritability, H^2**. Unlike h^2, the broad-sense heritability measures the proportion of the total variance that is genetic in origin, that is,

$$H^2 = \frac{V_G}{V_P}$$

without regard to the nature of the gene effects. It therefore has only limited usefulness in analyzing the components of variation, and further discussions of heritability will be confined to h^2.

▶ **FIGURE 5–14** The correlation coefficients for various hypothetical data plotted as y vs. x. (a) Perfect negative correlation. (b) Perfect positive correlation. (c) No correlation.

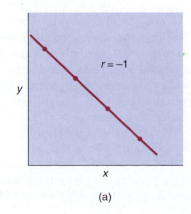

(a)

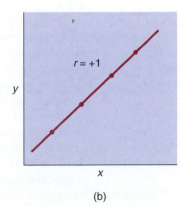

(b)

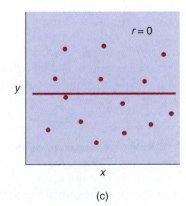

(c)

Genetics, Environment, and IQ

The intelligence quotient, or IQ, is defined as the ratio of an individual's mental age to his or her chronological age, multiplied by 100; it is determined by a standardized testing procedure. The IQ is actually a composite score that measures the average of a number of primary abilities, such as verbal ability, reasoning ability, and the abilities to memorize and visualize objects in space. Since primary abilities are not necessarily interrelated, their relative importance may vary from one individual to the next. A person's IQ is thus clearly a multidimensional character that includes several somewhat independent abilities; ideally, it should be expressed in terms of an individual's profile of primary abilities rather than by a single number. A case in point is the lowered IQ that is sometimes associated with Turner syndrome (the XO condition, where there are only 45 chromosomes, rather than 46, because a sex chromosome is missing). While first reports on individuals with Turner syndrome suggested a possible slight mental retardation, later studies revealed normal to superior verbal IQ. The reduction in overall score was largely caused by the difficulty that some individuals have with a form of space perception that requires right-left orientation.

When total IQ scores are plotted in a frequency distribution, they tend to follow the bell-shaped pattern that is expected for a normal curve. The distribution has a mean score of 100 for whites, 85 for African Americans, and 107 for those of Asian ancestry. The normal frequency distribution is often observed for quantitative traits, and while such curves cannot be taken by themselves as proof for a particular genetic mechanism, they are suggestive of a polygenic mode of inheritance.

Evidence for a genetic basis of variation in IQ scores has come from heritability estimates, which range from about 0.6 to 0.8, and from comparisons of IQ measurements in various groups of individuals reared together (unrelated individuals, parent and child, siblings, monozygotic (MZ) twins, and dizygotic (DZ) twins) and reared separately (unrelated persons, siblings, and MZ twins). The illustration to the right shows correlation coefficients for several of these IQ comparisons.

Each correlation coefficient measures the degree of correspondence in IQ scores between the persons concerned. The expected values are those predicted for strictly additive gene effects. For example, siblings are, on the average, identical in half their genes, giving an expected correlation coefficient of 0.5 if the variation in IQ scores is strictly genetic. Monozygotic twins are expected to show a correlation coefficient of 1, while no correlation is expected among unrelated individuals.

Correspondence of these expected values with the correlation coefficients actually observed is good in the case of siblings, but there are significant departures between expected and observed correlations for MZ twins and unrelated persons. The very high correlation in test scores between MZ twins supports the view that there is a significant degree of genetic determination for IQ, but note that rearing apart reduces the resemblance of MZ twins for this trait. This result and the fact that being raised together in the same orphanage or foster home significantly increases the correlation between unrelated persons provide evidence for a substantial influence of home environment on IQ score.

The value of h^2 can be determined from the strength of correlation between certain pairs of relatives. For example, if we measure the degree of correlation in a quantitative character between a parent and offspring (a mother-child or father-child correlation), the heritability can then be calculated from the relationship

$$h^2 = 2r$$

where r is the correlation coefficient. Since h^2 also equals V_A/V_P, estimates of V_A can then be obtained from independent measurements of r and V_P.

The inheritance of fingerprint patterns is an interesting human example to which a correlation analysis has been successfully applied. The patterns of fingerprint ridges are highly variable, and thus no two fingerprints are exactly alike, either on the right and left hands of the same individual or on corresponding hands of different individuals. However, the ridge patterns vary within limits that allow for systematic classification into three main types: arches, loops, and whorls (▶ Figure 5–15). This classification is based on the number of triradii—the points where three ridges join; the whorl has two triradii, the loop one triradius, and the arch none. Fingerprint patterns are also differentiated on the basis of their total ridge count (TRC),

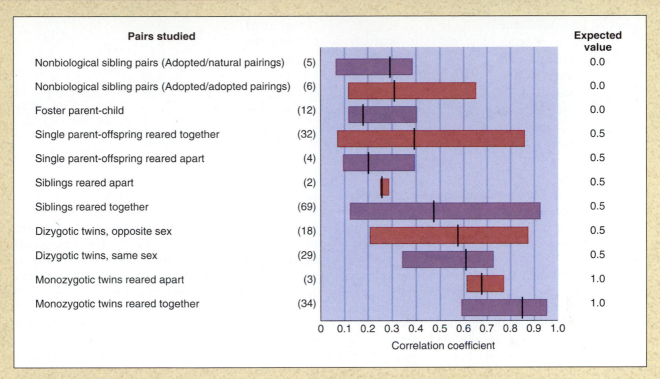

Pairs studied		Expected value
Nonbiological sibling pairs (Adopted/natural pairings)	(5)	0.0
Nonbiological sibling pairs (Adopted/adopted pairings)	(6)	0.0
Foster parent-child	(12)	0.0
Single parent-offspring reared together	(32)	0.5
Single parent-offspring reared apart	(4)	0.5
Siblings reared apart	(2)	0.5
Siblings reared together	(69)	0.5
Dizygotic twins, opposite sex	(18)	0.5
Dizygotic twins, same sex	(29)	0.5
Monozygotic twins reared apart	(3)	1.0
Monozygotic twins reared together	(34)	1.0

Correlation coefficient

The bar represents the range of correlation in each case, and the vertical line designates the median correlation coefficient. The number of studies included in each sample is given in parentheses. *Adapted from:* T. J. Bouchard, Jr., and M. McGue, "Familial studies of intelligence: A review," *Science* 212 (1981): 1055–1059. Copyright 1981 by the AAAS. Used with permission.

There has been extensive controversy over whether the differences in average IQ scores among white, African American, and Asian ancestry groups are genetic or environmental, or both, in origin. The high heritability estimate of 0.8 has been taken by some persons as evidence that the differences are mainly genetic. This reasoning is not valid, however, because heritability measures the genetic proportion of the variability *within* a measured population. It does not measure variation *between* populations. Thus, even if the heritability value measured in a particular population is high, this measurement is relevant only to that population and not to any comparison of it with a different population.

▶ **FIGURE 5–15** Fingerprint patterns. (a) Arch with no triradius. The ridge count is zero. (b) Loop with one triradius. The ridge count is 13. (c) Whorl with two triradii. The ridge count to the farthest triradius is 17. *Source:* John B. Jenkins, *Human Genetics* (Harper & Row, 1990), p. 412.

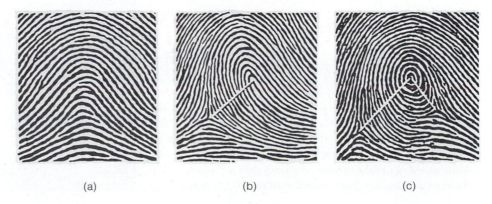

(a)   (b)   (c)

Relationship	Number of Pairs	Observed Correlation Coefficient	Theoretical Correlation Coefficient between Relatives
Mother—child	405	0.48 ± 0.04	0.50
Father—child	405	0.49 ± 0.04	0.50
Husband—wife	200	0.05 ± 0.07	0.00
Sibling—sibling	642	0.50 ± 0.04	0.50
Monozygotic twins	80	0.95 ± 0.01	1.00
Dizygotic twins	92	0.49 ± 0.08	0.50

Source: S. B. Holt, "Quantitative genetics of fingerprint patterns," Br. Med. Bull. 17 (1961): 247–250.

which is obtained by first calculating the number of ridges that lie along a line drawn from the core of the pattern to the farthest triradial point on each finger. These individual ridge counts are then summed to get the TRC (the total of all fingers).

Studies show that the TRC tends to vary from individual to individual according to a normal distribution; for British males, this distribution has a mean of 145 and a standard deviation of ±51. Moreover, the variation in the TRC is almost entirely a result of variation in the number of genes that have an additive effect on the trait. The importance of additive gene action is shown in ■ Table 5–1, in which the correlations for TRC observed between relatives are compared with the values that would be expected if the variation is strictly genetic in origin and is produced by additive gene effects ($h^2 = 1$). The theoretical genetic correlation between each pair of relatives is equal to the proportion of segregating genes that the members of the pair have in common. For example, first-degree relatives (siblings or a parent and child) are, on average, identical in half of their genes. Thus, for purely additive gene effects, first-degree relatives are expected to show a correlation coefficient of 0.5. Monozygotic twins have the same genotype and are therefore expected to show a correlation coefficient of 1, with members of each twin pair having identical total ridge counts. No correlation is expected between unrelated persons. The nearly perfect agreement that we observe between the theoretical and observed values in Table 5–1 indicates that variation in this trait is mainly genetic in origin and, moreover, conforms very well to an additive gene model.

Estimating the Number and Contributions of Polygenes

When a trait in experimental organisms conforms to an additive gene model, it is a relatively easy task to measure the effects of individual genes and to determine the number of segregating gene pairs through controlled breeding experiments. For a general example, suppose that a cross is made between two homozygous strains (AABBCC ... × aabbcc ...)

that differ in genotype at n gene pairs. Such matings give rise to a genetically uniform F_1 of genotype AaBbCc ... that has a mean phenotypic value intermediate to the values of the parental strains. Crossing the F_1s will then produce a highly variable F_2, which will show variation of both genetic and environmental origin. This crossing sequence is shown in ▶ Figure 5–16 in terms of the variation expressed by each generation. If we assume that all active (uppercase-letter) genes assort independently and contribute an equal and additive amount to the trait, it is possible to estimate the number of genes and contribution of each gene from the analysis of variance shown in ■ Table 5–2. Observe that by letting $R/2n$ represent the contribution of each active allele, where R is the difference between the parental means, the genotypic variance of the F_2 becomes $R^2/8n$. Since the phenotypic variance of the F_2 (designated V_{F_2}) is equal to the sum of the genetic variance and the environmental variance, it can be written as

$$V_{F_2} = \frac{R^2}{8n} + V_{F_1}$$

▶ FIGURE 5–16 A hypothetical series of crosses involving parental varieties of opposite phenotypic extremes. The distributions represent the variability expressed in each generation.

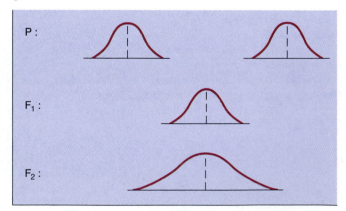

1. Start by considering only one gene pair. Contribution of the A locus:

Genotypes at the A Locus	Frequency among F_2	Contribution to Genotypic Value
AA	$\dfrac{1}{4}$	$2\left(\dfrac{R}{2n}\right)^*$
Aa	$\dfrac{1}{2}$	$\dfrac{R}{2n}$
aa	$\dfrac{1}{4}$	0

Mean contribution of A locus $= \dfrac{1}{4}\left(\dfrac{2R}{2n}\right) + \dfrac{1}{2}\left(\dfrac{R}{2n}\right) + \dfrac{1}{4}(0) = \dfrac{R}{2n}$

Contribution to $V_G = \dfrac{1}{4}\left(\dfrac{2R}{2n} - \dfrac{R}{2n}\right)^2 + \dfrac{1}{2}\left(\dfrac{R}{2n} - \dfrac{R}{2n}\right)^2 + \dfrac{1}{4}\left(0 - \dfrac{R}{2n}\right)^2$

$\qquad = \dfrac{1}{4}\left(\dfrac{R}{2n}\right)^2 + \dfrac{1}{4}\left(\dfrac{R}{2n}\right)^2$

$\qquad = \dfrac{R^2}{8n^2}$

2. Now extend the model to all n gene pairs.
   Computation of $\bar{P}$ and V_G for F_2 (take n times the contribution of a single locus): $\quad \bar{P} = n\left(\dfrac{R}{2n}\right) = \dfrac{R}{2}$

$$V_G = n\left(\dfrac{R^2}{8n^2}\right) = \dfrac{R^2}{8n}$$

*Note: $\dfrac{R}{2n} = $ the contribution of each active allele, where $R = $ the difference between the phenotypic means of the parental lines and $n = $ the total number of segregating pairs of polygenes.

In this expression, the phenotypic variance of the F_1 (V_{F_1}) is taken as an estimate of the environmental variance, since the F_1 individuals are genetically uniform. Solving for n, we get

$$V_{F_2} - V_{F_1} = \dfrac{R^2}{8n}$$

which gives

$$n = \dfrac{R^2}{8(V_{F_2} - V_{F_1})} \qquad (5.5)$$

We should emphasize that in deriving equation 5.5, we have assumed a number of simplifying conditions that, if not correct, can lead to errors in estimation. These simplifying conditions include independent assortment (unlinked genes) and an equal and additive contribution of all active genes. Most departures from these assumptions tend to inflate the variance of the F_2 and thereby lead to underestimates of the number of genes when equation 5.5 is applied.

Example 5.2

The average body weight of the Flemish breed of rabbits is 3600 g. The average weight of the Himalayan breed is 1875 g.

When crosses were made between these two breeds, the mean weights of the F_1 and F_2 were found to be intermediate between these extremes. The values of the variances were calculated as $V_{F_1} = 26,244\ g^2$ and $V_{F_2} = 52,900\ g^2$. Estimate the number of pairs of genes that contribute to variability in body weight in these crosses, and determine the average contribution of each active gene.

Solution: The number of gene pairs can be estimated using equation 5.5 as follows:

$$n = \dfrac{(3600 - 1875)^2}{8(52,900 - 26,244)} = 13.95, \text{ or } 14 \text{ gene pairs}$$

The average contribution of each active gene would be

$$\dfrac{R}{2n} = \dfrac{(3600 - 1875)}{2(14)} = 61.6\ g$$

Follow-Up Problem 5.2

If you were to examine the F_2 from the crosses in Example 5.2, within what range of body weights are 95% of the individuals expected to vary?

1. Because there is not a direct correspondence between genotype and phenotype in the case of quantitative traits, these traits are usually analyzed by statistical methods instead of by ratios among phenotypic classes. The mean value ($\overline{x}$) and variance (s^2) of the trait are calculated by using equations 5.2 and 5.3 for the sample mean and sample variance, respectively.

2. The statistical technique of analysis of variance can be used to analyze the underlying contributions of genes and the environment to the overall variation of a quantitative trait. In this type of analysis, the overall phenotypic variation is partitioned into three component parts: the genetic variance (V_G), the environmental variance (V_E), and the genotype-environment interaction variance (V_{GE}), so that $V_P = V_G + V_E + V_{GE}$.

3. Since the genetic variance (V_G) includes components associated with dominance (V_D), gene interaction (V_I), and additive gene effects (V_A), the phenotypic variance should actually be expressed as $V_P = V_A + V_D + V_I + V_E + V_{GE}$. In practical applications, however, V_{GE} is usually assumed to be zero, and V_D and V_I are combined into one term that represents deviations from strict additivity.

4. V_A and V_P also define the value of the narrow-sense heritability (h^2) for a trait ($h^2 = V_A/V_P$). The heritability gives the proportion of the total variation of a trait that is caused by additive gene effects.

5. In human populations, the heritability is usually estimated from statistical measurements of the degree of correspondence (as given by r, the correlation coefficient) in phenotypic values between various groups of relatives.

6. If a quantitative trait is strictly additive, the values of V_E and V_G can be determined from values of the variance of the F_1 and F_2 populations arising from a cross between homozygous lines that differ in genotype at all n gene pairs. The value of V_E is estimated by measuring the phenotypic variation within the homozygous populations or within the F_1 (where $V_G = 0$). Since the F_2 variance includes both environmental and genetic components, $V_G = V_{F_2} - V_{F_1}$. This analysis can also be used to estimate the number of genes (n) that determine the trait through the relationship

$$n = \frac{R^2}{8(V_{F_2} - V_{F_1})}$$

Chapter Summary

Quantitative traits such as height and weight vary in degree along a quantitative scale of measurement and thus exhibit a continuous variation pattern that lacks distinct phenotypic classes. The analysis of these traits therefore depends on statistical techniques that are concerned with the combined actions of several genes rather than on the conventional Mendelian analysis of phenotypic ratios.

Variation in quantitative traits is produced by both genetic and environmental factors. When different pure-breeding types are crossed, they tend to produce an intermediate F_1 that shows only environmental variation and a highly variable F_2 that incorporates both genetic and environmental variability. The additive polygene model offers the simplest explanation for the origin of this varia-tion: It assumes that genetic variation in a quantitative trait is the result of the combined action of several independently assorting polygenes, each contributing a small amount to the phenotype in an additive fashion.

The statistical analysis of quantitative traits is mainly concerned with estimating the components of phenotypic variation. One approach is to measure the heritability—the proportion of the phenotypic variance that is a result of additive gene effects—from the correlation in phenotype between various pairs of relatives. The additive genetic variance obtained in this way can then be used to estimate the number of polygenes and the average contribution of each active allele to the overall trait.

Questions and Problems

Polygenes and Quantitative Traits

1. Define the following terms and distinguish between paired terms:

   quantitative and qualitative traits
   continuous and discontinuous variation
   polygenes
   genetic and environmental variation
   threshold and meristic traits

2. Although the inheritance of eye color in humans is complex and incompletely understood, at least seven different eye colors can be identified: light blue, blue, blue-green, hazel, light brown, brown, and dark brown. (a) Assuming that eye color is a quantitative trait, propose genotypes for these classes of eye color. (b) Use your genetic model from part (a) to predict the distribution of eye color that you would observe among the offspring of hazel-eyed couples, where both the husband and wife are heterozygous for all of the contributing genes.

3. The number of toes on the hind legs of guinea pigs is thought to be determined by a polygenic system with four loci and a threshold effect. An individual having four or fewer contributing genes appears phenotypically as three-toed. However, an individual possessing at least five contributing genes exceeds

the threshold and has four toes. Predict the phenotypic ratio among the offspring from tetrahybrid crosses ($AaBbCcDd \times AaBbCcDd$).

4. Scott and Fuller have studied the mode of inheritance of various behavioral differences among different breeds of dogs. In one experiment, they attempted to measure the inheritance pattern for barking tendency. They used the African basenji breed, which has a high threshold of stimulation, and the American cocker spaniel breed, which has a low threshold. The tendencies to bark were assessed by measuring the percent of dogs that barked in 10-minute test periods when pairs of littermates were allowed to compete for a bone. The results obtained for the basenji and cocker spaniel breeds and their F_1 and F_2 offspring are presented below:

Dogs Barking in a Ten-minute Period

Basenji	19.6%
Cocker spaniel	68.2
F_1	60.1
F_2	55.5

Compare the F_1 and F_2 results with what is expected if the lower threshold (greater tendency to bark) of the cocker spaniel is inherited as the expression of a dominant allele of a single pair of genes.

Inheritance of Quantitative Traits

5. The number of days that elapse between the planting of wheat and the time that heads of grain appear (known as the heading date) is a quantitative trait. Crosses between two pure-breeding varieties of wheat with heading dates of 56 and 72 days produced a quite uniform F_1 with an average heading date of 64 days. An $F_1 \times F_1$ cross resulted in a highly variable F_2 with heading dates that varied symmetrically between the extremes of the early- and late-heading pure-breeding parental strains. Among 3000 F_2 plants examined, 12 had a heading date of 56 days. Estimate the number of gene pairs involved in determining the heading date, and calculate the average contribution of each active gene.

6. Two other pure-breeding strains of wheat with heading dates of 60 and 68 days were crossed. The average heading date of the F_1 was again 64 days, as was the average in the F_2. In this case, however, the F_2 distribution extended beyond the extremes in heading date that were exhibited by the pure-breeding parental strains. Out of 1000 F_2 plants, 3 had a heading date of 56 days, 5 had a heading date of 72 days, and the others had heading dates between these extremes. What genotypes are possible for the pure-breeding parental plants in this series of crosses? Use the first letters of the alphabet to designate genes.

7. Three pairs of independently assorting genes (R_1 and r_1, R_2 and r_2, and R_3 and r_3) are responsible for kernel color in wheat. Kernels are white in the $r_1r_1r_2r_2r_3r_3$ strain and red in all other strains. Active R genes contribute to red coloration in an additive fashion, so that kernel color forms an almost continuous gradation from very light red in strains with only one active gene to very dark red in the completely homozygous $R_1R_1R_2R_2R_3R_3$ strain. Suppose that we make the cross $R_1R_1r_2r_2R_3R_3 \times r_1r_1R_2R_2r_3r_3$ and carry it to the F_2 by intercrossing the F_1. Ignore the effects of environmental variation in answering the

following questions about the F_2 generation. (a) What proportion will have kernels of the same color as the $R_1R_1r_2r_2R_3R_3$ parental strain? (b) What proportion should breed true (will be pure-breeding) for the kernel color of the $R_1R_1r_2r_2R_3R_3$ strain? (c) What proportion will have kernels of the same color as the F_1? (d) What proportion should breed true for the kernel color of the F_1?

8. What proportion of the F_2 should be homozygous for all the contributing genes if the homozygous parental strains differ in genotype at (a) 2, (b) 3, (c) 4, (d) n gene loci?

9. What proportion of the F_2 should resemble a parental strain that is homozygous for completely dominant alleles at each of (a) 2, (b) 3, (c) 4, (d) n gene loci?

10. Suppose that height in a particular plant is determined by three independently assorting gene pairs (A,a, B,b, and C,c), with each active gene showing complete dominance and adding (in both heterozygous and homozygous combinations) 2 cm to a base height of 10 cm. The cross $AABBCC \times aabbcc$ is performed, and the F_1 are intercrossed to produce the F_2. Give the heights that are expected for the parents and for the F_1, and describe the distribution of heights expected in the F_2.

11. The average fruit weight of the Red Currant tomato is 1 g, while that of the Putnam's Forked variety is 58 g. When crossed, the two varieties produce an F_1 that has an average fruit weight of only 7.6 g, far less than the arithmetic average of the two parental strains. One plausible explanation for the low average of the F_1 is to assume that, say, 5 gene pairs determine fruit weight, with each active gene exerting a geometric effect by multiplying the value of the residual genotype by a constant amount of 1.5 g. Thus, genotypes $aabbccddee$, $AaBbCcDdEe$, and $AABBCCDDEE$ would have phenotypic values of 1, $(1.5)^5 = 7.6$, and $(1.5)^{10} = 58$ g, respectively. Use this model to predict the distribution of fruit weight among the F_2 of a cross between the Red Currant and Putnam's Forked strains.

Statistical Analysis of Quantitative Traits

12. R. A. Emerson and E. M. East found the F_1 from a cross between Black Mexican corn and a variety of popcorn to have the following ear lengths:

Ear length (cm):	9	10	11	12	13	14	15
Number of ears:	1	12	12	14	17	9	4

(a) Calculate the mean and variance of ear length in this sample. What is the probable cause of variation? (b) If ear length is assumed to follow a normal distribution, within what range of values can we expect 95.4% of the measurements to fall?

13. Indicate whether the following factors would serve to increase or decrease the heritability of a quantitative trait: (a) an increase in the homozygosity of the relevant genes, (b) a reduction in the environmental variability.

14. Data based on observations of 612 families by Miall and Oldham indicated that only 14.2% of the total variance in blood pressure (systolic pressure) could be ascribed to environmental variation. (a) Assuming a simple additive gene model, what is the heritability for this trait? (b) How much greater is the effect of genetic differences than environmental differences on the total variability?

15. The average weight of a mature male of the Bantam breed of chickens is 1.4 lb, while that of the Plymouth Rock male is 6.6 lb. Crosses between the Bantam and Plymouth Rock breeds produce an F_1 with a mean weight of 3.4 lb and a variance of 0.3 lb^2. The F_2 has a mean weight of 3.6 lb and a variance of 1.2 lb^2. (a) Ignoring dominance and interlocus interactions, estimate the number of gene loci involved in determining the difference in weight between these breeds of chickens. (b) Calculate the heritability for weight in these chickens.

16. In a cross between two diverse types of corn, Emerson and East obtained the distribution of ear length shown below for the parental strains, F_1, and F_2. (a) Calculate the mean and variance of each distribution. (b) Using equation 5.5, estimate the number of loci that determine ear length in these crosses. (c) How does the value calculated in part (b) compare with the number you would get if you based your estimate on the fraction of the F_2 with the longest (or shortest) length? Explain the reason for any difference.

Ear Length (cm)

	5	6	7	8	9	10	11	12	13	14	15	16	17	18	19	20	21
P	4	21	24	8					3	11	12	15	26	15	10	7	2
F_1					1	12	12	14	17	9	4						
F_2			1	10	19	26	47	73	68	68	39	25	15	9	1		

17. Suppose that variation in height in a particular plant is determined by three gene pairs (A,a, B,b and C,c) that exhibit complete dominance. Assume that each active gene, in either the homozygous or heterozygous state, contributes 4 inches of height to the base height of 10 inches. If we ignore variation owing to the environment, an $AABBCC$ plant would then be 22 inches high and an $aabbcc$ plant would be 10 inches. The cross $AABBCC \times aabbcc$ is made and is carried into the F_2 by intercrossing the F_1. (a) What is the height of the F_1? (b) Determine the height distribution in the F_2. (c) Calculate the mean and variance of height in the F_2. How do they compare in value to those of the F_1?

18. Reconsider problem 17, but now assume that the following interlocus interactions occur. Suppose that the cc allele pair exerts a masking effect over the B gene, so that whenever BB or Bb occurs with cc, the plant will exhibit the same height as a plant of the $bbcc$ genotype. A plant of genotype $AABBcc$ would then be 14 inches high, as would a plant of genotype $AAbbcc$. Any $C-$ plant will respond as it did in problem 17. Again, the cross $AABBCC \times aabbcc$ is made and is carried into the F_2. (a) Determine the height distribution in the F_2. (b) Calculate the mean and variance of height in the F_2. How do these values compare with the corresponding values computed in problem 17?

19. While performing a fingerprint analysis in a genetics class, ten students determined the ridge counts on their left and right hands and obtained the following results:

						Students					
		1	2	3	4	5	6	7	8	9	10
Ridge	left hand	73	59	79	83	17	102	88	61	50	34
Counts	right hand	67	65	97	70	26	109	74	47	58	45

(a) Calculate the product-moment correlation coefficient for these results. (b) How does the correlation coefficient calculated in (a) compare with the value reported for identical twins (see Table 5–1)? Why should these values be similar? Explain your answer.

THE CHEMICAL BASIS
OF INHERITANCE

In Part I we learned that genes are the fundamental units of heredity. They are both **units of expression** that act in the development of a particular trait by coding for proteins and **units of transmission** that are passed from parents to offspring according to demonstrable rules. We will now consider the chemistry of the genetic material and describe genes and their protein products in molecular terms.

The concept of the gene as a chemical substance has largely taken form since the identification of the genetic material in the early 1940s. Geneticists realized that any type of molecule that is to function as a gene must meet certain basic requirements (which were introduced in Chapter 1); it must be able to (1) store genetic information in a stable form, (2) transfer this information to other parts of the cell so that it can be expressed as a trait, (3) duplicate the information accurately during cell division, and (4) vary through mutation and recombination.

Of all the types of molecules in a cell, only proteins and the nucleic acids (DNA and RNA) are sufficiently complex to possibly qualify as candidates for

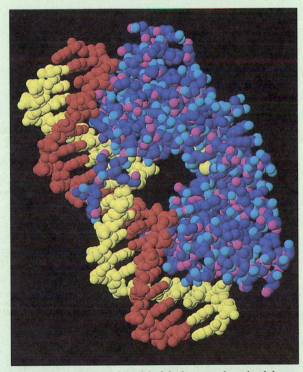

Computer-generated model of the bacteriophage lambda repressor protein bound to DNA.

genes. Both proteins and nucleic acids are **macromolecules**—large, complex polymers composed of different types of simple building-block molecules. The proteins are polymers of amino acids and the nucleic acids are polymers of nucleotides. Both classes of polymers are also **informational molecules**—each protein and each nucleic acid carries information essential for living processes within its characteristic building-block sequence. In the next two chapters, we will explore the nature of the proteins and nucleic acids, concentrating on the important structural features that enable these molecules to function in the transmission and expression of inherited traits. We will also consider some of the experiments that were instrumental in identifying the nucleic acid DNA (rather than a protein) as the primary carrier of the encoded instructions of each gene. DNA is thus directly responsible for the organism's genotype. RNA and proteins also carry genetic information, but do so for the purpose of carrying out the metabolic activities responsible for the organism's phenotype.

Proteins

In Chapter 1 we saw that genes are segments of chromosomal DNA that contain the information necessary for an organism to assemble various proteins. We will now focus on the exact relationship between genes and proteins. Proteins are the most abundant biological macromolecules— they make up about 50% of the dry weight of cells— so it is not at all surprising that proteins are so important in the determination of phenotypic traits. They are also the most functionally diverse of the biological molecules. Many proteins function as enzymes and catalyze specific metabolic reactions. Other proteins are antibodies, which function in the immune system to defend an organism against infectious disease. Some proteins are hormones, which act as chemical messengers in regulating a variety of cellular processes. Still others serve as transport proteins, (e.g., hemoglobin), contractile proteins (e.g., actin), or structural proteins. Proteins are therefore capable of controlling lit-

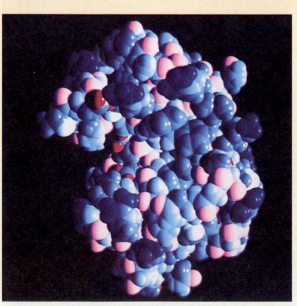

Computer simulation of the enzyme Ivozyme.

erally every aspect of cell chemistry and architecture through the multitude of activities that they perform.

The connection between genes and proteins was first recognized early in this century by the British physician Archibald Garrod. Garrod studied several human hereditary diseases and concluded from pedigree analysis that certain of these disorders are caused by defective recessive genes. He postulated that a defective gene produces a mutant phenotype because of a failure in a specific enzyme reaction, which leads to a block in a metabolic pathway in the cells. For example, in the case of alkaptonuria, which is one of the disorders studied by Garrod, the gene defect results in the inability to break down the benzene ring of homogentisic acid, also known as alcapton (▶ Figure 6–1). Failure to complete this reaction results in a buildup of homogentisic acid and the excretion of this metabolic intermediate in the urine. Garrod referred to these gene-controlled enzymatic failures that lead to human biochemical disorders as **inborn errors of metabolism.**

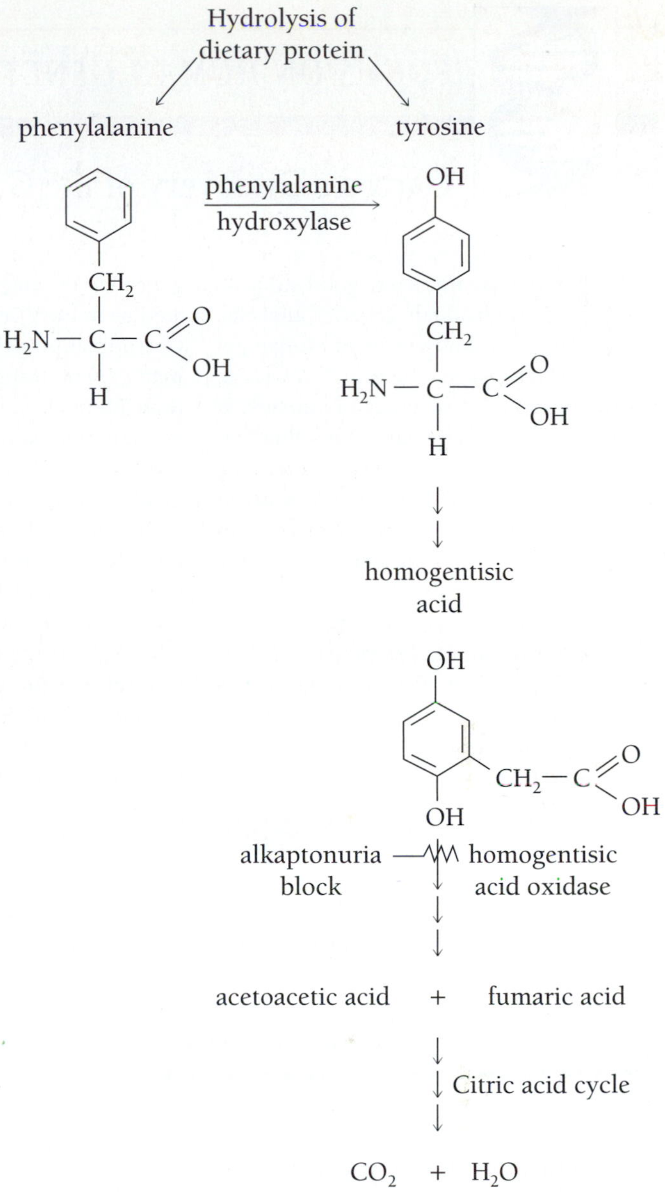

Garrod was the first person to make the connection between mutant genes and malfunctioning metabolic pathways, thereby establishing a relation between genes, proteins, and phenotypic traits. In the 1940s, through experimental work to be discussed in a later chapter, the **one gene–one enzyme hypothesis** was formulated; it states that the function of a gene is to control the synthesis of one enzyme. This idea was quickly extended to include proteins other than enzymes, establishing the concept that genes (at least most genes) determine traits by controlling the synthesis of proteins in the cells of an organism.

To understand how proteins can give rise to traits, it is important to realize that, despite their diversity of functions, proteins as a group perform their various metabolic activities in the same basic manner—by selectively binding to other molecules. For example, an enzyme binds specifically to a substrate, an antibody binds specifically to an antigen, a transport protein binds specifically to the molecule that it carries, and so on. Even structural proteins are involved in selective chemical binding; in this case, the binding typically joins identical molecules together as monomer units in a larger, more inclusive polymeric structure. The different binding specificities of proteins thus explain how this one group of molecules can carry out such diverse activities and give rise to the various traits of an organism. The differences in the binding specificities of proteins are the result of their structural differences, which we will now examine.

▦ COVALENT STRUCTURE OF PROTEINS

To carry out a particular gene-directed activity, each protein must have a specific, functional structure. Every protein is composed of at least one **polypeptide,** which is an unbranched polymer of **amino acids.** Polypeptides are typically large polymers that are composed of many amino acid residues. Moreover, each residue location in the polypeptide

▶ **FIGURE 6–1** Homogentisic acid is a tyrosine metabolite that is normally broken down by the enzyme homogentisic acid oxidase into products that enter the citric acid cycle. People affected with the recessive disorder alkaptonuria lack this enzyme, causing homogentisic acid to accumulate and be excreted in the urine. Thus a recessive mutant gene is associated with a metabolic block.

can be occupied by any of 20 different kinds of amino acids. If these amino acids were incorporated into the chain at random, the number of possible sequences is 20^n, where n is the number of residues. For a typical polypeptide with $n = 200$, the number of possible sequences is astronomical—equal to 20^{200} (or 10^{260}). Although there are thousands of different proteins in each species of organism and millions of different species, the number of polypeptides that actually exist on this planet is small when compared to the number theoretically possible. The reason for this discrepancy is that polypeptides are not just a random sample of 10^{260} possibili-

Garrod's Discovery of the Gene–Enzyme Connection

Humans do not make good subjects for traditional genetic analysis, and thus knowledge of human genetics has lagged behind knowledge of the genetics of other organisms. It is therefore somewhat remarkable that in 1903, just three years after the discovery of Mendel's work, Garrod postulated a link between Mendelian genes and certain human diseases like alkaptonuria.

Garrod's explanation that a recessive mutant allele is responsible for alkaptonuria showed that human genes obey the same laws of inheritance as the garden pea. (However, the significance of his work, in terms of relating genes to enzymes, was not appreciated until forty years later.) Of course, Garrod's approach to the study of human genetic traits was much different than Mendel's, since humans cannot be manipulated in controlled crossing sequences. Garrod, a British physician, instead analyzed the family histories of individuals with alkaptonuria. He observed a number of children who had the disease, and he noted that about 60% of them were born to parents who were first cousins. Since alkaptonuria and the other "inborn errors of metabolism" studied by Garrod are individually rather rare, the high frequency of the disease among first-cousin marriages suggested to Garrod that the causative gene is recessive. He postulated that the parents, who were unaffected, were heterozygous for the alkaptonuria allele, since they then would be much more likely to carry the same rare recessive gene than would unrelated individuals.

Garrod was the first person to use pedigree analysis for studying a human genetic trait. His work involved three disciplines that at the time were almost totally isolated from one another—biochemistry, genetics, and medicine. He was, like Mendel, far ahead of his time.

---

ties. Instead, each kind of polypeptide is the product of evolution, having a unique sequence of amino acids dictated by a particular gene that allows it to function efficiently in its environment. This characteristic sequence of amino acids in a polypeptide is called its **primary structure**; it is the most basic structural feature that distinguishes different proteins and, as we shall see in a later section, is ultimately responsible for their functional specificity.

Amino Acids and Peptides: Structure and Properties

The properties of proteins are determined by the chemical and physical properties of their constituent amino acids. The structures of the 20 amino acids that occur naturally in proteins are shown in ▶ Figure 6–2. Note that all amino acids with the exception of proline are α-amino acids, which have the following general structure:

$$\begin{array}{c} COO^- \\ | \\ {}^+H_3N - C - H \\ | \\ R \end{array}$$

where R is the side chain that is attached to the central carbon atom (called the alpha (α) carbon). At physiological pH, both the amino group (H_2N-) and the carboxyl group ($-COOH$) are present in their charged forms. The amino acid thus exists minimally as a *dipolar ion* and exhibits both acidic and basic properties.

The different amino acids owe their chemical individuality to the structure of their side chains. The different R groups range in complexity from a simple hydrogen ($-H$) in glycine and a methyl group ($-CH_3$) in alanine to the complex ring structures of phenylalanine, tyrosine, and tryptophan. The side chains are polar in some amino acids and nonpolar in others. The polar R groups can be further classified on the basis of the charge they possess at physiological pH. Two amino acids, aspartic acid and glutamic acid, have acidic R groups that carry a negative charge. Three amino acids, lysine, arginine, and histidine, have basic R groups and carry a positive charge. All other polar amino acids have side chains that are either uncharged or only slightly ionized at pH 7.0.

One important property of amino acids is their ability to form **peptide bonds** (▶ Figure 6–3). A peptide bond ($-CO-NH-$) is an amide linkage joining the α-amino group of one amino acid to the α-carboxyl group of another. Two amino acids joined together in this manner yield a **dipeptide** (Figure 6–3a); three amino acids joined together yield a **tripeptide**; and so on. When many amino acids are joined together, the result is a **polypeptide** (Figure 6–3b) that may contain up to a thousand or more amino acid residues. Because of the unbranched character of the polypeptide, all α-amino and α-carboxyl groups of its constituent amino acids are involved in peptide bonds ex-

Nonpolar (hydrophobic) R groups

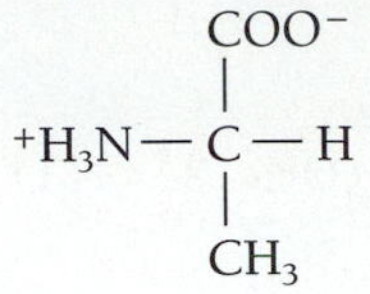

$$
\begin{array}{c}
COO^- \\
| \\
{}^+H_3N-C-H \\
| \\
CH_3
\end{array}
$$

alanine
(Ala or A)

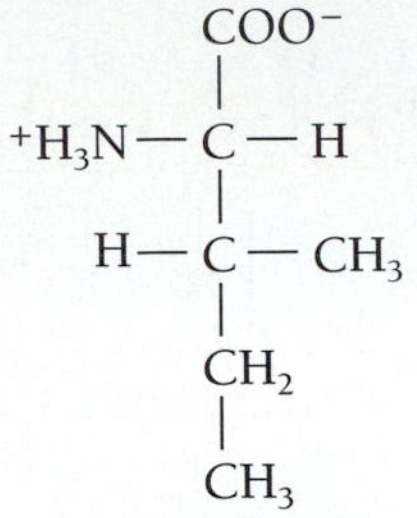

$$
\begin{array}{c}
COO^- \\
| \\
{}^+H_3N-C-H \\
| \\
H-C-CH_3 \\
| \\
CH_2 \\
| \\
CH_3
\end{array}
$$

isoleucine
(Ile or I)

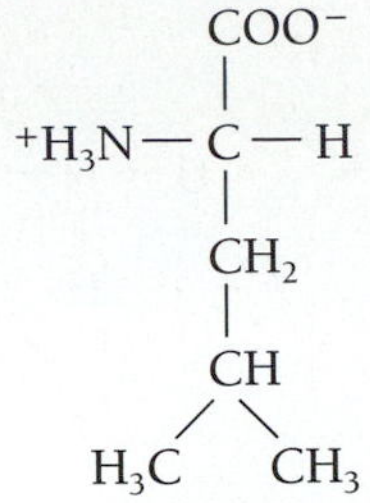

$$
\begin{array}{c}
COO^- \\
| \\
{}^+H_3N-C-H \\
| \\
CH_2 \\
| \\
CH \\
H_3C \quad CH_3
\end{array}
$$

leucine
(Leu or L)

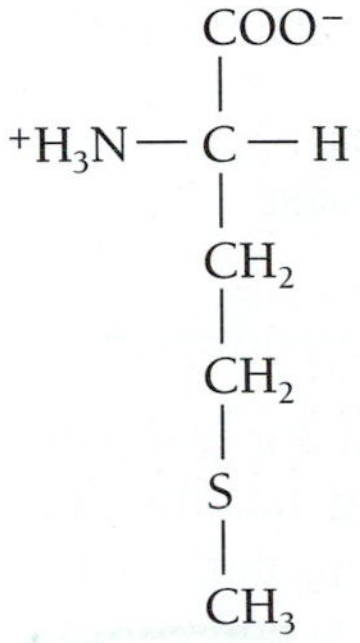

$$
\begin{array}{c}
COO^- \\
| \\
{}^+H_3N-C-H \\
| \\
CH_2 \\
| \\
CH_2 \\
| \\
S \\
| \\
CH_3
\end{array}
$$

methionine
(Met or M)

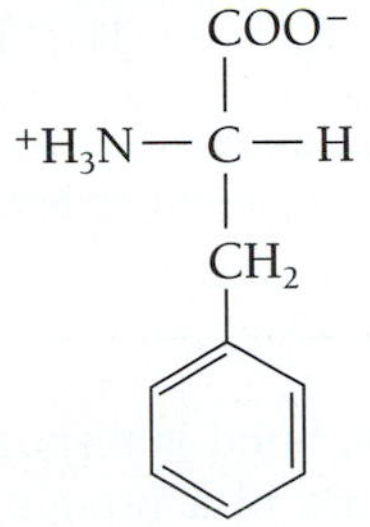

$$
\begin{array}{c}
COO^- \\
| \\
{}^+H_3N-C-H \\
| \\
CH_2
\end{array}
$$

phenylalanine
(Phe or F)

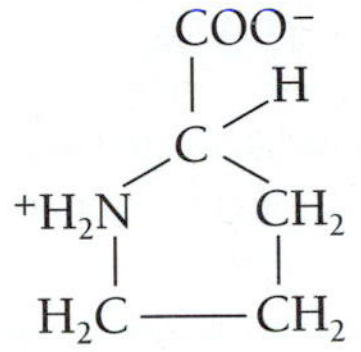

$$
\begin{array}{c}
COO^- \\
| \\
H \\
C \\
{}^+H_2N \quad CH_2 \\
H_2C-CH_2
\end{array}
$$

proline
(Pro or P)

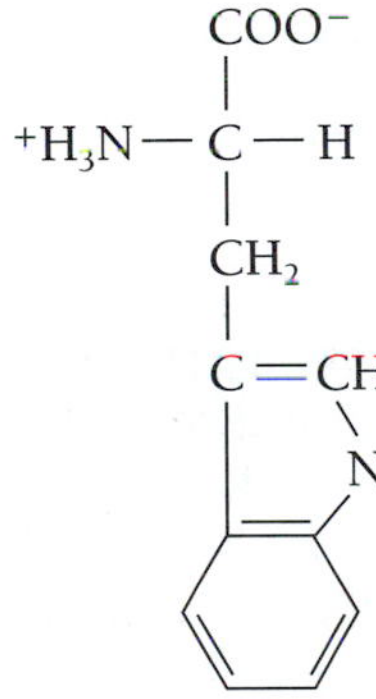

$$
\begin{array}{c}
COO^- \\
| \\
{}^+H_3N-C-H \\
| \\
CH_2 \\
| \\
C=CH \\
NH
\end{array}
$$

tryptophan
(Trp or W)

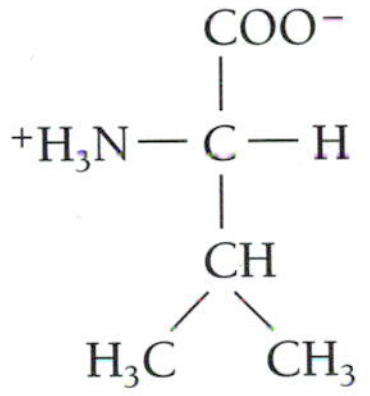

$$
\begin{array}{c}
COO^- \\
| \\
{}^+H_3N-C-H \\
| \\
CH \\
H_3C \quad CH_3
\end{array}
$$

valine
(Val or V)

Negatively charged R groups

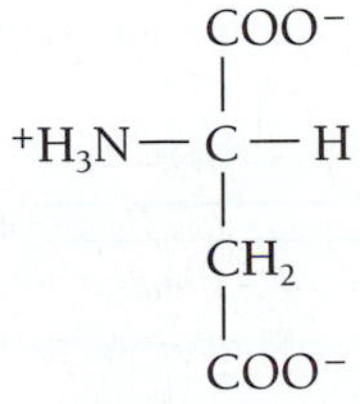

$$
\begin{array}{c}
COO^- \\
| \\
{}^+H_3N-C-H \\
| \\
CH_2 \\
| \\
COO^-
\end{array}
$$

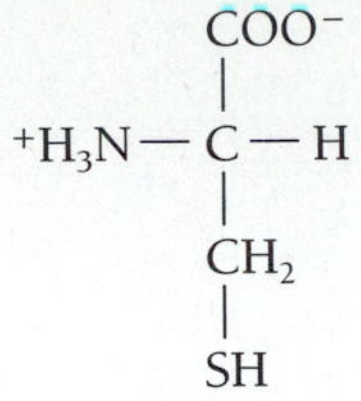

aspartic acid
(Asp or D)

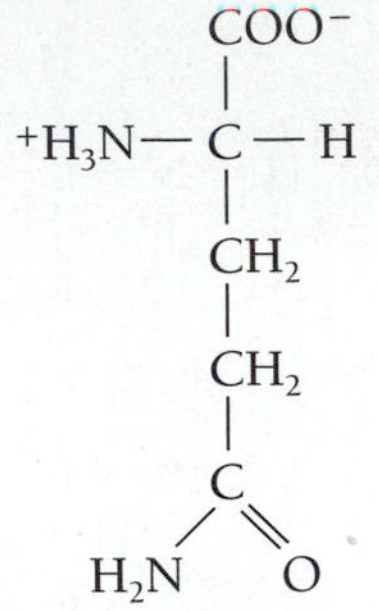

$$
\begin{array}{c}
COO^- \\
| \\
{}^+H_3N-C-H \\
| \\
CH_2 \\
| \\
CH_2 \\
| \\
COO^-
\end{array}
$$

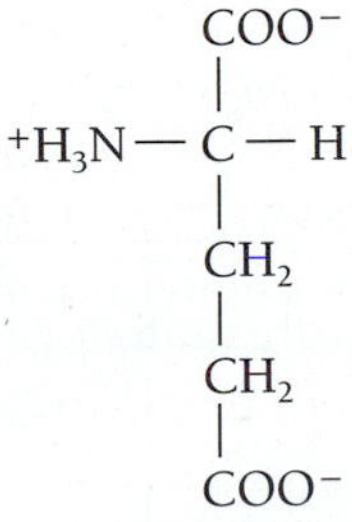

glutamic acid
(Glu or E)

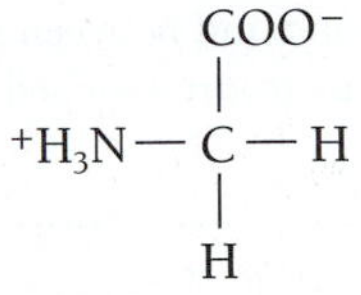

Polar but uncharged R groups

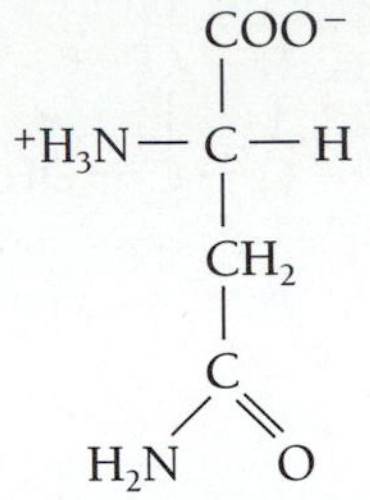

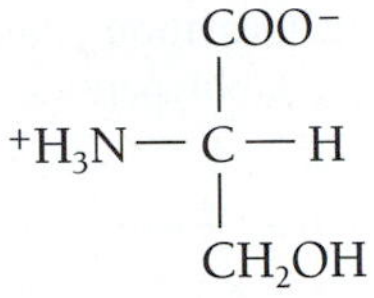

$$
\begin{array}{c}
COO^- \\
| \\
{}^+H_3N-C-H \\
| \\
CH_2 \\
| \\
C \\
H_2N \quad O
\end{array}
$$

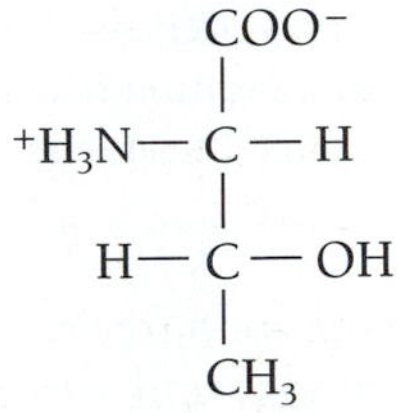

asparagine
(Asn or N)

$$
\begin{array}{c}
COO^- \\
| \\
{}^+H_3N-C-H \\
| \\
CH_2 \\
| \\
SH
\end{array}
$$

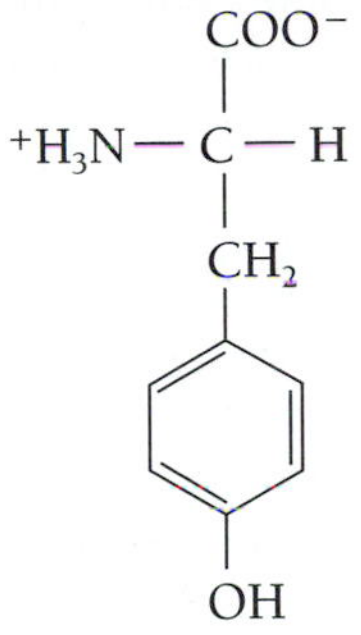

cysteine
(Cys or C)

$$
\begin{array}{c}
COO^- \\
| \\
{}^+H_3N-C-H \\
| \\
CH_2 \\
| \\
CH_2 \\
| \\
C \\
H_2N \quad O
\end{array}
$$

glutamine
(Gln or Q)

$$
\begin{array}{c}
COO^- \\
| \\
{}^+H_3N-C-H \\
| \\
H
\end{array}
$$

glycine
(Gly or G)

$$
\begin{array}{c}
COO^- \\
| \\
{}^+H_3N-C-H \\
| \\
CH_2OH
\end{array}
$$

serine
(Ser or S)

$$
\begin{array}{c}
COO^- \\
| \\
{}^+H_3N-C-H \\
| \\
H-C-OH \\
| \\
CH_3
\end{array}
$$

threonine
(Thr or T)

$$
\begin{array}{c}
COO^- \\
| \\
{}^+H_3N-C-H \\
| \\
CH_2 \\
| \\
\text{(benzene ring)} \\
OH
\end{array}
$$

tyrosine
(Tyr or Y)

Positively charged R groups

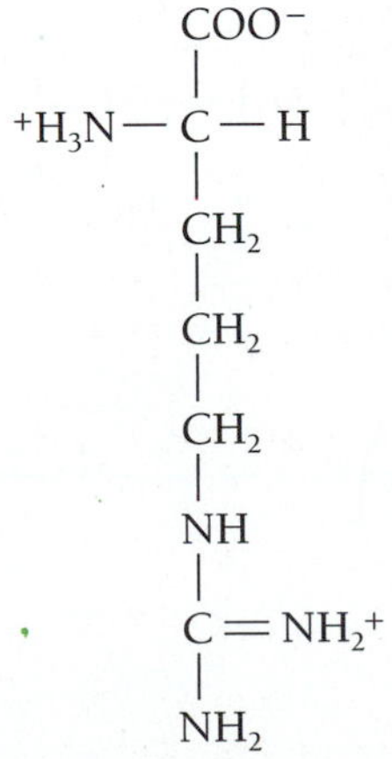

$$
\begin{array}{c}
COO^- \\
| \\
{}^+H_3N-C-H \\
| \\
CH_2 \\
| \\
CH_2 \\
| \\
CH_2 \\
| \\
NH \\
| \\
C=NH_2^+ \\
| \\
NH_2
\end{array}
$$

arginine
(Arg or R)

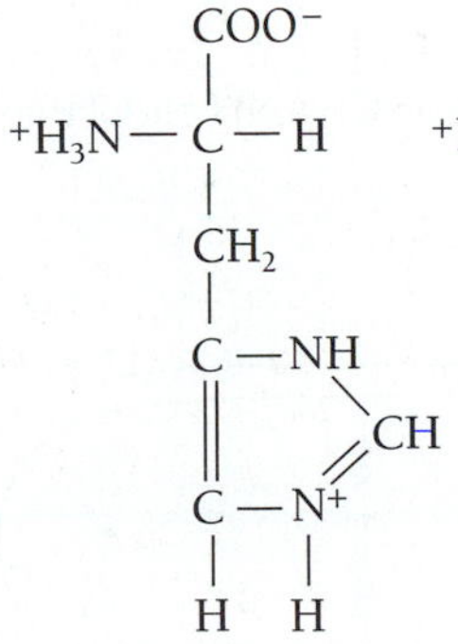

$$
\begin{array}{c}
COO^- \\
| \\
{}^+H_3N-C-H \\
| \\
CH_2 \\
| \\
C-NH \\
\| \qquad CH \\
C-N^+ \\
H \quad H
\end{array}
$$

histidine
(His or H)

$$
\begin{array}{c}
COO^- \\
| \\
{}^+H_3N-C-H \\
| \\
CH_2 \\
| \\
CH_2 \\
| \\
CH_2 \\
| \\
CH_2 \\
| \\
NH_3^+
\end{array}
$$

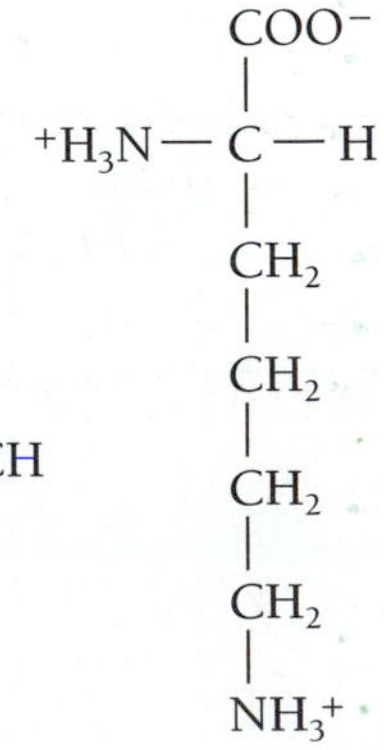

lysine
(Lys or K)

▶ **FIGURE 6–2** The structures of the 20 common amino acids with their standard one-letter and three-letter abbreviations.

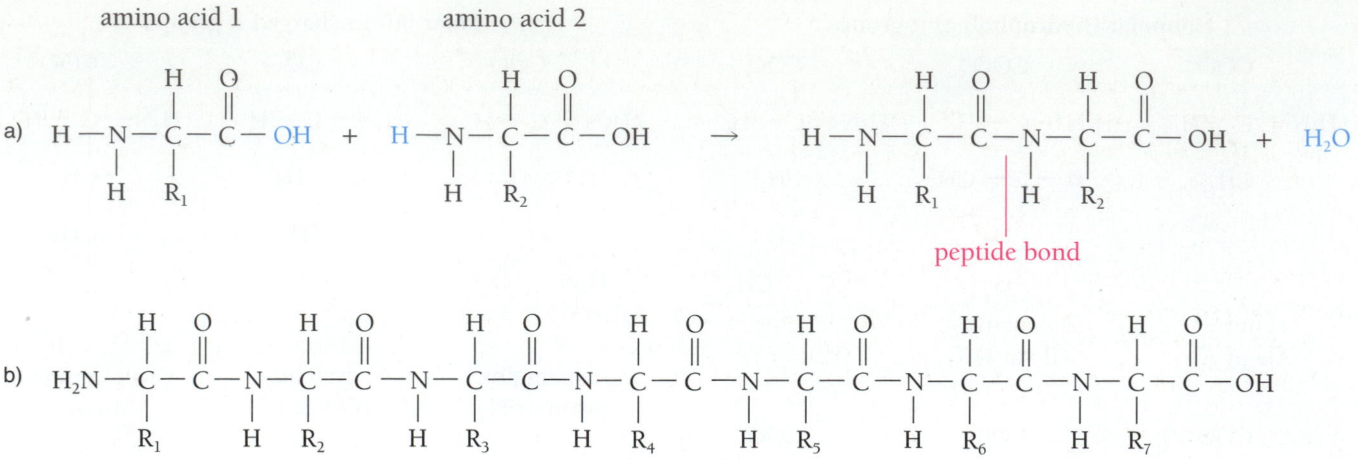

amino acid 1   amino acid 2

a)

b)

▶ **FIGURE 6–3** (a) The formation of a peptide bond between two amino acids occurs through the elimination of water in a condensation reaction. (b) A polypeptide is a linear, unbranched polymer of amino acids having a free amino terminus and a free carboxyl terminus.

cept for a free α-NH$_2$ group at one end (the **amino or N terminus**) and a free α-COOH at the other end (the **carboxyl or C terminus**). The R groups are not involved in joining the amino acids together in the covalent backbone of the chain, so they are free to interact chemically with one another or with the solvent. The amino acid side chains thus determine the specific chemical properties of a polypeptide.

Although the peptide bond is responsible for maintaining the primary structure of a protein, it is not the only covalent bond that can be formed between amino acids. Another is the **disulfide bond** ($-S-S-$), which forms between the *sulfhydryl* ($-SH$) *groups* in the side chains of two cysteine residues. As shown in ▶ Figure 6–4, disulfide bonds can involve cysteine residues of either the

▶ **FIGURE 6–4** The disulfide bond. (a) Formation of a disulfide bond between the sulfhydryl groups of two cysteine amino acids. (b) Intrachain disulfide bonds. (c) Interchain disulfide bonds.

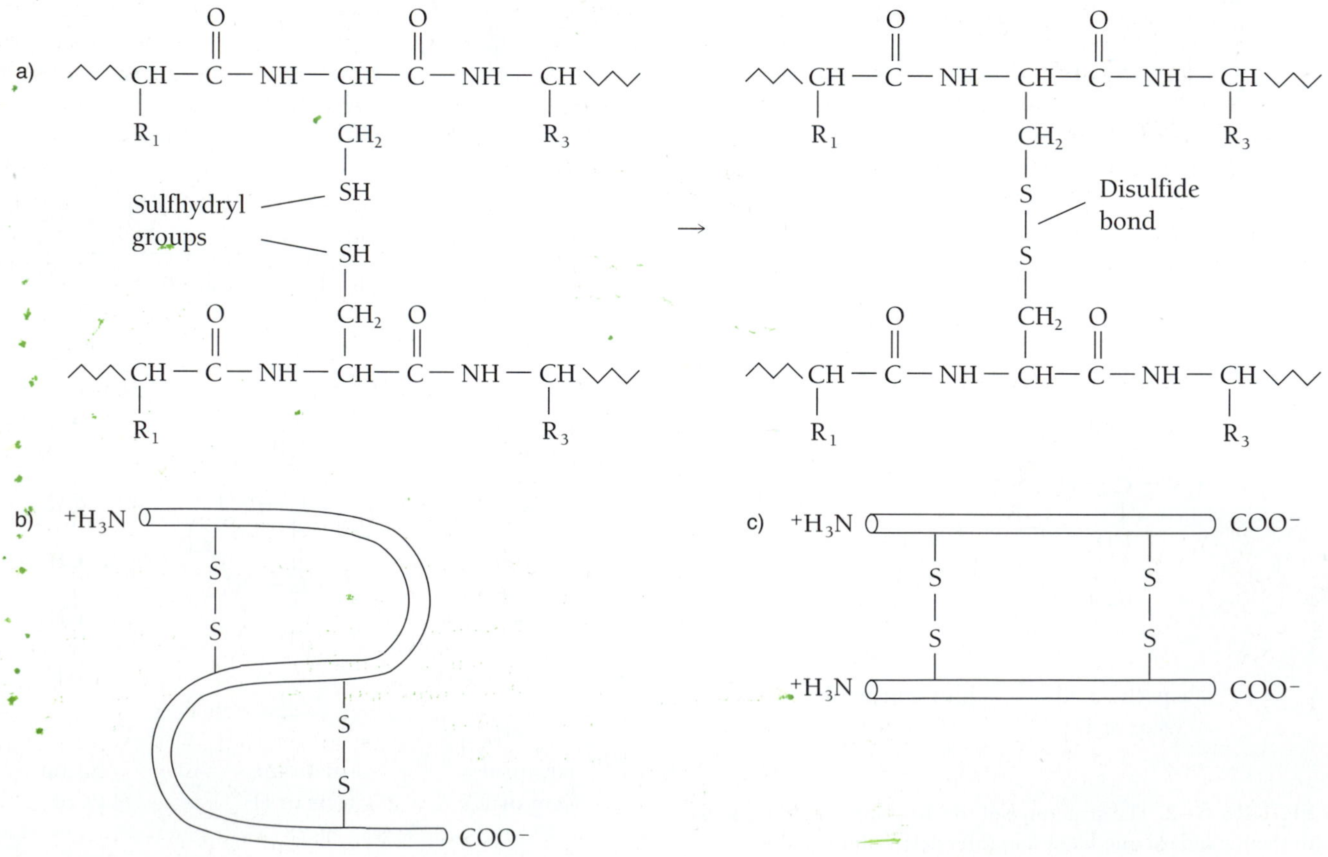

same polypeptide chain (Figure 6–4b) or different chains (Figure 6–4c).

Since several of the amino acid side chains have ionizable groups, a polypeptide tends to possess a net charge at a given pH. The strength and sign of the charge are determined by the polypeptide's particular mix of negatively charged (acidic) and positively charged (basic) amino acids. Differences in charge serve as the basis for separating polypeptides by **gel electrophoresis** (▶ Figure 6–5). Electrophoresis is the migration of charged molecules in response to an externally applied electric field. The molecules move in a direction determined by the sign of their charge and at a rate that depends directly on the magnitude of their charge and inversely on their size and shape. Molecules that differ in charge and conformation will therefore travel different distances in the imposed electric field and become separated from one another. In gel electrophoresis, such separations are typically carried out on thin slabs of agarose or polyacrylamide gels. A gel provides a porous supporting medium which differentially retards the movement of larger molecules as molecules move through it. The mobility of the molecules varies inversely with the gel concentration, so the amount of gelling agent used in a given experiment depends on the size of the polypeptides being separated. Following electrophoresis, strips can be sliced from the gel and stained with a dye to reveal the positions of the separated bands. The polypeptides in the remainder of the gel can then be eluted and analyzed.

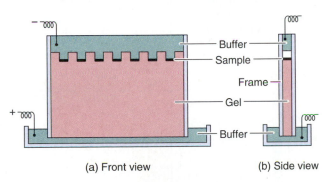
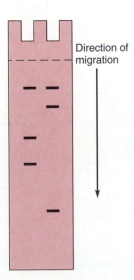

▶ **FIGURE 6–5** Gel electrophoresis. (a), (b) Apparatus. (c) The stained gel reveals the positions of the separated proteins.

(a) Front view
(b) Side view

Buffer
Sample
Frame
Gel
Buffer

−
+

Direction of migration

(c) Separation of components on the gel

To Sum Up

1. The connection between genes and proteins was first made in the early 1900s by Garrod, who postulated that certain human "inborn errors of metabolism" are due to blocks in metabolic pathways. He correlated these blocks with nonfunctional enzymes specified by mutant genes.

2. Proteins are large molecules that are composed of one or more polypeptide chains. Each polypeptide chain is a linear polymer of amino acids that are linked to one another by peptide bonds. The specific sequence of amino acids in a polypeptide chain is called its primary structure.

3. The positive and negative charges of certain amino acid side groups confer a net charge on a polypeptide chain and allow it to be separated from other polypeptides of different net charge and conformation using gel electrophoresis.

THREE-DIMENSIONAL STRUCTURE OF PROTEINS

Few polypeptides exist in a fully extended state. Instead they tend to be folded into a characteristic three-dimensional structure or **conformation**. This precise folding of the polypeptide chain provides a surface configuration that enables the protein to interact specifically with other molecules. The conformation of the polypeptide is therefore a major determinant of protein function, and any large perturbations of this structure can lead to a loss of biological activity.

Although the conformation of every kind of polypeptide is unique, proteins in general can be divided into two main classes, **globular** proteins and **fibrous** proteins, based on the overall shape of their folded structure. Globular proteins are tightly folded into compact spherical or ellipsoidal shapes. They are usually soluble in aqueous solutions, and include most enzymes, antibodies, and protein hormones, as well as many transport proteins. Fibrous proteins, on the other hand, are typically long filamentous structures that are insoluble in aqueous solutions and mainly serve structural or contractile roles. For example, α-keratin, the structural protein of hair, and collagen, the structural protein of tendons, are fibrous proteins, as are actin and myosin, the contractile proteins of muscle.

The conformation of a protein can be further complicated if the protein is conjugated with a nonpolypeptide substance called a **prosthetic group**. Conjugated proteins are usually classified according to the chemical nature of the prosthetic group. For example, glycoproteins contain sugar groups, lipoproteins contain lipids, and nucleoproteins contain nucleic acids (DNA or RNA), in addition to their polypeptide components. The overall three-dimensional con-

Characterizing and Purifying Proteins by Electrophoresis

Several variations of the basic technique of gel electrophoresis are useful for analyzing proteins. One variation is a method known as **SDS polyacrylamide gel electrophoresis,** or simply **SDS PAGE.** In this method, polypeptides are first treated with the detergent SDS (sodium dodecyl sulfate). Molecules of SDS associate with amino acid residues in the polypeptides and contribute to an overall negative charge that is proportional to the polypeptide chain length. As a result, all SDS-polypeptide complexes then have essentially the same charge-to-mass ratio. These complexes would tend to migrate toward the positive electrode at similar rates if they were free in solution. In the gel, however, the large complexes are significantly more retarded in their movements than the small complexes, and therefore their mobilities are inversely proportional to their molecular weights. The molecular weight of a polypeptide can then be determined by comparing its migration behavior with standards of known molecular weight, as shown in the following graph:

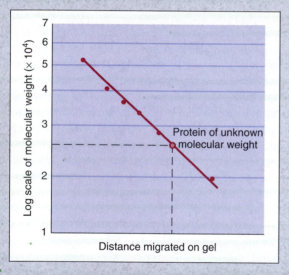

In this example, the line constructed through the points corresponds to distances traveled by the protein standards, and interpolation gives a molecular weight of 25,000 for the protein of unknown size (open circle).

SDS PAGE separates proteins on the basis of differences in their molecular weight. **Isoelectric focusing,** another useful variation of the electrophoretic technique, separates proteins on the basis of differences in their isoelectric point (i.e., the pH at which the net charge of a protein is zero). In isoelectric focusing, separations are carried out on gels containing a pH gradient. (The pH gradient is established first by electrophoresing a mixture of small multicharged polymers called polyampholytes.) When proteins are placed in such a gradient and subjected to an electric field, each protein will migrate to the pH on the gel where its net charge is zero. Since each kind of protein has its own unique isoelectric point (determined by its mix of acidic and basic residues), different proteins will cease to migrate at different points on the gel, thus enabling the proteins to become separated.

Especially high-resolution separations of proteins can be achieved by two-dimensional electrophoresis using a combination of isoelectric focusing and SDS PAGE. In this technique, the protein sample is first subjected to isoelectric focusing. The gel containing the partially separated proteins is then placed alongside an SDS gel and electrophoresed at a 90° angle to the first gel, as shown:

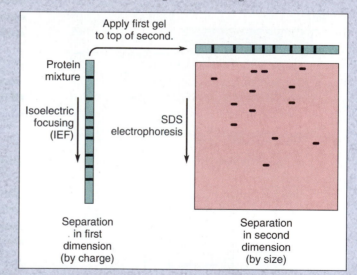

At the end of this procedure, the different proteins are revealed as a two-dimensional pattern of spots, having been separated by isoelectric point in one direction and by molecular weight at right angles to it. More than 1000 different protein components of a cell can be separated and revealed in a single experiment by this combination of electrophoretic methods.

figuration of a protein therefore depends not only on the spatial organization of its polypeptide chain but also on the structure of any prosthetic group that it may contain.

Because of the complex nature of proteins, it is helpful to visualize polypeptide conformation at different levels of organization. As we have seen, the primary structure is the actual sequence of amino acids making up the covalent backbone structure. The polypeptide chain is also folded into local, repeating conformational patterns or secondary structures, which together are organized into the complete polypeptide conformation or tertiary structure. We shall see that as one progresses up the structural hierarchy, each level carries the information necessary for the next, so that all higher-order structures are ultimately determined by the primary structure.

Example 6.1

The primary structures of polypeptides can sometimes be determined by differential peptide bond cleavage using specific cleavage reagents such as cyanogen bromide (CNBr) and the proteolytic enzymes trypsin and chymotrypsin. Cyanogen bromide cleaves peptide bonds on the carboxyl side of methionine, trypsin cleaves on the carboxyl side of arginine and lysine, and chymotrypsin cleaves on the carboxyl side of phenylalanine, tryptophan, and tyrosine. List the peptide fragments formed when the following polypeptide is treated with chymotrypsin:

Val-Ala-Lys-Glu-Glu-Phe-Val-Met-Tyr-Cys-Glu-Trp-Met-
Gly-Gly-Phe

Solution: Since chymotrypsin cleaves on the carboxyl side of Phe, Trp, and Tyr, cleavage by this enzyme would yield the following fragments:

Val-Ala-Lys-Glu-Glu-Phe

Val-Met-Tyr

Cys-Glu-Trp

Met-Gly-Gly-Phe

Follow-Up Problem 6.1

The overlapping patterns in fragments produced by different enzyme or reagent cleavages can be analyzed to piece together the primary sequence of a polypeptide chain. Suppose that a polypeptide chain is treated separately with chymotrypsin and trypsin and yields the peptide fragments shown below. Determine the primary sequence of the original chain.

Chymotrypsin Treatment	Trypsin Treatment
Met	Arg
Gly-Trp	Ala-Lys
Glu-Arg-Tyr	Phe-Glu-Arg
Arg-Ala-Lys-Phe	Tyr-Gly-Trp-Met

Secondary Structure

Secondary structure refers to the regular, local folding patterns of the polypeptide chain. The patterns that form this level of structure are largely determined by two properties of the peptide bond. First, electrons are shared among the atoms of the peptide bond in a way that gives it a partial double-bond character and a degree of torsional stiffness that makes the bond resistant to rotation about its axis. As a result, all the atoms of a peptide bond, along with the adjacent α-carbon atoms, lie in a rigid plane (▶ Figure 6–6). The polypeptide backbone therefore behaves much like a chain of linked plates that can fold by rotations around the α-carbon bonds into a relatively restricted number of conformations.

The second important property of the peptide bond is that its hydrogen and oxygen atoms are available for hydrogen bonding. A hydrogen bond is a weak electrostatic bond that forms between an electronegative (electron-seeking) atom (usually nitrogen or oxygen) and a hydrogen atom covalently bonded to another electronegative atom. An example of hydrogen bonding between the backbone peptide groups is shown below:

$$\overset{\diagdown}{\underset{\diagup}{\delta^-}}N-\underset{\delta^+}{H}\cdots\underset{\delta^-}{O}=C\overset{\delta^+}{\underset{\diagdown}{\diagup}}$$

The polarization of the N—H and C=O bonds leaves the hydrogen atom with a partial positive (δ^+) charge and the oxygen with a partial negative (δ^-) charge, thus allowing the formation of a hydrogen bond ($\cdots$) between the two groups. Note that a peptide bond includes both an H-bond donor group (N—H) and an H-bond acceptor group (C=O) extending in opposite directions. There is thus a tendency for the different peptide groups to form alternate H bonds between themselves in the polypeptide's folded state.

Linus Pauling and R. B. Corey predicted the existence of certain regular conformational patterns in proteins in the early 1950s. They sought patterns that would account for the earlier results of William Astbury, whose X-ray diffraction studies suggested that α-keratin, the structural protein of hair, and fibroin, the protein of silk fibers, consist of regular structural units with a periodicity of 0.54 nm and 0.70 nm, respectively. One important type of secondary structure that they proposed is the α-helix, which consists of a polypeptide wound in a helical configuration (▶ Figure 6–7). This structure was later found to be the basis of α-keratin, which is composed of a cablelike network of polypeptide chains, each in the form of a long α-helix. The α-helix has a pitch (vertical rise per turn) of 0.54 nm and a vertical rise per amino acid residue of 0.15 nm. There are thus 0.54/0.15 = 3.6 amino acids per turn of the helix. The structural integrity of the α-helix is maintained by hydrogen bonds formed between the C=O group of each peptide bond and the N—H group of the peptide bond located four amino acids away. The R groups of the amino

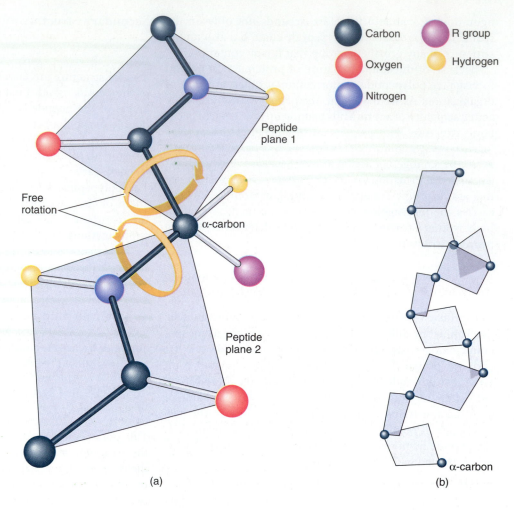

► **FIGURE 6–6** The planar character of the peptide bond. (a) The atoms of a peptide bond are coplanar, so each α-carbon atom connects two adjacent peptide planes. (b) The polypeptide backbone can be thought of as a chain of linked plates that folds by rotation of the α-carbon bonds.

Carbon
Oxygen
Nitrogen
R group
Hydrogen

Peptide plane 1

Free rotation

α-carbon

Peptide plane 2

(a)

α-carbon

(b)

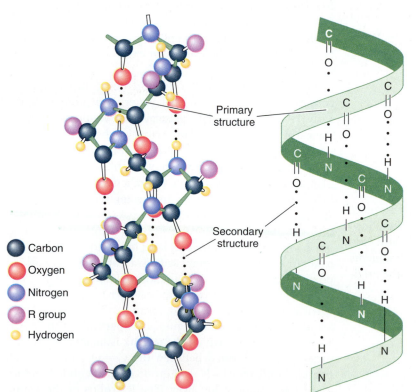

Carbon
Oxygen
Nitrogen
R group
Hydrogen

Primary structure

Secondary structure

► **FIGURE 6–7** The α-helical secondary structure of a polypeptide chain. The hydrogen bonds (····) between the amino and carboxyl groups of different amino acids maintain the helical arrangement.

acid residues extend outward from the helical backbone, leaving them free to interact with one another or with the solvent.

Example 6.2

Calculate the length of a polypeptide chain that contains 105 amino acids, if it exists entirely in the α-helical form.

Solution: A vertical rise per residue of 0.15 nm gives (105 amino acids) (0.15 nm/amino acid) = 15.75 nm.

Follow-Up Problem 6.2

An α-helical polypeptide is 30 nm in length. How many turns does the α-helix make over this 30 nm?

◆

In addition to the α-helix, Pauling and Corey proposed another type of secondary structure called the β-sheet or pleated sheet. This structure, which was later found to be the repeating unit in silk fibroin, consists of two or more polypeptide segments lying side by side in a sheetlike arrangement (▶ Figure 6–8). Once again, the structure is maintained by hydrogen bonds between the C=O and N—H groups of different peptide linkages, but instead of coiling into a helix, the backbone of each polypeptide segment is almost fully extended to form a zigzag structure resembling a series of pleats. The R groups of the amino acid residues project alternately above and below the pleated sheet. β-sheets can occur in two different arrangements: a parallel sheet, in which polypeptide segments are aligned in the same N-terminal-to-C-terminal direction, and an antiparallel sheet, in which adjacent segments are aligned in opposite directions. The antiparallel β-sheet has a repeat distance of 0.70 nm per residue pair, in agreement with that observed in fibroin, while the repeat distance of the parallel sheet is somewhat shorter.

The α-helix and β-sheet are common forms of secondary structure in both fibrous and globular proteins. The conformations of some proteins are, in fact, predominantly based on one or both of these structural forms. In fibrous proteins, such as α-keratin and fibroin, the polypeptides have elongated structures with a single secondary conformation throughout. In globular proteins, however, the polypeptides cannot exist as long chains, but must reverse direction several times in order to form the compact shapes of the protein molecules. A commonly observed and efficient way for a polypeptide to change direction is by forming a tight loop called the β-bend (▶ Figure 6–9). The β-bend involves four amino acid residues and is stabilized by a hydrogen bond. Proline is among the most common of the amino acids found in β-bends. Strictly speaking, proline is an imino acid, because the nitrogen atom is part of a rigid ring rather than an amino group. Because of its structure, proline tends to introduce a kink or bend wherever it is situated in a polypeptide chain. Glycine also occurs frequently in β-bends. Glycine is one of the more structurally flexible amino acids, since it has an R group that consists of a single hydrogen atom. It can therefore assume conformations in a polypeptide that are often impossible for other amino acids that are restricted by their more bulky side chains.

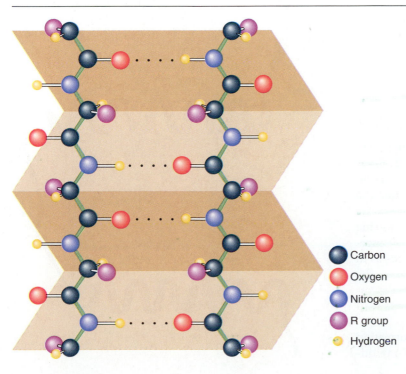

▶ **FIGURE 6–8** The β- or pleated sheet secondary structure of polypeptides. The structure is maintained by hydrogen bonds (····) between the amino and carboxyl groups of amino acid residues in different chains or in different regions of the same chain.

- ● Carbon
- ● Oxygen
- ● Nitrogen
- ● R group
- ● Hydrogen

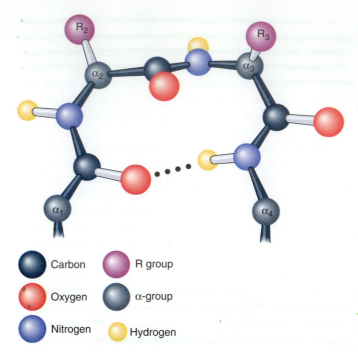

Carbon	R group
Oxygen	α-group
Nitrogen	Hydrogen

▶ **FIGURE 6–9** The β-bend. This conformation is a tight loop often formed when a polypeptide changes direction. The β-bend is stabilized by a hydrogen bond formed between the first and fourth amino acids in the loop.

Tertiary Structure

Tertiary structure refers to the spatial conformation of the entire polypeptide chain, including the R groups as well as the covalent backbone. This level of structure is the result of various chemical interactions between different parts of the polypeptide chain. There are two main sources for these interactions: (1) the N—H and C=O groups of the peptide linkages, which participate in hydrogen bonding, and (2) the R groups of the various amino acid residues. Recall that the R groups are not directly involved in forming the α-helix or β-sheet structures, so they are free to interact to produce additional folding. The principal types of R-group interactions are illustrated in ▶ Figure 6–10 and include: (1) hydrogen bonds—the example in Figure 6–10(b) shows a hydrogen bond between the side-chain hydroxyl group of serine and the side chain of histidine; (2) ionic interactions—Figure 6–10(a) shows the ionic attraction between the side-chain amino group of lysine and the carboxylate group of glutamic acid; (3) covalent cross-linkages, such as the disulfide bond between two cysteine residues in Figure 6–10(d); and (4) hydrophobic interactions between the side chains of nonpolar amino acids as in Figure 6–10(c). Hydrophobic interactions arise from the tendency of nonpolar R groups to escape from the polar environment and cluster together in the interior of the protein molecule, away from water.

Much of our knowledge about the overall conformation of polypeptide chains has been obtained from X-ray dif-

fraction studies on globular proteins. These studies, which began with the structural analysis of myoglobin (▶ Figure 6–11), the oxygen-carrying protein of animal muscle, have revealed that globular proteins tend to share certain fundamental structural similarities. First, globular proteins usually possess a polar surface and a nonpolar core. Most of the hydrophobic side chains of the nonpolar amino acids are clustered in the interior of these proteins, away from water, leaving mainly polar R groups exposed on the protein surface. Other polar groups that are buried within the protein structure, such as the N—H and C=O groups of the covalent backbone, tend to be tied up by hydrogen bonding. Second, globular proteins usually have very tightly packed structures, with the R groups occupying most of the open space in the interior of the molecules. The few internal cavities remaining are normally filled by water molecules. Because of this close packing, segments of secondary structure that are adjacent in the polypeptide are often in contact with each other. This interaction between sequentially linked segments of secondary structure can lead to regular patterns of substructure called **motifs** that are found

▶ **FIGURE 6–10** A segment of a globular protein showing different types of side-chain interactions: (a) ionic attractions, (b) hydrogen bonds, (c) hydrophobic interactions, and (d) disulfide bonds.

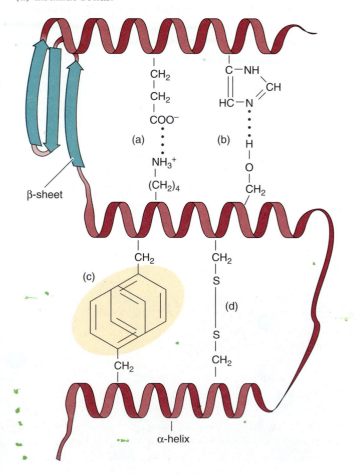

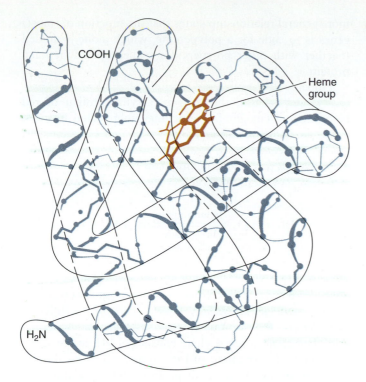

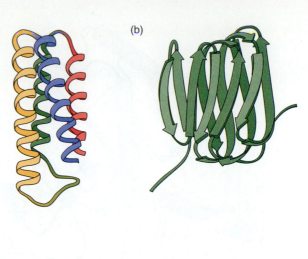

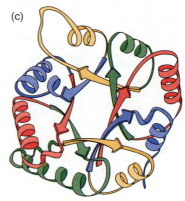

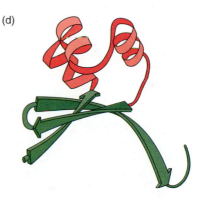

▶ **FIGURE 6–11** The tertiary structure of a myoglobin polypeptide chain from the sperm whale. The folding pattern is maintained by hydrogen and other noncovalent bonds that form between the amino acid side chains of the polypeptide. Each dot represents the α-carbon of an amino acid residue. *Adapted from:* R. E. Dickerson, *The Proteins,* vol. II, ed. H. Neurath (Academic Press, 1964), pp. 603–778. Used with permission.

▶ **FIGURE 6–12** Four common structural motifs in proteins. (a) α-motif—predominantly α-helices. (b) β-motif—predominantly β-sheets. (c) αβ-motif—predominantly alternating α-helices and β-sheets. (d) Helix-turn-helix motif—two short α-helical regions separated by a β-turn. *Source:* Used with permission of Jane Richardson.

in several different kinds of proteins. A few of the more common structural motifs are shown in ▶ Figure 6–12. A major area of current genetic research is investigation of the structural motifs of proteins that bind to DNA and thereby regulate the expression of certain genes. We will discuss these motifs and their significance in DNA binding in a later chapter.

In addition to the structural patterns that we have just considered, X-ray diffraction studies also reveal that many of the larger polypeptides are not merely folded into a single mass (as was the case with myoglobin) but are organized into two or more folded units called **domains**. Each domain is a polypeptide subregion that folds into a compact globular shape that appears separate from the rest of the chain. In certain cases, this modular domain structure allows a separation of functional activities. For example, enzymes frequently have binding sites on different domains, with each site showing specificity for a different substrate or cofactor in a reaction (▶ Figure 6–13). Domains also give the folded protein additional mobility. The domains are held together in a single aggregate (like beads on a string), enabling them to function together as a cooperative unit, but the segments of the chain connecting the domains are flexible enough to allow them to move in relation to one another and thus alter the shape of the entire protein.

Domain structure may also have implications for the evolution of proteins through a process called domain shuffling—the evolution of new proteins by recombining preexisting polypeptide domains. The structures of domains often resemble each other closely, and the remarkable simi-

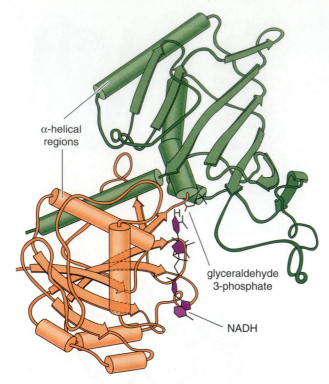

α-helical
regions

glyceraldehyde
3-phosphate

NADH

▶ FIGURE 6–13 Modular domain structure of the enzyme glyceraldehyde 3-phosphate dehydrogenase. The enzyme catalyzes the transfer of hydrogen and electrons from glyceraldehyde 3-phosphate to the enzyme cofactor NAD to form NADH. To facilitate this reaction, the enzyme has one substrate-binding domain that binds glyceraldehyde 3-phosphate and one NAD-binding domain. *Sources:* G. Biesecker, et al., "Sequence and structure of D-glyceraldehyde 3-phosphate dehydrogenase from *Bacillus stearothermophilus*," *Nature* 266 (1977): 331. Reprinted with permission from *Nature.* Copyright © (1977) Macmillan Magazines Limited. Used with permission.

larity of many domains has contributed to the belief that there are relatively few kinds of domain structures and that different combinations of them have served in the evolution of complex proteins. We will consider a possible mechanism for domain shuffling in Chapter 16.

Quaternary Structure

Although many proteins function as single polypeptides, many others are oligomeric (or multimeric) proteins (i.e., they consist of two or more polypeptide subunits) in their active states. The primary structure of the subunits can be identical or different, and the subunits bind to one another by the same types of forces that are involved in folding each peptide chain into its secondary and tertiary conformation. The spatial relationship between the folded subunit chains constitutes the quaternary structure of oligomeric proteins.

The coding of an oligomeric protein often requires more than just one gene. Oligomeric proteins can be included in the one gene–one protein relationship, however, by modifying it to the one gene–one polypeptide hypothesis. This

more general relationship states that the function of (most) genes is to code for a polypeptide, which alone or folded together with other polypeptides makes up a functional protein.

Probably the best known example of an oligomeric protein is hemoglobin, the oxygen-carrying protein of red blood cells. Hemoglobin, sometimes abbreviated Hb, is a conjugated protein that consists of a protein portion, called *globin*, associated with four iron-containing heme prosthetic groups that enable hemoglobin to combine reversibly with four molecules of oxygen in the following manner:

$$1 \text{ Hb} + 4 \text{ O}_2 \underset{\substack{\text{In capillaries of} \\ \text{other tissues}}}{\overset{\text{In lung capillaries}}{\rightleftharpoons}} 1 \text{ Hb} \cdot (\text{O}_2)_4$$

The globin portion of hemoglobin consists of four polypeptides—two identical α-chains (coded by a gene represented by Hb_α) and two identical β-chains (coded by a different gene, Hb_β)—that are linked together by various noncovalent interactions. Each α-chain consists of 141 amino acids, and each β-chain consists of 146 amino acids. X-ray diffraction studies reveal that each peptide subunit of the $\alpha_2\beta_2$ tetramer is folded into a roughly spherical tertiary structure and contains a single, partially buried heme group (▶ Figure 6–14). The subunits are arranged into the approximately tetrahedral quaternary structure of the complete protein. This structure is formed by the spontaneous association of the four subunits once the individual polypeptides have been formed.

Having a subunit structure confers several advantages on a protein. First, the possession of multiple subunits allows greater variation in form and function. The subunits of a multimeric protein can associate across complementary surfaces to form a variety of regular geometric arrangements. As a general rule, however, the preferred arrangement is the one that maximizes the number of subunit contacts. Therefore, in the case of a tetramer like hemoglobin, a tetrahedral arrangement with six subunit contacts is preferred over a square (four contacts) or straight chain (three contacts). When subunits differ, each arrangement also may vary in terms of its subunit composition. An interesting example of this type of variation occurs in the isoenzymes of lactate dehydrogenase. Isoenzymes (also called isozymes) are different forms of a protein that catalyze the same reaction in an organism. In the case of lactate dehydrogenase, the isoenzymes are tetramers that differ in their relative composition of two polypeptides designated A and B. In many vertebrate tissues, the polypeptides appear to associate at random, and thus the five isoenzymes, A_4, A_3B, A_2B_2, AB_3, and B_4, are observed at frequencies that correspond to the terms of the binomial distribution $(a + b)^4 = a^4 + 4a^3b + 6a^2b^2 + 4ab^3 + b^4$, where a and b are the respective average proportions of the two types of chains. Thus if the A and B chains were to occur in equal proportions, the five isoenzymes would appear in a ratio of $1:4:6:4:1$. In mammalian tissues, there is normally an imbalance in the proportions of the two poly-

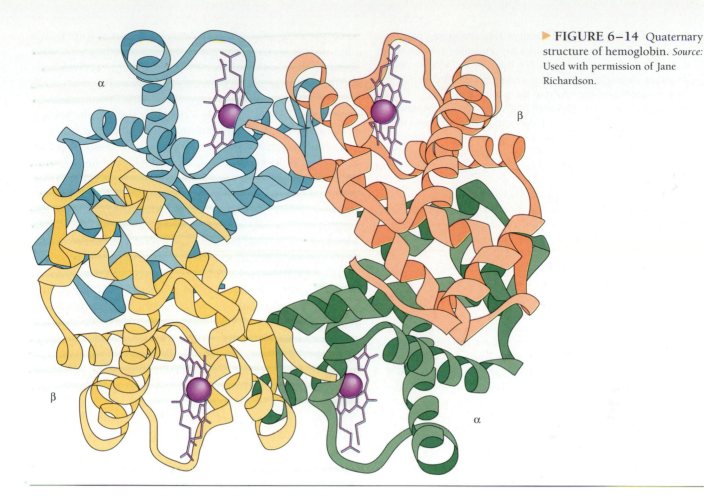

► **FIGURE 6–14** Quaternary structure of hemoglobin. *Source:* Used with permission of Jane Richardson.

peptides, with A_4 predominant in skeletal muscle and liver tissue and B_4 in heart tissue. This difference has been linked by some investigators to a difference in the catalytic properties of these enzyme forms, but despite a great deal of study, the different physiological roles of the lactate dehydrogenase isoenzymes is still not clear.

A second important advantage of a subunit structure is that it is economical to produce. By constructing a protein from multiple polypeptide chains, a large protein complex can be formed using only the genetic information required to specify a few different kinds of subunits. A subunit organization thus helps to minimize the amount of information needed to specify a complex structure. Subunits are also economical from the standpoint of quality control. Using a precise assembly of subunit chains in the construction of a protein allows for the rejection of faulty subunits before they become part of the final complex. The rejection process can conceivably occur at each stage of assembly, minimizing the chance of producing a nonfunctional product.

A third important advantage of a subunit structure is that it allows greater flexibility and movement. Like the folded domains of a single polypeptide chain, the subunits are held together in a single aggregate where they function together as a cooperative unit, yet the contact interactions between the subunit chains are loose enough to allow them to move in relation to one another and thereby alter the conformation of the entire protein. Such changes in confor-

mation can greatly affect the binding sites on the protein and thus its ability to combine with other molecules. For example, a DNA-binding protein may be able to attach to DNA in one conformational state but not in another. When such changes in conformation are induced in the protein by another molecule, they provide an important mechanism for regulation.

To Sum Up

1. The activity of a protein depends on its specific three-dimensional conformation, which is determined entirely by the information contained in the primary structure of each polypeptide chain.

2. Proteins can be divided into two main conformations based on their overall folded shape: globular proteins and fibrous proteins. Globular and fibrous proteins tend to have certain fundamental structural properties as a result of their higher-level folding patterns.

3. The three-dimensional structure of proteins has several different levels of complexity: (1) the secondary structure includes the regular folding patterns of localized regions of a polypeptide chain, (2) the tertiary structure refers to the spatial conformation of the entire chain, and (3) the quaternary structure refers to the aggregation of two or more polypeptide chains to form a single protein molecule.

4. The secondary structure of a polypeptide chain is maintained by hydrogen bonds that form between the C=O and N—H

groups of different peptide bonds within the same chain. The α-helix and β-sheet are commonly encountered secondary structures. In some polypeptides there is a single secondary structure throughout the entire chain; in others, different secondary structures occur within the same chain.

5. Tertiary structure involves both hydrogen bonding between the C=O and N—H groups of different peptide linkages and a variety of chemical interactions between the side groups of the different amino acids along the chain (hydrogen bonding, ionic interactions, covalent cross-linkages, and hydrophobic interactions).

6. The tight folding that characterizes globular proteins places segments of secondary structure in close juxtaposition, giving rise to several well-known substructural patterns, including the α-motif, the β-motif, the αβ-motif, and the helix-turn-helix motif.

7. The entire chain of some polypeptides is folded into a single unit. Many polypeptides, however, have a modular domain structure, where localized regions of a polypeptide chain fold into separate units or domains.

8. Proteins that function as oligomers, which are composed of two or more polypeptide chains, also have a quaternary structure. The advantages of an oligomeric structure include variation in subunit arrangement and composition, minimization of the genetic information needed to code for a large complex protein structure, and the ability to alter protein configuration by movement of the subunit polypeptide chains.

 ## FUNCTION OF PROTEINS AS ENZYMES

Nowhere is the importance of a specific three-dimensional structure to the biological activity of a protein better illustrated than among enzymes. Enzymes are proteins that function as catalysts. Most reactions in cells have a formidable energy barrier (a large activation energy) that prevents their spontaneous occurrence in the absence of a catalyst. Enzymes lower this energy barrier, thereby permitting the reactions to proceed under moderate conditions at rates that are compatible with life.

Catalysis and Specificity: The Active Site

Unlike other types of catalysts, an enzyme tends to show a high degree of specificity for the substrate that it acts on and for the reaction that it catalyzes. Most enzymes will combine with only a few different substrates, (often only one) and participate in only a single kind of reaction or a small set of closely related reactions. The structural basis for the substrate and reaction specificities of an enzyme is a particular region of the enzyme molecule known as its active site. The active site is formed by the specific folding of the polypeptide, which brings certain R groups into juxtaposition so that they can interact in a complementary fashion with the substrate molecule (▶ Figure 6–15). During catalysis, the enzyme first combines with the substrate to form an intermediate enzyme-substrate complex. Something akin to a "lock-and-key" fit is achieved through the complementarity of shape between the substrate and the active site. When the two molecules bind, functional groups of the amino acid side chains at the active site form (often weak) chemical bonds with complementary groups on the substrate. This combination of reactant and catalyst promotes a chemical change by increasing the probability of reaction. The enzyme may participate in the transfer, addition, or removal of groups in the substrate, or it may simply bring different reactants into close proximity and into the proper orientation for a favorable interaction. Following this conversion, the product is released, freeing the enzyme to function in another round of catalysis.

In those enzymes that have so far been subjected to X-ray analysis, the active site usually appears as a cleft or depression penetrating deep into the interior of the molecule. In many of these enzymes, the cleft lies between two globular domains. The domain structure acts as a flexible hinge that permits the enzyme to change shape when the substrate is bound (▶ Figure 6–16). In such cases, the cleft can exist as an open structure into which the substrate can fit or as a closed structure that provides an environment in which the substrate is protected from competing reactants such as water.

Because the ability of an enzyme to catalyze a reaction depends on the nature and arrangement of the R groups at the active site, any change in enzyme structure that affects the distribution of amino acids at the active site will tend to impair the function of the protein as a catalyst. Of major importance are those amino acids that play a direct role in catalysis; their distribution may be absolutely crucial, allowing no variation without a complete loss of enzyme activity. Changes at other locations can also destroy enzyme

---

▶ **FIGURE 6–15** The reaction of an enzyme with its substrates. The substrates specifically bind to the active site of the enzyme to form a highly reactive enzyme-substrate complex.

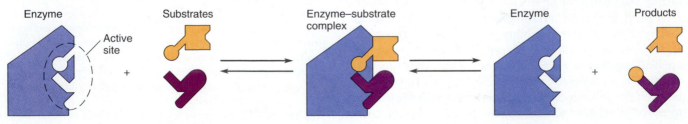

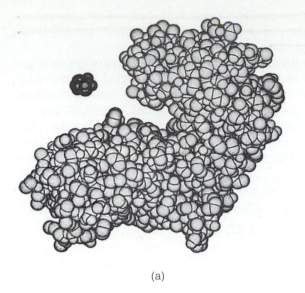

(a)

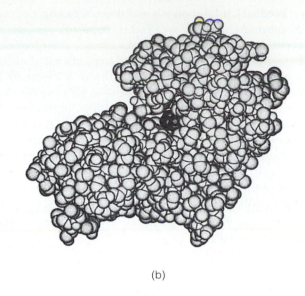

(b)

▶ **FIGURE 6−16** Structure of yeast hexokinase A. (a) Free hexokinase and its substrate, glucose. (b) Hexokinase bound to glucose at its active site, showing the change in enzyme structure that accompanies binding. *Source:* J. Bennett and T. A. Steitz, "Structure of a complex between yeast hexoleinage A and glucose," *J. Mol. Biol.* 140 (1980): 211. Used with permission.

function, however, by changing the conformation of the active site. For example, if mutation causes the replacement of a nonpolar amino acid with one that is charged, the shape of the enzyme might become so distorted that catalytic activity is lost, even though the mutational change occurred some distance from the active site. Many R groups are typically involved in the proper folding of a polypeptide chain, so that changes at any of the amino acid residue sites may lead to a loss or alteration of enzyme activity.

Enzyme Regulation: The Allosteric Site

Many enzymes possess not only an active site for the substrate but also a separate binding site called the **allosteric site**, which is involved in the regulation of enzyme activity. The allosteric site reversibly binds specific substances of comparatively low molecular weight called **effectors**. The binding of an effector molecule to the allosteric site either increases the activity of the enzyme or inhibits it (▶ Figure 6−17), that is, an allosteric effector can be an activator or an inhibitor. The allosteric effector achieves this regulation by altering the three-dimensional structure of the enzyme. The interaction of the effector with the allosteric site produces a local change in conformation that carries through to the active site, even though the allosteric site may be far (in molecular terms) from the site of substrate binding. If the effector is an activator, the binding affinity of the altered active site is enhanced; if it is an inhibitor, the affinity is reduced.

▶ **FIGURE 6−17** Conformational changes induced in an enzyme upon binding an effector molecule to the allosteric site. (a) Binding an activator to the allosteric site increases the affinity of an enzyme's active site for its substrate. (b) Binding an inhibitor to the allosteric site decreases the affinity of the active site for the substrate.

(a) Activation

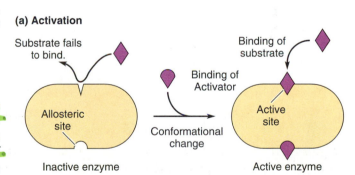

(b) Inhibition

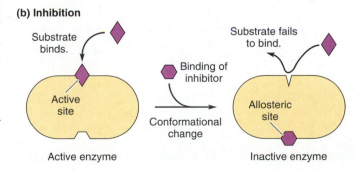

Feedback inhibition is a well-known example of control of enzyme activity by an effector. Feedback inhibition frequently involves inhibition of the enzyme that catalyzes the first step (the so-called committed step) of a metabolic pathway by the end product of that pathway. ▶ Figure 6–18 illustrates this form of regulation in the pathway involved in the synthesis of the amino acid histidine. In this case, histidine itself acts as an allosteric effector; when it is present in excess, histidine combines with and inhibits the enzyme that catalyzes the first step in the reaction sequence. Thus

the end product of a biosynthetic pathway can, when it accumulates, temporarily turn off the enzyme needed for its own formation.

The ability to undergo reversible changes in conformation on binding an effector is not restricted to enzymes but is characteristic of many other kinds of proteins that function in genetic and metabolic control. In general, proteins that can take on different conformational states with different functions are called **allosteric proteins**, and the ability of an effector to determine which conformation an allosteric

▶**FIGURE 6–18** The metabolic pathway for histidine biosynthesis in *Salmonella typhimurium*. Excess histidine binds allosterically to the enzyme that catalyzes the first step of the pathway. This binding inactivates the enzyme and temporarily shuts off further synthesis of histidine.

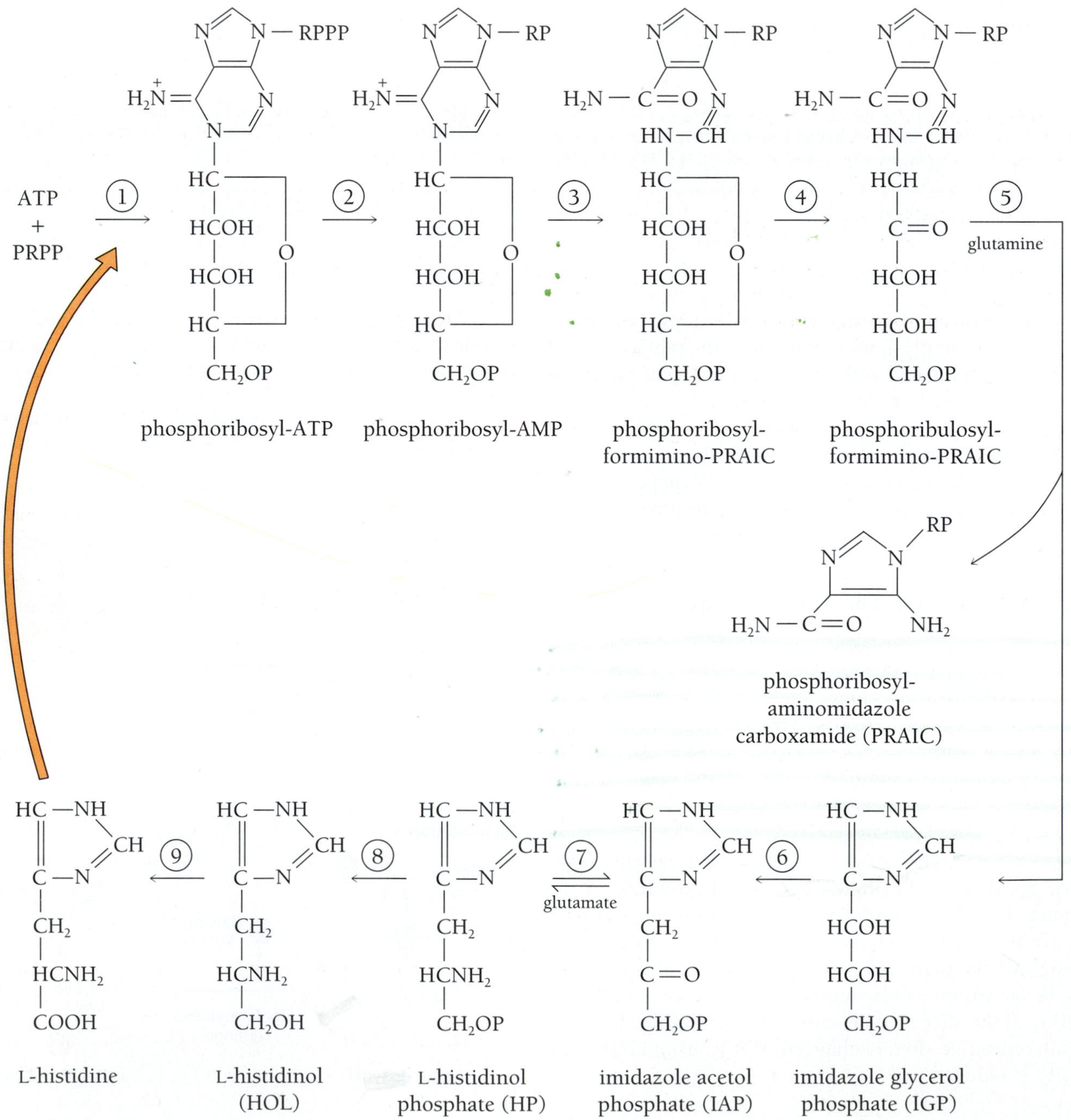

protein adopts is fundamental to the regulation of most biological processes.

To Sum Up

1. The activity of an enzyme depends on interaction with its substrate at the active site of the enzyme molecule. The active site is formed by specific folding of the polypeptide to yield a region with a shape that is complementary to that of the substrate. Because the primary structure is ultimately responsible for polypeptide folding, changes in the amino acid sequence caused by mutations can alter the spatial conformation of the protein, thereby modifying or destroying its catalytic ability.

2. Many enzymes possess an allosteric site that interacts with effector substances to alter the three-dimensional enzyme structure. The change in conformation serves to regulate the enzyme by either increasing or inhibiting its activity.

 ## DENATURATION AND RENATURATION OF PROTEINS

The biologically active conformations of a protein are stable only within a very limited range of environmental conditions. Heating or extremes in pH, for example, can disrupt the folded structure of the protein and lead to a loss of activity. Loss of the active three-dimensional structure of a protein is known as **denaturation**, and the factors or conditions that produce denaturation are known as *denaturing agents*. In addition to heating and extremes in pH, many other factors, including organic solvents, detergents, and high concentrations of urea, can act as denaturing agents. These agents destroy the folded structure of a protein by disrupting various bonds, such as hydrogen bonds and hydrophobic interactions, that maintain the levels of organization beyond the primary structure. However, denaturation does not break the covalent bonds of the polypeptide backbone. When a protein is completely denatured, it simply unfolds into a random structure, called a random coil, that still has its sequence of amino acids intact.

In certain cases, denatured proteins can undergo renaturation and regain their active state. A clear demonstration of this was given several years ago by Christian Anfinsen and his coworkers, who showed that denatured pancreatic ribonuclease will spontaneously fold into its active structure when the denaturing agent is removed. Pancreatic ribonuclease is a comparatively small enzyme that has a single polypeptide chain consisting of 124 amino acid residues. The folding of the polypeptide is stabilized by four disulfide bonds in addition to various weak interactions. To achieve denaturation, Anfinsen treated the enzyme with a concentrated urea solution in the presence of a reducing agent, mercaptoethanol, which cleaves the disulfide bonds (▶ Figure 6–19). Refolding took place, however, along with the formation of disulfide bonds, when the urea and reducing agent were removed by dialysis. The result-

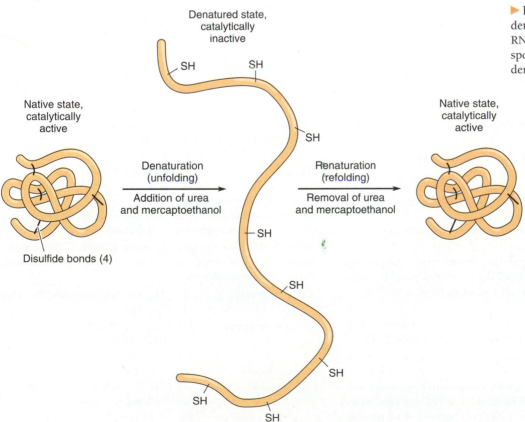

▶ **FIGURE 6–19** Reversible denaturation of pancreatic RNase. Renaturation is spontaneous once the denaturing agents are removed.

Denatured state, catalytically inactive

SH   SH

SH

Native state, catalytically active

Native state, catalytically active

Denaturation (unfolding)

Addition of urea and mercaptoethanol

Renaturation (refolding)

Removal of urea and mercaptoethanol

Disulfide bonds (4)

SH

SH

SH

SH

-ing enzyme was indistinguishable from the original ribo-nuclease in both its physical characteristics and catalytic ability. Interestingly, of the 105 possible combinations of disulfide bonds that could be created in this enzyme on a random basis, only the correct set was eventually formed. The refolding of the polypeptide chain is thus a very precise process. From these results and the results of similar experiments on other proteins, it is possible to conclude that *all the information required for the proper folding of a polypeptide chain resides in its amino acid sequence.* For some polypeptides, participation of other proteins *assists* the folding process; for example, **protein chaperones**, which will be discussed in Chapter 17, are thought to help in the folding process by preventing the formation of improperly folded structures. Chaperones do not, however, impart additional information to the folding polypeptide chain.

To Sum Up

1. Various environmental factors can cause a protein to unfold in a process called protein denaturation. Denaturation destroys the biological activity of the protein.
2. Because denaturation does not harm the covalent backbone of the polypeptide, some denatured proteins can spontaneously refold into their active three-dimensional pattern when the denaturing agent is removed. This renaturation directly shows that all the information required for the proper folding of a polypeptide chain resides in its amino acid sequence.

 # Chapter Summary

An individual's phenotype is a result of the combined activities of his or her proteins. Each protein is composed of at least one polypeptide chain, and each polypeptide consists of an unbranched polymer of amino acids. The one gene–one polypeptide relationship expresses the basic connection between the genotype and phenotype of an individual: It states that the function of a gene is to code for the structure of a polypeptide chain.

A gene specifically codes for the primary structure (the amino acid sequence) of a polypeptide. The sequence of amino acids enables the polypeptide to fold into a specific three-dimensional conformation or secondary structure that consists of regular, local folding patterns such as the α-helix and β-sheet. These structures are organized into a specific overall spatial arrangement or tertiary structure. In oligomeric proteins, two or more polypeptides combine to form a subunit arrangement or quaternary structure.

A protein's function is determined by its three-dimensional structure. In the case of enzymes, the catalytic activity of the protein depends on the proper folding of the polypeptide to form a functional active site that specifically binds the substrate and facilitates the conversion of substrate to product. Proper folding is also essential for the formation of an allosteric site, which specifically binds an effector molecule and functions in the metabolic regulation of certain (allosteric) enzymes. The biological activity of a protein is lost when it is exposed to factors that cause it to unfold and undergo denaturation.

 # Questions and Problems

Covalent Structure of Proteins

1. Fill in the blanks.
   (a) Amino acids are linked together into a chain by _____ bonds.
   (b) The amino acid that can stabilize protein structure by forming covalent cross-links within a polypeptide chain or between chains is _____.
   (c) The amino acid sequence of a polypeptide defines its _____ structure.

2. Assuming there are no restrictions on the relative frequencies of the 20 amino acids, calculate the number of possible amino acid sequences in a polypeptide consisting of (a) 2, (b) 3, (c) 4, and (d) n amino acids.

3. A polypeptide is known to have the amino acid sequence given by the following one-letter abbreviations (see Figure 6–2):

   A-T-E-C-N-C-P-K-L-C-A-R-R-C-Q-H

   The polypeptide was treated with a reagent that combines with free sulfhydryl groups. No reaction occurred. (a) How many disulfide bonds are present in the polypeptide? (b) Assuming that the polypeptide exists as a monomer in solution, draw a probable folding pattern for the polypeptide chain, indicating the location of any disulfide bonds.

4. The enzyme trypsin breaks a polypeptide into two peptide fragments, L-T-W-I-D-R and V-A-S. The enzyme chymotrypsin breaks it into two different peptide fragments, I-D-R-V-A-S and L-T-W. Determine the amino acid sequence of the polypeptide from which these fragments were derived.

5. Suppose that cleavage of a nanopeptide with CNBr, trypsin, and chymotrypsin yields peptide fragments with the following amino acid compositions:

CNBr treatment:	(1) (Ala, Arg, Gly, Lys, Met, Phe, Pro)
	(2) (Ile, Trp)
Trypsin treatment:	(1) (Ala, Arg, Phe, Pro)
	(2) (Ile, Met, Trp)
	(3) (Gly, Lys)
Chymotrypsin treatment:	(1) (Ala, Arg, Gly, Lys, Met, Trp)
	(2) (Phe, Pro)
	(3) (Ile)

(Note: the sequence of amino acids in each fragment is not known.) From these results, deduce the amino acid sequence of the nanopeptide.

*6. You are given a mixture containing the following proteins, listed with their molecular weights and isoelectric points:

Protein	Molecular weight	pI
α-antitrypsin	45,000	5.4
chymotrypsinogen	23,200	9.5
cytochrome c	13,400	10.6
lysozyme	13,900	11.0
myoglobin	17,000	7.0
serum albumin	69,000	4.8
transferrin	90,000	5.9

(a) If these proteins are separated by SDS PAGE, what will be their order from the top (point of sample application) to the bottom of the gel? (b) If these proteins are separated by isoelectric focusing, what would be their order from the negative (high pH) to the positive (low pH) end of the gel?

*7. Four proteins of known molecular weight were used to construct a standard curve for a molecular weight analysis by means of SDS PAGE. Protein 1 (mw = 17,000) had the highest mobility and moved the farthest on the gel. Protein 2 (mw = 45,000) moved 64% as far as protein 1. Protein 3 (mw = 66,000) moved 38% as far as protein 1. Protein 4 (mw = 132,000) moved 16% as far as protein 1. Construct the standard curve by plotting the logarithm of the molecular weight against the relative distance traveled, and determine the molecular weight of an unknown protein that moved 50% as far as protein 1.

Three-Dimensional Structure of Proteins

8. Fill in the blanks.
   (a) The folding pattern of amino acids that are near each other in the polypeptide chain is referred to as the _____ structure.
   (b) The secondary structure is maintained by _____ bonding.
   (c) _____ and _____ are the two major types of secondary structure characteristic of polypeptides.
   (d) The conformation of the entire polypeptide chain is referred to as its _____ structure.
   (e) In most globular proteins, the surface tends to be composed of _____ amino acids, while the interior (core) tends to be made up of _____ amino acids.
   (f) A protein that is an aggregate of two or more polypeptide chains is said to be a(n) _____ protein.
   (g) The folded configuration of an oligomeric protein is called its _____ structure.

9. A polypeptide contains an α-helical section that includes 180 amino acids. How many turns are there in this α-helical section?

10. An α-helical polypeptide is 45 nm in length. (a) How many

*An asterisk indicates that the question or problem is based on information in the Extensions and Techniques text.

amino acids make up this peptide? (b) What is its molecular weight? (Assume an average residue weight of 120 per amino acid.)

11. How long, in nm, is an antiparallel β-sheet that consists of 200 amino acid pairs? What is its molecular weight?

12. List five types of bonding that can be responsible for the tertiary structure of a polypeptide.

13. The N-terminal region of a particular protein is known to be in the form of an alpha helix and to have the following amino acid sequence: **Ile**-Gln-Asp-**Leu**-**Val**-**Cys**-Asn-**Leu**-Gln-Thr-Asp-**Phe**-His-Ser-**Ile**-Arg-. (The hydrophobic amino acids are shown in boldface.) Suggest a possible reason for the periodicity shown by the hydrophobic residues.

14. Suppose that the coat protein of a particular virus consists of 150 identical polypeptide subunits, each composed of 300 amino acid residues. If the probability of inserting the wrong amino acid into a peptide during its synthesis is 1 out of 3000 per residue, how many polypeptide subunits must be synthesized, on average, in order to produce a perfect viral protein coat? Compare this result with the number needed to produce a perfect viral coat that consists of one large polypeptide chain with the same total number of amino acid residues.

*15. Subjecting an oligomeric protein to SDS PAGE resulted in two bands corresponding to molecular weights of 5 and 13 kilodaltons. After treating the protein with a reagent that covalently cross-links subunits at their contact points, an SDS PAGE of the product yielded six bands corresponding to molecular weights of 5, 10, 13, 18, 23, and 36 kilodaltons. Explain these results in terms of a diagram of the quaternary structure of this oligomeric protein.

Function of Proteins as Enzymes

16. A particular enzyme contains an alanine residue in the cleft that makes up its active site. If a mutation substitutes glycine for this alanine, there is no effect on the activity of the enzyme. If, however, a different mutation substitutes glutamate for alanine, the enzyme loses its activity. Explain these results.

17. Predict, in a general way, the possible consequences of a mutation affecting an enzyme at the (a) active site, (b) allosteric site, (c) another location outside the active and allosteric sites.

18. A mutation that alters the allosteric site of an enzyme destroys the ability of the enzyme to respond to feedback inhibition. Would this mutation be dominant or recessive? Explain.

Denaturation and Renaturation of Proteins

19. Some proteins are extensively modified after they are synthesized. For example, the hormone insulin is converted to an active form by the removal of a central portion of its polypeptide chain after it is formed. When such proteins are denatured, they often fail to regain full biological activity by spontaneous renaturation. Why would such proteins fail to renature?

20. You have isolated a protein that exists in a biologically active conformation when in a nonpolar solvent, and it undergoes denaturation in a polar solvent such as water. In what part of the cell would this protein most likely function?

Nucleic Acids

The nucleic acids are macromolecules that consist of **deoxyribonucleic acid (DNA)** and **ribonucleic acid (RNA)**. This major class of bipolymers was discovered in 1868 by Friedrich Miescher, a Swiss physician, while he was undertaking a chemical study of the cell nucleus. His analysis revealed the presence of an acidic component with high nitrogen and phosphorus contents. Miescher called this substance *nuclein*—a name that was later changed to nucleic acid.

During the century following Miescher's discovery, a great deal of study was carried out on the basic chemistry of the nucleic acids. These polymers were shown to consist of monomer units called **nucleotides**, which are covalently joined into long, unbranched **polynucleotide chains**. Moreover, these polymers were found to include some of the largest molecules known. For example, the molecular weight of DNA varies from around 10^6 for the smallest virus

Computer model of the double-helical structure of DNA.

to about 2×10^9 for bacteria, and it is even larger for many eukaryotic organisms.

The great length of these molecules makes them especially susceptible to breakage by the shear forces of pipetting and mixing during laboratory studies, and thus the large DNA molecules are ordinarily fragmented into much smaller pieces during their isolation unless great care is taken. A typical isolation procedure is shown in ▶ Figure 7–1. In this example, the nucleic acids are extracted using a detergent and an organic solvent to lyse the cells and denature the proteins, a process that releases the DNA and RNA into solution in the aqueous phase. The DNA and RNA are then precipitated with ethanol, and the DNA is freed from the RNA (or vice versa) using a specific **nu-** **clease**—either DNase or RNase—to selectively degrade the unwanted nucleic acid.

The nucleic acids have a number of different roles in organisms, although not nearly as many as proteins. Some nucleic acids form complexes with proteins and act as important structural components within the cell. Certain other nucleic acids function as transport molecules. A few nucleic acids can even act as catalysts. However, the primary role of DNA and RNA is to act as informational molecules that provide the cell with a set of instructions in the form of a genetic code. Their encoded instructions enable cells to reproduce with near-perfect fidelity and to synthesize a large variety of specific proteins. This genetic code thus allows nucleic acids to store genetic information in a stable form,

▶ **FIGURE 7–1** A procedure for the isolation of DNA from bacteria.

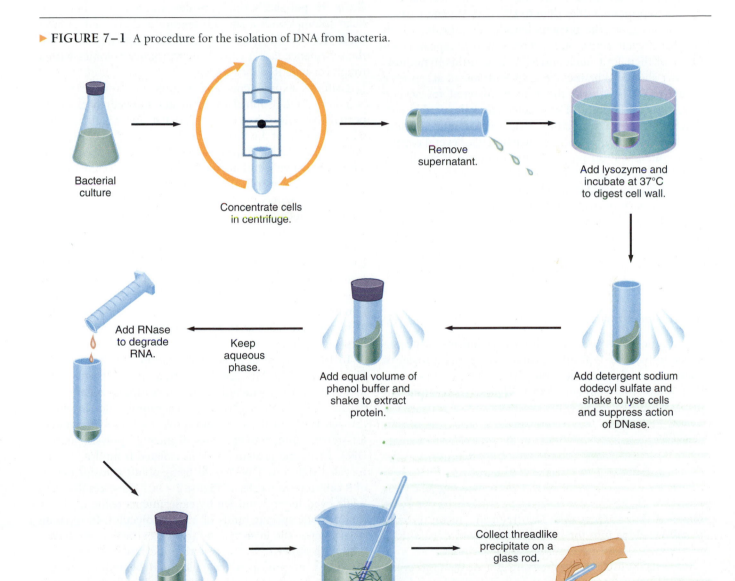

Bacterial culture

Concentrate cells in centrifuge.

Remove supernatant.

Add lysozyme and incubate at 37°C to digest cell wall.

Add RNase to degrade RNA.

Keep aqueous phase.

Add equal volume of phenol buffer and shake to extract protein.

Add detergent sodium dodecyl sulfate and shake to lyse cells and suppress action of DNase.

Shake with phenol to remove added enzyme.

Add two volumes of ethanol to precipitate DNA.

Collect threadlike precipitate on a glass rod.

to transfer this information to other parts of the cell so that it can be expressed as a trait (protein), to duplicate this information accurately during cell division, and to produce variation—the four requirements of the genetic material. In this chapter, we will examine the structural properties of nucleic acids that specifically allow them to meet each of these requirements.

DNA AS THE GENETIC MATERIAL

Although the concept of the gene was first developed at the beginning of this century, it took some fifty additional years to correctly identify the gene as DNA. During the 1930s and 1940s, misconceptions about the structure of the nucleic acids led scientists to ascribe the informational role of genes to proteins rather than to DNA. Most geneticists at the time accepted the **tetranucleotide hypothesis**, which held that nucleic acids have a monotonously repeating sequence of their four nucleotides. Although this hypothesis was later shown to be incorrect, its widespread acceptance obscured DNA's role as the genetic material for several years, even in the light of experimental evidence that correlated the known properties of genes with the physical and chemical properties of the nucleic acids. Several experiments conducted during the 1940s and early 1950s finally proved that genes are composed of DNA. Two experiments dealing with the genetic transformation of bacteria and the replication of bacterial viruses were particularly important in establishing the genetic role of DNA and will be considered in some detail in the following sections. We will see later that RNA is also capable of serving as the genetic material, but it does so only in certain viruses that lack DNA.

Genetic Transformation

Unlike gene transmission in higher organisms, gene exchange in bacteria typically involves a one-way transfer of genetic material from one strain (the donor strain) to another strain (the recipient strain). Genetic transformation was the first such mechanism of gene exchange to be discovered in bacteria. In genetic transformation, the genetic material from a donor cell is taken up directly by a recipient cell in a process that does not involve direct contact of the cells or mediation by any vector, such as a virus. Fragments of genetic material from a donor, each containing just a few genes, are adsorbed directly to receptor sites on the outer surface of the recipient cell. The fragments move through the cell envelope into the cell interior, where some are incorporated into the chromosome of the recipient.

The usual source of transforming genetic material in nature is the occasional release of such material from donor cells by autolysis (or perhaps by secretion). In the laboratory, large amounts of genetic material from a specified donor strain can be isolated and purified, thereby making

transformation more efficient than it probably is in nature. Even so, only certain kinds of bacteria have a cell wall structure that is permeable to the transforming agent. We will see in later chapters that modern genetic engineering techniques use special treatments to make cell walls more permeable to DNA. Without these treatments, however, the types of bacteria that are able to undergo transformation are limited, and most studies have involved only three species: *Streptococcus pneumoniae*, *Bacillus subtilis*, and *Haemophilus influenzae*.

Transformation was first discovered in 1928 by F. Griffith during his work with *S. pneumoniae*, commonly known as pneumococcus. The virulent form of this organism causes pneumonia in humans and is lethal when injected into laboratory mice. Virulence in pneumococcus is a genetically determined characteristic. Pneumococcal cells that are enclosed in polysaccharide capsules are virulent; they produce colonies with a smooth texture and are said to have the S (for smooth) phenotype. Mutant cells that have lost their virulence lack a capsule and produce colonies with a rough (or type R) appearance.

Griffith's experiments and results are summarized in ► Figure 7–2. Control experiments showed that, as expected, live cells of the S type are lethal to mice (Figure 7–2a), while live cells of the R type and dead cells of the S type are not lethal if only one type is injected (Figure 7–2b and c). If a mixture of live R and heat-killed S cells is injected into the same mouse, however, the mouse dies (Figure 7–2d). Furthermore, live pneumococcal cells extracted from the dead mouse produce smooth colonies and are lethal when used subsequently to infect other mice (Figure 7–2e). Griffith concluded that something in the debris of the dead S cells had somehow transformed the live R cells into virulent S cells, and that this transformation results in a permanently inherited change in genotype. Just what this substance was remained a mystery, since Griffith's experiment was not designed to give clues as to the chemical nature of the transforming agent.

In 1944, a team of three investigators—O. Avery, C. M. MacLeod, and M. McCarty—reported on results of experiments designed to identify the transforming agent. They isolated the different classes of macromolecules found in the debris of killed S cells and tested each class separately for transforming activity. Tests of purified polysaccharide, DNA, RNA, and protein fractions isolated from dead S cells revealed that only DNA could bring about transformation of R cells to type S cells (► Figure 7–3a). These results were made even more definitive by experiments using enzymes that degrade specific kinds of macromolecules. In separate experiments, the investigators treated extracts from dead S cells with either DNase (deoxyribonuclease), RNase (ribonuclease), or proteases (enzymes that degrade proteins); they then mixed each extract with live R cells and injected the mixture into mice. RNase and proteases had no effect on the transformation of R cells. In contrast, DNase eliminated all transforming activity (Figure 7–3b).

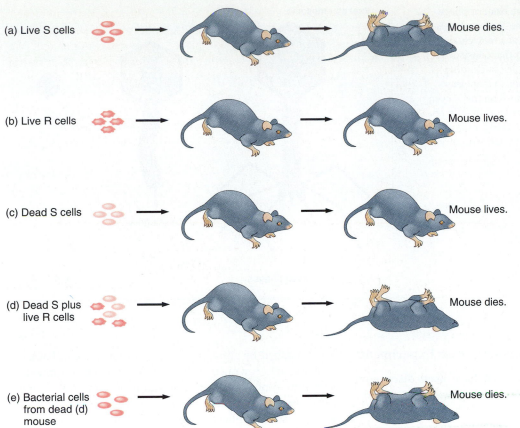

(a) Live S cells — → — → Mouse dies.

(b) Live R cells — → — → Mouse lives.

(c) Dead S cells — → — → Mouse lives.

(d) Dead S plus live R cells — → — → Mouse dies.

(e) Bacterial cells from dead (d) mouse — → — → Mouse dies.

▶ **FIGURE 7–2** Griffith's original demonstration of transformation. (a) Mice die after injection of virulent S bacteria. Mice survive infection by (b) nonvirulent R bacteria and (c) heat-killed S bacteria. (d) Mice die following injection with a mixture of dead S and live R bacteria, indicating that some of the R cells have been transformed to the virulent S condition. (e) Pneumococcal cells isolated from the dead mouse in (d) are lethal when injected into another mouse, indicating that the transformation event is a permanent genetic change.

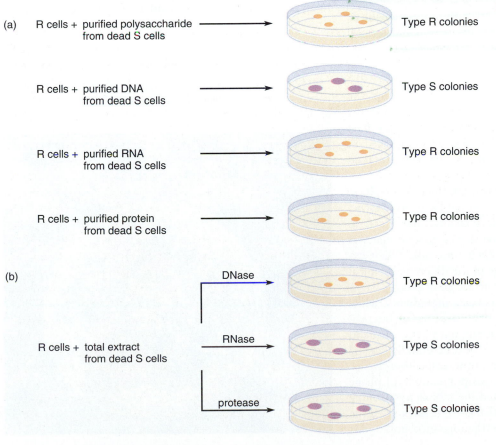

(a) R cells + purified polysaccharide from dead S cells — → Type R colonies

R cells + purified DNA from dead S cells — → Type S colonies

R cells + purified RNA from dead S cells — → Type R colonies

R cells + purified protein from dead S cells — → Type R colonies

(b) DNase — → Type R colonies

R cells + total extract from dead S cells — RNase → Type S colonies

protease — → Type S colonies

▶ **FIGURE 7–3** Identification of the transforming agent by Avery, MacLeod, and McCarty. (a) Purified classes of molecules isolated from heat-killed S cells were tested for transforming activity. Only DNA was active. (b) The transforming activity was destroyed only by DNase, confirming that the transforming agent is composed of DNA.

▶ **FIGURE 7–4** Representative bacteriophage strains of *Escherichia coli*. (a) The T-even phages (T2, T4, and T6) are composed of a molecule of DNA contained in a protein coat that consists of a polyhedral head and a complex tail ending in six tail fibers. (b) Phage lambda (λ), another DNA-containing phage, is smaller than the T-even phages and has a simpler tail that ends in only one tail fiber. (c) Phage φX174 consists of a protein head surrounding the DNA but no tail structure.

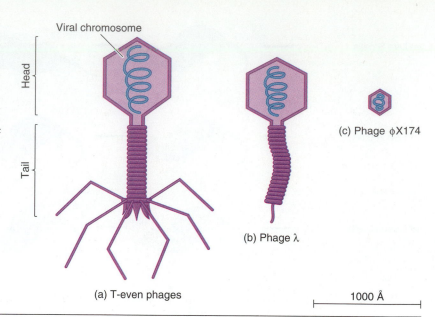

Viral chromosome

Head

Tail

(a) T-even phages

(b) Phage λ

(c) Phage φX174

1000 Å

Phage Replication: The Hershey-Chase Experiment

Of the many kinds of viruses that have been discovered, the viruses of bacteria, called bacteriophages (a term conventionally shortened to *phages*), have been used most extensively in genetic research. Bacteriophages have a comparatively simple structure, which consists of a molecule of DNA or RNA surrounded by a protein coat (▶ Figure 7–4). Phages can exist in this form indefinitely while they are stored in the laboratory, but in order to reproduce, phage particles must bind to receptor sites located on the bacterial cell wall and inject their genetic material into the cell (▶ Figure 7–5).

Bacteriophages were first described and named in 1917 by F. D'Herelle, a French bacteriologist, but not until the early 1950s did researchers learn that only the DNA of the phage, and not the protein coat, is injected into the host during the infection process. This discovery, reported in 1952 by A. D. Hershey and M. Chase, further substantiated that DNA is the genetic material.

Hershey and Chase studied the reproduction of phage T2, which infects the common intestinal bacterium *Escherichia coli*. In their experiment, Hershey and Chase used radioactive tracers to follow the transfer of genetic material from infecting T2 particles into their host cells. Their experimental design took advantage of the different chemical compositions of the DNA and protein coat of phages. DNA contains phosphorus as one of its basic ingredients, while most proteins do not. In contrast, DNA lacks sulfur, but proteins usually contain a substantial amount of this element. The DNA and protein can therefore be differentially labeled using isotopes of phosphorus and sulfur, respectively.

Hershey and Chase prepared two populations of phage particles. In one, they incorporated the radioactive isotope ^{32}P into the phage DNA; in the other, they incorporated the isotope ^{35}S into the phage protein. (Isotope incorporation is accomplished by allowing phages to reproduce in bacterial cells that are growing in a chemically defined medium whose sole source of phosphorus or sulfur is made up of radioactive isotopes of these elements.) Each culture of radioactively labeled phages was allowed to infect a separate population of *E. coli*. After the investigators had allowed sufficient time for the genetic material of the phages to be injected into the cells, they agitated the two cultures separately in an electric food blender to shear off any phage

▶ **FIGURE 7–5** Electron micrograph of adsorbed T4 phage particles. Numerous phages are attached by their tails to the wall of an *E. coli* cell.

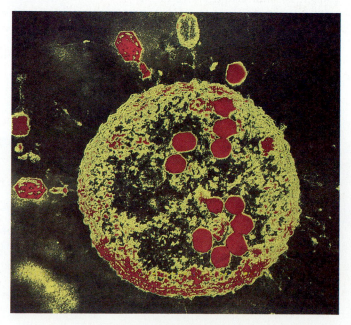

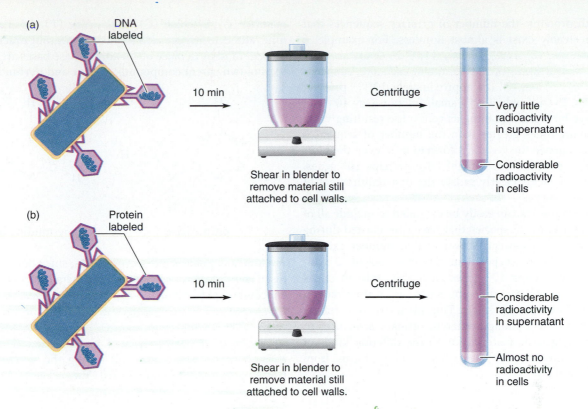

FIGURE 7-6 The Hershey-Chase experiment demonstrating that DNA is the genetic material of phage T2. (a) After infection, the labeled DNA of the phage remains with the cells after shearing. (b) The labeled protein coat of the phage is released into the supernatant after shearing.

parts that might still be attached to the cell walls. They then subjected the samples to low-speed centrifugation, which forced the infected bacteria toward the bottom of the tube, where they accumulated in the form of a pellet. Finally, the investigators used a radiation counter to determine the amount of radioactivity that remained in the supernatant and the amount that appeared in the pellet.

Hershey and Chase found that radiation emanating from the ^{32}P was localized in the pellet containing the infected cells, while most of the ^{35}S radioactivity remained in the supernatant (▶ Figure 7-6). In other words, shearing removed nearly all the ^{35}S from infected cells, but hardly any of the ^{32}P. From these results, Hershey and Chase concluded that the DNA of the phage enters the bacterium, while the protein coat remains on the cell surface following infection. The shearing force generated by the blender breaks the tails by which the empty phage capsids (phage ghosts) are held to the cell wall and releases them into the medium. These results clearly show that DNA is the material that enters the cells and directs the synthesis of phage progeny. DNA is thus the genetic material of the virus.

To Sum Up

1. Nucleic acids are large polymeric molecules in which the repeating monomeric units are nucleotides. The two types of nu-

cleic acids are DNA (deoxyribonucleic acid) and RNA (ribonucleic acid).

2. DNA is the genetic material in all prokaryotic and eukaryotic organisms that have been studied. Some viruses contain only RNA (no DNA); this RNA serves as the genetic material.

3. The transformation experiments of Avery, MacLeod, and McCarty, which were reported in 1944, were the first to show unequivocally that DNA is the genetic material. These experiments demonstrated that naked fragments of DNA isolated from a donor bacterium could be taken up by a recipient bacterium and could induce a permanently inherited change in the genotype of the recipient cells.

4. Hershey and Chase reinforced the conclusion that genes are composed of DNA with radioactive labeling experiments using bacteriophage T2, which they reported in 1952. Their work showed that the DNA of the phage, but not the protein coat, enters the bacterial cell to direct the synthesis of phage progeny.

COVALENT STRUCTURE OF NUCLEIC ACIDS

One of the requirements of the genetic material is that it store coded information (see Chapter 1). It is now known that the genetic information carried by a nucleic acid molecule is stored in its **primary structure**, that is, in the nucleotide sequence of the polynucleotide chain. As is the case with proteins, the number of primary sequences that can conceivably exist is almost limitless. For example, a polynucleotide consisting of only 1000 residue sites, with any one of four kinds of nucleotides at each site, can form $4 \times 4 \times 4 \ldots = 4^{1000}$ (or approximately 10^{600}) possible sequences. Even if only a very small fraction (say 10^{-6}) of these have biological meaning as genes, the resulting value is still many times greater than the number of sequences needed to encode the genetic information of even the most complex organism, which might have perhaps 10^5 genes. Thus, nucleic acids clearly satisfy the first requirement of the genetic material—the storage of information in a stable form. The argument can easily be extended to include all of the different kinds of species that have ever existed during the history of life on earth, even if that number exceeds 10^{10}, as some biologists postulate. The existence of 10^{10} different species having 10^5 genes each (certainly an upper estimate) requires 10^{15} different genes, still a tiny fraction of the 10^{600} sequences possible. Thus nucleic acids certainly possess more than enough different primary sequences to serve as the genetic material for all the different kinds of genes that constitute the blueprints for all life forms, from viruses to humans.

Example 7.1

If there were only two possible nucleotides (say one purine and one pyrimidine) instead of four, how many different primary sequences could be formed in a polynucleotide consisting of 1000 residue sites?

Solution: The number of different sequences would be 2^{1000}, which is still a very large number (about 10^{300}), much in excess of 10^{15}.

Follow-Up Problem 7.1

According to the now defunct tetranucleotide hypothesis, each nucleic acid chain was thought to consist of a particular repeating sequence of its four nucleotides. If any sequence of the four nucleotides could exist, how many different primary sequences would be possible under this hypothesis?

◆

Structure of Nucleotides

The nucleotides, which constitute the monomer units of a nucleic acid molecule, consist of three primary components: (1) a nitrogen-containing base, (2) a five-carbon, or pentose, sugar, and (3) a phosphate group. The main source of variation among nucleotides involves the nature of their nitrogenous base. The bases in DNA can be adenine (A),

guanine (G), cytosine (C), or thymine (T). Adenine, guanine, and cytosine are also found in RNA, but uracil (U) is present in RNA in place of thymine. These bases are derivatives of two parent compounds, **purine** and **pyrimidine**:

purine base   pyrimidine base

Adenine and guanine are purine bases, since both have the fused nine-member ring system that characterizes their parent compound (▶ Figure 7–7a). In adenine, an amino group is substituted for the hydrogen at position 6 in the purine ring, while guanine has an amino group at position 2 and an oxy (=O) group at position 6. Cytosine, uracil, and thymine are pyrimidine bases, all having the six-membered ring of pyrimidine (Figure 7–7b). Various substitutions at positions 2, 4, and 5 in the pyrimidine ring differentiate these bases.

The nitrogen atom at position 9 of the purine ring or position 1 of the pyrimidine ring is involved in linking the base to a pentose sugar to form a **nucleoside** (▶ Figure 7–8a). The carbon atoms of the sugar are then numbered $1', 2', 3', \ldots$, to distinguish them from the numbered ring positions of the base, with the $1'$ carbon joined to the purine or pyrimidine base by a **glycosyl linkage**. The nucleosides of RNA and DNA differ in the nature of their sugar residue. The sugar is **ribose** in the precursors of RNA (the ribonucleosides), and it is **deoxyribose** in the precursors of DNA (the deoxyribonucleosides). Both sugars are cyclic structures that have free hydroxyl (—OH) groups at the $3'$ and $5'$ carbon positions of their corresponding nucleosides. An additional hydroxyl group is present at the $2'$ carbon position in ribose. By convention, the nucleosides are named after their constituent bases using the suffix *-osine* for the purine nucleosides and *-idine* for the pyrimidine nucleosides. The common ribonucleosides are thus adenosine, guanosine, cytidine, and uridine, and the common deoxyribonucleosides are deoxyadenosine, deoxyguanosine, deoxycytidine, and deoxythymidine.

A nucleotide is formed when the sugar portion of a nucleoside is esterified to phosphoric acid (Figure 7–8b). In nucleotides with a single phosphate (the nucleoside monophosphates), the phosphate group can be attached at one of three carbon positions ($2', 3',$ or $5'$) of the ribonucleotides and at one of two carbon positions ($3'$ or $5'$) of the deoxyribonucleotides. Cyclic monophosphates in which the phosphate group is attached at two positions (e.g., adenosine $3', 5'$-cyclic phosphate, or cyclic AMP) also exist. However, the most common nucleotides are the nucleoside

(a) Purine bases

H—N—H

$\underset{1}{N}\ \underset{6}{C}\ \underset{5}{C}\ \underset{7}{N}$ ... adenine (A)

adenine (A)

O

guanine (G)

(b) Pyrimidine bases

H—N—H

cytosine (C)

O

uracil (U)

O

CH₃

thymine (T)
(5-methyluracil)

► **FIGURE 7–7** The principal purine and pyrimidine bases of nucleic acids. Adenine, guanine, cytosine, and thymine are present in DNA. Adenine, guanine, cytosine, and uracil are present in RNA.

(a) Nucleoside

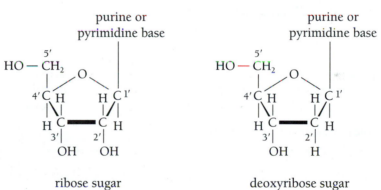

ribose sugar

deoxyribose sugar

(b) Nucleotide

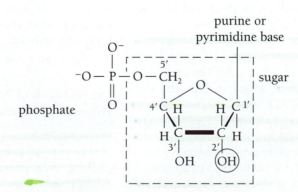

phosphate

► **FIGURE 7–8** (a) General structure of a nucleoside. A purine or pyrimidine base is bonded to C1′ of ribose (in RNA) or deoxyribose (in DNA). (b) General structure of a nucleotide. Here a nucleoside is bonded at C5′ to a phosphate group. In deoxyribonucleotides, the circled —OH group is replaced with —H.

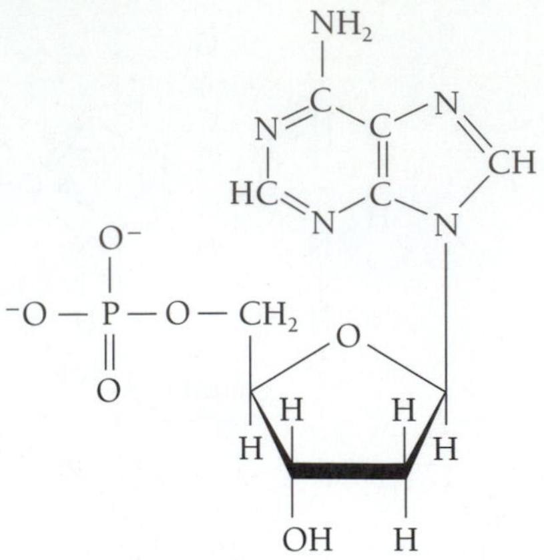

deoxyadenylate
deoxyadenosine 5′-monophosphate
Symbols: A, dAMP

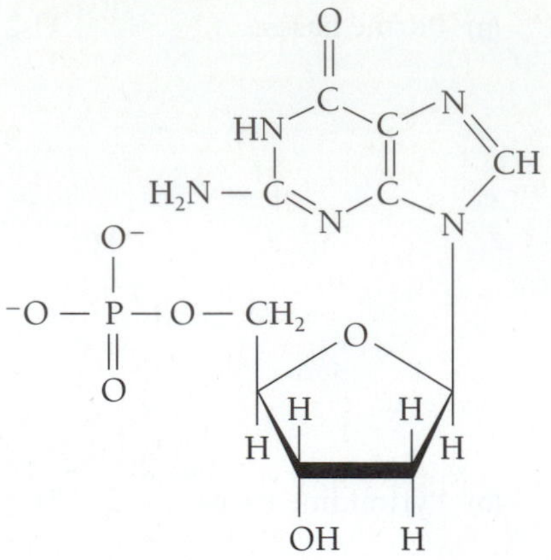

deoxyguanylate
deoxyguanosine 5′-monophosphate
Symbols: G, dGMP

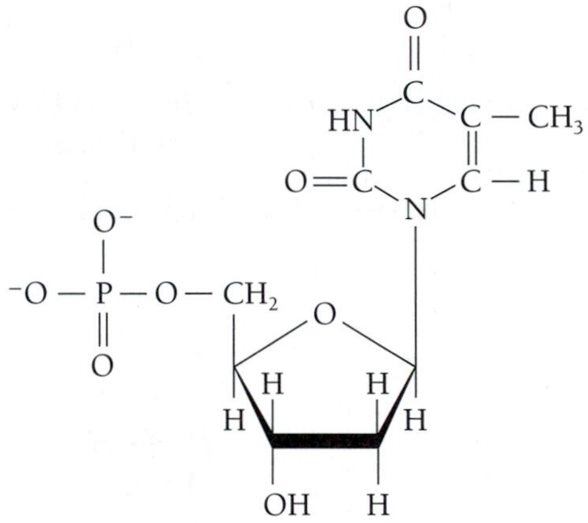

deoxythymidylate
deoxythymidine 5′-monophosphate
Symbols: T, dTMP

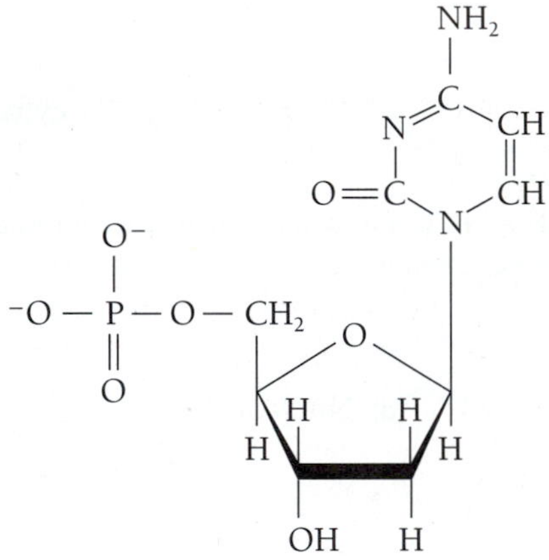

deoxycytidylate
deoxycytidine 5′-monophosphate
Symbols: C, dCMP

▶ **FIGURE 7–9** (a) The principal nucleoside 5′-phosphates in DNA.

5′-phosphates. The structures and names of the principal nucleoside monophosphates found in DNA and RNA are given in ▶ Figure 7–9.

All the common nucleosides can also exist in the form of 5′-diphosphates and 5′-triphosphates (▶ Figure 7–10), which play a variety of roles in living cells. The nucleoside diphosphates and triphosphates, which include the well-known examples ADP and ATP, are high-energy compounds that produce a large amount of free energy when their terminal phosphate residues are released upon hydrolysis; they therefore serve as important energy sources in various cellular activities. Of particular significance to genetics is the fact that the high-energy nucleoside triphosphates are precursors of DNA and RNA, making DNA replication and transcription possible. Recall that DNA replication and transcription of DNA to form RNA are the

processes that allow DNA to meet two of the requirements of the genetic material: inheritance of coded information and expression of the coded information into phenotypic traits, respectively.

Organization of Nucleotides into Polynucleotide Chains

DNA and RNA are polynucleotides—they consist of chains of covalently linked nucleotide residues. To form a polynucleotide, each phosphate group participates in two ester linkages, forming a bridge between the 3′ carbon of one nucleoside and the 5′ carbon of another. The structure of the polynucleotide is therefore maintained by repeating 3′ → 5′ phosphodiester bonds. The fundamental repeating struc-

(b)

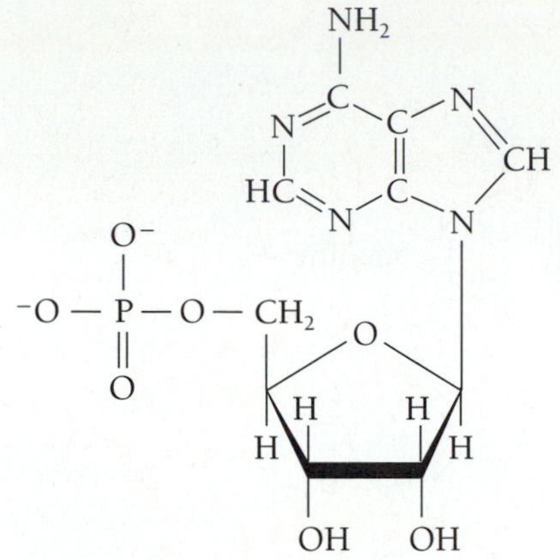

adenylate
adenosine 5'-monophosphate
Symbols: A, AMP

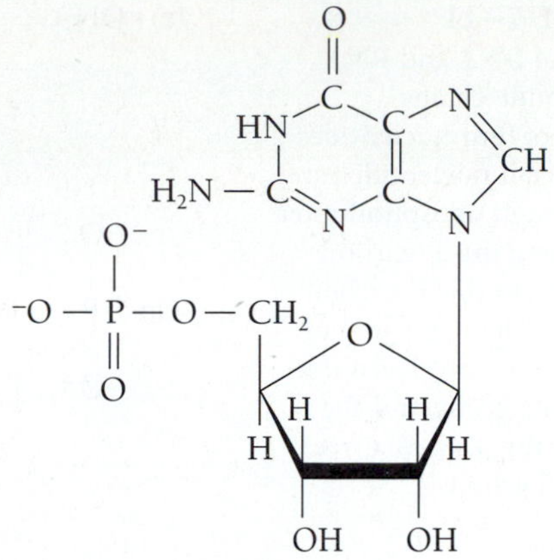

guanylate
guanosine 5'-monophosphate
Symbols: G, GMP

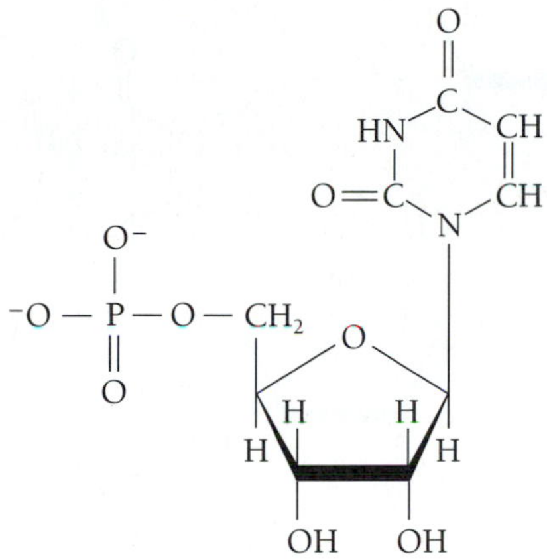

uridylate
uridine 5'-monophosphate
Symbols: U, UMP

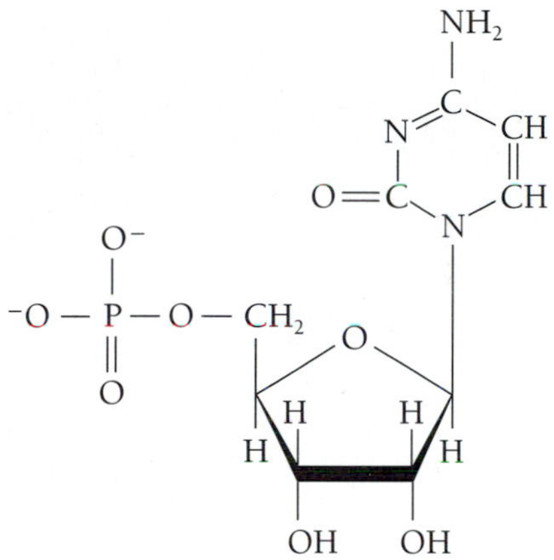

cytidylate
cytidine 5'-monophosphate
Symbols: C, CMP

► FIGURE 7–9 (*continued*) (b) The principal nucleoside 5'-phosphates in RNA.

(a)

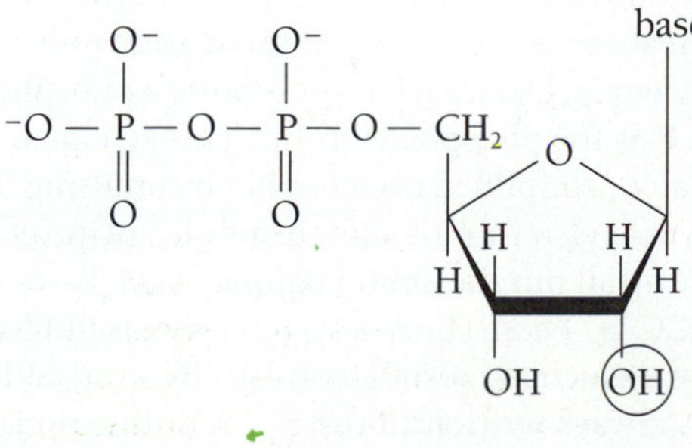

(b)

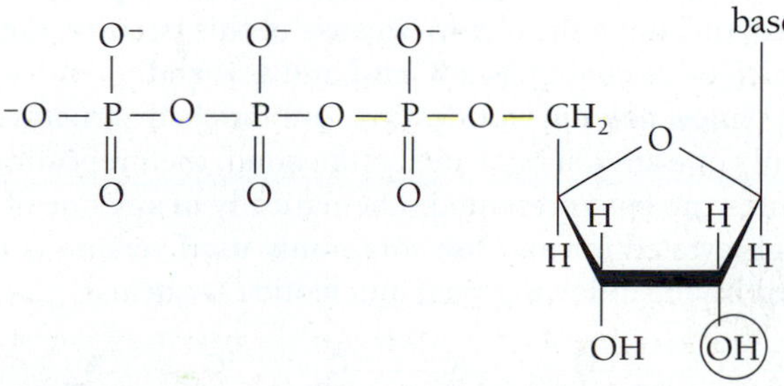

► FIGURE 7–10 (a) General structure of a nucleoside 5'-diphosphate. (b) General structure of a nucleoside 5'-triphosphate. In both cases, the circled —OH group is replaced with —H in the corresponding deoxyribonucleotides.

► **FIGURE 7–11**

Segments of DNA and RNA polynucleotide chains consisting of four nucleotides each. Adjacent nucleotides are linked through phosphodiester bonds joining the 3′ carbon of one sugar to the 5′ carbon of the next. Thus, one end of a chain (the 5′ end) has a free 5′ phosphate group and the other end (the 3′ end) a free 3′ hydroxyl group.

(a) DNA

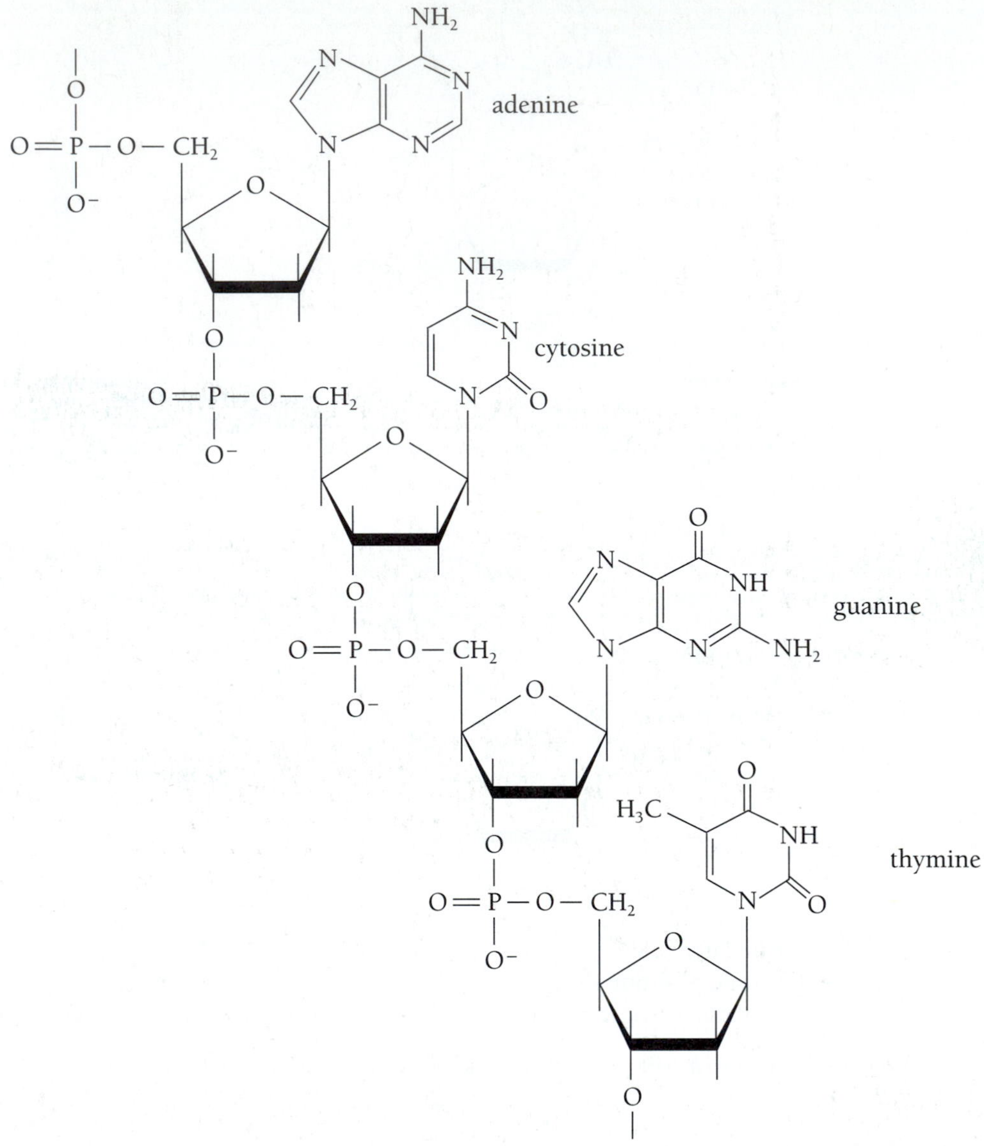

adenine

cytosine

guanine

thymine

---

ture of a polynucleotide is given in ► Figure 7–11. Note that the repeating 3′ → 5′ linkage gives the covalent sugar phosphate backbone of the structure a specific **polarity**, or direction, along the chain. Because of this polarity, each polynucleotide chain has a **5′ end** and a **3′ end**.

Since polynucleotides are unbranched structures with only one kind of internucleotide bond, their repeating structures can be represented schematically in any one of several abbreviated forms. One commonly used version is illustrated by the following pentanucleotide sequence:

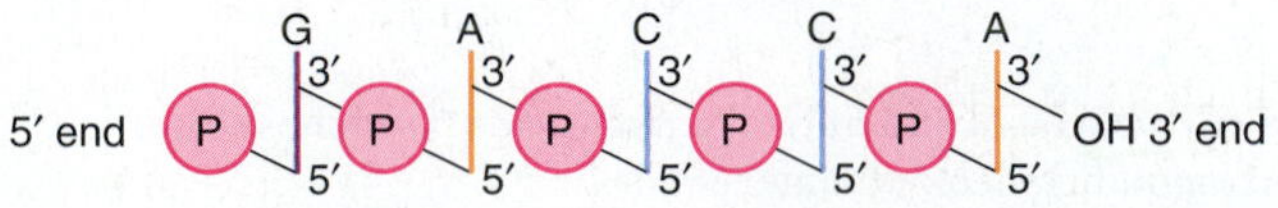

Each nucleotide residue in this chain is represented by a vertical line, ending in the 1′ position at the top and 5′ position at the bottom, with the single letter abbreviation of the base written above it. The different nucleoside units are connected by diagonal lines, each representing a phosphodiester bond, with P as the phosphate group. This schematic representation can be simplified even further by omitting the lines and the hydroxyl terminus (for example, pGpApCpCpA) or by omitting all but the abbreviations of the bases (for example, GACCA). These shorter systems are useful for designating long sequences of nucleotides. By convention, the sequence is always written in the 5′ to 3′ direction, so that the 5′ end is at the left. Thus, for example, pApG is a dinucleotide that has a phosphate group at its 5′ end and a phosphodiester bridge connecting the 3′ carbon of A to the 5′ carbon of G.

At physiological pH, the phosphate groups of the polynucleotide are fully ionized, contributing one negative charge per nucleotide residue. This negative charge means that nucleic acid chains of different length can be separated by gel electrophoresis in much the same manner as proteins. The fractionation of polynucleotides by this method is a widely used technique that is a crucial part of molecular genetic technologies. The procedure is usually carried out in an agarose gel as the matrix of choice, although a polyacrylamide gel is sometimes used. In either type of gel, nucleic acid chains travel in the same direction (toward the anode) but at rates inversely dependent on chain length.

For chains with somewhat similar base compositions, the molecular weight is directly proportional to the number of nucleotide residues. The polynucleotides therefore have essentially the same charge-to-mass ratio and tend to experience the same force pulling them through the electric field. As in the case of proteins, however, larger molecules have greater difficulty penetrating the gel matrix and consequently migrate at a slower rate than smaller molecules. For this reason, the distance traveled by a polynucleotide in a gel is often observed to vary inversely with its molecular weight.

When nucleic acids are fractionated by gel electrophoresis, the positions of the bands can be determined by staining with a fluorescent dye that interacts with the nucleic acid (e.g., ethidium bromide) or by using polynucleotides that have been previously labeled with a radioactive isotope (e.g., ^{32}P). Autoradiography is often employed to detect radiolabeled polynucleotides. In autoradiography, the sample is overlaid by a photographic film that is sensitive to radioactivity. The film-covered sample is stored in the dark while radioactive decay occurs, and then the film is developed. As shown in ▶ Figure 7–12, a series of bands appears where the film was exposed to radioactive decay, thus establishing the positions of the labeled polynucleotides in the gel.

To Sum Up

1. Each nucleotide consists of a nitrogenous base (either a purine or a pyrimidine), a five-carbon sugar, and an acidic phosphate group. In DNA the purine bases are adenine (A) and guanine (G), the pyrimidine bases are cytosine (C) and thymine (T), and the sugar is deoxyribose. The same nucleotides are found in RNA, except that the pyrimidine base uracil (U) occurs in

place of thymine, and the sugar in RNA is ribose rather than deoxyribose.
2. Nucleotides are linked by repeating $3' \rightarrow 5'$ phosphodiester bonds to form long polynucleotide chains. The directional nature of the phosphodiester bonds gives the covalent sugar-phosphate backbone of a chain structural polarity.
3. The first requirement of the genetic material—the ability to store information—is met by the huge variety of nucleotide sequences possible within DNA. Different nucleotide sequences contain the information that constitutes the different genes.

THREE-DIMENSIONAL STRUCTURE OF NUCLEIC ACIDS

The polynucleotide chain tends to fold into an ordered conformation that is crucial to its function. The specific folding of the chain is determined in large part by the interactions of the nitrogenous bases with each other and with the solvent. First, the purine and pyrimidine rings are planar structures with hydrophobic faces that give the bases a nonpolar character. Thus neighboring bases tend to stack on top of each other in parallel planes to limit their contact with water. This hydrophobic base stacking reduces the average distance between adjacent bases in the ordered state and gives a certain degree of rigidity to the polynucleotide chain. Furthermore, the edges of the bases carry certain weakly charged chemical groups that can act as hydrogen donors and acceptors for hydrogen bonding. These groups, which include the polarized N—H and C=O bonds, are located mainly at positions 1, 2, and 6 of the purine ring and positions 2, 3, and 4 of the pyrimidine ring, they provide certain bases with the ability to form H bonds with each other in the polynucleotide's folded state.

Double-Helical Structure of DNA

One of the first indications that nucleic acids have a regular structure came from the work of the American biochemist Edwin Chargaff, who by 1950 had analyzed the base composition of the DNA from several different sources. Chargaff discovered that while the overall percentages of the bases varied widely from source to source, the proportion of adenine was always equal to that of thymine (or very nearly so) and the proportion of guanine was always equal to that of cytosine. In terms of molar proportions, Chargaff's equivalence rules are

$$[A] = [T] \quad \text{and} \quad [G] = [C]$$

Observe that adding these two equalities gives $[A] + [G] = [T] + [C]$, which states that the proportions of the purine and pyrimidine bases in DNA are the same.

Another indication of the regular structure of DNA came from the X-ray diffraction studies of Maurice Wilkins and Rosalind Franklin. They beamed X-rays at a stretched DNA

▶ FIGURE 7–12 Detection of radioactively labeled materials by autoradiography. When an X-ray film is exposed to the radioactive compounds in an electrophoretic gel, the radioactive decay products strike the film, interacting with the silver halide contained in the emulsion and thereby producing a latent image. The image is made visible by standard photographic development techniques.

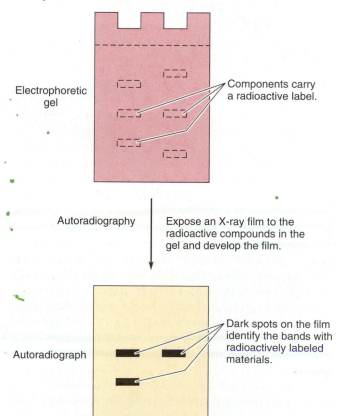

Electrophoretic gel

Components carry a radioactive label.

Autoradiography — Expose an X-ray film to the radioactive compounds in the gel and develop the film.

Autoradiograph

Dark spots on the film identify the bands with radioactively labeled materials.

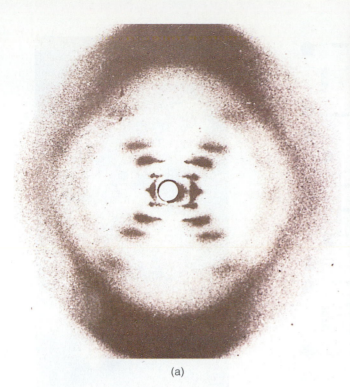

(a)

(b)

▶ **FIGURE 7–13** (a) Rosalind Franklin's first X-ray crystallography of DNA-B (1952). The pattern of discrete spots of varying spacing and intensity shows the sharply defined directions in which the X-rays were diffracted by the crystalline DNA structure. The central crosslike pattern is indicative of the molecule's helical structure. The upper and lower dark patterns provide an estimate of the periodicity of the base pairs, which occur every 0.34 nm. (b) Watson and Crick deduced the structure of DNA based on X-ray analysis.

fiber containing many similarly oriented DNA molecules and then measured the diffraction pattern that appeared on a photographic film. The X-ray diffraction diagram shown in ▶ Figure 7–13a has a central crosslike pattern, indicating that DNA has the form of a helix, and strong reflections with 3.4 nm and 0.34 nm spacings that give the pitch and rise per residue, respectively, of the helix. There are thus 3.4/0.34 = 10 residues per turn of the helix. Moreover, it can be inferred from the density of DNA that the DNA helix contains two polynucleotide chains.

James Watson and Francis Crick correctly interpreted these results in their now well-known model of DNA structure, which they published in 1953. (See Figure 7–13b.) They proposed that a molecule of DNA is a double helix consisting of two polynucleotide chains held together by hydrogen bonds between opposing bases on the two chains. ▶ Figure 7–14 shows the general form and dimensions of the double helix. Note that the model describes a right-handed helix, in which the chains are wound in a clockwise direction as they progress up and around the helix axis. The paired chains are slightly offset, creating two unequal grooves—the larger major groove and the smaller minor groove. Also note that the negatively charged phosphate groups are on the outside of the helix, in contact with the solvent, while the largely nonpolar base pairs are

stacked in a parallel array within the central core of the structure, roughly perpendicular to the helix axis. The interaction between the stacked bases is a major factor in stabilizing the helical form. Because of the hydrophobic character of the bases, the stacked base pairs tend to exclude water from the core of the molecule, contributing further to the stability of the helix in the aqueous environment.

In constructing the model, Watson and Crick recognized that the constant diameter of the helix required that a purine on one chain always be opposite a pyrimidine on the other chain. A base pair consisting of two purines or two pyrimidines, while theoretically possible, would alter the dimensions of the helix. They also noticed that only certain purine-pyrimidine pairs could be formed by hydrogen bonding. Model building revealed that under normal conditions, proper hydrogen bonding can occur only between adenine and thymine and between guanine and cytosine (▶ Figure 7–15). Pairing between adenine and cytosine and between guanine and thymine cannot take place, since the identical charges on both bases of each combination would cause them to repel one another. (Rare errors in base pairing do occur and are a source of mutations—the fourth requirement of the genetic material. Their consequences will be discussed in Chapter 22.) Thus, A normally pairs with T, and G normally pairs with C. These base pairing rules mean

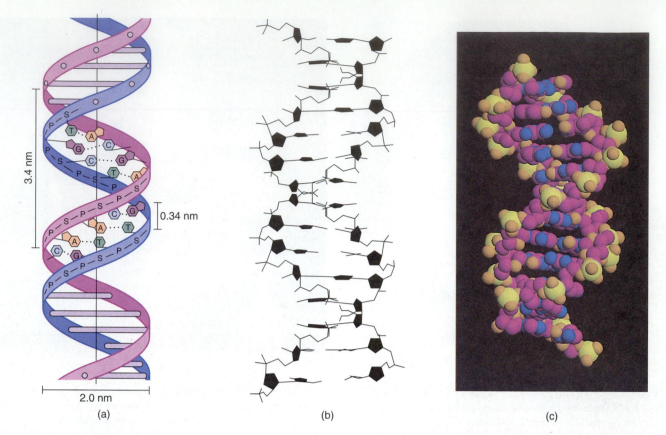

▶ **FIGURE 7–14** Different representations of the Watson-Crick DNA double helix. (a) The general form and dimensions of the double helix. (b) A skeletal model and (c) a space-filling model show the relationship of the different components of the helix.

that the nucleotide sequence of one chain of the double helix specifies the nucleotide sequence of the other. The two chains of a DNA duplex are thus said to be **complementary in primary structure**. The base pairing restrictions are also the basis of Chargaff's rules, since a double helix made up of only AT and GC base pairs must contain the same amount of A as T and the same amount of G as C. Note, however, that there is no restriction on the total proportions of AT versus GC pairs in the molecule; that is, the amount of A + T does not bear any set relationship to the amount of G + C. General agreement with this aspect of the model is shown in ■ Table 7–1, which lists the base compositions of the DNA from various organisms. With the exception of the phage ϕX174, all of these organisms conform to Chargaff's equivalence rules, but show a wide range of variation in the ratio of A + T to G + C. The reason why phage ϕX174 appears to violate Chargaff's rules is that the DNA of this phage is single-stranded and does not normally exist in the form of a double helix.

Example 7.2

The base composition of a certain nucleic acid is 30% A, 30% C, 20% G, and 20% T. Is this nucleic acid single-stranded DNA, double-stranded DNA, single-stranded RNA, or double-stranded RNA?

Solution: The nucleic acid is single-stranded DNA. The presence of thymine rather than uracil indicates DNA. However, [A] ≠ [T] and [G] ≠ [C], so it must be single-stranded DNA.

■ **TABLE 7–1** DNA base compositions in a variety of organisms.

Source	A	T	G	C	$\frac{A+T}{G+C}$	$\frac{A+G}{C+T}$
Human	31.0	31.5	19.1	18.4	1.67	1.00
Mouse	29.1	29.0	21.1	21.1	1.38	1.00
Chicken	28.0	28.4	22.0	21.6	1.29	1.00
Frog	26.3	26.4	23.5	23.8	1.11	0.99
Fruit fly	27.3	27.6	22.5	22.5	1.22	0.99
Corn	25.6	25.3	24.5	24.6	1.04	1.00
Tobacco	29.7	30.4	19.8	20.0	1.51	0.98
Bread mold	23.0	23.3	27.1	26.6	0.86	1.00
E. coli	24.6	24.3	25.5	25.6	0.96	1.00
Pneumococcus	30.3	29.5	21.6	18.7	1.48	1.08
Herpes simplex virus	13.8	12.8	37.7	35.6	0.36	1.06
Phage λ	26.0	25.8	23.8	24.3	1.08	0.99
Phage ϕX174	24.7	32.7	24.1	18.5	1.35	0.95

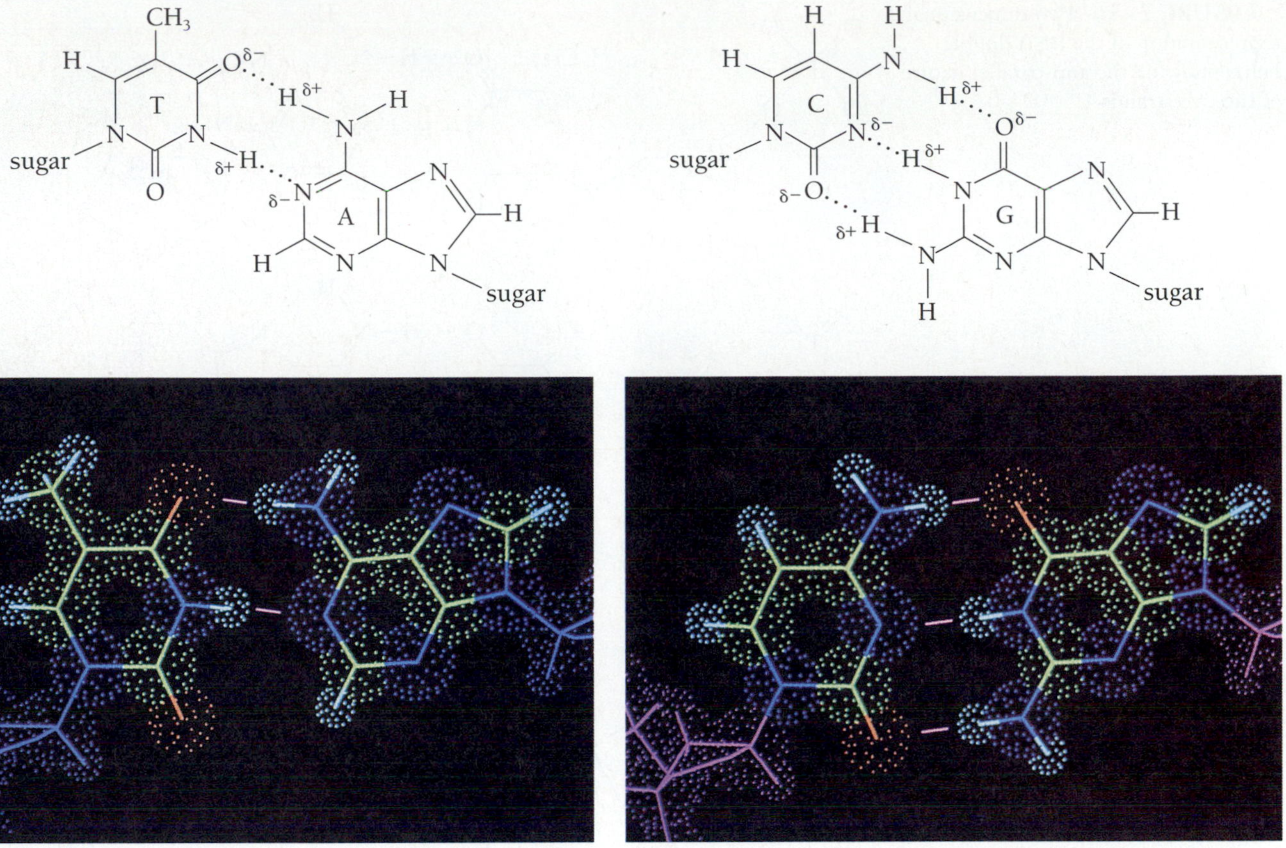

▶ FIGURE 7−15 Purine-pyrimidine base pairs in DNA. The structures show two possible hydrogen bonds between adenine and thymine and three between guanine and cytosine. (δ^- and δ^+ denote partial negative and partial positive charges.) Computer-generated models of (a) adenine-thymine base pair and (b) guanine-cytosine base pair show carbon in green, oxygen in red, nitrogen in blue, hydrogen in cyan, and hydrogen bonds in purple.

Follow-Up Problem 7.2

For the nucleic acid described in Example 7.2 above, the $(A + G):(C + T)$ ratio is 1. Does this finding contradict your conclusion that this is single-stranded DNA?

◆

A final feature of the Watson-Crick model is the opposite (**antiparallel**) polarity of the paired polynucleotide chains (see ▶ Figure 7−16). One chain runs in the 5′ to 3′ direction, while its partner runs in the 3′ to 5′ direction. Thus rotation by 180° does not alter the general appearance of the helix.

The genetic significance of the double helical structure can be understood by considering how DNA meets two of the requirements of the genetic material: information transfer and accurate duplication of information during cell division. Quoting from the paper on DNA structure of Watson and Crick (1953), "It has not escaped our notice that the specific pairing we have postulated immediately suggests a possible copying mechanism for the genetic material." Chapter 8 will discuss the mechanism by which the double helix replicates itself, showing how each of the two complementary strands serves as a template for the synthesis of a new strand complementary to it. In Chapter 16 we will describe the transfer of the genetic information from the nucleus to the cytoplasm, where it can be decoded into proteins. In this process, the genetic information carried by the double helix is transcribed onto messenger RNA molecules that are complementary to the DNA sequences that make up the genes. Complementarity is thus a crucial genetic feature of DNA.

Point to Ponder 7.1

Edwin Chargaff did not share in the Nobel Prize that was awarded to Watson, Crick, and Wilkins for their work in deducing the double-helical structure of DNA. (Franklin died before the Nobel Prize was given.) In light of his work on the base composition of DNA, why do you think the Nobel Prize Committee decided not to include Chargaff in the award? (Think about the various steps of the scientific method and the importance of interpreting and deriving meaning from data.)

► **FIGURE 7–16** Two-dimensional representation of the DNA double helix, showing the anti-parallel nature of the two strands.

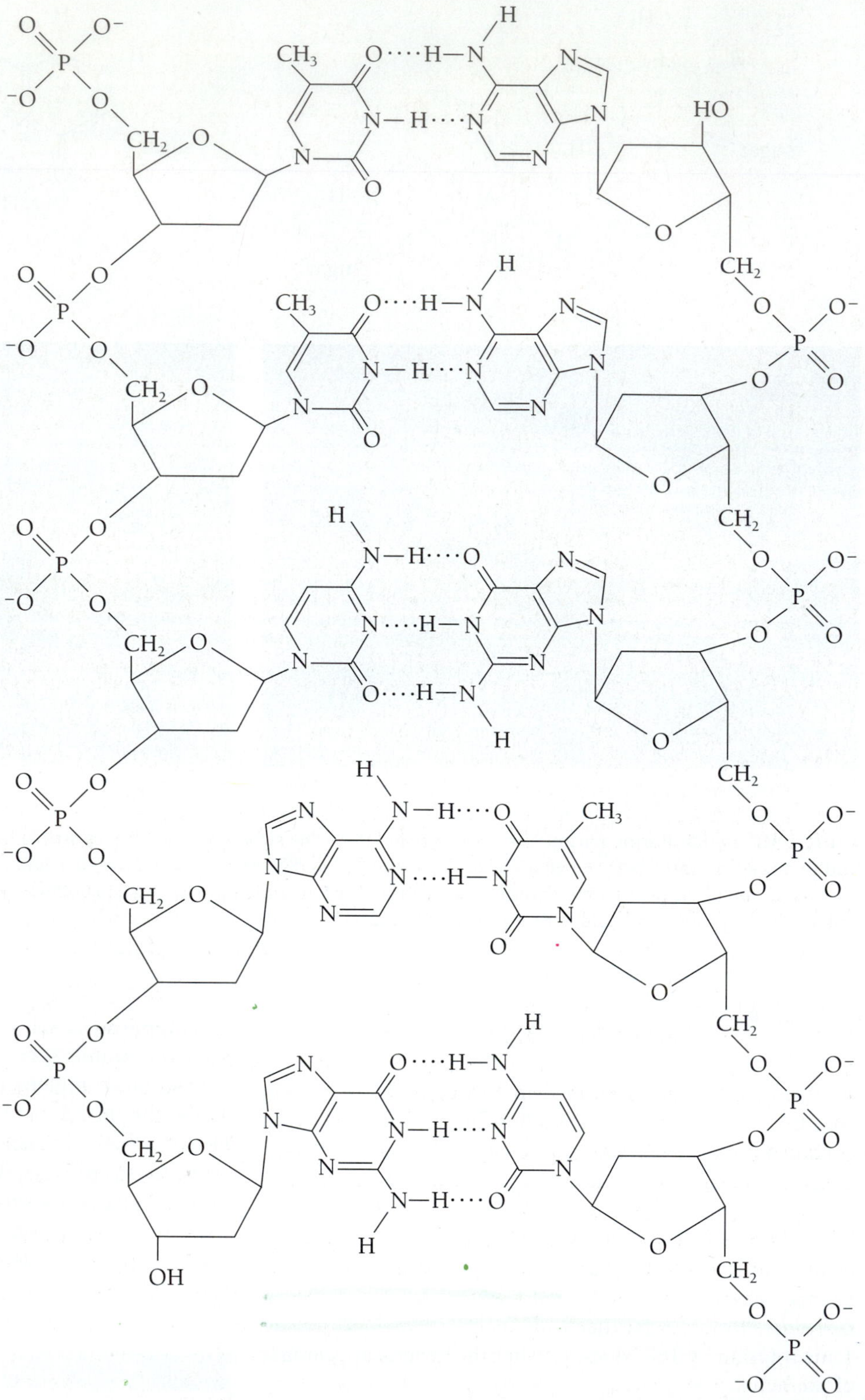

Conformational Variants of the Double Helix

In all nucleic acids that possess secondary structure, the three-dimensional form is a double helix based on the principle of complementary base pairing. In DNA, the normal base pairs are AT and GC, while in RNA, the base pairs are usually AU and GC. With the exception of certain viruses (e.g., φX174), most DNA is double-stranded and helical throughout much of its length. In contrast, most RNA is single-stranded and is helical only in regions where there

are complementary sequences within the same polynucleotide. The RNA chain then folds back upon itself to form a *hairpin* structure (▶ Figure 7–17a and b) consisting of a double-stranded helical stem with a single-stranded loop at one end.

Hairpin structures can also form in DNA in regions of an **inverted repeat** (a segment in which a nucleotide sequence is followed by its complementary sequence in reverse order) (Figure 7–17c). Because DNA is double-stranded, both chains have copies of the inverted repeat. Thus, if the strands should separate, two hairpin loops can form across from one another, creating a **cruciform** (crosslike) structure by base pairing in each of the separated strands. Note that inverted-repeat sequences in double-stranded molecules read identically when both strands are followed in the 5′ to 3′ direction (e.g., $\begin{smallmatrix} 5' \\ 3' \end{smallmatrix} \begin{smallmatrix} C\,T\,G\,.\,C\,A\,G \\ G\,A\,C\,.\,G\,T\,C \end{smallmatrix} \begin{smallmatrix} 3' \\ 5' \end{smallmatrix}$ where the dot indicates the axis of symmetry). Such sequences are called **palindromes**, in analogy to English phrases that read the same forward and backward. As we will see in later chapters, palindromes frequently serve as recognition sites for various proteins that bind specifically to DNA.

Since the proposal of the Watson-Crick model, studies have shown that there is considerable variation in the local conformation of the double helix. These variations are most clearly demonstrated by the X-ray diffraction patterns of native and synthetic DNAs, which indicate that the double helix is not a single structure consistent only with the Watson-Crick model but is actually capable of adopting more than one conformation. Three main conformational types have been identified, designated as the A, B, and Z forms, which differ in such properties as (1) the number of base pairs per turn of the helix, (2) the distance between base pairs, (3) the rotation angle per base pair, (4) the location of the helix axis with respect to a base pair, and (5) the orientation (inclination) of the base pairs relative to the helix axis (■ Table 7–2). Furthermore, each form can be thought of as a "family" of structures that can vary within a range of parameter values, with the preferred conformation depending on physical and chemical factors, such as degree of hydration and salt concentration, and on the local sequence of base pairs in the molecule.

The B form has a structure very similar to that deduced by Watson and Crick and is the predominant conforma-

▶ **FIGURE 7–17** Hairpin loops and inverted repeat sequences. (a) A hairpin structure is formed when a polynucleotide chain folds back on itself by base pairing in complementary regions to form a double-helical stem with a single-stranded loop at one end. (b) Formation of a hairpin structure in single-stranded RNA. (c) Formation of a cruciform structure in double-stranded DNA. Inverted-repeat sequences in the double-stranded molecule form a palindrome.

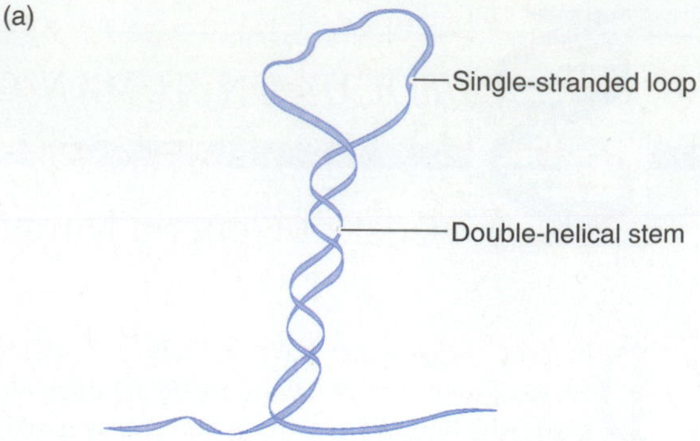

(a) Single-stranded loop

Double-helical stem

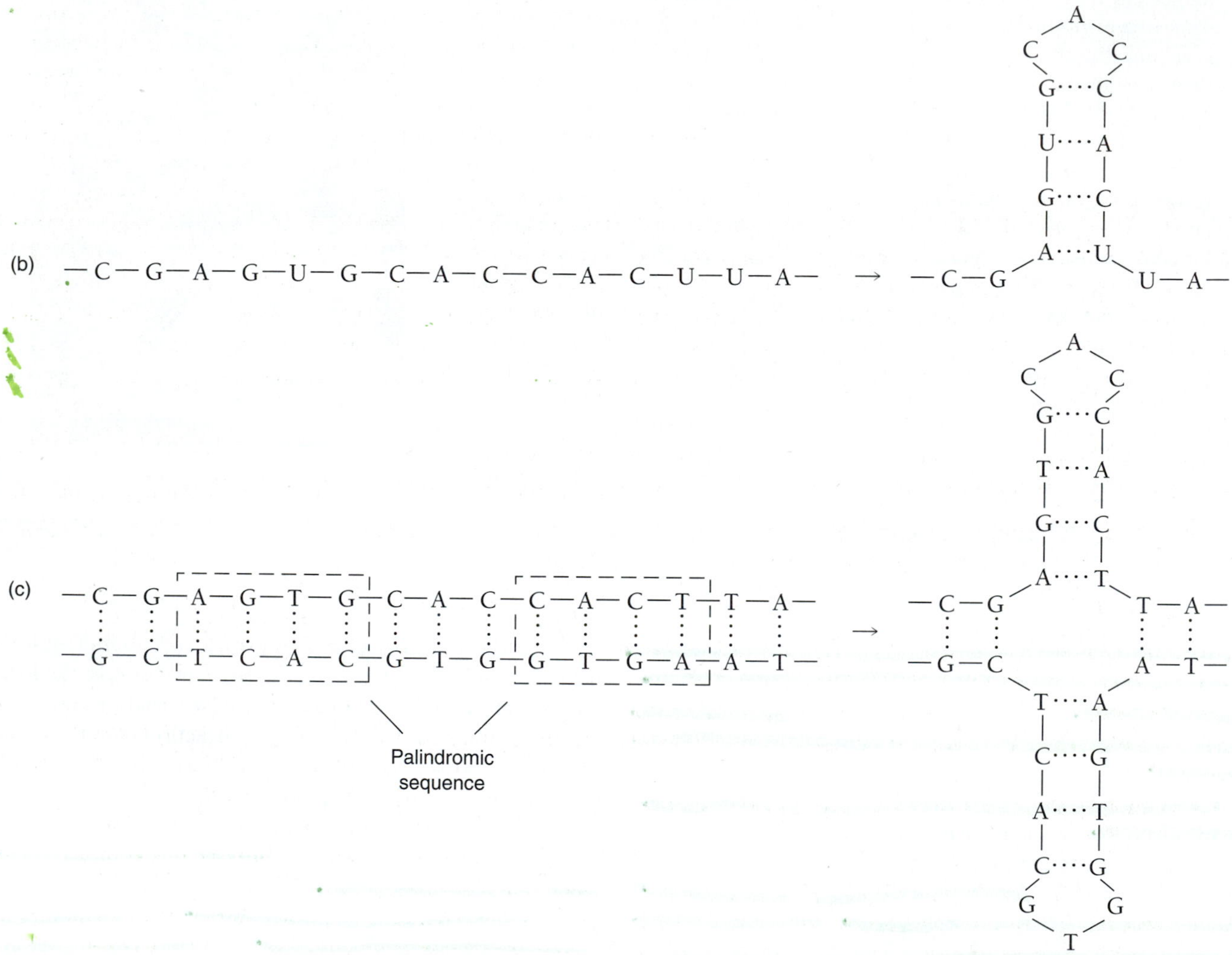

(b) $-$C$-$G$-$A$-$G$-$U$-$G$-$C$-$A$-$C$-$C$-$A$-$C$-$U$-$U$-$A$-$ $\rightarrow$ $-$C$-$G ... U$-$A$-$

(c) $-$C$-$G$-$A$-$G$-$T$-$G$-$C$-$A$-$C$-$C$-$A$-$C$-$T$-$T$-$A$-$ $\rightarrow$

$-$G$-$C$-$T$-$C$-$A$-$C$-$G$-$T$-$G$-$G$-$T$-$G$-$A$-$A$-$T$-$

Palindromic sequence

---

tion of DNA in solution. It has major and minor grooves of similar depth, and its base pairs are centrally arranged around the helix axis with almost no tilt or inclination (▶ Figure 7–18a). In oriented fibers of native DNA and crystals of synthetic DNA, the B form has an average of 10 base pairs per turn. In solution, however, its helical repeat appears to be slightly larger, on the order of 10.4 to 10.6 base pairs per turn.

The A form is thicker and more compact than the B form. It has 11 base pairs per turn, and they are tilted with respect to the helix axis (Figure 7–18b). The A conformation occurs in DNA only under relatively dehydrated conditions, so it is doubtful that any significant sections of A DNA exist in living cells. The A form is biologically important because it (or a very similar form) occurs in the double-stranded helical regions of RNA. Model building reveals that it is im-

■ **TABLE 7–2** A comparison of the structural properties of A, B, and Z DNA.

	Helix Type		
	A	B	Z
Base pairs (bp)/turn	11	10	12
Vertical rise/base pair	0.23 nm	0.34 nm	0.38 nm
Helix diameter	2.3 nm	1.9 nm	1.8 nm
Rotation/base pair	+33°	+36°	−30°
Helix axis location	major groove	through base pairs	minor groove
Tilt of base pairs	+19°	−1.2°	−9°

Adapted from: R. E. Dickerson, H. R. Drew, B. N. Conner, R. M. Wing, A. V. Fratini, and M. L. Kopka, "The anatomy of A−, B−, and Z−DNA," *Science* 216 (1982): 475–485. Copyright 1982 by the AAAS. Used with permission.

possible to accommodate the 2′-hydroxyl groups of the ribose residues of RNA in the Watson-Crick B conformation.

Both the A form of DNA and RNA and the B form of DNA are right-handed helices. In contrast, the Z form has a left-handed helix (Figure 7–18c). The Z helix is thinner and more extended than the Watson-Crick structure and has 12 base pairs per turn rather than 10. It gets its name from the zigzag path taken by the phosphate-sugar backbone of the Z-DNA chains about the periphery of the helix. In many respects, the structure of Z DNA is the inverse of A DNA. The Z form is tall and thin and has a deep minor groove and a flattened (essentially nonexistent) major groove; in contrast, the A form is short and compact, with a deep major groove and very shallow minor groove. These differences in structure lead to differences in reactivity to other molecules.

Some geneticists currently believe that all segments of DNA are in an equilibrium between the right-handed B helix and the left-handed Z helix. Under most physiological conditions, the Z conformation is much less stable, so that B DNA is greatly favored in the equilibrium between the B and Z forms. In B DNA, all the nucleotides are in the so-called anti conformation. To form Z DNA, the base pairs must flip upside down relative to their position in the B helix (▶ Figure 7–19a). To accomplish this transformation in an alternating purine-pyrimidine sequence, the purines rotate 180° about their glycosyl bond into the syn conformation (Figure 7–19b). Both the base and the sugar of the pyrimidine rotate, so they retain their anti conformation. This rotation of the sugar gives rise to the zigzag course of each phosphodiester chain and brings the negatively charged phosphate groups of opposite chains closer to one another,

▶ **FIGURE 7–18** Space-filling models of (a) the right-handed helix of B form DNA, (b) the right-handed helix of A form DNA, and (c) the left-handed helix of Z form DNA.

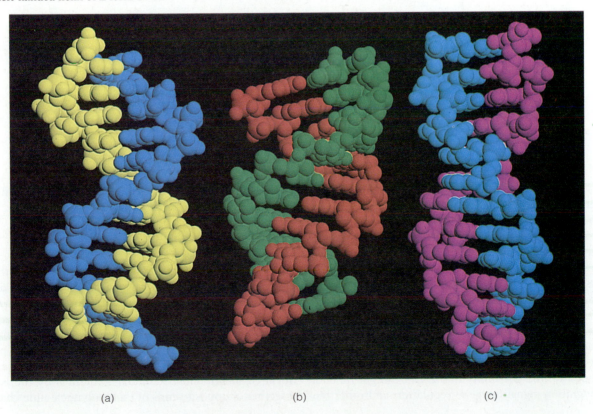

(a) (b) (c)

► **FIGURE 7–19**

(a) Conversion of a segment of B DNA to Z DNA by rotation of the bases. (b) Conformation of deoxyguanosine in B DNA and Z DNA. *Source:* Geoffrey Zubay, *Biochemistry*, 3d ed. (Wm. C. Brown, 1993), p. 704. Copyright © 1993 Wm. C. Brown Communications, Inc., Dubuque, IA. All Rights Reserved. Reprinted by permission.

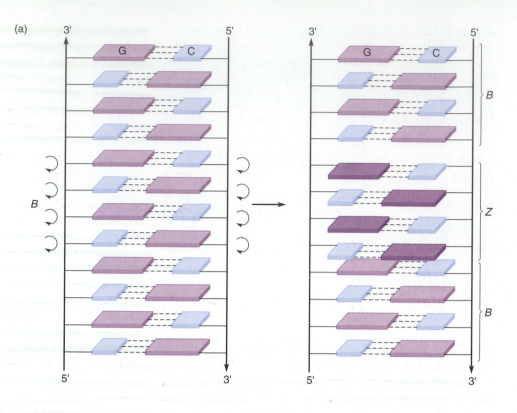

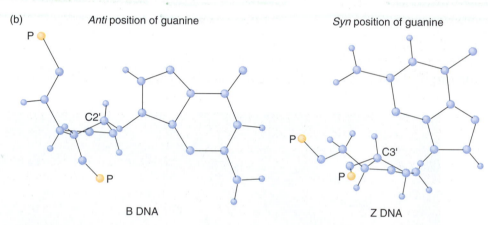

(b)   *Anti* position of guanine   *Syn* position of guanine

B DNA   Z DNA

thereby contributing to the lowered stability of the Z form by the enhanced electrostatic repulsion.

Several chemical factors are known to influence the transition from the B to Z form, including the effects of salts (e.g., high salt concentrations tend to favor the Z form) and various covalent modifications (e.g., methylation of the C5 position of cytosine helps to stabilize the Z form). The B-to-Z transition also tends to be favored in segments of DNA with alternating purine-pyrimidine sequences, with GpC sequences having the greatest effect. Several attempts have been made to identify proteins that bind to Z DNA and stabilize its conformation. The occurrence of such proteins would open up the possibility that Z DNA might function as a recognition signal for gene regulation and other bio-logical activities, but at this time the biological significance of Z DNA, if any, remains unknown.

Supercoiled DNA

The term **supercoiling** (or **superhelicity**) refers to the coiling of the double helix to form a higher-level spiral structure. The supercoiled or superhelical conformation is a common form of tertiary structure in DNAs that are topologically constrained so that their ends are not free to rotate (► Figure 7–20). For example, supercoiling is typical of DNA molecules in bacteria and many viruses; these DNAs form covalently closed circular structures by the bonding together of opposite ends of each polynucleotide chain. Su-

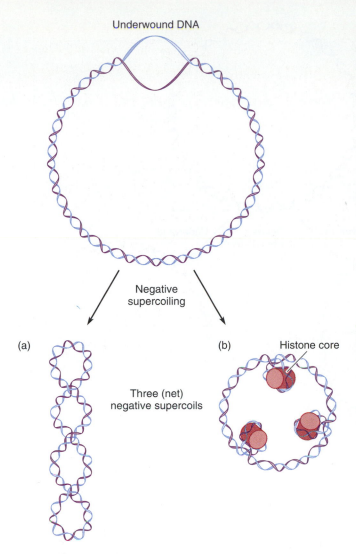

Underwound DNA

Negative supercoiling

(a)

Three (net) negative supercoils

(b)   Histone core

▶ **FIGURE 7–20** Supercoiled tertiary structures for topologically constrained DNA molecules. (a) Most circular DNA molecules in bacteria undergo negative supercoiling characterized by the presence of right-handed interwound coils. (b) Negative supercoils in eukaryotic DNA are introduced during chromatin assembly when DNA is wound around chromosomal (histone) proteins. The supercoils are introduced by enzymes (topoisomerases) by an underwinding of the DNA.

percoils are also present in the DNA of higher organisms, but in this case, supercoiling is induced by the winding of DNA around certain chromosomal proteins (see Chapter 9).

Supercoils are formed in DNA whenever the double helix is *underwound,* with fewer helical turns than if it were completely relaxed and under no torsional strain, or *overwound,* with too many turns. Possibly the simplest way to visualize the phenomenon of supercoiling is to consider a circular duplex molecule of DNA in which one of the strands is broken (nicked) at some point in the circle (▶ Fig-

ure 7–21). With the nicked strand free to rotate about the intact strand, the double helix will then adopt the least stressful state: the *relaxed* state, which is a B conformation with an average of 10.5 bp per turn. Now let us suppose that one of the free ends in the relaxed DNA is rotated about its complementary strand by 360° and subsequently joined to the other free end to again form a covalently closed circle. Since there is no longer a free end, the complementary strands cannot easily adjust to form hydrogen bonds in the conventional Watson-Crick manner. The only way the strands can reunite in the standard double-helical conformation is if the molecule twists as a whole and forms a supercoil (twisted circle), taking on the appearance of a figure eight.

The degree of supercoiling in the circular DNA molecule depends on the number of times the free end is rotated (one supercoil is formed for each 360° rotation), while the direction of supercoiling depends on which way the free end is rotated. If the free end is rotated in the same (right-handed) direction as the double helix is wound, the DNA circle becomes overwound compared to the relaxed state and reacts by forming a positive (left-handed) supercoil (Figure 7–21a). If the free end is rotated in the opposite (left-handed) direction, the DNA circle becomes underwound and forms a negative (right-handed) supercoil (Figure 7–21b). Most cellular DNA is underwound and exists in a negatively supercoiled state. Moreover, the degree of supercoiling in different DNA molecules is often similar. Typically circular DNA molecules isolated from bacterial cells have a supercoil density of about one negative superhelical turn for every 200 base pairs.

The extent of supercoiling in circular DNA can be determined by experimental means. The most direct approach is to observe the superhelical turns by electron microscopy (see ▶ Figure 7–22). Supercoiling also leads to changes in conformation that can be detected by gel electrophoresis. Because of their compact shape, supercoiled DNA circles have greater electrophoretic mobilities than non-supercoiled molecules of the same molecular weight. This shape-induced change in electrophoretic mobility is sufficient to distinguish DNA molecules that differ in only one superhelical turn.

In general, supercoiling is associated with a higher energy state, since a supercoiled molecule is under torsional stress. Processes that help to reduce the level of supercoiling are therefore energetically favored. One such process in negatively supercoiled DNA is the local unwinding of the double helix. Negatively supercoiled DNA is underwound in comparison to its relaxed counterpart and has a greater tendency to unwind. Negative supercoiling therefore facilitates processes that require the separation of DNA strands, such as replication, recombination, and RNA synthesis; it may also help to promote the transition of B DNA into Z DNA. The energy stored in superhelical turns may thus provide an important driving force for influencing the structure and function of the double helix.

Supercoiling of a circular DNA
molecule. (a) Overwinding
produces a left-handed
(positive) superhelix.
(b) Underwinding produces
a right-handed (negative)
superhelix.

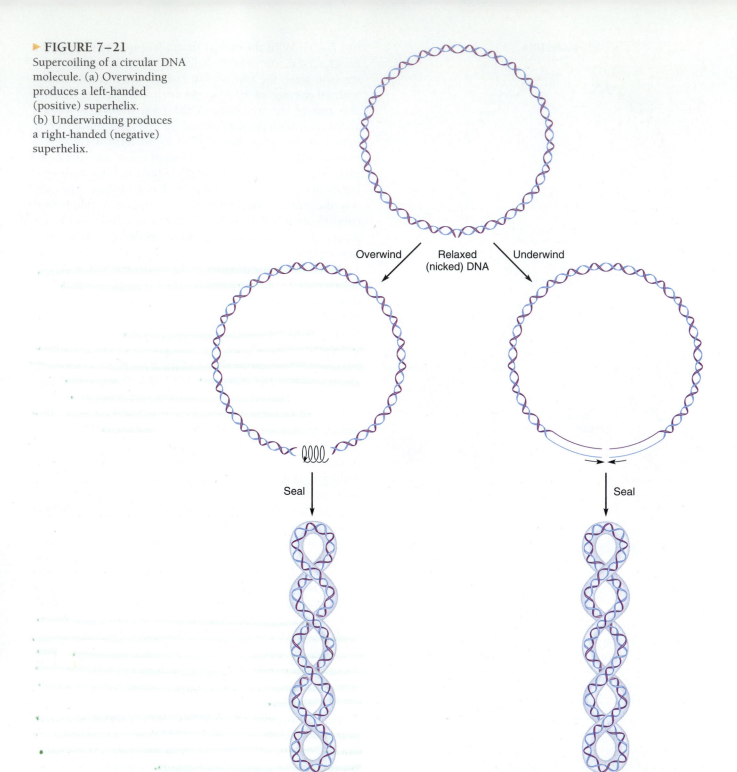

(a) Left-handed (positive) superhelix   (b) Right-handed (negative) superhelix

The amount of negative supercoiling in DNA can be altered by the action of enzymes known as **topoisomerases.**
There are two main types of topoisomerases, designated as
types I and II on the basis of whether they change the
number of superhelical turns in steps of one or steps of two

(► Figure 7–23). Both types can convert DNA into the relaxed state by catalyzing the removal of supercoils through
the breaking and rejoining of phosphodiester bonds. Type I
enzymes catalyze transient single-strand breaks in the DNA
duplex, while type II enzymes catalyze transient double-

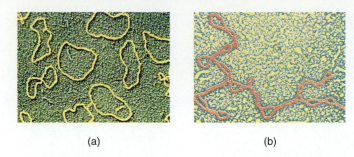

(a)   (b)

▶ FIGURE 7–22 Electron micrographs showing (a) relaxed circular DNA and (b) supercoiled circular DNA.

strand breaks in DNA. In neither case are the broken ends of a DNA strand free to rotate—they are always attached to the surface of the enzyme. All the bonds that are broken at any step of the uncoiling process are reformed once the step is complete. This enables the enzyme to reduce the level of supercoiling in a DNA molecule through a series of discrete cutting-unwinding-rejoining cycles.

Negative supercoiling in DNA can also be generated by enzyme catalysis. One special kind of topoisomerase II, called **DNA gyrase**, is able to introduce negative supercoils into DNA using the energy supplied by the breakdown of ATP. DNA gyrase is produced only in bacteria, and as we shall see in Chapter 8, it plays an important role in the replication of the bacterial chromosome.

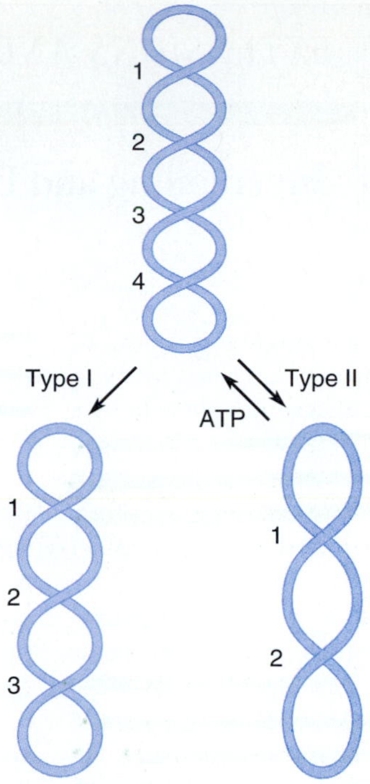

▶ FIGURE 7–23 Changes in the number of superhelical turns caused by type I and type II topoisomerases, which relax supercoiled DNA in steps of one and two, respectively. The type II topoisomerase, called DNA gyrase, can also introduce negative supercoils using the energy supplied by ATP cleavage. *Source:* Geoffrey Zubay, *Biochemistry,* 3d ed. (Wm. C. Brown, 1993). Used with permission.

To Sum Up

1. Most RNA molecules exist as single polynucleotide chains. In contrast, most DNA molecules take the form of a double helix consisting of two polynucleotide chains wound around each other in a spiral configuration; the two chains are antiparallel, that is, their $5' \rightarrow 3'$ polarity is reversed.

2. The chemical features of the purine and pyrimidine bases are critical in determining how the polynucleotide chains fold into helical secondary structures. The two chains of a DNA double helix are held together by hydrogen bonds between bases opposite each other on the complementary strands. Normally, only adenine can form hydrogen bonds with thymine, and only guanine can bond with cytosine. These restrictions on base pairing are the basis for Chargaff's rules, which state that the amount of adenine equals the amount of thymine and the amount of guanine equals the amount of cytosine. The two chains of a DNA duplex are thus complementary to one another.

3. The complementary nature of the two strands of a DNA molecule enables DNA to meet two of the requirements of the genetic material—information transfer and accurate duplication of information during cell division. Information is transferred by messenger RNA molecules (that are the complements of the DNA sequences) to locations in the cell where it is decoded into proteins. Duplication of DNA is also based on complementarity, with new chains of DNA being synthesized as complements of the strands of a DNA double helix.

4. A double-helical secondary structure can also form in single-stranded DNA and RNA regions that contain complementary sequences. The polynucleotide chain folds back upon itself to form a hairpin structure composed of a double-stranded duplex stem with a single-stranded loop at one end.

5. The Watson-Crick double-helical structure is known as B form DNA. The B form describes a family of duplex structures whose exact parameters vary depending on factors such as solvent conditions and the local nucleotide sequence. Although the B form is the predominant secondary structure of DNA, at least two other conformations have also been identified. Helical regions of RNA take on the A form, which accommodates the 2'-hydroxyl group of the ribose sugars. In contrast to the B and A forms, the Z form is a left-handed duplex found in certain localized DNA regions under particular in vitro conditions. Its function, if any, remains unknown.

6. Duplex DNA often coils into a higher-level spiral tertiary structure, a phenomenon known as supercoiling. Negative supercoiling occurs when the duplex helix is underwound relative to the relaxed state. It is thought that the higher energy level of the negatively supercoiled state helps to drive processes that require strand separation, such as DNA replication. Certain enzymes can either reduce or increase the number of superhelical turns in localized regions of DNA.

Supercoiling and DNA Topology

Supercoiling in circular DNA can be mathematically described in terms of a topological property known as the **linking number**, L, which is defined as the number of times one strand of the double helix encircles the other. For a closed circle, the linking number is necessarily an integer and can be changed only by physically opening one of the strands. Many experiments measure the **linking difference** $\Delta L = L - L_0$, where L_0 is the linking number of DNA in the relaxed state (L_0 equals the number of Watson-Crick helical turns in an unconstrained DNA molecule). The linking difference can be partitioned between changes in two variable quantities, the **twist** (T) and **writhe** (W), by the following equation:

$$\Delta L = \Delta T + \Delta W$$

In this equation, ΔT measures the change in the number of turns in the Watson-Crick double helix and ΔW measures the change in the number of turns of the superhelix. In DNAs from natural sources, which are normally underwound, the value of ΔW gives the change in the number of negative supercoils. A decrease in W (i.e., a negative ΔW) corresponds to an incorporation of negative supercoils into the DNA molecule, while an increase in W (i.e., a positive ΔW) corresponds to a loss of negative supercoils. For a relaxed DNA circle lying flat on a plane, $W = 0$, since there are no supercoils, and the linking number then equals the number of twists.

Since $\Delta W = \Delta L - \Delta T$, we see that there are two ways to change the number of supercoils (see the accompanying illustration). One approach is to change the linking number. DNA molecules that differ in linking number are **topoisomers**. Cells therefore change L by the action of *topoisomerases*—enzymes that interconvert topoisomers by introducing transient nicks into one or both strands of DNA. When the salt concentration and temperature are normal, the change in linking number is accompanied by a change in the amount of supercoiling without a change in twist. Under these conditions, the tendency of DNA to remain in the standard B conformation is so great that the molecule bends about its axis so as to minimize the amount by which the helical parameters (pitch and rise per residue) depart from the relaxed state.

A second approach to changing the amount of supercoiling is to change the twist of the DNA molecule. This approach is accomplished in the laboratory by adding ethidium bromide, a dye with a planar structure, to a solution of DNA circles. Ethidium bromide

---

PHYSICOCHEMICAL PROPERTIES OF THE DOUBLE HELIX

Many properties of a nucleic acid molecule result from its double-helical structure. Some of these properties have proven to be extremely useful in genetic research and will be the focus of discussion in the following sections.

Denaturation and Renaturation

The hydrogen bonds and base-stacking interactions that maintain the double helix are weak enough to be broken under conditions that leave the covalent bonds of the polynucleotide chains intact. Therefore, like proteins, nucleic acids tend to denature by unwinding into near-random, disordered coils when they are heated or exposed to chemicals such as urea or extremes of pH. In nucleic acids with highly ordered secondary structure, such as double-stranded DNA, denaturation usually results in an abrupt loss of native conformation over a relatively narrow range of denaturing conditions. For instance, when a solution of DNA is heated to a temperature of about 80° to 100°C, sufficient thermal energy is supplied to disrupt the forces holding the complementary chains together. The double helix then unwinds, allowing the intact strands to separate. This helix-coil transition often occurs over a temperature range of only 15°C. Since the disruption of the ordered helical structure occurs over such a limited temperature interval, the thermal denaturation of DNA is often referred to as **melting**, and the temperature at 50% denaturation is called the **melting temperature** (T_m).

Several changes in the physical properties of double-stranded nucleic acids can be used to measure the extent of denaturation. One change that accompanies denaturation is an increase in absorption of ultraviolet light. This increase in absorbancy, called the **hyperchromic shift**, is measured spectrophotometrically at a wavelength of 260 nm, where the absorbance of UV light is at a maximum. At 260 nm na-

binds to DNA by intercalating between the base pairs and thereby causes a local unwinding of the double helix. Increasing the concentration of ethidium bromide gradually decreases the amount of negative supercoiling, not by increasing the linking number L (which can be changed only by creating nicks in the DNA strands), but by reducing the twisting number T. Eventually a concentration of ethidium bromide will be reached where $T = L$ and thus $W = 0$, and the DNA circle will be completely free of superhelical turns.

Two approaches to changing the degree of supercoiling. (a) Change the linking number (L) by the action of topoisomerase enzymes. (b) Change the twisting number (T) by the addition of an intercalating agent such as ethidium bromide.

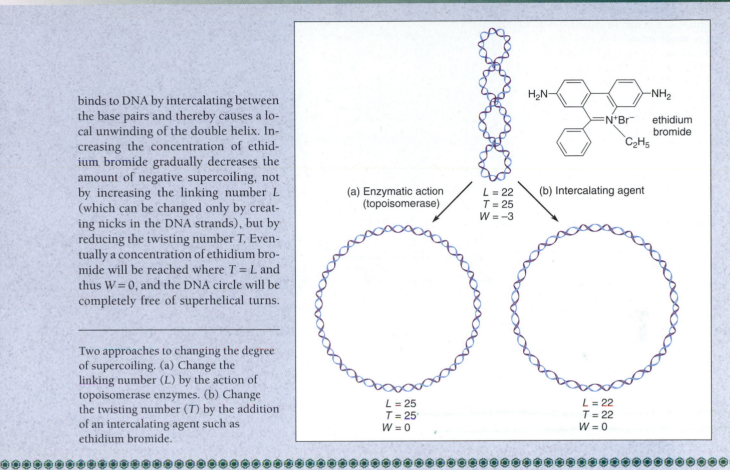

(a) Enzymatic action (topoisomerase)

$L = 22$
$T = 25$
$W = -3$

(b) Intercalating agent

ethidium bromide

$L = 25$
$T = 25$
$W = 0$

$L = 22$
$T = 22$
$W = 0$

tive DNA has a lower absorbance than denatured DNA because of the effect of the stacked bases. Much of this stacking effect is lost, however, with the disruption of the double helix, resulting in an increase in absorbancy of the bases in the separated chains. ▶ Figure 7–24 shows an example of the hyperchromic shift for a typical denaturation (melting) curve for DNA. Note that the melting temperature marks the midpoint of the transition from double-stranded to single-stranded DNA, that is, the point where the rise in absorbancy is half complete.

One interesting property of the melting temperature is its linear dependence on the GC content of the nucleic acid. Studies show that the relationship between T_m and the GC content of DNA (in a 0.15 M NaCl and 0.015 M sodium citrate solution at pH 7) is expressed by the equation

$$T_m = 69.3 + 0.41(\%\text{GC}) \qquad (7.1)$$

This empirical relationship results from the increased stability provided by the three hydrogen bonds of GC base pairs (compared with only two hydrogen bonds in AT pairs). It simply requires more thermal energy to break three hydrogen bonds than to break two.

Denaturation of double-stranded nucleic acids is a reversible process under certain conditions. If a solution of denatured DNA is held at about 20° to 25°C below T_m, complementary strands of DNA will rewind or renature upon collision, forming an intact double helix. To obtain maximum renaturation, the temperature must be kept sufficiently high to break hydrogen bonds that may have formed between complementary bases in the same chain but not high enough to interfere with the association of separate complementary chains.

The renaturation or **annealing** process is not restricted to nucleic acids of the same kind or from the same source but can occur between different nucleic acids as long as their chains exhibit a sufficient degree of complementarity. For example, double helices can be formed by the combination of complementary DNA and RNA chains, resulting in

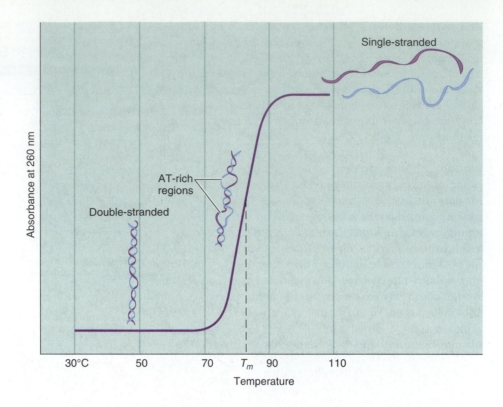

► **FIGURE 7–24** A typical denaturation (melting) curve for DNA, showing the sharp rise in absorbance at 260 nm that accompanies denaturation. During the conversion of double-stranded DNA to single-stranded DNA, regions rich in AT base pairs tend to denature first. The midpoint of the helix-coil transition corresponds to the melting temperature.

(Graph labels: y-axis "Absorbance at 260 nm"; x-axis "Temperature" with marks 30°C, 50, 70, T_m, 90, 110; curve labels "Double-stranded", "AT-rich regions", "Single-stranded")

DNA:RNA hybrid molecules. Such hybrids have a double-helical structure that is a mixture of the A and B forms. Denatured DNA chains from different strains or different but related species can also join to form DNA:DNA hybrids; however, sequence differences in such hybrid molecules give rise to single-stranded regions that cannot hydrogen bond because of mismatched nucleotides. Mismatched bases in a hybrid molecule lower its melting temperature, providing an experimental means of quantifying the extent of mismatching between the two nucleic acid chains. Nucleic acid hybridization has been extremely useful in genetic research; it is a crucial part of the current techniques of genetics and evolutionary biology that are described in later chapters.

Buoyant Density

Another property of the double helix that depends on GC content is buoyant density. In general, double helices with a high percentage of GC base pairs have a greater mass-to-volume ratio (density) than those with a lower percentage of GC pairs. For DNA in a CsCl solution, the buoyant density, ρ, varies according to the following empirical relationship:

$$\rho = 1.660 + 0.00098(\%GC) \qquad (7.2)$$

where the density is given in units of g/cm³. For example, *E. coli* DNA has a 51% GC content and a density of 1.710 g/cm³.

The buoyant density of a double helix is measured by a technique known as **equilibrium density-gradient centrifugation**. In this ultracentrifugal technique, the nucleic acid is centrifuged at high speeds (40,000–60,000 rpm) for several hours in a concentrated solution of a "heavy" salt, such as CsCl, which has a density close to that of DNA. When spun at such high speeds, the Cs⁺ and Cl⁻ ions tend to move toward the bottom of the centrifuge tube in response to the centrifugal force. Counterbalancing this migration is the tendency of these ions to move in the opposite direction through diffusion, as they tend to randomly disperse throughout the tube. The opposing processes of sedimentation and diffusion produce a stable concentration gradient of CsCl, and hence a **density gradient**, in which the lowest density is found at the top of the tube and the greatest density at the bottom, with a density equal to the density of the nucleic acid in between (► Figure 7–25).

A nucleic acid in such a density gradient is subjected to a centrifugal force that is proportional to the difference between the density ρ of the macromolecule and the density ρ_0 of the surrounding medium ($\rho - \rho_0$). The molecule therefore sediments in the direction of the centrifugal field only when its density is greater than that of the medium

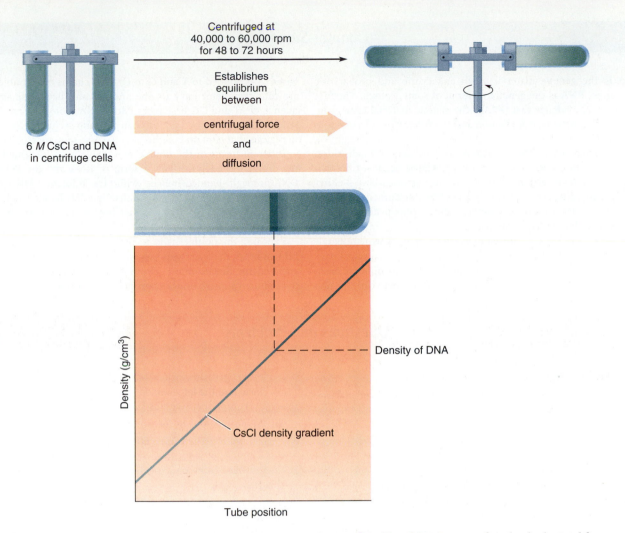

6 M CsCl and DNA in centrifuge cells

Centrifuged at 40,000 to 60,000 rpm for 48 to 72 hours

Establishes equilibrium between

centrifugal force

and

diffusion

Density (g/cm³)

Density of DNA

CsCl density gradient

Tube position

▶ **FIGURE 7–25** Equilibrium density-gradient centrifugation. A solution of CsC1 and DNA is centrifuged at high speed for several hours. The CsC1 forms a linear density gradient covering a range that includes the density of the DNA. The DNA moves to the position in the tube where its density is equal to that of the medium, forming a band at that location.

($\rho > \rho_0$); it floats upward against this field when $\rho < \rho_0$ and comes to rest at an equilibrium position when $\rho = \rho_0$. The nucleic acid molecules thus concentrate in a band centered at the point where their density is equal to the density of the medium (Figure 7–25). This concentrating tendency is opposed by diffusion, so that the macromolecules are distributed about their equilibrium position in a band whose width is inversely proportional to their molecular weight. Thus, large DNA molecules have a narrow band, whose position can be precisely determined by scanning the tube with ultraviolet light. Since the solution density can be measured throughout the tube, the buoyant density of the DNA can be established from the location of the band (which appears dark due to UV absorption) in the density gradient.

To Sum Up

1. Exposing DNA to high temperatures causes denaturation or unfolding of the double helix. The input of excessive thermal energy disrupts the hydrogen bonds between the paired bases, allowing the two strands of a double helix to separate from each other. Renaturation (rewinding) will occur spontaneously when temperatures are reduced below T_m. Hybridization of DNA strands from different sources or DNA and RNA strands can also occur if there is sufficient complementarity between the chains.

2. Both the melting (denaturation) temperature and the buoyant density of DNA depend on the GC content of the molecule and can therefore be used to determine the base composition of the DNA being investigated. The buoyant density of DNA can be measured by a technique known as equilibrium density-gradient centrifugation.

DNA is the genetic material of all eukaryotic and prokaryotic organisms; RNA is the genetic material of some viruses. The first unequivocal evidence that DNA is the genetic material came from studies on genetic transformation and the replication of viruses in bacteria.

The nucleic acids DNA and RNA are polymers of nucleotides. Each nucleotide consists of a phosphorylated sugar—ribose in RNA and deoxyribose in DNA—and a nitrogenous base—either adenine, guanine, cytosine, thymine (in DNA), or uracil (in RNA). The nucleotides are linked together by phosphodiester bonds to form long polynucleotide chains.

Most DNA molecules consist of a double helix of two antiparallel polynucleotide chains wound about each other in a spiral configuration. The two chains are held together by hydrophobic base-stacking interactions and by hydrogen bonds between specific

bases (A is bonded to T and G to C). The chains of a double helix are thus complementary to one another and have a specific sequence of AT and GC base pairs. Most RNA molecules are single-stranded, but they can fold into hairpin-loop secondary structures in regions that contain complementary base sequences.

Different forms of the double helix have been identified by X-ray diffraction analysis. The A form of RNA and the B form of DNA are right-handed helices, while the Z form of DNA (a less common variant of the double helix) is a left-handed helix. Duplex DNA can also coil into a higher-level spiral tertiary structure, a process known as supercoiling.

Exposing the double helix to high temperature or extremes of pH can break the hydrogen bonds, causing the strands to separate and undergo denaturation. The separated strands can be induced to renature (anneal) under appropriate conditions.

DNA as the Genetic Material

1. State whether each of the following statements is true or false. If it is false, explain why.
   (a) Genetic transformation refers to the transfer of DNA from a donor cell to a recipient cell by a bacterial virus.
   (b) The transformation of type R pneumococcus cells by an extract of dead S cells is not affected by DNase.
   (c) In the Hershey-Chase experiment, the discovery that most of the ^{35}S remains in the supernatant indicates that DNA is the genetic material.

2. You have two strains of a certain species of bacteria. One strain is penicillin-sensitive (pen^s), and the other is penicillin-resistant (pen^r). You would like to determine whether this particular species is capable of undergoing genetic transformation. Describe an experiment involving the pen^s and pen^r strains that could be used to test for transformation.

3. Explain why Chargaff's equivalence rules provided evidence against the tetranucleotide hypothesis.

Covalent Structure of Nucleic Acids

4. Fill in the blanks:
   (a) The three bases common to both DNA and RNA are _____ , _____ , and _____ .
   (b) Both _____ in DNA and _____ in RNA form hydrogen bonds with adenine.
   (c) The sequences of the two strands of a DNA double helix are _____ .
   (d) Bases across from each other on opposite DNA strands are held together by _____ bonds, while adjacent nucleotides along a DNA strand are connected by _____ bonds.
   (e) The specific carbon atoms that are connected by phosphodiester bonds between adjacent sugars are the _____ and _____ atoms.
   (f) The _____ form of DNA is its predominant structure; however, certain nucleotide sequences can coil into

other structures such as the right-handed _____ form and the left-handed _____ form.

5. A double-stranded DNA molecule contains 30% adenine. What are the proportions of the other three nucleotides?

6. One strand of a DNA molecule contains the nucleotide proportions 20% A, 30% C, 40% G, and 10% T. What proportions of the four base pairs are expected in the double-stranded form of this DNA?

7. A single-stranded DNA virus has an A:T base ratio of 2, a G:C ratio of $\frac{1}{2}$, and a (A + T):(G + C) ratio of $\frac{1}{4}$. (a) What is the (A + G):(T + C) ratio of this molecule? (b) If this single-stranded DNA forms a complementary strand, what are these four ratios in the complementary strand? (c) What are these four ratios in the resulting DNA duplex (the original and complementary strands together)?

8. Explain why the (A + G):(T + C) ratio in double-stranded DNA must equal 1, when the (A + T):(G + C) ratio can take on any value, depending on the species from which the DNA is derived. If the (A + G):(T + C) ratio is 1, can you conclude that the DNA of that organism is double-stranded, or would you need more information?

9. A bacteriumlike microorganism is brought back from Mars and found to contain an interesting nuclease enzyme that cleaves individual DNA strands into 5'-dinucleotides (pApT, pGpA, etc.). You decide to use the nuclease to test whether the complementary DNA strands of this organism are antiparallel or parallel by comparing the relative frequencies of the dinucleotides produced upon DNA digestion. (a) From your knowledge of DNA structure, predict which of the following equalities will exist among the resulting dinucleotide frequencies if the strands of this organism are antiparallel (assume that the (A + T):(G + C) ratio does not equal one in this case):

   (1) pApA = pTpT   (4) pCpG = pApT
   (2) pCpA = pTpG   (5) pApG = pTpC
   (3) pGpT = pCpA   (6) pCpT = pApG

   (b) Which of the above equalities will exist if the DNA of this organism has a random sequence of base pairs?

Three-Dimensional Structure of Nucleic Acids

10. The helical structures of three different nucleic acid molecules (call them P, Q, and R) gave X-ray diffraction patterns that revealed that P had a helix diameter of 2 nm, Q was shorter and more compact with a helix diameter of 2.3 nm, and R was tall and thin with a helix diameter of 1.8 nm. Upon seeing these patterns, a geneticist concluded that one of these molecules had the A form of secondary structure, another had the B form, and the third had the Z form (not necessarily in the order given). (a) Which of these molecules has a left-handed helix? (b) One of these molecules happens to be RNA. Which one?

11. How many twists (complete turns) are there in a DNA molecule that consists of 2,000,000 base pairs?

12. The chromosome of a particular bacteriophage is a linear duplex of DNA with a molecular weight of 6.93×10^6 daltons. For the following questions, assume that the molecule is in the B form with 10.5 bp per turn and that each nucleotide pair has an average molecular weight of 660 daltons. (a) Calculate the number of helical turns in the molecule. (b) Calculate the length (in nm) of the molecule. (c) If the molecule contains 20% G, what is the (A + T):(G + C) ratio of the DNA?

13. A DNA molecule that is isolated from a particular virus is 130 μm in length. (a) How many base pairs are contained in this molecule? (b) What would be its molecular weight?

*14. When the virus described in question 12 infects its host cell, the ends of the viral DNA are covalently joined, forming a supercoiled circular molecule. If electron micrographs reveal that the circular viral DNA has 50 supercoils, what is the linking number of the molecule?

*15. The linking number and twist of a negatively supercoiled DNA have the values $L = 26$ and $T = 28$. (a) What is the value of W? (b) What would be the values of L, T, and W for this DNA after one catalytic cycle with topoisomerase I? with topoisomerase II? with DNA gyrase and ATP?

Physicochemical Properties of the Double Helix

16. (a) Using equation 7.1, calculate the base composition of DNA that has a melting temperature of 86.5°C under standard conditions. (b) Using equation 7.2, calculate the buoyant density of this DNA in a CsCl density gradient.

17. The base compositions of nucleic acids isolated from four different sources are given below:

   (1) 30% A, 20% C, 20% G, 30% T
   (2) 40% A, 10% C, 40% G, 10% T
   (3) 30% A, 30% C, 20% G, 20% T
   (4) 20% A, 30% C, 30% G, 20% T

(a) In each case, identify the nucleic acid as DNA or RNA. (b) When the intact forms of these nucleic acids were heated to 100°C, only two of the four exhibited a hyperchromic shift. Identify these two in terms of their base compositions. (c) Which of the two forms identified in (b) will have the higher melting temperature? Why?

18. If heavy DNA (labeled with ^{15}N) that was isolated from phage λ is mixed with light λ DNA and subjected to equilibrium density-gradient centrifugation, two bands appear in the centrifuge tube. In contrast, if the DNA mixture is heated to 100°C and then slowly cooled and held at 65° prior to centrifugation, three bands appear in the centrifuge tube. Account for the difference in the results of centrifugation between the two experiments. What would be the position of the third band (formed by the heated and slowly cooled material) relative to the other two bands?

19. If the experiment in problem 18 is repeated using a mixture of DNA from heavy λ and light T2 bacteriophages, only two bands appear following the heating and slow cooling. Furthermore, these bands are found at the same positions as the bands formed by the unheated mixture. How would you account for the absence of a third band?

20. The DNA of a certain bacteriophage is broken into short double-stranded fragments and is then subjected to equilibrium density-gradient centrifugation. The fragments form one band at a density position that is identical to the band formed by unfragmented DNA. What does this tell you about the base pair composition along the DNA molecule of this organism?

21. The DNA of a newly discovered virus fails to yield an increase in absorption of ultraviolet light when gradually heated to temperatures over 110°C. Postulate an explanation for this result.

22. The DNA molecules of two different double-stranded DNA viruses are isolated. These molecules are found to have the same (A + T):(G + C) ratio but different lengths. In which of the following properties will the molecules differ? Explain your answer. (a) melting temperature, (b) molecular weight, (c) electrophoretic mobility (in an agarose gel), (d) buoyant density, (e) primary structure.

23. The following is part of a DNA molecule that was isolated from a given virus:

   5' CGTCATCGATGATGCAGCTC 3'
   3' GCAGTAGCTACTACGTCGAG 5'

(a) The above DNA contains a six-base-pair palindromic sequence that serves as a recognition sequence for an enzyme that cleaves double-stranded DNA at these sites. What is the recognition sequence for this system? (b) The DNA of this virus has 50% GC base pairs. Assuming that the base pairs are dispersed at random within the viral DNA, what is the average length of the fragments produced upon digestion of the DNA with this enzyme (i.e., what is the average spacing between adjacent recognition sites)?

*An asterisk indicates that the question or problem is based on information in the Extensions and Techniques text.

STRUCTURE AND REPLICATION OF CHROMOSOMES

Organisms have a packaging problem. The DNA chains that carry the genes are extremely large polymeric molecules; they are much too long to fit neatly into a cell or virus when fully extended or even when coiled into the Watson-Crick double helix. These molecules must therefore take on a higher level of structural organization to make themselves more compact. This higher-order compaction of the genetic material occurs in all organisms to varying degrees and is the basis of the structure of chromosomes.

To attain its compact in vivo state, each DNA molecule must undergo different levels of folding and supercoiling. This systematic condensation or **packaging** of DNA typically involves association with specific proteins. From a purely structural standpoint, the act of binding to specific proteins helps to stabilize DNA in a more tightly folded conformation than it would ordinarily attain on its own. Many of the chromosomal proteins are positively charged at physiological pH and can therefore neutralize the strong negative charges on the nucleic acid molecules. These proteins thus help to reduce the charge repulsion between different parts of a DNA chain when it is compressed into a small space.

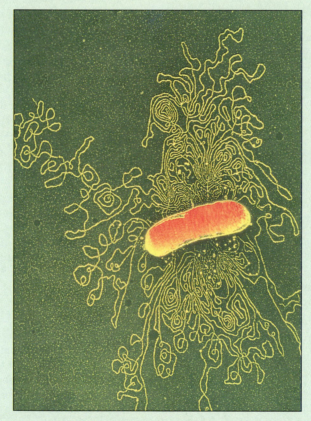

E. coli DNA released by gentle lysis of the cell.

In this part, we will examine the mechanisms of DNA packaging and replication in two fundamentally different types of organisms: prokaryotes and eukaryotes. Prokaryotes (organisms without a cell nucleus) include the bacteria and blue-green algae (cyanobacteria). They possess the simplest and smallest cells and are thought to have the most ancient evolutionary origins. In contrast, eukaryotes (organisms with cell nuclei) encompass a much broader variety of living systems, including plants and animals and both unicellular and multicellular forms. As we will see, there are basic differences in the ways that prokaryotes and eukaryotes package and replicate their DNA, with eukaryotes generally possessing larger and structurally more complex chromosomes, a greater amount of genetic information, and a more elaborate mechanism for DNA replication than prokaryotes.

In addition to looking at cellular organisms, we will also consider how intracellular genetic elements and viruses solve the problem of packaging and replication. Viruses are neither prokaryotic nor eukaryotic, but have a subcellular organization and rely on living cells in order to reproduce.

CHAPTER 8

Bacteria and Bacterial DNA

Bacteria (prokaryotes) are typically small, unicellular organisms, with dimensions that often border on the resolution of an ordinary light microscope. Bacteria have rapid reproductive rates and well-defined growth requirements, and these properties make them useful as experimental organisms. Their small size is also an advantage, since it makes them easy to handle and permits an investigator to work with large numbers of individuals. Because of these properties, bacteria have long been popular experimental subjects in genetics, and many of the concepts presented in this book were derived from studies of these organisms. The characteristics of some bacteria that have been especially popular as experimental genetic systems are given in ■ Table 8–1.

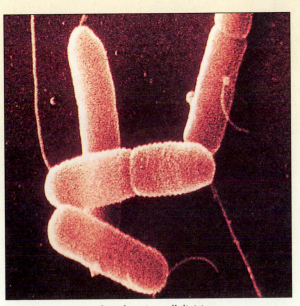

E. coli undergoing cell division.

THE STRUCTURE AND GROWTH OF BACTERIA

Many of the important structural features of prokaryotic cells are evident in the structure of the common intestinal bacterium, *Escherichia coli*—the most extensively studied prokaryote in genetics. (see ▶ Figure 8–1). An *E. coli* cell is a small rod-shaped structure with a diameter of 0.75 μm and a length that can vary between 1 and 3 μm, depending on the growth conditions and the stage of cell division. An average *E. coli* weighs approximately 2×10^{-12} g, of which 15% is protein, 6% is RNA, and 1% is DNA.

The DNA containing the complete set of genes or genome of *E. coli* is located in the nucleoid area of the cytoplasm. Unlike the nucleus of a eukaryotic cell, the bacterial nucleoid usually remains in a uniform state of compactness throughout the cell cycle and is not separated from the rest of the cell by a surrounding membrane.

In addition to the nucleoid, the cytoplasm of *E. coli* contains a number of smaller particulate elements; the most conspicuous of these are the ribosomes, which function as cellular sites of protein synthesis. The ten thousand or more ribosomes in *E. coli* are roughly spherical structures, each 20 nm thick, consisting of about one-third protein and two-thirds RNA. Also dispersed throughout the interior of the cell are numerous granules that contain stored nutrients such as starch and lipids, and various enzymes and smaller molecules that are dissolved in the cytosol, the aqueous phase of the cytoplasm.

The nucleoid and other cytoplasmic constituents of a bacterium are enclosed in a **cell envelope**. This structure protects the cell against osmotic rupture in dilute media and helps to regulate the passage of molecules into and out of the cell. (As we will see in later chapters, genetic engineering techniques must overcome the barrier imposed by the cell envelope to the uptake of foreign DNA.) The cell envelope is the principal structural basis for the classification of bacteria into two main groups: the **Gram-negative** and **Gram-positive** bacteria, which are named for their different responses to the Gram staining reaction. In *E. coli* and other Gram-negative bacteria, the cell envelope consists of three

■ TABLE 8–1 Some bacteria commonly used in genetic research.

Organism	Morphology	Characteristics
Bacillus subtilis	Gram-positive rods; often occur in chains; forms endospores	free-living, aerobic bacterium that inhabits the soil
Escherichia coli	Gram-negative; nonsporulating rods	facultative anaerobic bacterium that normally inhabits the intestinal tract of humans and animals; rarely pathogenic
Salmonella typhimurium	Gram-negative; nonsporulating rods; similar to *E. coli*	causes diseases of the intestinal tract in animals; causes gastroenteritis ("food poisoning") in humans
Streptococcus pneumoniae	Gram-positive spheres; cells often occur in pairs	infects the respiratory tract of humans and other mammals, where it causes pneumonia

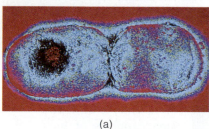

(a)

▶ **FIGURE 8–1** Structure of an *E. coli* cell shown as (a) an electron micrograph, and (b) with components labeled.

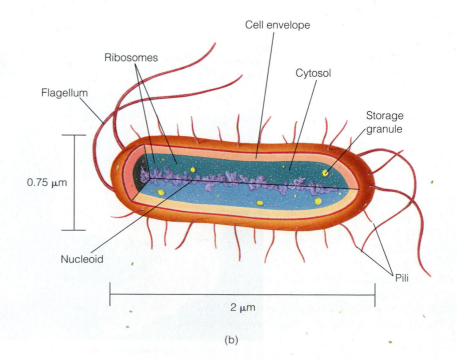

Cell envelope

Ribosomes

Cytosol

Flagellum

Storage granule

0.75 μm

Nucleoid

Pili

2 μm

(b)

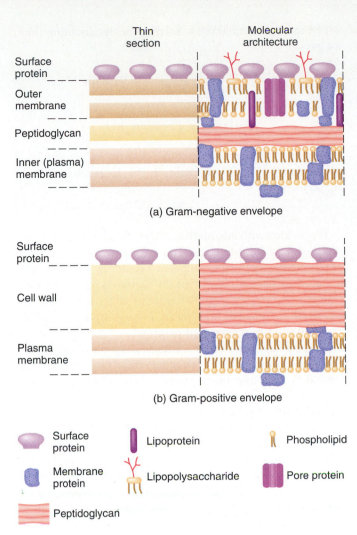

Surface protein
Outer membrane
Peptidoglycan
Inner (plasma) membrane

Thin section | Molecular architecture

(a) Gram-negative envelope

Surface protein
Cell wall
Plasma membrane

(b) Gram-positive envelope

Surface protein

Membrane protein

Peptidoglycan

Lipoprotein

Lipopolysaccharide

Phospholipid

Pore protein

▶ **FIGURE 8–2** Structure of the cell envelope of bacteria. Components of a typical envelope of (a) Gram-negative bacteria, (b) Gram-positive bacteria.

layers: a peptidoglycan layer sandwiched between inner (plasma) and outer lipoprotein membranes (▶ Figure 8–2a). The peptidoglycan layer consists of a network of polysaccharide chains cross-linked by short peptides to form a rigid cell wall that is separated from the inner and outer membranes by a space called the periplasm.

Gram-positive bacteria have a simpler cell envelope consisting of only two layers: a rather thick cell wall composed mainly of peptidoglycan and an inner plasma membrane (Figure 8–2b). Since Gram-positive bacteria lack an outer membrane, treatment of Gram-positive cells with the peptidoglycan-degrading enzyme lysozyme results in the complete digestion of the cell wall, converting the cells to osmotically sensitive spheres called **protoplasts**. Similar treatment of Gram-negative cells can also produce osmotically sensitive cells (now called **spheroplasts** because they retain remnants of the outer membrane), but only after damaging the outer membrane by some chemical or physical procedure to provide access of lysozyme to the peptidoglycan layer.

Bacterial Growth and Enumeration

For many genetic studies, bacteria are grown in the laboratory in a liquid medium. The growth medium often consists of some type of nutrient broth that contains many of the ingredients of cells in the form of amino acids, peptides, sugars, and so on, that have been extracted from meat or yeast. Such complex mixtures are generally cheaper to use than synthetic media and support a more rapid rate of growth, since the bacterial cells find many of the substances they require for growth ready-made in the broth. Despite the advantages of nutrient broth, many experiments require a medium of defined composition for selective or diagnostic purposes or simply to avoid complications resulting from one or more of the components in the nutrient media. If the synthetic growth medium contains no added organic compounds other than a carbon (energy) source such as a sugar, it is referred to as a **minimal medium**. For *E. coli*, a typical minimal medium contains the sugar glucose and the inorganic ions Fe^{3+}, K^+, Mg^{2+}, NH_4^+, HPO_4^{2-}, and SO_4^{2-}. For growth to occur in this medium, the bacterium must be capable of synthesizing all of its organic cellular components from the mixture of simple inorganic salts and glucose.

When bacterial cells are added to a liquid medium, the growth of the resulting cell population usually proceeds in three stages: the **lag phase**, a period of adjustment to the growth medium, the **log** (or **exponential**) **phase**, a period when cells are actively dividing, and the **stationary phase**, a period when the capacity to reproduce falls to zero and there is no further increase in the net number of cells. These phases are illustrated in the typical growth curve shown in ▶ Figure 8–3. Growth curves like Figure 8–3 are commonly used in the laboratory to determine the doubling time (generation time) of a culture during exponential growth; this information is useful when carrying out any experiments dealing with bacterial genetics. Doubling times determined in this way vary with the strain and species of bacteria, as well as with the temperature and type of growth medium. Typically the growth of bacterial cultures

▶ **FIGURE 8–3** Bacterial growth curve.

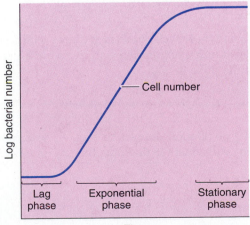

Log bacterial number

Cell number

Lag phase | Exponential phase | Stationary phase

Time

in a liquid medium is extremely fast. For example, *E. coli* growing under optimal conditions in nutrient broth can have a doubling time as short as 20 minutes. Progressing at this rate, it would take less than 7 hours for a single cell to produce a population of one million cells. The rapid growth of bacteria such as *E. coli* is one reason why transformed bacteria can be used as "factories" to produce large amounts of pure, cloned foreign DNA in a short time (see Chapter 18).

Example 8.1

The graph below depicts the growth of a bacterial culture following inoculation of a minimal medium with 1.5×10^4 cells/ml. The observed increase in log number of cells (log N) is shown as a function of time.

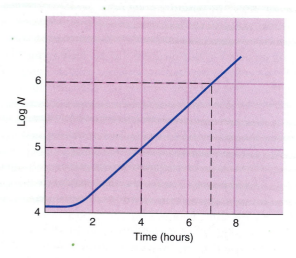

After a lag of about one hour, the culture entered the exponential growth phase. Determine the doubling time of the culture during exponential growth.

Solution: The straight-line portion of the graph shows the increase in log N during exponential growth. The slope of this line is

$$\frac{\log N_2 - \log N_1}{t_2 - t_1} = \frac{\log \frac{N_2}{N_1}}{t_2 - t_1}$$

where N_1 and N_2 are cell numbers at times t_1 and t_2, respectively. Substituting appropriate values from the graph, we find that the slope is

$$\frac{6 - 5}{7 - 4} = 0.33 \text{ per hour}$$

When the population doubles in size, N_2 will equal $2N_1$ and $t_2 - t_1$ will equal the doubling time, t_d. In this case the slope is

$$\frac{\log \frac{2N_1}{N_1}}{t_d} = \frac{\log 2}{t_d}$$

which must also equal 0.33. Solving for the doubling time, we get

$$t_d = \frac{\log 2}{\text{slope}} = \frac{0.301}{0.33} = 0.91 \text{ hour}$$

Follow-Up Problem 8.1

How many hours of exponential growth are required for the bacterial culture in Example 8.1 to reach a size of 10^9 cells/ml?

If provided with adequate nutrients, bacteria can also reproduce on a solid surface. The solid medium commonly used in the laboratory is prepared by adding agar (a solidifying agent derived from certain seaweeds) to a minimal medium or nutrient broth. After dissolving the agar during sterilization, the medium is poured into petri dishes and allowed to solidify at room temperature.

To grow bacteria on agar medium, a sample of liquid culture is deposited on an agar plate either by spreading the sample over the agar surface with a glass rod or by streaking with a wire loop dipped in the cell suspension. This process traps individual cells on the gel surface, where they divide to form separate macroscopic colonies (▶ Figure 8–4). Each colony becomes visible within 10–24 hours of incubation and constitutes a localized group of some 10^7 or more cells that originated from the same genetic ancestor. Such a group of cells is called a clone, because it was derived by asexual reproduction from a single parental cell. Because of their clonal origin, all cells in a colony are genetically identical (except for rare but inevitable mutants) and can therefore be used to establish a "pure" culture. Pure cultures are necessary for most genetic work and are usually prepared by the aseptic transfer of cells from a colony into liquid medium or to another agar plate.

Growth on agar medium is often used to determine

▶ **FIGURE 8–4** Bacterial colonies visible on the agar surface of an agar-filled petri dish.

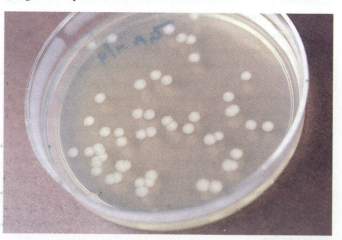

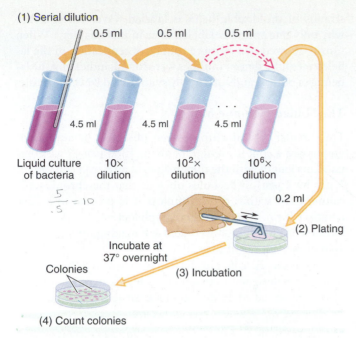

(1) Serial dilution

0.5 ml   0.5 ml   0.5 ml

4.5 ml   4.5 ml   4.5 ml

Liquid culture   10× 10^2× 10^6×
of bacteria   dilution   dilution   dilution

$\frac{5}{.5} = 10$

0.2 ml

(2) Plating

Incubate at
37° overnight

Colonies

(3) Incubation

(4) Count colonies

▶ **FIGURE 8–5** Growing bacterial colonies on solid medium for a viable count. A highly diluted sample of a bacterial culture is placed on nutrient agar medium. Each viable cell grows into a separate colony. The number of viable cells in the original culture is calculated by multiplying the number of colonies per sample volume plated by the dilution factor. In this example, if 100 colonies appear on the plate, the number of viable cells per ml is estimated to be $10^6 \left(\dfrac{100}{0.2} \right) = 5 \times 10^8$ per ml.

the number of viable (colony-producing) bacteria in a liquid culture. Viable counts are carried out by appropriately diluting a sample of the culture and plating a measured volume (spreading it over the surface of an agar plate) (▶ Figure 8–5). During a suitable incubation period, each bacterium on the agar surface will grow and produce a visible colony of descendants. The colonies are then counted, and the concentration of cells in the original culture is determined by multiplying the number of colonies formed per sample volume plated by the dilution factor, as shown in Figure 8–5.

To Sum Up

1. Bacteria have been useful experimental organisms in genetics because of their rapid reproductive rates, ease of handling, and well-defined growth requirements. Several fundamental concepts of molecular genetics have been derived from studies of bacteria, especially *E. coli.*

2. The bacterial genome of *E. coli* cells, in contrast to that of eukaryotic cells, is found in a prominent but ill-defined area referred to as the nucleoid, which is not separated from the rest of the cell by a membrane.

3. Bacteria can be grown in a nutrient-rich broth or in a minimal medium, which is a synthetic medium containing inorganic ions and a carbon (energy) source. In order to grow, bacteria must be able to synthesize for themselves all organic compounds needed for survival and growth. The advantage of a minimal medium is that it allows identification of the particular growth requirements of mutant cells.

4. Three stages characterize the growth of bacteria in a liquid medium. The lag phase is a period of adjustment by the cells to the growth medium. The log (or exponential) phase is the period of active cell division. It is followed by a stationary phase in which no further net increase in the number of cells occurs.

5. The doubling time of a bacterial population can be determined by examination of the log phase. The general equation from Chapter 1, $N_t = N_0 2^t$, can be used to describe the exponential growth, since each generation doubles the number of cells. The doubling time, t_d, can therefore be calculated from a graph of log N versus time.

6. Bacteria can also be grown on a solid agar-based medium. Each cell deposited on the agar plate forms a colony of 10^7 or more cells. Since the cells of any one colony all originated from a single ancestral cell, they are a clone—a group of genetically identical cells that can be used to establish a pure culture of cells.

STRUCTURE OF THE BACTERIAL CHROMOSOME

In bacteria, all essential genes are carried on a single chromosome that appears as a compact, irregularly-shaped body within the cell. This chromosome consists of a single molecule of DNA about 4×10^6 base pairs in length, complexed with small amounts of specific proteins. In rapidly dividing cultures, multiple (2–4) copies of the duplicating chromosome are usually present in each bacterial cell.

In addition to their essential genome, bacteria may also contain small, dispensable genetic elements (minichromosomes) called plasmids. Plasmids carry certain specialized genes that can be of benefit to the host cell and have the ability to replicate independently of the bacterial chromosome. Plasmids and their significance will be discussed in Chapter 10.

Circularity: The Cairns Experiment

The bacterial chromosome contains a circular molecule of DNA. The first direct evidence that bacterial DNA is circular in shape came from experiments conducted on *E. coli* by John Cairns in the early 1960s. Cairns grew cells for varying periods of time in a medium containing tritium-labeled thymidine, so that newly synthesized strands of DNA contained the radioactive isotope tritium (^{3}H). He lysed the cells very gently in order to keep the chromosomes intact and collected the chromosomal DNA on membrane filters. He then layered a photographic emulsion for autoradiography over the labeled chromosomes. The isotope tritium decays by emitting high-energy electrons called β particles. If a photographic emulsion is placed over a chromosome spread that contains ^{3}H, a chemical reaction will occur wherever a β particle strikes the film, thus exposing the film. After several weeks to allow decay of a significant portion of the tritium, the film is developed into a print that

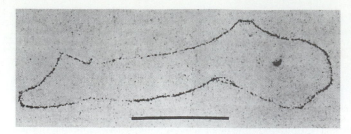

▶ **FIGURE 8–6** Autoradiogram of the *E. coli* chromosome replicating in tritium-labeled medium. The circular shape of the chromosome is apparent. The scale indicates 100 μm. *Source:* J. Cairns, "The form and duplication of DNA," *Endeavour* 22 (1963):144. Copyright © 1963, with kind permission from Elsevier Science Ltd., The Boulevard, Langford Lane, Kidlington OX516B, UK.

shows the emission tracks of the β particles as black spots or grains. The distribution of the radioactive atoms over the chromosomes can then be determined.

Cairns found that after one round of DNA replication in the labeled medium, the decay of the tritium produced a faint circular outline of the chromosomal DNA on the autoradiograph (▶ Figure 8–6). The outline of the chromosomal DNA is not very dense because only one of the two strands of the double helix is labeled. (We will consider why only one strand is labeled later in this chapter.) When Cairns allowed two or more rounds of replication in the labeled medium, denser outlines were found, indicating the labeling of both strands of a newly synthesized DNA molecule.

The Folded Genome Model

The circular *E. coli* chromosome observed in autoradiographs has a contour length of about 1360 μm. If we compare this length with the dimensions of an *E. coli* cell (about 1 μm by 2 μm), it becomes obvious that the chromosome cannot exist within the cell unless it is present in a very compact form. Electron micrographs show that the chromosome is indeed highly condensed, making up only about 10% of the volume of the cell.

When the *E. coli* chromosome is isolated by special procedures designed to maintain its highly compact in vivo state, the so-called **folded genome** structure is obtained. Studies on the folded genome structure suggest that the chromosome of *E. coli* consists of a circular molecule of DNA that is folded into about 40 to 100 supercoiled loops called **domains** (▶ Figure 8–7). This structure is stabilized

▶ **FIGURE 8–7** The folded genome model for chromosome structure in *E. coli*. Only six of the 40–100 loops or domains held together by an RNA-protein core are shown here. (a) The DNA within each loop is supercoiled. (b) DNase relaxes the supercoiling, while (c) RNase releases the loops. *Adapted from:* D. E. Pettijohn and R. Hecht, *Cold Spring Harbor Sympos. Quant. Biol.* 38 (1973): 31–41. Used with permission.

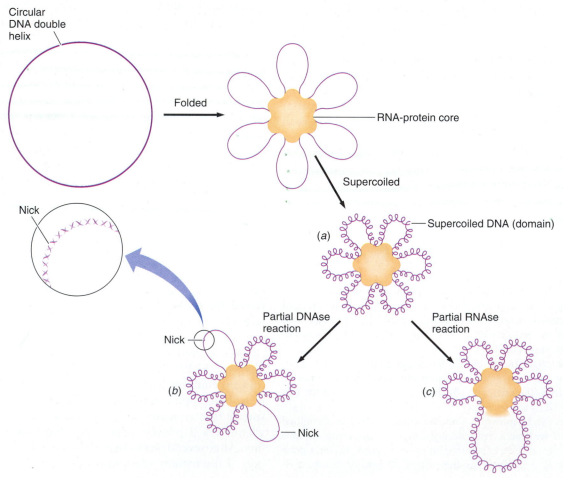

by attachment to a scaffold consisting of proteins (and possibly RNA). The DNA of each domain appears to be topologically constrained so that its ends are not free to rotate. Each domain therefore acts as a separate closed loop, maintaining its supercoiled conformation independent of the other loops in the overall structure. Thus, if a single nick is introduced into a domain by the limited action of DNase, the DNA of the domain is converted to the relaxed (non-supercoiled) state without interfering with the supercoiling of the other domains.

To Sum Up

All essential genes of a bacterium are contained in one main chromosome, which is tightly folded within the nucleoid area of the cell. The chromosome is composed of a supercoiled circular DNA molecule complexed with proteins (and possibly RNA). In addition to the main chromosome, bacterial cells may also harbor one or more small dispensable DNA molecules known as plasmids.

REPLICATION OF BACTERIAL DNA

The genetic material must be capable of faithfully copying its encoded instructions so that its information can be transferred essentially unchanged to different parts of a cell and from one cell generation to the next. In bacteria, the flow of information from parent to progeny cells is accomplished by a process of DNA replication that is closely synchronized with cell division. This replication process produces two complete daughter chromosomes that are separated and partitioned into progeny cells when the parent cell divides.

Semiconservative Replication:
The Meselson-Stahl Experiment

The first conclusive experimental data on the mechanism of DNA replication in bacteria was reported in 1958 by Matthew Meselson and Franklin Stahl, who studied replication in *E. coli*. Their experiment was designed to distinguish between three possible mechanisms that had been proposed (▶ Figure 8–8). One duplicative scheme called **semiconservative replication** had been suggested by Watson and Crick when they proposed the structure of the double helix. According to this model, the helix unwinds and each of the two separated chains serves as a **template** for the synthesis of a new complementary strand (Figure 8–8a). Thus the resulting daughter helices are each composed of one old polynucleotide chain (the template chain) and one new chain.

Two other proposed replication mechanisms—**conservative replication** (Figure 8–8b) and **dispersive replication** (Figure 8–8c)—lead to different arrangements of old and new material in the daughter molecules. In the conservative replication scheme, the two newly synthesized chains form a separate helix, leaving the parental molecule intact. In dispersive replication, old and new material is interspersed along each strand of a newly formed double helix. (The mechanisms of conservative and dispersive replication need not concern us here.) Experimentation was thus needed to establish which of the three models most accurately represents the replication process.

Meselson and Stahl realized that distinguishing among the three models required that the old parental strands be labeled in some way in order to differentiate them from the new strands synthesized during the replication process. To label the parental strands, they replaced the common isotope of nitrogen (^{14}N) that is normally present in the DNA base

---

▶ FIGURE 8–8   Three proposed models of DNA replication: (a) semiconservative, (b) conservative, and (c) dispersive. In each case, the parental strands act as templates for the synthesis of complementary new strands (shown by the darker shading). The arrangement of old and new material into daughter helices distinguishes the three models. The complementarity of the old and new strands enables DNA to satisfy one requirement of the genetic material—the ability of the genetic information to make an exact copy of itself during cell division.

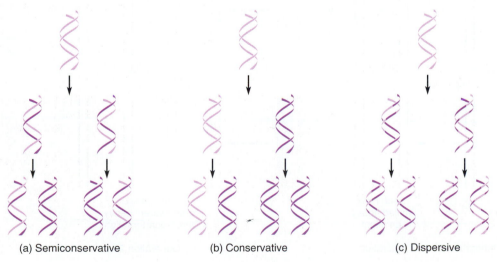

(a) Semiconservative   (b) Conservative   (c) Dispersive

pairs by another isotope (^{15}N); this replacement does not harm the cell. DNA labeled with the heavier ^{15}N has a greater buoyant density than ordinary DNA containing ^{14}N and thus bands at a lower position in the centrifuge tube when subjected to equilibrium density-gradient centrifugation.

The Meselson-Stahl experiment was performed as follows: They grew *E. coli* cells for many generations in a medium containing $^{15}NH_4Cl$ as the sole nitrogen source, until virtually all the bacterial DNA molecules were labeled with the heavy isotope. This point—the time at which the parental molecules are fully labeled—marks time zero. The investigators then transferred the cells to a medium containing ^{14}N, rather than heavy nitrogen, and allowed the bacteria to grow for controlled periods of time. All the DNA strands synthesized in this medium contained only ^{14}N. The researchers withdrew and lysed samples of bacteria from successive generations and extracted and centrifuged the DNA of the cells to determine its density.

The three replication schemes make very different predictions about the density profile following each generation of growth in ^{14}N medium (▶ Figure 8–9). According to the semiconservative model, the first generation grown in ^{14}N medium would have DNA molecules that are all of intermediate density, since each molecule would be composed of one heavy (H) and one light (L) chain (HL molecules). Only one band would form in the CsCl gradient, and it would be at a position of intermediate density. After two generations, however, two bands should form: one at the light density position, consisting of molecules with two light strands (LL molecules), and one at the intermediate position (HL molecules). The amount of DNA in each of these bands should be equal. In contrast, conservative replication would always yield two kinds of molecules: some light (LL molecules) and others heavy (HH molecules), representing the conserved parental helix. No DNA would band at an intermediate position. Finally, dispersive replication would always form molecules of intermediate density, composed of interspersed heavy and light material.

The ultraviolet scans from the Meselson-Stahl experiment are shown in ▶ Figure 8–10, along with the interpre-

▶ **FIGURE 8–9** Predictions of the three models of DNA replication. Heavy (old) strands are shown as solid lines, light (new) strands are shown as dashed lines. After one replication in ^{14}N medium, the semiconservative model predicts that all helices will be of intermediate density, the conservative model predicts half heavy and half light helices, and the dispersive model predicts helices of intermediate density. After two replications in ^{14}N medium, the semiconservative model predicts that half the helices will be of intermediate density and half light, the conservative model predicts one-fourth heavy and three-fourths light, and the dispersive model predicts that all helices contain one-fourth heavy segments, giving them a density between intermediate and light.

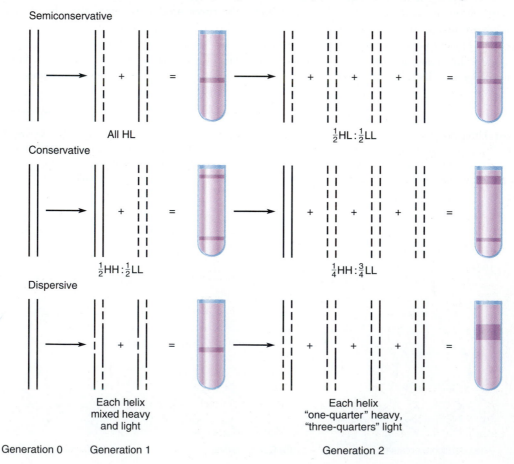

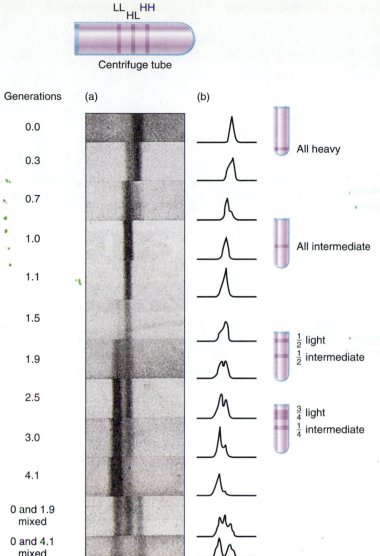

Generations (a)   (b)

0.0

0.3

0.7

1.0

1.1

1.5

1.9

2.5

3.0

4.1

0 and 1.9
mixed

0 and 4.1
mixed

LL   HH
   HL

Centrifuge tube

All heavy

All intermediate

$\frac{1}{2}$ light
$\frac{1}{2}$ intermediate

$\frac{3}{4}$ light
$\frac{1}{4}$ intermediate

► FIGURE 8–10 Results of the Meselson-Stahl experiment. (a) Photographs of the bands formed during centrifugation by scanning the centrifuge tubes with ultraviolet light. (b) Machine tracings of the photographs show peaks of UV absorption. The height of a peak is proportional to the amount of DNA. At time 0, all the DNA is heavy, giving a single band and a single peak at the heavy position. After one generation in ^{14}N medium, a single peak is again observed, but at an intermediate density. After two generations there are two peaks of equal height at the intermediate and light density positions, indicating equal proportions of DNA in these two bands. The results are clearly in agreement with the semiconservative model. *Source:* Matthew Meselson and F. Stahl, *Proc. Natl. Acad. Sci.,* U.S. 44 (1958): 671. Used with permission of Matthew Meselson.

tation of the band positions in the centrifuge tube. After one generation of growth in ^{14}N (unlabeled) medium, all of the DNA molecules band at the intermediate position; no heavy or light DNA is formed. Centrifugation after two generations yields equal proportions of molecules in each of two bands: one at the intermediate position and the other at the light position. These results are clearly in agreement with the semiconservative mechanism proposed by Watson and Crick; they are inconsistent with the conservative and dispersive replication models. They also explain why Cairns' autoradiographs showed only a faint circular outline of the bacterial chromosome after one round of replication in ^{3}H-thymidine medium but a more dense outline after two or more generations of DNA replication.

Example 8.2

In a Meselson-Stahl experiment, bacterial cells are grown in a medium containing ^{15}N as the sole nitrogen source. Once all the DNA is fully labeled, the cells are transferred to a medium containing ^{14}N as the only nitrogen source. If DNA replication is semiconservative, what proportion of the DNA molecules will be of intermediate density after two generations in ^{14}N medium?

Solution: According to the semiconservative scheme of DNA replication, each parental strand forms a double helix with the newly synthesized daughter strand that is copied from it. Therefore, the first-generation molecules will all be intermediate in density. Each strand of these molecules will then form a new molecule by pairing with newly synthesized light strands to form generation two. Of the four generation-two molecules, two will thus be intermediate in density and the other two will be fully light.

Follow-Up Problem 8.2

If DNA replication is conservative, what proportion of the DNA molecules in Example 8.2 would be of intermediate density?

◆

Topology of Replication

By the early 1960s, it was known that DNA replicated by a semiconservative mechanism. However, the relationship between the mode of replication and the circular DNA of the bacterial chromosome still remained obscure. It was not until Cairns performed his autoradiographic studies showing circular bacterial chromosomes in the act of replication that this relationship was clearly demonstrated. His studies revealed that replication begins from a single origin on the DNA molecule and then proceeds around the circular structure (▶ Figure 8–11), giving the appearance of an eye or a bubble. Thus when the DNA is caught midway through the act of duplicating, the replication intermediate is theta (θ)-shaped. More recent work has identified the replication origin as a fixed site on the chromosome; the site is designated *oriC,* and it consists of an AT-rich, 245 base-pair sequence that binds specific replication proteins. Once synthesis is initiated at this site, replication continues until the entire chromosome is duplicated. The bacterial chromosome thus acts as a replication unit that initiates and completes replication once every cell division. Units of DNA that replicate under the control of a single origin of replication are called replicons.

Autoradiograph   Interpretation

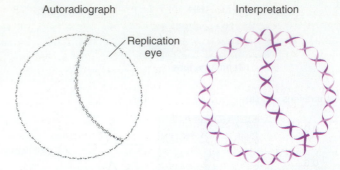

Replication eye

▶ **FIGURE 8–11** Drawing of an autoradiograph of a bacterial chromosome during the second round of replication in tritiated thymidine. One branch of the replication eye is labeled twice as strongly as the remainder of the chromosome, indicating that the DNA in this branch consists of two labeled strands.

Although the Cairns experiment provided evidence for a single replication origin, it failed to indicate whether replication from the starting point is **bidirectional** or **unidirectional** (see ▶ Figure 8–12). In bidirectional replication there are two growing points (Figure 8–12a). The two forks move in opposite directions and ultimately meet when each has

▶ **FIGURE 8–12** Possible modes of theta replication. (a) Bidirectional replication, in which replication proceeds in both directions around the circular replicon. (b) Unidirectional replication, in which replication proceeds in only one direction around the replicon. Both modes of replication lead to a θ-shaped intermediate.

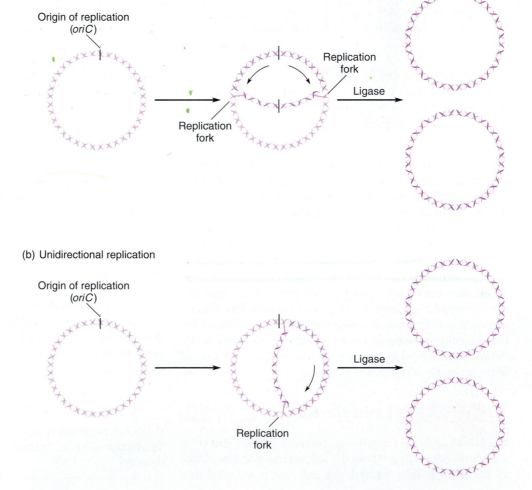

(a) Bidirectional replication

Origin of replication (*oriC*)

Replication fork

Replication fork

Ligase

(b) Unidirectional replication

Origin of replication (*oriC*)

Replication fork

Ligase

replicated about half of the DNA. In unidirectional replication there is only one replication fork at one end of the bubble, and it replicates the entire circle (Figure 8–12b). Both bidirectional and unidirectional replication can lead to θ-shaped intermediates that are indistinguishable in autoradiographs if they are uniformly labeled. To distinguish between these two possible modes of replication, geneticists have used a **pulse-labeling** technique in which the replicating chromosome is provided with two short bursts of a radioactive precursor (e.g., ^{3}H-thymidine), with the second pulse more intensely labeled than the first. In an autoradiograph, these pulses produce different densities of emission tracks that allow one to determine the direction(s) in which replication is proceeding. If replication is bidirectional, both replication forks will be more intensely labeled (▶ Figure 8–13a); if replication is unidirectional, only one replication fork will have the more intense labeling (Figure 8–13b). Studies of this type reveal that the bacterial chromosome replicates bidirectionally during cell division. Bacterial DNA can undergo a form of unidirectional replication, but it does so only under special conditions. We will consider the conditions for unidirectional replication in Chapter 10.

Enzymatic Synthesis of DNA: The DNA Polymerases

The first major breakthrough in understanding the enzymatic mechanisms by which bacterial DNA is replicated came when Arthur Kornberg discovered an enzyme that catalyzed the synthesis of DNA in vitro. The enzyme, now called **DNA polymerase I**, was isolated from *E. coli*. For activity, the enzyme requires the presence of a DNA molecule consisting of one partially synthesized **primer** strand and a longer **template** strand and the deoxyribonucleotides (in

(a) Bidirectional replication

(b) Unidirectional replication

—— Not labeled (invisible on autoradiograph)
—— Line from light density label
—— Line from heavy density label

▶ **FIGURE 8–13** Different densities of emission tracks in an autoradiograph can be used to distinguish (a) bidirectional and (b) unidirectional replication. These patterns are produced when a replicating chromosome is labeled with two short pulses of radioactivity, with the second pulse having greater intensity.

the form of the four high-energy nucleoside triphosphates, dATP, dGTP, dCTP, and dTTP). The reaction catalyzed by the enzyme is

$$(dNMP)_n + dNTP \xrightarrow{\text{Template strand}} (dNMP)_{n+1} + PP_i$$

Primer   Lengthened
strand   primer

where dNMP and dNTP denote deoxyribonucleoside monophosphate and triphosphate, respectively, and PP_i is inorganic pyrophosphate. The particular sequence of deoxyribonucleotides (dNTPs) in the reaction depends on the base sequence of the template strand.

The synthesis of each new strand of DNA occurs by the mechanism shown in ▶ Figure 8–14. The DNA polymerase

▶ **FIGURE 8–14** The 5' to 3' synthesis of a DNA strand under the direction of a template strand. The nucleotides are added in accordance with the Watson-Crick base-pairing rules. The energy-rich nucleoside triphosphates are the precursors for DNA synthesis; hydrolysis of the terminal phosphate residues provides the energy for extension of the strand in the 5' → 3' direction.

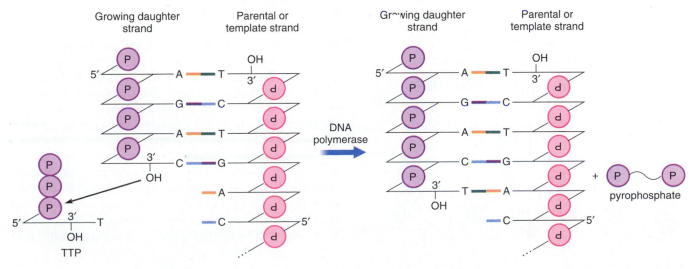

enzyme cannot start synthesizing a new DNA molecule by itself; it can only add nucleotides to a preexisting primer strand. The primer is lengthened nucleotide by nucleotide in a sequence that is complementary to the sequence of bases in the template strand, with each nucleotide being inserted in accordance with the Watson-Crick base-pairing rules. Note that the incoming nucleotides are added to the 3' end of the growing strand, so that DNA synthesis always proceeds in the 5' to 3' direction.

Since Kornberg's discovery of DNA polymerase I, two additional DNA polymerase enzymes that function in *E. coli* have been identified: **DNA polymerase II** and **DNA polymerase III**. These enzymes have the same synthetic ability as DNA polymerase I—they all catalyze DNA strand elongation in the 5' to 3' direction by basically similar mechanisms. In addition, they can all function as **exonucleases**. An exonuclease is an enzyme that sequentially excises (hydrolyzes away) nucleotides from an exposed end of a polynucleotide chain (▶ Figure 8–15). All three bacterial DNA polymerases are capable of the 3' → 5' excision of nucleotides in the reverse direction from DNA synthesis. This exonuclease activity is responsible for the proofreading ability of the bacterial DNA polymerases (see Chapter 23). With DNA polymerase I, strand digestion can occur in either direction, so this enzyme possesses 5' → 3' as well as 3' → 5' exonuclease activity.

▶ **FIGURE 8–15** The action of an exonuclease on a DNA strand. Nucleotides are excised sequentially from one end of the polynucleotide chain. Some exonucleases hydrolyze in the 5' → 3' direction, others in the 3' → 5' direction.

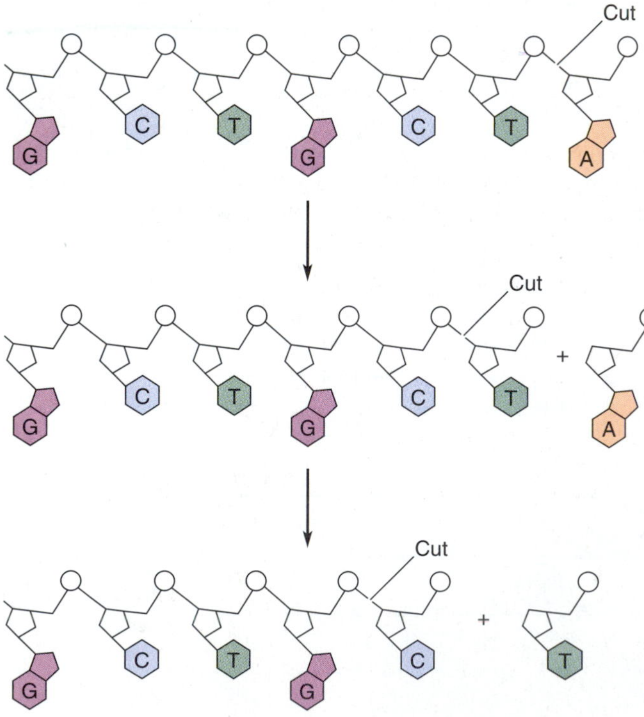

The function of DNA polymerase II is not well understood, but much more is known about the other two polymerase enzymes. DNA polymerase III is the principal enzyme catalyzing DNA strand elongation during the replication of the *E. coli* chromosome. It is an oligomeric enzyme with different subunit activities; for example, the polymerizing activity is located in one of its subunit chains, while another subunit carries the exonuclease activity. However, all subunits must be present together in the form of a **holoenzyme** for proper biological function. During replication, DNA polymerase III holoenzymes are positioned at each replication fork, where they act processively, without detaching from the template strand, to synthesize most of the new DNA.

DNA polymerase I, the enzyme isolated by Kornberg, functions in *E. coli* in a more general way. The Kornberg enzyme is a major **repair enzyme** that catalyzes repair synthesis, filling in gaps where they appear in the DNA structure. This multifunctional enzyme can use its 5' → 3' exonuclease activity to excise damaged (or otherwise incorrect) nucleotides from a polynucleotide segment and its 5' → 3' polymerizing activity to replace the cleaved segment with new DNA. In fact, these activities can go on simultaneously, working in concert in a process known as **nick translation** (▶ Figure 8–16). During nick translation, DNA polymerase I excises nucleotides from a nicked polynucleotide chain in the 5' to 3' direction and at the same time adds new nucleotides immediately behind it, using the exposed 3' end of the nicked strand as a primer. This combination of 5' → 3' exonuclease and polymerizing activities plays a key role in DNA replication during the removal of primer sequences, as will be discussed shortly. Unlike DNA polymerase III, the Kornberg enzyme functions as a monomer, so it carries its different enzymatic activities on a single polypeptide chain.

Discontinuous Replication of DNA

After the discovery of the DNA polymerases, it soon became apparent to geneticists that these enzymes are not the only proteins acting in DNA replication. The DNA polymerase enzymes can catalyze strand elongation only in the 5' to 3' direction, yet studies on the *E. coli* chromosome indicate that synthesis occurs simultaneously on both template strands at each replicating fork. This observation would suggest that DNA is being lengthened in the 3' to 5' direction on the other strand. Since no enzyme has been discovered that can catalyze 3' → 5' synthesis, how then does replication occur on both template strands at the replication fork? An answer to this question was provided in the late 1960s, when Reiji Okazaki discovered that DNA is synthesized discontinuously, in short segments, on at least one of the template strands. This discovery resulted from an experiment in which Okazaki administered shorter and shorter pulses of ^{3}H-thymidine to growing *E. coli* cells. The DNA was extracted from the bacteria and denatured, and the chain lengths of the newly synthesized labeled strands were measured by following their sedimentation rate in an

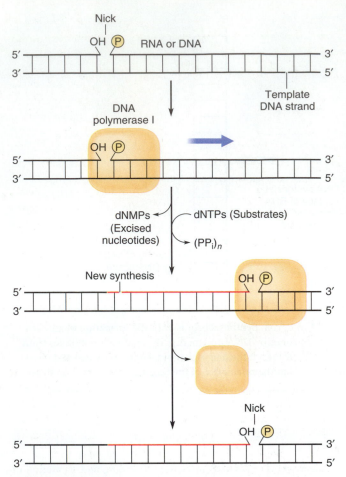

FIGURE 8–16 Nick translation catalyzed by DNA polymerase. The $5' \rightarrow 3'$ exonuclease activity at a single-strand break (a nick) can occur simultaneously with polymerization.

ultracentrifuge. Okazaki found that when the length of the pulses was shortened to less than one-tenth of the generation time, the label was incorporated primarily into short DNA pieces no more than 1 to 2 kilobases (kb) long. These pieces, now popularly known as Okazaki fragments, are joined together enzymatically during DNA synthesis. Thus when the pulse time was extended, the label appeared in progressively longer DNA segments.

▶ Figure 8–17 shows the involvement of Okazaki fragments in DNA synthesis at the replication fork. According to the current view of DNA replication, one growing strand (the leading strand) is synthesized continuously along its template from 5′ to 3′ while moving in the same direction as the replication fork. The other growing strand (the lagging strand), is synthesized discontinuously in the form of Okazaki fragments. These fragments are also synthesized from 5′ to 3′, but in a direction opposite to the movement of the replication fork. Each fragment will therefore exist only briefly, since its 3′ end is soon joined to the 5′ end of the previously synthesized fragment to form the lagging strand.

Another question arises regarding the requirement of DNA polymerase enzymes for a primer. If the DNA polymerases cannot start synthesizing a new DNA strand on their own, what initiates replication of the leading strand and of each Okazaki fragment? This question was answered when it was discovered that DNA is synthesized in vivo using a short RNA primer just a few ribonucleotides in length. The RNA primer is synthesized on the DNA template by an enzyme known as **primase**. The action of primase is similar to that of DNA polymerase, except that primase synthesizes the primer sequence from scratch using ribonucleoside triphosphates as substrates and can function effectively only in the presence of a group of proteins referred to collectively as a **primosome**. Like the DNA polymerase III holoenzyme, the primosome acts processively; it moves along the lagging strand template in the 5′ to 3′ direction, pausing occasionally for primer synthesis.

The sequence of events involved in the discontinuous synthesis of DNA in *E. coli* is illustrated in ▶ Figure 8–18. An RNA primer is first formed by the action of primase. Starting at the 3′ end of the RNA primer, deoxyribonucleotides are then added sequentially in the 5′ to 3′ direction by the action of DNA polymerase III. Elongation of the DNA fragment continues until the 5′ end of the previously synthesized portion of the lagging strand is reached. DNA polymerase I then removes the RNA primer of the lagging strand by nick translation and replaces the excised primer with the corresponding DNA sequence. Finally, the splicing (sealing) enzyme **DNA ligase** joins the 3′-hydroxyl group of the fragment that has just been synthesized to the 5′-phosphate group of the lagging strand.

▶ **FIGURE 8–17** Discontinuous synthesis of DNA. Short segments of new DNA (Okazaki fragments) are synthesized ($5' \rightarrow 3'$) in a direction opposite to the movement of the replication fork. The ends of the Okazaki fragments are joined to form the lagging strand. The leading strand is synthesized ($5' \rightarrow 3'$) in a continuous manner while moving in the same direction as the replication fork.

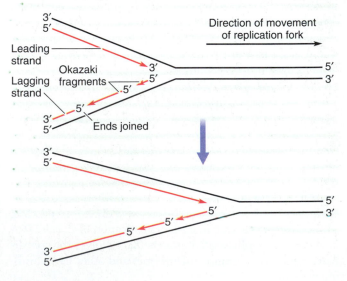

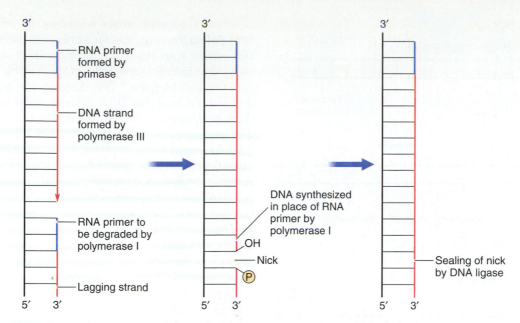

FIGURE 8–18 Steps involved in the synthesis of an Okazaki fragment in *E. coli*. Synthesis begins with the formation of an RNA primer. A segment of a DNA strand is then produced from the 3' end of the primer by DNA polymerase III. Synthesis continues until the 5' end of the previously synthesized portion of the strand is reached. The primer of the lagging strand is then removed and replaced with DNA by DNA polymerase I. The fragment is subsequently joined to the remainder of the lagging strand by DNA ligase.

Ongoing research reveals that the replication of DNA is even more complex than we have indicated so far. In addition to primase, DNA polymerases I and III, and DNA ligase, a variety of other proteins are required to unwind the double helix and keep the two strands apart. ▶ Figure 8–19 shows a current model of DNA replication that depicts the roles of some of these proteins at the replication fork. According to this model, a dimeric DNA polymerase III holoenzyme joins with the primosome to form a **replisome**, which jointly carries out the synthesis of both the leading and lagging strands. The template of the lagging-strand may make a full-turn loop to permit this joint synthesis. The polymerase making the lagging strand could then move in the same direction as the polymerase making the leading strand, even though the two strands are antiparallel. Acting ahead of primase and in concert with it are enzymes known as **helicases**, which unwind and separate the parental DNA strands. Energy for unwinding is provided by the breakdown of ATP. **Single-strand binding proteins (SSB)** temporarily bind to the single-stranded regions that are exposed just prior to replication to prevent the creation of base-pair hairpin loops and keep the parental helix from reforming. As the unwinding proceeds, the double-helical region ahead of the replication fork is prevented from twisting by the action of **DNA gyrase**. This topoisomerase enzyme introduces negative supercoils into the structure to compensate for the buildup of positive supercoils that form during unwinding of the DNA circle. In conclusion, replication of *E. coli* DNA is a complex, sophisticated process that involves the coordinated actions of a number of enzymes and other DNA-binding proteins.

To Sum Up

1. Replication of DNA requires the unwinding and separation of its polynucleotide chains. Each strand of the helix serves as a template for the synthesis of a new complementary strand. This mode of synthesis, in which the daughter molecules have one old parental strand and one new complementary strand, is called semiconservative replication.

2. Replication of the bacterial chromosome during cell division is bidirectional; it starts at the replication origin and proceeds around the circle in both directions.

3. Replication of one of the DNA strands is discontinuous; it proceeds in 1–2 kb segments called Okazaki fragments. DNA polymerase III makes each fragment separately by extending the synthesis in a 5' → 3' direction from an RNA primer. This primer is synthesized by a complex called a primosome, which is composed of primase and other proteins. After DNA polymerase I excises the RNA primer using its 5' → 3' exonuclease activity, it catalyzes repair synthesis to fill in the gap. The completed Okazaki fragment is then joined to the rest of the growing chain by DNA ligase.

4. Bacterial DNA replication requires a variety of other proteins in addition to the primosome, DNA polymerase III, DNA polymerase I, and DNA ligase. Together with DNA polymerase III, these proteins form a complex called a replisome that synthesizes both the leading (continuous) and lagging (discontinuous) strands at the same time. These proteins include helicases, single-strand binding proteins, and DNA gyrases; their functions are to unwind the two strands of the duplex, to keep the strands separated so that they can be read as templates by polymerase, and to prevent the twisting of the double-helical region ahead of the replication fork.

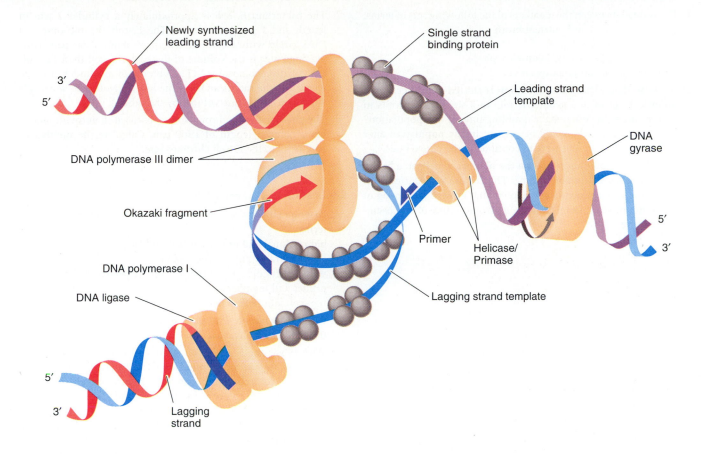

Newly synthesized
leading strand

3′
5′

DNA polymerase III dimer

Okazaki fragment

DNA polymerase I

DNA ligase

5′
3′

Lagging
strand

Single strand
binding protein

Leading strand
template

DNA
gyrase

5′
3′

Primer

Helicase/
Primase

Lagging strand template

▶ FIGURE 8–19 Current model of DNA replication showing some of the proteins that act in conjunction with the replicating enzymes. Looping of the template for the lagging strand enables a dimeric DNA polymerase III holoenzyme to synthesize both daughter strands.

Chapter Summary

Bacteria are small, single-celled organisms that lack a nucleus and the other membranous organelles that are characteristic of eukaryotic cells. Their ease of manipulation, rapid reproductive rates, and well-defined growth conditions have contributed to their widespread use in genetic research.

All essential genes of a bacterium are contained in a single chromosome. The chromosome consists of a circular duplex molecule of DNA that is complexed with small amounts of protein and RNA and tightly folded in the nucleoid of the cell.

Bacterial DNA undergoes semiconservative replication: the strands separate and serve as templates for the synthesis of complementary chains. Synthesis begins at a specific region (*oriC*) and proceeds bidirectionally around the DNA circle. DNA polymerases catalyze the 5′ to 3′ synthesis of each complementary strand

using deoxyribonucleoside triphosphates as substrates. 5′ to 3′ synthesis is continuous along one DNA strand; synthesis on the other strand is discontinuous and proceeds through the formation of Okazaki fragments. Each Okazaki fragment is initiated by the synthesis of a short RNA primer (catalyzed by the enzyme primase) and extended by DNA chain elongation (catalyzed by DNA polymerase III). The primer is ultimately degraded by hydrolysis (catalyzed by DNA polymerase I), and the Okazaki fragment is attached to the rest of the growing strand by the formation of a phosphodiester bond (catalyzed by DNA ligase). In addition to the enzymes that catalyze polynucleotide synthesis, several other enzymes, such as DNA gyrase and helicase, and DNA-binding proteins have roles in the DNA replication process.

The Structure and Growth of Bacteria

1. Distinguish between the members of the following sets of terms:
   minimal medium and nutrient broth
   protoplast and spheroplast
   lag phase, log phase, and stationary phase
   Gram-negative and Gram-positive

2. Suppose that a bacterial population is started with 2 cells and allowed to grow at a geometric rate to a size of 4096 cells in 275 minutes. (a) What is the doubling time of this population? (b) How many cells would be present in the population after 550 minutes, if growth were to continue at the above rate?

3. The graph below shows data on the number of cells in a bacterial culture as a function of time. The culture was started with 1.5×10^3 cells, which began exponential growth after a lag of about 30 minutes. Determine the generation time of this bacterial culture.

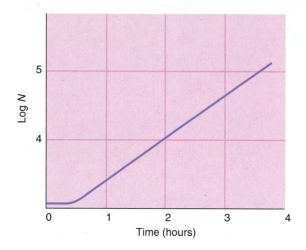

4. The data below were collected by measuring the turbidity of an exponentially growing bacterial population using a spectrophotometer. Determine the value of the doubling time, t_d.

Time (t, minutes)	Number of cells (N_t, per ml)
20	6.3×10^6
100	2.5×10^7
180	1.0×10^8
260	4.0×10^8

5. Suppose that you inoculate 10 ml of nutrient broth with 100 actively dividing cells each of strain A and strain B. Two hours later, you observe 6.4×10^2 per ml of strain A and 3.2×10^2 per ml of strain B. How many hours of exponential growth (after the time of inoculation) are required to obtain a culture in which the ratio of A:B cells is 99:1?

6. The smallest bacterial colony visible without magnification is around 0.1 mm in diameter. Assume the colony is a hemisphere containing cylindrical cells 2 μm in length and 1 μm in diameter. (a) How many cells would such a colony maximally contain? (b) Using the answer from (a) and assuming that an 8-hour period of incubation and exponential growth was required to reach this size, what is the doubling time of the cells?

Structure of the Bacterial Chromosome

7. The bacterium *E. coli* is approximately a cylinder 2 μm in length and 1 μm in diameter. The *E. coli* chromosome is a DNA circle with a contour length of about 1360 μm. How much greater is the volume of the bacterial cell than the volume of its chromosome? (The volume of a cylinder is $\pi r^2 h$, where r is the radius and h is the length. Note that π cancels when you divide one volume by the other.)

8. The circular *E. coli* chromosome observed in autoradiograms has a contour length of 1360 μm. Calculate the number of base pairs in the *E. coli* chromosome.

Replication of Bacterial DNA

9. Distinguish between the members of the following sets of terms:
   bidirectional and unidirectional replication
   leading strand and lagging strand
   semiconservative, conservative, and dispersive replication
   primer and template
   replicon and replisome

10. Postulate a reason why the hydrolysis of ATP is required for the action of helicases.

11. DNA biosynthesis in *E. coli* occurs at a rate of 50,000 base pairs per minute per replication fork. (a) How long does it take an *E. coli* cell to duplicate its chromosome? (b) The doubling time for *E. coli* cells can be as short as 20 minutes under optimum growth conditions. In terms of DNA replication, how is it physically possible for the cells to double in such a short period of time?

12. Consider a Meselson-Stahl experiment in which bacterial cells are grown in a medium containing ^{15}N until all of the DNA molecules are fully labeled. These cells are then transferred to a medium in which all of the nitrogen is ^{14}N and are permitted to duplicate further. (a) According to the semiconservative mechanism of replication, what proportion of the DNA molecules present after one round of duplication in ^{14}N medium will be of intermediate density? What proportion will be fully heavy? What proportion will be fully light? (b) What proportion of the DNA molecules present after three rounds of duplication in ^{14}N medium will be found in each density class? (c) Answer questions (a) and (b) if DNA replication were to follow the conservative mechanism. (d) Answer questions (a) and (b) if DNA replication were to follow the dispersive mode.

13. The Meselson-Stahl experiment cannot distinguish between semiconservative and dispersive replication if the cells are allowed just one act of duplication in ^{14}N medium. Outline an additional experimental procedure that could be used to distinguish between the first-generation molecules that are predicted by the semiconservative and dispersive replication models.

14. The following diagram shows density gradient profiles obtained from centrifuging ^{15}N-labeled DNA from a particular bacterium after 0, 1, and 2 replication cycles in ^{14}N-medium (as in the Meselson-Stahl experiment):

0   1   2

Which of the following choices could account for the two intermediate bands? Explain your choices. (a) The DNA has been sheared into two different lengths upon extraction. (b) The DNA is undergoing unidirectional replication. (c) The complementary DNA strands differ in their $(A + T):(G + C)$ ratios. (d) The complementary strands differ in their $(A + G):(C + T)$ ratios. (e) The DNA is undergoing conservative replication.

15. Suppose that four genetic markers are equally spaced on a bacterial chromosome, as shown in the following diagram:

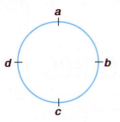

Also suppose that you could measure the number of copies of each marker per cell during cell growth. Given that a is the origin of replication, which of the following relationships would hold true for the average number of copies per cell of these markers in a growing population of bacteria (a) if DNA replication is bidirectional? (b) if DNA replication is unidirectional? Explain your answers. (1) $a > b > c > d$, (2) $d > c > b > a$, (3) $a > b = d > c$, (4) $c > d = b > a$, (5) $b = d > c > a$, (6) $a = b = c = d$.

16. Calculate the number of Okazaki fragments that are synthesized during the replication of an *E. coli* chromosome.

17. Suppose that you isolate two mutant strains of *E. coli*, designated x and y, that show defective DNA synthesis at high temperatures ($> 30°C$). In mutant strain x there is an accumulation of Okazaki fragments at the higher temperatures, while in mutant strain y there is an accumulation of short RNA segments. Give the likely cause of these mutant phenotypes, indicating which enzyme is affected in each of the mutant strains.

CHAPTER 9

Eukaryotes and Eukaryotic Chromosomes

All unicellular organisms other than bacteria and all multicellular organisms are classified as eukaryotes. Eukaryotic cells vary considerably in size and shape but are generally much larger than prokaryotes. For example, a human liver cell has a diameter of approximately 20 μm, compared with the 2-μm diameter of an *E. coli* cell; thus the eukaryotic cell is about 100 times greater in surface area and 1000 times greater in volume than the prokaryote.

Eukaryotic cells also have a considerably more complex internal organization than prokaryotes. In marked contrast to bacteria, the interior of eukaryotic cells is divided into a number of distinct membrane-bounded organelles that effectively compartmentalize the specialized cellular functions. For example, the nucleus of a eukaryotic cell separates the cell's genetic information from the rest of the cellular components. Other organelles in the cytoplasm carry out various specialized

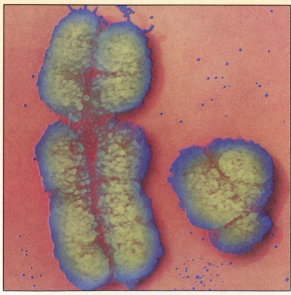

Scanning electron micrograph (SEM) of the
human X and Y chromosomes.

biochemical reactions such as those involved in the Krebs cycle and in oxidative phosphorylation. This separation of functions provides for greater versatility and control in the metabolic functioning of cells and enables cells to attain larger sizes than could otherwise be achieved.

In addition to membranous organelles, eukaryotic cells also contain various proteinaceous rods and filaments that function in cell motility and in the determination of cell shape. These protein structures collectively form a supportive network in the cytoplasm known as the **cytoskeleton**.

In this chapter, we will examine the complexities of eukaryotic chromosomes, focusing on the features of DNA packaging and nucleotide sequence organization that are fundamentally different from those in prokaryotes. We will then turn to the process by which eukaryotes replicate their DNA. As you will discover, the basic features of DNA replication are the same in bacteria and eukaryotes—replication is semiconservative, is initiated by RNA primers, and is discontinuous on one strand. However, the complex mechanism involved in packaging eukaryotic DNA and the fact that eukaryotic nuclear DNA is linear rather than circular mean that eukaryotic DNA replication has some additional complexities that are not present in bacteria.

 ## MORPHOLOGY AND FUNCTIONAL ELEMENTS OF CHROMOSOMES

Unlike prokaryotes, which carry their genes in a single circular chromosome, eukaryotes package their genome into several different linear (rod-shaped) chromosomes. The chromosomes of prokaryotes and eukaryotes also differ in another fundamental respect. While the prokaryotic chromosome tends to remain in a constant state of compaction throughout the cell cycle, eukaryotic chromosomes undergo periodic structural and biochemical changes in actively dividing cells. Recall from Chapter 1 that each eukaryotic chromosome exists in a **dispersed** (loosely coiled) state during interphase and in a **condensed** (tightly coiled) state during cell division. Just before cell division, eukaryotic chromosomes undergo extensive coiling and become visible with a light microscope. For this reason, most cytogenetic studies in eukaryotes have been done on dividing cells at mitotic metaphase, when the chromosomes are maximally condensed and can most easily be seen microscopically.

You should also recall from Chapter 1 that at metaphase, each eukaryotic chromosome has the appearance of a double rod (▶ Figure 9–1), because it consists of a pair of sister chromatids connected at the centromere. At this time, three morphological features are distinct enough to be used in identifying different chromosomes: (1) length, (2) position of the centromere (relative arm length), and (3) the presence of small terminal knobs called satellites. Depending on the position of the centromere, chromosomes are classified at this time as **metacentric**, if the centromere is in a median (centrally located) position; **submetacentric**, if the centromere is in a submedian (off-center) position; **acrocentric**, if the centromere is very near one end; or **telocentric**, if the

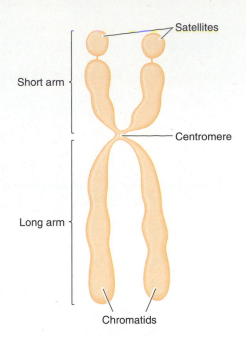

▶ **FIGURE 9–1** The basic structure of a nuclear chromosome when maximally condensed. Each chromosome is composed of identical sister chromatids held together at the centromere. Terminal knobs, called satellites, are seen on certain chromosomes.

centromere is at a terminal location and only one arm of each chromatid is visible (▶ Figure 9–2).

DNA Arrangement

Like the chromosomes of prokaryotes, each eukaryotic chromosome is thought to contain a single molecule of DNA running through its entire length. Evidence for a 1:1 relationship between a eukaryotic chromosome and a DNA molecule has come from studies employing a variety of physical and chemical techniques. For example, studies on the elastic properties of DNA in solution (known as **viscoelasticity**) give estimates of DNA size from the rate at which DNA recoils after being stretched into an extended conformation. In general, long DNAs take longer than short DNAs to relax after being stretched. Viscoelastometric measurements on the largest DNA molecules of *Drosophila* have yielded estimates of size that agree closely with the known DNA content of the largest chromosome of this species. This result is consistent with the belief that each chromosome contains a single, uninterrupted duplex of DNA.

Recently, researchers have conclusively demonstrated a 1:1 relationship between the number of DNA molecules and the chromosomes of yeast (*Saccharomyces cerevisiae*) using an electrophoretic technique known as **pulsed-field gel electrophoresis**. This technique uses pulses of current (rather than a constant current) to separate DNA molecules by alternating the direction of the electric field. The elongated DNA responds to each pulse by worming its way head-on in the direction of the positive electrode. When the direction of the field is suddenly changed, however, the

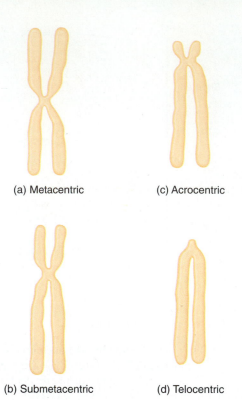

(a) Metacentric   (c) Acrocentric

(b) Submetacentric   (d) Telocentric

▶ **FIGURE 9–2** Different centromere locations in nuclear chromosomes. (a) Metacentric chromosomes have centromeres at a median position so that chromosome arms are about equal in length. (b) Submetacentric chromosomes have centromeres in a submedian position (off center). (c) Acrocentric chromosomes have centromeres close to one end. (d) Telocentric chromosomes have centromeres at one end so that only one arm of each chromatid is visible.

molecules must reorient themselves and move in a different direction. Since larger molecules require longer times to change their orientation, their net movement along the gel will be less than that of smaller molecules. With this technique, it is possible to separate very large pieces of DNA several orders of magnitude longer than can be separated by conventional (fixed-field) methods of electrophoresis. Thus researchers using this method have been able to separate the intact DNA of whole chromosomes (200–300 kb) of yeast, making it possible to study the organization of the DNA more directly.

Centromeres and Telomeres

Despite the uninterrupted nature of the underlying DNA, some differentiation of DNA structure is necessary for the chromosome to carry out its normal function. One essential feature of a nuclear chromosome is its centromere, a special DNA region that, with its associated kinetochore, provides an attachment site for one or more spindle microtubules (▶ Figure 9–3). In yeast, this attachment is made to individual microtubules, while in higher eukaryotes, it involves microtubular bundles. The centromere attaches to the mitotic spindle by means of a protein structure called a

kinetochore: one kinetochore is associated with each chromatid. The kinetochores are essential for an orderly distribution of chromosomes during cell division, since by attaching the centromere to microtubules, they provide for (1) the proper alignment of each chromosome at the spindle equator (Figure 9–3a), and (2) the exact disjunction of daughter chromosomes once centromere division has taken place (Figure 9–3b).

The DNA sequences that are essential for proper kinetochore organization and function have been designated *CEN* sequences, named for the centromere. Different *CEN* sequences have been isolated from the chromosomes of the yeast *S. cerevisiae* and their primary structures determined. These sequences are all very similar; they fall within a stretch of about 120 bp and are composed of three common sequence elements that are shown in ▶ Figure 9–4. These elements, which are essential for proper centromere function, consist of two short and highly conserved regions (elements I and III) flanking a region of high AT content (element II). Just how representative *CEN* sequences are of other eukaryotic centromeres remains to be seen; they are much smaller than the centromeric regions of higher organisms and lack the highly-repetitive DNA sequences typical of many other eukaryotes (to be discussed in a later section).

Telomeres, which are special DNA regions at the ends

▶ **FIGURE 9–3** Essential functions of kinetochores in attaching the centromeric regions of chromosomes to spindle microtubules. (a) Role of kinetochores in the alignment of chromosomes at the spindle equator. (b) Role of kinetochores in the disjunction of chromosomes following centromere division.

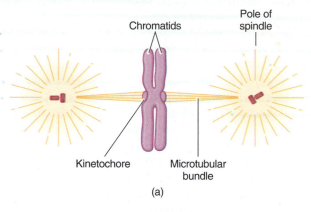

(a)

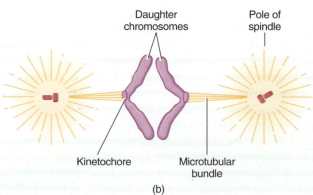

(b)

Taylor's Experiment: Unineme Structure and Semiconservative Replication of Eukaryotic DNA

Herbert Taylor's experiments with the broad bean, *Vicia faba*, in the 1950s gave us the first clear picture of the number of DNA molecules in a eukaryotic chromosome and the mode of segregation of the DNA strands during eukaryotic DNA replication. Taylor allowed interphase chromosomes of bean root tips to replicate once in the presence of ^{3}H-thymidine; autoradi-

ography of chromosomes during the following mitosis showed that both sister chromatids of each duplicated chromosome were labeled. When these labeled chromosomes were allowed to replicate again in the absence of ^{3}H-label, only one of the chromatids of each daughter chromosome remained labeled at the following mitosis. The autoradiographs for both replicated

generations are shown here, along with diagrams of chromosomal labeling and DNA replication. The simplest way to interpret these results is to assume the presence of a single DNA molecule in each chromatid and chromosomes that replicate in a semiconservative manner.

Source: J. H. Taylor, *Molecular Genetics,* Part I, J. H. Taylor, ed. Academic Press, 1963.

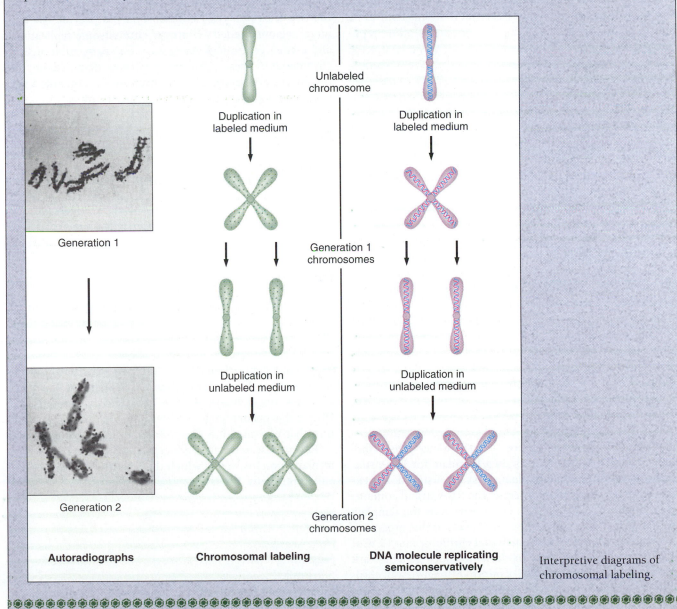

Interpretive diagrams of chromosomal labeling.

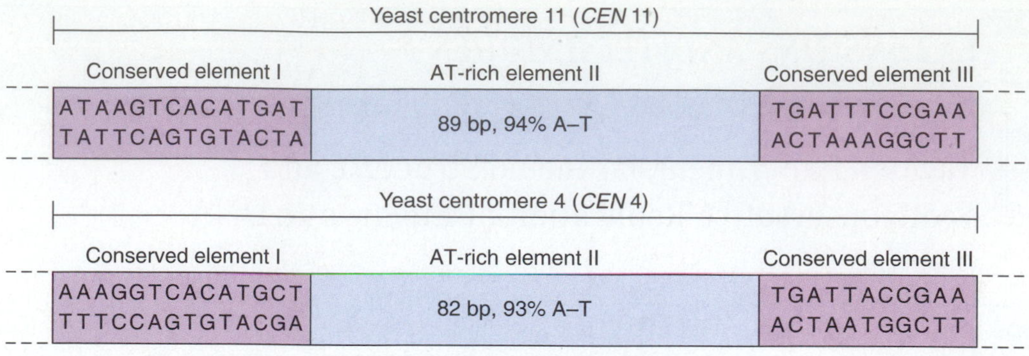

FIGURE 9–4 The DNA sequences of typical yeast centromeres. The centromere regions of *S. cerevisiae* chromosomes 11 and 4 are shown. The *CEN* regions of all 17 chromosomes of *S. cerevisiae* are very similar, composed of two highly conserved sequence elements (I and III) bracketing an AT-rich element (II).

of a chromosome, are another feature common to all linear eukaryotic chromosomes. Telomeric DNA solves a problem that occurs in the replication of linear DNA molecules—a few nucleotides at one end of each new strand tend to be omitted whenever DNA is copied. Telomeres allow replication of the ends of chromosomal DNA to be completed and help to preserve the integrity of linear DNA molecules. Without telomeric regions, the chromosomes in successive generations of cells would become progressively shorter. Telomeric regions also signal cells to maintain chromosomes as intact and distinct entities. For example, broken chromosome ends that have lost their telomeres become "sticky" and tend to combine with other chromosomes. If these broken ends are repaired, however, the chromosomes will return to their normally stable (unreactive) state.

The telomeric regions from a number of eukaryotes have been analyzed and have been found to consist of a series of short G-rich repeated sequences. For example, telomeres in the ciliated protozoan *Tetrahymena* consist of repeated TTGGGG sequences at the 3′ end that fold back and form an unusual G — G bonded hairpin loop at the chromosome terminus (▶ Figure 9–5a). This structure helps to avoid the replication problems and the susceptibility to degradation that are associated with free DNA ends. The removal of a primer from the 5′ end of a newly synthesized strand is expected to leave a 5′-terminal gap after each round of replication (Figure 9–5b). To prevent shortening of the ends of chromosomes, an enzyme called telomerase catalyzes the nucleotide-by-nucleotide addition of repeated sequences to the 3′ terminal overhang, and this extended region then forms the hairpin loop (Figure 9–5c). Telomerase is a ribonucleoprotein that carries its own template for DNA synthesis in the form of a complementary RNA strand. For example, in *Tetrahymena* telomerase, the RNA strand contains the sequence CAACCCCAA, which serves as the template for the TTGGGG repeats in the telomeres of this organism.

The construction of **yeast artificial chromosomes (YACs)** is an interesting example of modern molecular genetic work that provides a clearer picture of our traditional view of chromosomes. Through recombinant DNA techniques (to be described in Chapter 18), researchers can construct artificial

yeast chromosomes that replicate and segregate their DNA normally during cell division. These artificial chromosomes contain an origin of replication region, DNA segments from the two ends of a yeast chromosome (to provide the telomeric regions needed for proper chromosome replication), and a yeast centromere (necessary for segregation of sister chromatids). These artificial constructs demonstrate that the minimal eukaryotic linear chromosome contains a centromere, telomeres, and an origin of replication.

To Sum Up

Each eukaryotic nuclear chromosome contains a single linear molecule of duplex DNA. Specific DNA regions are important to normal chromosome function; they include the centromere (the attachment site for spindle microtubules), the telomeres (the ends of a chromosome), and several origins of replication.

◆

DNA PACKAGING

Eukaryotes have many orders of magnitude more DNA than a bacterial cell (■ Table 9–1). Even though the DNA of eukaryotes is divided into several chromosomes, the amount of DNA in a typical eukaryotic chromosome is still very large. For example, the DNA in the largest human chromosome is about 85 mm long when fully extended, more than 60 times longer than the DNA of an *E. coli* chromosome. How is this greater length of DNA organized within a structure that may measure only 10 μm at mitotic metaphase?

Recent research has produced a substantial body of information on the way in which the DNA is organized within eukaryotic cells. During interphase, most of the eukaryotic DNA is present in the form of chromatin, a complex of DNA and protein that forms a dispersed network of nucleoprotein fibers within the nucleus. When chromatin is isolated from interphase nuclei, it is a viscous, gelatinous material containing relatively fixed amounts of DNA and specific proteins called histones, along with varying amounts of RNA and nonhistone proteins (■ Table 9–2). The histones are the most prominent and best understood of the protein com-

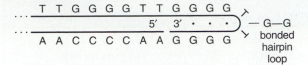

(a) *Tetrahymena* telomere

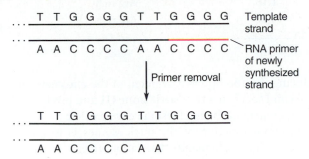

(b) Creation of 5'-terminal gap during DNA synthesis

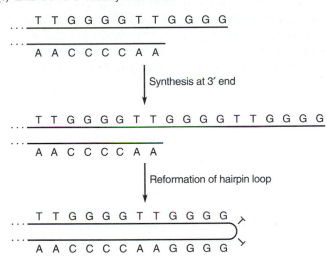

(c) Extension of 3' ends by telomerase

▶ **FIGURE 9–5** (a) Structure of the telomere of *Tetrahymena*, showing the G—G bonded hairpin at the end of the chromosome. (b) A gap is left at the 5' end of a newly synthesized strand when the RNA primer is removed during DNA replication. (c) The enzyme telomerase extends the 3' end of the top strand, using an RNA template (not shown). The extended region then reforms a hairpin loop, completing synthesis of the telomeric region.

■ **TABLE 9–1** Haploid DNA content of different species.

Organism	Molecular weight	Base pairs	Length (m)
Higher plants			
Lily	2.0×10^{14}	3×10^{11}	100
Maize	4.4×10^{12}	6.6×10^{9}	2.2
Vertebrates			
Human	1.9×10^{12}	2.8×10^{9}	9.4×10^{-1}
Mouse	1.4×10^{12}	2.2×10^{9}	7.5×10^{-1}
Frog	1.4×10^{13}	2.2×10^{10}	7.6
Invertebrates			
Fruit fly	1.2×10^{11}	1.8×10^{8}	6.0×10^{-2}
Sea urchin	5.0×10^{11}	8.0×10^{8}	2.7×10^{-1}
Fungi			
Neurospora	1.8×10^{10}	2.7×10^{7}	9.2×10^{-3}
Baker's yeast	1.2×10^{10}	1.8×10^{7}	6.0×10^{-3}
Bacteria			
Pneumococcus	1.2×10^{9}	1.8×10^{6}	6.0×10^{-4}
E. coli	2.8×10^{9}	4.1×10^{6}	1.4×10^{-3}
Viruses			
Lambda	3.3×10^{7}	4.6×10^{4}	1.6×10^{-5}
T2	1.2×10^{8}	1.8×10^{5}	6.0×10^{-5}
SV40	3.5×10^{6}	5226	1.7×10^{-6}

■ **TABLE 9–2** Composition of chromatin.

Component	Pea Embryo*	Liver*
Histone proteins	36%	33%
DNA	31%	31%
Nonhistone proteins	28%	18%
RNA	5%	17.5%

*Data expressed as percent of dry weight.

ponents of chromatin. Histones are comparatively small, basic proteins that contain a large proportion of the positively charged amino acids arginine and lysine. Five different types of histones are found in chromatin, H1, H2A, H2B, H3, and H4; they occur in a molar ratio of approximately 1 H1:2 H2A:2 H2B:2 H3:2 H4. The five types differ in structure and molecular weight and in arginine and lysine content (■ Table 9–3).

Chromatin Structure: Nucleosomes

Studies on isolated chromatin have revealed how eukaryotic DNA interacts with proteins. Chromatin has been found to have a highly ordered structure in which histone proteins are combined with DNA to form repeating subunits called **nucleosomes**. The structure and organization of nucleosomes depend upon the histone composition of chromatin and on the ionic strength of the medium. At low ionic strength, chromatin that has been stripped of histone H1 looks like a beaded filament when viewed with an electron microscope; the beads are the basic structural components of nucleosomes. Each bead or **nucleosome core particle** is formed from a protein core and a length of DNA consisting of 146 nucleotide pairs; the DNA is wrapped one and three-quarters times around the protein core in a helix (▶ Figure 9–6a). Each protein core is a histone octamer that is composed of two molecules of each of the core histones H2A, H2B, H3, and H4.

The addition of histone H1 to the complex yields what is called a **minimal nucleosome** or **chromatosome**. In this structure, a single molecule of H1 is thought to associate with another 20 base pairs of filament DNA (10 pairs on each

Histone	Relative Proportion	Molecular Weight*	Arg/Lys Ratio	General Composition
H1	1	21,000	0.0345	Lys rich
H2A	2	14,500	0.818	Slightly lys rich
H2B	2	13,700	0.375	Slightly lys rich
H3	2	15,300	1.18	Arg rich
H4	2	11,300	1.27	Arg rich

*Values for calf thymus chromatin.

side of the 146 core pairs) so that two full turns of DNA (166 base pairs) are wrapped around the histone core (Figure 9–6b). Adjacent nucleosomes are connected by approximately 34 base pairs of internucleosomal linker DNA (the number varies among species and among the tissues of a species). This 10-nm nucleosome fiber has a zigzag appearance when viewed with an electron microscope (Figure 9–6c).

There are three major sources of evidence for the nucleosomal model of chromatin: X-ray diffraction analyses, electron microscopy, and biochemical studies of unfolded chromatin fibers. The biochemical studies have included nuclease digestion experiments that have used a bacterial enzyme, micrococcal endonuclease, to break down chromatin into smaller fragments (▶ Figure 9–7). When chromatin is exposed to this enzyme for a very brief time and the cut DNA is then isolated and subjected to gel electrophoresis, the result is a ladder of discrete bands corresponding to DNA fragments whose lengths are multiples of approximately 200 base pairs (Figure 9–7a). This result indicates that each mononucleosome is associated with about 200 base pairs of DNA. Further digestion with the enzyme results in two additional sets of cuts (Figure 9–7b). The first set of cuts removes the exposed linker region of DNA,

producing the 166 bp DNA segment of the chromatosome. The second set of cuts releases histone H1 and produces the 146 bp segment of the nucleosome core particle.

One major function of the nucleosomes is to package DNA into compact filaments that can more easily fit within the confines of the eukaryotic nucleus. Coiling each 200-base-pair segment of DNA about a protein core of approximately 10 nm thickness reduces the length of each nucleosomal DNA segment from 68 nm (200 bp × 0.34 nm/bp) to 10 nm. The DNA packing ratio (the ratio of the DNA length to the length of its container) resulting from nucleosome formation is therefore approximately seven:

$$\frac{68 \text{ nm}}{10 \text{ nm}} = 6.8$$

Higher-Level Coiling: Solenoids and Domains

Further packing is needed to form the so-called unit chromatin fiber of about 30 nm thickness that is thought to be the predominant form of chromatin in interphase nuclei. F. Thoma and Th. Koller have shown that as the ionic strength of the solvent is increased, a gradual condensation of the nucleosome fiber takes place (▶ Figure 9–8). At in-

▶ **FIGURE 9–6** Organization of DNA into nucleosomes. (a) Each core nucleosome consists of 146 base pairs of DNA wrapped in a helical fashion around a protein core that consists of eight histone molecules. (b) The association of a core particle with a molecule of histone H1 yields a minimal nucleosome, which contains a total of 166 base pairs of DNA. The ends of the 146 base pairs of the core particle are marked by bars to show the addition of ten more base pairs at each end of the core DNA. Histone H1 is thought to bind to the DNA at the site at which the DNA enters and leaves the nucleosome. (c) Because the entry and exit points are on the same side of the nucleosome, the nucleosome fiber takes on a zigzag shape; adjacent nucleosomes are separated by approximately 34 base pairs of linker DNA.

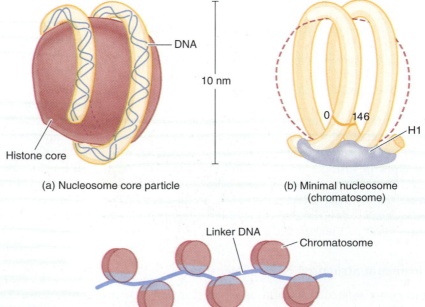

(a) Nucleosome core particle

(b) Minimal nucleosome (chromatosome)

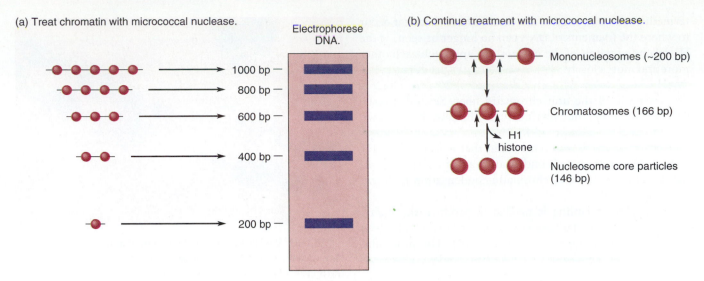

(a) Treat chromatin with micrococcal nuclease.

Electrophorese DNA.

1000 bp —
800 bp —
600 bp —
400 bp —
200 bp —

(b) Continue treatment with micrococcal nuclease.

Mononucleosomes (~200 bp)

Chromatosomes (166 bp)

H1 histone

Nucleosome core particles (146 bp)

► **FIGURE 9–7** Substructure of the chromatin fiber as revealed by partial digestion with the enzyme micrococcal endonuclease. (a) Increasing nuclease digestion of the polynucleosome and electrophoresis on an agarose gel results in a ladderlike sequence of DNA bands corresponding to multiples of the 200 bp associated with a mononucleosome. (b) Further trimming of the linker DNA produces chromatosomes with 166 bp of DNA, followed by nucleosome core particles with 146 bp of DNA. To produce nucleosome core particles, H1 histones must first be released.

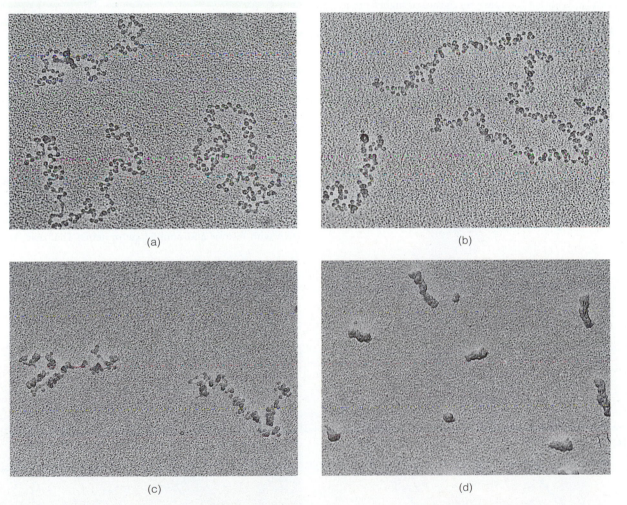

(a)

(b)

(c)

(d)

► **FIGURE 9–8** Electron micrographs of chromatin from rat liver cells, shown at increasing ionic strengths: (a) 1 mM NaCl, (b) 10mM NaCl, (c) 40 mM NaCl, and (d) 100 mM NaCl. The nucleosome filaments in (a) gradually condense to form closed, zigzag-shaped fibers in (b), then discontinuous, compact fibers in (c), and finally, continuous, compact fibers in (d). *Source:* Reprinted with permission from F. Thoma and Th. Koller, *J. Mol. Biol.* 149 (1981): 709–733. Copyright: Academic Press, Inc. (London) Ltd.

termediate ionic strengths, increases in fiber diameter occur to where the filaments of DNA can no longer be seen, as the H1-binding regions of the nucleosome approach each other more and more closely. High ionic strengths produce a cylindrical coil or **solenoid** with a diameter of 30 nm, which is characteristic of the unit chromatin fiber. Several investigators using different experimental techniques have shown that histone H1 is required for this condensation of chromatin into 30 nm fibers (chromatin that is stripped of H1 shows a less pronounced condensation and an irregular coiling pattern). A model of chromatin condensation is shown in ▶ Figure 9–9.

Much further coiling is needed to produce the highly compact state of metaphase chromosomes that can be seen in electron micrographs (▶ Figure 9–10). The molecular basis of the condensation of chromosomes from the inter-

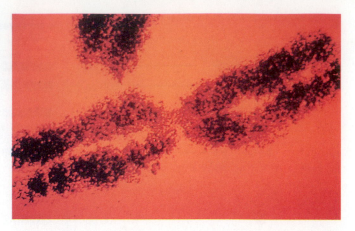

▶ **FIGURE 9–10** Electron micrograph of a human chromosome showing the centromere region located at the middle of the chromosome.

▶ **FIGURE 9–9** The condensation of chromatin containing H1 with increasing ionic strength. The open zigzag fiber closes up to form helices with increasing numbers of nucleosomes per turn (*n*). When H1 is absent (bottom), no zigzags or higher-order levels of organization are found. *Adapted from:* F. Thoma and Th. Koller, *J. Mol. Biol.* 149 (1981): 709–733. Copyright: Academic Press, Inc. (London) Ltd. Used with permission.

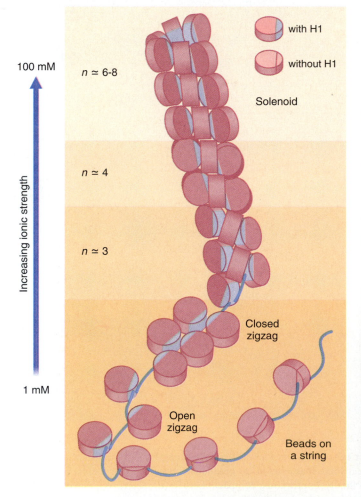

phase state to the metaphase state is still largely unknown. However, it is clear that at least part of this condensation is accomplished by the folding of the chromatin fiber into a very large series of looped **domains**. The looped structure is maintained by the attachment of the domains to some type of central **scaffold** consisting of nonhistone proteins. Evidence for this view has been provided by electron micrographs that reveal proteinaceous scaffolds remaining after the histones are removed from metaphase chromosomes (▶ Figure 9–11).

A model summarizing the structural hierarchy of nuclear chromosomes is shown in ▶ Figure 9–12.

To Sum Up

During interphase, nuclear chromosomes exist as dispersed chromatin fibers of about 30-nm diameter. They consist of a complex of DNA with histone and nonhistone proteins. The basic structural

▶ **FIGURE 9–11** Electron micrograph of a histone-depleted chromosome showing a proteinaceous scaffold to which loops of DNA are anchored.

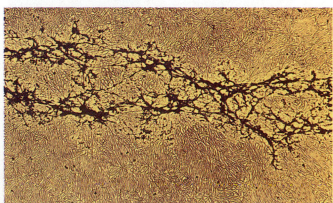

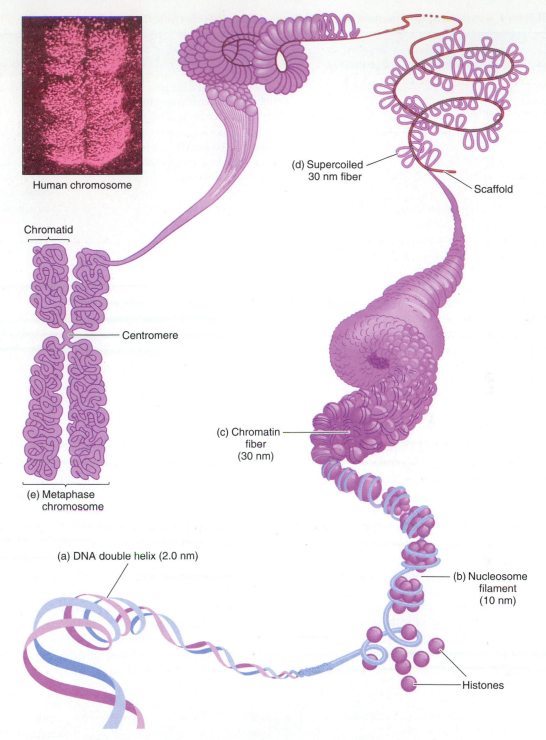

Human chromosome

Chromatid

Centromere

(e) Metaphase
chromosome

(d) Supercoiled
30 nm fiber

Scaffold

(c) Chromatin
fiber
(30 nm)

(a) DNA double helix (2.0 nm)

(b) Nucleosome
filament
(10 nm)

Histones

▶ **FIGURE 9–12** A model of the structure of a eukaryotic chromosome, showing the hierarchy of folding into a metaphase chromosome arm. The loops of 30 nm fiber are anchored to a nonhistone scaffold.

unit of nuclear chromatin is the nucleosome, a segment of DNA wrapped in helical fashion around a protein core of eight histone molecules. The 30-nm chromatin fiber is formed from condensation of a nucleosome fiber into a solenoid-type structure involving histone H1. Higher-level coiling of this solenoid structure produces a microscopically-visible mitotic chromosome.

▧ REPETITIVE DNA SEQUENCES

As we have seen in our discussions of centromeres and telomeres, some of the DNA of eukaryotic chromosomes is **repetitive**, that is, it consists of sequences that are repeated many times. These sequences vary in length and

■ TABLE 9–4 Characteristics of the three major renaturation classes of eukaryotic DNA.

Class	Percentage of Genome	Repetition Frequency	General Features	Function
Highly repetitive	0–50	10^5–10^7	5–300 bp per repeating unit; located mainly in centromeres and telomeres; often clustered in long tandem arrays	none known
Moderately repetitive	10–50	10^1–10^5	10^2–10^7 bp per repeating unit; located throughout the genome, often in a regular dispersion pattern; composed of sequence families	except for repeated genes, such as those coding for rRNA, tRNA, histones, and antibodies, function remains unclear
Unique sequence	10–80	1	> 10^8 bp per repeating unit	structural genes

base composition and are repeated from a few to several million copies per genome, either in tandem (one after another) at certain chromosomal locations (e.g., centromeres and telomeres) or dispersed through the DNA. The repeated sequences of eukaryotic DNA are in sharp contrast to the unique (single-copy) sequences that are the predominant form of prokaryotic DNA.

Sequence Organization

The repetitive sequences in eukaryotic DNA can be conveniently divided into two broad classes, **highly repetitive** and **moderately repetitive**, according to their repetition frequency. The characteristics of both types are given in ■ Table 9–4, along with those of unique-sequence DNA.

Highly repetitive DNA is composed of short (5–300 bp) sequences that are repeated about 10^5 to 10^7 times, often in large tandem arrays. In *D. virilis*, for example, sequences such as ACAACT, ATAACT, and ACAATT are each repeated over a million times within the genome. Because of differences in GC content such simple-sequence DNA can often be separated from the bulk of chromosomal DNA by centrifugation in a CsCl gradient (see Chapter 7). DNA fragments containing these sequences form **satellite bands**, which differ in buoyant density from the main-band DNA (▶ Fig-

ure 9–13). For this reason, such DNA segments are often referred to as **satellite DNA.**

The short, reiterated sequences that make up highly repetitive DNA are located mainly at the centromeres and, to a lesser extent, in the arms and telomere regions of chromosomes. The functions of these sequences are largely unknown. It is currently believed that the highly repetitive sequences at the centromere have a structural role, possibly in the binding of spindle proteins. This DNA has no known genetic function and is considered genetically inert.

Moderately repetitive DNA consists of sequences that are repeated 10 to 10^5 times within the haploid genome. This class of DNA differs from highly repetitive DNA in a number of important respects. First, moderately repetitive DNA does not just contain simple repetitions of one kind of sequence—it is instead composed of **sequence families**. A sequence family includes regions that are similar enough in base sequence to hybridize with one another under standard renaturation conditions. Members of different families do not hybridize. Even within a sequence family, hybridization occurs with a certain amount of mismatched base pairs. Members of the same family are therefore related, most likely having evolved from the same ancestral sequence, but are not identical. Some of these sequences are members of gene families that are related to each other by the criterion

▶ **FIGURE 9–13** Eukaryotic DNA separated into a main band and a satellite band by equilibrium density-gradient centrifugation.

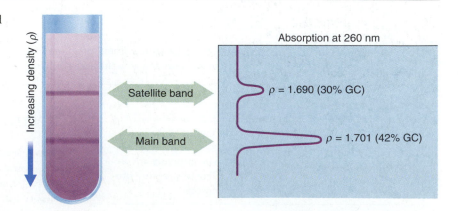

Fragile X Syndrome

Fragile X syndrome is the most common cause of inherited mental retardation. It draws its name from a chromosomal deformity in which the tip of the long arm of the X chromosome is attached by only a slender thread of DNA. In 1991, research groups in France, Australia, and the United States reported the discovery of the underlying defect in fragile X syndrome. A gene called *FMR-1*, which should contain fifty or fewer copies of the trinucleotide repeat sequence CGG/GCC, is altered in affected individuals. Healthy carriers of the trait have more repeats of this sequence (as many as 200) than do noncarriers, but their retarded children have hundreds or thousands of extra copies. In passing from carriers to affected persons, the fragile X site sometimes expands rapidly in its number of CGG/GCC repeats.

At least three other human diseases—myotonic dystrophy (the common form of muscular dystrophy in adults), spinal and bulbar dystrophy, and Huntington's disease—also appear to be caused by genes that expand rapidly between generations. For example, the severity of myotonic dystrophy correlates with massive duplication of CTG/GAC repeats at a particular chromosomal site.

It thus appears that asymptomatic individuals and people with very mild forms of these diseases carry genes that are functional but are unstable structurally. These genes seem to gain a few copies of each repeated unit with each generation, and there may be some kind of threshold number of units beyond which a sudden, rapid increase in the number of trinucleotide repeats results in the diseased state. Thus, some men who carry the fragile X chromosome are normal, and their children are also normal, because their trinucleotide sequence has expanded only minimally. The grandchildren of the original carriers are often retarded, however, because the repeat region has dramatically expanded in length. Inheritance patterns like this do not follow Mendelian rules, making predictions of the occurrence of the disease in a family very difficult if not impossible. The instability of these mutations may also provide an explanation for incomplete penetrance and variable expressivity (see Chapter 4).

The precise mechanism that causes an expansion in the number of triplet repeats from one generation to the next remains unknown (the regions also occasionally shrink, further complicating the picture). Nor do researchers know why the sex of the parent transmitting the gene affects the trinucleotide repeat expansions. The fragile X gene seems to expand most rapidly in individuals who have inherited the fragile chromosome from their mother, while in Huntington's disease, the enlargement usually occurs when the mutant gene is transmitted by the father. The discovery of this unprecedented type of mutation raises new questions about the role of genetics in human disease, about the influence of the sex of the transmitting parent on the mutation, and about how the human genome operates normally. It will be interesting to determine how widespread this mutational mechanism is and whether it operates in organisms other than humans.

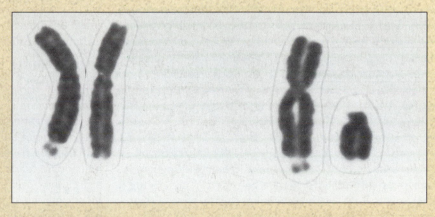

Mutation in fragile X syndrome makes the tip of the X appear as though it were fragile and hanging from a thread. (a) Heterozygous female. (b) Affected male.

(a)   (b)

that they encode a similar gene product. The genes coding for the chromosomal histone proteins fall into this group. However, many of the other moderately repetitive sequences do not appear to have a protein coding function. Moderately repetitive DNA thus consists of a complex collection of sequence families that, for convenience, are grouped together because they renature at somewhat similar rates.

A second important feature that distinguishes moderately repetitive from highly repetitive DNA involves the chromosomal locations of these sequence classes. Most moderately repetitive DNA families are dispersed throughout the genome rather than clustered in tandem arrays. In general, there is a certain degree of regularity in their distribution pattern. The most common pattern is short-period interspersion, in which short, interspersed, repeated sequences (*SINES*) of approximately 300 bp alternate with unique sequences of 1000 to 3000 bp. A less common pattern is long-period interspersion, which involves long, interspersed, repeated sequences (*LINES*) of 5000 to 7000 bp alternating with even greater lengths of unique-sequence DNA. The nature of *SINES* and *LINES* is currently a topic of intensive research. Studies reveal that some of these interspersed elements resemble DNA copies of RNA viral genomes, while others appear to be DNA copies of cellular RNA molecules. These elements serve no obvious functions within the cell and have been regarded by some geneticists as "molecular parasites" that infest the genome but rarely confer a selective advantage on the organism. Because of their similarities to known RNA forms, further discussion of these sequence elements will be postponed until Chapter 16, after we have considered more fully the interrelationships between RNA and DNA.

Renaturation Rate and Sequence Complexity

The presence of repeated sequences can be demonstrated experimentally by a study of the kinetics of renaturation of single-stranded DNA. When complementary strands of DNA are separated in solution and are then allowed to renature, they anneal together at a rate that depends on how rapidly the strands collide and remain attached by hydrogen bonding. To determine this rate, the time course of renaturation is generally followed experimentally by measuring the fraction of DNA still in the single-stranded form after different times of incubation. The results can then be graphed by displaying the data on semilogarithmic coordinates as a plot of C/C_0 against the logarithm of C_0t, where C is the concentration of single-stranded DNA remaining after t seconds of incubation and C_0 is the concentration of single-stranded DNA at time zero. The resulting curve, which is usually referred to as a **cot curve** ("cot" being a colloquialism for C_0t), thus describes the progress of the renaturation in terms of the product C_0t.

An idealized example of a cot curve is shown in ▶ Figure 9–14. The cot curve is an S-shaped curve showing that about 80% of renaturation occurs over a 100-fold range of C_0t (between 10^{-1} and 10 in this example). The inflec-

tion point at 50% renaturation is designated $C_0t_{\frac{1}{2}}$, since it marks the midpoint of the transition from single-stranded to double-stranded DNA. Note that a smaller value of $C_0t_{\frac{1}{2}}$ implies a faster rate of renaturation (a shorter time is required to achieve half-renaturation), while a larger value of $C_0t_{\frac{1}{2}}$ indicates a slower reaction. $C_0t_{\frac{1}{2}}$ is therefore inversely related to the renaturation rate, and its measured value can be used to compare the reassociation potential of different DNA samples.

Studies on the kinetics of renaturation provide information on the amount of sequence repetition through observations of the **DNA complexity**. The complexity of a DNA sample is defined as the sum of the lengths of its unique (single copy) sequences plus the sum of the unit lengths of its repetitive sequence classes. For example, if a DNA sample were to consist entirely of one repeating sequence, say AGCTAGCTAGCT . . . , the DNA complexity would be 4—the unit length of the repeated sequence. If, at the other extreme, there is no repetition of sequence elements, the complexity of the DNA sample would equal the genome length. Equality between DNA complexity and total genome length is typical of prokaryotes and their viruses, in which there is no appreciable sequence repetition. The DNA of these organisms consists of a single kinetic class that renatures according to the pattern shown in Figure 9–14. The cot curves of such DNA are all similar in shape (see ▶ Figure 9–15) but are located at different positions along the C_0t axis, each having a $C_0t_{\frac{1}{2}}$ value that is proportional to its genome length. Thus, for example, E. coli DNA, with a genome length of 4×10^6 bp, has a $C_0t_{\frac{1}{2}}$ that is approximately 20 times greater than that of phage T4, which has a genome length of about 2×10^5 bp.

Unlike prokaryotes, the DNA of eukaryotes is characterized by sequence repetition, so its complexity is less than the total genome length. For example, suppose that the genomic DNA of a eukaryote consists of 10^6 copies of a highly repetitive sequence X bp long, 10^2 copies of a moderately repetitive sequence Y bp long, and 1 copy of a (unique) sequence Z bp long. The complexity of this DNA is then $X + Y + Z$, compared with a total genome length of $10^6X +$

▶ **FIGURE 9–14** Idealized cot curve for nonrepetitive DNA. The curve shown has a value of $C_0t_{\frac{1}{2}} = 1$ when 50% of the single strands have renatured.

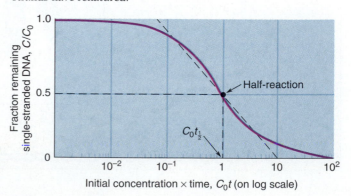

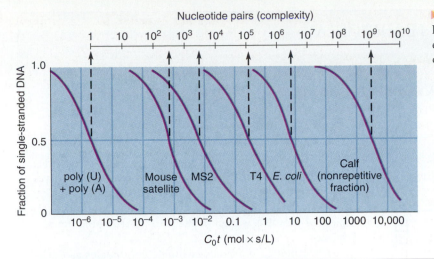

▶ **FIGURE 9–15** C_0t curves for various genome lengths show a direct relation between the experimentally determined $C_0t_{\frac{1}{2}}$ value and the complexity of the DNA.

$10^2Y + Z$. As this example shows, the disparity between DNA complexity and genome length depends on the complexities of the individual sequences X, Y, and Z. Sequence complexities such as these can be measured under appropriate conditions from the individual $C_0t_{\frac{1}{2}}$ values of the different sequence components using the following equation:

$$C_0t_{\frac{1}{2}} = AX \qquad (9.1)$$

where X is the sequence complexity (in base pairs) of a renaturing component and A is a proportionality constant that depends on reaction conditions. The direct dependence of $C_0t_{\frac{1}{2}}$ on sequence complexity reflects the fact that as the length (complexity) of a given sequence class increases, the renaturation rate decreases, since there are fewer copies of this sequence in the fixed initial quantity (C_0) of DNA.

The cot curve of human DNA shown in ▶ Figure 9–16 typifies the complex pattern of renaturation shown by eukaryotic DNA. This curve is much broader than would be expected for a kinetically pure sample of DNA, since it comprises the sum of the individual cot curves of various kinetic classes that renature at different rates. The simplest interpretation of these results is shown in ▶ Figure 9–17,

▶ **FIGURE 9–16** A cot renaturation curve for human DNA. The complex renaturation pattern indicates the presence of more than one renaturing class. *Adapted from:* C. W. Schmid and P. O. Deininger, "Sequence organization of the human genome," *Cell* 6 (1975): 345–358. Copyright © by Cell Press. Used with permission.

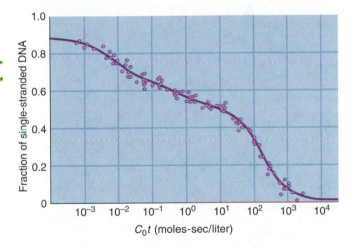

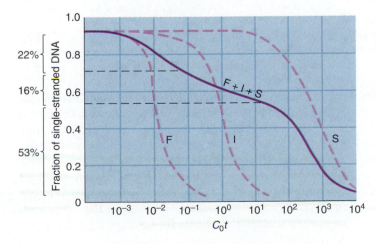

▶ **FIGURE 9–17** Simplest interpretation of the renaturation pattern in the cot curve for human DNA. The main curve (solid line) is the sum of the separate components (dashed lines), adjusted for their relative contributions. F, I, and S are the individual curves of the fast (hightly repetitive), intermediate (moderately repetitive), and slow (unique-sequence) renaturing components. The estimated plateaus of the individual cot curves give the percentage of each component class in the total DNA.

Measuring the DNA Renaturation Rate

The renaturation of DNA in solution is a two-step process involving the following two reactions:

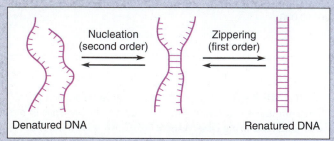

Nucleation (second order) — Zippering (first order)

Denatured DNA | Renatured DNA

The first step is the slow, bimolecular process of nucleation, which brings the nucleotides of complementary chains into register, and the second is a more rapid unimolecular "zippering" reaction that reforms the double helix. Under the conditions normally employed for renaturation, the bimolecular nucleation process is the rate-limiting step. Since this step proceeds through the random collision of complementary DNA sequences, the overall rate of renaturation depends on the product of the concentrations of the complementary chains. In mathematical terms, this means that renaturation occurs at a rate proportional to the square of the denaturated DNA concentration, and it is therefore said to follow second-order kinetics. Thus if we let C equal the concentration of single-stranded DNA at time t, the rate of decrease in C then follows the second-order rate equation:

$$-\frac{dC}{dt} = kC^2$$

This equation integrates to

$$\frac{C}{C_0} = \frac{1}{1 + kC_0t}$$

where C_0 is the concentration of single-stranded DNA at time zero and k is the second-order rate constant. Observe that when $t = 0$, $C/C_0 = 1$ and all DNA is still single-stranded. When t gets large, C/C_0 approaches zero, and most of the DNA is then in the form of renatured double helices.

When plotted on semilogarithmic coordinates as C/C_0 vs. log C_0t, the integrated form of the rate equation gives a relationship exemplified by the idealized cot curve in Figure 9–14. Note that at half-renaturation,

$$\frac{C}{C_0} = 0.5 = \frac{1}{1 + kC_0t_{\frac{1}{2}}}$$

which rearranges to give

$$k = \frac{1}{C_0t_{\frac{1}{2}}}$$

where $C_0t_{\frac{1}{2}}$ is the value of C_0t when renaturation is half-complete. It is thus possible to evaluate k from the experimentally determined value of $C_0t_{\frac{1}{2}}$.

Studies reveal that k and hence $1/C_0t_{\frac{1}{2}}$ depend on the temperature and ionic strength of the solution and on the length of the renaturing DNA. Measures of the rate of renaturation are therefore determined experimentally at a constant temperature and salt concentration from samples in which DNA has been sheared into fragments of more or less uniform length. To measure the time course of the reassociation process, samples are removed periodically from the reaction mixture and passed through a column of hydroxyapatite (calcium phosphate) crystals. Under appropriate ionic conditions, only double-stranded DNA binds to this column, so that the amount of renatured DNA formed in a given time interval can be readily determined.

in which the three component curves represent three renaturing classes of DNA: (1) a fast-renaturing class composed of highly repetitive sequences, (2) a heterogeneous intermediate-renaturing class composed of moderately repetitive sequences, and (3) a slow-renaturing class composed of nonrepetitive sequences (unique-sequence DNA). In this interpretation, the estimated plateaus of the three component cot curves give the percentage of each renaturing class in the total DNA. Observe that in humans, as in many eukaryotes, the overall curve does not extrapolate back to unity because of the presence of a fourth class of DNA that renatures faster than the smallest renaturation time (or C_0t value) that can be detected experimentally. This extremely rapid renaturing component (called foldback DNA) consists of inverted-repeat (palindromic) sequences that can renature by folding back on themselves to form hairpinlike structures (▶ Figure 9–18). This class comprises about 9% of human DNA.

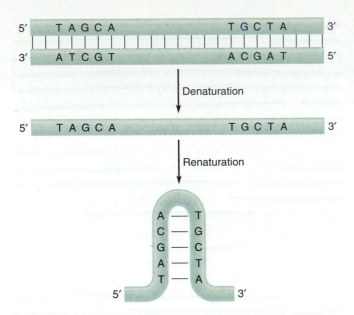

Denaturation

Renaturation

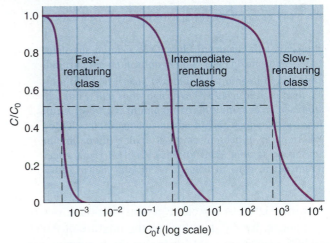

(and thus consist of 2×10^8 bp, 4×10^8 bp, and 4×10^8 bp, respectively), with respective $C_0 t_{\frac{1}{2}}$ values of 10^{-3}, 1, and 10^3. Interpretation of the cot curve in terms of the individual renaturation curves of the isolated components then gives:

Note that the $C_0 t_{\frac{1}{2}}$ values expected for the isolated sequences are smaller than those in the mixture because C_0 along the $C_0 t$ axis for the mixture is the total DNA concentration, whereas C_0 on the corresponding axis of the individual curves is the concentration of each isolated component. To obtain the $C_0 t_{\frac{1}{2}}$ values for the pure isolated samples, simply multiply the proportion of each component by its $C_0 t_{\frac{1}{2}}$ value in the mixture, that is, $(0.2)(10^{-3}) = 2 \times 10^{-4}$, $(0.4)(1) = 0.4$, and $(0.4)(10^3) = 4 \times 10^2$.

(a) To evaluate the sequence complexities, recall that the $C_0 t_{\frac{1}{2}}$ value is proportional to the sequence complexity and that the complexity of the nonrepetitive class (assumed to be the slow-renaturing component in this case) is equal to its total length. Thus, if we use the nonrepetitive component as the internal standard, we can write

$$\frac{\text{complexity of repetitive component}}{C_0 t_{\frac{1}{2}} \text{ of repetitive component}} = \frac{\text{complexity of nonrepetitive component}}{C_0 t_{\frac{1}{2}} \text{ of nonrepetitive component}}$$

Taking the complexity of the nonrepetitive component in this case to be $(0.4)(10^9) = 4 \times 10^8$ bp, the complexities of the highly repetitive and moderately repetitive sequences can then be calculated from the preceding relationship as

$$\frac{(4 \times 10^8)(2 \times 10^{-4})}{4 \times 10^2} = 2 \times 10^2 \text{ bp}$$

and

$$\frac{(4 \times 10^8)(0.4)}{4 \times 10^2} = 4 \times 10^5 \text{ bp}$$

respectively.

(b) These sequence complexities can now be used to determine the number of copies of each repeating sequence (repetition frequency) in the DNA as follows:

► **FIGURE 9–18** Renaturation of foldback DNA. Upon denaturation, inverted-repeat (palindromic) sequences in DNA can rapidly renature by forming base-paired hairpin loop structures instead of combining with a complementary strand and forming a duplex.

Example 9.1

The following graph is an idealized cot curve of the DNA from a hypothetical haploid eukaryote with a total genome size of 10^9 bp:

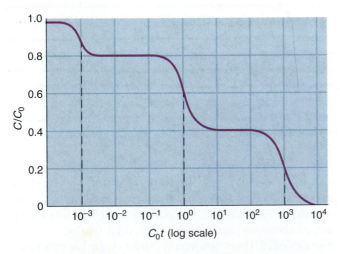

Suppose that you could isolate the DNA of each renaturing class in pure form and rerun cot curves on these isolated components. Construct a graph showing the cot curves of the isolated components, and determine (a) the sequence complexity of each kinetic class and (b) the number of copies (repetition-frequency) of each repetitious sequence.

Solution: First observe that the fast, intermediate, and slow components of this DNA occur in a ratio of $0.2 : 0.4 : 0.4$

$$\text{repetition frequency} = \frac{\text{total length of DNA component}}{\text{complexity of DNA component}}$$

Assuming that each kinetic class consists of only one repeating sequence, the DNA in the above example consists of

$$\frac{2 \times 10^8}{2 \times 10^2} = 10^6$$

copies of the highly repetitive sequence and

$$\frac{4 \times 10^8}{4 \times 10^5} = 10^3$$

copies of the moderately repetitive sequence.

Follow-Up Problem 9.1

The genome of a eukaryote consists of 10% fast-renaturing DNA, 30% intermediate-renaturing DNA, and 60% slow-renaturing DNA. A cot curve reveals that these renaturing components have $C_0 t_{\frac{1}{2}}$ values in the mixture of 0.002, 30, and 500, respectively. (a) Given that the sequence complexity of the fast-renaturing component is 10^2, what is the haploid genome size (in bp) of this eukaryote? (b) What is the sequence complexity of the intermediate-renaturing component? (c) Calculate the number of copies of the repeated sequence in the fast-renaturing component.

To Sum Up

1. Eukaryotic nuclear chromosomal DNA contains repeated nucleotide sequences. Different classes of repetitive DNA have different chromosomal locations and functions. Highly repetitive DNA sequences are repeated many times throughout the genome of an organism, tend to be clustered in tandem arrays, and are located in the nongenic regions of chromosomes, such as the centromere, where they might serve a structural role. Moderately repetitive DNA is composed of families of related sequences, some of which represent gene families (whose members code for similar proteins). In contrast to highly repetitive DNA, moderately repetitive sequences tend to be dispersed throughout the genome, some forming distribution patterns referred to as *SINES* and *LINES*.
2. The degree of repetitiveness of DNA, along with its sequence complexity, influences its rate of renaturation in cot curve analysis. Analysis of a cot curve can reveal the proportion of the genomic DNA in each renaturing class, the sequence complexity of each class, and the number of copies of each repeating sequence.

CHROMOSOME REPLICATION AND THE CELL CYCLE

Studies reveal that DNA replication in eukaryotes is in many respects similar to the process in prokaryotes. Like replication in bacteria, replication of DNA in eukaryotes requires many different enzymes. Five different DNA polymerases, referred to as α, β, γ, δ, and ϵ, have been identified in mammalian cells. The α and δ polymerases are thought to be the main enzymes catalyzing nuclear DNA replication, while the β and ϵ polymerases are believed to function in repair. The γ polymerase is localized primarily in the mitochondria, where it functions in the replication of mitochondrial DNA.

Current evidence also indicates that eukaryotic DNA replication, like that in bacteria, requires a variety of other protein activities. These activities apparently involve primase, ligase, helicases (to unwind and separate the parental DNA strands), single-strand binding proteins (to prevent the parental strands from reforming a helix or other folded configuration), and topoisomerases.

The eukaryotic DNA polymerases synthesize DNA in much the same manner as the corresponding prokaryotic enzymes. They all catalyze the template-directed 5' to 3' synthesis of DNA from an RNA primer. Some, including the δ and ϵ polymerases, also have associated exonuclease activity. Nevertheless, some important features of chromosome replication are unique to eukaryotic cells, and they are discussed in the following section.

Replicons and Replication in Eukaryotes

Eukaryotes differ most clearly from prokaryotes in their mode of DNA replication by having multiple replication origins. While each circular prokaryotic chromosome constitutes a single replicon, having only one replication initiation site, each linear eukaryotic chromosome usually consists of many replicons (▶ Figure 9–19). Each replicon in a eukaryotic chromosome extends over a length of DNA that has its own replication origin. As in prokaryotes, DNA replication is discontinuous on at least one of the template strands and is bidirectional, so that every replicon has two replication forks. However, the rate of movement of the replication forks is much slower than in prokaryotes (presumably because of the limiting effect of the attached histone proteins), and the Okazaki fragments that are formed are much shorter; each extending only to about 200 nucleotides in length.

The existence of multiple replication origins in eukaryotic chromosomes is not surprising, considering the vastly greater amount of DNA that must be replicated by higher organisms. For example, the human haploid genome consists of about 600 times more DNA than that of *E. coli*. If fully extended, this DNA would form a single bacteriumlike chromosome almost a meter in length. Large numbers of replication origins are therefore needed to complete the duplication of chromosomes within the normal time frame provided by the eukaryotic cell cycle.

Geneticists have been successful in identifying multiple replication origins in yeast. These origins are associated with AT-rich sequences of DNA called autonomously replicating sequences or ARS elements (ARS elements are used as the replication origins in constructing YACs). The ARS elements

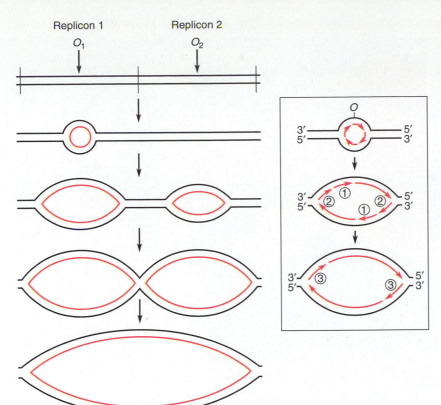

Replicon 1 Replicon 2
O_1 O_2

▶ FIGURE 9–19 A section of a replicating eukaryotic chromosome containing two replication origins (O_1 and O_2). The separation of the DNA strands at these points allows synthesis of new DNA (red lines) to proceed in both directions from each origin, forming replication "bubbles." Each replication unit is called a replicon. Different replicons do not necessarily replicate at the same time, as illustrated here. The insert shows the discontinuous nature of DNA synthesis within each bubble; each Okazaki fragment (numbered in order of replication) is about 200 nucleotides long.

are identified by inserting random segments of yeast DNA into bacterial plasmids using recombinant DNA technology (see Chapter 18). Bacterial plasmids are normally incapable of replicating in yeast, since they lack a yeast replication origin. However, when yeast DNA is inserted into the plasmid DNA, an integrated ARS region is sufficient to initiate DNA synthesis in a yeast cell, thus allowing the plasmid to replicate in the foreign host. The integrated ARS region must be free of nucleosomes in order to initiate replication; whether a nucleosome-free origin is a general necessity for the initiation of DNA replication is still unknown.

The Eukaryotic Cell Cycle

Another major distinction between chromosome replication in eukaryotes and prokaryotes involves the timing of DNA synthesis. Unlike prokaryotes, in which DNA replication continues throughout each cell-division cycle, most DNA synthesis in eukaryotic cells is restricted to a limited period during interphase. Time gaps occur between chromosome duplication and mitosis. It is therefore customary to divide the cycle of eukaryotic cells into four phases: (1) a presynthetic gap, G_1, (2) a period of DNA synthesis, S, (3) a postsynthetic gap, G_2 and (4) a period of nuclear division, M (for mitosis). The gaps, G_1 and G_2, are periods of growth and metabolic activity in which the cell prepares for subsequent DNA synthesis and mitosis. These two gaps along with S ($G_1 + S + G_2$) make up the period of interphase.

The cell cycle is qualitatively the same in different eu-

karyotes, but it can vary in length under optimum conditions from one cell type to another. In a typical mammalian cell growing in culture, G_1 lasts about 11 hours, S takes about 8 hours, G_2 lasts for 4 hours, and M is completed in 1 hour, for an average cell cycle of 24 hours (▶ Figure 9–20). Most of the variation in the length of the cell cycle can be attributed to differences in G_1. This phase may be very short or nonexistent, as in the rapidly dividing cells of certain embryos, or very long, as in the nondividing nerve and muscle cells of adult animals. In the G_1 phase the cell either becomes committed to completing the cycle and progressing to S, or else it ceases proliferation and enters a quiescent state called G_0. For this reason, differentiated cells that do not normally divide appear to be arrested at G_1. Similarly, starved cells and those restricted by density-dependent growth inhibition stop dividing in G_1 but will continue through the cycle once the proper growth conditions are restored.

Example 9.2

A certain line of mammalian cells growing in culture has an S phase of 8 hours. DNA synthesis in these cells during S occurs at a rate of 2000 base pairs per minute at each replication fork. If the DNA in each cell nucleus has a total length of 1.36×10^9 nm, what is the minimum number of replicons that must be operating in each cell during chromosome duplication?

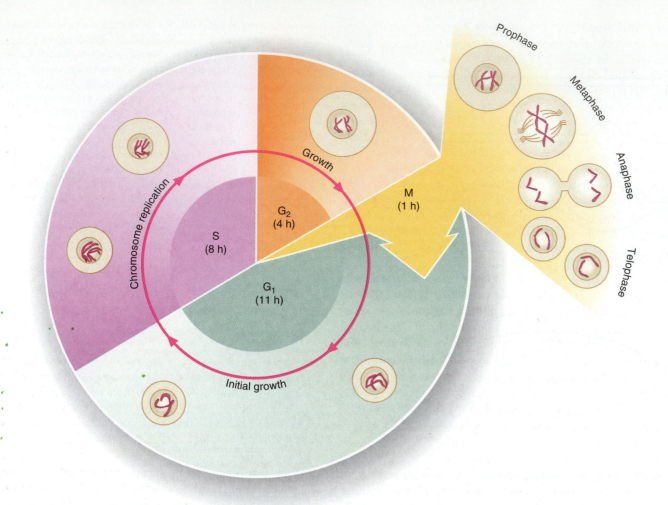

► **FIGURE 9–20** Phases of the life cycle of a dividing cell. The cell cycle is normally divided into four phases: G_1 (the period prior to DNA synthesis), S (the period of DNA synthesis), G_2 (the period between DNA synthesis and mitosis), and M (mitosis). The time spent by a typical mammalian cell in culture in each different stage is shown.

Solution: Each cell has a total of

$$\frac{1.36 \times 10^9 \text{ nm}}{0.34 \text{ nm/bp}} = 4 \times 10^9$$

base pairs of DNA. Since there are two replicating forks per replicon, each replicon can synthesize

$$(2)(2000 \text{ bp/min})(8 \text{ h})(60 \text{ min/h}) = 1.92 \times 10^6 \text{ bp}$$

during the 8-hour S phase. Therefore, the number of replicons needed to replicate all the DNA in a cell (assuming all operate simultaneously) is

$$\frac{4 \times 10^9}{1.92 \times 10^6} = 2083$$

Follow-Up Problem 9.2

Suppose that you discover a eukaryotic cell with one haploid chromosome that replicates its DNA bidirectionally from a single replication origin (as in prokaryotes). If the length of the DNA in the chromosome is 3.4×10^6 nm and DNA replication proceeds at 2000 bp per minute at each replicating fork, what is the minimum time required for S in this organism?

———————————————◆———————————————

If the cell does not enter a quiescent G_0 state, it will eventually reach a point in G_1 when it becomes irreversibly programmed to complete the S phase and subsequently divide. Uncovering the genes and gene products that govern the transition from G_1 to S and therefore control the cell's decision as to whether or not to divide has been a matter of intense research. One approach to this problem has involved **cell fusion experiments**, in which somatic cells in different phases of the cycle are induced to fuse so that their nuclei function together in one cell. (In Chapter 15 we will consider the procedures commonly used to induce cell fusion.) For example, when a cell in S is joined with one in G_1, the G_1 nucleus begins to synthesize DNA prematurely, suggesting that some activating signal for DNA synthesis is present in S-phase cells. G_2 nuclei are apparently refractory to this signal, since fusion of a cell in S with one in G_2 does not lead to DNA synthesis in G_2 chromosomes.

The transition of G_1 cells into the S phase is marked by the initiation of DNA synthesis at certain replicons, with some chromosomal regions undergoing replication earlier than others. Typically, functionally active genes replicate early, while the relatively inactive, tightly coiled regions of the genome (**heterochromatic regions**) replicate late.

Accompanying DNA replication is the assembly of new nucleosomes to form the daughter chromatin strands. It is still not known whether parental histones dissociate from DNA during replication. The fate of the parental histones in post-replicative chromatin is also unclear. The most recent evidence suggests that the core histones from the unreplicated region are transferred randomly to the newly synthesized leading and lagging strands (▶ Figure 9–21). It may be that core histones are not transferred to new DNA as complete ocatamers, however. Recent studies indicate that nucleosome formation is a sequential, two-step process in which deposition of H3/H4 tetramers onto the newly synthesized DNA molecules occurs first, followed by a random association of old and new H2A/H2B dimers with old and new H3/H4 tetramers. Thus, most of the histone cores on newly replicated DNA may actually be hybrids formed from a mixture of old and new histone proteins.

Chromosome replication in the S phase is ordinarily followed by a progressive condensation of the chromatin fibers. Chromatin condensation begins in G_2 and continues into metaphase of mitosis, when the chromosomes reach their highest level of compaction. Although little is known about the molecular basis of condensation, the phosphorylation of histone H1 is thought to be involved in the process. Cell fusion experiments show that condensation appears to be controlled by specific factors that appear in the nuclei of dividing cells. When an interphase cell is fused to an M phase cell, chromosome condensation begins to occur in the interphase nucleus, regardless of what particular stage (G_1, S, or G_2) it is in; the premature condensation of chromosomes does not require DNA synthesis, however, since the chromosomes that condense in G_1 are still unduplicated.

Another approach to the study of cell cycle control is to isolate temperature-sensitive *cdc* (*cell division cycle*) **mutants**. These mutants have specific genetic blocks at different points in the cycle and display their mutant phenotypes only at high (restrictive) temperatures. Typically, these mutant cells fail to complete the cycle at the restrictive temperature and are arrested at a point in the cycle where the functional form of their mutated gene would normally act. Through experiments with these *cdc* mutants, various genes that are essential to normal cell-cycle function have been identified in yeast and in other cell types.

Studies on *cdc* mutants have revealed that cell-cycle control is exerted at two main points in the cycle: One (designated **Start** in yeast) occurs toward the end of G_1 and controls entry into S; the other occurs at the end of G_2 and controls the completion of mitosis. Specific proteins called **cyclins** have been implicated in controlling progress past these points. One class of cyclins, the G_2-cyclins, is believed to act as a mitotic trigger, driving the cell through M phase when its concentration exceeds a threshold value. The intracellular concentration of the G_2-cyclins has been found to oscillate with the same periodicity as the cell cycle, rising steadily from zero during interphase and then suddenly dropping back to zero as the result of degradation during mitosis. G_2-cyclins appear to function as regulatory subunits of enzymes called **cyclin-dependent kinases (cdk)**. A cdk catalytic subunit alone is inactive, but when combined

▶ **FIGURE 9–21** Random distribution of parental histone cores during DNA synthesis. Parental histones are shown as purple spheres. (The nucleosomes on the replicated DNA molecules may actually be hybrids of old and new histones.)

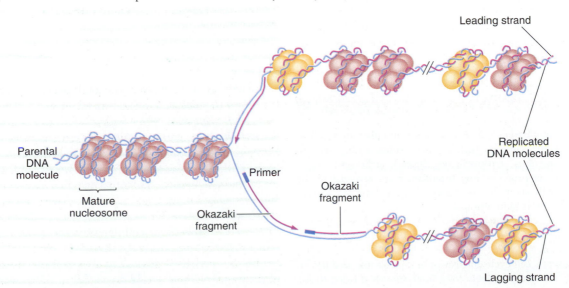

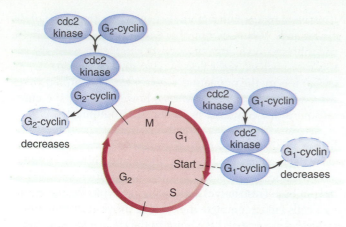

▶ **FIGURE 9–22** A model of the role of cyclins and the protein cdc2 kinase in the control of the cell cycle. In this model, a threshold level of cyclin regulates key points in the cell cycle. The G_1-cyclin–cdc2 complex functions at Start, the time at which cells become committed to division, and the G_2-cyclin–cdc2 complex triggers mitosis.

with a cyclin, it forms an active holoenzyme that can phosphorylate other proteins including histone H1.

A second class of cyclins, the G_1-cyclins, is involved in the decision to pass Start. Like G_2-cyclins, the G_1-cyclins appear to function as regulatory subunits of cdk's; they also oscillate in concentration with the same periodicity as the cell cycle, but they peak during G_1. ▶ Figure 9–22 gives a unified view of the role of the cyclins as regulatory subunits of a cdk (called **cdc2 kinase**) that functions in cell-cycle control in the fission yeast *Saccharomyces pombe*. As this model suggests, the mechanisms involved in the passage of Start and the induction of mitosis are fundamentally similar, with both occurring through the phosphorylation of proteins.

Although our knowledge of cell-cycle control is still incomplete, it is clear that the factors that control the structural transitions undergone by chromosomes during the cell cycle are the same factors that play major roles in the molecular control of that cycle. With the current research activity in this area, we should soon have a more detailed picture of the genetic control mechanisms operating in the eukaryotic cell cycle.

To Sum Up

1. The enzymatic basis of DNA replication in eukaryotes is fundamentally the same as that in prokaryotes, with primases, DNA polymerases, exonucleases, ligases, helicases, single-strand binding proteins, and topoisomerases involved in the process. Synthesis is discontinuous along at least one strand, and replication is bidirectional.
2. The DNA in a eukaryotic chromosome consists of several replicating units or replicons, each under the control of one origin of replication.
3. The eukaryotic cell cycle consists of interphase followed by cell division. The comparatively long interphase is divided into a presynthetic period (G_1), a period of DNA synthesis (S), and a presynthetic period (G_2). During G_1 the cell either becomes committed to division or it ceases proliferation.
4. The chromosomal DNA of dividing cells begins to condense at the end of G_2; this process continues into mitosis. Studies on *cdc* (cell division cycle) mutants have revealed cellular proteins, the cyclins, that are involved in the decision to divide, in the decision to initiate DNA synthesis, and in driving cells through mitosis. The cyclins appear to function in protein complexes with kinase enzymes, which phosphorylate other proteins, including histone H1. Thus, the structural transitions undergone by chromosomes during the cell cycle are coupled with the molecular events of that cycle.

◆

✦ ORGANELLES AND THEIR CHROMOSOMES

Until now our discussion of eukaryotes and their chromosomes has focused on the nucleus. Eukaryotic cells also possess several other membrane-limited structures that play an important role in the transmission and expression of genetic information. Two of these structures—**mitochondria** (found in both plant and animal cells) and **chloroplasts** (found only in plant cells)—are of special interest to geneticists because they are semiautonomous organelles that arise by growth and division and possess their own unique DNA and RNA.

Cytoplasmic Organelles

The cytoplasm of eukaryotic cells is characterized by the presence of complex systems of membranes that enclose specific regions and separate them from the rest of the cell (▶ Figure 9–23). Prominent among these systems is the network of membranous channels known as the endoplasmic reticulum (ER). Certain of these channels—the **rough ER**—have ribosomes attached to their outer (cytosolic) surface, while others—the **smooth ER**—are devoid of attached ribosomes. The ribosomes and membranes of the rough ER form a system for the synthesis and transport of gene-encoded proteins to various parts of the cell. Depending on their type, the proteins formed on the membrane-bound ribosomes are either inserted into the ER membrane or released into the lumen (interior space) of the enclosed ER channels.

Closely associated with the endoplasmic reticulum is the array of parallel, flattened membranous vesicles called the **Golgi apparatus**. This organelle plays a central role in the modification, sorting, and packaging of proteins for delivery to other regions of the cell and to the cell exterior. Many of the proteins that pass through the Golgi apparatus are destined to become lysosomal enzymes and are packaged into membrane-bound sacs called **lysosomes**. Lysosomes are small spherical vesicles containing various hydrolytic enzymes that break down cellular proteins, polysaccharides, and lipids that are no longer needed. The lysosomal enzymes are capable of degrading the structural components of the intact cell and would do so if they were not confined to these specialized organelles.

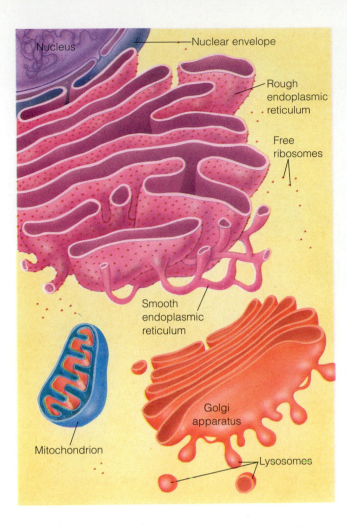

The organelles considered so far are important in genetics because of the special role they play in the expression of nuclear genes. Mitochondria and chloroplasts also function in the expression of the nuclear genome. In addition, these organelles transmit and express their own extranuclear genes (extranuclear genes were described briefly in Chapter 2).

Mitochondria are complex bodies about the size of bacteria. They are bounded on the outside by a double membrane whose two-unit membranes are separated by an intermembranous space (▶ Figure 9–24). The outer membrane of the mitochondrion appears smooth and continuous, while the inner membrane has folds known as cristae that extend into the interior or matrix of the organelle. The mitochondria are responsible for the bulk of the aerobic respiratory activities of a eukaryotic cell, including the biochemical processes of the citric acid (or Krebs) cycle, electron transport, and oxidative phosphorylation. The enzymes involved in the Krebs cycle are found mainly within the matrix, while the enzymes that participate in the coupled processes of electron transport and oxidative phosphorylation are located in the inner membrane.

Chloroplasts are present only in eukaryotic cells that can carry out the vital function of photosynthesis. Chloroplasts vary in size and shape, but they are frequently disk-shaped and larger than a typical mitochondrion (▶ Figure 9–25). Like mitochondria, chloroplasts are bounded by a double membrane that encloses an interior matrix or stroma. In chloroplasts, the inner membrane is organized into flattened sacs called thylakoids, which in higher plants are piled one on top of another in ordered stacks called grana. The thylakoids contain the chlorophyll of the plant and are formed during the development of mature chloroplasts from smaller, colorless proplastids. The membranes of chloro-

▶ **FIGURE 9–23** Diagram of the membranous organelles in the cytoplasm of a eukaryotic cell. The cell membrane is not shown.

▶ **FIGURE 9–24** The mitochondrion: (a) photomicrograph; (b) illustration of a mitochondrion, cut to reveal the internal organization. The outer membrane forms a continuous barrier around the organelle. The inner membrane contains a series of tubular extensions or folds, the cristae. The inner membrane, including the cristae, contains the biochemical machinery that functions in aerobic respiration. The internal aqueous matrix of the mitochondrion contains various solutes as well as ribosomes and DNA.

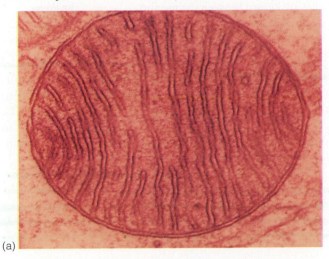

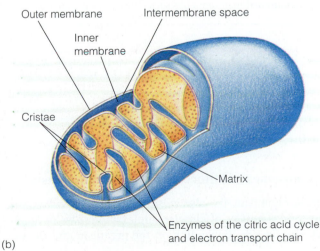

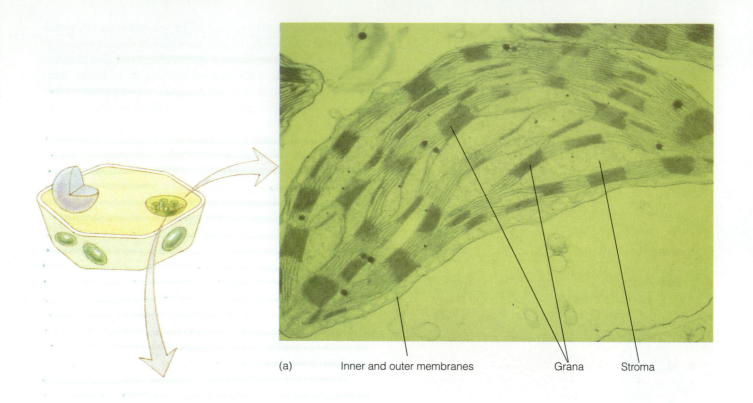

(a)   Inner and outer membranes   Grana   Stroma

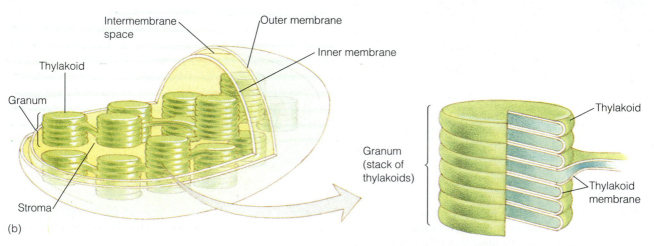

Intermembrane space

Outer membrane

Inner membrane

Thylakoid

Granum

Stroma

(b)

Thylakoid

Granum (stack of thylakoids)

Thylakoid membrane

▶ **FIGURE 9–25** The chloroplast: (a) Transmission electron micrograph; (b) three-dimensional drawing showing the cylindrical stacks of flattened membraneous sacs known as the grana. DNA molecules are located between the membrane layers of the grana.

plasts, like those of mitochondria, serve to compartmentalize functions. For example, the thylakoids house the light-sensitive pigments and enzymes that are involved in the light-capturing phase of photosynthesis. The stroma contains the enzymes that are involved in the conversion of atmospheric carbon dioxide to sugars.

Chromosomes of Mitochondria and Chloroplasts

In most organisms the chromosomes of mitochondria and chloroplasts consist of circular duplex molecules of DNA. They vary in size, depending on the organism and the or-

ganelle, but in general, they are circular like a bacterial chromosome, although they are smaller in size. For example, DNA circles in mitochondria average about 5 μm in contour length in most animals and 80–800 μm in plants. The chromosomes of chloroplasts are similar in structure to those of mitochondria but average about 45 μm in circumference in the higher plants.

Unlike nuclear chromosomes, the chromosomes of mitochondria and chloroplasts are not associated with histones, nor are they surrounded by a membrane comparable to the nuclear envelope. Instead they are organized into nucleoids within the matrix of each organelle. The DNA molecules of mitochondria and chloroplasts also differ in base composition from nuclear DNA, so they can often be isolated in separate (satellite) bands by equilibrium density-gradient centrifugation. The similarity of organelle chromosomes to prokaryotic (rather than nuclear) chromosomes (as well as many other similarities between organelles and prokaryotes that will be considered later in this text) has prompted several investigators, led by Lynn Margulis, to propose that mitochondria and chloroplasts were once free-living bacteria that were engulfed by ancient eukaryotic cells and established a symbiotic relationship with them.

The genes carried in the chromosomes of mitochondria and chloroplasts code for some of the components of these organelles, while nuclear genes code for others. For example, some of the genes in the mitochondrial chromosomes are known to code for essential components of the cellular respiration process and the protein-synthesizing system in mitochondria, but the enzymes of the Krebs cycle and many of the proteins involved in electron transport are specified by nuclear genes. Organelles are thus interactive genetic systems, depending in large part on the nuclear genome for their structural and functional parts.

Mitochondria and chloroplasts are self-replicating structures, that is, new organelles arise from existing ones rather than being assembled from scratch by the cell. The mechanism by which organelles reproduce themselves is not fully understood. However, it is known that they divide during the process of cell division and that the organelle chromosomes replicate prior to division through a process catalyzed by an organelle-specific DNA polymerase. In mitochondria the polymerase enzyme catalyzes a displacement-type synthesis in which both DNA strands are replicated continuously from different points of origin (▶ Figure 9–26). Initially only one of the parental strands is copied. The other parental strand is displaced during the process, forming a bubble called a **displacement** or **D loop**. Once the second replication origin is reached, the displaced strand is then copied in the opposite direction.

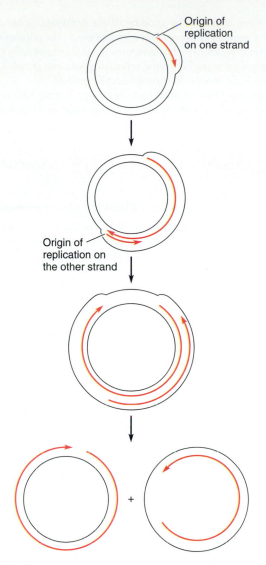

▶ **FIGURE 9–26** The displacement-loop (D-loop) mechanism of replication in mammalian mitochondrial DNA. The two DNA strands are replicated continuously from different points of origin.

To Sum Up

Mitochondria and chloroplasts contain their own chromosomes, which resemble those of prokaryotes in structure and appearance. These chromosomes contain genes that specify some of the components needed to carry out the functions of these organelles.

 ## Chapter Summary

Eukaryotic cells have several nuclear chromosomes, each containing a linear, duplex DNA molecule that extends from end to end. Each chromosomal DNA contains specialized regions that are essential for the normal replication and disjunction of chromosomes: the centromere (for attachment to the spindle microtubules), the telomeres (to replicate and preserve the integrity of the chromosome ends), and several origins of replication.

The DNA of each chromosome is combined with histone and nonhistone proteins to form a nucleoprotein complex called chromatin. The DNA of chromatin undergoes several levels of packaging to achieve the compact chromosome shape: The DNA is wrapped around octamers of histones to form nucleosomes, the nucleosome fiber is coiled into a helical solenoid, the solenoid structure is organized into many looped domains attached to a nonhistone scaffold, and the entire structure is supercoiled.

Nuclear chromosomes contain DNA sequences that are repeated many times in the genome of the organism. Some highly repetitive sequences are arranged in tandem in the telomeres and centromeric regions of chromosomes. Certain moderately repetitive sequences, including some repeated genes and families of related genes, are dispersed throughout the genome.

The nuclear DNA of eukaryotes replicates during a limited period (designated S) of the cell cycle. Replication in S is initiated at several replication origins and proceeds by a mechanism that is fundamentally the same as in prokaryotes.

The mitochondria and chloroplasts of eukaryotic cells also contain chromosomes. The chromosomes of these organelles resemble prokaryotic chromosomes and carry genes related to organellar function.

Questions and Problems

Morphology and Functional Elements of Chromosomes

1. The following sketch shows several eukaryotic chromosomes at mitotic metaphase. Indicate whether each chromosome is metacentric, submetacentric, acrocentric, or telocentric. Which, if any, of these chromosomes have satellites?

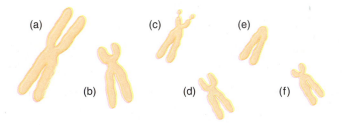

DNA Packaging

2. Calculate the packing ratio achieved by folding a nucleosome filament into a 30-nm solenoid that has six nucleosomes per turn and a pitch of 10 nm.

Repetitive DNA Sequences

3. The DNA of a particular eukaryote consists of 10^9 base pairs with a mixture of three renaturing components, as seen in the following cot curve:

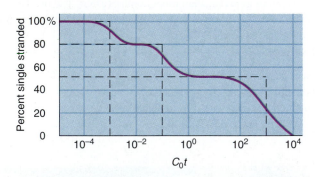

(a) What is the proportion of each renaturing component in the genome of this eukaryote? (b) Estimate the value of $C_0t_{\frac{1}{2}}$ for each renaturing component in the mixture. (c) What $C_0t_{\frac{1}{2}}$ values would you expect to obtain for the three components if each is isolated in pure form and renatured separately? (d) Determine the sequence complexity (number of base pairs per repeating unit) of each of the repetitive components. (e) Calculate the repetition frequency of each repetitive sequence. (f) What is the overall complexity of this DNA? Compare this figure to the total DNA content of 10^9 base pairs.

4. Suppose that you have three renaturing DNA components with $C_0t_{\frac{1}{2}}$ (pure) values of 0.002, 0.3, and 50, occurring in a ratio of 2:3:5, respectively. You conduct an experiment on the renaturation kinetics of this mixture. (a) Draw the cot curve that you would obtain from this experiment, indicating the appropriate proportions and $C_0t_{\frac{1}{2}}$ values of the components in the mixture. (b) If the three components have sequence complexities of 40, 6×10^4, and 10^7 bp, respectively, and the slow renaturing component is nonrepetitive, calculate the repetition frequency (copy number) of each of the sequences that make up the two repetitive components.

Chromosome Replication and the Cell Cycle

5. At G_1 there are 2.8 picograms (2.8×10^{-12} g) of DNA in the haploid set of chromosomes in humans. (a) What is the DNA content of a human somatic cell nucleus at G_2? at mitotic anaphase? at the close of mitotic telophase? (b) Compute the average DNA content per chromatid equivalent in each of these phases. (c) Compute the average DNA content per chromosome in each of these phases.

6. The rate of DNA replication in *E. coli* is 25 times greater than the average eukaryotic rate of 2000 base pairs per minute. Give at least two possible reasons for the large difference in the DNA replication rate in prokaryotic and eukaryotic cells.

7. When a sample of eukaryotic cells is grown in a culture, the cells tend to be randomly distributed with respect to the stage of the cell life cycle. At any given moment, some of the cells are in G_1, others are in S, and so on. The number of cells in each stage at any given moment depends solely on what fraction of the average cycle time is taken up by that stage. (a) Suppose that G_1, S, G_2, and M for a particular cell culture last for 11 hours, 8 hours, 4 hours, and 1 hour, respectively. What percentage of cells are expected to be in each of these stages at any given moment? (b) When cells from another population from this sample are examined for mitotic activity, it is found that of the cells undergoing mitosis, 40% are in prophase, 30% are in telophase, 20% are in metaphase, and 10% are in anaphase. How long (in minutes) is each of these mitotic stages in this cell population?

Organelles and Their Chromosomes

8. Distinguish between the D-loop replication mechanism of mitochondrial DNA and the semidiscontinuous replication mechanism of a bacterial chromosome.

9. Explain how some chloroplasts can have more DNA but fewer genes than an *E. coli* cell.

CHAPTER 10

Plasmids, Transposable Elements, and Viruses

Plasmids
Transposable Elements
Viruses
FOCUS ON HUMAN GENETICS:
Treating Viral Diseases

The living cell is the simplest natural environment capable of independent metabolism. It is also the habitat for a variety of even simpler subcellular genetic systems that are not part of the nuclear or organellar genetic systems and that must use the metabolic machinery of the cell to reproduce. These subcellular genetic systems include the largely extrachromosomal *viruses* and *plasmids,* which can replicate independently of the cell genome, as well as a group of mobile, intrachromosomal DNA segments known as **transposable elements (transposons).** In many respects, these subcellular systems are molecular parasites that depend on the cell for propagation. However, they are an important potential source of genetic variability, providing organisms with varied and novel mechanisms for gene transfer and recombination. Some, such as the transposable elements, can even act as mutagens.

In this chapter, we will consider three major types of subcellular

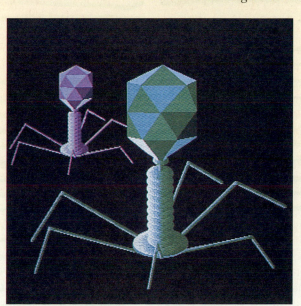

A virus: T2 bacteriophage.

genetic systems that replicate within cells. We will first look at those systems that normally exist only within the confines of a living cell. These so-called intracellular genetic elements include the plasmids and transposons. We will then consider viruses, which can exist in extracellular (infectious) as well as intracellular forms.

Although these subcellular genetic systems occur naturally in both prokaryotes and eukaryotes, they have been studied most extensively in bacteria. Therefore, most of the genetic systems described in the following sections are prokaryotic rather than eukaryotic in origin.

PLASMIDS

Plasmids are extrachromosomal genetic elements that are found in most prokaryotic species and in some eukaryotes. Almost all known plasmids are DNA molecules (the RNA killer plasmid in yeast is an exception). They are relatively small molecules, generally less than one-twentieth the size of the *E. coli* chromosome, and are usually isolated as double-stranded closed circles in the supercoiled state (▶ Figure 10–1). Plasmids can therefore be separated from similar-sized linear fragments of the host cell chromosome by taking advantage of the fact that supercoiled DNA molecules are more compact than nicked circles or linear duplexes and hence travel at a faster rate in an ultracentrifuge or electrophoretic gel. Plasmids also have a lower affinity for dyes such as ethidium bromide, which intercalate between base pairs and thus lower the buoyant density of the DNA (see Chapter 7). Since circular plasmids bind less of the dye than fragmented chromosomal DNA, they have a higher buoyant density in the presence of the intercalating agent and can be separated by equilibrium density-gradient centrifugation.

Plasmids are replicons; each plasmid has at least one replication origin and can replicate autonomously within the host cell. Replication is semiconservative and discontinuous along the lagging strand. Depending on the plasmid, replication is catalyzed by some or all of the host replication enzymes. Plasmids tend to differ in their pattern of replication during cell division, with some undergoing bidirectional replication and others unidirectional replication. However, most remain in a circular form throughout the replication cycle.

Many different kinds of plasmids are known. A large number of them are beneficial to the survival of the host cell under certain conditions and can provide the host cell with a selective advantage over plasmid-free cells. These benefits include resistance to drugs and heavy metals and the ability to produce certain types of toxic proteins. For example, **Col plasmids** (**colicinogens**) are toxin-producing plasmids that are carried by Col+ strains of *E. coli*. These plasmids carry the genes for toxic proteins called **colicins**, which prevent the growth of bacteria that lack a Col plasmid of the same type. The presence of the Col plasmid has survival value for its bacterial host, since Col+ cells can se-

▶ FIGURE 10–1 DNA conformations present during the isolation of plasmid DNA circles: (a) linear form, (b) open circle, (c) closed circle. The open circle has incurred a single-stranded break or nick (arrow), which relaxes it from the supercoiling of the unnicked (closed) circle. The three forms can be separated by centrifugation or gel electrophoresis because of differences in their hydrodynamic properties.

crete the colicins into the medium and thereby inhibit the growth of related Col– strains.

Conjugative Plasmids: The F factor

Conjugative plasmids have genes that code for the components of sex pili and certain other proteins required for conjugal gene transfer. In bacteria, **conjugation** is a mating process that involves a one-way transfer of genetic material from one bacterium (the donor) to another bacterium (the recipient) during contact between the cells. Donor cells possess special proteinaceous, hairlike appendages called **sex** or **F pili** (singular, **pilus**) that protrude from their cell walls. When a donor cell randomly comes in contact with a recipient cell, which lacks sex pili, some kind of recognition process occurs. A pilus physically connects the two cells and becomes modified into a **conjugation tube** (▶ Figure 10–2). The transfer of genetic material from donor to recipient can then take place through this tube.

In *E. coli*, a conjugative plasmid called the **F factor** (F for

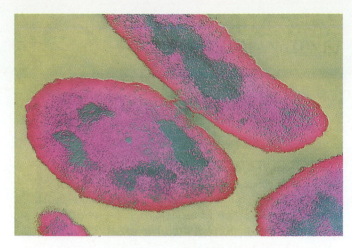

► FIGURE 10–2 Electron micrograph of conjugating *E. coli* bacteria. A slender bridge called the conjugation tube has been formed between a donor cell and a recipient cell. The DNA of the donor (elongated) cell was labeled (green) to demonstrate the transfer of DNA.

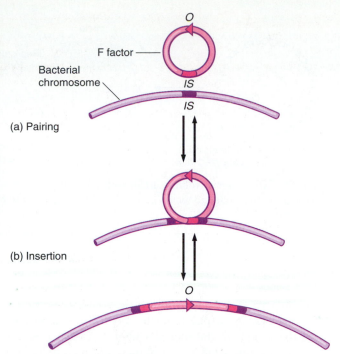

(a) Pairing

(b) Insertion

► FIGURE 10–4 Integration of the F factor into a bacterial chromosome. (a) Pairing between homologous regions of the F factor DNA and the bacterial DNA. The homologous regions are short sequences called insertion (*IS*) elements that are common to both the F factor and the bacterial chromosome (see Figure 10–3). (b) Insertion of the F factor DNA by a single crossover event. The *O* marker on the plasmid is the origin of replication (*oriT* site).

fertility) plays a special role in the conjugation process. A cell is a donor during conjugation if it contains the F factor; it is a recipient (designated F⁻) if the F factor is absent. The F factor consists of 94,500 bp of DNA that encode a variety of functions, including conjugal DNA transfer (encoded by the *tra* genes) and the ability of the plasmid to replicate itself (► Figure 10–3). The F factor is also occasionally in-

► FIGURE 10–3 Organizational map of the F factor showing the genes required for transfer of the F factor during conjugation (the *tra* genes) and the genes required for DNA replication. The insertion (*IS*) elements are responsible for the integration of the F factor into the bacterial chromosome.

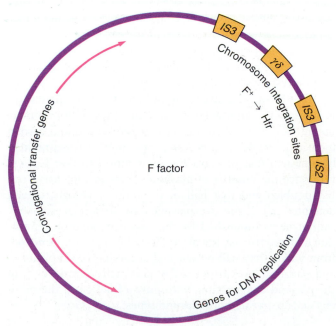

tegrated into the chromosome of its host. During the integration process, pairing occurs between one of several homologous sequences in the plasmid and the bacterial DNA, and a single crossover inserts the F factor into the bacterial chromosome (► Figure 10–4). The F factor can thus replicate either as an autonomous unit or as an integrated part of the bacterial chromosome. Plasmids with this dual replicative ability are known as **episomes.**

Since the F factor can replicate in one of two states, two types of donor cells exist. In F⁺ donor cells the F factor is present in the autonomous state, replicating independently of the cell chromosome. The other type of donor, **Hfr** cells, contain the F factor integrated into the cell chromosome. In F⁺ cells, replication of the F factor occurs bidirectionally from the *oriV* site during cell division and unidirectionally from the *oriT* site during conjugation. In conjugation it replicates through a process known as **rolling-circle replication** (► Figure 10–5). Rolling-circle replication occurs after one of the strands becomes detached from the origin, so that the replication intermediate is sigma-shaped (σ). This replication scheme is unidirectional, with only one replication fork per DNA molecule; it is initiated when a site-specific **endonuclease** cuts a DNA strand at the transfer origin (*oriT*). An endonuclease is an enzyme that cuts a single phosphodiester bond to produce a **nick** in one DNA strand;

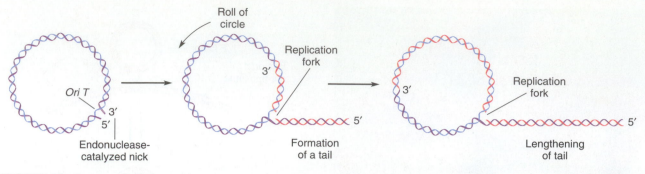

► FIGURE 10–5 Sigma (rolling-circle) replication. Replication occurs as one strand of the parental molecule is progressively displaced from the circle and moves in only one direction from the replication origin. The result is a sigma-shaped (σ) replication intermediate.

it does not cleave off any nucleotides. The unbroken circular strand then acts as a template for the synthesis of a new complementary strand, starting at the exposed 3′ end at the nick. Extension of synthesis in the 5′ to 3′ direction displaces the 5′ end of the nicked strand so that it forms a "tail" on the circle while synthesizing its own complementary strand. (Notice the sigma shape of the replication intermediate at this step.)

During an F⁺ × F⁻ mating, the displaced 5′ end of the replicating F factor is threaded through the conjugation tube and enters the recipient cell (► Figure 10–6a). Two copies of the F factor now exist. One copy is retained by the donor, which remains F⁺; the other copy is transferred to the recipient, which is converted to the F⁺ state in the process.

Rolling-circle replication also occurs in Hfr donor cells during conjugation, but in this case, the replicating DNA circle consists of both the F factor and the attached bacterial chromosome (Figure 10–6b). In an Hfr × F⁻ mating, replication is initiated at the same point (*oriT*) in the F factor as in F⁺ cells. Because of the fragile nature of the conjugation tube, the passageway between the conjugating cells is often broken before the entire replicating chromosome is transferred. The result is that usually only the *O* (*ori*-carrying) end of the F factor and a section of the bacterial genome enter the recipient cell. If the conjugation tube remains intact long enough, the remaining half of the F factor will be transferred last. Because the process is usually interrupted before the entire chromosome can be transferred, the recipient cell typically remains F⁻. The original Hfr cell retains its donor status, since replication accompanies transfer.

Copy Number and Plasmid Incompatibility

The process of conjugation, along with other forms of DNA transfer, makes it possible for several different kinds of plasmids to enter a single cell. For stability, each kind of plasmid is normally maintained at a characteristic copy number. Some, called high copy-number plasmids, are maintained in multiple (10–100) copies per cell, while others, called

low copy-number plasmids, occur in just 1–2 copies per cell. Large, conjugative plasmids, such as the F factor, are generally low-copy number plasmids, and small, nonconjugative plasmids are usually high-copy number plasmids, but exceptions do occur. Studies show that the copy number of a plasmid tends to be controlled at the origin of replication through regulation of the initiation of DNA synthesis. Control is exerted by a plasmid-encoded molecule that prevents the initiation of further replication once a particular copy number is reached. In the Col plasmid ColE1, for example, the regulatory molecule is an RNA chain that inhibits initiation by combining with a complementary sequence in the RNA primer.

Whether two plasmids can cohabit a cell depends on the effect each has on the other's copy number. In general, different plasmids that regulate their replication and assortment independently of each other are capable of coexisting in the same cell. In contrast, plasmids that have the same or very similar mechanisms of regulation cannot exist together in a cell line in the absence of selection pressure. Plasmids that cannot stably replicate together in a single cell line are said to be **incompatible**, and a collection of incompatible plasmids constitutes an **incompatibility group**. A variety of evidence suggests that members of the same incompatibility group are evolutionarily related, so such groups are often used as a system of classification.

A model of plasmid incompatibility (and compatibility) is shown in ► Figure 10–7; here incompatibility is interpreted as the net effect of random replication and assortment. This model assumes that copy number is maintained by a system that prevents the initiation of further replication once the number of plasmid copies in the parent cell has doubled and that half of the total copies go to each daughter cell. If the two plasmids are so similar that they cannot be distinguished by the system regulating replication and assortment, the plasmids will form a common pool from which they will be selected at random for replication and partitioning. Chance differences will then develop in the plasmid composition, with only a fraction of the progeny cells receiving both plasmid types.

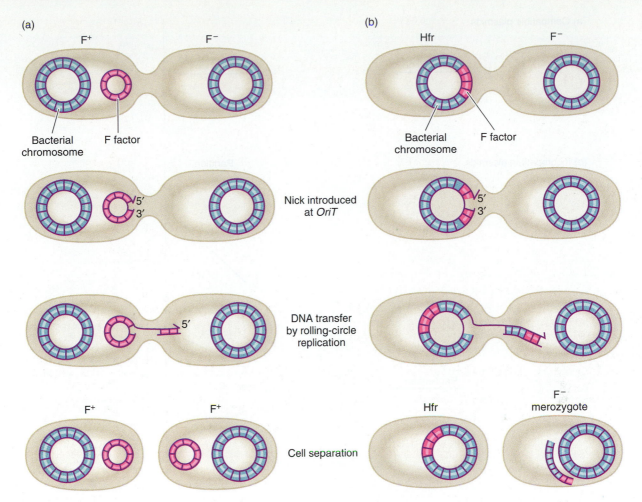

(a)

F+   F−

Bacterial   F factor
chromosome

Nick introduced
at *OriT*

5′
3′

DNA transfer
by rolling-circle
replication

5′

F+   F+

Cell separation

(b)

Hfr   F−

Bacterial   F factor
chromosome

5′
3′

DNA transfer
by rolling-circle
replication

Hfr   F−
merozygote

▶ FIGURE 10−6 Conjugation in *E. coli*. (a) In F+ cells, the F factor is free in the cytoplasm. (b) In Hfr cells, the F factor is integrated into the *E. coli* chromosome. In both types of matings, rolling-circle replication accompanies the transfer of DNA through the conjugation tube. In F+ × F− matings (a), only the F factor is transferred, to make both cells F+. In Hfr × F− matings, (b) the origin half of the F factor and the adjacent section of the bacterial chromosome are transferred, leaving the recipient cell F−.

Example 10.1

Consider a cell with two incompatible plasmids, A and B. Assume that both plasmids always replicate prior to cell division (unlike the mechanism in Figure 10−7) and that two of the four plasmids in the parent cell then enter each daughter cell at random, without regard to plasmid type. For this mechanism, compute the probability that a daughter cell will receive only one type of plasmid.

Solution: After plasmid replication and before cell division, the parent cell will contain four plasmid copies, two of A and two of B (2A,2B). Since any two of the plasmid copies can enter each daughter cell during cell division, there are

$$\frac{4!}{2!2!} = 6$$

possible ways to select two plasmids from two plasmids of one type and two of another. For a cell to receive two plasmids of the same type, both copies must be of type A or of type B. The probability that both are A, P(AA), equals $\frac{1}{2}$ for the first A times $\frac{1}{3}$ for the second, or $(\frac{1}{2})(\frac{1}{3}) = \frac{1}{6}$. (Note that after selecting the first A, only one of three remaining plasmids is of type A.) The probability that both are B, P(BB), is also $(\frac{1}{2})(\frac{1}{3}) = \frac{1}{6}$. Thus the probability that a cell will receive only one plasmid type is P(AA) + P(BB) = $\frac{2}{6} = \frac{1}{3}$.

Follow-Up Problem 10.1

Consider again the situation in Example 10.1. However, now suppose that the two plasmids are selected at random for both replication and partition, as in Figure 10.7(b). For this mechanism, what is the probability that a daughter cell will receive only one plasmid type?

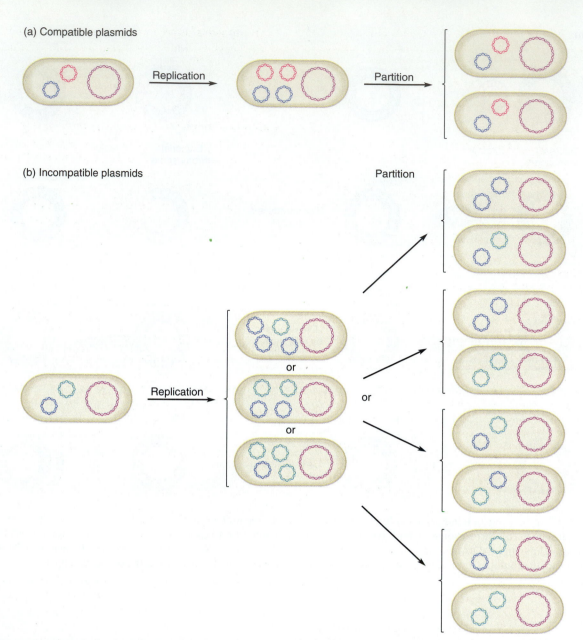

(a) Compatible plasmids

Replication

Partition

(b) Incompatible plasmids

Partition

Replication

or

or

or

or

▶ FIGURE 10–7 Replication and partition of compatible and incompatible plasmids (small DNA rings; the large ring is the host chromosome). (a) Unrelated plasmids are compatible and can stably replicate together in the same cell. (b) Closely related plasmids are incompatible. If the plasmids are chosen at random for replication and partition, cell lines that lack one plasmid or the other are produced.

To Sum Up

1. Plasmids are small, circular, supercoiled DNA molecules that can replicate autonomously within the host cell. There are many different kinds of plasmids, and they are found in most bacterial species as well as in some eukaryotes.

2. Conjugation is a bacterial mating process that involves a one-way transfer of genetic material during cell-to-cell contact. Conjugation is mediated by a specific plasmid, the F factor. Donor cells in the conjugation process are either F⁺ (with the F factor

replicating autonomously in the cytoplasm) or Hfr (with the F factor inserted into the bacterial chromosome). Recipient cells are F⁻ and lack an F factor.

3. Rolling-circle replication occurs during conjugation. The displaced DNA strand moves from the donor through the conjugation tube into the recipient cell. Only the F factor is transferred in this manner during F⁺ × F⁻ matings. In Hfr × F⁻ matings the bacterial chromosome is transferred, starting at the origin of transfer end of the integrated F factor.

4. Different kinds of plasmids maintain themselves at characteris-

tic copy numbers in bacterial cells. The number of copies of a particular plasmid is regulated by events that occur at the origin of replication of the DNA.

⊞ TRANSPOSABLE ELEMENTS

DNA consists of two types of sequences: those that remain at constant positions within the genome, changing only rarely on the time scale of evolution, and those that move by special mechanisms from one chromosomal location to another. Making up the second class of sequences—the mobile sequences—are the transposable elements, or simply transposons. Like plasmids, transposons are dispensable elements (cells can exist without them), but unlike plasmids, which are mainly extrachromosomal elements, transposons can exist only as part of other replicating DNA.

Transposons move to a different chromosomal location by a process called transposition, in which the transposon is inserted into a specific target sequence located in the same DNA or a different DNA. Such insertions are relatively rare, occurring at a frequency comparable with spontaneous mutation rates. Typically, a transposon-encoded enzyme called transposase first cuts the target DNA by making staggered nicks in the two strands of the target sequence (▶ Figure 10–8). Insertion of the transposon then follows, leaving short single-stranded gaps in the host DNA at the ends of the inserted element. These gaps are filled by DNA synthesis in order to restore the double helix, and thus the transposition creates direct repeats of the target DNA flanking the transposon.

Just as the target sequence serves as the site for transposase recognition in the recipient DNA, the ends of the transposon serve a corresponding role in the donor DNA. These ends are unrelated in sequence to the target DNA, so the donor and recipient sites involved in transposition are

---

▶ FIGURE 10–8 The insertion of a transposable element is accompanied by duplication of nucleotide pairs in the recipient DNA. The two strands of recipient DNA are cleaved at staggered sites in the target sequence. The transposable element then inserts itself, and single-stranded regions on either side of the inserted element are filled in with duplicate sequences to produce the DNA product shown at the bottom.

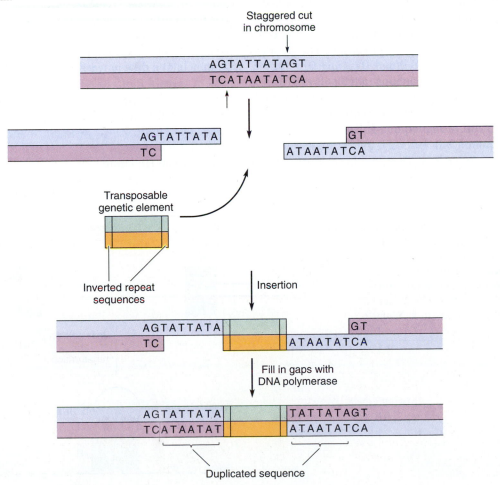

nonhomologous. The ends of almost all the transposable elements studied so far in Gram-negative bacteria have been found to be inverted repeats (IRs) ranging from 8 to 40 bp in length. For example, if one end of one of the DNA strands of a transposon were to begin with the sequence 5′ TAGC . . . , the other end of that strand would terminate in the sequence . . . GCTA 3′. Upon denaturation and renaturation, the ends of a transposon therefore tend to pair with each other to form a "lollipop"-shaped structure (▶ Figure 10–9).

Transposons as Mutagens

The movement of a transposon from one chromosomal site to another not only changes the location of the transposon but can also lead to various mutational alterations in the host DNA. Indeed, transposable elements are thought to be a major source of cellular mutations. Some mutations arise when the insertion of a transposon disrupts the normal function of a gene. Since the target site of a transposon can be situated within a gene as well as between genes, chance insertion of the transposon at a gene locus can alter the nucleotide sequence of the gene and destroy the gene's function.

A mutation of this type led Barbara McClintock to first propose the existence of transposable elements in the late 1940s. McClintock discovered the transposon Ds (for *dissociation*) while studying the genetics of kernel color in corn. The Ds element can induce the loss of kernel color when it becomes inserted in the color gene, C. The colorless mutation caused by the insertion of Ds is highly unstable, however, because of the high rate of excision of the Ds element from its target site. Many of the resulting kernels have a mottled or dotted appearance with spots of pigment occurring on an otherwise colorless background (▶ Figure 10–10). This high rate of excision is possible only in the presence of another transposon Ac (for *activator*). If the

Ac element is not present, the Ds element cannot be excised and the colorless mutation is stable.

With the advent of gene isolation and sequencing techniques, the molecular basis for these results has been revealed. Ds is actually a defective transposon that is derived from Ac by deletion (▶ Figure 10–11). A number of different forms of Ds have been discovered, and each has lost part or all of its transposase gene. Since these elements all lack a functional transposase, they are unable to excise themselves, but they can undergo excision in the presence of the transposase produced by Ac.

In addition to inducing mutations through insertions, transposons can also cause a variety of genetic rearrangements as a result of recombination. The mechanisms of some of these rearrangements are shown in ▶ Figure 10–12.

Bacterial Transposons

Bacteria typically contain a variety of transposons, which are conveniently divided into different classes. Three of the better-known classes of transposons in E. coli are the insertion sequences, the Tn3 family, and the transposing bacteriophages (see ■ Table 10–1 and ▶ Figure 10–13). The insertion sequences (ISs) are the simplest transposons. Each IS element consists of fewer than 2000 base pairs and ends in inverted repeats. Unlike larger transposons, ISs encode only transposition functions; that is, their only message is to insert themselves into the chromosome. A bacterial chromosome can contain several different types of ISs, with each having one to several copies. For example, E. coli has five prominent ISs, designated IS1 through IS5, that are usually present in multiple copies dispersed throughout the genome. Insertion sequences are also normal constituents of plasmids. The F factor, for example, contains four IS elements (see Figure 10–3), which provide homologous sites at which the F plasmid can be inserted into the E. coli DNA to produce an Hfr strain.

▶ FIGURE 10–9 Transposons possess inverted repeat (IR) sequences at their termini. (a) The transposon in its double-stranded form before denaturation. (b) The lollipop structure formed by intrastrand annealing after denaturation. (X′, Y′, and Z′ denote bases complementary to X, Y, and Z, respectively.)

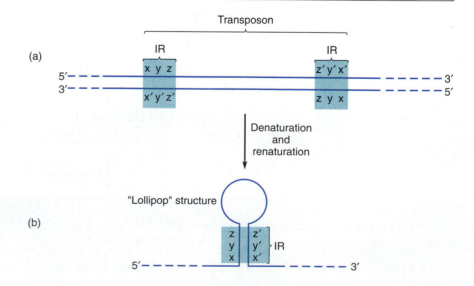

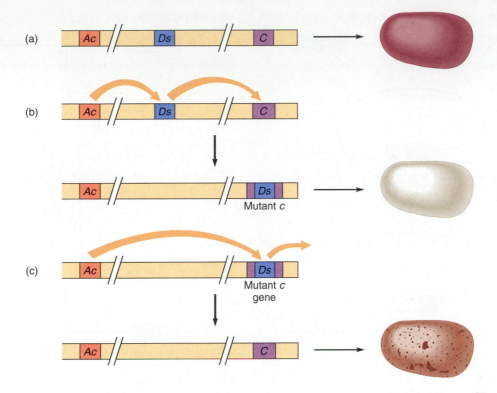

▶ FIGURE 10–10 Changes in kernel coloration caused by transposable elements in corn. (a) Normal solid pigmentation produced by an uninterrupted *C* gene. (b) Colorless mutation resulting from the insertion of the *Ds* element into the *C* gene. The *Ac* element activates this transposition of the *Ds* element. (c) Mottled pigmentation resulting from a high rate of excision of the *Ds* element from the *C* gene; *Ac* activates the transposition of *Ds* out of gene *C*.

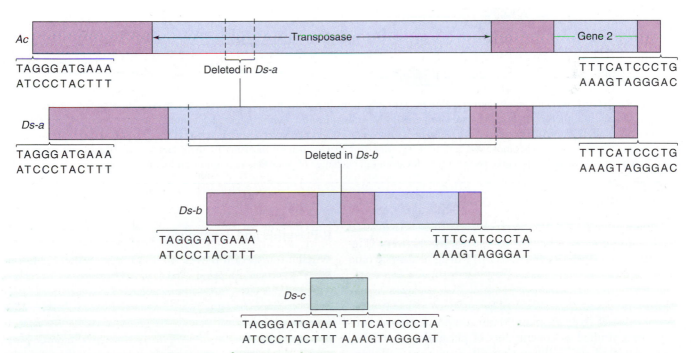

▶ FIGURE 10–11 Structural relationships between the *Ac* element and different forms of the *Ds* element. The *Ac* element contains two genes (one for transposase) and two imperfect IR sequences. The *Ds* elements (*Ds-a*, *Ds-b*, and *Ds-c*) are derived from *Ac* by progressively larger deletions. *Ds-c* is very short and retains only the inverted terminal repeats of *Ac* (see braces). *Adapted from:* N. V. Federoff, "Transposable genetic elements in maize," *Scientific American* 250 (1984): 84–98. Copyright © 1984 by Scientific American. All rights reserved. Used with permission.

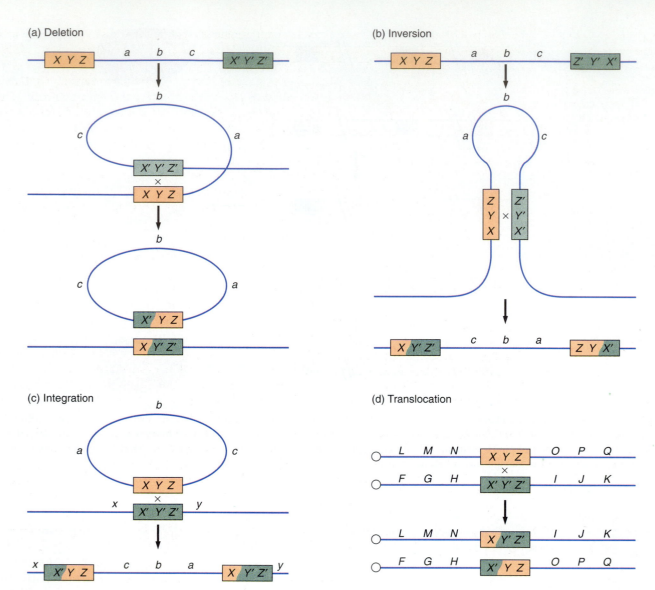

(a) Deletion

(b) Inversion

(c) Integration

(d) Translocation

▶ FIGURE 10−12 Genetic rearrangements due to recombinations involving homologous transposon sequences (shaded boxes). The repeats can be either separate transposons or parts of the same transposon. (a) Direct repeats along the same chromosome give rise to a deletion. (b) Inverted repeats along the same chromosome lead to an inversion (a reversal of the gene order). (c) Homologous repeats on different circular chromosomes lead to integration (the mechanism for insertion by the F factor). (d) Homologous repeats on different linear chromosomes give rise to a translocation (an exchange of parts between nonhomologous chromosomes).

Insertion sequences can exist alone or as constituents of larger mobile elements called **composite transposons** (Figure 10−13a). These transposons carry drug resistance (and other) genes as well as encoding transposition functions. The drug resistance genes are contained in a central region of the transposon that is flanked by identical or nearly identical copies of an *IS* element. Many composite transposons were first studied as constituents of plasmids, where they were detected as "lollipop-shaped" structures following strand separation and annealing of the plasmid DNA (▶ Figure 10−14). In this case, base pairing in the stem-and-loop occurs between the flanking *IS*s, which have inverted orientations relative to each other.

R Plasmids and Transposons

When a transposon moves from one DNA molecule to another, for example, from a bacterial chromosome to a plasmid or from one plasmid to another, the genes carried by the transposon become part of the recipient molecule. This transposon-mediated transfer of genes is widely thought to be responsible for the development of **drug-resistance plasmids**, (**R plasmids**). R plasmids carry genes that give the host cell resistance to a variety of antibiotics and heavy metals. Most R plasmids are conjugative plasmids that consist of two parts: a resistance-transfer segment, which carries the genes for conjugal transfer and replication, and a resistance-

■ **TABLE 10–1** Three classes of bacterial transposons.

Class	Characteristics
Insertion sequences (*IS*) and composite transposons	*IS*s vary in size from about 750 bp to 1600 bp and have a single transposase gene (*tnp*).Composite transposons consist of host-related genes (such as drug-resistance genes) flanked by two copies of an *IS*; the *IS*s may be in the same or inverted orientation.
Tn3 family transposons	Members of the *Tn3* family have genes for transposase (*tnpA*) and resolvase (*tnpR*) and have a site for cointegrate resolution (*res*). The remainder of the transposon may carry additional genetic markers, such as drug-resistance genes. Two subgroups are *Tn3* and *Tn501*.
Mu	Mu is a bacteriophage that also functions as a transposon. In addition to phage-related genes, Mu has the *A* and *B* genes, needed for efficient transposition and replication, an invertible *G* segment, and the *gin* gene, related to the *tnpR* gene of *Tn3*.

Source: N. D. F. Grindley and R. R. Reed, *Ann. Rev. Biochem.* 54 (1985): 863–896. Reproduced, with permission, from the *Annual Review of Biochemistry*, Volume 54, © 1985 by Annual Reviews Inc.

determinant segment, which carries the drug-resistance genes (▶ Figure 10–15). Resistance to many drugs can be involved; the genes for resistance to ampicillin (*amp*) chloramphenicol (*cam*), streptomycin (*str*), kanamycin (*kan*), and sulfonamide (*sul*) are among the most common. The genes conferring resistance are located on transposons, so an R plasmid can include several genes for drug resistance and several transposons. The resistance-determinant segment can thus develop (and continue to enlarge) within the basic plasmid structure through the insertion of additional transposons; transposition thus provides a mechanism for the localization of several drug-resistance genes in a single plasmid.

R plasmids are of considerable medical concern because of their ability to be transferred across species lines to various pathogenic bacteria. Drug resistance in a normally drug-sensitive strain can arise spontaneously as a result of gene mutation (see Chapter 11). Fortunately, most spontaneous mutations confer resistance to only a single kind of antibiotic, leaving the bacteria susceptible to the actions of other antibiotics. However, the acquisition of R plasmids can make bacteria simultaneously resistant to more than one antibiotic. This mechanism for transferring multiple resistance is particularly troublesome, since resistance spreads rapidly in the presence of antibiotics and is very difficult to control. For example, in one hospital in Japan, multiple drug resistance increased in *Shigella* (the bacterium that produces dysentery) from 0.2% in 1954 to 52% in 1964. Similar dramatic increases in multiple drug resistance have

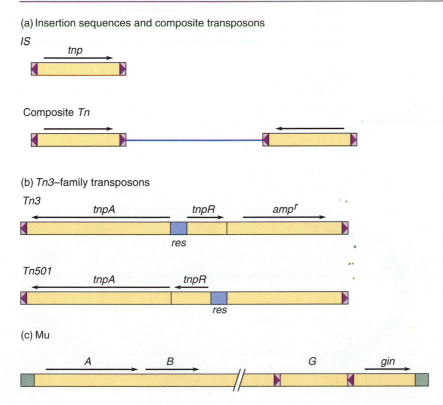

(a) Insertion sequences and composite transposons

IS

Composite *Tn*

(b) *Tn3*–family transposons

Tn3

Tn501

(c) Mu

▶ FIGURE 10–13 Three classes of bacterial transposons. (Also see Table 10–1.) *Adapted from:* N. D. F. Grindley and R. R. Reed, "Transpositional recombination in prokaryotes," *Ann. Rev. Biochem.* 54 (1985): 863–896. Reproduced, with permission, from the *Annual Review of Biochemistry*, Volume 54, © 1985 by Annual Reviews Inc.

► FIGURE 10–14 The lollipop structure of a transposon following strand separation and annealing of plasmid DNA. The helical stem of the stem-and-loop is formed by intrastrand annealing between two oppositely oriented copies of an insertion sequence.

(a)

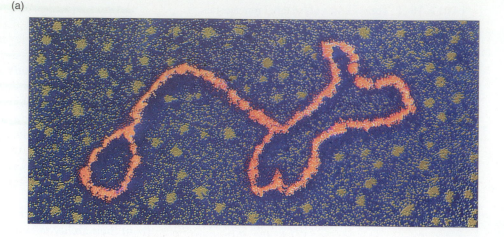

(b)

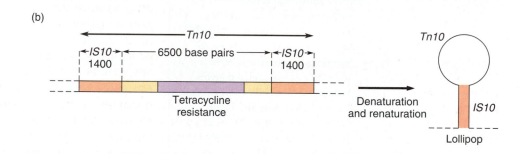

Mechanisms of Transposition

Transposable elements employ several different mechanisms of transposition. In *E. coli*, we can identify two main types of mechanisms: replicative and conservative (► Figure 10–16). In replicative transposition, the transposon duplicates itself during the transposition event, so that one copy remains at the old site as another copy inserts itself at the new site. This type of mechanism thus leads to a doubling of transposons. In conservative transposition, no extensive replication occurs—both parental strands of the transposon are excised from the donor DNA and inserted into the target site, without increasing the number of transposons.

been observed in other bacterial species. Domestic animals are one source of bacteria that carry R plasmids, since the widespread use of antibiotics as supplements in livestock feed has created a selective environment for these drug-resistant strains.

Replicative transposition is best understood in the bacterial transposon *Tn3* (see Figure 10–13b). This transposon has IR sequences of 38 bp at each end and is flanked by 5-bp direct repeats of target DNA. *Tn3* contains three genes: the *tnpA* and *tnpR* genes, which are necessary for

► FIGURE 10–15 A typical R plasmid. In this example, the genes encode resistance to ampicillin (*amp*), chloramphenicol (*cam*), kanamycin (*kan*), mercury (*Hg*), streptomycin (*str*), sulfonamide (*sul*), and tetracycline (*tet*). The resistance genes are clustered within transposons *Tn3, Tn4, Tn10,* and *Tn55*. Although transposon *Tn3* is within *Tn4*, each transposon can be transferred independently. *Adapted from:* S. M. Cohen and J. A. Shapiro, "Transferable genetic elements," *Scientific American* 242 (1980): 40–49. Copyright © 1980 by Scientific American. All rights reserved. Used with permission.

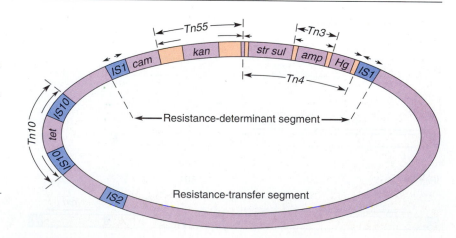

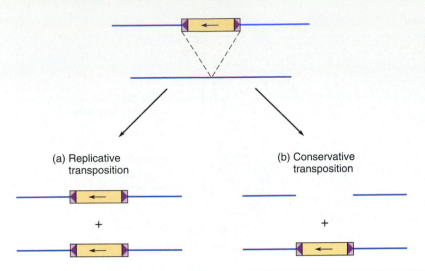

(a) Replicative
transposition

(b) Conservative
transposition

+

+

▶ FIGURE 10–16 Two major mechanisms of transposition in bacteria. (a) Replicative transposition involves doubling the transposon, with one copy remaining at the old site and the other copy inserting at the new site. (b) Conservative transposition involves the insertion of the transposon at a new site without extensive replication. The free ends remaining at the old site may or may not join.

transposition, and the ampicillin-resistance gene *amp^r*. The *tnpA* gene encodes transposase, the *tnpR* gene encodes another transposition protein called **resolvase**, and the *amp^r* gene encodes the ampicillin-inactivating protein β-lactamase. When transposition of *Tn3* occurs, it proceeds through the two-step pathway shown in ▶ Figure 10–17. Two circular DNA molecules are involved, one serving as donor DNA, the other as the target DNA. The first step (the replication step) is catalyzed by transposase and produces a monomolecular intermediate called a **cointegrate**, in which the donor and target DNAs are joined through a pair of *Tn3* copies. In the second step (the resolution step), which is catalyzed by resolvase, the cointegrate structure is converted back to the two separate molecules, each now containing a copy of *Tn3*.

A genetic analysis of mutations that prevent transposition of *Tn3* has revealed the two-step nature of its replicative pathway. Mutations in the *tnpA* gene that destroy the activity of transposase prevent the formation of the cointegrate structure. In contrast, mutations in the *tnpR* gene that destroy the activity of resolvase prevent cointegrate resolution and lead to a buildup of the cointegrate structure as the final product of transposition.

A model of the mechanism of replicative transposition and its possible relationship to conservative transposition is shown in ▶ Figure 10–18. After transposase makes staggered nicks at the ends of the transposon and at the target site, the two DNA molecules are joined by two of the four strands at the nicked transposon and target ends. What happens next depends on whether a replicative or a conservative pathway is followed. In transposons like *Tn3*, which use the replicative mode of transposition, a cointegrate is formed by replication of the joined transposon and target strands (Figure 10–18a). Cointegrate resolution then occurs through recombination of the two transposon copies. In transposons that use the conservative mode of transposition, a second set of cuts is made (Figure 10–18b). These

cuts allow insertion of the transposon at the target site but leave a potentially lethal gap in the donor DNA. No cointegrate is formed in conservative transposition.

Although bacterial transposons appear to undergo either replicative or conservative transposition, many eukaryotic transposons have evolved entirely different mechanisms. For example, some eukaryotic transposons move from site-to-site by forming an RNA intermediate. This mode of transposition is similar to the mechanism of replication of certain RNA viruses and is discussed in Chapter 16.

▶ FIGURE 10–17 The two-step process of transposition of *Tn3*. The donor and target DNAs are first fused in a replicative process catalyzed by transposase (the *tnpA* gene product) to form a cointegrate. The cointegrate is then resolved by the action of resolvase (the *tnpR* gene product) to regenerate the donor DNA and form a product DNA with an inserted transposon.

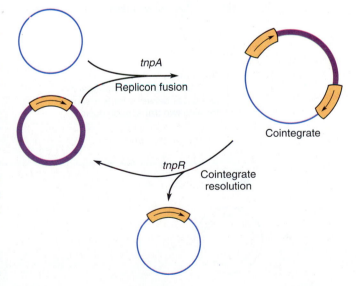

tnpA
Replicon fusion

Cointegrate

tnpR
Cointegrate
resolution

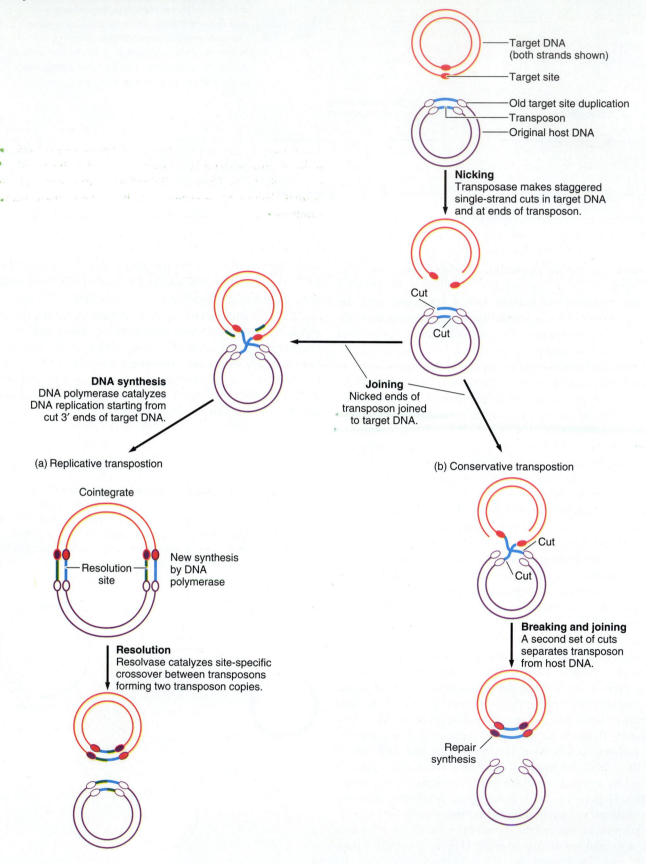

Target DNA (both strands shown)

Target site

Old target site duplication

Transposon

Original host DNA

Nicking
Transposase makes staggered single-strand cuts in target DNA and at ends of transposon.

Cut

Cut

DNA synthesis
DNA polymerase catalyzes DNA replication starting from cut 3′ ends of target DNA.

Joining
Nicked ends of transposon joined to target DNA.

(a) Replicative transpostion

(b) Conservative transpostion

Cointegrate

Resolution site

New synthesis by DNA polymerase

Resolution
Resolvase catalyzes site-specific crossover between transposons forming two transposon copies.

Cut

Cut

Breaking and joining
A second set of cuts separates transposon from host DNA.

Repair synthesis

1. Transposable genetic elements are DNA segments that move from one chromosomal location to another by a process called transposition. Insertion of a transposon into a target site is accompanied by the formation of direct repeats of the target DNA sequence that flanks the inserted element. The ends of most transposons are composed of inverted repeat sequences.

2. Transposable genetic elements were first discovered in corn in the late 1940s, when Barbara McClintock explained the unstable mutagenic effects of the *Ds* element on a color gene. Recombination between homologous transposons at different chromosomal sites can also induce mutations through genetic rearrangements such as inversions in the gene order and translocations (exchanges of segments between nonhomologous chromosomes).

3. Bacterial transposons include *IS* elements, which carry information only for transposition functions, and composite transposons, which carry genes for additional functions such as drug resistance. The conjugative R plasmids carry one or more drug-resistance genes on transposons that have become integrated into the plasmid DNA. The ability of R plasmids to carry and transmit drug resistance through conjugation makes them a serious medical concern.

4. In replicative transposition, the transposon replicates itself so that movement of the element is accompanied by a doubling in the number of copies of the transposon. In contrast, conservative transposition occurs with no increase in the number of copies of the transposon.

VIRUSES

Viruses are intracellular parasites that must use the energy sources and biosynthetic machinery of the host cell for their replication. They are divided into three main classes, animal viruses, plant viruses, and bacterial viruses (bacteriophages), which are defined by the type of host organism. Within each class, each virus is further restricted in the range of specific hosts that it can successfully infect.

A great variety of viruses are known, and they differ in both size and shape (▶ Figure 10–19). Electron microscopy

▶ FIGURE 10–19 Virus shapes and sizes (compared with an *E. coli* cell).

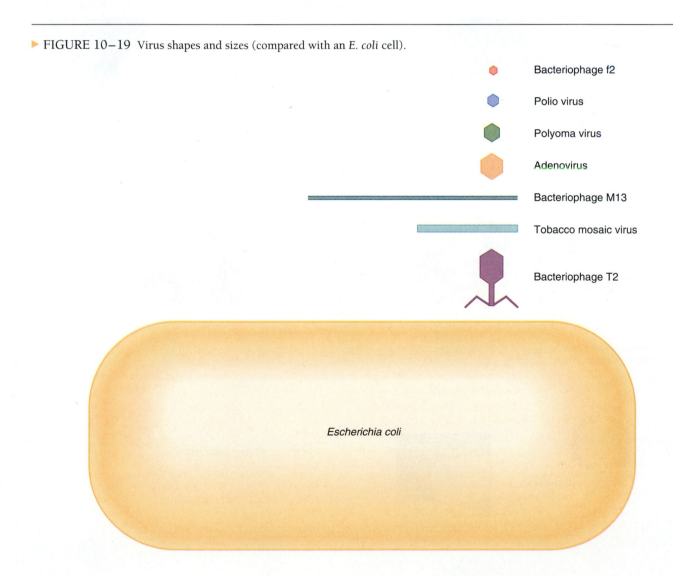

Bacteriophage f2

Polio virus

Polyoma virus

Adenovirus

Bacteriophage M13

Tobacco mosaic virus

Bacteriophage T2

Escherichia coli

reveals that the mature form of each virus, the **virion**, consists minimally of a protein coat, or **capsid**, that surrounds a genome of either DNA or RNA. The capsid and nucleic acid (DNA or RNA) together form the **nucleocapsid**. In certain viruses (the naked viruses), the nucleocapsid itself is the virion. In other viruses (the enveloped viruses), the virion also has a membranous covering, the **envelope**, which is derived from the membranes of the host cell.

The capsid is made up of many protein subunits called **capsomers**, which form a protective covering for the genome. The number and organization of the capsomers and the presence or absence of an envelope serve as a basis for viral classification. Five major morphological types are shown in ▶ Figure 10−20. The capsids of many viruses, including adenoviruses and herpesviruses, have icosahedral symmetry. (An icosahedron is a polyhedron with 12 corners and 20 triangular faces.) The size of the capsids in these viruses is determined by the number of capsomers they contain. In other viruses, such as tobacco mosaic virus and the viruses that cause influenza, the capsomers are organized about the nucleic acid component in the form of a helix. Both icosahedral and helical virions can be either naked or enveloped, depending on the type of virus involved.

Viruses also have diverse nucleic acid structures. In contrast to cells, in which the hereditary molecules are invariably double-stranded DNA, the genome of a virion can be made up of DNA or RNA, can be linear (rod-shaped) or circular (without ends), and can exist in either a single-stranded or a double-stranded configuration. There seems to be no general rule that relates the specific way in which the nucleic acid is organized to the category of virus. For example, there are both plant and animal viruses that contain RNA rather than DNA, either in a single-stranded

---

▶ FIGURE 10−20 Representative animal, plant, and bacterial viruses and their actual electron micrographs.

Naked icosahedral, human wart virus

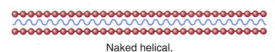

Naked helical, tobacco mosaic virus

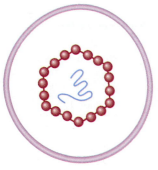
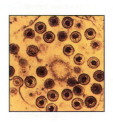
Enveloped icosahedral, herpes simplex virus

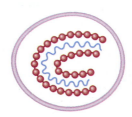

Enveloped helical, influenza virus

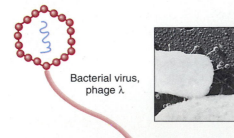
Bacterial virus, phage λ

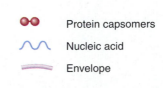
Protein capsomers

Nucleic acid

Envelope

linear form (e.g., the tobacco mosaic virus), in a single-stranded circular form (e.g., the endomyocarditis virus), or as a double helix. Mammalian reovirus is an example of a virus that contains RNA in a double-stranded configuration; in this case the genome is segmented into ten double-stranded RNA molecules coding for different proteins. Many types of RNA bacteriophages are also known, including the well-studied *E. coli* phages MS2, R17, f2, and Qβ. The RNA of these phages is single-stranded, but it assumes a flowerlike structure consisting of hairpin loops that result from hydrogen bonding between complementary regions of the same strand.

The DNA viruses also have a wide variety of chromosome structures, including single-stranded linear forms (parvoviruses), single-stranded circular forms (φX174), double-stranded linear forms (herpesvirus), and double-stranded circular forms (polyoma tumor virus). Other variations also exist. For example, the hepatitis B viral chromosome is about half double- and half single-stranded; the chromosome consists of a partially double-stranded, circular DNA where one of the strands extends over only part of the entire genome length.

Bacteriophage Structure and Growth

Of the many kinds of viruses that have been discovered, bacteriophages have been used most extensively in genetic research. The rapid growth of phages and the ease with which they can be manipulated in the laboratory have contributed to their use as model systems for viruses in general and for the study of cellular genetic processes. A DNA phage has a nucleic acid structure that is similar enough to that of its prokaryotic host to allow it to use many of the cellular enzymes for gene expression and replication. Intensive research on phages during the past several years has therefore contributed a wealth of information concerning the molecular biology not only of viruses but also of their host cells. In this section, we will consider some of the concepts that have emerged from these studies, focusing on the structure and growth of a few of the better-known DNA phages.

The DNA-containing phages can be divided into three morphological types: icosahedral, filamentous (helical), and a type that can be described as an icosahedral capsid (also called a "head" or "coat") attached to a tail. In the third type, the tail is the adsorption organ of the phage; it allows the phage to attach tailfirst to a receptor site on the surface of the bacterial host cell. The phage tail is a complex structure. For example, in the T-even phages (T2 and T4) that infect *E. coli*, the tail is composed of at least six different protein components (▶ Figure 10–21). Phage tails are highly variable in size, shape, and subunit composition, however, and many of the components present in the T-even phages may differ or be lacking altogether in the tails of other phage types.

Phages, like other viruses, do not grow and divide as cells do, but multiply instead by synthesis of their separate components followed by assembly of these components into infectious particles. Although considerable variation exists in the process of phage reproduction, there is, nevertheless, what may be called a basic infection cycle for bacteriophages. Any phage infection that ends in the release of

▶ FIGURE 10–21 The structure of phage T4. (a) Electron micrograph. (b) Schematic drawing.

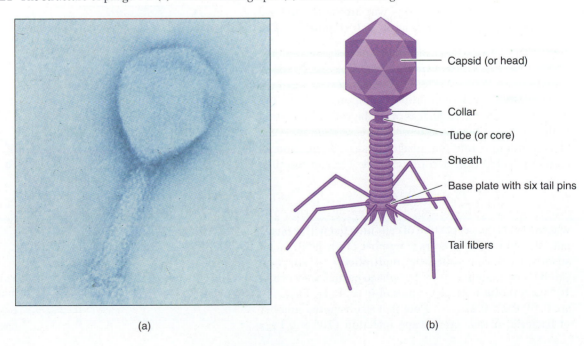

Capsid (or head)

Collar

Tube (or core)

Sheath

Base plate with six tail pins

Tail fibers

(a)   (b)

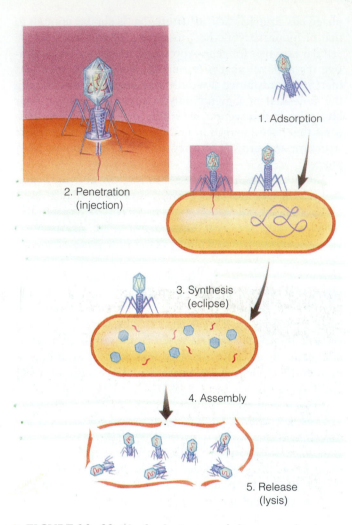

1. Adsorption

2. Penetration (injection)

3. Synthesis (eclipse)

4. Assembly

5. Release (lysis)

▶ FIGURE 10–22 The five basic steps of phage reproduction.

1.4×10^7), even though an average bacterium is infected by two phage particles.

2. *Penetration.* This second step of phage infection involves the passage of the phage genome into the host cell. In the T-even phages and in several other tailed phages, penetration occurs through an active injection process. The tail sheath of these viruses contracts upon infection, forcing the hollow inner tube through the bacterial cell wall like a hypodermic syringe (▶ Figure 10–23). These events trigger the release of the phage genome through the tube into the cell. Not all phages have contractile tails, however, indicating that injection must also occur by different mechanisms. Little is known about the mechanisms of penetration among the tailless phages.

3. *Synthesis.* Following the entry of the viral genome is a period when phage genes direct the synthesis of the protein and nucleic acid components of the virion. The genes are expressed in a temporal order that correlates closely with the timing of events in the phage growth cycle. The phage genes that are expressed early (the **early** genes) code for proteins involved in the replication of the phage genome and in diverting the cellular functions of the bacterium to the synthesis of progeny phages. The genes that specify late functions (the **late** genes) are expressed last in the sequence to form the lysis proteins and various protein subunits of the phage capsid.

4. *Assembly.* Once the component parts of the phage have been synthesized, they are combined in an ordered sequence of steps to form the intact virion. The steps involved in the assembly of phage T4 are shown in ▶ Figure 10–24. The formation of the mature phage is, in part, a self-assembly process that is assisted by the action of enzymes and scaffold proteins. Neither the enzymes nor the scaffold proteins form part of the final virion.

newly synthesized progeny virions will ordinarily proceed through this sequence of five basic steps (▶ Figure 10–22):

1. *Adsorption.* The first step of phage infection involves the binding of phage to specific receptors on the bacterial surface. Binding occurs between an **attachment protein on the phage** (such as a protein component of the tail fibers in certain tailed phages or of the capsid in certain tailless viruses) **and a receptor protein** on the host cell. Since binding is random, phages tend to be distributed among infected cells according to the Poisson formula:

$$P(x) = \frac{m^x}{x!} e^{-m} \qquad (10.1)$$

where $P(x)$ is the proportion of cells infected with x phage particles and m is the average number of phage particles adsorbed per cell, called the **multiplicity of infection** (**MOI**). For example, if 2×10^8 phage particles adsorb to 10^8 bacteria ($m = 2$), the values of $P(0), P(1), P(2), \ldots$ are $0.14, 0.27, 0.27, \ldots$. Note that a substantial number of bacteria in this case escape infection ($10^8 \times 0.14 =$

▶ FIGURE 10–23 Injection of the phage genome by a T-even phage.

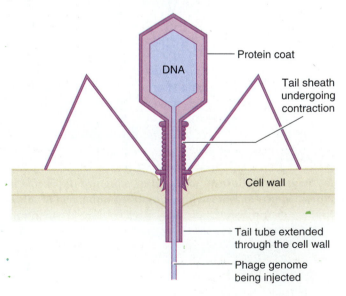

Protein coat

DNA

Tail sheath undergoing contraction

Cell wall

Tail tube extended through the cell wall

Phage genome being injected

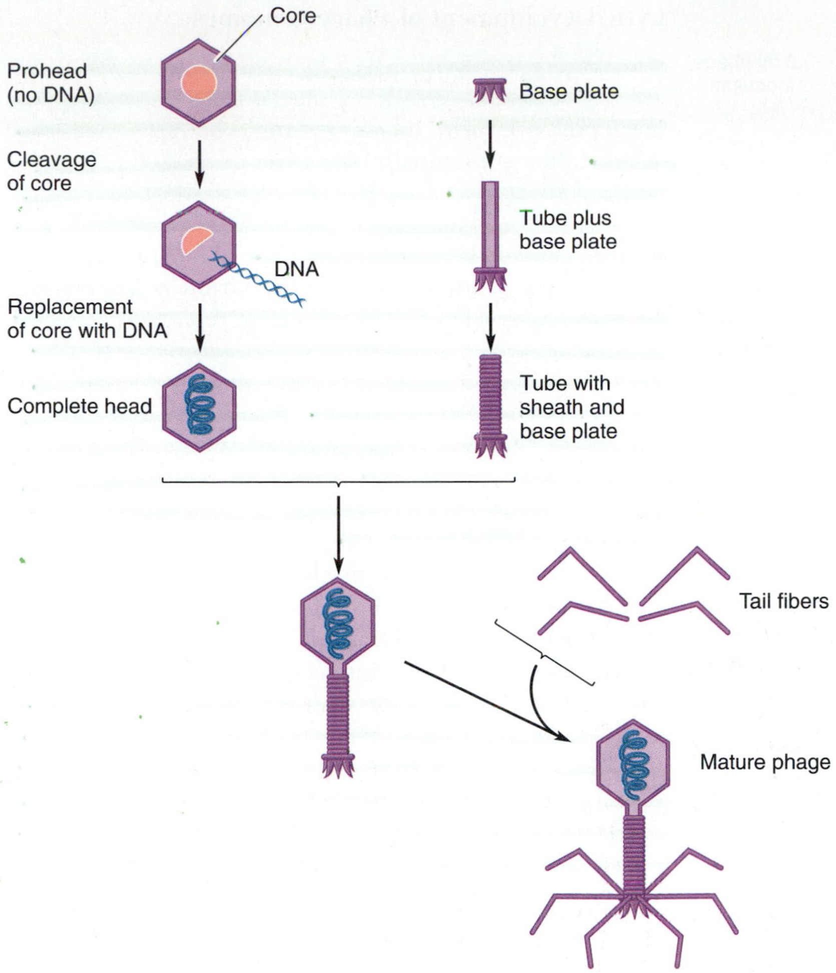

Core

Prohead (no DNA)

Cleavage of core

Replacement of core with DNA

Complete head

Base plate

Tube plus base plate

Tube with sheath and base plate

DNA

Tail fibers

Mature phage

► FIGURE 10–24 Maturation of phage T4 during the assembly period of intracellular phage growth.

but only ensure that the structural components are assembled in the proper sequence.

5. *Release*. The fifth and final step of the infection process is the release of newly synthesized phage from the confines of the infected cells (► Figure 10–25). In the T-even phages and most other phage types, release occurs as a result of cell lysis. The lytic process usually involves the destruction of the bacterial cell wall by a phage-coded lysis enzyme called a **lysozyme** or an **endolysin**. The cells then become sensitive to osmotic rupture and subsequently lyse, releasing perhaps as many as 50 to 1000 progeny phages, depending on the type of phage and the culture conditions.

The ability of a phage to lyse its host serves as the basis for the **plaque assay**, a widely used method for measuring phage concentration. Plaques are localized areas of lysis that develop when phage-infected bacteria are inoculated on an agar plate along with a high concentration of uninfected cells. The uninfected bacteria divide to form a lawn of growth over the agar surface, except where the infected cells were originally present. In these areas, progeny phages have lysed the bacteria to produce visible clear zones (plaques) that enlarge as a result of repeated lytic infec-

tions. Although the extent of clearing varies with the phage type, the bacterial strain, and culture conditions, plaques are generally limited in size to at most a few millimeters in diameter. They do not increase in size indefinitely, because most phages multiply only in growing cells.

► FIGURE 10–25 A bacterial cell undergoing lysis and releasing phage.

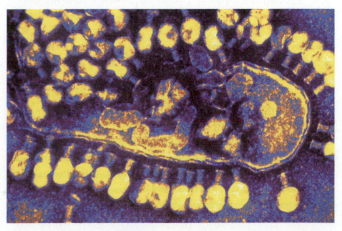

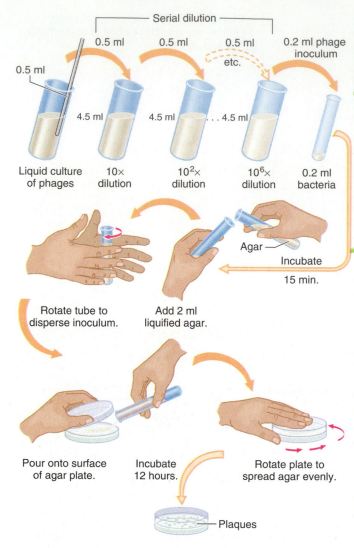

► FIGURE 10-26 Plaque assay for determining phage concentration.

In a typical plaque assay (► Figure 10–26), a small volume of a highly diluted phage suspension is added to a similar volume of a large concentration of indicator bacteria. Under these conditions of very low MOI, each plaque that develops originates from infection by a single phage particle. The concentration of phage particles in the original suspension can then be determined by multiplying the number of plaques formed per sample volume plated by the dilution factor employed.

Not all phages lyse their host at the completion of intracellular growth. For example, some filamentous phages produce a persistent infection in their host bacteria, which continue to release virus particles intermittently for very long periods of time. In a persistent infection, progeny phages are extruded through the cell membrane without causing major damage to the cell.

Lytic Development of Phage Genomes

When a phage infects its host by introducing its DNA into the host cell, the genome of the virion is converted into the replicative form. In many phages the replicative form differs (often substantially) from the corresponding molecule in the virion. For example, the replicative form of phage φX174 is double-stranded, while in the mature state the DNA molecule is single-stranded (the complementary strand is not synthesized when the virion is assembled). The formation of a double-stranded replicative form is common to all single-stranded DNA viruses and allows them to replicate in much the same manner as viruses with double-stranded DNA genomes. The genomes of double-stranded DNA phages may also differ from their corresponding replicative forms. The DNA of phage λ, for example, is linear in the mature state but is circular while replicating in the bacterial cell.

The precise replication mechanisms of phage genomes vary, but they all tend to show a similar developmental pattern within the cell. An idealized view of the cellular phage population during these intracellular stages is shown in ► Figure 10–27. The eclipse phase—the first phase of intracellular phage development—follows the injection of the phage genome into the bacterial host. It includes an initial growth lag, followed by a period of DNA replication. During the eclipse phase the phage DNA directs the synthesis of the structural components of the progeny phages.

► FIGURE 10–27 Idealized view of the changes in the number of replicative phages and mature assembled phages during intracellular growth. The time required for intracellular growth (the latent period) is divided into the eclipse and assembly stages. The eclipse period, during which no mature phages are present, is characterized by an initial lag followed by a phase of replication of phage chromosomes. When assembly of mature phages begins, the amount of phage DNA in the vegetative form reaches a steady state, as phage chromosomes are withdrawn for assembly into infective viral particles as fast as they are produced by replication.

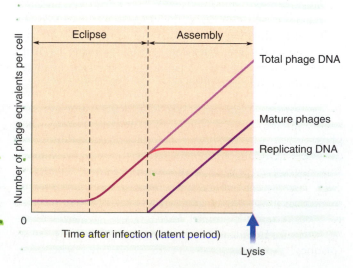

FOCUS ON HUMAN GENETICS

Treating Viral Diseases

Infectious diseases usually result from microbial growth or from toxins produced by microorganisms. The various chemicals and antibiotics used to combat microbial infections are characterized according to whether they are effective against viral, bacterial, fungal, protozoan, or metazoan infections. For example, bacteria can be inhibited or killed by chemicals and antibiotics that block DNA replication, RNA synthesis, or protein synthesis, and by agents that disrupt membranes, block cell wall synthesis, poison electron transport chains, or poison ATP synthases. Chapter 17 discusses some commonly used antibiotics used to combat bacterial infections and their modes of action.

Most bacterial infections can be much more successfully treated than can fungal or viral infections. Fungal infections are particularly difficult to treat because most chemicals that inhibit fungi also inhibit human cells, since both are eukaryotic organisms. In contrast, many bacteria can be effectively treated with antibiotics because the synthesis of bacterial nucleic acids and proteins and the formation of bacterial cell membranes and cell walls do not involve components of the human host cells. Antibiotics therefore act selectively against the bacterial cells and not against the host organism.

Viral infections, like those caused by fungi, are difficult to treat because many of the enzymes used in viral replication are host enzymes. However, some viral infections can be treated (often with limited usefulness) with drugs that inhibit virus uncoating or block viral protein synthesis or with nucleotide analogues or certain other chemicals that block DNA and RNA synthesis. Even though these drugs also affect host nucleic acid synthesis, blocking DNA and RNA synthesis can be effective against the virus because viruses replicate so much faster than eukaryotic cells. These drugs are often very toxic to the patient, however.

Most of the antiviral drugs that have been developed are used to combat herpes infections. **Acyclovir** has been used since 1982 to treat genital herpes as well as herpes infections of the eye. Acyclovir selectively inhibits herpes simplex II replication and thus is not toxic to the patient; it shortens the healing time and reduces the pain of the infection, but it does not eliminate the virus. **Idoxuridine**, a synthetic antiviral nucleoside (2'-deoxy-5'-iodouridine), is incorporated into the viral DNA in place of thymidine and inhibits viral DNA replication. It is used to treat herpes simplex I infec-

tions of the eye, but it is associated with side effects such as allergic reactions and edema of the eyes or eyelids, and since it causes birth defects, it cannot be given to pregnant women. **Vidarabine** is a naturally occurring antibiotic that is isolated from the bacterium *Streptomyces antibioticus*; it is a nucleoside that resembles adenine, and it acts by attaching to and inhibiting the DNA polymerase used by the virus to replicate its DNA. It is most widely used against the herpes virus that causes encephalitis.

Herpes simplex viruses in a cell (false-color TEM).

Mature phage assembly begins about midway through the reproductive or **latent** period. Phage assembly can be visualized as a steady-state process in which the rate of phage maturation keeps pace with the rate of DNA replication. Thus the number of DNA molecules in the replicating pool remains more or less constant.

The progression depicted in Figure 10–27 is accomplished by different replicative schemes in different phages. Since the intracellular stage of phage λ has been studied intensively, we will consider the details of its replication process as an example.

The course of DNA replication during a lytic infection by phage λ is shown in ▶ Figure 10–28. Immediately following injection of λ DNA into the host cell, the DNA molecule assumes a circular superhelical form (Figure 10–28c–e). The formation of a circular molecule is made possible by base pairing between complementary single-stranded (cohesive) ends of the injected DNA. These ends are joined by DNA ligase to form a **cos site** (for cohesive site). The circular molecule then replicates in a semiconservative manner. At first, replication is bidirectional, involving a theta-shaped intermediate (Figure 10–28f). Theta

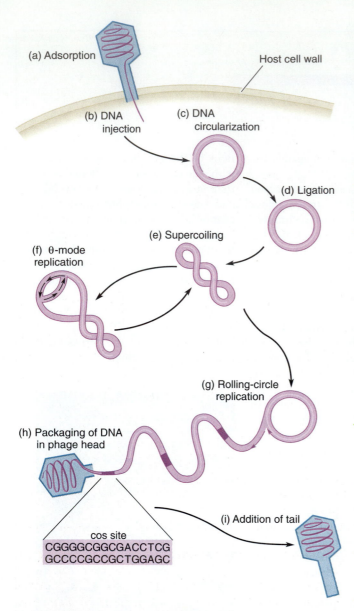

(a) Adsorption

Host cell wall

(b) DNA injection

(c) DNA circularization

(d) Ligation

(e) Supercoiling

(f) θ-mode replication

(g) Rolling-circle replication

(h) Packaging of DNA in phage head

cos site
CGGGGCGGCGACCTCG
GCCCCGCCGCTGGAGC

(i) Addition of tail

▶ FIGURE 10–28 DNA replication and packaging during a lytic infection by bacteriophage λ. *Adapted from:* M. E. Furth and S. H. Wickner, in R. W. Hendrix, J. W. Roberts, F. W. Stahl, and R. A. Weisberg (eds.), *Lambda II,* Cold Spring Harbor Laboratory, 1983, p. 146. Used with permission.

structures can often be observed at this stage if the replicating genomes are viewed in an electron microscope. Partway through the latent period, λ DNA switches to rolling-circle replication (Figure 10–28g). The linear daughter chromosomes that are generated by each "roll of the circle" remain attached end to end, forming a series of tandemly linked genomes called a **concatemer.** The genomes making up each concatemer are then cut at their joining cos sites into linear DNA molecules of the mature phage when they are packaged into the phage head.

It is important to note that the two replication schemes of phage λ contribute in different ways to the replicating

DNA pool of the infecting phage. Theta replication produces a geometric increase in the number of phage chromosomes, since each replicating DNA molecule forms two identical daughter molecules at each round of synthesis. This mode of replication is responsible for the initial rise in the number of phage chromosomes that begins midway through the eclipse period (see Figure 10–27). The switch from theta to sigma (rolling-circle) replication marks the beginning of the assembly period. During this stage, only one of the two strands of each parental DNA molecule continues to serve as a template for successive rounds of replication. The other strand replicates once to form a complementary strand in the duplex tail of the sigma structure and is then assembled into a mature phage. The amount of DNA in the replicating pool then remains constant, while the total intracellular phage DNA (replicating plus assembled DNA) increases linearly with time.

Lysogeny and Temperate Phages

Certain types of viruses sometimes establish a long-term relationship with their host, usually without harming the cell. In such cases, the viral DNA replicates more or less in synchrony with the host genome, so that whenever the host cell divides, at least one copy of the viral genome is retained in each daughter cell. Viruses capable of this form of replication are called **temperate viruses.** Phage λ is an example of such a virus. Usually, the formation of a circle by the infecting λ chromosome is followed by its lytic development, which includes the production of progeny phages and the killing of the host cell upon lysis. This mode of phage reproduction is known as the **lytic cycle.** Occasionally, however, λ DNA follows an alternate route of reproduction known as the **lysogenic cycle,** in which it replicates along with the host chromosome, as an integrated part of that chromosome. The integrated phage DNA, now called a **prophage,** is a latent genetic structure that expresses none of the functions that would lead to autonomous multiplication of the λ phage or to lysis of the cell.

To establish the lysogenic state, the circular λ DNA integrates into the bacterial host DNA through site-specific recombination between a specific attachment (*att*) site on the λ chromosome (designated *att POP′*) and corresponding site on the *E. coli* chromosome (designated *att BOB′*). A breakage and reunion event inserts the phage DNA into that of the bacterium (▶ Figure 10–29). The integration reaction is catalyzed by a phage-encoded enzyme, integrase (Int), which makes staggered cuts in the common core (*O*) sequence in the two *att* sites, and it also requires the involvement of an accessory host protein, the integration host factor (IHF). During λ integration, integrase, IHF, and the *att* site DNAs form part of a higher-order nucleoprotein complex called an **intrasome,** on which strand exchange is made.

The lysogenic condition is generally quite stable. However, on rare occasions, the prophage is spontaneously released from the bacterial chromosome to initiate a lytic

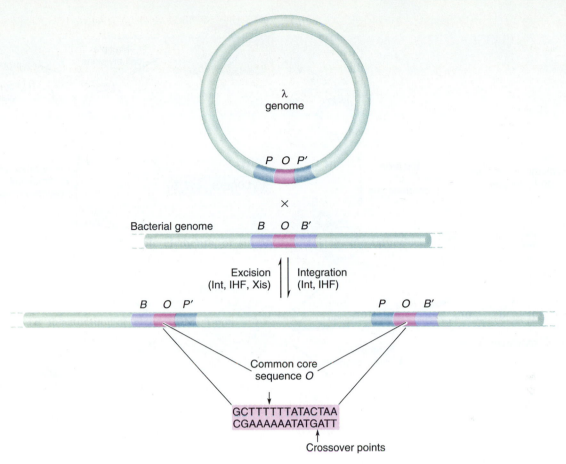

GCTTTTTTATACTAA
CGAAAAAATATGATT

↑
Crossover points

▶ FIGURE 10–29 Integration and excision of the λ prophage. The λ genome is inserted into the bacterial chromosome by a crossover that occurs between the *att POP′* site of the λ chromosome and the *att BOB′* site of the *E. coli* DNA. The crossover occurs at staggered nicks in the common core sequence, *O*, creating a seven-base overlap.

cycle. This process, which is called **prophage induction**, involves excision of the prophage from the host DNA by a reaction that requires another phage-encoded enzyme, excisionase (Xis), in addition to Int and IHF (see Figure 10–29). Prophage induction occurs spontaneously in only about one in every million cell divisions, although UV light, a potent inducing agent, can be used experimentally to greatly increase the chance of this event. Once free of its association with the host chromosome, the phage DNA undergoes autonomous replication and eventually lyses the cell. Because the bacterial host of a prophage is always potentially subject to such a lysis, it is called a **lysogenic cell**.

The lytic and lysogenic cycles of phage λ are shown in ▶ Figure 10–30. Unlike temperate phages (such as λ), virulent phages such as T2 and T4 can undergo only the lytic cycle.

Several temperate phages are used in genetic research, including λ and P1, which are both phages of *E. coli,* and P22, a phage of *Salmonella typhimurium,* as well as many others. Although many temperate phages have a lysogenic cycle similar to that of λ, some interesting variations occur. The lysogenic state of P1, for example, is established without integration into the bacterial chromosome. The phage replicates, instead, as a separate structure in a manner similar to a plasmid.

Another interesting variation involves the replication cycle of bacteriophage Mu. Mu, a temperate phage of *E. coli,* combines many of the characteristics of phage λ with those of bacterial transposons (see Figure 10–13c). Like phage λ, Mu replicates as a prophage while inserted in the *E. coli* chromosome, but unlike λ, integration of Mu DNA occurs at random target sites by transpositional recombination. Since Mu inserts its DNA by transposition, integration always leads to a duplication of the target sequence flanking the prophage, and it can also lead to mutation if the prophage is established within a host gene. In fact, the name Mu is derived from *mutator* because there is such a large percentage of mutants among Mu lysogens. The lytic cycle of phage Mu is also accomplished by a mechanism different from that of λ. Rather than replicating as an autonomous unit, Mu DNA remains integrated in the host chromosome, where it undergoes multiple rounds of replicative transposition in order to achieve lytic growth. Mature phage DNA is then cut from the host DNA during the assembly phase and used directly for packaging, so that Mu DNA is never found free in the cell.

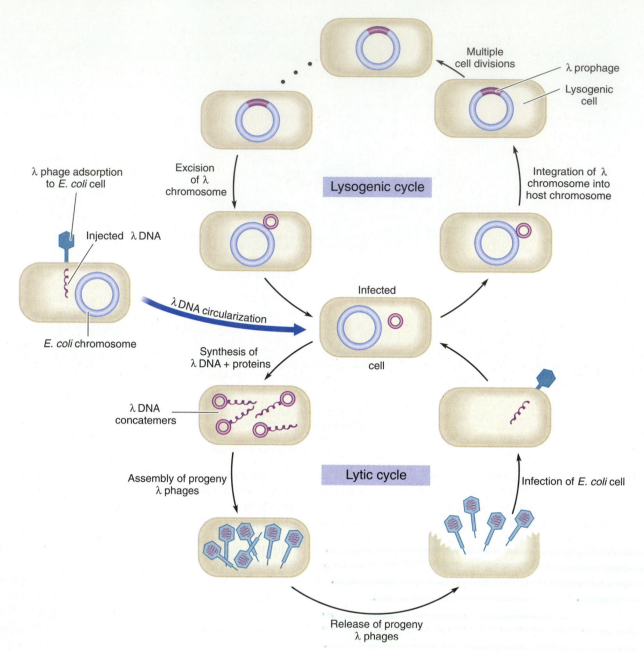

► FIGURE 10–30 The lysogenic and lytic growth responses of phage λ. Temperate phages, such as λ, are capable of both growth cycles. Virulent phages, such as T2 and T4, can only undergo the lytic cycle. (For convenience, the *E. coli* chromosome is not shown in the lytic cycle.)

Example 10.2

An interesting example of interaction between different viruses involves the growth of phages P2 and P4 in non-lysogenic bacteria. When phage P2 infects a cell, the cell normally lyses and produces P2 progeny. When phage P4 infects a cell, the cell survives and becomes lysogenic for P4. However, when both P2 and P4 infect the same cell, the cell lyses and produces only P4 progeny. Suppose that a sample containing 2×10^7 P2 phages and 2×10^7 P4 phages is added to a sample of 10^7 bacterial cells. (a) What percentage

of the bacteria will escape infection? (b) What percentage of the bacteria will become lysogenic for P4?

Solution: In this problem, both infecting phages occur at a multiplicity of infection (MOI) of

$$\frac{2 \times 10^7}{10^7} = 2$$

a. Since the distribution of infecting phages per bacterium is random and follows a Poisson distribution (see equation 10.1), the probability that a cell escapes infection (i.e., is not infected by either phage) is e^{-2} (for no P2) $\times e^{-2}$ (for no P4) $= e^{-4} = 0.0183$. Thus, 1.83% of the bacteria will escape infection.

b. To become lysogenic for P4, a cell must be infected by 1 (or more) P4 and no P2 phages. Therefore, the probability that a cell becomes lysogenic for P4 is $1 - e^{-2}$ (for 1 or more P4) $\times e^{-2}$ (for no P2) $= (1 - e^{-2})(e^{-2}) = (0.135)(0.865) = 0.117$. Thus, 11.7% of the bacteria becomes lysogenic for P4.

To Sum Up

1. The structure of viral chromosomes varies greatly, depending on the particular virus and its stage in the viral life cycle. In general, viral chromosomes can be either DNA or RNA, single- or double-stranded, and circular or linear.

2. The infection (lytic) cycle of most bacteriophages consists of five basic steps: adsorption of phage to cell surface receptors, penetration of the phage genome into the bacterial cell, synthesis of viral protein and nucleic acid (the eclipse stage); assembly of completed virions, and release of newly synthesized phages as a result of lysis of the infected cell.

3. Temperate phages can replicate autonomously in the cytoplasm of the host (the lytic cycle) or in association with the bacterial chromosome (the lysogenic cycle). For most temperate phages, the lysogenic cycle is initiated when the phage chromosome is inserted into the bacterial chromosome to become a prophage. Virulent phages, in contrast, are capable only of lytic growth.

Chapter Summary

Living cells provide a habitat for a variety of simpler genetic systems that must use the metabolic machinery of the cell in order to reproduce. These subcellular genetic systems include the largely extrachromosomal plasmids and viruses, which can replicate independently of the cell genome, and the mobile intrachromosomal DNA segments called transposons.

Plasmids are self-replicating DNA circles that often carry genes that are beneficial to the host cell. Some of these genes provide resistance to antibiotics, and some have the ability to produce certain types of toxic proteins. In *E. coli*, a plasmid called the F factor carries genes that code for proteins required for conjugation. In bacterial conjugation, DNA is transferred from an F factor–containing donor cell to an F⁻ recipient cell. F⁺ donor cells, in which the F factor replicates autonomously within the cytoplasm, donate only F factor DNA to F⁻ cells during conjugation, while Hfr donor cells, which carry the F factor inserted in their chromosome, also transfer bacterial DNA.

Transposable elements or transposons are DNA segments that can move from one chromosomal location to another by transposition, a process that often results in mutation when the transposon is inserted into a host gene. In bacteria, the insertion sequences (ISs), which are the simplest transposons, encode only transposition functions, while more complex transposons also encode other functions, such as resistance to antibiotics.

Unlike plasmids and transposons, viruses exist in both extracellular and intracellular forms. The extracellular form or virion consists minimally of a protein coat or capsid surrounding a genome of DNA or RNA. Bacteriophages also have a tail for injection of DNA into the host cell. In the virulent phages, such as T2 and T4, reproduction occurs through a lytic cycle that results in the lysis of the host cell and the release of progeny phages. Temperate phages, such as λ, can also reproduce through a lysogenic cycle by replicating as a prophage while inserted in the host chromosome.

Questions and Problems

Plasmids

1. Distinguish between the members of each of the following sets of terms:
   Hfr cell, F⁺ cell, and F⁻ cell
   conjugative plasmids, colicinogens, and R plasmids
   high copy-number and low copy-number
   sigma replication and theta replication

2. Fill in the blanks:
   (a) Some plasmids can exist either _____ in the host cell cytoplasm or _____ into the host chromosome; they are thus known as _____.
   (b) Because of their _____ physical state, plasmids can

be differentiated from bacterial DNA by separation techniques such as ultracentrifugation or electrophoresis.
   (c) During conjugation, replication of the F factor (in F⁺ × F⁻ matings) and Hfr chromosome (in Hfr × F⁻ matings) is _____ directional, proceeding by a process known as _____ replication.
   (d) The copy number of a plasmid is controlled by events occurring at the _____ site on the plasmid DNA.
   (e) The inability of certain plasmids to replicate together in a single cell is termed _____.

3. (a) Describe an experiment in which you might determine whether an *E. coli* strain from nature carries the F factor. (b) If

F⁻ cells are converted to F⁺ upon receiving an F factor, why aren't all *E. coli* cells in nature F⁺ or Hfr?

4. It takes 100 minutes for an Hfr donor cell of *E. coli* to transfer its entire chromosome during conjugation. (a) What chromosome length (in base pairs) can an Hfr cell transfer in 1 minute? Compare your answer with the rate of chromosome replication in *E. coli* (50,000 bp/min). (b) Because of random breakage of the conjugation tube, the probability $P(t)$ that two conjugating cells remain together t minutes after the start of conjugation follows the relationship $P(t) = e^{-kt}$, where k is a rate constant. Assuming that $k = 0.075$ per minute, what is the probability that an Hfr donor cell of *E. coli* will transfer its entire chromosome to an F⁻ cell during the conjugation process?

5. The laboratory you are associated with is interested in studying the mode of replication of a newly discovered plasmid. The plasmid is isolated during replication, treated with a site-specific endonuclease (which cuts both strands of the circular plasmid DNA at a unique site) and then viewed with an electron microscope. Some of the cut replicating structures that you observe are shown below. The structures are not arranged in any particular order, since you are unable to tell one end of the DNA from the other under the microscope. From these results, determine whether replication in this plasmid is bidirectional or unidirectional. How can you tell?

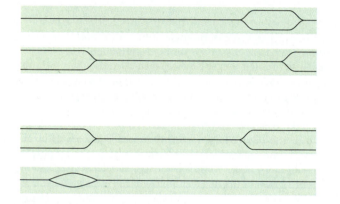

6. Suppose you have a bacterial population in which each parent cell contains one copy of each of three incompatible plasmids, A, A', and A'', at the start of generation 0. If these plasmids are transmitted in random assortment, with each daughter cell receiving half of the replicated plasmids of the parent cell, what proportion of the generation 1 progeny cells would you expect to have exactly the same plasmid makeup as the parents?

Transposable Elements

7. Distinguish between the members of each of the following sets of terms:
   plasmids and transposons
   conservative transposition and replicative transposition
   insertion sequence and composite transposons

8. Fill in the blanks:
   (a) The process of movement of a DNA segment from one chromosomal position to another is known as _____ .
   (b) The ends of nearly all transposons consist of _____ repeat sequences, whereas the target DNA sequences that flank the inserted transposon are _____ repeats.

(c) Unstable mutations can be the result of _____ of a transposon, allowing the target gene to return to its _____ state.
(d) Bacterial transposons that encode only transposition functions are called _____ .

9. Which of the following pairs of sequences qualify as the 3' and 5' terminal sequences of an *IS* element? Which qualify as target site duplications flanking the *IS* element? (a) 5' GAAGCCTAT and CTTCGGATA 3', (b) 5' GAAGCCTAT and ATAGGCTTC 3', (c) 5' GAAGCCTAT and GAAGCCTAT 3', and (d) 5' GAAGCCTAT and TATCCGAAG 3'.

10. When multiple copies of an *IS* element are present in a plasmid genome, these copies can recombine to form new plasmids with different gene arrangements (see Figure 10–12). If a plasmid has the gene arrangement shown below, how many different gene orders are possible after a single recombination act between the *IS* sequences? (*Hint:* Consider recombination between all pairs of the *IS* elements shown in this example.)

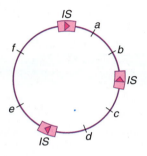

11. Describe how a multiply drug-resistant plasmid might be formed and how it might spread antibiotic resistance throughout a bacterial population.

Viruses

12. Distinguish between the members of each of the following sets of terms:
    capsid, nucleocapsid, and envelope
    eclipse period, latent period, and assembly period
    lysogenic cycle and lytic cycle
    temperate phage and virulent phage
    virion, replicative form, and provirus

13. Fill in the blanks:
    (a) The three morphological types of DNA phages are _____ , _____ , and _____ .
    (b) A bacteriophage particle (virion) is made up of a capsid composed of _____ , which encloses a chromosome made up of _____ , and a tail composed of _____ .
    (c) The average number of phage absorbed per bacterial cell is called the _____ .
    (d) The five basic steps of the bacteriophage infection cycle are _____ , _____ , _____ , _____ , and _____ .
    (e) The two stages of intracellular phage growth are the _____ and _____ phases, collectively referred to as the _____ period.
    (f) DNA replication during some phage infections results in the formation of long, tandemly linked phage genomes called _____ .

(g) During the eclipse period, the number of intracellular phage genomes increases _____, whereas during the assembly stage, intracellular phage DNA increases _____ over time.

(h) A _____ phage such as lambda can occasionally integrate its DNA into the host chromosome, thereby entering the _____ cycle; the integrated phage genome is known as a _____ .

(i) Release of a prophage from the host DNA initiates a lytic cycle and is termed prophage _____ .

14. The interior of the capsid (head) of phage T4 has a volume of about 2×10^{-16} cm^3. The DNA of this phage is 2×10^5 nucleotide pairs in length. Calculate the volume of this DNA (assume a cylindrical shape) and compare it with the space inside the capsid into which the DNA is packed.

15. A 0.1 ml sample is withdrawn from a T4 phage suspension (10^{10} phage/ml) and is added to a flask containing 10 ml of *E. coli* at a density of 10^8 cells/ml. (a) What is the ratio of phage added to bacterial cells in the flask? (b) Assuming that the number of phages that infect any given cell is determined randomly, calculate the percentage of cells that are not infected by phage. (c) What percentage of cells are infected by exactly one phage each? (d) What percentage of cells are infected with two or more phages each? (*Hint:* Use the Poisson probability formula.)

16. After performing a serial dilution on a sample of bacteriophage particles of unknown concentration, you remove 0.2 ml from the final dilution tube and add it to 0.2 ml of sensitive bacteria for the purpose of obtaining a plaque count (see Figure 10–25). This procedure yields 300 plaques on an agar plate. (a) What is the concentration of bacteriophage particles (per ml) in the final dilution tube from which you removed the 0.2 ml sample? (b) The initial sample of phage particles was diluted in steps of 1:10 in a buffer to a total dilution of 10^{-8}. What is the concentration of bacteriophage particles in the initial sample?

17. A lambda-like phage with a genome length of 50 kb has been discovered; it undergoes the lytic cycle in *E. coli* with an eclipse period of 15 minutes and a latent period of 30 minutes under optimal conditions. If the DNA of this phage replicates at a rate of 20 kb/min at each replicating fork, how many progeny will result from a single infecting phage particle? (*Hint:* Assume that DNA replication begins halfway through the eclipse period.)

18. A convenient way to study the lytic cycle of bacteriophages is by using a technique known as the one-step growth experiment. In such an experiment, an exponentially growing culture of bacteria is inoculated with phage particles to give an MOI of 5. After 2–3 minutes to allow the phages to adsorb to the cells, unadsorbed phages are inactivated by the addition of antiserum. The culture is then diluted 200-fold to reduce the antibody concentration to a level at which progeny phages will not be affected. The culture is then incubated at 37°C and sampled periodically to determine the number of infectious phage particles. Plotting the results as infectious units per ml versus time yields curves such as those shown below. In the example shown, curve I was obtained by assaying the culture after artificially lysing the cells; curve II was obtained by assaying the medium without lysing the cells. (a) What is the length of the latent period for the phage in this example? (b) Estimate the length of the eclipse period from these results. (c) What is the average burst size (phages produced per bacterium) under these conditions? (d) Why are infectious phage particles present in the culture during the eclipse period?

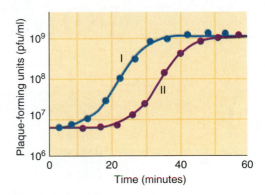

19. Virulent phages, like T2, form plaques with a clear center when plated on sensitive bacteria, whereas temperate phages, like λ, form plaques with a cloudy, or turbid, center. Explain the difference in plaque morphology in terms of the life cycles of these phages, and tell why the plaques of a temperate phage have a turbid appearance.

20. The DNA of a temperate phage with characteristics similar to phage λ has the gene arrangement *a b c d e* in the virion. The corresponding gene arrangement in the prophage of this temperate virus is *c b a e d*. Explain how these gene arrangements are related, describing how the prophage is derived from the infecting virion DNA.

21. Some of the λ phages released upon infection of a *tet^r* strain of *E. coli* are capable of conferring tetracycline resistance to a recipient *tet^s* strain after becoming inserted into the chromosome of the recipient strain as a prophage. Explain these results, describing how the λ phages could acquire this drug-resistance gene during lytic infection.

22. Compare the lengths of the DNA molecules from phage λ (49,000 bp), *E. coli* (4×10^6 bp), and the human genome (haploid content of 2.9×10^9 bp).

GENETIC VARIATION: MUTATIONS AND MUTANT STRAINS

Ordinarily, genes are copied exactly during DNA replication and are transmitted faithfully to the next generation. Considering the complex nature of living organisms, such fidelity requires remarkable accuracy in the reproductive process. On rare occasions, however, hereditary changes (**mutations**) occur that give rise to altered (**mutant**) forms. These rare, usually sudden changes in the genome of an organism can be **spontaneous** (occurring without any apparent external influence) or can be **induced** by some mutagenic agent. Mutations can affect any self-replicating system—they occur in viruses, in free-living cells, and in all cell types of multicellular organisms. In humans and other organisms with a distinct germ line, only mutations that occur in germ cells (**germinal mutations**) are heritable in the sense of having the potential for perpetuation to future generations. Mutations that occur in somatic cells (**somatic mutations**) are perpetuated only in the same individual and only within the clone of cells that descends from the original mutant cell.

This cat has polydactyly and thus more than the normal number of toes. Polydactyly is an inherited characteristic.

Mutations are the ultimate source of all genetic variation; they are thus essential to evolution, since they provide the genetic variability that allows organisms to adapt to different environments. Still, the mere appearance of genetic variants does not guarantee evolutionary success. Most mutant forms do not fare as well in their environmental settings as the forms that produce them. These less adapted variants are, nevertheless, important to genetics because of the variability they provide for genetic analysis.

Mutations can be classified as either **gene mutations** or **chromosome mutations** according to the size of the alteration. Gene mutations occur in single genes and often involve changes in single base pairs. Chromosome mutations, on the other hand, involve substantially larger aberrations that can lead to changes in the number and arrangement of genes, chromosomes, or even whole genomes. This part of the text will look at these major types of mutations, focusing first on gene mutations.

Gene Mutations

A gene mutation is any change in the sequence of nucleotides in a gene that is not a result of normal recombination. As in replication, the Watson-Crick model of DNA suggests a possible mechanism for mutation: A gene mutation can occur when a new base pair is erroneously substituted for the normal base pair at a given location. For example, an AT base pair at a particular site might be replaced with a GC pair, or vice versa. Since a gene is composed of many base pairs (often a thousand or more), many different mutations are possible, with each giving rise to a mutant gene that has a different sequence of nucleotides than its normal allele (▶ Figure 11–1). Mutations occurring at nucleotide sites thus have the potential to produce a large variety of mutant alleles.

Since many allelic forms of a gene can result from mutation, it is customary to define a standard or reference allele. The most commonly used standard is the wild-type allele, which is often designated by a

Navel oranges are products of somatic mutation.

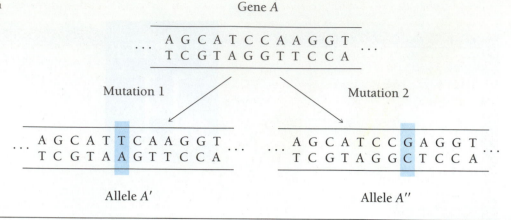

► **FIGURE 11–1** Gene mutation results in different allelic forms of a gene through substitution of different base pairs. (Only 12 pairs of the gene are shown.)

+ superscript (e.g., a^+). Usually, the wild-type allele is the form of the gene found most frequently in natural populations. For example, the ability to synthesize the amino acid threonine (designated Thr⁺) is the prevalent condition among *E. coli* cells in nature; this characteristic, known as the wild-type condition, is the expression of the allele *thr⁺*. The inability to synthesize threonine (designated Thr⁻) is the expression of the alternative allele *thr⁻*. In practice, all forms of a gene that do not produce the wild-type condition are considered mutant. Using the wild type as the reference, we call any change away from this standard (e.g., *thr⁺* → *thr⁻*) a **forward mutation** and any change toward this standard (e.g., *thr⁻* → *thr⁺*) a **reverse (back) mutation** or **reversion.**

The existence of reverse mutation implies that mutation is a reversible process, that is, a mutant allele can undergo a second mutation back to its original form. True reversion is the exact reversal of the DNA alteration that produced the original mutation. The individual formed by a reverse mutation (a **revertant**) is then indistinguishable from the original wild type. A different process, **suppression,** mimics the phenotypic consequences of reversion (► Figure 11–2). Suppression occurs when a second mutation at a different site in the DNA somehow compensates for the effects of the first, restoring the wild-type phenotype but leaving the genotype in a mutant condition. Since reversion and suppression yield the same wild-type phenotype, the distinction between them requires genetic and/or biochemical analysis.

TYPES OF MUTATION

Gene mutations, by definition, are changes in the genotype of an organism. Ordinarily, gene mutations must cause some change in phenotype in order to be recognized. The effects of mutation on the phenotype are varied, ranging from subtle differences from the wild type that are difficult to detect to gross abnormalities that make the organism ill-suited for survival in its environment. We generally associate mutation with some morphological change—vestigial

wings in *Drosophila* and wrinkled seeds in peas are two classic examples. However, many different types of mutations have been studied by geneticists, including a substantial number that can be detected only by biochemical means. A few of the many different types of mutation are considered in the following sections.

Lethal Mutations

Lethal mutations inactivate genes that are indispensable to survival and reproduction. In phages, such mutations lead to failure to complete infection and are measured by the loss of plaque-forming ability, while in bacteria, affected cells

► **FIGURE 11–2** Suppression occurs when a second-site mutation, called a suppressor mutation, at least partially restores the wild-type condition by somehow compensating for the effects of the first mutation.

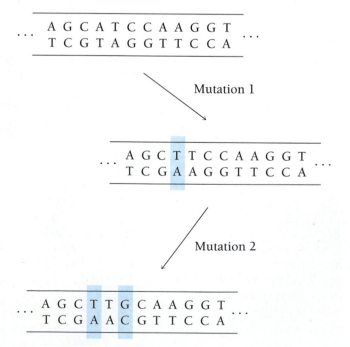

lose the ability to divide and produce colonies. In higher organisms with complex life cycles, lethal mutations manifest themselves through early mortality or the inability of an organism to complete some critical stage of the life cycle. Lethal mutations can be dominant or recessive, and the lethal effect can be expressed at any stage of development from the precleavage egg to a time after formation of adult structures.

Some of the better-known genetic disorders in humans are caused by lethal mutant genes that exert their effect at some point after birth. One example is Tay-Sachs disease, an untreatable, fatal condition that affects infants who are homozygous for the Tay-Sachs allele. Although these infants are normal at birth, they undergo rapid, progressive neurological deterioration leading to mental retardation, blindness, paralysis, and finally death by the age of three or four years. These symptoms are caused by a defect in hexosaminidase A, an enzyme that normally breaks down the lipid ganglioside GM_2. In the absence of this enzyme, ganglioside GM_2 accumulates, especially in brain tissue, resulting in the deterioration of the central nervous system.

Tay-Sachs disease is most common among Jewish people of eastern European origin (Ashkenazi Jews). It is estimated that as many as 1 out of 40 persons of such descent is a heterozygous carrier of the mutant gene. Since the condition is recessive and fatal before the reproductive age, most affected infants originate from matings between heterozygotes.

Example 11.1

The so-called creeper condition in chickens results from the heterozygous presence of the lethal mutant gene c. Heterozygous Cc chickens have short, crooked legs, giving them a squatty appearance. Homozygous cc embryos die on or about the fourth day of development. Only CC chickens are normal. From a mating of two creeper chickens, what phenotypes are expected among their live offspring and in what proportions?

Solution: The cross $Cc \times Cc$ produces a genotypic ratio of $1\ CC : 2\ Cc : 1\ cc$ among the zygotes. The cc embryos die, however, leaving a modified Mendelian ratio of 2 creepers : 1 normal (or $\frac{2}{3} : \frac{1}{3}$) among the live chicks produced by the cross.

Follow-Up Problem 11.1

A and A^Y are two alleles that affect coat color in mice. AA mice have wild-type coat color, AA^Y mice have yellow coats, and the $A^Y A^Y$ condition is lethal during early development. When yellow mice are bred together, what will be the effect of the lethal gene on (a) litter size and (b) the phenotypic ratio among the live offspring?

Conditional Mutations

Conditional mutations produce a mutant phenotype under certain **restrictive conditions** but have no effect (i.e., produce a wild-type phenotype) under other **permissive conditions.** The conditional lethal effects of **temperature-sensitive (ts) mutations** are among the most frequently studied. Some temperature-sensitive strains have a lower maximum temperature for growth than the wild-type strain. For example, a ts *E. coli* mutant will divide at 30°C but not at 40°C, while wild-type *E. coli* cells grow at both temperatures. In this example, 40°C is the restrictive temperature for the conditional lethal effect, while 30°C is a permissive temperature. In addition to heat-sensitive strains, cold-sensitive strains are also known; they require a higher minimum temperature for growth than the wild type. Such increased heat or cold sensitivity is almost invariably caused by amino acid changes at protein sites that are important in maintaining molecular conformation. The mutant protein, having less thermal stability than the wild-type form, undergoes denaturation and loses its biological activity at the restrictive temperature.

Conditional mutations are important in genetic research because they allow the identification and study of genes that might otherwise be unavailable for research. A nonconditional lethal mutation destroys the reproductive capability of the affected cell or virus, so the mutant passes from the scene without leaving a clue about the genetic alteration that produced it. In contrast, a conditional lethal mutation produces an altered cell or virus that can be maintained and cloned in the laboratory under permissive conditions. The effects of the mutation can then be studied by simply shifting a sample of the mutant strain to restrictive conditions.

Nutritional (Biochemical) Mutations

Nutritional mutations alter the normal **anabolic** (biosynthetic) or **catabolic** (degradative) capability of a cell. They are usually studied in microorganisms, where they often lead to a change in nutritional requirements. Wild-type strains of microorganisms such as *E. coli* are **prototrophs**, that is, they are capable of synthesizing all the organic molecules they require for metabolism and growth from a few inorganic salts and a carbon source, such as glucose, that also supplies energy. Prototrophic cells, being nutritionally self-sufficient, are capable of growing on a minimal medium. Biosynthetic mutants, however, are **auxotrophs**—they have an anabolic deficiency that may be as simple as a failure to produce a particular type of amino acid. Auxotrophic cells are thus nutritionally dependent and are unable to grow on a minimal medium unless the medium is supplemented with whatever compound they cannot synthesize for themselves. Auxotrophic mutants have been studied extensively in biochemical genetics, where they have been used to identify various chemical reactions involved in biosynthesis (see the discussion of pathway analysis in the following section).

Nutritional mutations can also lead to a loss of degradative (catabolic) functions. In *E. coli*, for example, wild-type cells can convert a variety of sugars, including lactose, galactose, and arabinose, to glucose. Any one of these sugars can ordinarily be used in place of glucose to support the growth of wild-type bacteria on a minimal medium. However, nutritional mutants that have lost the capacity to make one of the conversions occasionally arise. These mutants are characterized by an inability to utilize a particular sugar, other than glucose, as their sole carbon source.

Nutritional mutations can be regarded as a special class of conditional mutations—the absence of a suitable catabolite or growth requirement from a minimal medium constitutes the restrictive condition in these cases.

Resistance Mutations

Resistance mutations confer resistance to a drug such as a metabolic inhibitor or to a pathogen such as a virus. The development of resistance to the antimicrobial agents known as antibiotics is of particular medical interest. These drugs are produced by microorganisms that inhibit the growth of other microorganisms. Some of the better-known antibiotics and their mechanisms of action are listed in ■ Table 11–1. When antibiotics were first used in the treatment of infectious diseases, the development of bacterial resistance was rather infrequent. Resistance has become much more of a problem with the widespread and sometimes indiscriminate use of antibiotics, which has led to the selection of resistant strains. Normally, drug-resistant mutants are rare, but when an antibiotic is present, it selectively inhibits the growth of sensitive cells and allows any resistant cells to eventually dominate the population. Thus drug-resistant strains are now particularly common among people who work in hospitals where antibiotics are in constant use.

Drug resistance may arise through mutation or by the acquisition of an R plasmid (see Chapter 10), and it can be expressed in several ways. In certain strains, drug resistance is due to the presence of an inactivating enzyme (such as β lactamase in ampicillin-resistant bacteria) that destroys the activity of the antibiotic. In other strains, resistance may result from reduced permeability to the antibiotic or from an alteration that renders the target site less sensitive to the action of the drug. For example, ribosomes in streptomycin-resistant mutants have a reduced affinity for streptomycin that has been traced to a change in the ribosomal protein S12.

In addition to mutations that provide protection from the action of drugs, another class of resistance mutations decreases susceptibility to pathogens. For example, bacteria may become resistant to a particular bacteriophage through a mutational alteration of the attachment site for that phage, thus preventing the phage from binding to the cell wall, or resistance can result from the appearance of an enzyme that destroys the genetic material of the phage once it enters the

■ **TABLE 11–1** Some antibiotics and their mechanisms of action.

Antibiotic	Microbial Source	Mode of Action
ampicillin	*Penicillium* sp.	inhibits cell wall synthesis
bacitracin	*Bacillus subtilis*	inhibits cell wall synthesis
chloramphenicol	*Streptomyces venezuelae*	blocks protein synthesis
erythromycin	*Streptomyces erythreus*	blocks protein synthesis
gentamicin	*Micromonospora purpurea* and *M. echinospora*	blocks RNA synthesis
novobiocin	*Streptomyces niveus* or *speroides*	inhibits DNA synthesis
penicillin G	*Penicillium chrysogenum*	inhibits cell wall synthesis
polymyxin B	*Bacillus polymyxa*	destroys cell membrane
rifamycin B	*Streptomyces mediterranei*	blocks RNA synthesis
streptomycin	*Streptomyces griseus*	results in abnormal protein synthesis
tetracycline	*Streptomyces aureofaciens*	blocks protein synthesis

cell. Any increase in resistance in the host will tend to select for an increase in virulence in the pathogen. The change in genetic resistance to pathogens is therefore often accompanied by a counteracting change in the ability of the pathogen to infect its host. For example, the effects of mutations in bacteria that alter the attachment site for a phage are often offset by mutations in the phage that produce compensatory changes in the structure of the phage tail. This process of genetic change that occurs jointly in two or more interacting populations, such as a parasite and its host, is called coevolution.

To Sum Up

1. A gene mutation can occur when an error in base pairing substitutes one DNA base pair for another. The large number of nucleotide pairs making up each gene means that many mutant alleles are possible at any given locus.

2. Mutation is a reversible process. Mutant alleles occasionally revert back to the wild-type form through exact reversal of the original mutational change. Suppression phenotypically mimics the effects of true reversion. Suppression restores the wild-type phenotype through a second mutation that somehow compensates for the effects of the original alteration.

3. The phenotypic effects of mutant genes vary greatly and include recognizable alterations in morphology, survival, metabolism, and resistance to chemicals and viruses that are normally harmful. An organism with a conditional mutation expresses the altered phenotype only under restrictive environmental conditions.

MUTATIONS AND METABOLIC BLOCKS

Many of the mutations we have described produce a mutant phenotype by blocking a normal metabolic pathway within a cell. Recall from Chapter 6 that the connection between mutant genes and malfunctioning metabolic pathways was first recognized early in this century by the physician Archibald Garrod. Garrod's idea that genes specify enzymes and that the failure of a particular enzymatic reaction is the basis of a mutant phenotype was far in advance of its time. It was not until the studies of George Beadle and Edward Tatum in the early 1940s that any serious, productive steps were taken to correlate genes and metabolic functions. These researchers were awarded the Nobel Prize for experimentally demonstrating the relationship between genes and enzymes.

Beadle and Tatum worked with the common bread mold, *Neurospora crassa.* Wild-type *Neurospora,* like wild-type *E. coli,* has simple growth requirements and is able to reproduce on a minimal medium that contains a carbon and energy source and the vitamin biotin. By irradiating the asexual spores (conidia) of *Neurospora* with X rays, Beadle and Tatum were able to induce a variety of auxotrophic mutant strains that were deficient in different enzymatically controlled reactions. Further testing using genetic complementation tests revealed that the enzymatic deficiencies were the result of separate gene mutations: There was a one-to-one correspondence between a gene mutation and the loss of a particular enzymatic function.

On the basis of their findings, Beadle and Tatum proposed the one gene–one enzyme hypothesis, which states that each gene controls the synthesis of one enzyme. To include oligomeric enzymes as well as proteins other than enzymes in this relationship, the one gene–one enzyme hypothesis has since been refined to the one gene–one polypeptide hypothesis. This more general relationship holds that the function of (most) genes is to code for a polypeptide, which, alone or folded together with other polypeptides, makes up a functional protein.

Pathway Analysis

The one gene–one enzyme hypothesis and its successor, the more inclusive one gene–one polypeptide concept, have proved to be of great value in elucidating many of the details of metabolism in microorganisms, including several biosynthetic pathways. The main ideas underlying these investigations are illustrated for a hypothetical pathway in ▶ Figure 11–3 and can be summarized as follows:

1. Metabolism proceeds through a series of chemical reactions in which each step is catalyzed by a specific enzyme.

2. The enzymes that function in a metabolic pathway are controlled by different genes.

3. If one enzyme is rendered nonfunctional by a mutation, the reaction that it would otherwise catalyze is effectively blocked, and the substrate for that reaction undergoes no further transformation along the pathway.

Genetic defects that affect metabolic pathways are called metabolic blocks. Blockage at one step of a metabolic pathway affects only that step—the block is a failure to produce the next product in the sequence. The enzymes that catalyze the other steps of the pathway are present in normal amounts.

There are two direct consequences of a metabolic block: (1) an accumulation of precursor substances that are formed prior to the blocked step and (2) a failure to produce the end product of the pathway (because of the lack of the intermediate substance immediately following the block). Accumulation of precursors is the basis for elucidating pathways by means of accumulation studies. In the unbroken pathways of the wild-type strain, intermediate metabolites are seldom present in sufficient quantity to be easily detected, but when a metabolic block breaks the normal chain of reactions, the precursor substance at the blocked step is no longer transformed into its usual product and will therefore accumulate, often to be excreted by the cell. Thus the nature of the intermediate compounds in the pathway can sometimes be established by isolating and identifying the substances that accumulate in different mutant strains. For example, there is a different metabolic block after every precursor substance in the D-producing pathway in Figure 11–3, so theoretically each metabolite could be identified by its accumulation in a different mutant: compound A

▶ FIGURE 11–3 The one gene–one enzyme concept and its relationship to a hypothetical biosynthetic pathway consisting of three steps. Metabolites A–D are reactants and products in the pathway.

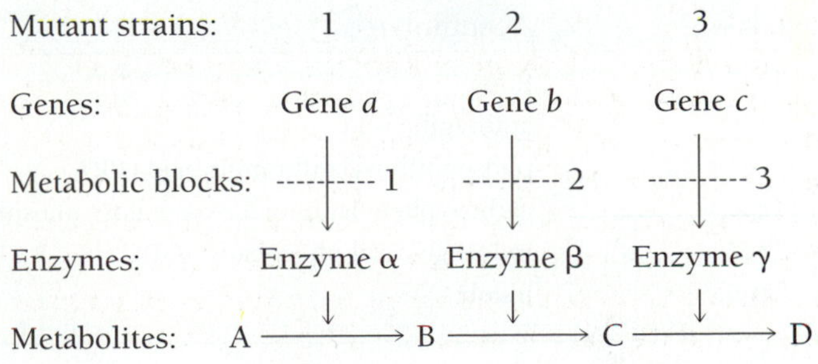

Mutant strains:	1	2	3
Genes:	Gene *a*	Gene *b*	Gene *c*
Metabolic blocks:	----┼---- 1	----┼---- 2	----┼---- 3
Enzymes:	Enzyme α	Enzyme β	Enzyme γ
Metabolites:	A ⟶ B	⟶ C	⟶ D

in mutant 1, compound B in mutant 2, and compound C in mutant 3.

The identification of accumulated metabolites in mutant strains does not by itself establish the sequence of these compounds in the pathway. To reconstruct the pathway, we need to determine the sequential order of the metabolic blocks. This can be done through the use of **growth studies** in which the mutant strains are tested for their ability to grow on minimal medium in the presence of each of the suspected intermediate metabolites. In a growth study, we assume that the end product of the pathway is essential for growth and that it can be formed in a mutant only from compounds that appear after the block. Therefore, only these intermediate substances or the end product itself will promote growth when added to a minimal medium. In the hypothetical pathway in Figure 11–3, for example, mutant 1, which has a block at the first step in the pathway, can grow if the minimal medium is supplemented with compounds B, C, or D; mutant 2, which has a block after B, can grow if the medium contains C or D; and mutant 3, which has a block in the last step of the pathway, can grow only in the presence of the end-product D.

An actual example of pathway analysis is presented in ■ Table 11–2, which lists the accumulated metabolites and growth responses of six tryptophan-requiring (Trp⁻) auxotrophs of *E. coli*. Accumulation studies suggest that the order of metabolites in this pathway is anthranilic acid → PRA → CDRP → IGP → indole → tryptophan. The reactions involved in the synthesis of tryptophan are shown for comparison in ▶ Figure 11–4. Observe that the growth responses to the phosphorylated intermediates PRA, CDRP, and IGP are not listed in Table 11–2—phosphorylated compounds of this type cannot enter the bacterial cell from the growth medium and so cannot be included in these tests. Also note that TrpE⁻ mutants fail to accumulate chorismic acid, the intermediate immediately preceding the metabolic block in this strain. Chorismate is an important precursor of other amino acids in addition to tryptophan, and thus it continues to be utilized by the cell in other pathways despite the block in tryptophan biosynthesis.

Example 11.2

Three auxotrophic mutant strains of *Neurospora* (strains 1, 2, and 3) are blocked at different steps of the same metabolic pathway, which involves the compounds F, G, H, and I. Each of these substances is tested for its ability to support the growth of the three mutant strains, with results as follows:

Mutant strain	Growth on minimal medium plus:			
	F	G	H	I
1	+	−	−	+
2	+	−	+	+
3	−	−	−	+

Determine the sequence of compounds making up the metabolic pathway, and determine the specific step in the pathway that is blocked in each mutant strain.

Solution: The end product of the pathway is the substance that supports the growth of all mutants; thus, compound I is the end product. Working backward, compound F supports the growth of the next largest number of mutants; therefore, it is the third substance in the pathway. Compound H supports the growth of just one mutant, and compound G no mutants. The sequence of substances in the pathway is therefore G → H → F → I.

Note that mutant 1 responds to compounds F and I; it is unable to utilize G and H, so it must be blocked at the second step (between H and F). Mutant 2 responds to all compounds except G; it is unable to utilize G and so must be blocked at the first step (between G and H). Mutant 3 responds only to compound I, so it must be blocked at the third step.

Follow-Up Problem 11.2

Set up a table showing the growth requirements of three mutants that are blocked in a certain metabolic pathway as follows:

$$C \xrightarrow{\text{Step 1}} A \xrightarrow{\text{Step 2}} T \xrightarrow{\text{Step 3}} S$$

■ **TABLE 11–2** Accumulated metabolites and growth responses of tryptophan-requiring (Trp⁻) auxotrophs of *E. coli*.; + shows that growth occurred; − indicates that no growth occurred.

Mutant Class	Substance Accumulated	Growth on Minimal Medium Plus:		
		Anthranilic Acid	Indole	Tryptophan
TrpE⁻	none	+	+	+
TrpD⁻	anthranilic acid	−	+	+
TrpC₁⁻	phosphoribosylanthranilic acid (PRA)	−	+	+
TrpC₂⁻	carboxyphenylamino deoxyribulose phosphate (CDRP)	−	+	+
TrpA⁻	indole glycerol phosphate (IGP)	−	+	+
TrpB⁻	indole	−	−	+

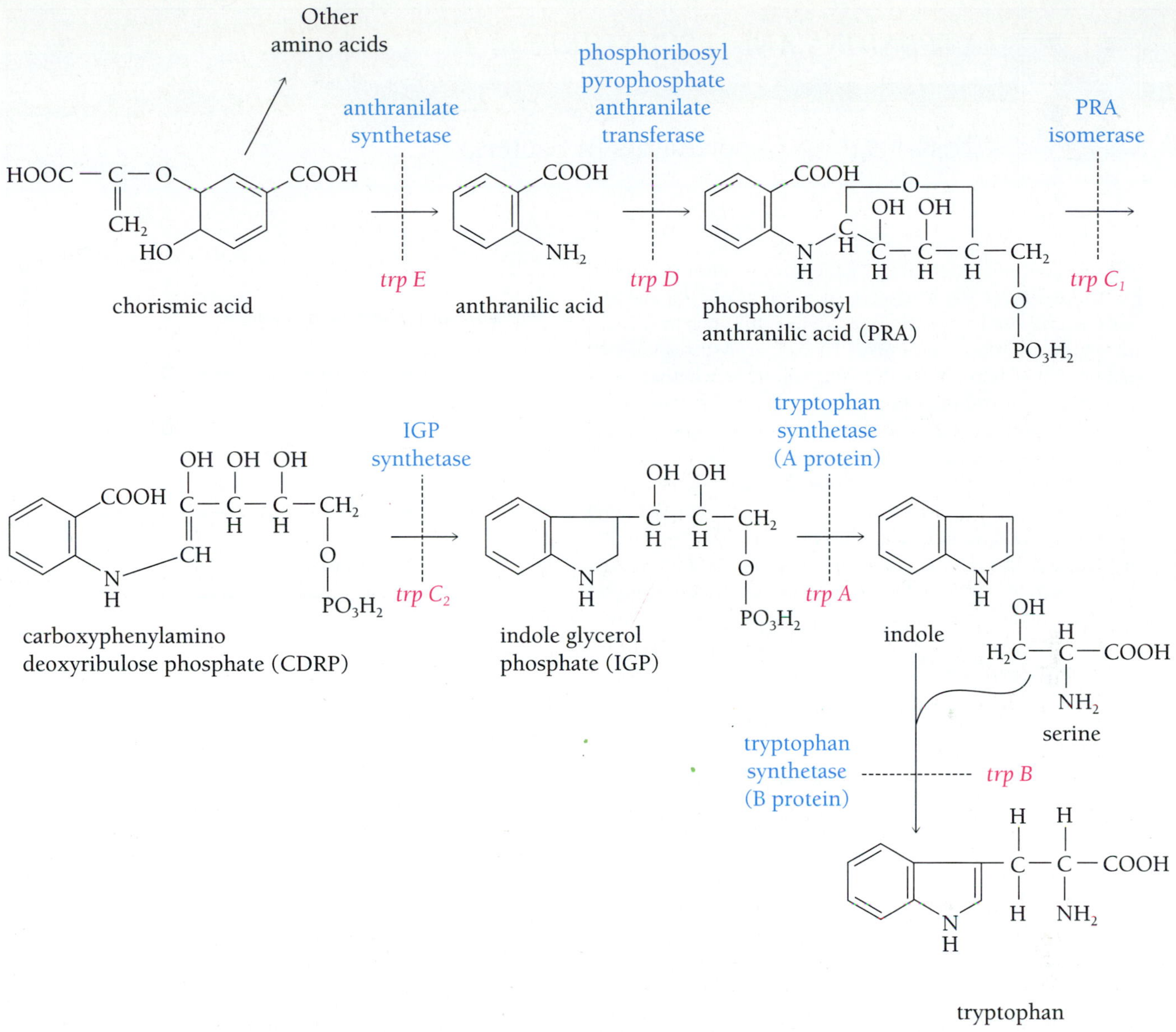

▶ **FIGURE 11−4** Metabolic pathway involved in the conversion of chorismic acid to tryptophan, showing the classes of *trp* mutations in *E. coli* that affect the function of the various enzymes in this pathway.

Mutant 1 is blocked at step 2, mutant 2 is blocked at step 3, and mutant 3 is blocked at step 1.

Of the various enzymes affected in tryptophan-deficient mutants, the one showing the most complex behavior is tryptophan synthetase, the final enzyme in the pathway. Tryptophan synthetase is a tetramer consisting of two α chains (the A protein) and two β chains (the B protein). It catalyzes the overall reaction

indole-3-glycerol phosphate (IGP) + serine →
     tryptophan + glyceraldehyde-3-phosphate

which is believed to consist of two parts:

indole-3-glycerol phosphate →
     indole + glyceraldehyde-3-phosphate

and

indole + serine → tryptophan

The ability of the enzyme to catalyze these steps and hence the overall reaction is affected in TrpA⁻ and TrpB⁻ mutant strains. The TrpA⁻ mutants have defective α chains and are unable to complete the first step of the reaction, whereas TrpB⁻ mutants produce defective β chains and lack the

Cross-Feeding Experiments in Bacteria

Cross-feeding experiments sometimes make it possible to determine the arrangement of metabolic blocks in bacteria without first identifying the intermediate metabolites. These experiments test the growth responses of nutritional mutants to determine quickly whether the accumulated substance that is excreted by one mutant strain will promote the growth of other strains. The theory behind cross-feeding is shown in part (a) of the figure. Note that one mutant can stimulate the growth of other mutant strains that are blocked at an earlier reaction in the pathway, but it will have no effect on those strains that are blocked at a later reaction. In the example, mutant 2, which excretes intermediate B, stimulates the growth of mutant 1, which is blocked at the preceding step and can use B as a precursor in the synthesis of the end product D. However, mutant 2 does not stimulate the growth of mutant 3, which, like mutant 2, has a block after B. Part (b) of the figure shows the results of a cross-feeding experiment involving these three mutant strains. In this test, the mutant strains are streaked on a plate of minimal agar that contains just enough of the required end product to permit a slight amount of growth. The streaks are patterned in such a way that the enrichment of growth caused by excreted metabolites diffusing from one strain to another (cross-feeding) can be readily assessed. Since mutant 3 cross-feeds mutants 1 and 2 in this example, steps 1 and 2 must precede step 3; and since mutant 2 cross-feeds mutant 1, step 1 must precede step 2. Therefore, the experimentally determined sequence of blocked steps in this pathway is 1–2–3.

(a) Blocked steps in metabolic pathway

A $\longrightarrow$ B $\longrightarrow$ C $\longrightarrow$ D   wild type

A $\longrightarrow$ B $\longrightarrow$ C $\vert$ - - - - D   mutant 3

C accumulates

A $\longrightarrow$ B $\vert$ - - - - C $\longrightarrow$ D   mutant 2

B accumulates

A $\vert$ - - - - B $\longrightarrow$ C $\longrightarrow$ D   mutant 1

A accumulates

(b) Results of cross-feeding tests

Mutants 1, 2, and 3 are streaked on minimal agar supplemented with limiting amounts of end product D.

Growth of mutant 1 is stimulated in the presence of mutants 2 and 3, and the growth of mutant 2 is stimulated in the presence of mutant 3.

Cross-feeding tests on a hypothetical series of mutant strains.

---

ability to complete step 2. Tryptophan synthetase is therefore an oligomeric protein with two enzymatic functions, one ascribable to each of the subunit chains.

Metabolic Basis of Inherited Disease

The relationship between genes and biochemical reactions has important applications in the diagnosis and treatment of metabolic disease. Many human biochemical disorders are now known to be the result of single gene mutations that lead to a nonfunctional enzyme and a metabolic block. For example, the metabolic blocks that lead to disorders in phenylalanine and tyrosine metabolism are identified in ▶ Figure 11–5, which shows the metabolic pathway for the

breakdown of these amino acids. In addition to the block that causes alcaptonuria (the disorder first studied by Garrod), blocks at other steps in this pathway result in albinism and phenylketonuria (PKU). In PKU, the failure to produce the liver enzyme phenylalanine hydroxylase causes a block in the conversion of phenylalanine to tyrosine. The resulting accumulation of phenylalanine and its derivatives, particularly phenylpyruvic acid, can reach levels that are toxic to the central nervous system, causing irreversible brain damage during the first few months of postnatal life.

Phenylketonuria and albinism illustrate the two major ways in which a metabolic block can cause disease. Phenylketonuria results from the accumulation of a substance that is toxic to cells at high concentrations, while albinism (the

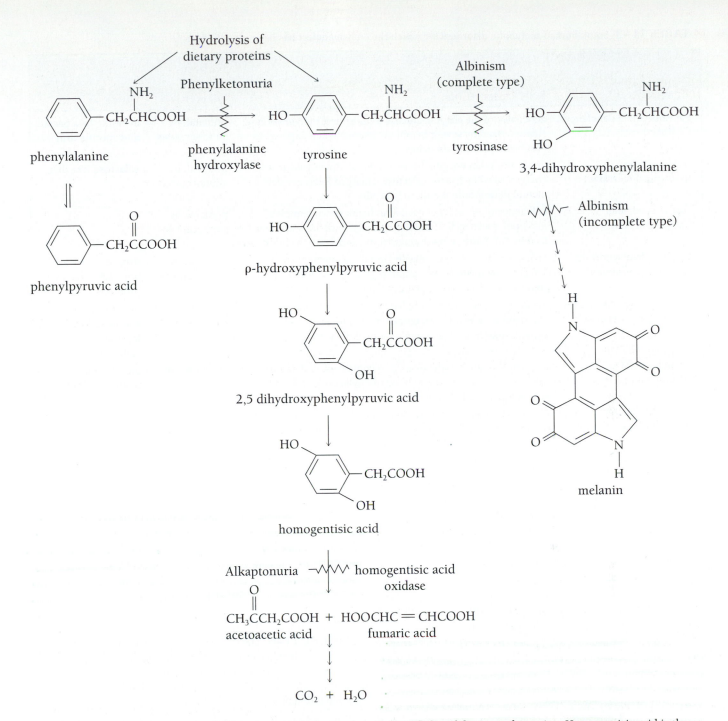

▶ **FIGURE 11–5** The metabolic pathway responsible for the degradation of phenylalanine and tyrosine. Homogentisic acid is the intermediate that accumulates in individuals with alkaptonuria. Phenylketonuria and albinism are also caused by blocks in this pathway. *Source:* Reprinted with the permission of Macmillan College Publishing Company from *Genetics* by Monroe W. Strickberger. © 1985 by Monroe W. Strickberger.

failure to produce melanin) results from the lack of the end product of a metabolic pathway. Most metabolic diseases can similarly be attributed to either an accumulation or a deficiency, although affected individuals might display some of the features of both. Individuals with phenylketonuria, for example, are often fair-skinned with blond hair and blue eyes because of their lower-than-normal levels of melanin.

■ Table 11–3 lists a few examples of metabolic diseases for which the enzyme defect has been identified. In many cases diet or drug therapy can alleviate certain symptoms of the disease. For example, if PKU is diagnosed shortly after birth, the affected infant can be placed on a controlled diet that minimizes the amount of protein containing phenylalanine. (Phenylalanine is one of the essential amino acids for

Disorder (and Genetic Basis)	Brief Description	Earliest Possible Diagnosis	Therapy
Favism (G6PD deficiency) (X-linked recessive)	deficiency of the enzyme glucose 6-phosphate dehydrogenase; characterized by severe hemolytic anemia when exposed to fava beans and certain drugs	prenatal, by amniocentesis	avoiding fava beans and specific drugs
Galactosemia (autosomal recessive)	deficiency of an enzyme needed to break down galactose; characterized by excess galactose, enlarged liver, eye defects, mental deterioration, and early death	prenatal, by amniocentesis	galactose-free diet
Hurler syndrome (autosomal recessive)	deficiency of an enzyme needed to break down complex mucopolysaccharides found in connective tissue; characterized by stiff joints, growth retardation, and death in childhood	prenatal, by amniocentesis	—
Lesch-Nyhan syndrome (X-linked recessive)	deficiency of an enzyme needed in purine metabolism; characterized by mental retardation, high uric acid levels, muscle spasms, and compulsive self-mutilation	prenatal, by amniocentesis	drugs for uric acid; none for behavioral problems
Maple syrup urine disease (autosomal recessive)	deficiency of an enzyme needed to break down branched-chain amino acids; characterized by maple syrup odor of urine, progressive degeneration, and early death	prenatal, by amniocentesis	diet low in the amino acids valine, leucine, and isoleucine
Phenylketonuria (autosomal recessive)	deficiency of an enzyme needed to convert phenylalanine to tyrosine; characterized by excess phenylalanine in blood and severe mental deterioration	newborns, by screening blood samples	diet low in phenylalanine
Tay-Sachs disease (autosomal recessive)	deficiency of an enzyme needed to break down a ganglioside lipid; characterized by progressive deterioration and death in infancy	prenatal, by amniocentesis	—

humans because it cannot be produced by the body's own metabolic machinery. Some phenylalanine must therefore be included in the diet so that proteins can be synthesized.) Some individuals with PKU who were placed on this diet within the first month of life have been able to discontinue it by the age of six years with no serious effects on brain function; however, in most cases, patients are encouraged to continue the diet for much longer.

Most human metabolic disorders are inherited as autosomal recessives. One normal allele is sufficient to produce a physiologically adequate amount of the enzyme. Although outward appearance of a heterozygote may suggest complete dominance by the normal allele, dominance is usually incomplete with regard to the level of enzyme concentration. Heterozygotes tend to have an intermediate amount of the enzyme—less than normal homozygotes but more than persons affected with the disease. This lack of complete dominance at the enzymatic level is the basis for tests to detect carriers of defective genes (▶ Figure 11–6). Carrier screening using enzyme analysis has been particularly successful for Tay-Sachs disease, since the screening test for heterozygosity is convenient and reliable and the disorder, although rare, is mainly restricted to a single population group.

Many metabolic disorders can also be detected by prenatal screening using amniocentesis (▶ Figure 11–7) or a newer method called chorionic villus biopsy (▶ Fig-

ure 11–8). In a chorionic villus biopsy, a catheter is inserted into the uterus to remove a small sample of cells from the outer fetal membrane or chorion, which is attached to the placenta. The cells can then be cultured for chromosome and enzyme analysis. A chorionic villus biopsy has two main advantages over amniocentesis, which is currently the more

▶ **FIGURE 11–6** The detection of heterozygotes for the recessive allele that causes Tay-Sachs disease. The serum concentration of the enzyme hexosaminidase A in parents of affected children is about 50% of the level in homozygous normal individuals.

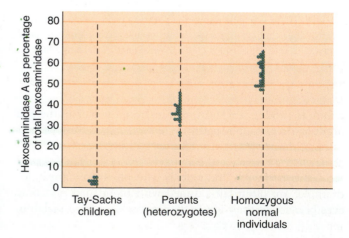

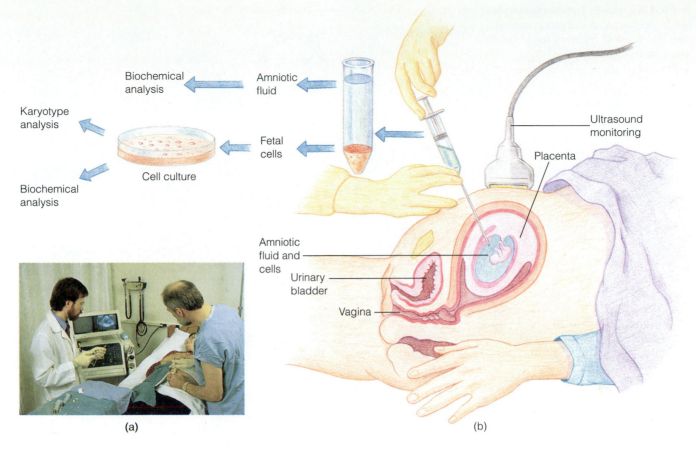

Biochemical analysis ← Amniotic fluid

Karyotype analysis

Fetal cells

Cell culture

Biochemical analysis

Ultrasound monitoring

Placenta

Amniotic fluid and cells

Urinary bladder

Vagina

(a)

(b)

► **FIGURE 11–7** Prenatal diagnosis using amniocentesis. (a) A patient undergoing amniocentesis. (b) A small sample of the amniotic fluid is removed by means of a syringe. Fetal cells that have been shed into the amniotic fluid can then be grown in culture and used for biochemical and chromosome studies.

common procedure: the results of chorionic villus biopsy are available overnight, compared with up to four weeks for amniocentesis, and the method can be used as early as the 7th to 9th week of pregnancy, while amniocentesis cannot be done until the 15th to 16th week.

Of course, not all diseases caused by metabolic blockage are common or severe enough to warrant the use of mass screening tests. Some of the factors that can act to alleviate the severity of metabolic blocks are summarized in ► Figure 11–9. In many cases, the excess metabolic intermediate can be excreted in the urine, and the cells can use an alternative pathway to compensate for the blocked one, thereby preventing a shortage of the end product (alcaptonuria is an example of this type of compensation). In other instances, the affected pathway is used so little under normal dietary conditions that the blockage has no severe effect unless an abnormally high amount of a particular dietary substance or an unusual substance is ingested. Certain drug reactions are now known to fall into this category—some individuals lack an enzyme that is required for the normal metabolism of a particular drug, and they can have a severe reaction to therapeutic doses of the drug, which would normally be metabolized with no ill effects.

To Sum Up

1. The phenotypic change associated with a mutant gene often results from the loss of function of the enzyme that is coded for by the wild-type allele of that gene. The lack of a functional enzyme creates a block in a metabolic pathway. Beadle and

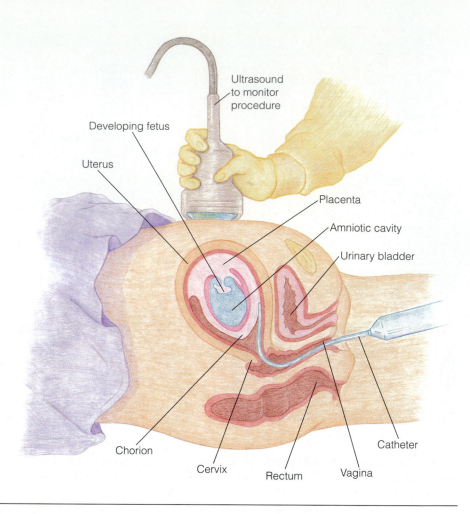

► **FIGURE 11–8** The chorionic villus biopsy technique. A catheter is inserted into the uterus to remove a tissue sample from the fetal chorion.

Ultrasound to monitor procedure

Developing fetus

Uterus

Placenta

Amniotic cavity

Urinary bladder

Chorion

Cervix

Rectum

Vagina

Catheter

Tatum's studies of metabolic blockage in *Neurospora* in the early 1940s led to experimental proof that the function of (most) genes is to code for a polypeptide.

2. The consequences of a metabolic blockage depend on the extent of the accumulation of the metabolic intermediates and the deficiency of the end product. A number of human metabolic diseases have been associated with particular enzyme defects, and prenatal screening is possible for a growing number of these disorders.

► **FIGURE 11–9** Factors that can reduce the severity of a metabolic block. A hypothetical pathway is blocked at the step that normally converts compound X to compound Y. The severity of the effects that result from the accumulation of compound X can be reduced by the excretion of excess compound X and by utilizing compound X in an alternative pathway. The lack of the end product might be alleviated by dietary sources or by producing it through an alternative pathway.

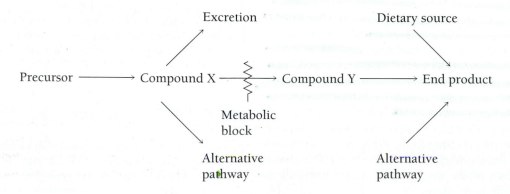

Excretion   Dietary source

Precursor ⟶ Compound X ⟶ Compound Y ⟶ End product

Metabolic block

Alternative pathway   Alternative pathway

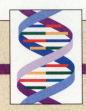

FOCUS ON HUMAN GENETICS

Dietary Therapy in the Treatment of Genetic Diseases

Several human metabolic diseases can be "treated" by dietary therapy. These modifications of the diet are essential for preventing severe phenotypic consequences such as mental retardation or death. Galactosemia, homocystinuria, phenylketonuria, tyrosinemia, and maple syrup urine disease are the major human metabolic disorders that can be controlled through dietary therapy.

The treatment that has been developed for phenylketonurics illustrates how a special diet can be devised to contain a low amount of the offending metabolite (in this case, phenylalanine). A commercial protein preparation such as casein (a protein extracted from milk) is enzymatically digested, and the resulting mixture of amino acids is charcoal-filtered. The filtering process removes the phenylalanine, tyrosine, and tryptophan. The latter two amino acids are added back to the filtrate, along with a source of fat (e.g., corn oil), carbohydrates (e.g., sugars, corn starch, or corn syrup), and vitamins and minerals. This powder or "formula" is then used at each meal as the main dietary protein (amino acid) source. The following example shows a typical daily menu for a young phenylketonuric child:

Breakfast

$\frac{2}{3}$ cup ready-to-eat rice cereal
$\frac{1}{2}$ banana
6 oz. formula

Lunch

$\frac{1}{2}$ can vegetable soup
3 crackers
1 cup fruit cocktail
4 oz. formula

Snack

$\frac{1}{2}$ cup popcorn with 1 tablespoon margarine

Dinner

2 cups low-protein noodles
$\frac{1}{2}$ cup meatless spaghetti sauce
1 cup lettuce with French dressing
4 oz. formula

Notice that an individual on this diet takes in a low amount of phenylalanine from other sources of protein in the diet to achieve the delicate balance needed between a phenylalanine level high enough to permit normal development of the nervous system but low enough to avoid causing mental retardation.

Even with this diet, an individual affected with phenylketonuria must be alert to seemingly inconspicuous sources of phenylalanine. For example, the artificial sweetener aspartame is composed of aspartic acid and phenylalanine. PKU individuals must therefore avoid all consumption of aspartame, which is the active ingredient in the most popular artificial sweetener on the market today.

Dietician working with a patient.

⊞ MUTATIONS AND PROTEIN STRUCTURE

Individual gene mutations alter the structure of a polypeptide chain by changing its amino acid sequence. The clearest demonstrations of how mutations affect protein structure have come from studies on hemoglobin, a protein whose primary structure is known. Researchers have identified more than 300 naturally occurring mutant forms of hemoglobin in the human population. Most of these variants result from single base-pair changes in the genes that code for this protein. Recall (from Chapter 6) that normal adult hemoglobin (HbA) is a tetramer consisting of two α- and two β-chains. The α-chains and β-chains are encoded by different genes, so a single gene mutation affects one chain or the other but not both. Some of the better-known mutations that affect these chains are shown in ▶ Figure 11–10. In each case the variant chain differs from HbA by a single amino acid residue: One mutation in a gene is responsible for a single amino acid replacement in the protein molecule.

Many of the amino acid replacements that occur in hemoglobin lead to significant changes in the properties of the protein. One of the more obvious effects is a change in charge. For example, the replacement of glutamic acid by valine in the sixth position of the β-chain results in sickle-cell hemoglobin (HbS), which has two fewer negative charges than HbA (one for each of the two chains). Such charge differences permit the identification of different hemoglobin variants by electrophoretic techniques.

Changes in conformation and oxygen affinity can also occur and can lead to serious clinical manifestations in individuals who are homozygous for a mutant hemoglobin gene. Sickle-cell anemia, for example, is a recessive hemo-

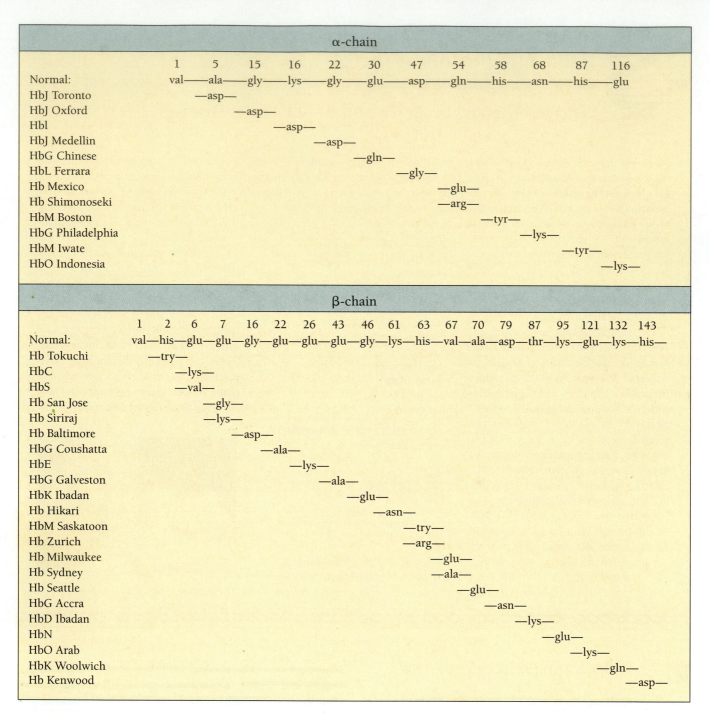

α-chain

	1	5	15	16	22	30	47	54	58	68	87	116
Normal:	val	ala	gly	lys	gly	glu	asp	gln	his	asn	his	glu
HbJ Toronto	—asp—											
HbJ Oxford		—asp—										
HbI			—asp—									
HbJ Medellin				—asp—								
HbG Chinese					—gln—							
HbL Ferrara						—gly—						
Hb Mexico							—glu—					
Hb Shimonoseki							—arg—					
HbM Boston								—tyr—				
HbG Philadelphia									—lys—			
HbM Iwate										—tyr—		
HbO Indonesia											—lys—	

β-chain

	1	2	6	7	16	22	26	43	46	61	63	67	70	79	87	95	121	132	143
Normal:	val	his	glu	glu	gly	glu	glu	glu	gly	lys	his	val	ala	asp	thr	lys	glu	lys	his
Hb Tokuchi		—try—																	
HbC			—lys—																
HbS			—val—																
Hb San Jose				—gly—															
Hb Siriraj				—lys—															
Hb Baltimore					—asp—														
HbG Coushatta						—ala—													
HbE							—lys—												
HbG Galveston								—ala—											
HbK Ibadan									—glu—										
Hb Hikari										—asn—									
HbM Saskatoon											—try—								
Hb Zurich											—arg—								
Hb Milwaukee												—glu—							
Hb Sydney												—ala—							
Hb Seattle													—glu—						
HbG Accra														—asn—					
HbD Ibadan															—lys—				
HbN																—glu—			
HbO Arab																	—lys—		
HbK Woolwich																		—gln—	
Hb Kenwood																			—asp—

▶ **FIGURE 11–10** A sample of the known hemoglobin mutations. The numbers above the normal amino acid sequences indicate the location of the specific amino acid that has been changed in each mutant type. *Source:* From L. T. Hunt, M. R. Souchard, and M. O. Dayhoff, in *Atlas of Protein Sequence and Structure,* vol. 5, ed. M. Dayhoff, National Biomed. Res. Found. (Washington, D.C., 1972): 67–87. Used with permission.

lytic disease characterized by anemia, jaundice, and episodes of severe pain (sickle-cell crises). Homozygotes for the sickle-cell gene produce only HbS hemoglobin. Unlike normal hemoglobin, the deoxygenated form of HbS has a tendency to polymerize through intermolecular hydrophobic interactions, forming bundles of insoluble fibers. These fibers distort the red blood cells of affected homozygotes into bizarre sickle shapes (▶ Figure 11–11). These cells are fragile and less able to pass through the smaller blood vessels, resulting in the rapid hemolysis (red blood cell destruction) and vascular obstruction that characterize this genetic disorder. Some of the symptoms that can occur in this disease are summarized in ▶ Figure 11–12. These many seemingly unrelated effects that can develop from a single genetic cause illustrate an important principle of genetics: A single genetic change can influence more than one

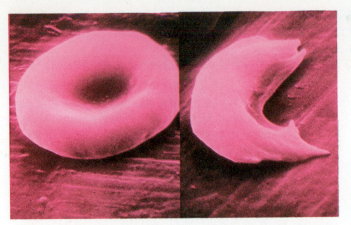

▶ **FIGURE 11–11** Normal red blood cells (disc shaped) and sickle cells (crescent shape). Cells from individuals homozygous for the sickle-cell gene have collapsed and have the shape of a sickle.

character. The manifestation of multiple phenotypic effects from a single mutation, a phenomenon called **pleiotropy**, reflects the involvement of a single gene product in several different processes during development.

⊞ RATES OF MUTATION

Gene mutations are rare events; newly mutated alleles at any given locus are typically found in fewer than 1 of every 10^5 cells or virus particles. However, gene mutations are also recurring events; the same kind of mutation can occur repeatedly within a large group of individuals, thereby leading to a change in the genetic makeup of the population over time. Because of the recurring nature of mutation, the probability that any given gene will undergo mutation is usually expressed in the form of a **mutation rate**.

There are two common methods for expressing how frequently a mutation occurs. For actively replicating cell (or

▶ **FIGURE 11–12** Some consequences of sickled red blood cells in individuals who are homozygous for the sickle-cell gene. Notice the pleiotropic effects at the molecular, cellular, and organ levels, caused by this single mutant gene.

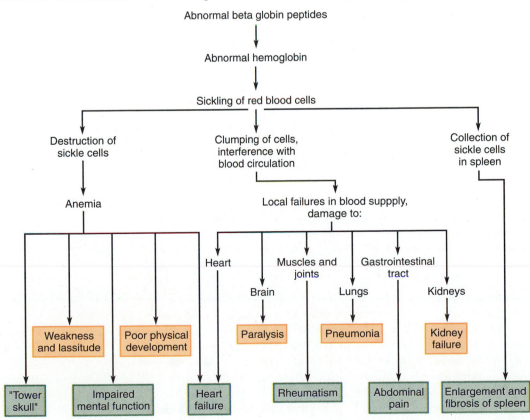

virus) populations, it is customary to express the mutation rate as the probability of mutation per cell division (or per act of gene duplication for viruses). Thus the mutation rate links the intrinsic tendency of a gene to mutate to the replicative state of the genome, rather than to some arbitrary unit of actual time such as hours or days. For continuously dividing cell populations, it is useful to define the mutation rate, μ, as:

$$\mu = \frac{m}{d} \qquad (11.1)$$

where m is the average number of mutations and d is the number of cell divisions. Although the number of cell divisions, d, cannot be measured directly, it can be estimated from the increase in the number of cells over the period of cultivation (see ▶ Figure 11–13a). The number of divisions then becomes $d = N - N_0$, where N is the final number of cells and N_0 the number of cells at time zero. For example, if an average of 2 mutations occurs during the growth of a culture from an initial number of 10^4 cells to a final number of 10^8 cells, there are a total of $10^8 - 10^4 \simeq 10^8$ cell divisions, so the mutation rate is

$$\frac{2}{10^8} = 2 \times 10^{-8} \text{ per cell division}$$

Implicit in equation 11.1 is the assumption that μ is constant per replicative act—this does not mean that the chance of mutation is genetically fixed independent of the surroundings, but rather that under uniform conditions replicative errors will tend to occur at a constant relative rate.

The mutation rate for multicellular organisms is generally expressed in the form of a **mutant frequency**, which is the ratio of mutant to total cells in a specified cell population. The cell population can theoretically consist of any type of cell. However, in sexually-reproducing organisms, mutant frequencies ordinarily refer to the proportion of mutants among gametes produced per sexual generation. In these organisms, therefore, the mutation rate is calculated as the probability of mutation per gamete per sexual generation. For example, suppose that for every 50,000 offspring,

▶ **FIGURE 11–13** Simple cell pedigrees showing (a) the relationship between the number of cells and the number of cell divisions, and (b) the combined effects of mutation and cell division on the proportion of mutant cells.

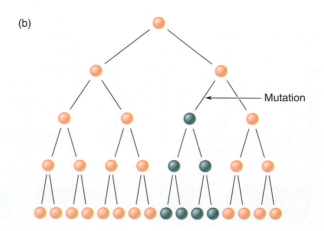

	Number of generations	Number of cells	Number of divisions
	0	$2^0 = 1$	0
	1	$2^1 = 2$	1
	2	$2^2 = 4$	3
	3	$2^3 = 8$	7
	4	$2^4 = 16$	15

Mutation

produced by normal (recessive) parents, one individual on the average is born with a dominant mutant trait. The formation of 50,000 diploid individuals requires 100,000 gametes, but because of the infrequent nature of the mutation, it is likely that only one of these gametes carries the mutant gene. The mutant frequency of the gene per gamete per generation is therefore

$$\frac{1}{100,000} = 10^{-5}$$

Some examples of mutation rates for various organisms are given in ■ Table 11–4. Genes in most eukaryotes appear to undergo mutation at a significantly higher rate than bacterial and viral genes; probabilities average between 10^{-6} and 10^{-5} per gamete per generation for higher organisms and around 10^{-8} to 10^{-7} per cell division for bacteria. It is not entirely clear why mutation rates for higher organisms are so much greater than those for bacteria and viruses; however, a significant part of this discrepancy is believed to be due to the different units used to measure these rates. By expressing the mutation rate in terms of the mutation frequency, as we do in higher organisms, we tend to overestimate the intrinsic probability that a gene will undergo mutation. In higher organisms, gamete-producing cells divide many times by mitosis before they ultimately form the gametes that are transmitted to the next sexual generation. Therefore, mutant frequencies determined for gametes include the effects of both mutation and the subsequent replication of mutant alleles during cell growth.

The simple cell pedigree in Figure 11–13b shows how the replication of mutant alleles can inflate estimates of mutation rates. In this pedigree, there are 16 cells in the final generation when the cells are examined. Of these 16 cells, 4 are mutant, yielding a mutant frequency of $\frac{4}{16} = \frac{1}{4}$. However, only one mutation occurred (two generations back), so the number of mutant cells in this example is greater than the number of mutations by a factor of four.

To Sum Up

The ability to mutate is an intrinsic property of the structure of DNA. Mutations occur at a very low rate per gene per cell division (in bacteria) or per gamete per sexual generation (in sexually reproducing organisms). The latter is actually a measure of the mutant frequency, which tends to overestimate the actual mutation rate because of possible replication of mutant alleles during mitotic reproduction of gamete-producing cells prior to their entering meiosis.

✳ RANDOMNESS OF MUTATION

Gene mutation is often described as a random genetic process. As we shall see in this section, there are several components of mutational randomness, including the following: randomness with respect to clone (cell lineage) and time of occurrence (cell generation), randomness with respect to the affected gene, and randomness with respect to the adaptive value of the mutation to the organism involved.

Random Distribution of Mutations

By defining the mutation rate as a probability of mutation per cell division, we are implying that we have no way of knowing in which cell line or cell generation a particular mutation will arise—we can only predict that such mutations will appear with some degree of regularity among the cells of a large population when considered over time. Mutation is thus a stochastic (statistically random) process that is characterized by variation in both space and time. Moreover, since mutations are exceedingly rare events, this vari-

■ **TABLE 11–4** Mutation rates in various organisms.

Organism	Mutation	Value	Units
Bacteriophage T2	lysis inhibition, $r \rightarrow r^+$	1×10^{-8}	rate, per gene duplication
	host range, $h^+ \rightarrow h$	3×10^{-9}	
Escherichia coli	lactose fermentation, $lac^- \rightarrow lac^+$	2×10^{-7}	
	phage T1 sensitivity, $ton^s \rightarrow ton^r$	2×10^{-8}	
	histidine requirement, $his^+ \rightarrow his^-$	2×10^{-6}	rate, per cell division
Chlamydomonas reinhardi	streptomycin sensitivity, $str^s \rightarrow str^r$	1×10^{-6}	
Drosophila melanogaster	eye color, $w^+ \rightarrow w$	4×10^{-5}	
Mus musculus	dilute coat color, $D \rightarrow d$	3×10^{-5}	
Homo sapiens	hemophilia A, normal $\rightarrow$ hemophilic	3×10^{-5}	frequency, per gamete
	albinism, normal $\rightarrow$ albino	3×10^{-5}	
Human bone-marrow tissue-culture cells	8-azaguanine resistance, normal $\rightarrow$ resistant	7×10^{-4}	rate, per cell division

Source: Ruth Sager and F. J. Ryan, *Cell Heredity* (New York: John Wiley and Sons, 1961), p. 55. Used with permission of Ruth Sager.

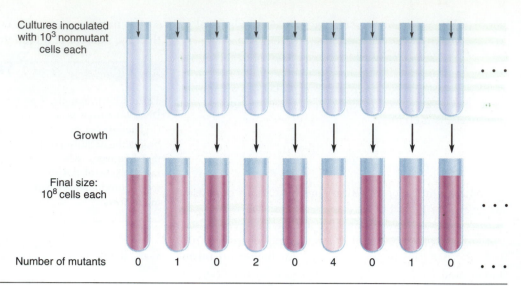

► **FIGURE 11–14**
Distribution of mutants in a series of independent but similar cultures. (Only a few of the cultures are shown.)

Cultures inoculated with 10^3 nonmutant cells each

Growth

Final size: 10^8 cells each

Number of mutants   0   1   0   2   0   4   0   1   0   ...

---

ation can lead to substantial departures from expected values in all but the largest of populations.

Consider an experiment in which several cultures are inoculated with the same number of nonmutant bacterial cells. After several generations of growth, each culture is examined for the presence of mutant cells of a specified type. When such experiments are performed, the results are often similar to those shown in ► Figure 11–14. At the time the cultures are examined, some contain no mutant cells, others contain one mutant cell, still others two mutant cells, and so on, with each culture differing from the others by what appears to be chance variation.

Two main sources of variation can be identified in this experiment. First, part of the variability can be ascribed to differences in the number of mutations that have occurred during growth. Some of the cultures (specifically those without mutant cells) have had no mutations, others have had one mutation, others two mutations, and so on, during the time that the cells were allowed to multiply. But if mutations are random events, how might we predict their precise distribution in this series of cultures? It is possible to do so if we recognize that the characteristics of mutation (a very small but equal chance of occurrence for each cell division) and the large number of opportunities for mutation (a large number of cell divisions in each culture) satisfy the requirements of the Poisson distribution. The distribution of mutations in different cultures is therefore described by the Poisson equation:

$$P(x) = \frac{m^x}{x!} e^{-m} \qquad (11.2)$$

where $P(x)$ is the fraction of cultures with exactly x mutations and m is the average number of mutations per culture. There is thus a probability of e^{-m} that no mutation (and, hence, no mutant cell) will occur in a culture and a probability of $1 - e^{-m}$ that at least one mutation will occur. For example, suppose that of 100 cultures examined, 85 contain one or more mutants. How many mutations, on the av-

erage, have occurred in each culture prior to examination? To solve this problem, we first note that the fraction of cultures with no mutations is $e^{-m} = 1 - 0.85 = 0.15$. The average number of mutations per culture can then be computed as $m = -\ln(0.15) = 1.9$.

► Figure 11–15 shows the results of an experiment demonstrating the Poisson distribution of mutations. In this experiment, a Lac⁻ (lactose-nonfermenting) strain of *E. coli* was grown on a nutrient agar medium containing the sugar lactose. Also incorporated in the medium was an

---

► **FIGURE 11–15** Poisson distribution of *lac*⁺ mutations in *lac*⁻ colonies. The mutations appear as red-colored outgrowths on the colony surface. *Source:* Ruth Sager and F. J. Ryan, *Cell Heredity* (New York: John Wiley and Sons, 1961), p. 50. Used with permission of Ruth Sager.

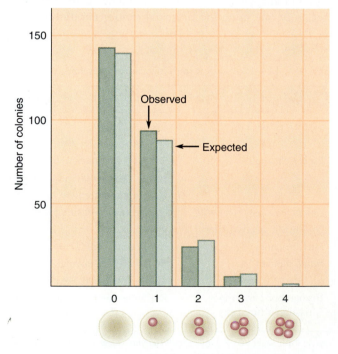

Number of colonies

Observed

Expected

Number of mutations

acid-base indicator dye that turned red whenever the acid by-products of lactose fermentation were produced. Therefore, when Lac+ (lactose-fermenting) revertants arose in a colony by mutation, they formed red-colored outgrowths (papillae) on the colony surface. Each colony (or clone) in this experiment is thus equivalent to a separate culture, and each papilla indicates a different mutation event. The close agreement between the observed results and those expected on the basis of the Poisson formula indicates that the mutations were "Poisson distributed" among the various colonies.

A second major component of variability in the number of mutants per culture results from differences in the time of mutation. ▶ Figure 11–16 illustrates how such variation occurs. Each clone in this series is assumed to multiply in perfect synchrony and to undergo zero or one mutation at some generation prior to the time the clones are examined. Observe that only one mutant appears if a mutation occurs in the last generation, two mutants appear if a mutation occurs one generation back, four mutants appear if a mutation occurs two generations back, and so on, yielding a simple geometric progression, 1, 2, 4, . . . , with a general term 2^i for a mutation occurring i generations back from

the time of examination. Moreover, since the total number of divisions that have occurred up to a given generation is equal to one less than the number of cells at that generation, the number of divisions (and thus the total number of chances for mutation) leading up to the cells produced i generations back is therefore

$$N_i - 1 = \frac{N}{2^i} - 1 \simeq \frac{N}{2^i}$$

where N_i is the number of cells i generations back and N is the number of cells in the clone at the time of examination (i.e., $N = N_i 2^i$). The probability $P(M \geq 2^i)$ that a clone contains 2^i or more mutants at the time of examination is then:

$$P(M \geq 2^i) = \mu \frac{N}{2^i} \qquad (11.3)$$

where M is the number of mutants and μ is the mutation rate. There are thus half as many clones with two or more mutants than with at least one mutant, one-fourth as many clones with four or more mutants, and so on. Equation 11.3 describes the expected distribution of mutant clones (called a **clonal distribution**) resulting from spontaneous mutation,

---

▶ FIGURE 11–16 Numbers of mutants found in independent but similar clones following a single mutation in each. Only a few of the clones are shown.

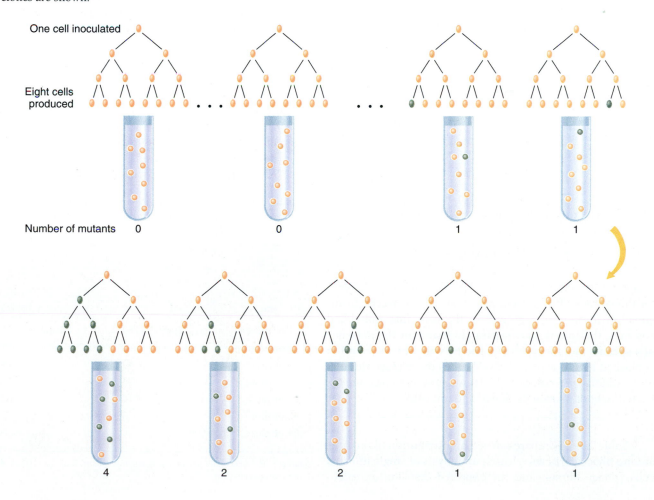

assuming a maximum of one mutation per clone. In practice, this assumption requires that mutation be very rare, so that it is extremely unlikely for two or more mutations to occur in the same clone. If this were not the case, we could not know, for example, whether two observed mutants arose from two mutations in the last generation or from a single mutation one generation back.

Example 11.3

Each of 50 identical tubes of nutrient medium was inoculated with exactly 10^3 tryptophan-requiring (Trp$^-$) bacterial cells. When the culture populations reached 6×10^8, they were tested for the presence of Trp$^+$ revertants; 42 of the cultures had at least 1 such cell. (a) From these results, calculate the average number of trp^- to trp^+ reversions per tube. (b) How many tubes are expected to contain exactly 1 Trp$^+$ cell?

Solution: (a) Since mutation follows a Poisson probability distribution, the proportion of tubes without any reversion is expected to equal

$$1 - \frac{42}{50} = 0.16 = e^{-m}$$

where m is the average number of reversions per tube. Thus, $m = -\ln(0.16) = 1.83$. (b) Only those tubes in which one mutation has occurred can potentially have just one mutant cell. From the Poisson distribution, the proportion of such tubes is $(1.83)e^{-1.83} = 0.294$. We see from equation 11.3, however, that half of all tubes with exactly one mutation should contain exactly one mutant cell:

$$\frac{P(M \geq 1) - P(M \geq 2)}{P(M \geq 1)} = \frac{\dfrac{\mu N}{2^0} - \dfrac{\mu N}{2^1}}{\mu \dfrac{N}{2^0}} = \frac{\mu N \left(1 - \dfrac{1}{2}\right)}{\mu N} = \frac{1}{2}$$

Therefore, $(\frac{1}{2})(0.294)(50) = 7.35$ (in other words, about 7 or 8) tubes are expected to contain exactly one revertant.

Follow-Up Problem 11.3

Suppose that 100 nonmutant (a^+) cells are placed randomly on the surface of a nutrient agar plate. After overnight incubation to allow the growth of the cells into individual colonies, the colonies are separately tested for the presence of a specific mutant type (a^-), and 87 colonies are found to contain at least one a^- cell. (a) What is the average number of mutations per colony? (b) How many colonies are expected to contain exactly one a^- mutant cell?

■ Table 11–5 compares an actual distribution of mutants among phage particles released upon lysis of single infected cells (phage bursts) and the expected distribution calculated

lated by assuming a single mutation event per clone. Each burst in this experiment is essentially a clone of phage that developed from a single infecting particle. Since the intracellular replication of phage genomes is not synchronized, not all bursts contain exactly 1, 2, 4, 8, . . . mutants. Synchronous growth is not essential to the analysis, however, as long as we treat the values of 2^i (1, 2, 4, . . .) as convenient although somewhat artificial class intervals. The reasonably close agreement between the experimental distribution and the distribution predicted by equation 11.3 indicates that there is, indeed, a clonal distribution of mutants among individual phage bursts.

Independence of Gene Mutations

As a general rule, each gene mutation can be regarded as a random event that occurs independently of other genes. This element of randomness applies to induced as well as spontaneous mutations. Although mutagens often show a certain degree of mutagenic specificity by acting on one kind of base more frequently than others (see Chapter 22), the presence of only four kinds of nucleotides within the genome tends to ensure that mutagens will not show specificity for the genes that they alter. However, this does not mean that all equal-sized segments of the genome undergo mutation at the same rate. Variations in the local structure of the genome make some chromosomal regions more prone to mutation than others, but these differences can be treated within the framework of a stochastic process, applying different probabilities to each.

Since gene mutations are independent events, the probability of the simultaneous occurrence of two different gene mutations can be calculated as the product of their individual probabilities. Thus if the mutation $a^+ \rightarrow a$ occurs at a rate of 10^{-6} per division and the mutation $b^+ \rightarrow b$ occurs at a rate of 10^{-7} per division, the two mutations are expected to occur together at a rate of $(10^{-6})(10^{-7}) = 10^{-13}$ per division. The extremely small probability of the joint occur-

■ **TABLE 11–5** A comparison between an actual distribution of mutants among phage bursts and the expected distribution calculated assuming a single mutation event.

r and w Mutants among Single Bursts of Phage T2	Observed Number of Clones	Expected Number of Clones
1 or more	Total = 183	Total = 183.0
2 or more	90	90.8
4 or more	41	45.1
8 or more	21	22.2
16 or more	10	10.8
32 or more	4	5.1
64 or more	0	2.2

Source: S. E. Luria, *Cold Spring Harbor Symp. Quant. Biol.* 16 (1951): 463. Used with permission.

rence of different gene mutations essentially ensures that such mutations will never occur together in the same cell during the same division cycle. Although it is possible to isolate double mutants, such as His⁻Met⁻ bacteria (which require the amino acids histidine and methionine), these mutants are produced by a two-step process involving events that are separated in time. For example, histidine-requiring mutants might later mutate to the methionine-requiring state (or vice versa) to cause the doubly auxotrophic condition.

Adaptive Randomness

The lack of gene specificity of environmental mutagens leads to another element of randomness in gene mutation: No particular mutation can be directed by the environment in which it occurs. In other words, particular mutations occur strictly by chance, not as an adaptive response to environmental conditions. For example, resistance to a specific antibiotic does not arise in bacterial populations in direct response to their exposure to the drug. Instead, it occurs spontaneously, regardless of whether the drug is present. When added to the environment, the antibiotic simply serves as a selective agent that permits only resistant bacteria to survive and reproduce.

Salvador Luria and Max Delbruck first demonstrated the undirected nature of gene mutation in the 1940s. In their experiments, Luria and Delbruck were interested in the origin of resistance to bacteriophage T1 among normally sen-

sitive (Ton^s) *E. coli* cells. To test for the appearance of the T1-resistant (Ton^r) character, it was necessary to treat the bacterial cells with the phage. There were two fundamentally different views regarding the origin of Ton^r clones: (1) the adaptation hypothesis, which states that the Ton^r character is induced in a small number of cells upon treatment with the virus, and (2) the preadaptation hypothesis, which states that the Ton^r character arises by spontaneous mutation, so that phage-resistant clones originate from the few Ton^r mutants that are already present in the culture at the time of exposure to the phage.

Luria and Delbruck devised a statistical approach called the **fluctuation test** to distinguish between these two hypotheses (see ▶ Figure 11−17). A nutrient medium in each of twenty 0.2-ml volumes (individual cultures) and in one 10-ml volume (bulk culture) was inoculated with 10^3 Ton^s cells per ml. The cultures were allowed to grow to a density of about 10^8 cells per ml, and then each of the twenty 0.2-ml individual cultures and ten 0.2-ml samples drawn from the bulk culture were spread on separate nutrient agar plates containing a high concentration of T1 phage. After overnight incubation, the plates were scored for number of T1-resistant colonies.

The hypothetical series of clones in ▶ Figure 11−18 show the results that the two hypotheses predict for this experiment. If the adaptation hypothesis is correct (Figure 11−18a), Ton^r cells would be induced at random in each of the plated samples, irrespective of their origin. We

▶ **FIGURE 11−17** The Luria-Delbruck fluctuation test showing the distribution of phage T1-resistant colonies on nutrient agar plates.

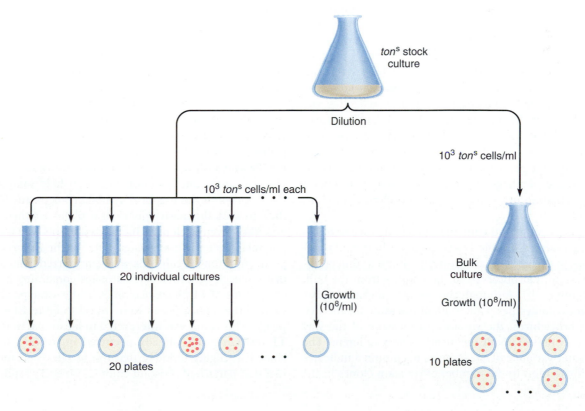

(a) Adaptation hypothesis

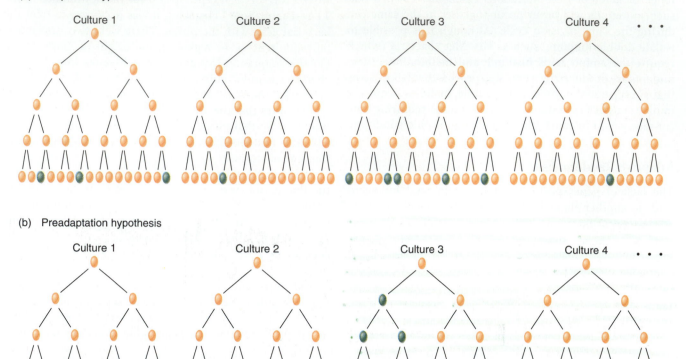

(b) Preadaptation hypothesis

▶ **FIGURE 11–18** Predicted distribution of mutants in a hypothetical series of clones for (a) the adaptation hypothesis and (b) the preadaptation hypothesis. *Adapted from:* G. S. Stent and R. Calendar, *Molecular Genetics,* 2d ed. (W. H. Freeman, 1978), p. 161. Copyright © 1978 by W. H. Freeman and Company. Reprinted with permission.

would therefore expect to observe the same Poisson distribution of resistant clones in the individual cultures and in the samples derived from the bulk culture, with only minor plate-to-plate variation. On the other hand, if the preadaptation hypothesis is correct (Figure 11–18b), we would again observe a Poisson distribution of resistant clones in the samples derived from the bulk culture, reflecting the process of random sampling, but the individual cultures would yield a clonal distribution of Ton^r cells, as a result of mutations occurring in different generations prior to treatment with the phage. Thus if spontaneous mutation is indeed the cause of phage resistance, we should observe a marked fluctuation in the number of Ton^r clones in the individual cultures.

The actual results of the fluctuation test that we have described are given in ■ Table 11–6. The distribution of resistant clones in the individual cultures (column a) differs significantly from that observed in the samples from the bulk culture (column b). Although the average (mean) number of resistant colonies per plate is about the same for the two sampling procedures, the variance (a measure of the way values are dispersed about the mean) is very different. The much greater variance of the samples in column a indicates marked fluctuation in the number of resistant clones in the

individual cultures. The results are thus in agreement with a spontaneous origin of phage resistance.

Although the fluctuation test provides an elegant confirmation of the undirected nature of gene mutation, the evidence rests on some very subtle statistical considerations. A more direct demonstration was given in the early 1950s by Joshua and Esther Lederberg, who used a technique known as **replica plating**. This technique involves transferring cells from colonies on a master plate to one or more replica plates (▶ Figure 11–19). The cells are transferred by pressing the agar surface of the master plate onto a sterile, velvet-covered block. The fibers of the velvet hold samples from the colonies in the identical spatial arrangement in which they grew on the master plate. Agar plates are then pressed onto the velvet. Each plate inoculated in this way becomes a potential replica of the master plate. In the Lederbergs' experiments, Ton^s *E. coli* cells were transferred from a master plate to each of several replica plates containing high concentrations of T1 phage. Colonies in the same positions on each of the replica plates were found to be resistant to phage. This result was interpreted to mean that mutation to T1 resistance had already occurred on the master plate, since it is highly unlikely that the phage could induce an identical pattern of resistant colonies on every replica plate.

■ **TABLE 11–6** Results of a Luria-Delbruck fluctuation test on the development of phage resistance in *E. coli*.

(a) Individual Cultures		(b) Bulk Culture	
Sample Number	Resistant Colonies	Sample Number	Resistant Colonies
1	1	1	14
2	0		
3	3	2	15
4	0		
5	0	3	13
6	5		
7	0	4	21
8	0		
9	5	5	15
10	0		
11	6	6	14
12	107		
13	0	7	26
14	0		
15	0	8	16
16	1		
17	0	9	20
18	64		
19	0	10	13
20	35		
	Mean = 11.3		Mean = 16.7
	Variance = 753.3		Variance = 18.2

▶ **FIGURE 11–19** The use of the replica-plating technique to show the adaptive randomness of mutation.

Furthermore, when the researchers tested the master plate colonies that gave rise to the resistant clones, these colonies were found to contain some resistant cells. In contrast, the other colonies on the master plate did not contain resistant cells. The conclusion from this experiment, like the conclusion of the fluctuation test, was that resistant colonies arise from preexisting mutant cells, not as a response to their exposure to the phage. Thus the mutations are not the result of any directed change.

To Sum Up

1. Spontaneous mutation is random with respect to both the cell line and the cell generation. Randomness with respect to cell lineage means that the number of mutational events that occur in different cell lines has a Poisson distribution. Randomness with respect to time means that the number of mutants per cell lineage follows a clonal distribution.

2. Mutation, whether spontaneous or induced by a mutagen, can also be regarded as random with respect to the affected gene. Thus gene mutations occur independently of one another. In addition, both spontaneous and induced mutations are random with respect to their adaptive value to the organism—no particular gene can be directed to mutate by any particular environmental condition.

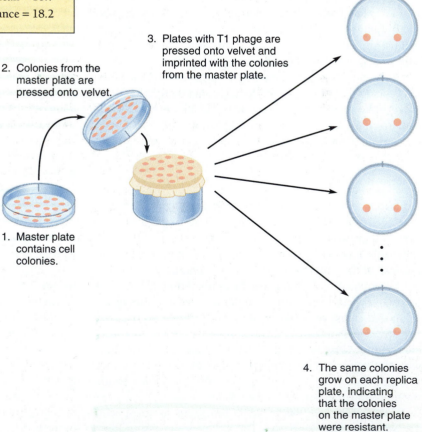

3. Plates with T1 phage are pressed onto velvet and imprinted with the colonies from the master plate.

2. Colonies from the master plate are pressed onto velvet.

1. Master plate contains cell colonies.

4. The same colonies grow on each replica plate, indicating that the colonies on the master plate were resistant.

MUTATION METHODOLOGY

The rare and unpredictable nature of mutation poses special difficulties in the detection and enumeration of mutant strains and in isolating these strains from predominantly wild-type populations. In this section we will look at some of the various techniques that geneticists have developed over the years as they have worked to overcome these difficulties. Most of the techniques that we will consider are important in the genetic analysis of microorganisms, where mutation methodology has been most extensively employed.

Isolation of Mutant Strains

Mutation, as it is ordinarily studied in genetics, is a process that occurs at the population level. The population may consist of millions of cells or virus particles, of which typically only a small fraction is mutant for a particular gene. Because of the infrequent occurrence of mutations, the initial step in most isolation procedures involving cell cultures or viruses is to expose the population to some mutagenic agent to increase the number of mutants. One chemical mutagen that is widely used for obtaining bacterial mutants is nitrosoguanidine (see Chapter 22). Treatment of a culture with this agent can increase the proportion of mutant cells in a population by as much as 100,000-fold under conditions permitting over 30% survival.

Nitrosoguanidine, like all other mutagens, is not specific for the genes that it affects. The initial mutagenesis must therefore be followed by specific enrichment procedures for preferentially increasing the proportion of the desired mutant strain over other strains. There are two main types of enrichment procedures. The simplest procedure, direct selection, is used to favor the growth of virus- and drug-resistant mutants in order to enrich the mutagenized culture with these mutant types. For example, phage-resistant bacterial cells can be isolated by simply treating a mutagenized culture with a high concentration of phage particles to kill all the phage-sensitive cells expressing the wild-type phenotype. Similarly, antibiotic-resistant clones can be isolated by plating the population on a medium containing the antibiotic, so that only the resistant cells develop into colonies.

Another set of methods serves as the basis for the second main enrichment procedure, counterselection, so named because the methods kill the nonmutant organisms that are capable of growing under the conditions employed. Counterselection is often used for the isolation of auxotrophs. Because auxotrophic cells will not grow on certain media that support the growth of prototrophic cells, the enrichment of auxotrophs on such media can be accomplished by killing only the growing wild-type strain.

Penicillin enrichment is one counterselective method that is used successfully for auxotroph isolation in bacteria. This method involves placing cells from a mutagenized culture into a minimal medium containing the antibiotic penicillin. Penicillin inhibits the synthesis of the peptidoglycan layer in the cell walls of bacteria that are sensitive to this antibiotic. The prototrophic cells that grow in the presence of penicillin thus become sensitive to osmotic rupture and lyse, but the nongrowing auxotrophs survive because the antibiotic does not affect preformed peptidoglycan. The auxotrophic cells are then removed from the antibiotic (by filtration or centrifuging and washing) and transferred to a nutrient, penicillin-free agar medium on which they can divide and form colonies. Some of the counterselective agents that are used for bacteria and cultured eukaryotic cells are listed along with their mechanisms of action in ■ Table 11–7.

Once enrichment has taken place, the next task in mutant isolation is to apply different screening procedures to identify specific mutant types and separate them from other mutants and the wild-type strain. Screening for mutant strains in microorganisms is facilitated by the use of various differential and selective media. For example, suppose that we want to identify a Gal⁻ mutant strain of E. coli, which is unable to metabolize the sugar galactose. One approach would be to grow the bacteria on a nutrient agar medium containing galactose and a pH indicator that consists of eosin and methylene blue (EMB). If the bacteria can metabolize galactose, they will excrete lactic acid as a metabolic byproduct and lower the pH to a point where the indicator acquires a violet color. Wild-type (Gal⁺) colonies therefore appear violet (▶ Figure 11–20a). The Gal⁻ bacteria, on the other hand, do not excrete lactic acid, so their colonies re-

Counterselective Agent	Mode of Action
Penicillin	kills growing bacterial cells by inhibiting the crosslinking of peptidoglycan chains during cell wall formation
8-azaguanine	kills growing eukaryotic and prokaryotic cells by becoming incorporated into DNA and rendering it nonfunctional
5-bromodeoxyuridine (5-BUdR)	kills growing eukaryotic and prokaryotic cells by becoming incorporated into DNA, which then undergoes photolysis and inactivates readily when exposed to near-ultraviolet light
Thymine deprivation	kills Thy⁻ cells that require thymine as a growth factor but are otherwise wild-type; used to enrich a culture with Thy⁻ cells that cannot grow in the particular medium because of the requirements imposed by a second mutation

main pale red in color (Figure 11−20b). Thus the mutant clones in this example can be identified using a single differential medium.

The cells of most conditional mutants, such as auxotrophs and temperature-sensitive mutants, must be transferred to different media in order to differentiate mutant clones from those of wild-type cells. The replica-plating technique is often used for this purpose. If the master plate contains nutrient medium and the replica plate minimal medium, any colony that does not appear on the replica plate can immediately be identified as one containing an auxotrophic mutant (▶ Figure 11−21). The mutant clones can then be tested on other supplemented media to determine the precise nature of their growth requirements. Alternatively, if the master plate is incubated at 30°C and the replica plate at 42°C, failure of a colony to grow on the replica plate identifies the clone as a temperature-sensitive mutant. The technique of replica plating therefore enables an investigator to retain the original set of clones on a master plate, while simultaneously testing them for various characteristics on replica plates.

Many of the steps involved in the isolation of mutant genetic strains are summarized in ■ Table 11−8, which shows a scheme suitable for isolating a temperature-sensitive auxotrophic strain of bacteria.

Detecting Recessive Mutations in Diploids

Although prokaryotic cells such as *E. coli* are haploid, the somatic cells of most higher organisms are diploid, i.e., they have two copies of genomic DNA. The diploid condition of somatic cells complicates the analysis of gene mutation, since the phenotype of a newly arising mutant allele will be expressed in the next cell generation only if the allele is dominant. Recessive mutations, in contrast, are not normally expressed in diploid cells, except in the very rare instance when two mutations of the same type make the cell homozygous for the mutant gene. A common approach for circumventing this problem is to study recessive somatic mutations in cell lines that are haploid for various genes as a result of chromosome aberrations. For example, there are various aneuploid cell lines whose chromosome sets have only one copy of one or more chromosomes (see Chapter 12 for a discussion of aneuploidy). Such cells are functionally haploid for the genes involved in the aberration and express all mutations of these genes when they occur.

Diploidy also complicates the analysis of germinal mutations. If a recessive mutation appears in the gametes of an individual and is then transmitted to the progeny, the offspring that receive the mutant allele will be heterozygous. The mutation may then go unnoticed for many generations before a chance mating between carriers of the mutant allele produces the homozygous condition. Various procedures have been developed to permit the detection of recessive mutations among gametes of a diploid organism.

▶ **FIGURE 11−20** (a) Wild-type Gal⁺ colonies and (b) mutant Gal⁻ colonies grown on EMB medium.

(a) Gal⁺

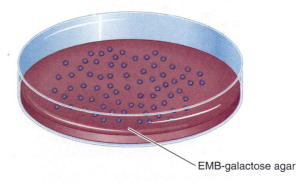

EMB-galactose agar

(b) Gal⁻

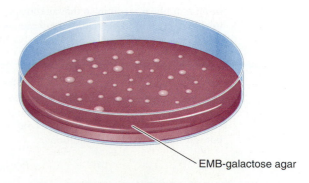

EMB-galactose agar

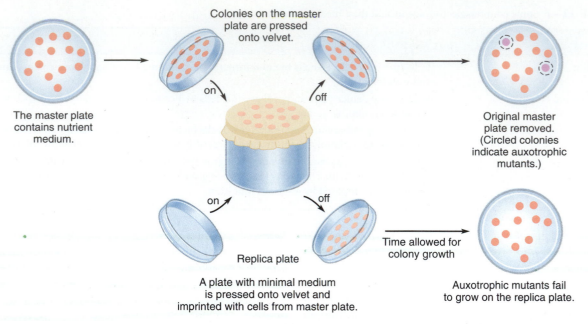

Colonies on the master plate are pressed onto velvet.

The master plate contains nutrient medium.

on

off

Original master plate removed. (Circled colonies indicate auxotrophic mutants.)

on

off

Replica plate

Time allowed for colony growth

A plate with minimal medium is pressed onto velvet and imprinted with cells from master plate.

Auxotrophic mutants fail to grow on the replica plate.

▶·FIGURE 11−21 The replica plating technique for identifying mutant clones.

We will consider one of the earliest of these procedures, the **ClB method**, which was devised by H. J. Muller in the 1920s to detect X-linked recessive lethal mutations in the gametes of normal male *Drosophila*. This method uses females that carry a specially constructed X chromosome called the ClB chromosome. This chromosome contains a

■ **TABLE 11−8** Summary of steps involved in the isolation of a temperature-sensitive auxotrophic (tryptophan-requiring) mutant strain of bacteria.

Wild-type culture

Treat with suitable mutagen (e.g., nitrosoguanidine) to increase the proportion of mutants among survivors.

Mutagenized culture

Grow for a few generations at permissive temperature (30°C) on minimal medium or at restrictive temperature (40°C) with tryptophan present to ensure expression of the mutant trait.

Phenotypically expressed culture

Initiate growth of culture at restrictive temperature on minimal medium so that auxotrophs are unable to grow. Add penicillin to kill metabolically active wild-type cells.

Mutant-enriched culture

Using the replica-plating technique, transfer cells to a medium under conditions that will restrict the growth of the desired mutants. Select colonies that grow under permissive conditions but not under restrictive conditions.

Mutant clone

large inverted segment that suppresses the recovery of viable recombinant progeny (*C* stands for "crossover suppressor," to be discussed in Chapter 12), a recessive lethal gene, *l*, and a dominant gene, *B*, which causes a narrowed eye (bar-eyed) phenotype.

The crossing sequence used in the ClB method is diagrammed in ▶ Figure 11−22. First, a parental (P) cross is performed in which the males to be tested are mated to tester females carrying the ClB chromosome. Next, the bar-eyed daughters, which carry the ClB chromosome of their mother and the potentially mutant X chromosome of their father, are selected and individually mated to normal males. Many such matings are performed, each in a separate vial, in order to sample a large number of X chromosomes derived from the gametes of the original male. Since the ClB/Y condition is lethal, viable males in the F_2 generation must carry the X chromosome from the original male. Therefore, the absence of viable male progeny in a vial would indicate that a recessive lethal mutation had occurred in a sperm of the original normal male.

Measuring the Mutation Rate

To determine the mutation rate, we must be able to detect and follow the results of mutation in a population over time. For microbial populations, researchers often measure the increase in the number of mutants (or mutant clones) in replicating cultures of cells or viruses and relate this increase to the mutation rate defined in equation 11.1. Three methods are commonly used.

Determining the mutation rate from the clonal distribution of mutants. One method that is appropriate for viruses is to calculate the mutation rate directly from the propor-

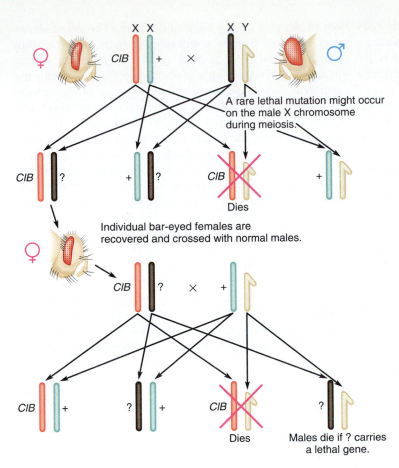

A rare lethal mutation might occur on the male X chromosome during meiosis.

Dies

Individual bar-eyed females are recovered and crossed with normal males.

Dies

Males die if ? carries a lethal gene.

tion of single cell bursts that contain one or more mutants. As long as mutation is very rare, so that no more than one mutation is likely to occur in the same clone, the proportion of bursts with one or more mutants (obtained by setting i equal to zero in equation 11.3) is then $P(M \geq 1) = \mu N$, and the mutation rate becomes

$$\mu = \frac{P(M \geq 1)}{N}$$

Let's look at a numerical example. Suppose that of 5000 single bursts examined, 50 had at least one mutant phage of a specified type. If the burst size (N) of this phage is 200, the mutation rate is then calculated as

$$\mu = \frac{50/5000}{200}$$

$$= 5 \times 10^{-5} \text{ per replication}$$

Determining the mutation rate from the Poisson distribution of mutations. In bacteria, where cultures generally get to be quite large and may undergo more than a single mutation, mutation rates are often calculated from the average number of mutations, m, in a series of identical cultures. The mutation rate can be related to m from equation 11.1 as

$$\mu = \frac{m}{N - N_0}$$

where $N - N_0$ is the total number of cell divisions that have occurred over the period of cultivation. Since the number of mutations cannot be directly observed, the value of m is usually computed from the $P(0)$ term of the Poisson formula (equation 11.2) as $m = -\ln P(0)$.

For a numerical example, suppose that 100 tubes of nutrient broth are each inoculated with 10^4 nonmutant bacterial cells ($N_0 = 10^4$) and then incubated over a period of time until a final size of 10^9 cells per tube is reached ($N = 10^9$). If at the end of this period all but 10 of the tubes are found to contain mutant cells of a specified type, so that $P(0) = 0.10$, the mutation rate is then calculated as

$$\mu = \frac{-\ln(0.10)}{10^9 - 10^4}$$

$$= 2.3 \times 10^{-9} \text{ per cell division}$$

Determining the mutation rate from the increase in mutants during growth. The mutation rate can also be determined experimentally from the increase in mutant frequency in the same bacterial culture over many generations of exponential growth. Since mutant cells arise by mutation, the mutation rate and the mutant frequency are obviously related. The relationship is not a simple one, however, because mutants are also formed from preexisting mutant cells by cell division. To avoid complications due to growth and back mutation, the cells are cultured in a nonselective medium (so that mutant and nonmutant cells divide at the same rate) and maintained under conditions where nonmutant cells greatly outnumber mutant cells. As long as the mutant frequency is small, we can conveniently ignore back mutation, because the number of mutants lost by reversion is negligible compared with the number gained by growth and forward mutation. Assuming a negligible rate of back mutation and equal viability of mutant and wild type, the only way that the mutant-to-nonmutant ratio can change is by mutation. Thus if we let q represent the proportion of mutants (the mutant frequency) and $p = 1 - q$ the proportion of nonmutants in the population, the increase per generation in the proportion of mutants (Δq) will then be

$$\Delta q = \mu(1 - q)$$

where μ is the mutation rate (i.e., the probability of mutation per cell generation). Since q is small under the conditions employed, $1 - q \to 1$, and Δq becomes approximately equal to μ. Therefore, after one generation, $\Delta q = q_1 - q_0 = \mu$, so that $q_1 = \mu + q_0$. After two generations, $q_2 = \mu + q_1 = 2\mu + q_0$. After three generations, $q_3 = \mu + q_2 = 3\mu + q_0$, and so on, so after t generations,

$$q_t = \mu t + q_0 \qquad (11.4)$$

We see that the mutant frequency depends not only on the mutation rate but also on the number of generations that have elapsed. The relationship is linear, so a plot of q versus t will yield a straight line with a slope equal to μ.

▶ Figure 11–23 gives the results of an experiment used to evaluate the mutation rate from the increase in the number of mutants during bacterial growth. The graph shows data obtained from a continuous growth experiment in which His$^-$ auxotrophs of *E. coli* were grown for several hundred generations on a medium supplemented with the required growth factor, histidine. Samples of the culture were tested periodically for the presence of His$^+$ revertants by plating on a minimal (histidine-free) agar medium. As these data show, the frequency of His$^+$ revertants increases linearly with the number of cell generations, in agreement with the model. Moreover, the rate of the $his^- \to his^+$ mutation can be calculated from the slope of the line to be approximately equal to 10^{-8} per cell generation.

Example 11.4

A liquid medium is inoculated with 3×10^5 thymine-requiring (Thy$^-$) cells. When the cell population reached 1.2×10^6, the mutant (revertant) frequency was measured and found to equal 10^{-5}. (a) What will the mutant frequency equal when the culture reaches 2.4×10^6 cells? (b) What is the mutation rate to the Thy$^+$ condition?

Solution: (a) Two generations of growth are involved in going from 3×10^5 to 1.2×10^6 cells. The mutant frequency will therefore increase from 10^{-5} when $t = 2$ to

$$q = \frac{3}{2}(10^{-5}) = 1.5 \times 10^{-5}$$

when $t = 3$. (b) Letting $q_0 = 0$, the mutation rate can be calculated using equation 11.4 as

$$\mu = \frac{10^{-5}}{2} = 5.0 \times 10^{-6} \text{ per cell generation}$$

Follow-Up Problem 11.4

Referring back to the bacterial culture described in Example 11.4, how many generations of growth will be required to attain a mutant frequency of 4.0×10^{-5}?

To Sum Up

1. Because of the rare and random nature of mutation, specialized techniques are required to detect and isolate particular mutant strains from a population. Various enrichment procedures can be used to increase the proportion of a desired mutant type in a population of microorganisms. Direct selection uses culturing conditions that favor growth of the mutant type. Counterselection uses culturing conditions that kill any nonmutant organisms that are capable of growing.
2. Once a bacterial culture has been enriched for mutant types, a screening procedure is used to identify the desired mutant type and isolate it from other mutants and from the wild-type cells. Replica plating can be used to identify and isolate conditional mutants.
3. The diploid nature of the somatic cells of most eukaryotic organisms greatly complicates the detection of recessive mutations. In some cases, somatic cell lines that are haploid for cer-

▶ FIGURE 11–23 Increase in the proportion of *his$^+$* revertants during the continuous growth of *his$^-$ E. coli* cells in the presence of histidine. *Source:* F. J. Ryan and L. K. Wainwright, *J. Gen. Microbiol.* 11 (1954): 376. Used with permission.

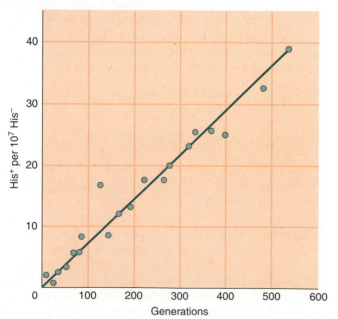

tain genes can be used to reveal recessive mutations in those genes. Recessive mutations in the germ line of *Drosophila* can be detected using the ClB technique, which reveals X-linked recessive lethal mutations.

4. Three methods can be used to measure the mutation rate of a particular gene in a microbial population. All three measure the increase in the number of mutants or mutant clones in replicating cultures and then relate this increase to the mutation rate using an appropriate mathematical relationship.

♦

 ## Chapter Summary

Mutations are sudden, heritable changes in the genetic material. They can be spontaneous or induced, and they are often classified according to their effect on an organism's phenotype.

Gene mutations can arise from a single base-pair alteration in the DNA sequence of a gene. Alterations of this type often lead to single amino acid changes in the encoded polypeptide that modify or destroy its normal function. If the polypeptide is an enzyme, the mutational loss of activity can produce a metabolic block. The phenotypic consequences of metabolic blocks depend on the extent of accumulation of metabolic intermediates and the level of deficiency of the end product, and are responsible for many human metabolic diseases.

Mutations are recurring events, so the same kind of mutation can occur repeatedly in a large group of individuals. They are also rare events, typically occurring at rates of less than 10^{-5} per cell division. Most mutations are reversible—a mutant allele can undergo a second mutation back to its original form. Mutations are random events with respect to cell line and cell generation and in their adaptive significance to the organism involved. The adaptive randomness of mutation means that no particular type of mutation can be directed by its environment—this fundamental concept was first conclusively demonstrated by the Luria-Delbruck fluctuation test.

Special techniques have been developed to isolate mutant strains and measure their rates of mutation. These techniques have been widely used in microbial genetics, where mutation methodology has been most extensively employed.

 ## Questions and Problems

Types of Mutation

1. Define the following terms, and distinguish between the members of paired terms:

   mutation and mutant
   reversion and suppression
   induced and spontaneous mutations
   germinal and somatic mutations
   prototrophs and auxotrophs

2. Matings between members of the Mexican Hairless breed of dogs produce smaller than usual litters consisting of both hairless and haired offspring in a ratio of 2:1. Occasional puppies with certain structural abnormalities are born dead. (a) Provide a genetic explanation for these results. (b) How might dog breeders produce the maximum number of hairless offspring without losses resulting from the birth of dead and abnormal pups?

3. Tay-Sachs disease is an untreatable, lethal abnormality that is inherited as an autosomal recessive trait. A husband and wife both have siblings who died of this disorder. (a) What is the probability that their firstborn child will have Tay-Sachs disease? (b) What would your answer in part (a) be if blood tests of the same husband and wife showed that both are carriers of this allele?

4. In a particular cross involving *Drosophila*, a female with short thoracic bristles is mated to a male with normal (long) thoracic bristles. The progeny produced by this cross occur in a ratio of 1 long-bristled female : 1 short-bristled female : 1 long-bristled male. Provide a genetic explanation for these results.

5. The navel orange and the Delicious apple both arose as the result of somatic mutation. In each case, a spontaneous mutation occurred in a single cell that, through successive mitotic divisions, produced an entire branch with the characteristics of the mutant type. Both the navel orange and the Delicious apple now enjoy widespread popularity. Why has it been possible to take advantage of the products of somatic mutation in plants, while such products in animals are uncommon?

6. In a pure-breeding (red-eyed) strain of *Drosophila melanogaster*, males are occasionally found that have one red eye and one white eye. (Females of this type are never found in this strain.) Explain the genetic basis for this anomaly.

7. You have discovered two conditional mutant strains of bacteria, *a* and *b*. Strain *a* is sensitive to infection by a particular phage (phage x) at temperatures below 30°C, and is resistant to infection at higher temperatures. Strain *b* is sensitive to infection by a different phage (phage y) at temperatures above 30°C, and is resistant to infection at lower temperatures. (The wild-type strain is normally sensitive to phage x and resistant to phage y at both high and low temperatures.) Explain these results.

Mutations and Metabolic Blocks

8. Consider the reaction sequence A $\xrightarrow{1}$ B $\xrightarrow{2}$ C $\xrightarrow{3}$ D $\xrightarrow{4}$ E, which is controlled by genes 1–4. Describe what effects mutations in each gene might have with respect to enzyme activity, accumulations, and growth tests. (Assume that the reactions occur in *E. coli* and that compound E is required for cell growth.)

9. Two auxotrophic mutant strains of *Neurospora* lack the ability to synthesize the amino acid leucine. Strain 1 can grow if the medium is supplemented with either α-ketoisocaproic acid or leucine. Strain 2 can grow on leucine but not on α-ketoisocaproic acid. Diagram the steps involved in the synthesis of leucine, and indicate which step in the pathway is blocked in each of the mutant strains.

10. Four auxotrophic mutant strains of *E. coli* (1, 2, 3, and 4) have different genetic blocks at different steps of the same metabolic pathway, which involves compounds A, B, C, D, and E. Each of these compounds is tested for its ability to support the growth of mutant strains, and the results are given below (+ = growth, – = no growth):

Mutant strain	Compound used as supplement				
	A	B	C	D	E
1	+	–	+	–	+
2	–	–	+	–	–
3	+	–	+	+	+
4	–	–	+	–	+

Give the order of the compounds in the metabolic pathway, and indicate the specific step that is blocked in each mutant strain.

11. The following table gives the growth responses (+ = growth, – = no growth) of four auxotrophic mutant strains of *Salmonella typhimurium* (1, 2, 3, and 4) to the addition of each of the compounds F through K to minimal medium:

Mutant strain	Compound used as supplement					
	F	G	H	I	J	K
1	–	+	–	–	+	–
2	–	+	–	–	+	–
3	–	–	–	–	+	–
4	+	+	–	–	+	–

Both strains 2 and 4 grow if both I and K are added to minimal medium. Diagram a metabolic pathway that is consistent with the data involving all six compounds, and indicate the specific step that is blocked in each mutant strain.

12. Among the many known eye colors in *Drosophila* are purple and scarlet. The genes responsible for purple (*pr*) and scarlet (*st*) are located on nonhomologous chromosomes and are recessive, so that wild-type flies of genotype $pr^+ - st^+ -$ have red eyes. These genes also interact—flies that are homozygous for both mutant genes (genotype *pr/pr st/st*) have white eyes. Construct a metabolic pathway for pigment synthesis that accounts for these phenotypes, and indicate which gene locus specifies the enzyme that catalyzes each step of the pathway.

13. The body of a certain species of extraterrestrial being normally has a red glow. Two variant types also occur, however, caused by recessive mutations in nonallelic genes located on different chromosomes. Mutant gene *s* gives a bright scarlet glow, and mutant gene *b* imparts a brown glow to the organism. The metabolic pathway for synthesis of glow pigment has recently been discovered:

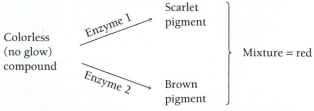

(a) Which enzyme is nonfunctional in scarlet individuals? in brown individuals? (b) One particular mating between scarlet and brown individuals gives only red offspring. Give the genotypes of the scarlet and brown parents and the red progeny. (c) If the red offspring described in part (b) intercross, what phenotypic classes and proportions are expected among their progeny?

14. A certain pigment is the metabolic end product of either of two separate reactions:

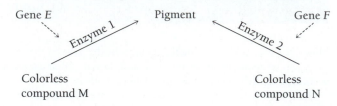

(a) If enzymes 1 and 2 each catalyze the production of exactly half the amount of pigment that the cell can use, what phenotypic ratio will result from a dihybrid cross? (Assume that the intensity of the coloration is directly proportional to the amount of pigment produced.) (b) If enzymes 1 and 2 each catalyze the production of a saturating amount of pigment (meaning that only one active gene is sufficient to obtain a full amount of the end product), what phenotypic ratio will result from a dihybrid cross?

15. Gene *A* codes for an enzyme that is needed to catalyze the formation of a flower pigment in a particular diploid plant. Its mutant allele, *a*, codes for an inactive enzyme that can no longer combine with its substrate. Gene *A* is said to be completely dominant, since the phenotype of *AA* homozygotes is indistinguishable from that expressed by *Aa* heterozygotes. When studies were conducted to determine the enzyme levels present in *AA* and *Aa* plants, it was discovered that under normal (physiological) conditions, the amount of active enzyme (estimated by the quantity of substrate converted to product per minute by tissue extracts) was the same in both genotypes. When large amounts of a particular effector molecule that is required by this enzyme for activity were added to enzyme samples taken from *AA* and *Aa* individuals, the enzyme activity increased in both samples, but the amount of substrate converted by the tissue extracts from *AA* plants was twice that converted by corresponding extracts from *Aa* plants. (a) Explain the possible molecular basis for complete dominance in this case. (b) Mutation in a different gene converts allele *A* to an incompletely dominant gene. Describe the probable molecular basis of this mutation in light of your previous explanation for complete dominance.

16. Referring to Figure 11–5, explain how it is possible that a couple could have a normally pigmented child if both parents are albino.

Mutations and Protein Structure

17. Suppose that two hemoglobin samples, one of HbS and another of HbC, lose their labels during storage. Referring to Figure 11–10, devise a method that you could use to determine which is which.

18. Suppose that you discover a new mutant form of hemoglobin, HbX, which, like HbS, causes sickling of red blood cells. Genetic analysis reveals that the offspring of marriages between homozygotes for HbS and homozygotes for HbX are normal in phenotype, expressing neither form of hemoglobin disease.

In which hemoglobin chain (α or β) is the mutation for HbX? Explain.

Rates of Mutation

19. Aniridia, a form of blindness caused by the absence of the iris, is an autosomal dominant trait that shows full penetrance. Of 4,664,799 persons reported born in the state of Michigan during the period 1919–1959, 41 were aniridic offspring of normal parents. (a) What is the mutation rate to aniridia? (b) Could the mutation rate of a recessive mutant gene be estimated by this same kind of study? Explain your answer.

20. If humans have 100,000 genes in a haploid complement of chromosomes, and an average mutation rate per gene of 10^{-5}, what would be the overall mutation rate per genome?

21. Explain why forward mutation rates are generally greater than reverse mutation rates.

Randomness of Mutation

22. An experiment was performed in which several cultures were inoculated with 10^3 penicillin-sensitive bacterial cells. When the bacterial population in each culture had grown to 10^8 cells, the cultures were tested for the presence of penicillin-resistant cells. Of 100 cultures examined, 85 contained at least one resistant cell. (a) How many mutations to penicillin resistance, on the average, have occurred in each culture during the period of growth? (b) From these results, calculate the mutation rate (per cell division) to penicillin resistance.

23. Bacteria were infected with exactly one each of wild-type phage T2. Single bursts were then tested for the presence of phage progeny with the host-range mutation h. Among 10^5 bursts examined, 400 contained at least 1 h mutant. Assuming that each burst contained 200 total phage particles, (a) how many bursts would you expect to find with exactly 1 h mutant? (b) 2 h mutants? (c) more than 2 h mutants?

24. Suppose that 300 Gal$^-$ bacterial cells (unable to utilize galactose) were deposited randomly upon the surface of a nutrient agar plate. The plate was incubated for several hours to allow cell growth, and each colony was then carefully removed and tested for the presence of Gal$^+$ revertants. Among the 300 colonies tested, 294 contained at least 1 Gal$^+$ cell. (a) Assuming that each colony contained a total of 10^7 cells, calculate the mutation rate per cell division from Gal$^-$ to Gal$^+$. (b) How many of the colonies are expected to contain exactly 1 Gal$^+$ cell?

25. When a bacterial culture is exposed to a high concentration of ethidium bromide (a potential mutagen), all but a few cells die. Further analysis reveals that the few surviving cells are highly resistant to the lethal effects of ethidium bromide. Design an experiment by which you can test whether resistance to ethidium bromide is induced by exposure to this chemical or is a result of spontaneous mutation (i.e., is preadaptive).

Mutation Methodology

26. Describe a procedure by which you could isolate a double-mutant strain of E. coli that is histidine-requiring (his$^-$) and also resistant to streptomycin (str^r).

27. We inoculated 100 ml of nutrient broth with 4.0×10^4 non-mutant bacterial cells at time zero. When the number of bacteria reached 1.6×10^5, the frequency of mutants of a specified type was measured and found to be 10^{-4}. Assuming a negligible rate of back mutation and equal viability of the mutant and wild types, how many mutant cells would we expect to find when the number of cells reaches 3.2×10^5? From these results, what is the mutation rate?

28. Liquid nutrient medium in a sterile tube was inoculated with 10^5 actively dividing His$^+$ cells. After 12 hours of growth, the number of His$^-$ cells was measured and found to be 7×10^3. If the generation (doubling) time of both His$^+$ and His$^-$ cells on this medium is 72 minutes, what is the mutation rate from His$^+$ to His$^-$?

29. A student inoculated 100 ml of nutrient medium with bacterial strains A and B. The ratio of A cells to B cells in the starting inoculum was 10^5. When A cells had increased to 64 times their initial concentration, the ratio of A to B had decreased to 10^4. The student proposed two alternative hypotheses that could account for the reduction in the A to B ratio: *Selection hypothesis:* Strain B has a faster rate of growth than strain A on this medium. Mutation between the two strains does not occur. *Mutation hypothesis:* Strain A cells give rise to strain B cells by spontaneous mutation. The two strains grow at the same rate.

    (a) If the selection hypothesis is correct, by what factor is the generation (doubling) time of strain A greater than that of strain B on this medium? (b) What would be the mutation rate from A to B under the mutation hypothesis? (c) The student also conducted the experiment using a 10^{-5} A-to-B ratio in the inoculum and was surprised to find that when A cells had increased to 64 times their initial concentration, the A-to-B ratio had increased to 10^{-4}. From these results, is selection or mutation the most likely basis for the change in the A-to-B ratio?

30. Suppose that 10 Arg$^-$ auxotrophic cells were added to each of 100 tubes and allowed to grow exponentially for several hours. At the end of this time, each tube contained a total of 10^7 cells. (a) Assuming that prototrophic cells are formed by spontaneous mutation during growth, by what factor is the number of Arg$^+$ cells produced in an average tube expected to exceed the average number of $arg^- \rightarrow arg^+$ mutations per tube? (b) Suppose that the rate of mutation from arg^+ to arg^- is four times greater than the reverse mutation rate. What percentage of Arg$^+$ cells must be present in the starting inoculum in order to maintain a constant average ratio of Arg$^+$ to Arg$^-$ cells?

Chromosome Mutations

So far we have implicitly assumed that chromosomes behave normally during cell division, so that the chromosome complement remains constant from one generation to the next. Irregularities do occur, however, and they can lead to changes in the number, size, and genetic organization of chromosomes. Chromosome anomalies have been studied most extensively in eukaryotes, where they often result in changes in the chromosome complement that are visible with the light microscope. In this chapter, we will look at these microscopically visible deviations from the norm, focusing on the chromosome mutations in higher organisms.

There are two major classes of chromosome mutations: (1) changes in the number of chromosomes and (2) changes in the structure of chromosomes. Changes in chromosome number are further divided into two main types: **aneuploidy**, which is the loss or gain of less than a complete set of chromosomes, and **polyploidy**,

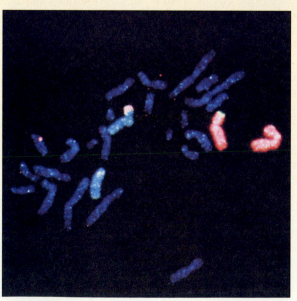

Normal human chromosome 7 (red), normal chromosome 12 (green), and translocation chromosomes (middle) showing an exchange of parts between chromosomes 7 and 12.

in which there is a gain of one or more entire sets of chromosomes.

ANEUPLOIDY

Aneuploid organisms have a chromosome complement that is not an exact multiple of the basic (genomic) set. Two of the most common forms of aneuploidy are **monosomy**, in which one copy of a chromosome is missing (chromosome number is $2n - 1$), and **trisomy**, in which there are three copies of one of the chromosomes rather than the normal two copies (chromosome number is $2n + 1$). When such conditions are not lethal, they are important as a source of variation because they are frequently heritable. In a trisomic individual, for example, the three copies of the affected chromosome often move in a regular 2:1 segregation pattern during meiosis, as shown in ▶ Figure 12–1. Segregation then results in two kinds of gametes, one containing n chromosomes and the other containing $n + 1$ chromosomes. Thus if the $n + 1$ gamete types are fertilized by gametes containing n chromosomes, offspring with the $2n + 1$ condition will be produced.

The various aneuploid conditions that affect all cells of an organism generally arise as a result of **meiotic nondisjunction**, which is the failure of homologous chromosomes to separate during meiosis. Nondisjunction can occur at either the first or the second meiotic division, as shown in ▶ Figure 12–2. Note that both $n + 1$ and $n - 1$ gamete types are produced; if these gametes unite with normal (n) gametes during fertilization, both trisomic and monosomic offspring will result.

Example 12.1

Geneticists have studied aneuploidy quite extensively in *Drosophila* and in a number of flowering plants, such as Jimson weed, corn, tomatoes, and wheat. In Jimson weed,

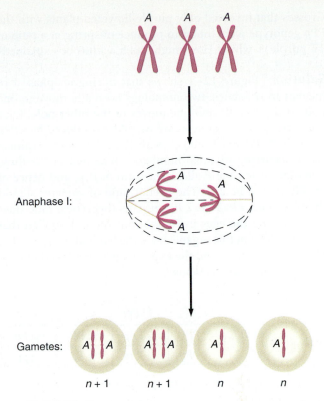

▶ **FIGURE 12–1** A 2:1 chromosome segregation pattern in a trisomic individual. During anaphase I of meiosis, two of the homologs migrate to one pole of the spindle and one migrates to the other pole, producing n and $n + 1$ gametes.

Datura stramonium, researchers have identified 12 different trisomics, corresponding to the 12 different chromosome pairs of this organism. Genetic studies on these trisomic strains reveal that crosses do not result in standard Mendelian ratios. Consider, for example, the purple locus on chromosome 9. At this locus, P (for purple color) is completely dominant to its recessive allele p (for white).

▶ **FIGURE 12–2** Chromosome nondisjunction at (b) the first and (c) the second meiotic divisions. All other chromosomes are assumed to segregate normally.

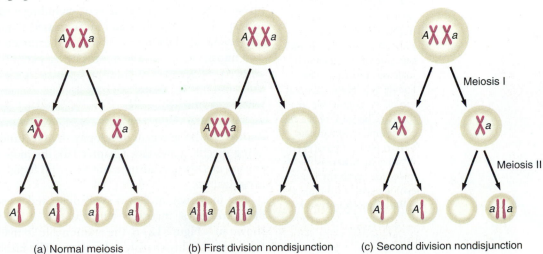

(a) Normal meiosis   (b) First division nondisjunction   (c) Second division nondisjunction

Crosses that involved only purple-flowered plants with the PPp genotype were found to produce offspring at a ratio of 17 purple:1 white. How might such a ratio be explained?

Solution: Figure 12–1 shows that during anaphase I of meiosis in trisomics, two homologs normally move to one pole of the spindle and one moves to the other pole. Four different classes of gametes are therefore produced by a PPp individual: PP, Pp, P, and p, with the n + 1 and n gamete types occurring in equal numbers. Since two of the three homologs carry the P allele, the ratio of P:p, and hence of Pp:PP, should be 2:1. The overall ratio of gametes should therefore be 1 PP:2 Pp:2 P:1 p. This theoretical ratio does occur among the eggs in *Datura*, but the only pollen that are functional in this organism are the haploid ones, 2 P:1 p. Therefore the cross and the expected genotypic and phenotypic outcomes are as follows:

		Eggs			
		$\frac{1}{6}$ PP	$\frac{2}{6}$ Pp	$\frac{2}{6}$ P	$\frac{1}{6}$ p
Pollen	$\frac{2}{3}$ P	$\frac{2}{18}$ PPP	$\frac{4}{18}$ PPp	$\frac{4}{18}$ PP	$\frac{2}{18}$ Pp
	$\frac{1}{3}$ p	$\frac{1}{18}$ PPp	$\frac{2}{18}$ Ppp	$\frac{2}{18}$ Pp	$\frac{1}{18}$ pp

Genotypic ratio: 2 PPP:5 PPp:2 Ppp:4 PP:4 Pp:1 pp
Phenotypic ratio: 17 purple (P–– and P–):1 white (pp)

Follow-Up Problem 12.1

Predict the ratio of purple-flowered (P–– and P–) and white-flowered (ppp and pp) *Datura* plants from the cross Ppp × Ppp.

◆

Aneuploidy can also occur in somatic cells after fertilization, resulting in individuals who are **chromosome mosaics** with a mixture of aneuploid and normal cell lines (▶ Figure 12–3). Chromosome mosaics can develop in one of two ways. First, they can result from the failure of chromosomes to separate during mitosis. **Mitotic nondisjunc-**tion (▶ Figure 12–4a) tends to give rise to three cell types: one normal, a second deficient, and a third with an extra chromosome. This process can result in a large amount of abnormal tissue if it occurs at an early stage of embryological development. A second mechanism for the formation of a chromosome mosaic is **anaphase lag** (Figure 12–4b), in which a chromosome fails to migrate properly during mitotic anaphase and is not incorporated into a daughter nucleus. In contrast to nondisjunction, anaphase lag produces only two cell types, of which one has the normal chromosome constitution and the other is missing a chromosome.

Aneuploid Conditions in Humans

Most forms of aneuploidy in humans result in a condition that is incompatible with life. Comparisons between the observed number and the expected number of aneuploids bear out this relationship. For example, 24 kinds of trisomics and 24 kinds of monosomics are possible in humans, 1 for each of the 22 different autosomes and 1 for each sex chromosome. Of these 24 possibilities, only 7 have been observed with any regularity among live-born infants (■ Table 12–1). The chromosomally defective fetuses that survive to term are predominantly trisomic for a sex chromosome or for one of the smaller autosomes and are usually born with a pattern of symptoms—a **syndrome**—characteristic of each abnormality. Trisomy and monosomy of the larger autosomes do occur, but these chromosomal mutations create such severe developmental abnormalities that they result in spontaneous abortion long before the time of normal birth.

One autosomal abnormality that has received intense study is **Down syndrome**, which occurs in humans with an extra chromosome 21 (47,+21) or trisomy of at least the bottom third of chromosome 21. Down syndrome is the most common cause of mental retardation in the United States, with an average frequency of 1 in 700 live births. Individuals with Down syndrome have certain characteristic features such as short stature, a rounded face and broad head, and epicanthic folds (folds of skin on the upper eyelid) (▶ Figure 12–5). They are also likely to have various physical and biochemical defects, including a susceptibility to respiratory infections and an increased risk of developing leukemia, Alzheimer's disease, and cataracts or other vision impairments. Biochemically, Down patients suffer from an overproduction of certain proteins encoded by genes on chromosome 21. These proteins function in diverse metabolic processes where their excess can lead to mental retardation and the other pathologies associated with this disease. With the recent development of techniques for isolating specific genes (see Chapter 18), the molecular mechanisms of Down syndrome are now the focus of intensive investigation.

The incidence of Down syndrome, and to some extent other trisomies, increases with the age of the mother. As shown in ▶ Figure 12–6, the relationship to maternal age is curvilinear, rising sharply at later ages. The mechanism that

▶ **FIGURE 12–3** A bilateral gynandromorph of *Drosophila*. Such individuals are female on one side of the body and male on the other. The female side of the bilateral mosaic (recognizable in this case because of the red compound eye) consists of normal XX (genotype w^+w) cells, whereas the male side (with the white eye) is derived from an XO cell lineage (genotype $w–$).

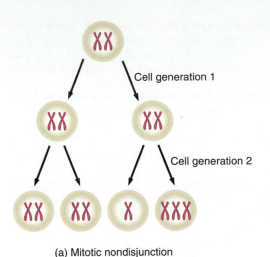

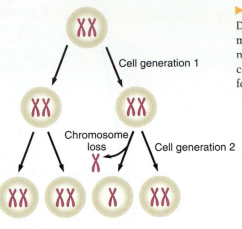

Cell generation 1

Cell generation 2

Cell generation 1

Chromosome loss

Cell generation 2

(a) Mitotic nondisjunction

(b) Anaphase lag (chromosome loss)

causes this increase is not yet known, but it is thought to be related to the fact that in females the developing egg begins meiosis before birth and remains in prophase I until just before ovulation. Because the length of the time delay before ovulation depends on age, a longer exposure to potential mutagenic agents (such as X rays and gamma rays) could contribute to a higher rate of nondisjunction in the eggs of older women.

Like Down syndrome, several sex chromosome aneuploids have also received intensive study, including the monosomic condition known as **Turner syndrome** (45,XO). Individuals with Turner syndrome are female, because of the absence of a Y chromosome, but have underdeveloped ovaries and are typically sterile. Physical characteristics that are often associated with this abnormality include short stature, infantile genitalia, sparse pubic hair, wide-spaced and underdeveloped breasts, and webbing of the neck. The low incidence of this disorder (1 in 2,500–10,000 female births) is a consequence of its high rate of fetal mortality. Over 90 % of 45,X fetuses spontaneously abort.

Among the various forms of sex chromosome aneuploids are two conditions that result from an extra X chro-

▶ **FIGURE 12–5** Down syndrome. (a) Chromosomes of a boy with Down syndrome. Note trisomy of chromosome 21. (b) A boy affected with Down syndrome.

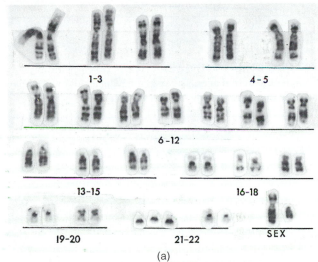

1-3 4-5

6-12

13-15 16-18

19-20 21-22 SEX

(a)

(b)

■ **TABLE 12–1** Human aneuploids occurring in live births.

Chromosome Abnormality	Syndrome	Estimated Frequency
trisomy 21	Down syndrome	1 in 700 births
trisomy 18	Edwards syndrome	1 in 4000 births
trisomy 13	Patau syndrome	1 in 5000 births
47,XXX	triple-X syndrome	1 in 1000 female births
47,XXY	Klinefelter syndrome	1 in 1000 male births
47,XYY	—	1 in 1000 male births
45,XO	Turner syndrome	1 in 5000 female births

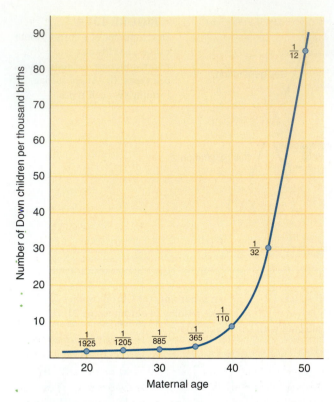

▶ **FIGURE 12–6** Estimated incidence of Down syndrome for different maternal ages. The risk of having a child with Down syndrome increases sharply in older mothers.

mosome; one is **Klinefelter syndrome** (47,XXY) and the other is **triple-X syndrome** (47,XXX). Individuals with Klinefelter syndrome are male, since they have a Y chromosome, but they frequently have poorly developed male secondary sex characteristics, as well as some feminine traits, such as breast development. Triple-X individuals are female, and the behavioral and physical expressions of this disorder tend to deviate little from 46,XX individuals. Many triple-X females are fertile and most are normal in IQ and appearance, indicating that an additional X chromosome does not necessarily lead to an abnormal condition.

These syndromes also occur among individuals who are chromosome mosaics, having a mixture of aneuploid and normal cell lines. Some interesting examples of sex chromosome mosaics that involve two and three cell lines include the forms X/XX, X/XX/XXX, and XY/XXY. Individuals with either of the first two forms tend to exhibit Turner syndrome, while individuals with the latter form show Klinefelter syndrome. The severity of the symptoms in each case depends on the relative abundance of aneuploid cells and hence on how early the mosaicism developed.

To Sum Up

1. Aneuploid organisms have a chromosome complement that is not an exact multiple of the basic chromosome set. Two of the most common types of aneuploids are monosomics (with one less chromosome than normal) and trisomics (with one additional chromosome).

2. Nondisjunction is the failure of chromosomes to separate during meiosis or mitosis. Meiotic nondisjunction can occur in either the first or the second meiotic division, and it leads to the production of gametes with one less or one more chromosome than normal. If an abnormal gamete unites with a normal gamete during fertilization, an aneuploid state can arise in all the cells of the offspring. In contrast, mitotic nondisjunction, which can occur in somatic cells after fertilization, leads to an individual who is a chromosome mosaic (has a mixture of aneuploid and normal cell lines). Mosaics can also be produced by anaphase lag.

3. Trisomy 21 (Down syndrome) is an example of an autosomal aneuploidy that occurs among live-born human infants. The incidence of Down syndrome increases dramatically with the age of the mother.

4. Unlike most autosomal aneuploids, aneuploids for the sex chromosomes in humans are usually viable. Turner syndrome (45,X) is the only human monosomy that is found in live births. Two trisomic conditions result from an extra X chromosome: Klinefelter syndrome (47,XXY) and triple-X syndrome (47,XXX).

◆

Points to Ponder 12.1

Suppose that prenatal testing reveals the presence of a fetus with Down syndrome. There is a great deal of variability in terms of the severity of this syndrome; some Down syndrome individuals respond well to intensive physical therapy and special education, while others are so severely retarded mentally and physically that they require institutionalization. The severity of any one particular case cannot be determined by prenatal testing. The parents must therefore make a decision as to whether to continue the pregnancy based only on the knowledge that the fetus is trisomy-21. The detection of a fetus with Down syndrome is legal grounds for therapeutic abortion.

What would you do if you were one of the parents of this fetus? Would you continue the pregnancy? This is a very complex decision, as demonstrated by the following questions.

Suppose the parents decide to terminate the pregnancy. Do you agree that they have the right to do this? Is this a matter of personal decision?

Suppose the parents choose to continue the pregnancy. Is society obligated to assume the responsibility for the specialized services required for a child with Down syndrome? Should society's responsibility outweigh that of the parents? How can we balance the rights of the parents with the role of society in such matters?

Suppose the child becomes unmanageable and must be institutionalized. Should the parents be held financially responsible?

Should knowledge that the fetus is affected with Down syndrome be private, or do health insurance companies have the right to know the condition of the fetus? Do insurance companies have the right to require prenatal testing? Do insurance companies have the right to demand abortion of an affected fetus? If an employer pays for insurance coverage for his or her employees, does that employer have any rights in this matter?

POLYPLOIDY

In contrast to aneuploids, the somatic cells of polyploid organisms have a chromosome number that varies above the diploid number by a simple multiple of the genomic set. Possible polyploid chromosome states are triploid (3n), tetraploid (4n), pentaploid (5n), and so on, each having one or more entire sets of chromosomes in addition to the diploid number.

One potential source of polyploids is the production of gametes with an unreduced chromosome number (▶ Figure 12–7a). For example, a meiotic irregularity such as the accidental disruption of the spindle in meiosis I can result in the formation of a diploid (2n) gamete. If this gamete is fertilized by a normal haploid gamete, the offspring would be triploid (3n). If two unreduced gametes happen to combine during fertilization, the offspring would be tetraploid (4n). Clearly, offspring of even higher levels of ploidy could be formed if such meiotic irregularities were to occur in polyploid individuals.

Polyploids may also arise from a failure in mitosis in which chromosome replication without cell division leads to chromosome doubling in somatic tissue (Figure 12–7b). This process, called **somatic doubling**, is usually accompanied by some form of asexual reproduction (such as stolons or runners in plants), which allows the resulting tetraploid somatic tissue to form an entirely new individual. Researchers have used drugs that interfere with spindle formation during mitosis to induce the polyploid condition in various plants. One drug that is often used for this purpose is the alkaloid **colchicine**, which is extracted from the autumn crocus *Colchicium*. Colchicine prevents the polymerization of spindle microtubules, thus blocking cell division at metaphase. The application of colchicine does not interfere appreciably with any other plant cell component or process, so continued treatment of plant tissue with the drug leads to a doubling of chromosome number in each affected cell. The resulting tetraploid tissue can then be propagated vegetatively in the absence of colchicine to produce a plant with a 4n chromosome number.

There are three important consequences of polyploidy. First, an increase in the number of chromosomes is often accompanied by an increase in cell size. This increase often leads to an increase in the size of the fruit and other plant parts that are commercially important. Consequently, many familiar, commercially grown food and crop plants are polyploids or at least have some polyploid varieties.

A second consequence of polyploidy is a reduction in fertility relative to diploid organisms of the same species; this reduction is particularly evident in triploids and other odd-numbered types (3n, 5n, 7n, . . .). Because of inherent reproductive difficulties such polyploid varieties are normally restricted to organisms that reproduce asexually. They usually encounter no difficulty during mitosis, since each chromosome behaves independently on the mitotic spindle. However, in meiosis, the extra chromosomes tend to be distributed to the gametes in a more-or-less random pattern, resulting in nonfunctional sex cells with unbalanced chromosome sets.

The chromosomal basis for sterility in polyploids with odd-numbered chromosome sets is most easily seen in triploids. Triploids have three homologs of each kind of chromosome, rather than two. In meiosis, two of these homologs move to one pole of the spindle and the remaining

---

▶ **FIGURE 12–7** Possible mechanisms of polyploidy. (a) Production of diploid gametes, resulting from a failure to complete meiosis I. (b) Somatic doubling, resulting from chromosome replication without cell division.

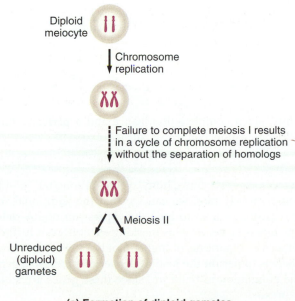

(a) Formation of diploid gametes

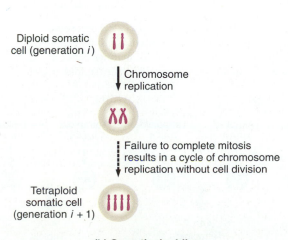

(b) Somatic doubling

(a)   (b)

homolog moves to the opposite pole, resulting in a 2:1 segregation pattern (see Figure 12–1). Each product of meiosis in a triploid organism thus has a 50% (1 out of 2) chance of receiving only one copy of a particular chromosome, and a 50% chance of receiving two copies of the chromosome. Since all n of the chromosome triplets assort independently, the chance that a fully balanced (n or $2n$) gamete is formed is equal to $(\frac{1}{2})^n$. Thus when n is large, so that $(\frac{1}{2})^n$ is very small, most of the gametes produced by the triploid organism are genetically unbalanced. The unbalanced products of meiosis are usually either incapable of fertilization (especially in plants) or produce nonviable aneuploid zygotes. In either case, the triploid organism is effectively sterile. Plant breeders have taken advantage of the sterile nature of triploid plants by developing seedless varieties such as the common banana and the seedless watermelon (► Figure 12–8).

Polyploid organisms with an even number of chromosome sets have a much better chance of being fertile. In certain tetraploids, for example, meiotic pairing results in a 2:2 segregation pattern in which two homologs of a particular chromosome move to one pole of the spindle and two move to the other pole. These tetraploids are interfertile—their matings produce fertile tetraploid offspring (► Figure 12–9). They are also reproductively isolated from their diploid ancestors, since matings between a tetraploid

and a diploid result in sterile triploid progeny. Such matings do have a practical use; for example, seeds from a tetraploid × diploid cross can be used to grow seedless watermelons.

A third important consequence of polyploidy is a decline in the frequency of recessive phenotypes, compared to the frequency in diploid varieties. Polyploidy thus reduces the chance that deleterious recessive genes are expressed in a population. To illustrate, consider the pair of alleles A and a in the tetraploid heterozygote $AAaa$. Assume a 2:2 segregation pattern, so that the alleles segregate in pairs in random assortment. What is the probability that a gamete of this tetraploid is fully recessive aa? To keep better track of the alleles, we will label them $A_1A_2a_3a_4$. Note that there are

$$\frac{4!}{2!2!} = 6$$

different combinations of alleles that can arise from 2:2 disjunction: A_1A_2, A_1a_3, A_1a_4, A_2a_3, A_2a_4, and a_3a_4. Of these combinations, only a_3a_4 includes both recessive alleles. The probability of a gamete of this type is therefore $\frac{1}{6}$. In the mating $AAaa \times AAaa$, we would expect the frequencies of recessive ($aaaa$) and dominant ($A{-}{-}{-}$) offspring to be:

$$P(aaaa) = P(aa)P(aa) = \left(\frac{1}{6}\right)\left(\frac{1}{6}\right) = \frac{1}{36}$$

and

$$P(A{-}{-}{-}) = 1 - P(aaaa) = \frac{35}{36}$$

Thus, given complete dominance and a perfectly random 2:2 segregation pattern of alleles, a monohybrid cross in tetraploids will yield a phenotypic ratio of 35:1, compared with the 3:1 ratio expected for the corresponding $Aa \times Aa$ cross in diploids. Since homozygotes among tetraploids are comparatively rare, we can see that the additional sets of chromosomes tend to hide the expression of any deleterious recessive genes. This masking effect is even further enhanced in organisms of higher ploidy. However, the overall effect is to permit the load of deleterious recessive genes in the population to build up, since these genes are so rarely expressed.

► **FIGURE 12–9** Consequences of 2:2 chromosome segregation in tetraploids. (a) Tetraploids produce diploid gametes and are interfertile, while (b) matings between tetraploids and diploids tend to yield sterile triploid offspring.

	(a)		(b)	
Parents:	Tetraploid $AAAA$	× Tetraploid $AAAA$	Tetraploid $AAAA$	× Diploid AA
Gametes:	AA	AA	AA	A
Offspring:		$AAAA$ Tetraploid		AAA Triploid

Example 12.2

Consider a pair of alleles A and A' that are lacking in dominance, so that each genotype expresses a different phenotype. Determine the number of phenotypic classes produced by a cross between tetraploids $AAA'A' \times AAA'A'$, and predict the expected frequency of each class. (Assume a perfectly random 2:2 segregation pattern of alleles.)

Solution: The cross will yield five different phenotypic classes, corresponding to the five different genotypes: $AAAA$, $AAAA'$, $AAA'A'$, $AA'A'A'$, and $A'A'A'A'$. We can deduce the relative proportion of each class by means of a Punnett square. First, recall that each tetraploid of this genotype can produce six combinations of alleles. These combinations form three genotypes in the gametes, in the ratio $1\ AA : 4\ AA' : 1\ A'A'$. Random fertilization produces the following possible zygotic combinations:

	1 AA	4 AA'	1 $A'A'$
1 AA	1 $AAAA$	4 $AAAA'$	1 $AAA'A'$
4 AA'	4 $AAAA'$	16 $AAA'A'$	4 $AA'A'A'$
1 $A'A'$	1 $AAA'A'$	4 $AA'A'A'$	1 $A'A'A'A'$

These results yield a genotypic (and thus a phenotypic) ratio of $1\ AAAA : 8\ AAAA' : 18\ AAA'A' : 8\ AA'A'A' : 1\ A'A'A'A'$. We see that with incomplete dominance, polyploids can exhibit greater phenotypic variation than can diploids.

Follow-Up Problem 12.2

Reconsider the A and A' alleles described in Example 12.2. What phenotypic ratio would be found among the offspring of a cross between tetraploids $AAA'A' \times AAAA'$?

Unlike higher plants, in which polyploids are relatively common, viable polyploids among animals are rare for several reasons. Two of the most important reasons involve differences in the nature of sex determination and reproduction. Many more animals than plants have a chromosomal mechanism of sex determination. Thus, animals have a greater dependence on the delicate balance between autosomes and sex chromosomes, which is severely disrupted by polyploidy. Moreover, the greater reliance on sexual reproduction among animal species makes it more difficult for newly formed polyploids to propagate themselves. The problems encountered by polyploid animals in sex determination and meiosis help to explain why many of the naturally occurring polypoid forms in animals are either hermaphroditic, such as earthworms, or can reproduce by parthenogenesis (that is, without fertilization).

Allopolyploids and Interspecific Hybridization

Polyploids are usually classified according to the origin and nature of their multiple chromosome sets. The polyploids that we have considered so far are classified as **autopolyploids**, since their chromosome sets are multiples of the same genome (i.e., homologous sets). Autopolyploids are thus characterized by having multiple chromosome sets that are derived from the same species. Polyploidy may also be accompanied by some form of interspecific hybridization, making it possible for the polyploid to acquire complete sets of chromosomes from different species (that is, when the sets are nonhomologous). When the multiple chromosome sets are initially derived from different species, the resulting polyploids are classified as **allopolyploids**.

Many plants and even a few animals are capable of producing hybrids from the mating of two different species. Such hybrids are usually sterile, because they lack the chromosome homology needed for proper pairing and accurate distribution of chromosomes to the gametes. However, if chromosome doubling should occur in these hybrids (either because of chance somatic doubling or from the application of colchicine in the laboratory), the result is a fertile allotetraploid called an **amphidiploid**, which has a normal diploid chromosome complement from both ancestral species. The origin of a fertile amphidiploid from a naturally occurring plant hybrid is shown in ▶ Figure 12–10.

The distinction between allopolyploids and autopolyploids is not always clear-cut. Studies reveal that multiple chromosome sets are not always strictly homologous (with identical pairing affinities at meiosis) or nonhomologous (with no pairing affinity). Varying degrees of similarity exist, depending on the relationship between the parent species. If the species are closely related in their evolution, the chromosomes they contribute to the allopolyploid will not have had sufficient time to diverge completely in genetic character and will be likely to show some degree of similarity. Chromosomes that are derived from different species but still retain some similarity are called **homeologous** (meaning "similar"), as opposed to *homologous* (identical). Genetic similarities in different sets of chromosomes can thus lead to pairing irregularities, in which homeologs of different sets compete with homologs for pairing partners during meiosis. Allopolyploids can therefore show a low level of fertility and could be mistakenly identified as autopolyploids.

Common bread wheat (*Triticum aestivum*) is an example of an allopolyploid species in which the extra sets of chromosomes exhibit some degree of similarity. Common wheat is a hexaploid of complex origin and exists in over 20,000 cultivated varieties. The total chromosome complement in this organism consists of two sets of chromosomes in each of three genomes (A, B, and D), so the formula for the genome of common wheat is AABBDD. Since each genome is made up of 7 different chromosomes, the somatic cells of this species contain 14 (for AA) + 14 (for BB) + 14 (for DD) = 42 chromosomes.

The A, B, and D genomes of wheat are homeologous, hav-

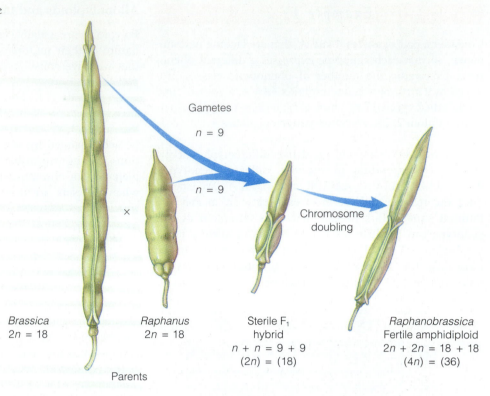

▶ **FIGURE 12–10** Formation of the amphidiploid *Raphanobrassica* by chromosome doubling in a sterile hybrid formed from a radish (*Raphanus*) and a cabbage (*Brassica*). *Adapted from:* A. M. Srb, R. D. Owen, and R. S. Edgar, *General Genetics*, 2d ed. Copyright © 1965 by W. H. Freeman and Company. Reprinted with permission.

Gametes

$n = 9$

$n = 9$

Chromosome doubling

×

Brassica
$2n = 18$

Raphanus
$2n = 18$

Sterile F₁
hybrid
$n + n = 9 + 9$
$(2n) = (18)$

Raphanobrassica
Fertile amphidiploid
$2n + 2n = 18 + 18$
$(4n) = (36)$

Parents

ing many genetic similarities. Chromosomes 1A, 1B, and 1D are homeologs, as are chromosomes 2A, 2B, and 2D, and so on. Despite the close relationship between the three genomes, chromosome pairing in wheat is like that of a diploid organism, with synapsis occurring only between true homologs. Homeologous pairing is suppressed by a gene named *Ph* (pairing *h*omeologous), which is present on the long arm of chromosome 5B. When this gene is present, only homologs pair during meiosis. When it is absent, homeologs also pair, but less frequently.

Researchers have demonstrated that the A, B, and D genomes are homeologous by developing viable **nullisomics** in wheat. Nullisomics are individuals that have lost both members of a pair of chromosomes. In diploids, this condition (designated $2n - 2$) is ordinarily lethal, because vital gene functions are lost. In the hexaploid bread wheat, however, homeologous pairs of chromosomes compensate for the loss to some extent because they have similar genes. Although the nullisomics are typically less fertile and vigorous than normal plants, at least their survival is assured because the partially duplicated genomes contain genes that are similar to the missing ones.

Evolution through Polyploidy

Polyploidy has been an important mechanism in the evolution of higher plants. Close to 40% of the flowering plant species appear to be derived in this manner. The most im-

pressive feature of polyploid evolution is the suddenness with which a new species can emerge, often occurring in just one or two generations without a long history of genetic change.

One example of polyploid evolution is the development of the present-day species of bread wheat. The genus *Triticum* contains a number of different species that can be subdivided into three groups having 14, 28, and 42 chromosomes. The group with 14 chromosomes contains, among others, the wild and cultivated forms of einkorn wheat, *T. monococcum* (genomic formula AA) The group with 28 chromosomes includes the wild and cultivated forms of emmer wheat, *T. turgidum* (genomic formula AABB), and the group with 42 chromosomes includes the different varieties of common bread wheat, *T. aestivum* (genomic formula AABBDD) Many geneticists currently believe that each group gave rise to the progenitors of the next higher group in the series through allopolyploidy.

One proposed evolutionary pathway for the development of the common bread wheat is shown in ▶ Figure 12–11. Two hybridization events have occurred, each of which is thought to involve a wild grasslike species with a chromosome number of 14. The first accidental cross-pollination may have taken place between einkorn wheat and the wild grass *T. searsii*, an inhabitant of wheat fields in southwestern Asia. The subsequent doubling of chromosomes in the F₁ produced an amphidiploid species of emmer wheat with 28 chromosomes. The second interspecies cross probably

Einkorn wheat
(*T. monococcum*)
14 chromosomes
(sets AA)

Wild grass
(*T. searsii*)
14 chromosomes
(sets BB)

Emmer wheat
(*T. turgidum*)
28 chromosomes
(sets AABB)

Wild grass
(*T. tauschii*)
14 chromosomes
(sets DD)

Bread wheat
(*T. aestivum*)
42 chromosomes
(sets AABBDD)

▶ **FIGURE 12–11** Evolution of
wheat (genus *Triticum*). This
sequence of hybridization events is
believed to be involved in the
evolutionary development of common
bread wheat. Each hybridization is
followed by chromosome doubling to
form fertile amphidiploids.

involved emmer wheat and the wild grass *T. tauschii*, which inhabits wheat fields in the Mediterranean region. After this second cross, the doubling of chromosomes in the F_1 gave rise to the progenitor of bread wheat, with 42 chromosomes. Thus, it appears that the AABBDD genomic constitution of bread wheat consists of chromosome sets derived from three diploid species: the AA of *T. monococcum*, the BB of *T. searsii*, and the DD of *T. tauschii*.

Two American scientists, E. S. McFadden and E. R. Sears, provided evidence to support this theory by successfully hybridizing emmer wheat (AABB) with *T. tauschii* (DD). Colchicine was used to artificially induce chromosome doubling in the hybrids. The 42-chromosome amphidiploid (AABBDD) that resulted had many of the characteristics of bread wheat. Moreover, it could readily cross with bread wheat to form fertile hybrids.

Example 12.3

Short oat (*Avena brevis*) has 14 chromosomes. Abyssinian oat (*Avena abyssinica*) has 28 chromosomes, and the common cultivated oat (*Avena sativa*) has 42 chromosomes. (a) How many chromosomes appear to make up the basic genomic set in the oat series? (b) How many genomic sets occur in the common oat?

Solution: (a) Short oat appears to be a diploid in this series, with $2n = 14$; its chromosome number is exactly divisible by 2 but not by any larger integer other than 7. Thus there are 7 chromosomes in the basic genomic set. (b) Common oat is a hexaploid, with $6 \times 7 = 42$ chromosomes.

Follow-Up Problem 12.3

Various species of raspberries have 14, 21, 28, 42, and 49 chromosomes. (a) How should the chromosome complements of these species be designated in relation to the basic number n? (b) How does evolution appear to be taking place in this plant group?

To Sum Up

1. Organisms with three or more complete sets of chromosomes are polyploid. Autopolyploids possess multiple chromosome sets derived from homologous genomes of the same species. They are formed from zygotes produced by unreduced diploid gametes or by a doubling of chromosomes in the somatic cells and the subsequent vegetative reproduction of the resulting tetraploid tissue.

2. Many autotetraploid plants have great commercial value because of their increased size and vigor. The inability of many autopolyploid plants to produce balanced gametes frequently results in infertility, however, so that many commercially useful polyploids must be propagated by asexual means.

3. Allopolyploids possess multiple chromosome sets derived from members of different species. Fertile amphidiploids, which are formed through hybridization and the subsequent doubling of the chromosome number, have the diploid set of chromosomes of each of the two parental species. They constitute an important source of new species in higher plants.

EXPERIMENTAL MANIPULATION OF CHROMOSOME NUMBER

In the decades ahead, an expanding food supply will be needed to accommodate the world's growing human population. A major part of any increase in food production will most likely come from using traditional breeding techniques to improve our current species of crop plants. The use of these techniques has already resulted in impressive gains in yield and nutritional quality during recent years, and there is every indication that these techniques will lead to further advances in the future. In this section, however, we will look at the creation of entirely new and possibly more productive genetic varieties through alterations in chromosome number. This area of research includes the use of monoploid (haploid) plants and the synthesis of polyploids by experimental techniques.

Experimental Production of Allopolyploids

Geneticists now routinely produce allopolyploids by using the drug colchicine to induce the doubling of chromosome number in hybrids of different species. This procedure was responsible for the production of the new genus *Triticale* from a hybrid of wheat (genus *Triticum*) and rye (genus *Secale*). First-generation seeds from such a cross are sterile, even though some germinate. If the tips of the growing sprouts from these seeds are treated with colchicine, however, the chromosome number doubles and the flowers that develop produce the fertile amphidiploid seeds of the *Triticale* plant.

The procedures used in the initial development of *Triticale* are shown in ▶ Figure 12–12. Notice that *Triticale* is a hexaploid like the common bread wheat but has a genomic formula of AABBRR. The A and B genomes are inherited from a tetraploid species of wheat (*T. turgidum*) and the R genome is inherited from rye.

The overall performance of *Triticale* was at first disappointing. Breeders have since produced improved genetic strains by selecting for traits such as high fertility, early maturity, greater insensitivity to day length, and greater resistance to lodging (collapse of the plants) due to weak straw and heavy head. Disease resistance has also been enhanced. For example, certain strains of *Triticale* appear to be more resistant than wheat to a fungal disease called rust. The improved varieties generally also combine the high protein content of wheat and the high lysine content of rye. Lysine is an essential amino acid that is present in only small amounts in the proteins of many cereal grains, so high lysine content thus improves the quality of the plant protein for human consumption.

It is also possible to produce amphidiploids artificially in the laboratory using the technique called **protoplast fusion** (▶ Figure 12–13). In this procedure, the somatic cells of different plant species are converted to protoplasts by removing their cell walls by enzyme treatment, and then they are cultured together in the presence of polyethylene glycol, a chemical that stimulates cell fusion. When two such protoplasts combine, they produce a hybrid cell with a single nucleus that contains a complete set of chromosomes from both parental species. Amphidiploids can then be produced by the regeneration of whole plants from the cultured hybrid cells, since plant cells, unlike animal cells, have the capacity to develop into a complete multicellular organism. Each hybrid protoplast is first induced to regenerate a cell wall and proliferate into an undifferentiated cell mass or callus. Adjustment of the plant hormone levels in the growth

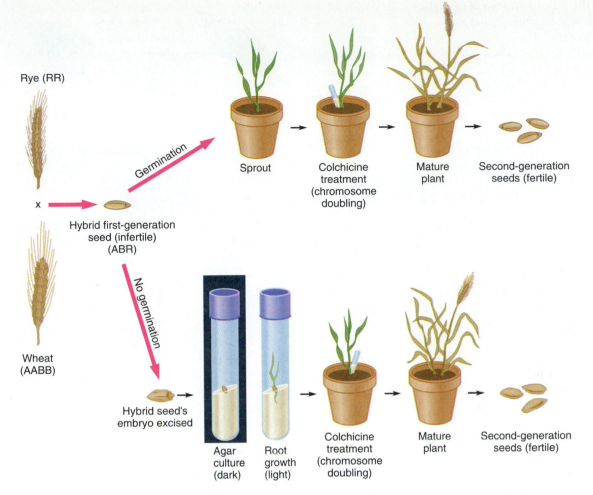

FIGURE 12–12 Experimental production of the amphidiploid *Triticale*. The cross of rye with wheat yields first-generation hybrid seeds that are usually sterile, although some may germinate (top). Sterility can then be overcome by treating the tips of growing sprouts with colchicine. This doubles the chromosome number, so the mature plant (with the genomic formula AABBRR) produces fertile second-generation seeds. In contrast, if the first-generation hybrid seed does not germinate (bottom), embryo culture can be used to produce the first-generation seedlings. Embryos are excised from immature seeds and cultured in the dark until the embryo sprouts.

medium then causes the callus tissue to develop into a small plantlet, which can be used to form the mature plant with roots, stems, leaves, and flowers. Protoplast fusion can thus overcome natural breeding barriers to produce hybrids between plant species that are incapable of cross-pollination.

Monoploids and Plant Engineering

Monoploids possess only the haploid set of chromosomes of a species that is normally diploid. Monoploids can arise naturally from unfertilized gametes (as do the monoploid males of bees, wasps, and ants) and are ordinarily smaller and less vigorous than their diploid counterparts.

Monoploids are of special interest to plant geneticists, who use them to isolate mutant strains and to develop more productive varieties. Since monoploid plants have only a haploid set of chromosomes, recessive mutations are imme-

diately expressed and can be studied in monoploid cell cultures in much the same way as in microorganisms. The cells derived from monoploid plant tissue are converted to protoplasts and induced to grow in broth or on agar medium. They can then be treated with a mutagen and screened for genetically useful characteristics in a manner similar to bacterial cells. For example, if the desired trait is resistance to the toxin produced by a certain plant pathogen, then screening could involve growing the cells in the presence of the toxin and selecting the resistant clones.

The procedures for the development and analysis of monoploid cell cultures are shown in ▶ Figure 12–14. In general, monoploid plants are produced artificially in the laboratory from the meiotic products of diploid plants by using the technique of anther culture (Figure 12–14a) or they are regenerated from the cultured somatic cells of other monoploids. The monoploid plants produced in this

▶ **FIGURE 12–13** The use of protoplast fusion to produce amphidiploid plants.

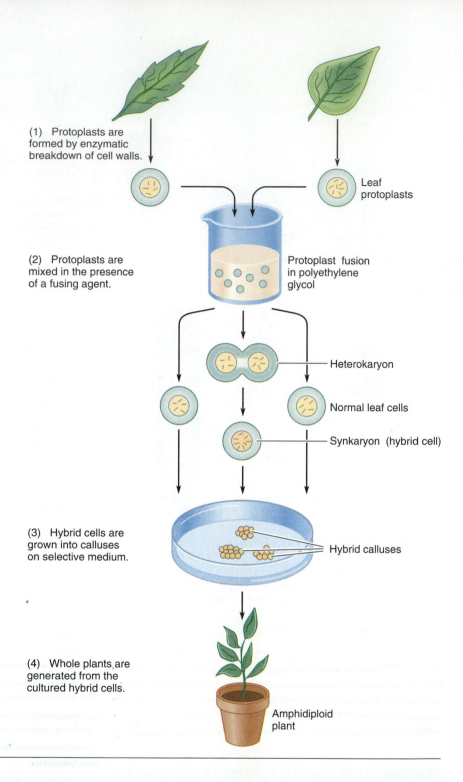

(1) Protoplasts are formed by enzymatic breakdown of cell walls.

Leaf protoplasts

(2) Protoplasts are mixed in the presence of a fusing agent.

Protoplast fusion in polyethylene glycol

Heterokaryon

Normal leaf cells

Synkaryon (hybrid cell)

(3) Hybrid cells are grown into calluses on selective medium.

Hybrid calluses

(4) Whole plants are generated from the cultured hybrid cells.

Amphidiploid plant

way are typically sterile; since each chromosome occurs singly at meiosis, gametes with less than a complete set of chromosomes are produced. However, these plants can be induced to form diploid tissue by the application of colchicine (Figure 12–14b). The treated tissue is then propagated vegetatively in the absence of the drug to produce a fertile diploid plant that is homozygous for all the genes of the monoploid parent. This procedure can therefore be used to establish a stable, true-breeding line in a very short time.

To Sum Up

1. Researchers can produce amphidiploids in the laboratory by using the drug colchicine to induce the doubling of chromosome number in sterile interspecific hybrids. Amphidiploids between two species that cannot cross-pollinate can also be formed by somatic cell protoplast fusion.

2. Monoploid individuals possess only the haploid set of chromosomes of a normally diploid species. Although monoploids are

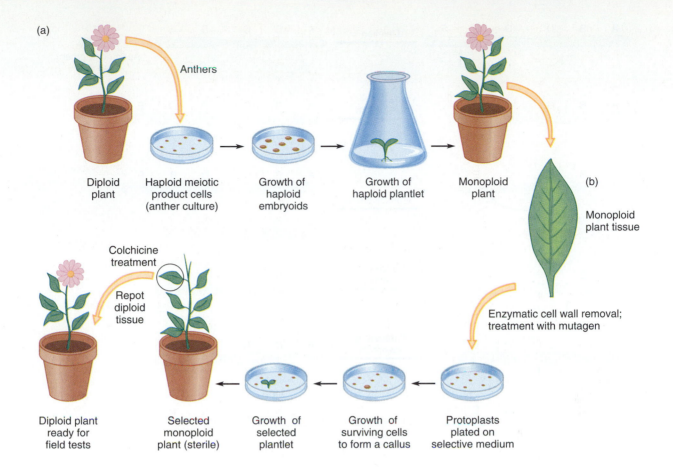

(a)

Anthers

Diploid
plant

Haploid meiotic
product cells
(anther culture)

Growth of
haploid
embryoids

Growth of
haploid plantlet

Monoploid
plant

(b)

Monoploid
plant tissue

Colchicine
treatment

Repot
diploid
tissue

Enzymatic cell wall removal;
treatment with mutagen

Diploid plant
ready for
field tests

Selected
monoploid
plant (sterile)

Growth of
selected
plantlet

Growth of
surviving cells
to form a callus

Protoplasts
plated on
selective medium

▶ **FIGURE 12–14** The development and use of monoploid cell cultures for plant engineering. (a) The development of a monoploid plant by tissue culture. Appropriately treated pollen grains can be induced by hormones to grow into haploid embryoids, which are then stimulated to develop into plantlets and whole plants. (b) Formation of a new diploid plant variety using monoploid cell cultures.

usually sterile, investigators can produce them in the laboratory by anther culture. Colchicine can then be used to produce a fertile diploid plant that is homozygous for all the genes of the monoploid parent.

◆

VARIATIONS IN CHROMOSOME STRUCTURE

When the disruption and rearrangement of genes occur on a more extensive scale than point mutations, they lead to a variety of structural alterations that we call **chromosome aberrations**. Such aberrations often involve more than one or two genes and may affect considerable portions of chromosomes.

Types of Structural Aberrations

There are four classes of chromosome aberrations (see ▶ Figure 12–15):

1. **Deletions** (or deficiencies) involve the loss of a segment from a chromosome. In eukaryotes, deletions are further classified as terminal deletions if the end of the chromosome is lost or interstitial deletions if an internal segment is missing.
2. **Duplications** involve the repetition of a segment, either in the same chromosome or on a different chromosome. Duplicated segments can be adjacent to each other (tandem duplication) or located at entirely different regions of the genome (transposition).
3. **Inversions** involve the reversal of a segment within a chromosome. In eukaryotes, further classification is based on whether the inverted segment includes the centromere (pericentric inversion) or is confined to an arm of the chromosome (paracentric inversion).
4. **Translocations** involve the transfer of a chromosome segment to an entirely different position within the genome, without a significant gain or loss of genetic material. The more common types of translocations in eukaryotes occur through the interchange of parts between nonhomologous chromosomes (reciprocal translocation) or by

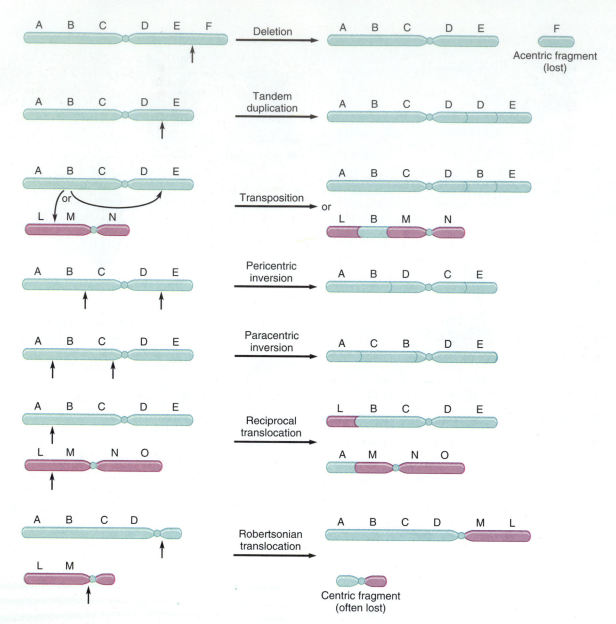

▶ FIGURE 12–15 The major classes of chromosome aberrations and possible origin of each. The arrows indicate points of breakage and alteration.

the fusion of the long arms of two nonhomologous chromosomes to a common centromere (**centric** or **Robertsonian fusion**).

These various structural aberrations have been studied in greatest detail in the large chromosomes of certain larval tissues, such as the salivary glands of the dipteran insects (flies, mosquitoes, etc.). These so-called polytene (many-stranded) chromosomes have enlarged by doubling their genetic material several times in succession without undergoing nuclear division. The many duplicate strands of each chromosome coil differentially along their lengths, resulting in a unique banding pattern that becomes evident upon

staining (▶ Figure 12–16). Moreover, the homologs of each chromosome remain tightly paired in a continuous state of somatic synapsis. These three properties—large size, differential staining pattern, and somatic pairing—enable geneticists to recognize each type of aberration and identify its precise location on the chromosome pair.

One aberration that is detectable in polytene chromosomes is the *Bar* duplication in *Drosophila*. This duplication is a tandem repeat of the X chromosome segment designated 16A in the banding pattern of the salivary gland chromosomes (▶ Figure 12–17). The *Bar* duplication has a distinct phenotypic effect—it narrows the compound eye by reducing the number of eye facets from an average of

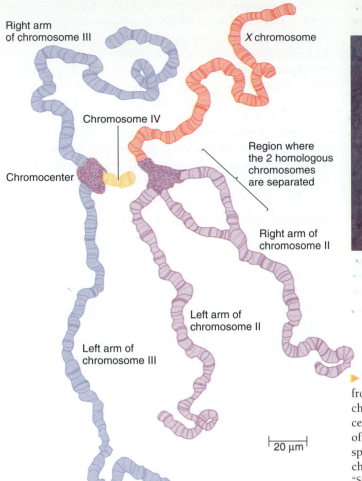

Right arm
of chromosome III

X chromosome

Chromosome IV

Chromocenter

Region where
the 2 homologous
chromosomes
are separated

Right arm of
chromosome II

Left arm of
chromosome II

Left arm of
chromosome III

20 µm

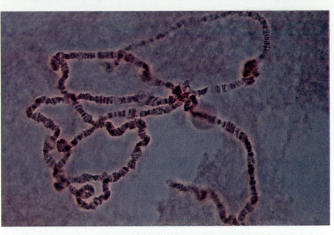

▶ FIGURE 12–16 The polytene salivary gland chromosomes from *Drosophila melanogaster,* showing banding patterns. Each chromosome is tightly paired with its homolog and the centromeres of all four chromosome pairs (X, II, III, and IV) are often fused at the chromocenter. (The chromocenter has been split by the squashing technique used to spread out the chromosomes on the microscope slide.) *Adapted from:* T. S. Painter, "Salivary chromosomes and the attack on the gene," *J. Hered.* 25 (1934): 466. Used with permission of Oxford University Press.

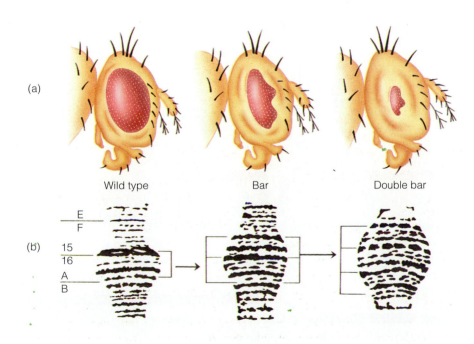

(a)

Wild type

Bar

Double bar

(b)

E
F

15
16

A
B

▶ FIGURE 12–17 The *Bar* mutation in *Drosophila.* (a) Comparison of the eye phenotypes of flies with 1 (wild type), 2 (Bar), and 3 (double-bar) doses of region 16A of the X chromosome. (b) Comparison of altered regions in the salivary gland chromosomes of Bar and double-bar mutants with that of wild type flies.

770 per eye in normal flies to about 358 in duplication heterozygotes and still further, to about 68, in duplication homozygotes.*

Once a chromosome contains a tandem duplication or two tandem regions with similar genetic sequences, additional duplications can arise when asymmetrically paired homologs cross over during prophase I (▶ Figure 12–18). Such unequal crossing-over leaves one homolog with a tandem repeat of a genetic segment and the other homolog with a deletion of the same region. For example, unequal crossing-over occasionally occurs in homozygotes for the *Bar* duplication in *Drosophila*, producing so-called double-bar offspring, in which region 16A is present in triplicate in the X chromosome (see Figure 12–17).

Many chromosome aberrations are associated with the loss of viability and/or sterility. Extensive deletions, for example, are usually lethal in the homozygous state because of the loss of essential gene functions. Translocations often result in reduced fertility, especially in plants, where chromosomes with reciprocal translocations typically give rise to nonviable gametes. In corn, for example, about half the meiotic products of translocation heterozygotes are nonfunctional. Consequently, half the pollen grains produced by these translocation carriers are aberrant (because they are abnormally small and deficient in starch content), and the ears produced have only about half the normal number of kernels.

The chromosomal basis for the reduced fertility of translocation heterozygotes is shown in ▶ Figure 12–19. Synapsis in translocation heterozygotes (Figure 12–19a) involves the formation of a cross-shaped configuration made up of two aberrant and two normal chromosomes. The chromosomes disjoin two by two at anaphase I in one of three different segregation patterns (Figure 12–19b). Two of these patterns yield meiotic products with extensive deletions and duplications; gametes with a complete haploid complement of genes can be produced only when the translocation figure forms a zigzag or figure-eight configuration during disjunction.

Structural Aberrations and Evolution

Despite the detrimental effects often associated with chromosome aberrations, changes in chromosome structure have undoubtedly played an important role in evolution. Duplications and deletions have been important mechanisms for the gain and loss of genetic material. Repeated doublings of DNA segments have been responsible for the development of **gene families**—groups of distinct but related genes that encode similar protein products. The hemoglobin genes provide an excellent example of how repeated gene duplications can give rise to a gene family derived from a single ancestral sequence. ▶ Figure 12–20 shows the phylogenetic scheme for the human hemoglobin

▶ **FIGURE 12–18** Production of duplications and deficiencies through unequal crossing-over. If homologs mistakenly pair in a region of partial homology and a crossover occurs between the loops, one of the products will carry a duplication of the region and the other will carry a deletion. (For simplicity, only one chromatid is shown for each chromosome.)

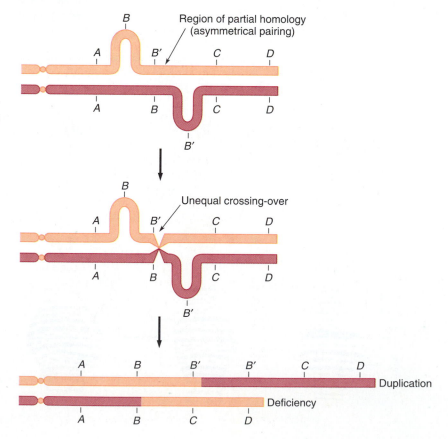

(a) Synapsis

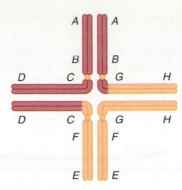

(b) Anaphase I

Adjacent

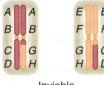

Alternate

Adjacent

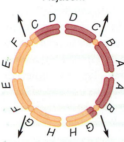

Gametes:

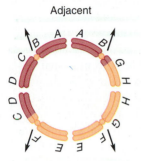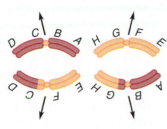

Inviable

Viable

Inviable

▶ FIGURE 12–19 The chromosomal basis for semisterility in translocation heterozygotes. (a) During prophase I of meiosis, the pairing arrangement of synapsed chromosomes forms a cross configuration in which two of the chromosomes are normal and two have undergone a reciprocal translocation. (b) During anaphase I, translocation heterozygotes show three segregation patterns: two known as **adjacent segregation**, in which each normal chromosome segregates with a translocated chromosome, and one known as **alternate segregation**, in which normal chromosomes (and translocated chromosomes) segregate together. Chromosomes with a complete haploid complement of genes can be produced only by the alternate segregation pattern.

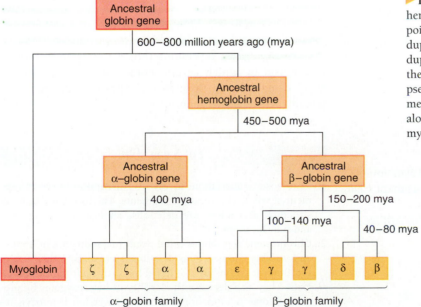

▶ FIGURE 12–20 Evolution of the human hemoglobin genes by duplication. Each branch point represents a presumed duplication step. The duplicated genes either diverged or remained as duplicates. In some cases, such as the ζ gene, one of the products of duplication is a nonfunctional pseudogene. Times of divergence for the various members of the hemoglobin gene family are shown along with the divergence of this family from myoglobin.

genes and the divergence of this gene family from myoglobin. The original α-β gene duplication occurred about 500 million years ago, with other divergences, especially within the β-globin cluster, being more recent. Once gene duplication has occurred, the duplicated copies can diverge in nucleotide sequence to become individually specialized for different functions. For example, in the β-globin cluster, distinct globins with different oxygen-binding affinities appear at different stages of development. The ε-chain combines with the ζ- (zeta) chain to form embryonic hemoglobin ($\zeta_2\epsilon_2$), the γ-chain and the α-chain form fetal hemoglobin ($\alpha_2\gamma_2$), and the β-chain (or in some cases the δ-chain) together with the α-chain make up adult hemoglobin ($\alpha_2\beta_2$).

Inversions and translocations have been important in evolution as mechanisms for the rearrangement of genes to form new linkage relationships. Such rearrangements sometimes lead to changes in phenotype as a result of a **position effect** (a change in the expression of a gene due to a change in chromosome location). For example, in *Drosophila*, female flies that are heterozygous for the recessive allele for white eyes ($X^{w^+}X^w$) normally have red eyes. However, when the wild-type allele (w^+) is moved by an inversion to a point adjacent to the centromere, the eye color of the heterozygotes becomes mottled or variegated, with red and white patches. In this case, the action of the gene is depressed by its movement to a heterochromatic (tightly coiled and genetically inactive) region of the chromosome.

Example 12.4

Four races of a particular species differ in their arrangement of genes *A* through *E* on a particular chromosome:

a. *ABCDE*   **b.** *ACDBE*   **c.** *ABEDC*   **d.** *ACBDE*

Race (a) is the most primitive and is believed to be the common ancestor of the others. Explain the origin of races (b), (c), and (d) from race (a) in terms of single inversion events.

Solution: A likely possibility is:

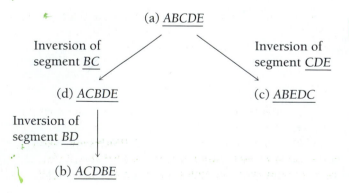

Follow-Up Problem 12.4

Diagram a sequence of single inversion events that can explain the evolution of a chromosome with the gene sequence *DGBAEFCH* from a starting ancestral sequence *ABCDEFGH*.

◆

The ability of inversions to act as **crossover suppressors** also has important evolutionary ramifications. The term crossover suppressor was originally proposed to describe the apparent inhibitory effect of inversions on the recombination of nonallelic genes. Geneticists at the time thought that the anomaly interfered with crossing-over within the inverted part of the chromosome. We now know that inversions do not necessarily inhibit crossing-over; instead, inversion heterozygotes form nonfunctional meiotic products when such crossovers do occur. The chromosomal basis of the effect of inversions on crossing-over is shown in ▶ Figure 12–21. Synapsis in inversion heterozygotes (Figure 12–21a) occurs through a double-loop pairing arrangement involving one aberrant and one normal chromosome. Problems develop when crossing-over occurs within the inversion loop. When a crossover occurs in the loop of a paracentric inversion (Figure 12–21b), it results in a fragment that lacks a centromere (an **acentric** fragment) and a chromosome with two centromeres (a **dicentric** chromosome). The acentric fragment fails to migrate on the meiotic spindle, and during anaphase the dicentric chromosome forms a bridge that subsequently breaks or interferes with the completion of meiosis. By contrast, a crossover in the loop of a pericentric inversion (Figure 12–21c) produces two chromosomes that contain large deletions and duplications. Both types of inversion have the effect of suppressing crossing-over. Gametes that have received a crossover chromosome are never recovered because they are nonfunctional or fail to produce viable zygotes. The genes in the inverted segment of a chromosome are thereby assured of joint inheritance and will always be transmitted together intact as a single unit or **supergene.** Inversions thus serve as an important mechanism by which certain combinations of genes can avoid being split up through crossing-over into new and possibly less adaptive genotypes.

To Sum Up

1. Chromosome aberrations involve the disruption and rearrangement of whole chromosome segments. The four basic types of structural aberrations are deletions, duplications, inversions, and translocations.
2. Chromosome aberrations often result in a reduction in fertility and viability. Homozygous deletions are usually lethal. Reciprocal translocations are not lethal but can result in position effects. They also decrease the fertility of heterozygotes by producing some nonfunctional gametes. Inversion heterozygotes

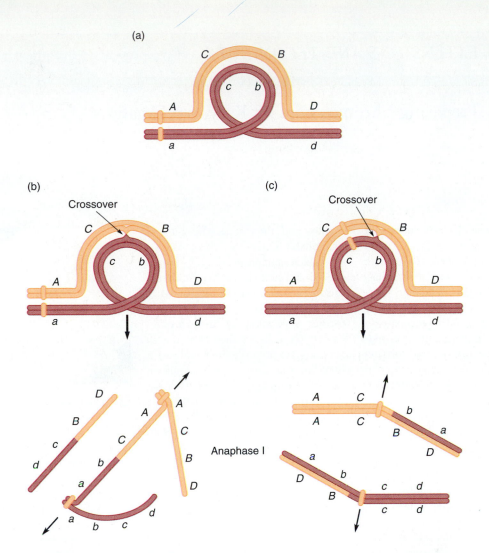

(a)

C   B
c   b
A   D
a   d

(b)

Crossover

C   B
c   b
A   D
a   d

(c)

Crossover

C   B
c   b
A   D
a   d

Anaphase I

▶ **FIGURE 12–21** (a) The double-loop pairing arrangement of synapsed homologs in a chromosome pair that is heterozygous for an inversion. The lower diagrams show the results of crossing-over within an inversion loop in heterozygotes for a (b) paracentric and (c) pericentric inversion. Crossing-over in a paracentric inversion loop gives rise to a fragment that lacks a centromere and to a chromosome with two centromeres that forms a bridge during meiosis. Crossing-over in a pericentric inversion loop yields two chromosomes with large deletions and duplications.

also form some nonfunctional meiotic products when crossovers occur within the inverted segment of an aberrant chromosome; this effect has led geneticists to describe inversions as crossover suppressors, since the results of such crossovers are never detected.

3. Despite the deleterious effects of many aberrations on viability and fertility, changes in chromosome structure have been important in evolution because they have provided a mechanism for the gain, loss, or rearrangement of genetic material.

✦ ANALYSIS OF MUTANT KARYOTYPES

Once a mutant strain that expresses a phenotype of interest has been isolated, it is often desirable to determine the exact nature and chromosomal location of the mutation. Gene mutations can usually be analyzed by genetic crosses and molecular techniques. However, if a eukaryotic cell has undergone a chromosome mutation, the nature and location of the mutation can sometimes be discovered by studying the **karyotype, or chromosome complement, of the mutant cell**. This section will discuss some of the methods used for the analysis and description of mutant karyotypes in human cells.

Identifying Abnormal Human Karyotypes

New staining methods developed in the past 30 years have revolutionized the study of human chromosomes. These methods, collectively referred to as **chromosome banding**, allow chromosome regions to be stained differentially so that each individual chromosome in the normal genome can

EXTENSIONS AND TECHNIQUES

Preparing Chromosomes for Karyotype Analysis

The karyotype of a cell can be analyzed from photographs of metaphase chromosomes in specially prepared smears called **metaphase spreads**. Ordinarily, lymphocytes or fibroblasts are used for karyotyping, but in theory any cell type capable of undergoing cell division in a nutrient culture medium can be used for such a study. A typical procedure used in the analysis of human chromosomes is shown here.

A small amount of blood is first added to a nutrient medium containing a plant extract called **phytohemagglutinin**, which acts as a mitogen (stimulates mitotic activity) for the lymphocytes. After sufficient time for cell division has elapsed, **colcemid** (a less toxic chemical derivative of colchicine) is added. Colcemid, like colchicine, interferes with spindle for-

mation and thus stops the cells in metaphase. Once the chromosomes are held at metaphase, the cells are removed and placed in a **hypotonic solution**, which causes water to enter into and swell the cells, resulting in further separation of the chromosomes. The cells are then treated with a fixative (to preserve them and retain their internal structures), spread on microscope slides, and stained so that the chromosome sets (metaphase spreads) can be located under a light microscope. Photographs are then taken, and individual chromosomes are cut out of each photograph and arranged in pairs in order of length (longest to shortest) on a standard karyotype form.

Automated analysis of chromosomes, in which computerized procedures are used for chromosome

Cytogeneticist with a human karyotype.

identification, is a more recent development in karyotyping. With an automated analysis, chromosomes can be stained in culture (rather than on a slide) and then automatically sorted by a computer-controlled detection system.

be identified. In some cases, the specific chromosomes or parts of chromosomes involved in an aberration can also be determined.

▶ Figure 12–22 summarizes some of the staining procedures used in chromosome banding. One of the more frequently used procedures is a differential staining method called **G banding** in which the chromosomes are treated with Giemsa stain (a mixture of basic dyes) after they have been heated in a salt solution and exposed to an enzyme (trypsin) that digests part of the chromosomal protein. This procedure causes the stain to bind preferentially to certain regions of the chromosome, producing a unique sequence of bands—the G (for Giemsa) bands—that can be studied under an ordinary light microscope (see ▶ Figure 12–23). With this staining method, investigators can unambigu-

ously match each chromosome with its homolog and distinguish it from other chromosomes with similar morphology.

In the conventional G-banding technique, the chromosomes are stained and viewed while in metaphase, when they are maximally condensed. Modifications of this procedure have now made it possible to observe even finer chromosomal detail by studying the chromosomes at earlier stages, such as early metaphase or late prophase. This high-resolution G-banding method permits about 2,000 specific bands of the human karyotype to be identified.

Cytogeneticists have adopted a convenient method for identifying the bands produced by the G-staining technique (see ▶ Figure 12–24). The short arm of each chromosome is designated by the letter p (for petite) and the long arm by the letter q. Each arm is divided into regions that are numbered

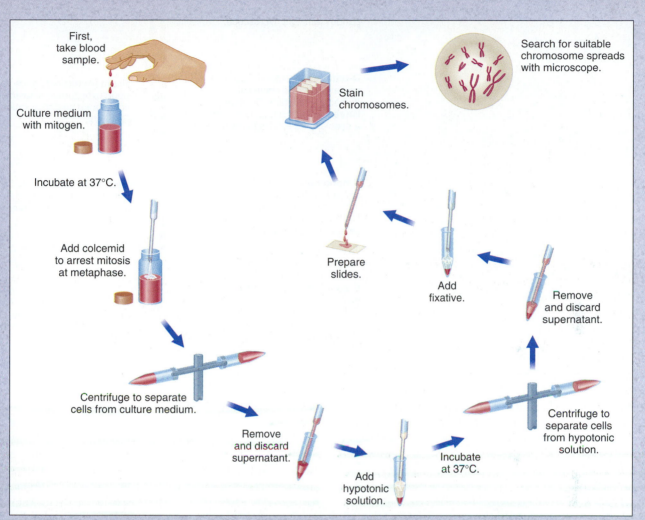

First,
take blood
sample.

Culture medium
with mitogen.

Incubate at 37°C.

Add colcemid
to arrest mitosis
at metaphase.

Centrifuge to separate
cells from culture medium.

Remove
and discard
supernatant.

Add
hypotonic
solution.

Incubate
at 37°C.

Centrifuge to
separate cells
from hypotonic
solution.

Remove
and discard
supernatant.

Add
fixative.

Prepare
slides.

Stain
chromosomes.

Search for suitable
chromosome spreads
with microscope.

A typical procedure used in the analysis of human chromosomes.

consecutively from the centromere. The boundaries of the regions are based on constant features of the chromosomes or major bands that are easily identified. It is thus possible to specify a particular band by a four-part code that gives: the chromosome number, the segment (p or q), the region number, and the band number within the region. For example, band 8q23 in a G-banded spread of human chromosomes refers to band 3 in region 2 of the long arm of chromosome 8.

The accumulation of information on chromosome mutations has made it necessary to devise a system for identifying human karyotypes. Under this system, karyotypes are described by indicating (1) the number of chromosomes, (2) the status of the sex chromosomes, (3) the presence or absence of a particular chromosome (indicated by a + or − placed before the chromosome symbol), and (4) the nature of any structural aberration. The symbols used for structural alterations include **del** for deletion, **dup** for duplication, **inv** for inversion, and **t** for translocation. For example, the karyotype of a human female with a deletion of the short arm of chromosome 5 would be 46,XX,del(5p).

With the advent of special staining techniques such as G banding, cytogeneticists can identify not only the particular chromosomes involved in structural rearrangements but also the specific band numbers of the precise break points within the chromosomes. When break points are known, they are given (with the segment—p or q) in parentheses following the chromosome designation. For example, if a male has a deletion in the long arm of chromosome 1 between G band 21 and G band 31, the karyotype would be 46,XY,del(1)(q21q31).

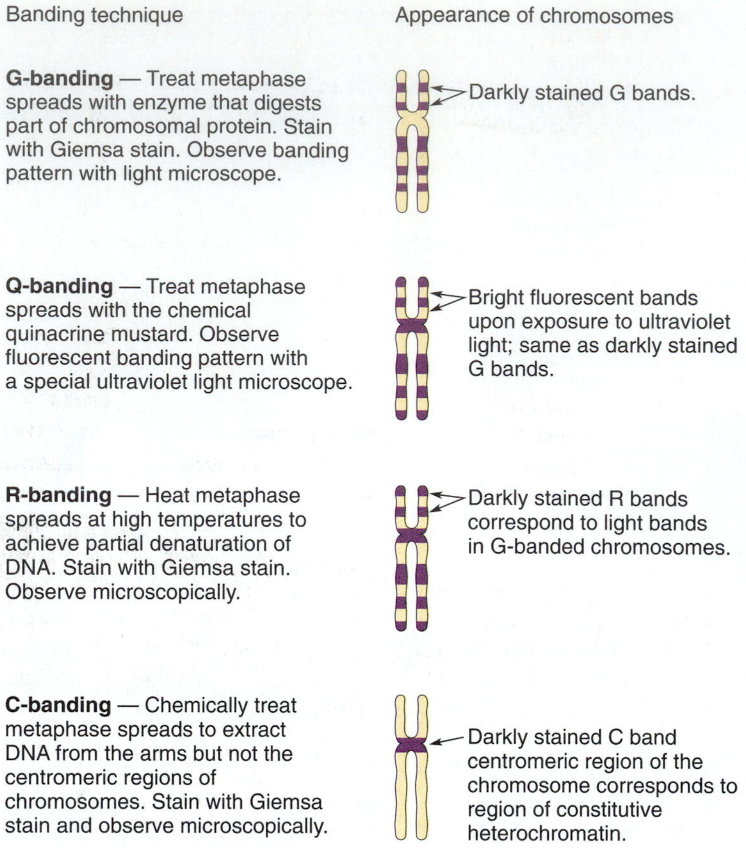

► **FIGURE 12–22** Highlights of four chromosome banding techniques commonly used in genetics.

Banding technique	Appearance of chromosomes
G-banding — Treat metaphase spreads with enzyme that digests part of chromosomal protein. Stain with Giemsa stain. Observe banding pattern with light microscope.	Darkly stained G bands.
Q-banding — Treat metaphase spreads with the chemical quinacrine mustard. Observe fluorescent banding pattern with a special ultraviolet light microscope.	Bright fluorescent bands upon exposure to ultraviolet light; same as darkly stained G bands.
R-banding — Heat metaphase spreads at high temperatures to achieve partial denaturation of DNA. Stain with Giemsa stain. Observe microscopically.	Darkly stained R bands correspond to light bands in G-banded chromosomes.
C-banding — Chemically treat metaphase spreads to extract DNA from the arms but not the centromeric regions of chromosomes. Stain with Giemsa stain and observe microscopically.	Darkly stained C band centromeric region of the chromosome corresponds to region of constitutive heterochromatin.

■ Table 12–2 lists some of the better-known chromosome aberrations in humans. One of the most extensively studied of these aberrations is the one associated with **translocation Down syndrome**, a chromosome abnormality that is responsible for 4% of the reported cases of Down syndrome. It differs from the common form of trisomy 21 in that individuals with translocation Down syndrome have 46 chromosomes rather than 47. However, they have one rather long chromosome that consists of a copy of chromosome 14 (or another medium-sized acrocentric chromosome) to which a copy of chromosome 21 has been translocated. (Thus, despite the presence of only 46 chromosomes, there are still three copies of chromosome 21.) In almost every instance of translocation Down syndrome, one of the parents is a translocation carrier who has only 45 chromosomes, including one copy of chromosome 21, one copy of chromosome 14, and a combined 14-21 chromosome produced by centric fusion. Down translocation carriers have a normal phenotype but are capable of producing six different gametes, as shown in ► Figure 12–25. In addition to having a high risk of producing offspring with Down syndrome, carriers can also produce another carrier with a normal phenotype. Thus, while the common form of trisomy 21 is usually not passed on to offspring, the translocation abnormality can be passed on by seemingly normal parents. Since translocation Down syndrome does not involve nondisjunction, its frequency does not increase with the age of the mother.

► **FIGURE 12–23** A normal human karyotype showing G-banded chromosomes. Each chromosome has a distinctive banding pattern.

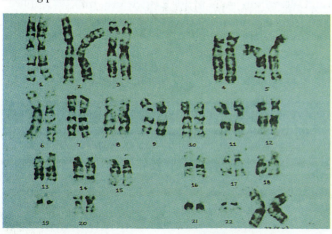

To Sum Up

1. Different staining methods produce identifiable bands in metaphase chromosomes. Modified Giesma staining (which produces G bands) permits each chromosome to be unequivocally identified.

Human chromosome 8

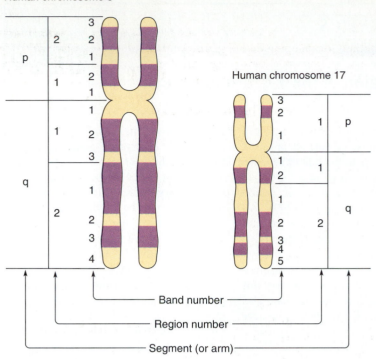

Human chromosome 17

▶ FIGURE 12–24 The method used to identify bands produced by the G (and Q) banding technique. Each band is identified by four items: the chromosome number, the segment (p or q), the region number, and the band number within the region.

Band number
Region number
Segment (or arm)

2. Human karyotype analysis of differentially stained chromosomes allows the identification of the particular chromosomes involved in a number of structural rearrangements and the affected positions along those chromosomes. One of the most extensively studied structural aberrations is translocation Down syndrome, which usually involves the fusion of the long arms of chromosomes 14 and 21. Unlike the usual trisomy 21, translocation Down syndrome is transmitted by a phenotypically normal carrier of the translocation chromosome. Since nondisjunction is not a factor, translocation Down syndrome does not increase in frequency with maternal age.

■ TABLE 12–2 Some well-defined chromosome aberrations in humans.

Chromosome Abnormality	Associated Disorder	Major Symptoms
46,del(5p)	Cri-du-chat (cat's cry) syndrome	mental retardation; microcephaly (small head), a broad "moon" face, cry in infancy that sounds like a cat's cry; severely reduced survival
46,del(4p)	Wolf-Hirschhorn syndrome	mental retardation; midline facial defects consisting of broad nose, wide-set eyes, small lower jaw, and cleft palate; heart, lung, and skeletal abnormalities common; severely reduced survival
46,del(15)(q11q13)	Prader-Willi syndrome	mental retardation; small stature, obesity, and poor muscle tone; underdeveloped gonads and genitals; behavior problems with wide mood swings
46,del(11)(p13)	WAGR syndrome	tumors of the kidney (Wilms tumor) and of the gonad (gonadoblastoma); aniridia (absence of the iris); ambiguous genitalia; mental retardation
46,−14,+t(14;21)	Translocation Down syndrome	same as trisomy Down syndrome
46,t(9;22)(q34q11)	CML (chronic myelogenous leukemia)	enlargement of liver and spleen; anemia; excessive, unrestrained growth of white cells (granulocytes) in the bone marrow having granules in the cell cytoplasm
46,t(8;14)	Burkitt's lymphoma	malignancy of B lymphocytes that mature into the antibody-producing plasma cells; solid tumors, typically in the bones of the jaw and organs of the abdomen

The Sex-Determining Region of the Y Chromosome

Ever since 1959, when the Y chromosome was shown to be male-determining, researchers have searched for the gene on the Y chromosome that codes for the "testis determining factor." (This factor acts on the primitive gonads of the early embryo, causing that tissue to differentiate into testis and setting in motion the sequence of events that leads to normal male development.) Since the mid-1980s this search has focused on individuals who carry X or Y chromosomes with some kind of translocation. For example, there are cases in which an individual is XX but is male because one X chromosome carries testis-determining DNA sequences inherited from the father's Y chromosome. Researchers have progressively narrowed down the precise region of the Y that must be carried on an X in order for the individual to be male, from the entire short arm (1966) to a specific 35 kb sequence in region 1A2 of the short arm (1990). Although it has not yet been conclusively proven that this sequence is the male-determining region, the evidence is strong. In addition to the finding that XX individuals who carry this region are male, researchers have also identified several XY females who have mutations within the 35 kb sequence.

The gene contained in this 35 kb sequence has been named the *SRY* gene (sex-determining region of the Y). Since the same gene is found in mice, experiments have been done to detect the time of expression of the gene. As would be expected, it is active in the embryonic gonad at the time of its differentiation.

Although the *SRY* gene appears to be necessary for sex determination, it is not sufficient by itself. It does appear to be the long-sought "master switch" that initiates the cascade of events that lead to maleness. However, there are other non-Y-linked genes that act at various steps along that cascade and play important roles in sex determination. Such genes are known to exist both on the X chromosome and on autosomes; identification of all of these genes and analysis of their functions will be necessary before the complex process of sex determination is fully understood.

► FIGURE 12–25

Chromosome segregation in a 14-21 translocation carrier. (a) Six types of gametes can be produced by the carrier. (b) Types of offspring produced when a carrier mates with an individual with a normal karyotype.

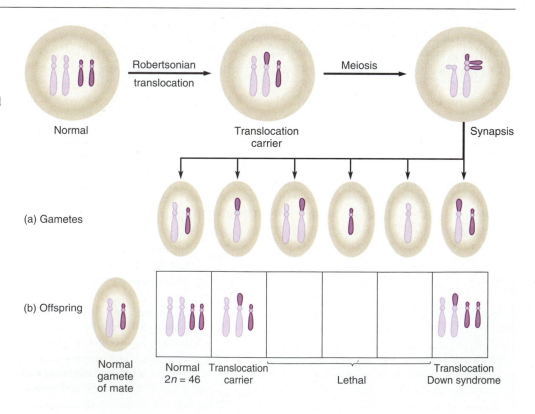

Chapter Summary

Chromosome mutations are departures from the wild-type condition in the number and/or structure of chromosomes. These deviations from the normal chromosome complement can arise spontaneously or can be induced by environmental mutagens.

Mutations that affect chromosome number can involve the gain or loss of partial sets of chromosomes (aneuploidy) or the gain of one or more complete sets of chromosomes (polyploidy). Aneuploidy usually affects only a single chromosome, producing a monosomic $(2n - 1)$ or trisomic $(2n + 1)$ condition as a result of chromosome nondisjunction or anaphase lag. Several important medical syndromes, including Down, Klinefelter, and Turner syndromes, are caused by aneuploidy.

Polyploidy in an organism can result from the production of diploid gametes or from a doubling of chromosome number in somatic cells. The multiple chromosome sets in polyploids may have identical or diverse genetic origins; autopolyploids have chromosome sets derived from the same ancestral species, while allopolyploids contain chromosome sets derived from different species. Allopolyploids can be made in the laboratory by using colchicine to induce chromosome doubling in sterile interspecies hybrids or through protoplast fusion. Polyploidy is widespread among plants, where it has served as an important mechanism in evolution.

Mutations that affect chromosome structure result from the breakage and rearrangement of chromosomes. The most common types of structural aberrations are deletions (loss of a chromosome segment), duplications (repeat of a chromosome segment), inversions (reversal of a chromosome segment), and translocations (transfer of a segment to another chromosomal location). These aberrations give rise to altered karyotypes that can be studied by special chromosome staining techniques.

Questions and Problems

Aneuploidy

1. Define the following terms and differentiate between the paired terms:

    aneuploidy and polyploidy
    monosomic and trisomic
    mitotic nondisjunction and anaphase lag

2. The haploid number of chromosomes in a certain organism is 12. How many chromosomes would be present in a monosomic of this species?

3. How many different kinds of trisomics are possible in an organism with 10 pairs of chromosomes in its somatic cells?

4. In corn, red aleurone is determined by the completely dominant gene *R*, and colorless aleurone is determined by the recessive allele *r*. (a) The pollen from a trisomic plant of genotype *Rrr* is used to fertilize a diploid plant of genotype *rr*. What phenotypic ratio is expected in the progeny, assuming that *n* + 1 pollen are nonfunctional? (b) How would you answer part (a) if the genotypes of the egg and pollen donor were reversed?

5. A certain woman has unusually prominent satellites on one of her copies of chromosome 21. Her mother also carries this unusual chromosome. (a) If the woman has a baby, what is the chance that her child will inherit this chromosome? (b) The woman has a child with Down syndrome. The child has one normal copy of chromosome 21 and two copies of the unusual chromosome 21. At what stage of meiosis did nondisjunction occur?

6. Down syndrome occurs in approximately 1 out of 700 live births in the general population. (a) The incidence of Down syndrome is about 22 times greater among children of 45-year-old women. What is the chance that a child with Down syndrome will be born to a woman in this age group? (b) A 45-year-old woman is about 60 times more likely to give birth to a child with Down syndrome than is a woman of 20. What is the incidence of Down syndrome among the offspring of 20-year-olds? (c) If women over 35 years of age give birth to 60% of the children with Down syndrome but produce only 15% of all children, what is the frequency of children with Down syndrome among the children born to women in this age group?

7. Two white blood cells from a normal human have abnormal chromosome numbers. One has 52 chromosomes, while the other has 69 chromosomes. One is aneuploid and the other is polyploid. Which is which?

8. Three different somatic cells of an animal were examined and were found to contain 59, 62, and 93 chromosomes. One of the cells is normal, one is polyploid, and the third is aneuploid (not necessarily in the order given). What is the genomic number of chromosomes in this species?

9. A pair of twins born to normal parents appear to be identical in every way (same blood types, can exchange skin grafts, and so on) except that one twin is a normal male and the other is a female with Turner syndrome. What is the likely explanation for this?

10. Defective tooth enamel is a dominant X-linked human trait. A person with Klinefelter syndrome whose parents have defective teeth has normal tooth enamel. In what parent and at what stage of meiosis did nondisjunction most probably occur?

Polyploidy

11. Define the following terms and differentiate between the paired terms:

    allopolyploid and autopolyploid
    homologous and homeologous
    amphidiploid and somatic doubling

12. An amphidiploid has 46 chromosomes, comprising chromosome sets derived from species 1 and species 2. If the haploid number of chromosomes in species 1 is 7, how many chromosomes make up the haploid set of species 2?

13. Emmer wheat has 28 chromosomes, comprising 14 chromosomes of genome A and 14 chromosomes of genome B. How many different nullisomics are possible in this species?

14. Consider the testcross *AAaa × aaaa*, in which the alleles show complete dominance and a perfect 2:2 segregation pattern. What phenotypic ratio is expected among the offspring?

15. Suppose that an autotetraploid that is heterozygous for two gene pairs (*AAaaBBbb*) is selfed. Assuming complete dominance and a 2:2 segregation pattern for each gene pair, what fraction of the offspring is expected to have the doubly recessive (*aaaabbbb*) genotype?

16. A 2:2 segregation pattern of alleles is possible in tetraploids only when the gene is closely linked to the centromere, so that crossing-over between the gene and the centromere is negligible. At the other extreme, when genes are separated from the centromere by a very large distance, crossing-over occurs so frequently that all eight chromatids of the four homologs assort independently in the region of the gene locus. Any two of the eight allelic genes can thus enter a gamete at random, resulting in

$$\frac{8!}{2!6!} = 28 \text{ possible allelic combinations}$$

(a) Suppose that a dominant gene *A* and its recessive allele *a* are located far from the centromere on their chromosome. What ratio of *AA:Aa:aa* will a tetraploid of genotype *AAaa* produce in the gametes? (*Hint:* treat each of the eight allelic genes as a distinguishable unit.) (b) If the *AAaa* tetraploid is selfed, what ratio of dominant:recessive phenotypes is expected among the progeny?

Experimental Manipulation of Chromosome Number

17. Suppose that you have an inbred (homozygous) plant that has many commercially desirable characteristics, but is sensitive to a particular virus. Describe a procedure that you might use to synthesize a new variety that retains the desirable characteristics of the original plant and is also resistant to the virus.

18. When pollen from an F_1 hybrid between plant species X (*n* = 9) and Y (*n* = 7) is used to fertilize species X, a few plants are produced, all with 25 chromosomes. Although highly sterile, continued selfing of these 25-chromosome plants eventually produces a few fertile offspring with 32 chromosomes. Outline the events that were probably involved in the origin of these 32-chromosome plants.

Variations in Chromosome Structure

19. Define the following terms and differentiate between the paired terms:

    interstitial and terminal deletion
    tandem duplication and transposition
    paracentric and pericentric inversion
    centric fusion and reciprocal translocation

20. The following diagram represents part of a pair of polytene chromosomes in the salivary glands of *Drosophila melanogaster*. The bands on the homologs have been arbitrarily designated by the letters a through f.

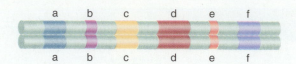

Draw a diagram showing the appearance of the chromosome pair when one member has (a) a deletion of band c, (b) a tandem duplication of band d, and (c) an inversion of the segment that includes bands c and d.

21. Five races of the fruit fly can be distinguished by the sequence of bands A through F on one of their polytene chromosomes. It is believed that these races were separated from the most primitive banding sequence, ABCDEF, by single inversion differences. Propose an evolutionary pathway that relates the banding patterns of the following five races: (a) ADCEBF, (b) ABCDEF, (c) BCDAEF, (d) ADCFBE, and (e) ADCBEF.

22. Certain closely related species have chromosome numbers that are not simple multiples of some basic integer but that vary almost continuously over a wide range of values. What mechanisms of chromosome mutation might be responsible for such variation?

Analysis of Mutant Karyotypes

23. Account for the following changes in karyotype: (a) seven pairs of small acrocentric chromosomes → five pairs of small acrocentric chromosomes plus one pair of large metacentric chromosomes. (b) seven pairs of small acrocentric chromosomes → six pairs of small acrocentric chromosomes plus one pair of small metacentric chromosomes.

24. Suppose that using X rays to irradiate a plant that is homozygous for the dominant *A* gene induces a deficiency of the chromosomal segment that includes the *A* gene on one of the homologs. Assuming that you have access to *aa* plants, outline a procedure that you could use to determine cytologically the location of the *A* gene locus on the chromosome.

25. Chimpanzees (*Pan troglodytes*), gorillas (*Gorilla gorilla*), and orangutans (*Pongo pygmalus*) have 48 chromosomes, compared to the 46 of humans. These apes possess two pairs of acrocentric chromosomes that are not present in humans but show similarities in G-banding patterns to the long and short arms of the large submetacentric human chromosome 2. Suggest a mechanism that could account for the difference in chromosome number between humans and apes.

26. A child with Down syndrome is found to have 46 chromosomes. The child has a normal 46,XY brother and an apparently normal sister with 45 chromosomes. The father of these children has a normal complement of chromosomes. How many chromosomes does the children's mother possess? Explain your answer.

27. Several human females have been reported who are 46,XY in chromosome makeup but are missing part of the short arm of the Y chromosome. These females have poorly developed (streak) gonads and are sterile. Individuals who are 46,XY and missing part of the long arm of the Y are also known, but these individuals appear to be normal males. What does this finding suggest concerning the location of the gene (or genes) for the factor that determines maleness?

LINKED GENES AND CHROMOSOME MAPPING

Now that we have studied the molecular organization of chromosomes and the nature of mutation, we can consider how to utilize mutations and other features of DNA to determine the chromosomal locations of genes. The association between chromosomes and the genes that they carry is highly specific—the genes are carried together in a specific sequence with each gene occupying its own particular position (locus) on the chromosome. Thus together the genes on a chromosome form a **linkage group** in which the relative order of loci corresponds to the arrangement of the genes in the chromosomal DNA.

The fact that different genes are physically linked on the same DNA molecule means that they tend to be transmitted together as a unit from parent to offspring. Linked genes are not joined together permanently, however. They can recombine with a homologous DNA segment, and the probability of recombination is proportional to the distance between the genes involved. The tendency for joint inheritance (the degree of linkage) therefore depends on the distance between the gene loci: the closer two genes are on the chromosome, the less likely they are to be separated by cross-

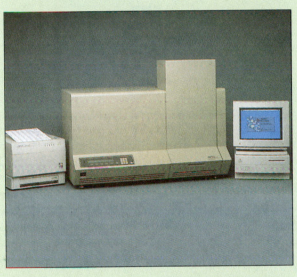

The 373A DNA Sequencer, a fully automated instrument for determining DNA sequences.

ing-over. This dependence on the distance between linked genes serves as the basis for the construction of **linkage maps** of chromosomes, which give the relative locations of genes in the chromosomal DNA.

In this part, we will be concerned with the methods used in mapping chromosomes, beginning with the conventional mapping procedures based on the probability of recombination between genes. These procedures give **genetic linkage maps** that reflect the arrangement of identifiable **genetic markers** along the chromosome. Traditionally these markers have been mutations that identify loci on the basis of mutant phenotypes. More recently, however, these markers have included differences in DNA that can be detected directly by enzyme digestion and other biochemical techniques, even without a change in phenotype.

In addition to the conventional mapping procedures, we will also discuss techniques for making **physical linkage maps** that directly reflect the arrangement of nucleotides in chromosomal DNA. These techniques include DNA sequencing and other procedures that can provide linkage information in the absence of a genetic cross.

CHAPTER 13

Mapping Genes in Eukaryotes

Genetic maps of eukaryotes are constructed from the linkage relationships established by genetic crosses. Crosses are performed and the offspring are examined to determine whether the genes in question are **linked** (carried on the same chromosome) or **unlinked** (carried on different chromosomes) and, if linked, to measure the amount of crossing-over between them. Each map derived in this way gives the relative locations of genes along a chromosome and the relative distances between successive mutant gene loci. Since the actual physical distance between gene loci is seldom known, the relative map distance used to measure length is expressed in proportion to the average number of crossovers occurring per nonsister chromatid pair between the markers involved. By expressing map distance in terms of crossover frequency, it is assumed that the number of potential sites for crossing-over will increase in proportion with the length of the interval between

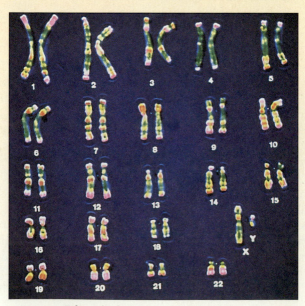

False-color karyotype of a human male.

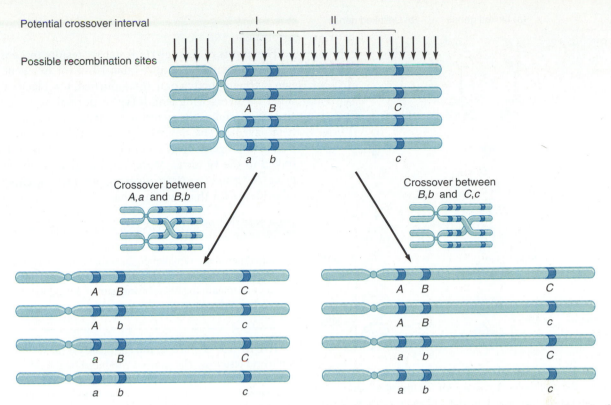

Potential crossover interval

Possible recombination sites

Crossover between
A,a and *B,b*

Crossover between
B,b and *C,c*

▶ **FIGURE 13–1** Genetic map distance. The frequency of crossing-over is assumed to be directly proportional to the length of the interval between gene loci. Thus, there is less crossing-over in a short interval (e.g., between genes *A,a* and *B,b*) than in a longer interval (e.g., between genes *B,b* and *C,c*). More recombination events therefore occur along the right-hand path, and fewer along the left-hand path.

gene loci. (▶ Figure 13–1). The probability of crossing-over should then vary directly with the physical distance separating the genes on the chromosome. Thus, in general, two genes that are far apart on the chromosome should have a proportionally higher average number of crossovers between them than two genes that are closer together.

 ESTABLISHING LINKAGE BETWEEN GENES

An essential first step in constructing a genetic map is to determine whether the genes in question are located on the same chromosome or on different chromosomes. This is usually accomplished by sampling the meiotic products of a heterozygote for two or more pairs of genes and comparing the frequencies of these products with the distribution expected for unlinked (independently assorting) genes (▶ Figure 13–2). Because of crossing-over, four types of gametes will be produced by a double heterozygote, regardless of whether the genes are linked. The difference between linked and unlinked genes lies in the frequencies with which the four gamete types are formed. When nonallelic genes are linked (Figure 13–2a) recombinant-type gametes are less frequent than parental types, but when genes assort independently (Figure 13–2b) the different gamete types occur with equal frequency. For linked genes, the proportions of recombinant-type and parental-type gametes differ because crossing-over does not occur with absolute certainty be-

tween most genes on a chromosome. The probability of crossing-over can actually be quite small when the genes lie close together, and thus the majority of meioses will produce gametes with the parental-type gene arrangement. A smaller percentage of meioses will involve a crossover in the interval between the loci and yield recombinant types. The net result is an excess of parental types in the total pool of gametes formed by a dihybrid parent. This outcome gives us one means of identifying linkage: *Linked genes yield greater than 50% parental-type gametes and less than 50% recombinant-type gametes.* The exact proportion of each type varies, depending on the distance between the genes.

Detecting Linkage from Testcross Results

The most straightforward way to demonstrate linkage in diploid eukaryotes is to perform a **two-point testcross**, in which a dihybrid is mated with a recessive homozygote. ▶ Figure 13–3 gives the general results expected for two testcrosses involving the same two pairs of linked genes and shows how the results depend on the linkage phase of the dihybrid. If the dominant alleles of both genes (*A* and *B*) are present on the same homolog of the dihybrid, linkage is said to be in **coupling**. We designate the coupling phase as

$$\frac{A \quad B}{a \quad b} \text{ (or } AB/ab) = \text{coupling phase}$$

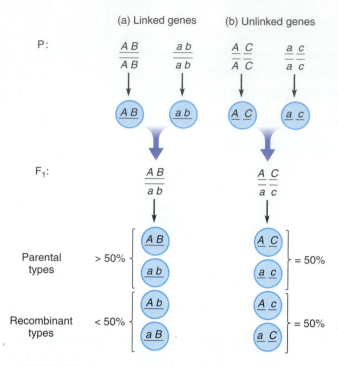

(a) Linked genes   (b) Unlinked genes

P:

$$\frac{A\ B}{A\ B} \qquad \frac{a\ b}{a\ b} \qquad \frac{A\ C}{A\ C} \qquad \frac{a\ c}{a\ c}$$

$\underline{A\ B}$ $\underline{a\ b}$ $\underline{A\ C}$ $\underline{a\ c}$

F_1:

$$\frac{A\ B}{a\ b} \qquad\qquad \frac{A\ C}{a\ c}$$

Parental types   > 50% $\underline{A\ B}$ $\underline{A\ C}$ = 50%
   $\underline{a\ b}$ $\underline{a\ c}$

Recombinant types   < 50% $\underline{A\ b}$ $\underline{A\ c}$ = 50%
   $\underline{a\ B}$ $\underline{a\ C}$

▶ **FIGURE 13–2** Comparison of gametes produced by an F_1 individual doubly heterozygous for (a) linked versus (b) unlinked genes. The same four types of gametes are produced in both cases. However, their frequencies differ depending on whether the genes are linked or unlinked. If they are linked (left), the original (P generation) arrangement of nonallelic genes tends to predominate in the gametes produced by the F_1, giving an excess of parental-type gametes (some frequency greater than 50%). The remaining gametes, which are nonparental or recombinant in their arrangement of nonallelic genes, are produced by crossing-over. In contrast, if the nonallelic genes are not linked (right), the gamete classes are equally frequent, with 50% total parental types and 50% total recombinant types; the recombinant gametes in this case are produced by independent assortment.

If the dominant alleles are on different homologs, linkage is said to be in **repulsion**.

$$\frac{A\quad b}{a\quad B} \quad (\text{or } Ab/aB) = \text{repulsion phase}$$

The dihybrid can be assigned a coupling or repulsion phase either from knowledge of its parental genotypes (e.g., $AB/AB \times ab/ab$ parents would produce a dihybrid in coupling, while $Ab/Ab \times aB/aB$ parents would form a dihybrid in repulsion) or from the testcross results. In the case of a testcross, the linkage phase determines which gametes are recombinant types and hence which testcross progeny classes will appear at a lower frequency. A dihybrid in the coupling phase can produce the Ab and aB gamete types only through crossing-over. Therefore, when genes are in coupling, Ab/ab and aB/ab are the recombinant offspring and will occur at a combined frequency of less than 50%. The situation is reversed for a dihybrid in the repulsion phase, where AB/ab and ab/ab are recombinants. In either case, linkage is revealed by the inability of the testcross to pro-

duce a 1:1:1:1 progeny ratio characteristic of independent assortment.

The effects of linkage on testcross results in *Drosophila* are shown by the data in ■ Table 13–1 for two gene pairs, one involving body color (b^+ = normal, b = black) and the other involving wing length (vg^+ = normal, vg = vestigial). In this example, the genes are linked in coupling in the heterozygote. The deviation from a 1:1:1:1 ratio in the testcross progeny clearly demonstrates that the b locus is linked to the vg locus. Moreover, the greater frequency of $b^+\ vg^+/b\ vg$ and $b\ vg/b\ vg$ offspring provides the experimental evidence for the coupling phase.

To Sum Up

1. Linked genes are nonallelic genes that are carried together on the same chromosome. They are characterized by their tendency to remain together during gamete formation, rather than assorting independently.

▶ **FIGURE 13–3** Effect of linkage phase on the results of a dihybrid testcross. The gametes produced by the dihybrid are the same in either case, except that their expected frequencies are reversed; the parental type gametes in the top diagram are the recombinant types in the bottom diagram, and vice versa.

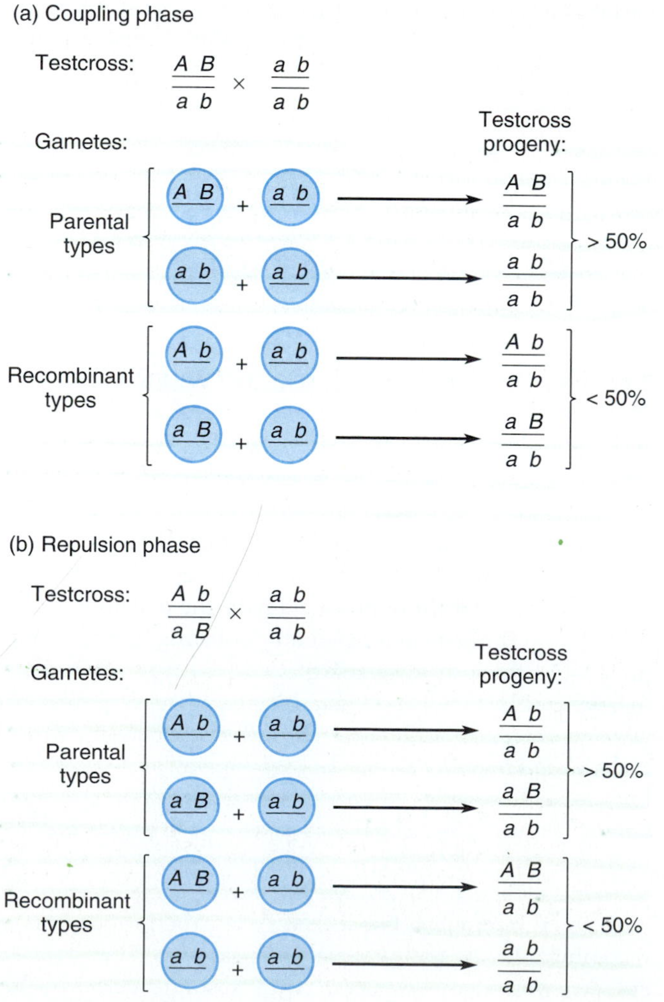

(a) Coupling phase

Testcross: $\dfrac{A\ B}{a\ b} \times \dfrac{a\ b}{a\ b}$

Gametes: Testcross progeny:

Parental types $\underline{A\ B}$ + $\underline{a\ b}$ ⟶ $\dfrac{A\ B}{a\ b}$ > 50%
  $\underline{a\ b}$ + $\underline{a\ b}$ ⟶ $\dfrac{a\ b}{a\ b}$

Recombinant types $\underline{A\ b}$ + $\underline{a\ b}$ ⟶ $\dfrac{A\ b}{a\ b}$ < 50%
  $\underline{a\ B}$ + $\underline{a\ b}$ ⟶ $\dfrac{a\ B}{a\ b}$

(b) Repulsion phase

Testcross: $\dfrac{A\ b}{a\ B} \times \dfrac{a\ b}{a\ b}$

Gametes: Testcross progeny:

Parental types $\underline{A\ b}$ + $\underline{a\ b}$ ⟶ $\dfrac{A\ b}{a\ b}$ > 50%
  $\underline{a\ B}$ + $\underline{a\ b}$ ⟶ $\dfrac{a\ B}{a\ b}$

Recombinant types $\underline{A\ B}$ + $\underline{a\ b}$ ⟶ $\dfrac{A\ B}{a\ b}$ < 50%
  $\underline{a\ b}$ + $\underline{a\ b}$ ⟶ $\dfrac{a\ b}{a\ b}$

■ **TABLE 13−1** Evidence of linkage among the progeny of a testcross $b^+b\ vg^+vg \times bb\ vgvg$ in *Drosophila*.

Phenotype	Gamete	Numbers		$\dfrac{(O-E)^2}{E}$
		Observed	Expected*	
Wild-type	b^+vg^+	975	600	234
Normal, vestigial	b^+vg	217	600	245
Black, normal	$b\ vg^+$	236	600	221
Black, vestigial	$b\ vg$	972	600	231
		2400	2400	$\chi^2 = 931$

*Expected numbers assuming independent assortment.

2. Linked genes form new (recombinant) gene arrangements through crossing-over. Even with crossing-over, however, linked genes yield greater than 50% parental-type gametes and less than 50% recombinant-type gametes. The exact frequency of parental- and recombinant-type gametes depends on the distance between the genes; genes very close together form very few recombinant-type gametes through crossing-over, while genes farther apart form a larger proportion of recombinant types.

3. The most straightforward way to detect linkage experimentally is to examine the progeny of a two-point testcross. The dihybrid parent in the testcross can have its genes arranged either in coupling or in repulsion. If the genes are linked, the offspring will be distributed in two high-frequency classes and two low-frequency classes, rather than in the 1:1:1:1 ratio expected for genes that assort independently. The lower-frequency classes have the recombinant gene arrangements.

RECOMBINATION FREQUENCY AS A MEASURE OF MAP DISTANCE

Crossover events cannot be accurately counted microscopically; they are usually inferred from the presence of recombinant-type gametes. Therefore, in practice, researchers typically measure the amount of crossing-over between linked genes by the frequency of recombinant-type gametes of a heterozygote. To establish a unit of map length, a recombination frequency of 1% is defined as being equal to one map unit, (also designated as one **centiMorgan, cM**, in honor of the geneticist T. H. Morgan). Map distance can then be evaluated directly from testcross data by calculating the proportion of testcross progeny with recombinant gene arrangements. For a testcross then, the recombination frequency (R) is given by

$$R = \frac{\text{number of recombinant testcross progeny}}{\text{total number of testcross progeny}}$$

Using this formula, the total map distance is 100R, and 1 map unit = 1% R. For example, the testcross results in Table 13−1 for the genes involved in body color and wing length in *Drosophila* give an R value of

$$R = \frac{236 + 217}{2400} = 0.189$$

The b and vg loci are therefore separated by a genetic distance of 18.9 map units. The following linkage map shows the relative locations of these genes:

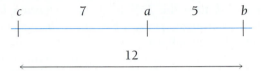

Once the distance between two genes is known, they can be mapped in relation to a third gene, a fourth, and so on within the linkage group. For example, suppose that testcross data show genes a and b to be separated by 5 map units and that a third locus, c, is found to be 7 map units from a and 12 map units from b. The linkage map is then

Other genes can be added to this map until all the genes in the linkage group have been assigned positions. The gene lying closest to one end of the linkage group is assigned the map position 0.0, and all other loci are then given map position values according to their distance from the first gene. In the hypothetical example just given, let us suppose that gene d is found to lie 2 units to the left of the c locus and that no gene is found to map closer to the left end of the chromosome than d. In terms of genetic distances, the map is

These relationships are more conveniently represented as

$$\begin{array}{cccc} d & c & a & b \\ | & | & | & | \\ 0.0 & 2.0 & 9.0 & 14.0 \end{array}$$

The crossover data in this example yielded a precise linear order of genes along the chromosome. Linear linkage maps are in fact the rule in a wide variety of plants and animals.

Example 13.1

When conducting linkage studies on a naturally self-fertilizing species of plant, it is usually more convenient to allow dihybrids to self-fertilize and analyze their offspring than it is to make a separate testcross. Suppose that you are interested in determining whether the genes for normal leaves (*M*) versus mottle (*m*) and smooth fruit skin (*P*) versus peach (*p*) in tomatoes are linked. A dihybrid *MmPp* plant is selfed and produces four phenotypic classes in the offspring, of which 0.16% are doubly recessive *mmpp*. (a) Are the genes *M,m* and *P,p* linked? If so, what is the linkage phase (coupling or repulsion) of the heterozygous parent? (b) Calculate the map distance between the *m* and *p* loci.

Solution: (a) The genes appear to be linked, since the frequency of doubly recessive offspring is significantly less than $\frac{1}{16}$ (6.25%), the proportion that would be expected for a dihybrid cross with independent assortment. We deduce that the *mp* gamete type is recombinant by noting that double recessives, *mp/mp*, are formed by the combination of two *mp* gametes. The expected frequency of this genotype is $f(mp/mp) = f(mp) f(mp) = (0.04)(0.04) = 0.0016$. Thus, the *mp* gamete arrangement occurs in only 4% of the gamete types produced by the heterozygous parent. Since this proportion is less than the 25% frequency expected for independent assortment, gametes that carry the *mp* chromosome must be recombinant and must have been formed through crossing-over. Therefore the genes in the heterozygous parent must be linked in the repulsion (*Mp/mP*) phase. (b) Since the *mp* chromosome is recombinant, its frequency (0.04) should equal $R/2$, where R is the frequency of recombination between the genes involved. Thus, the map distance is $(100)R = (100)(2)(0.04) = 8$ map units.

Follow-Up Problem 13.1

In Example 13.1, we concluded that the genes *M,m* and *P,p* are linked, with a genetic map distance of 8 units separating

them. From a dihybrid cross *Mp/mP* × *Mp/mP*, what proportion of the offspring are expected to be phenotypically (a) *M–pp*? (b) *mmP–*? (c) *M–P–*?

Maximum Frequency of Recombination

While the recombination frequency serves as a useful measure of the number of crossovers between closely linked loci, the relationship breaks down as the distance between the markers increases. As a rule, more crossing-over occurs between widely separated loci than is indicated by the frequency of recombinant gametes. The recombination frequency is thus an underestimate of the true genetic distance between widely separated loci. This discrepancy results from the occurrence of multiple exchanges within the longer intervals. These multiple crossovers affect the true map length (as measured by crossover frequency) but do not contribute to the observed recombination of genes that bracket the interval.

The eventual effect of multiple crossovers is to put an upper limit of 50% on the frequency of recombination. This upper limit holds for *all* multiple exchanges, whether they occur along the same chromatid pair or involve different combinations of nonsister chromatids. ► Figure 13–4 shows the effect of multiple crossovers along the same chromatid pair. In case 1, the genes are separated by a very short inter-

---

► **FIGURE 13–4** The effects of increased frequency of crossing-over on the frequency of observed recombination, when all crossovers involve the same two nonsister chromatids within the bivalent. As the length of the interval separating the two loci increases, more crossing-over takes place. However, the additional exchange events are not registered as additional recombination. Only odd-numbered events generate recombinants, and no matter how many events take place, only two recombinant types are formed. Even-numbered crossovers yield no recombinants at all.

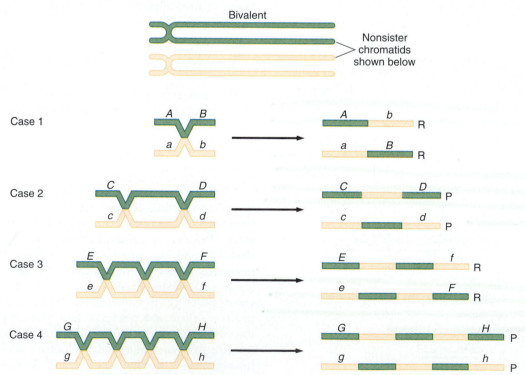

val, so crossing-over occurs only rarely, and multiple cross-over events do not occur at all; the result is very few (much less than 50%) recombinant gametes in the gamete pool. In case 2, the genes are farther apart, so that an occasional bivalent undergoes a double crossover, which re-creates the parental gene arrangement. No recombinant-type gametes are produced, since the second crossover nullifies the effect of the first. Case 3 shows two loci even farther apart, so that an occasional triple crossover is possible. Although two recombinants are produced by the triple crossover, the observed result is no different from that of a single crossover. In case 4, the loci are so widely separated that four crossovers can occur. Four crossovers, like a double crossover, yield no recombinant gametes, since they re-create the parental arrangement for the genes bordering the interval.

By now, you should see the general pattern: Any odd number of crossovers (1, 3, 5, . . .) between a pair of chromatids yields two recombinant gametes, while any even number of crossover events (2, 4, 6, . . .) yields two parental gametes. Because as many even-numbered as odd-numbered exchanges are expected to occur overall between widely separated loci, the recombination frequency between a pair of nonsister chromatids cannot exceed a maximum of 50%. The additional crossing-over allowed by a longer distance re-creates the parental types as often as it yields recombinants.

If different combinations of nonsister chromatids are involved in multiple crossing-over, the immediate results vary in terms of the gamete types produced by one meiotic event, but the average proportion of recombinant gametes is still not expected to be greater than 50%. The possible outcomes of double crossovers are shown in ▶ Figure 13–5, and the same general results hold for all higher-order events. If the distribution of crossovers is random among the four chromatids, four combinations of nonsister chromatids should be equally likely to occur in the double exchanges: (1) two-strand double exchanges, (2) and (3) two types of three-strand double exchanges, and (4) four-strand double exchanges. Case 1 yields four parental-type gametes. Cases 2 and 3 each yield two parental and two recombinant gametes—one parental type is derived from a chromatid that did not participate in crossing-over, and the other is derived from the re-creation of the parental arrangement by the second crossover event. Case 4 yields four recombinant gametes. The net result in the gamete pool is 8 (4 + 2 + 2) parental types and 8 (2 + 2 + 4) recombinant types, for a recombination frequency of 50%.

We can now see that even though the frequency of recombination tends to increase as the distance between linked genes increases, it will never exceed a maximum of 50%. Therefore, genes spaced very far apart on the same

▶ FIGURE 13–5 Alternative types of double crossovers involving different combinations of nonsister chromatids. The overall result is 50% parental and 50% recombinant gametes.

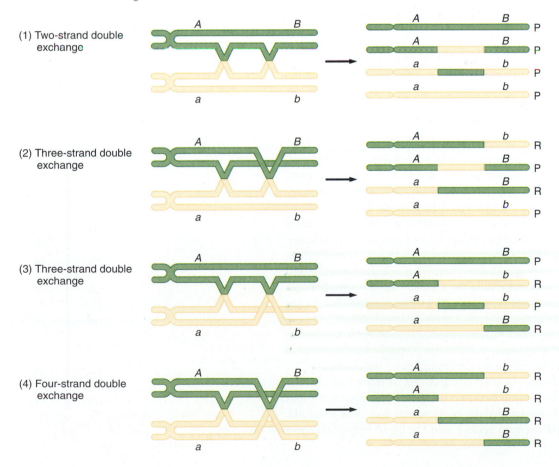

chromosome will appear to exhibit 50% recombination. In this situation, the four gamete classes produced by a dihybrid—whether in coupling or in repulsion—will occur in proportions of 25% each, the same result that we would expect for genes located on different chromosomes. Thus, genes can be located far enough apart on the same chromosome to assort independently in a testcross. Such loci can be said to be physically linked, in that they do lie on the same chromosome, but they do not appear to be genetically linked, since the data from genetic crosses fail to show ratios characteristic of linkage. How can we determine whether two genes are actually located on the same chromosome? The linkage of such genes can be discovered by finding that both show less than 50% recombination with some third gene. For example, suppose genes a and b yield 50% recombinants, but a and c yield 30% recombinants, and b and c yield 35% recombinants. Since both a and b are linked to the same common gene c, all three must be located on the same chromosome.

Points to Ponder 13.1

It has been estimated that humans have perhaps 100,000 genes, distributed over 23 pairs of chromosomes. Think about the implications of so many genes. Does it seem to you that there is a great deal of linkage among these genes? About how many genes are there per human chromosome? On the other hand, if you are studying two randomly chosen human traits caused by genes whose chromosome location is unknown, what is the chance that the two genes are linked? Consider the same questions for *Drosophila melanogaster*, which probably has about 20,000 genes.

Correcting for Multiple Crossovers

We have seen that a chromatid involved in any even number of exchanges will not be detected as a recombinant. Moreover, a chromatid involved in any odd number of crossovers, regardless of how many occur, will yield no more recombinants than one involved in a single exchange. Thus the increasing number of crossovers in longer intervals is simply not matched by a proportional gain in recombinants.

One way to avoid the inaccuracies that develop from multiple exchanges is to subdivide the linkage map into a series of shorter intervals, each small enough to include no more than a single crossover. In practical terms, this means restricting the size of each interval to a maximum of about 10 to 15 map units. Since recombination is now limited to single exchanges, recombination frequency and crossover frequency (per nonsister chromatid pair) will be equal.

To subdivide the map, we use one or more loci that are known to be present between the genes in question but that have not been studied in previous crosses. For example, suppose that we find genes e and f to be 16 map units apart

on the basis of the recombination frequency computed from testcross data. The preliminary map is then

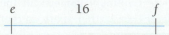

Because of multiple crossovers, however, this distance is probably an underestimate of the true distance. To obtain a more accurate value, we redo the mapping study, making use of gene h, which is located between e and f. (The h locus has always been present; we just did not score for its phenotype in the previous experiment.) We now make two testcrosses, one involving e and h and the other involving f and h. Assume that the crosses yield recombination frequencies of 8% and 9%, respectively. The improved map then becomes

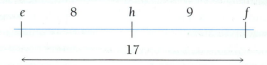

▶ **FIGURE 13–6** Linkage map of corn, showing locations of some known genes.

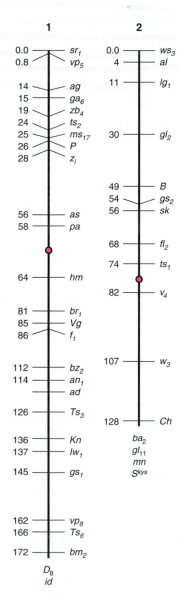

The distance between *e* and *f* is now greater than had been estimated earlier; it is 17 (8 + 9) map units.

We could extend this procedure to include the entire range of nonallelic genes known to be carried on a chromosome and thus construct a genetic map in which the total map distance (estimated by summing the intervals within it) is much greater than the maximum 50% recombination for widely separated loci. Genetic maps of the linkage groups for corn and *Drosophila* are shown in ▶ Figures 13–6 and 13–7, where the larger map position values have been determined by summing the shorter intervals.

The relationship between the recombination frequency and the true map distance (actual crossover frequency) obtained by the procedure we just described is illustrated in ▶ Figure 13–8. Observe that the recombination frequency is linearly related to the crossover frequency only over distances of up to about 15 map units. A discrepancy develops at longer distances, as the recombination frequency approaches its maximum value of 50% despite continued increases in the number of crossovers.

To Sum Up

1. Genetic maps show the relative locations of genes and the relative distances between them. A unit of map distance is defined as the equivalent of a 1% crossover frequency. In practice, genetic map distances are based on the percentage of recombination observed in testcross progeny.

2. The recombination frequency underestimates the true genetic (crossover) distance when the intervals between the genes are long. The discrepancy between crossover and recombination frequencies arises from the occurrence of multiple crossover events that yield either no recombinants or no more than the number produced by a single exchange.

3. The occurrence of multiple crossover events also limits the observed frequency of recombinant gametes to 50%, regardless of

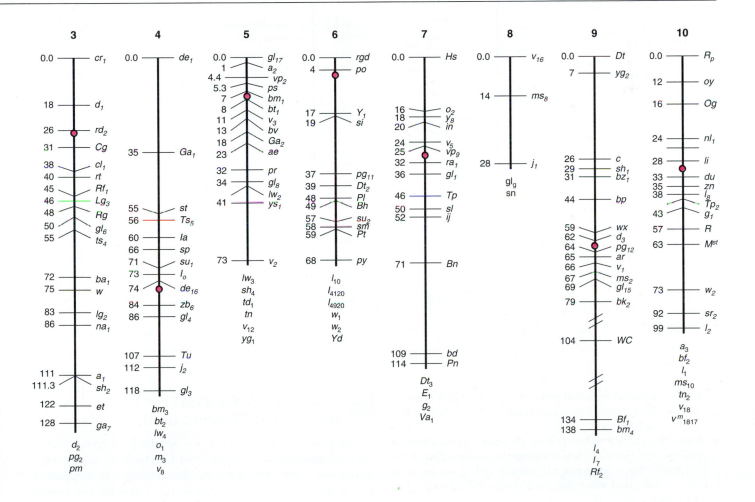

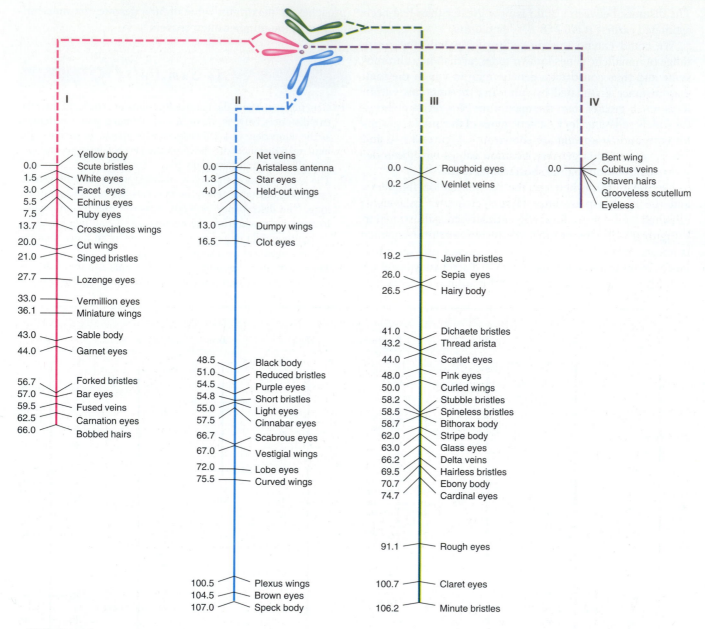

FIGURE 13–7 Linkage map of *Drosophila melanogaster,* showing the chromosomes corresponding to each linkage group.

the distance between linked genes, because multiple exchanges recreate the parental-type gene arrangement as often as they yield recombinants.

4. Genes that are spaced far apart on the same chromosome may not show the characteristics of linkage in testcross data. It can be shown that such genes are actually linked on the same chromosome by finding that each shows linkage with some third gene.

5. If two genes yield more than 15 to 20% recombinant gametes, a better estimate of the true distance between them can be obtained by subdividing the interval between genes into shorter regions, using other genes that are located between the original two loci. The true map distance is then best estimated as the sum of the shorter distances.

THE THREE-POINT TESTCROSS METHOD FOR MAPPING GENES

In the preceding discussions, we implicitly assumed that map distances are determined from the results of two-point testcrosses. There are two main disadvantages to mapping linked genes by this approach. First, by following the joint transmission of only two gene pairs at a time, we require a separate two-factor cross for each combination of two non-allelic genes—three separate crosses would be required just to arrange three gene loci unambiguously. Second, we cannot directly measure the genetic effects of multiple crossovers by two-point crosses; we can only infer the occurrence of double and higher-order crossovers from the failure

EXTENSIONS AND TECHNIQUES

The Haldane Mapping Function

Geneticists have developed various models to establish a mathematical basis for the graphical relationship shown in Figure 13–8. The **Haldane mapping function**, one of the simplest of these models, assumes that crossovers occur at random along the length of a chromosome pair. Given a large number of potential sites for crossing-over and a small probability of exchange at any one particular site, the probability of x crossovers per bivalent within an interval of interest should then follow the Poisson distribution:

$$P(x) = \frac{(2d)^x}{x!} e^{-2d} \quad (13.1)$$

where $2d$ is the average number of crossovers per bivalent (d being the crossover frequency per chromatid pair). For example, if there is an average of one crossover per bivalent ($2d = 1$) between the genes in question, the fraction of bivalents with single ($x = 1$), double ($x = 2$), triple ($x = 3$), . . . crossovers will be 0.368, 0.184, 0.061, . . . , respectively.

But as we have seen, only half the meiotic products of each crossover class will be recombinant. The recombination frequency must therefore equal one-half the proportion of bivalents with at least one crossover:

$$R = \frac{1}{2}[P(1) + P(2) + P(3) + \ldots]$$

$$= \frac{1}{2}[1 - P(0)]$$

Inserting the $P(0)$ term from equation 13.1, the Haldane mapping function then becomes

$$R = \frac{1}{2}(1 - e^{-2d}) \quad (13.2)$$

An analysis of equation 13.2 reveals that for widely separated loci (i.e., large values of d), the expression re-duces to $R = \frac{1}{2}$. The model therefore predicts a maximum of 50% recombination between genes. Moreover, for very closely linked markers (i.e., d values of about 0.1 or less), the expression becomes approximately $R = d$. Thus the function is linear for short genetic distances, demonstrating that R is a good measure of crossover frequency between loci that are closely linked.

The Haldane mapping function, given by equation 13.2, provides a convenient relationship for estimating crossover frequencies from recombination frequencies in organisms where the chromosomes have not been extensively mapped. The inaccuracies that result from multiple crossovers in longer distances can then be corrected directly without having to perform additional crosses.

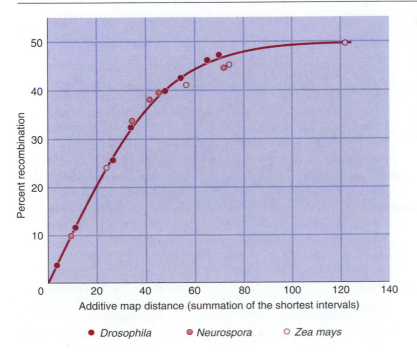

▶ **FIGURE 13–8** Relationship between the observed frequency of recombination and the actual frequency of crossing-over (the additive map distance). The observed recombination frequency is directly proportional to the crossover frequency only up to a map distance of about 15 units. *Source:* Perkins, D. D., "Crossing over and interference in a multiply marked chromosome of *Neurospora*," *Genetics* 47 (1962): 1253–1274. Used with permission.

● *Drosophila* ● *Neurospora* ○ *Zea mays*

of different pairs of matings to yield additive values of the recombination frequency.

Because of these drawbacks, a **three-point testcross** is the preferred method for mapping linked genes. In a three-point testcross, one parent is trihybrid, so three genetic markers can be followed in a single mating. Moreover, the presence of three linked loci spanning two adjacent crossover intervals permits the identification of recombinants that arise from double exchange events. The testcross

$$\frac{ABC}{abc} \times \frac{abc}{abc}$$

provides a particularly convenient example. (Any gene arrangement can be used—other possible trihybrid parents for the testcross are ABc/abC, AbC/aBc, and Abc/aBC). The trihybrid ABC/abc can produce eight kinds of gametes, which are listed in four classes in ■ Table 13–2. The simplest combinations of crossovers that can produce these gamete classes are shown in ▶ Figure 13–9. Class 1 includes parental-type gametes. For three linked genes, parental type gametes will occur in the highest frequencies. Class 2 and class 3 each have gamete types that result from a single recombination act that involves the a and b loci for class 2 and the b and c loci for class 3. Class 4 contains the double-recombinant gamete types and involves recombination events for both sets of markers. Since the probability that two crossovers will occur simultaneously is less than the chance that only one crossover event will occur, gametes of the double-recombinant type are present in the lowest frequency.

■ **TABLE 13–2** The eight kinds of gametes produced by a trihybrid. The genes are all in the coupling phase in this example. Parental vs. recombinant types are shown in the right column.

	Class	Frequency	Type
$\dfrac{ABC}{abc}$ ↓			
$\dfrac{ABC}{abc}$	(1)	$1 - \alpha - \beta - \gamma$	Parental types
$\dfrac{Abc}{aBC}$	(2)	α	Single recombinants $(a–b)$
$\dfrac{ABc}{abC}$	(3)	β	Single recombinants $(b–c)$
$\dfrac{AbC}{aBc}$	(4)	γ	Double recombinants $(a–b$ and $b–c)$

Determining Map Distance

We can use the frequencies of the various recombinant gamete classes (given in general as α, β, and γ in Table 13–2) to compute map distances for the different sets of loci. In the following discussion, we will use $f(\ )$ to mean "frequency of" and R_{a-b}, R_{b-c}, and R_{a-c} to denote the recombination frequencies for the designated loci (as determined from three separate two-point testcrosses, for example). Since all three genes are linked in the coupling phase in this example, the recombinant-type arrangements for each pair

▶ **FIGURE 13–9** The simplest possible origins of the various gamete classes produced by a trihybrid ABC/abc. More complex events are also possible, since any odd number of exchanges in an interval gives rise to a recombinant arrangement of genes, while any even number of crossovers produces a parental arrangement. (For simplicity, only the two chromatids involved in crossing-over are shown; a bivalent of course actually contains four chromatids.)

(a) No crossing-over

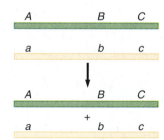

Parental for $a–b$; parental for $b–c$

(b) Crossing-over $a–b$; no crossing-over $b–c$

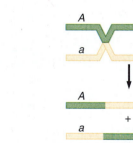

Recombinant for $a–b$; parental for $b–c$

(c) No crossing-over $a–b$; crossing-over $b–c$

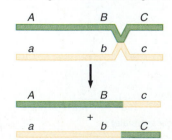

Parental for $a–b$; recombinant for $b–c$

(d) Crossing-over $a–b$; crossing-over $b–c$

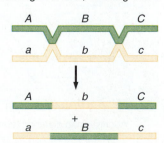

Recombinant for $a–b$; recombinant for $b–c$

of loci are *Ab* and *aB*, *Bc* and *bC,* and *Ac* and *aC*. For this example, $R_{a-b} = f(Ab) + f(aB)$, $R_{b-c} = f(Bc) + f(bC)$, and $R_{a-c} = f(Ac) + f(aC)$. Inspection of the gametes in Table 13–2 reveals that classes 2 and 4 are recombinant for the *a* and *b* loci, classes 3 and 4 are recombinant for the *b* and *c* loci, and classes 2 and 3 are recombinant for the *a* and *c* loci. Adding the frequencies of these pairs of gamete classes gives the recombination frequencies:

$$R_{a-b} = \alpha + \gamma$$
$$R_{b-c} = \beta + \gamma$$
$$R_{a-c} = \alpha + \beta$$

These relationships presuppose knowledge of the order of loci on the chromosome. How would we proceed when nothing is known about the gene arrangement? In this case we simply treat each pair of gene loci separately, without regard to the third locus, as indicated in the following equations by the slash through the locus not under consideration. We can then calculate the recombination frequencies directly from the frequencies of the trihybrid gametes ($f(Abc)$, $f(aBC)$, and so on):

$$R_{a-b} = f(Ab\cancel{c}) + f(aB\cancel{c}) + f(Ab\cancel{C}) + f(aB\cancel{C})$$
$$R_{b-c} = f(\cancel{A}Bc) + f(\cancel{a}bC) + f(\cancel{A}bC) + f(\cancel{a}Bc)$$
$$R_{a-c} = f(A\cancel{b}c) + f(a\cancel{B}C) + f(A\cancel{B}c) + f(a\cancel{b}C)$$

To calculate R_{a-b}, we ignore the *c* locus and add the frequencies of all gametes with recombinant gene arrangements for the *a* and *b* loci. In like manner, we determine the R values for *b–c* and *a–c* from the frequencies of their recombinant gene arrangements, ignoring *a* while calculating R_{b-c} and ignoring *b* while calculating R_{a-c}. By using this procedure, we are analyzing the data from a single three-point testcross as though we had conducted three separate two-point testcrosses, each involving a different pair of loci.

We are now ready to calculate the distances between loci from the results of an actual three-point testcross. Corn is an excellent subject for the evaluation of testcross data. Since each ear of corn contains several hundred seeds, the result of an entire cross can be obtained by simply recording the kernel phenotypes of some 10 to 20 ears. As an example, we will consider a three-point testcross involving the following three pairs of traits: colored kernels (*C*) vs. colorless (*c*), plump kernels (*Sh*) vs. shrunken (*sh*), and starchy kernels (*Wx*) vs. waxy (*wx*). For now, we will list the genes for these traits in alphabetical order, since neither the sequence nor the linkage phase of the genes is known.

The results of a testcross from an experiment of Claude B. Hutchison (1922) are given in ■ Table 13–3. The phenotypes of the offspring are listed in order of frequency, along with the genotype of the gamete received from the trihybrid parent. The lack of a $1:1:1:1:1:1:1:1$ ratio in the progeny clearly shows the linkage of the genes. Furthermore, *C sh Wx* and *c Sh wx* must be the parental gene arrangements (because of their higher frequencies), indicating that the linkage phase of the trihybrid parent is *C sh Wx/c Sh wx*. Remember, however, that we still do not know if this is the correct gene order.

The testcross results in Table 13–3 can be used to calculate map distances for the intervals between the different sets of genes. If we consider only two genes at a time, recombinant-type arrangements for the different sets of loci are *C Sh* and *c sh*, *Sh Wx* and *sh wx*, and *C wx* and *c Wx*. The map distances are then

$$100R_{c-sh} = 100[f(C\,Sh\,\cancel{wx}) + f(c\,sh\,\cancel{Wx}) + f(C\,Sh\,\cancel{wx}) + f(c\,sh\,\cancel{Wx})]$$
$$= 100\left(\frac{113 + 116 + 4 + 2}{6708}\right) = 3.5$$

$$100R_{sh-wx} = 100[f(\cancel{c}\,Sh\,Wx) + f(\cancel{C}\,sh\,wx) + f(\cancel{C}\,Sh\,Wx) + f(\cancel{c}\,sh\,wx)]$$
$$= 100\left(\frac{626 + 601 + 4 + 2}{6708}\right) = 18.4$$

$$100R_{c-wx} = 100[f(C\,\cancel{Sh}\,wx) + f(c\,\cancel{sh}\,Wx) + f(c\,\cancel{sh}\,Wx) + f(C\,\cancel{Sh}\,wx)]$$
$$= 100\left(\frac{113 + 116 + 626 + 601}{6708}\right) = 21.7$$

■ **TABLE 13–3** Results of a trihybrid testcross in corn involving alleles for colored (*C*) vs. colorless (*c*) aleurone, plump (*Sh*) vs. shrunken (*sh*) endosperm, and starchy (*Wx*) vs. waxy (*wx*) endosperm.

Kernel Phenotype	Gamete Received from Heterozygous Parent	Number
Colorless, plump, waxy	*c Sh wx*	2708
Colored, shrunken, starchy	*C sh Wx*	2538
Colorless, plump, starchy	*c Sh Wx*	626
Colored, shrunken, waxy	*C sh wx*	601
Colorless, shrunken, starchy	*c sh Wx*	116
Colored, plump, waxy	*C Sh wx*	113
Colored, plump, starchy	*C Sh Wx*	4
Colorless, shrunken, waxy	*c sh wx*	2
	Total	6708

Source: Claude B. Hutchinson, *Cornell Univ. Agr. Exp. Sta. Memoir* 60 (1922): 1419–1473.

Plump kernels.

Shrunken kernels.

Determining Gene Order

The next step in mapping genes by the three-point testcross method is to determine the proper gene order, since it may not be known to the investigator before the cross is performed. One approach is to calculate the various map distances, as we did in the previous analysis, and use these distances to arrange the genes in their correct order. For example, the preceding calculations gave map distances of 3.5, 18.4, and 21.7 map units for the $c-sh$, $sh-wx$, and $c-wx$ intervals, respectively. Since 21.7 is the longest of the three distances, we conclude that c and wx are the outside markers, with sh between them. The map is then

The sum $3.5 + 18.4 = 21.9$ is a more accurate measure of total distance than 21.7, since it compensates for the nullifying effects of those even-numbered exchanges between the outside markers that result from an odd number of exchanges in each of the two included intervals (see ▶ Figure 13–10).

It is also possible to determine the proper gene order without calculating map distances first. All we need to do is deduce the sequence of genes in the parental chromosomes that would produce the genotypes of the double-recombinant class by a double crossover. The double-recombinant class is produced by a single crossover event (more precisely, an odd-numbered event) in each of the included intervals, and therefore only the middle pair of alleles will be affected by the double exchange. For example, suppose that among the eight types of gametes produced by a trihybrid in a three-point testcross, AbC and aBc appear in the highest frequency, while Abc and aBC occur in the lowest frequency. What is the correct order of the a, b, and c gene loci

on the chromosome? It is apparent that gametes AbC and aBc, which occur in the highest frequency, are the parental types and that gametes Abc and aBC, which occur in the lowest frequency, are the double-recombinant types. Note that AbC differs from Abc and aBc from aBC only at the c locus. Thus, the double-recombinant gametes are formed from the parental gene arrangement by an exchange of chromosome parts involving the C,c allele pair. Since a minimum of two crossovers are required to achieve this result, the c locus must be in the middle:

$$
\begin{array}{ccc}
A & C & b \\
\hline
\times & \times & \\
\hline
a & c & B
\end{array}
\longrightarrow
\begin{array}{ccc}
A & c & b \\
\hline
\\
\hline
a & C & B
\end{array}
$$

Example 13.2

In *Drosophila*, the three gene pairs for normal bristles (f^+) vs. forked (bent or split bristles, f), red eye color (g^+) vs. garnet (pinkish eye color, g), and long wings (m^+) vs. miniature (m) are known to be present on the same chromosome. A trihybrid $f^+f\ g^+g\ m^+m$ female is crossed to a wild-type (normal-bristled, red-eyed, long-winged) male and produces the following offspring (in each case, only those genes that are known to be present from the phenotypes of the flies are listed):

Females (1,000):   All wild-type
Males (1,000): $f^+g^+m^+$ 408 $f^+g\ m$ 64 $f\ g\ m$ 399 $f\ g^+m^+$ 59
$\quad\quad\quad\quad\quad$ f^+g^+m 30 $f^+g\ m^+$ 4 $f\ g\ m^+$ 33 $f\ g^+m$ 3

(a) To which chromosome do these genes belong? How can you tell?
(b) Construct a linkage map of the f, g, and m loci.

Long wings vs. miniature wings.

Solution: (a) The three gene loci are present on the X chromosome. The female offspring all express the dominant wild-type genes that they received on the paternally-derived X chromosome. The males received an X chromo-

▶ **FIGURE 13–10** Double crossing-over along the interval separating the c and wx loci. (a) Because the outside markers are back in the parental arrangement, there is no way to detect the crossover event or know that the middle section of the chromosome is recombinant. (b) A gene marking the middle section of the chromosome allows detection of double crossovers as recombinants of the middle gene with each outside gene.

(a)
```
C                    Wx          C              Wx
  \  /        \  /                
   \/          \/                 
   /\          /\                 
  /  \        /  \                
c                    wx          c              wx
```

(b)
```
C    Sh    Wx                    C    sh    Wx
       \  /                       
        \/                        
        /\                        
       /  \                       
c    sh    wx                    c    Sh    wx
```

some from their maternal parent only. Since the males are hemizygous, they will express all the X-linked genes inherited from their mother. Therefore, the parental-type and recombinant-type gametes formed by the female parent can be observed directly in the male offspring. (b) It is first apparent that $f^+g^+m^+$ and $f\,g\,m$ are parental-type arrangements, since they occur in the highest frequencies, and $f^+g\,m^+$ and $f\,g^+m$ are the double recombinant types since they occur in the lowest frequencies. Therefore, $f^+g\,m$ and $f\,g^+m^+$ are single recombinant types for $f-g$, and f^+g^+m and $f\,g\,m^+$ are single recombinant types for $g-m$. The map distances can then be calculated as

$$100R_{f-g} = \frac{100(64 + 59 + 4 + 3)}{1000} = 13.0$$

$$100R_{g-m} = \frac{100(30 + 33 + 4 + 3)}{1000} = 7.0$$

$$100R_{f-m} = \frac{100(64 + 59 + 33 + 30)}{1000} = 18.6$$

The linkage map becomes:

Thus, mapping data for X-linked genes can be derived directly from the male offspring of heterozygous females even in the absence of a testcross.

Follow-Up Problem 13.2

The following sets of data were obtained from offspring of three-point testcrosses. Determine which gene locus is in the middle in each case. (Only the parental type and double-crossover type classes are given.)

Parental Classes	Double-Recombinant Classes
(a) $A\,B\,C$	$A\,b\,C$
$a\,b\,c$	$a\,B\,c$
(b) $d^+e\ f$	$d^+e\ f^+$
$d\ e^+f^+$	$d\ e^+f$
(c) $g\ H\,I$	$g\ h\,I$
$G\,H\,i$	$G\,H\,i$
(d) $j^+\,k\ l^+$	$j\ k\ l^+$
$j\ k^+l$	$j^+\ k^+l$

Interference and Coincidence

The previous analysis of the c, sh, and wx loci in corn gave recombination frequencies of $R_{c-sh} = 0.035$ and $R_{sh-wx} = 0.184$ for the two included intervals in the map. These numbers also represent the probabilities of an odd number of crossovers in the designated regions. If crossovers in the $c-sh$ region are independent of crossovers in the adjacent $sh-wx$ region, then the expected proportion of double-recombinant gametes would be $0.035 \times 0.184 = 0.00644$, or $0.00644 \times 6708 = 43$ double recombinants out of the total

number of offspring from the testcross. But note that the actual data contain only six (4 $C\,Sh\,Wx$ and 2 $c\,sh\,wx$) double recombinants. This outcome suggests that crossovers in adjacent intervals are not independent events but tend to inhibit each other.

The overall effect of a crossover in one region on the chance that another crossover will occur in an adjacent region is known as **interference**. The degree of interference (I) is measured as $I = 1 - C$, where C is the **coefficient of coincidence**, which is the ratio of the observed frequency of double recombinants to the frequency expected if the crossovers were truly independent:

$$C = \frac{\text{observed number of double recombinants}}{\text{expected number of double recombinants}}$$

For the preceding data, C has a value of

$$\frac{6}{43} = 0.14$$

giving $I = 0.86$.

When $C < 1$ (so that $I > 0$), adjacent crossovers are said to show **positive interference**. In this case, one crossover reduces the chance of another crossover nearby, causing double-recombinant types to be less frequent than would be expected from randomness alone. When $C > 1$ (so that $I < 0$), adjacent crossovers are said to show **negative interference**. This type of interference results in more double recombinants than expected for independent events.

Although the coefficient of coincidence is defined in terms of the frequency of the double-recombinant class, all recombinant classes are affected by interference. These effects are shown in ■ Table 13–4 for the general results of a

■ **TABLE 13–4** Effect of interference on the frequency of single- and double-recombinant classes from a three-point testcross. Since the value of γ is affected by interference (measured as the coefficient of coincidence, C), the frequency of the single crossover classes are affected as well.

Testcross	Recombinant progeny			
	Recombinant Class	Frequency	Type
$\dfrac{ABC}{abc} \times \dfrac{abc}{abc}$	$\dfrac{Abc}{abc}$		
	$\dfrac{abc}{abc}$	$\alpha = R_{a-b} - \gamma$	Single, $a-b$
	$\dfrac{aBC}{abc}$		
	$\dfrac{ABc}{abc}$		
	$\dfrac{abc}{abc}$	$\beta = R_{b-c} - \gamma$	Single, $b-c$
	$\dfrac{abC}{abc}$		
	$\dfrac{AbC}{abc}$		
	$\dfrac{abc}{abc}$	$\gamma = CR_{a-b}R_{b-c}$	Double, $a-b$ and $b-c$
	$\dfrac{aBc}{abc}$		

three-point testcross $ABC/abc \times abc/abc$. Observe that the frequency of each single recombinant class is equal to the frequency of recombination (R value) for the interval in question minus the frequency of double recombinants. For the example given, $\alpha = R_{a-b} - \gamma$ and $\beta = R_{b-c} - \gamma$. Moreover, the actual frequency of double recombinants is given from the definition of the coefficient of coincidence as

$$\gamma = CR_{a-b}R_{b-c} \tag{13.3}$$

We can therefore obtain a relationship between the frequency of recombination for the outside markers ($R_{a-c} = \alpha + \beta$) and C:

$$R_{a-c} = R_{a-b} + R_{b-c} - 2CR_{a-b}R_{b-c} \tag{13.4}$$

It is thus possible in principle to measure C either from the results of a three-point testcross, using equation 13.3, or from the pooled results of three two-point testcrosses, using equation 13.4.

Example 13.3

Suppose that three loci gave the following linkage map:

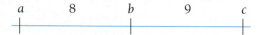

Assuming no interference between crossovers in adjacent intervals, predict the frequencies of the recombinant classes produced by the three-point testcross $ABC/abc \times abc/abc$.

Solution: The recombinant frequencies R_{a-b} and R_{b-c} are the probabilities of obtaining a recombinant gene arrangement for each of the two pairs of loci. Thus, the probability of getting an Ab or an aB arrangement is $R_{a-b} = 0.08$, and the probability of getting a Bc or bC arrangement is $R_{b-c} = 0.09$. The probabilities of obtaining parental-type arrangements are thus $1 - R_{a-b}$ and $1 - R_{b-c}$, respectively. Also note that without interference, crossovers that occur in adjacent intervals are independent events. Applying the multiplication rule of probability, the frequencies of the three recombinant classes can be calculated as follows:

Recombinant Class	Frequency	Type
$\underline{\underline{Abc}}$ abc $\underline{\underline{aBC}}$ abc	$R_{a-b}(1 - R_{b-c}) = 0.08 \times 0.91 = 0.0728$	single, $a-b$
$\underline{\underline{ABc}}$ abc $\underline{\underline{abC}}$ abc	$(1 - R_{a-b})R_{b-c} = 0.92 \times 0.09 = 0.0828$	single, $b-c$
$\underline{\underline{AbC}}$ abc $\underline{\underline{aBc}}$ abc	$R_{a-b} \times R_{b-c} = 0.08 \times 0.09 = 0.0072$	double, $a-b$ and $b-c$

Follow-Up Problem 13.3

Three gene loci give the following linkage map:

Assuming interference with a coefficient of coincidence of 0.8, determine the expected frequencies of the eight classes produced by the three-point testcross

$$\frac{RaT}{rAt} \times \frac{rat}{rat}$$

To Sum Up

1. The three-point testcross method is a convenient method for mapping three genes in one experiment and for subdividing a long interval on a chromosome map into two shorter, adjacent intervals. Not only can three genes be mapped using data from a single mating, but double-crossover events in the adjacent intervals can be detected, thus improving the estimate of the map distance for the entire interval.

2. The three-point testcross method also permits detection of the effects of interference, in which one crossover alters the chance that another crossover will occur in an adjacent interval. Interference is usually measured by the coefficient of coincidence (C), which is the ratio of the observed frequency of double recombinants to the expected frequency. Positive interference ($C < 1$) occurs when there are fewer double crossovers than expected; negative interference ($C > 1$) occurs when there are more double crossovers than expected.

MAPPING BY TETRAD ANALYSIS

Linkage analysis in diploid eukaryotes is a form of random chromatid analysis, since inferences made about crossing-over are based on the progeny derived from a random sampling of all products resulting from many pooled meiotic events. Linkage studies can also be conducted on haploid eukaryotes, such as certain fungi and yeast, by means of **tetrad analysis**. In a tetrad analysis, all four products of a single meiosis are recovered and tested individually to determine their genotypes. Tetrad analysis thus provides information about details of chromosome behavior during meiosis that are not ordinarily obtainable from studies on higher organisms.

The orange bread mold, *Neurospora crassa*, is an example of a haploid eukaryote with a reproductive cycle that lends itself to tetrad analysis, and thus this fungus has been used extensively in genetic research. The life cycle of *Neurospora* (see ▶ Figure 13–11) includes both asexual and sexual modes of reproduction. The asexual (vegetative) cycle of this fungus is initiated when a single reproductive cell or **conidium** germinates into a long threadlike structure called a **hypha**. The hypha branches repeatedly as it elongates to form a multinucleate mass of cytoplasm enclosed within

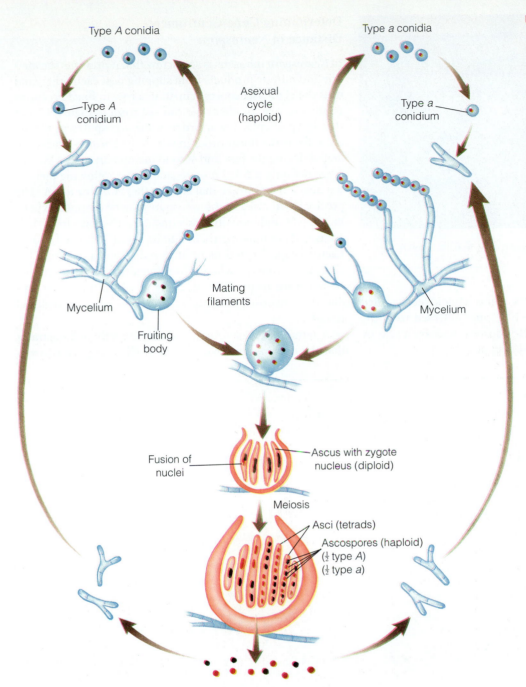

Type *A* conidia

Type *A* conidium

Asexual cycle (haploid)

Type *a* conidia

Type *a* conidium

Mycelium

Mating filaments

Mycelium

Fruiting body

Fusion of nuclei

Ascus with zygote nucleus (diploid)

Meiosis

Asci (tetrads)

Ascospores (haploid)
($\frac{1}{2}$ type *A*)
($\frac{1}{2}$ type *a*)

▶ **FIGURE 13–11** Asexual and sexual reproduction in *Neurospora.* In the asexual cycle, individual spores (conidia) germinate into hyphal systems that release masses of conidia, each of which repeats the germination process. The more complicated sexual cycle of reproduction begins when a conidium of one mating type (*A* or *a*) encounters a fruiting body of the opposite mating type (*a* or *A*, respectively). The conidium enters the fruiting body as it divides by mitosis and fertilizes several of the maternal nuclei. Each resulting zygote is enclosed in a saclike structure called the ascus. Meiosis and a subsequent mitotic division yield mature asci, each containing four pairs of ascospores referred to as a tetrad. Release of ascospores allows their germination into hyphae and a repeat of either the sexual or asexual cycle.

a branched system of tubes collectively referred to as the **mycelium.** The haploid mycelium releases masses of haploid conidia, which, upon germination, repeat the vegetative reproductive cycle. Note that in *Neurospora,* all vegetative reproduction is derived from and results in haploid cells.

Sexual reproduction involves the union of cells of opposite mating types. There are two mating types in *Neurospora,* designated *A* and *a.* Sexual reproduction is initiated when the mycelium produces fruiting bodies, each of which contains many haploid maternal nuclei. Specialized mating filaments that recognize the opposite mating type extend

from each fruiting body. If a hypha or conidium of one mating type contacts a mating filament of the opposite mating type, it moves into the fruiting body by successive mitotic divisions and fertilizes several of the maternal nuclei. Each fertilization yields a diploid zygote nucleus that is enclosed in an elongated sac called an **ascus.** A mature fruiting body may contain a hundred or more such asci.

The diploid phase of *Neurospora* is of short duration. The zygote within each ascus immediately undergoes meiosis, yielding the four haploid products, or **tetrad,** that will be analyzed. Meiosis in *Neurospora* is followed by a single

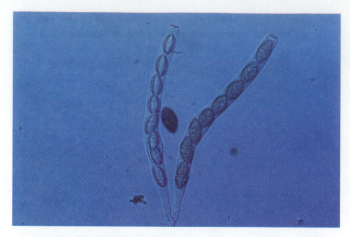

► FIGURE 13–12 The eight ascospores in a *Neurospora* ascus.

Determining Gene-Centromere Distance in *Neurospora*

In *Neurospora* the ascus is so narrow that it prevents the meiotic and mitotic products from slipping past each other, and thus the eight ascospores remain in a row in fixed positions relative to one another. We can see from ► Figure 13–13 that the particular linear order of the ascospores in the ascus reflects the linear order in which the chromosomes separated during the first and second meiotic divisions. We observe not only a 2 *A* : 2 *a* segregation ratio following meiosis but also a linear order that reads either *AAaa* or *aaAA*. The precise linear order of the ascospores can be detected experimentally by individually dissecting the ascospores from their ascus and germinating them into hyphae (► Figure 13–14). Each ascospore can thus be analyzed genetically on the basis of its phenotype when grown in culture. Because the ascospores are arranged in the order in which they were produced by meiosis, *Neurospora* is said to produce **ordered tetrads**.

A tetrad order of *AAaa* (or *aaAA*) directly demonstrates that the chromosome bearing the *A* allele segregated from

mitotic division that produces eight haploid cells called **ascospores** (see ► Figure 13–12). Once released from the ascus, the ascospores complete the reproductive cycle by germinating again into haploid hyphae.

► **FIGURE 13–13** Meiosis and subsequent mitosis in a *Neurospora* ascus. Because of the narrow structure of the ascus, meiotic products remain in a linear order that reflects the meiotic segregation events.

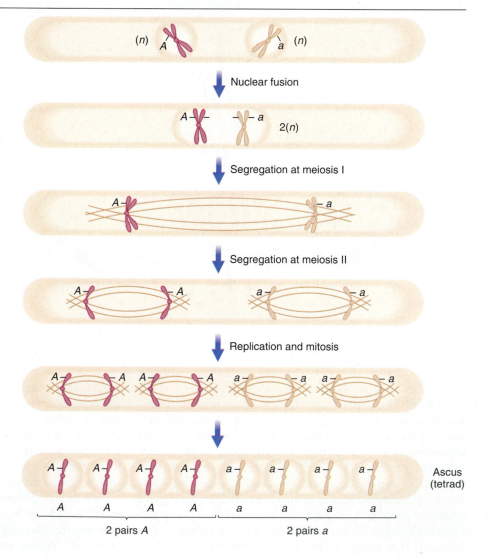

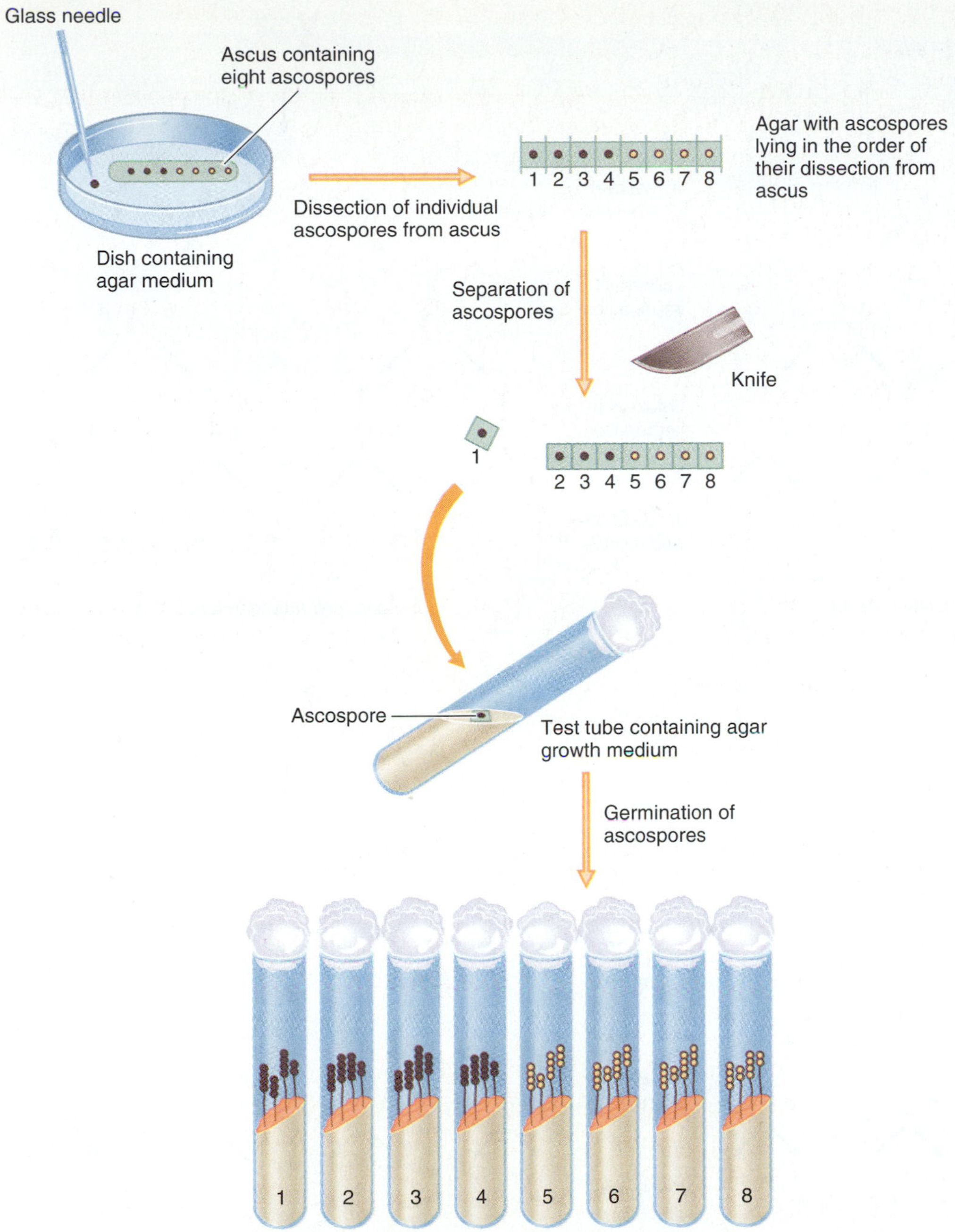

Glass needle

Ascus containing
eight ascospores

Dish containing
agar medium

Dissection of individual
ascospores from ascus

Agar with ascospores
lying in the order of
their dissection from
ascus

1 2 3 4 5 6 7 8

Separation of
ascospores

Knife

1

2 3 4 5 6 7 8

Ascospore

Test tube containing agar
growth medium

Germination of
ascospores

1 2 3 4 5 6 7 8

▶ FIGURE 13–14
Technique used to isolate
ascospores produced by a
Neurospora cross.

its homolog bearing an *a* allele at the first meiotic division (refer to Figure 13–13 to see why this is so). Thus the *AAaa* and *aaAA* arrangements are called **first-division segregation (FDS)** patterns.

In addition to the two FDS patterns, four other tetrad arrangements are also observed: *AaAa*, *aAaA*, *aAAa*, and *AaaA*. Each of these tetrad arrangements can be explained by crossing-over. ▶ Figure 13–15 shows that when the four different combinations of nonsister chromatids participate in a single exchange between the locus and the centromere, one of these four tetrad orders will result. In each case, a crossover brings the *A* and *a* alleles together on the same homolog and thereby delays the segregation of these alleles

until the second meiotic division. Hence, these four tetrad arrangements are referred to as **second-division segregation (SDS)** patterns.

Since an SDS pattern results from crossing-over, the total frequency of SDS asci produced by a cross in *Neurospora* can be used to estimate the distance between a gene and the centromere. However, because only two of the four chromatids are actually involved in the crossover that produces the SDS tetrad, the gene-centromere distance is equal to half the percentage of the SDS asci.

The results shown in ■ Table 13–5 for the *Neurospora* cross *AB* × *ab* are an example of the kind of data that can be used for evaluating gene-centromere distance. To determine

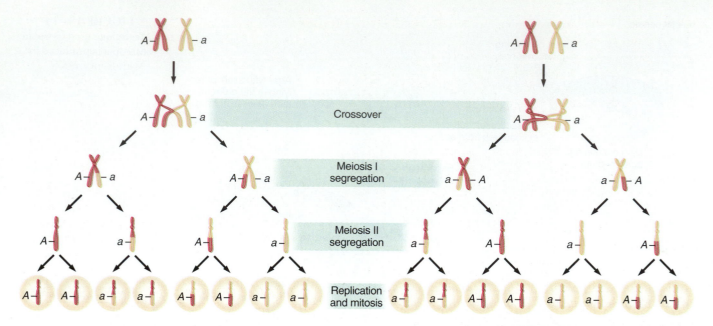

(a) Spore pair arrangement: *AaAa*

(b) Spore pair arrangement: *aAaA*

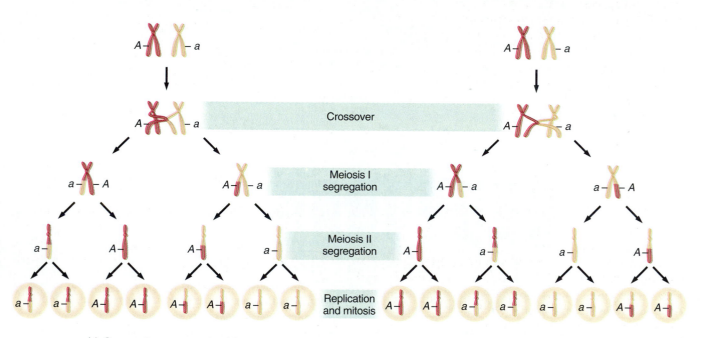

(c) Spore pair arrangement: *aAAa*

(d) Spore pair arrangement: *AaaA*

▶ **FIGURE 13–15** Second-division segregation patterns in a *Neurospora* ascus, each caused by a single crossover between the centromere and the gene *a* locus. Since the crossover can involve any pair of nonsister chromatids, four tetrad arrangements are possible: *AaAa*, *aAaA*, *aAAa*, and *AaaA*.

the segregation patterns, we consider one gene pair at a time by looking only at the spore-pair arrangements of the gene of interest. Thus, in the case of the *A/a* gene pair, only 100 of the 400 tetrads (class 3) show an SDS pattern, indicating

$$\frac{1}{2} \times \frac{100}{400} = 12.5 \text{ map units}$$

between the *a* locus and its centromere. In the case of the *B/b* allele pair, only 32 of the 400 tetrads (class 4) show an SDS pattern, indicating

$$\frac{1}{2} \times \frac{32}{400} = 4.0 \text{ map units}$$

between the *b* locus and its centromere.

■ TABLE 13–5 The classes and numbers of tetrads produced by the *Neurospora* cross *AB × ab*.

	Classes			
	1	2	3	4
Spore pair 1	*AB*	*Ab*	*AB*	*AB*
Spore pair 2	*AB*	*Ab*	*aB*	*Ab*
Spore pair 3	*ab*	*aB*	*Ab*	*aB*
Spore pair 4	*ab*	*aB*	*ab*	*ab*
	134	134	100	32 = 400 Total

Mapping Two Genes by Tetrad Analysis

Tetrad analysis can also provide information on the linkage relationships of two different gene loci. Linkage (or the lack of it) is established for genes by considering the combined segregation patterns of two pairs of alleles. In a two-point cross, such as *AB × ab*, the two gene pairs segregate together to form three types of tetrads: One type, called a **parental di-type (PD)** tetrad, contains ascospores with only the parental genotypes (in this case, *AB* and *ab*); a second type, called a **nonparental ditype (NPD)** tetrad, contains ascospores with only recombinant genotypes (in this case, *Ab* and *aB*); and a third type, called a **tetratype (T)** tetrad, contains ascospores with both parental and recombinant genotypes (in this case, *AB, ab, Ab*, and *aB*). Since all of the meiotic products in NPD tetrads are recombinant, but only half of the products in T tetrads are recombinant, the recombination frequency is given by

$$R = f(\text{NPD}) + \frac{1}{2}f(\text{T}) \qquad (13.5)$$

where $f(\text{NPD})$ and $f(\text{T})$ are the frequencies of the NPD and T tetrad classes, respectively.

When the genes under consideration are on different chromosomes, the PD and NPD tetrads result from the two possible chromosome alignments shown in ▶ Figure 13–16a. Thus, for independent assortment, the frequencies of the PD and NPD tetrads are equal, regardless of the frequency of T tetrads (in which half the spores are always recombinant and half are parental). This result is also shown by the data in Table 13–5, where the frequency of tetrad class 1 (PD tetrads) equals the frequency of tetrad class 2 (NPD tetrads). Tetrad classes 3 and 4 are T tetrads, which result from a crossover between a gene locus and its centromere (see Figure 13–16b). It should be stressed that the results

▶ **FIGURE 13–16** Origin of the three types of tetrads—PD, NPD, and T—for genes located on separate pairs of chromosomes. The cross is *AB × ab*. (a) The PD and NPD tetrads occur with equal frequency. (b) The T tetrads arise from crossing-over between a gene and its centromere. Only one of several possible T arrangements is shown here.

Cross: *AB × ab*
(a)

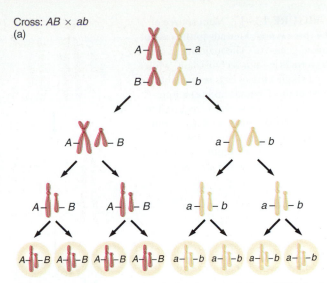

Spore pair arrangement: *AB, AB, ab, ab* parental ditype (PD) tetrad

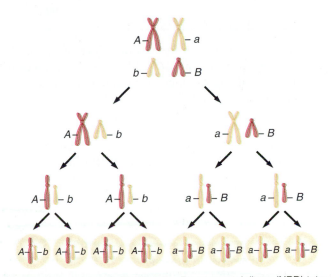

Spore pair arrangement: *Ab, Ab, aB, aB* nonparental ditype (NPD) tetrad

(b)

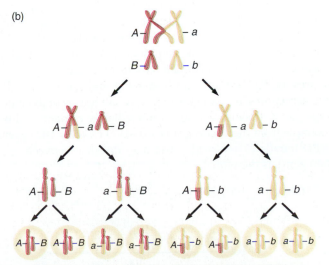

Spore pair arrangement: *AB, aB, Ab, ab* tetratype (T) tetrad

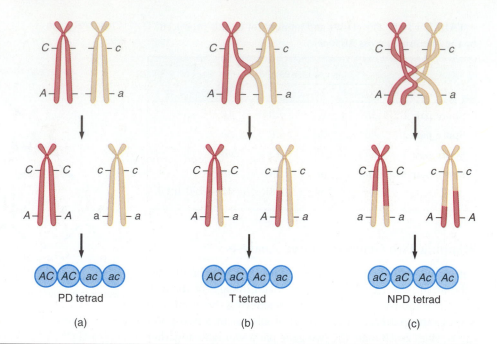

▶ **FIGURE 13–17** Main source of the three types of tetrads in the case of linked genes. (a) The majority of the PD tetrads are nonrecombinant, while (b) a single crossover is responsible for most of the T tetrads and (c) a four-strand double crossover is required to produce NPD tetrads. Linkage is thus manifested as an excess of PD over NPD tetrads. (For simplicity, only one member of each spore pair is shown.)

PD tetrad   T tetrad   NPD tetrad

(a)   (b)   (c)

in Table 13–5 do not differ from earlier observations on independent assortment, concerning the equal occurrence of all genotypes among the meiotic products. If we were to pool the ascospores from the different classes of asci, each of the four genotypes would make up one-fourth of the total number of ascospores present.

In contrast to independent assortment, linkage of two genes results in a large excess of PD over NPD tetrads. Consider the cross $AC \times ac$ in *Neurospora* in which the genes are linked. The most likely origins of the three major tetrad classes in this cross are shown in ▶ Figure 13–17. Observe that for linked genes, an absence of crossing-over will produce a PD tetrad, while a single exchange between the loci in question will result in a T tetrad; a relatively rare four-strand double exchange is now the simplest origin of a NPD tetrad. Thus, when genes are linked, NPD tetrads are expected to occur in the lowest frequency.

As an example of mapping by tetrad analysis, suppose that the cross $AC \times ac$ yields the classes and numbers of tetrads given in ■ Table 13–6. We see that classes 1 and 5, which contain only parental-type ascospores, are PD tetrads; classes 2 and 6, which contain only recombinant-type ascospores, are NPD tetrads; and classes 3, 4, and 7, in which half the ascospores are parental-type and half are recombinant-type, are T tetrads. Linkage is immediately evident in this example from the excess of PD tetrads ($175 + 7 = 182$) over NPD tetrads ($5 + 6 = 11$). The map distance between the two genes is therefore

$$100R_{a-c} = 100[f(NPD) + \frac{1}{2}f(T)]$$

$$= 100 \frac{5 + 6 + \frac{1}{2}(80 + 120 + 7)}{400}$$

$$= 28.6$$

▶ Figure 13–18 shows the origin of each of the tetrad classes listed in Table 13–6, along with the segregation pattern (FDS or SDS) of each pair of alleles. By combining the SDS asci for each gene pair, the distance between the genes and their centromere can be calculated as follows:

$$\begin{array}{ll} \text{centromere to} \\ \text{locus } a \text{ distance} \end{array} = \frac{1}{2}(\%SDS_{A,a})$$

$$= \left(\frac{1}{2}\right)\frac{(80 + 7 + 6 + 7)(100)}{400}$$

$$= 12.5$$

$$\begin{array}{ll} \text{centromere to} \\ \text{locus } c \text{ distance} \end{array} = \frac{1}{2}(\%SDS_{C,c})$$

$$= \left(\frac{1}{2}\right)\frac{(120 + 7 + 6 + 7)(100)}{400}$$

$$= 17.5$$

■ **TABLE 13–6** The classes and numbers of tetrads produced by the *Neurospora* cross $AC \times ac$.

	Classes						
	1	2	3	4	5	6	7
Spore pair 1							
AC	Ac	AC	AC	AC	Ac	AC
Spore pair 2							
AC	Ac	aC	Ac	ac	aC	ac
Spore pair 3							
ac	aC	Ac	aC	AC	Ac	Ac
Spore pair 4							
ac	aC	ac	ac	ac	aC	aC
175	5	80	120	7	6	7 = 400 Total

Tetrad class	Segregation pattern		Origin	Tetrad type
1	AC AC ac ac	A,a FDS	C,c FDS	No exchange	PD
2	Ac Ac aC aC	FDS	FDS	Four-strand double exchange — both in I or both in II	NPD
3	AC aC Ac ac	SDS	FDS	Two-strand single exchange — in I	T
4	AC Ac aC ac	FDS	SDS	Two-strand single exchange — in II	T
5	AC ac AC ac	SDS	SDS	Two-strand double exchange — one in I and one in II	PD
6	Ac aC Ac aC	SDS	SDS	Four-strand double exchange — one in I and one in II	NPD
7	AC ac Ac aC	SDS	SDS	Three-strand double exchange — one in I and one in II	T

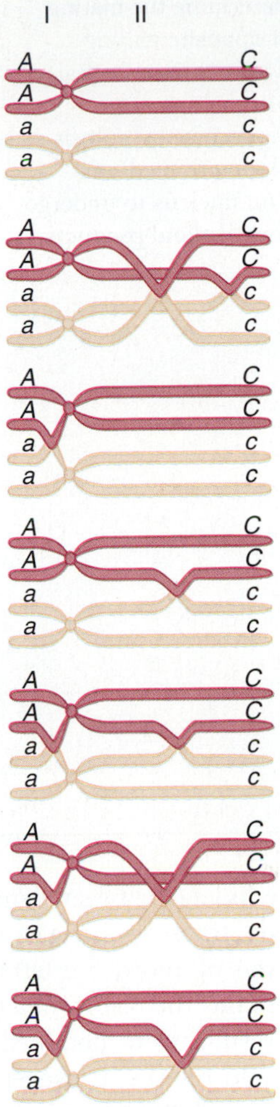

▶ **FIGURE 13–18** The origin of the seven classes of ordered tetrads arising from the *Neurospora* cross that generated the data given in Table 13–6. The recombination frequency between the genes is estimated as $R = f(\text{NPD}) + \frac{1}{2}f(\text{T})$. This formula assumes that the vast majority of NPD tetrads have undergone a double exchange, whereas nearly all the T tetrads result from a single exchange. The formula underestimates the distance somewhat, because the occasional triple-exchange NPD tetrads and some double-exchange T tetrads are not accounted for, and also because a few PD tetrads (class 5) have undergone crossing over but are not included in the calculation of the recombination frequency.

We can conclude that these genes lie on opposite arms of their chromosome from the fact that the distance between the *a* and *c* loci (determined to be 28.6) is compatible with the sum of the gene-centromere distances, rather than with their difference. The map is thus constructed as follows:

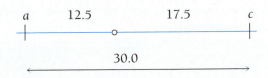

Unordered Tetrads

In contrast to *Neurospora*, sexual reproduction in many organisms, such as yeast and certain algae, produces **unordered tetrads**, in which the four meiotic products are jumbled together rather than occurring in a precise linear array. Common baker's yeast, *S. cerevisiae*, is an example (▶ Figure 13–19). The haploid cells of *S. cerevisiae* exist in either of two mating types (*a* and *α*) that divide mitotically by budding or undergo fusion to form *a/α* diploid cells. These diploid cells will also reproduce asexually as long as the medium is nutritionally suitable. Under conditions of nutritional deficiency, however, the *a/α* cells undergo meiosis to form a tetrad of haploid ascospores in a thick-walled, spherical ascus. In contrast to *Neurospora*, the yeast ascospores are arranged randomly within the ascus, so that the order in which the meiotic products are released is purely arbitrary.

The analysis of unordered tetrads is basically the same as previously described for *Neurospora*, except that FDS and SDS patterns cannot be (directly) determined and used to map the centromere. Since the tetrads need not be ordered for us to classify them as PD, NPD, or T, unordered tetrads

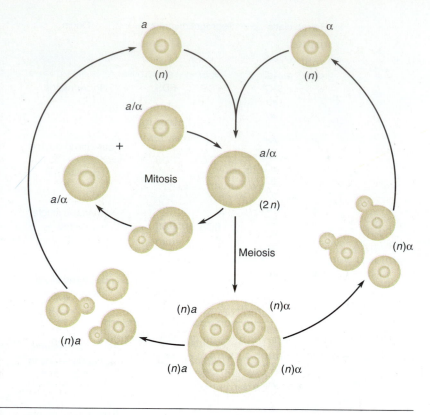

▶ **FIGURE 13–19** Life cycle of baker's yeast. The nuclear *a* and α alleles determine the mating type. Fusion between cells of opposite mating types produces a diploid cell that contains equal proportions of cytoplasm from each parental cell. Diploid cells normally reproduce by mitosis (budding). Plating on special growth medium, however, can induce the fusion nucleus to undergo meiosis instead, producing four haploid products.

can be used to estimate the linkage between genes in much the same manner as ordered tetrads, applying equation 13.5.

An example of the analysis of unordered tetrads is given in ■ Table 13–7, which contains the results of a three-point cross $ab^+c^+ \times a^+bc$. The results were obtained by allowing each ascus to mature and release its ascospores itself, thus avoiding the tedious task of dissecting individual asci. Table 13–7 shows 191 tetrads of nine different types that were recovered and analyzed genetically. Because of the wide variety of exchange events that can occur in the intervals that separate the genes, simultaneous analysis of all three loci in the tetrad is not possible. Instead, we analyze the genes in pairs, beginning in this case with the *a*–*b* region. Tetrad classes 1, 3, and 8 are PD with respect to the *a* and *b* loci (ignoring the *c* locus); these three classes contain only ab^+ and a^+b ascospores, which are the parental arrangements. Class 9 is the only NPD tetrad, and classes 2, 4, 5, 6, and 7 are tetratype with respect to these two loci. Thus, linkage of *a* and *b* is demonstrated by the excess of PD and

T tetrads over the number of NPD tetrads. We can now find the distance separating these loci by using equation 13.5:

$$100R_{a-b} = 100 \left[\frac{1 + \left(\dfrac{1}{2}\right)(54 + 2 + 6 + 3)}{191} \right]$$

$$= 19.1$$

The other two intervals (*a* to *c* and *b* to *c*) can be analyzed in a similar fashion and give the results shown in ■ Table 13–8.

To Sum Up

1. Tetrad analysis is a useful method for mapping genes in haploid eukaryotes. Linkage is determined from the relative frequencies of parental ditype (PD), nonparental ditype (NPD), and tetratype (T) tetrads. The frequencies of PD and NPD tetrads

■ **TABLE 13–7** The classes and numbers of unordered tetrads produced by the three-point cross $ab^+c^+ \times a^+bc$ in yeast. Since the tetrads are unordered, the order in which the spores are listed within each tetrad is purely arbitrary.

Classes								
1	2	3	4	5	6	7	8	9
$a\ b^+c^+$	$a\ b^+c^+$	$a\ b^+c^+$	$a\ b^+c^+$	$a\ b^+c^+$	$a\ b^+c$	$a\ b^+c$	$a\ b^+c$	$a\ b\ c^+$
$a\ b^+c^+$	$a\ b\ c$	$a\ b^+c$	$a\ b\ c^+$	$a\ b\ c$	$a\ b\ c^+$	$a\ b\ c$	$a\ b^+c$	$a\ b\ c$
$a^+b\ c$	$a^+b^+c^+$	$a^+b\ c^+$	a^+b^+c	a^+b^+c	$a^+b^+c^+$	$a^+b^+c^+$	$a^+b\ c^+$	$a^+b^+c^+$
$a^+b\ c$	$a^+b\ c$	$a^+b\ c$	$a^+b\ c$	$a^+b\ c^+$	$a^+b\ c$	$a^+b\ c^+$	$a^+b\ c^+$	a^+b^+c
14	54	103	2	6	3	6	2	1

EXTENSIONS AND TECHNIQUES

Estimating the Crossover Frequency by Tetrad Analysis

We have seen that the frequency of recombination (equation 13.5) can be used to measure the distance between gene loci. However, since tetrad analysis provides information on the frequencies of both single- and double-crossover events, it can also be used for directly estimating the crossover frequency (d in equation 13.2) to provide a more useful measure of map distance.

If we limit our analysis to the single- and double-crossover classes, the average number of crossovers per bivalent ($2d$) can be expressed as

$$2d = P(1) + 2[P(2)]$$

where $P(1)$ and $P(2)$ are the proportions of bivalents with one or two crossovers, respectively. Recall that bivalents with single crossovers give rise to only T tetrads, while bivalents with double crossovers give rise to PD, T, and NPD tetrads in a ratio of 1:2:1 (assuming that double cross-

overs are random between chromatids). By subtracting the T tetrads arising from double crossovers (equal to twice the frequency of NPD tetrads), the proportion of bivalents with one crossover can be estimated as

$$P(1) = f(T) - 2f(NPD)$$

Moreover, since one-fourth of all double crossovers yield NPD tetrads, the proportion of bivalents with two crossovers can be estimated as

$$P(2) = 4f(NPD)$$

Making the substitutions for $P(1)$ and $P(2)$ in the equation for $2d$, we get

$$2d = [f(T) - 2f(NPD)] + 2[4f(NPD)]$$
$$= f(T) + 6f(NPD)$$

The crossover frequency per chromatid pair (d) will then be

$$d = \frac{1}{2}f(T) + 3f(NPD) \qquad (13.6)$$

Comparing this expression with equation 13.5

$$R = \frac{1}{2}f(T) + f(NPD)$$

we see that the recombination frequency underestimates the crossover frequency measured by equation 13.6 by an amount equal to $2f(NPD)$. For example, suppose that $f(T) = 0.295$ and $f(NPD) = 0.0134$. Substituting these values into the expressions for R and d, we get $R = 0.16$ and $d = 0.187$— a difference in this case of $2f(NPD) = 0.027$.

Equation 13.6 is useful for estimating d from tetrad data. The value obtained by this procedure is accurate up to map distances of 35–40 cM. At greater distances, triple and higher-order exchanges become important and will then lead to underestimates of the true value of d.

---

are expected to be the same for unlinked genes, while linkage results in more PD than NPD tetrads.

2. In a tetrad analysis, the recombination frequency between genes is calculated as

$$R = f(NPD) + \frac{1}{2}f(T)$$

where $f(NPD)$ and $f(T)$ are the frequencies of nonparental ditype and tetratype tetrads, respectively. This relationship holds for both ordered and unordered tetrads. In ordered tetrads, the distance between a gene locus and the centromere is half of the percentage of second-division segregation (SDS) tetrads.

■ **TABLE 13–8** Evaluation of map distances from the three-point tetrad data given in Table 13–7.

	Intervals				
	$a–b$	$a–c$	$b–c$		
%PD	62.3	8.4	35.6		$100R_{a-b} = 0.5 + \frac{1}{2}(37.2) = 19.1$
%NPD	0.5	4.2	1.0		$100R_{a-c} = 4.2 + \frac{1}{2}(87.4) = 47.9$
%T	37.2	87.4	63.4		$100R_{b-c} = 1.0 + \frac{1}{2}(63.4) = 32.7$

Map: a 19.1 b 32.7 c

51.8

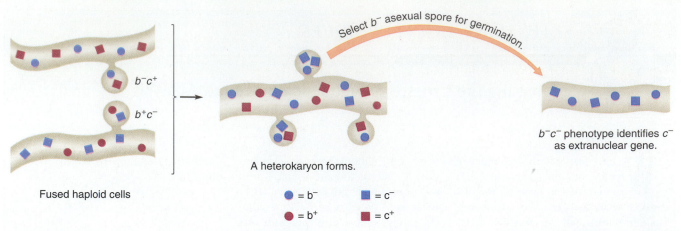

Select b^- asexual spore for germination.

b^-c^+

b^+c^-

A heterokaryon forms.

b^-c^- phenotype identifies c^- as extranuclear gene.

Fused haploid cells

● = b^- ■ = c^-

● = b^+ ■ = c^+

▶ **FIGURE 13–20** The heterokaryon test is used to detect cytoplasmic inheritance in filamentous fungi. A haploid strain carrying a known nuclear mutation (b^-) is fused with a strain carrying a mutation (c^-) of unknown location but suspected to be extranuclear. Uninucleate asexual b^- spores are then selected from the heterokaryon. Since there is no nuclear recombination in the heterokaryon, these spores can be c^- only if the c^- gene is inherited through the cytoplasm.

MAPPING ORGANELLAR GENES

Special techniques have also been developed for mapping genes on organellar chromosomes. However, the application of these techniques is complicated by the fact that both nuclear and extranuclear genes affect organelle function. Therefore, before a mutation can be assigned to an organellar chromosome, the possibility that it is present on a nuclear chromosome must be ruled out.

The **heterokaryon test** is useful for distinguishing between nuclear and cytoplasmic mutations in *Neurospora* and in other filamentous fungi. A **heterokaryon** is a cell with two genetically distinct nuclei. In *Neurospora*, two haploid cells of the same mating type will often fuse and produce stable dikaryons that are capable of undergoing cell division (▶ Figure 13–20). If the two haploid cells carry different genetic markers, the dikaryon is referred to as a heterokaryon. In the heterokaryon test, heterokaryons are formed by fusing cells having a known nuclear genetic marker with cells carrying a mutation of unknown location that is believed to be present on an organellar chromosome. Uninucleate asexual spores (conidia) are then selected from the heterokaryons and tested for the presence of both ge-

netic markers. In heterokaryons, the two cytoplasms mix but the nuclei remain separate, so that genes in the different nuclei cannot recombine. Therefore, the presence of both markers in a conidium rules out nuclear inheritance of the gene in question, since a uninucleate spore can acquire and express both markers only if this gene is indeed present in the cytoplasm.

Nuclear and organellar mutations can also be distinguished through analysis of their segregation ratios in a genetic cross. This approach is particularly helpful in those fungi where the non-Mendelian segregation ratios characteristic of organellar genes can be identified by tetrad analysis. One interesting example involves the **petite mutation**, which affects the mitochondria of yeast. When grown on agar medium, petite yeast form very small colonies compared with those of nonmutant or **grande** yeast. Petite mutants lack functional mitochondria (see ▶ Figure 13–21) and thus they are deficient in respiratory function because they are unable to produce ATP through oxidative phosphorylation. Some ATP is provided by cellular fermentation, but not enough to produce the larger colonies that are characteristic of the nonmutant cells. Genetic studies have distinguished three types of petite mutants. Two of them,

▶ **FIGURE 13–21** A comparison of the mitochondria of (a) colonies of grande (normal) yeast and (b) petite yeast.

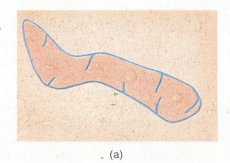

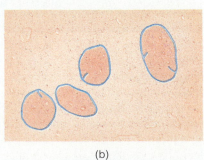

(a)

(b)

OXPHOS Diseases

A variety of neuromuscular diseases in humans have recently been shown to be associated with defects in oxidative phosphorylation (OXPHOS) due to mitochondrial mutations. Since OXPHOS is the primary source of energy for several organ and tissue systems, including the brain, heart, kidney, liver, pancreatic islets, and muscle, these diseases involve a variety of degenerative processes.

OXPHOS genetics is very complex, not only because of the large number of proteins involved in oxidative phosphorylation but also because the genes that code for these proteins are distributed throughout both nuclear and mitochondrial DNA (mtDNA). In addition, the mitochondrial genes are maternally inherited. Each cell contains thousands of copies of the mitochondrial genome and can harbor various mixtures of mutant and normal mtDNAs, a condition referred to as heteroplasmy. Since this mixture is randomly partitioned into daughter cells during cell division, the mtDNA genotype fluctuates from one cell division to the next. Eventually most of the cell lines drift toward either predominantly mutant or predominantly normal mtDNA. Because the mutation rate of mt genes is substantially higher than that of most nuclear genes, however, many cells are newly heteroplasmic. The severity of an OXPHOS defect therefore depends not only on the mt gene that is mutated but also on the proportion of newly mutant mtDNAs in particular cell types. The severity of these diseases usually also increases with age, a phenomenon that may be a consequence of accumulated damage to the mitochondria and the mtDNA by oxygen free radicals.

Despite these complexities, a number of diseases have been associated with particular mutational defects in certain proteins of OXPHOS. These diseases include Kearns-Sayre syndrome (KSS, which can include paralysis of the eye muscles, retinal degeneration, heart block, hearing loss, and dementia); Leber's hereditary optin neuropathy (LHON, a form of sudden-onset vision loss); neurogenic muscle weakness, ataxia, and retinitis pigmentosum (NARP, characterized by seizures and dementia); and Pearson's syndrome (which involves loss of blood cells, pancreatic fibrosis, and splenic atrophy). These diseases and others associated with mutant mitochondria are characterized by a maternal inheritance pattern (except in those cases where a nuclear gene is involved along with the mtDNA), an age-related expression of the symptoms, a correlation between the severity of the disease and the number of mutant mitochondria, and a high rate of occurrence in individuals with no family history of the disorder.

Researchers have recently formulated an OXPHOS paradigm as a hypothesis to explain age-relative degenerative diseases and the aging process itself. According to this paradigm, each individual is born with an initial OXPHOS capacity that declines with age; when the OXPHOS capacity falls below the energetic threshold of an organ, the symptoms of disease appear. Individuals who are born without mutations that affect oxidative phosphorylation reactions usually live out their lifespans before organ thresholds are reached, while those born with OXPHOS mutations start at lower OXPHOS capacities and therefore cross threshold levels in their lifetimes. This hypothesis is currently being applied to well-known disorders such as ischemic heart disease, late-onset diabetes mellitus, Parkinson's disease, Alzheimer's disease, and Huntington's disease in efforts to better understand their causes and clinical symptoms. However, no mutant OXPHOS genes that predispose individuals to these disorders have yet been identified. If such mutant genes are discovered, the metabolic therapies that are already being developed for the known OXPHOS mtDNA diseases may allow treatment and even prevention of the better-known disorders.

The OXPHOS hypothesis is: Individuals who are born without mutations that affect oxidative phosphorylation reactions are usually healthy in later years.

the cytoplasmic (rho⁻) petites, show cytoplasmic inheritance, while the third type, the nuclear or **segregational** (*pet⁻*) **petites**, shows nuclear inheritance. The two types of cytoplasmic petites are classified as **neutral** petites, which have lost essentially all mitochondrial DNA and thus retain no mitochondrial genes, and **suppressive petites**, whose mitochondrial DNA has large, random deletions.

The two types of cytoplasmic petites can be distinguished from segregational petites and from each other on the basis of cross results (see ▶ Figure 13–22). When

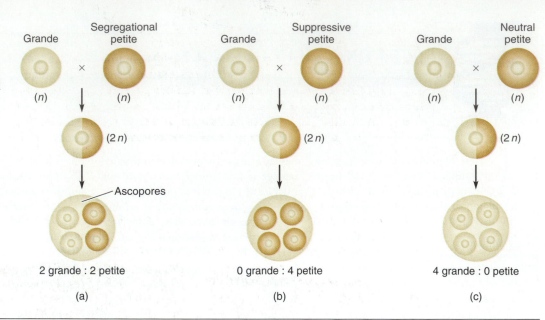

FIGURE 13-22
Crosses of grande (wild-type) yeast to various petite mutants. The cross involving the segregational petite yields a Mendelian ratio of 2 grande:2 petite offspring, which is a reflection of nuclear inheritance. In contrast, the crosses of grande × suppressive and grande × neutral petites yield progeny ratios that are more characteristic of cytoplasmic inheritance.

crossed to a grande cell, the segregational petite produces a tetrad of haploid ascospores with a Mendelian segregation ratio of 2 grande:2 petite. The cross of a cytoplasmic petite to a grande cell produces all grande ascospores in the case of neutral petite × grande and, when the diploid zygote undergoes meiosis immediately, all petite ascospores in the case of suppressive petite × grande. Thus when the mutant organellar genes are involved in the cross, the tetrads fail to give a 2:2 ratio and a non-Mendelian pattern is observed.

Mapping Mitochondrial Genes in Yeast

Mitochondrial genes have been mapped in yeast by taking advantage of various genetic markers on the mitochondrial chromosome. These markers include ant^R mutations, which provide resistance to a number of different antibiotics, syn^- mutations, which lead to an impaired system of mitochondrial protein synthesis, and mit^- mutations, which result in the loss of specific respiratory functions.

One approach to gene mapping has involved **recombination analysis**. In a cross of genetically dissimilar strains of yeast that differ in their mitochondrial DNA at two gene loci, both parental and recombinant types are observed among the progeny. These results indicate that mitochondrial chromosomes, like nuclear chromosomes, undergo recombination. It is thus possible to map the genes involved by computing the frequency of recombination. However, because of the high frequency of recombination shown by mitochondria, linkage is usually undetectable by this method unless the genes are very close to one another on the mitochondrial chromosome. This approach has therefore been limited to mapping closely linked markers.

Another mapping technique that has proved useful in yeast is based on **marker retention**. In this procedure, a mutagen such as ethidium bromide is used to induce random deletions in the mitochondrial DNA of a multiply drug-resistant strain of yeast, thus producing various suppressive petites. Each petite strain produced in this way is then tested to determine which resistant markers are retained. Since the deletions are randomly induced, it is likely that different drug-resistant genes will be lost in different strains. The deletions can be quite extensive, so the simultaneous loss of more than one resistant marker is possible. A genetic analysis of these strains might yield results similar to those given in ▶ Figure 13–23. Note that the marker

▶ **FIGURE 13–23** (a) Hypothetical data from a marker retention analysis. Each petite strain has retained a particular combination of resistance markers. (b) Combining the data from the different strains gives a group of overlapping chromosome segments that fit together into a circular map.

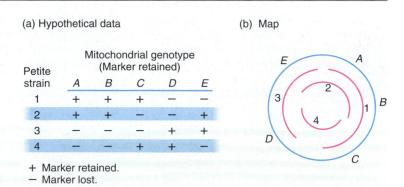

(a) Hypothetical data

Petite strain	Mitochondrial genotype (Marker retained)				
	A	B	C	D	E
1	+	+	+	−	−
2	+	+	−	−	+
3	−	−	−	+	+
4	−	−	+	+	−

+ Marker retained.
− Marker lost.

(b) Map

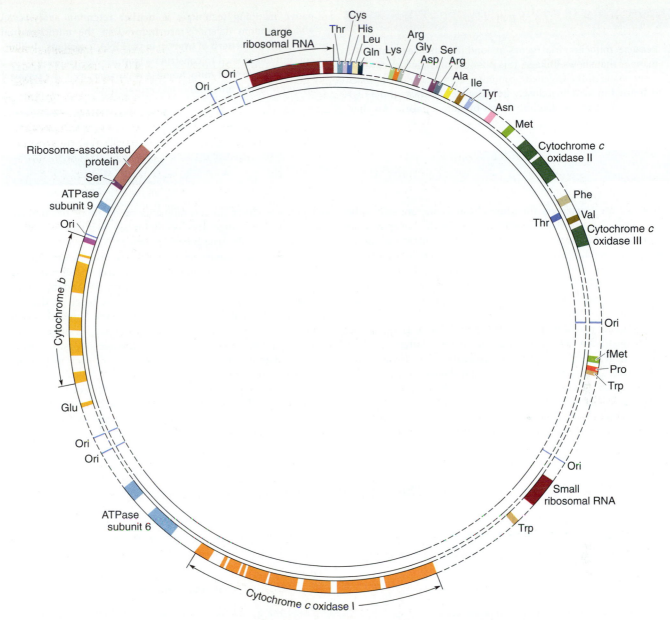

▶ **FIGURE 13–24** Map of yeast mitochondrial DNA showing genes coding for known proteins, for rRNA, and for tRNA, and unassigned genes, which presumably are genes coding for proteins that have not yet been identified. Dashed lines indicate portions of DNA that have not yet been mapped. *Adapted from:* L. A. Grivell, "Mitochondrial DNA," *Scientific American* 248 (1983): 78–89. Copyright © 1983 by Scientific American. All rights reserved.

retention data from the different strains can be combined to give a series of overlapping segments that can be arranged to form a map of the DNA molecule.

A map of yeast mitochondrial DNA is given in ▶ Figure 13–24. Unlike linkage maps of nuclear chromosomes, the map of the mitochondrial chromosome is circular, corresponding to the structure of the organelle DNA. This map is based on the results of traditional mapping techniques, including those just described, as well as the results of molecular methods that will be discussed in Chapter 15. The map shows the genes for ribosomal RNA and transfer RNA (RNAs involved in protein synthesis) and for a number of different proteins. Most RNAs are not free to move in and out of the mitochondrion, so the majority of the RNAs required by this organelle must be encoded by the mitochondrial DNA. In contrast, proteins, like many small molecules, can be transported across the mitochondrial membranes. As a result, only a few of the mitochondrial proteins have their genes in the mitochondrial DNA; most are encoded by nuclear genes and transferred to the mitochondria after synthesis.

To Sum Up

1. Because mitochondrial genes recombine and segregate in a manner that shows linkage, it is possible to construct chromosome maps of the organelle genomes. However, this approach is limited to mapping closely linked genes because of the high frequency of recombination shown by mitochondria. An alternative mapping technique is marker retention analysis, in which random deletions are induced in the mitochondrial DNA and the pattern of overlap in the DNA fragments of different mutants is used to construct a map.

2. The linkage map of mitochondria is circular, as is the physical structure of the mitochondrial chromosome.

Chapter Summary

Genes that are located on the same chromosome are said to be linked. Linked genes do not assort independently during meiosis but tend to remain together in the same arrangement they occupied on the parental chromosomes. When the alleles of two linked genes segregate, more than 50% of the gametes will be parental-type in gene arrangement and less than 50% will be recombinant-type. The recombinant gene arrangements are produced as a result of crossing-over.

The frequency of crossing-over between different genes can be used to construct a linkage map that shows the relative order and locations of genes on a chromosome. Since the actual frequency of crossing-over is seldom known, the map distance between linked genes is typically expressed in terms of their frequency of recombination. Recombination frequencies are linearly related to crossover frequencies only when genes are closely linked. Multiple crossovers tend to occur between more widely separated markers

and produce either no recombinants or no more recombinants than a single exchange; the recombination frequency is then an underestimate of the true genetic distance.

Recombination frequencies in diploid organisms are ordinarily calculated from the results of two-point or three-point testcrosses. Linkage analysis in these organisms is a form of random chromatid analysis, since inferences about linkage are based on the progeny formed from the products of many pooled meiotic events. Tetrad analysis, which is based on the products of single meiotic events, is another useful approach for linkage analysis in certain haploid eukaryotes that retain the products of each meiosis in a specialized structure over a period of time.

Special techniques have also been developed for mapping genes on organellar chromosomes. However, before a gene can be mapped to an organellar chromosome, the possibility of nuclear inheritance must first be ruled out.

Questions and Problems

Establishing Linkage between Genes

1. Define the following terms and distinguish between members of paired terms:
   linked and unlinked genes
   linkage map and linkage group
   parental-type and recombinant-type gametes
   coupling and repulsion linkage phase

2. A particular testcross involving genes a^+,a and b^+,b gives the following data among the offspring:

   $$882\ a^+ab^+b : 101\ a^+abb : 117\ aab^+b : 900\ aabb$$

   (a) Are the genes linked? Explain your answer.
   (b) Calculate the frequency of recombinant types.

3. The r and s loci in a certain organism are linked as follows:

   From a testcross of the dihybrid Rs/rS, determine the types and frequencies of the expected offspring.

4. In tomatoes, the gene for round fruit shape (O) is dominant to its allele for elongate shape (o), and the gene for smooth fruit skin (P) is dominant to its allele for peach (p). Two series of crosses involving these genes produced the following results:

P cross 1: round, smooth × elongate, peach
   F_1:   all round, smooth

P cross 2: round, peach × elongate, smooth
   F_1:   all round, smooth

When the F_1 of cross 1 were testcrossed (to doubly recessive plants), they produced F_2 offspring in the following proportions: 46% round, smooth; 4% round, peach; 4% elongate, smooth; and 46% elongate, peach. In contrast, testcrosses involving the F_1 of cross 2 gave the following results: 4% round, smooth; 46% round, peach; 46% elongate, smooth; 4% elongate, peach. Explain these results, giving the genotypes of both F_1s and their progeny.

5. Albinism in mice can arise as the homozygous expression of one (or both) of two recessive genes, c and d. Individuals must have the dominant alleles of both genes (C–D–) to express color. Suppose that testcrosses between a dihybrid male ($CcDd$) and albino $ccdd$ females produce 148 albino and 52 colored offspring. Are the two loci linked? Explain your answer.

6. Suppose that you are interested in determining whether the genes A,a and B,b are linked, and you have data on the phenotypic classes of offspring produced by a selfing of a particular dihybrid. Among these offspring, 4% have the doubly recessive phenotype ($aabb$). Are the gene loci linked? If so, what is the linkage phase of the heterozygous parent?

7. Refer to question 6. What would be the expected frequency of each of the other phenotypic classes of offspring produced by selfing the dihybrid *AaBb* plant?

Recombination Frequency as a Measure of Map Distance

8. Criticize the following statements:
   (a) Crossing-over always results in the formation of recombinant types.
   (b) Genes that are located on the same chromosome fail to assort independently.
   (c) A genetic map shows the dimensional relationships of genes along a chromosome.
   (d) The farther apart two genes are on a chromosome, the greater their frequency of recombination.
   (e) If 10% of the cells undergoing meiosis experience a crossover between loci *a* and *b*, then 10% of the gametes produced are expected to be recombinant for these genes.

9. Consider the following linkage data:

Gene loci:	a–b	b–c	c–d	d–e	c–e
Map distance:	8	6	2	4	6

   Construct a genetic map of these loci.

10. The following data were obtained from a series of two-factor testcrosses. Explain these results in terms of the relative locations of the genes on a linkage map.

Gene loci:	a–b	b–c	a–c	b–d	a–e	d–e	a–d
Percent recombination:	42	44	50	48	50	50	50

11. In rabbits, genes at two loci, *b* and *c*, interact to produce coat color; $B-C-$ individuals are black, $bbC-$ individuals are brown, and $--cc$ ($B-cc$ or *bbcc*) individuals are albino. Certain crossing data suggest that these loci are linked. In one series of testcrosses, matings between a dihybrid male and females of the doubly recessive albino strain resulted in 65 black, 34 brown, and 101 albino offspring. (a) Are the genes linked in the coupling or repulsion arrangement in the dihybrid parent? (b) Calculate the map distance between these loci. (c) The dihybrid male in this series of testcrosses was the product of a mating between members of two homozygous strains. What were the genotypes of these strains?

12. You discover a recessive mutant gene in *Drosophila* that is not reported in the scientific literature, and you want to determine on which chromosome this gene is located. Assume that you have access to two indicator strains of flies, one homozygous for a gene on chromosome II and another for a gene on chromosome III. Describe an experimental breeding procedure that would permit you to assign this mutant gene to one of the four chromosomes in this species.

13. In *Drosophila*, sable body color (*s*) and Bar eyes (*B*) are both sex-linked, separated by a genetic distance of 14 units. From a cross of heterozygous s^+B/sB^+ females to sable, wild males, what phenotypic classes are expected in the offspring? What are the proportions of these phenotypes?

14. In *Drosophila*, the *ec* and *y* gene loci (for echinus (rough) eye and yellow body color) are both sex-linked and are separated by a distance of 5.5 map units. The gene for dumpy wings (*dp*) is located on chromosome pair II. (a) What phenotypic ratio is expected among the offspring of a testcross of ec^+y/ecy^+ dp^+/dp females to fully recessive males? (b) What phenotypic ratio is expected among the offspring from a cross of the female parents in part (a) to ec^+y/Y dp^+/dp males?

15. Suppose that gene loci *s* and *t* are linked as follows:

    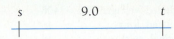

    What is the expected ratio of offspring produced by the dihybrid cross $s^+t/st^+ \times s^+t^+/st$?

16. In *Drosophila*, crossing-over occurs in females but not in males. Suppose that the dihybrid cross $pr^+ b/pr b^+ \times pr^+ b/pr b^+$ is made, in which the genes for the dominant red eyes (pr^+) vs. purple (*pr*) and the dominant wild-type body (b^+) vs. black (*b*) are in the repulsion phase. The genes for these traits are located six map units apart on chromosome II. (a) Predict the phenotypic ratio among the progeny of this cross. (b) Does this ratio directly indicate anything about how far apart the *pr* and *b* loci are on chromosome II? Explain your answer.

17. Three plants that are dihybrid for different pairs of gene loci are selfed. The frequency of double-recessive offspring produced by each mating is given below. In each case, determine whether the genes are linked, and if they are, determine the linkage phase in the dihybrid parent and calculate the map distance involved.

Genotype of self-pollinated plant:	AaBb	BbCc	CcDd
Frequency of double-recessive progeny:	1.00% *aabb*	6.25% *bbcc*	9.00% *ccdd*

18. The recessive genes for hemophilia A and red-green color blindness in humans are located about 10 map units apart on the X chromosome. The ABO locus is on chromosome 9. A normal man with type AB blood marries a normal, type AB woman whose mother is color-blind and whose father is hemophiliac. They plan on several children. (a) What is the chance that their first child is color-blind? (b) What is the chance that their second child is a boy with type A blood who is both hemophiliac and color-blind? (c) What is the chance that their third child will, like the parents, be normal in color vision and clotting time and have the AB blood type?

19. Nail-patella syndrome in humans is characterized by congenital abnormalities of the fingernails (and sometimes toenails) and the patellae (kneecaps). The gene for this disorder is dominant and is located on chromosome 9 about 10 map units away from the ABO locus. Suppose that a man with nail-patella syndrome and type A blood marries a normal woman with type B blood. The mothers of both the husband and wife are normal and have type O blood. (a) The husband and wife have two children, one with type A blood and the other with type B. What is the probability that both children have normal fingernails and patellae? (b) The couple are now expecting another child. What is the chance that the child will have nail-patella syndrome and type O blood? (c) Amniocentesis reveals that the fetus has type AB blood. What is the chance that it has nail-patella syndrome?

Three-Point Testcross Method for Mapping Genes

20. In *Drosophila*, the three gene pairs for red eyes (cn^+) vs. cinnabar (*cn*), normal bristle number (rd^+) vs. reduced (*rd*),

and long wings (vg^+) vs. vestigial (vg) are known to have their loci on chromosome II. Suppose that a three-point testcross yields the following offspring:

cinnabar, reduced, vestigial	406
cinnabar, reduced, long	46
cinnabar, normal, vestigial	28
cinnabar, normal, long	3
red, normal, long	438
red, normal, vestigial	45
red, reduced, long	33
red, reduced, vestigial	1

(a) Which progeny classes are the parental types? Which are the single-recombinant types? Which are the double-recombinant types? (b) Identify the most probable origin of each of the recombinant classes in terms of the number and location of crossovers. (c) Calculate the map distances between the genes, and construct a linkage map of these loci. (d) Determine the coefficient of coincidence for this set of loci.

21. In the seedling stage, a completely recessive corn plant was glossy (leaves have a shiny appearance), virescent (poor in chlorophyll), and liguleless (lacking certain appendages at the base of the leaves). This plant was crossed to a trihybrid that as a seedling had dull leaves, normal chlorophyll content, and ligules. Of the many seedlings produced in the next generation, a random sample of 1000 had the following characteristics:

dull, normal, with ligules	28
dull, normal, liguleless	179
dull, virescent, with ligules	69
dull, virescent, liguleless	250
glossy, normal, with ligules	198
glossy, normal, liguleless	70
glossy, virescent, with ligules	183
glossy, virescent, liguleless	23

(a) Calculate the map distances between the genes and construct a linkage map of these loci (glossy = gl, virescent = v, and liguleless = lg). (b) Give the genotype of the trihybrid parent, indicating the proper linkage phase on the pair of chromosomes.

22. A testcross was made using the trihybrid $PpRrSs$ as the dominant parent. The testcross progeny revealed that the trihybrid produced the following gametes:

11 PRS	15 pRS
238 PRs	240 pRs
242 PrS	230 prS
11 Prs	13 prs

(a) Which genes are linked and which assort independently? (b) Determine the arrangement (coupling or repulsion) of the linked genes in the trihybrid parent. (c) Calculate the map distances between linked loci.

23. Gene loci c and d are linked and are separated by a genetic distance of 10 map units. Gene h is located on a different chromosome. (a) What phenotypic ratio is expected among the offspring of the testcross

$$c^+d^+/cd \ \ h^+/h \ \times \ cd/cd \ \ h/h$$

(b) If the trihybrid parent in part (a) is selfed, what phenotypic ratio is expected among the offspring?

24. The genes a, b, and c are arranged as shown on the following linkage map:

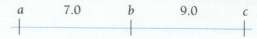

(a) Calculate the frequencies of the parental, single-recombinant, and double-recombinant classes expected among the progeny of the three-point testcross $ABC/abc \times abc/abc$, assuming that crossing-over occurs without interference. (b) Repeat your calculations, but now assume interference with a coefficient of coincidence of 0.5.

25. Suppose that genes a, b, and c are sex-linked in a certain mammal as follows:

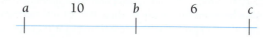

From a testcross of a trihybrid female of genotype a^+bc/ab^+c^+, what phenotypic classes are expected among the offspring and in what frequencies? (Assume no interference.)

26. Three two-point testcrosses gave the following recombination frequencies:

$$R_{p-q} = 12\% \qquad R_{q-r} = 8\% \qquad R_{p-r} = 5\%$$

(a) Construct a linkage map of the p, q, and r loci. Why are the distances not additive? (b) Calculate the frequencies of the parental, single-recombinant, and double-recombinant classes expected among the progeny of the three-point testcross $PpQqRr \times ppqqrr$, based on the recombination values given above. (c) Determine the coefficient of coincidence for this set of loci.

27. Consider the four-point testcross $ABCD/abcd \times abcd/abcd$. (a) How many recombinant classes can we possibly detect in the progeny of this cross? (b) Identify the most probable origin of each recombinant class in terms of the number and location of crossovers. (c) Suppose that data from previous two-point crosses indicate that the members of each of the pairs of gene loci (a–b, b–c, and c–d) are two map units apart. How many offspring must we examine from this four-point cross in order to observe about eight individuals in the lowest frequency class?

*28. Three gene markers in *Drosophila* give the following recombination frequencies:

$$R_{a-b} = 0.197 \qquad R_{b-c} = 0.316 \qquad R_{a-c} = 0.388$$

(a) Calculate the average number of crossovers per chromatid d between each pair of markers using the Haldane mapping function. Check the values for additivity. (b) Assuming independent crossovers, predict the percentage of bivalents in the trihybrid $+++/abc$ that involve a four-strand double crossover between the most distal markers.

Mapping by Tetrad Analysis

29. The cross $AB \times ab$ in *Neurospora* gives the following kinds and numbers of tetrads (spore pairs in each class of tetrads are listed vertically):

*An asterisk indicates that the question or problem is based on information in the Extensions and Techniques box.

Classes

	1	2	3	4
Spore pair 1:	AB	Ab	AB	AB
Spore pair 2:	AB	Ab	aB	Ab
Spore pair 3:	ab	aB	Ab	aB
Spore pair 4:	ab	aB	ab	ab
	243	242	120	55

Classes

6	7	8	9
+ + c	+ + c	+ b c	+ + c
+ b +	+ b c	+ b c	+ + c
a + +	a + +	a + +	a b +
a b c	a b +	a + +	a b +
3	2	1	5

(a) Identify each tetrad class as either PD, NPD, or T. (b) Show that the genes have assorted independently. (c) Calculate the percent SDS asci for each gene pair.

30. *Neurospora* of opposite mating types with the genotypes *CD* and *cd* are crossed. Analysis of the 389 ordered tetrads produced from this cross yields the following data:

Classes

	1	2	3	4	5	6	7
	CD	Cd	CD	CD	CD	Cd	CD
	CD	Cd	cD	Cd	cd	cD	cd
	cd	cD	Cd	cD	CD	Cd	Cd
	cd	cD	cd	cd	cd	cD	cD
	259	14	29	65	7	7	8

Determine whether the genes are linked. If they are linked, calculate the distance between the genes. What are the gene-centromere distances?

31. Gene pairs c^+,c (c = compact growth) and leu^+,leu (leu = requires the amino acid leucine) are observed in a mating of *Neurospora* strains $c^+ leu^+ \times c\ leu$. The following classes of ordered tetrads are obtained:

Classes

	1	2	3	4	5	6	7
	+ +	+ leu	+ +	+ +	+ +	+ leu	+ +
	+ +	+ leu	c +	+ leu	c leu	c +	c leu
	c leu	c +	+ leu	c +	+ +	+ leu	+ leu
	c leu	c +	c leu	c leu	c leu	c +	c +
	367	4	11	50	60	2	6

Determine the linkage relationship and the map distance between the genes.

32. Unordered tetrad analysis of a cross between $a^+b^+c^+$ and abc strains in yeast yields the following data. Determine the map distances (genes are in the order given) and explain the simplest origin of each tetrad class.

Classes

	1	2	3	4	5
	+ + +	+ + +	+ + +	+ + +	+ + +
	+ + +	+ b c	+ + c	+ b +	+ b c
	a b c	a + +	a b +	a + c	a + c
	a b c	a b c	a b c	a b c	a b +
	352	54	78	3	2

33. Unordered tetrad analysis of a cross between $d^+e^+f^+$ and *def* strains in yeast gives the following data (genes are linked but not necessarily in the order given). Determine the correct gene order and genetic distances.

Classes

	1	2	3	4	5
	+ + +	+ + +	+ + +	+ + +	+ + +
	+ + +	d + f	+ + f	d + +	d + f
	d e f	+ e +	d e +	+ e f	+ e f
	d e f	d e f	d e f	d e f	d e +
	299	42	90	10	11

	6	7	8	9	10
	+ + f	+ + f	d + f	+ + f	d + +
	d + +	d + f	d + f	+ + f	d + f
	+ e +	+ e +	+ e +	d e +	+ e +
	d e f	d e +	+ e +	d e +	+ e f
	10	12	5	20	1

34. Calculate the coefficient of coincidence for the data in problem 32.

Mapping Organellar Genes

35. Define the following terms and distinguish between members of paired terms:

cytoplasmic and nuclear inheritance
cytoplasmic and segregational petites
neutral and suppressive petites

36. There is a particular type of slow-growing mutant in *Neurospora* called *poky*. Crosses between wild-type females and poky males yield all wild-type offspring, while crosses between poky females and wild-type males yield all poky progeny. (a) What seems to be the inheritance mechanism of the poky characteristic? (b) Following the formation of a heterokaryon between arg^+poky and arg^-poky^+ cells, about one-fourth of all the uninucleate spores from the heterokaryon are totally wild-type (arg^+poky^+). Is this result consistent with your answer to part (a)? Explain your answer.

37. There are three genetic mechanisms that can cause reciprocal crosses to give different results among the progeny: (a) sex-linkage, (b) extranuclear inheritance, and (c) maternal effects. How could you distinguish experimentally among these three alternatives?

CHAPTER 14

Mapping Genes in Prokaryotic Systems

Because of their ease of handling and rapid reproductive rates, bacteria and their viruses have been used extensively for mapping studies. The genetic analysis of these prokaryotic systems is in many respects similar to that of eukaryotes. However, the different mechanisms of gene transfer in prokaryotes necessitate the use of special mapping techniques. Bacteria do not undergo the meiotic process characteristic of sexually reproducing eukaryotes, but many prokaryotic systems can be experimentally manipulated in such a way that a form of diploidy can occur, followed by synapsis and crossing-over. Recombination rates can then be measured and used to evaluate the relative positions of genes along the chromosome. In this chapter, we will consider various mapping techniques that have been developed for prokaryotic systems, focusing on the genetic analysis of bacteria and their viruses.

False-color TEM of newly synthesized and assembled T2 bacteriophages (black ovals) within the bacterium *Escherichia coli*.

MAPPING THE BACTERIAL GENOME

Bacteria have three main mechanisms of gene exchange (▶ Figure 14–1): **conjugation**, which involves the plasmid-directed transfer of genes during cell-to-cell contact; **transduction**, in which bacterial genes are transferred by a temperate phage; and **transformation**, the transfer of genes in the form of naked DNA. Of these mechanisms, conjugation has proved to be the most useful in mapping studies. A relatively large segment of the bacterial chromosome can be transferred during conjugation, so geneticists can map genes that are separated by large distances. A rough map of the entire bacterial chromosome can be obtained by combining data from several crosses involving different Hfr donors and many different genes. Such studies have revealed that the bacterial linkage map, like the structure of the bacterial chromosome, is circular.

Under normal circumstances, conjugation does not provide the resolving power needed to map genetic markers that are very closely spaced. Furthermore, it occurs in only a few bacterial species. These limitations have led to the use of transduction and, to a lesser extent, transformation for chromosome mapping. Like conjugation, both transduction and transformation involve the one-way transfer of a chromosome fragment from a donor cell to a recipient cell. Only relatively short lengths of chromosomal material are normally transferred by these methods, however, thus limiting their usefulness to the mapping of genes that lie very close together. In both cases, the recipient cell becomes a "partial diploid," since it has its own chromosome plus a fragment of the chromosome from the donor cell. Crossing-over between the recipient chromosome and the donor fragment can then yield recombinants whose proportions are related to the distance separating the genes involved.

Interrupted Conjugation

Conjugation (see Chapter 10) is a mating process in bacteria in which chromosomal DNA is transferred from an Hfr donor to an F⁻ recipient, following the initiation of replication of an integrated F factor in the Hfr cell. One of the strands of replicating DNA enters the F⁻ recipient through a cytoplasmic bridge—the conjugation tube—that physically connects the conjugating cells.

E. L. Wollman and F. Jacob first used conjugation for chromosome mapping in the late 1950s. They employed a method known as the **interrupted mating procedure**. Rather than allowing conjugation to proceed until the conjugation tube randomly breaks, the interrupted mating procedure controls the precise length of time allowed for the transfer of the donor chromosome by withdrawing samples from a population of conjugating cells at various time intervals and subjecting them to agitation in a blender. The shearing breaks the conjugation tubes that connect the conjugal pairs but does not irreparably damage the cells. By interrupting the mating process in this way, the proportion of

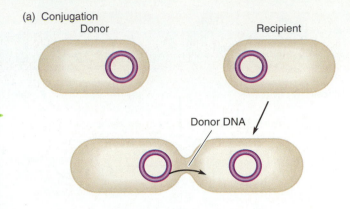

(a) Conjugation

Donor DNA

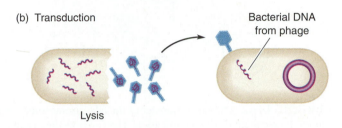

(b) Transduction

Bacterial DNA from phage

Lysis

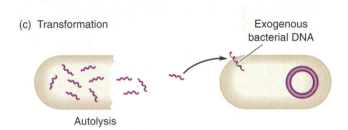

(c) Transformation

Exogenous bacterial DNA

Autolysis

▶ **FIGURE 14–1** Three major mechanisms of gene transfer in bacteria.

recombinants formed for each gene can be measured as a function of transfer time.

To illustrate the method, let us consider the cross

$$\text{Hfr } str^s\ a^+b^+c^+d^+ \times \text{F}^-\ str^r\ a^-b^-c^-d^-$$

which is shown in ▶ Figure 14–2. The str^r gene determines resistance to the antibiotic streptomycin and serves as a selectable marker in the cross. If the cells are placed in a growth medium that contains streptomycin immediately after the termination of conjugation, only F⁻ cells will survive. This is a convenient way to eliminate the Hfr cells from the sample after they have donated their genes, so that the recipient cells can be selectively studied.

In the interrupted mating procedure, samples are removed from the mixture of conjugating cells after defined time intervals and placed in a blender to terminate gene transfer. The separated cells are placed on a medium containing streptomycin to kill the Hfr cells, and the surviving F⁻ colonies are then tested individually for the presence of allelic characteristics from the donor strain. For example, samples of each colony might be transferred to various

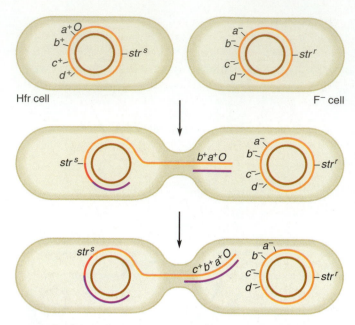

▶ FIGURE 14–2 The linear transfer of the Hfr chromosome into the F⁻ recipient during conjugation. The *O* marks the origin half of the integrated F factor, which leads chromosome transfer. The Hfr chromosome replicates itself as transfer proceeds. (Also see Figure 10–6.)

media to test for the acquisition of such donor traits as the ability to utilize certain compounds for growth, the ability to grow in the absence of certain nutrients, or resistance to different drugs and viruses.

▶ Figure 14–3 shows the results published by Wollman and Jacob. The genes *azi*ˢ (azide sensitivity), *ton*ˢ (sensitiv-

ity to T1 phage), *lac*⁺ (ability to metabolize lactose), and *gal*⁺ (ability to metabolize galactose) correspond to the *a, b, c,* and *d* markers used in our generalized description of the experiment. Notice that each donor gene first appears in the F⁻ recipients at a specific time after mating begins. For instance, the *azi*ˢ marker first makes its appearance about 9 minutes after the start of mating, while the *ton*ˢ marker does not appear until 11 minutes. Also notice that the genes that appear later yield lower maximal levels of F⁻ recombinants. These are the results we would expect if the Hfr chromosome is transferred to the F⁻ cell in a linear fashion, beginning at the *O* end (the origin of transfer or *oriT* site) of the F factor. Gene loci near the *O* site are likely to be transferred even during brief periods of conjugation. Genes farther away from the *O* site have an increasingly smaller chance of entering the recipient cell, since the conjugation tube often breaks by itself even before the sample is subjected to mechanical shearing. Thus, the greater the distance between the gene and the origin, the less likely it is to enter a recipient cell and to participate in recombination.

▶ Figure 14–4 further analyzes these relationships.

If we assume that the transfer of the chromosome occurs at a relatively constant rate, we can use the length of time that it takes each donor gene to first appear among the F⁻ cells as a measure of the distance of the gene from the origin. The unit of distance commonly used for this purpose is one minute of transfer time. A map of Wollman and Jacob's results is shown below:

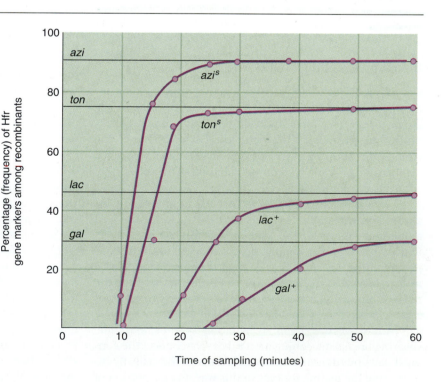

▶ FIGURE 14–3 The interrupted mating experiment of Wollman and Jacob. F⁻ *str*ʳ *azi*ʳ *lac*⁻ *gal*⁻ cells were allowed to take up genetic material from Hfr *str*ˢ *azi*ˢ *ton*ˢ *lac*⁺ *gal*⁺ cells. At specific time intervals, samples were withdrawn and the mating process was terminated by disruption in a blender. Cells were then placed in medium containing streptomycin to kill the Hfr. Surviving F⁻ cells were then tested to determine their genotype, and the results were plotted as percent of each Hfr characteristic among the surviving cells as a function of the time allowed for mating. *Adapted from: E. L. Wollman and F. Jacob, C.R. Acad. Sci., Paris 240 (1955): 2449.*

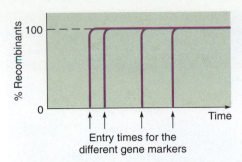

(a) Idealized results: All cells begin conjugation at the same time and continue transfer of their chromosomes until mating is artificially interrupted.

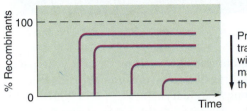

Probability of transfer decreases with distance of marker from the origin.

(b) Less idealized results: Mated cells separate spontaneously prior to and during artificial rupture.

(c) Actual results: Variability in the time of pair formation, as well as spontaneous separation of donor and recipient cells.

▶ **FIGURE 14–4** Explanation of Wollman and Jacob's graphical results. The shape of each curve is determined by three main effects: artificial rupture of mated cells, which delineates the time that each marker first enters the recipient; spontaneous rupture of mated cells, which decreases the probability that markers distal to the origin will eventually be received by the recipient; and variation in the time of pair formation.

Circularity of the Linkage Map

The circular nature of the bacterial chromosome was first suggested when mapping studies revealed that the linkage map of *E. coli* is circular. This circular map was generated from the combined results of interrupted mating experiments using different Hfr strains, which are formed by the insertion of the F factor at different sites in the bacterial chromosome (▶ Figure 14–5). There are several possible integration sites for the F plasmid during the conversion of an F⁺ cell to an Hfr strain, each site corresponding to an *IS* element (see Chapter 10). Moreover, the *ori* site of the F factor has polarity and can be oriented in either of two di-

rections during the chromosomal integration of F. The end result is that different Hfr strains will transfer genes from different starting points and in different directions during conjugation. For example, interrupted mating experiments using six different Hfr strains could give rise to six different maps, based on the order of transfer for the three hypothetical genes *a*, *b*, and *c*:

Hfr Strain	Relative Gene Order
1	$O-a-b-c$
2	$O-c-b-a$
3	$O-b-c-a$
4	$O-a-c-b$
5	$O-c-a-b$
6	$O-b-a-c$

The genes in the maps of these various strains differ in their relationship to the point of origin (*O*) and in their order of transfer. As we can see from Figure 14–5, the six gene orders are circular permutations produced by cutting a single circular map at different points and initiating transfer in either direction.

As we learned in Chapter 10, the entire Hfr chromosome is not usually transferred in any one mating. However, a variety of Hfr strains can be used to map different sections of the chromosome, and these maps can be pieced together to yield a composite map of the entire chromosome. One such map of the *E. coli* chromosome is shown in ▶ Figure 14–6.

▶ **FIGURE 14–5** Origin of different Hfr strains. The insertion of an F factor at one of many potential integration sites in the bacterial chromosome leads to the conversion of an F⁺ strain into a specific Hfr strain that transfers its chromosome from a unique starting point during conjugation. The direction of the arrow shows the direction in which the Hfr chromosome is transferred. The F factor has polarity and can thus orient itself in either of two directions during integration.

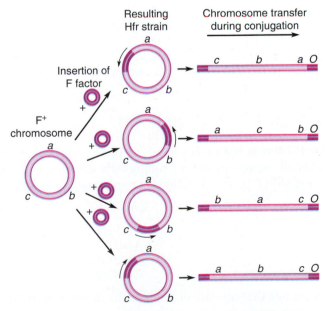

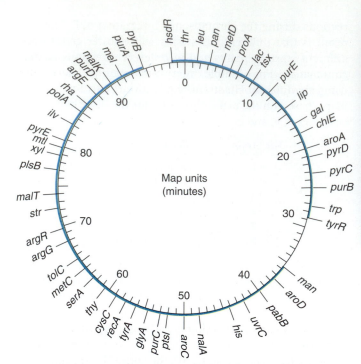

▶ **FIGURE 14–6** Circular map of the *E. coli* chromosome, showing the locations of a few of its genes as determined by conjugation analysis. The total map length is 100 minutes. Some of the better-known genes are *leu* = leucine synthesis, *pol B* = DNA polymerase II, *dna E* (*pol C*) = DNA polymerase III, *met* = methionine synthesis, *pro* = proline synthesis, *lac* = lactose operon, *tsx* = T6 resistance, *pur* = purine synthesis, *pyr* = pyrimidine synthesis, *gal* = galactose operon, *trp* = tryptophan synthesis, *tyr* = tyrosine synthesis, *uvr C* = ultraviolet sensitivity (repair), *his* = histidine synthesis, *rec A* = recombination and repair, *cys* = cysteine synthesis, *str* = streptomycin sensitivity, *pol A* = DNA polymerase I. The 1990 map contains more than 1400 genes and can be found by consulting the Bachmann reference in the Reading List for this chapter. *Source:* B. J. Bachmann, K. B. Low, and A. L. Taylor, *Bacteriol. Rev.* 40 (1976): 116. Used with permission.

The entire length of this map is 100 minutes, which is the time needed for the complete donor chromosome to be transferred to the recipient cell during the conjugation process. Observe that the map is circular, as is the bacterial chromosome.

Example 14.1

Four different Hfr strains of *E. coli* transfer their genes in different sequences during conjugation. After each strain mates with F⁻ cells, the following results are obtained:

Hfr Strain	Gene Order
1	P D X A L
2	A L U R T
3	P W Q C M
4	Q C M T R

Determine the sequence of genes on the chromosome of the F⁺ strain from which these Hfr strains were derived.

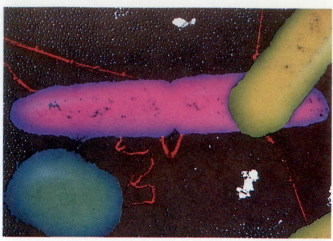

Electron micrograph of conjugating *E. coli* cells showing the sex pili.

Solution: Assuming only one copy of each gene per genome, the genes transferred by the different Hfr strains can be arranged along a circular map:

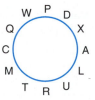

The circular arrangement of the loci is determined from the partial overlaps in gene order, which indicate that in this case the strains can differ in both the direction and origin of transfer.

Follow-Up Problem 14.1

Show where (and how) the F factor became integrated into the *E. coli* chromosome to produce the four different Hfr strains.

Generalized Transduction

The interrupted mating procedure can accurately map genes that are separated from one another by at least two minutes of transfer time. Genes that are closer together can be correctly mapped only by using other methods of gene transfer. One frequently used method is **generalized transduction,** which employs temperate phages such as P1 and P22 as vectors for the transfer of bacterial genes. In generalized transduction, virtually any locus in the donor chromosome can be transferred by the temperate phage to a recipient cell. During lytic multiplication of the temperate phage, a phage-encoded nuclease that causes fragmentation of the bacterial chromosome is produced. Since the maturation system of the phage cannot discriminate between bacterial and phage DNA, a small piece of the fragmented host chromosome is sometimes packaged into a mature phage by mistake in place

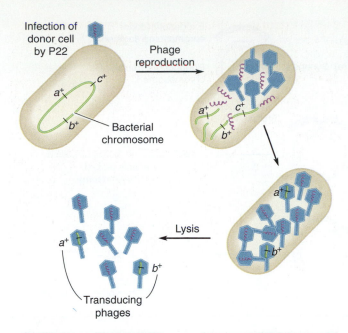

▶ **FIGURE 14–7** Generalized transduction by phage P22. Following phage infection of a donor cell, the phage DNA replicates, protein coats are synthesized, and the bacterial chromosome is fragmented. Occasionally, a piece of bacterial chromosome is accidently included in a phage head, producing a transducing phage.

of phage DNA (▶ Figure 14–7). The phage particle that contains the bacterial DNA is called a **transducing phage**.

If the transducing phage is subsequently involved in another infection, it will inject the donor fragment into its new host. The transferred block of donor genes then becomes available for expression in the recipient cell (▶ Figure 14–8). Two possible fates await the donor genes: Part or all of the donor fragment may become incorporated into the recipient chromosome in place of allelic genes—a result known as **complete transduction**. All the progeny of the transduced cell will then express the incorporated genes. Alternatively, the donor fragment may fail to be incorporated into the recipient chromosome, and therefore be unable to replicate—a result known as **abortive transduction**. The donor genes are then expressed only in the daughter cell that receives the unduplicated fragment. If the donor genes are needed for growth of the recipient on a solid medium, complete transduction would give rise to a normal-sized colony, while abortive transduction would result in a microcolony consisting of the transduced cell along with a comparatively small number of nontransduced cells that are able to grow on the selective medium as a result of cross-feeding.

In principle, the process by which a recipient cell is converted to a stable recombinant is the same in generalized transduction as it is in conjugation. In both processes, the recipient of chromosome transfer is initially a partial diploid or merozygote, since it has its own chromosome plus a chromosome fragment from the donor cell. The segment of donor chromosome in the recipient cell (the **exogenote**) may now synapse with the homologous section of the recipient chromosome (the **endogenote**) (see ▶ Figure 14–9). Crossing-over then yields a recombinant genotype. Although a double crossover is the simplest mechanism by which part or all of the exogenote can replace the corresponding segment on the

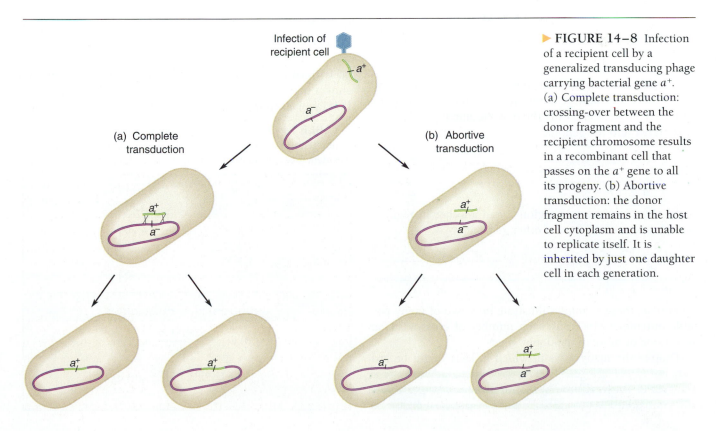

(a) Complete transduction

Infection of recipient cell

(b) Abortive transduction

▶ **FIGURE 14–8** Infection of a recipient cell by a generalized transducing phage carrying bacterial gene a^+. (a) Complete transduction: crossing-over between the donor fragment and the recipient chromosome results in a recombinant cell that passes on the a^+ gene to all its progeny. (b) Abortive transduction: the donor fragment remains in the host cell cytoplasm and is unable to replicate itself. It is inherited by just one daughter cell in each generation.

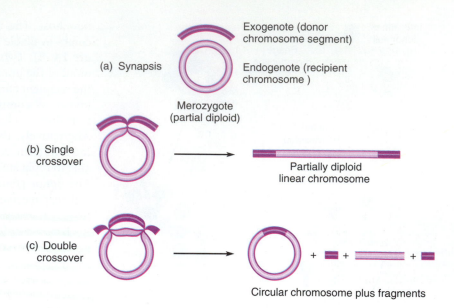

► **FIGURE 14–9** Crossing-over between exogenote and endogenote in a merozygote formed by the transfer of part of the donor chromosome during conjugation, transformation, or transduction. (a) Synapsis between exo- and endogenotes. (b) A single crossover leads to a partially diploid linear chromosome, which would be lethal. (c) A double (or other even-numbered) crossover event must occur in order to keep the circular chromosome intact. This recipient is now recombinant, having exchanged a section of its chromosome for the homologous section of the donor chromosome. The remaining chromosome fragments are lost from the cell upon division.

endogenote and thus be stably incorporated into the recipient chromosome, any even number of exchanges will suffice.

In transduction, only bacterial genes that are closely linked can be transferred on the same chromosome fragment. Still, physical proximity of the genes does not guarantee cotransfer, since linkage can be broken during fragmentation of the donor chromosome prior to packaging. Closely linked genes can also be separated by crossing-over during the incorporation of the donor fragment into the recipient chromosome. For example, suppose that homologous pairing of the donor fragment with the recipient chromosome yields the following temporary diploid condition:

a^+	b^+	c^+		Exogenote
(1)	(2)	(3)	(4)	Potential crossover intervals
a	b	c		Endogenote

If only two crossovers occur, one in region 1 and the other in region 3, the linkage between c and the other two loci would be broken, and only the $a^+ b^+$ segment would be incorporated into the genome:

$$a^+ \quad b^+ \quad c^+ \qquad\qquad a \quad b \quad c^+$$
$$\times\times \longrightarrow$$
$$a \quad b \quad c \qquad\qquad\quad a^+ \quad b^+ \quad c$$

Similarly, a crossover in region 2 and another in region 4 would separate a^+ from the other genes:

$$a^+ \quad b^+ \quad c^+ \qquad\qquad a^+ \quad b \quad c$$
$$\times\times \longrightarrow$$
$$a \quad b \quad c \qquad\qquad\quad a \quad b^+ \quad c^+$$

Of course, these examples illustrate only two of many possible outcomes, since any even number of crossovers can insert one or more of the donor genes.

Despite the many ways in which linkage might be broken, the closer two genes are on the bacterial chromosome, the greater the likelihood that they will both be present on the same donor fragment and will be incorporated together

into the recipient genome. The probability of simultaneous transduction of both markers (cotransduction) can thus be used to order gene loci on a linkage map. To illustrate how we can establish a linkage map from cotransduction frequencies, let us consider the results of generalized transduction as shown in ■ Table 14–1. In this example, phage P1 was grown on a wild-type $x^+ y^+ z^+$ donor bacterium, and the lysate was used to infect the mutant $x^- y^- z^-$ strain. Transductants for x^+ were selected and later tested for cotransfer of the other genes. The data in Table 14–1 give cotransduction frequencies for $x^+ y^+$ and for $x^+ z^+$ of 0.54 and 0.86, respectively, indicating that x and z are more closely linked than x and y. These frequencies are consistent with the orders $x-z-y$ and $z-x-y$. To distinguish between these possibilities, we apply reasoning similar to that used for three-point crosses in eukaryotes: the rarest recombinant class must represent the least likely crossover event. As shown in ► Figure 14–10, the least likely event is the one that involves four exchanges rather than two. Since a quad-

■ **TABLE 14–1** Results of a transduction experiment. The donor genotype is $x^+ y^+ z^+$ and the recipient genotype is $x^- y^- z^-$. The data give the genotypes of 500 x^+ transductants.

Genotypes Observed	Number of Colonies
$x^+ y^+ z^+$	265
$x^+ y^- z^+$	165
$x^+ y^+ z^-$	5
$x^+ y^- z^-$	65

Cotransduction Frequencies
$f(x^+ y^+) = \dfrac{265 + 5}{500} = 0.54$
$f(x^+ z^+) = \dfrac{265 + 165}{500} = 0.86$

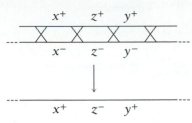
(a) Order x–z–y

x^+ z^+ y^+ — Donor fragment

x^- z^- y^- — Recipient chromosome

x^+ z^- y^+ — Lowest frequency class

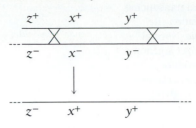
(b) Order z–x–y

z^+ x^+ y^+

z^- x^- y^-

z^- x^+ y^+

▶ FIGURE 14–10 Crossover origin of the lowest frequency class in Table 14–1, $x^+y^+z^-$. (a) If the gene order is x–z–y, four crossovers are needed to produce $x^+z^-y^+$ transductants. (b) If the gene order is z–x–y, only two crossovers are needed to yield $z^-x^+y^+$ transductants. The rarity of this class of transductants indicates that the gene order is x–z–y.

ruple exchange can produce the rarest class of transductants ($x^+y^+z^-$) only when z is the middle marker, the correct order must be x–z–y.

We can also use the results of transduction experiments to evaluate map distance—the problem is relating cotransduction frequencies to minutes of transfer time, the conventional unit of map distance in bacteria (see Figure 14–6). One mathematical equation that is commonly used for this purpose is:

$$\text{cotransduction frequency} = \left(1 - \frac{d}{L}\right)^3 \quad (14.1)$$

where d is the distance (in minutes) between a selected marker and another marker and L is the average length (in minutes) of transducing DNA. In phage P1, for example, L = 2.2 minutes. Note that the map distance and cotransduction frequency are inversely related. Thus, if two markers are very close together (d approaches 0), the cotransduction frequency will approach 1. On the other hand, if two markers are sufficiently separated that they are seldom carried on the same chromosome fragment (d approaches L), the cotransduction frequency will approach 0.

Example 14.2

A generalized transduction experiment was performed in which the donor E. coli cells had the genotype $C^+M^+R^+$ and recipient cells had the genotype $C^-M^-R^-$. P1-mediated transductants for C^+ were selected, giving the following data:

Genotypes Observed	Number of Colonies
$C^+M^+R^+$	548
$C^+M^-R^+$	555
$C^+M^-R^-$	93
$C^+M^+R^-$	4

a. Determine the order of the three genes.
b. Determine the map distances (in minutes) between the genes, using equation 14.1.

Solution: a. The cotransduction frequencies are

$$f(C^+M^+) = \frac{548 + 4}{1200} = 0.46$$

$$f(C^+R^+) = \frac{548 + 555}{1200} = 0.92$$

Thus, C and R are more closely linked than C and M, so the gene order is either C–R–M or R–C–M. To distinguish between these possibilities, note that the lowest frequency class, $C^+M^+R^-$, would be produced by two crossovers if the gene order is R–C–M but would require four crossovers if the order is C–R–M. (Diagram the crossovers for each case to convince yourself.) Thus, C–R–M is the correct order of the genes.

b. Substituting the cotransduction frequencies into equation 14.1, we obtain

$$d_{C-M} = L(1 - \sqrt[3]{0.46}) = 2.2(1 - 0.772) = 0.502 \text{ min}$$

$$d_{C-R} = L(1 - \sqrt[3]{0.92}) = 2.2(1 - 0.973) = 0.059 \text{ min}$$

Follow-Up Problem 14.2

Donor E. coli that are $a^-b^+c^+$ are used in P1-mediated transduction of recipient $a^+b^-c^-$ cells. Only a few colonies are observed on minimal medium. In contrast, the reciprocal experiment (i.e., $a^+b^-c^-$ donors and $a^-b^+c^+$ recipients) results in a large number of colonies that are able to grow on minimal medium. Given that c is to the right of the a and b markers, what is the order of the a and b genes relative to the c locus?

Specialized Transduction and Sexduction

While some temperate phages, such as P1 and P22, participate in generalized transduction, others, such as λ, mediate another related gene-transfer process called **specialized transduction**. Unlike generalized transduction, which can involve virtually any chromosomal locus, specialized transduction is restricted to the transfer of bacterial genes that are located adjacent to the site of prophage insertion. The mechanism of specialized transduction can be seen in the events that lead to the formation of a λ transducing phage (▶ Figure 14–11). The λ prophage is inserted between the E. coli gal (galactose) and bio (biotin) genes. During induction, excision of the prophage usually occurs without error, but rarely—in about one out of 10^6 or 10^7 cells—an aberrant excision occurs in which incorrect cuts are made, one in the host chromosome and the other in the phage DNA. These aberrant excisions can produce transducing particles that carry either the gal gene, if the cuts are displaced in

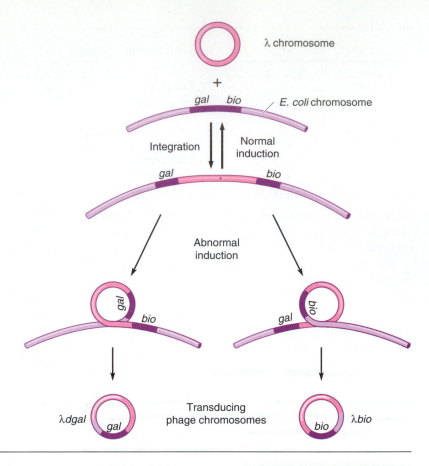

► **FIGURE 14-11** Production of transducing phage chromosomes by the specialized transducing phage lambda. The λ prophage integration site lies between bacterial genes *gal* and *bio*. Normal induction (top diagrams) yields an intact λ genome. However, an error in induction can produce chromosomes in which either of the bacterial genes replaces part of the phage chromosome, giving rise to transducing phages.

one direction from the normal cleavage sites, or the *bio* gene, if the cuts are displaced in the other direction.

For normal phage packaging to occur, the length of the transducing λ phage chromosome must be between 79% and 106% of normal λ DNA. Too little DNA leads to phage sensitivity to osmotic disruption, too much DNA will not fit into the phage protein coat. Therefore, the mature *gal-* and *bio*-transducing phages that are eventually released upon lysis are missing some λ genes. The missing genes in the *bio*-transducing particles are usually nonessential, so that λ *bio* can generally replicate upon further infection. In contrast, the *gal*-transducing phage is defective, since it lacks essential phage genes and is incapable of producing progeny by the lytic cycle. The *gal*-transducing particles are therefore known as λ*dgal* phage (*d* meaning defective).

When a specialized transducing phage, such as λ*dgal*, infects a suitable bacterium, transduction can occur by several mechanisms (► Figure 14-12). If the phage infects a nonlysogenic Gal⁻ cell, part of the *gal* region may replace a homologous segment in the host DNA by recombination and produce a stable haploid Gal⁺ transductant (Figure 14-12a). Alternatively, the entire phage chromosome (including the *gal* region) can become inserted as a prophage to produce a partial diploid Gal⁺ transductant called a **heterogenote** (Figure 14-12b). The heterogenote has two copies of the *gal* region: its own *gal*⁻ region and the *gal*⁺ region brought in with the phage DNA. If the phage infects a lysogenic Gal⁻ cell, the phage chromosome can become inserted as a prophage with high efficiency and produce a doubly lysogenic

Gal⁺ transductant (Figure 14-12c). The double lysogen, which is also a heterogenote, releases a mixture of normal λ phage and λ*dgal* phage upon lysis; the normal phage chromosomes act as "helpers" by supplying the defective phage with the missing gene products that it requires to synthesize progeny.

■ Table 14-2 compares the two forms of transduction, listing some of the characteristics of specialized transduction that set it apart from generalized transduction. Note that specialized transduction is strictly an episome-mediated form of gene transfer, since the phage vector must be capable of integrating into and excising from the host chromosome. The F factor, which also participates in an integration-excision cycle with the bacterial chromosome, can acquire bacterial genes in a manner similar to the phage λ mechanism, but unlike λ, the F factor rarely has any of its own genetic information deleted; it simply expands to include the extra piece of bacterial DNA. Since the F factor is never encased in a protein coat, the addition of a few bacterial genes does not create a packaging problem as it would in a phage. Such modified F factors, which carry bacterial genes in addition to their own, are called F′ **factors**.

Once it has acquired donor genes, the F′ factor can transfer them to a recipient cell during conjugation. This transfer, called **sexduction**, is a highly efficient process that occurs by the same mechanism as the transfer of the F factor itself. For example, on rare occasions an error in excision can lead to the inclusion of the *lac* (lactose) region of *E. coli* in the F factor. The *lac*-carrying factor, F′*lac*, can be rapidly

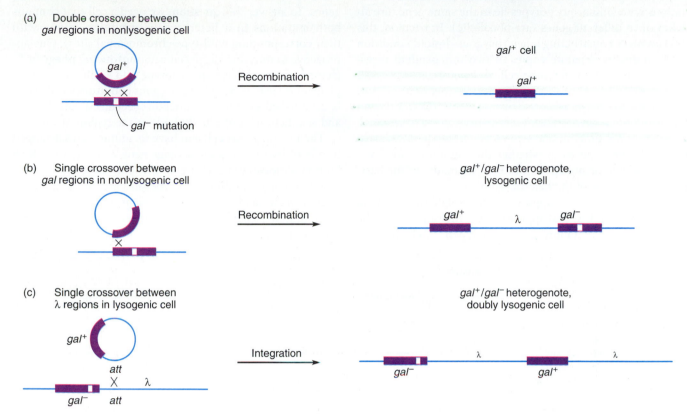

Event

Transductant

(a) Double crossover between *gal* regions in nonlysogenic cell

gal⁺

gal⁻ mutation

Recombination →

gal⁺ cell

gal⁺

(b) Single crossover between *gal* regions in nonlysogenic cell

Recombination →

gal⁺/*gal*⁻ heterogenote, lysogenic cell

gal⁺   λ   *gal*⁻

(c) Single crossover between λ regions in lysogenic cell

gal⁺

att

gal⁻   att   λ

Integration →

gal⁺/*gal*⁻ heterogenote, doubly lysogenic cell

λ   λ

gal⁻   *gal*⁺

▶ **FIGURE 14–12** Three possible mechanisms of transduction by a λ*dgal*⁺ phage infecting a *gal*⁻ recipient cell.

disseminated in the population by matings between F'*lac*-bearing donor cells and F⁻ cells. If the recipient cells are Lac⁻, they can thus be converted to partially diploid Lac⁺ cells with the genotype F'*lac*⁺/*lac*⁻.

An F' factor carrying a specific marker can be isolated initially by mating an appropriate Hfr strain having the gene of interest with a recombination deficient (Rec⁻) recipient. Since the F⁻ cells are deficient in recombination (unable to incorporate a donor DNA fragment into their own genome), they can stably carry the donor marker only if the gene is present on an F' factor. For example, by selecting for a rare Lac⁺ recipient from an Hfr *lac*⁺ × F⁻ *rec*⁻ *lac*⁻ mating, it is possible to isolate a strain with the F'*lac* factor.

Genetic studies of *E. coli* have made frequent use of specialized transducing phages and F' factors for performing **complementation tests**. Like the tests described for eukary-

■ **TABLE 14–2** A comparison of specialized and generalized transduction.

	Specialized Transduction	Generalized Transduction
Origin of the transducing phage	Bacterial genes are acquired as an error in the deintegration of the prophage during induction.	Bacterial genes are acquired as an error in the assembly of the mature phage during lytic multiplication.
Nature of the transducing phage	The phage particle contains a chromosome in which some phage genes are missing and have been replaced by genes of the bacterial host.	The phage particle contains mainly (or all) bacterial genes on a chromosome fragment incorporated into the capsid by mistake.
Nature of the donor genes transferred	Restricted to bacterial genes that are normally located close to the site of prophage attachment.	Any set of closely linked bacterial genes can be transferred without restriction as to chromosomal location.
Incorporation of the donor genes into the recipient chromosome	Donor genes are incorporated when the chromosome of the transducing phage is inserted as a prophage in the recipient genome.	Donor genes are incorporated by crossing-over between the donor fragment and a homologous region on the recipient chromosome.

otes in Chapter 4, these complementation tests provide an experimental basis for deciding whether two bacterial mutations with similar phenotypes lie in the same gene (are allelic) or in different genes (are nonallelic). In bacteria, this test involves constructing a heterozygous diploid condition by introducing separate copies of two independent recessive mutations into the same cell. If the two mutations are in the same gene, neither will provide normal function and a mutant phenotype will be observed. If the two mutations lie in different genes, then each will supply the function missing in the other and a wild-type phenotype will result. Thus, the decision as to whether the mutations are functionally allelic or nonallelic can then be made on the basis of the phenotype of the heterozygote.

For example, let us suppose that we isolate two bacterial auxotrophs, x and y, that appear to have the same metabolic block. Knowing that more than one polypeptide chain might be required to catalyze the normal reaction, we may ask whether these mutations are in the same gene or in different genes. To answer this question, we need a cell that contains both mutations in a heterozygous condition (along with their corresponding wild-type chromosomal sites). One approach is to introduce a specialized transducing phage or F′ factor with an appropriate genotype into a cell with a complementary genotype. A heterozygous condition would then be established, since the products of specialized transduction and sexduction would be diploid for the region of interest.

The heterozygous cell can have its mutant sites arranged in one of two ways: a **cis** arrangement, x^+y^+/xy, in which both mutations (x and y) are on one chromosome segment and their corresponding wild-type sites are on the homologous segment, and a **trans** arrangement, x^+y/xy^+, in which one mutation is present on each segment. As shown in ▶ Figure 14−13, the phenotypic results of these two arrange-

▶ **FIGURE 14−13** The theoretical basis of the complementation test. (a) When two mutations (x and y) are present in the same gene in a heterozygote, a wild-type polypeptide can be produced only in the cis arrangement. (b) When the mutations affect different genes, wild-type polypeptides are produced in both the cis and trans arrangements.

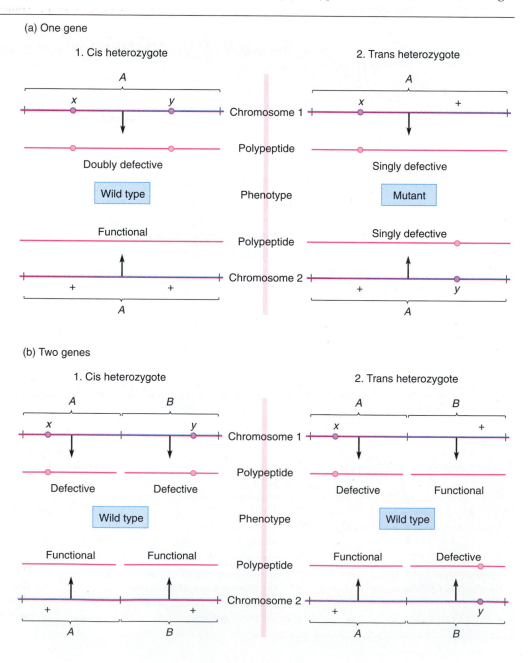

ments can be different. When the mutations are in the same gene (e.g., gene *A* in Figure 14–13a), neither allele can produce a functional polypeptide if the mutation sites are in the trans arrangement. However, if the sites are in the cis arrangement, one of the alleles can still produce a fully functional polypeptide chain. Therefore, the phenotype of the trans arrangement is mutant and the phenotype of the cis arrangement is wild type. Figure 14–13b shows the effects of mutations in different genes. The heterozygote has a normal allele of each gene to supply the missing function of its mutant allele, so functional polypeptides are formed by both genes, and a wild-type phenotype is produced in both the cis and trans arrangements.

Thus two recessive mutations in a cis arrangement should always give a wild-type phenotype. In contrast, two mutations in the trans arrangement should give a wild-type phenotype only if they are nonallelic, that is, in separate genes. Since the cis arrangement gives the same results whether the mutations lie in the same or in different genes, complementation tests are based in practice on the phenotype of the trans heterozygote; if it is wild type, the mutations are in different genes, but if it is mutant, the mutations are in the same gene. The cis arrangement, if used at all, serves only as a control to establish the validity of the test by demonstrating that the mutations are indeed recessive.

It should be stressed that complementation between different mutations depends only on the interactions of gene products within a cell; it does not require the direct contact of chromosomes. For example, it is possible to perform a complementation test on certain haploid eukaryotic cells by making heterokaryons of two different mutant strains. Since heterokaryon formation requires cell fusion, the chromosomes of the two mutants will share a common cytoplasm, but it is impossible for the chromosomes to interact directly, since they remain physically separated in different nuclei. As we will see later in this chapter, complementation is also possible in viruses when two different mutants infect the same host cell. In this case, each mutant virus must supply the missing function of the other so that both can grow and be released upon lysis.

Transformation

In bacterial transformation, which was described in Chapter 7 in terms of its role in identifying the genetic material, recipient bacterial cells acquire donor genes through the uptake of free DNA molecules from the surrounding medium. Transformation of a recipient cell proceeds through a series of three distinct phases (▶ Figure 14–14), beginning with the **binding** of DNA to specific receptor sites on the cell envelope. Several molecules can bind to a bacterium (a pneumococcal cell has about 80 DNA-binding sites on its surface), so a cell can be transformed by more than a single DNA molecule. DNA **penetration** then occurs, with bound DNA molecules actively transported into the cell. In Gram-positive bacteria, penetration is accompanied by nuclease digestion of one of the strands of the entering DNA, con-

verting it to a single-stranded form. DNA uptake is followed by **integration**, in which the entering DNA undergoes recombination with the recipient chromosome. During integration, the single strand of donor DNA displaces a homologous strand in the recipient DNA to form a heteroduplex recombinant structure (see Chapter 23).

The overall efficiency of transformation is usually rather low. For a given gene, the proportion of bacteria that are actually transformed is seldom greater than 10% and is often considerably lower. One reason for the low efficiency of transformation is that typically only a minority of cells are in a physiological state that permits them to undergo the transformation process—these cells are said to be **competent**. Competence represents the capacity of bacteria to take up DNA from the medium, the extent of which varies with growth conditions and the phase of growth of the recipient cells. In Gram-positive bacteria, such as pneumococcus, a small protein called the **competence factor** mediates the de-

---

▶ **FIGURE 14–14** The three phases of transformation: binding of donor DNA to the recipient cell, penetration of donor DNA into the cell, and integration of the donor gene into the host chromosome. In Gram-positive bacteria, the donor DNA is converted to a single-stranded form during the integration stage.

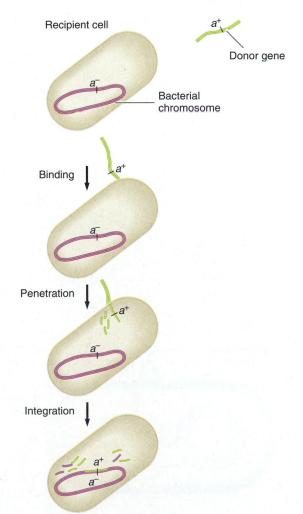

velopment of competence by noncompetent cells. This protein is produced and excreted into the medium continuously in growing cultures, but it causes a rapid rise in the competent state only when a critical cell concentration in the range of 10^7–10^8 cells per ml is reached (▶ Figure 14–15). In contrast, Gram-negative bacteria such as *Haemophilus* do not produce a competence factor; competence in these species appears to be internally regulated, influenced mainly by the composition of the growth medium.

In transformation, as in the other forms of gene transfer in bacteria, only fragments of donor DNA ordinarily enter a recipient cell. When researchers extract the DNA from donor cells and purify it, the donor chromosome breaks down into fragments that are usually less than 1% of its total length. Because of this chromosome fragmentation, linkage information is generally derived from the probability of simultaneous transformation (**cotransformation**) of different gene markers. Two closely linked markers will tend to be carried on the same DNA fragment and will therefore exhibit a large cotransformation frequency. On the other hand, widely separated markers will almost never be on the same fragment and will have a frequency approaching that expected for two independent transformations. For example, if transformation for each gene marker occurs with a frequency of 0.01 and the markers are so widely spaced that they are always carried on different fragments, we would expect their simultaneous independent transformation to occur at a frequency of $(0.01)^2 = 0.0001$. If such an expectation is fulfilled, the genes are regarded as being unlinked, even though they were originally present on the same chromosome.

Because of difficulties relating to competence, and to the fact that transformation frequencies tend to vary with DNA concentration, linkage relationships are usually analyzed by comparing two markers' joint transformation dependence on DNA concentration with that shown by each marker individually. An example is shown in ▶ Figure 14–16 for hypothetical gene markers that are assumed to transform with the same efficiency. At low DNA levels, single transformation by each gene shows a linear dependence on DNA concentration (C), so log (number of transformants) = log(C) + constant. A similar dependence is shown by the simultaneous transformation with two closely linked genes, because both are carried on the same DNA fragment. However, simultaneous transformation of two unlinked genes will increase as C^2; in this case log (number of transformants) = 2 log (C) + constant, so this line has a greater slope. Observe that in every case, a plateau is reached at very high concentrations where the number of transformants no longer changes with C. At this point, all competent cells that are capable of being transformed have undergone transformation for this set of loci.

▶ **FIGURE 14–15** Development of competence in a culture of *Streptococcus pneumoniae*. (a) Time course of competence development during growth. The competent state reaches a peak after a critical cell concentration is reached and then rapidly declines as the culture approaches its stationary phase. (b) Steps leading to the development of competence. (1) Cells produce a protein called competence factor (▲) that (2) binds to a cell-surface receptor (M) causing (3) the expression of certain gene products. Among these products is an autolysin (■) that exposes a DNA-binding protein and a nuclease on the cell surface. The DNA binding protein and nuclease participate in the uptake of the DNA into the cell. *Adapted from:* A. Tomasz, "Control of competent state in pneumococcus . . . ," *Nature* 208 (1965): 155–159. Reprinted with permission from *Nature.* Copyright © (1965) Macmillan Magazines Limited for part (a); (b) *Adapted from:* H. O. Smith, D. B. Danner, and R. A. Deich, "Genetic transformation." Reprinted with permission from the *Annual Review of Biochemistry,* Volume 50, © 1981 by Annual Reviews Inc.

(a)

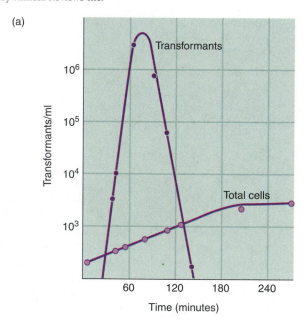

(b)

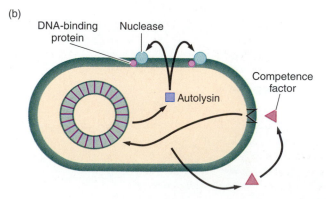

To Sum Up

1. The interrupted mating technique controls the duration of mating in bacteria by artificially separating conjugating cells. The resulting linkage map shows the minimum time for entry of each gene into the recipient cell. Linkage data from several different Hfr strains can be combined to give a circular composite map of the bacterial chromosome.

2. Transduction is the transfer of bacterial genes from donor to recipient cells by means of a temperate phage. Generalized transduction involves genes from any single area of the bacter-

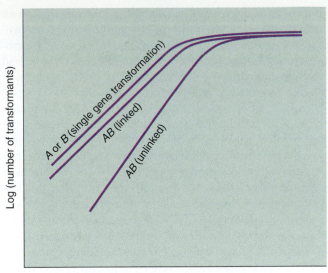

FIGURE 14–16 Test for cotransformation (linkage) of two genes, *A* and *B*. A plot of log (number of transformants) vs. log (DNA concentration) will be similar for single gene transformation (*A* or *B*) and double transformation (*AB*) if *A* and *B* are closely linked. These graphs reflect the linear dependence of transformation on DNA concentration. However, if *A* and *B* are carried on different donor DNA fragments (*A* and *B* are unlinked), then the frequency of double transformation varies as the square of the DNA concentration.

ial chromosome. Crossing-over between the donor fragment and a homologous region on the recipient chromosome incorporates the donor genes. Cotransduction frequencies are used to determine gene order and map distances.

3. In specialized transduction, only genes in the vicinity of the prophage are transferred. Insertion of the transducing phage as a prophage incorporates the donor genes in the recipient chromosome.

4. The complementation test determines whether two recessive mutations are allelic or nonallelic. Two independent mutations are introduced into the same cell (for example, via specialized transduction or sexduction) in a heterozygous condition. If the mutations are in separate genes, the mutant chromosomes complement each other and the normal alleles produce functional polypeptides, giving the heterozygote a wild-type phenotype. If the mutations are in the same gene, neither allele can produce a functional polypeptide and the heterozygote has a mutant phenotype.

5. In transformation, bacterial genes are transferred from one cell to another by the uptake of naked DNA by the recipient. Two genes are closely linked if their frequency of cotransformation is linearly dependent on the concentration of transforming DNA.

MAPPING PHAGE CHROMOSOMES

Bacteriophages, like their bacterial hosts, have been widely used as experimental subjects in mapping studies. In this section, we will concentrate on the results obtained with the T-even phages T2 and T4. The intracellular development of these phages has served as a model for many different processes in molecular biology and continues to provide new insights into the mechanisms of recombination.

The Phage Cross

The methods used for genetic mapping in bacteriophages superficially resemble those applied to eukaryotic organisms. The progeny produced by parental phages of contrasting genotype are analyzed for recombinants, and the frequency of recombinant types is then used as a measure of the distance between linked genes. A phage cross, however, is not a cross in the traditional sense but is a mixed infection in which a bacterium is infected by a mixture of phage particles of two (or more) different genotypes. These infecting phages serve as the "parents" in the cross. The concentration of each parental phage is kept high enough to ensure that most bacterial cells are simultaneously infected by both phage types. Since the infecting phages are distributed among host cells according to a Poisson distribution, a sufficiently high concentration usually means an average infection of five or so phage particles of each genotype per cell.

To illustrate the nature of a phage cross and the methods used in the analysis of phage recombinants, we will consider the following pairs of traits exhibited by bacteriophage T2: (1) plaque morphology: small plaques (r^+) vs. large plaques (r); and (2) host range: ability to infect *E. coli* strain B but not B/2 (h^+) vs. ability to infect both strains B and B/2 (h). *E. coli* B/2 is a T2-resistant (or Tto^r) strain that is derived by mutation from the normally sensitive (or Tto^s) *E. coli* B strain.

Now suppose that a cross is made between a mutant for plaque morphology (h^+r) and a mutant for host range (hr^+). *E. coli* B is infected with an average of five of each of these parental phages per cell. During the latent period that follows, the replicating phage chromosomes undergo recombination, participating in something akin to synapsis (called a mating by phage geneticists) and crossing-over. Following maturation and lysis, the progeny phages are analyzed for different phenotypes by spreading them on a lawn of mixed bacterial strains B and B/2. Phage progeny that carry the h^+ allele will produce cloudy plaques, since they are unable to infect the B/2 bacteria in the mixture, while those with the h allele infect both strains and produce clear plaques. The four possible phenotypes among the progeny are shown in ▶ Figure 14–17, and their inferred genotypes are listed here:

Appearance of Plaque	Genotype	
Cloudy and large	h^+r	Parental types
Clear and small	hr^+	
Clear and large	hr	Recombinant types
Cloudy and small	h^+r^+	

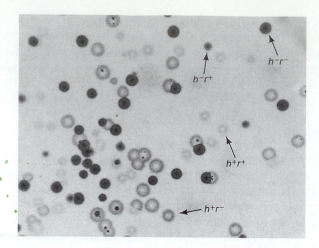

▶ **FIGURE 14–17** Four possible progeny phenotypes resulting from the phage cross $h^+r \times hr^+$. *Adapted from:* G. S. Stent, *Molecular Biology of Bacterial Viruses* (New York: W. H. Freeman, 1963), p. 185.

The recombination frequency is then calculated as

$$R = \frac{\text{number of } hr \text{ plus number of } h^+r^+}{\text{total number of plaques}}$$

Although the analysis so far seems straightforward, chromosome mapping in phages is subject to several complications. First, each parental phage is duplicated several times soon after infection, so multiple copies of the chromosomes of both parental types will exist in a single cell. Since the phage DNA molecules pair at random, paired matings will not be restricted to contrasting genotypes but can occur between two chromosomes of the same genotype as well.

Thus, in an equal-input cross, as in our example, half of all the matings in the vegetative pool would be unable to produce recombinants, since matings between two h^+r chromosomes and between two hr^+ chromosomes (which lead to no recombination) will occur as often, on the average, as matings between h^+r and hr^+ chromosomes.

Mapping in phages is further complicated by the repetitive acts of mating and gene exchange that occur in the DNA pool during phage development. When researchers examine the frequency of recombinants among the intracellular mature viruses (under conditions of lysis inhibition in order to prolong the latent period), they find that this frequency increases as long as phage multiplication is allowed to continue (▶ Figure 14–18). Genetic exchange must therefore occur throughout the lytic cycle—it is not restricted to any particular phase of phage growth. Moreover, since DNA molecules pair at random, a single phage chromosome may participate in several rounds of mating while it replicates within the host cell. Successive pairings of the same chromosome lineage will initially increase the chance of recombination, but once recombinant chromosomes are formed, they can be converted back to parental types by subsequent pairing and exchange events (▶ Figure 14–19). Matings between recombinant-type chromosomes will therefore counter the genetic effects of earlier matings between parental-type chromosomes. The net result in a two-factor equal-input cross is a drift toward a genetic equilibrium in which half of the viral genomes in the DNA pool would be recombinant for the gene of interest. The frequency of recombinants in a phage cross will therefore depend not only on the distance between the linked loci but also on the number of rounds of mating that occur by the time the phage genomes are withdrawn

▶ **FIGURE 14–18** Frequency of recombinant progeny phage as a function of time after infection. Lysis has been inhibited so that intracellular phage can be measured. The recombination frequency increases over time in the same manner as the number of progeny phages, showing that recombination occurs throughout the entire period of intracellular phage growth. *Adapted from:* B. D. Davis, R. Dulbecco, H. N. Eisen, H. S. Ginsberg, and W. B. Wood, *Principles of Microbiology and Immunology* (New York: Harper & Row, 1967), p. 705. Used with permission.

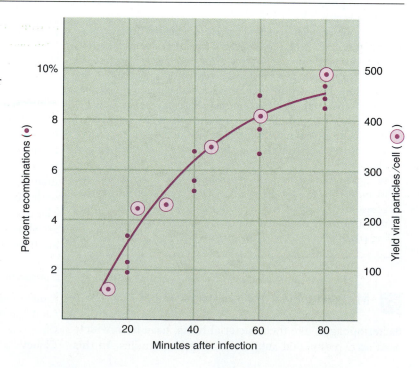

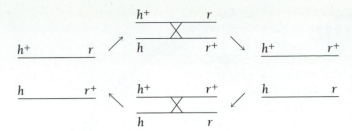

▶ **FIGURE 14–19** Random mating and recombination events between phage chromosomes. These events occur in the DNA pool during replication and would ultimately lead to an equilibrium between parental and recombinant genotypes if lysis were to be delayed long enough.

from the pool for irreversible maturation. In the T-even phages, an average phage chromosome undergoes approximately five rounds of mating during a normal lytic cycle. The resulting high rate of recombination makes it essential that two genetic markers be situated fairly close to one another on the T2 or T4 chromosome in order for them to appear as being linked in a two-point cross.

The Chromosome and Linkage Map of T4

Some of the earliest mapping experiments utilizing bacteriophages were carried out on phage T4. The T4 chromosome consists of a linear duplex molecule of DNA containing approximately 1.7×10^5 bp, and it differs from typical DNA by having a modified base, 5-hydroxymethyl cytosine (HMC), in place of cytosine. Moreover, the hydroxymethyl groups of HMC are covalently linked to various sugars (▶ Figure 14–20), which protect the T4 DNA from degradation by a host-encoded nuclease. Furthermore, the DNA of this phage exhibits a remarkably complex replication cycle in which recombination plays an essential role. If recombination is eliminated, DNA replication stops at an early stage and no mature virus particles are produced.

Despite the linear nature of the T4 chromosome, genetic studies reveal that the linkage map of T4 (and T2) is circular, with a contour length of 1500 map units (▶ Figure 14–21). How does a linear chromosome produce a circular linkage map? The answer lies in the terminal redundancy of the T4 DNA: Each DNA molecule has the same gene sequence at the end as it does at the beginning. (For example, the phage genome consisting of the sequence of gene markers *abcdefgab* has terminal redundancy because the *ab* sequence is repeated at both ends.) Terminal redundancy in phage T4 is a result of its DNA replication process. After the linear phage genome has infected the host cell, a recombination-dependent replication of this structure produces large, highly branched DNA concatemers that contain repeats of the basic genome (▶ Figure 14–22). These concatemers are regularly found when cells infected with T4 are prematurely lysed to study the intermediate stages of phage production. Many are several chromosomes in length.

Once maturation of the phage progeny begins, the T4 concatemers are processed by what is known as the **headful mechanism**. Branches of the concatemeric DNA are trimmed by an endonuclease and cut into unit-length chromosomes measuring approximately 103% of the genome—just the size that can be incorporated into the heads of the developing progeny. This cleavage process is not restricted to a single genetic site but can occur between any pair of markers. (Compare this process with the one that cuts λ phage chromosomes out of concatemers generated by rolling-circle replication—see Chapter 10.) For example, let us assume that processing according to the headful mechanism means a cut in the concatemeric DNA every nine genes, as shown:

$$\text{--- } a\,b\,c\,d\,e\,f\,g\,a\,b\,c\,d\,e\,f\,g\,a\,b\,c\,d\,e\,f\,g\,a\,b\,c\,d\,e\,f\,g\,a\,b \text{ ---}$$

Cutting the DNA into head-sized lengths that include a terminally redundant region results in a circularly per-

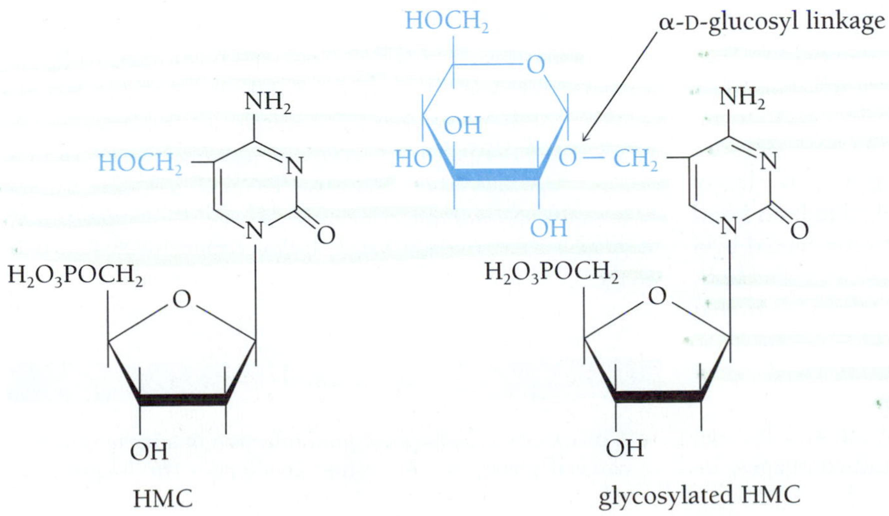

▶ **FIGURE 14–20** Structures of 5-hydroxymethyl cytosine, HMC (left) and glycosylated HMC (right).

HOCH₂   α-D-glucosyl linkage

HMC

glycosylated HMC

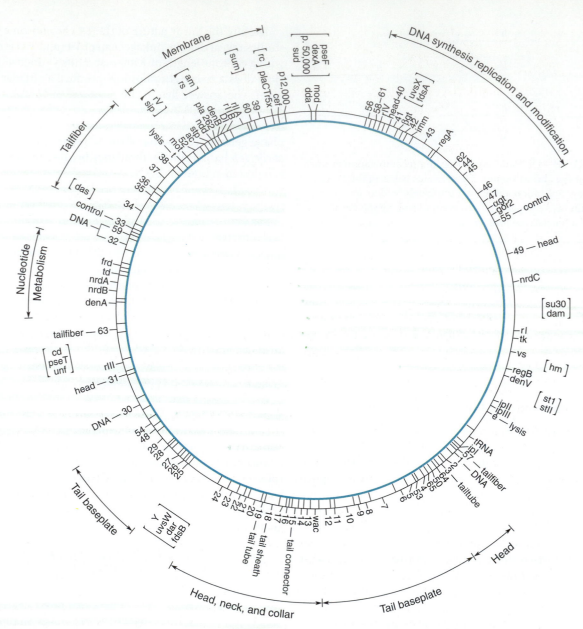

► **FIGURE 14–21** The linkage map of phage T4. Genes whose exact order is not known are enclosed in brackets in arbitrary order. Notice the clustering of related functions. *Source:* W. B. Wood and H. R. Revel, "The genome of bacteriophage T4," *Bact. Rev.* 40 (1976): 847–868. Used with permission.

muted collection of phage chromosomes, with some reading *abcdefgab*, others *cdefgabcd*, still others *efgabcdef*, and so on. Recombination analysis of these progeny would reveal that gene *a* is closely linked not only to *b* but also to *g*, the end marker in the sequence. Similarly, gene *b* is closely linked to both *a* and *c*, gene *c* is closely linked to both *b* and *d*, and so on. These relationships are best represented by a linkage map in the form of a circle. Thus, the circular linkage map of phage T4 is derived from a circularly permuted collection of linear progeny chromosomes. In general, a circular map is expected for any phage that uses the headful mechanism to package its genetic material.

■ Table 14–3 compares the chromosome and linkage map of T4 with those of two other well-studied phages, T7

and λ. Like T4, the chromosome of T7 has terminally redundant ends that are produced from the enzymatic cutting of unit chromosomes from concatemeric DNA; but unlike T4, all ends of T7 chromosomes are identical, since cuts are made at specific sites. Thus the DNA of T7, like that of λ, consists of a unique collection of molecules with regard to gene order, giving a linear linkage map for the mature phage.

To Sum Up

1. A phage cross involves the joint infection of a bacterium with parental phages of contrasting genotypes. The frequency of

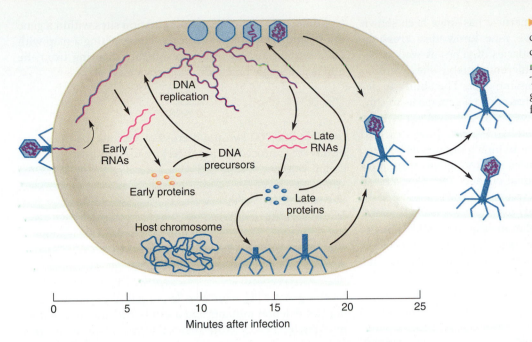

DNA
replication

Early
RNAs

DNA
precursors

Early proteins

Host chromosome

Late
RNAs

Late
proteins

0	5	10	15	20	25

Minutes after infection

recombinant-type phages among the progeny is used as a measure of distance between linked genes.

2. Some phages, such as λ, have linear maps, while others, such as T2 and T4, have circular maps. Circular maps in bacterial viruses do not normally indicate that the chromosome itself is circular; instead, they indicate a circularly permuted gene arrangement.

FINE-STRUCTURE MAPPING

Until the 1940s, a gene was regarded as a unit of chromosome structure that could not be subdivided by crossing-over. Crossing-over was thought to occur only between genes, never within the genes themselves. According to this view, different mutation sites within a gene could never recombine and would always map at the same point. This view

■ **TABLE 14–3** Chromosome and linkage maps of three well-studied phages.

Phage	Genome Size (kb)	Genome Characteristics	Linkage Map
T4	166	*b c d e* ⎯⎯//⎯ *z a b c* *d e f g* ⎯⎯//⎯ + *b c d e* + etc. Circularly permuted terminally redundant ends	circular
T7	40	*a b c d* ⎯⎯//⎯ *y z a b* Unique sequence terminally redundant ends	linear
λ	49	GGGCGGCGACCT ⎯⎯//⎯ CCCGCCGCTGGA Unique sequence single-stranded cohesive ends	linear

of the gene as an indivisible particle has since been shown to be completely wrong. We now know that crossing-over occurs within the boundaries of genes as well as between them. Furthermore, different mutable sites within a gene can be separated by crossing-over. The ability to detect intragenic crossing-over has allowed geneticists to construct fine-structure linkage maps, which show the relative positions of mutation sites within genes. Fine-structure mapping is difficult to perform in higher organisms because of the extremely short distances separating base pairs that undergo a detectable mutational change. The shortness of these intervals makes it necessary to analyze 100,000 or more progeny in order to obtain accurate recombination frequencies. Such large population sizes are readily obtainable with microorganisms, however; consequently most fine-structure mapping has been done on viruses, bacteria, and certain lower eukaryotes, such as yeast.

The procedures used in fine-structure mapping are basically similar to those employed in mapping the chromosomal locations of different genes. Two-point crosses are conducted between different (independently isolated) mutant strains that have undergone mutation in the same gene. The different mutants may be indistinguishable in phenotype, but crossing-over between the strains can be detected by the appearance of wild-type recombinants:

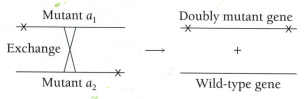

Since only half of the recombinant progeny are expected to have wild-type markers at both sites, the map distance in this case is twice the frequency of wild-type recombinants.

Completely mapping all the mutation sites within a gene is a formidable task. For example, constructing a map with 1000 mutant strains (each presumably differing from the others at a single base pair) requires a total of

$$\frac{1000!}{2!998!}$$

or about 500,000 two-point crosses! Various modifications of this basic procedure are therefore used to make the task easier. One useful approach is to cross the different mutant strains that have been isolated with deletion mutants from the same gene. Unlike single-site mutants, a deletion mutant is missing a segment of DNA that may include many bases. When a single-site mutant is crossed with a deletion mutant or when two deletion mutants are crossed, they will recombine and produce wild-type progeny only if their mutations do not share a common region. The theoretical basis for this genetic behavior for a hypothetical example involving two deletion mutants and a single-site mutant is shown in ▶ Figure 14–23. Note that wild-type offspring are possible only when the mutable site (base pair) affected in the single-site mutant is not one of those missing in the deletion mutant.

We can establish a deletion map of the two deletions in Figure 14–23:

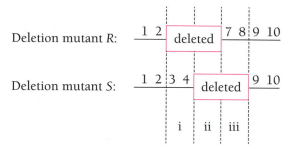

▶ FIGURE 14–23 (a) Three hypothetical mutants, a single-site mutant (left) and two deletion mutants, R (left) and S (right). (b) The effects of crossing-over between the single-site mutant and each of the deletion mutants. Wild-type recombinants are produced only by the crossover involving deletion mutant S (right), since the S deletion does not overlap the point of the single-site mutation, as deletion R does.

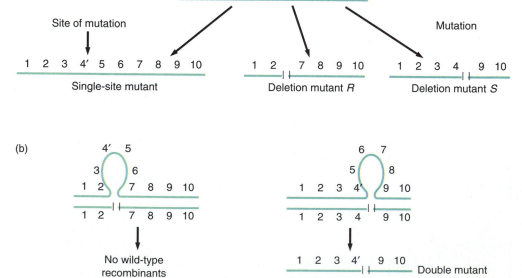

The overlap of the two deletions defines three regions of the gene (i, ii, and iii), and any single-site mutant can now be tested for its approximate location in one of these regions. There are three possibilities:

1. Wild-type recombinants are formed only by crosses between the single-site mutant and the deletion mutant *R*—this means that the single-site mutation lies in region iii.

2. Wild-type recombinants are formed only by crosses between the single-site mutant and deletion mutant *S*—this means that the single-site mutation lies in region i.

3. The single-site mutant gives no wild-type recombinants when crossed to either *R* or *S*—this means that the single-site mutation lies in region ii.

Once the approximate location of the single-site mutation has been established, its precise location can be pinned down more closely by crossing the mutant to a series of other known deletion strains that are missing progressively smaller amounts of DNA in the general region containing the mutation site (▶ Figure 14–24). The distances between mutation sites that have been localized in this manner can then be determined by a much smaller number of two-factor crosses.

Example 14.3

Deletion mutants can be intercrossed and their deletions mapped just like single-site mutations. As an example of how a deletion map is constructed, consider the following problem. Four deletion mutants are crossed in paired combinations to test for their ability to produce wild-type recombinants. The results are given in the table below, where + indicates that recombinants were observed. Draw the deletion map that corresponds to this group of mutations, and label the regions into which the deletion map subdivides the locus.

		Deletion Mutants			
		1	2	3	4
	1	–	–	+	+
Deletion	2		–	–	+
Mutants	3			–	–
	4				–

Solution: Deletions 1 and 3, deletions 1 and 4, and deletions 2 and 4 do not overlap, since they give wild-type recombinants when crossed. However, deletion 3 overlaps 4, and deletion 2 overlaps both 1 and 3. A topological representation (deletion map) of these results is:

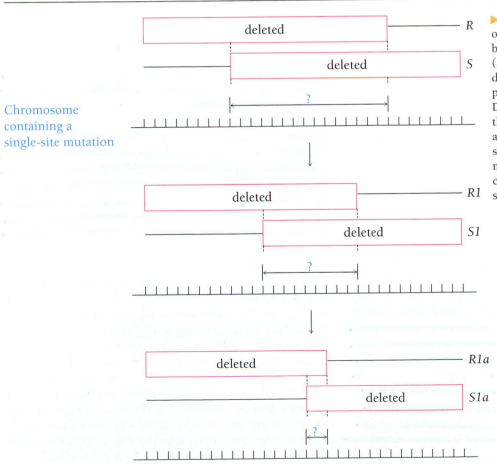

Chromosome containing a single-site mutation

▶ FIGURE 14–24 Localization of the point of a single-site mutation by using a series of known deletions (*R*, *S*, *R1*, *S1*, *R1a*, and *S1a*). The deletion mutants are missing progressively smaller amounts of DNA in the chromosomal region that contains the mutation site, so a series of crosses between a single-site mutant and these deletion mutants progressively reduces the chromosomal region in which the single-site mutation is known to lie.

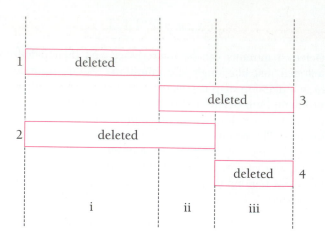

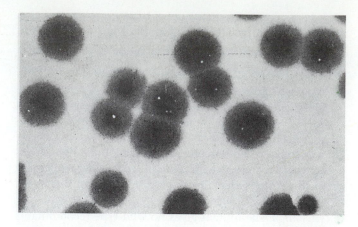

Follow-Up Problem 14.3

Three single-site (point) mutants (*a*, *b*, and *c*) were tested against the deletion mutants described in Example 14.3. The results of these crosses are given below (+ indicates that wild-type recombinants were formed). In what region of the deletion map (i, ii, or iii) is each of the point mutants located?

		Deletion Mutants			
		1	*2*	*3*	*4*
Point	*a*	+	−	−	+
Mutants	*b*	+	+	−	−
	c	−	−	+	+

The *rII* Region

Many of the methods used in fine-structure mapping were developed by Seymour Benzer in the 1950s. Benzer mapped the *rII* region of phage T4, which consists of two genes, *A* and *B,* that influence plaque morphology and host specificity. Wild-type (*r*⁺) phages form small plaques on both *E. coli* strains B and K12(λ) (the latter strain is lysogenic for λ) (see ▶ Figure 14–25). Mutant *rII* phages produce large plaques on strain B but are unable to infect strain K12(λ). Benzer began his experiments with a group of about 2400 independently isolated *rII* mutants, including 145 whose deletions spanned different lengths of the *rII* region. Benzer used these deletion mutants to construct a deletion map consisting of 47 smaller segments (▶ Figure 14–26).

To help speed up the mapping procedure, Benzer developed a special technique called a spot test that enabled him to determine whether two mutants could give a wild-type phenotypic response as a result of complementation or by the production of *r*⁺ recombinants (▶ Figure 14–27a). The test involved plating a mixture of *E. coli* strains B and K12(λ) on nutrient agar along with particles of an *rII* tester strain. Drops containing other *rII* mutants to be tested against the tester strain were then placed on the agar surface so that the

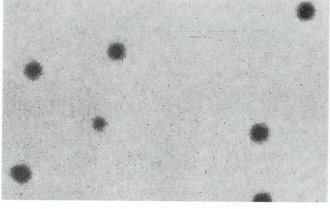

▶ **FIGURE 14–25** Mutant *rII* phages form large plaques on *E. coli* strain B (top) while wild-type (*r*⁺) phages form small plaques on both *E. coli* strains B and K12(λ) (bottom).

E. coli cells within the area of each drop would be infected by both *rII* mutants (a mixed infection). After incubation, three types of responses were observed (Figure 14–27b): (1) massive lysis produced as a result of complementation, which indicated that the pair of *rII* strains were mutant for different genes, (2) low-level lysis produced by occasional wild-type recombinants, which indicated that the pair of *rII* strains had different, nonoverlapping defects within the same gene, and (3) no lysis, which indicated that the pair of *rII* strains had overlapping defects and could neither complement nor recombine and produce wild-type recombinants.

With this test Benzer was able to divide the single-site *rII* mutants into two complementation groups or **cistrons** (a term now synonymous with genes) corresponding to the *A* and *B* genes. The approximate location of each mutation was then determined by spot-testing the single-site mutants with various deletion mutants to look for wild-type recombinants. This deletion mapping technique enabled Benzer to subdivide the mutations into even smaller intervals within the *rII* region. For more refined mapping, individual two-point crosses were finally performed between the mutant strains that had been localized within the same region of the deletion map. By employing this sequence of steps,

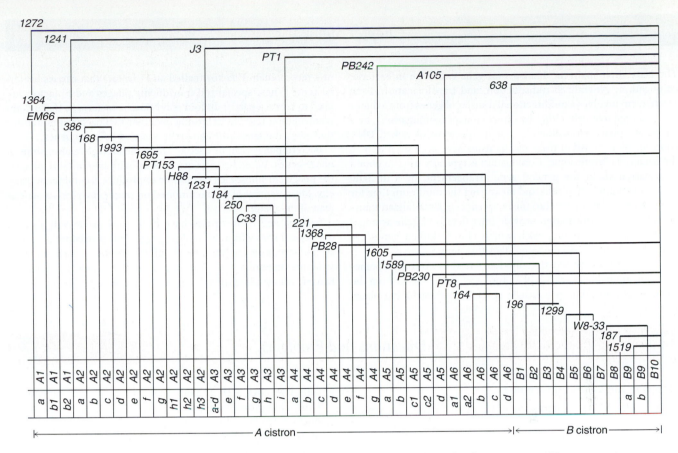

▶ **FIGURE 14–26** Benzer's deletion map of the *rII* region of bacteriophage T4 subdivides the region into 47 segments. *Source: Seymour Benzer, Proc. Natl. Acad. Sci., USA, 47 (1961): 403. Used with permission of Seymour Benzer.*

Benzer was able to identify over 300 mutation sites that were separable by recombination. The sensitivity of Benzer's mapping procedure was sufficient to detect one recombinant in 10^6 phage. However, the smallest experimental map distance that was observed in all of the crosses was 0.02 map units. Since the genome of T4 consists of 1.66×10^5 bp and has a total map length of 1500 map units, the smallest experimental map distance corresponds to roughly

$$1.66 \times 10^5 \left(\frac{0.02}{1500} \right)$$

or between 1 and 3 bp. Benzer was thus able to map the *rII* region down to its ultimate level of resolution—the nucleotides. As Benzer put it, he had "run the map into the ground."

▶ **FIGURE 14–27** The spot test technique. Three different mutants are crossed with a tester strain (rII_x) and spotted on a mixture of *E. coli* B and K12(λ).

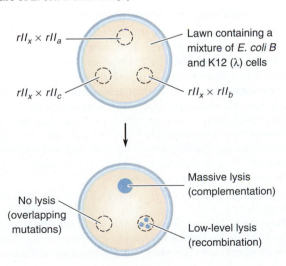

To Sum Up

1. Detailed maps can be made of the mutation sites within a gene locus. These fine-structure maps are derived from crosses between mutant strains having mutational changes at different base pairs within the same gene.
2. To facilitate the construction of a fine-structure map, geneticists subdivide a gene into smaller intervals based on the location, length, and overlap of the DNA segments missing from deletion mutants. Single-site mutations can then be quickly mapped into defined intervals by crossing the mutant strains with each of the overlapping deletion mutants.

There are three main mechanisms for gene exchange in bacteria: conjugation, generalized transduction, and transformation. Each mechanism involves a unidirectional transfer of genes from a donor cell to a recipient cell, either by direct contact (conjugation), by a temperate phage (transduction), or in the form of naked DNA (transformation). All of these mechanisms have been used to map the bacterial chromosome. Conjugation is typically used to obtain information about the general gene arrangement, by artificially separating conjugating cells and measuring the minimum time for entry of each donor gene into the recipient cell. Generalized transduction and transformation provide finer details of gene arrangement from the frequency with which closely linked genes are transferred together on the same DNA fragment.

Bacterial genes are occasionally transferred by episomes. Sometimes an anomalous excision of a prophage or an F factor from the bacterial chromosome produces a specialized transducing phage

or a recombinant F factor (called an F′ factor) that carries bacterial genes. These specialized transducing phages and F′ factors are used to create a partial diploid condition in bacteria so that a complementation test can determine whether two bacterial mutations that affect the same function are in same or different genes.

Bacteriophages can be used to study the linkage relationships of phage genes. When bacteria are infected with two or more phages, the phage chromosomes recombine to produce recombinant progeny. Measurement of the recombination frequency yields a genetic map of the phage.

Bacteria and their viruses have been used extensively to map mutational sites within individual genes. This procedure is known as fine-structure analysis. Many of the methods used for fine-structure mapping were developed in the 1950s by Seymour Benzer, who studied the *rII* region of bacteriophage T4.

 ## *Questions and Problems*

Mapping the Bacterial Genome

1. Define the following terms and differentiate between the paired terms:
   endogenote and exogenote
   transduction and sexduction
   generalized and specialized transduction

2. Suppose that in a series of crosses four different Hfr strains of *E. coli* transfer their genes in the order given below:

Hfr Strain	Gene Order				
1	L	D	P	H	Z
2	H	Z	C	O	Y
3	D	L	A	R	X
4	X	M	B	Y	O

Determine the sequence of genes on the chromosome of the F⁺ strain from which these Hfr strains were derived.

3. Suppose that in a series of crosses four different Hfr strains of *E. coli* transfer their genes in the order given below:

Hfr Strain					
1	Gene markers:	A	B	C	F
	Time (minutes):	4	54	72	79
2	Gene markers:	D	E	F	G
	Time (minutes):	7	16	67	80
3	Gene markers:	E	D	A	G
	Time (minutes):	11	20	35	47
4	Gene markers:	F	B	E	
	Time (minutes):	2	27	53	

Construct a linkage map of the *E. coli* chromosome from these results, using the entry times to calculate the distances (in minutes) between the adjacent pairs of loci.

4. When Hfr cells conjugate with F⁻ cells that are lysogenic for λ, the recipient cells usually survive. But when Hfr cells that are lysogenic for λ conjugate with F⁻ nonlysogens, the recipient

cells lyse after a short period of time. This phenomenon is known as zygotic induction. (a) Explain the basis of zygotic induction. (b) How could you determine the chromosomal locus of the integrated prophage?

5. You have discovered a form of gene transfer between two strains of bacteria. Assume that you have access to an enzyme that can catalyze the degradation of isolated (naked) bacterial DNA; you also have a device that allows you to physically separate the different bacterial strains without interrupting the free flow of smaller materials (e.g., viruses and chromosome fragments) between them. Explain how you could differentiate between conjugation, transduction, and transformation as the possible basis for this transfer of genetic material.

6. The following groups of genes can be cotransduced by phage P1:
   1. *A, N,* and *R*
   2. *A* and *B*
   3. *E* and *N*
   4. *E* and *Y*
   What is the relative order of the genes?

7. A P1-mediated transduction experiment was performed in *E. coli* using *trpA⁺trpC⁺pyrF⁺* cells as donors and *trpA⁻trpC⁻pyrF⁻* cells as recipients. The markers *trpA* and *trpC* are involved in tryptophan biosynthesis, and *pyrF* is involved in pyrimidine biosynthesis. Among the *pyrF⁺* transductants that were initially selected, 521 were *trpA⁻trpC⁻*, 131 were *trpA⁺trpC⁻*, 42 were *trpA⁺trpC⁺*, and 1 was *trpA⁻trpC⁺*. (a) Determine the correct order of these markers on the *E. coli* chromosome. (b) Calculate the cotransduction frequencies for *pyrF* and *trpA* and for *pyrF* and *trpC*. (c) Calculate the map distances (in minutes) between these markers using equation 14.1.

8. In pneumococci, genes *a, b, c,* and *d* give strain 1 resistance to drugs A, B, C, and D, respectively. Strain 2 is sensitive to all of these drugs because it carries the wild-type alleles of these genes. DNA from strain 1 is used to transform strain 2, and the recipient cells are then plated on nutrient agar media contain-

ing various combinations of the four drugs. The following results are obtained:

Drugs Added to Growth Medium	Number of Colonies
none	10,000
A	1,177
B	1,152
C	1,196
D	1,183
A,B	420
A,C	31
A,D	710
B,C	39
B,D	580
C,D	26
A,B,C	22
A,B,D	410
A,C,D	18
B,C,D	16
A,B,C,D	20

(a) Which three of the four genetic markers are so closely linked that they are normally present on the same DNA fragment? (b) Give the sequence of these loci based on the frequency of cotransfer.

9. Consider two closely linked genes in E. coli, designated b and c. To determine the order of these genes relative to the adjacent a locus (shown to the left of b and c by interrupted mating experiments), reciprocal crosses were made by generalized transduction. Each strain was alternatively used as a donor and recipient by growing the transducing phage on each mutant strain and then using the phage to transduce the other strain. The number of wild-type recombinants produced by each of these crosses is:

Genotypes of		Number of wild-
Donor strain	Recipient strain	type recombinants
$+\ +\ c$ $\times$	$a\ b\ +$	15
$a\ b\ +$ $\times$	$+\ +\ c$	292

(a) What is the correct order of the a, b, and c loci? Explain. (b) Suppose that you discovered another gene, d, that mapped closer to the a locus than either b or c. If you were to make the reciprocal crosses $+\ +\ d$ donor $\times$ $a\ b\ +$ recipient and $a\ b\ +$ donor $\times$ $+\ +\ d$ recipient, how would the relative number of wild-type recombinants formed in the two crosses compare? Explain.

Mapping Phage Chromosomes

10. Assume we have two mutant strains of bacteriophage T2: h is unable to synthesize head protein, and t is unable to form tail protein. Mature forms of both mutant strains can inject their chromosomes into E. coli, but neither strain by itself can complete the reproductive cycle and produce lysis. When phages of both strains infect the same cell, lysis occurs with the release of the same number of progeny viruses as would normally be produced by wild-type T2. Most of the progeny of such mixed infections are genetically mutant, but about 6% are wild type and can complete the lytic cycle by themselves to produce progeny when they subsequently infect sensitive bacterial cells. (a) Explain these results in terms of the life

cycle of bacterial viruses. (b) From the data given, estimate the distance between the h and t loci.

11. Three two-factor crosses were made using different strains of bacteriophage T2. The following results are obtained:

Cross 1: $+\ b \times a\ + \rightarrow$ 1773 $+\ b$, 1747 $a\ +$, 104 $+\ +$, 96 $a\ b$

Cross 2: $+\ c \times b\ + \rightarrow$ 1348 $+\ c$, 1312 $b\ +$, 124 $+\ +$, 108 $b\ c$

Cross 3: $+\ c \times a\ + \rightarrow$ 1443 $+\ c$, 1483 $a\ +$, 51 $+\ +$, 55 $a\ c$

(a) Construct a linkage map of these three loci. (b) Calculate the coefficient of coincidence from these results. What type of interference appears to be operating in this case? (c) What factors, in addition to multiple crossovers, influence the value of the coefficient of coincidence when evaluated from the results of two-point phage crosses as in (b)?

12. A series of two-point phage crosses involving the loci a, b, c, and d gave the following unique relationships for the recombination frequencies: $R_{a-c} = R_{b-d} > R_{a-b} = R_{b-c} = R_{c-d} = R_{a-d}$. Explain these results in terms of a suitable linkage map.

13. C. A. Thomas (1967) demonstrated that when denatured T2 DNA is allowed to renature and then is viewed with an electron microscope, the DNA molecules reassociate to form circular double helices. Explain how the linear DNA molecules of this phage can give rise to circular molecules upon renaturation. (Hint: Recall that base sequences are circularly permuted in this phage.)

14. Consider seven uniformly spaced genes a–g on the T4 chromosome. (a) If an E. coli cell is infected with a phage having the gene arrangement abcdefgab, how many arrangements could be present among the progeny? Give the arrangements. (b) Suppose that you isolate a T4 deletion mutant that is missing gene c. If the deletion mutant is used to infect an E. coli cell, what gene arrangements could be present among the progeny, assuming that the headful mechanism of assembly is correct?

15. How many different two-factor phage crosses are needed to map unambiguously (a) four mutant sites? (b) five mutant sites? (c) N mutant sites? (Hint: Consider the different combinations of genetic markers, taken two at a time.)

Fine-Structure Mapping

16. Four deletion mutants are crossed in pairwise combinations to test for their ability to produce wild-type recombinants. The results are given in the table below, where + indicates that recombinants were observed:

		Deletion Mutants			
		1	2	3	4
	1	–	+	+	–
Deletion	2		–	–	+
Mutants	3			–	–
	4				–

Show that the topological representation of these results (the deletion map) subdivides the locus into three regions.

17. Three point mutants were tested against the deletion mutants described in problem 16 through paired crosses. The results of these crosses are given below, where + again indicates that wild-type recombinants were observed. What is the relative order of the point mutants?

Deletion Mutants

		1	2	3	4
Point	a	+	+	-	-
Mutants	b	-	+	+	-
	c	+	-	-	+

18. Seven deletion mutants of bacteriophage T4 were crossed in paired combinations to test for their ability to produce wild-type recombinants. The results are given in the table below (+ indicates recombinants). Construct a deletion map based on these findings.

Deletion Mutants

		1	2	3	4	5	6	7
	1	-	+	-	-	-	-	-
	2		-	+	-	+	-	-
Deletion	3			-	+	-	+	-
Mutants	4				-	-	-	-
	5					-	+	-
	6						-	-
	7							-

19. Five point mutants were tested in paired crosses with the deletion mutants listed in problem 18 for their ability to yield recombinants. The results of these crosses are given below. What is the relative order of the point mutants?

Deletion Mutants

		1	2	3	4	5	6	7
	a	+	-	+	-	+	-	-
	b	-	+	-	+	-	+	-
Point	c	-	+	+	-	-	+	-
Mutants	d	-	+	+	-	+	-	-
	e	-	+	-	+	-	+	+

20. Five *rII* mutant strains of phage T4 give the following results in complementation tests (+ = complementation, - = no complementation, blank = not tested):

Mutant Strain

		1	2	3	4	5
	1	-	+		+	-
	2		-	+		
Mutant	3			-	-	
Strain	4				-	+
	5					-

Arrange the mutant loci into cistrons.

21. Six mutant loci are known to reside in three cistrons. They give the following results in complementation tests. From these results, fill in the seven indicated blank spaces in the following table using + for complementation and - for no complementation.

Mutant Loci

		1	2	3	4	5	6
	1	-	+	___	+	+	+
	2		-	+	___	___	+
Mutant	3			-	-	___	___
Loci	4				-	___	___
	5					-	+
	6						-

22. The following experiments relate to a fine-structure analysis of the *his* locus of *Salmonella typhimurium* using P22-mediated transduction.

(a) The results of complementation tests between eight single-site His⁻ mutants during abortive transduction are shown in the following table. Complementation in these studies is indicated by the growth of microcolonies on minimal medium, designated by +. Lack of complementation is indicated by the failure to grow, designated by -.

Single-Site Mutants

		1	2	3	4	5	6	7	8
	1	-	+	+	+	-	+	+	-
	2		-	-	+	+	-	+	+
Single-	3			-	+	+	-	+	+
Site	4				-	+	+	-	+
Mutants	5					-	+	+	-
	6						-	+	+
	7							-	+
	8								-

(b) The results in the following table were obtained when seven nonreverting (deletion) His⁻ mutants were crossed with each other in all pairwise combinations. The + sign indicates that wild-type recombinants are formed by complete generalized transduction; the - sign indicates that no wild-type recombinants are formed.

Deletion Mutants

		a	b	c	d	e	f	g
	a	-	+	+	+	+	-	+
	b		-	+	+	-	+	+
Deletion	c			-	+	+	+	-
Mutants	d				-	+	-	-
	e					-	+	+
	f						-	+
	g							-

(c) The eight single-site mutants in (a) gave the following results when crossed with the deletion mutants in (b) (+ indicates wild-type recombinants, - indicates no wild-type recombinants).

Single-Site Mutants

		1	2	3	4	5	6	7	8
	a	-	+	+	+	+	+	+	-
	b	+	+	+	-	+	+	-	+
Deletion	c	+	-	-	+	+	+	+	+
Mutants	d	+	+	+	+	-	-	+	+
	e	+	+	+	-	+	+	+	+
	f	+	+	+	+	-	+	+	-
	g	+	-	+	+	+	-	+	+

(1) Using the data to support your conclusion, how many cistrons do these mutants represent? Which cistrons, if any, are contiguous? (2) Map the *his* region in the *S. typhimurium* genome as completely as possible, indicating any ambiguities.

CHAPTER 15

Cellular and Molecular Approaches to Mapping

The traditional methods for chromosome mapping have led to the construction of genetic linkage maps for a variety of experimental organisms. The opportunities to apply these methods to humans have been rather limited, however, because of their long reproductive cycle and the obvious impossibility of designing genetic crosses. Until recently, human geneticists have had to rely solely on family pedigrees to deduce the linkage of genes for various traits. Such data are useful for genes located on the X chromosome, since sex linkage is relatively easy to demonstrate, but pedigree information alone is usually inadequate for assigning autosomal genes to particular chromosomes.

Only in the past few decades have comparatively rapid methods for localizing genes on human chromosomes become available. These methods are of two main types: (1) cellular methods, which involve tissue culture techniques to assign genes to individual chromosomes and

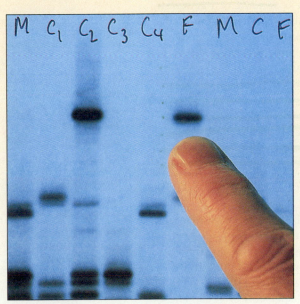

DNA fingerprinting can be used to show paternity.
The finger is pointing to one of the father's DNA bands that
matches a band in the child marked C2.

specific chromosome segments, and (2) **molecular methods**, which use various enzymatic and nucleic acid hybridization techniques to cleave DNA and analyze the fragments. These new technologies have revitalized human gene mapping and have provided geneticists with the tools needed to directly examine the gene itself without first having to identify the phenotype or isolate the gene product.

The techniques discussed in this chapter (mapping, genetic testing, and forensics) form the basis of modern human genetic analysis. In addition, the molecular methods have found broad application in a variety of other biological areas. In fact, the technical developments described in this chapter, along with those discussed in Chapter 18, have made possible major leaps forward in our knowledge of the molecular biology of all organisms. From ecology to medicine, from the study of the expression of a single gene to the study of evolutionary relationships among organisms, all of biology has been revolutionized by the development and application of these technologies.

 ## CELLULAR METHODS: SOMATIC CELL GENETICS

Multicellular organisms, such as humans, have rather long life spans and small numbers of progeny. In contrast, their somatic cells may have rapid reproductive rates when cultured outside the body, and can undergo several division cycles in a short period of time. The large reproductive potential of somatic cells has made them extremely useful tools in genetic research, since geneticists can examine several generations consisting of large numbers of cells within a relatively brief interval of time. Moreover, the ability to culture one or just a few cell types under carefully controlled conditions has provided a refinement of genetic analysis that would be impossible with the intact organism. The many advantages of using somatic cells in a variety of genetic investigations, including the identification of genetic defects in cultured cells, have led to the development of an area of cell research known as **somatic cell genetics.**

This section provides a brief introduction to the methods of somatic cell genetics, with a particular focus on techniques that are useful in linkage analysis. Many of these methods are used in conjunction with a procedure called **somatic cell hybridization**, which was developed in the early 1960s. Scientists use this procedure to form hybrid cells by fusing the somatic cells of different species. Although the hybridization of somatic cells was not purposely developed for the study of linkage, it has been adapted by geneticists for use in determining the chromosomal location of genes.

Somatic Cell Fusion

When somatic cells, such as fibroblasts, are grown together in a culture medium, they are capable of undergoing fusion. Under normal circumstances, fusion occurs quite rarely, but its incidence can be greatly enhanced by adding certain chemicals (such as polyethylene glycol) or ultraviolet-inactivated Sendai virus to the suspension of cells. These fusing agents alter the surfaces of cells in a way that holds cell membranes in close contact, allowing them to coalesce and form a single cell.

To begin the procedure for somatic cell fusion, researchers mix two different lines of cells in the presence of a fusing agent (▶ Figure 15–1). Fusion of the cytoplasm oc-

▶ **FIGURE 15–1** Cell fusion techniques applied to human and mouse cells. Cells are fused on a medium containing a fusing agent, such as inactivated Sendai virus. Cytoplasmic fusion occurs first, producing a heterokaryon with a nucleus from each parental cell line. Subsequent fusion of the nuclei gives rise to hybrid cells, which are isolated on a selective medium. The photo shows G-banded human and mouse chromosomes in a hybrid cell.

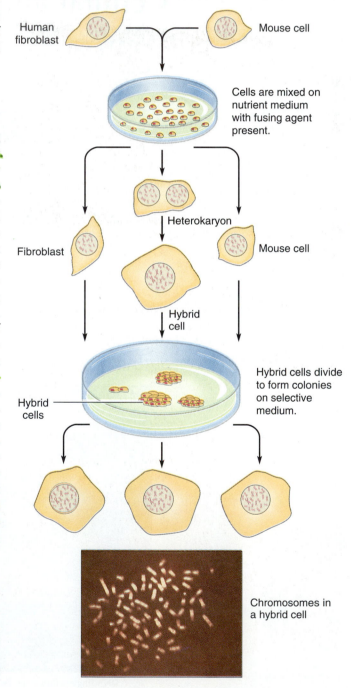

curs first, forming a heterokaryon with two different nuclei. The nuclei may then fuse to form a **synkaryon** (or hybrid cell) with a single nucleus containing the chromosomes of both parental cell lines. Hybrid cells, which may express genes of both parental cell types, are capable of dividing on their own to yield hybrid clones. The nuclei that fuse should contain complete sets of chromosomes from both parents, but if they are from separate species, the hybrids tend to lose chromosomes during the first few subsequent mitoses. Interestingly, chromosomes are usually lost from only one of the sets in an interspecific hybrid cell. For example, in hybrid cells made from mouse and human cells, the human chromosomes are lost preferentially. In this case, the loss of human chromosomes can be followed microscopically, since mouse and human chromosomes are fairly easy to distinguish.

Detection of Hybrid Cells

Once cell fusion occurs, the geneticist must be able to distinguish the hybrid cells from all the other cells in the medium and selectively isolate those hybrids whose genotypes are to be studied. Two readily identifiable cell characteristics are commonly used for this purpose: (1) the ability of certain cells to grow in the absence of specific nutrients and (2) the resistance of certain cells to the actions of various drugs. The **HAT method** is an example of a technique that selects for hybrid cells on the basis of both of these characteristics. The selective medium for this technique contains hypoxanthine and thymidine (H and T), which are normal precursors of DNA, and aminopterin (A), a drug that specifically blocks one pathway leading to DNA synthesis. Normal cells can synthesize DNA by two metabolic pathways: the major pathway, which is blocked by aminopterin, and a salvage pathway (▶ Figure 15–2). The major pathway produces DNA enzymatically from scratch, starting with simple sugars and amino acids. The salvage pathway synthesizes DNA from more complex precursors (nucleosides) supplied by the breakdown of RNA and by dietary sources. This alternative (salvage) pathway thus enables cells to utilize substrates that are already supplied in the diet, without having to break them down completely and resynthesize them from simpler starting materials. Two functioning enzymes are required for completion of the salvage pathway: thymidine kinase (TK) and hypoxanthine-guanine phosphoribosyl transferase (HPRT). Cells that are unable to produce either TK (TK$^-$ cells) or HPRT (HPRT$^-$ cells) cannot convert precursors into DNA along the salvage pathway and must depend on the major pathway for DNA synthesis. Thus, if the major pathway is blocked by the presence of aminopterin in the medium, such cells would be unable to grow.

The application of the HAT method is shown in Figure 15–2b. Note that when TK$^-$HPRT$^+$ mouse cells are fused with TK$^+$HPRT$^-$ human cells and plated on HAT medium, both parental strains die. Only hybrid cells that have both the mouse HPRT$^+$ character and the human TK$^+$ character will survive, since they are now capable of producing the functional HPRT and TK enzymes that are needed for the synthesis of DNA along the salvage pathway. The amounts of hypoxanthine and thymidine in the selective medium provide sufficient substrate for the salvage pathway alone to support growth.

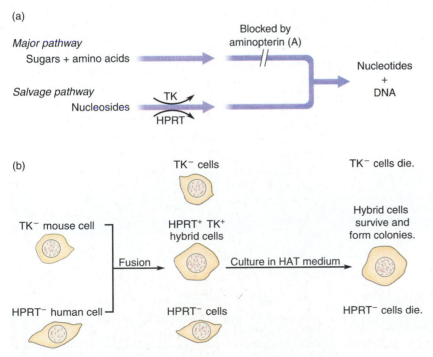

(a)

Major pathway
Sugars + amino acids → Blocked by aminopterin (A)

Salvage pathway
Nucleosides → TK, HPRT

Nucleotides + DNA

(b)

TK$^-$ cells → TK$^-$ cells die.

TK$^-$ mouse cell ⎤
                    ⎥ Fusion → HPRT$^+$ TK$^+$ hybrid cells → Culture in HAT medium → Hybrid cells survive and form colonies.
HPRT$^-$ human cell ⎦

HPRT$^-$ cells → HPRT$^-$ cells die.

▶ **FIGURE 15–2** (a) Two metabolic pathways used in the synthesis of DNA: the major pathway, which is specifically blocked by aminopterin, and a salvage pathway, which requires the enzymes thymidine kinase (TK) and hypoxanthine-guanine phosphoribosyl transferase (HPRT). (b) Cells incapable of producing the TK or HPRT enzymes will die when placed in HAT medium, since it contains aminopterin. Hybrid cells that possess both the TK and HPRT enzymes will survive and grow.

Assigning Genes to Chromosomes

Once a number of hybrid cell lines (clones) have been isolated, they can be used to assign genes to chromosomes. In the case of rodent-human hybrid cells, for example, the preferential loss of human chromosomes is the basis for this procedure. The human chromosomes are lost in a random order as the cells divide until only one or just a few human chromosomes remain in the hybrid nucleus. At this point, stable hybrid clones are produced that no longer lose human chromosomes. If these hybrid clones, which retain different partial human chromosome sets, can be analyzed for the presence of human genes (or gene products) and have sufficient karyotypic variability, then these genes can be readily assigned to specific human chromosomes with great accuracy.

In a typical assignment test, several different hybrid clones are selected to form a **clone panel** in which each clone has a unique subset of human chromosomes. The clones are then analyzed for the retention or loss of specific human genes, either by testing for the genes themselves (to be described in a later section) or by testing for the enzyme or other protein product of each of these genes. Chromosome assignments can then be made by comparing the pattern of retention (+) or loss (−) of each gene tested in the different hybrid clones with the presence or absence of specific human chromosomes (■ Table 15–1). If the presence (or absence) of a particular human chromosome is always associated with the retention (or loss) of a specific human gene, then it can be concluded that the gene tested for is located on that chromosome.

The number of different clones needed on a clone panel in order to uniquely assign a human gene to its chromosome is not very large. In general, the number of different patterns of + and − that can be generated in a clone panel is 2^c, where c is the number of hybrid clones. To produce a different pattern for every chromosome, 2^c must be greater than 24 (22 autosomes plus two different sex chromosomes). Thus, a total of five different clones ($2^5 = 32$) is sufficient. In practice, however, somewhat larger clone panels are generally used to help avoid problems of false positive and false negative tests resulting from experimental error.

When many different genes are tested with a clone panel, some are likely to exhibit identical patterns of retention and loss. Such genes would be assigned to the same chromosome, and to distinguish them from genes that are determined to be linked in family studies, they are said to be **syntenic** (from Greek words meaning the "same ribbon"). Although syntenic genes have their loci on the same chromosome, they may or may not show genetic linkage. (Recall that two widely separated markers can recombine so readily

■ **TABLE 15–1** Examples of using human chromosome panels to assign genes to specific chromosomes. Each panel consists of several different hybrid clones, each containing a unique subset of human chromosomes. + means that the chromosome is present in the clone; − means that it is absent in that clone. The right-hand column gives the results of tests for 6-phosphogluconate dehydrogenase (6PGD, panel 1) and galactokinase (GALK, panel 2) activity; + means the enzyme is present, − means it is not. Comparison of the pattern of retention of the chromosomes and the enzyme in each panel shows that the 6PGD gene is located on chromosome 1 and the GALK gene on chromosome 17.

Hybrid Clone	Panel 1 Human Chromosomes										6PGD Activity

Hybrid Clone	X	Y	1	4	7	9	11	17	21	6PGD Activity
1	+	+	+	−	−	+	+	+	+	+
2	+	−	+	−	+	+	−	+	−	+
3	+	−	−	+	+	−	−	+	−	−
4	−	−	−	−	−	−	+	+	+	−
5	+	−	−	−	−	+	−	−	−	−
6	−	−	+	+	−	−	+	−	−	+

Panel 2 — Human Chromosomes

Hybrid Clone	X	Y	1	4	7	9	11	17	21	Galactokinase Activity
1	+	+	−	−	−	+	+	+	+	+
2	+	−	+	−	+	+	−	+	−	+
3	−	−	+	+	+	−	−	+	−	+
4	+	−	+	−	−	−	+	+	+	+
5	+	+	−	−	−	+	−	−	−	−

that they will appear to assort independently in genetic crosses.) The term syntenic is therefore a useful description of genes that are known to be physically linked, regardless of whether genetic data show their linkage.

Making Regional Assignments

The techniques just described can assign genes to certain chromosomes, but they do not give the precise location of any one gene on a chromosome map. More detailed analysis is necessary to pinpoint the location of genes and to establish the relative order and degree of linkage of the different loci along a chromosome.

One useful approach for localizing genes on chromosomes is to use hybrid clones that carry a chromosome aberration, such as a deletion or translocation. Aberrations of this type can be induced or can occur spontaneously, either in the hybrid clone itself or in the original parent cell. When such hybrid clones are included in a clone panel, researchers are able to make regional assignments by associating the retention (or loss) of a gene with the presence (or absence) of the chromosomal segment involved in the aberration.

As an example, ▶ Figure 15–3 shows a regional mapping test that was used to assign the human genes for the enzymes HPRT, PGK (phosphoglycerate kinase), and G6PD (glucose 6-phosphate dehydrogenase) to different regions on the X chromosome. In this study, clones were used that were missing progressively longer sections of the long arm of the X chromosome as a result of chromosomal aberrations. By comparing the cellular activities of the enzymes with the particular chromosomal segments that were missing and retained, it was possible to demonstrate that the order of the three genes was centromere-*PGK-HPRT-G6PD*,

and to determine the approximate location of each of these genes on the chromosome.

To Sum Up

1. The genes on human chromosomes present a problem to the traditional methods of mapping since controlled testcrosses cannot be ethically performed. An alternative method, such as somatic cell hybridization, must therefore be used to determine the chromosomal location of human genes.
2. The usual procedure in somatic cell genetics is to form human-mouse hybrid cell lines through cell fusion and then isolate the hybrid clones on a selective medium. As the hybrid cells divide, they preferentially lose the human chromosomes. Proper chromosome assignments can then be made by correlating the loss of certain genes or gene products with the loss of specific chromosomes.
3. More precise assignments of genes to particular chromosome regions can be made by relating the presence or loss of a gene to defined chromosomal deletions or translocations.

MOLECULAR METHODS: DNA RESTRICTION AND HYBRIDIZATION

Methods developed since the early 1970s have made it possible to construct detailed maps of large genomes using markers that are independent of gene function. The positions that are mapped by these methods are enzyme-cleavage sites, that is, they are short DNA sequences that are recognized for cleavage by site-specific endonucleases. Unlike gene loci in conventional linkage maps, the recognition sites for these nucleases can exist anywhere along the length of a DNA molecule, and thus it is possible to map regions on a chromosome where no other markers can be found.

DNA Restriction and Modification: Restriction Endonucleases

Underlying many of the new techniques for chromosome mapping is the discovery of a group of bacterial enzymes known as **restriction endonucleases.** Certain of these enzymes catalyze the double-stranded cleavage of DNA at specific base sequences, thus acting as precise molecular scalpels. Because they cut DNA at specific target sites, these enzymes have proven to be indispensable not only for DNA mapping but also for isolating, amplifying, and manipulating genes—procedures that will be discussed in Chapter 18.

In nature, the restriction endonucleases defend bacteria from the DNA of an infecting phage or other potentially harmful genetic element by cleaving foreign DNA into fragments while leaving their own DNA intact. These enzymes were discovered (without any preconceived notion of the great contributions they would make to current biotechnology) through research on an interesting biological phenomenon known as **host-controlled restriction and modifica-**

▶ **FIGURE 15–3** Regional mapping of the human X-linked genes for hypoxanthine-guanine phosphoribosyl transferase (HPRT), phosphoglycerate kinase (PGK), and glucose 6-phosphate dehydrogenase (G6PD). The genes were assigned regionally to the X chromosome in the order centromere-*PGK-HPRT-G6PD*, based on the enzymes present in a series of hybrid clones (1–4) missing different parts of the long arm of the human X chromosome due to chromosomal aberrations.

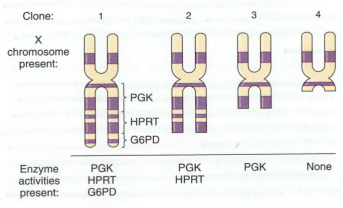

Clone:	1	2	3	4
X chromosome present:	PGK HPRT G6PD	PGK HPRT	PGK	
Enzyme activities present:	PGK HPRT G6PD	PGK HPRT	PGK	None

tion. This phenomenon is most readily observed when phages are transferred from one bacterial host strain to another. For example, if phage λ is grown in the K12 strain of *E. coli,* the progeny phages (designated here as λ.K) also grow well in subsequent infections of the K12 strain. Thus when a limited number of λ.K particles are spread on an agar plate with a large number of *E. coli* K12 cells, each phage will initiate an infection and produce a plaque. However, when the λ.K phages are spread on an agar plate with the B strain of *E. coli* (instead of the K12 strain), they grow poorly, with only 0.01% of the λ.K phages giving rise to a plaque. This phenomenon is summarized in ▶ Figure 15–4, where we see that the same kind of results are also obtained when the experiment is performed in reverse: If phage λ is grown on *E. coli* B, the resulting progeny (λ.B) grow well on strain B but poorly on strain K12.

How can we explain these results? In the 1960s, work by Werner Arber and his colleagues showed that the molecular basis of this phenomenon is a host-induced restriction and modification system that catalyzes structural changes in the phage DNA. Arber's group was able to demonstrate that most phage DNA molecules are restricted to a particular host strain because they are modified in a way peculiar to that host. By means of isotopic labeling, the researchers found that when the phages infect a host strain with a different modification system (such as λ.K in strain B or λ.B in strain K12), the phage DNA is recognized as being foreign and is degraded by a strain-specific restriction endonuclease. Restriction is not absolute, since a few phages slip by and acquire the modification pattern of the new host.

Arber and his colleagues were also able to show that modification involves a host-specified pattern of **DNA methylation.** In addition to the usual bases (adenine, cytosine,

5-methylcytosine (m⁵C)   N⁶-methyladenine (m⁶A)

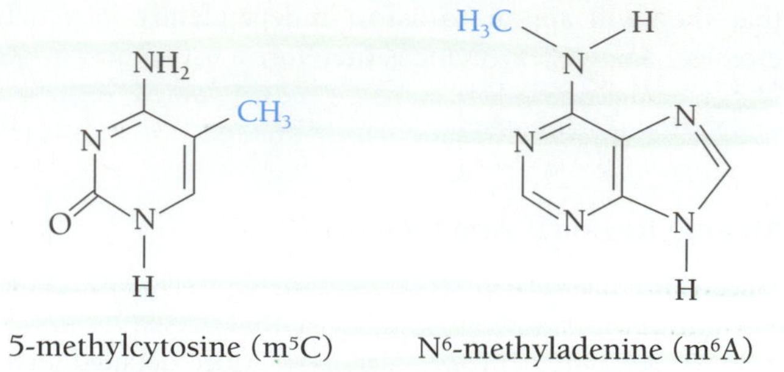

▶ **FIGURE 15–5** Two methylated bases found infrequently in DNA.

guanine, and thymine), DNA ordinarily contains a minor proportion of methylated bases, such as 5-methylcytosine and N⁶-methyladenine (▶ Figure 15–5). In the K12 and B strains of *E. coli,* methyl groups are enzymatically attached to adenine groups in a strain-specific pattern. The adenines that are methylated are found in the nucleotide sequences that are recognized for cleavage by the cell's restriction endonuclease—thus the modification and restriction specificities of a bacterial strain are the same. If the DNA lacks the methylation pattern peculiar to the host strain, it is broken down by the restriction endonuclease of the cell (▶ Figure 15–6a). However, if the phage DNA entering a particular bacterial strain has methylated adenine at the host-specified restriction (nuclease recognition) sites, it is permitted to replicate and carry out a successful infection (Figure 15–6b). The restriction and modification enzymes therefore provide bacteria with a primitive immune system that protects against the uptake of foreign DNA.

Properties of Restriction Enzymes

We now recognize three main classes of restriction endonucleases (■ Table 15–2). Type I and type III enzymes are bifunctional, acting as both endonucleases and methylases. Each type I or type III enzyme carries the catalytic activities for DNA restriction and modification on different subunits of a single protein. Both the type I and the type III enzymes cut DNA at some distance from the recognition sequence— the recognition and cleavage sites are not the same. In the type I system (found, for example, in the *E. coli* K12 and B strains), cleavage occurs at a random site at least 1 kb away from the recognition site. In the type III system, cutting occurs at a distance of 24–26 bp on one side of the recognition site.

Unlike the other two types of restriction systems, the type II enzymes are monofunctional, possessing only endonuclease activity. Prokaryotes that rely on the type II system for DNA restriction and modification also possess a separate DNA methylase with the same sequence specificity as the restriction enzyme. The type II endonucleases also differ from their type I and type III counterparts by having a cleavage site that is in most cases the same as the recognition site. By cutting DNA within the recognition site, each

▶ **FIGURE 15–4** Host restriction of phage growth. Phage particles grown on *E. coli* strain K12 or strain B are used to infect lawns of *E. coli* strain K12 or strain B.

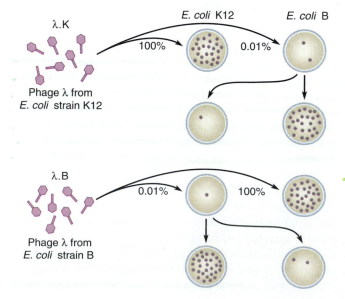

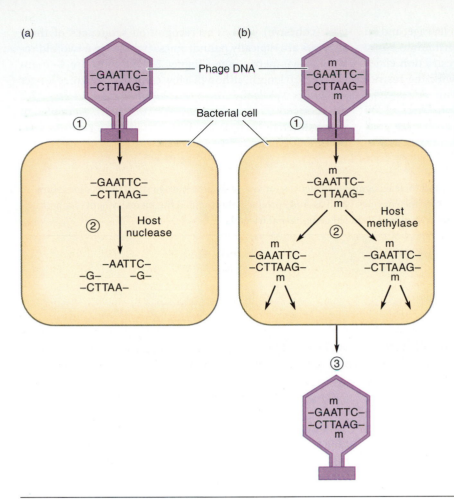

(a)

(b)

Phage DNA

Bacterial cell

—GAATTC—
—CTTAAG—

① Host
② nuclease

—AATTC—
—G—   —G—
—CTTAA—

m
—GAATTC—
—CTTAAG—
m

①

m
—GAATTC—
—CTTAAG—
m

② Host
methylase

m
—GAATTC—
—CTTAAG—
m

m
—GAATTC—
—CTTAAG—
m

③

m
—GAATTC—
—CTTAAG—
m

▶ **FIGURE 15–6** The molecular basis of host-controlled modification of DNA. (a) Phage DNA that is not methylated in a restriction site ① is cut by the restriction enzyme that recognizes the sequence ②, terminating the infection process. (b) DNA that has been methylated ① (m = methyl group) is able to replicate within the host cell ②, producing progeny phage ③.

type II restriction enzyme generates a consistently reproducible set of DNA fragments whose terminal sequences are part of the recognition sequence. These proteins are therefore indispensable tools for DNA manipulation, and all further discussions concerning restriction endonucleases will be limited to the type II enzymes.

Several hundred type II endonucleases with more than 100 different specificities have been isolated. These endonucleases are designated by the first letter of the genus name followed by the first two letters of the species name of the bacterium from which the enzymes are obtained; in many cases, a letter follows to indicate the strain or antigenic type of the bacterium, and a Roman numeral is included if the cell contains more than one type of restriction enzyme. For example, *Hind*II *and Hind*III are restriction enzymes isolated from *Haemophilus influenzae*, serotype d. Each of these enzymes cuts both strands of the DNA double helix, but the cuts are made in different sequences of six base pairs:

$$\textit{Hind}\text{II:} \quad 5' \text{ G T Py} \overset{\downarrow}{\text{ Pu A C }} 3'$$
$$3' \text{ C A Pu Py} \underset{\uparrow}{\text{ T G }} 5'$$

$$\textit{Hind}\text{III:} \quad 5' \overset{\downarrow}{\text{ A}} \text{ A G C T T } 3'$$
$$3' \text{ T T C G A} \underset{\uparrow}{\text{ A}} 5'$$

■ **TABLE 15–2** The three major classes of restriction endonucleases.

	Type I	Type II	Type III
Protein structure	bifunctional protein with three subunits for DNA recognition, cleavage, and methylation	monofunctional protein for cleavage only (separate protein for methylation)	bifunctional protein with two subunits, one for cleavage and one for methylation and recognition of DNA
Recognition site	bipartite and asymmetrical (e.g., TGAN$_8$ TGCT for the *E. coli* B enzyme)	4–6 bp sequence, often palindromic*	5–7 bp sequence, asymmetrical
Cleavage site	nonspecific, more than 1000 bp from recognition site	same as or close to recognition site	24–26 bp downstream (to the 3' side) of recognition site

*See Table 15–3 for examples.

The arrows indicate the points of enzymatic cleavage, and Pu and Py designate unspecified purine and pyrimidine bases.

■ Table 15–3 lists a few of the type II restriction endonucleases used in genetic research, along with the restriction sequence (recognition site) of each enzyme. Some of the enzymes, such as *Hind*II, cleave the DNA into blunt-ended fragments. Other enzymes, such as *Hind*III, make staggered cuts, creating fragments with single-stranded complementary (cohesive) ends. The recognition sequences of these enzymes are typically palindromes, (which have twofold rotational symmetry—see Chapter 7), and most are 4–6 nucleotides in length, although a few enzymes recognize longer sequences. While the restriction sequence tends to vary with the bacterial source, there are several known examples in which different bacteria produce enzymes, called **isoschizomers**, that recognize the same site.

■ **TABLE 15–3** Some restriction endonucleases and their recognition sequences (arrows indicate cleavage sites). The colors show the nature of the DNA ends resulting from cleavage. Some enzymes (such as *Alu*I) produce blunt-ended fragments; others (such as *Eco*RI) make staggered cuts and produce fragments with single-stranded complementary ends.

Microorganism	Enzyme Abbreviation	Target Sequence
Arthrobacter luteus	*Alu*I	5′ A G C T 3′ / 3′ T C G A 5′
Bacillus amyloliquefaciens	*Bam*HI	5′ G G A T C C 3′ / 3′ C C T A G G 5′
Bacillus globiggi	*Bgl*II	5′ A G A T C T 3′ / 3′ T C T A G A 5′
Escherichia coli	*Eco*RI	5′ G A A T T C 3′ / 3′ C T T A A G 5′
Haemophilus aegyptius	*Hae*II	5′ Pu G C G C Py 3′ / 3′ Py C G C G Pu 5′
Haemophilus aegyptius	*Hae*III	5′ G G C C 3′ / 3′ C C G G 5′
Haemophilus haemolyticus	*Hha*I	5′ G C G C 3′ / 3′ C G C G 5′
Haemophilus influenzae	*Hind*II	5′ G T Py Pu A C 3′ / 3′ C A Pu Py T G 5′
Haemophilus influenzae	*Hind*III	5′ A A G C T T 3′ / 3′ T T C G A A 5′
Haemophilus parainfluenzae	*Hpa*I	5′ G T T A A C 3′ / 3′ C A A T T G 5′
Haemophilus parainfluenzae	*Hpa*II	5′ C C G G 3′ / 3′ G G C C 5′
Serratia marcescens	*Sma*I	5′ G G G C C C 3′ / 3′ C C C G G G 5′
Streptomyces albus	*Sal*I	5′ G T C G A C 3′ / 3′ C A G C T G 5′
Xanthomonas oryzae	*Xor*II	5′ C G A T C G 3′ / 3′ G C T A G C 5′

Restriction Mapping

Researchers have used the specificity shown by the type II restriction endonucleases to construct **restriction maps.** Each restriction map of a DNA molecule shows the locations of the recognition sites for particular endonucleases; these sites thus serve as the markers on the map. To obtain a restriction map, various restriction enzymes are used both separately and in combination to cleave the DNA into fragments of different sizes. The fragments are then separated according to size by gel electrophoresis (see Chapter 7) and analyzed to determine the number and kinds of restriction sites occurring in the DNA. By measuring the chain length of each fragment upon electrophoresis, it is possible to express map distance in base pairs, thus establishing map lengths with complete additivity. Like conventional linkage maps, restriction maps locate regions of genetic importance on a chromosome. Once a restriction fragment that carries all or part of a particular gene is found, the location of this fragment on the restriction map identifies the chromosomal position of the gene relative to the various restriction sites. Restriction mapping has thus been used (in combination with conventional mapping and DNA sequencing) to construct detailed maps of both mitochondrial and chloroplast genomes (▶ Figure 15−7). Another example of its application is mapping regions of tumor virus DNA that correspond to particular viral gene functions. Because tumor viruses are difficult to work with using traditional genetic techniques, restriction mapping has been extremely useful in providing geneticists with information about the locations of genes on tumor virus chromosomes.

Different approaches can be taken in developing a restriction map. In the construction of the first restriction map in 1971, the enzyme HindII was used to cut the circular DNA of the animal tumor virus SV40 into 11 fragments (▶ Figure 15−8). The order in which these fragments occur in the DNA was deduced by **partial digestion.** Partial digests are obtained under conditions (such as limited hydrolysis time) in which an enzyme is not able to cleave all of its target sites on the DNA. The pattern of enzyme action can then be followed over time. The first cut breaks the circular DNA molecule into a linear one. The linear molecule is then cut into progressively smaller fragments that are overlapped by the fragments of previous cuts.

The experimental results in ▶ Figure 15−9 illustrate the use of partial digestion for constructing a restriction map. In this example, the DNA is labeled at its 5′ ends with the radioactive isotope ^{32}P (Figure 15−9a). One of the labels is subsequently removed using a different endonuclease, so that the DNA to be analyzed has only one labeled end. By combining **end-labeling** with partial digestion, researchers obtain an autoradiograph (see Chapter 7) that reveals only

▶ **FIGURE 15−7** (a) Restriction map and (b) gene map of the 195-kb chloroplast DNA of *Chlamydomonas*. On the restriction map, the three circles from the inside to the outside represent, respectively, *Bgl*II, *Bam*HI, and *Eco*RI fragments. The order of fragments in parentheses has not been determined. The two inverted repeats containing the rRNA genes are shown on the outside. Designations on the gene map are as follows: *atp* = genes for subunits of coupling factor; *ori* = replication origins; *psa* = genes for chlorophyll a apoproteins; *psb* = genes for chlorophyll a-binding proteins; *rbc, rpl,* and *rsp* = genes for chloroplast ribosomal proteins; and *tufA* = gene for elongation factor EF−Tu. Asterisks indicate genes whose positions are imprecise; transfer RNA genes are not shown. Filled boxes represent exons; open boxes, introns. Arrows indicate direction of transcription. *Source:* Restriction map from J. D. Rochaix, *J. Mol. Biol.* 126 (1978): 597−617. Used with permission. Gene map from J. D. Palmer, "Comparative organization of chloroplast genomes." Reproduced with permission from the *Annual Review of Genetics,* Volume 19, © 1985 by Annual Reviews Inc.

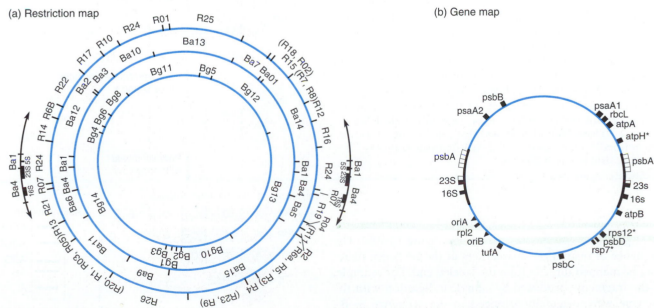

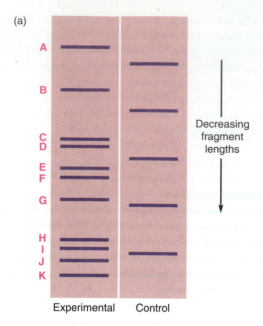

(a)

A
B
C
D
E
F
G
H
I
J
K

Decreasing
fragment
lengths

Experimental   Control

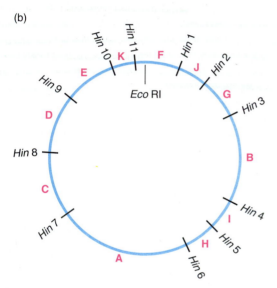

(b)

FIGURE 15–8 Restriction map constructed for the circular SV40 chromosome. (a) Gel electrophoresis separates 11 fragments of different size, labeled A through K, produced by cleavage with *Hind*II. (The original enzyme actually turned out to be a mixture of *Hind*II and *Hind*III.) The length of any particular fragment is determined by calibrating the gel; a mixture of standard markers of known lengths are run in one lane of the gel as a control. (b) The order of these fragments on the DNA. *Source:* O. Nathans, "Restriction endonucleases, Simian virus 40, and the new genetics," *Science* 206 (1979): 903–909. Copyright © 1979 by the AAAS. Used with permission.

the fragments containing the labeled end, arranged in order of length along the electrophoretic gel (Figure 15–9b). The sequence of restriction site markers in the DNA can therefore be mapped directly from the labeled end. For example, if the fragments produced by complete digestion with the enzyme were originally arranged in the sequence A–B–

C–D–E–… from the labeled end, the smallest fragment to appear on the autoradiograph after partial digestion would be A, the next smallest would be A–B, the third smallest A–B–C, and so on, up to the largest fragment, which consists of the uncut labeled DNA.

The **double-digestion** technique is another useful method for restriction mapping and is simpler technically. In this procedure, the DNA is cut at all available sites by a mixture of two enzymes, and the fragments produced are compared with those formed by each of the enzymes separately. A restriction map is then constructed on the basis of the observed overlaps.

To see how a restriction map can be derived from the results of double digestion, consider an example in which a 10-kb DNA molecule is cut at two sites by *Eco*RI, producing three fragments with lengths of 2.2, 3.7, and 4.1 kb. The same molecule is cut at one site by *Bam*HI, producing two fragments measuring 4.8 and 5.2 kb. When the DNA is cut by both enzymes, double digestion releases four fragments measuring 1.1, 2.2, 3.0, and 3.7 kb. The results of this ex-

▶ FIGURE 15–9 Use of end-labeling and partial digestion to construct a restriction map. (a) Each DNA strand of the double helix is labeled at its 5′ end (certain enzymes can add phosphate moieties specifically to 5′ ends), and the label at one end of the duplex is subsequently removed by enzyme x. (b) A second restriction enzyme (enzyme y) breaks down the DNA into overlapping fragments, all of which are labeled at the same end. Electrophoresis separates the fragments, whose positions on an autoradiogram reveal the order of the fragments relative to the labeled end.

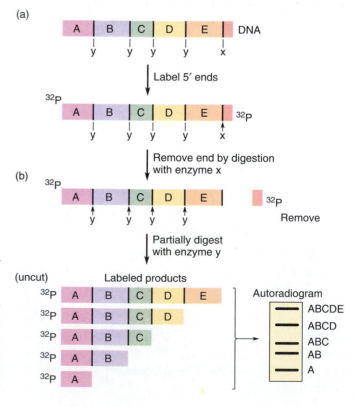

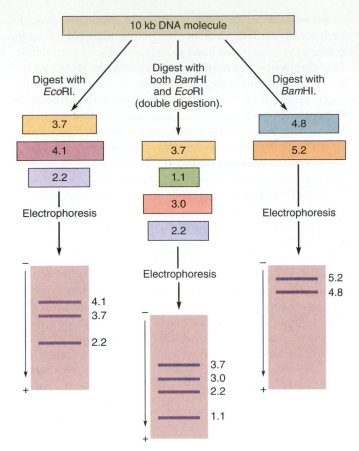

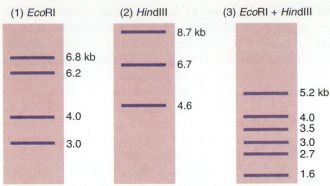

Construct a restriction map based on these data.

Solution: The analysis can be started at any of several places where smaller fragments can be added together to generate larger gel fragments. For example, the *Hind*III 8.7-kb fragment can be seen from the double digest to be divided by an *Eco*RI site into smaller fragments of 3.5 and 5.2 kb. Similarly, the *Eco*RI 6.8 kb fragment contains a *Hind*III site that divides it into fragments of 5.2 and 1.6 kb. Thus, a partial map is

Continuing on to complete the left end of the map, notice that the *Eco*RI 6.2 fragment is divided by a *Hind*III site into fragments of 2.7 and 3.5 kb, and the *Hind*III 6.7 fragment is divided by *Eco*RI into 4.0- and 2.7-kb fragments. At the right end, the *Hind*III 4.6-kb fragment is subdivided by *Eco*RI into 1.6- and 3.0-kb fragments. Adding this information to our map, we obtain

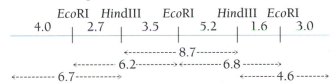

► FIGURE 15–10 Use of the double-digestion technique in restriction mapping. The results of single digests with different restriction endonucleases (left and right) are compared with the fragments produced when the DNA is digested by both enzymes together (center). Fragments that are produced by the *Eco*RI single digest but not by the double digest contain cleavage sites for *Bam*HI, and vice versa. These results allow construction of the restriction map shown in the text.

periment are summarized in ► Figure 15–10. Observe that the 2.2 and 3.7 kb fragments of *Eco*RI are contained in the double digest. Therefore, the single *Bam*HI site must lie in the 4.1 kb *Eco*RI segment, which is missing from the double digest. By relying on the additive nature of the fragment lengths, we can construct the following map:

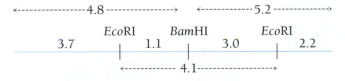

Example 15.1

A 20-kb DNA molecule is subjected to the double digestion technique using the enzymes *Eco*RI and *Hind*III. The resulting fragments are separated by electrophoresis, with these results:

Follow-Up Problem 15.1

A 4.1-kb DNA molecule undergoes double digestion with the restriction enzymes *Eco*R1 and *Alu*I, giving the following data: (1) Digestion with *Eco*R1 gives gel bands of 1.8 kb, 1.4 kb, and 0.9 kb; (2) digestion with *Alu*I gives gel bands of 2.1 kb, 0.8 kb, 0.7 kb, and 0.5 kb; (3) the double digest gives gel bands of 0.3 kb, 0.4 kb, 0.5 kb, 0.6 kb, 0.8 kb, and 1.5 kb. Construct the restriction map.

DNA Probes and Southern Blots

The number and size of the fragments produced by a restriction enzyme depend on the size of the recognition sequence of the endonuclease. The shorter the sequence, the greater the number of such sequences in the DNA and the shorter the lengths of the fragments that are produced by enzymatic digestion. For example, suppose that the GC content of a molecule of DNA is 50% so that any given base has a one-fourth chance of occurring at any nucleotide position. A recognition sequence consisting of n base pairs is then expected to occur at a frequency of $(\frac{1}{4})^n$; in other words, each sequence is expected to be present, on the average, once in every 4^n base pairs. Thus an endonuclease with a recognition sequence of 4 bp (for example, *Hae*III) is expected to produce fragments averaging $4^4 = 256$ bp in length, while those with a recognition sequence of 6 bp (for example, *Bam*HI and *Eco*RI) are expected to produce fragments averaging $4^6 = 4096$ bp in length. A mammalian genome comprised of 3×10^9 bp could therefore contain over a million copies of a restriction sequence and could generate about the same number of different fragments upon digestion by a restriction enzyme. If we were to subject such a digest to electrophoresis, the result would be a continuum of fragment sizes on the gel—a smear, rather than distinct bands.

Because there are so many possible restriction sites in high-molecular-weight DNA, restriction mapping of large genomes is usually confined to analyses of selected regions of a chromosome. These regions are usually identified by a DNA transfer-hybridization technique that uses specific single-stranded DNA (or RNA) sequences called **probes**. A probe is a radiolabeled or chemically labeled DNA (or RNA) strand, ranging from 15 to thousands of nucleotides long, that is complementary to the DNA sequence of the chromosomal region one wishes to detect. The labeled probe detects this region by selectively hybridizing with it. (The methods used to obtain probes will be discussed in Chapter 18.)

In order to detect a specific sequence with a DNA probe, it is first necessary to transfer the DNA segments from the agarose gel, where they have been separated by electrophoresis, onto the surface of a membrane filter where nucleic acid hybridization can take place. This transfer is accomplished by a process known as **Southern blotting** (named after its inventor, E. Southern). In this process, which is shown in ▶ Figure 15–11, the gel is treated with alkali to

▶ **FIGURE 15–11** The Southern blotting technique. Electrophoresis separates the DNA fragments from a restriction digest. The fragments are then denatured and "blotted" onto a membrane filter—capillary action draws the transfer buffer through the gel, transferring the denatured DNA fragments to the membrane, where they adhere strongly. A labeled probe and autoradiography are then used to detect the band (or bands) of DNA containing sequences that are complementary to the probe.

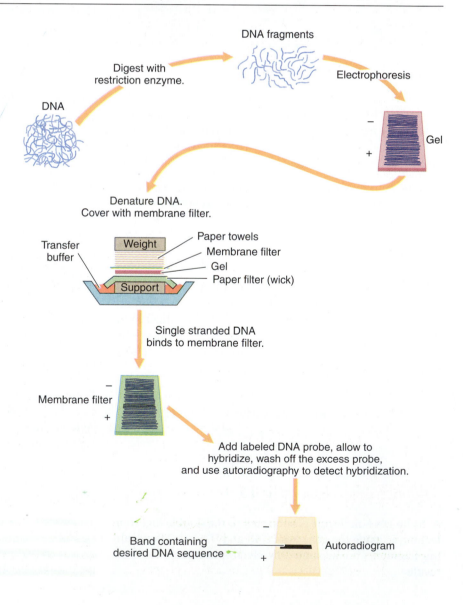

denature the double-stranded fragments into single strands, and then it is covered with a membrane filter. A buffer is drawn through the gel by capillary action, forcing the single-stranded DNA fragments onto the filter, where they bind in the same relative positions they occupied on the gel. Once the DNA fragments are immobilized on the filter, they are exposed to the DNA probe, which hybridizes with the band(s) of DNA containing the complementary sequence. Since the probe is radioactive, the band(s) containing the sequence can be detected by autoradiography, permitting equivalent DNA fragments in the original gel to be identified for restriction analysis.

The Southern blotting technique with a labeled probe is an extremely sensitive procedure for identifying DNA sequences. The hybridization reaction is so selective that it can detect a single-copy sequence in the genome of a eukaryotic cell. A similar procedure called **northern blotting** (so-named because it was patterned after the Southern blotting technique) has also been developed for analyzing RNA that has been separated by size on an electrophoretic gel. The RNA molecules in the gel are blotted onto membrane filters as described above (omitting the alkaline treatment) and detected using appropriate probes. Northern blotting thus provides for the identification of specific sequences in RNA, just as Southern blotting does for DNA.

In Situ Hybridization

Labeled DNA probes can also be used to assign DNA sequences to specific chromosome regions. The technique most often used is **in situ hybridization** (▶ Figure 15–12), which is a cytological procedure that combines nucleic acid hybridization and a hybrid detection procedure, such as autoradiography or fluorescence microscopy, to determine the chromosomal location of a labeled sequence. The chromosomes are first mounted on a slide, so that they can be viewed by light microscopy, stained, and photographed to reveal their banding patterns. The mounted chromosomes are then denatured by alkali and incubated with an appropriate labeled probe after the pH and temperature are adjusted to permit hybridization. The probe can be labeled with a radioisotope or with a substance such as biotin. After the excess unhybridized probe is removed by washing, the hybridized probe can be detected by autoradiography if a radioisotopic label was used. If a nonisotopic label such as biotin was used, a biotin-binding protein that is complexed with a yellow-green fluorescent dye is added, and the probe is detected by its fluorescence. When the autoradiograph or fluorescent micrograph is compared with the chromosome spread, the location of the spots or fluorescence reveals the chromosomal location(s) of the DNA sequence to which the labeled probe is bound (▶ Figure 15–13).

Historically, in situ hybridization was devised to locate certain repetitive chromosomal DNA sequences that bind large amounts of the radioactive probe. However, improvements in the resolving power of this technique now make it

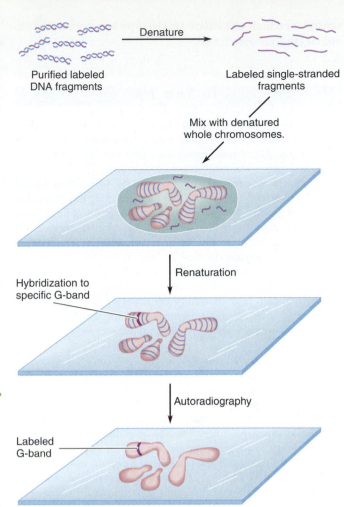

▶ **FIGURE 15–12** Localization of a DNA sequence to a specific chromosomal band by in situ hybridization.

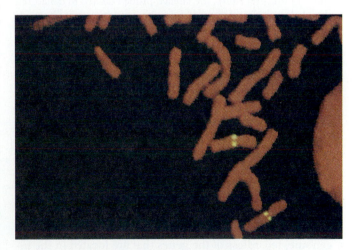

▶ **FIGURE 15–13** Photograph of human chromosomes probed in situ with a biotin-labeled fluorescent probe specific for a single gene that encodes a muscle protein. The gene locus is revealed as a fluorescent spot on each member of a pair of homologous metaphase chromosomes. *Source:* P. Lichter and David Ward, "High-resolution mapping of human chromosome 11 by in situ hybridization with cosmid clones," *Science* 247 (1990): 64–69. Copyright © 1990 by the AAAS. Used with permission.

sensitive enough to localize single-copy sequences on individual chromosome bands.

To Sum Up

1. Restriction endonucleases are bacterial enzymes that catalyze the breakdown of DNA. Many restriction enzymes cleave DNA at specific palindromic recognition sites that are often four or six base pairs in length.

2. Restriction enzymes are found in bacteria, where their natural role is to defend against the DNA of an infecting phage or other harmful genetic element. Each bacterial strain modifies its own DNA through a specific pattern of methylation to protect it from the action of its own restriction enzymes.

3. Restriction enzyme cleavage is used in restriction mapping, in which recognition sites for various endonucleases are ordered in relation to one another along a DNA molecule. If the length of each restriction fragment is known (from gel electrophoresis), then the resulting map gives the distance between sites in base pairs, and the individual distances are completely additive.

4. A probe is a labeled, single-stranded DNA (or RNA) sequence that is complementary to the DNA sequence of a particular gene or other chromosomal region of interest. The probe hybridizes to its complementary sequence on a Southern blot (a membrane filter onto which denatured fragments from a restriction digest have been bound). A suitable detection technique, such as autoradiography, can be then used to reveal the location on the membrane (and therefore on the original gel) of the fragment(s) to which the probe has hybridized. A similar technique called northern blotting is used to identify RNA molecules that have been bound to a nylon membrane.

5. Radioactively or chemically labeled probes can also be used to determine the chromosomal location(s) of particular sequences of interest through the technique known as in situ hybridization. In this procedure, the labeled probe is added to denatured banded chromosomes on a slide; the resulting autoradiograph or fluorescent micrograph is compared to the chromosome spread to detect the precise chromosomal regions containing the hybridized probes.

RESTRICTION SITE POLYMORPHISM AND GENE MAPPING

A restriction map is a *physical* map of a chromosome or chromosomal segment that directly reflects chromosomal distances in terms of DNA base pairs. Since the information for a restriction map is derived from the cleavage of DNA, a map of this type can be constructed for any portion of a genome, regardless of the genes that are located in the included interval.

Restriction analysis of DNA can also provide the information needed to construct a *genetic* map that correlates the restriction sites and gene loci of a particular region according to their frequencies of recombination. To obtain a genetic map from an analysis of restriction fragment lengths,

we must determine the crossover distance between the restriction sites and the neighboring gene loci. Maps of this type can be made when genetic variations in restriction sites are present in a population. For example, if a particular restriction sequence is absent in some individuals as a result of mutation, two allelic forms of DNA will exist in the population: a nonmutant form (+) having the restriction site and a mutant form (−) lacking the site (▶ Figure 15–14a). Even though the alleles may not express a difference at the phenotypic level (as would be the case, for example, if the restriction sequence is located between genes), they can still be distinguished at the molecular level by a difference in restriction fragment lengths on an electrophoretic gel (Figure 15–14b). Such variation in fragment lengths is referred to as **restriction fragment length polymorphism** (**RFLP**).

Detecting an RFLP in a eukaryotic genome requires the use of a DNA probe (▶ Figure 15–15). The probe is chosen, often at random, from an array of sequences in the genomic DNA; it is labeled with a radioactive isotope, denatured, and applied to Southern blots of restriction fragments derived from different individuals. The probe and the RFLP that it detects form a highly sensitive marker system that can be used to define a convenient point of reference on the genetic map. Usually, the RFLP detected in this way is inherited as a simple Mendelian marker with two codominant alleles, + and −. Occasionally, however, two polymorphic sites occur sufficiently close together to be detected by a single probe; the result is a marker with four possible forms: ++, +−, −+, and −− (▶ Figure 15–16). Each form is a **haplotype**—a particular combination of the restriction site polymorphisms at the two sites.

Mapping a Human Disease Gene

One important use of RFLP analysis in human gene mapping is locating polymorphic markers that are associated with defective (human disease) genes. If a restriction marker is closely linked to a defective site, with very little chance of recombination between them, the presence of the marker can then be used to identify potential carriers of the disease.

An example of the use of RFLPs in mapping a human disease gene is shown by the systematic attempts of James Gusella and his colleagues in the 1980s to locate the dominant autosomal gene for Huntington's disease (HD). Huntington's disease is a neurodegenerative disorder marked by spasmodic uncontrolled movements and mental deterioration. The age of onset is highly variable. Symptoms most often begin to appear between 30 and 40 years, but they sometimes do not appear for the first time until after 60 years of age. Thus most individuals who have the causative gene do not manifest any symptoms until middle age or older, when many of them have already passed on the gene to an average of half their children.

In the attempt to locate the HD gene, the investigators first looked for a RFLP marker that would show evidence of

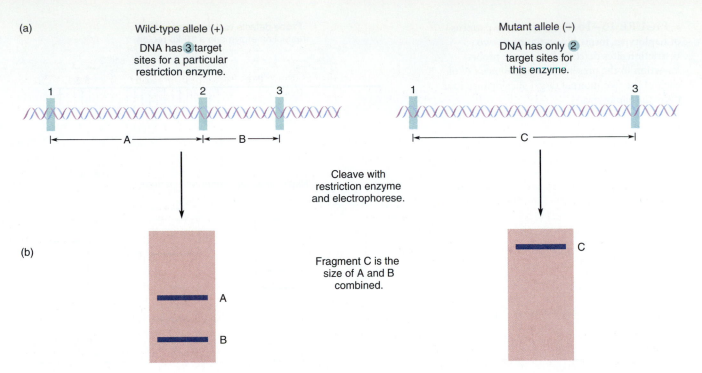

▶ FIGURE 15–14 Restriction fragment length polymorphism is produced if a mutation eliminates a restriction site along a section of a DNA molecule. This mutation alters the pattern of fragments revealed by electrophoresis. (RFLP would also occur if a mutation created a restriction enzyme target site.)

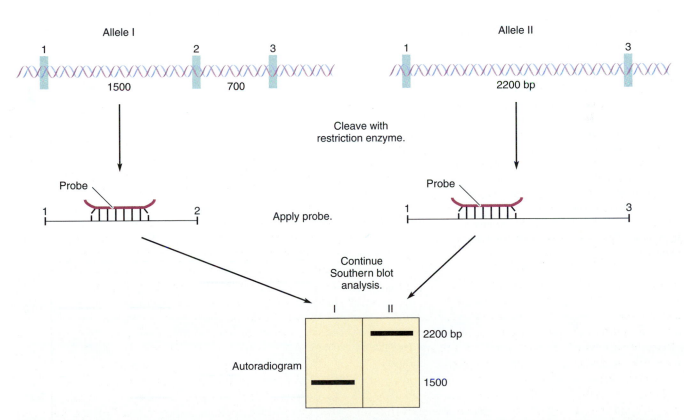

▶ FIGURE 15–15 Detection of a restriction fragment length polymorphism. Allele I is identified by the DNA probe as a 1500-bp-long fragment. Since the middle restriction site is missing in allele II, the same probe now identifies that allele as a 2200-bp-long fragment. Autoradiography of the Southern blot reveals the different alleles detected by the labeled probe.

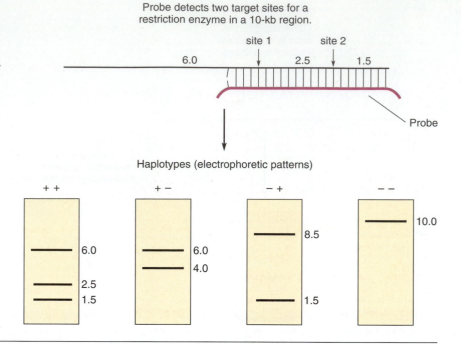

▶ **FIGURE 15–16** Electrophoretic patterns of haplotypes formed by variation at two restriction sites detected by a single probe. Variation in the presence (+) or absence (−) of each of the two internal target sites defines four different haplotypes.

close linkage to this genetic disease. After testing a number of random probes, they were successful in finding one, called G8, that detected polymorphism at two *Hind*III restriction sites on chromosome 4 that are close to the Huntington gene (▶ Figure 15–17). These restriction sites are separated by 18.7 kb, which corresponds to about 0.02 map units (cM) in humans. The G8 locus therefore consists of four haplotypes, designated A through D, that would normally

▶ **FIGURE 15–17** The human chromosome 4 RFLP that is detected by the action of *Hind*III and the G8 probe. An autoradiograph of a Southern blot reveals four alleles: A (−+), B (−−), C (++), and D (+−). (Cleavage at a third *Hind*III site located between sites 1 and 2 always occurs in this case and therefore plays no role in generating polymorphism.) The 2.5 kb fragment cannot be detected on the autoradiogram because the labeled probe is not long enough to hybridize to it. *Adapted from:* J. F. Gusella et al., "A polymorphic DNA marker genetically linked to Huntington's disease," *Nature* 306 (1983): 234–238. Reprinted with permission from *Nature.* Copyright (1983) Macmillan Magazines Limited.

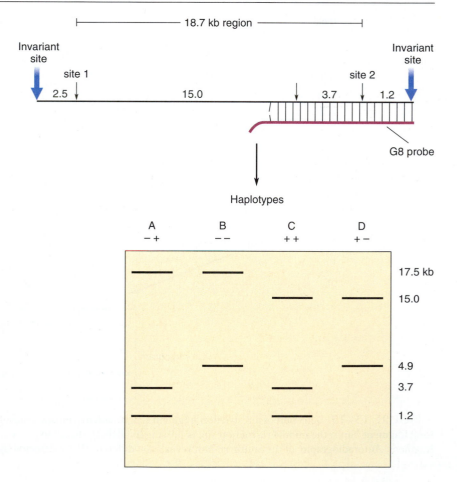

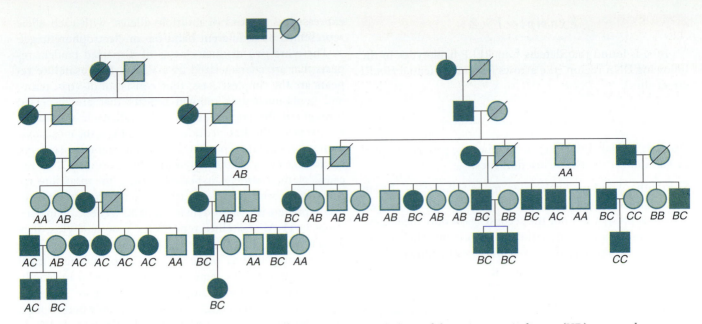

▶ **FIGURE 15–18** The pedigree of a large Venezuelan family indicates linkage of the Huntington's disease (HD) gene to the chromosome 4 RFLP described by Figure 15–17. In this family, the HD gene always segregates with allele C; however, the HD gene shows linkage to the A or B allele in other families. *Source:* J. F. Gusella et al., "A polymorphic DNA marker genetically linked to Huntington's disease," *Nature* 306 (1983): 234–238. Reprinted with permission from *Nature.* Copyright (1983) Macmillan Magazines Limited.

behave as alleles. Studies on two families, one in America and a much larger one in Venezuela, indicated that the HD gene segregated with the A haplotype in the American family and with the C haplotype in the Venezuelan family (see ▶ Figure 15–18). At first, the data suggested that the RFLP and the HD gene were very closely linked, providing a diagnostic test with nearly 100% accuracy. Later studies revealed, however, that the gene and its marker are approximately 4 map units apart. The 4% chance of recombination, coupled with the fact that many more people have the A or C haplotype than carry the disease gene, would indicate that the G8 locus is of limited diagnostic value. Researchers looked for additional polymorphic sites to provide better presumptive screening, and in March of 1993 Gusella's group announced the discovery of the Huntington gene itself (as a result of extending RFLP analysis with additional techniques that will be discussed in Chapter 18). Thus, the screening test for the HD gene can now employ the mutant DNA itself as the probe.

In addition to Huntington's disease, several other human disorders have been linked to RFLP markers, including those listed in ■ Table 15–4. In some cases, the mutation responsible for the loss of the restriction site is also the one that causes the disease. It is possible in these cases to provide a completely accurate diagnosis, even without a family history, since the disease locus and the diagnostic marker are the same. In most cases, however, the change in the restriction site does not result in the disease, so crossing-over is possible between the loci. If recombination should occur between these sites, the disease locus could segregate with a + marker allele in one family and a – marker allele in another family (as has happened with Huntington's disease),

making diagnosis difficult without a family history. It would then be necessary to follow the marker through a family pedigree in order to determine which marker allele is associated with the disease state.

■ **TABLE 15–4** Some human disorders that have been linked to RFLP markers.

Disorder	Chromosome
Adenomatous polyposis (familial colon cancer)	5
Alpha-1 antitrypsin deficiency	14
Alzheimer's disease	21
Cleft palate	X
Cystic fibrosis	7
Fragile X syndrome (X-linked mental retardation)	X
Hemophilia A (Factor VIII deficiency)	X
Hemophilia B (Factor IX deficiency)	X
Huntington's disease	4
Lesch-Nyhan syndrome	X
Muscular dystrophy (Duchenne)	X
Muscular dystrophy (myotonic)	19
Nevis basal cell carcinoma (Gorlin syndrome)	9
Ornithine transcarbamylase deficiency	X
Phenylketonuria	12
Polycystic kidney disease	16
Retinoblastoma	13
Sickle-cell anemia	11
Thalassemias	11
Wilms' tumor	13

Example 15.2

A probe is found that detects four RFLP haplotypes in the following DNA region (the arrows indicate potential *Hind*II target sites):

Diagram and label the four electrophoretic patterns formed by variation at the two restriction sites.

Solution: The four haplotypes are as follows: ++ if both restriction sites are present on the DNA; +− if the first target site is present but not the second; −+ if only the second site is present; and −− if neither site is present. The electrophoretic gels produced by these haplotypes are shown below:

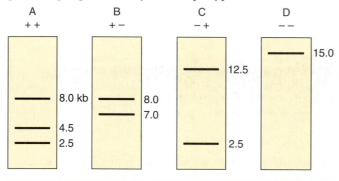

Follow-Up Problem 15.2

The four haplotypes of Example 15.2 are found to be associated with the pedigree of a certain family as follows (A, B, C, and D indicate the RFLP haplotypes; darker symbols indicate individuals affected by a particular genetic disorder):

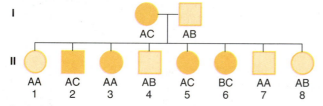

Which haplotype is the best marker for the gene causing the disorder? Does this haplotype show complete linkage to the gene locus? What is the significance of individual II-3?

DNA Fingerprinting and the Polymerase Chain Reaction

So far in our discussion of RFLPs, we have considered DNA polymorphisms that arise due to the presence or absence of a restriction enzyme recognition site. Another type of RFLP is characterized by a VNTR—a variation in the number of tandem repeats. Each VNTR unit in a DNA molecule consists of a variable number of short repeated sequences. In a VNTR, the restriction sites are unaltered, but the fragments generated by cuts at these sites vary in length because of differences in the number of tandem repeats between the sites (▶ Figure 15–19). This variation in fragment length is

expressed as a series of multiple alleles, with each allele represented by a different band on an electrophoretic gel.

There are two distinct classes of dispersed tandem repeats that are characterized by a VNTR. **Microsatellite repeats** are the simplest class; they consist of di-, tri-, tetra-, and penta-nucleotide tandem repeats that are dispersed throughout the arms of most mammalian chromosomes. For example, the dinucleotide repeat $(GT)_n$ is the most common microsatellite repeat in the human genome, occurring on average every 30,000 bp in the chromosomal DNA. Because of their widely dispersed nature, microsatellite repeats have been used as markers in the construction of a comprehensive, high-density human genetic map (see Table 1–1). **Minisatellite repeats** are more complex; they form a second class of dispersed tandem repeats, in which the repeating unit is about 30 to 35 bp in length, each with a common core sequence of 11 to 16 bp in length. In human DNA, for example, the 16-bp core sequence AGAGGTGGGCAGGTGG can occur in anywhere from a few to hundreds of copies at dozens of different loci. We do not know why these **hypervariable regions** differ so much in number of tandem repeats; however, it is thought to be the result of unequal crossing-over (see Chapter 12).

Because of the large number of tandem repeat (VNTR) alleles that can occur in a population, researchers have used the discriminating power of RFLP analysis to develop a technique called **DNA fingerprinting** (or **DNA typing**) for making individual DNA identifications. DNA fingerprinting involves the use of Southern blotting and labeled DNA probes. Two types of probes are used: a **single-locus probe**, which hybridizes with a tandem repeat sequence at a unique hypervariable region, and a **multilocus probe**, which hybridizes with a tandem repeat sequence occurring in numerous regions throughout the genome. When used alone, a single-locus probe can identify at most two alleles (bands) in a heterozygote and only one in a homozygote. A precise identification therefore requires several single-locus probes, each specific for a different hypervariable region. With a multilocus probe, on the other hand, a much larger pool of variability can be sampled in a single test, since the number of tandem repeats detected by such a probe is variable both within a locus and between loci. Thus, when DNA is isolated from a single individual, cleaved with a single restriction enzyme, and hybridized (following electrophoresis and Southern blotting) with a single multilocus probe, a complex ladder of DNA fragments can be detected on an autoradiograph (▶ Figure 15–20).

DNA fingerprinting has applications in a broad range of disciplines where identity testing is involved, including forensics. Unlike blood testing, which can yield only exclusionary evidence (i.e., it can exclude a suspect of a crime or an individual in a paternity suit), DNA fingerprinting can provide positive evidence of a person's identity, just as conventional fingerprinting does. When conducted with suitable probes, the test is so sensitive that the probability of two individuals having the same DNA fingerprint can be extremely low (less than 10^{-6}).

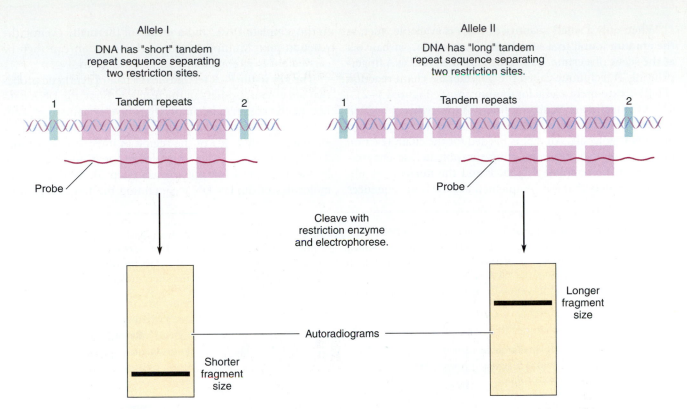

▶ **FIGURE 15–19** RFLP due to a variation in the number of tandem repeats (VNTR). Since the length of the span separating restriction enzyme cutting sites 1 and 2 differs for the two alleles, different fragment sizes appear on the autoradiograms of the Southern blots. (Only two alleles of a multiple allelic series are shown.)

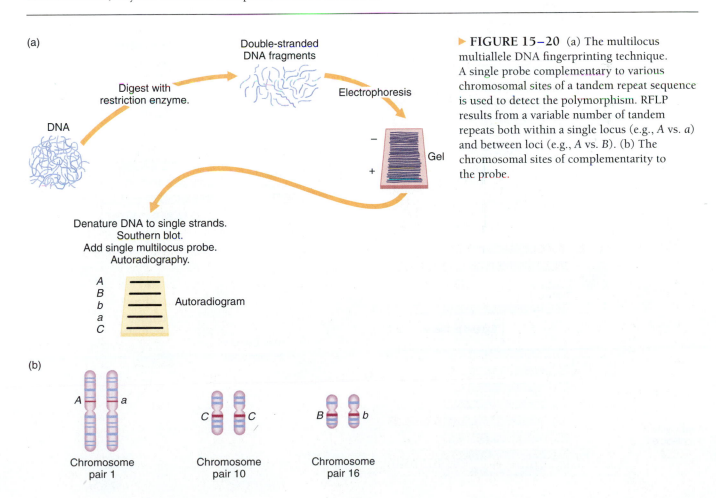

▶ **FIGURE 15–20** (a) The multilocus multiallele DNA fingerprinting technique. A single probe complementary to various chromosomal sites of a tandem repeat sequence is used to detect the polymorphism. RFLP results from a variable number of tandem repeats both within a single locus (e.g., *A* vs. *a*) and between loci (e.g., *A* vs. *B*). (b) The chromosomal sites of complementarity to the probe.

When only a small quantity of DNA is available, such as the amount found in a sample of blood, semen, or hair left at the scene of a crime, it must be amplified for DNA fingerprinting. A technique called the **polymerase chain reaction** (**PCR**) is extremely useful for this purpose (▶ Figure 15–21). The PCR technique is an amplification procedure in which a DNA polymerase-catalyzed reaction replicates portions of a DNA molecule. All that is needed for the chain reaction to occur are a buffer containing the polymerase enzyme, a copy of the template sequence and the nucleotide substrates, primers that are complementary to short sequences in the template DNA, and a means of thermally cycling the reaction mix. Multiple cycles of this reaction can then be carried out to exponentially amplify the DNA.

The key to the PCR technique is a pair of synthetic probes that are used as primers in the reaction. First, the DNA containing the region to be amplified is heat-denatured and annealed to the primers, which attach to the complementary strands on either side of the region of interest (steps 1 and 2). The primers are then extended in the 5′ to 3′ direction by the action of DNA polymerase (step 3), producing two molecules of duplex DNA containing the target region. A

▶ **FIGURE 15–21** Amplification of a selected target DNA sequence (gray) using the polymerase chain reaction (PCR). Each cycle consists of denaturing the DNA that includes the sequence, annealing a primer (colored stripes) on either side of this sequence, and extension of those primers by a DNA polymerase. At each cycle, the number of DNA molecules doubles. As the number of cycles increases, more and more of the DNA molecules precisely represent the selected target sequence (the short molecules that begin to appear after 3 cycles), until eventually these molecules predominate.

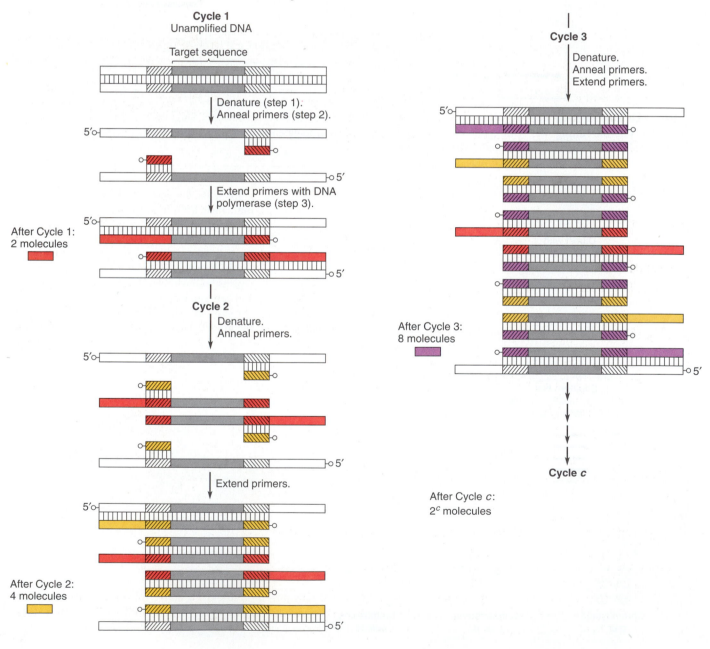

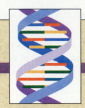

FOCUS ON HUMAN GENETICS

Practical Uses of DNA Fingerprinting

Perhaps the best-known practical application of DNA fingerprinting is in forensic work. When hypervariable (VNTR) DNA was discovered in 1980 and found to have a wide range of variation in humans, forensic scientists recognized its potential for identifying a criminal from DNA samples (in blood, hair, or semen) left at a crime scene. There has been some debate over just how variable human populations are in VNTR loci. Not all scientists agree on the probability of two individuals sharing the same RFLP. Some researchers think that there is considerably less variability within certain ethnic groups, so that RFLP matches within such groups may be much more common than among randomly chosen individuals. Even if humans are not as variable as was originally thought, however, most scientists agree that it is essentially certain that a wrongly accused individual will be cleared by his or her RFLP pattern and that there will be little chance of a coincidental match between the DNA profile of a suspect and that of a sample. Courts in the United States and England have therefore admitted DNA evidence in thousands of criminal and civil litigations.

In addition to its use in forensics, DNA fingerprinting has a wide variety of uses, for example, in establishing (or ruling out) parentage, in animal and plant sciences, and in cases of possible wildlife poaching. In zoology, for instance, the degree of evolutionary relationship within different populations of a given species (bird and wolf populations are examples) are being studied to determine classification into genus, species, or subspecies. This kind of work is particularly important in identifying endangered species and in determining whether different populations of a given species are similar enough that individuals from an abundant population can be successfully transplanted to another population. DNA profiles are also being used to avoid excessive inbreeding in captive-breeding programs. Mitochondrial DNA has proved to be a rich source of RFLP for these kinds of applications.

A new use of DNA fingerprinting involves the identification of big-game animals that have been illegally poached. In almost all instances of poaching, a large quantity of tissue is left at the kill site. Matching the DNA profile of such tissue with tissue recovered from a freezer, pelt, or mounted trophy head identifies the poaching incident. As data accumulate on DNA profiles for legally hunted versus protected animals (and plants), this type of nonhuman forensic work will become important in controlling the illegal killing of big-game animals and the illegal collecting of endangered plants.

Modoris Ali and his wife Aygun Bibi, photographed with four of their five children in London in 1988. Modoris Ali applied in 1975 for his family to join him in Britain from Bangladesh, but the British authorities did not believe Aygun was his wife or that the children were his. DNA fingerprinting proved that Modoris Ali and Aygun Bibi were the parents of all the children. Aygun and the younger children were allowed entry to Britain in 1988. Entry was refused to Boshir, the eldest son, seen by Aygun's forefinger in the photo the parents are holding, because he was over 18—even though he was 13 when they applied for entry.

DNA analysis. Autoradiograph of DNA.

second cycle of heat denaturation, annealing, and primer extension follows the first, then a third cycle is carried out, and so on, with each cycle doubling the amount of DNA present. The result is an exponential accumulation of DNA by a factor of 2^c, where c is the number of cycles involved. It is thus possible to amplify picogram starting quantities of DNA (the amounts present in a single hair or sperm cell) to microgram amounts in a matter of hours. After the first and even the second cycles, the amplified DNA molecules are longer than just the desired target region. After three cycles, however, DNA duplexes having the precise length of the target region are produced, and as the number of cycles increases, these molecules quickly begin to predominate. By the end of the amplification procedure, virtually all of the DNA molecules are the desired precise replicas of the short target sequence originally identified by the primers.

To avoid denaturation of the polymerase enzyme, the PCR uses a highly heat-stable enzyme called *Taq* DNA polymerase, which is isolated from the thermophilic bacterium *Thermus aquaticus*. This enzyme is stable at the temperatures used in the PCR procedure, so that all reactions can occur continuously in a single reaction vessel without the need to add more DNA polymerase after each heat-denaturation step.

Because of the simplicity of the polymerase chain reaction and the speed with which it can be carried out, sequence amplification by the PCR method has become a standard procedure in a large variety of research and diagnostic applications. Essentially any DNA can be amplified by this technique, depending only on the availability of primers. DNA fragments from bone, mummified tissue, charred wheat, and leaves embedded in clay—some of these millions of years old—have been amplified by this procedure, permitting the study of samples previously refractory to molecular analysis (▶ Figure 15–22). Recently, 120-million-year-old weevil DNA entombed in amber was amplified by the PCR reaction. These accomplishments, coupled with DNA sequencing, open up the possibility of "molecular archaeology."

the cleavage site. The two alleles can be detected molecularly by their variation in fragment lengths, known as restriction fragment length polymorphism (RFLP). If two polymorphic sites occur close enough together to be detected by a single probe, there are four combinations of allelic states for the two sites; each state is referred to as a haplotype.

2. Treating restriction cleavage sites as genetic markers allows the construction of genetic maps that measure distances between cleavage sites and gene loci by recombination frequency. One aspect of this type of gene mapping is finding the chromosomal position of a gene by identifying its close linkage to a particular haplotype.

3. A probe that identifies a particular haplotype then identifies the allelic state of the closely linked gene. This technique is used to diagnose the presence of certain mutant genes in humans. The accuracy of such a diagnosis of course depends on the frequency of recombination between the haplotype and the mutant gene.

4. Restriction fragment length polymorphism also arises from variation in the number of tandem repeat sequences found between two given DNA cleavage sites. The large number of VNTR alleles that can exist in a population means that the fragment pattern detected by a probe that is specific for a certain tandem repeat sequence constitutes a DNA fingerprint for each individual.

5. When only small quantities of DNA are available, the polymerase chain reaction is used to amplify the DNA so that fingerprinting (or some other procedure of interest) can be performed. Amplification is achieved in the PCR by using a DNA polymerase to replicate a segment of DNA that is bracketed by a pair of synthetic primers.

To Sum Up

1. The presence of a certain restriction site at a particular location in a DNA molecule can be deduced by comparing its restriction digest with one obtained from DNA that lacks this site due to mutation. Thus, a restriction sequence can be treated as a genetic marker having two allelic forms: presence or absence of

▶ FIGURE 15–22 DNA fragments can be obtained from this 38 million-year-old cranefly which was found embedded in amber.

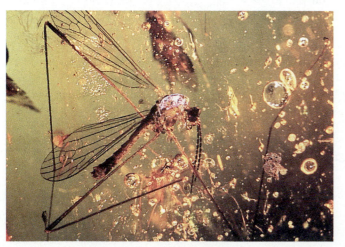

Points to Ponder 15.1

Suppose that your grandmother has just been diagnosed as having Huntington's disease. Her son (your father) is 40 years old and shows no symptoms of the disease. You are contemplating marriage and children. (a) What is your probability of carrying the HD gene? Would you be worried about the possibility of passing the gene to your children? Would you worry so much that you would decide not to have children? (b) Would you want to be screened using the RFLP test to determine whether you have the HD gene? Consider the various "costs" of having this genetic knowledge: For example, the psychological and social costs and the implications regarding medical insurance, employment, and privacy—do other family members, prospective employers, or insurance companies have the right to know the results of the genetic testing? (c) Suppose your father does not want to know if he carries the HD gene. Would that be a factor in your decision whether or not to have the screening test done? (d) Suppose you decide not to be tested. You marry and your spouse demands that prenatal testing be done to determine whether your fetus carries the gene. What would your reaction be? (e) Suppose you decide to be tested. Would finding that you carry the HD gene change your answers to the second and third questions in part (a)?

DNA SEQUENCING

The ultimate level of resolution of a molecularly based linkage map is the nucleotide sequence of the DNA in each of the chromosomes. Early approaches to sequencing nucleic acids were similar to those applied to proteins: obtain the monomer sequence of small isolated fragments and then deduce the overall sequence of the chain from the overlapping character of the fragments. Such analyses are generally limited to short polynucleotide chains less than 100 nucleotides in length. The main difficulty in sequencing longer nucleic acids with this approach was that in the absence of modified bases, only four different nucleotides make up each molecule. This meant that many of the fragments produced by cleaving a long polynucleotide tend to have similar base compositions and, despite differences in sequence, will be virtually impossible to separate. Thus while several short RNA molecules were sequenced by these methods, a different approach was clearly needed to determine the primary structures of the longer DNA molecules.

In the late 1970s, two efficient techniques were developed for DNA sequencing: the **chemical cleavage** method of Allan Maxam and Walter Gilbert and the **chain-terminator** (or **dideoxy**) method of Fred Sanger. Both sequencing methods use single-stranded DNA and generate a series of oligonucleotide fragments of increasing length, each one nucleotide longer than the last. The fragments, which are radioactively or chemically labeled, are separated on the basis of length by gel electrophoresis and identified according to their position on the gel. The two methods differ primarily in the way the fragments are produced and in the manner of incorporating the radioactive or chemical label into the DNA chain. In the chemical cleavage method, the label is added to one end of the DNA to be sequenced. Fragments of the labeled chain are then produced by using different chemical reagents to cleave the DNA at specific bases. In the chain-terminator (dideoxy) method, a polymerase enzyme is used to synthesize labeled DNA in vitro. Labeled fragments of different lengths are produced by terminating the synthesis of the chain at different sites with specific inhibitors.

To describe the details of DNA sequencing, we will look at Sanger's chain-terminator (dideoxy) method. This is currently the most widely used method for DNA sequencing, and it is the easiest to automate.

Chain-Terminator Method

The chain-terminator (dideoxy) method is based on the synthesis of duplex DNA using the strand with the sequence of interest as a template. Synthesis is carried out in each of four reaction vessels containing the usual nucleotide substrates and DNA polymerase. At least one of the nucleotide substrates (usually dATP) is radioactively or chemically labeled in each case so that the product strand can be detected. To produce fragments ending in specific bases, the synthesis of DNA in each reaction vessel is terminated by the addition of a small amount of a 2',3'-**dideoxynucleoside triphosphate (ddNTP)**

$$HO-\overset{\overset{O}{\|}}{P}-O-\overset{\overset{O}{\|}}{P}-O-\overset{\overset{O}{\|}}{P}-O-CH_2 \quad base$$

which can be either ddATP, ddTTP, ddCTP, or ddGTP, depending on the base at which termination is to occur. When the dideoxynucleotide is present in the reaction mixture, it will be incorporated into the DNA in place of the corresponding normal nucleotide, but once it is present in the growing chain, further elongation is halted by the absence of a 3' hydroxyl group. The result is the formation of partially completed product chains, each with a particular dideoxynucleotide incorporated at its 3' end.

To better visualize the Sanger (dideoxy) method of sequencing, let us assume the following DNA chain:

Template   3'———CGAGGTGCAT . . . 5'
Primer   5'———

in which only the first ten nucleotides in the template strand are shown. A primer has been combined with the template strand just upstream to the 5' side of the sequence of interest to provide a 3'-OH group for chain growth. Suppose that a small amount of ddATP is added to reaction vessel 1. Letting A* represent the dideoxynucleotide, synthesis in this vessel would then produce the following products:

3'———CGAGGTGCAT . . . 5'   3'———CGAGGTGCAT . . . 5'
5'———GCTCCA*   , 5'———GCTCCACGTA*   ,
                                                  etc.

When these products are denatured and the strands are separated, we are left with complementary sequences of varying lengths (———GCTCCA*, ———GCTCCACGTA*, etc.), each having a 3' dideoxy terminus that corresponds in location to a T in the template strand. Similarly, the reactions occurring in the other three vessels will produce partially synthesized chains ending in dideoxy derivatives of either T, C, or G, corresponding to the positions occupied by A, G, and C in the template.

The results obtained from the four reaction vessels in our example are summarized in ▶ Figure 15–23. A different dideoxy derivative is radioactively labeled in each of the four vessels. The partially synthesized chains are then separated by gel electrophoresis and detected on the gel by autoradiography. Since the shortest chains travel the greatest distance in the electric field, the base sequence complementary to the template can be read directly from the autoradiograph, starting at the bottom of the gel with the fastest-moving chains from the 5' end of the DNA strand. An actual autoradiograph of a dideoxy sequencing gel is shown

▶ **FIGURE 15–23** DNA sequencing using the Sanger chain-terminator (dideoxy) method. Extension of the primer occurs until a dideoxynucleotide is incorporated. The base sequence complementary to the template can be read directly from the autoradiograph, starting at the bottom.

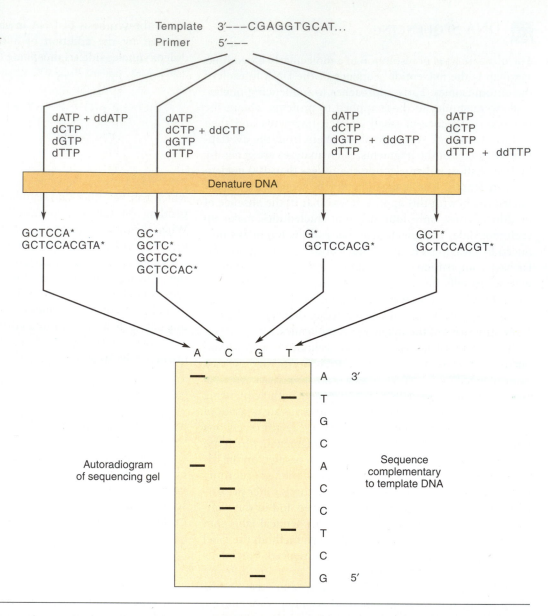

in ▶ Figure 15–24. Note the ladderlike appearance of the bands in each lane.

One of the more recent adaptations of the chain-terminator method has been attaching different fluorescent dyes to the dideoxynucleotides. The presence of different dyes imparts a characteristic color to the chains that end in A, T, C, or G, so that the chains can be produced in a single reaction vessel containing all four dideoxynucleotides and then identified in a single lane of an electrophoretic gel.

To Sum Up

The most widely used method for DNA sequencing is the Sanger chain-terminator (dideoxy) method. This technique involves the synthesis of radioactively labeled DNA fragments of increasing length, each ending in a specific dideoxynucleotide that terminates synthesis of the chain. The partially synthesized DNA chains are separated by electrophoresis and the base sequence is read directly from the gel.

Point to Ponder 15.2

The use of RFLP markers to detect the genes for Huntington's disease and certain other disorders (see Table 15–4) is an example of the application of new molecular techniques to the diagnosis of human genetic diseases. The accuracy of these tests depends on several factors, including the distance between the disease gene and the RFLP marker and the number of different mutational changes within the gene locus that can lead to the disease. For example, screening for the genetic disorder cystic fibrosis is complicated by the fact that at least 60 different mutations within the CF gene have been discovered, and probes have been developed to detect only about 75% of these mutations. Of course, any time a screening test is not completely accurate, there is a chance that it will incorrectly fail to reveal the presence of the mutant gene. How accurate do you think a test must be before it is allowed to become available as a routine genetic testing procedure? What factors must be taken into consideration in deciding on an acceptable level of test accuracy?

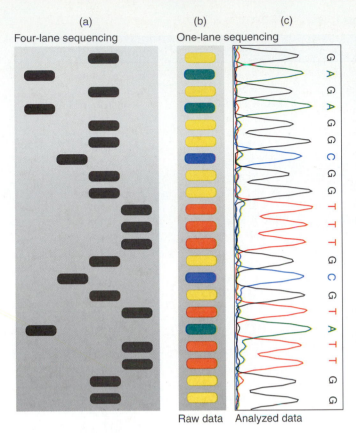

(a) Four-lane sequencing

(b) One-lane sequencing

(c)

G A G A G G C G G T T T G C G T A T T G G

Raw data   Analyzed data

▶ **FIGURE 15–24** (a) Autoradiograph of a dideoxy sequencing gel with 4 lanes, one for each base. (b) One lane sequencing made possible by the use of fluorescent labels to differentiate the bases. (c) Analyzed sequencing data. *Source:* Applied Biosystems Division of Perkin-Elmer. Used with permission.

MAPPING THE HUMAN GENOME

Human genome mapping (as in the **Human Genome Project**) is a multidisciplinary endeavor involving scientists with diverse backgrounds ranging from biochemistry to anthropology. This collaborative effort, which involves many of the newer technologies described in this chapter and in Chapter 18, has produced a variety of map and sequence information that must be processed in order to construct a unified linkage map. The diverse nature of the information is exemplified by the mapping data for chromosome 1, which is listed in ■ Table 15–5. These data are summarized in ▶ Figure 15–25, which shows a genetic map of RFLP markers and confirmed gene assignments and a cytological (physical) map showing certain regional assignments derived from somatic cell studies. Identification of the genes in all the bands of this chromosome is estimated to be over 0.5% complete.

In general, the results of these studies have shown that the order of loci is the same in both the physical and genetic maps. However, because of regional variation in recombination, there is no simple relationship between the physical distances of the chromosome regions and recombination frequencies. To complicate matters further, sex variation in recombination is also observed (▶ Figure 15–26). Overall, meiosis tends to produce a proportionally greater amount of recombination in females than in males.

Possession of a complete map of the human genome will have many practical benefits, particularly in medicine. For example, a genetic map can facilitate diagnosis of certain genetic diseases: A disease gene that cannot be detected prenatally or at least before it is expressed may be revealed if it is closely linked to a marker that can be detected. Furthermore, a genetic map giving the approximate position of a gene locus can aid in the isolation of that gene once closely linked markers have been identified. If a gene can be isolated, its function can be studied. For example, isolation of the cystic fibrosis gene has permitted the characterization of its previously unknown protein product, thereby giving insight into the primary causes of the disease and hopes for improved treatments for people affected with CF (see Chapter 18).

Linkage Detection: The Lod Score Method

Obtaining a complete genetic map requires the determination of recombination frequencies between all linked genes. Such evaluations are hindered by the limited (usually very small) size of human families and the researcher's lack of control over matings and the linkage phase of the parental genes. Because of the difficulties in determining linkage from human pedigrees, researchers are forced to collect data from many families demonstrating similar segregation patterns and estimate the recombination frequency from the pooled data. To increase their accuracy in determining whether the genes in question are linked, human geneticists commonly use a probability technique known as the **lod score method**. This method involves calculating the value of the linkage odds ratio L, which is the ratio of the probability that a particular distribution of births came from linked loci with a certain frequency of recombination to the probability of that same birth distribution from unlinked genes. By convention, linkage between two loci is demonstrated when the *logarithm* of the *od*ds ratio (lod score) is 3.0 or greater, which corresponds to an odds of at least 1000:1 in favor of linkage (but see the accompanying Extensions and Techniques on the lod score method).

To illustrate the calculation of L, consider the pedigree in ▶ Figure 15–27, which shows the joint inheritance of the genes for Rh factor (D for Rh$^+$ vs. d for Rh$^-$) and for elliptocytosis (E for oval erythrocytes vs. e for normal erythrocytes). Assuming these loci are linked, the heterozygous

■ TABLE 15–5 Some confirmed gene assignments to chromosome 1.

Gene Symbol	Gene Name	Mode*	Regional Assignment†
ACTA, ASMA,	Actin, skeletal α chains	REa	p21–qter
AK2	Adenylate kinase-2, mitochondrial	F, S	p34
ALPL, HOPS	Alkaline phosphatase, liver/bone form	A, F, Fd, S	p36.2–p34
AMY1	Amylase, salivary	A, F, REa	p21
AMY2	Amylase, pancreatic	A, F, REa	p21
ANF, PND	Atrial natriuretic factor	A, REa	p36.2
APCS, SAP	Amyloid P component, serum	A, Fb, REa	q12–q23
APOA2	Apolipoprotein A-11	REa	q21–q23
AT3	Antithrombin III	A, D, F, REa	q23.1–q23.9
C1QB	Complement component C1q, β-chain	REa	p
C4BP, C4BR	Complement component 4 binding protein	A, F, REa	q32
CAE	Cataract, zonular pulverulent	F	q2
CFH, HF	Complement factor H	F, REa	q32
CMT1	Charcot-Marie tooth disease	F, Fd	p22–q23
CR1, C3BR	Complement component-3b, receptor	A, F, REa	q32
CRP	C-reactive protein	A, REa	q12–q32
DAF	Decay-accelerating factor of complement	A, REa	q32
DIZ1	Satellite DNA III	A	q11
EL1	Elliptocytosis-1 (Rh-linked)	F	p36.2–p34
ENO1, PPH	Enolase-1	F, S	pter–p36.13
F3, TFA	Clotting factor III	Fd, S	p22–p21
FH	Fumarate hydratase	D, S	q42.1
FTHP	Ferritin, heavy chain	A, D, REa	p
FUCA, FUCA1	α-L-fucosidase-1	F, S	q34
FY	Duffy blood group	F, Fd	q12–q21
GALE	UDP-galactose-4-epimerase	S	pter–p32
GBA	Acid β-glucosidase	A, D, S	q21
GDH	Glucose dehydrogenase	F, S	pter–p36.16
GUK1	Guanylate kinase-1	D, S	q32.1–q42
GUK2	Guanylate kinase-2	D, S	q32.1–q42
HLM2	Oncogene HLM2	F, REa	pter–p36
LCA, T200	Leucocyte common antigen	A, S	q31–q32
NGFB	Nerve growth factor, β	A, Fd, RE, REa	p13
NRAS1	Oncogene NRAS1	A, REa	p22
PEPC	Peptidase C	S	q42
PGD	6-phosphogluconate dehydrogenase	F, S	p36.2–p36.13
PGM1	Phosphoglucomutase-1	F, S	p22.1
PK1, PKR	Pyruvate kinase, red cell type	A, REa	q21–q22
RD	Radin blood group	F	p36.2–p34
REN	Renin	A, D, REa	q32
RH	Rhesus blood group	D, F, Fd	p36.2–p34
RN5S	5S ribosomal RNA genes	A	q42–q43
RNU1	Small nuclear RNA U1	A, REa	p36.3
SC	Scianna blood group	F	p36.2–p34
TSHB	Thyroid stimulating hormone, β-subunit	Fd, RE, REa	p22
UGP1	UDP-glucose pyrophosphorylase-1	S	q21–q23
UMPK	Uridine monophosphate kinase	F, S	p32
UROD	Uroporphyrinogen decarboxylase	A, S, REa	p34

Source: V. A. McKusick, *Mendelian Inheritance in Man: Catalogs of Autosomal Dominant, Autosomal Recessive, and X-linked Phenotypes,* 10th ed. (Baltimore, MD: Johns Hopkins Press), 1992.

*The methods by which assignments have been made: A = in situ hybridization, D = deletion mapping, F = family linkage studies, Fd = family linkage studies combined with restriction endonuclease techniques, RE = restriction endonuclease techniques, REa = restriction endonuclease techniques combined with somatic cell genetics, S = somatic cell genetics.

†Numbers represent band designations: p = short arm, q = long arm, ter = terminus.

▶ **FIGURE 15–25** Mapping data for human chromosome 1. A recombination map showing sex-averaged genetic distances is on the left, and a cytological map with G band designations for each arm of the chromosome is shown on the right. Regional assignments of some genes from the recombination map to the cytological map are noted in the middle. The entire genetic map is 390 cM long and includes 58 markers that have been uniquely located (with odds exceeding 1000:1) and another 41 that have not yet been uniquely placed. The genetic intervals of these 41 markers are shown by the vertical lines to the left of the genetic map. HGM (human genome map) gene designations are given, along with D segment markers where available. (A D number represents a piece of DNA that has been assigned to a given chromosomal region but has not been characterized with respect to gene content. *Source:* NIH/CEPH Collaborative Mapping Group, "A comprehensive genetic linkage map of the human genome," *Science* 258 (1992): 67–86. Copyright © 1992 by the AAAS. Used with permission.

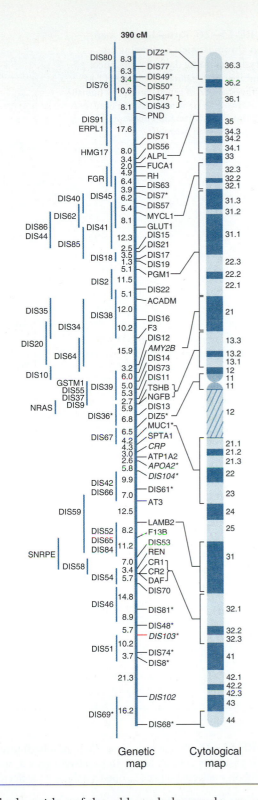

female in generation II must have her genes in coupling, since she received a *DE* chromosome from her father and a *de* chromosome from her mother. The mating in generation II is therefore

$$\frac{DE}{de} \times \frac{de}{de}$$

Of the children produced in generation III, only individual III–2 (who received a *dE* from his mother) and individual III–6 (who received a *De* from her mother) are recombinant. These results give an estimate of $\frac{2}{8} = \frac{1}{4}$ for the recombination frequency.

To determine the odds in favor of linkage for these data, we first calculate the conditional probability of the observed (*O*) combination of births (*O* = 3 *DE/de*, 3 *de/de*, 1 *De/de*, and 1 *dE/de*), given a recombination frequency of $\frac{1}{4}$. For the assumed conditions, each recombinant type (*De/de* and *dE/de*) should occur at a frequency of $R/2 = 1/8$ and each parental type (*DE/de* and *de/de*) should occur at a frequency of $(1 - R)/2 = 3/8$. Therefore, the conditional probability for the observed birth combination, given $R = \frac{1}{4}$, can be computed from the appropriate term of the multinominal distribution (see Chapter 3) as

$$P(O \mid R = \tfrac{1}{4}) = \frac{8!}{3!3!1!1!}\left(\frac{3}{8}\right)^3\left(\frac{3}{8}\right)^3\left(\frac{1}{8}\right)^1\left(\frac{1}{8}\right)^1 = 0.04866$$

In contrast, if the two loci are unlinked, R must equal $\frac{1}{2}$, and every genotype has a probability of $\frac{1}{4}$ of occurring. The conditional probability of the observed birth sequence, given $R = \frac{1}{2}$, can then be calculated as

$$P(O \mid R = \tfrac{1}{2}) = \frac{8!}{3!3!1!1!}\left(\frac{1}{4}\right)^3\left(\frac{1}{4}\right)^3\left(\frac{1}{4}\right)^1\left(\frac{1}{4}\right)^1 = 0.01709$$

Thus, the odds ratio in favor of linkage is

$$L = \frac{P(O \mid R = \tfrac{1}{4})}{P(O \mid R = \tfrac{1}{2})} = 2.8473$$

and the logarithm of the odds, or lod score, becomes

$$\log(2.8473) = 0.4544$$

Linkage is indicated here because the lod score is greater than zero (the odds ratio is greater than one). However, the lod score is substantially less than 3, the conventional threshold for linkage. Additional testing is therefore needed to provide a stronger likelihood of linkage. The standard

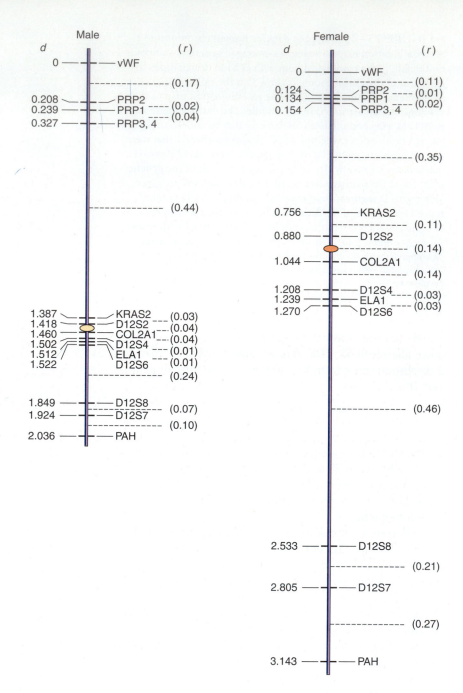

▶ **FIGURE 15–26** Partial linkage map of human chromosome 12, showing sex differences in mapping data (d = distance, r = recombination). *Source:* R. White and J. M. Lalouel, "Investigation of genetic linkage in human families," *Advances in Human Genetics* 16 (1987): 121–288. Copyright © Plenum Publishing Corporation. Used with permission.

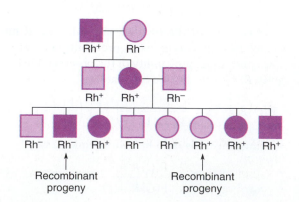

▶ **FIGURE 15–27** Pedigree showing the joint inheritance of the Rh trait (Rh⁺ = Rh positive = gene *D*, Rh⁻ = Rh negative = gene *d*) and elliptocytosis (solid symbols = affected = gene *E*, open symbols = unaffected = gene *e*).

FOCUS ON HUMAN GENETICS

The Human Genome Project

The development of gene amplification and restriction mapping techniques led researchers to suggest in 1980 that the entire human genome could be mapped. In the decade that followed, researchers from various laboratories in several different countries realized that an organized, international collaborative effort would be needed to accomplish this task, and in October of 1990 the **Human Genome Project** was officially initiated, with James Watson as its head in the United States. This ambitious project is designed around three related goals: (1) establishing a genetic map of the linkage relationships of known genes to each other and to physical markers such as restriction sites, (2) establishing physical maps based on restriction sites and other physical landmarks such as chromosome bands, and (3) determining the nucleotide sequence of all 23 chromosome pairs. The immediate objectives are to produce a linkage map with a resolution of 2 cM (that is, identify genetic markers separated by a recombination frequency of 2%) and generate physical maps with markers spaced at approximately 100,000 bp intervals.

Although the genetic and physical maps will not correspond over all chromosomal regions, the combination of the two mapping techniques allows researchers to determine the chromosomal location of genes that are at the root of human diseases. The genes that cause disorders such as cystic fibrosis and juvenile muscular dystrophy have already been identified and isolated, and genes that influence certain polygenic traits, such as Alzheimer's disease, breast and colon cancers, diabetes, heart disease, and affective disorders, such as manic depression, have been identified. Once a gene is identified and isolated, it can be amplified and its expression and regulation can be studied, leading to identification of the primary cause of the disease (as in the case of cystic fibrosis) and hopes for more effective treatments. Isolation of genes also allows probes to be constructed for prenatal testing and genetic screening for heterozygotes (carriers). Identification of genes involved in polygenic traits is particularly exciting, since it could lead to insight into the role of specific environmental factors causing these disorders and development of preventative measures to control their incidence.

What at first seemed like an overwhelming undertaking has been considerably simplified by tremendous technical advances that will be described in Chapter 18. Nucleotide sequencing—the ultimate goal of the Human Genome Project—is the most difficult task. Current automated methods allow 100,000 bp per person per year to be sequenced, so 30,000 person-years will be required to complete the task using existing technology.

Just as research conducted in connection with the space program has greatly impacted our lives in many areas that are not directly concerned with space, so has the Human Genome Project impacted genetic research. For example, one important result of sequencing is the ability to compare the nucleotide sequences of different species. Such comparisons already reveal that many genes (and proteins) are evolutionarily conserved over a wide range of organisms, a discovery with important implications for the understanding of the evolution of organisms (see Chapter 26).

It will take 30,000 person-years to determine the nucleotide sequence of the entire human genome using existing technology.

procedure is to locate and test other informative families and combine the lod scores for the different family units. By examining more families, and then pooling the data, we can refine our estimate of the map distance as well as improve the accuracy of our results.

To Sum Up

1. The Human Genome Project combines older techniques, such as somatic cell hybridization, and newer techniques, such as restriction mapping, in an ambitious long-range mapping

project aimed at eventually identifying all human genes and their chromosomal locations.

2. As more and more human genes are discovered, it becomes crucial to establish linkage relationships among certain genes. Since humans cannot be subjected to traditional genetic analysis to detect linkage, the lod score method has become an important aspect of the human genome project. This mathematical technique uses observed data on the joint inheritance of two traits in families to calculate the odds in favor of linkage of the two gene loci. The lod score is the logarithm of the odds ratio in favor of linkage; its value must be at least 3 to establish that the genes are linked.

 ## Chapter Summary

Genetic mapping has been facilitated by recent developments in cellular and molecular technologies. Cellular techniques include the use of interspecific cell hybrids (formed by somatic cell fusion) for assigning genes to chromosomes. In rodent-human hybrid cells, most of the human chromosomes are lost during the divisions following fusion. Proper chromosome assignments are made by correlating the loss of a particular chromosome from these hybrid cells with the loss of a specific gene function.

Molecular approaches to mapping include the use of restriction endonuclease enzymes that cut DNA at specific palindromic sequences called restriction sites. Restriction enzymes occur naturally in bacteria, where they protect the organism against infection by foreign DNA. Restriction mapping techniques have provided detailed physical maps of DNA that show the locations of restriction sites. Allelic variations in the locations of restriction sites

(called restriction fragment length polymorphisms, or RFLPs) can be detected using genetic probes. The use of RFLPs as genetic markers permits the identification and mapping of genes and provides a tool for the diagnosis of human genetic disease. DNA fingerprinting is a special type of RFLP analysis that detects variable tandem nucleotide repeats; it has proven valuable in forensics for dealing with questions of identity. Rapid DNA sequencing methods and PCR (polymerase chain reaction—a technique for amplifying specific DNA sequences) now make it possible to efficiently obtain sequence information from relatively small starting quantities of DNA for use in mapping, genetic testing, and forensics. The molecular methods described in this chapter (and in Chapter 18) are now being used to map the human genome and make it possible to establish genetic and physical maps based on molecular markers instead of visible phenotypic differences.

Cellular Methods: Somatic Cell Genetics

1. Define the following terms and distinguish between members of paired terms:
   heterokaryon and synkaryon
   syntenic and linked genes
   somatic cell hybridization
   HAT method
   clone panel

2. Human cells are fused with mouse cells, and the resulting hybrid cells are placed into HAT medium. After extensive chromosomal loss, six different cell lines are selected and tested for the presence of human chromosomes as well as for the presence of four human enzymes, one of which is thymidine kinase (TK). In the results tabulated below, the presence of a chromosome or enzyme is indicated by a +, its absence by a −.

		Cell Lines					
		(a)	(b)	(c)	(d)	(e)	(f)
Human enzymes	E_1	+	+	−	−	−	−
	E_2	−	+	+	−	+	+
	E_3	+	+	−	−	−	+
	E_4	+	+	+	+	+	+
Human chromosomes	1	+	+	−	−	−	+
	11	+	+	−	−	−	−
	17	+	+	+	+	+	+
	X	−	+	+	−	+	+

(a) Identify the chromosome that carries the gene for each enzyme. (b) Which of the enzymes (E_1, E_2, E_3, or E_4) is thymidine kinase? How can you tell?

3. The following data are derived from Owerback et al. (1980)*, who used a labeled probe and Southern blotting to examine different hybrid cell lines for the presence (+) or absence (−) of DNA sequences complementary to the insulin gene.

Hybrid Clone	Insulin Sequences Detected	Human Chromosomes Present
1	+	3 5 11 14 15 17 18 21
2	−	2 5 6 10 12 18 20 21 X
3	−	8 10 12 15 17 21 X
4	+	4 5 10 11 12 17 18 21
5	+	6 7 10 11 14 17 18 20 21 X

From these data, identify the chromosome that carries the human insulin gene.

Molecular Methods: DNA Restriction and Hybridization

4. In the following section of a single DNA strand, identify the potential restriction endonuclease target sites in the double-helical form of the DNA:

5′...CATGAACGTTAAGATCAAAAGTCGACTGCAGAGAGA...3′

*Owerback, D., et al. "The insulin gene is located on chromosome 11." *Nature* 286 (1980): 83. Reprinted with permission from *Nature*. Copyright (1980) Macmillan Magazines Limited.

5. A preparation of DNA is subjected to partial and complete digestion by the restriction enzyme *Alu*I. The results are shown below, where end-labeled fragments are marked by asterisks. Construct a restriction map of this DNA.

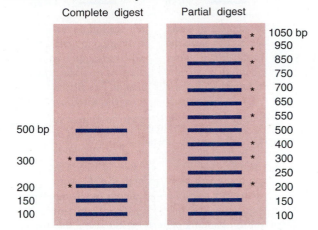

6. A preparation of circular plasmid DNA is digested to completion with the restriction enzymes listed below. Using the data given below on fragment sizes derived from gel electrophoresis, construct a restriction map of the plasmid DNA.

Enzymes	Fragment Sizes (kb)
*Bgl*II	12
*Hga*I	4, 8
*Sma*I	4, 8
*Bgl*II + *Hga*I	2, 8
*Bgl*II + *Sma*I	4
*Hga*I + *Sma*I	2, 4

7. A circular plasmid DNA is digested to completion with the restriction endonucleases listed below. Using the fragment sizes given in the table below, construct a restriction map of the plasmid DNA.

Enzymes	Fragment Sizes (kb)
*Bam*HI	14.1
*Eco*RI	5.5, 8.6
*Hind*III	2.1, 5.6, 6.4
*Bam*HI + *Eco*RI	1.5, 4.0, 8.6
*Bam*HI + *Hind*III	1.4, 2.1, 5.0, 5.6
*Eco*RI + *Hind*III	2.1, 2.6, 2.9, 3.0, 3.5
*Bam*HI + *Eco*RI + *Hind*III	1.4, 1.5, 2.1, 2.6, 3.0, 3.5

8. The linear segment of DNA shown below is labeled with ^{32}P at its 5′ ends and has four restriction sites for the enzyme *Bam*HI, indicated by the arrows. Diagram the gel banding pattern you would obtain following electrophoresis of the fragments produced by (a) complete and (b) partial digestion with *Bam*HI. (Lengths are given in kb.)

9. Two organisms have mutations that affect the DNA segment in problem 8. DNA of each is end-labeled with ^{32}P and digested to

completion with *Bam*HI. The gel banding patterns of the two mutants are shown below, with asterisks denoting end-labeled fragments. From these results, indicate the probable mutational alterations in each of the strains.

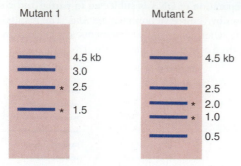

Mutant 1 | Mutant 2

Mutant 1:
4.5 kb
3.0
* 2.5
* 1.5

Mutant 2:
4.5 kb
2.5
* 2.0
* 1.0
0.5

10. A DNA segment in a certain gene has the following *Eco*R1 and *Bam*H1 restriction map (distances given in kb):

EcoRI | BamHI
2.0 | 4.5 | 3.0

Draw the gel banding pattern expected if a mutation (a) destroys the *Eco*R1 site, (b) adds another *Bam*H1 site precisely in the middle of the 3.0 section.

11. Suppose that complete digestion of a DNA molecule by the restriction enzyme *Hind*II produces eight segments, designated A through H in order of decreasing size. A partial digest of the DNA molecule was then made with *Hind*II, and the fragments produced were separated by gel electrophoresis, recovered from the gel, and redigested to completion to determine their *Hind*II-segment composition. The five fragments that were recovered were composed of the following segments: (A, B, D, F, G), (A, D, E, F), (B, G, H), (C, E, H), and (B, F), where the parentheses indicate that the order of the enclosed segments is unknown. Construct a restriction map of the DNA molecule from these data, giving the correct order of the segments A through H.

12. A preparation of circular plasmid DNA was digested to completion with three different combinations of restriction enzymes. The patterns of the fragments following electrophoresis are shown below:

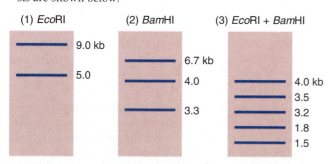

(1) *Eco*RI
9.0 kb
5.0

(2) *Bam*HI
6.7 kb
4.0
3.3

(3) *Eco*RI + *Bam*HI
4.0 kb
3.5
3.2
1.8
1.5

(a) Construct a restriction map of this plasmid DNA. (b) A Southern blot was prepared from gel 3. Which fragments would hybridize to a ^{32}P-labeled probe of the 5.0-kb fragment in gel 1?

Restriction Site Polymorphism and Gene Mapping

13. A particular DNA region is 10 kb long, includes two *Hind*III restriction sites, and is recognized by DNA probes as follows:

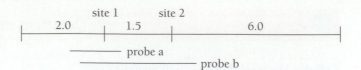

site 1 | site 2
2.0 | 1.5 | 6.0
probe a
probe b

(a) Draw the bands expected on an electrophoretic gel of the DNA after complete digestion with *Hind*III. (b) How many RFLP haplotypes are detectable on a Southern blot using probe a? Draw them. (c) How many RFLP haplotypes are detectable on a Southern blot using probe b? Draw them.

14. Four possible RFLP haplotypes can be generated by restriction enzyme digestion of the following DNA region:

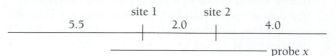

site 1 | site 2
5.5 | 2.0 | 4.0
probe x

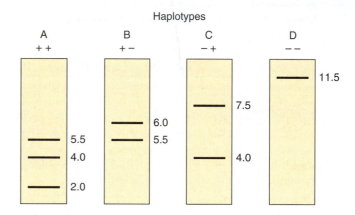

Haplotypes

A (+ +): 5.5, 4.0, 2.0
B (+ −): 6.0, 5.5
C (− +): 7.5, 4.0
D (− −): 11.5

Two pedigrees showing the inheritance of a particular autosomal dominant disorder are available for two different families. The haplotypes of living individuals in the pedigrees were determined, and the results are shown below:

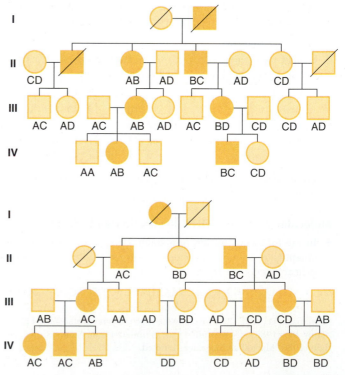

362   CHAPTER 15

(a) What conclusions can you draw about the relationship between the RFLP and the disease-causing gene? (b) What appears to be the approximate genetic (recombination) distance between the RFLP detected by probe *x* and the mutant gene? (c) Suppose somatic cell hybridization analysis reveals that probe *x* hybridizes to DNA in hybrid cells only if they contain the short arm of chromosome 5. What can you conclude about the chromosomal location of the gene causing the disorder? (d) Suppose that this disorder does not show its symptoms until middle age, even though it is caused by a dominant gene. Explain how you could use probe *x* to identify "affected" individuals at an early age, before symptoms appear. How accurate would your genetic test be? Explain your answer.

15. The techniques of amniocentesis and chorion biopsy make it possible to diagnose many genetic disorders prenatally by looking for protein deficiencies in the fetal cells, but several important disorders cannot be detected by this type of analysis. One such condition is phenylketonuria, which cannot be detected by protein analysis because the enzyme whose absence causes the disease is produced only in liver cells; the gene that is responsible for this enzyme is inactive in other types of cells. Other examples include sickle-cell anemia and other hemoglobin disorders, because cells obtained by amniocentesis and chorion biopsy do not include fetal blood cells. However, restriction enzyme analysis now makes it possible to detect hemoglobin disorders prenatally. The gene that encodes the β-polypeptide chain of normal hemoglobin contains an *Mst*II restriction site, but 80% of individuals with hemoglobin S lack this cleavage site. Use this information to outline a procedure that will allow the prenatal detection of sickle cell anemia.

16. RFLP analysis can be used for prenatal diagnosis of sickle-cell anemia because the mutation destroys an *Mst*II restriction site:

    *Mst*II site

    . . . GGACTCCTC . . . $\xrightarrow{\text{Mutation}}$ . . . GGACACCTC . . .
    Normal allele   Sickle-cell allele

    The recognition sequence for *Mst*II is GGANTCC, where N can be any base. Hemoglobin C disease is another disorder that is caused by a mutation in the same sequence of bases. In hemoglobin C the sequence is changed to . . . GGATTCCTC . . . . Can RFLP analysis with *Mst*II be used for prenatal diagnosis of hemoglobin C? Explain your answer.

17. One application of DNA fingerprinting technology is the identification of the biological parents of children. Suppose that there is reason to suspect that child X is actually the biological child of couple Y, even though couple X have been presumed to be his parents. DNA fingerprinting is done, using a probe that recognizes a certain hypervariable region. The results are as follows:

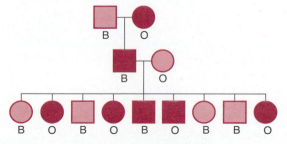

| Couple X | | | Couple Y | |
| Man X | Woman X | Child X | Man Y | Woman Y |

What is your conclusion regarding the parentage of child X?

18. Amplification of a desired DNA sequence by the PCR technique cannot continue indefinitely. After a certain number of cycles, the sequence stops increasing at an exponential rate (i.e., stops doubling in concentration in every cycle) and enters a linear or stationary phase in which the rate of increase is constant. Explain why the exponential growth of DNA cannot continue, identifying the factors that are responsible for limiting the amplification process.

DNA Sequencing

19. A dideoxynucleotide sequencing gel obtained from the beginning of a certain DNA segment is shown below. What is the nucleotide sequence of this section of the DNA?

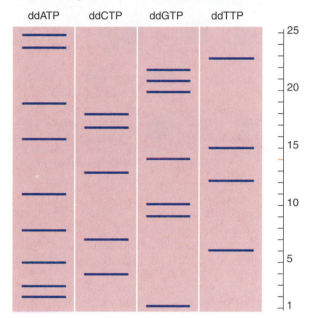

20. A short DNA fragment with the sequence 3′ CTGATAAG-GCTTTG 5′ is used as a template for sequencing by the dideoxynucleotide method. Each of the four reaction vessels contains labeled deoxynucleotide triphosphates plus a small amount of either ddATP, ddTTP, ddCTP, or ddGTP. Draw the dideoxy sequencing gel expected for the DNA strand complementary to the template.

Mapping the Human Genome

21. In Chapter 13, problem 19 discussed the linkage between the ABO group and nail-patella loci. The pedigree below shows the joint inheritance of these loci in three generations. Individuals with nail-patella syndrome are represented by darker symbols, and the blood type is given below each symbol.

(a) Determine the linkage phase (coupling or repulsion) of the heterozygote in generation II. (b) Identify which of the progeny in generation III received a recombinant gamete from the heterozygote, and determine the frequency of recombination. (c) Calculate the probability of the birth distribution in generation III, assuming linkage with the frequency of recombination computed in (b). (d) Calculate the odds ratio in favor of linkage and the logarithm of the odds, or lod score.

22. The pedigree below shows the joint inheritance of two linked pairs of alleles for three generations. One pair is for myotonic dystrophy (D) vs. normal (d), and the other pair is for secretor (S) vs. nonsecretor (s). (A secretor is an individual whose body secretions contain certain blood group antigens.) Individuals with myotonic dystrophy are represented by darker symbols; the secretor status is given below the symbol for each individual.

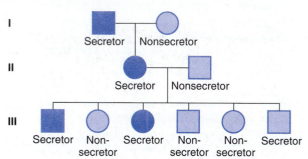

(a) Determine the linkage phase (coupling or repulsion) of the heterozygote in generation II. (b) Identify which of the progeny in generation III received a recombinant gamete from the heterozygote, and determine the frequency of recombination. (c) Calculate the probability of the birth distribution in generation III, assuming linkage with the frequency of recombination computed in (b). (d) Calculate the odds ratio in favor of linkage and the logarithm of the odds, or lod score.

23. If one more child were born to the generation II parents in the pedigree in problem 22 and the child is a secretor affected with myotonic dystrophy, what would be the odds ratio in favor of linkage of the d and s loci?

VI

GENE EXPRESSION AND CLONING

Although the genes of an organism carry genetic information, they do not directly express inherited traits. Instead, genes direct the synthesis of proteins, which in turn serve as the molecular instruments for expressing the genotype.

An organism may possess thousands of genes, and it can produce about as many different kinds of proteins. Although each kind of protein has its own functional role in an organism, all are synthesized by a similar process. To produce a protein, a gene first serves as a template for the synthesis of an RNA molecule; the RNA then moves into the cytoplasm carrying the instructions for the structure of a polypeptide chain. This flow of information in biological systems is outlined by a scheme that Crick called the **central dogma of molecular biology.**

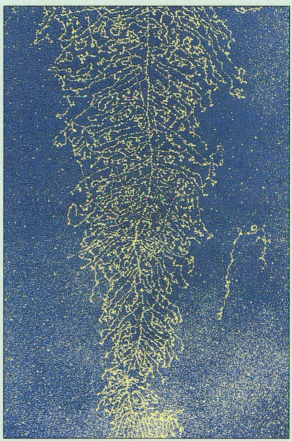

Transcription of DNA showing a feather-like structure. Numerous mRNA molecules extend in clusters from the DNA strand (backbone of feather).

The solid arrows trace the major directions of information flow, while the dashed arrows represent allowable but less common transfers that are essential mainly for certain viruses that contain RNA in place of DNA. Two types of information transfer are shown: **homocatalytic** or **replicative transfer** (represented by the circular arrows), which occurs between successive generations of the same molecule, and **heterocatalytic transfer,** (represented by the straight arrows), which occurs between different molecules of the same generation.

If we include only the major heterocatalytic transfers that are required for protein synthesis, the diagram simplifies to

$$\text{DNA} \xrightarrow{\text{transcription}} \text{RNA} \xrightarrow{\text{translation}} \text{Protein}$$

which now depicts the flow of information that occurs when genes express themselves phenotypically. The DNA of the genes first undergoes **transcription** to form complementary RNA strands. Some of this RNA—the messenger RNA—carries information from the structural genes on the primary structure of proteins, while other components of this RNA—the ribosomal RNA and transfer RNA—function in the **translation** of each structural gene message into the amino acid sequence of a polypeptide chain.

In this part of the text, we will consider the processes involved in the transfer of information from genes to proteins, focusing on the two main steps in gene expression, transcription and translation. We will also describe some of the techniques for manipulating DNA that make up the new arsenal of methods now available for recombinant DNA research. The key techniques in this arsenal are gene isolation and cloning methods that use the natural reproductive ability of cells to isolate and replicate specific gene sequences. This new arsenal of methods has revolutionized the way researchers study gene expression by making it possible to isolate and specifically modify genes from complex genomes and transfer these genes back into an organism for further analysis.

CHAPTER 16

RNA Synthesis and the Genetic Code

There are two major forms of RNA synthesis: **DNA-dependent RNA synthesis,** in which RNA is formed on a DNA template, and **RNA-dependent RNA synthesis,** in which RNA is formed on an RNA template. DNA-dependent RNA synthesis is the more widely used of these forms, since it serves as both the mechanism of transcription during protein synthesis and the means of forming RNA primers during DNA replication. In contrast, the RNA-dependent process has a much more restricted role in nature, functioning principally in the replication of certain RNA viruses.

In this chapter, we will consider both forms of RNA synthesis, concentrating on the DNA-dependent process of transcription first. We will also examine the genetic structure and processing of messenger RNA and the nature of the genetic code.

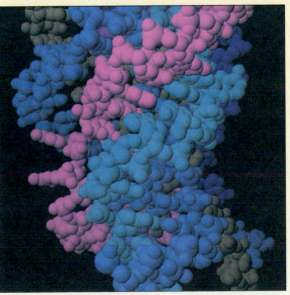

Computer-generated model of ribosomal RNA
undergoing self-splicing.

DNA-Dependent Synthesis of RNA

In cells that are actively synthesizing protein, individual genes and groups of genes are selectively transcribed to produce the three major classes of RNA: **messenger RNA (mRNA)**, **ribosomal RNA (rRNA)**, and **transfer RNA (tRNA)**. In the regions of DNA that carry these genes, a small portion of the double helix is unwound to expose the bases of the separated strands in the immediate area of transcription (▶ Figure 16–1). A complementary single-stranded chain of RNA is then synthesized from its nucleoside triphosphate precursors (ATP, GTP, CTP, and UTP), using one of the exposed strands of DNA as a template.

The idea that DNA can serve as a template for the synthesis of RNA was adopted as a working hypothesis in the early 1950s. At this time RNA synthesis was known to be correlated with protein synthesis, and since it was also known that genes code for proteins, this correlation suggested that RNA might function as a "messenger," transferring genetic information from the chromosomes to the ribosomes on which proteins are synthesized. The first convincing evidence for DNA's template role in RNA synthesis came from studies on the RNA produced in *E. coli* cells after infection by phage T2. First, work done by Elliot Volkin and Lawrence Astrachan in the mid-1950s showed that the newly synthesized RNA had a GC content similar to that of T2 DNA, in agreement with results expected for a template mechanism. Even more persuasive, however, were the results of later experiments conducted by Benjamin Hall, Saul Spiegelman, and others, who showed that the RNA produced after T2 infection could hybridize specifically with denatured T2 DNA. Thus, not only are this DNA and RNA alike in overall base composition, but they also exhibit base-sequence complementarity. It was therefore apparent that shortly after infection, the infecting T2 DNA served as a template for producing the phage-specific RNA.

DNA–RNA hybridization also provided evidence that in vivo, only one DNA strand serves as a template for a given RNA chain. For example, J. Marmur and his colleagues separated the DNA strands of the *Bacillus subtilis* phage SP8 by equilibrium density gradient centrifugation, taking advantage of differences in the buoyant densities of the strands. They attempted to hybridize each of the strands to the RNA produced upon phage infection, and discovered that the RNA combined exclusively with the heavier of the two strands. Thus, only the heavier DNA strand of this phage serves as a template for the synthesis of RNA.

It is apparent that all of the information for RNA synthesis in phage SP8 is restricted to just one DNA strand, but this is not the case for cells or for most other DNA viruses. In these systems, the template strand is gene-specific, so the strand that is copied during transcription can vary from one RNA-coding region to another. Thus, one DNA strand may code for RNA in one region of the genome, while the other strand may serve as the template strand in another region. However, in any given region, only one of the two DNA strands normally serves as the template for RNA synthesis.

▶ **FIGURE 16–1** Transcription of RNA from a DNA template. (a) A short region of the DNA is unwound. (b) One of the DNA strands in this region serves as a template for the synthesis of a complementary RNA transcript.

(a)

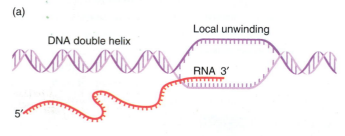

(b)

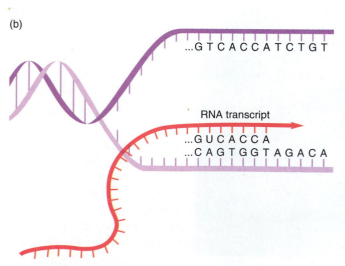

Example 16.1

If you had samples of pure RNA and duplex DNA, how could you tell if the RNA was synthesized using part of the DNA as a template?

Solution: If the RNA was transcribed from the DNA in question, it would be complementary to one or more single-stranded regions of that DNA. Therefore, the first step in the analysis would be to denature the DNA. Following separation and purification of the DNA strands (e.g., by equilibrium density-gradient centrifugation), the next step would be to determine if the RNA hybridizes with one or more portions of one of the DNA strands. Hybridization of the RNA with a DNA strand would establish their complementarity and would mean that the RNA was indeed synthesized using that DNA strand as a template.

Follow-Up Problem 16.1

If the RNA referred to in Example 16.1 is complementary to one or more regions of the DNA strands, will its base com-

Subunit	Number	Molecular Weight	Possible Function
α	2	40,000 (each)	promoter binding(?)
β	1	155,000	nucleotide binding (catalytic site)
β'	1	160,000	DNA binding
σ	1	85,000	promoter recognition, initiation

position reflect that of those DNA regions? Explain your answer.

RNA Polymerases

RNA synthesis that occurs on the DNA template strand is catalyzed by an enzyme called **RNA polymerase**. This enzyme catalyzes the stepwise addition of each ribonucleotide substrate to the growing RNA chain by the following reaction:

$$(NMP)_n + NTP \xrightarrow{\text{DNA template strand}} (NMP)_{n+1} + PP_i$$

$$\text{RNA strand} \qquad\qquad\qquad \text{Lengthened RNA}$$

where NMP and NTP denote ribonucleoside monophosphate and triphosphate, respectively, and PP_i is inorganic pyrophosphate. Unlike the DNA-replicating enzymes, RNA polymerase does not require a preexisting primer chain—it can initiate synthesis de novo, beginning with the first ribonucleotide residue in the growing chain. The resulting RNA therefore consists entirely of newly incorporated residues.

In bacteria, all genes can be transcribed by a single type of RNA polymerase enzyme. This enzyme is an oligomeric protein that is encoded by several structural genes. In *E. coli,* for example, the complete molecule of RNA polymerase (referred to as the **holoenzyme**) consists of four different polypeptide subunits, α, β, β', and σ, in a ratio of 2:1:1:1 (■ Table 16–1). The catalytic site for RNA synthesis is thought to reside on the β subunit, which is a binding site for the antibiotic rifampicin, a potent inhibitor of the initiation of RNA synthesis in prokaryotes.

The *E. coli* holoenzyme can be reversibly separated into the **core enzyme** and the **sigma factor** according to the following reaction:

$$\alpha_2\beta\beta'\sigma \rightleftharpoons \alpha_2\beta\beta' + \sigma$$

$$\text{holoenzyme} \quad \text{core} \quad \text{sigma}$$

The polymerase core enzyme can catalyze the synthesis of RNA, but it cannot initiate transcription at the proper sites. The σ factor allows the polymerase holoenzyme to recognize appropriate DNA regions to begin transcription. Several different σ factors are known—each recognizes a different initiation region in DNA. However, the majority of cellular transcription involves a single predominant σ factor (designated σ^{70} in *E. coli*).

Unlike prokaryotes, eukaryotic cells have at least four different kinds of RNA polymerases. Three of these polymerase enzymes (I, II, and III) are nuclear, and a fourth is mitochondrial. The general properties of the nuclear RNA polymerases are given in ■ Table 16–2. Note that each of these enzymes is responsible for synthesizing a different class of RNA. RNA polymerase I transcribes rRNA genes into ribosomal RNA; RNA polymerase II transcribes structural genes into mRNA precursors (hnRNA); and RNA polymerase III synthesizes tRNA and an rRNA species designated 5S, as well as other small RNAs. All three enzymes are large complex proteins containing multiple subunits. They are distinguishable from one another by their nuclear location and by their sensitivity to the drug α-amanitin (a highly toxic peptide from the poisonous mushroom *Amanita phalloides*). RNA polymerase I resides in the nucleolus and is insensitive to α-amanitin. RNA polymerases II and III are both located in the nucleoplasm; both are inhibited by α-amanitin, but only the polymerase II enzyme shows high sensitivity to the drug.

Transcription in Prokaryotes

The transcription process in prokaryotes occurs in three phases: **initiation, chain elongation,** and **termination.** The steps in each phase are summarized in ▶ Figure 16–2.

Initiation begins with the noncovalent binding of RNA polymerase to a specific DNA region known as the **promoter.** Each transcription unit in DNA (a transcription unit can include from one to several genes) has a promoter region that selectively binds the polymerase molecule. The polymerase holoenzyme with its attached σ factor recognizes the promoter site through certain base sequences that are common to this region. For example, in *E. coli* there are

■ TABLE 16–2 Properties of eukaryotic RNA polymerases.

Polymerase	Location	RNA Products	α-amanitin Sensitivity
I	Nucleolus	rRNA	not inhibited
II	Nucleus	hnRNA (protein coding); snRNA	strongly inhibited
III	Nucleus	small RNA (tRNA, 5S rRNA, snRNA)	species specific (vertebrates weakly inhibited, yeast and insects not inhibited)

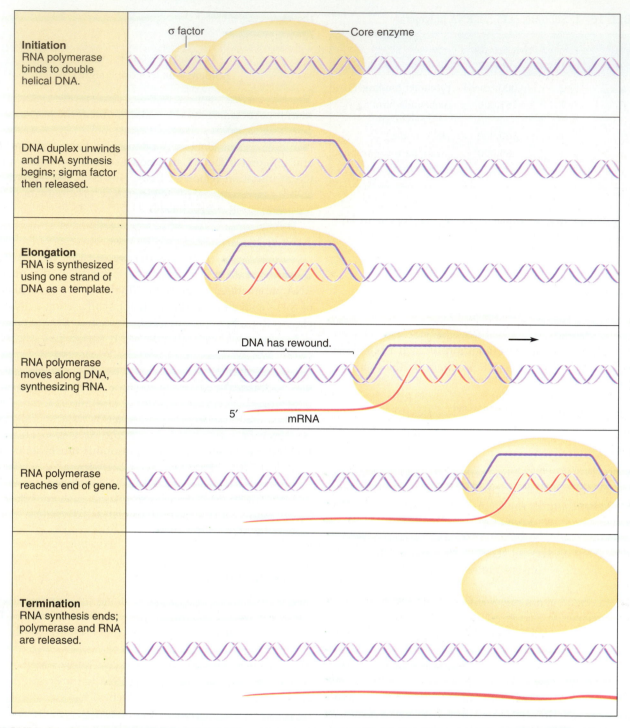

Initiation RNA polymerase binds to double helical DNA.	σ factor — Core enzyme
DNA duplex unwinds and RNA synthesis begins; sigma factor then released.	
Elongation RNA is synthesized using one strand of DNA as a template.	
RNA polymerase moves along DNA, synthesizing RNA.	DNA has rewound. 5′ mRNA
RNA polymerase reaches end of gene.	
Termination RNA synthesis ends; polymerase and RNA are released.	

▶ **FIGURE 16–2** A summary of the steps in prokaryotic transcription.

two relatively invariant sequences that bind and orient the σ^{70}-containing holoenzyme: one is centered about 35 bp and the other about 10 bp before the point where RNA synthesis starts. By convention, bases on the 5′ side (upstream) of the transcription start site are assigned negative values, so these promoter sequences are commonly designated as the **−35 sequence** and the **−10 sequence** (also called a **Pribnow box,** after its discoverer). The −35 and −10 regions are both six base pairs long and have the following **consensus** sequences (based on comparisons of many different promoters):

$$
\begin{array}{ccc}
-35 & -10 & +1 \\
5'\text{----TTGACA----TATAAT----start} \\
& & \text{site}
\end{array}
$$

These consensus sequences are highly conserved regions that have been identified by aligning different promoter sequences so as to achieve maximum homology. (▶ Fig-

lac	A C C C C A G G C T	T T A C A C	T T T A T G C T T C C G G C T C G	T A T G T	T G T G T G G Ⓖ Ⓐ A T T G T G A G C G G
lacP115	A C A C T T T A T G	C T T C C G	G C T C G T A T G T T G T G T G G	T A T T G T	G A G C G G Ⓐ T A A C A A T T T C A
lacI	C C A T C G A A T G	G C G C A A	A A C C T T T C G C G G T A T G G	C A T G A T	A G C G C C C Ⓖ G A A G A G A G T C
galP₁	T T C C A T G T C A	C A C T T T	T C G C A T C T T T G T T A T G C	T A T G G T	T A T T T C Ⓐ T A C C A T A A G C C
galP₂	A T T T A T T C C A	T G T C A C	A C T T T T C G C A T C T T T G T	T A T G C T	A T G G T T Ⓐ T T T C A T A C C A T
araBAD	G G A T C C T A C C	T G A C G C	T T T T T A T C G C A A C T C T C	T A C T G T	T T C T C C A T Ⓐ C C C G T T T T T
araC	G C C G T G A T T A	T A G A C A	C T T T T G T T A C G C G T T T T	T G T C A T	G G C T T T Ⓖ G T C C C G C T T T G
trp	A A A T G A G C T G	T T G A C A	A T T A A T C A T C G A A C T A G	T T A A C T	A G T A C G C Ⓐ A G T T C A C G T A
bioA	T T C C A A A C G	T G T T T T	T T G T T G T T A A T T C G G T G	T A G A C T	T G T A A Ⓐ C T A A A T C T T T T
bioB	C A T A A T C G A C	T T G T A A	A C C A A A T T G A A A A G A T T	T A G G T T	T A C A A G T C Ⓣ A C A C C G A A T
bioP98	T A A T T C G G T G	T A G A C T	T A T A A A C C T A A A T C T T T T	T A A A T T	T G G T T T Ⓐ C A A G T C G A T T A
tet	A T T C T C A T G T	T T G A C A	G C T T A T C A T C G A T A A G C	T T T A A T	G C G G T A Ⓖ T T T A T C A C A G T
tRNA_tyr	C A A C G T A A C A	C T T T A C	A G C G G C G C G T C A T T T G A	T A T G A T	G C G C C C C Ⓖ C T T C C G A T A
strep	T G T A T A T T T C	T T G A C A	C C T T T T C G G C A T C G C C C	T A A A A T	T C G G C Ⓖ T C C T C A T A T T G T
Spc	T T A T T T T T T C	T A C C C A	T A T C C T T G A A G C G G T G T	T A T A A T	G C C G C Ⓖ C C C T C G A T A T G G
rrnD₁	C A A A A A A A T A	C T T G T G	C A A A A A A T T G G G A T C C C	T A T A A T	G C G C C T C C Ⓖ T T G A G A C G A
rrnE₁	C A A T T T T T C T	A T T G C G	G C C T G C G G A G A A C T C C C	T A T A A T	G C G C C T C C Ⓐ T C G A C A C G G
rrnX₁	C A T T T T T C C G	C T T G T C	T T C C T G A G C C G A C T C C C	T A T A A T	G C G C C T C C Ⓐ T C G A C A C G G
rrnD₂E₂X₂	G A A A T T C A G G	G T T G A C	T C T G A A A G A G G A A A G C G	T A A T A T	A C G C C A C Ⓒ T C G C G A C A G T
rrnA₁	T A A A T T T C C T	C T T G T C	A G G C C G G A A T A A C T C C C	T A T A A T	G C G C C A C C Ⓐ C T G A C A C G G
rrnA₂	A A A A T A A A T G	C T T G A C	T C T G T A G C G G G A A G G C G	T A T T A T	G C A C A C Ⓒ Ⓒ Ⓒ G C G C C G C T G

TTGACA ← 16 to 19 bp → TATAAT ← 6 to 9 bp → ◯ ⟶

▶ **FIGURE 16–3** Promoter sequences in 21 different genes of *E. coli*. The Pribnow (TATAAT) boxes have been aligned, and the consensus sequence is shown at the bottom. Circled nucleotides denote the transcription start sites. (By convention, DNA sequences are written for the nontemplate strand, that is, the strand that matches the mRNA rather than the strand that serves as template for mRNA synthesis.)

ure 16–3). A typical bacterial promoter therefore has three components: the −35 and −10 consensus sequences and the start site.

Following binding, the polymerase enzyme unwinds nearly two turns of template DNA in the region of the start point to form an **open promoter complex.** Initiation of the RNA chain then begins at the start site with the insertion of the first base (usually either A or G) in the form of its nucleoside triphosphate. Unlike the other nucleotide components of the RNA chain, the first (initiating) nucleotide does not lose its terminal phosphates as PP_i. Every RNA transcript therefore starts with pppN—usually either pppA or pppG—at its 5′ end.

Once the RNA chain is initiated, chain elongation proceeds through the sequential linking of nucleotides by phosphodiester bonds. At first, chain elongation is catalyzed by the RNA polymerase holoenzyme. However, once a few (8–10) nucleotides have been added to the growing chain (enough to form a stable pairing with the DNA template), the σ factor is released, leaving the core enzyme to direct the continued chain growth. Nucleotide-by-nucleotide lengthening of the chain progresses at a rate of about 50 nucleotides per second in a 5′ to 3′ direction. Throughout the elongation phase, DNA is unwound just ahead of the enzyme and rewound at the same rate behind it, creating an enzyme-associated "transcription bubble" in which active synthesis takes place.

The elongation process ends once a **terminator** sequence is reached. The enzyme and the completed RNA chain are then released through a process that may require certain pro-

tein factors as well as polynucleotide signals. There are two main types of termination sites: type I terminators are self-terminating sites requiring no known auxiliary factor for termination to occur in vitro, while type II terminators require one or more protein-termination factors to facilitate RNA chain release.

Type I terminators can be relatively long (up to around 50 nucleotides) and function through the formation of hairpin loops in the RNA transcripts. A typical type I terminator sequence consists of a palindromic GC-rich region followed by a run of AT base pairs (▶ Figure 16–4a). When the RNA polymerase reads through this sequence, the nascent RNA chain folds into a hairpin secondary structure through complementary base pairing (Figure 16–4b). Although the mechanism is still unclear, it is believed that the hairpin structure causes the polymerase to pause or stop elongating, leaving the transcript attached to its template by only a short length of AU base pairs. Since the strength of pairing between A and U is weak, the completed RNA chain dissociates from the DNA template and then from the enzyme, terminating transcription (Figure 16–4c).

Type II terminators may also contain palindromic sequences, but they generally lack the string of AT base pairs preceding the point of termination. The strength of pairing between the completed transcript and template is greater in this case, so termination requires protein factors that combine with the transcript or enzyme (or both) and help unwind the RNA from the DNA strand. Many different protein-terminating factors have been identified in prokaryotes. One such protein, the **rho** (ρ) **factor,** has been shown to

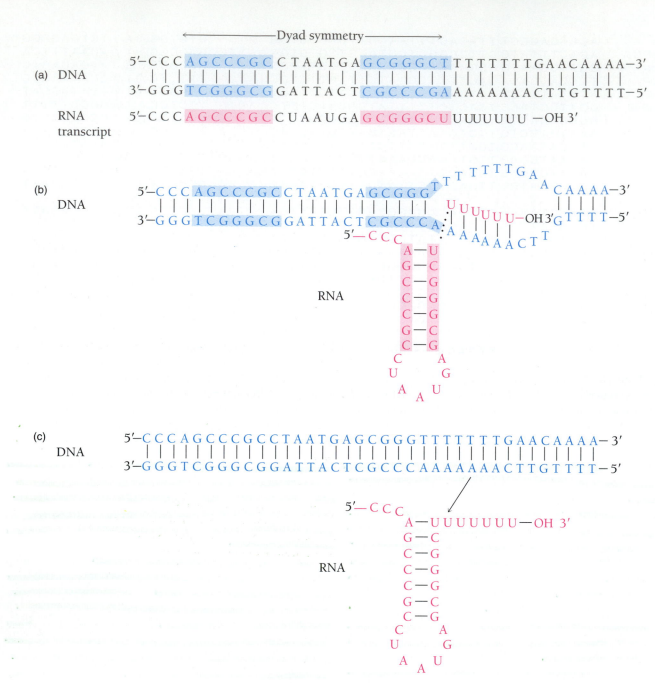

(a) DNA

5′–C C C AGCCCGC C T A A T G A GCGGGCT T T T T T T G A A C A A A A–3′

3′–G G G TCGGGCG G A T T A C T CGCCCGA A A A A A A A C T T G T T T T–5′

RNA transcript

5′–C C C AGCCCGC C U A A U G A GCGGGCU U U U U U U –OH 3′

(b) DNA

5′–C C C AGCCCGCCTAATGAGCGGG T T T T T T G A A C A A A A–3′

3′–G G G TCGGGCGGATTACTCGCCCA A A A A A A C T T G T T T–5′

5′–C C C

RNA

A—U
G—C
C—G
C—G
C—G
G—C
C—G
C   A
U   G
A   A U

(c) DNA

5′–C C C A G C C C G C C T A A T G A G C G G G T T T T T T T G A A C A A A A–3′

3′–G G G T C G G G C G G A T T A C T C G C C C A A A A A A A C T T G T T T T–5′

5′–C C C

RNA

A—U U U U U U U U—OH 3′
G—C
C—G
C—G
C—G
G—C
C—G
C   A
U   G
A   A U

▶ **FIGURE 16–4** (a) Sequence of a type I terminator and its terminated RNA transcript. (b) Stem-and-loop formation by the transcript at the terminator site leaves the RNA attached to the template by a short stretch of AU base pairs. (c) The unstable AU hybrid dissociates, releasing the completed RNA transcript from the terminator site. *Source:* Part (a) adapted from C. Yanofsky, "Attenuation in the control of expression of bacterial operons," *Nature* 289 (1981): 751. Reprinted with permission from *Nature.* Copyright (1981) Macmillan Magazines Limited. Parts (b) and (c) from P. J. Farnham and T. Platt, "A model for transcription termination suggested by studies on the *trp* attenuator in vitro using base analogs," *Cell* 20 (1980): 739–748. Copyright © by Cell Press. Used with permission.

be an important termination factor in *E. coli*. The ρ factor functions at ρ-dependent terminators by combining with the RNA transcript and using the energy provided by the breakdown of ATP to dissociate the RNA and the polymerase from the DNA template. A model of ρ-dependent termination is shown in ▶ Figure 16–5. According to this model, ρ binds to the transcript and follows the RNA polymerase; it finally catches up when the polymerase pauses at the ρ-dependent terminator, allowing ρ to destabilize the RNA–DNA bonds and dissociate the RNA and polymerase from the template.

Another important termination factor, NusA protein, also aids in the release of RNA, but it combines directly with the *E. coli* RNA polymerase. The NusA protein binds to the core enzyme once the σ factor has been released and stimulates transcription termination at various terminator sites.

(a) ρ pursues polymerase.

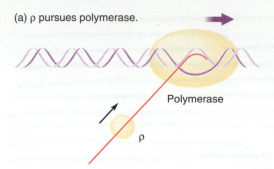

Polymerase

ρ

(b) Polymerase pauses at hairpin; ρ catches up.

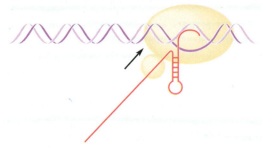

(c) ρ causes termination.

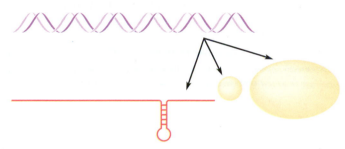

▶ **FIGURE 16–5** A model for the function of the ρ factor at a ρ-dependent terminator. The ρ factor binds to the transcript and moves along the RNA in the 5′ to 3′ direction, using the hydrolysis of ATP to provide energy for the movement. When the hairpin forms and the polymerase pauses, the ρ factor interacts with RNA polymerase and causes termination.

Transcription in Eukaryotes

Transcription in eukaryotic cells is similar in several respects to the process in prokaryotes. Chain elongation occurs in a 5′ to 3′ direction in much the same manner as the elongation process catalyzed by the bacterial core enzyme. Various promoterlike signals are also associated with eukaryotic genes, but they occur at different locations than in prokaryotes. For example, most promoters for RNA polymerase II have a short consensus sequence of A and T nucleotides (TATAAAA) centered about −25; this region is called the **TATA** or **Hogness box** (after its discoverer), and it is similar to the −10 sequence in bacterial promoters. Many promoters for RNA polymerase II also contain a CCAAT sequence (or **CCAAT box**) and a GGGCGG sequence (or **GC box**), and they may have a variety of other conserved sequence elements located upstream from the TATA box (▶ Figure 16–6a). In contrast, consensus sequences for most promoters for RNA polymerase III are located within the genes transcribed by the enzyme. In tRNA genes, these intragenic sequences, called **internal control regions** (**ICR**), consist of two sequence elements, box A and box B, which have approximate coordinates of 8–19 and 52–62, respectively (Figure 16–6b). The B box has some sequence similarities to the −35 sequence of prokaryotes.

The promoterlike signals in eukaryotes serve as binding sites for various stimulatory proteins called **transcription factors**. Some of these factors are more or less ubiquitous, such as those recognizing the TATA and CCAAT boxes, while others are tissue- or developmental-stage–specific. Unlike the bacterial holoenzyme, eukaryotic RNA polymerases cannot recognize their cognate promoters by themselves in the absence of auxiliary protein factors. A larger transcription complex must therefore be involved. RNA polymerases isolated from eukaryotes are thus more like core enzymes than holoenzymes—all are capable of catalyzing the same polymerization reaction, but each class responds to a different set of transcription factors and promoterlike signals for the initiation of transcription. We will continue our discussion of transcription factors and their DNA-binding sites in Chapter 20, when we consider their role in eukaryotic gene regulation.

(a) mRNA

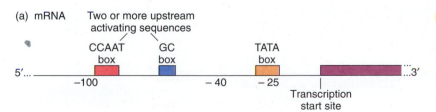

Two or more upstream activating sequences

CCAAT box

GC box

TATA box

5′...   −100   − 40   − 25

Transcription start site

...3′

(b) tRNA

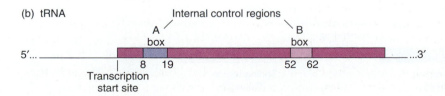

Internal control regions

A box

B box

5′...   8 19   52 62   ...3′

Transcription start site

▶ **FIGURE 16–6** General features of promoters for (a) mRNA and (b) tRNA synthesis (by RNA polymerases II and III, respectively) in eukaryotes.

The eukaryotic RNA polymerases may also require specific signals for the termination of transcription. Termination by RNA polymerase III has the simplest sequence requirement—a DNA segment containing a run of at least four T residues surrounded by GC-rich sequences. Little is known about the termination signals for RNA polymerase II. The transcripts of this enzyme have defined 3' ends, but they result from a site-specific endonucleolytic scission of the RNA chain rather than from a transcription termination event.

To Sum Up

1. Information flows from DNA to RNA in a process known as transcription. A single strand of DNA serves as a template for the synthesis of a complementary strand of RNA. This process is catalyzed by a DNA-dependent RNA polymerase. In bacteria, all three major classes of RNA (mRNA, rRNA, and tRNA) are synthesized by one RNA polymerase. In eukaryotes, different RNA polymerases synthesize different kinds of RNA. Like the synthesis of DNA, all RNA synthesis occurs in a 5' → 3' direction.

2. To initiate transcription, RNA polymerase selectively binds to DNA at a promoter region. Bacterial promoter regions have certain common base sequences that define the three components of a promoter: the −35 sequence, the −10 sequence, and the transcriptional start site, whose position is designated +1.

3. Transcription of eukaryotic genes also begins with recognition of a promoter region by an RNA polymerase. Some eukaryotic promoter sequences, such as those making up the Hogness and B boxes, are similar to sequences of prokaryotic promoters. In contrast to prokaryotes, however, eukaryotic RNA polymerases require specific proteins called transcription factors to recognize the appropriate promoters.

◆

▓ MESSENGER RNA AND THE GENETIC CODE

Messenger RNA (mRNA) carries the encoded instructions of structural genes to the ribosomes, where proteins are synthesized. Monocistronic mRNA molecules code for a single polypeptide while polycistronic mRNA codes for two or more polypeptide chains. Monocistronic messengers are the RNA products of single genes (the word *cistron* is a synonym for gene), and they are the predominant form of mRNA in eukaryotes. Polycistronic messengers are multigenic in origin—each is transcribed from two or more contiguous genes in the DNA—and they represent the major form of mRNA in prokaryotes. Both types of mRNA tend to have noncoding sequences at their ends—a leader sequence at the 5' end and a trailer sequence at the 3' end (▶ Figure 16–7). In addition, polycistronic mRNA usually contains noncoding sequences interspersed between the coding regions.

Each coding region in mRNA carries the information for the primary structure of a polypeptide chain in a sequence of code "words" called codons. Each codon is a specific combination of three adjacent nucleotides (for example, AGU) that specifies a particular amino acid in the polypeptide. Because three nucleotides compose each codon, the code is called a triplet code.

The codons that make up the genetic code are nonoverlapping, so each nucleotide participates in the specification of only one amino acid. The codons are read consecutively without punctuation and are translated into a linear sequence of amino acids. For example, if a coding region of mRNA has the nucleotide sequence AUGAGAUCU . . . , then AUG would code for the first amino acid in the polypeptide chain, AGA would code for the second amino acid, and so on. There is thus a linear correspondence between the nucleotide sequence and the amino acid sequence—a coding relationship known as colinearity.

Deciphering the Code

The genetic code was deciphered by work in several laboratories beginning in the 1950s. Researchers initially deduced the triplet nature of the code by comparing the number of naturally occurring amino acids with the number of code words possible for different numbers of nucleotides. Because there are 20 different kinds of amino acids but only four different bases, it is apparent that a codon cannot consist of a single nucleotide. A doublet code, in which two nucleotides would specify each amino acid, would also be in-

▶ **FIGURE 16–7** General features of mRNA structure. (a) Monocistronic mRNAs have a single coding region carrying the information of a single gene. (b) Polycistronic mRNAs have multiple coding regions transcribed from two or more adjacent genes. Both kinds of mRNA have noncoding regions at their ends: a leader sequence at the 5' end and a trailer sequence at the 3' end.

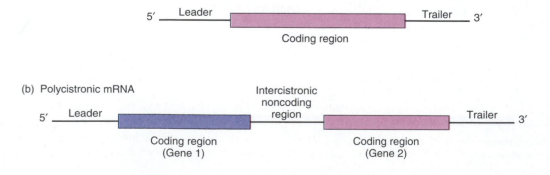

(a) Monocistronic mRNA

5' ——— Leader ——— [Coding region] ——— Trailer ——— 3'

(b) Polycistronic mRNA

5' ——— Leader ——— [Coding region (Gene 1)] ——— Intercistronic noncoding region ——— [Coding region (Gene 2)] ——— Trailer ——— 3'

adequate, since there would be only 16 (4 × 4) possible sequences. A triplet code, on the other hand, provides more than enough code words to specify all 20 amino acids since reading nucleotides in groups of three would give 64 (4 × 4 × 4) possible codons. Investigators working on the code therefore expected that it would be triplet.

The base composition of the codons was established in the early 1960s after the development of an in vitro system that permitted researchers to produce polypeptides under the direction of synthetic mRNAs. In the first experiments designed to crack the code, the synthetic mRNAs were synthesized with the enzyme polynucleotide phosphorylase. This enzyme catalyzes the formation of single-stranded RNA molecules from nucleoside diphosphates, and it has the interesting property of not using a template to specify the nucleotide sequence. The proportion of each base in the RNA molecules formed in the reaction therefore depends solely on its proportion in the reaction mixture of nucleoside diphosphates. Thus, if ADP is the only nucleoside diphosphate in the reaction mixture, the RNA made by polynucleotide phosphorylase will be a homopolymer composed only of A. If the reaction mixture consists of 75% ADP and 25% CDP, the RNA that is formed will be a heteropolymer that contains $\frac{3}{4}$ A and $\frac{1}{4}$ C. Notice that although we know the overall base composition of the RNA that is produced in this manner, we have no way of knowing the specific sequence of nucleotides.

The RNA codon for phenylalanine was the first to be identified using synthetic mRNA. This identification was made by M. Nirenberg and H. Matthei in 1961, when they discovered that the synthetic RNA polyuridylic acid (poly U) promoted the synthesis of only the polypeptide polyphenylalanine, despite the presence of all 20 amino acids in the reaction mixture. Nirenberg and Matthei correctly interpreted this result to mean that the RNA codon for phenylalanine must be UUU. Shortly after this discovery, they found that poly A and poly C yield polylysine and polyproline, respectively. Thus AAA is the codon for lysine, and CCC is the codon for proline.

Codons containing more than one kind of nucleotide were identified using mRNAs composed of two types of bases. The nucleotide sequences in the copolymers are essentially random, so it is possible to predict the frequency of each codon in the synthetic mRNA. By relating the frequencies of the various amino acids in the polypeptides that the RNA codes for to the expected frequencies of the various triplet sequences in the random RNA copolymer, it is possible to deduce the base compositions of the codons for all 20 amino acids. For example, a reaction mixture consisting of $\frac{3}{4}$ ADP and $\frac{1}{4}$ CDP should produce a random copolymer in which the codons AAC, ACA, and CAA each occur at a frequency of $(\frac{3}{4})(\frac{3}{4})(\frac{1}{4}) = \frac{9}{64}$. The occurrence of asparagine at about this frequency in the resulting polypeptide indicates that a codon consisting of 2 As and 1 C codes for asparagine.

Experiments with randomly ordered synthetic RNAs provided information about the base composition of the various codons, but they did not establish the actual sequence of the nucleotides within any codon, except for the trivial cases of AAA, CCC, GGG, and UUU. The subsequent development of other experimental procedures led to the assignment of base sequences within codons. One such procedure was the synthesis of polypeptides using artificial mRNAs of defined rather than random nucleotide sequence. For example, H. Khorana synthesized the RNA copolymer poly UG, which contains the alternating codon sequence −UGU−GUG−UGU−GUG−. Using this copolymer as an mRNA produced a polypeptide chain with the alternating amino acid sequence -cysteine-valine-cysteine-valine-. Since it was already known that the triplet UG_2 codes for valine but not for cysteine, this experiment showed that the codons for cysteine and valine must be UGU and GUG, respectively. Using this and other procedures, investigators soon established the code words for each of the 20 amino acids.

Example 16.2

Several of the first experiments designed to crack the genetic code were conducted in the laboratory of S. Ochoa. In one experiment reported by Ochoa's group in 1963, an artificial mRNA was synthesized by polynucleotide phosphorylase from a reaction mixture consisting of $\frac{5}{6}$ ADP and $\frac{1}{6}$ CDP. When used in a cell-free extract, this mRNA coded for the synthesis of a polypeptide with the following amino acid composition:

Amino Acid	Observed Frequency Relative to Lysine
lysine	100%
threonine	26%
asparagine	24%
glutamine	24%
proline	7%
histidine	6%

From these data, postulate codon assignments for each of the amino acids listed.

Solution: Since AAA is the most frequently occurring codon in the artificial mRNA, we will express the expected codon frequencies relative to AAA:

Possible Codons in mRNA		Expected Codon Frequency	Codon Frequency Relative to AAA
AAA		$(\frac{5}{6})^3 = \frac{125}{216}$	100%
AAC			20%
ACA	each	$(\frac{5}{6})^2(\frac{1}{6}) = \frac{25}{216}$	20%
CAA			20%
ACC			4%
CAC	each	$(\frac{5}{6})(\frac{1}{6})^2 = \frac{5}{216}$	4%
CCA			4%
CCC		$(\frac{1}{6})^3 = \frac{1}{216}$	0.8%

The codon AAA specifies lysine. This codon assignment was known before the experiment was conducted and prompted the investigators to express their data relative to lysine. Threonine, with a relative frequency of 26%, would appear to be coded for by one of the A_2C triplets (either AAC or ACA or CAA) and one of the AC_2 triplets (either ACC or CAC or CCA). Its expected relative frequency would then be 20% + 4% = 24%. Asparagine would be coded by another of the A_2C triplets, for an expected relative frequency of 20%. The final A_2C triplet would be assigned to glutamine, again giving a frequency of 20%. We would postulate that proline is coded for by another of the AC_2 triplets and by the CCC triplet, for an expected relative frequency of 4.8%. Finally, we would assign the remaining AC_2 triplet to histidine, for an expected relative frequency of 4%. Note that some of the actual frequencies must be inferred by looking at the group of amino acids produced as a whole. For example, we might guess that asparagine is coded for by one A_2C triplet and one AC_2 triplet, to give an expected frequency of 24% (the same as the observed frequency), but we would then be short one arrangement of triplets to account for the observed frequencies of proline and histidine.

Follow-Up Problem 16.2

Suppose that an artificial mRNA is synthesized using a reaction mixture containing equal amounts of ADP and UDP. What proportion of its codons would code for (a) isoleucine, (b) leucine, (c) phenylalanine, (d) tyrosine?

◆

Characteristics of the Code

The genetic code dictionary for RNA is given in ■ Table 16–3. It has several important features. First, of the 64 possible triplet codons, all but three (UAA, UAG, and UGA) code for specific amino acids. As we will see later, these three code words function as termination (or stop) codons during translation. They are found either singly or in combination at the end of a coding region, where they signal the completion of the gene message.

Second, the 61 remaining sense codons code for 20 different amino acids (an average of approximately 3 codons per amino acid). We therefore say that the genetic code is degenerate, meaning that more than one codon can specify the same amino acid. For example, six different code words specify leucine. There are also six codons for serine and six for arginine. Several amino acids have four codons each, one (isoleucine) has three codons, and several others have two. Only methionine and tryptophan are specified by a single codon each.

The degeneracy of the genetic code raises some interesting questions concerning its evolution. The code seems to have evolved in a way that provides a buffer against the effects of mutation. For instance, a nucleotide change in the third position of the codon GUU would not alter the amino acid sequence of the polypeptide, because all codons that start with GU (GUU, GUC, GUG, and GUA) specify the same amino acid. The same pattern occurs for a number of other codons. Furthermore, if we compare the genetic code with the amino acid structures given in Figure 6–2, we can see that amino acids with similar chemical properties often have codons that differ in only one nucleotide. Thus, even if a nucleotide change results in an amino acid change, the new amino acid may be similar enough to the original to conserve protein function.

A third major feature of the genetic code is that it is very widespread, but not quite universal. The same "standard" code has been found in all the prokaryotes and nuclear

■ **TABLE 16–3** The mRNA codon assignments.

First (5') Letter	Second Letter								Third (3') Letter
	U		C		A		G		
U	UUU UUC	Phe	UCU UCC	Ser	UAU UAC	Tyr	UGU UGC	Cys	U C
	UUA UUG	Leu	UCA UCG		UAA UAG	terminator terminator	UGA UGG	terminator Trp	A G
C	CUU CUC	Leu	CCU CCC	Pro	CAU CAC	His	CGU CGC	Arg	U C
	CUA CUG		CCA CCG		CAA CAG	Gln	CGA CGG		A G
A	AUU AUC	Ile	ACU ACC	Thr	AAU AAC	Asn	AGU AGC	Ser	U C
	AUA AUG	Met	ACA ACG		AAA AAG	Lys	AGA AGG	Arg	A G
G	GUU GUC	Val	GCU GCC	Ala	GAU GAC	Asp	GGU GGC	Gly	U C
	GUA GUG		GCA GCG		GAA GAG	Glu	GGA GGG		A G

genes of eukaryotes that have been studied, with the exception of a mycoplasma and certain species of protozoa. While exceptions to the code are rare in nonorganellar genomes and affect only the termination codons, several differences have been documented in the code utilized by mitochondria. For example, in mammalian mitochondria, AGA and AGG are stop codons, while AUA (normally a codon for isoleucine) codes for methionine, and UGA (normally a chain-terminating codon) codes for tryptophan. The basis of these (and other) differences in the genetic code of mitochondria is considered in Chapter 17.

Points to Ponder 16.1

There has been a great deal of speculation about the evolution of the genetic code. For example, one hypothesis is that there was an earlier code that was, in effect, a doublet code, with only the first two of the three nucleotide positions having a distinct meaning. Study the genetic code in Table 16–3, especially noting its degeneracy. Does the code have features that suggest a possible evolutionary past as a doublet code? Also note that the same code is used by all prokaryotes and the nuclear genes of nearly all eukaryotic organisms. What does this suggest to you about the evolutionary origin of the genetic code?

To Sum Up

1. Messenger RNA carries the message of the structural genes. Prokaryotic mRNAs are usually polycistronic, while mRNAs of eukaryotes are usually monocistronic. Noncoding sequences (the leader and trailer) bracket the coding regions in both prokaryotes and eukaryotes.

2. The coding regions of mRNA molecules consist of triplet codons. Each codon specifies a particular amino acid at a residue site in the polypeptide. The relationship between an mRNA molecule and the polypeptide for which it codes is colinear, that is, a linear sequence of codons on the mRNA molecule corresponds to a linear sequence of amino acids in its polypeptide product.

3. The genetic code was deciphered using in vitro systems that produced polypeptide chains under the direction of synthetic mRNAs. Initially, the synthetic mRNAs were polyribonucleotides of random base sequence; later, synthetic mRNAs of defined sequence were used. These experiments established that the code is degenerate—usually more than one codon specifies each amino acid. Bacteria, viruses, and nearly all nuclear genes of eukaryotes use the same genetic code.

GENE-PROTEIN RELATIONS AND RNA PROCESSING

Colinearity of Gene and Protein

We have noted that the sequence of codons in mRNA is colinear with the sequence of amino acids in the polypeptide. Since mRNA is transcribed from a DNA template, colinearity should also exist between the polypeptide and the template strand of the structural gene. Research has shown that this is indeed the case in prokaryotes.

The first strong evidence for gene-protein colinearity came from studies on the *trpA* gene in *E. coli* (see Chapter 11) by Yanofsky and others. Recall that the *trpA* gene codes for one of the polypeptides (peptide α) that make up the *E. coli* enzyme tryptophan synthetase. Yanofsky's group worked with a collection of auxotrophic mutants that produced an inactive form of this enzyme. Mapping studies revealed that these mutants differed from one another in the precise positions of their mutation sites within the *trpA* gene. Furthermore, sequence analysis of each mutant polypeptide showed that many of the mutants had amino acid changes at different residue locations. Comparison of the linkage map and the peptide sequences revealed a close correlation between the sequence of mutant sites on the linkage map (reflecting the sequence of nucleotides in the gene) and the amino acid substitutions in the polypeptide chain (▶ Figure 16–8). Not only did the order of the positions match, but the relative distances between these positions were also very similar, thus confirming the idea of a direct correspondence between the sequence of a gene and its protein product.

▶ **FIGURE 16–8** Colinearity between the α peptide of tryptophan synthetase in *E. coli* and the gene that codes for this enzyme. There is a linear correlation between the mutation sites mapped along the gene and the relative positions of the amino acid changes in the peptide. *Source:* C. Yanofsky, "Gene structure and protein structure," *Scientific American* 216 (1967): 80–94. Copyright © 1967 by Scientific American. All rights reserved. Used with permission.

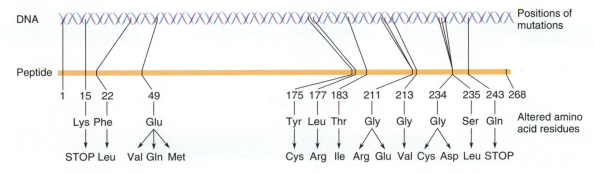

Overlapping Genes

So far, we have drawn a comparatively simple picture of a structural gene as a unique segment of DNA that contains exactly three times as many nucleotide pairs between the beginning of its start and stop codons as there are amino acids in the polypeptide chain it codes for. This view, which was first deduced from genetic studies, has been confirmed for a number of bacterial genes through DNA, RNA, and protein sequencing. However, a more complex picture began to emerge in the 1970s, when other studies demonstrated that genes can depart significantly from this simple paradigm both in their structure and in their coding relationships with other genes (▶ Figure 16–9). One interesting departure is shown by the **overlapping genes** (Figure 16–9b–d), in which the same DNA sequence is used by more than one gene.

The existence of overlapping genes was first discovered in the *E. coli* phage φX174. This phage has long been a favorite subject of experiments in molecular genetics because of its small size, the ease with which it can be manipulated, and the fact that its DNA can be replicated in a test tube. The chromosome of this phage is a single-stranded DNA molecule consisting of only 5386 nucleotides. Assuming that an average gene in this phage is 1000 nucleotides long (a typical size for a gene in bacteria) this chromosome should have enough information to encode 5 or 6 proteins. However, studies show that this phage makes 11 proteins containing a combined total of more than 2300 amino acids (which would require more than 6900 nucleotides). In brief, the φX174 chromosome appears to contain an insufficient amount of genetic material for all of its genes!

This paradox was finally resolved when the complete nucleotide sequence of the phage DNA was determined. When the nucleotide sequence was compared with the amino acid sequences of the proteins, two of the genes (*B* and *E*) were found to lie entirely within the boundaries of other genes, and one gene (*K*) overlapped three others (▶ Figure 16–10a). In this case, each pair of overlapping genes can specify entirely different proteins by having different reading frames, as illustrated in Figure 16–10b for the *D* and *E* genes. The overlapping *A* and *A** genes, on the other hand, have the same reading frame; they represent a simpler situation in which one gene (gene *A** in this case) is a shorter version of the other.

Gene overlap to one degree or another has been observed in the DNA of a number of viruses, cellular organisms, and mitochondria. It thus raises many interesting evolutionary questions. For example, a single nucleotide change in an overlapping sequence will cause both genes in the sequence to mutate. If both mutant genes are advantageous or both deleterious, the effect of the change is clear. However, if one change is beneficial and the other is harmful, how will selection proceed?

Split Genes

In most prokaryotes, there is a one-to-one relationship between the nucleotide sequence of a structural gene and the complementary sequence of its mRNA product. For this

---

▶ **FIGURE 16–9** Structure and coding relationships of genes. (a) The nonoverlapping prokaryotic gene. (b–d) Overlapping genes; in (c), different genes share a common boundary, while in (d) one gene is completely within the boundaries of another. (e) The split or interrupted gene, common in eukaryotes. Split genes contain one to several noncoding sequences between the start and stop codons. The crosshatched areas represent noncoding intergenic spacers.

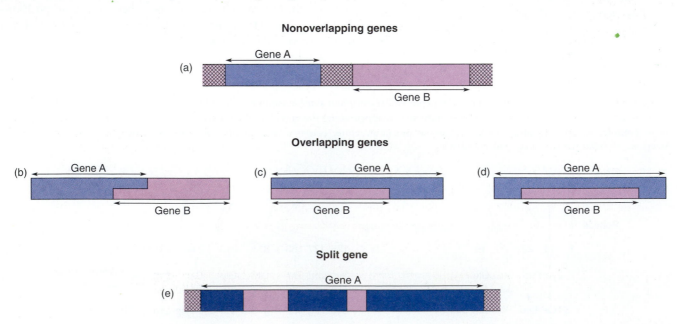

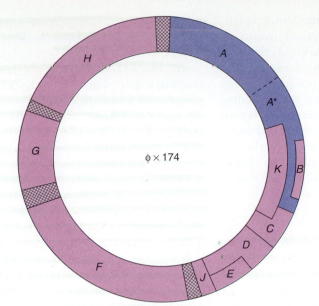

(a)

φ×174

► **FIGURE 16–10** Overlapping genes in phage φX174. (a) Genetic map of φX174, showing boundaries of protein products. Crosshatched regions are untranslated spacers. (b) The nucleotide and amino acid sequences of the D and E genes and their respective polypeptides. The nucleotides are numbered from the start of the D gene. Gene E begins at nucleotide 179, with its codon sequence offset from that of gene D. The D and E genes thus code for different amino acid sequences. *Source:* C. Yanofsky, "The nucleotide sequence of a viral DNA," *Scientific American* 237 (1977): 54–67. Copyright © 1977 by Scientific American. All rights reserved. Used with permission.

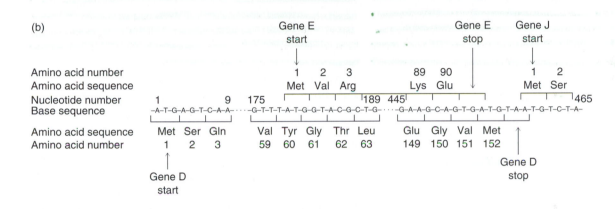

(b)

reason, prokaryotes do not ordinarily modify their mRNA after transcription; they can use the RNA transcript to synthesize proteins immediately, even while it is being formed. This is not generally the case in eukaryotic cells, however. Many (probably most) genes in eukaryotes are discontinuous or split (Figure 16–9e). In these organisms, the DNA that codes for a single polypeptide is interrupted at various points by noncoding sequences that must be removed from the RNA transcript before translation can occur.

The interrupted nature of a eukaryotic structural gene is illustrated by the ovalbumin gene shown in ► Figure 16–11. This gene is responsible for producing the major egg-white protein in chickens. It is composed of eight segments that

are represented in its mRNA product, separated by seven segments that are missing from the mRNA. The segments that specify mRNA are called **exons**, and the noncoding intervening segments are called **introns**. The exons in the ovalbumin gene vary considerably in length, ranging from 47 to 1043 nucleotides, and specify the untranslated 5′ and 3′ sequences in addition to the protein-coding region in the mRNA. Some eukaryotic genes are split into as many as 50 or more exons. It is not uncommon for a coding region containing 1000 nucleotides to be split into exons that are distributed over a length of DNA ten times that size.

Visual evidence of noncoding intervening sequences in eukaryotic structural genes can be obtained by forming

► **FIGURE 16–11** The interrupted nature of the ovalbumin gene. The gene is composed of eight segments called exons (*L*, 1–7) that are represented in its mRNA. The size of the introns (intervening sequences) varies from 251 bp (intron *B*) to about 1600 bp (intron *G*).

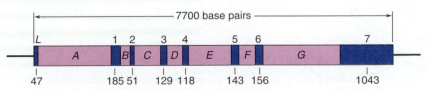

RNA Synthesis and the Genetic Code   379

DNA–RNA hybrids between mRNA and its DNA template. If DNA containing a structural gene is partially melted in the presence of its RNA transcript (pre-mRNA), the RNA will hybridize with its complementary sequence in the template strand of the DNA, displacing the noncoding strand in the entire region of the gene and leaving the remaining DNA intact (▶ Figure 16–12a). But if mature mRNA is used in place of the pre-mRNA transcript in this experiment, electron micrographs of these hybrids reveal the existence of unhybridized DNA loops that correspond to the introns within the interrupted genes (Figure 16–12b).

Modification and Splicing of Pre-mRNA

The intervening sequences that are missing from mRNA are removed from the RNA chain by a process known as **RNA splicing** sometime after the initial transcript is formed. Splicing is one of several molecular changes involved in the maturation of the RNA chain. These changes are collectively referred to as **posttranscriptional modification** or simply **processing**. Since many eukaryotic structural genes are split, transcription normally produces long RNA molecules called **heterogeneous nuclear RNA (hnRNA)**. Each molecule of hnRNA contains all the alternating exon and intron segments of the original template DNA strand in a complementary base sequence. The hnRNA (pre-mRNA) is then processed into mRNA.

The reactions involved in the processing of hnRNA are summarized in ▶ Figure 16–13. One of the first steps is the addition of a 7-methylguanosine residue or **cap** to the 5′ end of the molecule. The cap is attached to the RNA chain by an unusual 5′—5′ triphosphate linkage (▶ Figure 16–14). The two adjacent riboses may also be methylated at their 2′ positions. Capping occurs through a series of enzymatic reactions shortly after the initiation of transcription, and it forms an important structural feature that aids in the binding of eukaryotic mRNA to ribosomes during protein synthesis. The 5′ cap may also help to protect eukaryotic mRNA from digestion by 5′ exonucleases within the cell.

In conjunction with the termination of transcription, the distal portion of the pre-mRNA transcript is cleaved, and some 200–250 adenylate residues are added to the newly generated 3′ end to form a **polyadenylate (poly A) tail**. Both the cleavage and the polyadenylation steps are catalyzed by ribonucleoprotein complexes and occur at a site 10–30 residues downstream from a highly conserved sequence in the RNA chain, AAUAAA. This sequence is the predominant element for signaling the location of the poly A site. The function of the poly A tail is still somewhat of an enigma. It does not appear to be essential for the transport of mRNA from the nucleus or its subsequent translation, since some eukaryotic mRNAs (some mRNAs coding for histones, for example) do not have a poly A terminus. Regardless of its function, the poly A tail is a convenient experimental feature. The fact that the poly A tail is unique to mRNA transcripts has been used by researchers to isolate mRNA from the remaining cellular RNA by hybridizing it to chromatographic columns containing immobilized poly T chains (see Chapter 18).

Polyadenylation is followed by the splicing reactions that convert the modified pre-mRNA transcript into the final mature mRNA. Splicing involves the removal of the intron segments from hnRNA and the end-to-end joining of exons in the proper arrangement to code for a specific polypeptide chain. When RNA is experimentally extracted from nuclei at different times, the pre-mRNA transcripts appear to change in length, indicating that the splicing is done in a step-by-step manner. These changes occur in association with multicomponent splicing complexes called **spliceosomes**, which are ribonucleoprotein complexes that contain the various factors involved in splicing. In addition to proteins, these factors include small RNA molecules less than 300 nucleotides in length, which are called **snRNA** (for small nuclear RNA). In splicing, the role of the snRNAs, which normally exist in ribonucleoprotein particles called **snRNPs** ("snurps"), is to recognize certain regions in the intron-containing transcript. These regions include three consensus sequences common to all introns—the 5′ and 3′ splice sites at the ends of each intron and an internal site that is called a branch point for reasons that we will see shortly. Studies of a large number of splice junctions reveal

▶ **FIGURE 16–12** Demonstrating the existence of introns by DNA–RNA hybridization.

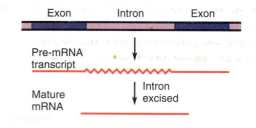

Exon Intron Exon

Pre-mRNA
transcript

Intron
excised

Mature
mRNA

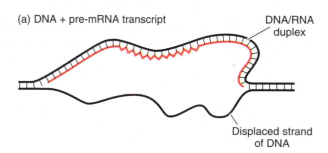

(a) DNA + pre-mRNA transcript

DNA/RNA
duplex

Displaced strand
of DNA

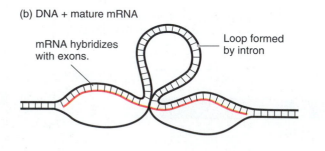

(b) DNA + mature mRNA

mRNA hybridizes
with exons.

Loop formed
by intron

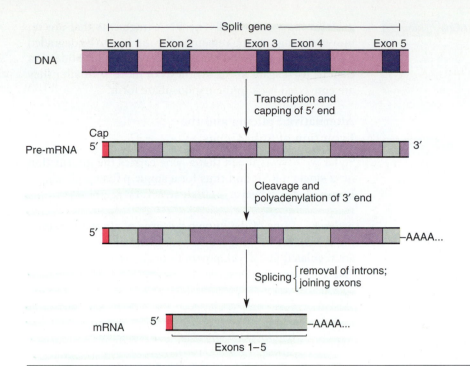

▶ **FIGURE 16–13** Steps involved in the processing of the pre-mRNA from a split gene. The pre-mRNA transcript is capped at the 5′ end, and a poly A tail is added to the 3′ end. The intron regions are then excised from the transcript, and the exons are spliced together to produce the mature mRNA.

---

that the 5′ end of each intron almost always begins with GU and the 3′ end almost always terminates with AG, preceded by a run of pyrimidines. Moreover, the branch point nearly always involves a specific adenine-containing residue. These features are:

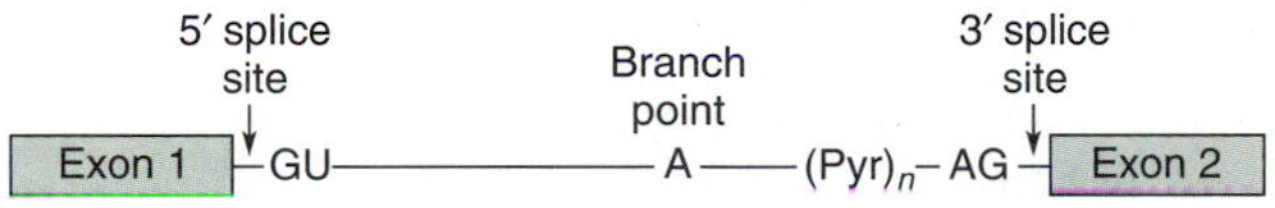

5′ splice site		Branch point		3′ splice site	
Exon 1	⌐GU———	— A ——	(Pyr)$_n$– AG ⌐	Exon 2	

The role of these regions in the pre-mRNA splicing pathway is shown in ▶ Figure 16–15. The intron is excised by two **transesterification reactions**. The first step involves cleavage at the 5′ splice site and the subsequent formation of a phosphodiester bond between the 5′-terminal phosphate of the intron and the 2′-hydroxyl of the adenylate residue at the branch point. This 2′, 5′-phosphodiester bond at the branch site produces a loop structure commonly referred to as a **lariat intermediate**. The second step is cleav-

---

▶ **FIGURE 16–14** Structure of the 5′ methylated cap of eukaryotic mRNA. A 7-methylguanosine is attached by a triphosphate linkage between its 5′-OH and the 5′-OH of the transcript terminus. The 2′-OH groups of the first two residues in the transcript may also be methylated, as is shown in this case.

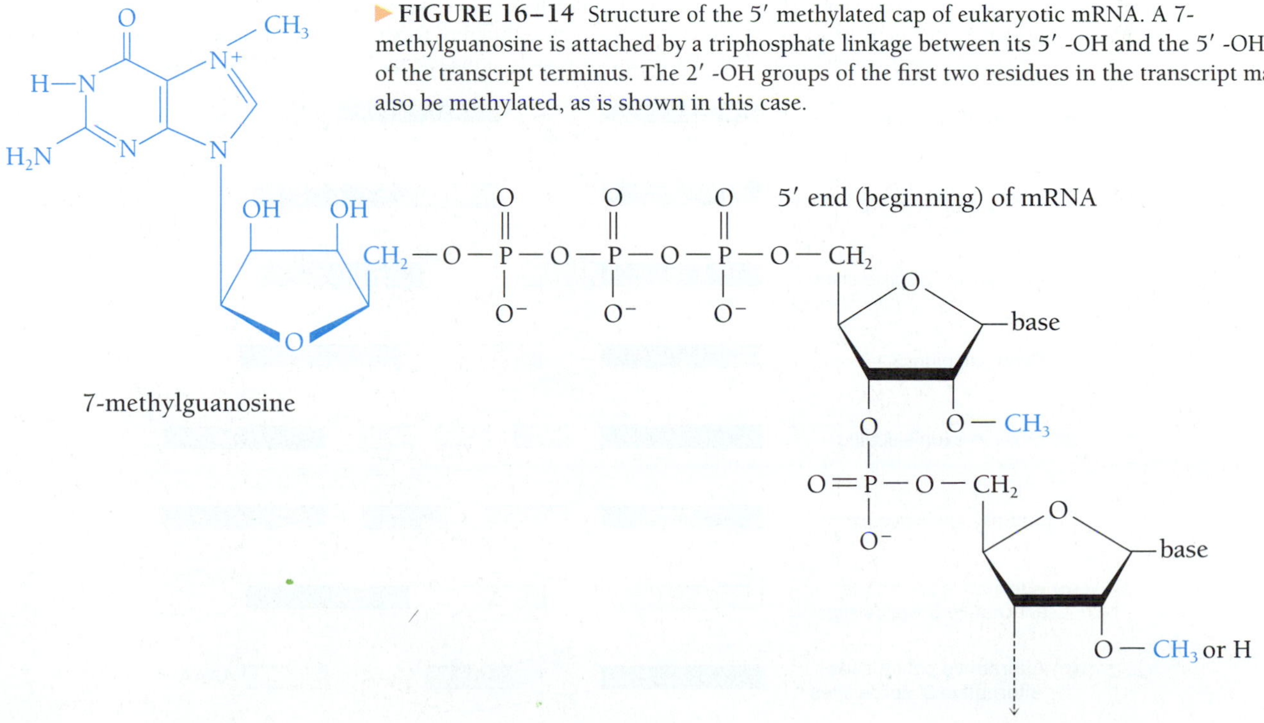

7-methylguanosine

5′ end (beginning) of mRNA

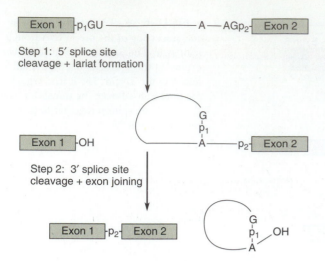

Exon 1 ⊣p₁GU ─────── A ── AGp₂⊢ Exon 2

Step 1: 5' splice site
cleavage + lariat formation

Exon 1 ⊢OH

Step 2: 3' splice site
cleavage + exon joining

Exon 1 ⊢p₂⊣ Exon 2

▶ **FIGURE 16–15** The pre-mRNA splicing pathway. In this schematic representation, boxes represent exons, the line is an intron, and GU, AG, and A indicate the 5' splice site, the 3' splice site, and the branch point, respectively; p indicates a phosphodiester bond.

age at the 3' splice site and the joining of the two exons by another phosphodiester bond. The result is the excision of the intact intron in a lariat configuration.

All the reactions involved in the processing of hnRNA occur in the nucleus of the cell. The RNA molecule is thus in mature mRNA form before it enters the cytoplasm, where

proteins are synthesized. The RNA fragments that are removed during splicing remain in the nucleus to be degraded by enzymes into ribonucleotides. These ribonucleotides can then be used again in the synthesis of more RNA after they are converted back to the triphosphate form.

Alternative Splicing and the Evolution of Split Genes

So far we have assumed that a gene carries the information for a single mRNA and thus for a single polypeptide chain. In eukaryotes, however, a single structural gene can produce more than one kind of mRNA through alternative splicing pathways. Some of the known patterns of alternative splicing are shown in ▶ Figure 16–16; many of these pathways are regulated in a developmental- or tissue-specific manner (see Chapter 20). Observe that differential recognition of splice sites makes it possible for particular exons to be included in some mRNA chains but not in others: one transcript's exon can be another transcript's intron, thus blurring the conventional distinction between exons and introns.

One of the functions of alternative splicing is to produce multiple proteins from the information encoded in a single gene. An interesting example of alternative splicing involves the gene for calcitonin in the rat (▶ Figure 16–17). In the thyroid, three of the four potential coding regions are spliced together in the mRNA to produce calcitonin, a protein that controls blood calcium levels. In the brain, however, the 3' calcitonin-coding exon is replaced by another

▶ **FIGURE 16–16** Patterns of RNA splicing. The dashed lines represent alternative pathways for the splicing of RNA sequences. (a) Simple splice/don't splice alternatives. (b) and (c) The possible alternatives if different 3' or 5' splice sites are recognized; a partial or entire exon sequence could be deleted from the mRNA. (d) and (e) Similar results occur if one or more exons are skipped or (f) if the inclusion of exons into mRNA is mutually exclusive. In the last two cases additional nonsplicing events also occur: (g) alternative 5' ends with alternative 5' splice sites, and (h) alternative 3' ends with alternative 3' splice sites. Alternative splicing mechanisms may be an important genetic regulatory mechanism affecting protein production in different tissues. *Source:* After M. McKeown, "Alternative mRNA splicing," *Annual Review of Cell Biology* 8 (1992): 136. Reproduced with permission from the *Annual Review of Cell Biology*, Volume 8, © 1992 by Annual Reviews Inc.

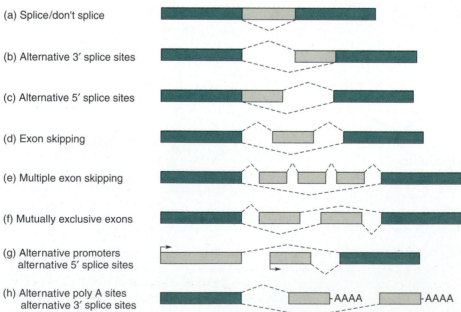

(a) Splice/don't splice

(b) Alternative 3' splice sites

(c) Alternative 5' splice sites

(d) Exon skipping

(e) Multiple exon skipping

(f) Mutually exclusive exons

(g) Alternative promoters
alternative 5' splice sites

(h) Alternative poly A sites
alternative 3' splice sites

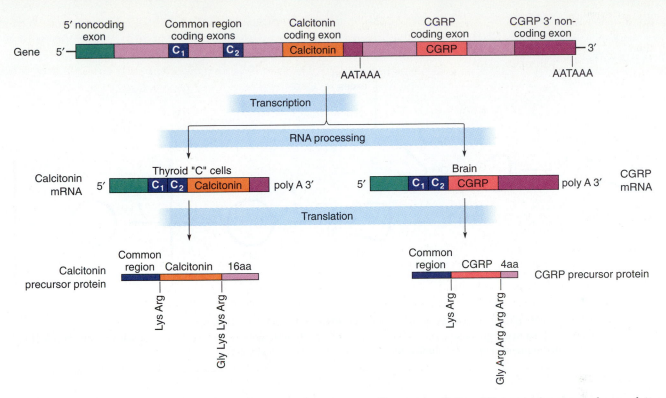

► FIGURE 16–17 Different splicing patterns in the rat calcitonin gene. Alternative splicing of 3′-terminal exons produces calcitonin precursor protein in thyroid cells and CGRP precursor protein in the brain. The precursor proteins are further processed by proteolytic cleavage at the basic residues indicated to yield the final hormone products: calcitonin in the thyroid and a neuropeptide called calcitonin gene-related peptide (CGRP) in the brain. *Source:* J. D. Watson, et al. *Molecular Biology of the Gene* (Menlo Park, CA: Benjamin Cummings, 1987), p. 721. Used with permission.

exon through alternative splicing, and the mRNA then specifies the structure of a neuropeptide.

The split gene arrangement that is so essential for alternative splicing may also be responsible for the rapid rise in protein diversity that has occurred over evolutionary time. It is becoming increasingly clear that DNA is subject to a surprising degree of movement within the genome, with small fragments of DNA being excised from one location and inserted into another. Such rearrangements would be harmful if they disturbed an existing gene whose products are needed, but they could be beneficial if they formed a new (additional) combination of exons and created a new gene with a different function. Evidence of the importance of exon shuffling in evolution comes from the remarkable structural similarity observed in parts of proteins that function in a similar manner. These so-called **functional domains** are coded for by specific exon sequences that have been largely conserved during evolution. Thus the joining of exons that code for functional domains could result in the rather rapid evolution of new genes.

Self-Splicing RNA: RNA as a Catalyst

Studies on RNA processing have shown that a few RNAs can undergo splicing on their own, in the absence of any protein or spliceosome complex. These results were surprising at first, since it had generally been expected that the removal of introns, like other cellular reactions, needed to be catalyzed by proteins. It was later demonstrated, however, that the process of self-splicing is **autocatalytic**. The catalytic activity needed for intron removal is provided by the RNA molecules themselves. Since RNA molecules that have this catalytic activity behave like enzymes in a chemical reaction, they are referred to as **ribozymes**.

Self-splicing was discovered in a precursor of ribosomal RNA in the ciliated protozoan *Tetrahymena*. In this organism, pre-rRNA processing includes the excision of a 413-nucleotide intron to form the mature rRNA molecule. The removal of the intron proceeds through a series of transesterification reactions that requires only the 3′-hydroxyl group of a free guanosine nucleoside to initiate the process. The self-splicing reactions involving the intron are shown in ► Figure 16–18. Note that once the intron is released, it undergoes two more rounds of self-splicing through repeated autocyclization and hydrolysis. The result is the formation of two small oligonucleotide fragments (one containing the initiating guanosine residue) and a 395-nucleotide intron core, which still has the catalytic site of the RNA. Studies on the intron core revealed that it could still function as both an RNA polymerase and a ribonuclease by joining and cleaving shorter oligonucleotide fragments in a specific manner. Moreover, it could catalyze these reactions without being consumed in the process. The intron core thus exhibits all the major catalytic properties that we normally associate with enzymes.

Since the discovery of self-splicing in pre-rRNA introns,

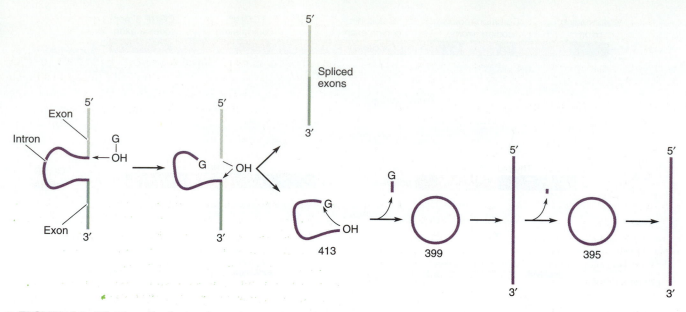

► **FIGURE 16–18** The self-splicing of *Tetrahymena thermophila* rRNA. A 413-nucleotide intron is excised through a transesterification reaction requiring guanosine (G) as a cofactor. The intron undergoes two more rounds of self-splicing, producing two small oligonucleotide fragments and a 395-nucleotide core that is still catalytically active.

self-splicing introns have been found in the RNAs of some bacteriophages as well as in an assortment of eukaryotic RNAs located principally in the mitochondria of fungi and yeast and the chloroplasts of unicellular algae. These self-splicing introns fall into two main categories (► Figure 16–19): group I introns, which require a free guanosine molecule to initiate splicing, and group II introns, which excise themselves in the absence of a guanosine co-factor. Group II introns differ from group I introns in another important respect: after self-splicing they are released in a lariat configuration by a process that resembles the excision of hnRNA introns. Because of the similarities in their splicing mechanisms, the spliceosome-catalyzed splicing of hnRNA introns is thought to have evolved from the RNA-catalyzed self-splicing of group II introns. Group II splicing may thus represent an evolutionary intermediate between

► **FIGURE 16–19** Groups of self-splicing introns. Splicing of group I introns is initiated by a guanosine nucleoside. Group II introns, which do not require guanosine, produce a lariat-shaped intermediate.

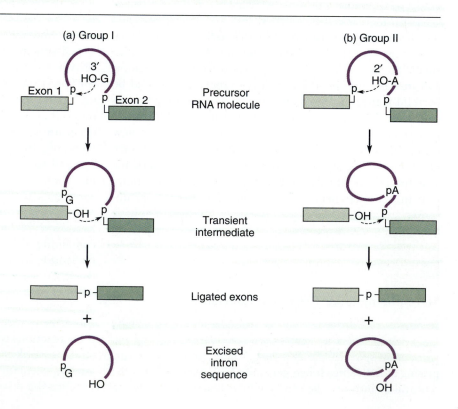

group I splicing (as seen in the pre-rRNA introns of *Tetrahymena*) and the splicing of nuclear mRNA precursors.

To Sum Up

1. In prokaryotes, the gene and the polypeptide it encodes are co-linear. In eukaryotes, however, many (if not most) genes are split into alternating coding (exon) and noncoding (intron) regions. A gene in a eukaryote may therefore extend over a length of DNA that is many times longer than the length needed for coding. Intron regions are cleaved out of the pre-mRNA, and the exons are spliced together.

2. In addition to removing introns and splicing together exons, pre-mRNA processing involves the addition of a cap to the 5′ end of the molecule and a poly A tail to its 3′ end. Spliceosomes are nuclear ribonucleoprotein complexes that catalyze the pre-mRNA splicing reactions. These reactions involve the recognition of splice sites at the 5′ and 3′ ends of each intervening region and a two-step transesterification process that excises the intron in a lariat configuration.

3. Some RNAs can undergo self-splicing—the removal of introns and splicing together of exons without the involvement of any protein or spliceosome complex. In these cases, the RNA itself contains the catalytic site for splicing. RNA molecules that act as enzymes are called ribozymes.

RNA-DEPENDENT RNA SYNTHESIS AND REVERSE TRANSCRIPTION

Information transfer can also proceed from RNA to RNA and from RNA to DNA, but these transfers occur on a much more limited scale than transfers from DNA to RNA. In the remainder of this chapter, we will consider the special cases in which RNA, rather than DNA, serves as the template for the production of RNA or DNA transcripts.

Replication Strategies of RNA Viruses

The RNA viruses include a heterogeneous group of infectious particles that contain RNA rather than DNA as their genetic material. Because of their RNA genomes, these viruses cannot utilize the host's DNA-directed replication and transcription systems; instead, they must encode their own replicative enzymes, in some cases even carrying the enzymes with them when invading the host cell. A list of some of the better-known RNA viruses is given in ■ Table 16–4.

▶ Figure 16–20 outlines the different replication strategies used by RNA viruses. Notice that the precise replication scheme depends on whether the RNA genome is single-stranded or double-stranded and on its relationship to its mRNA. In single-stranded RNA viruses (which constitute

■ **TABLE 16–4** Examples of RNA viruses.

	Type of Virus	Structure of Virion
	Animal viruses	
	(+) RNA viruses	
	polio virus	naked, icosahedral
	rubella virus	enveloped, icosahedral
	common cold virus	enveloped, helical
	(−) RNA viruses	
	rabies virus	enveloped, helical
	mumps virus	enveloped, helical
	influenza virus	enveloped, helical
Influenza virus	**dsRNA viruses**	
	reovirus	naked, icosahedral
	Retrovirus	
	HIV (AIDS) virus	enveloped, icosahedral
	Plant viruses	
	(+) RNA viruses	
	cucumber mosaic virus	icosahedral
Tobacco mosaic virus	tobacco mosaic virus	helical (rod-shaped)
	Bacteriophages	
	(+) RNA viruses	
	Qβ	icosahedral
	dsRNA viruses	
A bacteriophage	φ6	enveloped, icosahedral

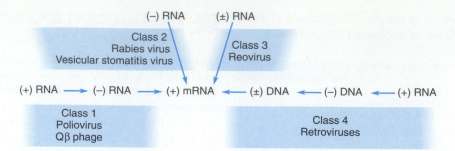

Class 2
Rabies virus
Vesicular stomatitis virus

Class 3
Reovirus

(+) RNA ⟶ (−) RNA ⟶ (+) mRNA ⟵ (±) DNA ⟵ (−) DNA ⟵ (+) RNA

Class 1
Poliovirus
Qβ phage

Class 4
Retroviruses

▶ **FIGURE 16–20** Modes of replication of RNA viruses. The + and − notations refer to the nature of the single-stranded viral genome with respect to the polarity and sequence of the viral mRNA. Viral + strand RNA is the same as the viral mRNA, and viral − strand RNA is the complement of the mRNA. Four classes of RNA viruses can be defined on the basis of their replication strategy. Class 1 and class 4 viral genomes are + stranded. In class 1 viruses, either the viral genome itself serves as mRNA, or mRNA is produced from a − strand RNA that is synthesized by an RNA replicase. In class 4 viruses, mRNA is produced after the virion RNA serves as a template for the synthesis of − strand DNA, which is followed by formation of a DNA duplex. Class 2 viral genomes are − strands that serve directly as templates for the synthesis of mRNA. Class 3 viruses are composed of double-stranded RNA whose − strand is the template for mRNA production.

by far the largest proportion of RNA viruses), the virion chromosome is either a **plus (+) strand** or a **minus (−) strand** in reference to mRNA. The viral RNA is a + strand if it has the same polarity and sequence as mRNA and can be translated directly into viral proteins within the host. Such chromosomes in animal viruses typically have a poly A tail as well as other features needed for translation in the eukaryotic cell. In order to replicate, some + strand RNA genomes encode the structure of an RNA-dependent RNA polymerase called **RNA replicase**, which is produced by translation shortly after infection occurs. The replicase enzyme copies the viral RNA to make − strands, which then serve as templates for + strand synthesis.

The replication of a + strand RNA virus is illustrated in ▶ Figure 16–21, which shows the life cycle of phage Qβ. Note that translation occurs immediately after entry of the phage RNA, forming the polypeptide product of the replicase (*rep*) gene. Once an active RNA replicase is formed (in Qβ, the active replicase consists of one replicase polypeptide combined with three host proteins—the ribosomal protein S1 and two elongation factors for protein synthesis, EF-Tu and EF-Ts), it participates in the 5′ to 3′ synthesis of RNA, a two-step process in which a − strand is first synthesized from the + strand and then copied into more + strands. Observe that no double-stranded replicative form is produced during RNA replication; the product RNA is hydrogen-bonded to the template only in the region of polymerase activity.

If the viral genome is a − strand (class 2), the RNA is opposite in polarity to mRNA, and the infecting genome must first be copied into a complementary + strand in order to function as a messenger during translation. Since cells are normally incapable of this conversion, − strand viruses carry an RNA-dependent RNA polymerase (in this case called **transcriptase**) in association with their chromosomes. The transcriptase catalyzes the synthesis of + strands on the − strand template. This enzyme is encoded by the virus and packaged along with the genome when virions are assembled in the preceding cycle.

▶ **FIGURE 16–21** The life cycle of the RNA phage Qβ.

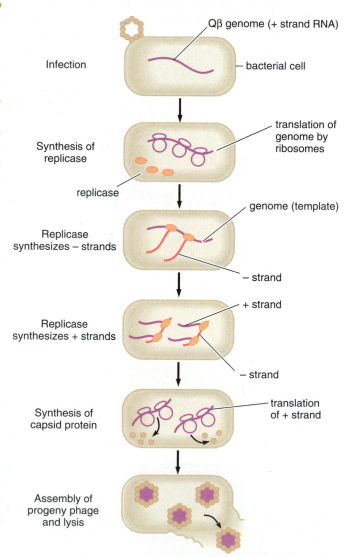

Qβ genome (+ strand RNA)

Infection — bacterial cell

Synthesis of replicase — translation of genome by ribosomes

replicase

Replicase synthesizes − strands — genome (template)

− strand

Replicase synthesizes + strands — + strand

− strand

Synthesis of capsid protein — translation of + strand

Assembly of progeny phage and lysis

Viruses with double-stranded (ds) RNA have genomes composed of both + and − strands, but like − strand RNA, these viruses must also carry a transcriptase enzyme into the cell during infection. Double-stranded RNA is inactive as a messenger, so the transcriptase is needed for transcription of the − strand of the duplex molecule into mRNA. The + strands that are formed by this process also function later as replication intermediates, serving as templates for the production of − strands to produce the double-stranded virion chromosomes.

Retroviruses and Reverse Transcription

Another mode of replication is exhibited by the + strand RNA viruses known as **retroviruses**. Retroviruses include many **oncogenic** (cancer-causing) viruses in animals as well as the virus that causes acquired immunodeficiency syndrome (AIDS) in humans. Unlike other viruses, each retrovirus virion contains two identical RNA molecules. Each RNA molecule normally has three distinct protein-coding regions: *gag*, which encodes the structural proteins of the nucleocapsid, *pol*, which encodes the catalytic proteins required for replication and integration, and *env*, which encodes the envelope glycoproteins (▶ Figure 16−22). In ad-

| R | U_5 | PBS | gag | pol | env | U_3 | R | $-A_n$ |

▶ **FIGURE 16−22** Genetic makeup of retroviral RNA. Three protein-coding regions—*gag, pol,* and *env*—are essential for productive infection. Untranslated repeated (R) sequences are needed for dimerization of RNA genomes, and unique (U_3 and U_5) sequences at the 5′ and 3′ ends contain signals for the promotion and regulation of virus replication. PBS is the primer binding site.

dition, each RNA molecule has untranslated repeated (R) and unique (U_3 and U_5) sequences at its 3′ and 5′ ends. The untranslated terminal sequences contain signals for the promotion and regulation of virus replication.

The retroviruses derive their name from their ability to form a DNA copy of their RNA chromosome in an infected cell by using an enzyme called **reverse transcriptase**. This enzyme, which is brought into the cell by the infecting virus, is capable of catalyzing three different reactions: RNA-dependent DNA synthesis, RNA degradation, and DNA-dependent DNA synthesis (▶ Figure 16−23). The enzyme

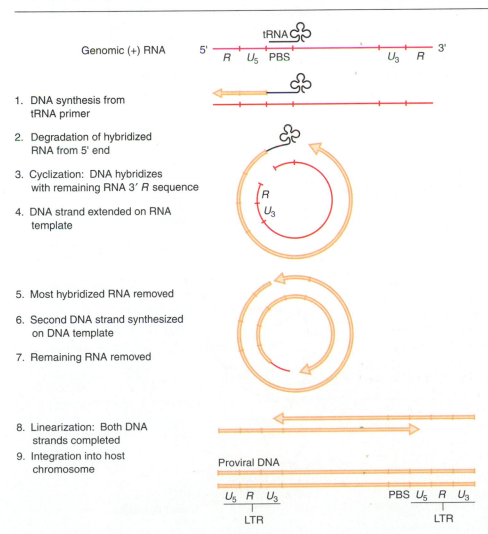

Genomic (+) RNA

1. DNA synthesis from tRNA primer

2. Degradation of hybridized RNA from 5' end

3. Cyclization: DNA hybridizes with remaining RNA 3′ R sequence

4. DNA strand extended on RNA template

5. Most hybridized RNA removed

6. Second DNA strand synthesized on DNA template

7. Remaining RNA removed

8. Linearization: Both DNA strands completed

9. Integration into host chromosome

Proviral DNA

▶ **FIGURE 16−23** Conversion of retroviral RNA into double-stranded DNA by the enzyme reverse transcriptase. A tRNA attached to the PBS site serves as the primer. LTR denotes the long terminal repeat at the ends of the proviral DNA.

first carries out the 5' to 3' synthesis of DNA on the viral RNA template, forming an RNA–DNA hybrid intermediate. Acting as a primer in this reaction is a molecule of tryptophan tRNA that was packaged with the viral genome in the preceding infection cycle. The enzyme then degrades the viral RNA, temporarily leaving a segment to serve as a primer for the synthesis of the complementary DNA strand. The resulting DNA duplex molecule is longer than the viral genome due to the creation of **long terminal repeat** (**LTR**) sequences (U_3-R-U_5 . . . U_3-R-U_5) at its ends.

The life cycle of a retrovirus known as Rous sarcoma virus is shown in ▶ Figure 16–24. This virus, like other retroviruses, reproduces by inserting a double-stranded DNA copy of its RNA genome into the host chromosome to form a **provirus**. While in the integrated state, the provirus replicates in synchrony with the cell's DNA and expresses itself using the transcriptional apparatus of its host. RNA polymerase II can transcribe the provirus, producing both viral mRNA, which codes for the viral proteins, and progeny RNA, which is packaged into nucleocapsids at the cell membrane. The cell infected by this virus is usually not killed but continues to produce mature enveloped progeny viruses over an extended period of time by **budding.**

Retroposons

Studies now indicate that many of the basic features of retroviruses are shared by a number of eukaryotic transposons. The best-characterized of these retrovirus-like transposons are the *Ty* elements of yeast and the *copia* and related transposons (*copia*-like elements) of *Drosophila*. The structure of these elements is similar to that of a retroviral provirus, including the LTRs and other retrovirus-like sequences. More important, these elements appear to transpose by way of an RNA transcript that uses reverse transcriptase to form the DNA of the newly inserted element. The first conclusive evidence of an RNA intermediate in transposition came from studies on the *Ty* elements. When an intron is inserted into a *Ty* transposon (using techniques that will be described in Chapter 18), the intron does not appear in the newly transposed copy of the element. This result is inconsistent with the mode of transposition in bacterial transposons, where the initial and transposed copies are identical. The simplest way to explain the excision of an intron, in light of the other information known about the transposon, is to assume that an RNA intermediate was formed and underwent splicing before being copied into the DNA of the transposed element.

The similarity of the *Ty* and *copia*-like elements to retroviruses has led to their classification as **retroposons**. In recent years, the classification of retroposons has been extended to a variety of transposed elements that include inserted DNA copies of nonviral cellular RNAs (■ Table 16–5). Certain of these retroposons form the **SINES** (short interspersed elements) and **LINES** (long interspersed elements) that make up a major portion of the moderately repetitive sequences in eukaryotic DNA (see Chapter 9). In general, SINES are DNA copies of RNA polymerase III transcripts and therefore have internal promoters, while LINES are DNA copies of RNA polymerase II transcripts and lack promoters. Because LINES are derived from RNA polymerase II

▶ **FIGURE 16–24** Life cycle of the Rous sarcoma virus.

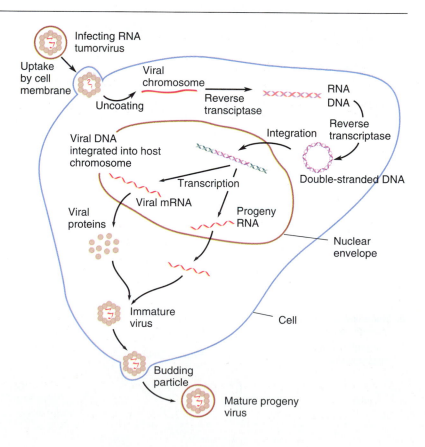

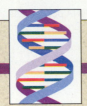

FOCUS ON HUMAN GENETICS

The AIDS Virus

There are a number of categories of diseases that affect the human immune system. Some of these diseases, such as lupus and juvenile diabetes, are the result of a failure to distinguish "self" from "nonself" that leads to an immunity to oneself or **autoimmunity**. Others, such as toxic shock syndrome and allergies, are caused by an overstimulation of the immune system in response to a foreign antigen. Still others are due to an inability to respond to foreign antigens (an **immunodeficiency**) and thus result in a decreased resistance to infection. Some of these disorders are genetic; others, such as **AIDS** (**acquired immunodeficiency syndrome**) are acquired. AIDS is caused by infection of the human immunodeficiency virus, **HIV**.

HIV is a retrovirus consisting of about 10,000 base pairs of RNA. The virus infects a particular type of immune cell, the helper T cells. Helper T cells are one of three kinds of lymphocytes that are involved in the immune response: **cytotoxic T cells** destroy foreign cells (including cancer cells)

or viruses through cell-to-cell contact; **B cells** make and secrete antibodies into the blood plasma; and **helper T cells** modulate the activities of both B cells and other T cells. Thus, helper T cells play a central regulatory role in the immune system, and their destruction by the human immunodeficiency virus leaves AIDS victims susceptible to a wide variety of infections and cancers.

Following infection of a helper T cell, the HIV reverse transcriptase copies the viral RNA into DNA that is integrated into the host cell genome. The transcription of the proviral DNA and the subsequent synthesis of viral proteins leads to the production of progeny viruses and the death of the cell. As helper T cells die over a period of months or years, opportunistic infections by fungi and other microbes take advantage of the weakening immune system. Pneumonia, fever, diarrhea, and weight loss are common, and cancer (especially the otherwise rare Karposi sarcoma) often develops. Most patients die within five to ten years after being infected with HIV.

Drugs that inhibit the viral reverse transcriptase, such as AZT (azidothymidine) and DDI (dideoxyinosine) are commonly given to AIDS patients; however, they are neither completely effective nor without serious side effects.

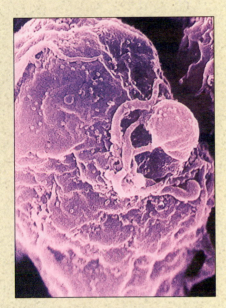

HIV budding from T-cell leukocyte.

■ **TABLE 16–5** Classes of retroposons.

Class	Characteristics of Superfamily
Viral Superfamily Ty (yeast) Copia family (Drosophila) L1 family (mammals)	Long terminal repeats (LTRs) and target site duplications of 4–6 bp; dispersed in genome; code for reverse transcriptase
Nonviral Superfamily Alu family (human) Processed pseudogenes of RNA polymerase II transcripts	Lack terminal repeats but have target duplications of 7–21 bp; dispersed in genome; do not code for reverse transcriptase

transcripts, they have many of the features common to mRNA, including a poly A tail and the absence of any recognizable introns; they are consequently referred to as **processed pseudogenes.** Like transposons, both SINES and LINES are flanked by short direct repeats, indicating that they were inserted into genomic DNA at some time in the past by a transposition act.

The best-characterized retroposons in mammals belong to a family of SINES known as the *Alu* family, so named because the sequences contain a site for the restriction endonuclease *Alu*I. In humans, for example, *Alu* sequences are repeated about 10^5 times and constitute about 5% of the total DNA. These sequences are approximately 300 bp long and are derived from internally deleted DNA copies of 7SL RNA, which is an RNA component of the signal-recognition particle involved in protein synthesis (see Chapter 17). Because the SINES contain an RNA polymerase III promoter,

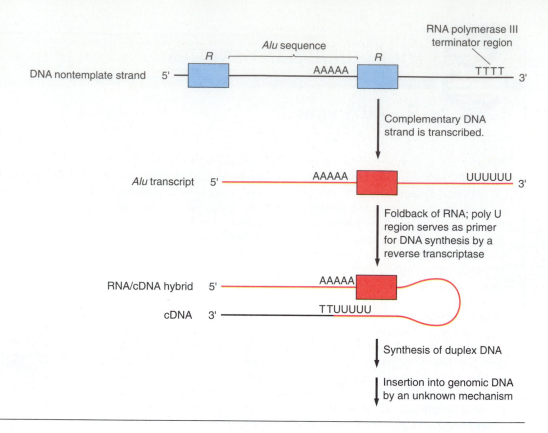

► FIGURE 16–25 A proposed mechanism for the RNA-mediated transposition of *Alu* sequences. The *Alu* transcript is hypothesized to use a borrowed reverse transcriptase to form a double-stranded DNA copy from an original DNA molecule. (*R* regions are the short direct repeats that flank an *Alu* region.)

at least some members of the *Alu* family can be transcribed in vivo into RNA. The abundance of *Alu* and other SINE families is usually attributed to their ability to undergo transcription. A proposed mechanism for the RNA-mediated transposition of *Alu* sequences is shown in ► Figure 16–25. To transpose by this mechanism, each sequence must use a borrowed reverse transcriptase (possibly from a retrovirus) to recreate the DNA insert.

In addition to the *Alu* family, mammalian genomes also contain several thousand copies of a family of LINES known as the *L1* family. These sequences are approximately 5–7 kb in length, have poly A tails, and are flanked by short direct repeats. Many are also truncated at their 5′ end, possibly as a result of premature termination by reverse transcriptase. Unlike typical processed pseudogenes, some members of the *L1* family are thought to be capable of RNA-mediated transposition. The ability to transpose could account for the relative abundance of this group of retroposons.

To Sum Up

1. In some viruses, RNA rather than DNA is the genetic material. These viruses must therefore encode special enzymes for replication and transcription of the viral genome inside a host cell.
2. Retroviruses are RNA viruses that form a double-stranded DNA copy of their genome inside an infected cell using a viral enzyme known as reverse transcriptase. The duplex viral DNA then inserts itself into a host chromosome, forming a provirus. The host RNA polymerase is then used to transcribe the provirus, resulting in the release of progeny viruses from the cell through budding.
3. Retroposons are transposable genetic elements that transpose by means of an RNA transcript that uses reverse transcriptase to form the DNA of the transposed element. Retroposons include both viral and nonviral cellular RNAs. The latter include a large portion of the moderately repetitive sequences that are found in the genomes of most eukaryotes.

Chapter Summary

Information flows from DNA to RNA through transcription; in this process, one of the strands in a transcription unit of double-stranded DNA serves as a template for the synthesis of a complementary strand of RNA. In prokaryotes, a single RNA polymerase catalyzes the transcription of all structural genes. In eukaryotes, different RNA polymerases are needed to synthesize the three major classes of RNA: mRNA, rRNA, and tRNA. Like DNA synthesis, all RNA synthesis occurs in a 5′ to 3′ direction.

The end product of transcription of a structural gene is a molecule of messenger RNA, which is usually polycistronic in prokaryotes and monocistronic in eukaryotes. Each coding region in an mRNA molecule consists of a sequence of triplet codons that specifies the sequence of amino acids in a polypeptide. The genetic code is degenerate: 61 of the 64 possible codons encode the 20 common amino acids (the three exceptions are stop or termination codons). The genetic code is essentially universal for all organisms.

Most structural genes, at least in prokaryotes, are nonoverlapping and colinear with their protein products; that is, a gene has three times as many nucleotide pairs between the beginning of its start and stop codons as there are amino acids in the polypeptide it codes for. Many genes depart from this simple paradigm, however. Several genes (found mainly in small viruses) are over-

lapping, so that the same DNA sequence is used by more than one gene. Moreover, many genes (mostly in eukaryotes) are split, containing both coding and noncoding regions. The noncoding (intron) regions in the RNA transcripts of split genes must be removed by splicing reactions before the mature mRNA product is formed.

Information transfer can also proceed from RNA to RNA and from RNA to DNA, but these processes occur on a much more limited scale. RNA-dependent RNA synthesis is catalyzed by enzymes encoded by most RNA viruses, while RNA-dependent DNA synthesis (reverse transcription) is restricted to retroviruses and certain retroposons.

 ## Questions and Problems

DNA-Dependent Synthesis of RNA

1. Fill in the blanks:
   (a) The transfer of genetic information from DNA to RNA is called _____ .
   (b) The specific DNA regions where an RNA polymerase attaches and initiates RNA synthesis are called _____ .
   (c) Promoter regions contain highly conserved _____ sequences that function in polymerase binding and the initiation of transcription at the proper startpoint.
   (d) The *E. coli* RNA polymerase holoenzyme consists of a _____ factor, which permits recognition of appropriate promoter regions, and the _____ , which catalyzes the synthesis of RNA.
   (e) Eukaryotic _____ transcribes structural genes, _____ transcribes rRNA genes, and _____ transcribes tRNA genes.

2. A short section of a particular gene includes the following sequence of nucleotide pairs:

   5′ G T C T T A C G C T A G 3′
   3′ C A G A A T G C G A T C 5′ Template strand

   What will be the base sequence of the mRNA transcribed from this gene?

3. Certain DNA nucleases degrade double-stranded DNA unless the DNA is "protected" by bound protein. Suppose you wanted to determine the location within a certain DNA sequence of a promoter region. How could you use the DNA, RNA polymerase, and a nuclease to make this determination?

4. The core enzyme of bacterial RNA polymerase covers approximately 70 bp of DNA and catalyzes chain elongation at a rate of about 50 nucleotides per second. (a) If a gene is 2.1 kb in length, what is the maximum number of polymerase molecules that can bind to the gene at one time? (b) Under conditions where chain elongation limits the rate of transcription, what is the maximum number of RNA chains that can be produced per second from this gene?

5. A consensus sequence can be used to describe a number of related but nonidentical sequences by representing each position in the sequence by the most commonly occurring nucleotide. Propose a suitable consensus sequence for the following:

   5′ T C T C T C T T A 3′
   A A A C T G T A A
   T A T G G G T A A
   T A T T T A T A T
   T A T C T G T A T
   T A T G T G A A T
   T A T C T G T A T

Messenger RNA and the Genetic Code

6. A double-stranded molecule with the following sequence produces in vivo a polypeptide that is five amino acids long.

Left end   Right end
Strand x
T A C A T G A T C A T T T A A G G G A A T T C T A G C A T G T A
A T G T A C T A G T A A A T T C C C T T A A G A T C G T A C A T
Strand y

   (a) Which strand (x or y) serves as the template strand for mRNA? (b) Which is the 5′ end of the template strand (the left or right end)? Why? (c) What is the amino acid sequence of the polypeptide?

7. Refer to the genetic code in Table 16–3. (a) In how many cases would you be able to identify the amino acid specified by a particular codon if you knew only the first two nucleotides of that codon? (b) Conversely, in how many instances would you be able to identify the first two nucleotides of a codon if you knew the amino acid that it specifies?

8. The amino acid sequence of a certain polypeptide is Leu-Tyr-Arg-Trp-Ser. (a) How many nucleotides are necessary in the DNA to code for this pentapeptide? (b) How many different nucleotide sequences in the DNA could possibly code for this pentapeptide?

9. Suppose that life is detected on another planet and it is based, as is our own, on nucleic acids and proteins. Instead of 20 different amino acids, however, there are 432, and they are coded by six nucleotides instead of four. (a) What is the minimum number of nucleotides required to make up a codon in this system? (b) How much degeneracy is there in this genetic code compared with our own?

10. If an organism synthesizes 150,000 different proteins but con-

tains 4×10^9 nucleotide pairs of DNA, what fraction of the DNA is actually coding for protein? (Do this calculation first assuming 1000 base pairs per gene and then assuming 10,000 base pairs per gene.)

11. Suppose you use polynucleotide phosphorylase and a reaction mixture consisting of GDP and UDP to synthesize an artificial mRNA. If the RNA is produced from a mixture that contains an excess of UDP, more valine than glycine is incorporated into protein, but when GDP is in excess, more glycine than valine is incorporated. Without looking at Table 16–3, list the codons containing G and U that could possibly be coding for Val and Gly.

12. Determine the relative frequencies of all possible triplet codons in the RNA produced by a reaction mixture consisting of 60% UDP and 40% CDP and polynucleotide phosphorylase.

13. Suppose that the RNA of the preceding problem is used as a messenger in a protein-synthesizing system. Using Table 16–3, determine what percentages of the various amino acids are expected to be incorporated into polypeptides under the direction of this RNA. Express each percentage relative to the amount of phenylalanine.

14. Suppose that synthetic mRNAs were made in the presence of polynucleotide phosphorylase in a solution that contained $\frac{2}{3}$ ADP and $\frac{1}{3}$ UDP. The mRNAs were then used in a cell-free system for in vitro protein synthesis. (a) Calculate the expected frequencies of the A_3, A_2U, AU_2, and U_3 triplet sequences in the synthetic mRNAs; express the frequencies relative to the most commonly occurring sequence. (b) The polypeptides produced under the direction of these mRNAs were found to contain amino acids in the following proportions (expressed relative to the most commonly occurring amino acid): Lys = 100%, Ile = 75%, Asn = 50%, Leu = 25%, Tyr = 25%, and Phe = 12.5%. On the basis of these data and your knowledge of the genetic code, deduce the most probable codon assignments for these amino acids. (Account for all eight codons, and give the correct base sequence of each codon when it is possible to do so.)

15. In an Ochoa-type experiment, an artificial mRNA was synthesized using a reaction mixture consisting of $\frac{3}{4}$ UDP and $\frac{1}{4}$ GDP and polynucleotide phosphorylase. This mRNA, when used in a cell-free extract, coded for the synthesis of a polypeptide with the following amino acid composition:

Amino acid	Observed frequency relative to phenylalanine
phenylalanine	100%
valine	44%
leucine	36%
cysteine	35%
glycine	14%
tryptophan	12%

From these data, postulate codon assignments for each of the amino acids listed.

16. Using Table 16–3, give the amino acid sequences in the polypeptides produced in an in vitro protein-synthesizing system under the direction of artificial mRNAs with the following defined sequences:
(a) A U A U A U A U A U A U . . .
(b) G A U G A U G A U . . .
(c) U A A U A A U A A . . .

Assume that under the conditions used in the in vitro system, translation is initiated at randomly selected codons in the artificial mRNAs. How would your results differ if the genetic code were actually a doublet code?

Gene-Protein Relations and RNA Processing

17. Fill in the blanks:
(a) Most eukaryotic genes are split into coding regions called _____, separated by noncoding regions called _____ .
(b) The RNA polymerase II transcripts in the nucleus are known as _____ molecules, because of the heterogeneity in their sizes.
(c) The large ribonucleoprotein complex that catalyzes splicing of the hnRNA is called the _____ .
(d) The two acts of cleavage and reformation of phosphodiester bonds that take place during intron removal and ligation of exons are known as _____ reactions.
(e) Mature eukaryotic mRNA has a _____ that has been added to its 5′ end and (usually) a _____ that has been added to its 3′ end.
(f) The posttranscriptional modifications that operate on hnRNA molecules to produce mature mRNA are collectively referred to as _____ .

18. A certain viral DNA molecule contains 5250 nucleotide pairs. (a) If the genes of this organism are nonoverlapping sequences of nucleotide pairs along the DNA, how many proteins of molecular weight 30,000 could be coded for by this DNA? (Assume the average molecular weight of an amino acid to be 120.) (b) This viral chromosome actually codes for 10 proteins of molecular weight 30,000 each. How many codons are needed to encode these 10 proteins? (c) Compare this figure with the length of the viral DNA. Explain how the DNA can contain enough codons for all the viral genes. (d) Would you expect there to be similarity in amino acid sequence among any of these proteins? Explain your answer.

19. Explain how the genetic code is nonoverlapping, yet there are known cases of overlapping genes.

20. At least three modifications occur during the processing of hnRNA into mRNA. List them and describe each one.

21. The following eukaryotic pre-mRNA sequence contains an intron:

Left end
5′ A G G U C C G U U C A A U G C C U C G A C U G G U —
Right end
C A C U A C U A A C U A A C U U C C U A G U C U C U A C 3′

Identify the intron sequence including the probable 5′ splice site, 3′ splice site, and branch point.

22. During the formation of a spliceosome complex, a snRNP designated U1 binds to the 5′ splice site on the intron, covering about 16 nucleotides; a snRNP designated U2 binds to the branch site on the intron, covering about 40 nucleotides; and a snRNP designated U5 binds to the 3′ splice site on the intron, covering about 15 nucleotides. Based on the reactions of these three snRNPs, estimate the minimum size of an intron needed to form a spliceosome.

23. Suppose that the *Tetrahymena* rRNA gene, which contains an intron, is isolated and introduced into *E. coli*. The bacterial RNA polymerase recognizes the promoter and transcribes the

gene. Would you expect to find mature rRNA species within the bacterial cell? Answer this question again for the case in which a split eukaryotic structural gene is introduced into *E. coli* cells.

RNA-Dependent RNA Synthesis and Reverse Transcription

24. Define the following terms and distinguish between members of paired terms:

    − strand and + strand
    retroposons and retroviruses
    RNA replicase and transcriptase
    reverse transcriptase
    SINES and LINES

25. RNA viruses are classified into four types based on the manner in which they replicate. Describe the general features of replication of each of these viral classes (see Figure 16–20).

26. Actinomycin D is an antibiotic that blocks DNA-dependent RNA synthesis, but not RNA-dependent replicases. Suppose you isolate a new virus and find that its growth is not inhibited by actinomycin D. (a) Is the genome of this virus composed of DNA or RNA? (b) What would your answer have been if it was found that actinomycin D does inhibit viral growth?

Protein Synthesis and the RNA Decoding System

Ribosomes and Ribosomal RNA
Transfer RNA
Decoding the Message: Translation
Protein Folding: Assisted Self-Assembly
Posttranslational Modification of Proteins

During protein synthesis, the codon sequence in each messenger RNA is translated into a corresponding sequence of amino acids in a polypeptide. Since mRNA cannot directly transfer its encoded instructions to the amino acid precursors of the protein, the translation process requires a complex RNA decoding system to recognize the mRNA codons and guide the amino acids into their proper positions in the growing polymer chain. Two classes of RNA are involved in decoding the message: ribosomal RNA (rRNA), a constituent of ribosomes, and transfer RNA (tRNA), which serves as an adaptor molecule to mediate the transfer of information from the mRNA to the polypeptide. Both rRNA and tRNA are coded for by multiple copies of specific genes and, like eukaryotic mRNA, are initially formed as large transcripts that require extensive processing in their maturation.

Color-enhanced scanning electron micrograph (SEM) of ribosomes.

In this chapter, we will consider the process of protein synthesis. We will begin with a description of the RNA decoding system and proceed to the mechanisms of translation and protein processing.

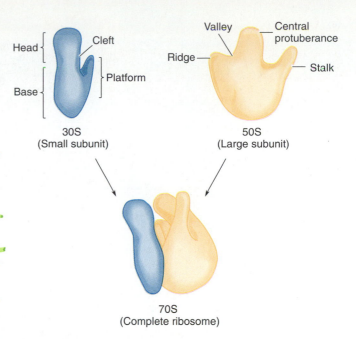

► FIGURE 17–1 The subunit structure of a ribosome in *E. coli*. The 30S subunit, which is the smaller of the two, associates with the larger 50S subunit to make up the complete 70S ribosome.

✳ RIBOSOMES AND RIBOSOMAL RNA

Ribosomes are the cellular sites where mRNA is translated into proteins. Electron micrographs reveal that each ribosome consists of two subunits, one about half the size of the other (► Figure 17–1). The size of the ribosome is often expressed in terms of its **sedimentation coefficient**, which measures the rate at which a particle will sediment in a unit centrifugal field of force. In general, the larger the size of the particle, the faster it will move in response to high-speed centrifugation. Thus the complete *E. coli* ribosome has a sedimentation coefficient of 70S, while the coefficient of the smaller subunit is 30S and that of the larger subunit is 50S. The subunit values are not additive (30S + 50S ≠ 70S) because the velocity of sedimentation depends not only on size but also on shape, which differs in the three forms. The slightly larger eukaryotic ribosomes vary in size in different species, but on the average they are composed of 40S and 60S subunits, with a sedimentation coefficient of 80S for the complete structure.

Ribosomes are ribonucleoprotein particles that consist of proteins in a complex with rRNA. The rRNA components of ribosomes have been sequenced, and their primary structures reveal that they can all be tightly folded into compact secondary structures. Moreover, sequence comparisons of corresponding rRNAs from different species indicate that many major structural features, such as complex loops and base-paired stems, are shared by the rRNAs of prokaryotes and eukaryotes, despite differences in nucleotide sequence (► Figure 17–2). These results suggest that evolution has conserved the folded structure, rather than the primary structure, of these RNA molecules.

Ribosomes are formed from their rRNA and protein components by a self-assembly process. In *E. coli*, the primary transcripts for rRNA are produced as large 30S precursor (pre-rRNA) molecules from each of seven transcription sites. Each rRNA precursor is then processed into three types of rRNA: 16S (about 1540 nucleotides), 23S (about 2900 nucleotides), and 5S (about 120 nucleotides), as shown in ► Figure 17–3a. Once formed, the 16S rRNA molecule combines with 21 ribosomal proteins (known as S1, S2, etc.) to yield the 30S ribosomal subunit. A 23S and a 5S rRNA molecule combine with 31 ribosomal proteins (L1, L2, etc.) to yield the 50S subunit.

In eukaryotes, rRNA is formed at numerous transcription sites, most of which are located in tandem repeats in **nucleolar organizer regions** of chromosomes. In these regions, the tandemly arranged and transcriptionally active rRNA genes loop out from the remaining chromosomal DNA to help form the nucleoli, where ribosomes are as-sembled. In humans, nucleolar organizers are found on the short arms of chromosomes 13, 14, 15, 21, and 22. In contrast, the nucleolar organizers of *Drosophila* and corn are present on a single pair of chromosomes. Transcription in the nucleolar organizer regions yields a large 45S pre-rRNA molecule, which is later processed into 18S, 28S, and 5.8S rRNAs (Figure 17–3b). Transcription from scattered sites outside of the nucleolar organizer regions produces the 5S rRNA. Union of the 18S rRNA molecule with ribosomal proteins forms the 40S subunit of the ribosome. The 28S, 5.8S, and 5S molecules unite with proteins to yield the 60S subunit.

Several of the steps involved in ribosome biosynthesis and assembly have been elucidated with the aid of genetic techniques, such as using temperature-sensitive mutants with specific genetic blocks. Various physical and chemical methods have also been helpful. For example, the structural relationships of the ribosomal components have been studied by introducing chemical cross-links to indicate which components are in close proximity (► Figure 17–4a). In another approach, the components are bound to specific antibodies, and then the location of the bound antibodies is observed with an electron microscope (Figure 17–4b). **Partial reconstitution experiments** are another biochemical method that has been useful in the study of ribosome assembly. In this method, ribosomes are assembled in vitro in the absence of individual ribosomal proteins. Omitting different proteins from the reaction mixture makes it possible to construct assembly maps for the subunits according to which ribosomal components still form a partially

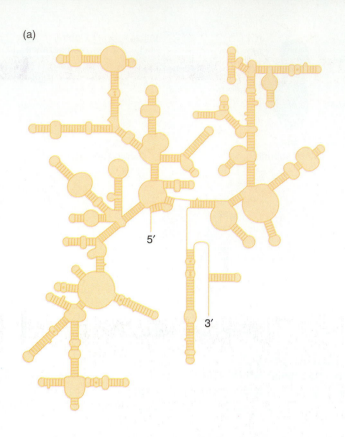

(a)

5′

3′

▶ **FIGURE 17–2** Similarities in the predicted secondary structures of the 16S rRNA of eubacteria and 16S-like rRNAs of distantly related species. (a) Eubacteria (*E. coli*). (b) Archaebacteria (*Halobacterium volcanii*). (c) Eukaryotes (baker's yeast). *Adapted from:* R. R. Gutell, B. Weiser, C. R. Woese, and H. F. Noller, *Prog. Nucleic Acid Res. Mol. Biol.* 32 (1985): 183. Used with permission.

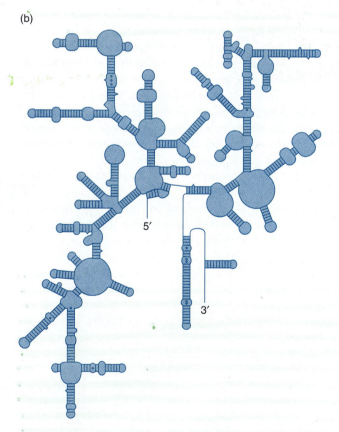

(b)

5′

3′

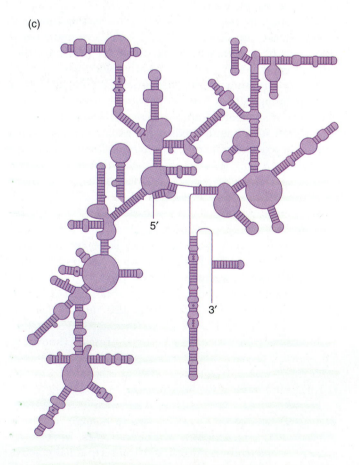

(c)

5′

3′

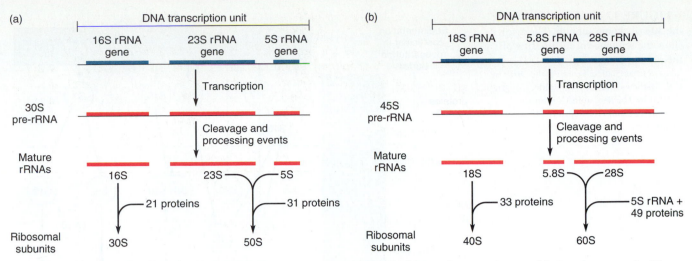

(a) DNA transcription unit
16S rRNA gene 23S rRNA gene 5S rRNA gene

Transcription

30S pre-rRNA

Cleavage and processing events

Mature rRNAs
16S 23S 5S

21 proteins 31 proteins

Ribosomal subunits
30S 50S

(b) DNA transcription unit
18S rRNA gene 5.8S rRNA gene 28S rRNA gene

Transcription

45S pre-rRNA

Cleavage and processing events

Mature rRNAs
18S 5.8S 28S

33 proteins 5S rRNA + 49 proteins

Ribosomal subunits
40S 60S

▶ **FIGURE 17–3** Biosynthesis of ribosomal subunits (a) in prokaryotes and (b) in eukaryotes (mammals). In eukaryotes, the 5S rRNA is coded for by a separate transcription unit.

assembled product and which components fail to bind (▶ Figure 17–5). These studies reveal that ribosome assembly is a highly ordered process that occurs in a series of sequential steps.

To Sum Up

The rRNAs of both prokaryotes and eukaryotes are cut out of larger precursor molecules synthesized from rRNA transcription units. The individual rRNAs fold with ribosomal proteins in a self-assembly process that forms the ribosomal subunits.

 ## TRANSFER RNA

The role of transfer RNA during protein synthesis is to carry amino acids to the ribosome in a sequence dictated by the mRNA. Many different tRNA molecules make up the normal tRNA set of a given cell, including several different types that carry the same amino acid. tRNAs that differ in structure but combine with the same amino acid are called isoacceptors.

Despite differences in their primary structure, all tRNAs have a number of features in common. All tRNA molecules

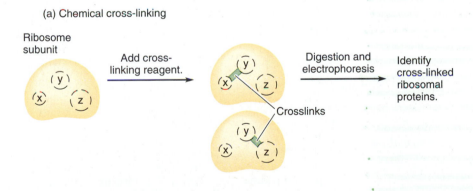

(a) Chemical cross-linking

Ribosome subunit

Add cross-linking reagent.

Crosslinks

Digestion and electrophoresis

Identify cross-linked ribosomal proteins.

▶ **FIGURE 17–4** Biochemical methods for analyzing ribosome structure.

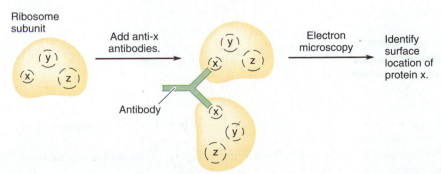

(b) Immunoelectron microscopy (antibody labeling)

Ribosome subunit

Add anti-x antibodies.

Antibody

Electron microscopy

Identify surface location of protein x.

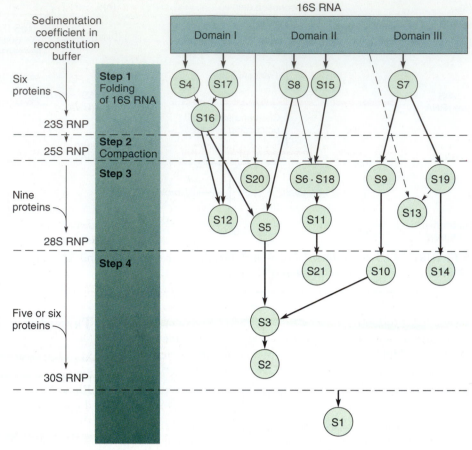

► **FIGURE 17-5** Assembly map for the 30S ribosome subunit. The arrows show the sequence of the binding of the protein components (S1–S21). *Adapted from: Source: From Biochemistry by* C.K. Mathews and K.E. vanHolde. Copyright © 1990 by The Benjamin/Cummings Publishing Company. Reprinted by permission.

are comparatively small, single-stranded nucleic acids ranging in length from 73 to 93 nucleotide residues. In addition, all tRNA chains have a substantial proportion of modified or "unusual" nucleosides, many containing methylated derivatives of the principal bases. These unusual nucleosides also include **dihydrouridine** (D), which has a saturated double bond in the uracil ring, **inosine** (I), which contains the purine hypoxanthine, **ribothymidine** (T), in which ribose is bonded to thymine, and **pseudouridine** (Ψ), which has an unusual carbon-to-carbon bond between the base and the sugar. The structures of some of the unusual nucleosides are given in ► Figure 17–6. All of these structures are formed by the enzymatic modification of the standard nucleosides, A, C, G, and U, after the initial tRNA chain is synthesized.

The primary structures of several hundred different tRNAs from a variety of bacterial and eukaryotic sources have been determined. These studies reveal that the different tRNA chains can all be folded into a common secondary structure in which the bases are arranged by hydrogen bonding in a **cloverleaf** configuration (see ► Figure 17–7). By convention, the 76 nucleotide residues of the most common tRNA structure are numbered starting from the 5′ end. This structure is sufficiently general that any extra nucleotides that may be present in a given tRNA (most commonly following positions 17, 20, and 47) are simply num-

► **FIGURE 17-6** Some modified or unusual nucleosides present in tRNAs. Inosine has a purine not found in other types of RNA. Color regions on the nucleosides point out differences from their standard counterparts.

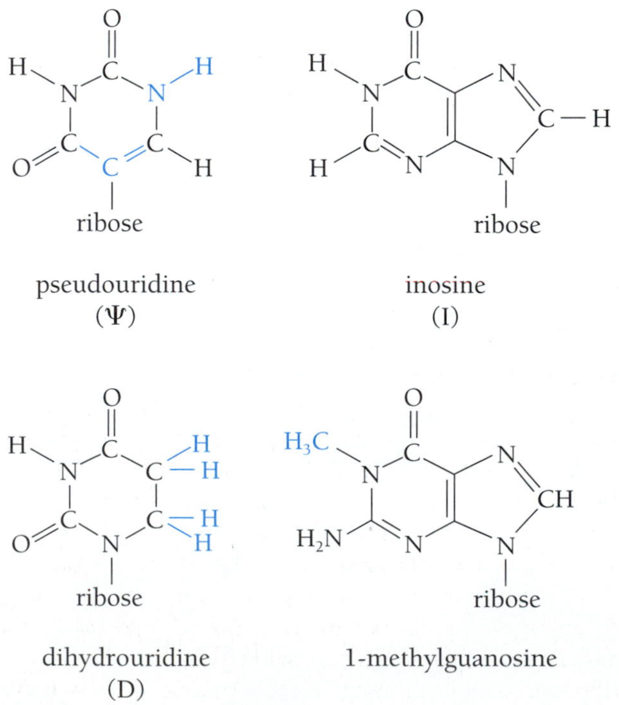

pseudouridine (Ψ)

inosine (I)

dihydrouridine (D)

1-methylguanosine

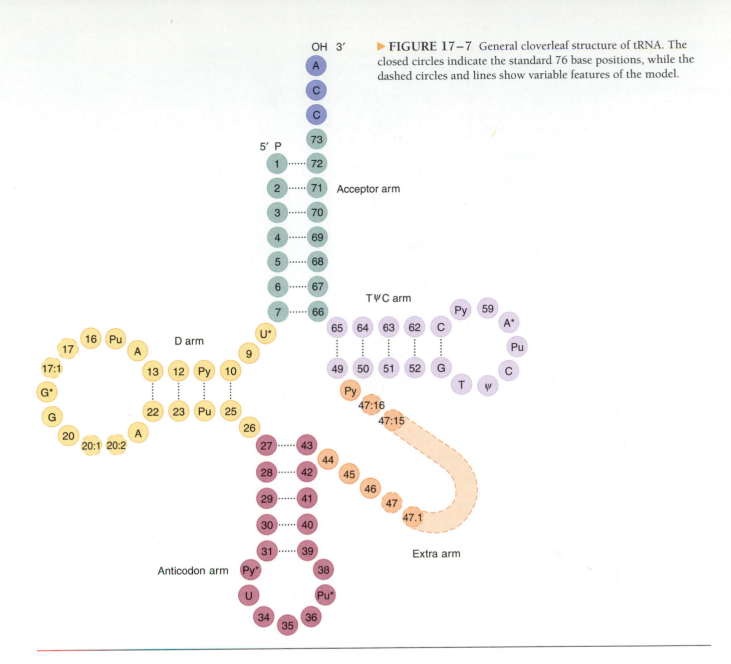

▶ **FIGURE 17–7** General cloverleaf structure of tRNA. The closed circles indicate the standard 76 base positions, while the dashed circles and lines show variable features of the model.

bered as additions to the standard sequence (such as 17.1, 20.1, 47.1, 47.2, and so on).

Note that all tRNAs have four main hydrogen-bonded arms, a D arm containing a loop of 8–12 unpaired bases, an anticodon arm containing a loop of 7 unpaired bases, a TΨC arm also containing a loop of 7 unpaired bases, and an acceptor arm containing an unpaired -C-C-A sequence at the 3′ end. Sequence comparisons demonstrate that several of the unpaired bases in the single-stranded loops are invariant (conserved). Each arm of the cloverleaf structure has an important function. The anticodon arm serves as the attachment site for a particular mRNA codon, the acceptor arm serves as an attachment site for an amino acid, and the D and TΨC arms are involved in the three-dimensional folding of the molecule. In addition to these four main arms, there is also a highly variable fifth arm, the so-called extra arm, which can range up to 21 nucleotides in length.

Although the primary and secondary structures of different tRNA molecules have been known for some time, more recent X-ray diffraction studies of crystallized tRNA have shown that the tertiary structure of this nucleic acid looks more like an L than a cloverleaf (▶ Figure 17–8a). The L-shaped tertiary structure is formed by the folding of the molecule into two double-helical branches—one consisting of the anticodon arm and the D arm and the other consisting of the acceptor arm and the TΨC arm—at approximately right angles to each other. This three-dimensional arrangement is stabilized by so-called tertiary hydrogen bonds, several of which involve invariant bases that are unpaired in the cloverleaf structure (Figure 17–8b). The end result of this three-dimensional folding is a tRNA molecule that has the anticodon loop at one end and the -C-C-A acceptor stem at the other end.

Like rRNA molecules, the different tRNAs are derived

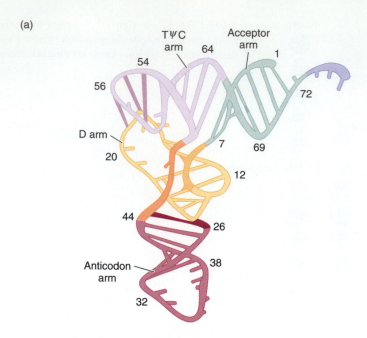

(a)

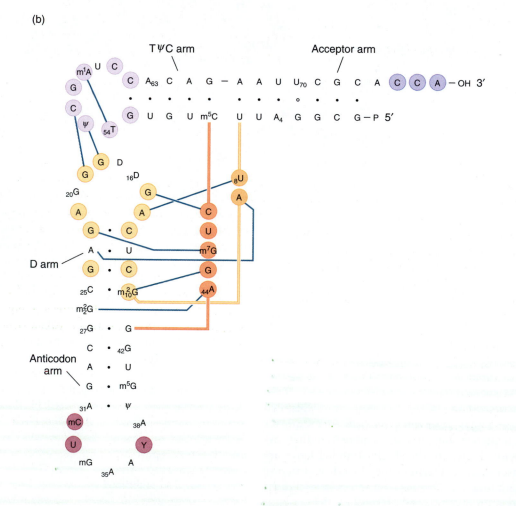

(b)

▶ **FIGURE 17–8** Tertiary structure of yeast phenylalanine tRNA. (a) The three-dimensional L-shaped conformation. (b) The nucleotide sequence in the L arrangement; the tertiary hydrogen bonds are shown as blue lines. *Source:* (b) From B. J. Lewin, *Genes* (New York: John Wiley & Sons, 1983). Used with permission of Benjamin Lewin.

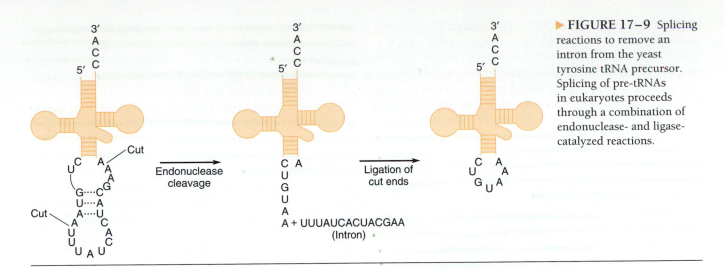

▶ FIGURE 17–9 Splicing reactions to remove an intron from the yeast tyrosine tRNA precursor. Splicing of pre-tRNAs in eukaryotes proceeds through a combination of endonuclease- and ligase-catalyzed reactions.

from larger, short-lived precursors. In *E. coli*, these precursors are transcribed from clusters of repeated tRNA genes that are scattered throughout the genome and may encode the structures of rRNAs as well as tRNAs. Following transcription, the tRNA precursors are cleaved and trimmed by different enzymes to form the mature tRNAs. It is interesting to note that one of these enzymes, the bacterial endonuclease RNase P, is able to function in tRNA processing by virtue of its catalytically active RNA component; this is another example of RNA acting as a biological catalyst. Some eukaryotic tRNA precursors may also contain noncoding intervening sequences that must be spliced out (▶ Figure 17–9). However, introns are removed from these nuclear pre-tRNAs by specific protein-based endonucleases rather than by spliceosome-catalyzed or RNA-catalyzed splicing.

Amino Acid Recognition: Charging tRNA

Each tRNA must exhibit two types of specificity in order to perform its role in protein synthesis. It must be able to (1) combine with a specific amino acid and then (2) combine with the specific mRNA codon for the amino acid. Specific **amino acid recognition** is needed so that the tRNA will bond chemically to only one type of amino acid and transfer that amino acid to the ribosome. Specific **codon recognition** is needed so that the tRNA will bind to the surface of the ribosome under the direction of mRNA, thus ensuring the insertion of the amino acid at its proper position in the polypeptide chain.

Amino acid recognition is accomplished by a class of enzymes known as **aminoacyl-tRNA synthetases**. There are (at least) 20 of these enzymes, each specific for a particular amino acid. The enzyme catalyzes the bonding of the amino acid to the 3' acceptor arm of the appropriate tRNA. The energy for this reaction is provided by the hydrolysis of ATP.

The overall reaction catalyzed by each synthetase enzyme proceeds through a two-step process. In the first step, the amino acid is activated by forming a complex with AMP:

$$\text{amino acid} + \text{ATP} \xrightarrow[\text{synthetase}]{\text{aminoacyl-tRNA}} \text{aminoacyl-AMP} + \text{PP}_i$$

The inorganic pyrophosphate (PP_i) released in this reaction is broken down by hydrolysis, so that a total of two high-energy phosphate bonds are consumed in order to convert the amino acid to an activated state.

The aminoacyl-AMP complex then undergoes the second step of the reaction, which results in the formation of a covalent bond between the amino acid and an appropriate tRNA:

$$\text{aminoacyl-AMP} + \text{tRNA} \xrightarrow[\text{synthetase}]{\text{aminoacyl-tRNA}} \text{aminoacyl-tRNA} + \text{AMP}$$

The product of this reaction is an **amino acid ester** formed by bonding between the carboxyl group of the amino acid and a hydroxyl group in the terminal adenosine of tRNA (▶ Figure 17–10). This structure, which is referred to as either an **aminoacyl-tRNA** or a **charged tRNA**, functions as a high-energy intermediate in protein synthesis.

Amino acid recognition by the synthetase enzyme must proceed with high fidelity; failure to do so would lead to the production of faulty proteins. Since errors inevitably occur, however, many aminoacyl-tRNA synthetases have proofreading capabilities and are able to hydrolyze the ester bond formed when an incorrect amino acid is mistakenly attached to a tRNA. This type of proofreading is important when there is a problem in discriminating between structurally similar

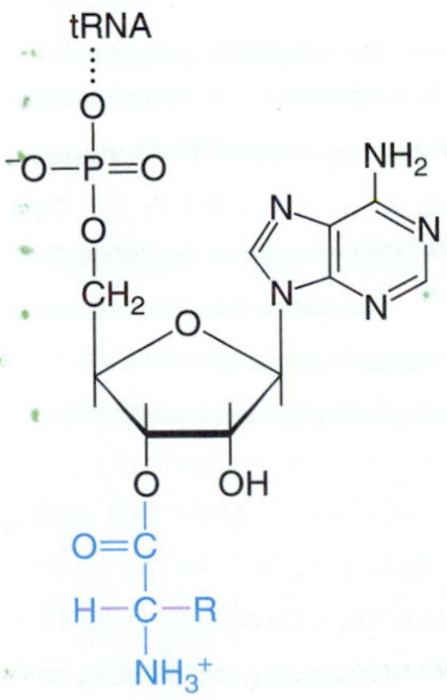

▶ FIGURE 17–10 Structure of the terminal 3' adenosine of a charged tRNA that is bonded to an amino acid.

amino acids. For example, valine differs from isoleucine only by having one fewer methylene ($-CH_2-$) group, and it is occasionally attached to isoleucine tRNA (tRNA^Ile). If the mischarged tRNA were free to function in protein synthesis, it would incorporate the wrong amino acid into proteins at an unacceptably high rate. The high error rate is avoided in this case by the ability of the synthetase to recognize and hydrolyze valyl-tRNA^Ile. This editing capability appears to be a common feature among the aminoacyl-tRNA synthetases, and the presence of a special hydrolytic site for proofreading permits this class of enzymes to maintain an overall error rate of less than 1 in 10^4.

In addition to binding the proper amino acid, it is also essential that each aminoacyl-tRNA synthetase combine with the appropriate tRNA. Numerous studies, including some that use mutationally altered tRNAs containing substituted bases, have attempted to determine the structural elements that a synthetase recognizes in identifying its cognate tRNAs. The results to date indicate that no single common feature is used for recognition by all of the synthetases. Instead, each synthetase appears to recognize its cognate tRNAs by just a few bases that can be located in the acceptor, D, and/or anticodon arms of the tRNA molecule. Thus, despite their common enzymatic function, the aminoacyl-tRNA synthetases seem to comprise a rather diverse group of proteins.

Codon Recognition: Wobble in the Anticodon

The second form of tRNA specificity, the ability to recognize a particular codon in the mRNA strand, is made possible by a specific sequence of three nucleotides called the **anticodon**, found in the anticodon loop of the tRNA. The anticodon of each different tRNA is complementary in sequence to one codon in the mRNA. This complementarity allows an aminoacyl-tRNA to hydrogen bond specifically to one (or in some cases, more than one) mRNA codon and thereby bring its attached amino acid to the ribosome surface as dictated by the genetic instructions in the mRNA.

Since there are 61 mRNA codons that specify amino acids (64 minus 3 stop codons), we might expect a cell to contain at least 61 tRNA molecules with 61 different anticodons. The evidence now indicates that fewer tRNAs are actually needed because of a lack of pairing specificity between the bases in the third (3′) position of the codon and the first (5′) position of the anticodon. Francis Crick, in his so-called **wobble hypothesis**, has proposed that the anticodon bonds with the first and second bases of the codon according to strict base-pairing rules (i.e., A with U and G with C) but that some degree of flexibility, or wobble, is allowed in the pairing at the third codon position. This flexibility is derived in part from the conformation of the anticodon loop and results in a less stringent set of rules for third-position pairing (see ■ Table 17–1). For example, an anticodon that consists of the nucleotide sequence GGG will pair not only with the exact complement CCC in mRNA, but, with a certain degree of wobble, it can also recognize and bond to the mRNA codon CCU. The first two bases of each codon

therefore confer most of the codon-anticodon specificity, since the wobble rules dictate that mRNA codons ending in either C or U (or even A in some cases) can specify the amino acid carried by a single type of tRNA. ▶ Figure 17–11 shows examples of wobble base pairings.

Not all decoding systems obey Crick's wobble rules. Recent studies indicate that codon-anticodon pairing in mitochondria obeys a simpler set of wobble rules, requiring a minimum of only 22 anticodons, as opposed to 32 in other proteinsynthesizing systems. This economic use of the genetic message in mitochondria seems to have evolved separately from nonmitochondrial systems and may account for the apparent differences in the genetic code in these organelles.

Example 17.1

In an experiment designed to test whether the amino acid portion of an aminoacyl-tRNA is involved in recognition of an mRNA codon, the cysteine moiety of a cysteinyl-tRNA (anticodon ACA) is chemically converted to alanine. It is found that the charged tRNA still recognizes the UGU codon on the mRNA, incorporating alanine into the growing polypeptide chain in place of cysteine. What does this result mean in terms of the roles of the amino acid and anticodon in recognizing the codon of the mRNA?

Solution: This experiment demonstrates that once a tRNA has been charged, the amino acid plays no role in recognition of the mRNA codon. The specificity for recognition of an mRNA codon resides with the tRNA anticodon and not with the amino acid carried by the tRNA.

Follow-Up Problem 17.1

Consider the wobble pairings given in Table 17–1. Which of the following mRNA codons could be recognized by the same tRNA anticodon? (a) GCA (b) GCC (c) GCG (d) GCU What is the minimum number of different tRNA codons required to recognize this set of codons?

■ **TABLE 17–1** Pairing at the third (3′) position of the codon and the first (5′) position of the anticodon.

Nucleoside on Anticodon	Bases Recognized on Codon
U	A
	G
C	G
A	U
G	U
	C
I*	U
	C
	A

*I = inosine (the purine hypoxanthine bonded to ribose).

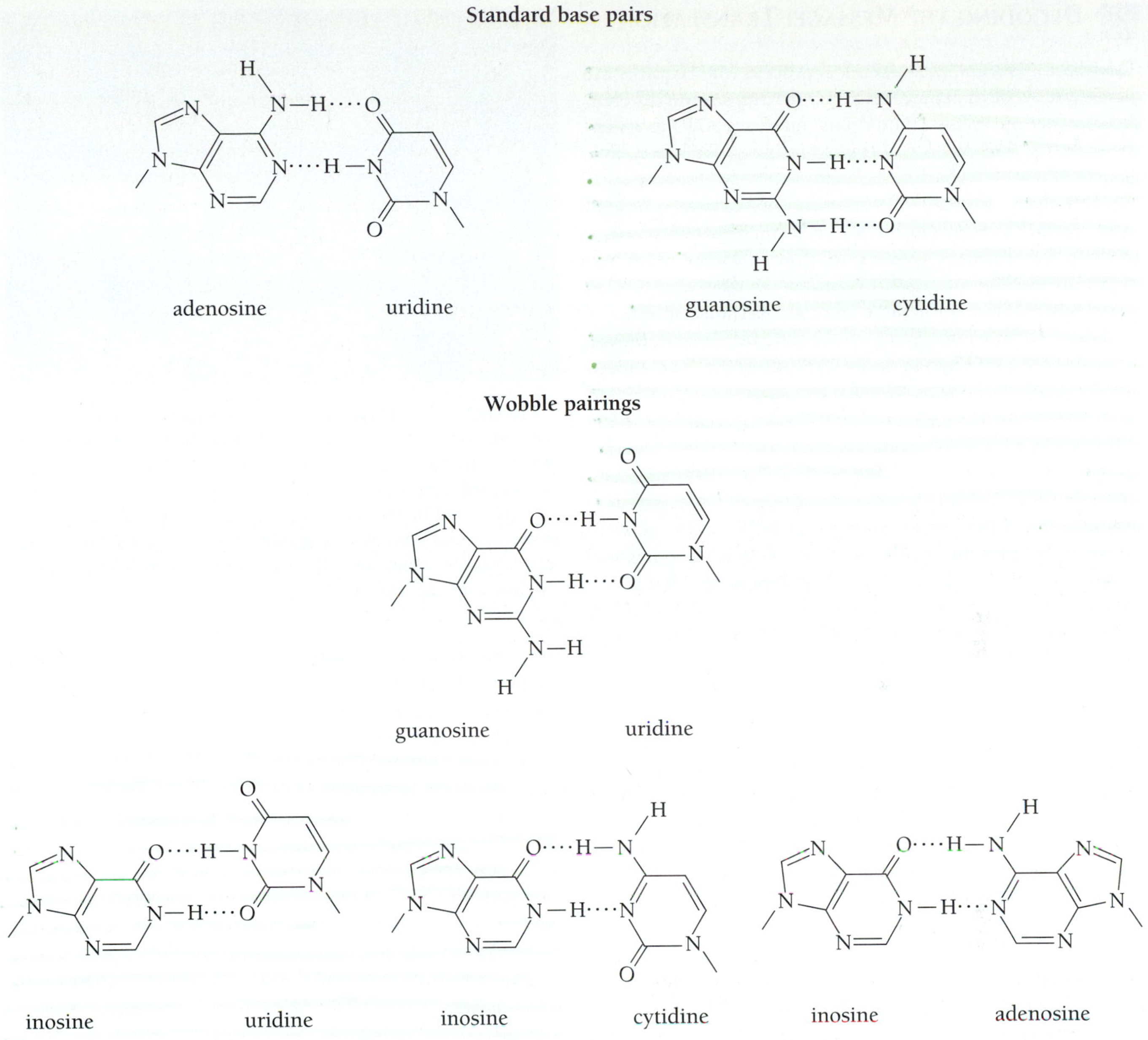

Standard base pairs

adenosine   uridine   guanosine   cytidine

Wobble pairings

guanosine   uridine

inosine   uridine   inosine   cytidine   inosine   adenosine

▶ **FIGURE 17–11** Wobble in base pairing allows some bases at the first (5′) position of an anticodon to recognize more than one base in the third (3′) position of an mRNA codon. G-U, I-U, I-C, and I-A wobble pairings are shown.

To Sum Up

1. Many tRNAs are cut out of larger transcription products. There are many different tRNA primary structures, but all of them fold into the same cloverleaf secondary structure. Two arms of the cloverleaf fold at a right angle to the other two arms to form an L-shaped tertiary structure that exposes the amino acid acceptor stem at one end and the anticodon loop at the other end.

2. A tRNA becomes active when it binds chemically to an amino acid; this process is called charging of the tRNA. The amino acid is esterified to the terminal adenosine of the 3′ end of the acceptor stem in a two-step reaction catalyzed by an amino-acyl-tRNA synthetase. There are 20 of these enzymes, each one specific for a particular amino acid and the corresponding group of isoacceptor tRNAs.

3. In addition to its specificity for a particular amino acid, a tRNA molecule also exhibits specificity in recognizing a particular mRNA codon. This specificity resides in the anticodon of the tRNA, a triplet nucleotide sequence that is complementary to the mRNA codon. Recognition of a codon by an anticodon follows strict base pairing rules for the first and second codon positions but tends to "wobble" in allowable pairings at the third position of the codon.

DECODING THE MESSAGE: TRANSLATION

Translation is a complex chemical process that requires the participation of ribosomes, charged tRNAs, several different protein factors, and GTP (and in some cases ATP) as an energy source. The process of translation begins when a ribosome becomes attached to the beginning of the gene message in the mRNA. The mRNA molecule then acts as a tape of instructions; the instructions are read consecutively, codon-by-codon, as the mRNA moves along the ribosome. The reading of the message proceeds in the 5′ → 3′ direction, so that translation occurs in the same direction as transcription.

Although only one polypeptide can be formed during the translation process on any single ribosome, many identical copies of the encoded polypeptide can be synthesized in a short time through tandem translation of the mRNA molecule by several ribosomes spaced about 90 nucleotides apart (▶ Figure 17–12). The complex of ribosomes attached to the same mRNA is referred to as a **polyribosome** or **polysome**. Electron micrographs of cells that are actively engaged in protein synthesis show extensive polysome complexes in which the longest polypeptide chains are found on ribosomes located toward the 3′ end of the mRNA molecule. In bacteria, polysomes are formed at the same

▶ **FIGURE 17–12** (a) A single mRNA molecule associates with many ribosomes to form a polysome. Each ribosome is in a different stage of polypeptide synthesis, so the lengths of the polypeptide being synthesized form a gradation from the shortest at the 5′ end of the mRNA to the longest at the 3′ end. (b) An electron micrograph showing ribosomes associated with a long molecule of mRNA. In the micrograph, we can actually see the gradation in length of the nascent polypeptide chains.

(a)

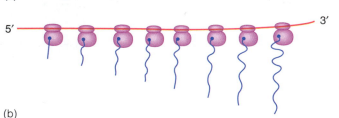

(b)

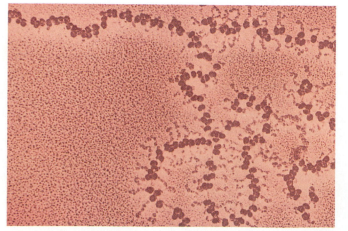

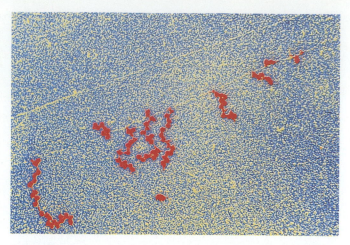

▶ **FIGURE 17–13** The action of a bacterial gene in protein synthesis: false color transmission electron micrograph (× 32,600) of a structural gene from the bacterium *Eschericia coli*, showing the coupled transcription of DNA into messenger RNA (mRNA) molecules and the translation of this messenger code into protein molecules. The DNA fiber runs down the image (in yellow) with numerous ribosomes (in red) attached to each mRNA chain. The longer chains are called polysomes.

time that mRNA is synthesized (▶ Figure 17–13). Such coupling of transcription and translation is possible only in prokaryotes, where no nuclear envelope separates the DNA from the protein-synthesizing apparatus.

To accomplish protein synthesis, every ribosome has two binding sites that are capable of combining with tRNA (▶ Figure 17–14a), the **peptidyl** or **P site**, and the **aminoacyl** or **A site**. During translation, the P site holds the tRNA that is attached to the growing peptide chain, while the A site binds the tRNA charged with the next amino acid to be added (Figure 17–14b). A ribosome in this state is in the proper form for further elongation of the polypeptide chain.

The decoding of the gene message in mRNA occurs in three distinct phases, with different protein factors catalyzing the various steps in the process. As in transcription, the three phases of translation are called **initiation, chain elongation**, and **termination**. Initiation brings the N-terminal amino acid of the polypeptide into position on the ribosome by decoding the **initiation codon**, AUG (or, less frequently in prokaryotes, GUG), which occurs at the beginning of every gene message. Elongation then extends the growing peptide chain from the N-terminal to C-terminal residue by decoding all the internal codons in the message. Termination finally releases the completed polypeptide from the ribosome when a termination codon is reached. These phases occur through similar mechanisms in both prokaryotes and eukaryotes; the few differences in detail mainly involve the initiation phase, the nature of the protein factors, and the structure of mRNA and its interaction with ribosomes.

Initiation of Translation

Translation is initiated when the 5′ end of the mRNA molecule combines with the smaller subunit of the ribosome and

(a)

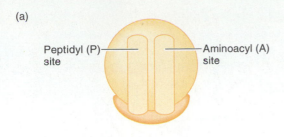

Peptidyl (P) site ———————————— Aminoacyl (A) site

(b)

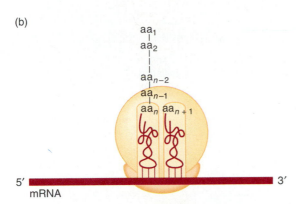

aa₁
aa₂
|
aa_{n-2}
aa_{n-1}
aa_n aa_{n+1}

5′ ———————————————————————————— 3′
mRNA

▶ **FIGURE 17–14** (a) A ribosome has two binding sites capable of combining with tRNA. (b) During polypeptide synthesis, a peptidyl-tRNA is bound to the P site and an aminoacyl-tRNA is bound to the A site. A ribosome in this state is in the proper form for further elongation of the polypeptide chain.

with the **initiator tRNA**, which carries the first amino acid, to start polypeptide chain growth. The process requires an energy source (mainly GTP) and several protein **initiation factors**—three called IF1, IF2, and IF3 in prokaryotes and as many as 10 or more, labeled eIF1, eIF2, and so forth, in eukaryotes. The initiator tRNA is charged with the amino acid **methionine** in eukaryotes. In prokaryotes, the methionine in the initiator tRNA (tRNA$_f^{Met}$) is formylated to produce **formylmethionyl-tRNA** (**fMet-tRNA**). Only methionine residues attached to the initiator tRNA receive a formyl group; those attached to the tRNA that recognizes internal AUG codons (tRNA$_m^{Met}$) do not (▶ Figure 17–15). Thus methionine is the initial amino acid of each newly synthesized polypeptide in eukaryotes, and formylmethionine is the initial amino acid in prokaryotes.

The initiation of translation in prokaryotes proceeds in three steps (see ▶ Figure 17–16a). The first step involves the binding of the initiation factors and GTP to the 30S ribosomal subunit. One of these factors, IF3, prevents the association of 30S and 50S subunits, while the other two factors, IF1 and IF2, aid in the binding of mRNA and initiator tRNA.

In the second step of the initiation process, initiator tRNA and the 5′ end of mRNA combine with the IF-GTP-30S ribosomal complex to form the 30S initiation complex. The binding of messenger RNA to the initiation complex involves specific pairing between a purine-rich sequence (called the **Shine-Dalgarno sequence**) located upstream of the initiation codon and a complementary sequence in the

16S rRNA of the 30S subunit. The binding of the initial transfer RNA also involves specific base pairing, but in this case with the mRNA AUG (or GUG) codon. When the AUG codon occurs in the interior of the gene message, it normally specifies the amino acid methionine and binds to Met-tRNA$_m^{Met}$. However, when AUG occurs at the beginning of the gene message in prokaryotes, it binds only to fMet-tRNA$_f^{Met}$.

Following the release of IF3, the 50S subunit of the ribosome combines with the 30S complex in the third step of initiation. This step requires the breakdown of GTP and results in the release of the two remaining initiation factors as well as formation of the complete ribosomal structure, the 70S initiation complex. The charged initiator tRNA is now bound at the P site, and the ribosomal complex is now ready to carry out the process of protein synthesis.

Figure 17–16b shows the corresponding events during initiation in eukaryotes. Note that the main events in eukaryotes are very similar to those in prokaryotic systems. The principal differences involve the nature of the initiation factors (many more are needed in eukaryotes) and the binding of mRNA. Eukaryotic messengers do not contain a ribosome binding sequence analogous to the Shine-Dalgarno sequence of prokaryotes. Instead, the 5′ cap of eukaryotic mRNA (see Chapter 16) seems to be the major structural feature needed for efficient binding. During initiation, a cap-binding protein combines first with the cap

▶ **FIGURE 17–15** Formation of initiator tRNA in prokaryotes. (a) Bacteria have two tRNAs that can accept methionine, but only one can be formylated by the enzyme transformylase to function as initiator tRNA. (b) The structure of the 3′ terminus of fMet-tRNA, showing the attached formylmethionine residue.

(a)

$$tRNA_f^{Met} \xrightarrow{Synthetase} methionyl\text{-}tRNA_f^{Met} \xrightarrow{Transformylase} fMet\text{-}tRNA_f^{Met}$$

$$tRNA_m^{Met} \xrightarrow{Synthetase} methionyl\text{-}tRNA_m^{Met} \xrightarrow{Transformylase} \text{no reaction}$$

(b)

tRNA$_f$

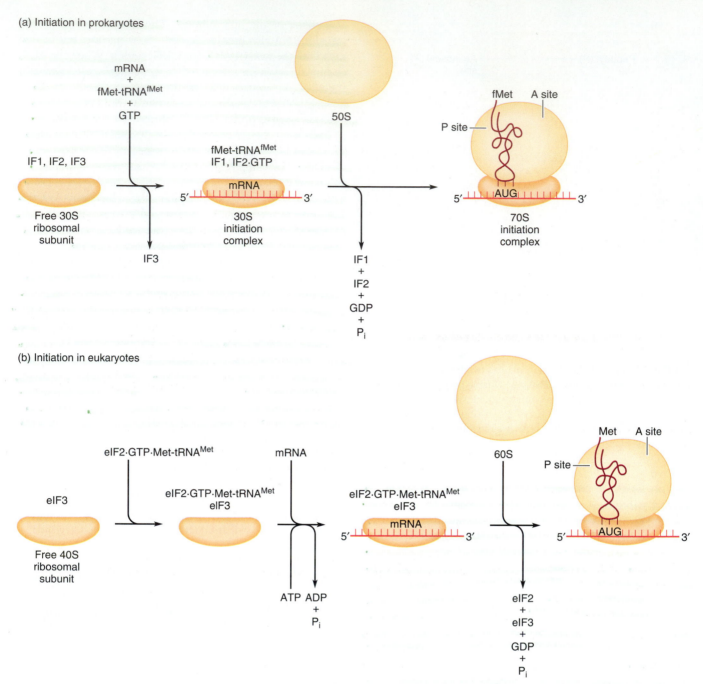

(a) Initiation in prokaryotes

IF1, IF2, IF3

Free 30S
ribosomal
subunit

mRNA
+
fMet-tRNA^{fMet}
+
GTP

IF3

fMet-tRNA^{fMet}
IF1, IF2·GTP

5′ 3′
mRNA
30S
initiation
complex

50S

IF1
+
IF2
+
GDP
+
P$_i$

fMet   A site

P site

5′ AUG 3′
70S
initiation
complex

(b) Initiation in eukaryotes

eIF3

Free 40S
ribosomal
subunit

eIF2·GTP·Met-tRNA^{Met}

eIF2·GTP·Met-tRNA^{Met}
eIF3

mRNA

ATP ADP
+
P$_i$

eIF2·GTP·Met-tRNA^{Met}
eIF3

5′ 3′
mRNA

60S

eIF2
+
eIF3
+
GDP
+
P$_i$

Met   A site

P site

5′ AUG 3′

▶ **FIGURE 17–16** Initiation of translation in (a) prokaryotes and (b) eukaryotes. The process requires various initiation factors and involves a stepwise binding of the initiator tRNA and mRNA to the ribosome subunits. In both cases, the initiator tRNA attaches first to the smaller ribosomal subunit and binds to the P site upon association with the larger ribosomal subunit. *Adapted from: Geoffrey Zubay, Biochemistry, 3d ed. (Wm. C. Brown, 1993), pp. 845 and 846. Copyright © 1993 Wm. C. Brown Communications, Inc., Dubuque, Iowa. All Rights Reserved. Reprinted by permission.*

and then with initiation factors to promote binding of the mRNA to the 40S ribosomal subunit.

Chain Elongation

Following initiation, polypeptide chain growth proceeds from its amino terminus to its carboxyl terminus. We can view the elongation of the polypeptide chain as a cyclic process that occurs through the repetition of the three steps

summarized in ▶ Figure 17–17: (1) **aminoacyl-tRNA binding**, (2) **transpeptidation**, and (3) **translocation**.

In the first step, an aminoacyl-tRNA associates with the vacant A site of the ribosome. This step is mediated by specific pairing between the codon and the anticodon and is accompanied by the breakdown of GTP. Aminoacyl-tRNA binding requires the action of the paired elongation factors EF-Tu and EF-Ts (EF1$_\alpha$ and EF1$_{\beta\gamma}$ in eukaryotes). As shown in ▶ Figure 17–18, EF-Tu in the form of an EF-Tu-

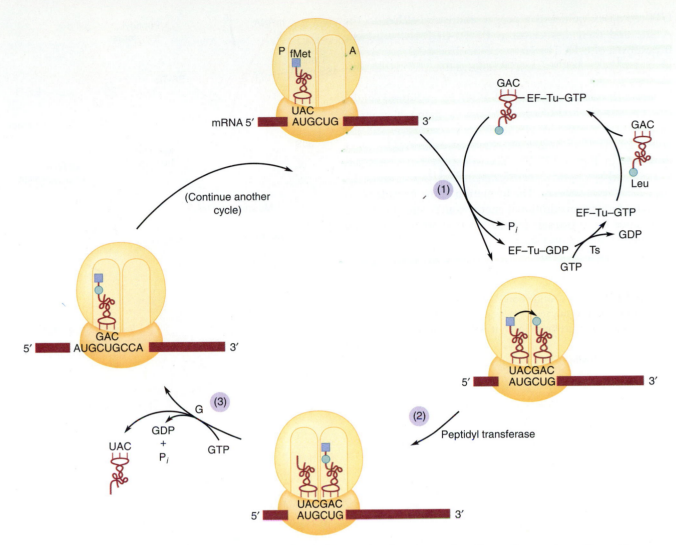

► **FIGURE 17–17** Elongation of the polypeptide chain. The addition of each amino acid residue to the growing polypeptide involves three steps: (1) aminoacyl-tRNA binding, (2) transpeptidation, and (3) translocation. The figure shows the addition of the second amino acid residue (in this case, leucine) to the initial formyl-methionine residue of a hypothetical prokaryotic polypeptide.

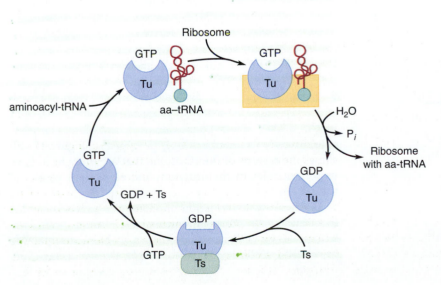

► **FIGURE 17–18** The reaction cycle of the elongation factors EF-Tu and EF-Ts in prokaryotes. The EF-Tu-GTP complex guides the proper aminoacyl-tRNA to the A site of the ribosome. GTP is subsequently hydrolyzed, and the GDP form of EF-Tu is released from the ribosome. EF-Ts then catalyzes the regeneration of EF-Tu-GTP to prepare it for interaction with another aminoacyl-tRNA.

GTP complex guides the proper aminoacyl-tRNA to the A site and then dissociates as EF-Tu-GDP following the breakdown of GTP. The factor EF-Ts then catalyzes the exchange of bound GDP for GTP to form a functional EF-Tu-GTP complex.

Once the incoming charged tRNA is stabilized at the A site, a peptide bond forms between the carboxyl group of the amino acid carried by the tRNA at the P site and the amino group of the amino acid carried by the tRNA now at the A site (▶ Figure 17–19). The transpeptidation reaction is catalyzed by **peptidyl transferase**, which is present in the large ribosomal subunit. The formation of the peptide bond does not require an additional energy source such as ATP or GTP, since the substrates for the reaction are already in an activated (high-energy) state.

In the final step in the cycle, translocation, the peptidyl-

▶ **FIGURE 17–19** Peptide bond synthesis on the ribosome. Bond formation is catalyzed by peptidyl transferase. "R" represents a formyl group in fMet; it is an H in Met.

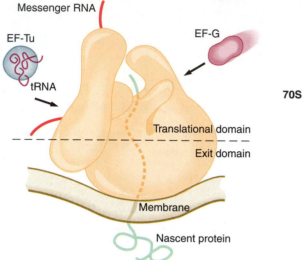

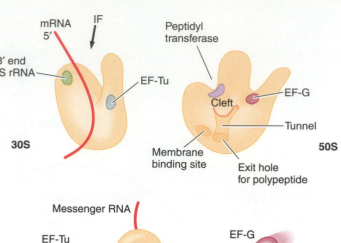

▶ **FIGURE 17–20** The functional sites of a bacterial ribosome. The polypeptide that is being synthesized is secreted through the cell membrane. A cleft in the 50S subunit leads to a tunnel that passes through the subunit. The polypeptide chain is thought to pass through this tunnel as it exits the cell. *Adapted from:* C. Bernabeu and J. A. Lake, *Proc. Natl. Acad. Sci. (U.S.)*, 79 (1982): 3111–3115. Used with permission of J. A. Lake.

tRNA at the A site moves to the P site, thus vacating the A site for further aminoacyl-tRNA binding. During translocation, the tRNA that was formerly bound to the P site is released for reuse (since it is no longer peptidyl tRNA). The transfer of the peptidyl-tRNA from the A site to the P site is accompanied by the movement of the mRNA along the ribosome by one codon, and requires the elongation factor EF-G (EF2 in eukaryotes) and the breakdown of another molecule of GTP. Translocation is believed to involve a change in the three-dimensional conformation of the entire ribosome in order to produce the one-codon shift in the mRNA. Once translocation is complete, the ribosome is ready for another elongation cycle. ▶ Figure 17–20 shows the regions of the ribosome that are thought to have important roles in the initiation and elongation phases of translation.

Much of our information on chain elongation has come from inhibiting different steps in this process with antibiotics. Many antibiotics are known to block polypeptide growth at specific stages. For example, the antibacterial ac-

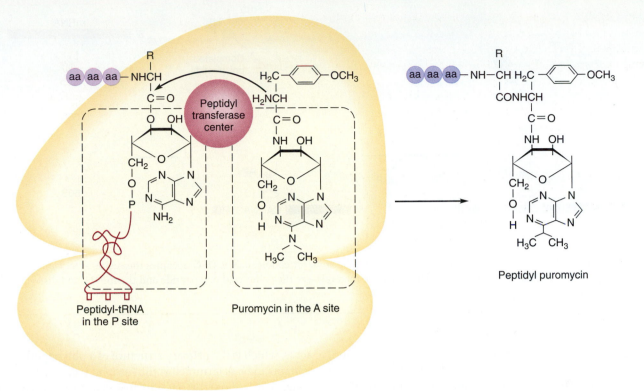

▶ **FIGURE 17–21** The antibiotic puromycin causes premature termination of polypeptide chain elongation because it resembles the aminoacyl terminus of an aminoacyl-tRNA. Its amino group joins to the carboxyl group of the growing polypeptide to form a peptidyl-puromycin chain that dissociates from the ribosome.

tion of puromycin lies in its ability to interrupt chain elongation by acting as a structural analog of aminoacyl-tRNA. Puromycin resembles the aminoacyl terminus of a charged tRNA and is mistaken for it during protein synthesis (▶ Figure 17–21). Puromycin binds to the A site and combines with the growing polypeptide chain, but it then dissociates from the ribosome, causing premature termination of protein synthesis.

■ Table 17–2 lists the modes of action of several of the better-known antibiotics that affect protein synthesis. All of the antibiotics listed, except for streptomycin, specifically inhibit the elongation phase of the process.

Termination of Translation

The elongation cycle of protein synthesis continues until a termination codon is encountered. Termination then occurs through the process shown in ▶ Figure 17–22. Termination requires the presence of certain termination (or release) factors that can recognize the termination codons UAA, UAG, and UGA in mRNA. In bacteria, one release factor (RF1) recognizes UAA and UAG, while another (RF2) recognizes UAA and UGA. Eukaryotes have only a single release factor (eRF) to recognize these codons. Once a termination codon arrives opposite the A site on the ribosome, it is recognized by the appropriate release factor, which then

blocks further chain elongation. The completed polypeptide chain then dissociates from the ribosome, followed by tRNA and mRNA. Finally, the ribosome separates into its two subunits, to be used again in the synthesis of another polypeptide chain.

■ **TABLE 17–2** Antibiotic inhibitors of protein synthesis.

Antibiotic	Mode of Action
chloramphenicol	inhibits peptidyl transferase in the 70S ribosomes of prokaryotes
cycloheximide	inhibits peptidyl transferase in the 80S ribosomes of eukaryotes
erythromycin	inhibits translocation in prokaryotes by binding to the 50S ribosomal subunit
puromycin	causes premature chain termination by acting as an analogue of charged tRNA
streptomycin	inhibits initiation and causes misreading of mRNA in prokaryotes by binding to protein S12 of the 30S ribosomal subunit
tetracycline	inhibits binding of charged tRNAs to the ribosomes of prokaryotes by binding to the 30S ribosomal subunit

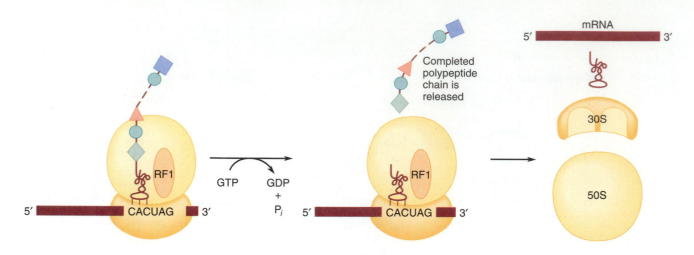

▶ **FIGURE 17–22** Termination of polypeptide synthesis. A terminator codon such as UAG occupies the A site. A termination factor (in this case, RF1) combines with the A site. The ribosome then releases the completed polypeptide and the mRNA and dissociates into its subunits.

All three phases of translation—initiation, elongation, and termination—are illustrated in ▶ Figure 17–23, which shows a ribosome reading the complete length of an mRNA molecule.

To Sum Up

1. Translation is the process in which the message in the mRNA is decoded into an amino acid sequence. Translation occurs on the ribosomes and is mediated by various protein initiation, elongation, and termination factors. Polypeptide synthesis is initiated by a special initiator tRNA, which brings in a methionine residue (formylmethionine in prokaryotes) in response to the mRNA initiator codon AUG. An initiation complex consisting of the AUG codon and the charged initiator tRNA forms at the P site of the ribosome; this complex then begins the synthesis of a polypeptide.
2. The elongation cycle of protein synthesis consists of three steps: aminoacyl-tRNA binding, in which the anticodon of an incoming charged tRNA is paired with a complementary codon in the mRNA at the ribosomal A site; transpeptidation, in which peptidyl transferase catalyzes the synthesis of a peptide bond between the amino acid residues; and translocation, which involves the transfer of a peptidyl-tRNA from the A site to the P site of the ribosome, accompanied by the movement of the mRNA along the ribosome by one codon.
3. Translation is terminated when the ribosome encounters any of the three mRNA terminator codons, UAA, UAG, and UGA.

◆

PROTEIN FOLDING: ASSISTED SELF-ASSEMBLY

To fully code for the structure of a protein, a gene must provide all the information needed for the polypeptide to fold into its biologically active conformation. We have seen how the sequence of codons determines the sequence of amino acids, which is the primary structure of a polypeptide. What remains to be determined is the "second half" of the genetic code, that is, how the one-dimensional information of the genetic message produces the specific three-dimensional protein structure needed for biological activity.

Until recently, most of our knowledge of protein folding came from studies on the refolding of denatured proteins in vitro in the absence of other cellular components. These studies indicated that refolding proceeds spontaneously along a pathway that involves (1) increasing compaction of the polypeptide due to the increasing association of nonpolar segments of the chain, (2) the creation of stable secondary structures (helices and sheets) through hydrogen bonding, and (3) the establishment of covalent interactions, such as disulfide bonds. The fact that certain proteins can refold on their own when placed in a suitable (nondenaturing) environment led to the hypothesis of protein self-assembly. According to this hypothesis, all information needed for proper folding is present in the amino acid sequence. Therefore, once the primary structure is formed, the polypeptide should automatically fold into its proper three-dimensional structure without requiring the assistance of other molecules.

▶ **FIGURE 17–23** Translation in *E. coli*. Initiation of translation occurs when the fMet-initiator tRNA and the 30S ribosome subunit combine with the mRNA (1). Elongation cycles add successive amino acids as the ribosome reads the mRNA in the 5′ to 3′ direction. The first elongation cycle consists of Leu-aminoacyl-tRNA binding at the A site (2), peptide bond formation at the P site (3), and translocation (4). Cycle two (5–7) adds Pro to the growing peptide chain; cycle three (8–10) adds Trp, and so on. Termination of peptide synthesis occurs when the stop codon UAA is encountered at the A site (11). The completed polypeptide chain is then released from the ribosome (12).

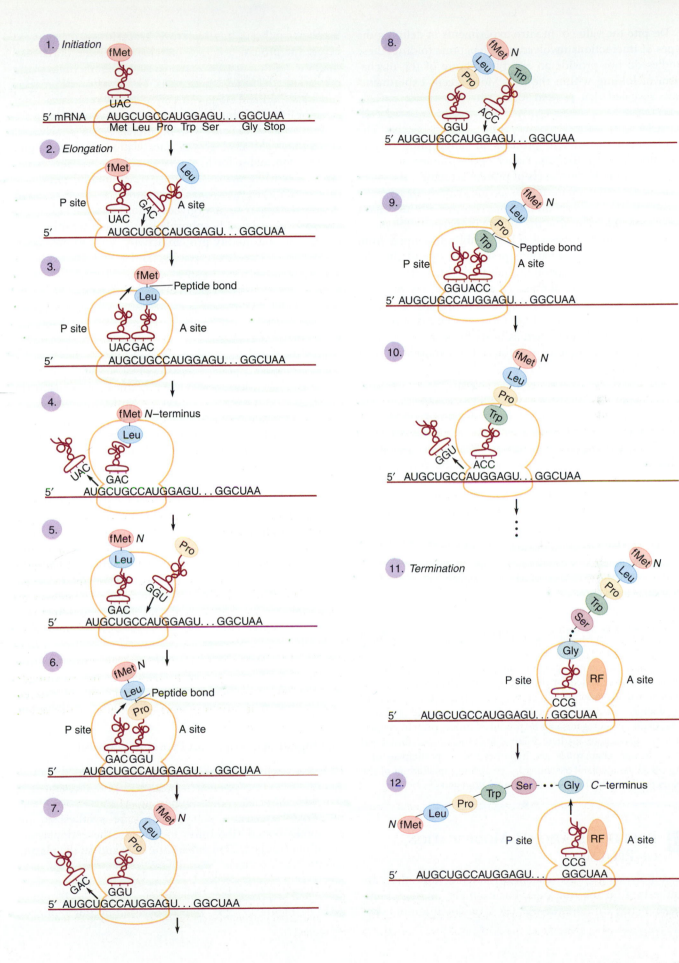

Despite the value of in vitro experiments in defining the types of interactions involved in polypeptide folding, these studies do not provide an accurate picture of the mechanism of folding within the cell. More recent experiments have revealed that protein folding in vivo is not entirely a self-assembly process but rather is one of assisted self-assembly, requiring the participation of other proteins. The involvement of different proteins in polypeptide folding does not diminish the importance of the primary structure in determining higher levels of protein structure; the function of the other proteins is to assist the primary structure in achieving the final functional conformation, not to impart additional information to the folding polypeptide chain.

The concept of assisted self-assembly developed from studies that implicated various proteins in modulating the state of folding of polypeptide chains in the cell. These proteins are produced in abundance under a variety of stress conditions that result in unfolded or malfolded polypeptides. For example, some of the proteins that assist in folding were originally identified as heat-shock proteins, since they increase in amount when a cell is exposed to high temperatures.

Two main types of proteins participate in protein folding: Conventional enzymes which catalyze specific steps (e.g., formation of disulfide bonds) that limit the rate of folding, and protein chaperones, which recognize and bind to folding polypeptides and thus mediate their correct assembly. When proteins with complex subunit structures are denatured, they often tend to form nonfunctional aggregates during refolding. Many of the protein chaperones that have been studied interact with and stabilize these denatured proteins and prevent the formation of aggregates. Their ability to prevent aggregation has contributed to the present hypothesis that chaperones normally assist in protein folding by preventing misfolding and the formation of inappropriate structures.

To Sum Up

The sequence of amino acids in a polypeptide chain contains all the information needed for the folding of the chain into its three-dimensional structure. However, it appears that there are a number of alternative ways in which a polypeptide can fold, and thus nonfunctional aggregates can be produced unless protein chaperones are present to mediate folding into the proper functional form. Protein chaperones are hypothesized to participate in a process of assisted self-assembly by preventing misfolding of the polypeptide chain(s) into nonfunctional aggregates.

POSTTRANSLATIONAL MODIFICATION OF PROTEINS

Most polypeptides are chemically altered after they are synthesized on the ribosome. For example, all peptide sequences start with methionine (or formylmethionine) during their synthesis, but the formyl group is usually removed before the protein becomes functional, and the initial methionine residue may be removed as well, sometimes along with a few additional amino acids. Thus functional proteins in prokaryotes and eukaryotes often do not contain methionine (or formylmethionine) as their N-terminal amino acid.

Protein structure may also be altered by the attachment of different chemical groups (acetylation, hydroxylation, methylation, and so forth) to amino acid residues after translation. Two of the more common covalent modifications are the phosphorylation of serine, threonine, and tyrosine and the attachment of sugars (glycosylation) to asparagine, serine, and threonine. Phosphorylation is often used by cells to regulate and modify protein activity. For example, protein synthesis in eukaryotes is inhibited by the phosphorylation of the initiation factor eIF2, which brings Met-tRNA to the 40S ribosomal subunit. Phosphorylation is catalyzed by an enzyme known as a protein kinase and uses ATP as a phosphate source. The factor eIF2 loses its ability to initiate protein synthesis when it is phosphorylated; however, it becomes functional again when the phosphate is removed by another enzyme, a phosphatase, by hydrolysis.

Glycosylation is generally used by eukaryotic cells to aid in targeting proteins for export or for delivery to certain cellular destinations, such as lysosomes. The oligosaccharide units are first acquired by the proteins in the endoplasmic reticulum (ER) and are later altered and elaborated in the Golgi apparatus. The role of the Golgi apparatus in the maturation of glycoproteins in a eukaryotic cell is shown in ▶ Figure 17–24. The membranous sacs of the Golgi apparatus are functionally polarized. The sacs at the cis end of the stack are specialized for receiving the glycoproteins arriving from the ER, while the sacs at the trans end sort and package the modified proteins for delivery to their intracellular and extracellular destinations. During their transit from the cis end to the trans end of the Golgi stack, the glycoproteins are processed by a sequence of reactions (including phosphorylation) that mainly alter the oligosaccharide chain. Some of the proteins that bud off from the Golgi stack are packaged into lysosomes to function as lysosomal enzymes, while others enter secretory vesicles to be released outside the cell. Glycosylation is thus a characteristic feature of all proteins that transit the ER and Golgi system.

The Signal Sequence and Protein Secretion

The proteolytic cleavage of a peptide chain is the most general of all covalent modifications of proteins, and it is a modification that is common to most secretory and membrane proteins. Such proteins are synthesized on membrane-bound ribosomes located on the endoplasmic reticulum (rough ER) in eukaryotic cells and on the plasma membrane in bacteria. Virtually all of these proteins contain a relatively hydrophobic N-terminal sequence called the signal sequence, which consists of about 15–35 amino acids (▶ Figure 17–25). During translation, this sequence

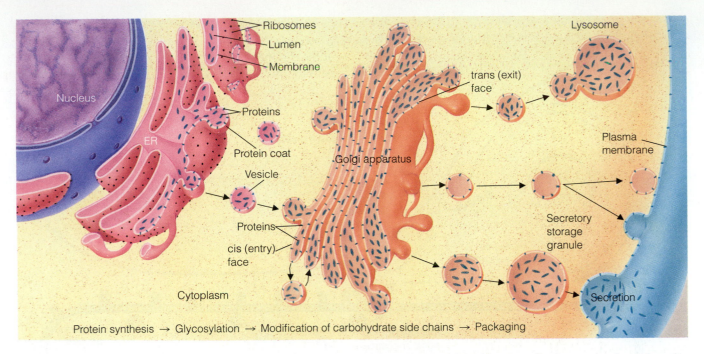

▶ **FIGURE 17–24** The sorting and maturation of proteins in the membrane system of a eukaryotic cell. The proteins formed on the ribosomes are inserted into the ER membrane or released into its lumen, where they are glycosylated. Certain glycoproteins are enclosed in vesicles and transported to the cis compartment of the Golgi apparatus. The glycoproteins are then modified as they pass through the stacks, eventually reaching the trans compartment, where they are sorted and packaged for delivery. Some glycoproteins are secreted, others are inserted into the plasma membrane, and still others are incorporated into lysosomes.

directs the ribosome to the membrane surface and initiates the passage of the protein through the membrane. The signal sequence is eventually cleaved from the growing polypeptide chain by a membrane-bound protease (called a **signal peptidase**) before the passage of the protein through the membrane is complete. Therefore, the protein that exits a bacterial cell by this mechanism or that enters the lumen of the rough ER in a eukaryotic cell will contain fewer

▶ **FIGURE 17–25** Role of the signal sequence in protein secretion. An N-terminal signal sequence on secretory proteins aids in the attachment of polyribosomes to the rough ER and in the subsequent passage of the protein through the ER membrane. This sequence is cleaved by a membrane-bound peptidase (the signal peptidase) before translation is complete.

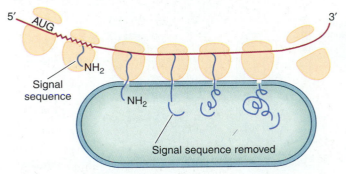

amino acids at its N terminus than coded for by its structural gene.

The stages of translation of a secretory protein on the endoplasmic reticulum are illustrated in ▶ Figure 17–26. In eukaryotes, ribosomes are delivered to the rough ER in combination with a ribonucleoprotein component called the **signal recognition particle (SRP)**, which binds to the newly formed signal sequence of the growing polypeptide, halting translation as it does so. The SRP-ribosome complex then diffuses to the membrane surface, where it combines with an **SRP receptor** (**docking** protein). Once at the membrane surface, the ribosome is attached to special receptor proteins, the SRP dissociates into the cytosol, and polypeptide chain elongation resumes, threading the protein across the ER membrane. The protein is finally released into the lumen of the rough ER following the cleavage of its signal sequence.

The proteolytic modification of proteins is not always limited to the removal of a signal sequence. Some secretory proteins undergo further cleavage en route to the plasma membrane during their release from the cell. One example is the proteolytic processing that occurs during the maturation of the hormone insulin (▶ Figure 17–27). Preproinsulin, the insulin molecule coded for by mRNA, consists of 107 amino acids. A signal peptide of 23 residues is cleaved off during the protein's passage through the ER membrane, leaving a molecule called proinsulin containing 84 amino

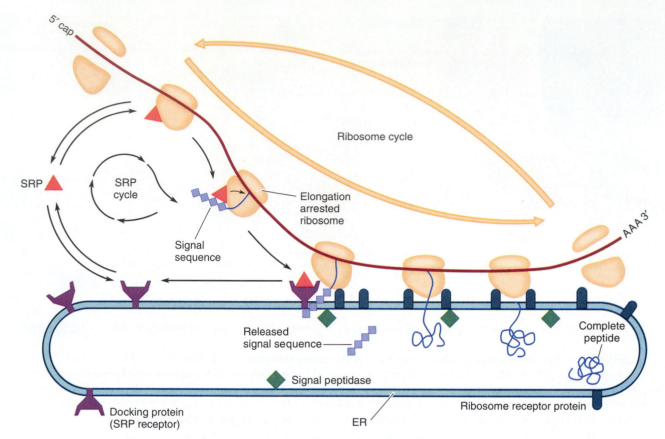

Ribosome cycle

SRP

SRP cycle

Signal sequence

Elongation arrested ribosome

AAA 3'

Released signal sequence

Complete peptide

Signal peptidase

Docking protein (SRP receptor)

ER

Ribosome receptor protein

5' cap

▶ **FIGURE 17−26** The translation of a secretory protein on the rough ER. The signal recognition particle (SRP) guides a ribosome to a receptor on the endoplasmic reticulum, arresting translation as it does so. Translation is then resumed, the growing polypeptide is transported across the membrane, and the completed chain is released into the lumen of the rough ER after the signal sequence is removed by the signal peptidase. *Adapted from:* P. Wallace, R. Gilmore, and G. Blobel. "Protein translocation across the endoplasmic reticulum," *Cell* 38 (1984): 5−8. Copyright © by Cell Press. Used with permission.

acids. Following this cleavage, two disulfide bonds are formed, and the molecule is transferred to the Golgi apparatus, where it is packaged into a secretory vesicle. While in the secretory vesicle, a connecting or C peptide is removed by a cleaving enzyme to yield the functional insulin molecule of 51 amino acids. The hormone that is finally secreted by the insulin-producing cell thus consists of two polypeptide chains, the A and B chains, linked by disulfide bridges.

To Sum Up

1. Most polypeptides are chemically modified after they are synthesized, for example, by phosphorylation of serine, threonine, or tyrosine, and/or by glycosylation of asparagine, serine, or threonine. Phosphorylation is often used as a means of regulating the activity of a protein once it is synthesized, while glycosylation in eukaryotic cells is generally part of the process of

▶ **FIGURE 17−27** The processing of insulin. Preproinsulin, the peptide coded for by the mRNA, is first converted to proinsulin in the endoplasmic reticulum by cleavage of the signal sequence and formation of disulfide bonds. Proinsulin is converted to insulin in the secretory vesicle by the removal of a connecting (C) peptide.

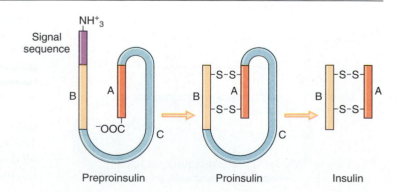

Signal sequence

NH_3^+

B

A

$^-$OOC

C

Preproinsulin

B

A

S–S

S–S

C

Proinsulin

B

A

S–S

S–S

Insulin

directing proteins to various intracellular and extracellular destinations.

2. Proteolytic cleavage is the most common posttranslational modification of proteins. It is part of the process of transporting a polypeptide chain through the cell membrane, and it is also important in the maturation process that some proteins must undergo to become functional. Functional proteins are often formed from longer precursor polypeptides by one or more cleavage events.

3. Almost all secretory and membrane proteins contain a signal sequence at their N-terminus. This hydrophobic segment directs the ribosome to the cell membrane and initiates the passage of the growing polypeptide chain through the membrane. The signal sequence is then cleaved from the polypeptide chain as it is transported through the membrane.

Chapter Summary

Once mRNA is formed, translation of its codon sequence into the amino acid sequence of a polypeptide occurs on the surface of a ribosome. Each ribosome is composed of two subunits, each containing both rRNA and protein components. The rRNA components are cut out of larger precursor molecules synthesized from rRNA transcription units.

During translation, tRNAs carry amino acids to the ribosome in a sequence dictated by the mRNA. The tRNAs thus serve as adaptor molecules that mediate the transfer of information from the mRNA to the polypeptide. There are several different kinds of tRNAs—at least one for each of the 20 common amino acids. All tRNAs are similar in length and three-dimensional structure and, like rRNAs, are cut out of larger transcripts. Every tRNA molecule shows two types of specificity: It binds chemically to only one kind of amino acid and recognizes one (in some cases, more than one) mRNA codon. Codon specificity resides in the anticodon—a sequence of three nucleotides that is complementary to a codon in mRNA.

Translation is a complex chemical process that requires the participation of ribosomes, charged tRNAs, an energy source (mainly GTP), and several different protein factors. Protein synthesis begins when an mRNA and the first charged tRNA are bound by a ribosome. The mRNA then acts as a tape of instructions; charged tRNA molecules are successively bonded to the codons in mRNA by anticodon-codon interactions as the mRNA moves along the ribosome. The message is read from 5′ to 3′ and ends once a stop codon in the mRNA is reached.

The growing polypeptide folds with the aid of various protein factors called chaperones and assumes a characteristic three-dimensional structure upon the completion of translation. Most polypeptides are also chemically modified (for example, by phosphorylation, glycosylation, and/or proteolytic cleavage) before they attain their final biologically active conformation.

Questions and Problems

Ribosomes and Ribosomal RNA

1. Compare the structures of prokaryotic and eukaryotic ribosomes. List the structural features that are common to both types of ribosomes and the features that are unique to each type.

2. For many RNA molecules, the sedimentation coefficient is directly proportional to the square-root of the molecular weight. Knowing that the 18S rRNA of eukaryotes is 1.9 kb in length and assuming that the ratio of the number of bases to the molecular weight is the same for the other rRNA components, estimate the lengths of 28S rRNA and 5.8S rRNA.

3. What ensures that there will be equal amounts of 5S, 16S, and 23S rRNA for prokaryotic ribosomes?

4. Which major activities of protein synthesis take place on the small ribosomal subunit? Which take place on the large subunit?

5. Suppose that bacteria are grown in a medium containing the "heavy" isotopes ^{13}C and ^{15}N until the bacterial ribosomes are fully labeled. The labeled ribosomes are then isolated and added with an equal quantity of unlabeled (^{12}C- and ^{14}N-containing) ribosomes to an in vitro system for protein synthesis. After several hours, the ribosomes are removed from the protein synthesizing system and subjected to equilibrium density-gradient centrifugation. How many bands would you expect to obtain in your gradient? Explain.

6. You are studying a ribonucleoprotein and want to determine the structure and organization of the RNA component. A primary structure analysis reveals that the RNA has the following sequence:

5′ CUCAGGCCUUAGCCGGAGGCCUUGGCUGGCA—
GGGCGUUGGCGACCCUGACC 3′

When you treat the ribonucleoprotein with an RNase that degrades single-stranded RNA, you are left with two RNase-resistant sequences, 5′ UAGCCG 3′ and 5′ GUUGGCG 3′, that were protected by the protein, and two base-paired double-stranded sequences,

5′ AGGCC 3′ and 5′ CAGGG 3′
3′ UCCGG 5′   3′ GUCCC 5′

Diagram the base-paired structure of this RNA, showing how it is attached to the protein.

Transfer RNA

7. What would happen to the amino acid sequence of a polypeptide if the anticodon of tRNA^His were changed by a single nucleotide substitution so that it recognized an Arg codon on the mRNA?

8. Look back at problem 2 in Chapter 16. (a) List the anticodons on the tRNA molecules that recognize each mRNA codon (assume no wobble in this case). (b) Give the amino acid sequence of the polypeptide made by translation.

9. The following table provides just enough information about a section of a particular gene to allow you to determine (a) the sequence of base pairs along the DNA, (b) which DNA strand serves as the template for mRNA transcription, (c) the mRNA nucleotide sequence, (d) the tRNA anticodons, and (e) the amino acid sequence of the polypeptide. Complete the table.

DNA		G		T		
DNA			T T			T G A
mRNA		U C	U			
tRNA				A U G		
amino acids	Met					

10. Because of wobble, many tRNA molecules can often recognize two (or more) mRNA codons (see Table 17–1). Referring to the mRNA codon assignments given in Table 16–3, determine the minimum number of different tRNA molecules required to recognize the leucine codons.

11. Inspection of the genetic code reveals that there are 2 amino acids with one codon each, 9 amino acids with two codons each, 1 amino acid with three codons, 5 amino acids with four codons each, and 3 amino acids with six codons each, accounting for the 61 codons that code for amino acids. Calculate the minimum number of tRNAs needed to decode all 61 codons, considering wobble.

12. For the three major classes of RNA that participate in translation, make a table summarizing the distinguishing features of (a) their origin, abundance, and processing and (b) their structure and function. Compare prokaryotic and eukaryotic systems where possible.

Decoding the Message: Translation

13. Describe the events involved in the three phases of translation (initiation, elongation, and termination) as they are known to occur in *E. coli*.

14. List seven differences between eukaryotic and prokaryotic protein synthesis.

15. The accompanying diagram of the relationships between tRNAs and mRNA at the ribosome is incorrect in many respects. (a) Redraw the illustration, correcting the errors.

(b) Draw and label appropriate diagrams to show how these (correct) relationships change during the remaining two steps in the same cycle of chain elongation.

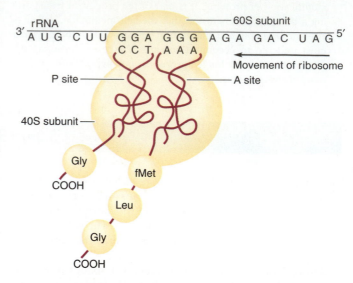

16. Suppose that you give a short pulse of the 20 amino acids (each labeled with a radioactive isotope) to cells that are actively synthesizing a particular protein. Immediately after the pulse, completely finished molecules of the protein are isolated and analyzed for the location of the radioactive label. In what part of the finished molecules would you expect to find the label: (a) toward the N terminus, (b) toward the C terminus, (c) toward the center, or (d) randomly incorporated, distributed over all parts of the chain? Explain your choice.

17. Calculate the number of high-energy ATP and GTP phosphate bonds that must be hydrolyzed in order for an *E. coli* cell to synthesize a protein containing 50 amino acid residues. (Include all $PP_i \rightarrow 2P_i$ reactions in your calculations.)

18. Despite intensive efforts, no one has been able to isolate a ribosomal protein(s) with peptidyl transferase activity. Can you speculate on possible reasons why this search has so far been fruitless?

Protein Folding: Assisted Self-Assembly

19. What major steps are believed to take place during protein folding? What functions do enzymes and protein chaperones perform in the process?

Posttranslational Modification of Proteins

20. List the major modifications that occur during the posttranslational processing of proteins.

21. Describe the process by which a eukaryotic cell targets proteins for lysosomes and for release outside the cell.

22. Distinguish between the signal sequence, the signal recognition particle (SRP), and the SRP receptor.

CHAPTER
18

Gene Cloning and the Analysis of Gene Function

The genetic technologies discussed in Chapter 15—restriction mapping, DNA probes, DNA sequencing, and DNA amplification techniques such as the polymerase chain reaction—make it possible to map and study the structure of many eukaryotic genes at the DNA level. Although these techniques have revolutionized the study of eukaryotic molecular genetics, they comprise only one part of the vast array of new technology available. An approach known as "reverse genetics" has been developed to study the function of eukaryotic genes: instead of using the traditional methods of working from a phenotype back to the gene responsible for the phenotype, researchers work forward, from the gene to the function that gene specifies. In reverse genetics, a gene is first isolated by a procedure known as **gene cloning**. Specific mutations are then directed to precise locations in the gene by the technique of **site-directed mutagenesis** (to be described in Chapter 22), and the altered gene

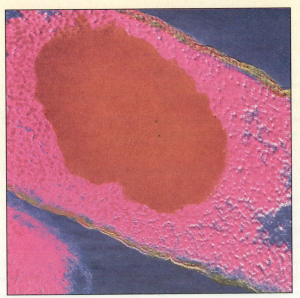

Transmission electron micrograph of *E. coli* cells genetically engineered to produce human insulin (orange).

is incorporated into a suitable host cell or organism where its expression can be monitored. Not only has this methodology been crucial to determining the molecular functions of many genes, it also has been the key to identifying the nucleotide sequences important in regulating eukaryotic gene expression.

This chapter will describe the basic techniques used to clone eukaryotic DNA and to identify and isolate the gene that has been cloned. It will also focus on the use of cloning technology to study the expression of genes inserted into bacterial and eukaryotic cells. Finally, genetic engineering of cells and organisms and its implications to medicine, agriculture, and industry will be discussed.

GENE CLONING TECHNIQUES

The development of **recombinant DNA** techniques in the early 1970s revolutionized the study of eukaryotic molecular genetics by providing the tools needed for **molecular cloning**, including the cloning of specific genes. DNA cloning is the amplification of a defined DNA fragment by linking it to a self-replicating vector (a plasmid or viral genome) and propagating the resulting recombinant DNA molecule in a host organism (often a bacterium). Large amounts of the DNA fragment are then produced by the host. Although newer technologies, such as the polymerase chain reaction, are now routinely used to amplify known DNA sequences for purposes such as sequencing or fingerprinting, gene cloning remains the method of choice for broader analyses of gene function.

Cloning Vectors

In 1972, investigators at Stanford University found that any two DNA fragments produced by the *Eco*RI restriction enzyme, regardless of their origin, would form hydrogen bonds with one another at their complementary ends. The fragments can then be joined permanently by the action of a DNA ligase, creating a recombinant DNA molecule. To create a recombinant DNA molecule that is also capable of replicating itself within a host cell, the DNA of a plasmid or virus is used in its construction. The plasmid or virus serves as a **vector** for the transfer of the recombinant DNA molecule into a host organism and for its subsequent replication. The DNA of the vector, which contains one or more different restriction sites, is cleaved at a unique site by a restriction enzyme, such as *Eco*RI (▶ Figure 18–1). Having a unique site for a particular restriction enzyme means the vector will be linearized but not cut into separate pieces. The linear DNA of the vector is then mixed with DNA fragments from another source (referred to here as the donor or foreign DNA), which also has cohesive ends generated by the same restriction enzyme. The donor DNA can be obtained from any microbial, animal, or plant source. Hydrogen bond formation between complementary ends and the subsequent action of DNA ligase yields the hybrid vector

chromosome that constitutes the recombinant DNA molecule. This recombinant DNA can then be used to transform competent bacterial cells from which it can subsequently be obtained in large quantities after replication as a plasmid or viral chromosome.

Many different plasmids and viruses are currently being used as vectors in recombinant DNA work; some of the commonly utilized ones are described in ■ Table 18–1. The most widely used cloning vectors have two major desirable features: (1) single cutting sites for each of a large number of restriction endonucleases and (2) the ability to confer readily selectable phenotypic markers on transformed host cells. Ideally, one of the restriction sites lies in a marker that is inactivated by the insertion of donor DNA, so that recombinant vectors can be distinguished from nonrecombinant ones (those that have self-annealed or annealed with another copy of the vector).

Certain plasmids that are widely distributed in prokaryotes and carry one or more genes for antibiotic resistance offer particular advantages as cloning vectors. The latter feature makes the selection of bacteria that have taken up a recombinant plasmid a simple matter. After being incubated with recombinant plasmids, the recipient bacteria (which previously carried no plasmids of their own) are placed on a medium containing one or more antibiotics. Only those cells that have acquired a plasmid are able to survive.

Plasmids that carry two different genes for antibiotic resistance, such as pBR322 (▶ Figure 18–2), have been constructed to allow specific selection of cells that have acquired a hybrid plasmid carrying the donor DNA. Each resistance gene contains a different set of restriction enzyme target sequences, so investigators can select an appropriate restriction enzyme to splice the donor DNA into one of the resistance genes (the tetracycline resistance gene in the example in Figure 18–2). This insertion inactivates the gene, and thus any bacteria that receive the hybrid plasmid will be resistant only to the other antibiotic. The replica-plating technique (see Chapter 11) can then be used to distinguish cells that maintain both resistance markers (that is, contain nonrecombinant plasmids) from those that have received a recombinant vector. This highly specific system ensures that only the bacterial cells that have taken up a recombinant DNA molecule of defined construction will be selected.

Because plasmid pBR322 has been sequenced, the exact locations of all of its restriction sites are known. Several restriction enzymes have unique sites that insertionally inactivate either the ampicillin or the tetracycline resistance genes, so this plasmid and its derivatives are versatile, widely used cloning vehicles.

In some derivatives of pBR322, one of the antibiotic resistance genes has been replaced by a selective marker whose phenotype is based on the ability of bacterial cells to metabolize the sugar lactose. ▶ Figure 18–3 shows one of these vectors, the pUC18 member of the pUC series. This plasmid retains the ampicillin resistance gene and replica-

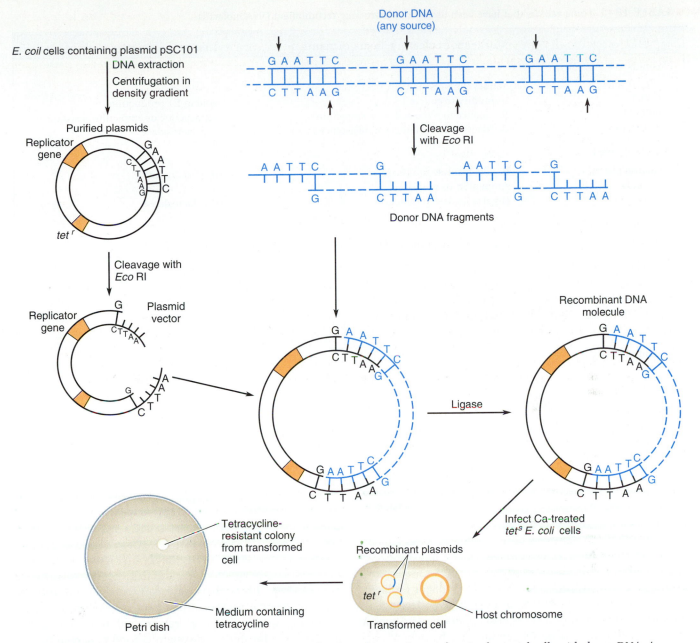

E. coil cells containing plasmid pSC101

DNA extraction

Centrifugation in density gradient

Purified plasmids

Replicator gene

tet ^{r}

Cleavage with Eco RI

Replicator gene

Plasmid vector

Donor DNA (any source)

Cleavage with Eco RI

Donor DNA fragments

Ligase

Recombinant DNA molecule

Infect Ca-treated tet ^{s} E. coli cells

Recombinant plasmids

tet ^{r}

Transformed cell

Host chromosome

Tetracycline-resistant colony from transformed cell

Petri dish

Medium containing tetracycline

▶ **FIGURE 18−1** Construction of a recombinant DNA molecule and its use in transforming bacterial cells with donor DNA. A restriction enzyme (for example, *Eco*RI) that cleaves DNA at 6-bp target sites generates donor fragments of average length 4,096 bp. Ligation into the plasmid gives the recombinant DNA, which is used to transform *E. coli* cells. Transformed cells are selected by their resistance to an antibiotic (in this case, tetracycline), which indicates that they have taken up a plasmid.

tion origin of pBR322, but 40% of its DNA, including the tetracycline resistance gene, has been deleted and replaced by the amino-terminal section of the bacterial gene *lacz* (denoted *lacz'*). A segment of so-called **polylinker DNA** containing **multiple cloning sites (MCS)** has been spliced into *lacz'*. The MCS segment, shown in detail in Figure 18–3b, contains several different restriction sites, so any of a large number of enzymes can be used to construct recombinant plasmids.

The *lacz'* region provides a means of distinguishing bacterial cells that have taken up a hybrid plasmid from those

that have not. This vector is used with a special strain of bacteria that carry the carboxyl-terminal portion of the *lacz* gene. When intact N-terminal and C-terminal regions are present in the same cell (as would be the case if a non-recombinant pUC were taken up by the bacterium), intra-genic complementation occurs and the cell produces the gene z product, the enzyme β-galactosidase, which begins the metabolism of the sugar lactose. When a lactose ana-logue is added to the bacterial growth medium, it is hy-drolyzed and causes the growth medium to turn blue. In contrast, when the N-terminal portion of the *lacz* region

■ TABLE 18–1 Some vectors that have been useful in constructing recombinant DNA molecules.

	Marker for Selecting Transformants	Insertional Inactivation of . . .
Plasmids		
pSC101	Tetracycline-resistance	——
ColE1	Immunity to colicin E1	colicin E1 production
pBR322	Resistance to tetracycline or ampicillin	tetracycline or ampicillin resistance
pUC series	β-galactosidase production; resistance to ampicillin	β-galactosidase
Phage λ-derived		
Charon phage	β-galactosidase production	β-galactosidase
cosmids	Ampicillin resistance	——
phasmids	β-galactosidase production	β-galactosidase
Filamentous coliphages		
f1, M13	β-galactosidase production	β-galactosidase
Yeast shuttle vectors		
2μ-based	Synthesis of histidine,	
ARS-based	leucine, lysine,	——
ARS-CEN-based	tryptophan, or uracil	
YAC		
Animal vectors		
SV40	Several dominant markers, e.g., neomycin	viral structural genes
retrovirus	resistance	
Plant vectors		
Ti plasmid	Synthesis of opines; several dominant markers, e.g., neomycin resistance	T region

has been destroyed by the insertion of donor DNA in recombinant pUCs, recipient cells are unable to metabolize lactose and their colonies remain white.

Although plasmids are perhaps the most convenient cloning vectors, they do have one disadvantage compared to some other vectors—they can accommodate only a rather limited length of donor DNA. This can be a major problem, for example, if one wants to clone a whole gene. This length limitation on the inserted DNA also applies to wild-type λ phage, which can increase its genome size by only 5% before exceeding the DNA capacity of its head. However, λ is so well known genetically that researchers have been able to construct derivatives of this phage that can incorporate much larger amounts of DNA. Donor DNA is inserted into these vectors and the recombinant molecules are then introduced into recipient cells in the form of either naked DNA (transfection) or as reconstituted particles (infection). The cloned DNA is then recovered in the form of a plaque on a lawn of susceptible bacteria. Because transfection is a very inefficient process, yielding only about 10^3–10^4 plaques per μg of recombinant DNA, an **in vitro packaging system** has been devised to package recombinant molecules into phage coats. Infection, which is far more efficient (up to 10^{12} plaques per μg of recombinant DNA), can then be used to introduce the DNA into host cells for cloning.

The phage λ derivatives that have become popular cloning vectors are called **replacement or insertional vectors,**

because a section of phage DNA is deleted to make room for insertion of the donor DNA. In the first phage vectors to be constructed in this way, the so-called **Charon phages,** the middle region, which contains no λ genes important to lytic growth, is replaced by a *lacz'* insert (▶ Figure 18–4). The capacity of the phage head dictates that the insert can be up to 20 kb long. Researchers have constructed 16 of these λ phage insertional vectors. Some have restriction (cloning) sites lying within the *lacz'* region and thus the simple blue/white color indicator system is used to screen for recombinant bacterial cells. Cells that receive a recombinant phage genome fail to metabolize lactose and yield colorless plaques.

Perhaps the most extreme examples of replacement vectors are the **cosmids** (▶ Figure 18–5), which are basically plasmid vectors that contain the λ cos sites (the cohesive termini of the chromosome; see Chapter 10) to allow efficient packaging in vitro. Very large donor DNA fragments of 37–52 kb (almost the entire length of a λ chromosome) must be inserted between the cos sites in order for packaging to occur. In vitro packaging yields phage particles that are used to infect a host cell. The cos sites cause the recombinant DNA to circularize within the cell, where the plasmid origin of replication allows the DNA to replicate as a plasmid rather than as a phage. Transformed cells are selected on the basis of antibiotic resistance.

The **phasmids** are another class of vectors that combine

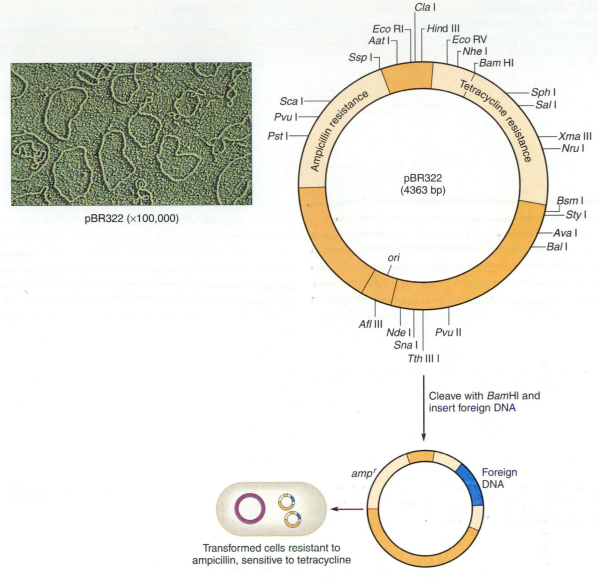

Cla I
Eco RI
Aat I
Ssp I
Hind III
Eco RV
Nhe I
Bam HI

Ampicillin resistance
Tetracycline resistance

Sca I
Pvu I
Pst I

Sph I
Sal I

Xma III
Nru I

pBR322
(4363 bp)

Bsm I
Sty I
Ava I
Bal I

ori

Afl III
Nde I
Sna I
Tth III I
Pvu II

Cleave with *Bam*HI and
insert foreign DNA

amp^r

Foreign
DNA

Transformed cells resistant to
ampicillin, sensitive to tetracycline

▶ **FIGURE 18–2** The structure of plasmid pBR322, which carries two genes for antibiotic resistance. Each antibiotic resistance gene contains several unique restriction sites. Insertion of foreign DNA into one of these restriction sites inactivates one of the antibiotic resistance markers. In this example, the *tet^r* gene is inactivated, leaving the transformed bacteria resistant only to ampicillin. *Adapted from:* R. W. Old and S. B. Primrose, *Principles of Gene Manipulation* (Oxford: Blackwell Scientific Publications, 1989), p. 51. Used with permission.

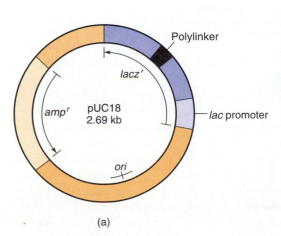

Polylinker

lacz'

amp^r

pUC18
2.69 kb

lac promoter

ori

(a)

▶ **FIGURE 18–3** (a) The widely used vector pUC18, a derivative of pBR322. pUC18 retains the *amp^r* marker and *ori* of pBR322, but the *tet^r* region has been replaced by part of a bacterial gene that codes for the N-terminal portion of an enzyme that metabolizes lactose (*lacz'*) and the *lac* promoter. An artificially synthesized polylinker region that contains multiple cloning sites (MCS) has been inserted into the partial *lacz'* gene, providing a metabolic selection mechanism for detecting transformed cells. (b) Some of the different restriction sites in the MCS of pUC18. Other members of the pUC series contain different MCS inserts.

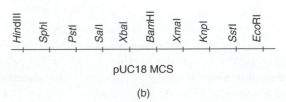

HindIII
SphI
PstI
SalI
XbaI
BamHI
XmaI
KpnI
SstI
EcoRI

pUC18 MCS

(b)

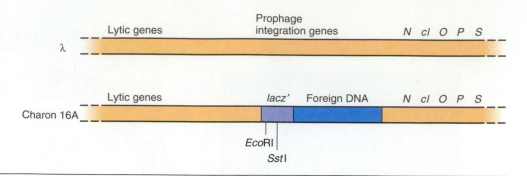

FIGURE 18–4 Structure of a Charon phage vector (bottom), compared to wild-type λ (top). The central portion of the λ genome is not involved in lytic growth. In the Charon phage vector, it is replaced by the selective *lacz'* marker, which contains unique *Eco*RI and *Sst*I sites, and foreign DNA.

λ and plasmid sequences and accommodate large inserts. These vectors are λ phage genomes that contain one (or more) plasmid molecules (▶ Figure 18–6). They are constructed using a plasmid that carries an *att BOB'* site, which allows it to insert itself into the phage genome by means of site-specific recombination (see Chapter 10). This reversible process is known as "lifting" the plasmid onto the phage genome. Donor DNA can either be inserted into the plas-

mid before it is lifted onto the phage genome, or it can be inserted into the phasmid, as in Figure 18–6. The recombinant molecules are then packaged into phage particles in vitro for efficient delivery into recipient cells. After the bacterial cells are infected, the recombinant DNA molecules can be propagated as either plasmids or phage, since the phasmid vector contains replication origins for both.

The growing emphasis on DNA sequencing by the chain-

▶ **FIGURE 18–5** (a) Formation of a cosmid vector. The λ cohesive ends (cos sites) are included in a vector that contains a plasmid *amp^r* gene and *ori* as well as a restriction enzyme target site. (b) Structure of a recombinant DNA molecule, and cloning the foreign gene in *E. coli*. Cosmids form long concatemers containing the donor DNA. The foreign DNA must be long enough to separate the cos sites by at least 37 but no more than 52 kb of DNA, the minimum and maximum distances required for packaging a phage-sized chromosome. In vitro packaging gives recombinant cosmids that are used to infect *E. coli* cells.

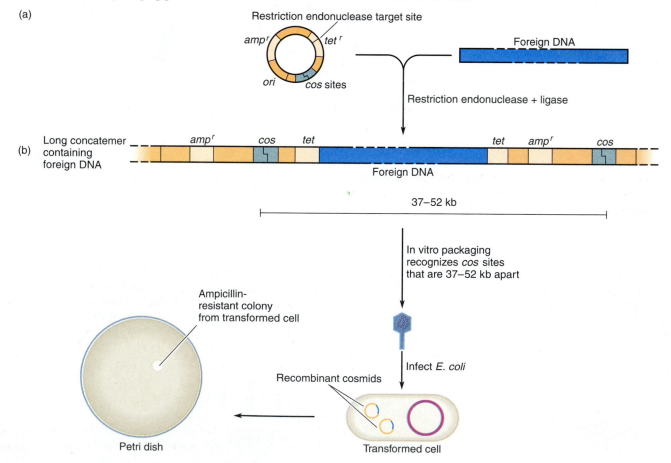

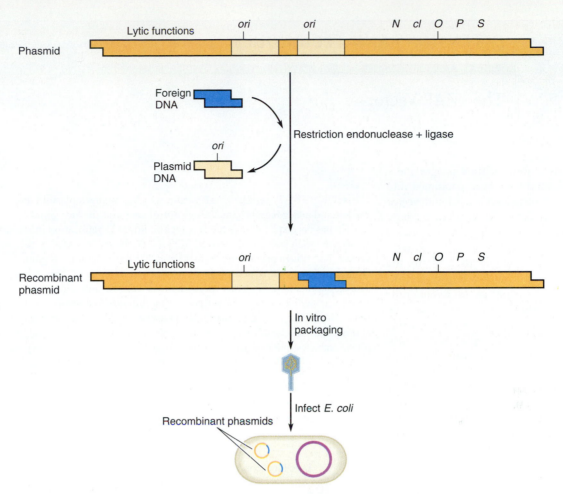

▶ **FIGURE 18–6** Use of a phasmid vector to transform *E. coli*. The phasmid contains all λ genes necessary for lytic growth, as well as two sections of plasmid DNA sequences containing plasmid *ori* sites. To construct a recombinant phasmid, foreign DNA is inserted in place of one of the plasmid segments. In vitro packaging of the recombinant phasmids is followed by infection of *E. coli*, where the phasmids can replicate from either the plasmid or the λ *ori* (located in gene *O*).

terminator (dideoxy) method and on site-directed mutagenesis has led to the development of another group of vectors that yield recombinant DNA in a single-stranded form; phage M13 and the phasmid λZAP are particularly useful examples of this type of vector. M13 and the closely related phages f1 and fd are filamentous coliphages containing circular, single-stranded DNA. These phages infect only F⁺ *E. coli* cells by absorbing to an F-pilus. As the phage DNA enters the host cell, it is converted to a double-stranded replicative form (RF); after a period of replication of duplex molecules, replication switches to the rolling-circle mechanism, and single-stranded phage genomes are produced for packaging into mature progeny phage. The double-stranded RF is used as the vector for cloning foreign DNA and is treated just like a plasmid—the RF is isolated and manipulated in vitro and then used to transform recipient bacterial cells. Unlike phage λ, however, packaging of the M13 genome is independent of the length of the DNA, so very large inserts of donor DNA can be cloned and recovered from phage plaques.

One disadvantage of M13 as a vector is that its DNA has very few unique restriction sites in its limited nonessential regions. For this reason, a series of artificial cloning vectors (the M13mp series) have been constructed by incorporating a *lacz'*-MCS region into the M13 genome. The blue/white screening assay is then used to select for recipient cells that have been transfected by a recombinant vector.

These few general types of vectors demonstrate the specific uses, advantages, and disadvantages of various cloning systems. New kinds of vectors are being designed daily to facilitate the cloning of particular kinds of donor inserts. It is beyond the scope of this text to describe them all. However, these examples should be sufficient to illustrate researchers' sophistication and cleverness in vector design. (The accompanying box describes one additional example, a vector for preparing single-stranded nucleic acid to be used as a probe.)

Our current discussion has been limited to the cloning of DNA in *E. coli* cells. Cloning can also be done in yeast,

The λZAP Vector

Parts of four different phages and a plasmid have been combined into a sophisticated insertional vector called λZAP, which is particularly suitable for preparing single-stranded DNAs or RNAs that are to be used as probes. ▶ Figure 18–7 shows the structure and use of this vector. To form the vector, an MCS (multiple cloning site) is inserted into the middle portion of the λ chromosome, where it is flanked by T3 and T7 promoters oriented in opposite directions and f1 replication initiator and terminator sites. The region bounded by these two sites is known as the **Bluescript phasmid**; it can replicate as a plasmid conferring ampicillin resistance or it can be packaged into M13-like particles.

The latter process is used to obtain cloned recombinant DNA. Donor DNA to be cloned is inserted into one of the sites within the MCS region, giving a recombinant λZAP molecule with an inactivated *lacz* region. F⁺ *E. coli* cells are transfected with the recombinant

▶ **FIGURE 18–7** (a) The λZAP vector, which contains a plasmid *ampʳ* marker and a *lacz* region (containing control sites and the *lacz* gene). A MCS has been spliced into the *lacz* region. Phage f1 replication initiation and termination sequences, derived from the f1 *ori*, are included in the vector. Two phage promoters (from T3 and T7) flank the MCS region. (b) Recombinant λZAP accepts foreign DNA inserts up to 10 kbp in length. The recombinant λZAP is incorporated into F⁺ *E. coli* cells, along with f1 or M13 helper phage to provide trans-acting proteins that recognize the f1 replication sequences. Single-stranded phasmid DNA is then synthesized and automatically excised from the vector. It circularizes and is packaged into phage particles that can then be used to infect *E. coli* cells. Cells carrying recombinant DNA are detected as white, *amp*-resistant colonies.

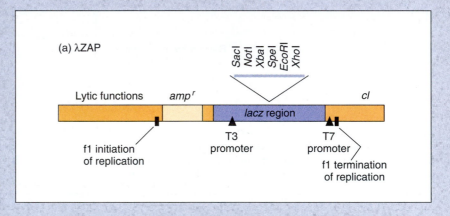

using yeast artificial chromosomes (YACs; see Chapter 9) as vectors. These vectors accommodate very large inserts, so they have been particularly useful in the Human Genome Project (see Extensions and Techniques under that title later in this chapter). Cloning in other species is much more difficult. Bacteriophages cannot infect eukaryotic hosts, and very few plasmids have so far been found to be indigenous to eukaryotic cells. This is not a major problem as long as cloning is the goal, since *E. coli* faithfully replicates animal and plant DNA. However, if the intent is to study the expression of a eukaryotic gene or to **genetically engineer** a eukaryotic organism, then vectors that can be incorporated by eukaryotic cells and integrated into a host chromosome are a necessity. The vectors used for these kinds of analyses are specially constructed expression vectors that will be described later in this chapter.

Cloning Strategies

We have discussed the general nature of recombinant DNA molecules and have given examples of some widely used cloning vectors and their properties. We now turn our attention to the variety of strategies that must be considered in planning how to clone a particular DNA segment. ▶ Figure 18–8, which outlines some alternative cloning strategies, shows that choices can be made at four levels: (1) the preparation of the DNA fragment to be cloned, (2) the mechanism for joining the fragment to the vector, (3) the means by which the recombinant DNA is introduced into the host cell, and (4) the method of selecting or identifying cells that have incorporated recombinant molecules.

DNA fragments to be cloned can be prepared in several ways. As we have discussed, digestion of donor DNA with a

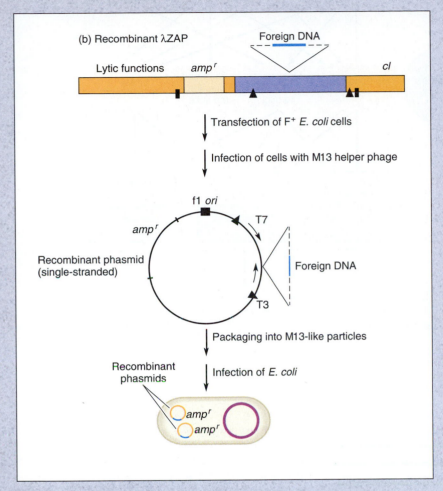

(b) Recombinant λZAP

Foreign DNA

Lytic functions   amp^r cl

Transfection of F$^+$ E. coli cells

Infection of cells with M13 helper phage

f1 ori

amp^r T7

Recombinant phasmid
(single-stranded)   Foreign DNA

T3

Packaging into M13-like particles

Recombinant
phasmids   Infection of E. coli

amp^r

amp^r

phage and then infected with M13 helper phage. The M13 phage supplies trans-acting factors that activate the f1 replication sites, leading to DNA synthesis in the region bounded by those sites and production of single-stranded recombinant phasmid molecules that are packaged into M13-like phage particles. Infection of sensitive E. coli by these particles confers antibiotic resistance, and white colonies (containing recombinant phasmids) can be selected.

There are two major advantages of this cloning system. First, the single-stranded phasmid region containing the donor DNA is automatically excised from the λZAP vector in vivo, providing a convenient way to clone single-stranded donor inserts, despite the complexity of the vector. Second, the T3 and T7 promoters that flank the inserted donor DNA allow in vitro transcription from either DNA strand. The resulting RNAs can be very useful as probes for northern blot analysis (see Chapter 15), in screening cDNA libraries (to be discussed later), or in studying the type of protein coded by the donor DNA using cell-free translation assays.

restriction enzyme leaves cohesive ends that can be ligated to the complementary cohesive ends of the vector. These reactions need only slight modifications to permit the use of restriction endonucleases such as HindII, which leaves blunt-ended fragments, in the construction of recombinant molecules. One approach, the homopolymer tailing technique, uses the enzyme called terminal deoxynucleotidyl-transferase to synthesize single-stranded tails at the 3′ ends of the blunt-ended fragments. ▶ Figure 18–9 shows the addition of oligo-A to the 3′ ends of the vector molecules and oligo-T to the 3′ ends of the donor fragments, thereby creating complementary single-stranded ends on the two populations of fragments. In more recently developed techniques, cohesive termini are synthesized and ligated to the blunt ends of fragments, making it possible to design whatever cohesive ends are convenient.

Researchers have also developed a technique in which artificially synthesized linker oligonucleotide segments are joined to blunt-ended fragments through blunt-ended ligation, a reaction catalyzed by T4 DNA ligase. ▶ Figure 18–10 illustrates this procedure. If T4 ligase joins a linker containing one or more restriction sites to each end of a donor fragment, a restriction enzyme can then be used to generate cohesive ends from the linkers.

Restriction enzyme digestion is not the only way to obtain donor DNA fragments for cloning. As Figure 18–8 shows, fragments can also be generated by mechanical shearing (although termini generated by this method have to be modified before any ligation to a vector is possible) or by direct chemical synthesis. The latter is an expensive and time-consuming procedure, but it has been used to obtain certain human genes that can be expressed by E. coli. Donor

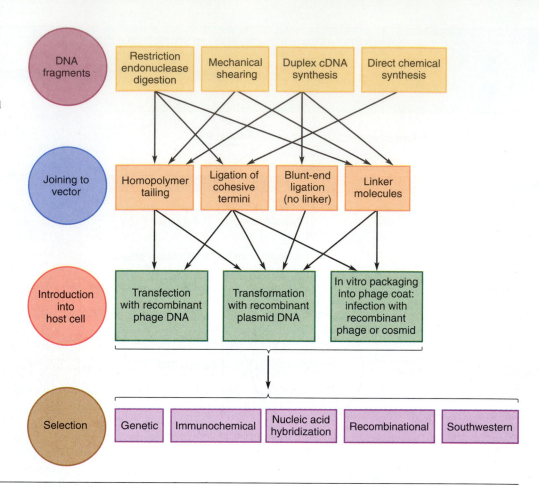

▶ **FIGURE 18–8** Strategies for DNA cloning in *E. coli*. Choices can be made at four different levels. The arrows show the possible choices at each point. *Source:* R. W. Old and S. B. Primrose, *Principles of Gene Manipulation* (Oxford: Blackwell Scientific Publications, 1989), p. 109. Used with permission.

DNA can also be obtained as **complementary** or **cDNA**. cDNA is single- or double-stranded DNA that is complementary to the messenger RNA of a particular gene, and therefore contains only the exon regions of that gene. cDNA molecules have many important applications, including use as DNA probes. We have already described (in Chapter 15) some of the uses of probes in gene mapping and DNA fingerprinting. The construction of probes and some of their other uses will be described shortly.

To Sum Up

1. A recombinant DNA molecule can be artificially created by linking a DNA segment to a self-replicating vector. The two DNA molecules can be spliced in several ways; one way is to join the fragments at their complementary cohesive ends, which have been formed by treatment with the same restriction enzyme.

2. A variety of vectors are used to construct recombinant DNA molecules; most are plasmids or phage chromosomes that occur naturally or have been specifically constructed for particular cloning purposes. Two desirable features of a vector are possession of unique cutting sites for a variety of restriction enzymes and the ability to confer a selectable phenotypic marker on cells that have incorporated a recombinant molecule.

3. Plasmids and wild-type phage λ, two widely used kinds of cloning vectors, tend to be limited in the length of donor DNA they can accommodate. However, derivatives of λ in which some of the nonessential DNA is replaced by donor inserts can accommodate up to a 20 kb insert. Vectors such as cosmids and phasmids, which incorporate DNA sequences of both plasmids and phage λ, can accommodate even longer inserts.

4. Phage vectors for cloning single-stranded DNA have also been developed. One particularly useful one is the single-stranded coliphage M13. However, M13 DNA has few unique restriction sites and selectable markers, so artificial M13 derivatives are created as cloning vectors. These vector constructs contain artificially synthesized polylinker DNA containing multiple unique cloning sites (MCSs) and a gene that provides a marker phenotype for transformed cells. MCS segments are also inserted into some plasmid and other phage DNAs to increase their effectiveness as cloning vectors.

5. Donor DNA can be prepared in a variety of ways—restriction enzyme digestion, mechanical shearing, chemical synthesis, and the preparation of complementary DNA (cDNA). cDNA carries only the exon regions of a particular gene; it has a variety of uses, including use as a DNA probe.

6. There are several means for joining donor fragments to vectors and introducing them into host cells, and several techniques can be used to select or identify host cells that have incorporated the desired recombinant DNA.

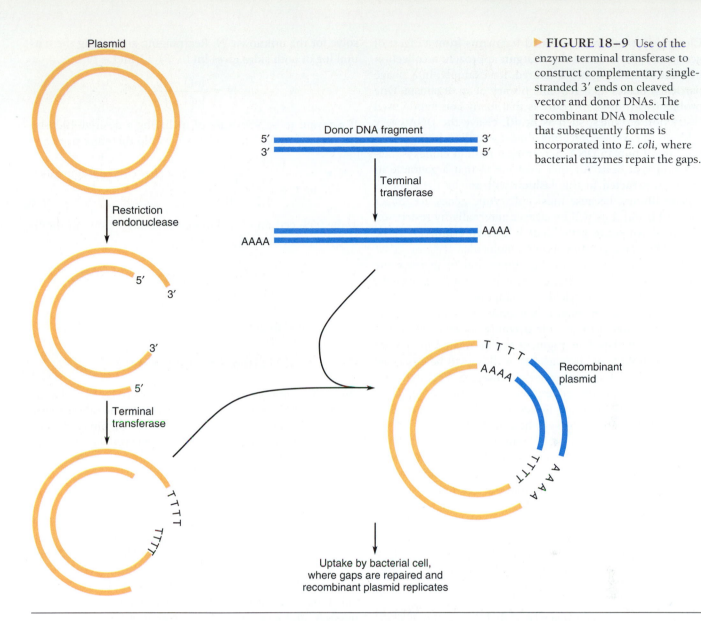

Plasmid

Restriction
endonuclease

Donor DNA fragment

5' ──────────────── 3'
3' ──────────────── 5'

Terminal
transferase

AAAA ──────────────── AAAA

5'
3'

3'

Terminal
transferase

5'

TTTT

T T T T

A A A A

Recombinant
plasmid

T T T T

A A A A

Uptake by bacterial cell,
where gaps are repaired and
recombinant plasmid replicates

► FIGURE 18–9 Use of the
enzyme terminal transferase to
construct complementary single-
stranded 3′ ends on cleaved
vector and donor DNAs. The
recombinant DNA molecule
that subsequently forms is
incorporated into E. coli, where
bacterial enzymes repair the gaps.

► FIGURE 18–10 The ligation of linker DNA segments to
blunt-ended restriction fragments. The linkers are synthesized to
contain one or more restriction sites that generate cohesive ends
upon cleavage.

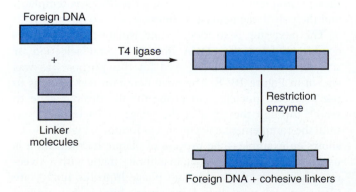

Foreign DNA

+

Linker
molecules

T4 ligase

Restriction
enzyme

Foreign DNA + cohesive linkers

GENE ISOLATION

Before the development of DNA cloning, the huge amount
of DNA in a typical eukaryotic nucleus made it impossible
to isolate any specific individual gene. Isolating a typical
mammalian gene that consists of just 0.00005% of the haploid genome would literally be like looking for a needle
in a haystack. Now, however, the isolation of a particular
gene is possible. By coupling DNA cloning technology with
selection/identification methods that can detect the desired
gene among a population of various recombinant clones, it is
possible to pinpoint which clone contains the desired gene.

Recombinant Selection and Screening

One approach to isolating a particular gene is to clone all
the unique-sequence DNA of the organism and then screen
each of the clones for the presence of the gene of interest.

Cloning all randomly generated fragments from a digest of genomic DNA is called the **shotgun** approach; a collection of these cloned fragments (stored, for example, in λ phage particles) makes up a **genomic library** of an organism. One problem with this procedure is that many commonly used restriction enzymes, such as *Eco*RI, cleave the DNA every 4 kb on the average, creating almost a million fragments in the case of human DNA. Screening a million clones is a tedious task at best! Another problem is that a genomic library constructed in this fashion will not be the same as a **gene library**, because most eukaryotic genes are longer than 4 kb and thus will be cleaved internally by restriction enzymes. An entire gene locus is therefore unlikely to be carried by one recombinant DNA molecule.

To decrease the size of the library and to increase the chance that it is a gene library, researchers use a **partial digestion** technique that yields overlapping fragments averaging about 20 kb in length. Not only are such fragments much more likely to include a complete gene locus, but since they overlap, one fragment can be related to another by the technique of chromosome walking (to be described shortly). These fragments are then inserted separately into high-capacity cloning vectors, such as an appropriate λ derivative, and the resulting library is screened to determine which clone contains the specific gene of interest.

In humans, the ratio of the genome size (2.8×10^6 kb) to the size of a single fragment (~ 20 kb) is about $n = 1.4 \times 10^5$. However, more than 1.4×10^5 fragments must be included in the library to ensure that the digest actually contains all genomic sequences. In practice, the number of clones required in order to be at least 95% certain of including any particular sequence in the library is about 3 to 4 times the n value (see Example 18.1); for the human genome, it would be about 5×10^5 clones. Thus, screening even a partial digest library remains a tedious procedure.

Example 18.1

How many cloned DNA fragments of average length 4,000 bp would have to be included in an *E. coli* genomic library to be 95% certain that the library contains a given gene of interest? Repeat this calculation for 99% certainty.

Solution: The probability that a given cloned fragment is included in the library depends on the length of the fragment and on the number of cloned fragments (the library size). If we let f represent the average length of a cloned fragment relative to the length of the genome and let N be the size of the library, we can treat this as a binomial situation, with f being the probability per clone of containing the desired gene. Thus, the probability that a clone does not contain the desired gene fragment is $(1 - f)$ and the chance that no clone in the library contains this fragment is $(1 - f)^N$. The probability that the desired fragment is included in the library is therefore $P = 1 - (1 - f)^N$.

Since we have a given probability in mind (95% for the first question), we rearrange this equation so that we can

solve for the unknown, N. Rearranging and taking the natural log of both sides gives $\ln(1 - P) = N\ln(1 - f)$, so

$$N = \frac{\ln(1 - P)}{\ln(1 - f)}$$

If we want to be 95% sure of including a desired 4000-bp fragment in the *E. coli* library, we would therefore need

$$N = \frac{\ln(0.05)}{\ln\left(1 - \dfrac{4 \times 10^3}{4.2 \times 10^6}\right)} \simeq 3 \times 10^3 \text{ cloned fragments}$$

To be 99% certain of including the desired clone in the library, we would need

$$N = \frac{\ln(0.01)}{\ln\left(1 - \dfrac{4 \times 10^3}{4.2 \times 10^6}\right)} \simeq 4.8 \times 10^3$$

clones in the library.

Follow-Up Problem 18.1

How many *Drosophila melanogaster* DNA fragments of average length 15 kb must be cloned in order to be (a) 90% sure and (b) 99% sure that a genomic library contains a desired DNA sequence? (The *Drosophila melanogaster* genome contains about 1.5×10^5 kb of DNA.)

One can come much closer to constructing a gene library (as opposed to a genomic library) by making a **cDNA** or **expression library**, which contains only those genes that are expressed by a certain type of cell. This library will be much smaller than a genomic library (and therefore easier to screen) because only a small proportion of the DNA of a eukaryotic organism actually consists of protein-coding sequences, and an even smaller proportion forms the subset of sequences expressed by any particular type of cell. The procedure for constructing a cDNA library is illustrated in ▶ Figure 18–11. Messenger RNA from a eukaryotic cell can be isolated by virtue of its poly A tail—the poly A tail binds to a cellulose resin on which poly T segments have been fixed, while other types of RNA do not recognize the poly T segments. Reverse transcriptase is then used to synthesize complementary DNAs using the RNAs as templates and the poly T segments as primers.

The next step is to identify and isolate the desired recombinant from the cDNA library. Several selection or screening methods can be used for this purpose, as was shown in Figure 18–8. The most common technique is to use a labeled nucleic acid probe for the desired gene to search through the DNA of the phage or bacterial clones until the complementary sequence is found. ▶ Figure 18–12 shows an example—the process of plaque hybridization is used to detect a desired gene in a library made with a λ vector. Thousands of recombinant phage from the library are plated on each petri dish. A membrane filter is then applied

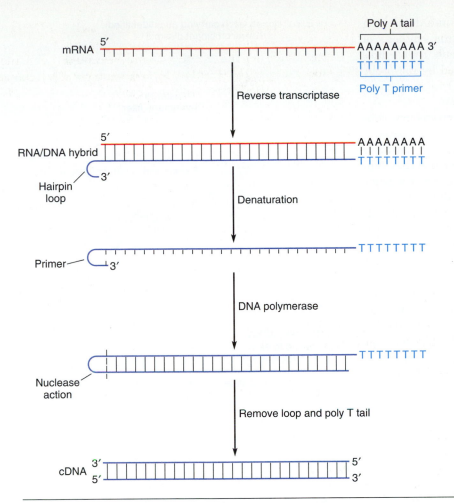

Poly A tail

mRNA 5'

AAAAAAA 3'
Poly T primer

Reverse transcriptase

RNA/DNA hybrid 5'

AAAAAAA

Hairpin loop
3'

Denaturation

Primer
3'

TTTTTTT

DNA polymerase

Nuclease action

TTTTTTT

Remove loop and poly T tail

cDNA 3'
5'

5'
3'

▶ **FIGURE 18–11** Construction of double-stranded cDNA. An mRNA molecule is bound to a cellulose resin by hybridization of its poly A tail to the poly T segment. The enzyme reverse transcriptase is then used to synthesize a DNA strand that is complementary to the mRNA, using the poly T segment as a primer. After denaturation of the RNA/DNA duplex, the hairpin loop left by reverse transcriptase on the 3' end of the DNA serves as a primer for DNA polymerase, which synthesizes a DNA strand that is complementary to the first one. Nucleases remove the loop and the poly T tail, leaving a DNA molecule that is complementary to the original mRNA. A collection of such cloned DNA molecules constitutes the cDNA library of the cell type from which mRNA molecules were isolated.

to the surface of the dish, where it makes direct contact with the plaques. Some of the phage from each plaque adhere to the filter; alkaline conditions lyse these phage, and the covalently bound, denatured DNA on the filter is then hybridized with the radioactive probe, as described in Chapter 15. Autoradiography reveals the position of the plaque containing the gene of interest, and that plaque is then isolated.

Construction of Probes

The limiting factor in screening a genomic or cDNA library in order to isolate a DNA sequence from a specific gene is the availability of a probe to detect that sequence. Because nucleic acid probes are also widely used in Southern and northern blots and in DNA fingerprinting analyses (see

▶ **FIGURE 18–12** The so-called plaque-lift method. A library contained in λ particles is being screened for a desired gene. A membrane filter is placed over the surface of a petri dish that contains plaques. Some phage from each plaque adhere (are lifted) to the filter, where they are then lysed and their DNA denatured. A labeled nucleic acid probe detects the desired gene via autoradiography.

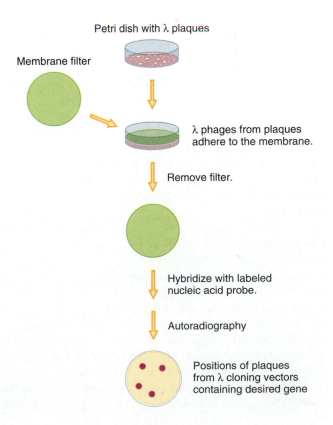

Petri dish with λ plaques

Membrane filter

λ phages from plaques adhere to the membrane.

Remove filter.

Hybridize with labeled nucleic acid probe.

Autoradiography

Positions of plaques from λ cloning vectors containing desired gene

Gene Cloning and the Analysis of Gene Function   429

Chapter 15), considerable effort has gone into developing techniques for constructing them. While the synthesis of nucleic acid probes is not particularly difficult, the identification of the specific gene sequence carried by the probe (i.e., the identification of the probe itself) is much more difficult.

To make pure DNA probes, a cDNA library is first constructed as shown in Figure 18–11. The cDNA library contains as many different types of cDNA molecules as there are types of mRNA (with poly A tails) in the eukaryotic cell. Identification of the desired cDNA can therefore be a major problem, because it is not a simple task to differentiate between the various mRNA or cDNA molecules. Identification is easiest if the cells from which the mRNA was isolated are specialized ones in which particular mRNAs are abundant. Reticulocytes (in which 90% of the protein made is hemoglobin), plasma cells (which synthesize mainly antibody protein), and hormone-stimulated chicken oviduct cells (in which over 50% of the protein is ovalbumin, the protein in egg whites) are examples of such specialized cells. As an illustration, let's assume we are using mRNA that was extracted from chicken oviduct cells. Reverse transcriptase generates a mixture of cDNA probes, of which about 50% are copies of the ovalbumin gene. We now need to specifically identify these cDNAs.

There are a variety of strategies for identifying these probes. The **HRT** (**hybrid released translation**) method works well with cDNAs derived from abundant RNAs. In this method the cDNA molecules are first cloned to provide sufficient amounts of material to work with. The cloned DNA is then fixed to a membrane filter and hybridized with the mRNA extract from the oviduct cells (▶ Figure 18–13). Only those mRNA molecules that are complementary to a given cDNA are retained on the filter; washing removes the rest. The hybridized mRNAs are individually eluted from the filter and placed in separate test tubes containing a cell-free translation system. Radiolabeled amino acids are added to the tubes so that the proteins that are made will be radiolabeled and can be analyzed by gel electrophoresis and autoradiography to identify those cDNAs that direct the synthesis of ovalbumin. In our example, we would expect to find that about 50% of the hybrid-selected mRNA samples produce ovalbumin. Once the cDNA molecules that correspond to these mRNAs are identified, they can be spliced into plasmids and cloned in E. coli. The recombinant plasmids can then be isolated from the bacterial cells and the cDNA isolated and labeled, yielding a large amount of known, labeled cDNA probe.

Because not all probes can be derived from abundant mRNAs like ovalbumin, alternative approaches for identifying particular probes have been devised. One strategy that has been applied to cDNAs derived from nonabundant mRNAs is to use a probe that has already been developed to detect a particular gene in one species to detect a homolog of that gene in a second species. For example, the mouse HPRT cDNA has been used as a probe to detect the homologous human DNA in molecular screening for the gene that

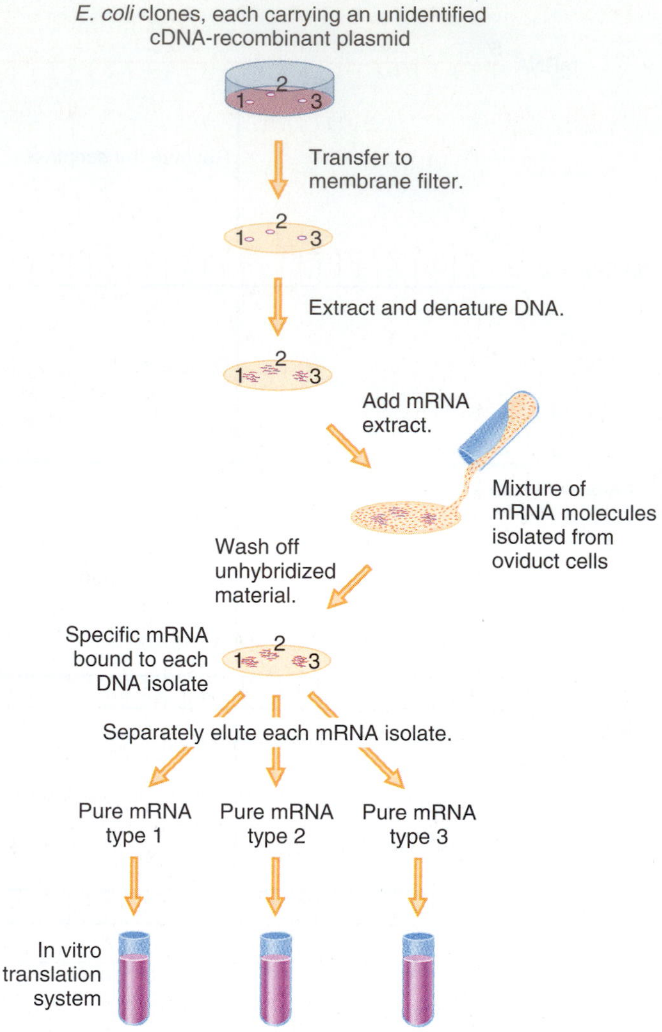

▶ **FIGURE 18–13** Using the HRT method to identify cDNA molecules in the cDNA library. The denatured DNA extracted from each clone is exposed to the mRNA extract from the specialized cell type (in this case, oviduct cells). The mRNA molecules hybridize to the DNA of clones that contain their DNA templates. The mRNA that is retained on the membrane filter at the position of each isolate is then eluted and placed in a test-tube protein-synthesizing system. The protein that is made is identified by biochemical procedures, in turn identifying the mRNA and the specific cDNA of each original clone.

causes Lesch-Nyhan syndrome. The same principle can be applied to members of protein families (for example, the serine protein kinases), which contain structurally similar domains—a probe that detects one member of the family can often be used to detect other members. Recent advances in the technology of amino acid sequencing provide still another method by which specific cDNA probes can be identified: Investigators determine the partial amino acid sequence of the protein coded for by the desired gene and then, working with the genetic code, they chemically synthesize an oligonucleotide that codes for that sequence. This oligonucleotide can be used as a primer in a PCR reaction to amplify a cDNA or genomic sequence, which can then be used as a probe, or it can be directly used as a probe to screen a genomic or cDNA library.

Example 18.2

Design a DNA probe that would allow you to identify the cloned gene that encodes a protein with the following N-terminal amino acid sequence:

N-Ala-Pro-Arg-Thr-Trp-Tyr-Cys-Met-Asp-Trp-Ile-Ala-Gly-Gly-Pro

The probe should be 18 to 20 nucleotides long, a size that provides adequate specificity to determine whether a gene is homologous to the probe.

Solution: A probe of 18–20 nucleotides corresponds to only 6–7 amino acids. We need to identify the codon assignments for the amino acids in the protein and then search through the sequence to find the stretch of 6–7 amino acids having the least codon degeneracy (the numbers below indicate the number of codons that code for each amino acid):

Ala-Pro-Arg-Thr-Trp-Tyr-Cys-Met-Asp-Trp-Ile-Ala-Gly-Gly-Pro
  4   4   6   4   1   2   2   1   2   1   3   4   4   4

A quick inspection shows that the 6–amino acid section bounded by the Trp residues has the least overall degeneracy. We would therefore need to synthesize eight different 18-nucleotide probes ($1 \times 2 \times 2 \times 1 \times 2 \times 1 = 8$) to be sure of including the sequence that exactly matches the sequence of the cloned gene. Of the 8 synthesized, the best probe is the one that shows the most homology to the cloned gene (as judged by hybridization).

Follow-Up Problem 18.2

If you started your probe one amino acid to the left of the Trp residue selected in Example 18.2, how many different 18-nucleotide probes would you have to make to be sure of including the matching one?

◆

Chromosome Walking

In some cases, a gene close to a second gene of interest has already been cloned. The precise location of the second gene can be found by using the first gene to probe a genomic library for an overlapping adjacent fragment. That fragment can in turn be used to identify a fragment adjacent to it, and so on, until the set of overlapping clones spans a long stretch of DNA. This technique is called **chromosome walking** and is illustrated in ▶ Figure 18–14. Hundreds of kilobases of contiguous DNA can be walked in this manner.

How does the researcher know when the desired gene has been reached? A case in point involves the search for the human dystrophin gene, in which a mutation causes Duchenne muscular dystrophy (DMD). Researchers knew the approximate location of the DMD gene, since most DMD patients have large deletions in the region of the X chromosome known as Xp21. When it was suspected that the chromosome walk was close to the gene, the investiga-

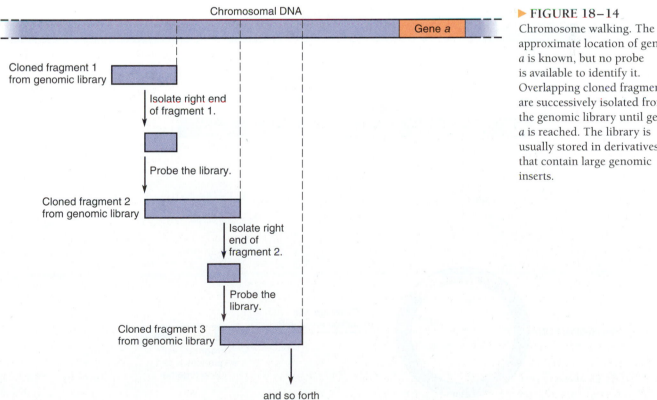

▶ **FIGURE 18–14**
Chromosome walking. The approximate location of gene *a* is known, but no probe is available to identify it. Overlapping cloned fragments are successively isolated from the genomic library until gene *a* is reached. The library is usually stored in derivatives that contain large genomic inserts.

tors began to sequence the cloned DNA fragments, looking for **open reading frames (ORFs)**. (An open reading frame is a long sequence of nucleotides that begins with a start codon and is free of stop codons until its terminus; such a sequence is indicative of a protein-coding gene.) The open reading frames that were found were then searched for the large deletions known to be characteristic of the mutant DMD gene. An alternative approach would have been to use DNA fragments with ORFs as probes to look for mRNA with the same sequence and thereby identify transcribed genes.

Both of these methods were used in finding the cystic fibrosis gene. Because of the extremely long length of the region that needed to be screened, a variation of the walking technique known as **chromosome jumping** was used to approach this gene. Jumping (▶ Figure 18–15) in this case involved circularizing a long genomic DNA fragment to bring together DNA sequences that originally were far apart, and then excising and cloning the DNA from the region that covered the closure site of the circle. DNA clones suspected of containing the CF gene were identified by a number of criteria, including homology to probes from other organisms and the presence of open reading frames. These clones were screened for a common mutational lesion; DNA sequencing showed that 70% contained the same 3-bp

deletion, identifying not only the gene but also its specific defect. The wild-type allele of this mutant gene was found to transform cultured CF cells, conclusively showing that the CF gene had been isolated. Northern blotting with clones that carried the gene identified mRNA in tissues affected by CF; the mRNA sequence indicates that the gene encodes a protein domain known to be common in membrane ion-transport proteins, thus identifying the protein coded by the CF gene.

To Sum Up

1. When a recombinant DNA molecule replicates within a bacterium, the foreign DNA is said to be cloned, and the self-replicating vector is called a cloning vector. Cloning is especially useful for creating a genome library, in which DNA fragments representing the entire genome of an organism are stored as recombinant DNA molecules. It is equally useful in constructing a cDNA library, which represents the set of genes expressed by a given type of cell.

2. To isolate a particular gene from a genome or gene library, geneticists usually use a probe that is specific for the gene of interest. Probes can be made using the enzyme reverse transcriptase to synthesize DNA molecules that are complementary to

▶ **FIGURE 18–15** Chromosome jumping. Partial digestion with restriction enzyme A yields very long, overlapping genomic fragments, one of which is shown here. Sequences x and y are separated by 100 kb in this example. The fragment is spliced into a vector, and a different restriction enzyme (B), which does not cut the vector but cuts often within genomic DNA, is used to generate a small fragment that spans the closure site of the recombinant DNA. Sequences x and y are close together on this fragment, which is then cloned. The procedure is repeated for the various overlapping fragments to create a "jumping" library in which the chromosome walking procedure can be performed over much more substantial distances than in a normal genomic library.

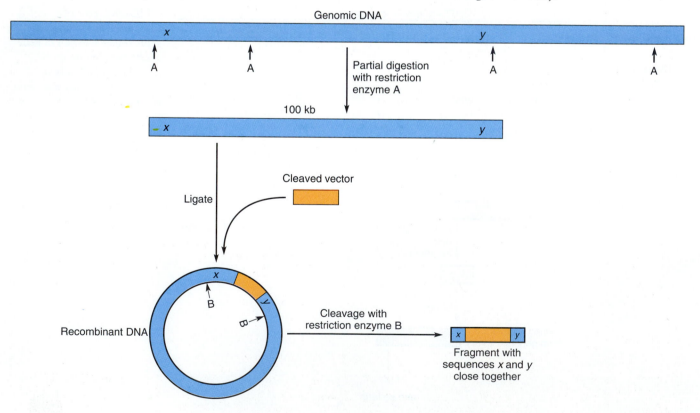

EXTENSIONS AND TECHNIQUES

The Human Genome Project: Physical Mapping with Sequence-tagged Sites

Technical advances have now made it possible to isolate individual human chromosomes; in fact, particular regions of a specific chromosome can be microdissected for analysis. The development of YAC vectors that accommodate huge inserts containing up to a million or more base pairs allows the cloning of fragments long enough to contain an entire gene locus plus its regulatory regions. Researchers have used these vectors to clone particular regions of particular human chromosomes (thus constructing chromosome libraries) as well as the entire human genome. However, the identification of the cloned DNA in chromosomal and genome libraries is still a time-consuming process that requires walking and jumping long regions of chromosomes.

Two laboratories have recently produced detailed physical maps of human chromosomes Y and 21 using

YAC cloning vectors and physical landmarks called **sequence-tagged sites (STSs)**. Sequence-tagged sites are short (200–500 bp), naturally-occurring unique sequences from known chromosomal locations; they can be used to determine the order of DNA clones spanning a particular long chromosomal region. More and more of these sequences are being identified, and unique sequences complementary to them are being synthesized for use either as probes to screen a YAC library or as primers for the polymerase chain reaction. To illustrate their use in mapping, consider a YAC library containing a DNA section that we wish to map. First, various pairs of primers complementary to specific STSs known to lie within the region are used to selectively amplify clones containing the DNA of interest from the YAC library. Hypothetical clones resulting from such a proce-

dure are shown below (the symbols represent six different STSs):

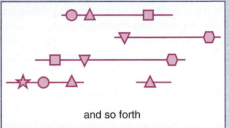

and so forth

Next, the clones are screened for overlapping STSs, using the cSTSs as probes. The pattern of overlaps yields the physical map shown below.

To map chromosomes Y and 21, STS markers were used to screen the YAC Y- and 21-chromosomal libraries, and the pattern of overlap of PCR-amplified DNA sections bounded by particular STS sites then gave the order of the cloned sections along the chromosomes.

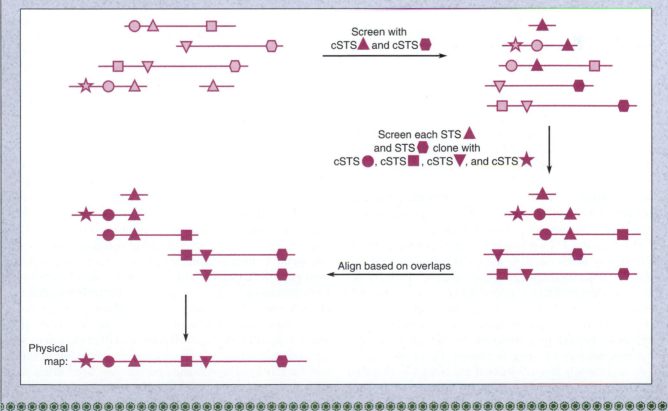

the mRNA molecules expressed by a cell. Identification of the desired probe can be accomplished by a variety of strategies, including techniques such as hybrid released translation. Probes can sometimes be artificially synthesized using the amino acid sequence of the polypeptide encoded by the gene as a guide to the nucleotide sequence of the probe.

3. When the approximate chromosomal location of a gene is known but no probe is available to directly isolate it, chromosome walking and chromosome jumping can be used to find the gene in a genome library. Chromosome walking and jumping move along a chromosome by identifying a series of overlapping adjacent fragments that eventually reach the gene of interest. Numerous human disease genes have been isolated in this manner, allowing researchers to investigate their expression in different tissues and the nature of their protein products.

EXPRESSION CLONING

So far, we have not considered whether foreign DNA can be expressed by the bacteria or other recipient cells into which it is cloned. In the mid-1970s it was discovered that some genes from yeast and *Neurospora* can be expressed when placed into *E. coli* cells. Although gene expression by cloned DNA may not, at first thought, be that surprising, more careful reflection reveals the complexities faced by a eukaryotic gene in a bacterial host. As we have seen in previous chapters, promoters, RNA polymerases, and ribosome binding sites all differ between bacteria and eukaryotes; even gene structure itself may be a complicating factor, since so many eukaryotic genes are split. However, special types of vectors and manipulations of cloned eukaryotic DNA have allowed some cloned genes to be expressed.

Expression of Foreign DNA in Bacteria

Bacteria are not able to remove introns from eukaryotic genes, so the foreign DNA in bacteria must either be prepared as cDNA or be chemically synthesized according to the amino acid sequence of the encoded protein. The latter approach is practical only for very short genes, but it has been used to obtain the coding sequences of the human insulin and somatostatin genes to allow their expression in *E. coli*.

Neither cDNA nor artificial genes contain the control regions that are normally adjacent to the coding sequences on chromosomes. Without a promoter region that will combine with bacterial RNA polymerase, an inserted eukaryotic gene cannot be transcribed. Researchers have overcome this problem by constructing **expression vectors**, which are derivatives of cloning vectors with added bacterial transcriptional control signals. Although replication of the recombinant vector is not a prerequisite for expression of the foreign gene, its high concentration does boost the level of mRNA templates and protein produced. Therefore, expression vectors usually are constructed in such a way that they retain their *ori* sites, and high copy-number plasmids, such as pBR322 derivatives, are used.

As an example, ▶ Figure 18–16 shows the insertion of the somatostatin gene into a pBR322-derived expression vector that contains the promoter and first seven codons of the *E. coli lacz* gene, including the Shine-Dalgarno sequence required for binding *E. coli* ribosomes to mRNA. The recipient cells synthesize somatostatin as a polypeptide attached to the first seven amino acids of the *lacz* gene product, from which it can be cleaved with cyanogen bromide and isolated in pure form. The production of a hybrid **fusion protein**, as in the synthesis of somatostatin, is the usual outcome of eukaryotic gene expression by *E. coli*; however, researchers have also developed special vectors that express the protein product from its own N-terminus, to save the time and expense of the chemical cleavage step.

Another problem with bacterial synthesis of eukaryotic gene products is that bacteria often degrade foreign protein. For this reason, researchers frequently modify the host cells by mutation so that their intracellular protease activity is greatly reduced.

The expression of a foreign gene of known function in *E. coli* can be detected in a number of ways, for example, by an assay for a particular enzyme activity or an immunochemical method. In the latter case, lysed colonies can be exposed to a radioactively labeled antibody probe specific for the protein product of the cloned gene. Autoradiography will then reveal the locations of colonies producing the protein; this procedure is known as **western blotting**. Another valuable screening technique is used to detect clones in which the foreign gene encodes a DNA-binding protein that specifically binds to a particular DNA sequence. These proteins are critical to the regulation of expression of prokaryotic and eukaryotic genes and will be discussed in more detail in Part VII. In this method, the lysed colonies are exposed to a radioactively labeled duplex DNA oligonucleotide containing the sequence for which the DNA-binding protein is specific, and the presence of clones expressing the binding protein is again detected by autoradiography. This procedure, in which a labeled DNA is used as a probe to detect the presence of a particular protein on the membrane filter is known as **southwestern blotting**.

The practical uses of engineering *E. coli* to express foreign proteins are well known. The pharmaceutical industry was the first to make widespread use of recombinant DNA techniques to express foreign proteins in *E. coli*. Bacteria are now used to synthesize a wide variety of pharmacologically active proteins, including **humulin**, a version of human insulin that has replaced insulin isolated from cattle and swine (which frequently causes allergic reactions), human **growth hormone**, which formerly was obtained only from human cadavers, several types of **interferon**, and vaccines against the hepatitis and foot-and-mouth viruses.

Another important application of recombinant DNA technology is in the detoxification and degradation of sewage and industrial wastes by microorganisms. For example, one bacterium, *Pseudomonas putida*, has been engineered with a plasmid carrying genes that code for enzymes that break down the octane, hexane, decane, xylene, toluene,

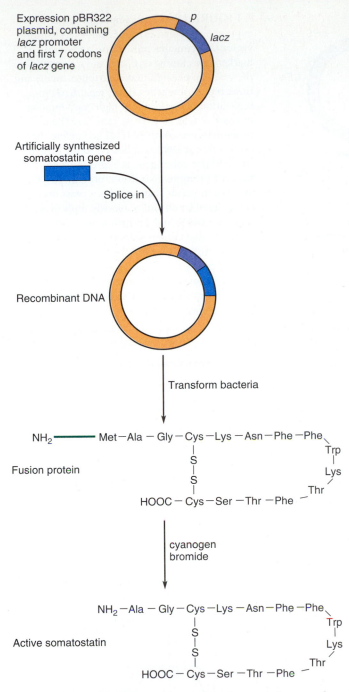

Expression pBR322 plasmid, containing *lacz* promoter and first 7 codons of *lacz* gene

p

lacz

Artificially synthesized somatostatin gene

Splice in

Recombinant DNA

Transform bacteria

Fusion protein

NH₂——Met—Ala — Gly—Cys—Lys—Asn—Phe—Phe
                              | Trp
                              S |
                              | Lys
                              S   /
                              | Thr
        HOOC—Cys—Ser —Thr —Phe

cyanogen bromide

Active somatostatin

NH₂—Ala — Gly—Cys—Lys—Asn—Phe—Phe
                    | Trp
                    S |
                    | Lys
                    S   /
                    | Thr
HOOC—Cys—Ser —Thr —Phe

▶ **FIGURE 18–16** Production of the polypeptide hormone somatostatin by *E. coli*.

camphor, and naphthalene components of oil. This particular bacterium made legal history in 1980, when M. Chakrabarty, its genetic engineer, was issued the first patent ever granted for a genetically manipulated microorganism.

Use of Yeast to Study Gene Function

To avoid some of the problems associated with expression of eukaryotic genes in bacteria and to more accurately assess the activity of the encoded proteins, researchers are increasingly using eukaryotic cells, instead of bacteria, as re-

cipients for foreign DNA. Yeast, which share some bacterial features such as rapid growth and ease of culturing, are popular organisms for this purpose. The ability to be propagated in both the haploid and the diploid state is another feature that makes yeast well suited for genetic analysis.

Baker's yeast (*Saccharomyces cerevisiae*) is one of several kinds of fungi that are transformable; protoplasts are easily made, and plasmids occur naturally, providing a basis for the construction of cloning and expression vectors. Like *E. coli*, yeast degrade foreign proteins, so strains with defective protease genes must be used to study the expression of foreign genes. This disadvantage is outweighed by the fact that yeast carry out the posttranslational protein modifications that are ubiquitous in eukaryotes but virtually absent in prokaryotes. Expression of foreign genes in yeast therefore gives a much more accurate picture of their normal expression, so gene regulation and the effects of posttranslational modifications on protein function can be studied.

The most efficient yeast expression vectors are **shuttle vectors**, which can replicate in both *E. coli* and yeast. The donor gene can be conveniently cloned in *E. coli* and then transferred to yeast to study gene expression. ▶ Figure 18–17 shows some typical yeast expression vectors. Most consist of pBR322 (or pUC) segments into which a yeast marker (such as *ura*⁺) linked to sequences from a yeast replicating plasmid (such as the **2μ plasmid** in Figure 18–17a) has been spliced at the *tet*ʳ gene. The yeast plasmid contributes its own *ori* and three genes essential for its maintenance in yeast cells at high copy-number (*rep1–3*). The *ura*⁺ gene codes for an enzyme involved in the synthesis of uracil and thus allows selection of transformed *ura*⁻ cells growing in the absence of uracil. The vector shown in Figure 18–17a also contains a complex transcriptional regulatory region consisting of the promoter and associated regulatory regions of a structural gene (in this case, the *cyc1* gene); this region is adjacent to a *Bam*H1 site into which donor DNA is spliced. The donor DNA is then expressed with the structural gene as a fusion protein.

For commercial applications, the secretion of the foreign protein by the yeast is desirable for ease of isolation. Yeast have been engineered to secrete the foreign protein by linking the protein-coding sequence to a signal peptide sequence (see Chapter 17). Production of interferon, hepatitis B vaccine, interleukin-2 (a white blood cell growth hormone), and β-endorphin (a pain-killing neuropeptide) have been achieved in this manner.

Other yeast expression vectors have been constructed using the same shuttle vector principle. A yeast chromosomal *ori* called the **autonomously replicating sequence** (**ARS**) can be used instead of the 2μ *ori*; the resulting vector is high copy-number, but it is not stable in dividing yeast cells unless a **centromere** (**CEN**) sequence is added to mediate partitioning of the vector at mitosis (Figure 18–17b). However, the addition of a centromere sequence maintains the vector at a low copy-number, with just one or two copies per cell, thus sacrificing high expression of the donor gene. Yeast artificial chromosome expression vectors (Fig-

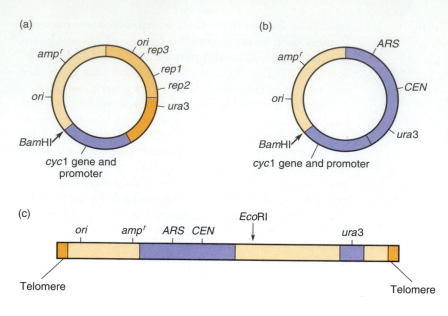

(a)

amp^r
ori rep3
rep1
rep2
ori
ura3
BamHI
cyc1 gene and promoter

(b)

ARS
amp^r
CEN
ori
ura3
BamHI
cyc1 gene and promoter

(c)

ori   amp^r ARS CEN   EcoRI   ura3

Telomere   Telomere

▶ **FIGURE 18–17** Examples of yeast expression vectors. (a) An *E. coli*–yeast shuttle vector that contains pBR322 *ori* and *amp^r* markers for cloning in *E. coli* and a section of the yeast 2μ plasmid that contributes an *ori* and three replication genes (*rep1, rep2, rep3*) for replication of the vector at high copy-number in yeast. A selectable yeast gene (*ura3*) and a promoter-gene complex (*cyc1*) complete the vector. Donor DNA is spliced into the *Bam*HI site, and the donor gene is expressed from the *cyc1* promoter as a fusion protein. (b) An expression vector that contains a yeast *ori* contributed by an autonomously replicating sequence (*ARS*) and a centromere (*CEN*) sequence that confers stability upon cell division. (c) Yeast artificial chromosome expression vector. These vectors accommodate very large foreign DNA inserts at the *Eco*RI site, which is especially useful for studying large human genes.

ure 18–17c), which contain both the *CEN* sequence and telomeres, are sometimes used in order to accommodate very large donor DNA inserts, such as the Duchenne muscular dystrophy gene, which spans more than one million base pairs.

Yeast are increasingly being used to study the expression of eukaryotic genes, especially the expression of their own cloned (in *E. coli*) genes in so-called surrogate yeast genetics. The availability of shuttle vectors allows geneticists to manipulate yeast genes in a fashion previously reserved for *E. coli* and phage. A mutation can be introduced into a cloned gene in vitro, and its effects can be studied by returning the altered gene to yeast and then directing its integration into a particular region of the genome. This process is called **gene targeting**, and it requires recombination between homologous regions on the donor and host DNAs.

An example of gene targeting is shown in ▶ Figure 18–18. After yeast protoplasts are transformed by a linearized recombinant vector carrying a mutated gene of interest, the free ends of the transforming DNA can pair with the homologous region (the target) on a host chromosome, and

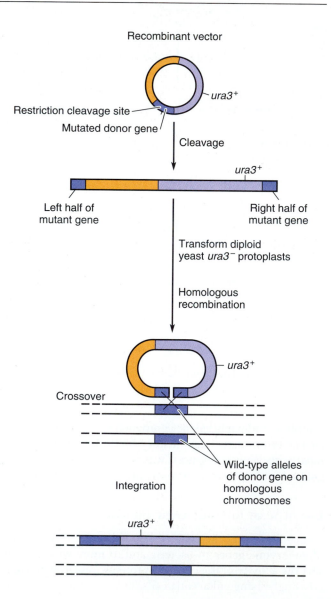

Recombinant vector

Restriction cleavage site
Mutated donor gene
ura3$^+$

Cleavage

ura3$^+$
Left half of mutant gene   Right half of mutant gene

Transform diploid yeast *ura3$^-$* protoplasts

Homologous recombination

ura3$^+$

Crossover

Wild-type alleles of donor gene on homologous chromosomes

Integration

ura3$^+$

▶ **FIGURE 18–18** Gene targeting by homologous recombination in yeast. A recombinant vector carrying a mutant donor gene of interest is linearized by a restriction enzyme that cleaves within the gene. After transformation of diploid yeast protoplasts, homologous recombination integrates the mutant donor gene into a protoplast chromosome, thus adding a third copy of the gene to the genome. In this example, gene targeting will have a phenotypic effect if the mutant gene is dominant to its wild-type alleles.

a single crossover event integrates the gene at the target site. The result is essentially gene duplication—the genome contains the two gene copies that were originally present and the mutated gene. Any dominant phenotypic effects of the mutation can then be observed in appropriately cultured cells.

In a specific form of gene targeting known as **targeted gene replacement**, a wild-type gene is replaced by a mutant allele. ▶ Figure 18–19 illustrates this technique. Restriction enzymes are used to generate a linear fragment of recombinant DNA carrying an inactivated gene. Homologous recombination between the transforming DNA and the endogenous DNA is again stimulated by the free ends of the transforming DNA, but now two crossovers are needed to mediate the recombination event, which results in the replacement of one copy of the normal gene with its mutant allele. If the mutant gene is completely inactive, this type of experiment is a so-called "gene knockout" experiment. Gene replacement experiments have proven particularly useful in studying the effect of recessive mutations in yeast. Diploid yeast are transformed and then allowed to sporulate—half of the haploid spores they generate will have the knocked out gene. The idea is that the function of the wild-type gene can be determined by seeing what cells grown from these spores are unable to do. The use of transformed yeast to study gene expression has given the first insight into eukaryotic genetic processes such as transcriptional activation, posttranslational modifications of proteins, protein secretion, control of the eukaryotic cell cycle, and other cellular signaling pathways.

To Sum Up

1. An expression vector is required to express the foreign DNA carried by a recombinant DNA molecule. Such vectors are derived from cloning vectors by adding transcriptional control signals that allow expression of the foreign DNA in a specific type of host cell.
2. Bacteria can be used to express foreign genes only if the recombinant DNA molecule contains an intronless version of the inserted gene. This insert can be either cDNA or artificially synthesized DNA.
3. In many cases, the expression of a eukaryotic gene is best studied in a eukaryotic host such as yeast rather than a bacterial host. The most efficient vectors in this situation are shuttle vectors, which can replicate (and therefore be easily cloned) in bacteria as well as in yeast.
4. Gene targeting in yeast is an important example of expression cloning with a shuttle vector. Gene targeting involves cloning a yeast gene in *E. coli,* introducing a mutation into that gene, returning the altered gene to yeast cells, where it recombines with the homologous region (target) of a yeast chromosome, and studying of the phenotypic effects of the mutation. By targeting specific genes in this way, researchers learn the function

of a wild-type gene by noting what the transformed yeast cells are unable to do after the mutation.

TRANSFERRING GENES INTO EUKARYOTES

Of course, the best way to study the expression of any gene is to transform cells of the organism from which the gene was obtained. (The term *transformation* is widely used to mean conversion of an animal or plant cell to the cancerous state as well as to refer to the uptake of foreign DNA by a recipient cell, but in this chapter the latter definition is always used.) One of the areas of current research in molecular biology involves the analysis of gene expression in animal and plant cells. The roles of genes in the immune response, in development, and in cancer are finally beginning to be revealed through gene targeting. This section will describe the methods used to introduce foreign DNA into mammalian cells in culture and then will discuss the production of genetically engineered animals and plants.

Transferring Genes into Mammalian Cells

The first successful transformation of mammalian cells growing in culture occurred in the early 1960s; recipient cells that were deficient in some selectable biochemical function were exposed to a calcium phosphate precipitate of DNA fragments carrying the wild-type gene, and the wild-type transformants were selected. Although this technique has been widely used, it is limited by the very low efficiencies with which cells take up the naked DNA and stably integrate that DNA into the genome (approximately one in 10^5 to 10^6 cells becomes stably transformed). Cellular enzymes in the transformed cells act on the incorporated DNA to form large concatemers composed of several copies of the original fragment and the vector DNA. A concatemer then recombines with a site on a recipient chromosome and is inserted into the genome of the host cell.

Direct **microinjection** of donor DNA into cultured cells is an alternative approach to the low-efficiency transformation process just described; it virtually ensures transformation, but it is too technically demanding to be applied to a large number of cells. A technique called **electroporation** has recently been used to transform cultured cells; it involves subjecting the cell-DNA mixture to a high-voltage electric pulse that apparently creates temporary holes in the plasma membrane through which DNA can pass. In another technique, **lipofection**, the donor DNA is included in positively charged liposomes that fuse with the plasma membranes of cultured cells to allow the DNA to pass through.

Because many genes do not convey a selectable phenotype, experiments can be designed to screen recipient cells

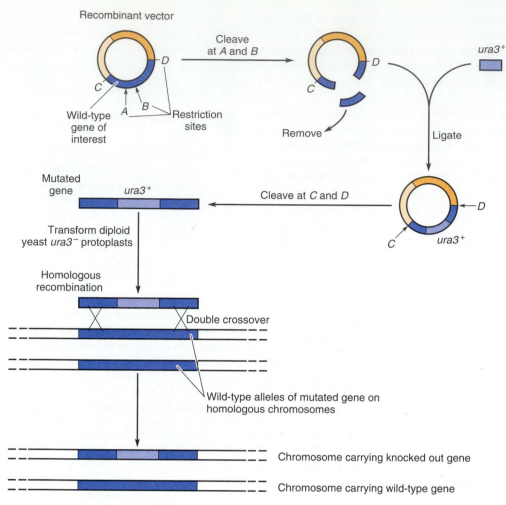

FIGURE 18–19 Targeted gene replacement (the replacement of a wild-type chromosomal gene by its mutant allele). A gene of interest is mutated by removing a short section of its DNA and replacing it with a selective marker, in this case ura3⁺. The deletion yields a completely nonfunctional gene. The mutant gene fragment is then cleaved from the vector and incorporated into diploid yeast protoplasts through homologous recombination involving two crossovers. This replacement of a chromosomal wild-type gene with the mutant allele is called gene "knockout" because the wild-type gene is replaced by a completely inactive allele. Recessive mutations are studied by allowing yeast that have undergone targeted gene replacement to form haploid spores. Since half of these spores carry the knocked out gene, the consequences of its loss can be determined.

for **cotransformation** by a selectable gene. The formation of concatemers in transformed cells is the basis of cotransformation. For example, suppose that cultured TK⁻ cells are exposed to a calcium phosphate precipitated DNA mixture containing a tk⁺ gene and a biochemically unselectable gene, say the β-globin gene, and that TK⁺ transformants are selected. Because the transforming DNA has gone through the concatemer stage, it is quite likely that the TK⁺ transformants have also been transformed by the β-globin (Hb_β) gene. Southern blot analysis of DNA extracted from recipient cells shows that this is indeed the case; nearly all the transformed cells have acquired both genes (► Figure 18–20).

Any cloned gene can be stably introduced into cultured mammalian cells through cotransformation. The original procedure was severely limited by the need for a mutant TK⁻ cell line, but this limitation has now been overcome by the development of dominant selectable markers that can be used with nonmutant cells. The most widely used marker is the bacterial gene *neo,* which confers resistance to G418, a neomycin-related antibiotic that kills mammalian (and other animal and plant) cells by blocking protein synthesis. Transformants of nonmutant cultured mammalian cells that have acquired this gene can be easily selected by their resistance to the antibiotic.

Our discussion so far has focused on mammalian cells that can be transformed without using a vector. Using certain animal viruses as vectors improves the efficiency of transformation of those cells and also extends gene transfer to cells that normally are not transformable (e.g., human cells). Simian virus 40, a DNA tumor virus, is an example of a virus that can be used as a vector; its genome structure is shown in ► Figure 18–21.

Because packaging strictly limits the size of DNA that

can be carried by SV40, several derivatives of the viral genome have been developed. Two of these derivatives are shown in ▶ Figure 18–22. **SV40 plasmid vectors** are designed solely for replication; they lack the late genes necessary for packaging into virions. These vectors thus transfect cells with low efficiency, but they are maintained in recipient cells as high copy-number, plasmid-like molecules that express a high level of donor gene product. Alternatively, if the region carrying the early genes is replaced by a fragment of foreign eukaryotic DNA of equivalent size (Figure 18–22b), the result is an **SV40 transducing vector** that can be packaged into virions in cells that contain both the recombinant vector and "helper" SV40 functions. In these cells, which are called COS cells, functional SV40 early genes are integrated into the host genome. Because these genes supply the functions missing from the vector, prog-

eny viral particles can be produced, and some will contain the foreign gene. These recombinant particles can very efficiently introduce the foreign DNA into recipient cells through the normal viral infection process. The injected genome is defective as a virus and cannot replicate and produce progeny, but if it has been constructed as an expression vector, the foreign gene will be expressed in the recipient cells.

The frequent use of SV40 derivatives as expression vectors is based on thorough knowledge of SV40's genetics and life cycle. However, the capacity of SV40 for foreign DNA is limited, and the virus infects only monkey cells. Other viruses have therefore been developed as vectors for transferring genes into other animal cells. The bovine papillomavirus can be used as a vector for transporting large DNA inserts into mouse cells, where it does not kill the cells but

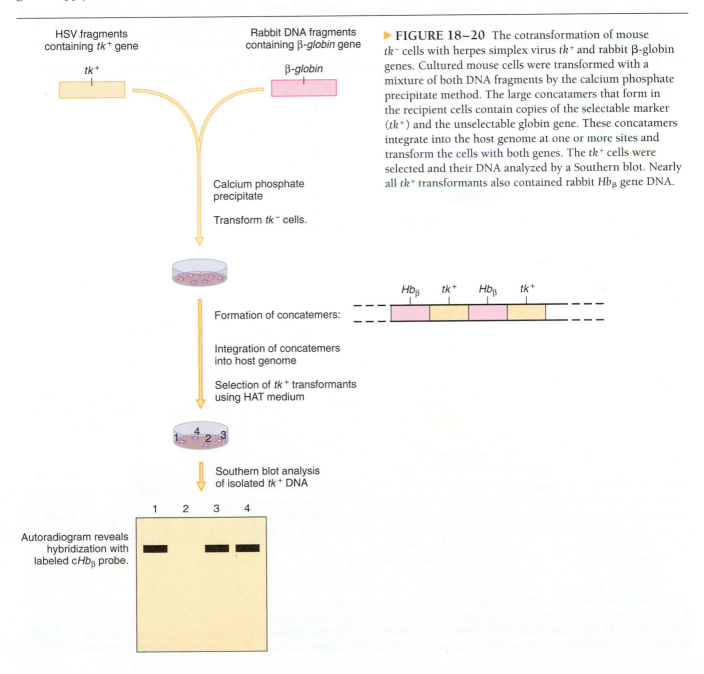

▶ **FIGURE 18–20** The cotransformation of mouse tk^- cells with herpes simplex virus tk^+ and rabbit β-globin genes. Cultured mouse cells were transformed with a mixture of both DNA fragments by the calcium phosphate precipitate method. The large concatamers that form in the recipient cells contain copies of the selectable marker (tk^+) and the unselectable globin gene. These concatamers integrate into the host genome at one or more sites and transform the cells with both genes. The tk^+ cells were selected and their DNA analyzed by a Southern blot. Nearly all tk^+ transformants also contained rabbit Hb_β gene DNA.

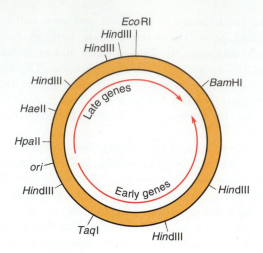

▶ FIGURE 18–21 The genome of simian virus 40. Early genes code for proteins essential to viral genome replication, late genes for capsid and packaging proteins. The circular arrows denote the directions of transcription of early and late viral genes.

instead replicates as a plasmid and expresses foreign protein at a high level. Another virus, the vaccinia virus, can be engineered to express a number of important genes, producing vaccines such as the hepatitis B surface antigen, an influenza haemagglutin, and the rabies antigen. Although these vaccines have shown promise in rabbits, hamsters, and foxes, they are rarely used in humans, since naturally occurring vaccinia is closely related to the virus that causes smallpox. However, vaccinia vectors do hold promise as expression vectors in human cells. Recently a specially constructed vaccinia vector was shown to express the wild-type *CF* gene in cells derived from a patient with cystic fibrosis, correcting the ion-transport defect of these cells.

The RNA tumor viruses (the retroviruses) have been extensively studied genetically (see Chapter 16), and shuttle vectors derived from retroviruses have been transfected directly from *E. coli* to animal cells. These vectors are constructed in vitro from a plasmid such as pBR322 (▶ Figure 18–23a) and consist of the plasmid *ori*, a selectable

(a) SV40 plasmid vector

(b) SV40 transducing vector

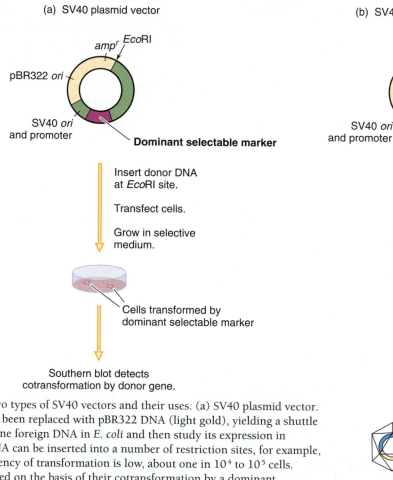

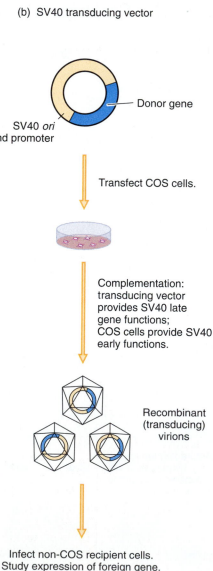

▶ FIGURE 18–22 Two types of SV40 vectors and their uses. (a) SV40 plasmid vector. The late gene region has been replaced with pBR322 DNA (light gold), yielding a shuttle vector that is used to clone foreign DNA in *E. coli* and then study its expression in monkey cells. Donor DNA can be inserted into a number of restriction sites, for example, the *Eco*RI site. The efficiency of transformation is low, about one in 10^4 to 10^5 cells. Transformants are selected on the basis of their cotransformation by a dominant selectable marker. (b) SV40 transducing vector. A eukaryotic gene (blue) is inserted in place of the early gene region, and the recombinant vector is used to transfect COS cells. These cells produce sufficient viral early gene product to support the replication of the SV40 recombinant vector, which subsequently is packaged into virions. Cultured non-COS cells are very efficiently transformed through infection by these defective virions, and the expression of the transduced gene can then be studied in the transformed cells.

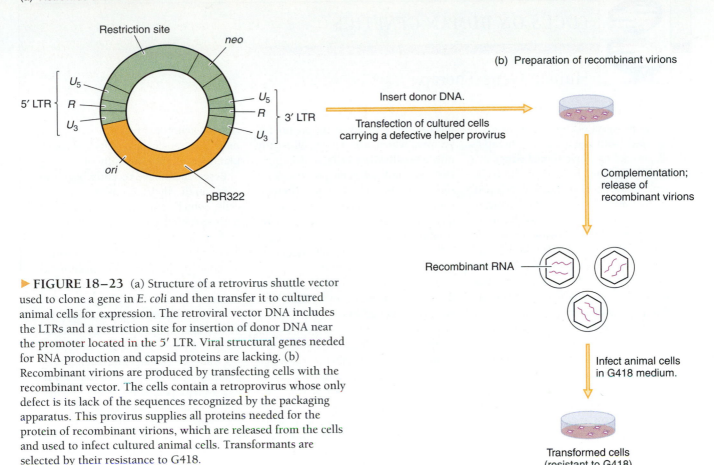

(a) Retrovirus shuttle vector

Restriction site

neo

5′ LTR { U_5 / R / U_3 }

3′ LTR { U_5 / R / U_3 }

ori

pBR322

(b) Preparation of recombinant virions

Insert donor DNA.

Transfection of cultured cells
carrying a defective helper provirus

Complementation;
release of
recombinant virions

Recombinant RNA

Infect animal cells
in G418 medium.

Transformed cells
(resistant to G418)

▶ FIGURE 18–23 (a) Structure of a retrovirus shuttle vector used to clone a gene in *E. coli* and then transfer it to cultured animal cells for expression. The retroviral vector DNA includes the LTRs and a restriction site for insertion of donor DNA near the promoter located in the 5′ LTR. Viral structural genes needed for RNA production and capsid proteins are lacking. (b) Recombinant virions are produced by transfecting cells with the recombinant vector. The cells contain a retroprovirus whose only defect is its lack of the sequences recognized by the packaging apparatus. This provirus supplies all proteins needed for the protein of recombinant virions, which are released from the cells and used to infect cultured animal cells. Transformants are selected by their resistance to G418.

marker such as the neo^r gene, and retroviral DNA sequences including the LTRs. A donor gene of interest is inserted into the shuttle vector near the 5′ LTR region in place of a section containing one or more viral genes. To prepare recombinant virions (Figure 18–23b), cells are transfected by the recombinant vector using the calcium phosphate method. A promoter region in the 5′ LTR drives transcription of the recombinant viral genome; transcription terminates at a termination signal located in the 3′ LTR, so plasmid DNA is not transcribed. Helper viral functions are needed to make up for the vector's lack of one or more viral genes. These functions can be provided if the recipient cells contain a provirus that lacks the sequences recognized by the packaging system but is otherwise normal. From the pool of helper and recombinant RNAs, only the latter are packaged into virions, which bud from the cell and are collected for use as infectious recombinant viruses. Cultured cells infected by these viruses express the donor gene but never produce progeny viruses because they lack the necessary viral helper genes.

Retroviruses are very useful vectors for a number of reasons: They can infect virtually any kind of animal cell; integration of the recombinant vector into the host genome means that the recipient cells permanently express the donor gene; the 5′ LTR promoter is very strong, leading to a high level of donor gene expression; and infection is not lethal to the host cells. Retroviruses are proving to be especially useful for transfecting human cells and for engineering the cells of very early animal embryos.

Medical researchers are currently working to adapt the transformation of cultured cells for therapeutic use in treating some human genetic diseases, a procedure known as **gene therapy**. For example, cells extracted from the bone marrow can be transformed in culture with a retrovirus vector carrying a wild-type allele of the defective gene and then reintroduced into the patient in the hope that they will produce the normal gene product. Several problems have so far prevented such gene therapy from being successful on a widespread basis. A major problem is the difficulty of working with cell types other than those from the bone marrow. Most genes are expressed only in particular types of tissue; for gene therapy to be successful, the specific cell types in which a given gene is normally expressed must be identified and manipulated. Another difficulty arises from the random integration of the transforming DNA into the recipient cell genome, since gene expression also depends on specific regulatory signals at specific chromosomal locations. However, the successful transformation of cultured

Human Gene Therapy

The prospect of treating human disease with gene-based therapies instead of conventional drugs is a tantalizing one. Several methods for gene delivery have been developed, and research on new methods continues. Efficient delivery has been attained by using retroviral vectors to transfer the desired DNA into a culture of relevant target cells, where it is integrated into the host cell nuclear genome. Some retroviral vectors are able to transform nearly 100% of the target cells; even in these cases, however, replication of the cells seems to be necessary for integration of the recombinant vector into the host genome. This requirement is a problem with human cells, which are very difficult to grow in culture. Another problem is that retrovirus particles are unstable and are rapidly inactivated in some types of primate cells in vivo; the transferred genes and therefore gene expression are transient. In addition, even though retroviral vectors have been developed to contain no genetic information for viral propagation and are thus defective as infectious viruses, several outbreaks of wild-type viruses from transformed cell lines have been reported, and some transformed primates have developed lymphomas.

Because of these problems, other types of viruses are being investigated for potential use in gene therapy. The most promising may be the adenoviral vectors, which are stable, efficiently infect nondividing cells, and express large amounts of gene product for a significant period of time. There is some evidence, however, that a low level of replication of the defective recombinant viruses occurs in target cells, so there may be safety issues associated with these viruses.

In addition to problems involving the vectors, technical difficulties have been encountered in isolating (and/or in growing in culture) large amounts of appropriate target cells and transplanting them back into the individual in a state where they will continue to manufacture the needed protein. Hematopoietic stem cells have been a center of focus because their transformation and transplantation would ensure a continuous supply of engineered hematopoietic cells during the patient's lifetime. The technology for extracting and transforming stem cells has been very difficult to develop, however, mainly because they are found only in small numbers in bone marrow, and, being relatively quiescent, they are usually not susceptible to infection by a recombinant retrovirus. So little is known of the biology of stem cells or of bone marrow transplantation that it has been difficult to determine how to get stem cells to develop an increased competence for transformation, and how to get them to persistently express a desired gene once the transformed cells are transplanted back into the individual.

Researchers have recently begun to concentrate their attention on liver cells, which can be transformed by both retroviral and adenoviral vectors, and on lung cells, which adenoviral vectors have successfully transformed in large proportions for meaningful amounts of time (e.g., transformation of lung cells with the *CF* gene).

A recent report on gene therapy for hemophilia suggests one way to avoid the problems of extracting and growing cells in culture and then transplanting them back into an individual: Researchers placed the canine IX factor gene (the mutant form of this gene causes hemophilia B) directly into the livers of dogs. Because they used a retroviral vector, they also removed about two-thirds of the liver to get the liver cells to divide (liver cells start dividing to regenerate the organ when part of the liver is removed). Although the level of factor IX produced by the dogs in these experiments was quite low, gene expression seemed to be long-term. There are obvious dangers that must be overcome before such a procedure could be adapted to humans, but because human cells are so extremely difficult to grow in culture, this method of gene therapy may hold promise for the future.

CF cells by the normal *CF* gene provides encouragement that gene therapy may eventually be useful in the treatment of some single-gene disorders.

Producing Transgenic Animals

It is important to distinguish between the transformation of eukaryotic cells in culture and the introduction of foreign DNA into early embryos. In the latter case, the foreign DNA integrates into the recipient's genome early enough in development to transform all cell lineages, both somatic and germ line. The result is a **transgenic** organism that passes the foreign gene on to progeny in a Mendelian fashion. In contrast, the transformation of cells growing in culture does not introduce heritable changes into an organism.

A transgenic animal is created when foreign DNA is incorporated into a single-celled embryo (usually the paternal pronucleus of a fertilized egg) or into embryonic stem (ES) cells (primitive undifferentiated cells capable of giving rise to all cell types of the organism). Transgenic frogs

(*Xenopus laevis*), *Drosophila*, and mice (and a few other mammals) have been produced. ▶ Figure 18–24 shows how transgenic mice were created in the first reported example of the expression of a human gene in another animal. A pBR322-derived expression vector carrying the human growth hormone gene adjacent to a mouse gene promoter was microinjected into the pronuclei of fertilized mouse eggs. The eggs were then implanted in a foster mother, and the DNA of the resulting progeny was screened by the Southern blot technique to detect which offspring carried the donor gene. In the original experiment, 23% of the progeny mice that stably incorporated the donor gene produced high levels of human growth hormone, which

caused them to grow at two to three times the rate of normal mice and to attain an unusually large size. Intermatings of the transgenic mice showed that the trait is passed on in a Mendelian fashion.

The ultimate goal of research on eukaryotic gene expression is to study how genes function in whole animals. The recent development of methods for gene targeting in mice have been a major step toward this goal. Researchers have been able to target specific genes in cultured mouse ES cells to create so-called knockout mice. In contrast to gene knockout in yeast, in which specific genes are inactivated through homologous recombination, transforming DNA in animal ES cells tends to be integrated at random chromo-

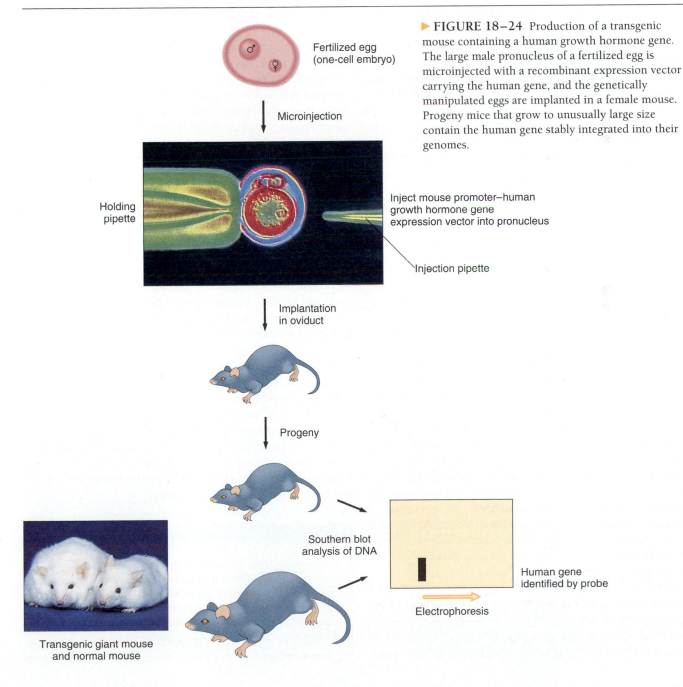

▶ **FIGURE 18–24** Production of a transgenic mouse containing a human growth hormone gene. The large male pronucleus of a fertilized egg is microinjected with a recombinant expression vector carrying the human gene, and the genetically manipulated eggs are implanted in a female mouse. Progeny mice that grow to unusually large size contain the human gene stably integrated into their genomes.

Fertilized egg (one-cell embryo)

Microinjection

Holding pipette

Inject mouse promoter–human growth hormone gene expression vector into pronucleus

Injection pipette

Implantation in oviduct

Progeny

Southern blot analysis of DNA

Human gene identified by probe

Electrophoresis

Transgenic giant mouse and normal mouse

somal locations. Thus, gene targeting in mice is not as efficient as it is in yeast, and a selectable marker gene must be used to detect a rare homologous recombination event.

The procedure for the creation of knockout mice is shown in ▶ Figure 18–25. The transforming DNA consists of a selectable tk^+ marker gene and the gene of interest, which has been inactivated by insertion of the neo^r gene to serve as a marker for stably transformed stem cells. Although the DNA can integrate randomly anywhere in the genome of the recipient cells, those rare (perhaps one in 10^3 or so) integration events that occur through homologous recombination can be selected by the absence of the TK⁺ phenotype. The combination of positive selection (for G418 resistance) and negative selection (for lack of TK function) enriches the concentration of stem cells containing a targeted mutation over 2000-fold. When these cells are introduced into mouse embryos, the mixture of altered and unaltered stem cells develops into chimeric mice that are identified by their variegated coats. The variegation occurs because the engineered stem cells are derived from dark mice while recipient embryos come from light-colored mice. Of course, only a very few of these chimeras have, by chance, had the gene altered in the germ line, so a large number of them must be created and mated together to obtain progeny mice that have the knocked-out gene in all their cells.

The effect that the loss of the targeted gene has on these mice gives clues about the gene's normal function. Significant advances in understanding the role of genes in the immune system, in development, and in cancer have already emerged from these knockout experiments (see Chapters 20 and 21). Researchers are now developing mouse models for the study of human gene disorders by engineering mice to contain the normal human gene and then studying the consequences of knocking it out. Clever new techniques are also allowing researchers to achieve the long-sought goal of inactivating mouse genes specifically in selected cell types.

The study of the regulation of eukaryotic gene expression has greatly benefited from the development of techniques that introduce DNA into whole organisms. As we will see in Chapters 20 and 21, the activity of a gene in different tissues of an organism and during the various stages of embryonic development is regulated by several kinds of DNA sequences (such as promoters) and other regulatory factors. Biologists have long been interested in elucidating these regulatory mechanisms. For example, one method of analyzing the influence of a regulatory DNA region on gene expression is to couple it to a gene that encodes an easily assayed protein. The expression (or lack of expression) of that gene can then serve as a "reporter" for the functioning of this promoter in different tissues and at different developmental stages. Genes that are used in this fashion are called **reporter genes**.

Identification of the specific tissue(s) and developmental stage(s) in which a particular regulatory DNA region exerts control opens up a variety of ways to study embryonic development. For example, if a promoter has been found to activate gene transcription only in a particular tissue, re-

searchers can attach a reporter gene that encodes a toxin in order to study the developmental effects of destroying that tissue. Studies can be even more precisely focused by using **inducible promoters**—promoters that do not stimulate transcription until activated by an inducing substance. Special expression vectors that contain inducible promoters can be used to precisely time the effects of the toxin on a specific stage of development.

One example of a study involving an inducible promoter uses the diptheria A (*dipA*) toxin gene as a reporter with a pBR322 expression vector described above. This vector carries a tissue-specific mouse metallothionein (*MMT*) gene promoter that triggers expression of the adjacent foreign gene only when mice are treated with a heavy metal such as cadmium or zinc (the metallothionein gene product is involved in detoxification of heavy metals). Female mice implanted with eggs that have been injected with pBR322-*MMT* promoter–*dipA* recombinant DNA can be treated with cadmium at various stages of embryonic development to study how destruction of a particular tissue at different developmental stages affects the developing embryo. Experiments of this type have studied the expression of more than 30 tissue-specific genes that are active in a dozen types of tissue ranging from the brain to the liver.

Commercially important agricultural animals have also been the subjects of genetic engineering, and transgenic pigs, rabbits, and sheep containing the human growth hormone gene have been produced. Pigs, for example, have been found to grow faster, have a greater weight gain per unit of feed, and produce leaner meat when injected with human growth hormone produced by recombinant *E. coli*. A more efficient means of achieving the same result is to engineer the pigs to produce the growth hormone themselves. Although this produces the desired effects on growth and meat production, transgenic pigs do not properly regulate the expression of the foreign growth hormone gene and as a result suffer severe physiological problems. (Transgenic mice suffer similar problems.)

Researchers are also working to engineer animals to produce pharmaceutical proteins, for example, to develop transgenic cows that secrete such proteins in their milk, where they can be conveniently harvested. This work has been successful with mice, but so far the levels of secreted proteins in cows have been too low to be of commercial value. The regulatory elements affecting expression of these introduced genes need to be elucidated before transgenic livestock become commercially valuable.

Points to Ponder 18.1

If gene therapy techniques could be adapted to the "correction" of gene defects in the developing human fetus, would this affect your opinion of prenatal diagnosis? Does gene therapy pose any ethical dilemmas for you? What important ethical differences do you perceive between gene therapy and the production of transgenic organisms?

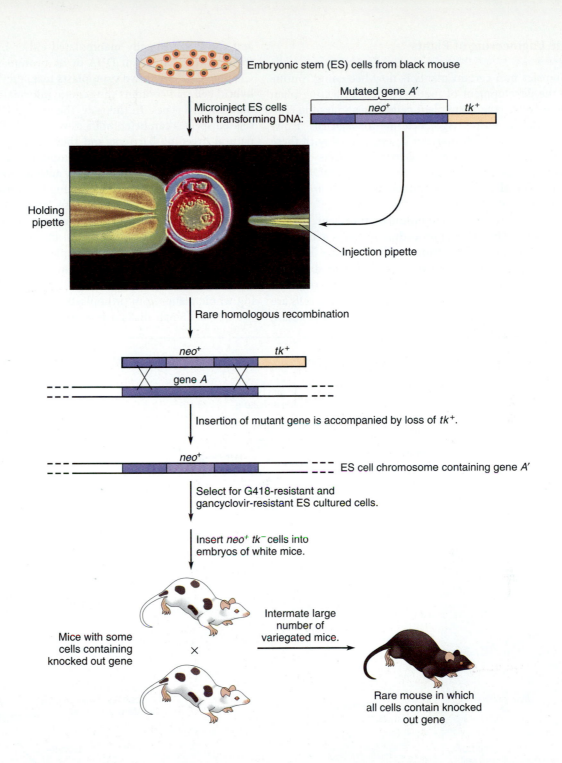

Embryonic stem (ES) cells from black mouse

Microinject ES cells with transforming DNA:

Mutated gene *A′*

neo⁺   *tk⁺*

Holding pipette

Injection pipette

Rare homologous recombination

neo⁺   *tk⁺*

gene *A*

Insertion of mutant gene is accompanied by loss of *tk⁺*.

neo⁺

ES cell chromosome containing gene *A′*

Select for G418-resistant and gancyclovir-resistant ES cultured cells.

Insert *neo⁺ tk⁻* cells into embryos of white mice.

Mice with some cells containing knocked out gene

×

Intermate large number of variegated mice.

Rare mouse in which all cells contain knocked out gene

▶ **FIGURE 18–25** Production of knockout mice by gene targeting. Embryonic stem (ES) cells from a black mouse are microinjected with DNA containing a mouse gene (*A′*) that has been mutated by insertion of a *neo* gene and a *tk⁺* gene from herpes simplex virus. Most of the transforming DNA is integrated into the ES genome at random locations, but integrations at the gene *A* target site occur through rare homologous recombination events. These targeted integrations are selected for by the resistance of the transformed ES cells to G418 and to the nucleoside analogue gancyclovir (to which TK⁺ cells are sensitive). The selected ES cells are microinjected into embryos of white mice,which subsequently develop a variegated coat. Many variegated mice must then be intermated in hopes of finding two that have incorporated the knocked out gene into the germ line and therefore can produce progeny having the knocked out gene in all their cells.

Genetic Engineering of Plants

Gene transfer into certain plants is now becoming routine due to the development of methods for transfecting plant protoplasts or regenerating plant cells using a plasmid contained in the bacterium *Agrobacterium tumefaciens*. However, the range of plants that can be engineered with these methods is limited, because *Agrobacterium* infects only dicotyledonous plants. Many important crop plants (for example, the cereal grains) are monocots, which are much more difficult to transform.

Two methods can be used to produce a transgenic dicotyledonous plant. The original procedure involves generating whole plants from single cultured cells (▶ Figure 18–26). Single undifferentiated cells can be induced to divide in broth or on agar medium to produce clones that consist of millions of genetically identical descendants. These cells are then converted to protoplasts by treatment with cellulase

and can be genetically manipulated either by transfecting them with recombinant DNA or by protoplast fusion (see Chapter 12). When two protoplasts fuse, they form a single hybrid cell that contains the genetic information of the two different cell types. In this way, the genes of even distantly related species can be combined without the restrictions of natural breeding barriers, and there is a potential for the creation of novel plants. The protoplasts are then induced to regenerate cell walls and grow into an undifferentiated cell mass called a callus. By adjusting the hormone levels in the medium, researchers can cause the callus tissue to develop into small plantlets with immature roots, stems, and leaves. Eventually, the plantlets are transferred to soil, where they can grow to form complete plants.

A recently developed procedure that is simpler and more rapid than the difficult, time-consuming task of growing whole plants from protoplasts is summarized in ▶ Figure 18–27. Small disks a few millimeters in diameter are

▶ **FIGURE 18–26** The four-step process used to produce a transgenic plant derived from genetically manipulated cells growing in culture.

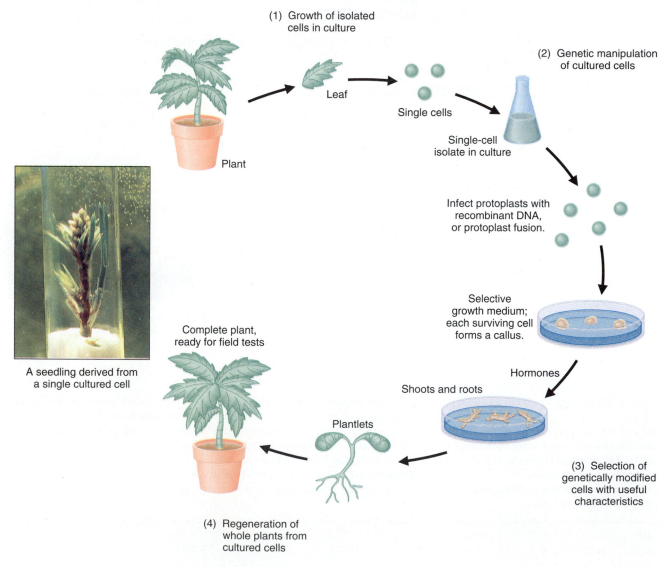

(1) Growth of isolated cells in culture

Leaf

Single cells

(2) Genetic manipulation of cultured cells

Single-cell isolate in culture

Plant

Infect protoplasts with recombinant DNA, or protoplast fusion.

Selective growth medium; each surviving cell forms a callus.

A seedling derived from a single cultured cell

Complete plant, ready for field tests

Hormones

Shoots and roots

Plantlets

(3) Selection of genetically modified cells with useful characteristics

(4) Regeneration of whole plants from cultured cells

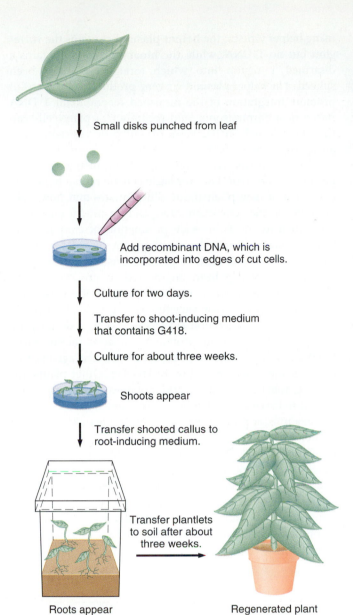

Small disks punched from leaf

Add recombinant DNA, which is incorporated into edges of cut cells.

Culture for two days.

Transfer to shoot-inducing medium that contains G418.

Culture for about three weeks.

Shoots appear

Transfer shooted callus to root-inducing medium.

Transfer plantlets to soil after about three weeks.

Roots appear   Regenerated plant

▶ **FIGURE 18–27** The leaf disk method for producing a transgenic plant.

punched from the leaves of a dicot (petunia, tobacco, or tomato plants are commonly used) and incubated in a medium containing the recombinant DNA, which includes a foreign gene and an antibiotic resistance marker such as neo^r. The leaf disks are a source of regenerating cells, which are efficiently transformed by the recombinant DNA molecules. After two days in this culture, the cells are transferred to a medium containing an antibiotic such as G418 to select for transformed cells. Culturing in the appropriate media yields rooted plantlets that can be transferred to soil within two months.

The leaf disk method provides tremendous opportunities to genetically engineer plants. The major obstacle has been a paucity of vectors that will allow the donor material

to be replicated in recipient plant cells. Plant DNA viruses have strict size requirements for packaging and contain almost no nonessential regions into which foreign DNA can be substituted. RNA viruses of plants have promise as vectors but have not yet been developed to the stage where they can be widely used. Thus plant geneticists have so far focused their attention on a bacterial plasmid called **Ti** (**tumor inducing**), which is found in *Agrobacterium tumefaciens*. A Ti gene map is shown in ▶ Figure 18–28.

A. tumefaciens causes crown gall tumors in wounded dicotyledonous plants by transferring a segment of the Ti plasmid (called **T DNA**) into plant cells by a conjugative process. Transfer of the plasmid and integration of T DNA are controlled by the plasmid *vir* genes. The T DNA inserts itself into a randomly selected position in the plant genome. Oncogenes that are carried on the T DNA prevent cell differentiation, causing the infected cells to grow in an uncontrolled manner to form a tumor. Other T DNA genes, such as *nos,* cause the infected plant cells to produce enzymes that catalyze the synthesis of a special class of nitrogen-rich amino acids called opines. The opines are required by *A. tumefaciens* as a source of nitrogen.

Because they cause tumorous growth of recipient cells, wild-type Ti plasmids are not suitable as vectors for transferring foreign DNA. However, because the oncogenes of the Ti plasmid lie entirely within its T region, they can be "disarmed" without affecting the ability of the plasmid to transfer T DNA to recipient cells. To disarm the T DNA, researchers replace its oncogenes with pBR322 sequences, leaving only the left and right border regions (which are

▶ **FIGURE 18–28** Gene map of a nopaline Ti plasmid. The *vir* region contains genes that are activated by specific signal molecules produced by wounded plant cells. The *vir* gene products control the transfer of T DNA into the wounded cells and its integration into the cell genome. Genes in the T region suppress the differentiation of cells, causing them to grow in the uncontrolled manner characteristic of a tumor. The *nos* gene codes for nopaline synthetase, an enzyme that synthesizes an opine required for the growth of *Agrobacterium.*

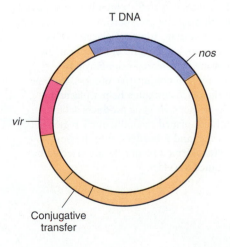

T DNA

nos

vir

Conjugative transfer

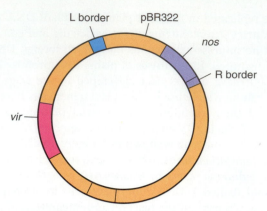

► **FIGURE 18–29** Structure of the Ti plasmid pGV3850, which has been disarmed by replacing most of the T region with pBR322 DNA. The *nos* gene and the left and right border regions of the T region remain; the border regions are required for integration of the T region into the host cell genome.

necessary for the integration of T DNA into a host chromosome) and the *nos* gene of the T DNA (see ► Figure 18–29). Although opines are still produced only by cells transfected by this construct (and could therefore serve as markers), researchers place a neomycin resistance gene next to the *nos* promoter, so that antibiotic resistance can more conveniently be used to select for transfected cells.

Although the pGV3850 vector can be used to transfer foreign DNA, the vector is so large that the pBR322-T region does not contain any unique restriction sites to ensure proper placement of foreign DNA. For this reason, researchers recently developed a **binary Ti system** using *Agrobacterium* that have incorporated two different plasmids (► Figure 18–30). The binary system is analogous to

using helper viruses; the **helper plasmid** contains the *vir* region but no T DNA, while the **binary plasmid** contains a disarmed T region into which foreign DNA has been spliced. The helper plasmid *vir* gene products act in trans to promote integration of the disarmed recombinant T DNA into a host chromosome, and recombinant plant cells can then be selected on the basis of antibiotic resistance and grown into whole plants.

These methods make the engineering of plants a very promising endeavor, both for basic genetic research into the questions of how plant tissue differentiates and how light induces specific gene expression, and for practical applications such as the large scale production of commercially valuable proteins and the development of plant varieties with desirable traits such as disease resistance. Transgenic plants have already been engineered to produce several pharmaceutical proteins. The development of disease- and pesticide-resistant plant varieties is currently the leading area of research. Recombinant DNA carrying a donor gene that codes for the coat protein of tobacco mosaic virus (TMV) has been used to "vaccinate" tobacco against this troublesome virus (see ► Figure 18–31). Other plants, including tomatoes, cotton, and potatoes, are being engineered to carry genes that confer resistance to insect pests. For example, a protein produced by the natural microbial pesticide *Bacillus thuringiensis* is toxic to the larvae of a number of insect species, and researchers are working to develop transgenic plants carrying the gene that encodes this protein. Some crop plants that have been engineered to carry this toxin gene (for example, cotton) have been field-tested and are ready for commercial release. Transgenic plants that are resistant to commonly used herbicides are also being developed.

► **FIGURE 18–30** The binary Ti system. A binary vector (left) contains the left and right border sequences of T DNA, a dominant selectable marker (in this case *neo*), and a promoter-*lacz'* region (for detection of a donor gene by the blue/white color assay) into which has been placed a multiple cloning site for insertion of donor DNA. These vectors contain bacterial plasmid DNA (dark gold), which allows them to replicate in *E. coli*; they are transferred from *E. coli* to *Agrobacterium* by a conjugative process. If these *Agrobacterium* also contain a helper plasmid (right), the helper's *vir* gene products act to integrate the disarmed recombinant T region (between the L and R borders) into the host genome. Transformed plant cells are selected by their resistance to G418.

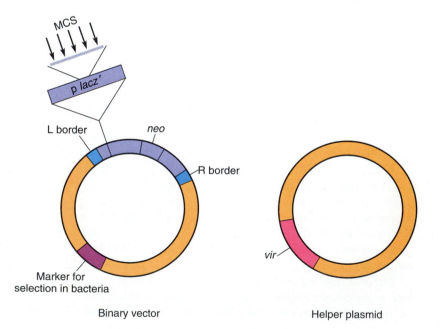

▶ **FIGURE 18–31** "Vaccinated" tobacco plant (left) and regular tobacco plant (right) exposed to TMV.

A different application of recombinant DNA technology has produced tomatoes that carry a polygalacturonase gene with its transcriptional template strand in the wrong 5'–3' orientation. In its normal orientation, this gene codes for an enzyme that breaks down the cells walls of tomatoes, causing them to soften. In reverse orientation, the gene produces **antisense** mRNA, which hybridizes with the mRNA produced from the normal gene, thus preventing the latter from being translated into protein. Commercial tomatoes with this modification do not have to be picked and shipped while they are still green; instead, growers can allow them to remain on the vine longer, greatly improving their taste (▶ Figure 18–32).

Although these uses of transgenic plants are quite impressive, there remains the problem of engineering monocots such as the cereal plants. Several plant viruses have been studied for possible use as vectors, but the plant DNA viruses studied to date are unable to replicate if foreign DNA is inserted into them, and research on plant RNA viruses has not yet provided usable vectors. Protoplasts of corn and rice have been successfully transfected using electroporation, but there has been little success in regenerating whole plants from protoplasts of these species. Researchers are now working, with some success, on techniques to "shoot" minute metal beads coated with DNA into whole cells using a particle gun designed specifically for this purpose. Genetic engineering of cereal grains in this fashion promises to have dramatic implications for worldwide food production.

To Sum Up

1. Several different techniques have been developed to transfer foreign DNA into mammalian cells. The efficiency of transformation can often be increased by using certain viral chromosomes or their derivatives as vectors. These vectors can be designed for cloning and/or expression purposes and have a wide range of possible applications. Retroviral vectors are particularly useful because virtually any kind of animal cell can be nonlethally and stably infected by retroviruses.

2. Transgenic organisms arise from transformation of early embryos. In contrast to transformed somatic cells, transformed embryos give rise to organisms that are genetically altered in all their cells and therefore pass on the acquired genes to their offspring.

3. Transgenic animals are being used for a variety of research purposes. Gene targeting experiments with mammals, such as mice, are elucidating the roles of genes involved in the immune system, in embryonic development, and in cancer. Transgenic animals are also being produced for practical purposes; for example, meat- and wool-producing farm animals are being engineered to grow faster and yield a better quantity and quality of product.

4. Genetic engineering of plants has been most successful with dicots, which can be transformed using the *Agrobacterium* plasmid as a vector. Like transgenic animals, transgenic plants are used for research into gene expression and genetic control of development as well as for practical applications, such as the production of disease- and pesticide-resistant varieties.

Point to Ponder 18.2

In the near future, recombinant DNA techniques will make it possible to engineer plants to have increased resistance to herbicides. This will make it possible for spraying to kill weeds more effectively without harming the crop plants. Although this might seem like a good idea on the surface, it has been intensely criticized by some scientists and environmentalists.

Consider the application of this technology—weigh its benefits against its various costs. Do you think engineering for increased herbicide resistance is a good idea?

▶ **FIGURE 18–32** MacGregor's® tomatoes grown from FLAVR SAVR™ seeds. These tomatoes were first introduced to customers in California and Illinois in May of 1994 by Calgene, Inc. Because their natural softening process has been slowed, they can stay on the vine longer for extra flavor and a desirable texture.

DNA cloning technology has greatly facilitated the study of gene structure and function. To clone a gene, restriction endonucleases and DNA ligase are used to insert foreign DNA into a self-replicating vector, such as a virus or plasmid. The recombinant vector with its foreign DNA is then transferred into a suitable host, where it can be propagated to form clones. Bacteria were the first organisms to be used as hosts for DNA cloning. Bacterial cloning vectors include plasmids such as pBR322 and its derivatives, bacteriophages such as λ, and vectors such as phasmids and cosmids, which incorporate DNA sequences of both plasmids and phage λ. Single-stranded phage vectors such as M13 and its derivatives are used for cloning DNA in a single-stranded form.

Often the first step in cloning a gene is the construction of a genomic library, which is a collection of DNA fragments representing most of the genome of an organism separately inserted into various vectors. Clones containing a specific gene from a genomic library can be identified by using a labeled probe whose base sequence is complementary to the gene of interest. To facilitate the isolation of a gene, cloning is often limited to the complementary DNA copies of mRNAs; this results in the construction of a cDNA or expression library, which contains only those genes that are expressed by a certain type of cell.

Expression vectors are designed to express the foreign DNA in a specific type of cell, and they permit the production of large amounts of cloned proteins for research and commercial purposes. In many cases, the use of a shuttle vector, which can replicate in bacteria as well as in a eukaryotic cell, allows the expression of a eukaryotic gene to be studied in a eukaryotic host such as yeast. Various eukaryotic vectors, such as yeast plasmids, modified tumor viruses in animals, and modified Ti (tumor inducing) plasmids in plants, also permit cloning and expression of eukaryotic genes directly in eukaryotic cells.

Gene Cloning Techniques

1. The plasmid pBR322 (see Figure 18–2) is used as a cloning vector. pBR322 DNA that has been cleaved with *Pst*I is mixed with eukaryotic DNA that has also been cleaved by *Pst*I; DNA ligase is also present. The DNA mixture is then used to transform competent *E. coli* cells, and tetracycline-resistant clones identify plasmid-containing bacteria. In addition to the desired recombinant plasmid, what other types of plasmids might be found among the tetracycline-resistant bacterial clones?

2. The target sequence and cleavage sites of the restriction enzymes *Sau*3A and *Bam*HI are shown below:

(a) What fraction of the *Bam*HI sites in a DNA molecule can be cut with *Sau*3A? What fraction of the *Sau*3A sites can be cut with *Bam*HI? (b) Suppose you ligate a *Bam*HI fragment to a *Sau*3A fragment. Can the hybrid joined ends be cut with *Bam*HI? with *Sau*3A?

3. Two DNA fragments were generated by restriction enzyme X digestion of a certain DNA segment; one fragment is 300 nucleotide pairs long; the other is 800 bp long. You want to join them together to form a hybrid gene, so you mix solutions containing the two fragments in the presence of DNA ligase and incubate the mixture for six hours. After 30 min of incubation, gel electrophoresis reveals a complex pattern of fragments, instead of just the 1.1 kb hybrid of interest (see diagram that follows). After the full six hours, electrophoresis reveals a somewhat different pattern, consisting of fewer short fragments and more long ones. When these fragments are cut with enzyme X, only the original two fragments are obtained:

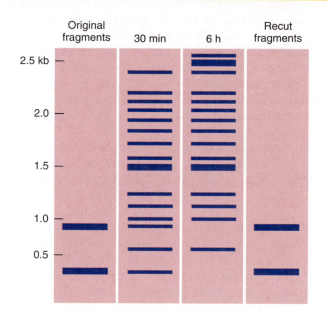

(a) Propose an explanation for the presence of so many bands on the gel after 30 min and after six hours of incubation. (b) Given the following restriction map of the hybrid region, what gel pattern would you expect to obtain if you isolate the 1.1 kb fragment from the gel and digest it with enzyme Y?

4. Suppose you cut two different DNAs, one with *Bam*HI and one with *Bgl*II, then ligate them together through their compatible sticky ends. Once joined, could you separate these two DNAs again with either restriction enzyme? Why or why not?

5. Draw a general map of the genetically engineered plasmid pUC18, showing the features that permit the insertion, replication, and retrieval of specific donor DNA sequences. Describe how each feature serves its particular role in a cloning experiment.

6. You want to use a Charon vector to clone donor DNA segments of length 10 kb. Describe how you would construct a Charon phage vector with a *Bam*HI cleavage site within a *lacz'* insert, and how you would then insert the donor DNA, replicate it, and select for cells that have been infected with a recombinant phage genome. (You have an in vitro packaging system available.)

7. Could you clone a DNA segment of length 45 kb in a Charon vector? Explain why or why not. What two kinds of lambda-based vectors described in the text could be used to clone such a large foreign DNA fragment? Describe how you would use each one to clone the DNA in question.

8. Under what specific circumstances might you use a member of the M13mp series of cloning vectors?

Gene Isolation

9. Define the following terms and distinguish between members of paired or grouped terms.
chromosome walking and chromosome jumping
gene library and genomic library
open reading frame

10. Suppose that you want to clone a particular DNA fragment cut from the lambda genome by *Eco*RI. You are given *Eco*RI-cut lambda DNA and the plasmid shown below, which has two *Eco*RI sites, to use as a cloning vector. Outline a set of procedures that would allow you to insert, replicate, and retrieve the DNA fragment in a cloning experiment. (Emphasize the selection methods you would use to isolate a recombinant plasmid that contains the lambda DNA fragment.)

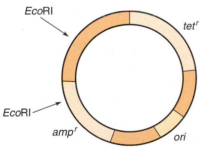

11. Suppose that you want to clone a particular DNA fragment cut from the lambda genome by *Bam*HI. You are given *Bam*HI-cut lambda DNA and the plasmid (pBR322) (see Figure 18–2) to use as a cloning vector. Outline a set of procedures that would allow you to insert, replicate, and retrieve the DNA fragment in a cloning experiment. (Emphasize the selection methods you would use to isolate a recombinant plasmid that contains the lambda DNA fragment.)

12. Would a plasmid like pUC18 be a good choice as a vector for constructing a genomic library of a complex organism? Why or why not?

13. How many yeast DNA fragments of average length 5 kb must be cloned in order to be (a) 90% certain and (b) 99% certain that a genomic library contains a particular segment? (The yeast genome contains 13,500 kb of DNA.)

14. How large a genomic library should you construct in order to be 90% sure of including a particular 15-kb human gene?

15. You have constructed a human cDNA library from an individual suspected of carrying the Lesch-Nyhan mutant gene, and you are using mouse HPRT cDNA as a probe to detect the homologous human gene. When hybridizing the mouse probe to the human library, would you use a higher, a lower, or the same temperature as for hybridizing the mouse probe to a mouse cDNA library? Explain your answer.

16. The sequence of part of a protein whose gene you wish to clone is as follows:

Phe-Pro-Arg-Leu-Met-Trp-Ile-Cys-Gln-Met-Lys-Val-Ser

(a) What sequence of amino acids would give an 18-oligonucleotide sequence with the least codon degeneracy? (b) How many different 18-nucleotide probes would you have to synthesize to be sure that the one that perfectly matches the cloned gene is included? (c) Could you extend the probe to 20 nucleotides without increasing the number of different probes you would have to synthesize? Explain your answer. (d) If you started your probe one codon to the left of the optimal one, how many different 18-nucleotide probes would you have to synthesize?

17. A certain animal virus contains a circular DNA that has five recognition sites for the restriction enzyme *Eco*RI. One of these fragments contains a viral gene that you would like to sequence; the gene codes for a viral surface protein with a known primary structure. Describe the steps you would take to identify the gene and isolate it for sequencing.

18. Identify the open reading frame in the following cloned DNA fragment, which has been sequenced by the dideoxy method:

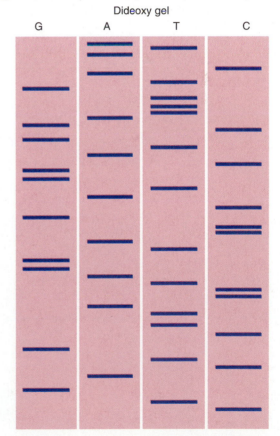

Expression Cloning

19. Suppose you are interested in studying a particular protein encoded by a mouse gene. You have isolated the mouse gene and intend to splice it into a vector for expression in *E. coli*. What sequences or sites are necessary to include on that vector to get this gene transcribed and translated in *E. coli*?

20. Distinguish between Southern blotting, northern blotting, western blotting, and southwestern blotting in terms of the techniques involved and what each is used for.

Transferring Genes into Eukaryotes

21. Figure 18–23 shows the structure of a retroviral vector commonly used to transform mammalian cells growing in culture. Describe how the vector can be assembled into infectious recombinant virions, even though it is defective in viral replication and packaging genes. Why is it critical that this vector be defective in the replication and packaging gene?

22. Explain why the binary Ti system (see Figure 18–30) is more useful in transferring foreign DNA into plant cells than would be a strategy employing a single vector such as pGV3850.

23. Describe the different methods that can be used to transfer genes into mammalian cells.

REGULATION OF GENE EXPRESSION

Life as we know it depends on gene control. Organisms must continually turn genes on and off in response to signals from their external and internal environments. The absence of such control would mean metabolic anarchy. If all the structural (protein-coding) genes of an organism were to function at the same time in an unregulated fashion, more energy would be consumed for RNA and protein synthesis than all the other metabolic processes could provide.

Gene regulation not only controls metabolism, but it is also required in multicellular organisms for **cell differentiation** (the divergence of different cell types that occurs during an organism's development). These developmental changes in cell structure and function occur in response to differences in the kinds, amounts, and activities of the proteins in the cells. Since all of the cells of an organism have the same complement of genes, such differences can only result from differential gene activity—turning various genes on and off at particular times during development.

In the next three chapters, we look at the mechanisms of gene control. The foundations of our under-

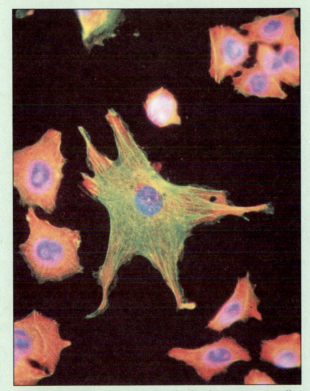

Immunofluorescent light micrograph of melanoma cancer cells and fibroblasts cultured from a human tumor. A star-shaped fibroblast (center) forms connective tissue to support other cells. Actin fibers (green) around it help provide this supportive network. Cytoplasm is orange. Smaller melanoma cells have a large nucleus (blue); derived from melanin-forming skin cells, they are highly malignant and divide rapidly (upper right).

standing of gene control have come from studies on prokaryotes, so we will begin our discussion of gene regulation by considering the genetic control processes that operate in bacteria. Most of the mechanisms of gene control in prokaryotic cells regulate metabolism by controlling enzyme synthesis. Some of these mechanisms are also found in eukaryotes. However, the greater complexity of gene structure and organization in eukaryotic cells provides additional opportunities for gene regulation. In the final two chapters of this section, we will examine several of the mechanisms for controlling gene expression in eukaryotes; many of these mechanisms involve complicated signal-generating, signal-transducing, and signal-receiving events not found in prokaryotes. The details of most of these regulatory mechanisms have emerged only since the mid-1970s, with the advent of molecular cloning techniques and the ensuing studies of eukaryotic genes. We will also look at some genetic control elements that affect development in different eukaryotes as well as some genes that have been linked to cancer, a group of diseases in which cell growth and development go awry.

CHAPTER
19

Regulation in Prokaryotes

I n bacteria, genetic control regulates the metabolic activities of the cell so that it can respond quickly to environmental changes. Bacteria are often exposed to highly fluctuating environments, so they need to be able to adjust their metabolic processes rapidly in order to optimize cell growth and reproduction. These adjustments are typically made at the level of transcription by turning the transcription of specific genes "on" and "off" to regulate the levels of the encoded proteins. The transcription of specific genes occurs in response to environmental signals and enables the cell to produce certain proteins only when they are needed. Often an environmental signal will affect the transcription of more than one gene. For example, when several enzymes act in the same metabolic pathway, usually either all or none of these enzymes are produced in response to an environmental change. This phenomenon, known as *coordinate regulation*, is the result of transcribing all of the genes for these en-

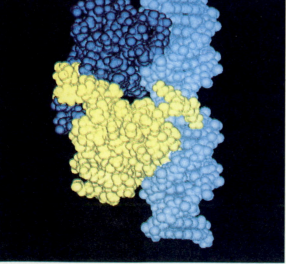

λ repressor protein bound to DNA.

zymes into a single polycistronic message or mRNA. The functionally related genes encoding these enzymes are thus regulated as a unit by means of coordinate transcriptional control. The coordinate regulation of the transcription of genes with related functions occurs in bacteria and their viruses, and is a fundamental principle of the genetics of prokaryotes.

OPERONS AND TRANSCRIPTIONAL CONTROL

Genetic studies on bacteria and their viruses reveal that genes with related functions tend to be clustered together in the same general region of the linkage map. In bacteriophage T4, for example, the genome is roughly divided into two parts: one containing the genes that are necessary for replication (the early genes), and the other containing the genes needed for maturation and lysis (the late genes). Furthermore, the related genes within each cluster are often regulated as a unit—they are transcribed from a single promoter by RNA polymerase and form a single polygenic (polycistronic) mRNA. Such a group of related genes under the joint transcriptional control of a single promoter is known as an operon.

▶ Figure 19–1 shows a hypothetical operon consisting of three adjacent structural genes (S_1, S_2, and S_3) and their shared promoter and transcriptional terminator sites. Since the three structural genes share a single promoter region, they constitute a single unit of transcription that codes for a single polygenic mRNA chain. Also included in the operon are various cis-acting and trans-acting control elements that play roles in determining whether transcription will occur. The cis-acting elements (the operator as well as the promoter) are specific sequences that affect the activity of genes in adjacent regions of the same DNA molecule. These cis-acting sites usually function as protein-binding sequences; for example, the promoter serves as the binding site for RNA polymerase. The trans-acting elements, on the other hand, are generally genes that encode the structures of diffusible regulatory products that may affect the activity of genes on different DNAs (located in trans) as well as genes on the same DNA. Among the trans-acting elements are regulatory genes, which encode regulatory proteins that exert transcriptional control by binding to either the promoter or operator sequence.

Regulatory Proteins: Positive and Negative Control

Regulatory proteins can control the transcription of the operon in either a positive or a negative manner, depending on the operon. In operons subject to positive control, the regulatory gene product stimulates transcription; in operons subject to negative control, the regulatory protein inhibits transcription. ▶ Figure 19–2 shows some of the better-known mechanisms for transcriptional control of operons. Positive control by the regulatory protein involves either enhancement of RNA polymerase binding at the promoter or prevention of early termination of RNA chain growth at a terminator site preceding the structural genes (Figure 19–2a). Negative control by the regulatory protein involves blocking access of RNA polymerase to critical promoter sequences (Figure 19–2b). It is not uncommon for an operon to be affected by two different regulatory proteins, one that exerts positive control and one that exerts negative control.

Most bacterial operons contain genes that code for enzymes. Transcription of these operons is indirectly regulated by the interaction of certain of the enzymes' substrates and products with the regulatory proteins. These regulatory proteins are allosteric proteins (see Chapter 6); in addition to binding to a specific regulatory region of the DNA at their active (DNA-binding) site, they also interact with certain key metabolites at their allosteric site (the key metabolite is often the first substrate in a degradative pathway or the end product of a biosynthetic pathway). These metabolites act as effector molecules by promoting conformational changes in the regulatory proteins, thus altering the capacity of these proteins to bind to the DNA and indirectly activating or repressing transcription.

Inducible and Repressible Operons

Bacterial operons fall into two major categories based on the actions of the effector molecule on the regulatory gene product. If the effector molecule alters a regulatory protein in such a way that transcription of the structural genes is "turned on," the operon is said to be inducible. Conversely, if the effector substance modifies the regulatory protein so that transcription of the structural genes is "turned off," the operon is termed repressible.

The structural genes in a bacterial operon ordinarily code for enzymes involved in the same metabolic pathway—the genes of the *lac* operon code for enzymes that are necessary for the degradation of lactose, those of the *trp*

▶ **FIGURE 19–1** Basic components of an operon. Transcription of the three structural genes (S_1, S_2 and S_3) is initiated at a single promoter site (*p*). The *t* site indicates the end of transcription of the operon. Also shown are the operator (*o*) and regulatory gene (*R*) control elements that determine whether transcription occurs. The operator region contains a nucleotide sequence that binds a particular regulatory protein; both *p* and *o* are cis-acting elements that affect only adjacent structural genes. The regulatory gene encodes a DNA-binding protein that acts on either the promoter or the operator, depending on the type of operon and control system. The regulatory gene is trans acting; its protein product can act on the *p* or *o* sites shown or on the same regions of a different DNA molecule. (Gene *R* has its own promoter and transcriptional termination sites which are not shown on the diagram.)

DNA ... —|—R—|—//—|—p—|—o—|—S_1—|—S_2—|—S_3—|—t—|—

Polygenic mRNA

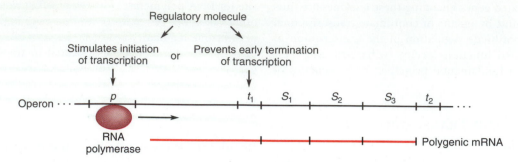

(a) Positive control

Regulatory molecule

Stimulates initiation of transcription · or · Prevents early termination of transcription

Operon ···

p t_1 S_1 S_2 S_3 t_2

RNA polymerase

Polygenic mRNA

(b) Negative control

Regulatory molecule prevents initiation of transcription

Operon ···

p t_1 S_1 S_2 S_3 t_2

No transcription

RNA polymerase

▶ **FIGURE 19–2** Basic components of operons subject to (a) positive control and (b) negative control. For simplicity, the locations of the regulatory gene and operator element are not shown. The t_1 site is an early terminator found in some operons; t_2 signifies the end of transcription of the operon. In positive control systems, the product of the regulatory gene either enhances the binding of RNA polymerase to the promoter region or it overcomes early termination of transcription. In negative control systems, the regulatory protein interferes with the access of RNA polymerase to the promoter region.

operon code for enzymes that catalyze the biosynthesis of tryptophan, and so on. Inducible operons are usually involved in the synthesis of enzymes that act in degradative (catabolic) pathways, such as those that break down lactose and other sugars (see ▶ Figure 19–3). The effector molecule is called the **inducer**, and it is often the initial substrate in the degradative pathway. For example, lactose (actually the isomeric form, allolactose) serves as the inducer of the *lac* operon—the operon is turned on in the presence of lactose, and enzymes are produced that catalyze the sugar's degradation. Repressible operons, in contrast, are mainly associated with anabolic pathways, which synthesize required substances, such as tryptophan and other amino acids. In this case the effector molecule is called the **corepressor**, and it is frequently the end product of the biosynthetic pathway. For example, the amino acid tryptophan serves as the corepressor of the *trp* operon. In the presence of tryptophan, the operon is turned off and enzymes that catalyze the synthesis of this amino acid cease to be produced by the cell.

▶ Figure 19–4 shows a general model of an inducible system. Note that in the absence of the inducer, the operon is under the negative control of a regulatory gene. The regulatory gene product, or **repressor**, binds to the operator region and prevents transcription. When the inducer is present, it combines with the allosteric site on the repressor,

causing the repressor protein to undergo a change in conformation. Transcription and subsequent enzyme synthesis are greatly enhanced, since the ability of the modified repressor to bind to the operator is much reduced. Induction thus results from the allosteric inhibition of the repressor protein.

In contrast to an inducible system, which is turned off until an effector substance activates transcription, a repressible system is on until an effector turns it off. In repressible operons, the allosteric binding of the effector to the regulatory gene product increases the affinity with which the repressor binds to DNA. For example, the enzymes that are involved in tryptophan synthesis are produced by the *trp* operon when the concentration of tryptophan in the cell is low, and the synthesis of enzymes by this operon is repressed when the concentration of tryptophan is high. The general model shown in ▶ Figure 19–5 shows the molecular basis for the action of tryptophan as a corepressor. Again, the corepressor or effector molecule (tryptophan) exerts its influence by combining allosterically with the regulatory protein. In this case, however, the unaltered product of the regulatory gene is an inactive **aporepressor** that is unable to combine with the operator site. The structural genes of a repressible operon are thus transcribed in the absence of the corepressor. When the corepressor is present,

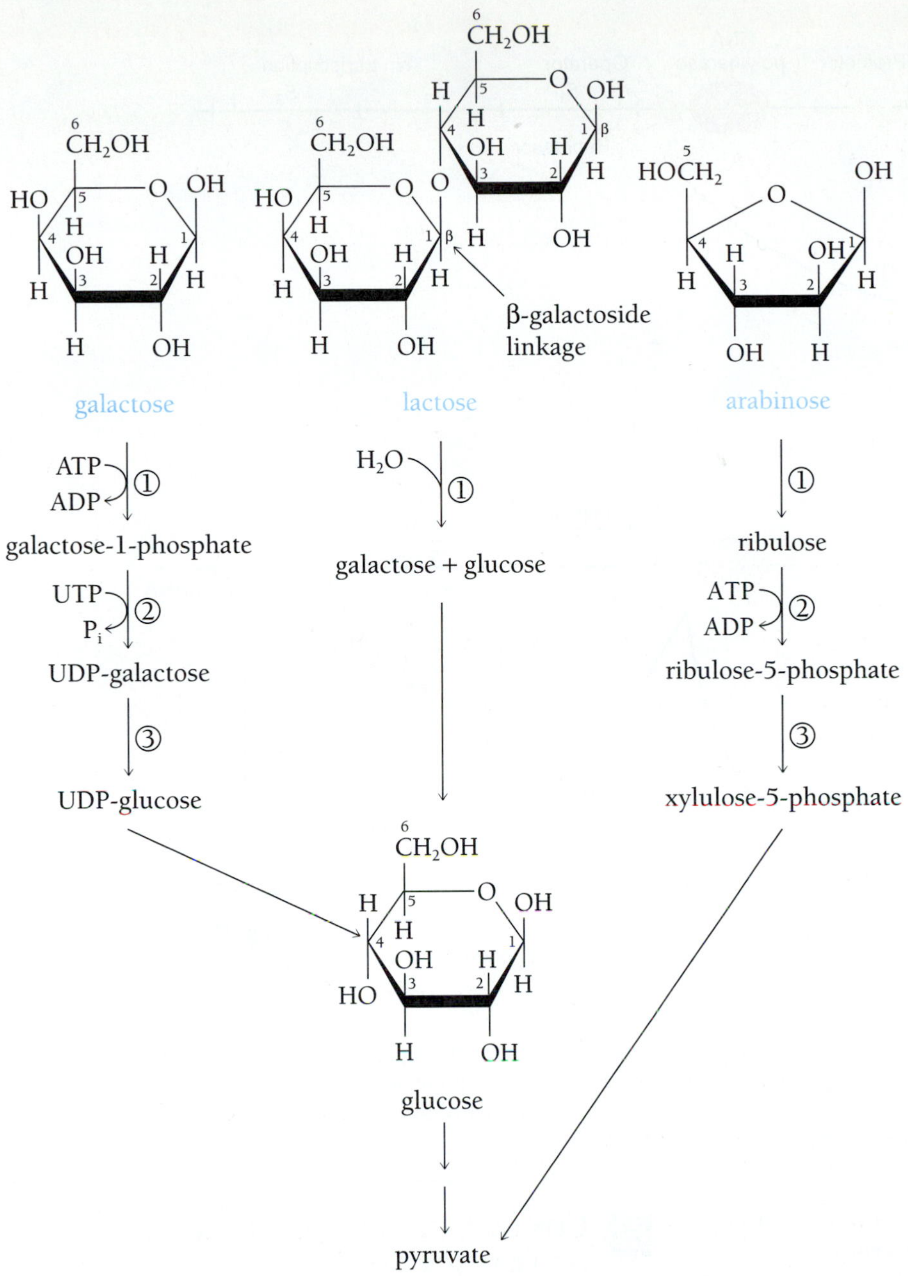

▶ FIGURE 19–3 Steps in the metabolism of the sugars galactose, lactose, and arabinose. The enzymes that catalyze the numbered steps of each pathway are coded for by genes that are organized into inducible operons.

it combines with the aporepressor, converting it to an active repressor that binds to the operator, shutting off transcription and enzyme synthesis.

Inducible and repressible operons exemplify the general economy of protein synthesis. The bacterial cell produces significant quantities of the enzymes that catalyze lactose degradation or tryptophan synthesis only when lactose is present or when tryptophan is in short supply. Because the synthesis of proteins requires the expenditure of large amounts of cellular energy, it is clearly advantageous for a cell to produce proteins only when they are definitely needed. Of course, not all genes are subject to regulation. **Constitutive genes** are continuously expressed; these genes encode proteins that are essential for growth and are re-quired by the cell in relatively constant amounts, regardless of the presence or absence of various metabolites in the growth medium.

To Sum Up

1. Many bacterial genes are organized into functionally related groups known as operons. An operon consists of a group of structural genes under the control of a single promoter, various other cis-acting regulatory sites, and one or more trans-acting sites. The trans-acting sites often code for regulatory proteins that control the transcription of the operon.

2. The structural genes in an operon are regulated as a group by being transcribed into a single polygenic mRNA. Coordinate

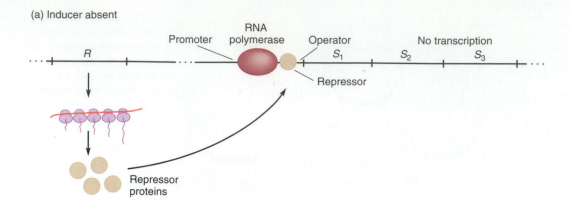

(a) Inducer absent

Promoter RNA polymerase Operator No transcription

R S_1 S_2 S_3

Repressor

Repressor proteins

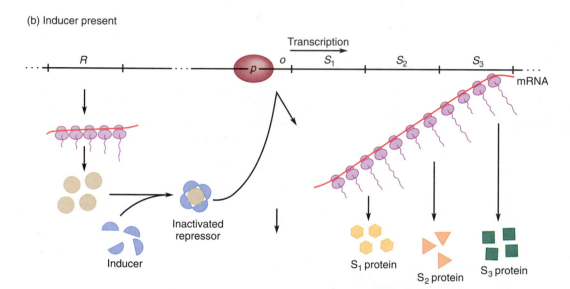

(b) Inducer present

Transcription

R p o S_1 S_2 S_3

mRNA

Inactivated repressor

Inducer

S_1 protein

S_2 protein

S_3 protein

▶ **FIGURE 19–4** General model of an inducible operon. Regulatory gene *R* codes for the repressor protein. (a) In the absence of the inducer, the repressor binds to the operator region, blocking transcription of the adjacent structural genes by RNA polymerase. (b) When the inducer is present, it allosterically inactivates the repressor, leaving it unable to bind to the operator. The RNA polymerase molecules can now transcribe the structural genes into mRNA.

regulation of the transcription of genes with related functions is a fundamental principle of prokaryotic genetics.

3. If a regulatory protein inhibits transcription, the operon is said to be under negative control. If the regulatory protein stimulates transcription, we say that the operon is subject to positive control.

4. Most bacterial operons contain genes that code for enzymes. In such cases the regulatory proteins interact at their allosteric sites with key metabolites called effectors. Interaction of a regulatory protein with an effector alters the DNA-binding site of the regulatory protein and indirectly activates or represses transcription.

5. Inducible operons respond to an inducer by turning on the transcription of the structural genes. Inducible operons are usually involved with the synthesis of enzymes that function in degradative pathways.

6. Repressible operons respond to a corepressor by turning off the transcription of the structural genes. Repressible operons usually synthesize enzymes that function in biosynthetic pathways.

CONTROL OF THE INITIATION OF TRANSCRIPTION

Transcription of an operon can be regulated by events occurring at the promoter or at an early transcriptional terminator (such as t_1 in Figure 19–2a). Examples of termination control will be discussed later in this chapter. Regulation of transcriptional initiation is more common and more straightforward, so we will first look at two major operons that are controlled by events in the vicinity of the promoter. These operons are concerned with the initial steps of metabolism of the sugars lactose and arabinose.

The *lactose* Operon: Inducible Negative Control

The negative control mechanism of the *lac* operon in the bacterium *E. coli* was first discovered by the French microbiologists François Jacob and Jacques Monod, and it has served as the prototype for the operon model, which these

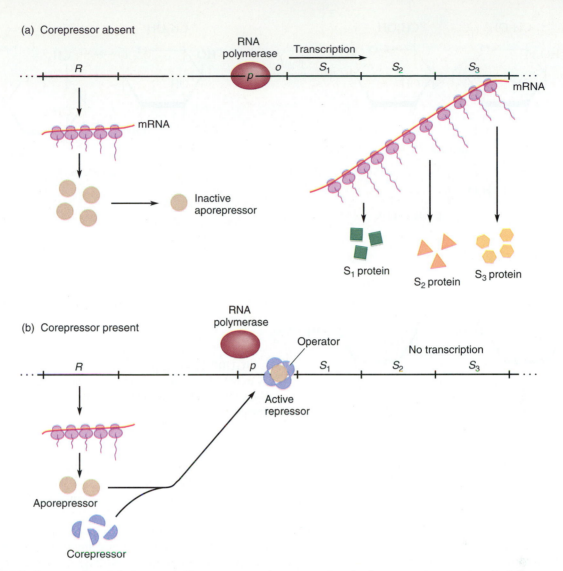

(a) Corepressor absent

RNA polymerase

Transcription

R

p o S_1 S_2 S_3

mRNA

mRNA

Inactive aporepressor

S_1 protein

S_2 protein

S_3 protein

(b) Corepressor present

RNA polymerase

Operator

R

p S_1 S_2 S_3

No transcription

Active repressor

Aporepressor

Corepressor

▶ **FIGURE 19–5** General model of a repressible operon. Regulatory gene *R* codes for an inactive protein called the aporepressor, which becomes an active repressor only if it first forms a complex with the corepressor. (a) In the absence of the corepressor, the aporepressor is unable to bind to the operator, allowing RNA polymerase to transcribe the adjacent structural genes. (b) The aporepressor is allosterically activated by binding with the corepressor; it can then bind to the operator, blocking transcription.

investigators first described in 1961. The *lac* operon is an inducible system with three structural genes that code for proteins involved in the metabolism of lactose (or lactose analogues). Gene *z* codes for the enzyme β-**galactosidase**, which catalyzes the conversion of lactose to galactose and glucose (▶ Figure 19–6). (β-galactosidase can also convert lactose to allolactose, the actual inducer of the *lac* operon.) Gene *y* encodes **galactoside permease**, a protein that is located in the bacterial cell membrane and facilitates the diffusion of lactose molecules into the cell. Gene *a* codes for a **galactoside transacetylase**, an enzyme that transfers an acetyl group from acetyl-CoA to β-galactosides. The acetylation reaction is not actually part of the metabolism of lactose. Its role is not completely understood, but it may be involved in detoxification of analogues of β-galactosides that cannot be metabolized.

Jacob and Monod found that *E. coli* cells growing in the absence of lactose contain only a very few molecules of β-galactosidase, galactoside permease, and galactoside transacetylase. That any of these molecules are present in an uninduced cell is the result of a **basal level** of transcription of the operon (0.1% of the full induced level) that takes place in the absence of lactose (when the operon is said to be turned off). When lactose is added to the growth medium of the cells, a few of the sugar molecules are able to enter the cell because of the basal level of permease, and they start the operon induction process. Jacob and Monod discovered that within just two or three minutes, protein production increases the cellular concentrations of β-galactosidase, permease, and transacetylase to as much as 5000 molecules each. When lactose is removed from the medium, the synthesis of the three proteins ceases just as quickly as

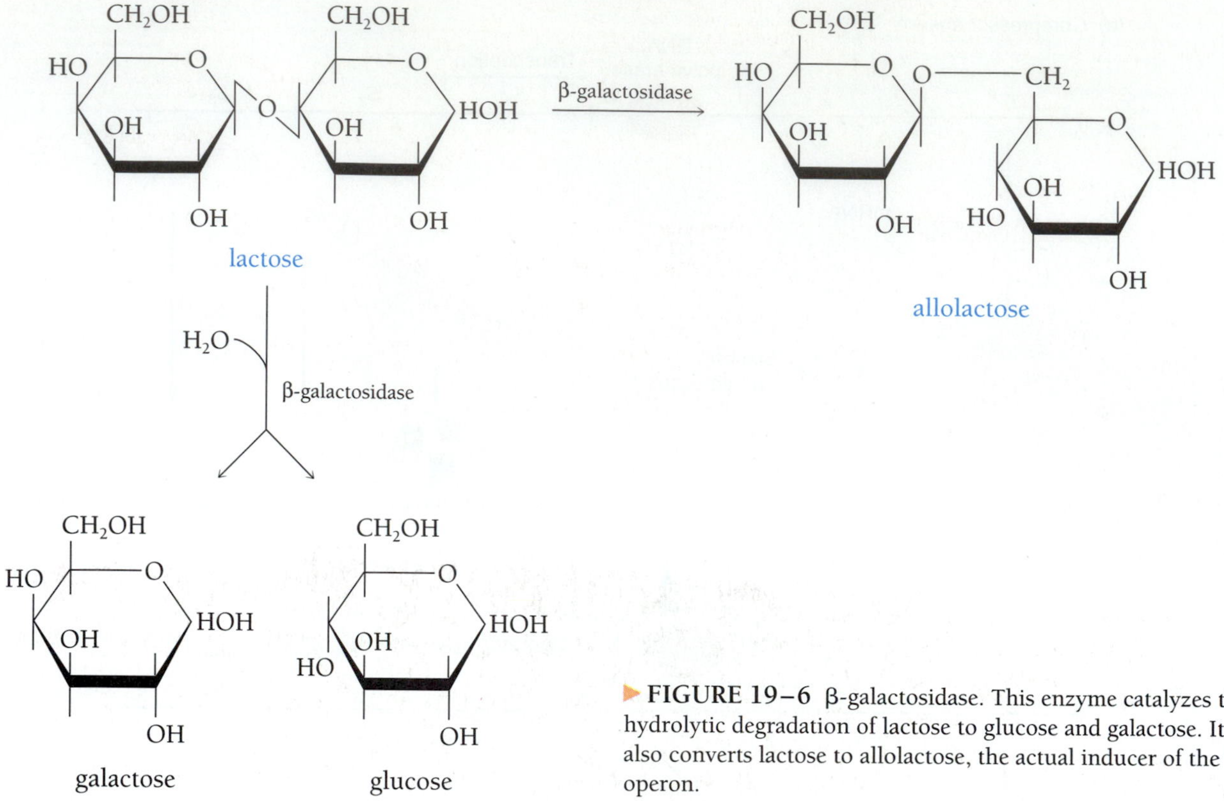

► **FIGURE 19−6** β-galactosidase. This enzyme catalyzes the hydrolytic degradation of lactose to glucose and galactose. It also converts lactose to allolactose, the actual inducer of the *lac* operon.

it began. The *lac* mRNA is very unstable, with a half-life of only about three minutes; thus the levels of the newly synthesized proteins fall rapidly when the effector is removed.

► Figure 19−7 shows the order of the structural genes on the *E. coli* chromosome, the single cis-acting promoter and operator sites, and the trans-acting negative regulatory gene *i*. Gene *i* codes for a repressor protein whose active form is a tetramer made up of four identical polypeptide chains. Each cell typically contains 10 to 20 repressor molecules. The repressor is an allosteric protein that undergoes a conformational change when bound to the inducer; the

resulting distortion of the repressor's shape leaves it unable to bind to the *lac* operator site.

Researchers have used techniques such as DNA footprinting (see Extensions and Techniques) to define the locations in the *lac* promoter and operator regions that are recognized by RNA polymerase and by the repressor. ► Figure 19−8 shows that the two regions are not entirely distinct. When bound to promoter DNA, RNA polymerase protects a region extending from about 50 nucleotides upstream (−50) to 5 nucleotides downstream (+5) of the transcriptional startpoint (+1) from nuclease digestion. The op-

► **FIGURE 19−7** The *lac* operon in *E. coli*. Structural genes *z*, *y*, and *a* code for proteins that are involved in the utilization of lactose by the cell. The coordinate transcription of these genes is controlled by regulatory gene *i*, which codes for a repressor protein. The binding of the repressor molecule to the operator blocks transcription of the adjacent structural genes (see Figure 19−4a). Lactose acts as the inducer, allosterically inactivating the repressor to allow transcription of the structural genes (see Figure 19−4b).

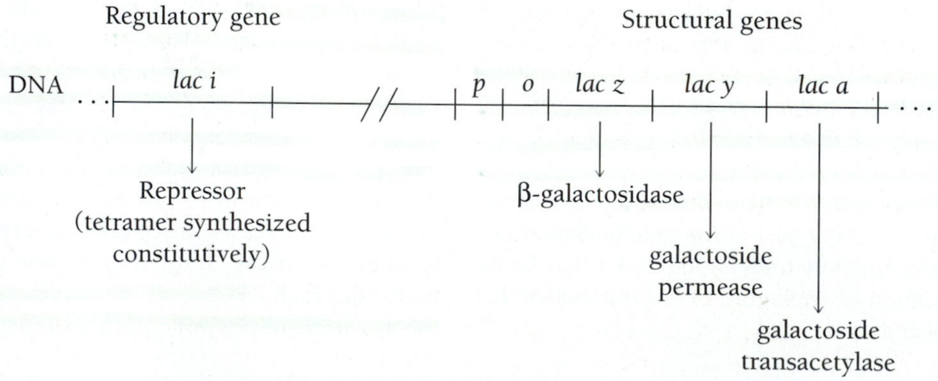

EXTENSIONS AND TECHNIQUES

DNA Footprinting

The region of DNA to which a protein, such as the *lac* repressor, binds can be identified by a technique called DNA footprinting. The principle behind this procedure is the same as in DNA sequencing. The DNA in question is labeled at one end and subjected to partial digestion with a DNase enzyme. The DNA region where the protein is bound is protected from nuclease action, so DNA fragments that would ordinarily be produced by cleavage in this region will not be present in the digest. Since all fragments are radioactively labeled at one end, autoradiography of an electrophoretic gel reveals the fragments produced and allows identification of the nucleotide positions protected by the bound protein (see illustration at right).

A DNA sequencing ladder can be simultaneously carried out on a gel run alongside the footprinting ladder to determine the exact nucleotide sequence covered by the bound protein.

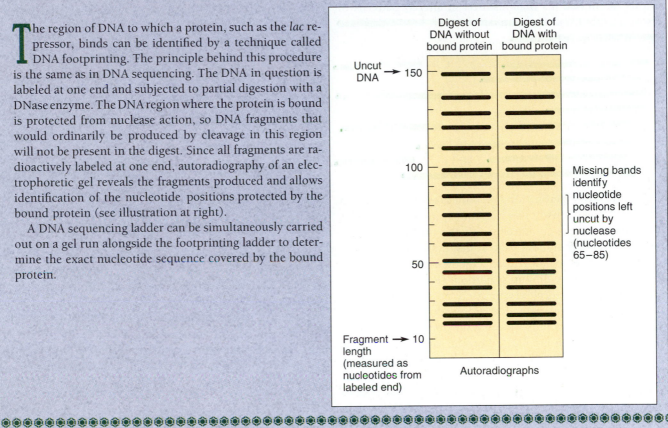

erator region defined by repressor binding and nuclease digestion extends from position −5 to position +21. The two control regions thus overlap by about 10 nucleotides. Because of this overlap, researchers have long speculated that polymerase and repressor binding to the promoter-operator DNA are mutually exclusive, i.e., that binding by the repressor prevents binding by RNA polymerase. Recent studies, however, indicate that both the polymerase and the repressor bind to the DNA at the same time. The most critical contact points between the bound polymerase and the

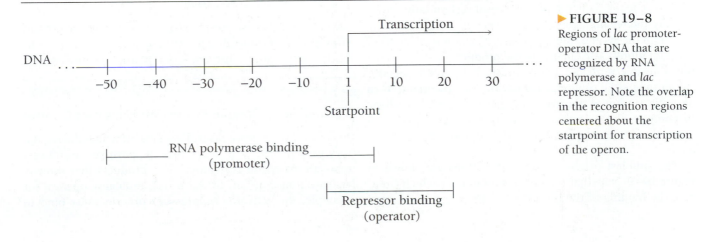

▶ **FIGURE 19–8**
Regions of *lac* promoter-operator DNA that are recognized by RNA polymerase and *lac* repressor. Note the overlap in the recognition regions centered about the startpoint for transcription of the operon.

DNA lie just upstream of the transcriptional startpoint, and those between the repressor and the DNA lie just downstream of the startpoint. The binding of the repressor within the region of overlap is thus thought to prevent RNA polymerase from gaining access to the mRNA startpoint. The addition of inducer releases the repressor from the operator, allowing immediate transcription by bound RNA polymerase.

Jacob and Monod based their operon model on studies of various *lac* mutations that resulted in altered patterns of lactose metabolism. These mutations can be divided into the following major classes:

1. Structural gene mutations (z^-, y^-, a^-): affect the structure of the enzymes produced.
2. Operon control mutations: affect only the regulation of enzyme synthesis.
   a. Constitutive mutants (i^c and o^c): always produce enzymes, even in the absence of inducer.
   b. Super-repressed mutants (i^s and p^-): never produce enzymes, even in the presence of inducer.

Examples of operon control mutations are shown in ▶ Figure 19–9. Constitutive mutations can occur in either the regulatory gene (i^c mutants) or the operator (o^c mutants). o^c and i^c mutations map at distinctly different sites, and they can also be differentiated by their properties when present in a partially diploid condition. (Recall from Chapter 14 that partial diploids in bacteria are usually constructed using either sexduction or transduction.) The constitutive influence of the o^c mutation extends only to those structural genes that are located on the same chromosome as the mutant o^c locus. Thus, a partial diploid of genotype $o^+z^-y^-a^-/o^cz^+y^+a^+$ would show constitutive synthesis of all three enzymes, making it appear that o^c is dominant to o^+. When the genes are arranged differently, however, as in a cell with genotype $o^+z^+y^+a^+/o^cz^-y^-a^-$, the three enzymes are synthesized only in the presence of the inducer. The o^c mutation is thus said to have a **cis dominant** effect. In comparison, i^c mutations do not show a cis dominant effect; these mutations are recessive to i^+ regardless of whether the arrangement is cis or trans in relation to the alleles at the structural gene loci.

Mutations of the super-repressed variety can occur in either the regulatory gene or the promoter region of the operon. The regulatory super-repressed (i^s) mutation is especially interesting, since it shows complete dominance to i^+ in both the cis and trans positions. The dominant effect of this mutation is attributable to the production of an altered but diffusible repressor that can attach to either operator site in the partial diploid, thus having the potential to turn off transcription of the structural genes on both homologs. In contrast, the p^- mutation affects only those structural genes that are located on the same chromosome as the p^- locus.

The simplest origins of these mutations are shown in Figure 19–9. Note that constitutive mutations, which eliminate the binding of the repressor to the operator site, may result from an altered operator locus (as in the o^c mutation) or from an altered repressor molecule (as in the i^c mutation). The super-repressed expression of the i^s mutation results from a defect in the allosteric (inducer) binding site on the repressor molecule. The defective repressor is still able to bind to the operator locus, turning off gene activity, but it does not respond to the effects of the inducer. The p^- mutation eliminates the ability of RNA polymerase to bind to the promoter and prevents transcription of genes located in cis with p^-.

Example 19.1

For each of the following situations involving operons and their effector molecules, determine whether transcription of the adjacent structural genes will occur. (R = the regulatory gene of the operon.)

Inducible Operon	Repressible Operon
(a) $R^+p^+o^+$, inducer absent	(d) $R^+p^+o^+$, corepressor absent
(b) $R^-p^+o^+$, inducer absent	(e) $R^+p^+o^c$, corepressor present
(c) $R^+p^+o^c$, inducer absent	

Solution: In an inducible system like the *lac* operon, the R^+ gene specifies an active repressor. Without an inducer to inactivate this repressor, transcription of the structural genes remains turned off (a). In (b), however, no active repressor is produced because of the mutant active site of the repressor (R^-), so transcription occurs. Transcription occurs in (c) as well, even though the R gene is wild-type and the inducer is absent, because the mutant operator site (o^c) prevents the repressor from binding to the DNA to block transcription.

In a repressible system, the corepressor is needed to activate the aporepressor (the product of gene R) in order to turn off transcription. Hence, transcription occurs in (d). Transcription also occurs in (e) because of the mutant condition of the operator region.

Follow-Up Problem 19.1

In the following *E. coli* cells, determine whether transcription of *lac* operon gene *z* occurs.

(a) $i^+p^+o^+z^+$, lactose present
(b) $i^cp^+o^+z^+$, lactose absent
(c) $i^cp^+o^+z^+$, lactose present
(d) $i^cp^-o^+z^+$, lactose present
(e) $i^sp^+o^+z^+$, lactose present
(f) $i^+p^+o^cz^+$, lactose absent

The *arabinose* Operon: Positive and Negative Control

In contrast to negative control systems, in which the regulatory gene product acts as a repressor, positive control systems are based on regulatory gene products that directly activate transcription. In the *E. coli arabinose* operon, for example, the **activator (expressor) protein** must bind to

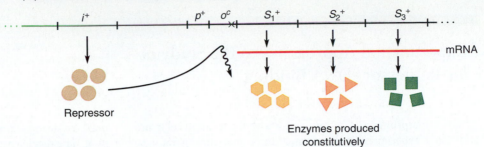

(a) Operator constitutive mutation
(repressor cannot bind to altered operator site):

Repressor

mRNA

Enzymes produced
constitutively

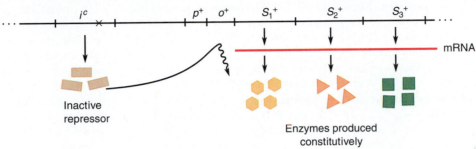

(b) Regulatory constitutive mutation
(altered repressor cannot bind to operator):

Inactive
repressor

mRNA

Enzymes produced
constitutively

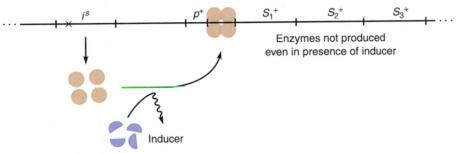

(c) Super-repressed regulatory mutation
(altered repressor cannot bind with inducer):

Enzymes not produced
even in presence of inducer

Inducer

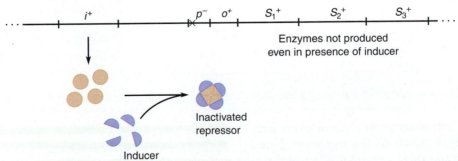

(d) Super-repressed promoter mutation
(RNA polymerase cannot bind to altered promoter site):

Enzymes not produced
even in presence of inducer

Inactivated
repressor

Inducer

▶ **FIGURE 19–9** Four classes of mutations that affect the control of the inducible *lac* operon. (a) A mutation in the operator region destroys its ability to bind the repressor, resulting in constitutive transcription and enzyme synthesis. (b) A mutation in the regulatory gene *i* affects the DNA-binding site on the repressor protein. The result is again constitutive transcription of the operon. (c) A different mutation in the regulatory gene *i* affects the site on the repressor that normally recognizes the inducer. The repressor cannot be allosterically inactivated by the inducer, and transcription of the operon is blocked by the continued binding of the repressor to the operator. (d) A mutation in the promoter region destroys its ability to bind RNA polymerase, so no transcription of the operon can occur.

Experimental Approaches to the Study of *lac* Repressor–DNA Binding

In addition to DNA footprinting, a number of additional experimental approaches such as DNA sequencing, genetic analysis of mutations within the operator region, and physical studies of the *lac* repressor, have been used to study the nature of the repressor and the DNA region to which it binds. These studies illustrate how integration of the results from a variety of experimental approaches is often necessary to elucidate all the details of a particular process.

DNA sequence analysis has precisely defined the nature of the repressor-binding region identified by DNA footprinting. The nucleotide sequence of the *lac* operator region is shown below, along with the locations of o^c mutations. Notice that the operator sequence is an imperfect palindrome (the inverted repeats are in blue) with an axis of rotational symmetry at position +11. Operator-constitutive mutations occur at eight sites in the region, thus establishing that these nucleotide pairs are crucial contact points for repressor binding. (The asymmetric distribution of o^c mutation sites suggests that the left side of the operator is more important to repressor binding than the right side.) Note that these contact points all lie downstream of the startpoint, showing that the major recognition points for polymerase and repressor are indeed adjacent, rather than overlapping. The dyad symmetry of the operator sequence is thought to reflect symmetrical binding of the repressor, with one repressor dimer contacting each inverted repeat.

Digestion of the tetrameric repressor with the protease trypsin reveals that each peptide consists of 360 amino acid residues. At the N-terminal end is a 59-amino-acid **headpiece** that can be cleaved from the rest of the protein; this is followed by a hinge region and a long C-terminal trypsin-resistant core. The domain structure of the *lac* repressor is shown below:

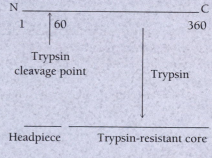

The C-terminal core retains the ability to form a tetrameric molecule and bind to the inducer. In contrast, only the headpiece retains the DNA binding specificity; the headpiece is thought to protrude from the body of the protein as it contacts the DNA, perhaps in the following manner:

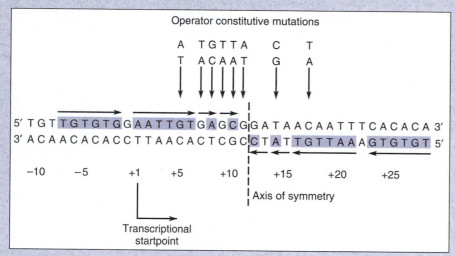

the promoter region for transcription of the adjacent structural genes. ▶ Figure 19–10 shows the regulatory gene *C*, the promoter region, and three structural genes of the *ara* operon, along with the three enzymes coded by the structural genes. These enzymes act in the following pathway:

$$arabinose \xrightleftharpoons{isomerase} L\text{-}ribulose \xrightleftharpoons{kinase}$$

$$L\text{-}ribulose\ 5\text{-}P \xrightleftharpoons{epimerase} D\text{-}xylulose\ 5\text{-}P$$

The *ara* operon, like the *lac* operon, is an inducible system. The difference between the two systems is that the regulatory C protein both activates and represses transcription, depending on the availability of arabinose. The C protein also represses transcription of its own gene; this **autoregulation** maintains a constant level of C protein. Both the structural genes and gene *C* are regulated at the region labeled *p* in Figure 19–10, which has been found to be a **multipartite** structure consisting of three separate regulatory sites—o_2, o_1, and *I*. ▶ Figure 19–11a shows that in the absence of C protein, only gene *C* is transcribed. A model explaining induction of the operon is shown in Figure 19–11b and c. In the absence of arabinose, the C protein binds co-

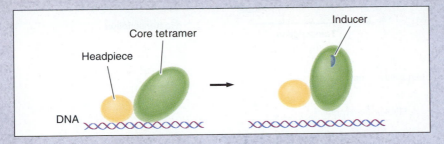

In this hypothetical arrangement, the headpiece contacts the crucial operator sites and an allosteric change removes the repressor from the DNA.

The nature of the binding between the *lac* repressor and the DNA has recently been investigated through crystallographic studies of repressor bound to half of an operator dyad. The DNA-binding domain of the repressor contains a short element called the **helix-turn-helix (HTH)** structural motif. This motif is a recurring substructure seen in a number of otherwise different phage, bacterial, and eukaryotic protein domains. It consists of a region of approximately 20 amino acids that forms two α-helices linked by a β-turn:

One helix (the recognition helix) is thought to lie in the major groove of the DNA, where it makes important base pair contacts through hydrogen bonding. The other helix lies across the DNA in a position that helps lock the recognition helix in place. The interaction between an HTH protein and DNA looks like this (the recognition helix is green):

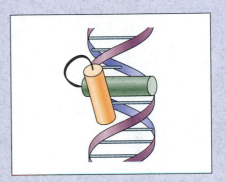

In DNA, the major groove H-bonding patterns are distinctive for each of the four base pairs and are thought to provide part but not all of the specificity of recognition for such DNA-binding proteins. Other polar and nonpolar interactions between bases and amino acid side chains may also be involved.

As is evident from this discussion, the exact structure of the *lac* repressor and the relationship between its structure and its DNA-binding function are not precisely known. Other regulatory DNA-binding proteins, such as those of certain phage, are better understood and will be discussed later in this chapter.

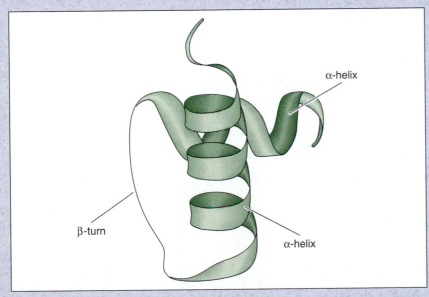

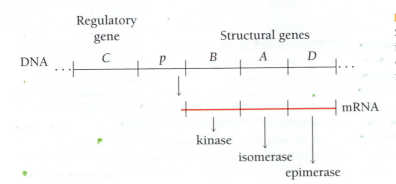

▶ **FIGURE 19–10** The *arabinose* operon of *E. coli.* Structural genes *B*, *A*, and *D* code for enzymes that are involved in the utilization of arabinose by the cell. The coordinate transcription of these genes is controlled by regulatory gene *C*.

▶ **FIGURE 19–11** Inducible control of the arabinose operon by gene *C*. The o_2, o_1, and *I* regions make up a complex multipartite promoter region. (a) In the absence of C protein, only gene *C* is transcribed, starting from the o_1 promoter region. (b) In the presence of C protein but with no arabinose, both gene *C* and the structural genes are repressed by the binding of C protein. (c) When arabinose is present, gene *C* remains repressed, but transcription of the structural genes is activated by the binding of allosterically-altered C protein to the o_2-o_1-*I* region.

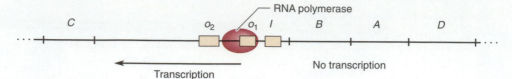

(a) No C protein

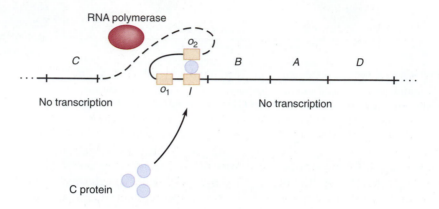

(b) C protein present, no inducer

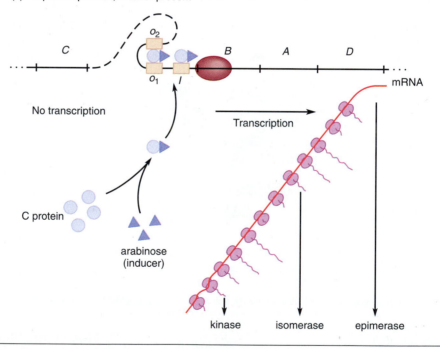

(c) C protein present, inducer present

operatively to both o_2 and *I*, generating a DNA loop that represses both gene *C* and the operon's structural genes. In the presence of arabinose, however, the allosterically altered C protein binds to all three sites, o_2, o_1 and *I*. The cooperative binding shifts from o_2 and *I* to o_1 and o_2, generating a DNA loop that continues to repress gene *C*. The allosterically altered C protein also exhibits a heightened affinity for site *I*, where it binds as an activator (expressor)

for transcription. In some unknown way, the binding of the activator protein improves the access of RNA polymerase to the transcriptional startpoint region, and the adjacent structural genes are transcribed. Thus, the enzymes for arabinose degradation are synthesized by the cell only when arabinose is present in the cellular growth medium.

Researchers are just beginning to recognize the possible significance of multipartite regulatory regions and DNA looping in transcriptional regulation. Regulatory regions with multiple sites for binding the regulatory protein have been identified in several phage and bacterial operons. Two weak secondary operator sites have even been identified in the *lac* operon. Although the exact mechanism by which binding of the repressor protein prevents transcription is not known for either the *lac* or the *ara* operons, one might speculate that a looped DNA structure denies access of bound RNA polymerase to the transcription initiation site. DNA bending or looping may turn out to be a general mechanism for transcriptional regulation, especially in situations where regulatory regions are located at some distance from the promoters they control.

Catabolite Repression

When *E. coli* cells are grown in the presence of both glucose and another sugar, such as lactose or arabinose, the cells will preferentially utilize glucose as their carbon and energy source. The *lac* and *ara* operons will not be induced, even though the appropriate effector substances are present. This inhibition by glucose of the transcription of a number of glucose-sensitive operons, including the *lac, ara,* and *gal* operons, is known as **catabolite repression.**

Catabolite repression is an indirect result of the effect of glucose on the cyclic AMP (cAMP) levels within a cell. The effect of glucose on cAMP is part of a separate positive control system that acts on glucose-sensitive operons such as the *lac* operon, which we will consider as an example. In addition to inducible negative control, the *lac* operon is also subject to an inducible positive control mechanism. The regulatory gene for positive control is the *cap* gene, which codes for the catabolite activator protein. In its activated form, this protein binds to a palindromic site in the *lac* promoter region, immediately adjacent to the site of RNA polymerase binding. To become activated, the CAP protein must be allosterically bound to its effector (inducer) substance, cAMP. This key metabolite is the effector for all glucose-sensitive operons. ▶ Figure 19–12 shows the induced *lac* operon. Transcription of the three structural genes occurs only in the presence of both effectors—cAMP (to activate the CAP protein) and lactose (to inactivate the repressor protein).

Although it is not entirely clear why CAP must be bound to the *lac* promoter in order for a significant amount of transcription to occur, several hypotheses can be advanced. One might speculate that CAP affects the conformation of the promoter region DNA, making it suitable for polymerase binding, or that the CAP and RNA polymerase molecules recognize each other and form a protein-protein complex that strengthens polymerase binding. CAP binding is known to induce a bend in the DNA, and this local distortion of the double helix is assumed to somehow be related to transcriptional activation.

The role of cAMP in the positive control of an operon such as *lac* explains how glucose exerts catabolite repression. The cellular concentration of cyclic AMP is inversely related to the glucose concentration. Thus, when glucose

▶ **FIGURE 19–12** Induced state of the *lac* operon, showing both inducible positive and inducible negative control. Transcription requires an activated CAP protein bound at the promoter as well as inactivation of the repressor protein.

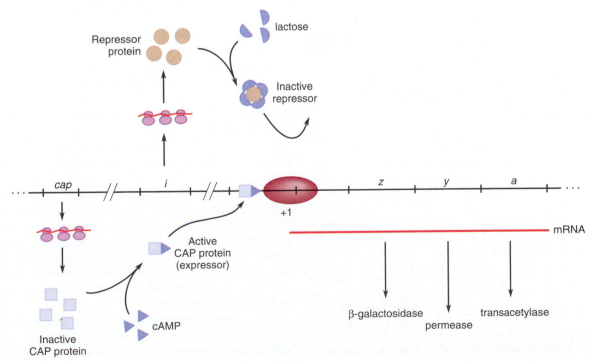

(a)

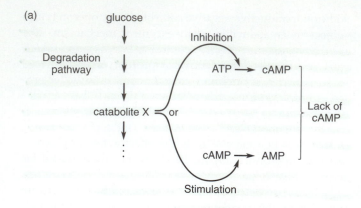

(b)

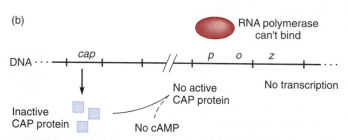

▶ **FIGURE 19–13** Catabolite repression. (a) A catabolite produced during glucose metabolism lowers cAMP levels, possibly by inhibiting the enzyme that converts ATP to cAMP or by stimulating the breakdown of cAMP into AMP. (b) Low cAMP levels in the cells mean the CAP protein cannot be activated. RNA polymerase cannot bind to the *lac* promoter region, so the *lac* structural genes are not transcribed, even in the presence of inducer (lactose).

levels are high, cAMP levels are low and the *lac* operon cannot be induced, no matter how much lactose is present (▶ Figure 19–13).

To Sum Up

1. The transcription of an operon is usually regulated by control of the initiation of mRNA synthesis at the promoter region of the operon. The control can be positive and/or negative, depending on the operon. The *lac* and *ara* operons are examples of inducible systems regulated by the initiation of transcription.

2. The *lac* operon consists of three structural genes, *z*, *y*, and *a*, which code for proteins involved in the metabolism of lactose. The cis-acting promoter-operator region regulates the initiation of transcription of these genes. The regulatory protein is a repressor that binds to the operator DNA to block transcription. Lactose binds allosterically to the repressor, inactivating its DNA-binding site and thereby inducing transcription indirectly—by blocking repression.

3. The three structural genes of the *ara* operon (*B*, *A*, and *D*) code for proteins involved in the degradation of arabinose. Initiation of transcription is regulated by a complex multipartite promoter-operator region. The regulatory gene (*C*) encodes a protein that activates transcription when it is allosterically bound to arabinose. The regulatory gene also controls its own transcription in response to the level of C protein, a phenomenon known as autoregulation.

4. In many bacterial operons, the initiation of transcription is inhibited in the presence of glucose through a phenomenon known as catabolite repression. These operons are subject to positive regulation by the CAP protein, which must allosterically bind the effector cAMP in order to activate transcription. Therefore, these operons cannot be induced when cAMP levels are low. High glucose levels somehow lower the cAMP level to cause catabolite repression. As a result of catabolite repression, cells preferentially utilize glucose when that sugar is present, thus avoiding the energetically expensive process of synthesizing enzymes that metabolize other sugars.

◆

CONTROL OF THE TERMINATION OF TRANSCRIPTION

Transcription terminates within a terminator sequence, as discussed in Chapter 16. There are, however, transcriptional control mechanisms that allow RNA polymerase to bypass the terminator region. Positive control systems that regulate transcriptional termination (rather than initiation) are called **antitermination** mechanisms. Antitermination is an important aspect of genetic regulation in some phages, where a regulatory protein prevents transcription from being terminated at a ρ-dependent terminator sequence. We will discuss this type of transcriptional control in the last section of this chapter. At this point, we are more interested in **attenuation**, a complex antitermination process that involves the availability of aminoacylated tRNAs.

Attenuation Control: The Tryptophan Operon

Several operons encoding enzymes that catalyze the biosynthesis of certain amino acids are subject to a form of transcriptional termination control called attenuation. In each of these systems, the ability of RNA polymerase to transcribe the structural genes of the operon is linked to the supply of the amino acid synthesized by the enzyme products of that operon. Hence, as the intracellular level of the amino acid increases through biosynthesis, transcription of the genes encoding enzymes for further synthesis decreases, following exactly the same principle of repressible control that we have discussed. However, in the case of attenuation, the issue is whether RNA polymerase will stop transcription at an early terminator that precedes the structural genes (see Figure 19–2), rather than whether transcription will be initiated. The decision is based on the supply of the relevant aminoacyl-tRNA. If this aminoacyl-tRNA is in short supply, RNA polymerase will read past the terminator region and into the structural genes, thereby increasing biosynthesis of that amino acid. If the aminoacyl-tRNA level is high, transcription will stop at the early terminator region, and no more enzymes for synthesis of that amino acid will be produced by the cell.

The *trp* operon provides an excellent example of attenuation control. It has long been known that the repressed state of the *trp* operon (which is a repressible system—see Figure 19–5) is not fully repressed. As in the case of the *lac* operon, a basal level of transcription takes place even when active repressor is bound to the operator, but repression of the *trp* operon is ten times less efficient than repression of the *lac* operon. Tryptophan itself, however, is able to partially compensate for the inefficient blocking of RNA polymerase at the operator. The *trp* operon contains an early terminator region called the **attenuator region** (▶ Figure 19–14) between the *trp* operator and the first structural gene. This region is part of a leader sequence and is composed of a GC-rich palindrome followed by a stretch of AT base pairs (recall from Chapter 16 that these are the typical features of a Type I (ρ-independent) transcriptional terminator region). Most of the RNA polymerase molecules that manage to initiate transcription of the repressed operon are terminated at this region by a mechanism that is activated by high tryptophan concentrations. Termination is not completely effective, however; 10% of the polymerase molecules still manage to read through the terminator region and transcribe the adjacent structural genes. Hence, this mechanism is viewed as a weakening or attenuation of transcription.

Events that occur during transcription and translation of the leader sequence are the key to attenuation control. ▶ Figure 19–15 shows the sequence of codons in the *trp* operon leader mRNA. Three important features of this mRNA should be noted. First, nucleotides 27 through 68 (beginning with the initiation codon AUG and ending with the translational stop codon UGA) are translated into a leader peptide. Second, there is a great deal of internal complementarity within the leader sequence: sequences in region 1 make it complementary to region 2, which in turn is complementary to region 3, which is also complementary to region 4. Several different hairpin structures can therefore be formed by this mRNA. Third, region 1 contains two successive trp codons (UGG). Whether or not the ribosomes can translate these two codons during translation of the mRNA into leader peptide determines whether a terminator hairpin will be formed.

▶ Figure 19–16 illustrates the various hairpin structures that can be formed by the leader mRNA. The Type I structure terminates transcription, while the other types are alternative structures that do not have the necessary properties to stop transcription. When the tryptophan level is high (Figure 19–16a), there is plenty of trp-tRNA^{Trp} and the leader mRNA is easily translated. A ribosome will move past the successive trp codons to the translational stop codon, where it physically covers a large portion of region 2 (just before it disassociates from the mRNA); thus region 3 can pair only with region 4. The result is a Type I terminator hairpin at the attenuator region, so transcription is stopped before it reaches the adjacent structural genes. In contrast, if the tryptophan level is low (Figure 19–16b), the scarcity of trp-tRNA^{Trp} causes the ribosomes that are translating the leader mRNA to stall at the two trp codons. This stalling leaves region 2 uncovered by a ribosome, allowing it to pair with region 3 into an alternative (nonterminating) hairpin structure. RNA polymerase then continues transcription through the adjacent structural genes. Figure 19–16c shows that if translation of the leader mRNA does not occur at all, a terminator hairpin is formed, halting transcription.

For attenuation to function correctly, transcription by RNA polymerase and translation by ribosomes must be very precisely coordinated. The ribosome must approach the trp codons just as RNA polymerase reaches the potential terminator region. Recent studies have shown that RNA polymerase pauses at about position 90 on the leader DNA until a ribosome reaches the trp codons on the mRNA. Polymerase then continues transcription, so that by the time the ribosome has or has not stalled at the trp codons, a hairpin secondary structure is being formed.

Attenuation has been found to be involved in regulating the transcription of at least seven operons concerned with amino acid biosynthesis. In each case, a leader sequence on the mRNA contains successive codons for the amino acid in

▶ **FIGURE 19–14** The *trp* operon of *E. coli*. The structural genes (*E* through *A*), which code for enzymes of the tryptophan biosynthetic pathway, are under the control of a promoter-operator-leader-attenuator region. Tryptophan serves as a corepressor to block the operator in a typical repressible system. RNA polymerase molecules that are still able to initiate transcription in spite of repression are terminated at the attenuator region, a ρ-independent transcriptional termination site.

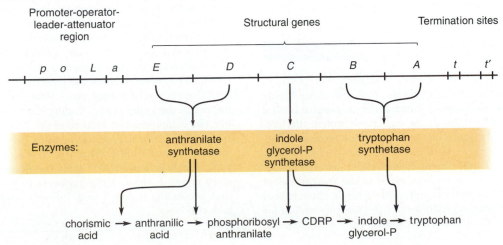

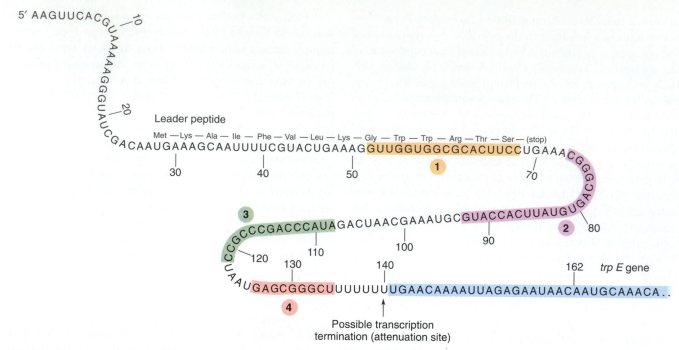

▶ **FIGURE 19-15** The nucleotide sequence of the *trp* operon leader mRNA, which extends from nucleotide position +1 to the beginning of structural gene *E*, at position 162. The leader peptide is encoded by nucleotides 27 through 68, with a translational stop codon at position 70. This region contains two consecutive tryptophan codons. The attenuator region extends from nucleotide position 115 to 140. The leader mRNA possesses the internal complementarity necessary to form a Type I terminator hairpin, since GC-rich regions 3 and 4 are complementary and lie adjacent to a poly U region. However, other internal complementarities in the leader sequence also make it possible for alternative secondary structures to form. Region 1 is complementary to region 2, which in turn is complementary to region 3. Thus, if 2 pairs with 3, then 3 cannot pair with 4 and transcriptional termination will not take place. *Adapted from:* J. D. Watson, N. H. Hopkins, J. W. Roberts, J. Steitz, and A. M. Weiner, *Molecular Biology of the Gene,* 4th ed. (Benjamin Cummings, 1987): 487. Used with permission.

question. In some operons, such as the *trp* operon, attenuation seems to act as a fine-tuning mechanism—it backs up operator-mediated repression of transcription when trp levels are high, and it also allows transcription to be turned on gradually as the level of trp in the cell decreases. It is thought that attenuation occurs even as repressor is released from the operator, so that transcription increases progressively as the trp level falls. In other cases, attenuation is the only form of transcriptional regulation of the operon; one example is the *histidine* operon, whose leader mRNA contains seven consecutive his codons.

To Sum Up

1. The transcription of some operons is regulated by control of the termination of mRNA synthesis. Antitermination is one such regulatory mechanism; it can occur in positive control systems and operates through a regulatory protein that prevents transcription from being terminated at a terminator site.

2. Attenuation is another type of transcriptional termination control; it has been found in many bacterial operons whose structural genes code for enzymes involved in the biosynthesis of certain amino acids. In these repressible systems, the supply of an amino acid controls production of the enzymes needed for its biosynthesis. Control is exerted at an early transcriptional terminator site that precedes the structural genes encoding these enzymes; whether transcription proceeds past this site depends on the level of the relevant aminoacyl-tRNA.

3. The *trp* operon is a negative repressible system with two means of transcriptional control. In the presence of tryptophan, the initiation of transcription is inefficiently repressed by the binding of the repressor. Any transcription that is initiated is then subject to termination through attenuation, a second transcriptional regulatory mechanism.

4. The *trp* operon contains an attenuator region between the operator and the first structural gene. Transcription can be terminated in this region if tryptophan levels are high. The transcriptional termination mechanism is based on the movement of ribosomes along the operon's leader mRNA, the initial portion of which codes for a leader peptide. The leader mRNA contains complementary sequences that can fold into either terminator or nonterminator hairpins. In the presence of ample trp-tRNA[Trp], the ribosomes move along the leader mRNA to the end of the leader peptide coding region, thus covering the first two complementary sections of the mRNA. The attenuator region then forms a Type I terminator hairpin from complementary regions 3 and 4. When trp levels are low, however, the short supply of trp-tRNA[Trp] stalls the ribosomes before they reach region 2, allowing that region to pair with region 3 and thus precluding region 3 from forming a terminator hairpin with region 4.

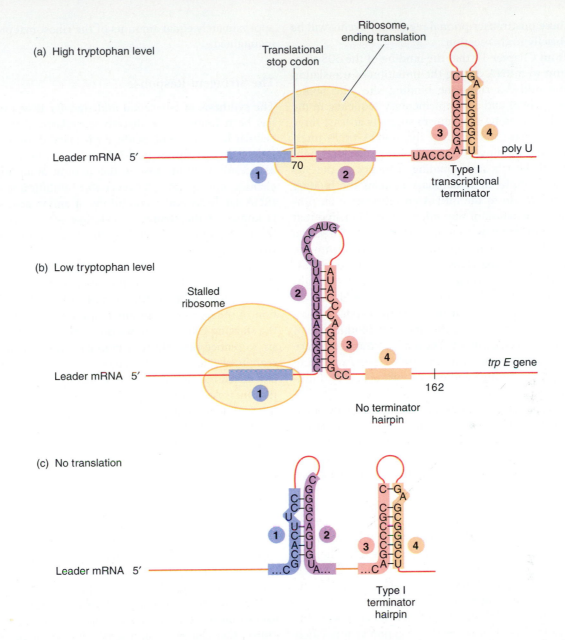

▶ **FIGURE 19–16** Effect of ribosome position on transcription of the *trp* operon. (a) When the intracellular tryptophan level is high, there is ample trp-tRNA^Trp, so translation of the leader mRNA continues past the two trp codons. Although the ribosome stops at the UGA codon, it physically covers part of region 2, so that 2 cannot pair with region 3. As a result, region 3 pairs with 4, generating a Type I terminator hairpin, which stops RNA polymerase from transcribing the structural genes of the operon. (b) In cells starved for trp, the low level of trp-tRNA^Trp causes the ribosome to stall in region 1 at the two trp codons. This allows regions 2 and 3 to pair, thereby preventing 3 and 4 from forming a terminator hairpin. RNA polymerase then continues transcription through the structural genes of the operon. (c) If no translation of the leader mRNA occurs, regions 1 and 2 pair, as do 3 and 4. The latter generates a terminator hairpin, halting transcription.

▦ POSTTRANSCRIPTIONAL CONTROL

So far, our discussion of regulation has focused on transcription, since that is the predominant level at which gene activity is controlled in prokaryotes. However, regulatory opportunities do not cease once a gene has been transcribed. Essentially all mRNAs in prokaryotes have the same rela-tively short half-life, so there is little opportunity to regulate gene activity through differential stability of mRNAs (in contrast to eukaryotic systems). However, the steps in translating an mRNA can be influenced by its accessibility to ribosome binding and by the availability of intracellular ribosomal and transfer RNAs. Thus a gene's activity can be controlled by regulating the rate of synthesis of its encoded peptide. These posttranscriptional regulatory events will be discussed briefly in this section.

Recall from Chapter 17 that the binding of the 30S ribosome subunit to mRNA during the initiation of translation involves the mRNA's ribosome binding site (the Shine-Dalgarno sequence) and a complementary sequence in the 16S rRNA. This binding mechanism suggests at least three possible events that could regulate this mRNA-rRNA interaction. First, the binding of a protein to the mRNA might limit access of the ribosome-binding sequence to the 30S subunit. For example, in some phage systems the binding of one phage protein to the mRNA encoding another protein prevents translation of that mRNA, thus signaling that enough of that protein has accumulated in the cell. In such a case, the regulatory protein might simply compete with the 30S subunit for the Shine-Dalgarno sequence of the mRNA. Second, the ribosome-binding sequence of the mRNA might be recognized by a small complementary RNA molecule. For example, the bacterial transposon *Tn10* encodes a small RNA that is complementary to the ribosome-binding site of the transposase enzyme mRNA. The double-stranded nature of the bound mRNA region prevents translation from being initiated, thus preventing synthesis of transposase and limiting the frequency of *Tn10* transposition events. This example is a type of **antisense** regulatory mechanism, in which gene expression is blocked by an interaction between the messenger RNA (sense) and a complementary sequence (antisense). A third possible mechanism for controlling translational initiation involves the formation of hairpin loops in mRNA and its consequences for ribosome binding. Polygenic mRNAs are often translated one gene at a time, with each gene having its own ribosome binding and termination sites. In some phages, progression of ribosomes along a polygenic mRNA is necessary to open up the secondary structure of adjacent regions of the mRNA, which would otherwise be folded in such a way as to "hide" the ribosome binding region of an adjacent gene.

Other posttranscriptional regulatory mechanisms involve the availability of ribosomes, elongation factors (such as EF-Tu and EF-G), and tRNAs. Ribosome availability depends on the intracellular amounts of rRNA and ribosomal proteins. The majority of the genes that code for ribosomal proteins are organized into just a few operons. For example, ■ Table 19–1 shows six operons that contain most of the genes that encode ribosomal proteins in *E. coli*; they also contain genes for elongation factors and the subunits of RNA polymerase. The primary control of these operons is at the level of translation rather than transcription. One ribosomal protein generated by each operon acts as a repressor for translation of the polygenic mRNA transcribed from that operon. For example, Table 19–1 shows that ribosomal protein L4 represses the translation of S10 operon mRNA starting at gene *S10* and ending at gene *L17*. L4 binds to the mRNA at a translational initiation point, preventing ribosomes from binding and thus preventing synthesis of proteins S10 through L17. Translation of the mRNA from the other operons is regulated in a similar manner. In this way,

approximately equal amounts of the ribosomal proteins are maintained.

The Stringent Response

The synthesis of ribosomal and transfer RNAs in bacteria has been found to be directly correlated with the intracellular level of amino acids. Cells starved for amino acids reduce their synthesis of rRNA and tRNA by 10- to 20-fold as a result of an increase in the amount of an unusual nucleotide, ppGpp. ppGpp's reversible inhibition of rRNA and tRNA synthesis under conditions of amino acid starvation is known as the **stringent response**—it prevents the cell from wasting energy synthesizing translation components that it cannot use.

Although the mechanism of ppGpp inhibition is not understood, it is known that this nucleotide also inhibits the transcription of many other genes, enhances the transcription of others (whose products are presumably needed by the starving cell), and slows down general protein synthesis to reduce possible translational errors during times of scarcity of charged tRNAs. The synthesis of macromolecules is reduced in starved cells, and the rate of protein degradation increases; these are logical responses to conditions of nutrient deprivation.

The ppGpp nucleotide is made in the following way. Starvation for amino acids means that an uncharged tRNA will occupy the A site of a ribosome. This halts peptide elongation, and an "idling" reaction occurs on the ribosome. During idling, the **stringent factor** (a protein found in about 1 in 20 ribosomes) catalyzes a chemical reaction in which ATP donates a pyrophosphate group to the 3′ position of either GTP or GDP, yielding pppGpp or ppGpp, respectively. The pppGpp is converted to ppGpp, the nucleotide that mediates the stringent response. ▶ Figure 19–17 shows the reactions that produce ppGpp.

We have now encountered two examples of bacteria using unusual nucleotides to respond to nutrient conditions—the stringent response is mediated by ppGpp, while catabolite repression is mediated by cAMP. Whether these examples are part of a more general phenomenon in prokaryotes or in eukaryotes remains an interesting question.

To Sum Up

1. Although regulation at the transcriptional level is the main control mechanism in prokaryotes, several types of posttranscriptional regulation of gene activity have also been discovered. Control of ribosome binding to mRNA at the Shine-Dalgarno sequence is one example. The availability of ribosomes, elongation factors, and tRNAs are other possible posttranscriptional regulatory mechanisms.

2. The availability of ribosomal and transfer RNAs is directly correlated with the intracellular levels of amino acids. When cells are starved for amino acids, the so-called stringent response inhibits rRNA and tRNA synthesis. The stringent response en-

■ **TABLE 19–1** Six operons containing genes for ribosomal proteins in *E. coli*. The ribosomal protein that regulates (represses) translation of the mRNA of each operon is listed on the right, an arrow indicates its site of action, and the segment of mRNA affected by its action is bracketed. (L and S refer to proteins of the large and small ribosomal subunits, respectively.)

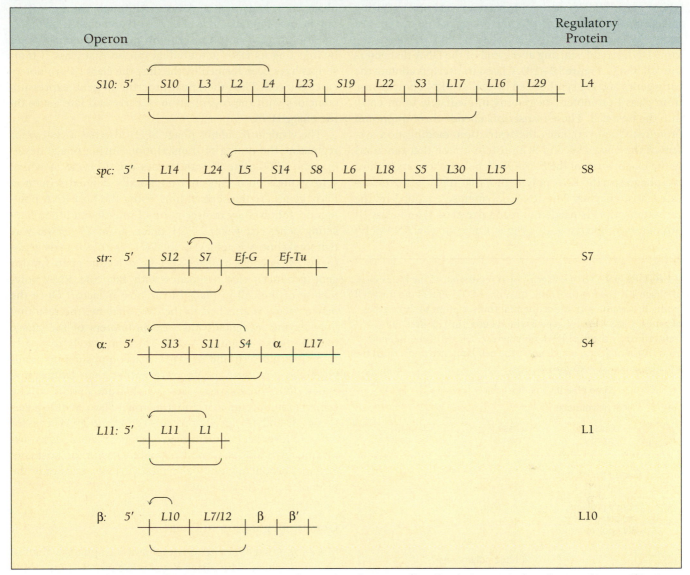

Operon		Regulatory Protein
S10: 5′ S10 L3 L2 L4 L23 S19 L22 S3 L17 L16 L29		L4
spc: 5′ L14 L24 L5 S14 S8 L6 L18 S5 L30 L15		S8
str: 5′ S12 S7 Ef-G Ef-Tu		S7
α: 5′ S13 S11 S4 α L17		S4
L11: 5′ L11 L1		L1
β: 5′ L10 L7/12 β β′		L10

Source: M. Nomura, R. Gourse, and G. Baughman, *Ann. Rev. Biochem.* 53 (1984): 82. Reproduced, with permission, from the *Annual Review of Biochemistry*, Vol. 53, © 1984 by Annual Reviews, Inc.

sures that bacterial cells do not waste energy synthesizing ribosomes and tRNAs that cannot be used.

◆

▶ **FIGURE 19–17** Reactions on idled ribosomes that lead to the production of ppGpp, the mediator of the stringent response.

$$pppG + ATP \xrightarrow{\text{Stringent factor}} pppGpp + AMP$$

$$P_i \longleftarrow \Big| \text{Ribosomal proteins}$$

$$ppG + ATP \xrightarrow{\text{Stringent factor}} ppGpp + AMP$$

GENETIC REGULATION OF PHAGE DEVELOPMENT

Like bacterial genes, phage genes are generally organized into operons, facilitating their coordinate control; but unlike bacterial operons, which allow the cell to respond efficiently to fluctuations in available nutrients, phage operons primarily ensure that genes are expressed in an orderly temporal sequence during phage development. When a virulent phage infects a bacterium, one or more products of the first set of genes in the sequence turn on the genes of the second set, one or more products of the second set of genes may then turn on the genes of a third set, and so on, until all pertinent sets of genes have been transcribed. The life cycle of a bacteriophage thus follows a programmed developmental sequence that is controlled at the level of transcription.

One example of temporal control during phage development is the regulation of gene expression in the virulent *E. coli* phage T7. Phage T7 is a double-stranded DNA phage whose entire 39,936-nucleotide sequence is known. During infection, the genome of this phage makes three types of transcripts (designated as classes I, II, and III) from the same DNA strand, starting at the end of the DNA that enters the cell first (▶ Figure 19–18). Class I transcripts contain the early genes and are synthesized by the host RNA polymerase from class I promoters, beginning immediately upon entry into the host cell. These transcripts encode three important proteins: One overcomes the restriction/modification system of the host, a second is a protein kinase that functions in the inactivation of the *E. coli* RNA polymerase, and a third is a phage-specific RNA polymerase that replaces the inactivated host enzyme. The RNA polymerase encoded by the early transcripts is needed to recognize class II and class III promoters for the remaining genes. Once the new RNA poly-merase is formed, it produces the class II transcripts, which encode proteins involved in DNA replication, followed by the class III transcripts, which encode proteins needed for phage morphogenesis and lysis. Although the new RNA polymerase can recognize both class II and class III promoters, the phage delays transcription of the late-acting genes by injecting its DNA at a gradual rate so that class II transcripts have been formed by the time the class III promoters enter the cell. Thus the linear ordering of gene expression coincides with the organization of early and late genes on the phage DNA.

The virulent *B. subtilis* phage SPO1 also has a system for regulating the timing of transcription during phage development. SPO1, like T7, undergoes a lytic cycle in which gene expression occurs in a temporally defined sequence. Early genes are transcribed first, using the host RNA polymerase, followed by transcription of middle- and then late-acting genes. However, SPO1 differs from T7 in the way temporal control is achieved. In SPO1, one of the early genes (gene *28*) encodes a phage-specific σ factor, gp28 (gp for gene product), that associates with the core RNA polymerase of the cell in place of the host σ factor. Once the new σ factor is attached to the core enzyme, the resulting holoenzyme recognizes only the promoters of the phage middle genes (▶ Figure 19–19). Thus the switch from early to middle transcription in SPO1 is accomplished by modifying the host RNA polymerase, rather than by forming a new RNA polymerase. The switch from middle to late transcription occurs in a similar manner. Two middle genes (genes *33* and *34*) specify proteins (gp33 and gp34) that interact to form yet another phage-specific σ factor. After displacing gp28, this latest σ also combines with the host core enzyme, producing a holoenzyme that recognizes only the promoters for the late genes.

Genetic Regulation of Phage λ

More is known about genetic regulation in phage λ than in any other phage. The phage λ genes are grouped into three main operons, as shown in ▶ Figure 19–20. There are two groups of early genes, reflecting the fact that an infecting λ chromosome can follow either of two growth cycles. The genes in the **early right operon** are concerned with DNA replication during the lytic cycle, while the genes in the **early left operon** are concerned with recombination and lysogeny. The genes coding for head, tail, and lysis proteins are found in the **late operon**. Each operon is transcribed from its own promoter site; the promoter sites are designated p_L (early left), p_R (early right), and $p_{R'}$ (late).

Six genes, *N*, *Q*, *cro*, *cI*, *cII*, and *cIII*, are important in the regulation of phage development. The functions of these genes and a number of other phage λ genes are given in ■ Table 19–2. Genes *N*, *Q*, and *cro* are involved in the lytic phase of multiplication. The products of the *N* and *Q* genes, gpN and gpQ, are regulatory proteins that exert positive transcriptional control. Both gpN and gpQ are **antiterminators** that allow RNA synthesis to proceed past certain tran-

▶ **FIGURE 19–18** Transcriptional regulation of the lytic cycle of phage T7. During infection, the phage DNA enters the host cell gradually, resulting in the sequential transcription of three types of genes. Class I (early) genes, which enter the cell first, are transcribed by the host RNA polymerase. Class II (middle) and class III (late) genes, which enter the cell later, are transcribed by a phage-encoded RNA polymerase.

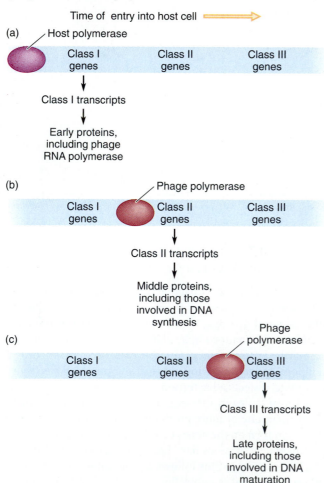

► **FIGURE 19–19** Transcriptional regulation of the lytic cycle of phage SPO1. The sequential transcription of early, middle, and late genes in the phage DNA is achieved by successive substitutions of the sigma factor that modify the initiation specificity of the host RNA polymerase.

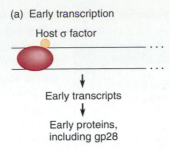

(a) Early transcription

Host σ factor

↓
Early transcripts
↓
Early proteins,
including gp28

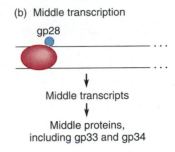

(b) Middle transcription

gp28

↓
Middle transcripts
↓
Middle proteins,
including gp33 and gp34

► **FIGURE 19–20** The genes of phage λ are grouped into three main operons: two are concerned with early functions and one with late functions. The early operons are each controlled by a promoter-operator complex ($o_L p_L$ and $o_R p_R$) and are transcribed from different strands in different directions. The late operon is transcribed beginning with the promoter $p_{R'}$. The promoters p_{RE}, p_{RM}, and p_I are concerned with lysogenic functions. *Adapted from:* Voet, Donald, and Judith G. Voet, *Biochemistry* (New York: Wiley & Sons, 1990): 1002. Copyright © 1990 by John Wiley & Sons, Inc. Reprinted by permission.

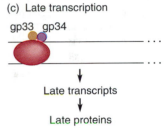

(c) Late transcription

gp33 gp34

↓
Late transcripts
↓
Late proteins

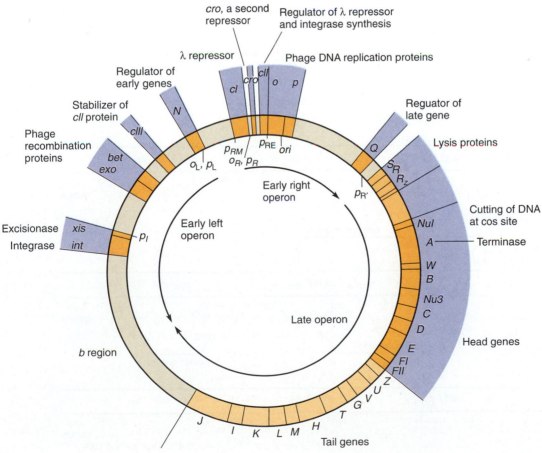

■ **TABLE 19–2** Selected genes and regulatory sites for bacteriophage λ.

Gene or Site	Function
cI	repressor of lytic response; establishment and maintenance of lysogeny
cII, cIII	establishment of lysogeny
cro	repressor of lysogenic response
N	antiterminator of early gene transcription
O	antiterminator of late gene transcription
int	prophage integration and excision
xis	prophage excision
B–E, W, Nu3, F1, F2	head assembly
G–M, U, V, Z	tail assembly
A, Nu1	DNA packaging
R, S	lysis of host cell
O, P	DNA replication
o_L, o_R	operator sites
p_I, p_L, p_R, p_{RM}, p_{RE}, $p_{R'}$	promoter sites
t_{L1}, t_{R1}, t_{R2}, t_{R3}, $t_{R'}$	transcriptional termination sites
nutL, nutR	N utilization sites
qut	Q utilization sites

scription terminators—they bind to specific DNA sequences called *nut* and *qut* sites (for Q utilization and Q utilization) and modify the host RNA polymerase as it passes these sites so that it no longer recognizes the early transcription terminators that follow. The *cro* gene product, gpcro, is a repressor protein that combines with operator sites to block transcription essential to the lysogenic response.

The control of gene expression in the lytic pathway of phage λ is summarized in ▶ Figure 19–21. Transcription is carried out by the host RNA polymerase in a temporally defined sequence consisting of three phases designated as early, delayed early, and late. The early phase produces three short transcripts starting from the promoters p_L, p_R, and $p_{R'}$ and ending at the ρ-dependent terminators t_{L1}, t_{R1}, and $t_{R'}$ (Figure 19–21a). The transcript starting at p_L encodes gpN, while the transcript starting at p_R encodes gpcro. The synthesis of gpN marks the transition between the early and delayed early phases. gpN allows transcription to proceed past three early terminators, including t_{L1} and t_{R1}, thus extending the transcribed region to include gene Q in addition to the various other genes found in the early right and early left operons (Figure 19–21b). The production of gpQ provides the switch from delayed early to late transcription. This regulatory protein allows RNA synthesis to proceed past the $t_{R'}$ early terminator and results in the transcription

▶ **FIGURE 19–21** Transcriptional regulation of the lytic pathway of phage λ. Sequential control is exerted by the actions of two antiterminators, gpN and gpQ, and the repressor gpcro, which turns off transcription of the early left and early right operons. Lytic control involves three phases of transcription: (a) early transcription for the synthesis of gpN and gpcro, (b) delayed early transcription, promoted by gpN and needed for the synthesis of DNA replication proteins, and (c) late transcription, promoted by gpQ and needed for the synthesis of head, tail, and lysis proteins. The transcripts produced at each phase are represented by wavy arrows pointing in the direction of synthesis. *Adapted from:* Voet, Donald, and Judith G. Voet, *Biochemistry* (New York: Wiley & Sons, 1990): 1004. Copyright © by John Wiley & Sons, Inc. Used by permission.

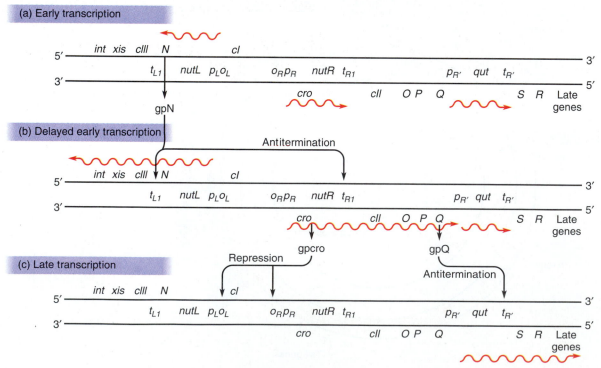

of the late operon (Figure 19–21c). By this time, gpcro has accumulated in sufficient quantity to bind to operators o_L and o_R and shut off transcription of the early left and early right operons. Thus, the lytic cycle of phage λ, like those of the virulent phages, exhibits a programmed developmental sequence that is controlled at the level of transcription, but unlike the control circuits in T7 and SPO1, the developmental switch operating in phage λ is based on antitermination.

The three remaining genes involved in λ regulation—*cI*, *cII*, and *cIII*—are essential to lysogeny and, along with gene *cro*, play an important role in "choosing" between the lysogenic and lytic pathways. The conditions in the bacterial host cell heavily influence the decision. The lysogenic response is favored in starved cells and in cells that have been infected with λ at a high ratio of phage to cells. Under either of these conditions, the concentration of the *cII* gene product, gpcII, is high. gpcII is a positive transcriptional regulator that stimulates the transcription of *cI* and the integrase (*int*) gene. The cII protein is metabolically unstable and requires the presence of gpcIII to protect it from degradation by the host. The concentration of gpcII is thought to be the determining factor in the choice of growth pathway: If the concentration of gpcII is high, then the lysogenic pathway is followed; if the concentration of gpcII is low, then lytic growth ensues.

The transcription of the *int* and *cI* genes promoted by gpcII is critical for the establishment of the lysogenic state. The *int* gene specifies integrase, which is needed for prophage integration (see Chapter 10). The *cI* gene specifies a protein repressor, gpcI, which (like the gpcro repressor) binds to both operators o_L and o_R and prevents transcription of the early operons. In preventing early transcription, gpcI turns off the synthesis of all λ gene products that are needed for the lytic response, thus stopping the lytic cycle before it can proceed beyond the early stages. In its active form, gpcI is a dimer. Each of its monomer units is a single polypeptide that folds into two globular domains, giving the monomer the appearance of a dumbbell (▶ Figure 19–22a). The two domains have different functions. The N-terminal domain has a helix-turn-helix motif that enables the dimer to bind to the approximate palindromic symmetry of the operator DNA (Figure 19–22b). The C-terminal domain has no direct role in DNA binding; it is responsible for the association of the two chains into a dimeric form.

Two promoters initiate transcription of the *cI* gene (▶ Figure 19–23). The promoter p_{RE} (RE for repressor establishment; Figure 19–23a) is used during the establishment of lysogeny and requires gpcII. The mRNA initiated at p_{RE} is translated very efficiently, resulting in a buildup of gpcI that further suppresses the lytic response. Once the cascade of events has tipped the balance in favor of lysogeny over lysis, the resulting prophage must continue to produce a sufficient amount of repressor to maintain a stable lysogenic state. In this stage the *cI* gene is transcribed from a promoter designated p_{RM} (RM for repressor maintenance; Figure 19–23b). Transcription from the p_{RM} promoter is activated not by gpcII, which is no longer produced at this time, but by the

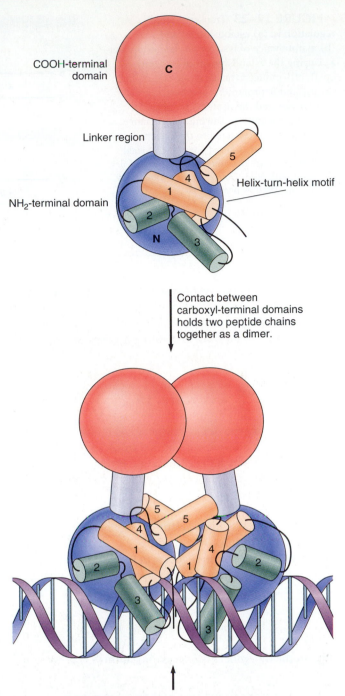

The two amino-terminal domains bind to the two halves of the palindromic recognition sequence in λ DNA.

▶ **FIGURE 19–22** The structure and action of the λ cI repressor molecule. (a) The monomer unit of the repressor is a polypeptide chain that is coded for by gene *cI*. The polypeptide folds into two domains, a carboxy-terminal domain needed for dimer formation and an amino-terminal domain needed for DNA binding, linked by a region that is susceptible to cleavage by proteases. (b) A functional repressor molecule is a dimer. Each amino-terminal domain of the dimer has a helix-turn-helix motif that binds to half of the palindromic recognition sequence in the o_L and o_R regions of λ DNA. *Adapted from:* M. Ptashne, *A Genetic Switch: Phage λ and Higher Organisms* (Cambridge, MA: Cell Press and Blackwell Scientific Pub., 1992): 38. Used with permission.

► **FIGURE 19–23** Transcriptional regulation in (a) establishment and (b) maintenance of lysogeny in bacteriophage λ. During the establishment of lysogeny, gpcII, cIII-induced transcription occurs from the p_{RE} and p_I promoters, synthesizing the cI repressor and the proteins required for prophage integration. During maintenance, transcription is restricted to the *cI* gene from the p_{RM} promoter. *Adapted from:* Voet, Donald, and Judith G. Voet, *Biochemistry* (New York: Wiley & Sons, 1990): 1011. Copyright © by John Wiley & Sons, Inc. Used by permission.

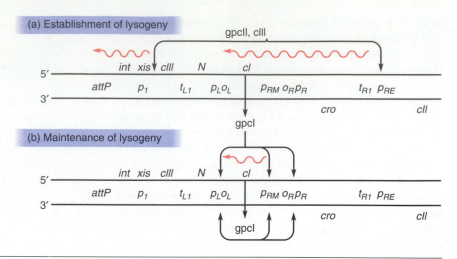

binding of gpcI to the overlapping o_R region. Thus, in a lysogenic cell, the cI repressor is a positive regulator of its own synthesis, acting to ensure that the cell has an adequate supply of gpcI to keep the remaining λ genes in a repressed state.

Example 19.2

Consider the *cI* gene of bacteriophage λ. If a bacterial cell is simultaneously infected with both a wild-type (cI^+) and a *cI* mutant of λ (in which the *cI* mutation renders the repressor nonfunctional), can this cell become stably lysogenic? Explain why or why not.

Solution: The *cI* gene is a trans-acting regulatory element, analogous to the repressor-encoding gene *i* of the *lac* operon. Therefore, the wild-type *cI* gene will encode enough repressor to allow a cell infected with both cI^+ and *cI* phages to become stably lysogenic.

Follow-Up Problem 19.2

Explain the differences in the actions of the λ promoters p_{RE} and p_{RM}.

◆

The cI repressor formed in a λ lysogen not only prevents the transcription of prophage genes but also blocks the transcription of DNA from other λ phages that might later infect the cell. For this reason, a lysogenic cell is said to be immune to superinfection by phages of the same type as the phage that is carried as a prophage (i.e., phages having the same control or **immunity** region). Lysogenic cells that are formed during λ infection are therefore capable of growing in the presence of superinfecting particles. Such growth is responsible for the **turbid** plaques that characterize this temperate phage (► Figure 19–24). While most cells lyse during plaque development, a few of the cells become lysogenic for λ and produce a cloudy or turbid area of growth

within the plaque. Immunity to superinfection is not the same as phage resistance. In immunity, the superinfecting phage DNA enters the cell but is unable to replicate; it is eventually diluted out during bacterial growth. Phage resistance, in contrast, typically involves the loss of functional receptor sites on the bacterial cell wall, so that phage particles are unable to attach themselves to the resistant cell.

In the race between the lytic and lysogenic pathways in a λ-infected cell, the competition between the cI and Cro repressors is perhaps the most pivotal. Both repressors bind to the operators o_L and o_R but with different results. Binding of the cI repressor, as we have previously mentioned, blocks transcription from p_L and p_R and stimulates it from p_{RM}. Binding of the Cro repressor, on the other hand, represses all λ genes, including *cI*. The two repressors therefore have opposing functions, with gpcI promoting transcription of *cI* during lysogeny and gpcro repressing transcription of *cI* during the lytic cycle. The Cro repressor is the smaller of the two repressor proteins, but like gpcI, gpcro binds to DNA

► **FIGURE 19–24** Plaque morphology of phage λ. The wild-type (c^+) plaque has a turbid center due to the growth of occasional cells that become lysogenic for the λ phage. Plaques produced by *c* mutants are clear, because such mutants are unable to form stable lysogens. The dotted appearance of *c* plaques is caused by the growth of a few λ-resistant mutant cells.

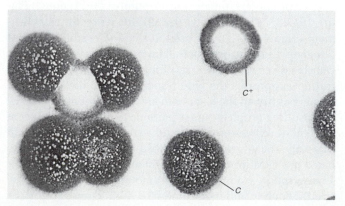

as a dimer. Each monomer unit folds into a single domain with a helix-turn-helix motif that mediates both binding to the DNA and dimerization.

The explanation of how two regulatory proteins can bind to the same operator regions of DNA and yet have different effects on gene expression is based on the organization of the o_R control region (▶ Figure 19–25a). The o_R region is an 80-base pair segment of DNA that is bracketed by the *cI* gene on the left and the *cro* gene on the right and is overlapped on the sides by two promoter regions, p_{RM}, which reads to the left, and p_R, which reads to the right. Three 17-bp palindromic sequences in the o_R region (o_R1, o_R2, and o_R3) serve as binding sites for the protein repressors. These palindromic sequences are similar enough that gp*cI*

and gpcro can bind to all three, but they bind with characteristically different affinities for each site. Chemical and nuclease protection experiments show that the affinity of gp*cI* dimers for the binding sites decreases in the order $o_R1 > o_R2 > o_R3$. The cI repressor is attracted most strongly to o_R1, but once a gp*cI* dimer is bound to o_R1, the bound dimer cooperatively binds a second dimer to o_R2 through associations between their C-terminal domains. The o_R3 site ordinarily remains free of repressor unless there is an abnormally high concentration of gp*cI* in the cell. The cooperative binding of gp*cI* dimers at o_R1 and o_R2 blocks access of RNA polymerase to the rightward promoter (p_R) and thereby prevents the transcription of *cro* and other early right genes (Figure 19–25b). At the same time, the cooperative binding of gp*cI* dimers stimulates synthesis of more repressor. By a mechanism that is still not completely clear, the N-terminal domain of the dimer bound at o_R2 interacts with RNA polymerase and promotes transcription from p_{RM}.

The affinity of gpcro dimers for the palindromic sequences in the o_R region is essentially opposite to that of gp*cI*, namely, $o_R3 > o_R2 = o_R1$. Once the Cro repressor is synthesized, it binds first to o_R3, preventing activation of p_{RM} by gp*cI* (Figure 19–25c). However, following the buildup in the gpcro concentration that normally occurs during the lytic cycle, o_R2 and o_R1 gradually become occupied, preventing rightward transcription from p_R and enabling the Cro protein to repress its own synthesis at high concentrations.

Induction of the prophage occurs spontaneously at a low rate (see Chapter 10), although exposure of lysogenic cells to UV light greatly enhances this rate. The switch from the lysogenic to the lytic state occurs when the cI repressor is inactivated. UV exposure causes repressor inactivation as a result of a repair mechanism known as the SOS response (see Chapter 23). In the SOS response, UV-induced damage of DNA triggers the activation of a protein called RecA, which induces the autocatalytic cleavage of the cI repressor. (The RecA protein normally has other roles in *E. coli* cells, most noticeably in general recombination—see Chapter 23). The self-cleavage of gp*cI* occurs in the polypeptide segment linking the repressor's two domains (▶ Figure 19–26). The loss of the C-terminal domain from the repressor monomers greatly reduces the efficiency of dimerization, which in turn affects the ability of the repressor to bind stably to DNA. Thus, the concentration of active repressor in the cell falls, and the o_R1 and o_R2 sites are left open, allowing *cro* and other early genes to be transcribed. In the absence of any corresponding mechanism to inactivate the Cro repressor, the cell will irreversibly enter the lytic cycle.

▶ **FIGURE 19–25** (a) Organization of the o_R region and the adjacent *cI* and *cro* promoters and genes in the DNA of phage λ. The o_R region is divided into three recognition sites: o_R1, o_R2, and o_R3. The *cI* promoter overlaps o_R3 and o_R2, and the *cro* promoter overlaps o_R1 and o_R2. (b) Transcription of the prophage during the lysogenic state. The cI repressor preferentially binds to o_R1 and o_R2, blocking the transcription of *cro* and other genes of the early right operon. RNA polymerase is able to bind to the *cI* promoter region, however, so that transcription of *cI* occurs. The repressor bound to o_R2 is thought to make contact with the RNA polymerase molecule, and this contact is hypothesized to aid the binding of polymerase to the *cI* promoter region. In this manner, *cI* stimulates its own synthesis. (c) Transcription of the λ genome during early lytic growth. The lack of functional cI repressor molecules leaves o_R1 and o_R2 open; RNA polymerase binds to the rightward promoter and transcribes *cro* and other early lytic genes. The Cro protein binds to o_R3, blocking further transcription of *cI*.

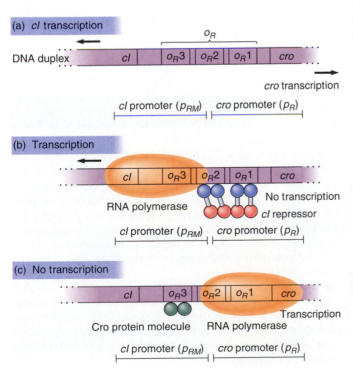

To Sum Up

1. Functionally related genes in bacteriophages also tend to be organized into operons. Phage operons ensure that genes are transcribed in an orderly temporal sequence during phage development.

(a) Lysogenic mode

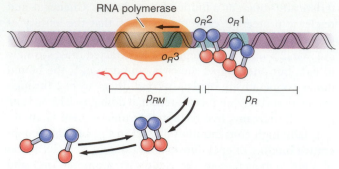

(b) Induction

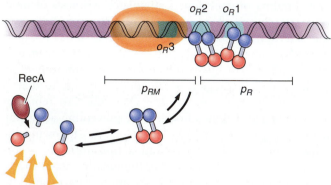

▶ **FIGURE 19–26** (a) Maintenance of lysogeny. (b) Induction of the lytic state by UV radiation. During induction, the RecA protein induces the autocatalytic cleavage of the cI repressor at its linker region, relaxing repression and permitting lytic multiplication to occur. *Adapted from:* Voet, Donald, and Judith G. Voet, *Biochemistry* (New York: Wiley & Sons, 1990), p. 1017. Copyright © John Wiley & Sons, Inc. Used by permission.

2. The genes of phages are typically organized into groups that are expressed at the same time during phage development. For example, early genes (such as those that control replication of the phage genome) are organized into one or two groups that are transcribed immediately or shortly after entry of the phage genome into the host cell. Late genes are grouped separately from early genes and are transcribed later in phage development. One or more products of the early set(s) of genes are required to turn on transcription of the later set(s) of genes.

3. Bacteriophage λ provides an example of the temporal sequencing mediated by phage operons. The lytic genes of λ are organized into two early and one late operon. Each operon is transcribed from a separate promoter, and each contains an early transcriptional terminator site. The major regulatory mechanism is antitermination; transcription of each of the early operons is halted at its early terminator until enough early gene product is available to allow RNA polymerase to proceed past these sites into the remaining genes of the operon (the delayed early genes). The production of gene product from the delayed early genes then allows transcription to proceed past the early terminator of the late operon.

4. The chromosomes of temperate phage, such as λ, also include another group of genes involved in lysogenic growth. Establishment of lysogeny depends on a positive transcriptional regulator (the cII-cIII gene product) that activates transcription from two promoters. One of these promoters, p_{RE}, controls the gene (cI) that encodes repressor protein, the other (p_I) controls genes needed for prophage integration. Maintenance of lysogeny depends on the repressor protein; it binds to operator regions that are adjacent to the promoters of the two early operons and thereby prevents the gene activity needed for lytic growth. During the maintenance of lysogeny, transcription of the cI gene initiates at the p_{RM} promoter, rather than at p_{RE}, allowing it to be positively regulated by the repressor protein itself (another example of autoregulation).

5. A bacterial cell harboring a λ prophage is immune to infection by additional phages of the same type as that carried as a prophage. The cI repressor protein mediates this immunity by blocking transcription of the early operons of the superinfecting phage genomes.

6. Induction of a lysogenic cell occurs when the cI repressor is inactivated, allowing the early operons to be transcribed and thus initiating lytic growth.

Chapter Summary

In prokaryotes, the genes that code for enzymes involved in the same metabolic pathway are typically clustered together into functional units called operons. The genes of an operon are coordinately transcribed from a single promoter and are coordinately controlled by the protein products of one or more regulatory genes. The regulatory proteins control the genes of an operon by binding to cis-acting control sites, such as the operator or the promoter, where they govern the initiation or early termination of transcription. The binding of the regulatory protein can affect the operon either positively, if it stimulates transcription of the structural genes, or negatively, if it inhibits transcription.

Operons can be either inducible or repressible. Inducible operons, such as the *lac* operon in E. coli, mainly regulate the synthesis of enzymes involved in degradative pathways. The transcription of an inducible operon is turned on in the presence of an inducer (often the first substrate in the pathway). In contrast, repressible operons, such as the *trp* operon in E. coli, tend to be involved in the synthesis of enzymes that function in biosynthetic pathways. The transcription of a repressible operon is turned off in the presence of a corepressor (usually the end product of the pathway). The *trp* operon, like several other operons involved in amino acid biosynthesis, is also subject to regulation by attenuation, an antitermination process that depends on the availability of aminoacylated tRNAs.

Functionally related genes in bacteriophages are also organized into operons. During lytic growth, this organization and regulation of genes ensures that they are expressed in an orderly temporal sequence. Temperate phages, such as λ, also have regulatory genes and cis-acting elements that regulate the establishment and maintenance of lysogeny.

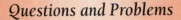

Questions and Problems

Operons and Transcriptional Control

1. Define the following terms and distinguish between members of paired terms:

    aporepressor and corepressor
    inducer and repressor
    inducible and repressible operon
    negative and positive control
    operon

2. The following two models have been proposed to account for the molecular basis of transcription in an inducible operon system.

    Model I   The regulatory gene (R) produces a repressor protein that in its active form combines with the operator-promoter site (o) of the operon to block transcription of the adjacent structural genes (S). The inducer acts by allosterically inactivating the repressor protein.

    Model II   The regulatory gene (R) produces an expressor protein that in its active form combines with the operator-promoter site (o) to promote transcription of the structural gene (S). The inducer acts by allosterically activating the expressor protein.

    Suppose that you have access to three kinds of mutations: R^i is a mutation that destroys the ability of the regulatory gene product to combine with the inducer; R^o is a mutation that destroys the ability of the regulatory gene product to combine with the operator-promoter site; and o^- is a mutation that destroys the ability of the operator-promoter site to combine with the regulatory gene product. For each of the following genotypes, indicate whether enzyme synthesis according to Models I and II will be (a) constitutive and independent of the concentration of inducer, (b) inducible and occurring in substantial amounts only in the presence of the inducer, or (c) repressed, with enzyme synthesis lacking or occurring only in trace amounts regardless of the level of inducer.

Genotype	Phenotype expected for	
	I	II
$R^i\,o^+\,S^+$	Repressed (c)	Repressed (c)
$R^o\,o^+\,S^+$		
$R^+\,o^-\,S^+$		
$R^i\,o^+\,S^+\,/\,R^+\,o^+\,S^+$		
$R^o\,o^+\,S^+\,/\,R^+\,o^+\,S^+$		
$R^+\,o^-\,S^+\,/\,R^+\,o^+\,S^-$		

Control of the Initiation of Transcription

3. In the *lac* operon, what is the probable effect of each of the following on gene expression:
    (a) mutation in the operator region,
    (b) mutation in gene *i* that inactivates the repressor's active site,
    (c) mutation in gene *i* that inactivates the repressor's allosteric site,
    (d) mutation in the promoter region?

4. Are the following *E. coli* cells constitutive or inducible for the transcription of the *z* gene?

(a) $i^+\,o^+\,z^+$ (c) $i^+\,o^c\,z^+$ (e) $i^s\,o^+\,z^+$
(b) $i^c\,o^+\,z^+$ (d) $i^c\,o^c\,z^+$ (f) $i^s\,o^c\,z^+$

5. Give two possible mutant genotypes for the *lac* operon corresponding to each of the following phenotypes:
    (a) The genes of the operon are always transcribed, even in the absence of lactose.
    (b) The genes of the operon are not transcribed (above basal level), even in the presence of lactose.

6. Consider the four mutant genotypes in your answer to question 5. State whether or not *lac* operon mRNA will be made in partially diploid cells in which one chromosome is the mutant genotype and the other is wild-type for the entire operon.

7. Several different haploid and diploid genotypes for the *lac* operon are given below. For each genotype, determine whether β-galactosidase and galactoside permease will be produced under conditions of noninduction (no lactose present) and induction (lactose present). Fill in the table with a + if the enzyme is produced (above basal level) and a − if it is not.

Genotype	β-galactosidase		Permease	
	No lactose	Lactose	No lactose	Lactose
Example $i^+\,p^+\,o^+\,z^+\,y^+$	−	+	−	+
(a) $i^c\,p^+\,o^+\,z^+\,y^+$				
(b) $i^+\,p^-\,o^+\,z^+\,y^+$				
(c) $i^+\,p^+\,o^c\,z^+\,y^-$				
(d) $i^+\,p^+\,o^+\,z^-\,y^-\,/\,i^c\,p^+\,o^+\,z^+\,y^+$				
(e) $i^+\,p^+\,o^c\,z^+\,y^-\,/\,i^+\,p^+\,o^+\,z^-\,y^+$				
(f) $i^+\,p^+\,o^c\,z^-\,y^+\,/\,i^c\,p^+\,o^+\,z^+\,y^-$				
(g) $i^s\,p^+\,o^+\,z^-\,y^+\,/\,i^+\,p^+\,o^+\,z^+\,y^-$				
(h) $i^s\,p^+\,o^+\,z^+\,y^-\,/\,i^c\,p^+\,o^+\,z^-\,y^+$				
(i) $i^c\,p^-\,o^c\,z^+\,y^-\,/\,i^+\,p^+\,o^+\,z^-\,y^+$				
(j) $i^+\,p^+\,o^c\,z^+\,y^+\,/\,i^s\,p^+\,o^+\,z^-\,y^-$				

8. A bacterial species contains three linked loci, *a*, *b*, and *c*. These loci are involved in the synthesis of a group of coordinately controlled enzymes in an inducible operon. In the table below, + means that the enzymes are synthesized, and − means that they are not. One of these loci is a regulator, one is an operator, and one is a structural gene. Which is which?

Genotype	Enzyme synthesis	
	Inducer absent	Inducer present
$a^+\,b^+\,c^+$	−	+
$a^+\,b^+\,c^-$	+	+
$a^-\,b^+\,c^+$	−	−
$a^+\,b^-\,c^+$	+	+
$a^+\,b^-\,c^+\,/\,a^-\,b^+\,c^-$	+	+
$a^-\,b^+\,c^+\,/\,a^+\,b^-\,c^-$	+	+
$a^+\,b^+\,c^-\,/\,a^-\,b^-\,c^+$	−	+

9. Four haploid strains of *E. coli* (1, 2, 3, and 4) carry *lac* operon mutations involving one or more of the loci *i*, *o*, and *z* (the promoter is wild-type). Analysis of β-galactosidase production yields the following results:
   (a) Strains 1 and 4 constitutively synthesize β-galactosidase.
   (b) Strains 2 and 3 do not synthesize β-galactosidase even in the presence of lactose.
   (c) Strain 3 synthesizes a protein that shares a common structure with β-galactosidase, although it has no enzymatic activity. It synthesizes this protein only in the presence of lactose. Strain 2 does not make this protein at all.
   (d) When a donor chromosome fragment with the markers $i^+ o^+ z^-$ is introduced into a strain 1 cell to make a partial diploid, the cell synthesizes β-galactosidase, but only when induced. When this same donor chromosome is introduced into strain 4, that strain still synthesizes β-galactosidase constitutively.
   (e) When the *lac* operon of strain 1 is introduced into strain 2, the resulting partial diploid does not synthesize β-galactosidase even in the presence of lactose.
   Give possible genotypes of the four strains with respect to the loci *i*, *o*, and *z*.

10. Describe the role of cAMP in transcriptional control in *E. coli*.

11. Researchers have isolated *E. coli* mutants that are simultaneously uninducible for the *ara*, *gal*, *lac*, and *mal* operons, even when there is no glucose present. Give an explanation for the nature of these mutations.

12. Consider the positive control of the *lac* operon by the CAP protein. How would a cap^- mutation affect transcription of the operon? Would a cap^- mutation be dominant or recessive to cap^+?

13. Consider both aspects of control (negative and positive) of the *lac* operon. Is enzyme production by each of the following genotypes constitutive, uninducible, or inducible? (Give separate answers for β-galactosidase and permease if necessary.)
   (a) $cap^-\ i^+\ p^+\ o^c\ z^+\ y^+$
   (b) $cap^-\ i^+\ p^+\ o^c\ z^-\ y^-\ /\ cap^+\ i^c\ p^+\ o^+\ z^+\ y^+$
   (c) $cap^-\ i^c\ p^+\ o^+\ z^+\ y^-\ /\ cap^-\ i^c\ p^+\ o^c\ z^-\ y^+$
   (d) $cap^+\ i^+\ p^+\ o^c\ z^+\ y^-\ /\ cap^+\ i^s\ p^+\ o^+\ z^-\ y^+$
   (e) $cap^+\ i^s\ p^+\ o^+\ z^+\ y^-\ /\ cap^-\ i^+\ p^+\ o^+\ z^-\ y^+$

14. The tryptophan operon is a repressible system; it is on in the absence of tryptophan and off in the presence of tryptophan. In the table below, + means that the enzymes coded by the operon are synthesized, and − means that they are not. Let the symbols *a*, *b*, and *c* represent three different loci of this operon. One of these loci is a regulator, one is an operator, and one is a structural gene. Which is which?

Enzyme synthesis

Genotype	Tryptophan absent	Tryptophan present
$a^-\ b^+\ c^+$	+	+
$a^+\ b^-\ c^+$	−	−
$a^+\ b^+\ c^-$	+	+
$a^-\ b^+\ c^+\ /\ a^+\ b^-\ c^-$	+	+
$a^+\ b^-\ c^+\ /\ a^-\ b^+\ c^-$	+	+
$a^+\ b^+\ c^+\ /\ a^-\ b^-\ c^-$	+	−
$a^+\ b^+\ c^-\ /\ a^-\ b^-\ c^+$	+	−

Control of the Termination of Transcription

15. Consider attenuation control of the transcription of the *trp* operon. Predict the effect of each of the following mutations on regulation of the operon.
   (a) The entire leader region is deleted.
   (b) The region that codes for the leader peptide is deleted.
   (c) The leader region does not contain an AUG codon.
   (d) The distance (number of nucleotides) between the leader peptide gene and region 2 is increased.
   (e) The distance between regions 2 and 3 is increased.
   (f) Region 4 is deleted.

16. Why wouldn't you expect to find attenuation control in any eukaryote?

17. In some operons involved in amino acid biosynthesis, the leader region that precedes the structural genes contains codons for several different amino acids. For example, the leader region of an isoleucine operon contains multiple codons for valine and leucine as well as for isoleucine. Explain why codons for different amino acids might be located in the same leader region.

Posttranslational Control

18. The genes that encode the subunits of RNA polymerase and those that encode translational elongation factors are located in the same *E. coli* operons that contain the genes for ribosomal proteins. What might be the significance of such an organization of genes?

19. Some antibiotics can inhibit protein synthesis by binding to ribosomes without increasing the level of ppGpp. Provide an explanation.

Genetic Regulation of Phage Development

20. Both bacteria and bacteriophages organize genes into operons for transcriptional control. Describe the principle differences between regulatory systems in bacteria and phages.

21. Why is the antitermination activity of gpN and gpQ regarded as a form of positive control?

22. What will be the effects on phage λ development of mutations that inactivate the following genes or sites?
   (a) *N* (b) *Q* (c) *nutL* (d) *nutR* (e) *qut* (f) *cro* (g) *cI* (h) *cII* (i) *int* (j) p_{RE} (k) p_{RM} (l) o_R1 (m) o_R3

23. When a certain Hfr cell that is lysogenic for phage λ conjugates with a nonlysogenic F⁻ cell, the F⁻ cell is subsequently lysed, releasing progeny λ phages. Explain what has happened.

CHAPTER

20

Regulation in Eukaryotes I: Transcriptional Activation

Many of the mechanisms of gene control that operate in prokaryotes are also important in eukaryotes. However, the greater complexity of eukaryotic cells requires more diverse regulatory mechanisms. In principle, gene activity in eukaryotes can be regulated at any of four fundamental levels: (1) at transcription (through control of transcriptional initiation, elongation, or termination), (2) posttranscriptionally (for example, through RNA processing, alternative splicing, or mRNA stability), (3) at translation (through control at the initiation, elongation, or termination of translation), and (4) posttranslationally (through modifications to proteins or protein-protein associations). Regulation at all of these levels has been identified, including several examples that we will consider. However, transcriptional regulation is thought to be the primary level of control in eukaryotes, and is therefore the major focus of this chapter.

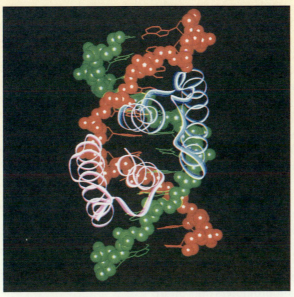

A basic helix-loop-helix protein bound to DNA.

TRANSCRIPTIONAL REGULATION IN EUKARYOTES

As in prokaryotes, the initiation of transcription is a key step in the regulation of many eukaryotic genes. While some aspects of transcriptional control are similar in prokaryotes and eukaryotes, others are not. One major difference is that in eukaryotes, functionally related genes are not organized into operons. Although there are some examples of clusters of related genes (for example, α-globin, β-globin, and histone genes) most related genes are not located together in one region of the chromosome. Even in cases where related genes are clustered, the individual genes within each cluster have their own promoters and are transcribed separately. Thus, the coordinate regulation of eukaryotic genes (when it occurs) nearly always reflects a common process occurring at separate promoters.

Transcriptional regulation in eukaryotes is also more complex. Unlike prokaryotes, which utilize a single RNA polymerase to synthesize all RNAs, eukaryotic cells contain three different nuclear RNA polymerases (see Table 16–2). RNA polymerase I transcribes 18S, 28S, and 5.8S rRNA genes, RNA polymerase II transcribes structural genes and genes for snRNA, and RNA polymerase III transcribes tRNA genes and several other genes encoding small cellular RNAs. The eukaryotic RNA polymerases are large, multimeric enzymes that consist of two large subunits and a number of smaller subunits. With the exception of a carboxyl-terminal tailpiece on the larger subunit of RNA polymerase II, the two major subunits share extensive structural and functional homology with the β and β′ units of *E. coli* RNA polymerase, indicating that all four enzymes are probably derived from a common ancestral protein.

In addition to the protein subunits that make up each functional holoenzyme, the eukaryotic RNA polymerases also require several proteins called **transcription factors** (**TFs**) to activate transcription of a eukaryotic gene. These protein factors bind to regulatory regions of the DNA as part of a **transcription initiation complex** located near the startpoint for RNA synthesis, and they signal the holoenzyme to begin transcription. In the sections that follow, we will limit our discussion to the transcription factors that are associated with RNA polymerase II (see ■ Table 20–1). The experimental study of these TFs was significantly advanced by the isolation of cDNA clones that encode yeast transcription factors. Subsequent isolation of cDNA clones encoding homologous proteins in a wide variety of organisms has made it possible to identify a number of RNA polymerase II TFs and to investigate and identify their key structural features.

RNA polymerase II transcription factors are often divided into two broad (and sometimes overlapping) categories: **general TFs**, which are essential for initiation and respond to specific DNA sequences that are common to most if not all promoters recognized by RNA polymerase II, and **regulatory TFs**, which are not required for initiation but react to specific sequences in certain promoters and/or are active in certain types of cells. We will first describe the general factors and their role in establishing a transcription initiation complex.

TRANSCRIPTION FACTORS: RECOGNITION SEQUENCES

Initiation of transcription of a eukaryotic structural gene requires RNA polymerase II and several general transcription factors. These factors together with the RNA polymerase II holoenzyme form a transcriptional apparatus that can recognize a highly diverse set of promoters and respond to a wide variety of regulatory signals. Since the RNA polymerase II TFs can recognize (nearly) all eukaryotic promoters, one would expect to find common DNA sequences in all promoters of structural genes. One such element was described in Chapter 16—the eight-base pair region that includes the so-called **TATA** or **Hogness box** (▶ Figure 20–1), located approximately 25 base pairs upstream of the transcriptional startpoint in higher eukaryotes and at a more variable position farther upstream in yeast. In most cases, mutations in the TATA box yield RNA products with vari-

■ **TABLE 20–1** Selected RNA polymerase II transcription factors and their functions.

Factor	Function
General factors	
Basal	
TFIID	binds to TATA box in first step of initiation complex assembly
TFIIB	binds to TFIID-TATA complex
TFIIF	recruits RNA polymerase II into initiation complex
TFIIE	binds to DBPolIIF complex
TFIIH	binds to DBPolIIFE complex; helicase
TFIIA	stabilizes assembly of complex (?)
Upstream	
C/EBP	binds to CCAAT box
CP family	bind to CCAAT box
CTF/NF-1 family	bind to CCAAT box
Sp1 family	bind to GC box
OCT-1	binds to octamer sequence
ATF	binds to ATF module (GTGACGT)
H2-TF1	binds to κB module (GGGACTTTCC)
Regulatory factors	
CBF	binds to TATA and CCAAT boxes
CDP	binds to CCAAT box as repressor
NF-E	binds to CCAAT box as repressor
Sp1 family	binds to GC box
OCT-2 (NF-A2/OTF-2)	binds to octamer sequence
NF-κB	binds to κB module
AP-1 (Jun, Fos, GCN4)	binds to TGANTNA (AP-1 element)
CREB	interacts with TFIID

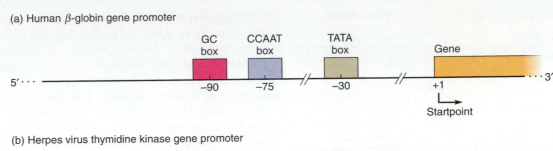

(a) Human β-globin gene promoter

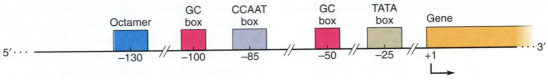

(b) Herpes virus thymidine kinase gene promoter

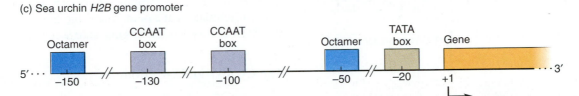

(c) Sea urchin *H2B* gene promoter

▶ **FIGURE 20–1** Promoters of representative eukaryotic structural genes, showing three common conserved sequence modules (TATA, CCAAT, and GC) and a fourth octameric sequence found in some promoters. All modules except TATA can occur in either orientation and at variable positions.

able 5′ ends, indicating that this sequence is usually required for initiation at the proper startpoint.

Most structural genes are also regulated by a second conserved sequence, the **CCAAT** ("cat") **box**, which is located upstream of TATA at approximately position −75. Mutations in this sequence usually affect the frequency of initiation but not its startpoint. A third consensus sequence, the **GC box**, is often found in multiple copies and at variable positions upstream of the startpoint; it also seems to affect the frequency of transcription. The promoter can thus be viewed as a modular structure consisting of short conserved sequences. Although not all promoters have all three modules, the structures in Figure 20–1 seem to represent promoters in a very wide variety of tissues and organisms.

Because native DNA structure is so different from that taken on in vitro, it is difficult to determine exactly how the general TFs and RNA polymerase II use the promoter recognition modules to initiate transcription. Some general TFs appear to be DNA-binding proteins that interact directly with promoter modules. Other TFs seem to function by complexing with each other or with the RNA polymerase holoenzyme through protein-protein interactions. We begin by discussing the general TFs that bind close to the startpoint and direct the binding of RNA polymerase II.

Initiation of Transcription: Events at the TATA Box

Initiation of transcription at the proper site is dependent on a series of **basal** general TFs, designated TFIIA, TFIIB, etc.,

which together with RNA polymerase II constitute the **basal transcription apparatus.** These factors assemble near the startpoint in a prescribed order to form a complex with RNA polymerase II (the initiation complex) that is needed to transcribe any eukaryotic structural gene. ▶ Figure 20–2 diagrams our current understanding of the events leading to the formation of the initiation complex; it is based on experiments using transcription inhibitors, gel filtration, gel electrophoresis, and DNA footprinting.

The first step in forming the initiation complex involves binding of TFIID at the TATA box. TFIID (also called the TATA factor) is a multisubunit protein that contains a TATA-binding protein (TBP), which binds to DNA at the TATA box in the minor groove (unlike most other DNA-binding proteins that bind in the major groove), and a number of other subunits called TAFs (for *TBP associated factors*).

Once TFIID has bound to the TATA box, TFIIA and TFIIB then join the complex, followed by RNA polymerase II in association with TFIIF. Three other factors, TFIIE, TFIIH, and TFIIJ, bind to the complex after RNA polymerase II has bound. TFIIH has been shown to have helicase activity and is thought to be responsible for unwinding the DNA around the startpoint during initiation. As Figure 20–2 shows, most of the basal TFs and the polymerase enzyme are joined near the startpoint by protein-protein interactions. No direct interaction of RNA polymerase II with the TATA box has ever been detected.

Following the formation of the complete preinitiation complex, an energy-dependent step, perhaps catalyzed by

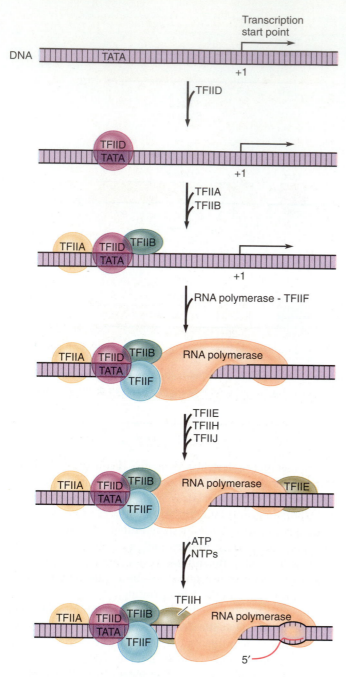

DNA

Transcription start point

+1

TFIID

+1

TFIID
TATA

+1

TFIIA
TFIIB

TFIIA | TFIID | TFIIB
TATA

+1

RNA polymerase - TFIIF

TFIIA | TFIID | TFIIB | RNA polymerase
TATA | TFIIF

TFIIE
TFIIH
TFIIJ

TFIIA | TFIID | TFIIB | RNA polymerase | TFIIE
TATA | TFIIF

ATP
NTPs

TFIIH

TFIIA | TFIID | TFIIB | RNA polymerase
TATA | TFIIF

5′

▶ **FIGURE 20–2** Assembly of the transcription complex for transcription initiation by RNA polymerase II.

recognizes this module. However, the transcription factor itself is not responsible for positioning the startpoint. Experiments in which the yeast analogue of TFIID is inserted into mammalian cells have shown that transcription starts at the normal mammalian startpoint rather than 40 to 120 base pairs downstream of TATA, as is characteristic of yeast transcription. Thus the distance separating the TATA box from the startpoint seems to be important for positioning the start correctly, rather than the binding by TFIID per se. The primary importance of TFIID must be its role in recognizing other TFs and the polymerase enzyme to form the initiation complex. The geometry of the complex then determines where initiation occurs, since the initiation complex stretches a defined distance downstream. This view is supported by experiments in which the startpoint is deleted from the DNA—transcription still occurs at a near normal rate, with initiation at a point that is the same distance from the TATA box as was the original startpoint.

The CCAAT and GC Modules

The CCAAT and GC recognition sequences, located upstream of the TATA box, can dramatically increase (or decrease) the rate of transcription. These modules can therefore be thought of as cis control elements, and research has revealed that they are binding sites for certain **upstream general TFs**. The TFs that bind to the CCAAT and GC modules seem to influence the transcription rate indirectly by interacting with the basal transcription factors of the initiation complex rather than with RNA polymerase itself. Since CCAAT and GC sequences can be as far as 90 or more base pairs upstream from the startpoint, RNA polymerase would not physically be able to extend over such a long DNA region (at least not in vitro, where DNA is not in its native coiled state). Thus indirect influences on RNA polymerase through various protein-protein interactions seem to be a logical explanation for the effects of CCAAT-and GC-binding factors, although how such interactions occur is mainly a matter of speculation.

Many general upstream transcription factors recognize the CCAAT module. These TFs can be grouped into several families, including the **CTF/NF-1** and **CP** families (Table 20–1). In addition to being a target for general transcription factors, the CCAAT box may also be a major target for tissue-specific gene regulation. For example, the histone *H2B* gene in sea urchins is expressed only during spermatogenesis, when the *H2B* promoter is bound by various factors, including the regulatory TF, CBF, at the TATA and CCAAT boxes. CBF is also present in embryonic tissue, but it is not bound to the *H2B* promoter. Instead, the CCAAT displacement protein (CDP) is bound to the *H2B* CCAAT box, preventing CBF from binding. These two proteins apparently compete for the CCAAT region, with CDP having the properties of a eukaryotic repressor. An analogous situation occurs in the case of the human γ-globin gene promoter. An erythroid cell-specific repressing factor, NF-E,

TFIIH, activates the complex to make it capable of initiation. The requirement for ATP hydrolysis is not understood; it may be needed to phosphorylate the C-terminal tailpiece of RNA polymerase II, and thus release the polymerase enzyme from the promoter so that it can enter the elongation phase of transcription.

So far we have developed a picture of transcription in which the TATA box is responsible for initiation at the proper startpoint, and TFIID is the basal general TF that

binds at the CCAAT box and inhibits binding of positive factor CP1, thereby preventing the expression of the γ-globin gene in adults. Mutation at the NF-E binding site in or near the CCAAT box is the cause of a hemoglobin disorder in which fetal hemoglobin persists in the adult. Other CCAAT-binding factors, both positive and negative, are now being identified. Some of these factors are present only in certain types of tissue, reinforcing the idea that this module is involved in both general and regulatory transcriptional events.

The human Sp1 factor is a general upstream TF that binds to the GC box (also called the Sp1 box). The exact mechanism by which this binding activates transcription is unknown (as is also the case for CCAAT-binding factors). Although we have classified Sp1 as a general transcription factor, recent studies suggest that there may be several related forms of Sp1 rather than just one general form. Some of these forms may exert promoter-specific regulatory influences, perhaps by recognizing either the number of copies of the GC box or the flanking sequences.

A few promoters contain a fourth recognition module, an **octamer sequence** of consensus ATGCAAAT. This sequence is recognized by certain upstream transcription factors, including both the general factor OCT-1 and several regulatory factors. For example, OCT-1 binds the octamer sequence of the promoter of the human histone gene *H2B*, which is ubiquitously expressed during the S phase of the cell cycle. A related factor, OCT-2, also binds to the octamer module but only in lymphoid cells, where it activates expression of the antibody genes. It is interesting that two related factors binding to the same sequence element can exert general transcriptional activation on the one hand and regulatory influence on the other. The situation is especially difficult to explain when in vitro studies reveal that OCT-1 has the ability to bind to the promoter of the *H2B* gene in lymphoid cells. The mechanism for this kind of

discrimination is undoubtedly a fundamental aspect of embryonic development, and its explanation is therefore a major challenge to current research.

Regulatory Transcription Factors

In our discussion of general transcription factors, we have seen that the recognition elements can also be recognized by regulatory TFs. More examples of regulatory TFs will be discussed later in this chapter and in Chapter 21. An increasing number of these proteins are being identified as **gene-specific** or **tissue-specific**. Regulatory TFs selectively activate (or repress) certain genes or groups of genes through their interactions with the upstream promoter boxes and with other recognition sequences located outside of the promoter (to be discussed shortly). In the example in ▶ Figure 20–3, TFs interact with regulatory modules of the same gene in different tissues. The primary targets of these TFs and their mechanisms of action are, for the most part, still a mystery. In some cases a certain sequence is recognized only by one specific TF; in other instances several TFs recognize the same sequence. One TF can sometimes recognize more than one DNA sequence, suggesting the presence of multiple DNA-binding domains.

To Sum Up

1. Much less is known about the mechanisms of gene control in eukaryotes than in prokaryotes. Functionally related genes in eukaryotic organisms are not organized into structural units for coordinate transcription—no operons have been discovered in eukaryotes.

2. Transcription of eukaryotic genes is activated by transcription factors. All three eukaryotic RNA polymerases require TFs for the initiation of RNA synthesis. RNA polymerase II transcrip-

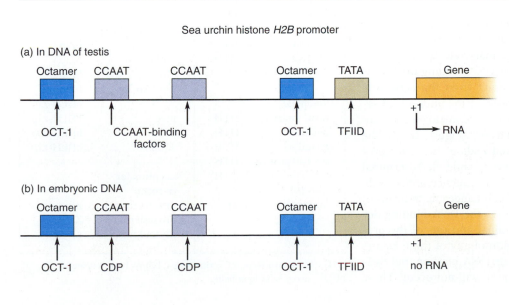

Sea urchin histone *H2B* promoter

(a) In DNA of testis

▶ **FIGURE 20–3** Tissue-specific regulation of transcription. (a) A particular *H2B* gene is transcribed only during spermatogenesis in the sea urchin, when CCAAT-binding factors interact with the promoter. (b) These factors are displaced from the DNA in embryos by the CDP protein (CCAAT-displacement factor), preventing transcription.

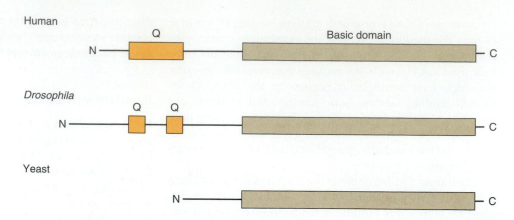

► **FIGURE 20–4** Comparison of TFIID from humans, *Drosophila,* and yeast. Q domains are stretches of uninterrupted glu residues. *Source:* M. G. Peterson, N. Tanese, B. F. Pugh, and R. Tjian, "Functional domains and upstream activation properties of cloned human TATA binding protein," *Science* 248 (1990): 1625–1630. Copyright © 1990 by the AAAS. Used with permission.

Human

Drosophila

Yeast

tion factors include general TFs (which recognize most if not all structural genes) and regulatory TFs (which exhibit promoter-, gene-, and/or cell-type specificity). However, transcription factors cannot be absolutely classified as general or regulatory—some TFs function in both capacities.

3. Several short DNA sequences are recognized by various polymerase II transcription factors. The TATA (Hogness) box seems to be required for initiation of RNA synthesis at the proper startpoint. The CCAAT and GC boxes, which are located upstream of the TATA sequence, are hypothesized to affect the frequency of transcriptional initiation. The promoter is thus a modular structure consisting of regions of short conserved nucleotide sequences.

4. How RNA polymerase II and its transcription factors interact with each other and with promoter modules to initiate transcription is not well understood. The interaction of RNA polymerase II and various basal general TFs with the TATA box generates a transcriptional initiation complex. The rate of transcription is mediated by CCAAT- and GC-bound upstream general TFs, probably through their direct or indirect interactions with factors that are part of this initiation complex.

TRANSCRIPTION FACTORS: STRUCTURAL MOTIFS

Transcription factors have two important functions: DNA binding and the subsequent activation of transcription. Isolating and sequencing the genes that code for these factors clearly demonstrates that the DNA-binding domain is structurally and functionally distinct from the activating domain. For example, a comparison of the amino acid sequences of mammalian TFIID with the analogous *Drosophila* and yeast proteins is shown in ► Figure 20–4. The 180 amino acids at the C-terminal ends of the three proteins are highly (80–90%) conserved, while the N-terminal region has diverged considerably in sequence and length. Since the three factors are fully interchangeable in terms of binding to each other's TATA boxes, the DNA-binding domains must lie within the C-terminal portions of the proteins. In fact, the N-terminal region of any of these factors can be deleted without affecting its DNA binding, but Sp1-dependent activation of transcription will not occur. These

results clearly show that DNA binding and transcriptional activation are independent processes that involve different protein domains of the same or possibly even different transcription factors. We will first discuss structural motifs that are characteristic of DNA-binding domains. Although we will focus on transcription factors, these motifs are fundamental to DNA-binding proteins in general.

DNA-Binding Domains

Even though there are many different kinds of protein-DNA interactions, just a few common protein structural motifs appear to be responsible for them. Four in particular stand out: the **helix-turn-helix, zinc finger, leucine zipper,** and **helix-loop-helix** motifs. ■ Table 20–2 gives examples of regulatory proteins in each of these structural categories.

The helix-turn-helix (HTH) motif was originally pro-

■ **TABLE 20–2** Structural motifs of selected DNA-binding regulatory proteins.

Protein	Motif	Protein	Motif
Yeast		Vertebrates	
GAL4	ZF	OCT-1	HTH
HAP1	ZF	NF-κB	ZF
MATa1	HTH	Sp-1	ZF
MATα1	HTH	C/EPB	bZip
MATα2	HTH	CREB	bZip
SWI5	ZF	Fos	bZip
		Jun (Ap-1)	bZip
Drosophila		AP-4	bHLH-Zip
fushi tarazu	HTH	myc	bHLH-Zip
paired	HTH	OCT-2	bHLH-Zip
engrailed	HTH	TFEB	bHLH-Zip
ANT-C proteins	HTH		
BX-C proteins	HTH	Receptors for	
kruppel	ZF	estrogen	ZF
knirps	ZF	glucocorticoid	ZF
		progesterone	ZF
		thyroid hormone	ZF

HTH = helix-turn-helix; ZF = zinc finger; bZip = basic leucine zipper; bHLH = basic helix-loop-helix.

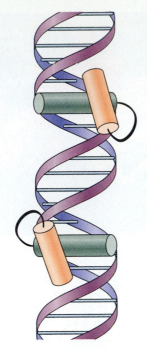

► **FIGURE 20–5** Helix-turn-helix motif of a protein dimer bound to a dyad-symmetric site on DNA. The recognition helix (darker) is situated in the major groove of each half of the binding site, while the other helix helps lock the first into place. (Only the helix-turn-helix region of the protein is shown.)

posed for some phage and bacterial regulatory proteins. Analysis of these protein crystals by X-ray crystallography and NMR reveals two short α-helical regions separated by a β-turn (see "Extensions and Techniques: Experimental Approaches for Studying *lac* Repressor–DNA Binding" in Chapter 19). One α-helix is thought to lie in the major groove of the DNA, making important base pair contacts, while the other lies at an angle across the DNA in a position that helps lock the recognition helix in place (► Figure 20–5). Although the first helix is the most important for recognition, the entire motif participates in binding; nearly all the protein-DNA contacts have been found to contribute to binding specificity.

Zinc finger motifs are characteristic of TFIIIA, Sp1, and a variety of other eukaryotic and viral proteins, including the steroid hormone receptors that we will discuss shortly. Coordinate binding of zinc atoms to properly spaced cysteines and histidines is proposed to give a tetrahedral zinc-Cys/His complex from which amino acid chains loop out in fingerlike projections (► Figure 20–6a). This finger motif is repeated a variable number of times, depending on the protein, with seven or eight amino acids linking adjacent fingers. The Cys and His positions, along with certain others, are highly conserved among different proteins, with classic CH (Cys$_2$His$_2$) fingers having the following consensus sequence:

$$-\text{Cys}-\text{X}_{2-4}-\text{Cys}-\text{X}_3-\text{Phe}-\text{X}_5-\text{Leu}-\text{X}_2-\text{His}-\text{X}_{3-4}-\text{His}-$$

where X represents any unspecified amino acid. The finger "tips" contain a large proportion of polar amino acids capable of forming hydrogen bonds. Several regulatory proteins have been found to possess CC (Cys$_2$Cys$_2$) fingers, a similar motif in which two cysteines replace the two histidines (Figure 20–6b).

The amino acid sequences in the fingers and at their bases are essential for DNA-binding specificity; a single amino acid change can be enough to change the DNA sequence recognized. ► Figure 20–7a shows the secondary structure of a zinc finger protein, based on NMR data. Various studies suggest that the second half of each finger assumes an α-helical structure that lies in the major groove of the DNA, where it forms hydrogen bonds with specific bases. The first half of each finger seems to form a β-sheet located on the outside of the DNA, contacting its backbone. One finger contacts three or four base pairs. In studies of TFIIIA binding to RNA polymerase III promoters, it was found that a C-terminal sequence of 68 amino acids following the last finger does not bind to DNA but is required for activation of transcription, suggesting that it contains the second domain, which is crucial to the assembly of the initiation complex.

The leucine zipper (Zip) and helix-loop-helix (HLH) motifs are found in many proteins whose functional units are dimers. ► Figure 20–8 shows the primary sequences of several Zip transcription factors; leucine occurs at about every seventh position in a series of five repeats in the dimerization domain. This domain forms an amphipathic (containing both strongly polar and strongly nonpolar residues) α-helical structure that puts all the leucines on one face of the helix. When two complementary helices are aligned in parallel, their hydrophobic faces form a "zipper"

► **FIGURE 20–6** Two-dimensional zinc finger motif. (a) CH finger, in which the zinc is bound by two cysteines and two histidines. (b) CC finger, in which all four zinc-binding residues are cysteines. Different transcription factors have different numbers of fingers in a series. (F is phenylalanine, L is leucine.)

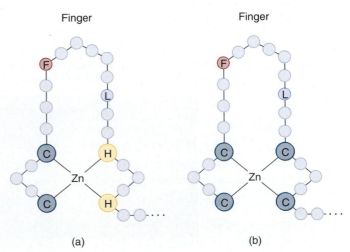

Finger   Finger

(a)   (b)

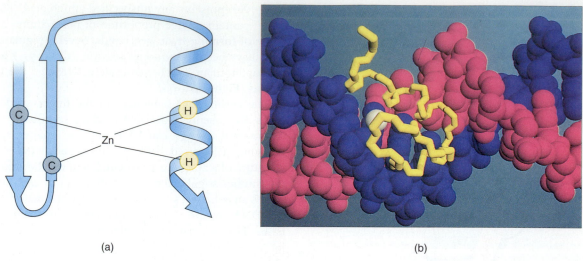

(a)   (b)

▶ **FIGURE 20–7** (a) The secondary structure of a zinc finger. The two cysteines are proposed to lie in a β-sheet hairpin region, while the histidines lie in an α-helical portion of the finger. (b) Computer-generated model of a zinc finger binding to the major groove of DNA. Zinc fingers constitute the DNA-recognition domains of many DNA regulatory proteins and are so-named for their resemblance to fingers projecting from the protein. (Zn finger carbon skeleton = yellow; Zn ion = white).

interlock as the leucines interdigitate with one another (▶ Figure 20–9).

In transcription factors of the basic-Zip (bZip) family, a DNA-binding domain adjacent to the leucine zipper contains a cluster of highly conserved polar amino acids bounded by Lys and Arg residues. Mutations that prevent dimerization also eliminate DNA binding, showing that the zipper arrangement is necessary to generate a functional DNA-binding element. Computer modeling studies suggest that the dimerized helices of the transcription factor wrap around each other to form a coiled coil that stops at the junction of the zipper and the basic domains (▶ Figure 20–10). The basic domains are assumed to also be α-helical; they diverge bilaterally from the junction, forming a fork that binds DNA in the major groove through hydrogen bonds and van der Waals contacts with the nucleotide bases.

The helix-loop-helix is another transcription factor structural motif that involves dimerization. ▶ Figure 20–11a shows that the dimerization motif consists of two amphipathic α-helices separated by a loop region of variable length. As in Zip proteins, interactions between hydrophobic faces of these helices are thought to mediate dimerization. A subclass of these proteins, bHLH, consists of DNA-binding proteins that contain a region of *basic* amino acids located to the N-terminal side of the first helix of each monomer. Again, dimerization is required to generate a functional DNA-binding element. Models of the bHLH structure are shown in Figure 20–11b and c.

A number of proteins have been found to contain a bHLH motif combined with a C-terminal leucine zipper. These proteins therefore contain two separate, presumably independent dimerization domains, raising the interesting pos-

▶ **FIGURE 20–8** The leucine zipper domains of five transcription factors (the one-letter abbreviations of the amino acids are shown). In addition to the leucine zipper domain, all contain a basic domain that is involved in binding the protein to the DNA. *Source:* S. J. Busch and P. Sassone-Corsi, "Dimers, leucine zippers and DNA-binding domains," *Trends in Genetics* 6 (1990): 36–40. Used with permission.

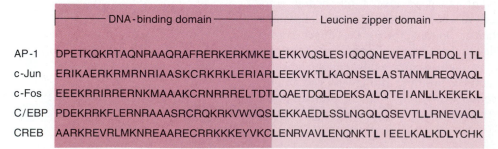

	DNA-binding domain	Leucine zipper domain
AP-1	DPETKQKRTAQNRAAQRAFRERKERKMKE	LEKKVQSLESIQQQNEVEATFLRDQLITL
c-Jun	ERIKAERKRMRNRIAASKCRKRKLERIAR	LEEKVKTLKAQNSELASTANMLREQVAQL
c-Fos	EEEKRRIRRERNKMAAAKCRNRRRELTDTL	QAETDQLEDEKSALQTEIANLLKEKEKL
C/EBP	PDEKRRKFLERNRAAASRCRQKRKVWVQSL	EKKAEDLSSLNGQLQSEVTLLRNEVAQL
CREB	AARKREVRLMKNREAARECRRKKKEYVKC	LENRVAVLENQNKTLIEELKALKDLYCHK

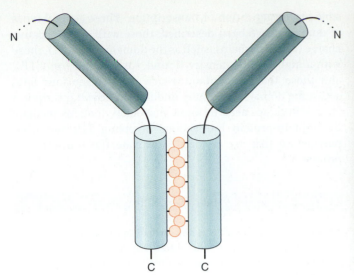

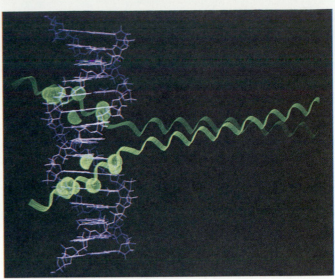

▶ **FIGURE 20–9** Dimerization of two peptides into a leucine zipper structure. The light green cylinders represent parallel α-helical regions containing five leucine repeats. Hydrophobic interactions between protruding leucines (spheres) lock the two helices together. The dark green cylinders represent basic domains that directly contact DNA in a region of dyad symmetry.

▶ **FIGURE 20–10** Computer model of the leucine zipper motif. The supercoiled dimerized α-helices form a forked junction with the basic DNA-binding domains. The basic domains are shown oriented in the major groove of the DNA. *Source:* K. T. O'Neil, R. H. Hoess, and W. F. DeGrado, "Design of DNA-binding peptides based on the leucine zipper motif," *Science* 249 (1990): 774–777. Copyright © 1990 by the AAAS. Used with permission.

▶ **FIGURE 20–11** (a) Amino acid sequences of two DNA-binding proteins that contain a basic domain followed by a helix-loop-helix structure. Both of these proteins also contain a leucine zipper motif. *Source:* S. J. Busch and P. Sassone-Corsi, "Dimers, leucine zippers and DNA-binding domains," *Trends in Genetics* 6 (1990): 36–40. Used with permission. (b) Proposed secondary structure of a bHLH protein, showing possible antiparallel (left) and parallel (right) dimerizations that expose the positively charged DNA-binding domain. (c) Computer-generated model of a bHLH dimer bound to DNA. A view perpendicular to the DNA axis is shown on the right, one looking down the DNA on the left. *Source:* b and c from S. J. Anthony-Cahill et al., "Molecular characterization of the helix-loop-helix peptides," *Science* 225 (1992): 979–986. Copyright © 1992 by the AAAS. Used with permission.

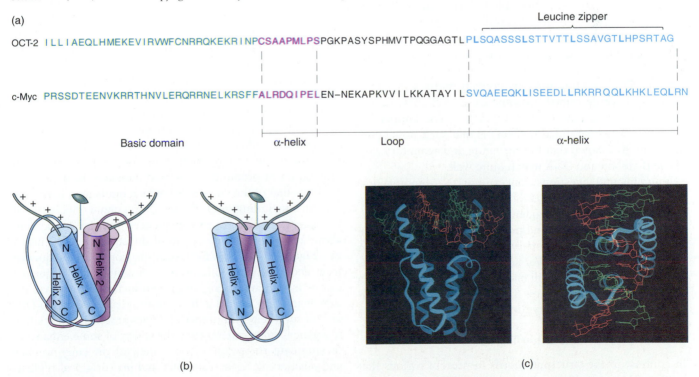

(a)

Leucine zipper

OCT-2 ILLIAEQLHMEKEVIRVWFCNRRQKEKRINPCSAAPMLPSPGKPASYSPHMVTPQGGAGTLPLSQASSSLSTTVTTLSSAVGTLHPSRTAG

c-Myc PRSSDTEENVKRRTHNVLERQRRNELKRSFFALRDQIPELEN–NEKAPKVVILKKATAYILSVQAEEQKLISEEDLLRKRRQQLKHKLEQLRN

Basic domain   α-helix   Loop   α-helix

(b)

(c)

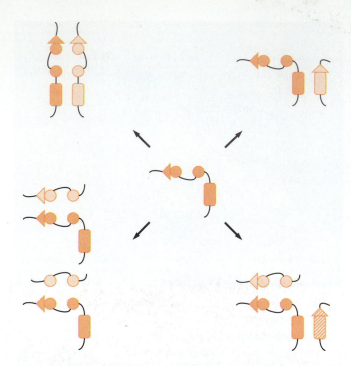

▶ **FIGURE 20–12** Possible bHLH-Zip complexes. At the center is a bHLH-Zip structure. Triangles represent the basic domain, circles joined by connectors the HLH domain, and rectangles the Zip domain. Since the bHLH-Zip structure has two dimerization domains, various combinations of bHLH-Zip-containing structures are possible. *Source:* M. A. Blanar and W. H. Rutter, "Interaction cloning: Identification of a helix-loop-helix zipper protein that interacts with cFos," *Science* 256 (1992): 1014–1018. Copyright © 1992 by the AAAS. Used with permission.

sibility that heterocomplexes involving one or both of these dimerization regions might be able to form. Various combinations of these motifs are illustrated in ▶ Figure 20–12. Such dimerization is an intriguing idea that might help explain the diversity of functions displayed by some transcription factors.

Some DNA-binding proteins contain motifs other than those we have described. In some cases, a β-sheet, rather than an α-helix, may contact the DNA. The antiparallel β-ribbon (a two-stranded β-sheet) is one such possibility; modeling has shown that the curvature and symmetry axes of the β-ribbon and DNA match quite well.

Fundamental questions concerning DNA-binding proteins remain. Does each class of structural motif recognize a particular class of DNA sequences? Are there any rules relating AT/GC base pairs and particular amino acids within motifs? Are there evolutionary relationships among DNA recognition sequences and/or among structural motifs? Answering these questions will make it possible to formulate a more general model of how DNA-binding proteins transcriptionally control eukaryotic gene expression.

Transcriptional Activation Domains

In contrast to the structural motifs in protein regions that bind to DNA, very little is known about TF motifs respon-

sible for the activation of transcription. Three general types of domains have been described: those with a net negative charge (acidic domains, such as the domain on GAL4), those with a high proline density (found, for example, on CTF), and those that are glutamine-rich (two of the four Sp1-activating domains). To fully understand how transcription factors work, we need to know how such domains mediate the protein-protein interactions involving TFs and RNA polymerase that are needed to activate the transcription process.

To Sum Up

1. Transcription factors contain separate structural domains for DNA binding and initiation of transcription. Different TFs that bind to the same promoter module tend to have quite similar DNA-binding domains, while their activating domains differ considerably.
2. Four main structural motifs characterize the DNA-binding domains of most proteins: helix-turn-helix, zinc finger, leucine zipper, and helix-loop-helix. Helix-turn-helix and zinc finger proteins can bind to DNA as monomer units. Leucine zipper and helix-loop-helix motifs are found in proteins that function as dimers; in DNA-binding proteins, a basic DNA-binding domain is located adjacent to the dimerized region of each protein.

ENHANCERS AND SILENCERS

Enhancers are sequence elements that are located outside of promoters and are able to greatly increase the rate of transcription. Gene manipulation experiments have shown that some enhancers can stimulate transcription from any promoter in their vicinity, while others are more specific, enhancing transcription of only certain genes. Most enhancers are tissue-specific—they function only in certain specialized cells.

Enhancers can be viewed as another group of recognition modules. They are located at variable (sometimes quite long) distances from their promoters, in either orientation, and either upstream or downstream of the transcriptional startpoint. In other words, there seem to be fewer restrictions on enhancer placement than on the placement and orientation of promoters. The first enhancer to be discovered was that of the virus SV40, which binds to at least four proteins, including OCT-1. The SV40 enhancer consists of two identical 72-base-pair sequences located in tandem about 200 base pairs upstream of the startpoint, as shown in ▶ Figure 20–13. However, only one copy of the sequence is needed for normal enhancement function. This sequence can be moved anywhere, and it will increase the transcription of any gene placed in its vicinity at least 200-fold. (This lack of specificity in number of copies and position relative to a gene is not a general rule; the ability of some enhancers to stimulate transcription does depend on copy-number and positioning.) Like promoter and *ori* (origin of replication) regions, the SV40 enhancer is extremely sensitive to

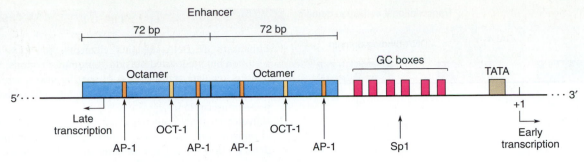

FIGURE 20–13 Transcriptional control regions of the virus SV40. The enhancer is shown in blue. Several general and regulatory transcription factors bind to the control regions, including Sp1, OCT-1 and AP-1, as well as others not shown here.

nuclease digestion, suggesting that enhancer regions consist of a loose chromatin structure that allows binding of regulatory proteins.

Yeast Enhancers

Yeast provide a good experimental system in which to study eukaryotic gene regulation, since their regulatory gene networks have been extensively investigated. Yeast DNA contains sequences called **upstream activator sequences (UAS)**, which are analogous to enhancers and can function in either orientation and at varying distances upstream (but not downstream) of promoters. When a UAS is bound by a regulatory protein, it activates one or more genes downstream. Gene activation is abolished if the UAS is bound by another protein or if a transcriptional termination site is introduced between UAS and the promoter. These results suggest that RNA polymerase slides along the DNA from the enhancer to the promoter; however, this is not the general mechanism by which all enhancers act, since some are located downstream of the startpoint.

An example of a transcriptional regulatory system in yeast which involves the metabolism of galactose is shown in ▶ Figure 20–14. Several structural genes, including *GAL1* and *GAL10,* have promoters that are positively regulated by protein GAL4 and negatively regulated by protein GAL80. GAL4 is a transcription factor that binds to a single UAS (*UAS$_G$*) located between these two structural genes. In the absence of galactose, the GAL80 protein complexes with the DNA-bound GAL4 in such a way that the latter can't activate transcription. When galactose is present, GAL80 changes its association with GAL4 and exposes the GAL4 transcriptional activation domain, thereby activating transcription.

The GAL4 protein has three structural domains corresponding to its three functions (DNA binding, activation of transcription, and binding of GAL80). The DNA-binding domain of GAL4 consists of the 98 N-terminal amino acids, which form a single CC zinc finger motif; if the other domains are deleted, this domain still binds DNA but it cannot activate transcription. The transcription-activating domain consists of residues 147–238 and 768–881; both

regions are needed for full activity. The 30 amino acids at the C-terminus bind GAL80; if they are deleted, transcription is constitutive. The transcriptional activation domain actually overlaps the GAL80-binding domain. The region of DNA that is covered when GAL80 binds to GAL4 may be long enough to block access to the point where the transcription initiation complex is supposed to form.

To determine how GAL4 activates transcription, researchers constructed a hybrid gene that produces a protein consisting of a LexA DNA-binding domain and a GAL4 tran-

---

▶ FIGURE 20–14 (a) Metabolism of galactose by yeast involves structural genes *GAL1* and *GAL10* and the *UAS$_G$* enhancer-like regulatory region. (b) In the absence of galactose, the GAL80 protein binds to the GAL4 protein, preventing the GAL4 transcriptional activation domain (black region) from functioning. (c) When galactose is present, an as yet unidentified induction signal is transmitted to the GAL4-GAL80 complex, exposing the activation domain of GAL4 and allowing transcription to occur.

(a) Structural genes and regulator region

(b) No galactose

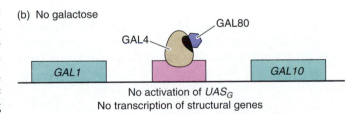

(c) Galactose

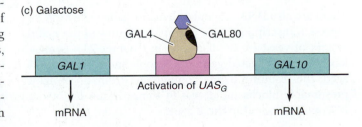

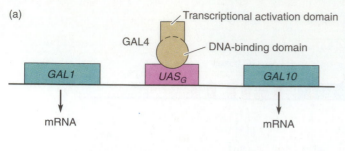

(a)

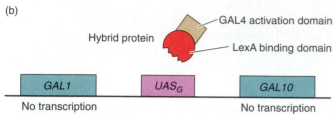

(b)

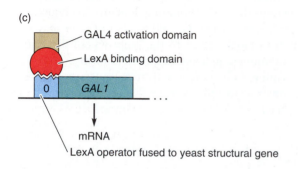

(c)

▶ FIGURE 20–15 A "swap" of yeast and bacterial DNA-binding domains. Yeast GAL4 will activate transcription from any gene as long as the DNA-binding domain is able to recognize a specific control region on the DNA.

scriptional activation domain. (LexA is a prokaryotic repressor protein that binds to several operators and blocks transcription.) As ▶ Figure 20–15 shows, the hybrid protein cannot activate a gene with a UAS_G, but it can activate a gene with a LexA operator. This result demonstrates that the main functional role of the DNA-binding domain is to locate the regulatory factor at a specific DNA module. Once the factor is bound to the DNA, it performs a separate function—transactivation. The transcriptional activation domain itself acts nonspecifically. It activates transcription from any promoter that the binding domain places it close to, suggesting that the responsibility for recognition specificity lies solely with the DNA-binding domain.

Silencers

There are also DNA sequence elements called silencers that repress transcription. One example is found in the mating system of yeast (to be discussed in Chapter 21). Some silencers cause affected genes to become more tightly packaged; the resulting highly condensed chromatin structure presumably blocks the access of transcription factors and RNA polymerase II to these genes.

RESPONSE ELEMENTS AND HORMONES AS ACTIVATING SIGNALS

Certain enhancer sequences have been found to activate transcription in response to particular extracellular signals, such as light, hormones, and environmental stress. These DNA sequences are collectively referred to as **response elements** (RE). One example of a system that responds to an external signal is the **heat shock system.** In a wide variety of prokaryotes and eukaryotes, a drastic increase in temperature alters the transcription of a group of genes called the **heat shock genes.** These genes are regulated by heat shock response elements located upstream of the genes' promoters. Heat causes a protein called the heat shock transcription factor (HSTF) to become phosphorylated. In this state it binds to the heat shock response elements and initiates transcription of the genes. The function of these genes is only partially understood; they seem to code for proteins that somehow help the cell or organism survive the environmental stress. Some heat shock proteins are chaperones (see Chapter 6), which mediate the folding of other proteins into their active forms. The chaperones that are produced during heat shock may stabilize the proteins in a cell and prevent their denaturation by heat.

The Steroid Hormones

Response elements are best known for their interactions of certain hormones, including the steroid hormones (glucocorticoids, mineralcorticoids, estrogen, progesterone, and androgens) and thyroid hormones, and for their role in differentiative processes such as those involving vitamin D_3 and retinoic acid (a derivative of vitamin A). These substances all belong to a superfamily of ligands that initiate transcription of specific target genes by binding to specific receptor proteins and thus initiating a sequence of activation events that leads to the binding of transcription factors at the appropriate RE of the DNA.

As an example of this process, we will consider the action of glucocorticoids—adrenal hormones that increase blood sugar level and exert an anti-inflammatory effect. The cytoplasmic glucocorticoid receptor (GR) is bound to another protein that masks its activity, as illustrated in ▶ Figure 20–16. When the hormone binds to this complex, it

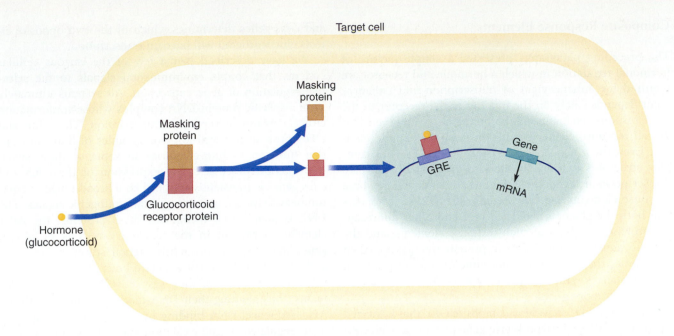

▶ **FIGURE 20−16** Mechanism of action of the steroid hormone glucocorticoid. The hormone interacts with a masked receptor protein, disassociating the receptor from the masking protein. The hormone-receptor complex then enters the nucleus and binds to a glucocorticoid response element (GRE) on the DNA, thereby activating transcription of a target gene.

dissociates the receptor from the masking protein, and the ligand-receptor complex then enters the nucleus and binds to glucocorticoid response elements (GRE) that are located near each gene that responds to the hormone. Binding to the GRE activates the nearby promoter, and a transcription initiation complex is formed. Experiments with hybrid genes have shown that any gene artificially linked to a GRE will be activated by glucocorticoid, again demonstrating the generality of transcriptional initiation once binding to a specific DNA target sequence has occurred. Similar mechanisms explain the influence of other steroid hormones, thyroid hormones, vitamin D_3, and retinoic acid on transcription. Each interacts with a specific receptor protein (sometimes located in the nucleus) that binds to the DNA at an appropriate response element.

Similar promoter and enhancer recognition sequences are involved in responses to various extracellular stimuli in viral, plant, and animal systems. For example, similar DNA sequences (the so-called Hex sequence and the G, ABRE, and E boxes) are involved in the responses of plants to certain hormones, to light, and to wounds and in the responses of viruses and mammals to a variety of signals. These sequences suggest that there may be a highly conserved family of eukaryotic proteins that recognize similar sequences in the regulatory regions of various genes involved in a number of different response pathways.

Molecular analysis of the genes encoding several hormone receptors has provided some information on their amino acid sequences. Each receptor has a modular structure (see ▶ Figure 20−17), in which a DNA-binding domain is flanked by separate independently functioning domains for ligand binding and transcriptional activation. The

DNA-binding motif is highly conserved among members of the ligand superfamily and in most cases consists of two CC zinc fingers. The ligand-binding domains of these receptors are also somewhat similar in amino acid sequence, but they do not have predictable structural motifs.

▶ **FIGURE 20−17** Domain organization of various proteins belonging to a superfamily of related receptors. The DNA-binding domains of these receptors are highly conserved, being 42−94% similar in amino acid sequence (depending on the proteins being compared). The ligand-binding domains are less similar to one another. *Source:* B. Lewin, *Genes V* (Oxford University Press, 1994), p. 895. Used with permission.

Receptor for:

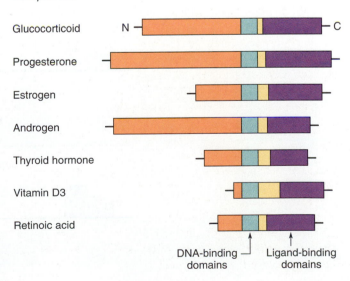

Composite Response Elements

The preceding paragraphs presented a simple picture of hormonal regulation in which a hormone and receptor are required for enhancement of transcription and enhancement depends solely on the binding of the receptor to the DNA response element. It is becoming clear, however, that not all response elements function in such a simple manner. Even the glucocorticoid system has GREs with complex structures that enable them to differentially regulate transcription in various genetic and cellular contexts. These response elements are called **composite GREs**. It appears that both the glucocorticoid receptor and other nonreceptor proteins can bind at such response elements, and the specific nature of their protein-protein interactions then determines the effect of the hormone on transcription. In some instances, glucocorticoids repress (rather than activate) one or more genes.

As an example, consider a composite GRE that is located upstream of the mouse proliferin gene promoter. This GRE can provide either positive or negative regulation, depending on the relative activities of two nonreceptor proteins, c-Jun and c-Fos. The possible results are shown in ▶ Figure 20–18. Glucocorticoid exerts positive control of transcription when c-Jun alone is present; however, in the presence of c-Jun and a relatively high level of c-Fos, it exerts negative control. Thus, the level of expression of the *c-jun*

and *c-fos* genes determines which of the two opposite effects glucocorticoid will have on transcription.

We have described just a few of the various cellular systems that couple environmental signals to the selective regulation of gene expression. All systems ultimately form a specific protein-DNA complex whose conformation either activates or represses transcription. The c-Jun and c-Fos proteins are well-known examples of inducible TFs that function as intermediaries in some of these signal transduction processes. The expression of the *c-jun* and *c-fos* genes is transiently induced by a broad range of environmental stimuli, including certain mitogenic signals. The DNA sequence that recognizes these proteins, the AP-1 element, is present in the regulatory regions of several genes that respond to environmental stimuli. The c-Fos and c-Jun proteins form a heterodimer, the AP-1 factor, through interaction of their respective leucine zipper domains, thus aligning their DNA-binding domains with the AP-1 sequence. DNA binding by the heterodimer appears to be regulated by the oxidation state of a single cysteine residue in each of the DNA-binding domains. When an as yet unidentified nuclear protein reduces these Cys residues, binding is stimulated. Regulation of the Fos and Jun transcription factors thus occurs at both the transcriptional and translational levels, providing a cascade of possible regulatory effects on the structural genes under their control.

▶ FIGURE 20–18 Responses of a composite GRE to multiple regulatory signals. For simplicity, the hormone-receptor complex is diagrammed just as the hormone. *Source:* M. T. Diamond, J. N. Miner, S. K. Yoshinaga and K. R. Yamemoto, "Transcription factor interactions: Selectors for a positive or negative regulation from a single DNA element," *Science* 249 (1990): 1266–1271. Copyright © 1990 by the AAAS. Used with permission.

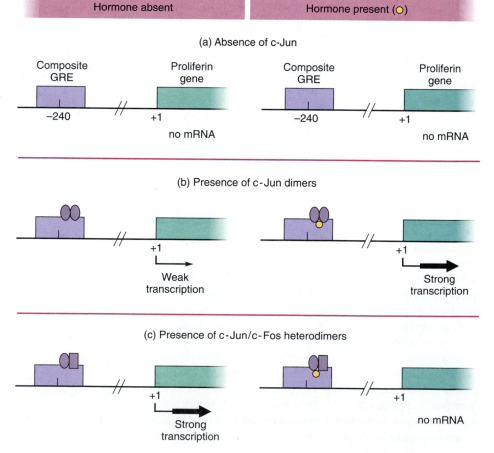

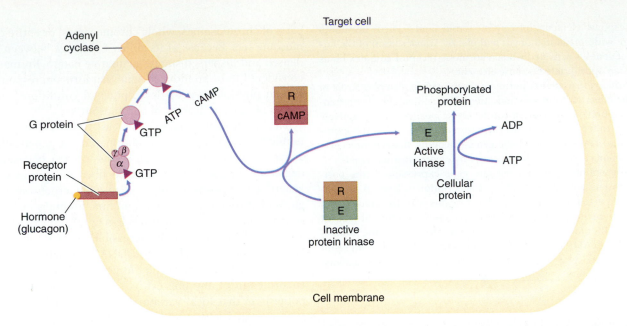

FIGURE 20–19 Action of a polypeptide hormone in exerting posttranslational control over the activity of a protein kinase. The hormone binds to a receptor on the target cell. The hormone-receptor complex causes the signal-transducing G protein to exchange GDP for GTP, activating the G protein. The GTP-bound α subunit of the G protein then dissociates and in turn activates a membrane-bound enzyme, adenyl cyclase, which is responsible for the conversion of ATP to cAMP. The cAMP then serves as a second messenger by allosterically interacting with a protein kinase. Protein kinases are composed of a regulatory subunit (R) and an enzymatically active subunit (E). Interaction with cAMP frees the enzymatically active subunit, allowing it to phosphorylate various cellular proteins and thereby control their activity.

Polypeptide Hormones

In contrast to the steroid hormones, which regulate gene expression directly at the level of transcription, the **polypeptide hormones** regulate gene activity indirectly, primarily through the posttranslational modification of proteins. The polypeptide hormones include insulin and glucagon, which are involved in the control of glucose metabolism, and somatotropin (growth hormone), which affects the rate of skeletal growth and gain in body weight. These hormones act through a signal-transducing pathway involving a **G (guanine nucleotide-binding) protein**. G proteins are general signal-transducing cell membrane phosphoproteins that react with transmembrane proteins to transduce signals from outside a cell. G protein–mediated transmembrane systems have been identified in organisms as divergent as yeast and humans, and they transduce a variety of signals, including hormones, neurotransmitters, odorants, pheromones, and light.

The effect of a polypeptide hormone (glucagon, for example) is initiated when it binds to a receptor protein on the outside of the target cell membrane (▶ Figure 20–19). The receptor activates the G protein by causing it to exchange bound GDP for GTP. The GTP-bound subunit of the G protein dissociates and diffuses along the membrane, binding to and activating adenyl cyclase. This activated enzyme then converts ATP to cAMP, whose major role is to allosterically activate a protein kinase called the cyclic-AMP-dependent protein kinase (A-kinase). (Protein kinases are regulatory enzymes that alter the activity of various proteins in the cell by phosphorylating them. Several different proteins are affected by these phosphorylation reactions, including the enzymes glycogen synthetase and phosphorylase, which are involved in glycogen metabolism. The proteins of membranes, microtubules, and ribosomes are also affected by protein kinases, as are the histones.) cAMP combines with a regulatory subunit (R) of the A-kinase protein. The cAMP-R complex then dissociates from the catalytic subunit (E, the active part of the enzyme) which is now free to participate in the phosphorylation reactions. Polypeptide hormones thus regulate genes by controlling the activity of their protein products.

An increase in the cAMP level can also activate the transcription of genes controlled by promoters that contain the cAMP response element (CRE). This regulatory region is recognized by the cAMP-response element binding protein (CREB), which is a transcription factor that is activated when phosphorylated by the A-kinase.

To Sum Up

1. Response elements are enhancer sequences that regulate transcription in response to extracellular signals. For example, heat shock or stress serves as an external signal for the phosphorylation of a specific transcription factor that binds to heat shock response elements to stimulate transcription of the heat shock genes.

2. Response elements for certain steroid (and related) hormones are widespread in animals and plants. Transcription of the target genes is activated when a hormone-receptor complex binds to an appropriate response element.

3. Proteins that are not hormone receptors can also play a role in the effect of hormones and other extracellular signals on transcription. Transcription can be either positively or negatively controlled, depending on the nature and amount of these proteins. Transcription of the genes encoding these nonreceptor proteins is sometimes influenced by environmental signals.

4. Polypeptide hormones regulate gene expression at the post-translational, rather than the transcriptional, level. A complicated signal-transducing pathway begins with the binding of a hormone to its membrane-bound receptor and the subsequent activation of a G protein; subsequent steps involve the production of cAMP and activation of protein kinases. The end result is the phosphorylation of target proteins to alter their activity.

 ## CHROMATIN STRUCTURE AND TRANSCRIPTION

Some recent ideas on the structure of chromatin were discussed in Chapter 9. The basic interphase 30 nm chromatin fiber is thought to consist of a 10 nm nucleosome fiber wound into a solenoid structure. This solenoid fiber coils even more tightly into a highly condensed state during cell division. The solenoid is thus a relatively loosely coiled or dispersed state compared to mitotic chromatin. However, variation still occurs in both the degree and pattern of chromatin coiling in this dispersed condition. In fact, variation in the extent to which chromatin is coiled is believed to be a mechanism in the regulation of transcription in eukaryotes.

Degree of Coiling of Chromatin

Interphase chromatin fibers show a greater degree of coiling in certain parts of chromosomes, especially in regions adjacent to the centromere and at the tips, and these regions tend to be stained more heavily by dyes that bind to DNA. These areas of tightly packed, dark-staining chromatin are referred to as **heterochromatin**; the more loosely packed and less intensely staining chromatin in other regions is called **euchromatin**. Euchromatin cycles between a dispersed state (interphase) and a condensed state (cell division), while the degree of condensation of heterochromatin remains relatively unchanged throughout the cell cycle.

Geneticists recognize two different types of heterochromatin: constitutive and facultative. **Constitutive heterochromatin**, which is commonly found in chromosome centromeres and tips, remains in a permanently condensed, dark-staining condition throughout the life cycle of the organism. In contrast, **facultative heterochromatin** may appear as heterochromatin at some stages of the life cycle and as euchromatin at others. For example, cells that are highly specialized have much more heterochromatin than do embryonic cells. Facultative heterochromatin may involve

parts of chromosomes, whole chromosomes, or entire chromosome sets. Because of its ability to change between condensed and dispersed states, facultative heterochromatin is considered to be an induced state that forms in response to certain physiological or developmental conditions, rather than a permanent feature of chromosomes.

Heterochromatin makes up at least 10 to 20% of total genomic DNA, and like mitotic chromatin, it is genetically inactive. A negative correlation seems to exist between the degree of coiling of the chromatin fiber and the extent of genetic activity—the highly condensed heterochromatic regions of chromosomes contain no functioning genes. Transcriptionally active chromatin is less tightly coiled, presumably to make the DNA more accessible to RNA polymerase. However, not all euchromatic DNA is genetically active. Only a small proportion of the genes (about 7%) in the euchromatin of any one cell is ever transcribed. One hypothesis is that local variations in euchromatin structure are associated with gene activity—active regions are less coiled than nontranscribed sections. Similarly, the chromatin structure of promoters and enhancers is relatively loose, while the DNA of silencers is very tightly coiled, as we have already noted.

One well-known example of the association between transcriptional activity and the level of euchromatin coiling is gene regulation by certain steroid hormones in fruit flies. Studies of *Drosophila* and other dipterans reveal that ecdysone, the molting hormone in insects, induces specific puffing patterns of the chromosomes in the larval stage. **Chromosome puffs** are regions that are enlarged or ballooned, as shown in ▶ Figure 20–20. They appear and regress in characteristic tissue-specific patterns during larval development. Since the puffed areas are actively engaged in synthesis of RNA, it would seem that they indicate a loosening of the chromosome coils in transcriptionally active regions.

Experiments reveal that certain puffs called "early" puffs

▶ **FIGURE 20–20** Photomicrograph of a polytene chromosome of the fly *Chironomus tentans*. Three puffs, also known as Balbiani rings, characterize this chromosome.

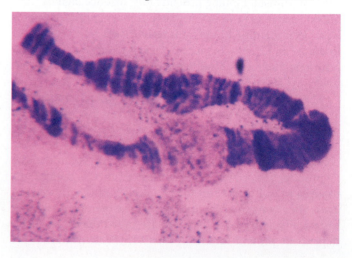

Imprinting and DNA Methylation

According to basic genetic principles, mammals inherit two complete sets of autosomal genes from their parents and express these genes according to the dictates of dominance/recessiveness, gene interactions, and environmental factors. However, researchers are now finding examples of a phenomenon known as **genomic imprinting**, in which the expression of a gene varies according to its maternal or paternal origin. That is, there seems to be some molecular mechanism that can leave a gene with an "imprint" from its mother or father, and this imprint determines whether that gene will be expressed.

Studies with mice have revealed that a small number of genes that act in embryonic development are imprinted. In some cases, the maternally inherited gene is imprinted, and in other cases, it is the paternally inherited gene. Thus, imprinting probably occurs during gametogenesis, since this is when the DNA of the maternal and paternal genomes can be separately subjected to modification.

The current model of imprinting proposes that a DNA sequence element known as an **imprinting box** is connected with the gene locus and that modification of this sequence is caused by an **imprinting factor**. According to this model, any gene that is influenced by an imprinting box is subject to parent-specific modification by the imprinting factor during ga-

metogenesis. It is not yet known why only some genes seem to be subject to imprinting; the imprinted genes that have been identified so far do not have common functions.

Because of its suspected association with lack of gene activity, DNA methylation has been proposed to be the mechanism by which imprinting occurs. Investigations of the methylation profiles of imprinted genes support this idea, although the findings are not clear-cut. In several mouse genes, for example, parent-specific methylation is present in the mature gamete and is maintained in the diploid embryo and adult as long as imprinted gene expression persists. However, in some cases methylation is associated with repression of a gene, and in other cases it is associated with expression. One explanation might be that transcription of the gene depends on methylation-sensitive repressors or methylation-sensitive activators, so that it is not the structural gene itself that is being imprinted but rather a regulatory gene that controls its expression.

Even if DNA methylation does turn out to be the imprinting factor, several questions remain. What is the nature of the imprinting box? What is the connection between methylation of DNA in general and imprinting? What function does imprinting play in regulating gene expression? Are genes that act in embryonic development the

only ones that are affected? What, if anything, do imprinted genes have in common? What is the evolutionary significance of imprinting? You may recall that bacteria protect themselves against foreign DNA by cleaving it with restriction endonucleases, and that their own DNA is protected by methylation of the restriction enzyme target sites. Could the imprinting box be a nucleotide sequence that for some reason is recognized as foreign DNA? Is this evidence of an evolutionary link between a host defense mechanism and the regulation of gene activity? Finding the answers to these questions will reveal a great deal about the vast complexities of gene regulation and its significance.

Mammals inherit a complete set of autosomal genes from each parent.

---

develop rapidly in response to molting hormone. These early puffs form even when protein synthesis is inhibited by drugs that block translation. Several hours later, the so-called "late" puffs appear, but they form only if protein synthesis can occur. Proteins encoded by the loci in the early puffs are needed to induce transcription at the sites where the later puffs develop. The genetic regions that are transcribed in the early puffs thus behave as regulatory genes.

The Structure of Active Chromatin

Studies comparing transcriptionally active chromatin to inactive chromatin have revealed interesting chemical modifications, including phosphorylation and methylation of histone and nonhistone proteins as well as methylation of DNA. **DNA methylation** is highly correlated with the transcriptionally inactive state in some organisms. In vertebrates, for example, CpG dinucleotides are occasionally found to be modified by methylation at the C5 position of cytosine:

$$5' \quad \frac{^mCpG}{GpC^m} \quad 5'$$

Although some CpG sites are either methylated or unmethylated in all tissues of an organism, others vary from one cell type to another. In those cases, the transcriptional control regions of genes that are expressed in a given cell type tend to be undermethylated relative to the same sequences in tissues where the genes never become active. The most striking evidence that demethylation may induce gene activity comes from studies using 5-azacytidine, an analogue of cytosine whose incorporation into DNA inhibits methylation (▶ Figure 20–21). It has been found that 5-azaC induces differentiation in cultured mouse cells. There thus appears to be a correlation between undermethylation and gene expression in tissue-specific regulation of gene activity. One outstanding question regarding methylation at specific genes is how the particular CpG sequences that are to be demethylated are determined.

The addition of a methyl group to carbon 5 of cytosine may seem like a small modification that is not likely to induce gross conformational changes in the DNA. This indeed may be the case; however, even subtle, local changes in the conformation of the DNA sugar-phosphate backbone might be important in the recognition of regulatory DNA sequences and the binding of regulatory proteins. Methylation may also alter the DNA structure more dramatically than might be anticipated. The C5 methyl group of ^mC protrudes into the major groove of B-form DNA, destabilizing

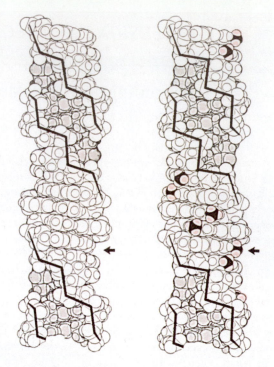

▶ **FIGURE 20–22** The crystal structures of Z-DNA in its unmethylated (left) and methylated (right) forms. The methyl group carbons are shown in black. The hydrophobic methyl groups stabilize the Z structure by filling in depressions (arrow) in the unmethylated form. *Source:* A. H. Wang, *Cold Spring Harbor Sympos. Quant. Biology* 47 (1982): 41. Used with permission.

the molecule, and some researchers have hypothesized that this destabilization aids in switching the DNA to the Z form. (Recall from Chapter 7 that Z form DNA has been observed in vitro for poly GC/CG DNA.) Methylating the cytosines in these polymers stabilizes the Z form because the hydrophobic methyl groups fill in depressions in the molecule (▶ Figure 20–22). Whether local regions of DNA take on a Z form in vivo is not known. Negatively supercoiled DNA is more likely than relaxed DNA to take on the Z form, and negative supercoiling also assists in strand unwinding, which is necessary for transcription. However, if the Z form were relevant to gene activation through its relationship to supercoiling, one would expect methylation to be associated with enhanced gene activity rather than with inactivity.

There is still no direct experimental evidence that adding methyl groups inactivates a gene, although 80% of the methylated cytosines in mammals are found in nucleosomes that contain histone H1, which indicates a tightly coiled chromatin structure. In vitro experiments that methylate DNA and then study its activity show inhibition of gene expression in some cases but not in others. (One complicating factor in these studies is that only a few of the possible DNA methylation sites may be critical for control, and these sites have not been identified.) Furthermore, there are many known cases in which genes are expressed even when their promoters are extensively methylated. It seems obvious that methylation is not a general regulatory mecha-

▶ **FIGURE 20–21** (a) Cytosine can be methylated without affecting its base-pairing properties. The methyl group is added by specific enzymes, the DNA methylases, after DNA replication. (b) The synthetic nucleotide 5-azacytosine cannot be methylated because position 5 is blocked.

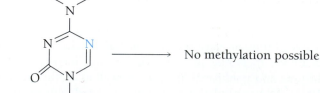

(a) cystosine → Methylation → 5-methyl cytosine

(b) 5-azacytosine → No methylation possible

nism; it appears to be important only for some genes, and it is not known what distinguishes these genes from others. Even in cases where methylation is associated with gene regulation, most evidence indicates that the DNA regions in question acquire their methylated state only after events involving regulatory proteins and/or chromatin structure have taken place. This suggests that DNA methylation reinforces regulatory decisions once they have been made, rather than actually making the decisions.

Transcriptionally active genes also exhibit a heightened sensitivity to nucleases. Experiments using extremely low concentrations of nucleases have identified particular DNA regions called **hypersensitive sites** that are as much as 100 times more susceptible to nuclease action than are the transcriptionally active genes themselves. Every active gene studied so far has at least one hypersensitive site upstream in its 5′-flanking region; inactive counterparts of these genes (in cells in which these genes are not expressed) do not exhibit hypersensitive regions. Hypersensitivity to nuclease action is therefore a distinctive feature of the regulatory regions (promoters/enhancers) of genes whose chromatin is in a state suitable for transcription.

Hypersensitivity to nucleases can be explained by the hypothesis that histones and transcription factors are mutually exclusive occupants of certain promoter regions. Investigations with the general transcription factor TFIID support this idea. TFIID forms a transcription initiation complex in vitro only if it is added to DNA prior to the addition of histones; it is unable to function on nucleosomal DNA. Thus some regulatory proteins apparently prevent nucleosomes from forming in transcriptional initiation regions (or perhaps displace nucleosomes that have already formed). A nucleosome-free region of DNA presumably is much less compacted than a region containing nucleosomes and is thus highly sensitive to nuclease action. Because RNA polymerase requires TFIID for assembly of an initiation complex, it cannot initiate RNA synthesis on nucleosomal DNA in vitro. The absence of transcriptional activation on nucleosomal DNA thus supports the idea that histones are general repressors of gene activity.

Since RNA polymerases are larger than nucleosomes and bind tightly to DNA lengths of approximately 50 base pairs, it is difficult to imagine the elongation phase of transcription occurring without unwinding the DNA from the nucleosomes. However, experiments designed to determine whether transcription displaces nucleosomes have given conflicting results. The DNA of heavily transcribed rRNA genes appears to be almost completely unwound from nucleosomes. In contrast, pictures of the transcription of SV40 viral DNA show the normal nucleosome "beaded" structure. In most cases, genes that are being transcribed contain the same frequency of nucleosomes as nontranscribed DNA regions. Thus, either nucleosomes are retained during transcription or they are transiently displaced and immediately reform once RNA polymerase passes by.

X Chromosome Inactivation

Somatic cells of a female mammal contain only one completely functioning X chromosome. The other X chromosome becomes heterochromatic and remains condensed and, for the most part, genetically inactive throughout the cell cycle. The condensed X chromosome, which consists of facultative heterochromatin, forms a dark-staining structure referred to as the **sex chromatin body** or **Barr body** (after its discoverer). Since the number of X chromosomes differs between the cells of males and females, the inactivation of one of the X chromosomes in females equalizes the dosage of functional X chromatin in the sexes.

A simple staining technique enables geneticists to detect the presence of X chromatin in cell nuclei at interphase. Barr body analysis is performed on cells from a buccal smear, a procedure in which epithelial cells are scraped from the lining of the mouth. The cells are spread on a microscope slide, fixed in ethanol, and stained with a basic dye. The Barr body appears as a small, compact structure (1 μm or less in diameter), located close to or touching the inner surface of the nuclear envelope, as shown in ▶ Figure 20–23. Barr body analysis is extremely valuable for identifying abnormalities in the number of X chromosomes.

▶ **FIGURE 20–23** (a) The Barr body in the nucleus of an epithelial cell from a human female. When epithelial cells from the buccal mucosa are spread on a microscope slide, fixed in ethanol, and stained, the Barr body appears as a dark-stained structure closely associated with the inner surface of the nuclear envelope. (b) The nucleus of an epithelial cell from a human male. This cell does not have a Barr body. (c) An abnormal XXX female nucleus, with two Barr bodies.

(a)

(b)

(c)

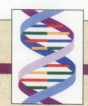

Why Two X Chromosomes?

In humans, the XO chromosome abnormality is almost always an embryonic lethal condition. Nearly all XO embryos are spontaneously aborted, and the few that develop to live birth have Turner syndrome (which includes sterility; see Chapter 12). At first glance, the severe consequences of XO may seem surprising—if one X chromosome in each embryonic cell is to be inactivated, why is having only one X chromosome lethal? Two possible reasons can be given.

First, inactivation occurs several days into gestation. There may be certain key X-linked genes whose expression in double dose is needed during this very early embryonic period. Second, recent studies of gene expression have identified several genes on the human X chromosome that are not inactivated. These genes may be required for normal development and sexual maturation. One such gene, located near the tip of the short arm of the X chromosome, encodes steroid sulfatase. A deficiency of steroid sulfatase results in a condition called ichthyosis (from the Greek *ichtyhos,* meaning fish), which is characterized by large dark brown scales covering much of the body, the result of a failure to remove cholesterol sulfate from the outer layer of skin. Heterozygous females may show mild symptoms of scaling.

It has been difficult to experimentally determine what embryonic roles are played by X-linked genes that are not inactivated, because other animal models are not applicable to humans. For example, X-chromosome inactivation in mice is complete; all genes on the inactivated chromosome are repressed, including the mouse homologs of the human genes that remain active. As you would expect, XO mice are normal and fertile.

All X chromosomes except one become inactivated, so individuals have one less Barr body than they have X chromosomes in each cell. Therefore, if the staining technique reveals the presence of more than one Barr body in a female or a Barr body in a male, it indicates an abnormal number of X chromosomes.

The inactivation of an X chromosome is referred to as **Lyonization,** after Mary Lyon, an early investigator of the phenomena. Today, more than 30 years after her work, the inactivation process remains, for the most part, a mystery. It has been studied in humans, marsupials, monotremes, and mice; these organisms show some similarities and some differences in how inactivation takes place and is maintained. In all organisms that have been studied, both X chromosomes are euchromatic in the early development of the female embryo. At a certain point in development, however, one of the X chromosomes in each embryonic cell becomes condensed, through a two-step process: the **initiation of inactivation** by events that occur at the X-inactivation center and **spreading of inactivation** from that center to other regions of the X chromosome. A gene that maps to the human and mouse X-inactivation centers has recently been cloned; this gene is expressed from the inactive X chromosome only, not from the active one. While the exact function of this gene has not yet been elucidated, the spreading of inactivation is presumably one of its roles.

A third step—stabilization of the inactivated state—occurs in most organisms. Stabilization means that the X chromosome that was originally inactivated will show up as the sex chromatin body in all of the descendants of the cell. In organisms (such as marsupials) where the stabilization step does not take place, individual genes on the X chromosome can become reactivated.

5-azaC has been found to activate the expression of some genes on an inactive X chromosome, suggesting that methylation of DNA may be connected to its genetic inactivity. However, methylation is more likely to be important in the stabilization step than in the initiation or spreading of inactivation. This conclusion is based on the discovery that differential methylation between the active and inactive X chromosomes is found in humans but not in marsupials.

Example 20.1

Hemophilia is an X-linked recessive disorder in humans. Heterozygous females are not affected with hemophilia, but they can show a wide range of clotting times. Suggest an explanation for the longer clotting times observed in some heterozygous females.

Solution: The variation in clotting times in heterozygous females presumably reflects which X chromosome is inactivated more often in stem cell lines. Since the X chromosome that is inactivated in each cell line originated from a random choice of an X in each early embryonic cell, heterozygous women will have different proportions of cells with the h^+ allele on the active X chromosome. The more cells having an h^+ allele on the active X, the faster the clotting time.

Explain what X chromosome inactivation has to do with the calico (tortoiseshell) coloration observed in female cats.

The calico coloration, shown by a cat, that is heterozygous for a sex-linked pair of alleles encoding yellow and black.

To Sum Up

1. Differential chromatin coiling can result in regulation of eukaryotic gene expression at the level of transcription. Highly condensed chromatin (heterochromatin) is genetically inactive. Euchromatin, which is more loosely coiled and becomes condensed only during cell division, contains potentially active genes. Most chromatin is euchromatic; however, only a small portion is ever transcribed in any one cell.

2. The degree of methylation of chromatin and its sensitivity to nucleases also seem to be associated with its transcriptional state. A lack of methylation in CpG dinucleotide sequences is correlated with transcriptional activity. Transcriptionally active DNA is also more sensitive to nucleases than transcriptionally inactive regions. In fact, promoters and other regulatory elements located near active genes are unusually hypersensitive to nuclease action. Whether this hypersensitivity reflects an absence of histones in regulatory sequences has not yet been determined.

3. One of the X chromosomes in each cell of a female mammal becomes heterochromatic early in embryological development and remains, for the most part, genetically inactive in all descendant cells, at all remaining stages of the life cycle. Heterochromatic X chromosomes can be microscopically detected in interphase cells as dark-staining Barr bodies.

 ## Chapter Summary

All three eukaryotic RNA polymerases require a number of transcription factors for initiation of transcription. These factors can be divided into two broad, overlapping categories—general and regulatory transcription factors. Regulatory TFs control transcription from particular genes, promoters, and/or tissues, while general TFs are required for transcriptional initiation at all promoters and respond to sequences that are common to promoters recognized by a particular RNA polymerase. Sequences recognized by RNA polymerase II general TFs include the TATA (Hogness), CCAAT, and GC boxes. General TFs can be further divided into basal factors, which interact with the TATA box along with RNA polymerase II and form complexes that initiate transcription, and upstream TFs, which interact with CCAAT and GC boxes to influence the frequency of initiation of transcription.

Transcription factors contain separate domains for DNA binding and transcriptional activation. Most DNA binding domains exhibit either the helix-turn-helix, zinc finger, leucine zipper, or helix-loop-helix structural motif.

Recognition sequences located outside of promoters also influence transcription. Enhancer sequences increase the rate of transcription, while silencers repress transcription. Enhancers that respond to extracellular signals are called response elements; they are best known for their interactions with certain hormones that regulate gene activity at the transcriptional level.

In contrast, polypeptide hormones regulate gene expression at the posttranslational level through an indirect cellular signaling process that ultimately phosphorylates target proteins coded by the gene and thereby alters the activity of those proteins.

Chromatin coiling is inversely related to genetic activity. Euchromatin cycles between a dispersed state (interphase) and a condensed state (cell division); however, even in the dispersed state, variations in the degree of coiling occur. Euchromatic regions that are more loosely coiled are genetically active; these regions also tend to be undermethylated and have a heightened sensitivity to nucleases. Heterochromatin is highly coiled throughout the cell cycle and is genetically inactive. The sex chromatin or Barr body is an X chromosome that has become heterochromatic; the inactivation of one X chromosome in a female mammal achieves dosage compensation.

 ## Questions and Problems

Transcriptional Regulation in Eukaryotes

1. Compare and contrast the mechanisms of transcriptional regulation in prokaryotes and eukaryotes.

2. List and describe the various levels at which gene activity can be regulated. Speculate on the advantages and disadvantages of regulation at each level.

Transcription Factors: Recognition Sequences

3. Is the binding of transcription factors to their DNA modules sufficient to initiate transcription? What else is necessary? Describe the current view of transcriptional initiation in eukaryotes.

4. The idea of transferring genes into the cells of eukaryotic organisms (for example, in gene therapy) was discussed in Chap-

ter 18. Would you expect an introduced gene to be functional if it is randomly inserted into a chromosome, or would it have to be targeted to a specific location? Explain your answer.

5. Compare and contrast eukaryotic transcription factors and prokaryotic sigma factors. Describe the features they have in common and how they differ.

6. Suppose that eukaryotic microbes are isolated from the depths of the ocean; they have no TATA boxes but do have another conserved consensus, GTGT, in about the same position. How could you experimentally determine whether the GTGT sequence is a transcriptional regulatory module?

Transcription Factors: Structural Motifs

7. Describe four major types of structural motifs found in DNA-binding proteins. Describe the role of each motif in binding transcription factors to DNA.

8. What experimental evidence supports the idea that transcription factors bind DNA and activate transcription through independent domains?

9. How many kinds of dimers could be formed using the bHLH-Zip protein shown in the center of Figure 20–12, if the HLH and Zip domains act independently in the formation of heterocomplexes with other HLH-Zip, HLH, and Zip proteins?

Enhancers and Silencers

10. In what ways are enhancers and promoters similar? In what ways are they different?

11. Hypothesize how an enhancer sequence located thousands of base pairs away from a promoter is able to stimulate transcription of the associated gene, even if its orientation is reversed.

12. A *GAL4^c* mutant of yeast exhibits constitutive galactose fermentation in haploid cells. What is the most likely molecular nature of this mutation?

13. You want to determine what specific regions of a particular DNA-binding protein are able to bind to an enhancer that the complete protein recognizes. You have several cloned partial cDNAs representing regions toward the 3′ end of the gene that encodes the protein:

Peptides are made by in vitro transcription and translation of these partial cDNAs, and these peptides are mixed with radioactively labeled DNA containing the enhancer region. Binding of peptide to DNA is judged by the resulting retardation of the movement of the labeled DNA upon electrophoresis:

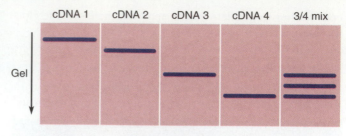

(a) Why are the bands at different positions on the gel?
(b) Where in the protein is the DNA-binding domain?
(c) When cDNAs 3 and 4 are mixed together prior to their transcription and translation, three bands appear on the gel, as shown above. Interpret this result.

Response Elements and Hormones as Activating Signals

14. Describe how steroid hormones control the transcription of their target genes. How important are the hormone receptors? How important are the response elements?

15. What functional domains and structural motifs are common to the different steroid hormone receptors?

16. What is the fundamental difference in the ways the steroid and the polypeptide hormones regulate gene activity? Describe the signal-transducing pathway that is involved in regulation by a hormone such as glucagon.

Chromatin Structure and Transcription

17. Describe the kinds of evidence that link the degree of chromatin coiling to its transcriptional activity.

18. What are chromosome puffs? What is the relationship between early and late puffs?

19. What sequences in DNA are methylated? What effect does methylation seem to have on gene activity?

20. The transcription of some eukaryotic genes can be induced by treatment with 5-azacytidine. Suppose that tissue culture cells are exposed to labeled dUTP for five minutes in the presence and in the absence of 5-azaC. Radioactivity in RNA is 1,800 counts per minute in the culture without 5-azaC but 27,500 in the presence of 5-azaC. Explain these results.

21. The inactivation of an X chromosome is strictly random, so each cell in an early embryo has an equal chance of having the maternally or the paternally derived X chromosome become heterochromatic. Suppose that X chromosome inactivation occurs in a certain mammal at the eight-cell stage of embryonic development. (a) What is the probability that the maternally derived X is inactivated in exactly half the cells? (b) What is the chance that the female that develops from this embryo is not mosaic (has either all of its maternally derived or all of its paternally derived X chromosomes inactivated)? (c) Suggest a way to determine the embryonic stage when X inactivation occurs in a certain species of mammal.

CHAPTER 21

Regulation in Eukaryotes II: Development and Cancer

All eukaryotic cells constantly respond to a complex environment in which gene expression is mediated by ligands such as hormones (see Chapter 20), growth factors, neurotransmitters, morphogens, and other substances that bind to cell-surface transmembrane receptors. The ligand-receptor information is transported across the membrane and relayed to the nucleus, where transcription of selected genes is activated or repressed.

Intracellular mechanisms that relay information to the nucleus can be very complex and may involve a series of signal-transducing cytoplasmic events. Once a signal has reached the nucleus and selectively activated one or more genes, these genes are often involved in the regulation of still other genes. The result is a cascade of regulatory events that lead to particular physiological responses to the original ligand. Researchers are

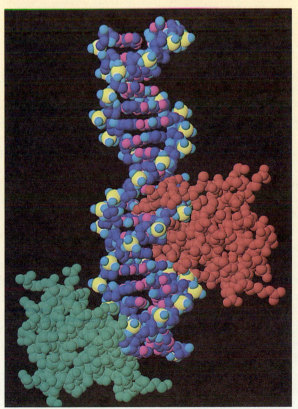

Computer-generated model of a homeodomain protein complexed with DNA. These regulatory proteins control the development of body architecture in the embryo.

now trying to identify the specific signaling mechanisms that lead to gene expression and clarify the role of gene products in events further along the pathway.

To illustrate what is currently known about transcriptional responses to external and internal signals, we will now describe three fundamental regulatory processes in eukaryotes. First, we will focus on developmental genetics and the connection between genes and embryonic development. Much of the current research on gene regulation in eukaryotes is based on studies of embryology and developmental biology. Next, we will describe specific types of developmental processes in which gene activity is associated with rearrangements of the organism's genome in certain types of specialized cells. Finally, the relationship between genes and cancer will be discussed, with emphasis on the role of genes that normally control cell division and are also involved in oncogenesis (cancer). Exciting experimental work over the last decade has revealed that common cellular signaling mechanisms underlie both normal development and the processes of cell growth and cancer.

▶ FIGURE 21–1 In *Antennapedia* mutants, antennal structures are transformed into legs. Occasionally, the entire antenna becomes leglike, as shown, but usually only a portion of the antenna is transformed.

CONTROL ELEMENTS AFFECTING DEVELOPMENT

The development of an adult organism from a zygote involves long-term, usually irreversible changes in gene expression that are directed by a large set of genetic regulatory events. The precise developmental program of an organism reflects the exact sequence, timing, and spatial pattern of expression of its genes. Initially, the zygote is **totipotent** (uncommitted to any particular fate). As cell division proceeds, cells gradually commit themselves irreversibly to a particular fate, a phenomenon known as **determination** or commitment. Cells often commit themselves to a particular fate long before they express any of the associated differentiating characteristics. Although this time lag would seem to make it difficult to associate the activation or inactivation of a particular gene with a particular determination event, the advent of genetic manipulation technologies (see Chapter 18) has made it possible to identify key developmental genes that code for regulatory proteins and the sites of action of these proteins. Complex, multistep signal-transducing processes involving cis- and trans-acting regulatory factors are being discovered as the genetic picture of development begins to be revealed.

The crucial question in understanding development is how eukaryotic organisms coordinately control the genes involved in developmental processes. Key regulatory genes have been identified through mutational analysis; mutation in these genes can cause abnormal development of major body parts, reflecting the disruption of entire sets of genes. For example, in the *Antp* mutation in *Drosophila* (see ▶ Figure 21–1), legs develop in place of antenna.

The mechanism that underlies the formation of the body plan of insects has been a major focus of investigation in developmental biology. Genetic analysis of *Drosophila* has been extremely valuable in identifying genes that regulate

this process, because its genetics is among the best known of any eukaryotic organism. The development of the segmentation pattern in *Drosophila* is thus a useful illustration of what has been learned about the genetic basis of embryogenesis. In order to understand how segmentation and the body plan arise, and the context in which regulatory genes act, we begin with a very brief description of fruit fly development.

Overview of *Drosophila* Development

Following fertilization, the nucleus of a *Drosophila* zygote undergoes its first nine divisions without corresponding cytoplasmic division. It is not until the blastoderm stage (after 12–13 division cycles) that cell membranes finally separate the 6000 or so nuclei into individual cells. At cycle 14, the level of transcription increases dramatically, the cells become motile, and gastrulation begins. The characteristic pattern of body segments becomes apparent by about 10 hours of development (▶ Figure 21–2). After the egg hatches (at about 24 hours), this segmentation pattern persists throughout larval development and is also apparent in the adult stage following metamorphosis.

During the larval stage, the chromosomes of most cells become polytenized through repeated DNA replication without separation of the daughter chromatids (see Chapter 12). One consequence of polytenization is that larval cells can't divide—they just grow larger as development continues. At the end of larval life, metamorphosis forms the adult fly. Polytene larval tissues are broken down and discarded during metamorphosis. Thus adult structures are

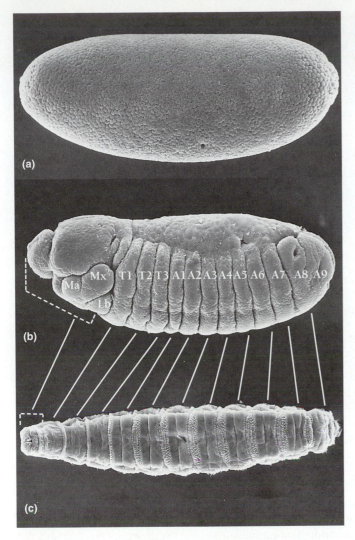

first event divides the cells into two groups, one of which produces the anterior portion of the wing and most of the dorsal thorax, while the other forms the posterior part of the wing and the rest of the dorsal thorax. Another determination event then separates the wing disk from the leg disk, retaining the anterior-posterior boundaries created earlier. The third determination event distinguishes the dorsal and ventral surfaces of the wing, the fourth separates the central elements that form the wing proper from the cells that generate the wing blade, and so on. Thus, as determination occurs, the wing disk becomes progressively subdivided into discrete, nonoverlapping compartments whose cells do not mingle. These determinative events are controlled by regulatory genes that are essential for informing cells of their proper spatial arrangement.

Genetics of Pattern Formation in *Drosophila*

Determination, which is the major event of early development, proceeds by establishing gradients in the distribution of cytoplasmic **effector substances** (determinants) along the anterior-posterior (A/P) and dorsal-ventral (D/V) axes of the fertilized egg. At least some of these determinants are the products of **maternal genes**, the first major class of genes involved in embryonic development. These genes are expressed by the mother during oogenesis, and the maternal RNA or protein is transported into the egg. Between the time of fertilization and the blastoderm stage, various maternal and embryonic gene products become differentially distributed throughout the embryo's cytoplasm in very precise concentration gradients.

The *bicoid* (*bcd*) **gene** is an example of a maternal gene whose product is differentially distributed in the embryo. Bicoid mRNA forms a concentration gradient within the oocyte and the early cleavage embryo, with most *bcd* mRNA near the anterior end. Mothers mutant in the *bcd* gene produce embryos lacking heads and thoraxes, regardless of the paternal or embryo genotype; this defect can be overcome by injection of cytoplasm from the anterior pole of a wild-type embryo. Thus, the maternal *bcd* gene codes for a product that acts in the determination of anterior body parts. The analogous gene product specifying the posterior parts is produced by the maternal gene *nanos* (*nos*).

▶ **FIGURE 21–2** Scanning electron micrographs of *Drosophila* embryos, showing the development of segmentation along the anterior-posterior axis. (a) Three-hour-old embryo. (b) Ten-hour-old embryo. (c) Newly hatched larva. (Ma, Mx, and Lb are the mandibular, maxillary, and labial segments, respectively; T1–T3 are the thoracic segments; A1–A9 are abdominal segments.)

not derived from remodeling larval structures. Instead they develop from groups of undifferentiated cells called **imaginal disks**, which have been set aside during embryogenesis. Although present throughout larval life, the imaginal disks play no role in larval development.

Even though imaginal disks consist of undifferentiated cells, each disk is "determined" during the larval stage, so that each one forms a particular element of the adult structure during metamorphosis. Consider the imaginal disk that gives rise to the wing, for example. The cells that will form the disks that give rise to the mesothorax begin to undergo a series of successive determination events just after blastoderm formation. These events create compartments that will eventually form specific parts of adult structures. The

Example 21.1

The 3′ untranslated regions of the *bcd* and *nos* mRNAs and their poly A tails are responsible for localizing these mRNAs to their respective poles. Suppose you perform an experiment that exchanges the 3′ end and poly A tail of *bcd* mRNA with the 3′ end and poly A tail of *nos* mRNA. You then transform doubly mutant *bcd/bcd nos/nos Drosophila* with the altered mRNAs. If one of these transformed female flies is bred, what will be the effect on A/P axis development in her embryos?

Solution: The only bicoid and nanos mRNAs that this doubly mutant female fly can give to her eggs are the altered mRNAs that she was transformed with. Since the specificity for A/P determination resides in the 3' untranslated and poly A tail regions of the *bcd* and *nos* mRNAs, the A/P axes will be reversed in her embryos.

Follow-Up Problem 21.1

If drugs that inhibit RNA synthesis, such as actinomycin D, are injected into fertilized frog eggs, will early cellular divisions and protein synthesis still occur? Explain your answer.

◆

Once maternal gene products have determined the basic polarity and spatial organization of the embryo, determinants in the early embryo allow it to form segments. Some of these determinants act as transcription factors that activate expression of the **segmentation** genes. These genes establish the segmentation pattern by partitioning the A/P axis of the egg into a defined number of segments with specific positions and polarities. Thus, a longitudinally distributed set of transcription factors (including both activating and repressing determinants) acting in early development controls not only gene expression but also the physical boundaries of that expression along the A/P axis.

Segmentation genes appear to be activated in a defined order; expression of one group is required for expression of the next, as successively smaller regions of the embryo are determined. ■ Table 21–1 lists the three groups of segmen-

■ **TABLE 21–1** *Drosophila* segmentation genes.

Group	Gene
Gap	*giant (gt)*
	hunchback (hb)
	knirps (kni)
	Kruppel (Kr)
Pair rule	*barrel (brr)*
	even-skipped (eve)
	fushi tarazu (ftz)
	hairy (h)
	odd-paired (opa)
	odd-skipped (odd)
	paired (prd)
	runt (run)
Segmentation polarity	*armadillo (arm)*
	cubitus interruptus[D] *(ci*[D]*)*
	engrailed (en)
	fused (fu)
	hedgehog (hh)
	paxh (pat)
	wingless (wg)

Source: S. F. Gilbert, *Developmental Biology* (Sunderland, MA: Sinauer Associates, 1988): 637. Used with permission.

tation genes and some representatives in each group. The segmentation genes express themselves in early embryonic units called **parasegments**, which eventually give way to the larval segmentation pattern. ▶ Figure 21–3 shows the rela-

▶ **FIGURE 21–3** Relationship between early embryonic parasegments and larval segmentation. P and A represent the posterior and anterior compartments of the larval segments. The segmentation pattern genes express themselves in regions based on parasegmental divisions. Because the parasegments are one compartment out of phase with the segmentation pattern, the parasegments start with their posterior segmental compartments toward the head. (Ma, Mx, and Lb are the mandibular, maxillary, and labial segments, respectively; T1–T3 are the thoracic segments; A1–A9 are abdominal segments.) *Source:* S. F. Gilbert, *Developmental Biology* (Sunderland, MA: Sinauer Associates, 1988): 638. Used with permission.

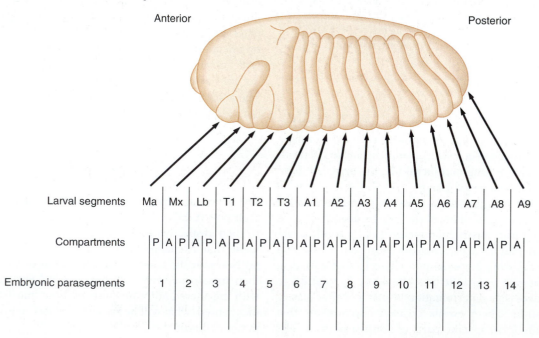

tionship of the developmental territories in the larva to those in the early embryo. The parasegments are one compartment out of phase with the segmental pattern; each parasegment is made up of the posterior and anterior compartments of adjacent segments.

The first set of segmentation genes, the **gap genes**, encode transcription factors. The expression of these genes interprets the early positional information and divides the embryo into nonoverlapping regional domains of cells. Different genes are expressed in these different domains, as shown by ▶ Figures 21–4a and 21–5a. The anterior gap gene *hunchback* (*hb*) is directly activated by *bicoid*, while the posterior gap gene *knirps* (*kni*) is indirectly activated (through elimination of a repressor) by *nanos*. *Kruppel* (*Kr*) is required for the formation of thoracic and anterior abdominal segments. Its expression is apparently regulated in a complex fashion involving competition between *bcd*-mediated activation and *kni*-mediated repression of a *Kr*

▶ **FIGURE 21–4** Probable sequence of expression of the *Drosophila* segmentation genes. Brackets indicate the domains of action of the various genes. *Adapted from:* S. F. Gilbert, *Developmental Biology* (Sunderland, MA: Sinauer Associates, 1988): 639.

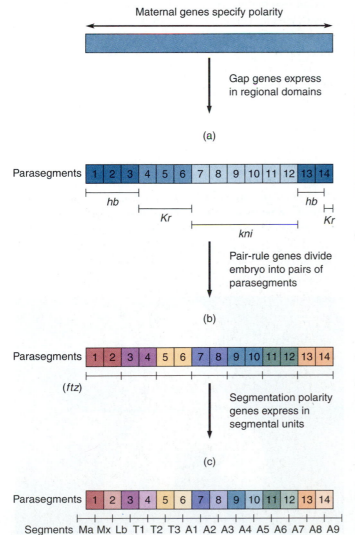

(a)

(b)

(c)

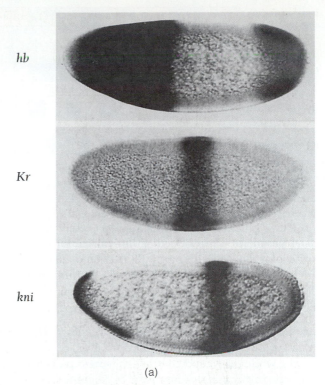

(a)

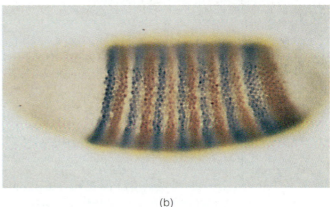

(b)

▶ **FIGURE 21–5** (a) Early blastoderm expression of three gap gene proteins: *hb* (top), *Kr* (middle), and *kni* (bottom). (b) In situ hybridization reveals the expression patterns of proteins from the pair rule genes *eve* and *ftz* during late blastoderm. Pair rule expression yields seven "zebra stripes," one for every two segments.

regulatory region. (The *kni* gene has recently been placed in the steroid hormone receptor superfamily.) Gap mutations eliminate whole regions of contiguous segments, because of the resulting inability of those regions to react to maternally-supplied information.

These events are followed by expression of the **pair rule genes**, which divide the embryo into a series of regions that each contain a pair of parasegments (Figures 21–4b and 21–5b). Like *Kr*, these genes are also regulated by interplay between transcriptional activators and repressors. It has recently been discovered, for example, that the *eve* promoter/enhancer contains binding sites for the products of the

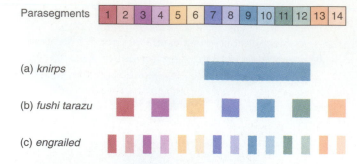

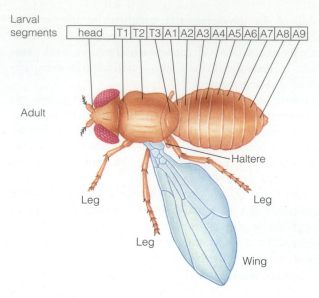

FIGURE 21–6 Three types of larval segmentation mutations. (a) The *knirps* gene normally expresses itself in parasegments 7 through 12; these are deleted in the mutant. (b) The *ftz* gene product is normally found in every other parasegment; these are deleted in the *ftz* mutant. (c) The *engrailed* mutant lacks the posterior compartment of each parasegment, which is replaced by the mirror image of the anterior compartment.

FIGURE 21–7 Comparison of larval and adult segmentation in *Drosophila,* showing the appendages that originate from each thoracic segment. *Adapted from:* S. F. Gilbert, *Developmental Biology* (Sunderland, MA: Sinauer Associates, 1988): 636. Used with permission.

maternal *bcd* gene and gap genes *hb, Kr,* and *gt.* The maternal morphogen and the Hb protein are transcriptional activators, while the Kr and Gt proteins repress transcription of *eve.* Pair rule mutations fuse pairs of adjacent segments, resulting in half the normal number of segments, each twice the usual width.

The transcription factors encoded by the pair rule genes regulate the expression of the **segmentation polarity genes** in each segment. The result is the determination of units that correspond to the segments that will appear along the larva and adult (Figure 21–4c). Segmentation polarity mutants have a normal number of segments, but the posterior compartment of each one is lost and replaced by the mirror image of the anterior compartment. Each segment thus consists of mirror images of two anterior halves. ▶ Figure 21–6 illustrates all three types of segmentation gene mutations.

Once the segmentation pattern is established, a third major class of developmental genes, the **homeotic** (structure-determining) **genes**, determines what body part develops from each segment during metamorphosis. The homeotic genes are activated by the gap genes, which thus execute two separate functions as transcriptional activators. External adult structures are derived from the imaginal disks, as described earlier. ▶ Figure 21–7 shows the normal structures that develop from the larval segments. While mutations in the segmentation genes result in the deletion of body parts, mutations in homeotic genes cause one part of the body to develop a structure appropriate for a different segment. For example, in one of the bithorax mutants, the third thoracic segment becomes transformed into the tissue type of the second segment. The fly then has four wings, because a second pair of functional wings replaces the normal halteres (▶ Figure 21–8). Drastic phenotypic changes like these suggest that the homeotic genes are "master genes" that control the activities of the many genes that dictate development of major body structures.

Most (but not all) homeotic genes in *Drosophila* are found in two major gene complexes located on separate regions of chromosome III. ▶ Figure 21–9 shows these gene complexes and the structures they define. Each complex contains several loci, and each of these genes is expressed in a restricted region of the developing fly and specifies the identity of that region. The **Antennapedia-complex** (*ANT-C*) consists of the *Lab* (*Labial*), *Pb* (*Proboscapedia*), *Dfd* (*Deformed*), *Scr* (*Sex-comb reduced*), and *Antp* (*Antennapedia*) loci, which are responsible for the development of anterior structures from the head through the anterior compartment of the third thoracic segment. If the entire complex is deleted, the labial and maxillary head segments and the T2 and T3 segments are all transformed into T1. The **Bithorax complex** (*BX-C*) defines the structures that develop from

FIGURE 21–8 A bithorax mutant in which the third thoracic segment is changed into the tissue type of the second segment, causing the development of a second pair of functional wings.

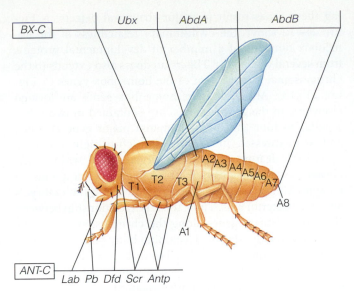

▶ **FIGURE 21–9** The two homeotic gene complexes of *Drosophila* and their functions. The Antennapedia complex (*ANT-C*) controls the development of anterior structures as far back as the wings and second pair of legs. The third pair of legs and abdominal structures are defined by the genes of the Bithorax complex (*BX-C*). *Adapted from:* S. F. Gilbert, *Developmental Biology* (Sunderland, MA: Sinauer Associates, 1988): 642. Used with permission.

the middle of the thoracic segments through the eighth abdominal segment. It contains the *Ubx* (*Ultrabithorax*), *AbdA*, and *AbdB* genes. Deletion of *BX-C* yields normal heads and first and second thoracic segments, but T3 and the abdominal segments are all transformed into T2.

Since there are eight abdominal segments, plus T3, but only three *BX-C* genes, researchers have searched for other regions of the bithorax complex that might be involved in sequentially activating the development of segments T2 through A8. Several cis-regulatory sites that probably function as enhancers have been identified; they include *abx, bx, bxd, pbx,* and *iab-2* through *iab-9*. ▶ Figure 21–10 shows the correspondence between these genes and regulatory sites and the abdominal segments.

The activity of the homeotic genes is also influenced by trans-acting factors. For example, the segmentation genes, which divide the fly into segments, also act early in embryogenesis to limit expression of the *BX-C* (and *ANT-C*) genes to broad domains along the A/P axis. Various homeotic genes also regulate each other's expression through enhancer elements that contain multiple binding sites for regulatory proteins. Furthermore, some homeotic genes contain multiple promoters and transcriptional startpoints, and some RNA transcripts are differentially spliced in different tissues. Thus, even though there is only a handful of homeotic genes, cellular variations in the levels of homeotic proteins, in the combinations of proteins produced, and in the proteins resulting from differential splicing and transcription startpoints provide the diversity needed for tissue-specific developmental differences.

Homeobox Genes

More than 20 known *Drosophila* genes, including some of the gradient-establishing, segmentation, and homeotic genes, are now known to contain a sequence called the **homeo box**. The homeo box consists of approximately 180 nucleotide pairs located near the 3′ end of the genes.

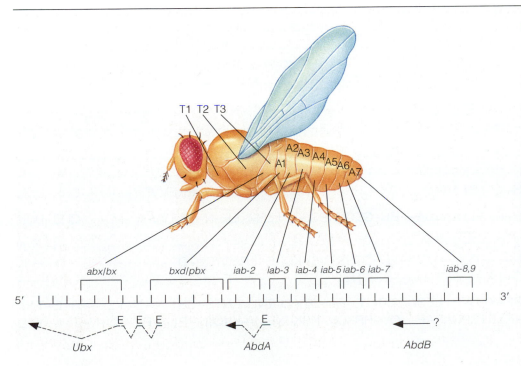

▶ **FIGURE 21–10** The *Drosophila* Bithorax complex, showing the three structural genes (transcription units), various regulatory regions, and the corresponding abdominal segments. The complex is 300,000 base pairs long; vertical scale marks on the DNA represent 10,000-base-pair increments. Transcription of the structural genes yields RNA consisting mainly of introns (dashed lines); exons are marked as E. *Adapted from:* S. F. Gilbert, *Developmental Biology* (Sunderland, MA: Sinauer Associates, 1988): 645. Used with permission.

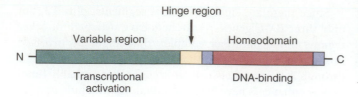

FIGURE 21–11 General domain structure of a homeobox protein, showing the presumptive functions of the two domains.

Homeobox genes encode proteins with a **homeo domain**, a stretch of 60 highly basic amino acids (▶ Figure 21–11) that appears to bind to DNA through a helix-turn-helix motif. The homeo boxes in different *Drosophila* genes range from about 50% to over 80% similar, depending on the genes being compared. This conservation of sequence, along with a structural motif known to occur in proteins that bind to DNA, suggests that homeogene-encoded proteins control major developmental events by acting as transcription factors.

Since their discovery in *Drosophila* in the early 1980s, homeobox genes have been found in a wide range of organisms, from the primitive jellyfish and hydra to humans. While mammals do not have the same kind of segmentation that *Drosophila* does, these genes do specify structure formation in higher animals. They seem to be active only in certain restricted embryonic regions, and mutations affect-

ing these genes result in major structural defects. ▶ Figure 21–12 shows the evolutionary relatedness of the homeobox domains of a number of developmental proteins from several organisms. This relatedness also extends to the chromosomal arrangement of the homeobox genes (▶ Figure 21–13). Although some homeobox genes are located elsewhere in the genome, most are contained in two (*Drosophila*) to four (mice and humans) major gene clusters, and are expressed in a particular left-to-right order. A remarkable feature of the gene organization within each cluster is that the position of each gene corresponds to its approximate location of expression along the embryo's anterior-posterior axis. Thus, there is a relationship between the order of these genes along the chromosome and their order of activation in the developing body plan. How these genes are coordinately regulated so that some are active and others silent at a particular developmental stage remains an area of intense experimental investigation.

It has been suggested that the homeo box is universally involved in the regulation of development and that homeobox genes evolved from a single gene complex (HOM-C) in a common, primitive metazoan ancestor. This is an exciting idea, because it suggests that the same basic developmental signals exist in all organisms.

Structural motifs other than the homeodomain helix-turn-helix motif are found in some of the proteins that regulate development in *Drosophila* and other organisms. Work with transcription factors GHF-1/Pit-1 (mouse pitu-

▶ **FIGURE 21–12** Comparison of amino acid sequences of homeo domains in developmental proteins from a variety of organisms. Amino acids that differ from the consensus are indicated by letter.

```
Consensus:    RKRGRTTYTRYQTLELEKEFHFNRYLTRRRRIEIAHALCLTERQIKIWFQNRRMKWKKEN

Drosophila Antp    -----Q------------------------------------------------------
Drosophila ftz     S--T-Q---------------------I-------D--N--S-S---------------S--DR
Drosophila Dfd     P--Q--A---H-I---------Y----------------T-V-S-----------------D-
Drosophila AbdB    VRKK-KP-SKF----------L--A-VSKQK-W-L-RN-Q-----V-----------N--NS
Drosophila Ubx     -R---Q--------------------T-H--------M-Y--------------------L---I

Frog AC1           -R---QIYS-------------------------N-------------------------R
Frog MM3           ------N-----------------------------V-----------------------

Mouse Hox-2.3      -----Q---------------Y---------------T-----------------------
Mouse Hox-2.5      SRKK-CP--K-----------L--M----D--H-V-RL-N-S---V-----------M--L-
Mouse GHF-1 (Pit-1)  KRKR---ISIAAKDA--RH-GEHSKPSSQEIMRM-EE-N-EKEVVRV--C---QRE-RUK

Bovine GHF-1 (Pit-1)  KRKR---ISIAAKDA--RH-GEQNKPSSQEILRM-EE-N-EKEVVRV--C---QRE-RUK

C. elegans Mab-5    S--T-Q--S-S------------Y------K--Q--SET-H-----V-----------H---A
C. elegans Ceh-11   S-K------Q----SV--AK-QQSS-VSKKQ-E-LRLQTQ--D--------------A---K
C. elegans Ceh-15   E--Q-TA---N-V----------THK----K----V--S-M-----V-----------H----

Yeast MATα2        -GHRF-KENVRILESWFAKNIE-P--DTKGLENLMKNTS-SRI---N-VS---R-E-TIT
Yeast MATa1        SPK-KSSISPQARAF--QV-RRKQS-NSKEKE-V-KKCGI-PL-VRV-VC-M-I-L-YIL
```

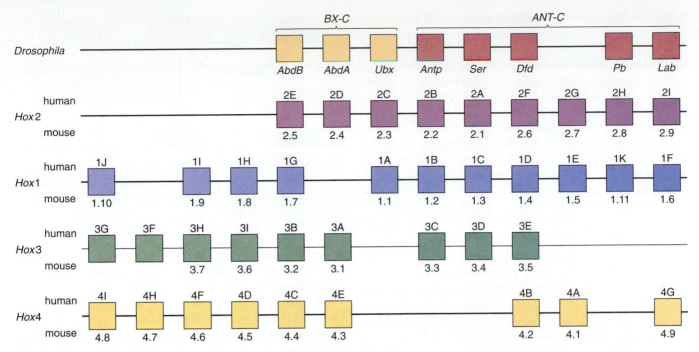

► **FIGURE 21–13** The homeobox gene clusters of *Drosophila*, mice, and humans. *Drosophila* has two clusters containing approximately 10 genes, while mice and humans have at least 40 such genes, in clusters of about 10 each. (Human *Hox* genes are designated by letters, mouse genes by numbers.) The order of homeobox genes on a chromosome reflects the location along the embryo's anterior-posterior axis where the genes are expressed. (*BX-C* and *ANT-C* are homeotic gene complexes.) *Adapted from:* R. Krumlauf, *BioEssays* 14 (1992): 245. © ICSU Press.

itary growth hormone factor), OCT-1 and -2, and the *C. elegans unc-86* gene product has identified a class of proteins with domains related to the homeo box. These proteins consist of a homeo domain plus an adjacent domain on the N-terminal side containing a high content of either basic or hydroxyl-group–containing residues; these two domains constitute the so-called POU box (Pit/OTF/unc). Some developmental proteins have no relation to the homeo box. For instance, the gap proteins contain zinc finger domains, and some segmentation and homeotic genes encode still other less well known motifs. We cannot conclude, therefore, that all genes regulating development contain homeo boxes. However, the widespread occurrence of homeo boxes in many different kinds of eukaryotes obviously gives them a great deal of importance in studies of developmental genetics.

To Sum Up

1. Development requires precise coordination in the expression of many genes. Since functionally related eukaryotic genes are not organized into operons, complex signal-transducing pathways mediate this coordination. Many of the genes involved in these pathways code for transcription factors; a hierarchy of activation and/or repression of gene transcription at successive levels mediates determination and differentiation.
2. Pattern formation in *Drosophila* illustrates the precisely coordinated gene expression that is required for development. The

anterior-posterior pattern is established through a hierarchy of protein-gene interactions that establish broad domains of gene expression which are progressively specialized into a highly defined set of subdivisions.

3. Expression of three levels of *Drosophila* segmentation genes generates the correct number and organization of body segments. A parallel cascade of expression establishes the identity of each segment. These cascades are initiated by the maternal *bcd* and *nos* gene products, which activate gap gene expression by the early embryo. The gap genes (1) activate the pair rule genes, eventually leading to the correct number of segments, and (2) activate the homeotic genes, which establish the identity of each segment. Misexpression of a homeotic gene causes cells to adopt a different segmentation identity.
4. The development of body plan and structure in a wide variety of animals appears to involve homeobox genes. Homeobox proteins contain a DNA-binding domain that interacts with nucleotide sequences through a helix-turn-helix motif and are thought to act as transcription factors. The homeo box may be universally involved in the regulation of development.

⧉ REGULATION BY GENE REARRANGEMENT

The chromosomal position of a gene can affect its expression. For example, chromosomal translocations (see Chapter 12) sometimes cause a change in phenotype, even though genes are only moved, not gained or lost. Such mutational changes appear to be important aspects of certain kinds of cancers (to be discussed in the next section). In addition to mutational events, there are also several known situations in which normal recombinational events reorganize certain DNA sequences. This section describes two examples in which gene expression changes as the result of genetically programmed rearrangements of DNA. The first involves yeast mating type, the second, the recombination of various DNA coding sequences to generate active antibody genes. In both cases, transcription of the relevant genes occurs only after DNA rearrangement.

Mating Type in Yeast

The common baker's yeast *Saccharomyces cerevisiae* does not have sexes; instead, there are two mating types, designated a and α. When haploid cells of opposite mating type fuse, they produce a diploid *a/α* zygote. The zygote then matures and may undergo meiosis to produce four haploid spores, completing the life cycle (see Figure 13–19).

Yeast mating type is determined by a single gene locus on chromosome 3, the *MAT* locus, which has two possible allelic states, *MATa* and *MATα*. *MATa* cells are of mating type a; *MATα* cells are mating type α. These alleles control the expression of specific proteins (pheromones) that are secreted by the cells and are recognized by cell membrane receptors of the opposite mating type. The interaction between the pheromone and its receptor triggers cell and nuclear fusion events through a signaling pathway involving a G protein (see Chapter 20). The GDP bound to the G protein is exchanged for GTP, activating the G protein. This step sets in motion a signaling pathway that ultimately activates transcription of certain genes involved in the establishment of the diploid state. Mutations in any of a variety of genes involved in the signaling pathway (including those coding for the a and α receptors, for the subunits of the G protein, for certain protein kinases, and for certain DNA-binding proteins) eliminate the ability of yeast cells to form the *a/α* diploid state. Thus mating and the formation of the diploid state are complex processes that require the transduction of a large number of regulatory signals.

Because mating and sporulation require cells of opposite mating type, we would expect populations composed of only a or only α cells to be unable to enter the diploid part of the yeast life cycle. Indeed, this is the case in some yeast strains. In other strains, however, cells can easily switch their mating type. As soon as one generation after initiating a population with pure a or pure α cells, mating begins to occur, as some cells switch their mating type from a to α or vice versa. A gene at the *HO* locus on chromosome 4 determines whether a yeast strain has a *MAT* locus that can't switch allelic states or a *MAT* locus that can switch; if a *MAT* locus can switch allelic states, it can do so as often as every generation. Most strains in nature repeatedly change the state of the *MAT* locus to switch mating type.

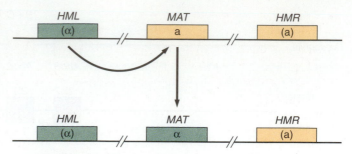

▶ **FIGURE 21–14** The cassette model of mating-type switches. A switch from a to α mating type is shown here. DNA replication yields a copy of the silent α cassette, which is then transposed into the active *MAT* locus.

Spontaneous mutation cannot explain mating type switches, since the switching frequency is so high. Instead, the switch is caused by a replicative transposition of DNA from an unexpressed mating type gene to the *MAT* locus. As shown in ▶ Figure 21–14, the active *MAT* locus is flanked by two silent mating type genes, designated *HML* (containing an unexpressed α allele) and *HMR* (with an unexpressed a allele). A switch in mating type occurs when a copy of one of the silent genes replaces the original *MAT* sequence. The silent cassettes do not move randomly with respect to the allelic state of *MAT*, however. *MATa* is preferentially invaded by *HML* (switching a to α), and *MATα* is preferentially replaced by *HMR* (switching α to a), so the probability of switching is over 80%.

▶ Figure 21–15 shows the structure of the *MAT*, *HML*, and *HMR* loci. The a and α alleles differ only in their Y regions. The Y regions of the silent loci contain the same nucleotide sequence as do *MATa* and *MATα*, including the

▶ **FIGURE 21–15** Structures of the *MAT*, *HML*, and *HMR* loci. The W, X, and Z segments are the same in all loci (except that *HMR* lacks a W region). There are two types of Y segments: identical Yα regions are present in *HML* and *MATα*, and identical Ya segments are present in *HMR* and *MATa*. The silent regions thus contain exactly the same genetic information as their corresponding active alleles.

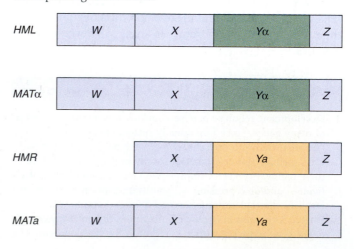

same potential promoter transcription startpoints, but their transcription is repressed by the action of the **SIR** (silent information regulator) gene products at two silencer sites, *HMLE* and *HMRE*. These silencers are located at least 1000 base pairs upstream of *HML* and *HMR*, but gene manipulation experiments have shown that they can silence transcription from either orientation and at long distances upstream or downstream of their targets. (Recall that enhancers have similar properties, but they stimulate rather than repress transcription.)

▶ Figure 21–16 illustrates a recombination process in which different cassettes are placed at the *MAT* locus to facilitate mating type switching. The mechanism of cassette movement involves *HO*, which encodes an endonuclease that recognizes a 24-base-pair sequence at the *Y-Z* junction of *MAT*. This enzyme makes a double-stranded cut that is thought to initiate a two-strand double crossover process between the recipient *MAT* region and the donor *HMR* region. In this model, a *MAT Y* region is degraded, and crossing-over within homologous regions on either side of Y then allows the gap to be filled in by a repair DNA poly-

merase that uses the homologous *HML* or *HMR Y* region as a template. Note that the silent cassettes are not affected by this process.

The timing of cassette recombination is carefully controlled. The HO endonuclease is made at the G_1 stage of the cell cycle, prior to DNA replication and cell division, so that both daughter cells are of the same mating type. However, a budded daughter cell contains no endonuclease, so it cannot switch mating type until the next generation, when it becomes a mother cell. Regulation of *HO* transcription appears to be quite complex, involving both repressor and activator proteins.

The *MAT* gene products regulate a large number of mating type-specific and haploid-specific genes, as shown in ▶ Figure 21–17. The *MAT* locus controls the expression of a-specific or α-specific genes (for example, genes encoding pheromones and receptors) to give the a or α phenotypes. *MAT*α encodes the α1 and α2 proteins. All cells constitutively express a-specific genes; the α2 protein represses transcription of these genes in a haploid cell by targeting the operator regions of the various a-specific genes. The α1 protein, in contrast, positively regulates expression of α-specific genes. Thus, to be of the α phenotype, a cell must repress its a-specific genes and activate its α-specific genes. Without α1 and α2, the cell expresses a-specific genes only.

In diploid cells, certain haploid-specific genes such as *HO* and *RME* (repressor of meiosis and sporulation) must be repressed. A third regulatory protein, a1, which is encoded by *MATa*, is thought to complex with α2, changing the operator specificity of the latter so that it represses transcription of *HO* and *RME*.

Comparisons of protein sequences have revealed an interesting homology between homeo domains and the a1 and α2 proteins, placing these proteins in the broad category of factors involved in developmental processes. The α2 repressor contains a C-terminal DNA-binding domain and an N-terminal region that must interact with another protein, PRTF, in order for transcription of a-specific genes to be repressed. PRTF binds to a short DNA consensus sequence (the P box) found in a variety of operators, where it mediates a repressive effect when interacting with certain other proteins (such as α2). If a1 instead of PRTF interacts with α2, the complex recognizes different operators—those of the haploid-specific genes that must be repressed in diploid cells.

The α1 protein activates transcription of α-specific genes by binding to their upstream activator sequences. Binding at these *UAS*s also requires interaction with PRTF: while PRTF binds to a P box, α1 binds to an adjacent Q box. Neither protein alone can activate transcription of the α-specific genes, although there is some evidence that PRTF alone is responsible for activating the a-specific genes that are constitutively expressed in haploid a cells. PRTF is thus a very interesting protein that seems to be able to function in haploid cells in three different ways: (1) independently, as a transcriptional activator of a-specific genes, (2) in

▶ **FIGURE 21–16** Model for a homologous recombination process that switches a *MATα* locus to a *MATa* locus. The *HMR* DNA has moved across from the homologous *MAT* region. The initial event is a double-strand break in DNA in the region that will receive new sequences. A two-strand double crossover is shown in this model, but the crossover products are both *Ya* due to degradation of the *Y* region and its replacement by *Ya* through DNA repair synthesis.

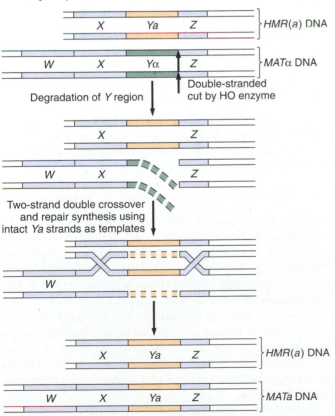

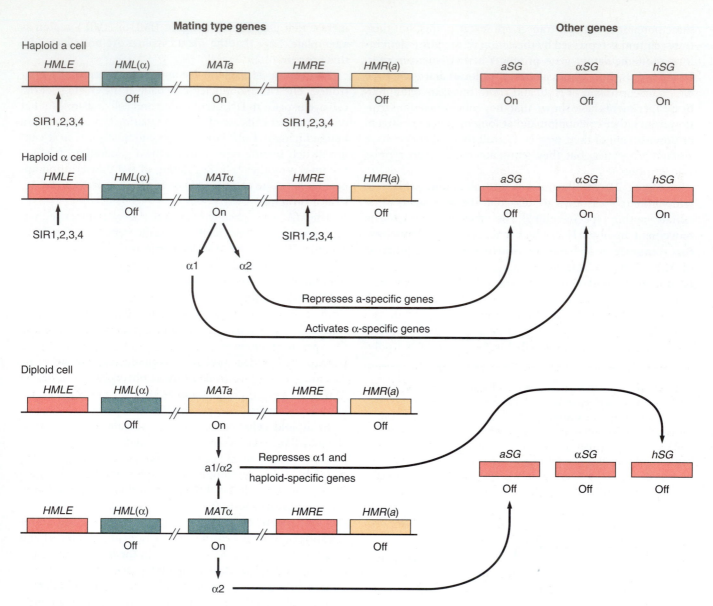

Mating type genes

Other genes

Haploid a cell

SIR1,2,3,4 (pointing to HMLE)
SIR1,2,3,4 (pointing to HMRE)

Haploid α cell

SIR1,2,3,4 (pointing to HMLE)
SIR1,2,3,4 (pointing to HMRE)

α1   α2

Represses a-specific genes

Activates α-specific genes

Diploid cell

a1/α2   Represses α1 and haploid-specific genes

α2

▶ **FIGURE 21–17** Gene expression in haploid and diploid cells. Essential regions adjacent to both *HML* and *HMR* repress their transcription through the concerted action of the *HMLE* and *HMRE* regulatory sequences. Haploid a cells constitutively express a-specific genes (*aSG*) needed for various mating functions. Haploid α cells express α-specific genes (*αSG*), and repress a-specific functions. In diploid cells, production of α1 and the haploid-specific genes (*hSG*) are repressed by α2, and α-specific genes are not expressed because of the lack of α1.

cooperation with α1 as an activator of α-specific genes, and (3) in cooperation with α2 as a repressor of a-specific genes.

Immunoglobulin Genes

Immunology deals with the way the vertebrate body responds to substances that are foreign to it. The immune response involves the integrated action of many different cell types that circulate freely throughout the body, moving in and out of the circulatory and lymphatic systems. The production of **antibodies** is one of the body's major lines of defense. Antibodies, which belong to a family of proteins called immunoglobulins, are complex molecules that recognize, combine with, and neutralize the effects of **antigens**.

Antigens are substances that are capable of inducing cells to produce antibodies against them; often they are foreign proteins or protein derivatives. For example, a protein on the surface of bacterial cells that have invaded the bloodstream of an animal would elicit an antibody response. The combination of antibody with antigen brings about a conformational change in the antigen and ultimately leads to its destruction. The antigen may be immediately scavenged by macrophages (specialized white blood cells) or it may be exposed to the action of **complement**, a heterogeneous group of proteolytic enzymes that degrade the antigen in a stepwise fashion.

An individual might encounter any of tens of millions of antigenic substances during a lifetime—each will elicit a different, specific antibody response. Bacteria, viruses, and toxins act as antigens, but the antibody response is not limited to fighting disease. The well-known phenomena of blood group incompatibilities (see Chapter 4) are antibody-antigen reactions in which a sugar on the surface of the red blood cells is the antigen. In the autoimmune diseases (in which an individual produces antibodies against itself), the antigens are normal components of the body.

The antibody-mediated or **humoral response** is just one possible way the immune system can respond to an antigenic invader. The antibodies are secreted by B lymphocytes, which are specialized white blood cells produced in the bone marrow. The other major arm of the immune system is the **cell-mediated response**, which is directed by receptor sites on the surface of T lymphocytes, which differentiate in the thymus. This system includes various kinds of T cells that are involved in the destruction of foreign tissues or host cells infected by a foreign agent. Rejection of tissue grafts and organ transplants occurs through T cell–mediated responses. The two immune systems can act separately or together to provide a diverse spectrum of ways to attack foreign antigens. Both systems involve a large number of genes, and DNA sequences must be rearranged to produce the gene complexes that encode the specific proteins needed for recognition of specific antigens. In this section we will focus our attention on the DNA reorganizations that are necessary for generating the functional antibody genes. The T cell receptor gene family, which is evolutionarily related to the immunoglobulin family, has essentially the same kind of genetic system.

Each particular kind of antigen elicits production of a unique kind of antibody molecule by the B lymphocytes. Each B lymphocyte has a single kind of membrane immunoglobulin on its surface, and contact with an antigen results in the multiplication of B cell clones specific for that antigen and in their differentiation into plasma cells capable of producing and secreting large amounts of antibody of the same specificity. Since any foreign macromolecule can, under appropriate conditions, stimulate an immune response, a tremendous diversity of antibody types must be genetically programmed into any individual. It has been estimated that humans, for example, can produce more than 10 million different antibody molecules. Each of these is encoded by information on the germ-line DNA, but only one unique antibody that binds to a particular stimulating antigen is actually produced by a given B lymphocyte. Two of the major questions of immunology are how the vertebrate genetic system can contain enough genes to specify the almost endless diversity of antibodies that might be needed and how any one B cell can express only one of these genes to the exclusion of all others.

The ability of antibodies to combine specifically with antigens depends on their three-dimensional structure. The structures of the peptides that form the major class of antibodies, IgG (immunoglobulin G), are shown in ▶ Figure 21–18. Each IgG molecule is composed of four polypeptides—two identical light (L) chains of about 220 amino

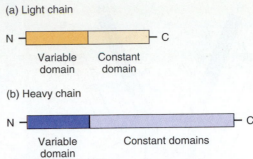

(a) Light chain

(b) Heavy chain

▶ **FIGURE 21–18** An IgG molecule is composed of four peptides: (a) two identical light chains and (b) two identical heavy chains. Each chain is divided into an N-terminal domain of variable amino acid sequence and a C-terminal domain whose amino acid sequence is the same among all variants of the same type of IgG molecule.

acids and two identical heavy (H) chains of about 450 amino acids. Differences among IgG antibody molecules are the result of variable amino acid sequences in the N-terminal domains of the L and H chains. These **variable domains** in both the L and the H polypeptides contain about 110 amino acids. The amino acid sequences in the C-terminal portions of the L and H chains do not change significantly from one type of IgG molecule to another and are referred to as the **constant domains** of the antibody.

The subdivision of the variable domains of the L and H peptides is shown in ▶ Figure 21–19. The L chain variable domain includes the V region (approximately 95 amino acids) and the short J region, which joins the constant and variable sections. The H chains have a similar arrangement, with a third region, D, added between V and J. The D (diversity) region consists of a short hypervariable sequence of amino acids. Each L chain is linked to one of the H polypeptides by a single disulfide bond, and the H chains are joined to each other by two disulfide linkages. Both the L and the H peptide chains also contain intrachain disulfide bonds, which contribute to the specific folding pattern shown in Figure 21–19b. The variable domains of the two pairs of chains form the business end of the bivalent molecule—the sites that combine with a specific antigenic determinant.

The specificity of the binding reaction is a result of the complementary nature of the antigen and antibody shapes. The variable amino acid sequences of different IgG molecules cause their variable domains to have different chemical properties and to assume slightly different folding patterns, each of which recognizes only one antigen conformation. Researchers have used antibody **Fab fragments** to study the nature of binding specificity and strength. These fragments consist of the entire L chain, the variable H domain, and the first constant H domain. Several Fab structures, both free and complexed with ligands, have been investigated by crystallography. ▶ Figure 21–20 shows a space-filling model of the Fab fragment of IgG as it contacts an antigen, illustrating the complementary contacting surfaces of the two molecules.

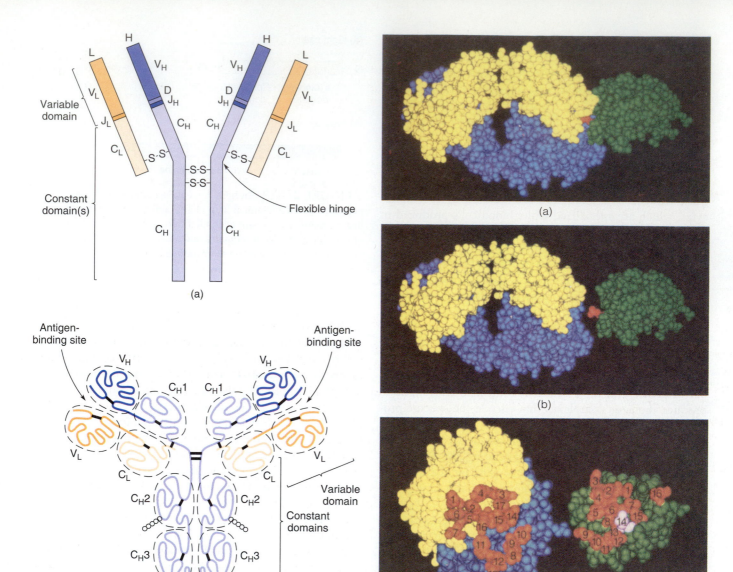

(a)

(b)

(a)

(b)

(c)

▶ **FIGURE 21–19** (a) The two-dimensional structure of an IgG antibody molecule. The letters L and H indicate the light and heavy chains, respectively. V, J, and C refer to the variable, joining, and constant regions of each chain. D is the diversity region, which is found only on the heavy chain. The variable domain of each chain includes the V, D (in heavy chains only), and J regions. Chains are held together by disulfide bonds. The flexible hinge allows the geometry of IgG molecules to adjust to fit a variety of spacings between antigenic sites. (b) The folding pattern of the L and H chains, showing the interchain grooves where the antigen is bound. Also shown is the subdivision of the constant region of the H chain into repeating domains of similar amino acid sequence (C_H1, C_H2, and C_H3).

▶ **FIGURE 21–20** Model of the three-dimensional structure of an antibody-antigen complex based on X-ray diffraction studies of the Fab fragment of IgG. The antibody variable H domain is shown in blue and the L chain in yellow. The antigen (lysozyme) is shown in green. Amino acids that contact the other molecule are colored red. (a) The antibody is complexed to the antigen; protuberances (including the glutamine residue shown in red) and depressions in complementary surfaces of the two molecules fit together. (b) The antibody and antigen have been pulled apart to better illustrate the complementary features of the surfaces of the two molecules. (c) The molecules have been rotated 90° to give an end-on view. L chain residues that contact antigen and lysozyme residues that contact antibody are numbered. The glutamine that was shown in (a) and (b), residue number 14, fits into the antibody surface pocket surrounded by V_L and V_H residues 2, 5, 6, 7, and 16. *Source:* A. Admit et al., "Three dimensional structure of an antigen–antibody complex at 2.8 Å resolution," *Science* 233 (1986): 747–753. Copyright © 1986 by the AAAS. Used with permission.

Class	Concentration in Blood Serum (mg/ml)	Molecular Weight	Type of Heavy Chain	Type of Light Chain	General Formula
IgM	1	950,000	μ	λ or κ	$(\mu_2\kappa_2)_5$ or $(\mu_2\lambda_2)_5$
IgD	0.03	180,000	δ	λ or κ	$\delta_2\kappa_2$ or $\delta_2\lambda_2$
IgG	13	150,000	γ	λ or κ	$\gamma_2\kappa_2$ or $\gamma_2\lambda_2$
IgA	2	160,000*	α	λ or κ	$(\alpha_2\kappa_2)_n$ or $(\alpha_2\lambda_2)_n$**
IgE	0.001–0.01	190,000	ϵ	λ or κ	$\epsilon_2\kappa_2$ or $\epsilon_2\lambda_2$

*Polymeric forms of greater molecular weight are known.

**$n = 1$, 2, or 3.

Data from J. T. Barrett, *Basic Immunology and Its Medical Applications* (St. Louis: C. V. Mosby, 1980).

Studies of the amino acid sequences of light chains from a variety of organisms have revealed that there are two types of polypeptides, the lambda (λ) and kappa (κ) chains. The two kinds of L peptides differ in amino acid sequence in both their variable and their constant domains. The kappa chains constitute about 60% of all the light chains of humans and about 95% of those of mice. Similar studies of heavy chains have found five different classes that differ in the amino acid sequence of their constant domains. These differences in the H chain constant region define the five major classes of antibodies recognized by immunologists: IgM, IgD, IgG, IgA, and IgE (■ Table 21–2). IgG is the dominant antibody in most sera, representing 75–85% of the total. IgM forms about 9% of human serum antibody content. IgG and IgM are the two major serum agents that activate complement to combat disease-carrying microorganisms. IgA is found in the highest concentration in external secretions such as milk, nasal mucus, saliva, and intestinal and respiratory mucus. IgE is important in the development of hay-fever type allergies, and the function of IgD is not yet known.

Three gene loci that code for antibody molecules have been identified—gene *H* for the heavy gene and two loci for the light chains, λ and κ, all of which are located on separate chromosomes. Either λ or κ (but not both) can be genetically active in any given plasma cell. Thus, only two genes in a plasma cell function in antibody production (*H* and either λ or κ). What, then, is the basis of the great diversity of antibodies that are produced by different B cells? Much of this variation originates from rearrangement of the DNA in the immunoglobulin genes during the development of an undifferentiated B lymphoblast into a mature cell capable of producing antibodies. ▶ Figure 21–21 illustrates the structures of the light and heavy chain genes and the joining of coding segments into functional variable domain exons. The λ and κ genes each contain three regions—one coding for the V_L portion of the light peptide, one for the J_L segment, and one for the C_L domain. The *H* gene is similarly split into four regions that code for the V_H, D, J_H, and C_H segments of the heavy chain. The *V*, *D*, and *J* regions consist of many repeated copies of coding segments, symbol-ized by V_1, V_2, V_3, . . . , V_m for the V segment, J_1, J_2, J_3, . . . , J_n for the J segment, and D_1, D_2, D_3, . . . , D_p for the D segment. Each copy differs slightly in nucleotide sequence from the other copies in that region. As an antibody-producing cell develops, one V copy is joined to one J copy (and to one D copy in the case of the *H* gene) by somatic chromosomal rearrangements. This *V–J* or *V–D–J* joining forms the complete variable domain exons of the light and heavy genes, respectively.

The molecular mechanisms that join V, J, and D segments are just beginning to be elucidated. Recombination proceeds through a different mechanism than is found in other DNA rearrangements such as mating type switching in yeast, transposition (Chapter 10), prophage integration (Chapter 10), and general recombination (Chapter 23). Specific conserved DNA sequence elements—the recombination signal sequences—that flank the V, J, and D copies mediate the joining reactions (▶ Figure 21–22). On the 3' side of each V_L copy lies a 28-bp structure composed of conserved heptamer and nonamer sequences separated by 12 bp that differ among the various V_L copies. The 5' end of each J_L copy is similarly flanked by the same heptamer and nonamer sequences (in reverse orientation), now separated by 23 varying bp. Recombination can occur only between repeats with 12-bp and 23-bp spacers. This "12–23" rule ensures that a V_L copy joins to a J_L copy rather than to another V_L. The same general arrangement dictates V_H–D and D–J_H joining (Figure 21–21b).

The recombination activator genes *RAG-1* and *RAG-2* are required for V(D)J rearrangement. However, it is not clear whether the proteins they encode function as recombination enzymes or regulators of these enzymes. A regulatory role is perhaps favored in light of the results of studies in which cultured cells were transfected with transcriptionally active *RAG-1* and *RAG-2* genes and the V(D)J joining processes occurred in the DNA of the transfected cells. In fact, evidence suggests that the *RAG-1* and *RAG-2* genes determine not only the lymphoid cell specificity of V(D)J joining but also the precise developmental stage of these cells when the joining reactions occur. A model for V_L–J_L joining is shown in ▶ Figure 21–23; the recombination

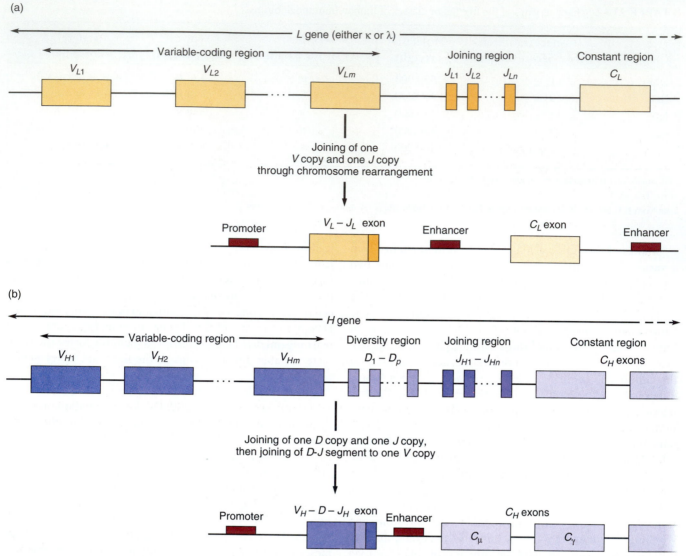

▶ **FIGURE 21–21** Generalized diagrams of the chromosomal regions that contain coding sequences for the variable portions of the (a) L and (b) H peptides. The variable-coding, joining, and diversity regions of the chromosomes each consist of repeated copies of coding segments, with each copy differing slightly in nucleotide sequence. Rearrangement of chromosomal material during the development of an antibody-producing cell joins one V copy to one J copy to form the L gene variable portion exon, and one V, one D, and one J copy to form the H gene variable portion exon. The exons coding for the constant regions of the peptides remain separate from the variable exons. After the rearrangements, the promoters are located just upstream of the variable domain exons, and the enhancers are located in the introns separating the variable and constant domains. The constant exon of the H gene is selected after the DNA is transcribed onto RNA.

sites are presumably brought together by the specialized proteins, double-stranded breaks in the DNA are made, and the copies are ligated together.

The rearrangements illustrated in Figure 21–21 bring promoters into close proximity to the variable domains and to enhancers located between the variable and constant regions of both the light and the heavy genes. The enhancers are very complex; each contains two or more sequence modules that recognize distinct regulatory transcription factors. For example, the octamer sequence described in the last chapter has been found in every Ig promoter and enhancer studied so far. As mentioned in Chapter 20, however, interaction between the octamer module and TFs such as OCT-1 and OCT-2 does not seem to fully account for the B cell specificity of Ig gene transcription. The transcription factor NF-κB and other proteins have been implicated in a signal transduction process that is important for modulating transcription in response to events associated with B cell development.

If the choice of V, J, and D copies is made randomly, the number of possible V–J (or V–D–J) combinations, which is equal to the product of $m \times n$ (or $m \times n \times p$), can be quite large. Thus, if there are 300 V_L copies to choose from and 5 J_L copies, there would be 300×5 or 1500 possible light chain variable domains. Heavy chain variable domains would be even more numerous because there are multiple

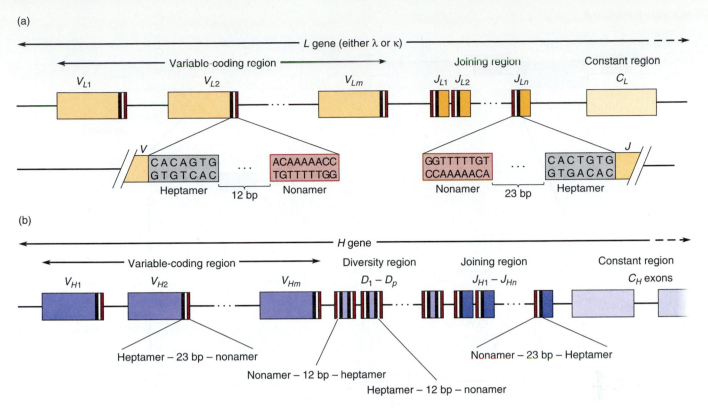

(a)

L gene (either λ or κ)

Variable-coding region

Joining region

Constant region

V_{L1} V_{L2} V_{Lm} J_{L1} J_{L2} J_{Ln} C_L

V

| CACAGTG | | ACAAAAACC |
| GTGTCAC | | TGTTTTTGG |

Heptamer   12 bp   Nonamer

| GGTTTTTGT | | CACTGTG |
| CCAAAAACA | | GTGACAC |

Nonamer   23 bp   Heptamer   J

(b)

H gene

Variable-coding region   Diversity region   Joining region   Constant region

V_{H1} V_{H2} V_{Hm} $D_1 - D_p$ $J_{H1} - J_{Hn}$ C_H exons

Heptamer – 23 bp – nonamer

Nonamer – 12 bp – heptamer

Heptamer – 12 bp – nonamer

Nonamer – 23 bp – Heptamer

▶ **FIGURE 21–22** Recombination signal sequences that mediate *V–J* and *V–D–J* joining. (a) In the light chain genes each V_L copy is flanked on its 3′ end by a heptamer–12 bp–nonamer sequence that is complementary to the nonamer–23 bp–heptamer sequence flanking the 5′ end of each J_L copy. Recombination occurs only between copies with 12-bp and 23-bp sequences, ensuring that a V_L copy joins to a J_L copy. (b) A similar mechanism generates the variable portion of a heavy chain gene. Each *D* copy is flanked on both sides by the heptamer–12 bp–nonamer signal sequence: the 12–23 spacer rule thus dictates that the two recombination events needed to generate the variable part of the H gene will involve *V–D* and *D–J*. *Adapted from:* J. D. Watson, M. Gilman, J. Witkowski, and M. Zoller. *Recombinant DNA*, 2d ed. (New York: W. H. Freeman and Company), p. 301. Used with permission.

copies of *D*. If there are 10 copies of *D*, 1500 × 10 or 15,000 heavy chain variable domains could be constructed. These numbers are still far short of the number needed to account for normal antibody diversity. However, since any pair of L chains can combine with any pair of H chains and each pair has acquired its variable domain independently, the poten-tial number of different IgG antibodies increases to greater than 10^7 (1500 × 15,000), a number more in line with the variability observed to exist. The joining reactions them-selves generate additional diversity, because base pairs can be inserted or deleted at the *V–J* and *D–J* junctions. Recent evidence indicates that somatic mutation in variable domain

▶ **FIGURE 21–23** Model of *V–J* recombination. The conserved signal sequences are brought together, and recombination takes place at the border between heptamer and coding regions. The spacer length defines two types of joining signals, and recombination occurs only with one signal of each type. The result is a fusion of the coding segments, as well as a heptamer-heptamer fusion of the signal sequences. Any one B cell is therefore left with just one *V–J* active coding region.

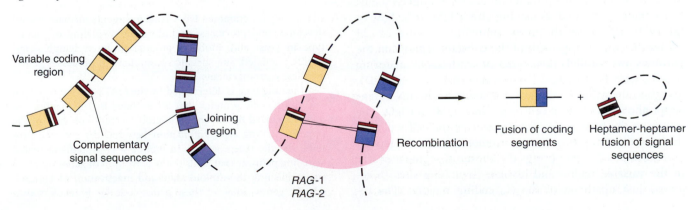

Variable coding region

Complementary signal sequences

Joining region

RAG-1
RAG-2

Recombination

Fusion of coding segments   +   Heptamer-heptamer fusion of signal sequences

Immature immune cell:

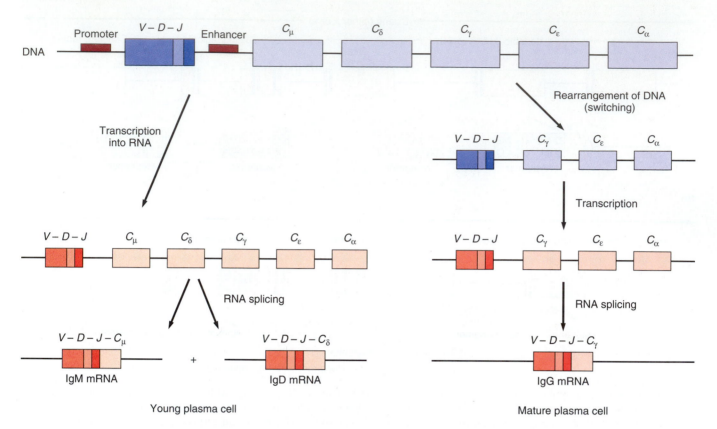

▶ **FIGURE 21−24** Production of messenger RNA from the antibody *H* gene. A young plasma cell makes either IgM or both IgM and IgD heavy chains (left) through differential splicing of the RNA transcript. The switch to the final class of heavy chain that will be produced and secreted by the cell involves a gene rearrangement (right). In the case illustrated, the cell will make an IgG heavy chain, and the C_μ–C_δ segment has been deleted to bring the completed variable region exon closer to the C_γ exon. The rearranged DNA is then transcribed and the RNA is spliced to form the mature mRNA. *Note:* The *C* coding regions are longer than shown here, and the division of each *C* coding region into domains is not shown. There are actually several related copies (not shown here) of the gamma coding sequence.

exons even further amplifies antibody gene diversity, and mutation rates in developing lymphocytes are quite high. All these ways of producing diversity give the possibility of several billion different B cell types producing IgG.

The particular class of antibody produced by a cell also depends on chromosomal rearrangement. Different classes of antibodies are produced in a set pattern during the development of a B lymphocyte. First, the cell makes IgM, followed by simultaneous production of IgM and IgD. Finally, it becomes committed to making either IgM, IgG, IgA, or IgE, with IgG being the most common. ▶ Figure 21−24 shows that the arrangement of the constant regions on the chromosome parallels their order of synthesis by maturing B cells, with C_μ (for IgM) first, followed by C_δ (for IgD), and then the others. Differential splicing of the RNA transcript allows the young cell to produce IgM and IgD. The switch to the final heavy chain class that the cell will produce permanently involves a chromosome rearrangement (▶ Figure 21−25) that brings the chosen *C* segment nearer to the variable region and involves switching sites (*S* regions) just upstream of the C_H coding regions. The re-

arranged DNA is then transcribed and the RNA is spliced to form the final H chain messenger. Notice that the same variable domain is maintained throughout this sequence of events, indicating that even though the class of antibody being secreted by the cell changes, the specific antigen recognized remains the same.

To Sum Up

1. Genetically programmed DNA rearrangements mediate several developmental processes, including the switching of mating type in yeast and antibody production by B lymphocytes. Highly regulated, site-specific recombination processes generate these rearrangements.

2. Yeast mating type is determined by the *MAT* gene locus, which encodes specific proteins that are secreted by the haploid cell and recognized by membrane-bound receptors on cells of the opposite mating type. Interaction between the secreted proteins and the receptors sets in motion a G-protein–mediated signaling pathway that leads to the activation of genes involved in establishing the diploid state and inactivation of haploid-specific genes. Some of these genes code for homeobox pro-

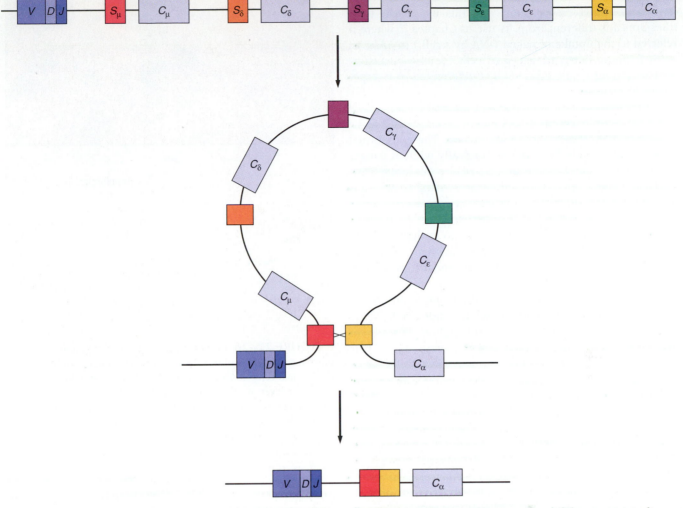

▶ **FIGURE 21–25** Possible mechanism of chromosomal rearrangement that joins the V–D–J region to one of the constant coding sequences. Each constant exon is preceded by a switching signal (S) that bears some homology to the S region of other C sequences. Loop formation is followed by deletion of the DNA within the loop and joining of the variable and constant regions. The S regions can be spliced out following transcription. Specialized but as yet unidentified recombination enzymes are hypothesized to mediate the rearrangement process.

teins. MAT proteins also help to maintain the diploid state by activating a gene that represses meiosis.

3. Mating type switching in yeast involves a site-specific homologous recombination event in which mating type-specific information at the *MAT* locus is replaced by alternative information copied from one of two unexpressed donor DNA sequences.

4. Development of a B lymphocyte involves rearrangement of DNA segments within the gene loci that encode the polypeptide chains that make up antibody molecules. Thus by the time a B cell is developmentally mature, it has been programmed to produce just one particular antibody.

5. The immune systems of vertebrates can produce millions, if not billions, of different antibodies to react with virtually any substance. They generate this enormous diversity through three basic mechanisms involved in B cell maturation: (a) the assembling of genes for antibody light chains and heavy chains from two or three component parts that are randomly selected from a heterogeneous pool of parts, and the random union of light and heavy chains to form antibody molecules, (b) joining the parts through an imprecise mechanism that can add or

delete nucleotides from the variable-coding region, and (c) a high rate of somatic mutation in the variable-coding regions during the proliferation of B cell clones.

6. The final class of antibody produced by a B cell also depends on DNA rearrangement, which links one constant-coding region to the already rearranged variable-coding region.

CANCER AND CELL PROLIFERATION

Many cells of the body, including skin and blood cells, must continue to divide throughout an individual's life in order to replace cells that are lost naturally or destroyed. Since the body can maintain a constant number of cells only by balancing the rate of cell production with the rate of cell death or loss, cell division in an organism must be under strict control (see Chapter 9). On occasion, abnormal immortalized cells escape the usual constraints and proceed to multiply at the expense of neighboring cells. Cells that no longer respond to normal constraints on growth are said to have been **transformed.** (Note that this use of the term *transformed* is different from its use in Chapter 7, where it referred to the uptake of naked DNA by a cell.) In addition to their immortality, transformed cells are also characterized, to various degrees, by a loss of differentiation and specialized function.

When transformed cells remain localized in their place of origin and do not invade the surrounding tissues, the resulting mass is called a **benign tumor.** These abnormal growths can usually be removed surgically without danger of recurrence. A **malignant tumor,** on the other hand, is capable of invasive growth into the surrounding tissues; immortalized cells shed from a malignant tumor can spread (metastasize) to distant sites via the lymph or blood circulation. Thus, in addition to their ability to proliferate without normal regulation, the cells of a malignant tumor are also able to spread to and produce secondary growths in other parts of the body.

Tumor cells grown in culture exhibit important differences from normal cultured cells and thus indicate the kinds of effects that cancer has on cell proliferation. Changes in the surface properties of cancer cells (compared to their normal counterparts) make the transformed cells unresponsive to the growth restraints that result from contact between cells (**contact inhibition**) (▶ Figure 21–26). These alterations also cause the cells to be unresponsive to the regulatory actions of various growth factors that normally exert their influence by binding to receptor sites on the plasma membrane to initiate a signal-transducing pathway. In addition, modifications of the cell surface result in loss of the normal affinities that cause cells of the same kind to stick only to each other. Cancer cells instead have aberrant interactions with each other and form aggregates with disorganized arrangements of cells.

Approximately 200 distinct varieties of human cancer are recognized. Most of these types belong to three major groups. The **carcinomas,** which include about 90% of all types of cancer, constitute the largest group. These tumors originate from epithelial tissues (e.g., skin and the linings of the respiratory and digestive tracts). Cancers of the lung, large intestine, and breast make up a large percentage of this group (40 to 50%). Most of the remaining types of cancer are either **sarcomas,** which are derived from the supporting tissues of the body (bone, cartilage, muscle, and fat), or **leukemias,** which originate in the progenitors of the white blood cells. In humans, sarcomas and leukemias make up about 8% of all types of cancers; in other mammals that have been studied, however, these two major groups are not nearly so rare.

Mutations and Cancer

One of the most important developments of the last 15 years has been the proof that cancer is a disease that involves gene and/or chromosome mutations, that is, that cancer is a genetic disease. However, cancer differs from "typical" ge-

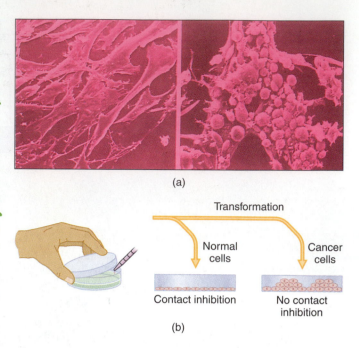

(a)

(b)

▶ **FIGURE 21–26** (a) Scanning electron micrographs of normal cells (left) and transformed cancerous cells (right). (b) Normal cells exhibit contact inhibition and stop growing after forming a layer of cells. Transformed cells lack contact inhibition and "pile up" on one another. The lack of adherence to the culture dish allows them to assume a rounder appearance and a disordered arrangement.

netic disorders in several respects. Most cancers are caused, at least in part, by somatic rather than germ-line mutations, so they are not inherited in any obvious Mendelian manner; each tumor is basically a clone whose origin is a single abnormal cell (the **monoclonal origin** of cancer). Furthermore, multiple mutational events are usually required for transformation. In a few cancers, a single mutation may be sufficient, but in most cases, several mutational events are necessary before a cell becomes cancerous. It is currently thought that at least three and perhaps as many as six separate mutations are required.

The clonal basis of cancer can be explained by assuming that the multiple mutational events are cumulative in their effects. Tumors thus develop by a process of clonal evolution driven by mutation. The first mutation affects the proliferation of a single cell, and later a second mutation transforms one of the progeny of the original cell. This cell then outgrows its sister cells to form a benign tumor. One cell from this tumor could undergo further mutation that favors its selection over the other cells, and so on. Finally, a mutation in one of the clonally selected cells could yield malignancy. This multihit model helps explain why most human cancers, for example, occur in older people—time is required to accumulate enough mutations to achieve malignancy.

Even though cancer is considered the result of somatic mutation, it is still possible for a genetic predisposition to a certain cancer to be inherited through successive generations of a family. In such cases, one of the mutations is inherited through the germ line. Cancer would still occur only when additional somatic alterations accumulate.

Mutations that cause tumors in somatic cells may arise spontaneously or from exposure to **carcinogens** (cancer-producing agents). Mutation can cause uncontrolled cell proliferation by affecting either events that normally stimulate cell division or events that normally inhibit it. A mutated gene that overly stimulates division is called an **oncogene**; its normally functioning allele is referred to as a **proto-oncogene**. Since proto-oncogenes are normal cellular genes, they are designated by the prefix *c*, for example, *c-myc* for the *myc* proto-oncogene. Genes that normally inhibit cell proliferation are sometimes referred to as **tumor suppressor genes**.

Functions of Proto-Oncogenes

Before we try to understand how oncogenes cause cellular transformation, we need to consider the normal functions of the alleles from which the oncogenes are derived. Theoretically, any gene that functions in the regulation of cell division is a potential proto-oncogene. More than 60 proto-oncogenes have so far been discovered. The 3T3 assay is one experimental technique used to identify proto-oncogenes; it involves determining the response of cultured mouse cells to uptake of DNA from tumor cells. Transformation of the cultured cells indicates that they contain an oncogene (▶ Figure 21–27). The DNA from transformed cultured cells can then be isolated, cleaved with restriction enzymes, ligated to vectors, and cloned. There are several selective cloning techniques that can be used to identify the particular oncogene involved.

The 3T3 assay detects only oncogenes that behave as dominant alleles (that is, oncogenes that deregulate cell proliferation even when normal alleles are present in the recipient cell). The recessive behavior of proto-oncogenes identified in this way indicates that they are involved in the normal stimulation of cell division, rather than in its suppression.

■ Table 21–3 lists some proto-oncogenes, their normal functions, and the cancers induced by their oncogenic derivatives. A number of these proto-oncogenes encode enzymes that phosphorylate proteins at their tyrosine residues (the tyrosine kinases). (The significance of phosphorylation in the cell cycle was initially discussed in Chapter 9.) These proteins, some of which are growth factors or their receptors, comprise an integral part of the signal-transducing pathway that regulates cell division.

During the cell cycle, a cell is advanced into G_1 by growth (competence) factors, such as EGF (epidermal growth factor), PDGF (platelet-derived growth factor) or FGF (fibroblast growth factor). It then becomes committed to DNA synthesis through the action of other growth (progression) factors, such as IGF-1 (insulinlike growth factor 1). In most, if not all, major tissue types, these growth factors ex-

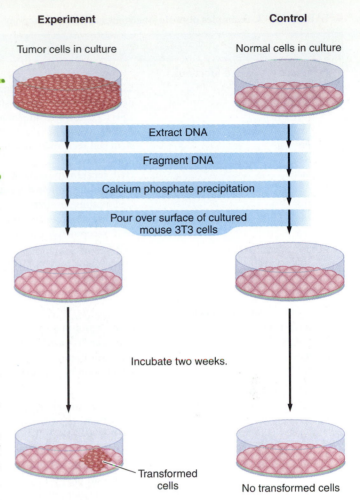

Experiment   **Control**

Tumor cells in culture   Normal cells in culture

Extract DNA

Fragment DNA

Calcium phosphate precipitation

Pour over surface of cultured mouse 3T3 cells

Incubate two weeks.

Transformed cells   No transformed cells

▶ **FIGURE 21–27** Assay for the presence of an oncogene in tumor cells. The DNA from cancerous cells is incubated with mouse 3T3 cultured cells under conditions in which the cells will take up the DNA, a process called transfection. Transformed cells are identified by their growth on a selective medium in which normal cells are unable to grow. Transformation of the cells is an indication that the DNA extracted from the tumor contains an oncogene.

ert their effects through receptor proteins that have tyrosine kinase activity. Binding the appropriate ligand (such as a growth factor, a hormone, or a neurotrophin) to the extracellular domain of the membrane-bound receptor activates its intracellular domain, which contains the kinase activity. The various receptor kinases then phosphorylate specific sets of substrates, including themselves, other membrane-bound and intracellular kinases (such as the cyclin-dependent kinases discussed in Chapter 9), and G proteins, all of which are encoded by various other proto-oncogenes. These so-called second messengers may then be involved in still another set of intracellular kinase activations, as a cascade of signals results in the sequential activation and repression of the various genes involved in cell division. (This general type of signaling process was described in Chapter 20 for the action of polypeptide hormones; see, for example, Figure 20–19.) In the absence of continued stimulation by the

Proto-oncogene	Protein Function	Cancer Caused by Oncogene
Membrane-associated functions:		
erbB	EGF receptor/tyrosine kinase	carcinoma, erythroleukemia, fibrosarcoma
fgr	nonreceptor protein/tyrosine kinase	sarcoma
fms	CSF receptor/tyrosine kinase	fibrosarcoma
hst	FGF	carcinoma
kit	stem cell receptor/tyrosine kinase	sarcoma
KS	unknown	Karposi's sarcoma
met	HGF receptor/tyrosine kinase	osteosarcoma
neu	receptorlike protein/tyrosine kinase	neuroblastoma
ras	GTP binding protein	carcinoma, erythroleukemia, sarcoma
ros	IFG receptor/tyrosine kinase	sarcoma
sea	HGF receptor/tyrosine kinase	leukemia, sarcoma
sis	PDGF, β-chain	sarcoma
src	nonreceptor protein/tyrosine kinase	sarcoma
trk	NGF receptor/tyrosine kinase	
yes	nonreceptor protein/tyrosine kinase	sarcoma
Cytoplasmic signaling functions:		
abl/bcr-abl	nonreceptor protein/tyrosine kinase	pre–B cell leukemia, sarcoma
fes/fps	nonreceptor protein/tyrosine kinase	sarcoma
mos	serine-threonine kinase	sarcoma
raf	serine-threonine kinase	sarcoma
Nuclear functions:		
erbA	thyroid hormone receptor	supplements action of *erbB*
ets	DNA-binding protein	supplements action of *myb*
fos	nuclear transcription factor/part of AP-1	osteosarcoma
jun	part of AP-1 nuclear transcription factor	fibrosarcoma
myb	DNA-binding protein	carcinoma, myeloblastic leukemia
myc	DNA-binding protein	carcinoma, myelocytoma, sarcoma
rel	NF-κB-related protein	reticuloendotheliosis
ski	TF?	carcinoma

critical factors, the cell will revert to its resting (G_0) state or perhaps even die. In contrast, mutations that convert proto-oncogenes to oncogenes cause either mutant expression, overexpression, or expression at the wrong time of growth factors or subsequent components of their signaling pathways.

Some proto-oncogenes code for nuclear rather than membrane-bound or cytoplasmic proteins. If these nuclear proteins function as transcription factors (as the c-Fos and c-Jun proteins do), then the oncogenic state results directly from altered transcriptional regulation. ▶ Figure 21–28 shows a possible scheme for a complex control network in which proto-oncogenes function.

Conversion of Proto-Oncogenes into Oncogenes

Proto-oncogenes can be converted into their oncogenic derivatives through either **quantitative** or **qualitative** alterations in gene expression (see ▶ Figure 21–29). In a quantitative change, the proto-oncogene itself is not altered—the oncogenic response is the result of overproduction of the normal protein due to gene amplification or to a chromosome rearrangement that brings the gene under the control of a different promoter or enhancer. In contrast, a qualitative alteration involves a mutation within the proto-oncogene to yield an altered protein that is hyperactive or unregulatable.

Some proto-oncogenes can become oncogenic through both quantitative and qualitative alterations. The *ras* gene family, whose members are found in many species ranging from yeast to humans, is a good example. Ras proteins are essential transducers in the signaling pathway through which gene activity responds to extracellular signals. Oncogenic versions of *ras* have been implicated in many types of cancer. Qualitative alterations, such as point mutations, cause some *ras*-mediated transformations, and *ras* oncogenes have also been discovered in tumors induced by certain viruses. These viruses integrate into the host chromosome in the vicinity of a *ras* proto-oncogene, disrupting its normal regulation and linking it to viral regulatory elements. The result is an overly expressed *ras* gene and malignant transformation of the cell. Similar results have been obtained by integrating multiple copies of the normal human H-*ras* gene into host DNA. The combination of quantitative and qualitative alterations is thought to explain the unusually wide diversity of phenotypes induced by *ras*

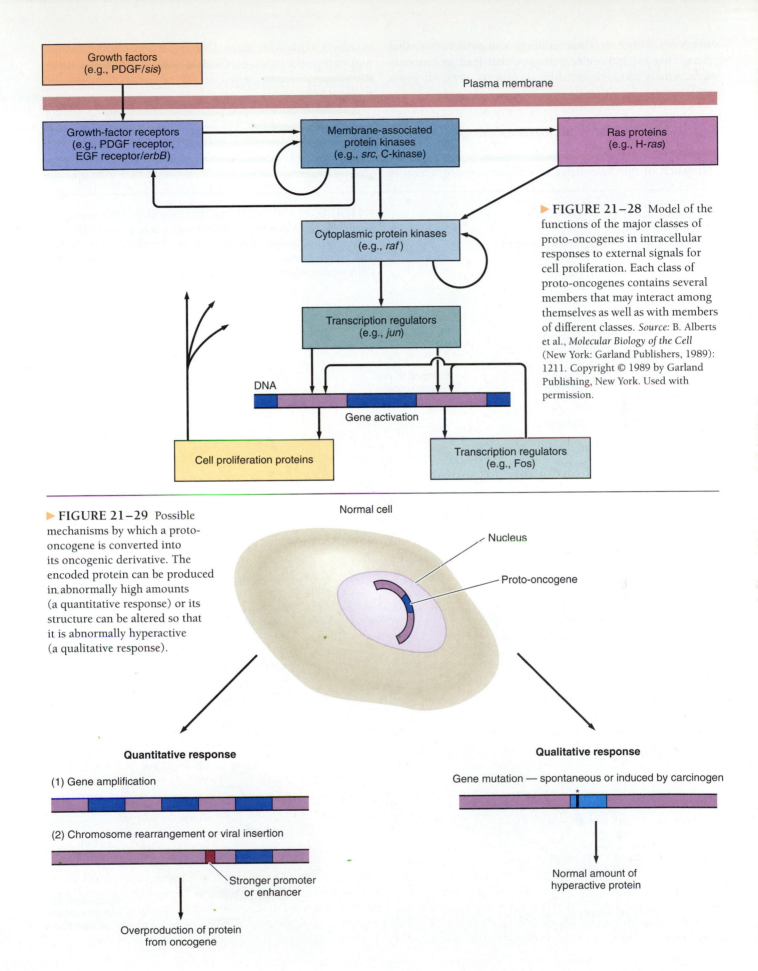

► **FIGURE 21–28** Model of the functions of the major classes of proto-oncogenes in intracellular responses to external signals for cell proliferation. Each class of proto-oncogenes contains several members that may interact among themselves as well as with members of different classes. *Source:* B. Alberts et al., *Molecular Biology of the Cell* (New York: Garland Publishers, 1989): 1211. Copyright © 1989 by Garland Publishing, New York. Used with permission.

► **FIGURE 21–29** Possible mechanisms by which a proto-oncogene is converted into its oncogenic derivative. The encoded protein can be produced in abnormally high amounts (a quantitative response) or its structure can be altered so that it is abnormally hyperactive (a qualitative response).

oncogenes. However, these findings also demonstrate that uncovering the molecular pathways that lead to cancer is an extremely complex and difficult process, even for just a single gene.

Oncogenic Viruses and Retroviral Oncogenes

Oncogenes were first discovered in the oncogenic viruses, and much of our knowledge of cellular transformation is based on work with them. Oncogenic viruses are divided into two general classes, the DNA tumor viruses and the RNA tumor viruses, based on the nature of their genetic material.

The life cycle of the DNA tumor virus SV40 (a simian virus) is illustrated in ▶ Figure 21–30. In general, infection with a DNA tumor virus can have one of two possible consequences: (1) productive infection, in which progeny virus particles are formed and released and eventually de-

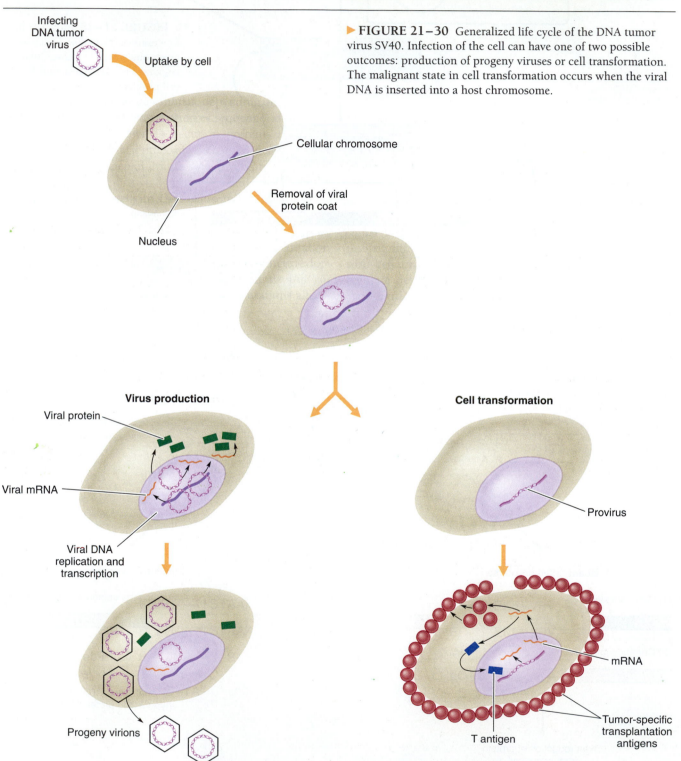

▶ **FIGURE 21–30** Generalized life cycle of the DNA tumor virus SV40. Infection of the cell can have one of two possible outcomes: production of progeny viruses or cell transformation. The malignant state in cell transformation occurs when the viral DNA is inserted into a host chromosome.

Infecting DNA tumor virus

Uptake by cell

Cellular chromosome

Removal of viral protein coat

Nucleus

Virus production

Viral protein

Viral mRNA

Viral DNA replication and transcription

Progeny virions

Cell transformation

Provirus

mRNA

T antigen

Tumor-specific transplantation antigens

stroy the host cell, or (2) **cell transformation, in which the malignant state develops with little or no viral production. In cell transformation, the viral DNA is inserted into the host chromosome, where it replicates in synchrony with the host DNA and transforms the cell by a mechanism that involves a viral protein called the T antigen. This interesting protein functions both in productive infection and in transformation of the cell to a tumorigenic state, and it is essential to the maintenance of the transformed state.**

The RNA tumor viruses are retroviruses (see Chapter 16), that use the enzyme reverse transcriptase to form a DNA copy of their RNA chromosome within the infected cell. The life cycle of the Rous sarcoma retrovirus was shown as an example in Figure 16–24. Transformation by this virus occurs when the DNA copy of the viral RNA is inserted into the host chromosome. RNA polymerase then transcribes the integrated viral chromosome, producing both viral mRNA, which codes for viral proteins, and progeny RNA, which is encapsulated to form new tumor viruses. In contrast to infection by the DNA tumor viruses, a cell infected by an RNA tumor virus usually is not killed but continues to produce progeny virus particles while in the transformed state.

Some retroviruses induce cell transformation because of the presence of an oncogene on the viral chromosome. In fact, the first oncogene to be discovered was found in the Rous sarcoma virus. Unlike the DNA tumor virus oncogenes, which encode proteins that function in normal viral multiplication as well as in transformation, RNA tumor virus oncogenes do not play a role in the production of progeny viruses—they are not normal viral genes at all, but instead are modified versions of host proto-oncogenes that have been accidently incorporated into the viral chromosome in place of part of the viral genome. For example, the cellular *src* gene (*c-src*) codes for a membrane-associated tyrosine kinase that functions in cell division. This proto-oncogene has an oncogenic counterpart carried by the Rous sarcoma virus, *v-src*, that transforms the cell following viral infection.

Proto-oncogenes are usually split into exons and introns,

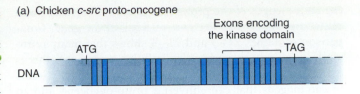

(a) Chicken *c-src* proto-oncogene

(b) Rous sarcoma viral genome

▶ **FIGURE 21–31** Structures of (a) the *src* proto-oncogene and (b) the Rous sarcoma virus genome containing the *v-src* oncogene. The dark blue regions in (a) and the red region in (b) represent exons; *gag, pol,* and *env* are viral genes that are essential for the viral life cycle. In addition to lacking introns, the *v-src* gene contains mutations that make the encoded protein kinase unregulated and hyperactive. Rous sarcoma virus is unusual among oncogene-containing retroviruses in that it is not rendered defective by acquisition of the *src* gene; all three essential viral genes are still present on its chromosome.

as are most eukaryotic genes. However, retroviral oncogenes do not contain introns, suggesting that they originated from processed proto-oncogene mRNA. ▶ Figure 21–31 compares the chicken *c-src* gene to the Rous sarcoma virus genome; the viral genome contains only *src* exons. Although it is not known how the proto-oncogene is incorporated into the viral genome, one attractive possibility is that the virus enters the proviral state at a site adjacent to the *src* gene. If the proviral gene section is transcribed as a unit under the control of a single viral promoter, processing of the resulting RNA could yield a viral genome that contains the oncogene exons only.

■ Table 21–4 lists some of the 20 or so presently known retroviral oncogenes and their normal cellular counterparts. The homology between the encoded cellular and viral proteins is striking. In some cases, the virus has acquired

■ **TABLE 21–4** Comparison of retroviral oncogenes and the cellular genes from which they are derived.

Gene	Codons in *c-onc*[1]	Codons in *v-onc*[2]	Number of Amino Acid Differences	Region Missing or Replaced in *v-onc*
erbB	1210	600	99	N-terminal half and C-terminus
fms	980	930	20	C-terminal 50 amino acids
mos	369	369	11	none
myb	640	372	11	N- and C-termini
myc	417	417	2	none
H-*ras*	189	189	3	none
sis	220	220	18	none
src	533	514	16	C-terminal 19 amino acids

[1]The *c-onc* codons represent the length of the coding region (exons only).

[2]The *v-onc* codons represent the length of the viral region that is homologous to *c-onc*.

Adapted from: B. Lewin, *Genes V* (Oxford: Oxford University Press, 1994): 1192. Used with permission.

the entire length of the proto-oncogene, so that only point mutations distinguish the two versions of the gene. Comparison of Tables 21–3 and 21–4 shows that many retroviral oncogenes correspond to cellular oncogenes; this is expected, since retroviruses are simply one of many kinds of carcinogens that can convert a proto-oncogene into an oncogene.

Retroviral oncogenes are also thought to transform cells by both quantitative and qualitative alterations. For example, loss of the C-terminal domain of the c-Fms protein during retroviral uptake increases its activity; this quantitative response suggests that this domain is a regulatory region. A qualitative response would occur if incorporation of the proto-oncogene into the retrovirus changes its coding sequence and results in a modified protein with altered activity.

We should also note that some retroviruses transform cells even though the viral genome does not contain an oncogene. In these cases, integration of the viral genome near a proto-oncogene of the host cell activates the proto-oncogene, causing transformation (as described earlier for some *ras* oncogenes).

Chromosome Aberrations and Human Cancer

Because viral cancers are known to exist in animals, and because viruses can transform cultured human cells in the laboratory, scientists have searched for viruses that cause cancer in humans. However, despite the vast amount of work in this area in recent years, the search for human cancer viruses has been largely unsuccessful. It seems unlikely that viruses will ever prove to be a major cause of human cancer, except in the case of certain leukemias (retrovirus HTLV-I), certain lymphomas (Epstein-Barr virus), liver cancer (hepatitis B virus), and uterine carcinoma (papillomavirus). Even in these cases, the viruses do not carry oncogenes and only a small proportion of the people infected with the virus develop cancer, perhaps indicating that additional factors must act in conjunction with the virus for cellular transformation.

Several nonviral oncogenes, however, have been detected in various lines of human cancer cells. For example, the *ras* cellular oncogenes, a ubiquitous eukaryotic gene family, have been the focus of especially intense study since 1982, when they were first identified by the 3T3 transfection assay. These genes were initially discovered as retroviral oncogenes, but the cellular oncogene members of the family are more prevalent. They have been found in many human cancers, including two sarcomas, several leukemias, and more than half a dozen carcinomas, such as bladder, breast, colon, liver, lung, ovarian, and pancreatic cancers. The *abl, erb, met*, and *myc* cellular oncogenes have also been detected in human cancers.

In only a few cases is it known exactly what kind of event converts the proto-oncogene into its oncogenic form. Karyotype analysis of tumor cells reveals particularly interesting associations between chromosomal translocations and certain cancers. ▶ Figure 21–32 shows the conversion of the *c-myc* proto-oncogene into an oncogene found in patients with Burkitt's lymphoma, a B cell cancer. A translocation between chromosome 8 and any of the three chromosomes (2, 14, or 22) that contain antibody-encoding genes positions the *c-myc* gene near an antibody locus. The *c-myc* protein product is a nuclear phosphoprotein that is expressed in actively dividing B cells. Repositioning of the *c-myc* gene enhances its transcription, producing cancer through a quantitative response. (As mentioned earlier, most cancers cannot be attributed to a single cause. In this case, the Epstein-Barr virus also seems to play a role in cellular transformation, but the nature of its effect is not

▶ **FIGURE 21–32** Reciprocal translocation that positions the *c-myc* gene near the constant region coding sequence of the heavy chain antibody gene. The translocation increases the level of expression of the *myc* gene by as much as 10 times.

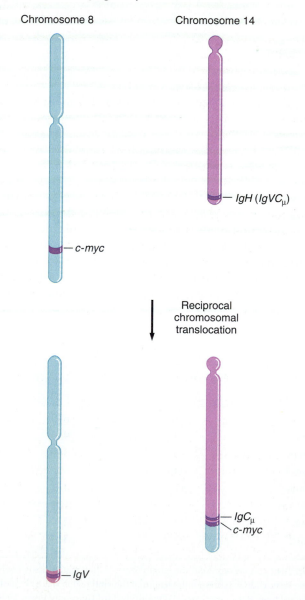

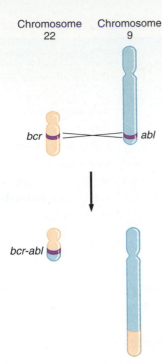

FIGURE 21–33 The reciprocal translocation between human chromosomes 9 and 22 gives rise to the Philadelphia chromosome (lower left), which is associated with chronic myelogenous leukemia. The Philadelphia chromosome carries a hybrid *bcr-abl* gene, whose protein product is abnormal and causes overproduction of hemopoietic bone marrow cells.

known. There also seems to be an association between the incidence of Burkitt's lymphoma and malaria, whose role in transformation is unknown.)

A chromosomal translocation is also responsible for chronic myelogenous leukemia. Reciprocal translocation between chromosomes 9 and 22 joins the *bcr* gene (function unknown) on chromosome 22 to the *c-abl* proto-oncogene on chromosome 9 (▶ Figure 21–33). The result is the so-called Philadelphia chromosome which contains a hybrid *bcr-abl* gene that produces an abnormal protein and causes excessive proliferation of hemopoietic bone marrow cells. In addition to the two cancers just described, at least a dozen others are known to be associated with other chromosomal translocations, showing how strongly the expression of a gene can be influenced by its position in the genome.

In addition to the translocations, a number of deletions have been found in tumor cells. A well-known example involves retinoblastoma, an eye tumor that occurs in children and has both hereditary and nonhereditary forms. In the hereditary form, multiple tumors arise in both eyes, while in the nonhereditary form, only one tumor arises in one eye. Karyotype analysis of tumors from both forms of the disease reveals deletion of a specific portion of the long arm of chromosome 13. The association of the loss of a chromo-

somal region with the development of tumors suggests that some cancers might be caused by the failure of a suppression of cell proliferation.

Tumor-Suppressor Genes

Although oncogenes have been the primary focus of most of the research on the cell cycle and on cancer during the past decade, it has become clear that cell proliferation is regulated not only by growth-promoting proto-oncogenes but also by growth-constraining genes, the tumor-suppressor genes. Just as mutations that cause constitutive expression of proto-oncogenes yield unrestrained cell growth, so do mutations that inactivate tumor-suppressor genes.

The first hint of the existence of tumor-suppressor genes came from somatic cell hybridization experiments. For example, the Ehrlich cell line produces a fatal malignancy when injected into mice. However, when investigators fuse the malignant cells to normal cells and then inject the hybrid cells into mice, no malignancy develops as long as the normal chromosome number is maintained in the hybrid cells (see ▶ Figure 21–34). The malignant state thus acts as a recessive character, but it reappears if an Ehrlich/normal hybrid cell loses many of its chromosomes during prolonged culture, as hybrid cells tend to do. These results suggest that the chromosomes of normal cells carry specific suppressors of the malignant state and that malignancy can

▶ **FIGURE 21–34** Results of fusing normal and Ehrlich cancer cells. When the hybrid cells are injected into mice, no malignancy develops. The cancerous state acts as a recessive character, but genetic suppression of the malignant state is relaxed over time with the loss of chromosomes from the hybrid cells.

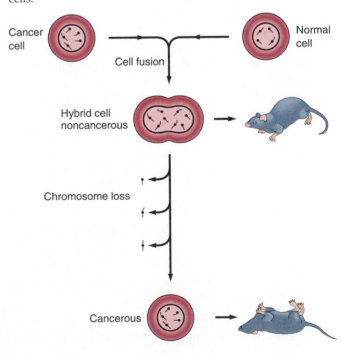

Gene	Cancer Type	Mode of Action	Hereditary Syndrome
APC	colon carcinoma	unknown	familial adenomatous polyposis
DCC	colon carcinoma	cell adhesion molecule	——
NF1	neurofibromas	GTPase-activator	neurofibromatosis type 1
NF2	Schwannomas and meningiomas	links membrane to cytoskeleton?	neurofibromatosis type 2
p53	colon cancer; many others	transcription factor	Li-Fraumeni syndrome
rb	retinoblastoma	transcription factor	retinoblastoma
RET	thyroid carcinoma; pheochromocytoma	tyrosine kinase receptor	multiple endocrine neoplasia type 2
VHL	kidney carcinoma	unknown	von Hippel-Lindau disease
WT-1	nephroblastoma	transcription factor	Wilms tumor

Source: J. Marx, "Learning how to suppress cancer," *Science* 261 (1993): 1385. Copyright © 1993 by the AAAS. Used with permission.

be the result of inactivation or loss of such a suppressor. In contrast to the dominant expression of the oncogenes we have discussed so far, cancers caused by the loss of tumor-suppressor genes are expected to behave as recessive traits, since both copies of the suppressor gene have to be inactivated before transformation occurs.

Wild-type genes whose normal function is the suppression of cell proliferation are identified by finding their mutant nonsuppressive alleles in tumors. ■ Table 21–5 lists some tumor-suppressor genes implicated in particular human cancers. The retinoblastoma gene is an example of a recessive mutant oncogene, *rb-1*, whose wild-type allele (*rb-1*⁺) is a tumor-suppressor gene. ▶ Figure 21–35 outlines the genetic events that occur in cases of retinoblastoma. In the heritable form of the disease (Figure 21–35a), the child at birth is heterozygous for the *rb* locus, having inherited a mutant (deleted) *rb-1* allele from one parent. A subsequent somatic mutation in retinal cells causes loss of the *rb-1*⁺ suppressor gene, leading to tumor formation. In contrast, the sporadic form of the disease (Figure 21–35b) begins with a homozygous *rb-1*⁺ child, and both alleles are lost, just by chance, through separate, independent somatic mutations.

Example 21.2

Consider a newborn of genotype *rb-1*⁺/*rb-1* (susceptible to hereditary retinoblastoma). Individuals affected with this cancer develop an average of three retinal tumors each. What is the chance that this newborn will never develop a retinal tumor?

Solution: If we assume that the occurrence of *rb-1* mutations in the retinal cells of this individual can be treated as a Poisson process with a mean of $m = 3$, then the probability of no mutations in the retina can be estimated from the Poisson probability formula as

$$P(0) = \frac{(3)^0 e^{-3}}{0!} = e^{-3} = 0.05$$

Thus, there is only a 5% chance that the newborn will not develop retinal tumors.

Follow-Up Problem 21.2

A somatic cell hybridization experiment fuses a retinoblastoma cell and a normal cell. Based on the preceding description of the function of the *rb-1*⁺ gene, what do you think will be the phenotype of the hybrid cell?

◆

Mutations in tumor-suppressor genes are among the cumulative factors that contribute over a long time period to the development of most human cancers. Colorectal cancer is a classic example of such development. Most cases of this cancer are the result of multiple mutations that lead to a progression of tumorigenesis. These alterations include mutations of *c-ras* to an oncogenic form (K-*ras*) and the inactivation of genes on at least five other chromosomes (2, 5, 8, 17, and 18). One of the inactivated chromosome 17 genes is *p53*, a tumor-suppressor gene that is being intensively studied because of its role in many common human cancers, including those involving the bladder, liver, brain, breast, and lungs, as well as the colon and rectum. (The *p53* gene is the most frequently mutated gene in human cancers—its mutated form is found in about half of all cases.) The chromosome 5 and 18 genes are also tumor-suppressor genes—*APC* (adenomatous polyposis coli) and *DCC* (deleted in colon carcinoma), respectively. It has been shown that malignant cells in a single tumor have the same set of mutations as is found in benign portions of the tumor, but they have at least one additional mutation that is not found in benign precursor cells. Thus, these cancers are indeed clonally derived, and a series of separate, independent events involving spontaneous and/or induced mutation in many different genes has to occur before carcinoma and subsequent metastasis develop. ■ Table 21–6 summarizes the current state of knowledge of the genes underlying most colon cancers.

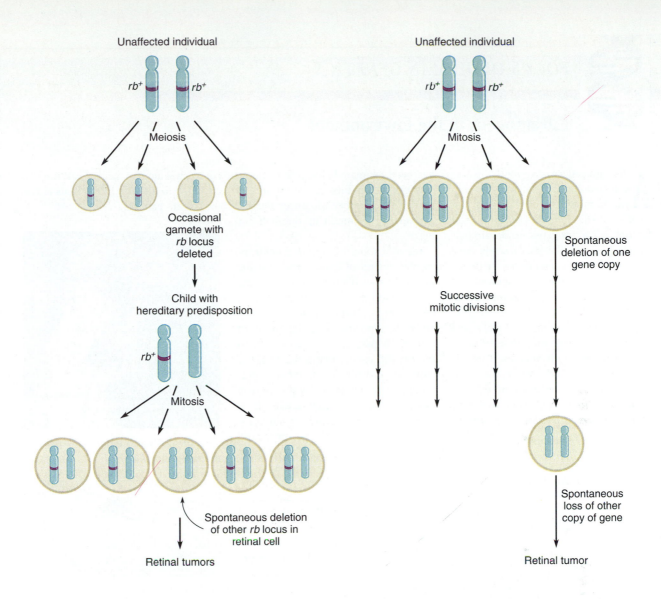

▶ FIGURE 21−35 Hereditary and nonhereditary forms of retinoblastoma. The *rb*⁺ gene is a tumor-suppressor gene. If both copies of the gene are lost, the cell becomes cancerous. In the hereditary form of retinoblastoma, an individual inherits one defective gene; the other is later lost through somatic mutation. In the nonhereditary form of the disease, both copies of the gene are lost through somatic mutation.

■ TABLE 21−6 Genes that are altered in human colon cancer.

Gene	Chromosome	Tumors with This Mutation	Type	Function
APC	5	>70%	tumor suppressor	unknown
cyclins	various	4%	oncogene	regulate cell cycle
DCC	18	>70%	tumor suppressor	cell adhesion molecule
FCC	2	~15%	unknown	maintains accuracy of DNA replication
K-ras	12	~50%	oncogene	intracellular signaling molecule
myc	8	2%	oncogene	regulates gene activity
neu	17	2%	oncogene	growth factor receptor
p53	17	>70%	tumor suppressor	regulates gene activity

Source: J. Marx, "New colon cancer gene discovered," *Science* 260: 752 (1993). Copyright © 1993 by the AAAS. Used with permission.

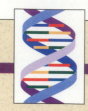

Carcinogens in the Environment

Although cancer is a genetic disease associated with gene and chromosomal mutations, it is important to clarify the role of environmental factors as carcinogens. Epidemiological studies over the past decade have clearly shown correlations between certain environmental factors and particular cancers. For example, occupational exposure to some substances, such as soot and coal tars, polyvinyl chloride, asbestos, and pesticides, is correlated with an increased likelihood of cancer. The American Cancer Society estimates that about 80% of lung cancer cases are related to smoking, and tobacco has also been implicated in cancers of the oral cavity, larynx, and esophagus. Almost all skin cancers are the result of repeated ultraviolet exposure from sunlight or tanning lamps. Diet has also been associated with cancer. In addition, many food plants manufacture their own toxic substances as a defense mechanism against predators, and a number of these substances are potentially carcinogenic to animals. Although some of these naturally occurring carcinogens cannot be avoided, it has been estimated that a few simple changes in our lifestyle could greatly reduce the number of cancer deaths, perhaps by as much as one-half (or more).

Identification of specific environmental causes of cancer and specific people at risk from exposure to various carcinogens is usually difficult. One complication is the long time lag between exposure and the onset of cancer, in part due, as mentioned in the text, to cancer being the result of cumulative effects of a number of mutations. Another factor is the genetic basis of cancers—individuals have different susceptibilities to environmental agents that cause cancer, and these differences are genetic. Whether a person will develop cancer as the result of exposure to a known carcinogen depends on that person's genetic makeup, that is, on the activity and efficiency of the metabolic processes that affect cancer susceptibility. Researchers need to identify the nature of oncogene and tumor-suppressor gene products and correlate these gene products with metabolic processes to better understand how environmental factors act as carcinogens and how to identify individuals at risk for certain cancers.

Cigarette smoke contains carcinogenic compounds.

Repeated exposure to the ultraviolet (UV) radiation in sunlight can result in skin cancer.

The Relationship between Development and Cancer

Even though embryogenesis seems, on the surface, to be quite different from cancer, both development and cell proliferation involve complex intra- and intercellular signal-transducing networks, so it seems only logical that there might be some genetic relationship between the two. Indeed, researchers have recently reported that certain genes in *Drosophila* and *C. elegans* (and perhaps in mice) function in a wide variety of developmental events as well as in cell growth and cancer.

As developmental biologists begin to isolate and molecularly characterize the particular genes that function in certain differentiation processes, the common features of the signal-transducing pathways are being revealed. Investigations of the development of neurons in the *Drosophila* eye and the development of sex organs in *C. elegans* have dis-covered genes that encode growth factor receptors with tyrosine kinase activity and other genes that function in the signaling pathways initiated by these kinases. Of particular interest are the developmental genes that are similar in sequence to activators or repressors of *ras* gene activity. Oncogene research has clearly established the Ras protein as an intermediate in the cell proliferation signal-transducing pathway that is initiated by the growth factor receptor kinases. Taken together, these discoveries suggest that the initial steps of a fundamental signaling pathway are common to both development and cell proliferation. At some point in this pathway, beyond Ras, divergence must yield cell growth on the one hand and differentiation on the other. Elucidation of the steps beyond Ras that separate differentiation from cell growth is a major challenge for researchers.

To Sum Up

1. Cancer involves uncontrolled cell proliferation and the loss of the differentiation and specialized functions of cells. Cancer is now recognized as a genetic disease that is caused by an accumulation of (mainly) somatic mutations. These mutations can affect genes that normally stimulate cell division or those that normally inhibit it. An oncogene, a mutant gene that overly stimulates cell proliferation, is formed by alteration of a proto-oncogene. Genes whose normal role is to inhibit cell division are called tumor-suppressor genes.

2. Proto-oncogenes can be converted into oncogenes in a number of ways. A mutation within the gene can lead directly to its hyperactivity (a qualitative response), or an alteration in the regulatory elements of the gene can indirectly cause its over-expression (a quantitative response).

3. The first oncogenes were discovered in tumor viruses. Both DNA and RNA tumor viruses have been studied, but the mechanism for transformation of the host cell is best understood in the retroviruses. Oncogenes carried by retroviruses lack introns, suggesting that they originated from proto-oncogenic mRNA. Uptake of a proto-oncogene by a retroviral genome can alter the gene and its expression in either a quantitative or a qualitative way.

4. Not all retroviruses carry oncogenes; some transform cells because integration of the proviral genome affects the regulation of a nearby locus.

5. The quantitative response of a proto-oncogene placed near a positive regulatory element of another gene has been associated with several human cancers. For example, certain chromosomal translocations are known to mediate genome rearrangements that lead to such cancers.

6. Tumor-suppressor genes are normal components of the genome. They are involved in the delicate balance of regulatory processes that control cell proliferation. Mutation of both copies of a tumor-suppressor locus is required for transformation of a cell; these genes thus behave as recessive oncogenes.

7. Developmental processes involve controlled cell proliferation and differentiation. Cancer involves uncontrolled cell proliferation and dedifferentiation. Thus, development and cancer are correlated, yet unique, phenomena.

Chapter Summary

Because eukaryotic genes are not organized into units of coordinate control, complex cellular signaling pathways coordinate the expression of the many genes involved in developmental processes and in the control of cell division. The complexity of this coordinate control of gene expression is illustrated by the genetics of pattern formation in insects such as *Drosophila*. The basic polarity and spatial organization of the embryo are determined by gradients in the distribution of maternal and embryonic gene products (determinants) along the anterior-posterior and dorsal-ventral axes. Some of these determinants are transcription factors that activate sequential expression of the various groups of segmentation genes as the embryonic segmentation pattern gives way to that of the larva. Once the larval segmentation pattern has been established, the homeotic genes determine what body structure develops from each segment during metamorphosis. Adult structures develop from imaginal disks, which are undifferentiated groups of cells that have been determined (committed to their fate) during the larval stage.

Many of the gradient-establishing, segmentation, and homeotic genes of *Drosophila* contain the homeo box, a sequence of nucleotides that encodes a stretch of highly basic amino acids known as the homeo domain. These gene products appear to act as transcription factors, with the homeo domain binding to DNA through a helix-turn-helix configuration. Homeobox genes have been found in a wide range of organisms, suggesting that they are universally involved in the coordinate regulation of gene activity during development.

The switching of mating type in yeast and the activation of antibody-encoding genes are two developmental processes that involve DNA rearrangements. During mating, an interaction between a MAT protein and its receptor sets in motion a signal-transducing pathway that activates the transcription of genes involved in the establishment of diploidy. Cells of most yeast strains repeatedly switch their mating type through events at the *MAT* locus. Each *MAT* locus is flanked by silent loci containing unexpressed *MAT*

alleles; replicative transposition of DNA from one of these silent loci into the active *MAT* locus switches the mating type of that cell if the transposition changes the allelic state of the *MAT* gene.

The DNA rearrangements that generate antibody-producing genes occur during the development of B lymphocytes. As each lymphocyte develops, it undergoes a DNA rearrangement that brings randomly chosen, variable encoding segments together into functional variable domain exons for both the light- and heavy-chain genes. Thus, a developmentally mature B cell can produce only one type of antibody, which is encoded by its rearranged variable region. However, because the variable encoding segments that are brought together to generate the active antibody gene are chosen randomly and because any pair of light chains can combine with any pair of heavy chains, the number of different antibody molecules that an organism can potentially produce is enormous. Imprecise joining of variable segments and a high rate of somatic mutation in the variable encoding regions of B cells add to this diversity.

Cancer is a genetic disease that usually involves the cumulative effects of multiple somatic mutations in genes that normally stimulate cell division (proto-oncogenes) or in genes that normally inhibit division (tumor-suppressor genes). A proto-oncogene can be converted into an oncogene when a mutation in the gene itself leads to its hyperactivity (a qualitative alteration) or when a mutation such as gene amplification or translocation brings the gene under the control of a different regulatory region that overstimulates its expression (a quantitative alteration).

Certain DNA and RNA tumor viruses also carry oncogenes. The oncogenes of RNA tumor viruses are thought to be derived from cellular proto-oncogenes that have been altered as a result of uptake by the virus.

Mutations in genes that normally restrain cell division (the so-called tumor-suppressor genes) can also give rise to unregulated cell proliferation. Most human cancers have been linked to altered tumor-suppressor genes.

Control Elements Affecting Development

1. Define the following terms and distinguish between members of paired terms:

   differentiation and determination
   homeo domain
   homeotic genes and homeobox genes
   totipotent

2. Summarize the effects of the *bicoid, nanos,* gap, pair rule, segmentation polarity, and homeotic genes on the development of *Drosophila.* How do these genes interact with each other?

3. In the sea urchin, development up to the gastrulation stage (the 15th hour of development) occurs even in the presence of actinomycin D. If actinomycin D is removed at the end of blastula formation, gastrulation still does not proceed. In fact, if this drug is present only between the 6th and 11th hours of development, gastrulation is arrested. What can you conclude about the role of gene transcription during hours 6 through 11?

4. *Drosophila* females homozygous for the recessive *bicoid* allele produce embryos that lack heads and thoracic structures. Females homozygous for the recessive *oskar* gene produce embryos that lack all abdominal segments. Both of these effects occur regardless of the paternal or embryo genotype. (a) What can you conclude about the normal functions of these genes? (b) Are these genes active before or after oogenesis?

5. The *Hox* genes of humans are strikingly similar in DNA sequence to the homeobox genes of *Drosophila.* What does this similarity suggest with regard to the genetic control of human development?

6. Why does the occurrence of identical quintuplets in humans prove that human embryonic cells remain totipotent for at least three cell division cycles?

Regulation by Gene Rearrangement

7. Describe the pattern of gene expression necessary to make a yeast cell mating type a. Describe the gene expression pattern required for the cell to be type α.

8. Describe the genes expressed and the events that occur during the switch of mating type in yeast from α to a.

9. What effects would each of the following mutations have on mating type in yeast? (a) $MAT\alpha1^-$, (b) $MAT\alpha2^-$, (c) $MATa1^-$, (d) deletion of *HMLE.*

10. How many different antibodies can a B lymphocyte produce? How many can it potentially produce before it differentiates?

11. How many polypeptide chains form an antibody molecule? Describe their characteristics. How many different antigens are recognized by an antibody molecule?

12. Describe the process that generates a complete variable domain coding region. Describe the process that determines the class of antibody made by a B cell, including the mechanism of class switching.

13. If the genome of a certain species contains 500 V_L, 10 J_L, 1 C_L, 300 V_H, 10 D, 10 J_H, and 5 C_H copies, which can join in all possible combinations during the differentiation of B lympho-

cytes, what is the minimum number of different antibodies that can be produced by this organism?

14. A radioactive cDNA probe is made from the mRNA for the light chain of an antibody molecule. DNA isolated from embryonic B cell clones and from mature B lymphocytes is digested with a restriction enzyme, and the resulting fragments are separated by gel electrophoresis. A probe of the gel yields the following results:

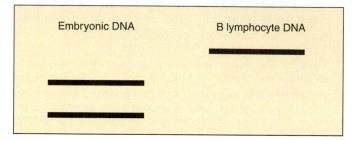

Explain these results.

Cancer and Cell Proliferation

15. Changes that can be detected immunologically occur on the surface of many types of cancerous cells. However, the cell surface antigens present on a cancer cell are often identical to those present on normal embryonic cells. Suggest an explanation of why surface antigens on many cancer cells are the same as those of normal embryonic cells.

16. Defend or criticize the following statements: (a) All retroviruses cause cancer. (b) All RNA tumor viruses carry an oncogene. (c) All RNA tumor viruses that cause cancer carry an oncogene. (d) The vast majority of cancers are preventable. (e) Cancer is a disease of aging; predominantly older people get cancer.

17. The methionine analogue L-ethionine causes liver cancer in rats. L-ethionine is a nonmutagenic carcinogen that inhibits DNA methylation. How might this analogue cause cancer?

18. The DNA sequences of proto-oncogenes are highly conserved in a large number of species. What does this tell you about the functions of these genes?

19. Describe the ways in which a proto-oncogene differs from an oncogene. What are some of the ways in which a proto-oncogene can be converted to an oncogene?

20. Can the amplification of a proto-oncogene (without mutation of that gene) lead to oncogenesis? Explain your answer.

21. What evidence is there that proto-oncogenes are not simply integrated retroviral oncogenes?

22. What is some of the evidence that the *v-src* gene arose after the *c-src* gene?

23. Describe a procedure you might use to determine whether particular cancer cells (say, cells of the colon) are expressing a certain proto-oncogene.

24. The genome of a retrovirus that lacks an oncogene is found to be integrated into a host chromosome about 3 kb away from a proto-oncogene. Cells harboring this provirus are found to produce 10 times more oncogene-specific mRNA than uninfected cells. Explain what could account for this increase in mRNA synthesis.

25. Is it correct to view the mutant *rb-1* allele as a recessive oncogene? Explain your answer.

26. The AIDS (HIV) virus can cause a variety of types of cancer, including Karposi's sarcoma, B-cell lymphoma, and cancer of the rectum and tongue. The HIV virus does not carry an oncogene. Postulate how this virus might be able to transform cells.

VIII

MECHANISMS OF MUTATION, RECOMBINATION, AND REPAIR

A recurring theme of genetics is that the chemical structure of DNA (and in some cases, RNA) must allow it to meet certain basic requirements that are essential to its role as the genetic material. These requirements include the capacity to store and transfer genetic information and the ability to replicate, mutate, and recombine. A valuable feature of the Watson-Crick model is that it predicts how the DNA molecule can fulfill certain of these requirements. We have already seen how DNA's seemingly endless variety of base-pair sequences enables it to code for the structure of proteins. We have also seen how the hydrogen-bonded chains can separate to permit the encoded information to be transferred directly to a complementary RNA strand during transcription and to complementary DNA strands during replication. We will now consider the features of the DNA double helix that allow it to undergo mutation and recombination.

Mutation and recombination both lead to genetic

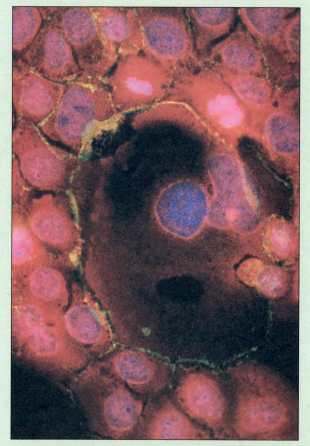

Cancerous skin cells, the result of failure of the DNA repair system to correct damage caused by exposure to sunlight.

change. Mutation is the ultimate source of variation, since it introduces changes in the base-pair sequence of individual genes. Recombination reshuffles the mutant genes in each generation and thereby serves as an immediate source of variation. Acting to moderate these changes (but also to protect them in some cases) is the process of repair. The enzymatic repair mechanisms of living cells serve as an adjunct to both mutation and recombination by controlling the rate at which variation is generated. These repair mechanisms are essential for maintaining the stability of the genome, since without them, organisms could not survive the natural rate of damage to their DNA.

The following two chapters describe the processes of mutation, recombination, and repair in terms of what is known about their mechanisms at the molecular level. We will begin by considering the molecular events involved in mutation and follow this with a discussion of the enzymology of recombination and repair.

CHAPTER 22

Molecular Basis of Mutation

DNA is subject to a surprising variety of chemical alterations in each generation. Replication errors occur resulting in the gain or loss of nucleotides and the incorporation of incorrect bases. Structural changes may also result from hydrolytic reactions that remove amino groups by deamination and cause base loss by depurination and depyrimidation. Data suggest that as many as 10,000 depurinations occur in each mammalian genome per day, a much higher rate than generally recognized. DNA is also the target of numerous physical and chemical agents. Heat can induce base loss, forming apurinic and apyrimidinic (AP) sites; UV radiation produces cross-linking; and ionizing radiation (for example, X-rays) can cause single- and double-strand breaks (▶ Figure 22–1). In addition, various chemicals can react with DNA to modify bases or otherwise alter its structure. In the following sections, we will consider several of these effects on DNA structure, concentrating

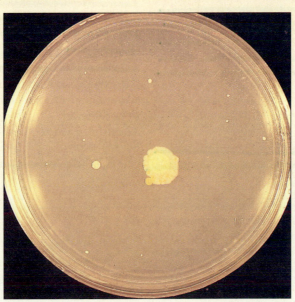

The Ames test. Bacterial growth surrounding the disk in the center of the dish signifies that the disk contains a mutagenic substance.

► **FIGURE 22–1** Various ways of damaging DNA that can subsequently give rise to mutation.

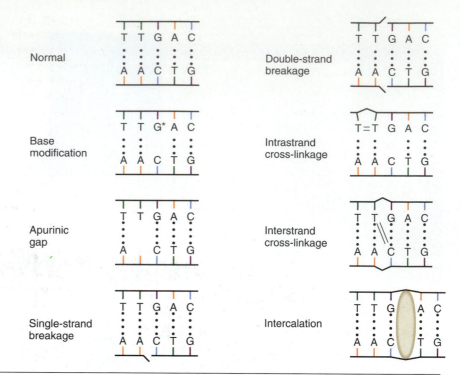

on changes that lead to point mutations, that is, alterations of a single base pair.

SPONTANEOUS POINT MUTATIONS

Point (single-site) mutations in DNA can occur through two possible mechanisms: (1) the substitution of one base pair for another and (2) the addition or deletion of a nucleotide pair. Both mechanisms give rise to a mutant gene that has a different sequence of nucleotides than its normal allele. The results differ in one fundamental respect (► Figure 22–2): Exchanging one base pair for another will alter only one codon within the mutant gene; the codons on either side of the mutation remain the same. In contrast, adding or deleting a base pair scrambles the genetic message downstream from the mutation, affecting all codons that are subsequently read.

Base Substitutions

Simple base substitutions are the most common form of base-pair change. They fall into two general classes: transi-

tions, in which a purine in one strand of DNA is replaced by another purine and the pyrimidine in the complementary strand is replaced by another pyrimidine, and **transversions**, in which a purine replaces a pyrimidine and a pyrimidine replaces a purine (► Figure 22–3). We can represent the different transition events by the general notation AT ⇌ GC, and the different transversions as AT ⇌ TA, AT ⇌ CG, GC ⇌ CG, and GC ⇌ TA.

Transitions are the most frequently occurring class of base substitution. Spontaneous transitions may occur during DNA replication because of isomeric changes in base structure, called **tautomerism**. These changes result from a shift in the location of a proton from one position to another in a purine or pyrimidine ring. Such a shift in proton position (called a **tautomeric shift**) can change the hydrogen-bonding properties of the base in which it occurs. For example, adenine normally pairs with thymine, but if adenine undergoes a tautomeric shift in which a hydrogen from its amino group at the number 6 position moves to its nitrogen at the number 1 position (► Figure 22–4), its hydrogen-bonding properties will be changed to those of guanine. Adenine in this rare **imino** form then pairs with

► **FIGURE 22–2** Effects of base substitution and nucleotide deletion on the coding sequence of a gene.

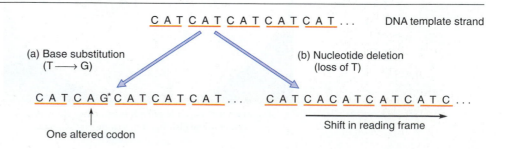

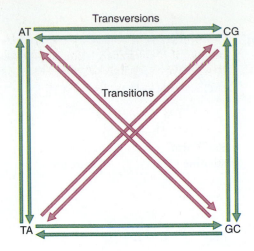

FIGURE 22–3 The possible base pair substitutions in DNA. The substitutions shown on the diagonals are transitions, while those on the periphery are transversions.

cytosine, as shown in ▶ Figure 22–5a. Note that the reverse situation also occurs: A tautomeric shift in guanine produces its rare **enol** form and changes its base-pairing properties to those of adenine (Figure 22–5b). All four bases are subject to these structural changes and can exist, at least briefly, in their rare tautomeric states, in which they have altered base-pairing properties. In general, therefore, a purine in its rare state can pair with the "wrong" pyrimidine, and a pyrimidine in its rare state can pair with the "wrong" purine.

For such a pairing mistake to occur and produce a transition, the altered base must appear in its rare tautomeric form at the moment of DNA replication. Only then can a mismatched base pair be incorporated into a DNA molecule. Tautomeric shifts can induce base-pairing mistakes in two ways (see ▶ Figure 22–6). One involves a **template error**, in which the rare tautomeric form of the base occurs in the template strand during replication (Figure 22–6a). The

▶ **FIGURE 22–4** Rare forms of the four DNA bases, caused when a proton shifts from one position to another in a purine or pyrimidine ring. The usual forms of the bases are shown on the left, the rare tautomeric forms on the right.

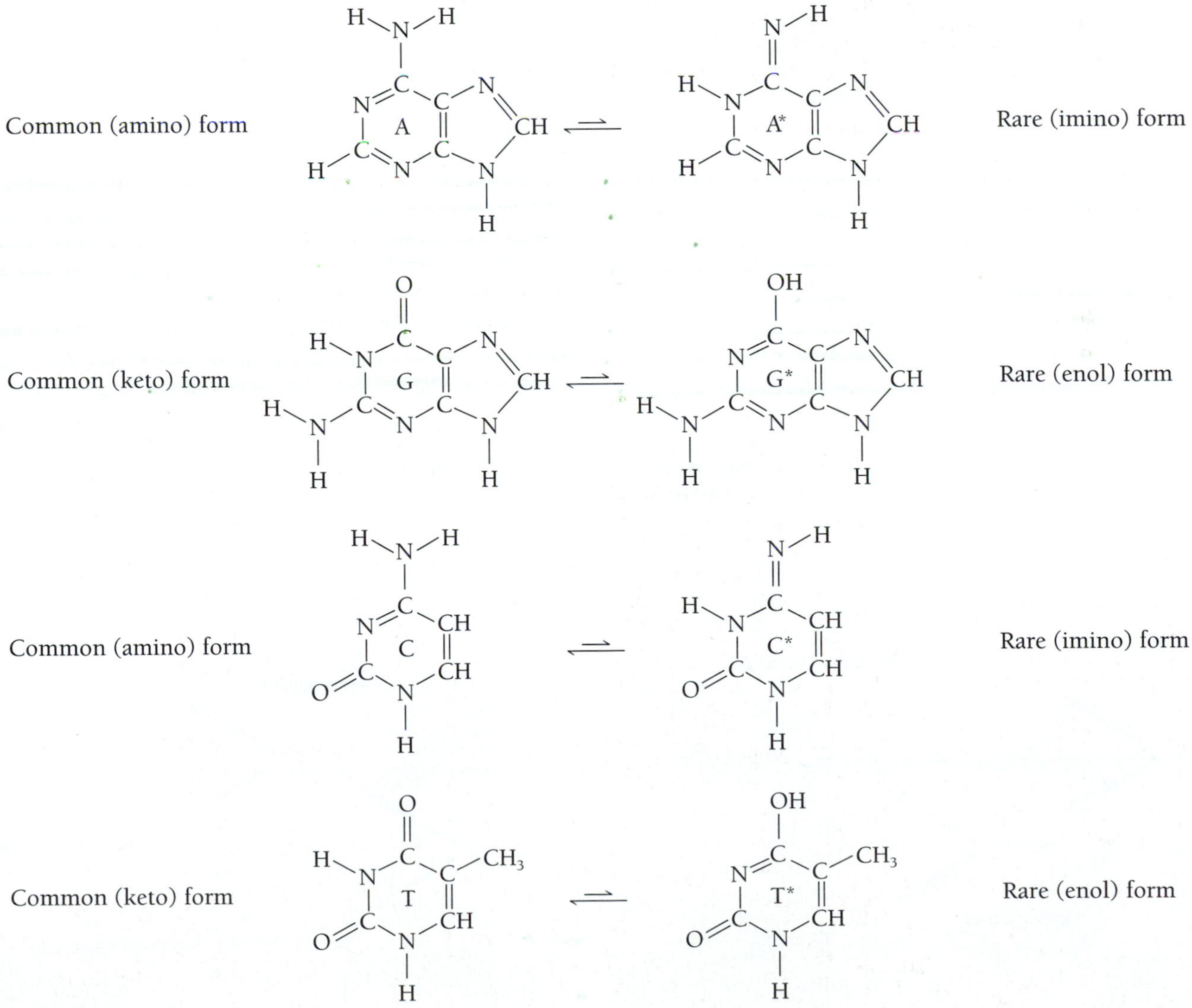

Common (amino) form ⇌ Rare (imino) form

Common (keto) form ⇌ Rare (enol) form

Common (amino) form ⇌ Rare (imino) form

Common (keto) form ⇌ Rare (enol) form

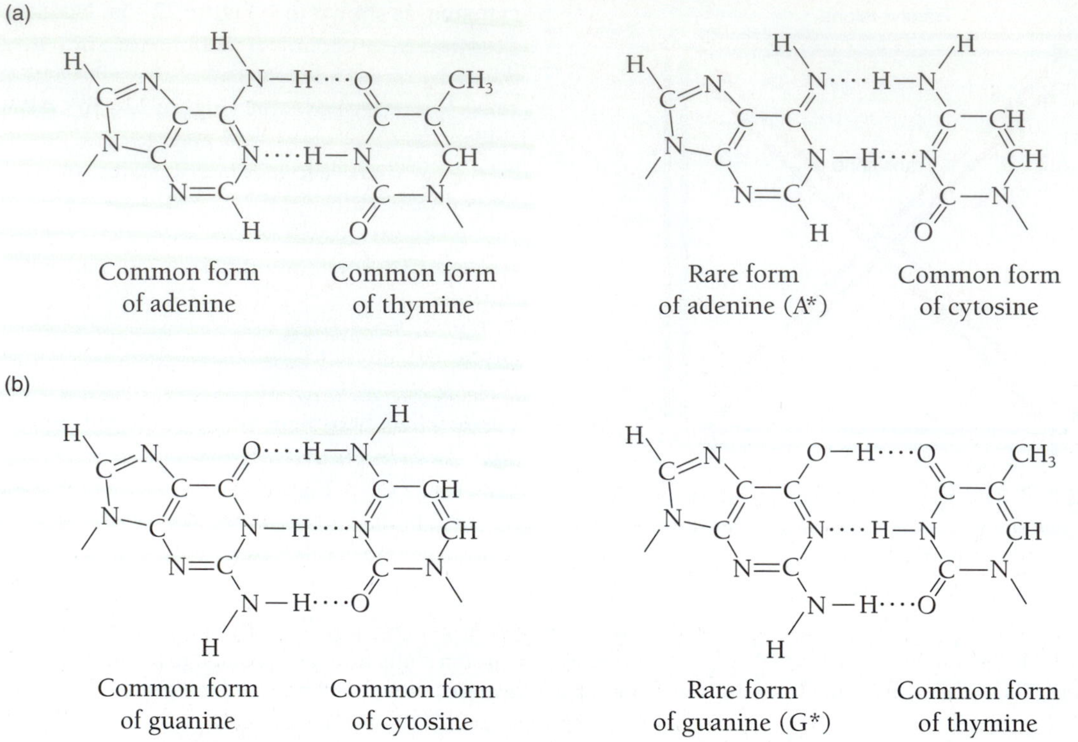

(a)

Common form   Common form
of adenine   of thymine

Rare form   Common form
of adenine (A*)   of cytosine

(b)

Common form   Common form
of guanine   of cytosine

Rare form   Common form
of guanine (G*)   of thymine

▶ **FIGURE 22–5** Errors in base pairing as a result of tautomerism. (a) The common form of adenine forms hydrogen bonds with thymine (left), while in its rare tautomeric form, it forms hydrogen bonds with cytosine (right). (b) Guanine in its common form bonds to cytosine (left); in its rare tautomeric form, it bonds to thymine.

other involves an **incorporation error**, in which the rare tautomeric form of the base exists in a deoxyribonucleotide precursor that is inserted into the growing chain (Figure 22–6b). In either event, the result is a DNA molecule that contains an incorrectly matched base pair (a **heteroduplex**). The rare base is highly unstable and is expected to return to its normal state. A second round of DNA replication is then required to segregate the mismatched base pair and pro-

duce a fully mutant double helix. Thus after two rounds of DNA replication, a replication error in the parental DNA molecule will result in the replacement of one base pair—either AT or GC—with a different base pair—either GC or AT—in the progeny DNA molecules.

In addition to the replication errors that we have just described, another potential mechanism for the origin of transitions is **deamination** of 5-methylcytosine (a rare, but nor-

▶ **FIGURE 22–6** Spontaneous mutations involving base substitutions caused by the mispairing of adenine. (a) Tautomerism of adenine in the template strand leads to the base pair change AT → GC. (b) Incorporation of the tautomeric form of adenine into a daughter strand leads to the base pair change GC → AT.

(a) Template error

(b) Incorporation error

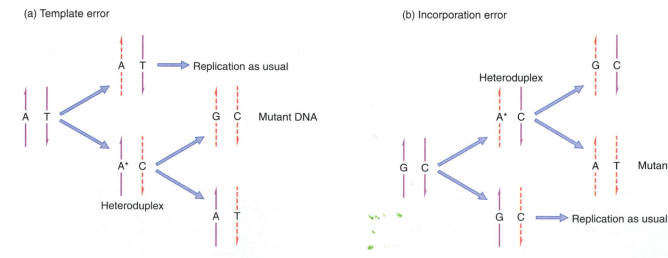

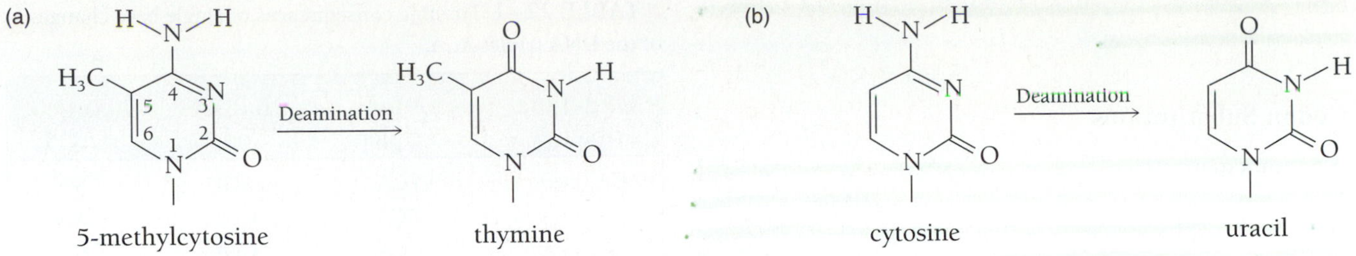

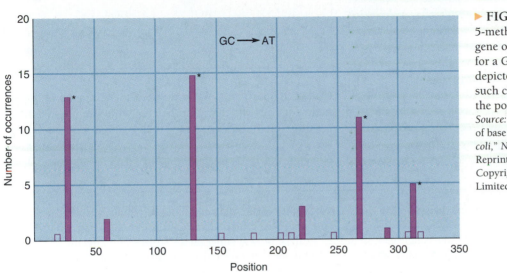

(a)

5-methylcytosine → Deamination → thymine

(b)

cytosine → Deamination → uracil

► **FIGURE 22–7** Deamination of (a) 5-methylcytosine and (b) cytosine.

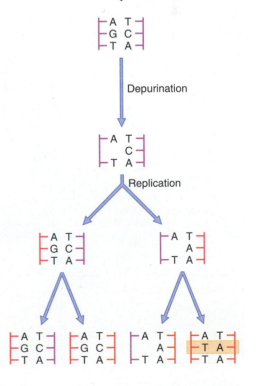

► **FIGURE 22–8** Distribution of 5-methylcytosine hot spots in the *laci* gene of *E. coli*. Of the 15 sites examined for a GC → AT transition, only the sites depicted by the closed bars showed such changes. The asterisks mark the position of 5-methylcytosines. *Source:* C. Coulondre et al., "Molecular basis of base substitution hotspots in *Escherichia coli*," *Nature* 274 (1978): 775–780. Reprinted with permission from *Nature*. Copyright 1978 Macmillan Magazines Limited.

mal constituent of DNA) and, to a lesser extent, cytosine (► Figure 22–7). Both of these bases normally pair with guanine. But when deaminated, 5-methylcytosine becomes thymine and cytosine forms uracil, which are bases that pair with adenine. The result is a GC → AT transition after a round of replication. Since uracil is not a normal constituent of DNA, there exists a repair system (to be considered later) that replaces uracil with cytosine. Thus most deamination sites involving cytosine are corrected. In contrast, the deamination product of 5-methylcytosine goes unrecognized, since it is not an unusual base; such sites often form **hot spots** of spontaneous transitions within genes (► Figure 22–8).

Tautomerism and deamination can explain the origin of many spontaneous transitions, but the origin of most spontaneous transversions is less clear. **Depuration** is one mechanism that is known to produce transversions in experimental systems; it results from breaking the glycosyl linkage that connects a purine base to the deoxyribose sugar. The gap left in the DNA by the loss of the purine (an apurinic or AP site) tends to be mutagenic if not repaired. During replication, the AP site can be bypassed and a nucleotide (usually an adenine-containing nucleotide) can be incorporated into the daughter strand exactly opposite the gap (► Figure 22–9). Since the AP site originally contained

► **FIGURE 22–9** A transversion caused when DNA polymerase inserts an adenine across from an apurinic site.

either A or G, the insertion of a purine across from the site leads to a transversion.

Codon Substitutions

The immediate effect of a base substitution in a structural gene is a change in a codon specifying a particular amino acid. Each triplet codon in a wild-type gene can undergo nine possible base substitutions. The possible mutations of the DNA triplet ACA are shown as an example in ■ Table 22–1. Note that not all of the mutations necessarily affect the protein product. Because of the degeneracy in the code, the base substitution ACA → ACG, which affects the third position of the mRNA codon, does not change the amino acid specification of the affected code word. Such substitutions are called **silent mutations**, since they have no effect on the amino acid sequence of the polypeptide.

Most single base changes in a DNA triplet lead to single amino acid substitutions. These mutations are called **missense mutations** because they make the wrong kind of sense to the ribosomal translational apparatus, that is, the codon codes for a different amino acid than the one originally in-

■ **TABLE 22–1** Possible consequences of single base changes of the DNA triplet ACA.

Wild-Type Triplet	Mutant Triplet	mRNA Codon	Amino Acid
ACA (cys)	TCA	AGU	Ser
	GCA	CGU	Arg
	CCA	GGU	Gly
	AAA	UUU	Phe
	AGA	UCU	Ser
	ATA	UAU	Tyr
	ACT	UGA	Term
	ACG	UGC	Cys
	ACC	UGG	Trp

tended. Inspection of the amino acid assignments of the genetic code reveals that amino acids with similar structures are more often than not coded for by similar codons (▶ Figure 22–10); thus a missense mutation will quite often involve amino acids with similar chemical properties. If the

▶ **FIGURE 22–10** The mRNA codons and the side-chain characteristics of the amino acids they code for. External amino acids, which are located on the surface of the protein, are normally charged or highly polar. Internal amino acids, which are buried inside a protein, are normally hydrophobic (nonpolar). Ambivalent amino acids can be present either on the surface or on the inside of the protein. Notice that a missense mutation tends to replace an amino acid with one that has similar side-chain characteristics.

Second Base

First (5') base		U	C	A	G	Third (3') base
U	UUU UUC	Phe (nonpolar, internal)	UCU UCC UCA UCG Ser (polar, ambivalent)	UAU UAC Tyr (bulky, polar, ambivalent)	UGU UGC Cys (polar, ambivalent)	U C
	UUA UUG	Leu (nonpolar, internal)		UAA UAG Stop	UGA – Stop / UGG – Trp (bulky, nonpolar, ambivalent)	A G
C	CUU CUC CUA CUG	Leu (nonpolar, internal)	CCU CCC CCA CCG Pro (nonpolar, ambivalent)	CAU CAC His (+ charge, external) / CAA CAG Gln (highly polar, external)	CGU CGC CGA CGG Arg (+ charge, external)	U C A G
A	AUU AUC AUA Ile (nonpolar, internal) / AUG – Met (nonpolar, internal)		ACU ACC ACA ACG Thr (polar, ambivalent)	AAU AAC Asn (highly polar, external) / AAA AAG Lys (+ charge, external)	AGU AGC Ser (polar, ambivalent) / AGA AGG Arg (+ charge, external)	U C A G
G	GUU GUC GUA GUG Val (nonpolar, internal)		GCU GCC GCA GCG Ala (nonpolar, ambivalent)	GAU GAC Asp (– charge, external) / GAA GAG Glu (– charge, external)	GGU GGC GGA GGG Gly (small, polar, ambivalent)	U C A G

substituted amino acid can participate in the appropriate intrachain and interchain bonding, the normal folding pattern and protein function will be maintained. (Recall that the amino acid sequence of a polypeptide chain determines its folded structure and consequently its activity.) It is thus quite possible that an alteration in the amino acid sequence will not manifest itself by altering the activity of the protein. A missense mutation that does not change the function of the protein is sometimes referred to as a **neutral mutation**.

On the other hand, some missense mutations severely affect protein function. Such mutations usually involve substituting an amino acid whose chemical properties differ from those of the normal residue at that position in the polypeptide. This alters the primary sequence of the polypeptide to such an extent that a functional conformation cannot be attained. A missense mutation of this type is the molecular basis of sickle-cell anemia, as shown by ▶ Figure 22–11.

The final type of mutational effect on the amino acid sequence is a **nonsense mutation**. This type of mutation converts a codon for an amino acid to a termination (stop) codon. As a result, polypeptide synthesis is prematurely halted, producing a shortened and almost certainly nonfunctional protein. Nonsense mutations usually have a very severe effect on protein function. Since there are three different termination codons, there are three kinds of nonsense mutations that give rise to the codons UAG, UAA, and

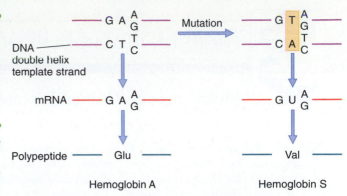

▶ **FIGURE 22–11** The base substitution thought to give rise to sickle-cell hemoglobin (HbS). Normal hemoglobin (HbA) contains glutamic acid (Glu) at the sixth position of the β-polypeptide, while HbS has valine (Val) at this position. The Glu codon that changes to a Val codon by a single base substitution could be either CTT or CTC, so the base pair substitution that changes the normal gene to its mutant allele would be AT → TA, a transversion.

UGA in the mRNA chain. These nonsense mutations are called *amber, ochre,* and *opal,* respectively (their names reflect the history of their discovery). A comparison of silent, missense, and nonsense mutations is shown in ▶ Figure 22–12.

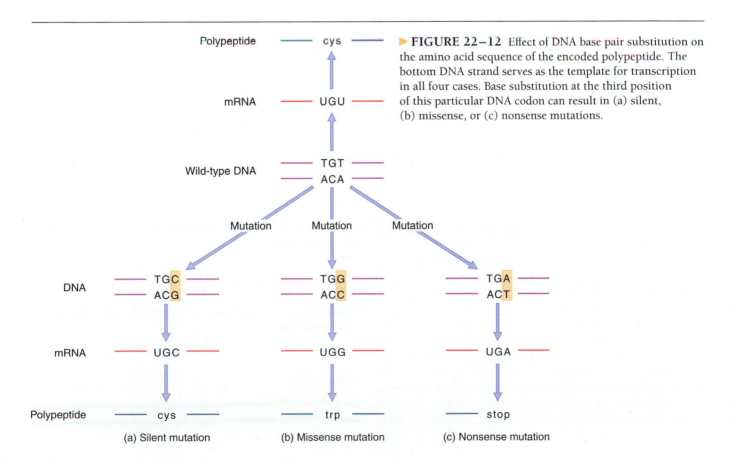

▶ **FIGURE 22–12** Effect of DNA base pair substitution on the amino acid sequence of the encoded polypeptide. The bottom DNA strand serves as the template for transcription in all four cases. Base substitution at the third position of this particular DNA codon can result in (a) silent, (b) missense, or (c) nonsense mutations.

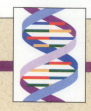

FOCUS ON HUMAN GENETICS

Mutations and Human Diseases

The use of molecular techniques to detect mutations is now uncovering the specific kinds of mutations that are responsible for certain human diseases. Two interesting discoveries are: (1) many of these mutations are not of the single-site (point) variety and (2) different kinds of alterations in DNA can lead to the same disease. In addition to base-pair substitutions and frameshift mutations, deletions, insertions, and errors in mRNA splicing reactions have also been identified as causes of certain genetic disorders.

Examples of genetic disorders caused by nucleotide-pair substitution mutations of the missense variety include sickle-cell anemia and some cases of hemophilia. Some instances of neurofibromatosis are caused by a base-pair substitution that generates a nonsense mutation. Frameshift mutations underlie most cases of Tay-Sachs disease, although the addition usually involves four base pairs and results in a premature stop codon (a nonsense mutation). Other cases of Tay-Sachs, as well as some cases of phenylketonuria, are the result of mRNA splicing errors.

Cystic fibrosis, the most common single gene disorder in the North American Caucasian population, has been linked to at least 20 different mutations in the *CF* gene. The *CF* gene is very long (250,000 nucleotide pairs); it codes for a 6.5-kb mRNA and a protein containing 1480 amino acids. About 70% of CF mutations are of the type designated ΔF508, which involves the deletion of one triplet codon in the DNA and the loss of one amino acid in the protein.

Other large human genes are also characterized by many different kinds of mutations and mutational sites. For example, in the dystrophin gene, which includes more than 2 million base pairs, various lengths and locations of large deletions account for about 60% of the cases of juvenile (Duchenne) muscular dystrophy. Some cases of Lesch-Nyhan syndrome are caused by large deletions in the *HPRT* gene. Insertions of several nucleotide pairs can also cause gene mutations; for example, fragile X syndrome is due to repeated extra copies of the triplet CGG in the *FMR-1* gene. It is becoming increasingly clear that gene muta-

tion can occur in numerous ways and that single-site base substitutions and frameshifts represent only two of many possibilities.

Pulmonary function in cystic fibrosis. A physiotherapist is performing a lung function test on a patient using a Vitalograph. This device measures various aspects of lung function such as the vital capacity, forced vital capacity, and forced expiratory rate. The age, height, and sex of the patient is entered into the computer which calculates the theoretical values for the test. The device then displays the patient's performance as a percentage of the theoretical value.

Example 22.1

Suppose that a transition type mutation occurs in the code word that specifies the amino acid methionine. What amino acids could be substituted for Met?

Solution: The mRNA codon for Met is AUG, corresponding to the DNA triplet base pair sequence ATG/TAC. An A/T → G/C transition at the first position would give GTG/CAC on the DNA, GUG on the mRNA, and Val as the amino acid. A transition mutation at the second position gives ACG/TGC on the DNA, ACG on the mRNA, and Thr as the amino acid. At the third position, a transition gives

ATA/TAT on the DNA, AUA on the mRNA, and yields Ile as the amino acid.

Follow-Up Problem 22.1

Repeat Example 22.1 for a transversion mutation in the code word that specifies Met.

◆

Frameshifts

Although the majority of spontaneous point mutations are a result of base substitutions, a surprising proportion of the

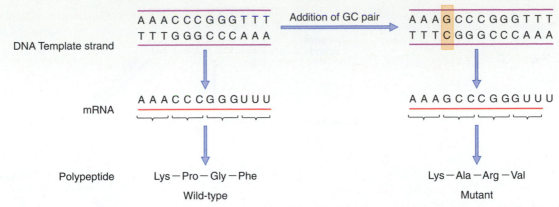

FIGURE 22-13 Frameshift mutation resulting from the addition of a nucleotide pair between the third and fourth base pairs of a wild-type gene. The reading frame of the mutant gene is shifted by one base pair, beginning at the point of addition. This shift changes the amino acid sequence of the polypeptide chain.

mutations studied in bacteria and phages are caused by the addition or deletion of a single nucleotide pair. The addition or deletion causes the reading frame of the triplet code words along the DNA sense strand to be shifted by one nucleotide, as illustrated in ▶ Figure 22-13. Nucleotide additions and deletions are therefore known as **frameshift mutations**. The shift in the reading frame originates at the point where the nucleotide has been added or deleted, and it changes the entire remaining sequence of codons. Some of the new codons are silent mutations, others are missense mutations, and still others are nonsense mutations. Since the primary structure of the encoded polypeptide is drastically altered, this kind of mutation generally abolishes the function of the protein.

Frameshift mutations are thought to result from a slippage error during DNA replication when one of the strands at the growing point "loops out" from the double helix (▶ Figure 22-14). Slippage would occur preferentially in a region containing a run of identical bases, where the failure of a base to pair with its normal partner could lead to the displacement of a DNA strand along its complementary strand by a single residue. This displacement would produce an addition if the growing strand loops out or a deletion if the template strand loops out.

Suppression and Suppressor Mutation

In the various base-pair alterations considered thus far, the wild-type sequence can be restored by a simple reversal of the original mutation. Total or partial restoration of the

▶ **FIGURE 22-14** Nucleotide additions and deletions caused by a slippage error during DNA replication.

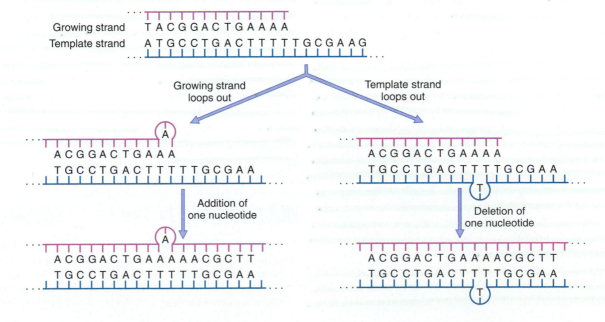

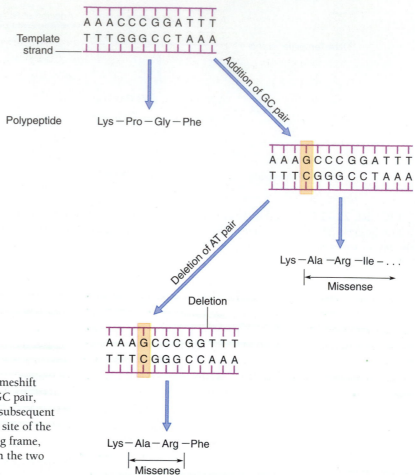

FIGURE 22–15 Intragenic suppression of a frameshift mutation. The original mutation, the addition of a GC pair, causes a frameshift. The suppressor mutation is the subsequent deletion of an AT pair six nucleotides away from the site of the original mutation. This mutation restores the reading frame, although it leaves two missense amino acids between the two sites of mutation.

wild-type phenotype is also possible through **suppression**. Recall that the term *suppression* refers to a second-site reversion that occurs when a second mutation at a different site on the DNA (a **suppressor mutation**) somehow compensates for the effects of the first. This compensation restores the wild-type phenotype, at least in part, but it leaves the genotype in a mutant condition.

Both **intragenic** and **intergenic suppression** can occur. In intragenic suppression, the original and the suppressor mutation occur in the same gene. The frameshift mutations in ▶ Figure 22–15 illustrate this type of suppression. In this example, the original mutation is the addition of a base pair, which shifts the reading frame one nucleotide out of phase. If a second frameshift mutation involving a deletion of one nucleotide should occur near the point of the first mutation, the reading frame will be restored, leaving mutant codons only between the points of addition and deletion. The effect of this stretch of mutant codons cannot be predicted, but it is likely to be less severe than if the second mutation had not taken place.

Intergenic suppression involves a suppressor mutation that occurs in a different gene than the first mutation. The best-studied example of intergenic suppression involves

nonsense suppressors. In this case, the suppressor mutation alters a tRNA anticodon so that it recognizes a termination or stop codon in a strand of mRNA. The mutant tRNA thus reads an mRNA codon containing a nonsense mutation as an amino acid and avoids the serious consequences of premature polypeptide termination. One example is the **amber suppressor** that changes the 5′ GUA 3′ anticodon of tyrosine tRNA to 5′ CUA 3′ in one of the multiple copies of tyrosine tRNA genes (▶ Figure 22–16). The wild-type tRNA^{Tyr} (the major species of tyrosine tRNA) continues to recognize the codon UAC in mRNA, while the suppressor tRNA (the minor species) recognizes the nonsense triplet UAG and inserts a tyrosine when this triplet appears in the mutant gene message.

To Sum Up

1. The ability to mutate is an intrinsic property of DNA that results in a low rate of spontaneous mutation at each gene per generation. There are two major mechanisms that cause point mutations: nucleotide-pair substitution, and addition or deletion of a nucleotide pair.

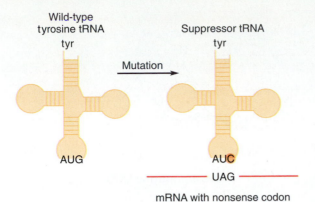

Wild-type
tyrosine tRNA
tyr

Suppressor tRNA
tyr

Mutation →

AUG

AUC

——— UAG ———

mRNA with nonsense codon

▶ **FIGURE 22–16** Nonsense suppression of UAG by a mutation in the anticodon of a tRNA for tyrosine. Tyrosine can now be inserted into a polypeptide at the position of the nonsense codon.

2. Base-pair substitutions can be classified as either transitions or transversions. Many transitions result from errors in base pairing during DNA replication. A base-pairing mistake can result from a tautomeric shift, that is, a shift in the proton distribution of a base, which causes a change in its hydrogen-bonding properties. Deamination of 5-methylcytosine is also important in causing transitions. The mechanism by which transversions occur is not well understood. One possibility is depurination followed by incorporation of a purine in the newly synthesized strand opposite the depurinated site.

3. Nucleotide-pair substitution may or may not affect the encoded peptide chain. If the changed DNA code word does not specify an amino acid change, the alteration in DNA is called a silent mutation. Missense mutations lead to an amino acid change. The effect of a missense mutation on protein function depends on the nature of the amino acid change and on the position of the affected amino acid in the polypeptide. Neutral mutations are missense changes that do not affect protein function. Nucleotide pair substitutions that produce a termination code word are called nonsense mutations.

4. Adding or deleting a nucleotide pair causes a frameshift mutation. The reading frame of the DNA triplet code words is shifted, severely altering the resulting protein structure.

5. Suppression restores the wild-type phenotype without reverting the original mutation. Suppression occurs through a second mutation (the suppressor mutation) that somehow compensates for the effects of the first mutation. The suppressor mutation can occur in the same gene as the original mutation (intragenic suppression) or in a different one (intergenic suppression).

◆

CHEMICAL MUTAGENESIS

Today's concerns about the effects of chemicals on our environment and on human health have focused attention on the mutagenic properties of such substances as cosmetics, drugs, food additives, pesticides, and synthetic industrial compounds like plastics and other petrochemical materials. Hundreds of chemicals that we encounter in our daily lives are now known to have mutagenic effects ranging from very slight to very potent. Closely tied to studies of the mutagenic properties of these chemicals have been studies of their carcinogenic (cancer-causing) properties. Nearly all known carcinogens have also been shown to be mutagenic, indicating that most forms of cancer are at least partially due to changes in DNA.

In addition to the obvious practical benefits of understanding the action of mutagenic agents, studies with chemical mutagens have been invaluable in probing the nature of the spontaneous mutation process. For example, consider a mutagen that is known specifically to cause $GC \rightarrow AT$ transitions in DNA. If we find that this mutagen is able to revert a spontaneous mutant allele back to its wild-type sequence, we can conclude that the original spontaneous mutation event was an $AT \rightarrow GC$ transition.

Much of our knowledge of the molecular action of chemical mutagens is based on studies using in vitro test systems in which viral or transforming DNA is treated with a direct-acting mutagen in vitro and then tested in vivo. In such systems, the chemical effects that might be produced when whole cells are exposed to a mutagen, such as the activation (or inactivation) of repair enzymes and the disruption of normal metabolism, are not major considerations. The mutagen-induced effects in the in vitro systems are restricted to DNA and can often be related to a specific chemical modification of a base pair.

The structures of several of the mutagens that are discussed in this chapter are given in ▶ Figure 22–17. These mutagens can be divided into two groups: those that are effective only on DNA that is actively replicating (base analogues) and those that do not require concurrent DNA replication for their primary action (hydroxylamine, alkylating agents, and acridines). The base analogue mutagens and alkylating agents cause two-way transitions ($AT \rightleftharpoons GC$). Alkylating agents can also cause transversions, and acridines induce frameshift mutations. Hydroxylamine is the most specific of the mutagens that we will consider—it induces only $GC \rightarrow AT$ transitions. A summary of these mutagens and the major types of mutations that they induce is given in ■ Table 22–2.

Base Analogues

A **base analogue** is a substance that is similar enough in structure to a normal purine or pyrimidine residue to be incorporated into a strand of DNA in place of the base it resembles. Two of the most widely cited base analogues are 5-bromouracil (BU) and 2-aminopurine (AP). The former is a pyrimidine analogue of thymine, while the latter is a purine analogue of adenine. When incorporated into DNA, each of these analogues pairs like the base that it replaces. Each analogue can also undergo tautomeric shifts that result in the same alteration of hydrogen-bonding properties. Thus

Base analogs

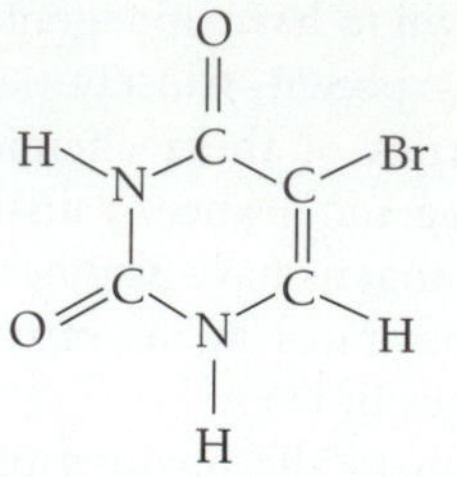

5-bromouracil (BU)

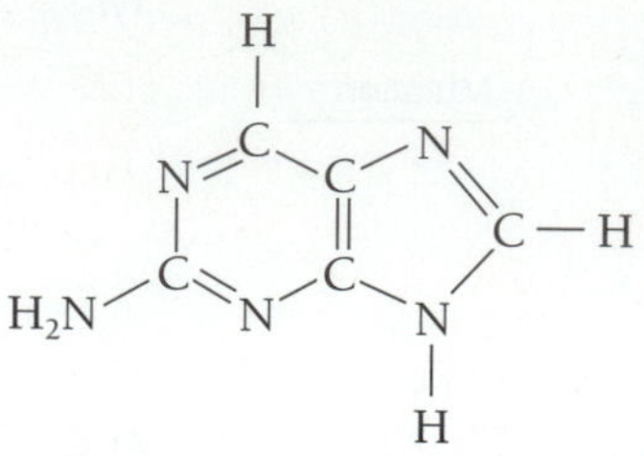

2-aminopurine (AP)

Alkylating agents

$CH_3—CH_2—O—SO_2—CH_3$

ethylmethane sulfonate (EMS)

$CH_3—CH_2—O—SO_2—CH_2—CH_3$

ethylethane sulfonate (EES)

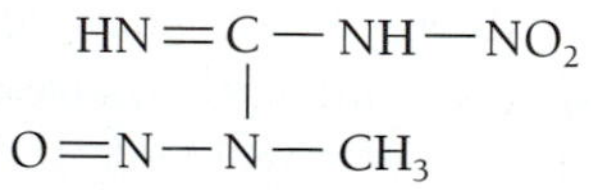

$HN{=}C—NH—NO_2$
$O{=}N—N—CH_3$

nitrosoguanidine (NG)

$Cl—CH_2—CH_2—S—CH_2—CH_2—Cl$

sulfur mustard (mustard gas)

Acridines

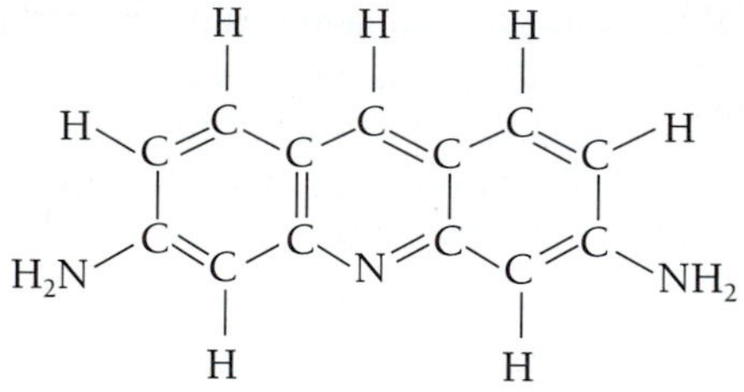

proflavin

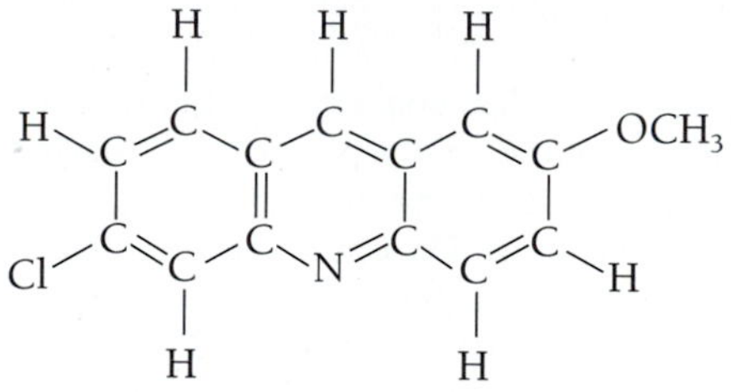

quinacrine

Miscellaneous

HNO_2

nitrous acid (NA)

NH_2OH

hydroxylamine (HA)

▶ **FIGURE 22–17** Some of the better-known chemical mutagens.

■ **TABLE 22–2** Summary of the action of selected chemical mutagens.

Mutagen	Damage to DNA	Mutagenic Effect
Base analogues (BU and AP)	base-pairing errors through tautomerism	two-way transitions (AT $\rightleftharpoons$ GC)
Hydroxylamine (HA)	base-pairing error through modification of C	one-way transition (GC $\rightarrow$ AT)
Alkylating agents (EMS, EES, NG, mustards)	base-pairing errors through modifications of G and T	two-way transitions (AT $\rightleftharpoons$ GC)
	depurination	transversions
	nucleotide deletion	frameshift mutations
	structural distortion of double helix	
Acridines (proflavin, acridine orange)	intercalation into double helix	frameshift mutations

G   BU*

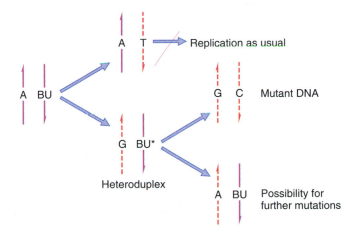

Tautomeric form of BU   Pairing of BU* with G

▶ FIGURE 22–18 The tautomeric form of bromouracil (BU*) has the base-pairing properties of cytosine, and thus it forms hydrogen bonds with guanine.

the rare enol form of BU (BU*) pairs with guanine (▶ Figure 22–18), just as the enol form of thymine does.

The base analogues act as mutagens because they undergo a higher rate of tautomerism than their normal counterparts and thus cause a higher rate of replication errors. Like the spontaneous mutations caused by tautomerism in normal bases, a mutation induced by a base analogue can occur through either an incorporation error or a template error. ▶ Figure 22–19 shows the base substitutions that can be induced by BU. Note that an error in the template strand gives rise to an AT → GC transition, while an error in incorporation yields a GC → AT transition.

Geneticists have studied the mechanism of action of base analogues in order to decipher the molecular basis of spontaneous mutation. The fact that an increased incidence of tautomerism can explain the mutagenic effects of base ana-

(a) Tautomerism of BU in template strand

▶ FIGURE 22–19 Base substitution mutations induced by BU. (a) The tautomerism of BU in a template strand leads to the transition AT → GC. (b) Incorporation of the tautomeric form of BU in a daughter strand gives the transition GC → AT.

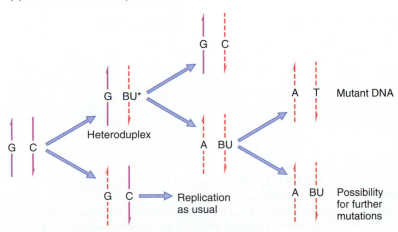

(b) Tautomerism of incorporated BU

logues supports the idea that spontaneous mutation can occur through tautomerism and errors in pairing of normal bases.

DNA-Reactive Mutagens

Hydroxylamine (HA) and the **alkylating agents** act by chemically modifying the DNA. Hydroxylamine, which is thought to be an intermediate in nitrate reduction in vivo, reacts specifically with cytosine, adding a hydroxyl group to yield hydroxylaminocytosine (▶ Figure 22–20). This chemical alteration favors a tautomeric shift that enables the modified base to pair with adenine. The overall result is the base-pair change GC → AT.

The alkylating agents, in contrast, are a diverse group of mutagens that include ethylmethane sulfonate (EMS), ethylethane sulfonate (EES), nitrosoguanidine (NG) and other N-nitroso compounds (which are also highly carcinogenic), and nitrogen and sulfur mustards (see Figure 22–17). Sulfur mustard, commonly known as mustard gas, was the first chemical mutagen to be discovered. Its development as an agent for chemical warfare prompted investigations into its mutagenic properties.

The alkylating agents are electrophilic substances that add alkyl ($-CH_3$, $-CH_2CH_3$, . . .) groups to centers of negative charge in other molecules. Among the sites for alkylation in DNA are the N-7 and O^6 positions of guanine and the O^4 position of thymine. These alkylations are potentially mutagenic. For example, EMS ethylation of the O^6 of guanine and the O^4 of thymine changes the hydrogen bonding properties of the affected bases to those of adenine and cytosine, respectively (▶ Figure 22–21a and b). The reaction with guanine results in the transition GC → AT, and the reaction with thymine produces the transition TA → CG. In contrast, ethylation of the N-7 of guanine induces depurination, creating an apurinic site in the DNA (Figure 22–21c). If the AP site is present during replication, the gap tends to be bypassed and an adenine nucleotide is often added in the daughter strand across from it, thus yielding a GC → TA transversion.

In addition to their direct mutagenic effects, it is believed that alkylating agents also act indirectly as mutagens by causing errors in the repair processes of living cells. The addition of bulky chemical groups by these mutagens can distort the helical structure of the DNA molecule, and nitrogen and sulfur mustards produce distortions by causing intra- and interstrand cross-linkage. Certain cellular repair processes recognize such lesions and are usually efficient enough to restore the damaged section of DNA to its original condition, but if the repair systems are unable to keep up with the magnitude of damage done by a mutagen, or if for some other reason a mistake is made in repair (that is, a misrepair occurs), then gene mutation might result. The idea that mutation can result from the misrepair of DNA damage will be discussed in Chapter 23.

Example 22.2

Nitrous acid (NA) is a mutagen that acts by deaminating cytosine and adenine. What kinds of base-substitution mutations will be caused by the action of NA?

Solution: Deamination of cytosine yields uracil, and deamination of adenine produces the base hypoxanthine.

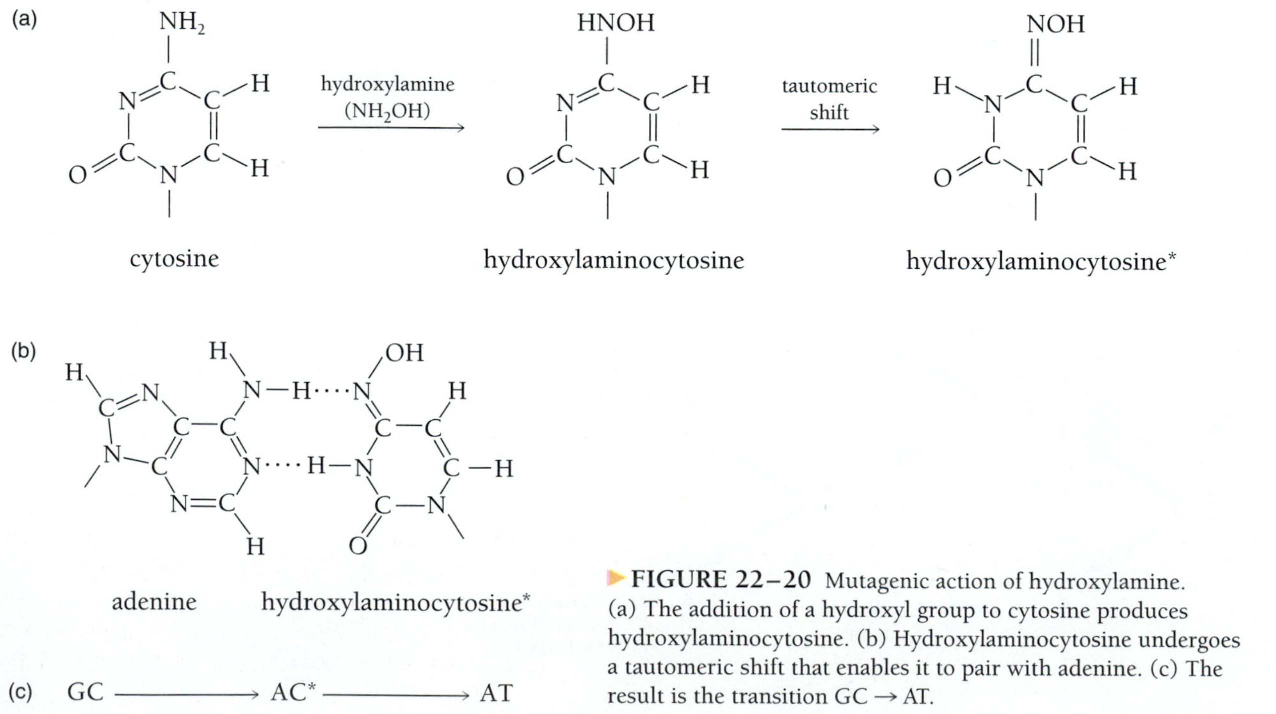

▶ **FIGURE 22–20** Mutagenic action of hydroxylamine. (a) The addition of a hydroxyl group to cytosine produces hydroxylaminocytosine. (b) Hydroxylaminocytosine undergoes a tautomeric shift that enables it to pair with adenine. (c) The result is the transition GC → AT.

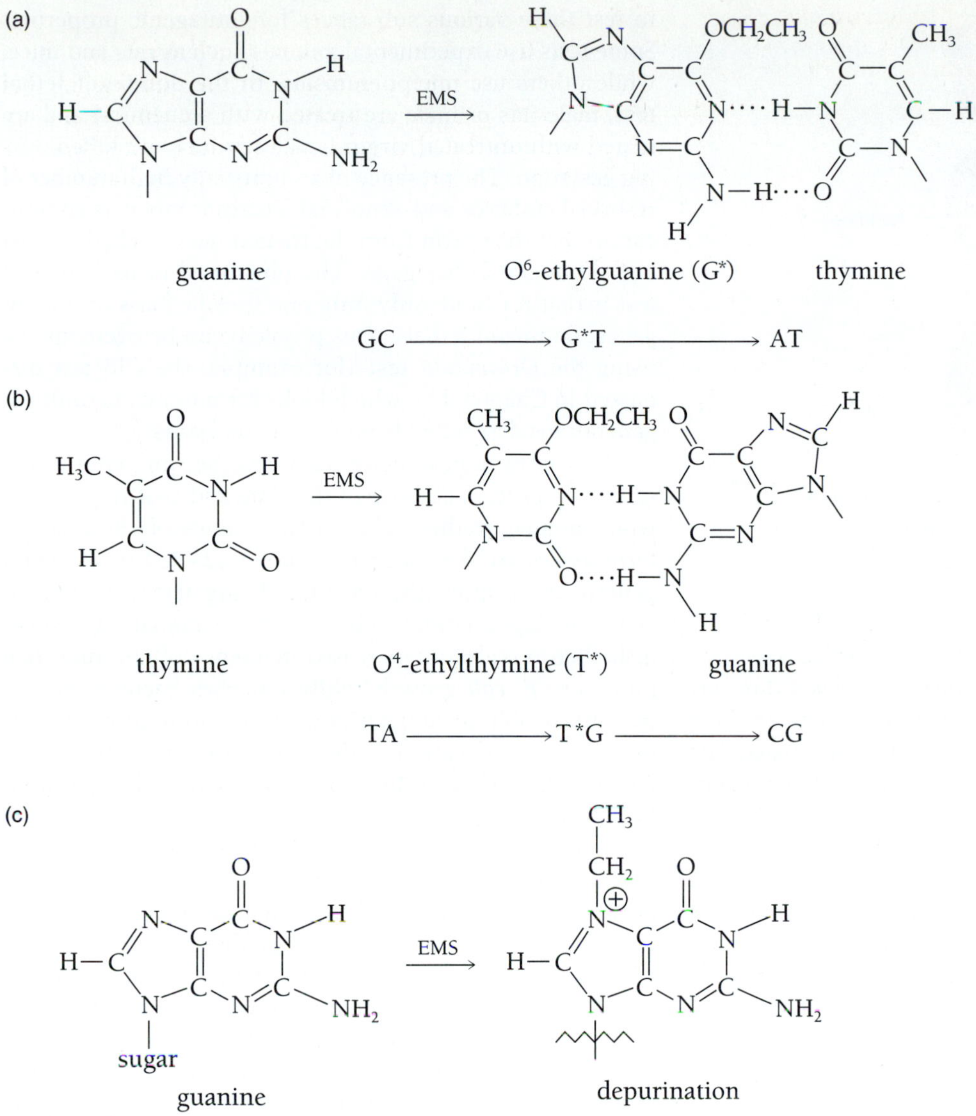

FIGURE 22–21 Direct mutagenic actions of the alkylating agent EMS. (a) The alkylation of the 6-oxy group of guanine yields a guanine derivative that pairs with thymine. The transition GC → AT results. (b) The alkylation of the 4-oxy group of thymine yields a thymine derivative that pairs with guanine. The transition TA → CG results. (c) The alkylation of the nitrogen at the seven position of guanine can lead to depurination, and transversion can result.

(a)

guanine O^6-ethylguanine (G*)   thymine

GC ⟶ G*T ⟶ AT

(b)

thymine O^4-ethylthymine (T*)   guanine

TA ⟶ T*G ⟶ CG

(c)

guanine   depurination

During DNA replication, uracil behaves like thymine rather than like cytosine (it pairs with adenine instead of with guanine) and hypoxanthine behaves like guanine rather than like adenine (it pairs with cytosine instead of with thymine). After DNA synthesis, the net result (shown below) is two different transition type mutations: GC pairs are converted to AT pairs, and AT pairs are converted to GC pairs, as shown:

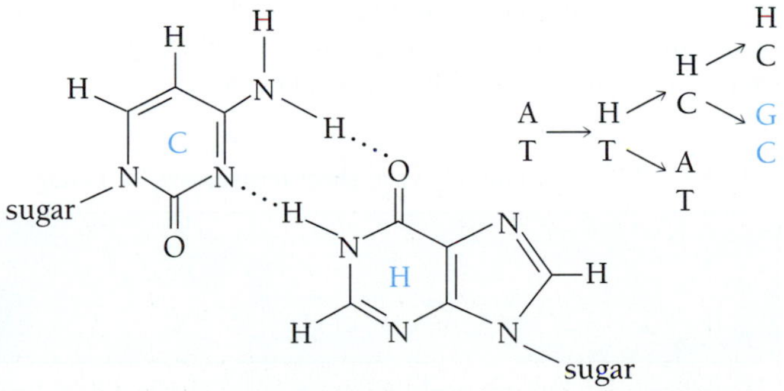

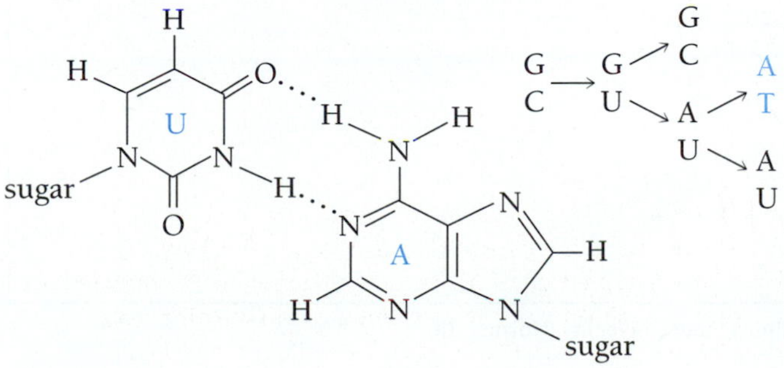

Follow-Up Problem 22.2

In a certain experiment, about half of the *rII* mutations induced with ultraviolet light were revertible using base analogues. However, almost none of these were revertible using hydroxylamine. What is the most likely nature of the original transition induced in the DNA by the UV light?

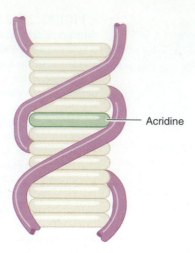

Acridine

▶ FIGURE 22–22 Intercalation of an acridine molecule between stacked base pairs of the DNA double helix.

Intercalating Agents

The **acridines**, which include proflavin and acridine orange, have large, flat hydrophobic surfaces that allow them to insert between the base pairs of a DNA double helix. The dimensions of an acridine approximate those of a base pair, making the **intercalation** possible (▶ Figure 22–22).

The acridines act as frameshift mutagens by causing the addition or deletion of a nucleotide pair. Their mutagenic activity is most potent during times of DNA replication or recombination. The mechanism by which the mutation occurs is not yet understood; however, the intercalated acridine may help to stabilize the looping-out process that appears to be important in frameshift mutations (see Figure 22–14).

Screening Tests for Mutagens

Several thousand natural and synthetic chemicals are now in use in this country, and many new ones are introduced each year. A number of different systems have been devised to test these various substances for mutagenic properties. Some tests use experimental animals such as rats and mice, while others use microorganisms. In the **dominant lethal test**, male rats or mice are treated with a chemical and are mated with untreated virgin females, which are killed during gestation. The presence of an unusually high number of resorbed embryos and abnormal fetuses is taken as an indication that the sperm from the treated male carried at least one dominant lethal gene. The obvious drawback of this test is that it can identify only one specific class of mutant genes: dominant lethals. This problem can be overcome by using the *Drosophila* **test** (for example, the ClB test discussed in Chapter 11), which looks for mutants in multiple generations descended from treated ancestors.

Short-term mutagenicity tests have become vital to the screening process because of the time and cost required by whole-animal studies. The characteristics of these short-term assays are described in ■ Table 22–3. In vitro **cytogenetic tests** apply the chemical being tested to human cells growing in culture. Microscopic examination of these cells then reveals any increased incidence of chromosomal mutation. **E. coli growth inhibition** uses bacterial strains that are unable to repair DNA lesions; inhibition of their growth is an indication of the mutagenic properties of the chemical being tested. **Eukaryotic DNA replication inhibition** is based on the induction of damage to DNA (which prevents its replication) as the basis for detection of mutagenicity. Other tests evaluate mutagenicity by observing the occurrence of **unscheduled DNA synthesis** in treated cells. One disadvantage of these tests is that specific kinds of point mutations are not revealed. Another is that a chemical that is mutagenic to a whole organism may not be mutagenic to cells growing in culture or to bacterial cells, since cultured cells and prokaryotic systems do not mimic the metabolic system of the whole eukaryotic organism. For example, cultured cells do not respond to substances that become mutagenic only after they have been chemically modified by certain metabolic enzymes. Nitrates and nitrites, which have been used as food preservatives, are examples of a group of mutagens that are metabolically ac-

■ **TABLE 22–3** Commonly used short-term assays for mutagenicity.

Assay	Principle upon which Mutagenicity Is Evaluated	
	Gene Mutation	DNA Damage
Cytogenetics (chromosomal aberration, sister chromatid exchange, etc.)		+
E. coli growth inhibition (using repair-deficient strains)		+
Eukaryotic DNA replication inhibition		+
Unscheduled DNA synthesis		+
Eukaryotic microorganisms (S. cerevisiae, N. crassa, A. nidulans)	+	
Cell culture and mutagenesis	+	
Ames/Salmonella	+	

Source: Courtesy of Dr. Ralph J. Rascati, Department of Biological Sciences, Illinois State University, Normal, Ill.

tivated. These substances are harmless until they are metabolically converted by the liver into highly mutagenic and carcinogenic nitrosamines.

The short-term tests that we have just described evaluate the mutagenic potential of a substance by its ability to break or otherwise damage DNA. Several other more direct tests use the induction of point mutations as a criterion for the mutagenic activity of a substance. The most commonly used of these test systems is the **Ames test**, in which the bacterium *Salmonella typhimurium* is the test organism. The Ames test combines the ease of handling and large population size of microorganisms with a eukaryotic metabolic system. Rat liver enzymes are isolated by sedimenting an extract of liver cells into what is called the microsomal fraction. The liver enzymes are added to the medium of the bacteria, along with the chemical being tested for mutagenic activity. The enzymes convert the chemical to the same types of metabolic intermediate compounds as would the metabolic system of the whole organism, and thus activate a potentially mutagenic substance.

The bacteria used in the Ames test are histidine auxotrophic mutants (his^-), so they cannot grow on minimal medium. But when placed on agar-based minimal medium, the bacterial cells do not die—they remain dormant unless they happen to mutate back to the wild-type state:

$$his^- \xrightarrow{\text{Reverse mutation}} his^+$$

$$\text{no growth} \qquad\qquad \text{growth}$$

The Ames test is conducted by spreading the his^- cells over the surface of a petri dish containing minimal medium and the rat liver enzyme extract. A disk saturated with the test chemical is placed in the center of the plate (▶ Figure 22–23), and the chemical diffuses outward from the disk, forming a concentration gradient. The number of bacterial colonies at various distances from the disk measures the effectiveness of the chemical in inducing back mutation. Because the strains of *Salmonella* used in the Ames test are very sensitive to mutagens, very weak mutagenic agents can be detected. The Ames test can even distinguish between a mutagen that causes base substitutions and one that induces frameshift mutations. Discrimination at this level is

▶ **FIGURE 22–23** The Ames test screens chemical substances for mutagenic properties. Two tests are shown: Experiment A utilizes a his^- base substitution mutant strain as the indicator, while Experiment B uses a his^- frameshift mutant strain. After incubation, all the plates are scored for revertant (his^+) colonies. The control plates show the background level of spontaneous reversion. In these experiments, it is clear that chemical 1 induces base substitution mutations but not frameshift mutations, while chemical 2 induces frameshift mutations but not base substitution mutations.

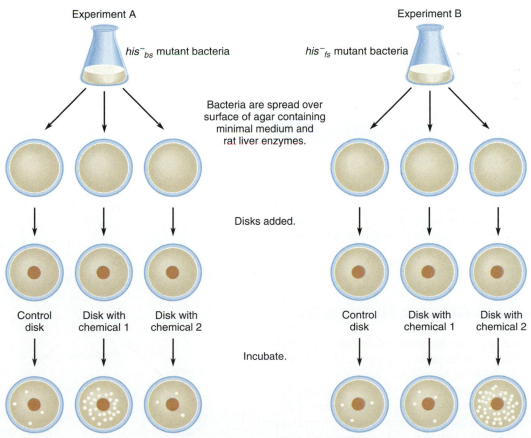

possible because various *his*⁻ strains whose DNA lesions are known are available for use as indicator bacteria. For example, if a substance is found to cause reversion in $his_{\overline{bs}}$ bacteria (which have a mutation of the base substitution type) but not in $his_{\overline{fs}}$ bacteria (which have a frameshift-type mutation), then we can conclude that the substance in question specifically induces base-substitution mutations. The requirement of some mutagens for DNA replication can be accommodated in an experiment with *his*⁻ mutants by adding a trace of histidine to the minimal medium, just enough to allow a few generations of growth, but not enough to allow the formation of colonies.

To Sum Up

1. Base analogues are incorporated into DNA in place of the bases they resemble. These substances have a high rate of tautomerism, which increases the frequency of base-pairing errors during DNA replication and thus leads to base substitutions.

2. DNA-reactive mutagens, such as hydroxylamine and alkylating agents, modify DNA bases chemically. Some modifications increase the frequency of base-pairing errors, while others cause depurination. The result is base substitution. Some DNA-reactive mutagens cause mutation by DNA cross-linkage and subsequent misrepair.

3. Chemical substances that act as intercalating agents, such as the acridines, cause addition or deletion of a nucleotide pair, resulting in frameshift mutations.

4. Chemical substances can be screened for mutagenic activity by a variety of tests. The Ames test is currently the most widely used of these tests; it combines the metabolic enzyme system of a eukaryote with a bacterial test organism. Because of the high correlation between mutagenicity and carcinogenicity, the Ames test is widely regarded as a screening test for carcinogens.

Point to Ponder 22.1

We usually don't think about our everyday exposure to a wide range of potential mutagens such as certain food additives, medical radiation, and therapeutic drugs. Now that we have discussed gene mutations and mutagenesis, do you have any reservations about the widespread use of such substances? How can we balance the benefits of their use against their potential costs?

IN VITRO SITE-DIRECTED MUTAGENESIS

Mutagenesis resulting from the chemical treatment of DNA ordinarily produces random point mutations. With the development of gene cloning techniques, however, it is now possible to selectively mutate a base pair at a particular location within a length of DNA. There are a number of different technical approaches to making specific changes in a DNA sequence, but the one that is most commonly used

is **oligonucleotide-directed site-specific mutagenesis.** This method produces mutant copies of a gene of interest by in vitro DNA replication after introducing a specific point mutation into the daughter strands. To introduce the mutation, a chemically synthesized oligonucleotide between 7 and 20 nucleotides in length serves as a primer for the replicative process. The primer is constructed so that it is complementary in sequence to a particular region in the gene, except for a single mismatched base.

One experimental approach that can be used for oligonucleotide-directed mutagenesis is shown in ▶ Figure 22–24. To obtain the gene in a single-stranded form for DNA replication, the gene is typically cloned using an M13-based vector (see Chapter 18). The synthetic oligonucleotide containing the mismatched base is then annealed to the single-stranded template, and a duplex form of the recombinant vector is produced by the actions of DNA polymerase and DNA ligase. When this duplex vector is introduced into *E. coli* cells by transformation, it will replicate on its own and produce a population of progeny DNA molecules, half of which carry the mutant gene.

Once the gene has been cloned, the recombinant molecules containing the mutation can be identified by DNA hybridization, using the same synthetic oligonucleotide (now radiolabeled) as a probe. Under suitable conditions of stringency (temperature and ionic strength), only perfectly matched DNA strands will form a duplex, thereby limiting

▶ **FIGURE 22–24** Oligonucleotide-directed mutagenesis on a cloned gene using a synthetic oligonucleotide with a mismatched base pair.

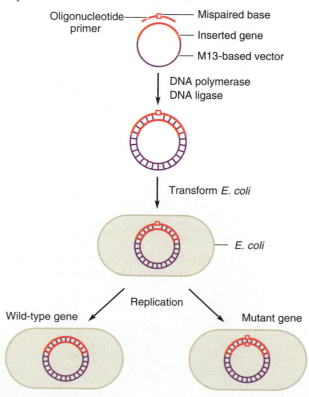

the detected response on an autoradiograph to the mutant sequence.

Oligonucleotide-directed mutagenesis can also be used to form additions, deletions, and multiple point mutations (▶ Figure 22–25). For example, a mutant gene that is missing a particular codon can be produced by using an oligonucleotide with this deletion. When the oligonucleotide is annealed to the gene of interest, the corresponding three nucleotides in the normal gene loop out, and the newly synthesized strand will lack this looped-out section.

To Sum Up

1. Site-directed mutagenesis refers to the selective introduction of a mutation at a particular location in a gene; one commonly used method is oligonucleotide-directed site-specific mutagenesis. To induce a single base pair substitution, a chemically synthesized oligonucleotide containing a single mutant nucleotide is used to initiate replication of a gene; half the progeny DNA molecules will then carry the mutation.

2. Site-directed mutagenesis can also be used to induce specific additions, deletions, and multiple point mutations in selected genes, thus permitting the study of the effects of many different kinds of mutation on gene function.

(a) Addition mutagenesis — Mutant oligonucleotide carrying an added sequence that loops out in hybrid

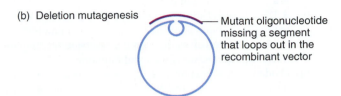

(b) Deletion mutagenesis — Mutant oligonucleotide missing a segment that loops out in the recombinant vector

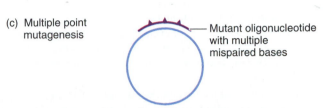

(c) Multiple point mutagenesis — Mutant oligonucleotide with multiple mispaired bases

▶ **FIGURE 22–25** Oligonucleotide-directed mutagenesis for the specific introduction of (a) additions, (b) deletions, and (c) multiple point mutations.

Chapter Summary

A spontaneous point mutation can occur through the substitution of one base pair for another or by the addition or deletion of a nucleotide pair. Base-pair substitutions are the most common type of point mutation; they are classified as transitions if they involve purine-for-purine and pyrimidine-for-pyrimidine substitutions and as transversions if a purine is replaced by a pyrimidine and a pyrimidine by a purine. Transitions are the best-understood class of substitution mutation; they often result from base-pairing mistakes that arise spontaneously from base tautomerism and deamination.

The immediate effect of a base-pair substitution is a change in a codon specifying a particular amino acid. Codon substitutions are classified as missense mutations if they cause the replacement of one amino acid by another in the encoded polypeptide, as nonsense mutations if they result in premature termination of protein synthesis, or as silent mutations if they do not affect the encoded peptide chain. Adding or deleting a nucleotide pair causes a frameshift mutation, which shifts the reading frame of the DNA triplet code words and severely alters the resulting protein structure.

The molecular mechanisms of point mutations have been studied using chemical mutagens. Base analogues, alkylating agents, and intercalating agents are common types of chemical mutagens. Base analogues are incorporated into DNA during replication and induce base-pairing mistakes; alkylating agents often cause depurination; and intercalating agents give rise to frameshift mutations. Chemical substances can be screened for mutagenic activity by a variety of tests; the Ames test is currently the most widely used for this purpose.

Researchers are now able to create specific alterations in DNA through site-directed mutagenesis, thus facilitating the study of the effects of different mutations on gene function.

Questions and Problems

Spontaneous Point Mutations

1. Define the following terms and distinguish between members of paired terms:

base substitution and codon substitution
intragenic and intergenic suppression
missense, nonsense, and frameshift mutations
silent mutation and neutral mutation
tautomerism
transitions and transversions

2. Fill in the blanks.
(a) If adenine undergoes a tautomeric shift, it will pair with _____ instead of with thymine.

(b) If thymine undergoes a tautomeric shift, it will pair with _____ instead of with adenine.

(c) A DNA base-substitution mutation that yields no change in the encoded amino acid is known as a _____ mutation.

(d) A DNA base substitution that changes the encoded amino acid is called a _____ mutation; if that mutation does not significantly alter the function of the protein, it is a _____ mutation.

(e) A DNA base-substitution mutation that generates a premature stop codon is called a _____ mutation.

(f) _____ involves the seeming reversion of a mutant phenotype as a result of a second mutation.

(g) Oxidative deamination is a common mutational mechanism that converts the base _____ to the base _____ .

3. Figure 22–6 shows spontaneous mutations caused by base-pair substitutions resulting from the tautomerism of adenine. Diagram the sequence of events that would lead to mutation as a result of tautomerism in each of the other three DNA bases. Include both template and incorporation errors in your diagram.

4. Explain why all base-pair substitutions in DNA do not result in an amino acid substitution. Explain why amino acid substitutions do not always produce a protein with altered activity.

5. In general, forward mutation rates are higher than reverse mutation rates. Explain this difference in terms of DNA structure and the molecular basis of mutation.

6. If a transition mutation occurs in a codon that specifies Trp, what amino acids could replace Trp? Answer this same question for a transversion.

7. Hemoglobin C was the next hemoglobin variant to be discovered after HbS. It carries lysine in place of glutamic acid at the sixth position of its β-chains. What mutational change at the DNA level would give rise to the HbC allele?

8. Which of the following amino acid changes can result from a single base pair substitution? (a) Arg → Leu, (b) Cys → Glu, (c) Ser → Thr, (d) Ile → Ser, (e) Arg → Asp, (f) Pro → His.

9. List the mRNA codons that could cause chain termination during protein synthesis if altered by a single base change.

10. Categorize each of the following base-substitution mutations in the DNA strand that serves as the template for transcription (consider its effects on the amino acid sequence of the encoded polypeptide): (a) CCT → CCA, (b) CCT → CAT, (c) CCT → ACT, (d) CTA → CTC, (e) AAC → GAC, (f) AGT → GGT, (g) ACC → ATC.

11. Suppose you are studying a protein in which part of the amino acid sequence is Trp-Lys-Ala-Arg-Thr-Val. Several mutants for the gene that specifies this protein are isolated, and the amino acid sequences of their proteins are found to be as follows:

wild-type: Trp-Lys-Ala-Arg-Thr-Val
mutant 1: Arg-Lys-Ala-Arg-Thr-Val
mutant 2: Trp-Met-Ala-Arg-Thr-Val
mutant 3: Trp-Lys-Val-Arg-Thr-Val
mutant 4: Trp-Lys-Ala
mutant 5: Trp-Lys-Cys-Ser-Asn-Gly
mutant 6: Trp-Lys-Cys-Ser-Thr-Val

(a) What mutational event at the DNA level would give rise to

each mutant allele? (b) Give the most probable nucleotide sequence for the DNA strand that is the template for mRNA in the wild-type gene. (c) How many base-substitution mutations are possible in the DNA that codes for this polypeptide?

12. Three mutant forms of *E. coli* tryptophan synthetase show single amino acid substitutions at position 23 in the α-chain. One form has threonine, another has serine, and a third has isoleucine at this position. The wild-type tryptophan synthetase has arginine at this position. If the mRNA codons for Thr, Ser, and Ile are ACA, AGU, and AUA, respectively, and each form arose through a single base-pair change in the normal DNA sequence, what is the mRNA codon for Arg?

13. Diagram the mutational consequences of rare AG pairing. Which of the transversion events have we not been able to explain by purine-purine pairing?

14. Some auxotrophic mutants are "leaky," that is, they are able to grow very slowly in minimal medium with no supplements. Postulate an explanation for the molecular basis of leaky mutants.

Chemical Mutagenesis

15. Show that the base analogue 2-aminopurine causes $AT \rightleftarrows GC$ transitions.

16. Aflatoxin B_1 is a powerful carcinogen (when metabolically activated) that binds to DNA at guanine bases and generates apurinic sites. Studies have shown that adenine is preferentially inserted across from an apurinic site during DNA replication. Predict the nature of the mutation caused by aflatoxin B_1.

17. The Ames test can be used to analyze a spontaneous mutation for a particular type of mutational change. An investigator exposes each of four spontaneously mutant bacterial strains (1, 2, 3, and 4) to a series of mutagens to see if reversion will occur. The results are given below, where + indicates reversion and − indicates no reversion:

		Mutagen			
		HA	BU	EMS	Acridine
	1	+	+	+	−
Bacterial	2	−	−	+	−
Strain	3	−	−	−	+
	4	−	+	+	−

Considering the known modes of action of these mutagens, describe the nature of each spontaneous mutational event. Be as specific as possible.

18. Suppose you want to use a mutagen to produce nonsense mutations. If you had to choose among BU, HA, and NA, which one would you use? Why?

19. Suppose you want to use a mutagen to revert nonsense mutations. If you had to choose among BU, HA, and NA, which one would you use? Why?

20. Consider the various mutagen screening tests discussed in this chapter. Discuss the strengths and weaknesses of each test. Which seem(s) to you to be best suited for screening products ultimately designed for human use? Why?

21. The wild-type coat protein of tobacco mosaic virus (TMV, an RNA virus) contains proline at position 20. Treatment of TMV RNA with hydroxylamine produces mutant forms with the following amino acid substitutions at this position (ar-

rows indicate the sequence of the various induced changes in amino acids at this location):

If the codon for Ser in this situation is UCU, what are the most likely codons for the other amino acids shown? (Give the correct base sequences based on the patterns to the genetic code and the mechanism of action of hydroxylamine.)

22. When an *rII* mutant strain of phage T4 is treated with an increasing dose of hydroxylamine, the number of mutants induced to revert by this chemical mutagen is given by the following dose-response curve:

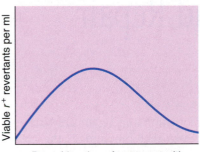

Dose (duration of treatment with constant HA concentration)

(a) Explain why there is a reduction in the number of revertants at the higher doses. (b) Why are there revertants at zero dose? (c) Assuming that killing and mutagenesis are independent events, construct a plot of the ratio of revertants to survivors as a function of dose.

In Vitro Site-Directed Mutagenesis

23. In site-directed mutagenesis, one often designs the oligonucleotide so as to add a new restriction site or eliminate a preexisting restriction site as a byproduct of the directed mutation. Why is this additional feature desirable?

24. You have cloned the gene for an *E. coli* enzyme in an M13-based vector and want to determine whether a lysine residue (encoded by the following 18-nucleotide sequence) is essential for enzyme activity, by replacing it with another amino acid.

Segment of gene in vector: GGC TCT AAA TTC GCC ACG
Amino acid sequence: Gly Ser Lys Phe Ala Thr

You also want to introduce a new restriction site (or eliminate a preexisting restriction site) without making any additional changes. Using the restriction sites in Table 15–3 and the genetic code assignments in Figure 22–10, design an oligonucleotide, 18 nucleotides in length, that can accomplish these objectives.

CHAPTER 23

Mechanisms of Recombination and Repair

Recombination and repair are both enzymatic processes that function in the creation and maintenance of genetic diversity. They are also closely related processes, which share a number of common enzymes, and for this reason, we will consider them together in the same chapter.

RECOMBINATION AT THE MOLECULAR LEVEL

Recombination, like mutation, serves as a mechanism for genetic change. But unlike mutation, recombination produces genetic change by rearranging existing DNA sequences; the creation of new DNA sequences is not involved.

There are three main types of recombination (▶ Figure 23–1). **General recombination**, in which genetic information is exchanged between DNA regions that share extensive sequence homology (Figure 23–1a), is the most common type. The exchange can involve any pair of such regions and can occur at essentially any point in the regions, although certain DNA

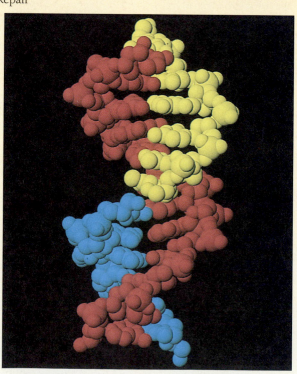

Computer-generated model of nicked (damaged) DNA. The severed strand is shown in blue/yellow.

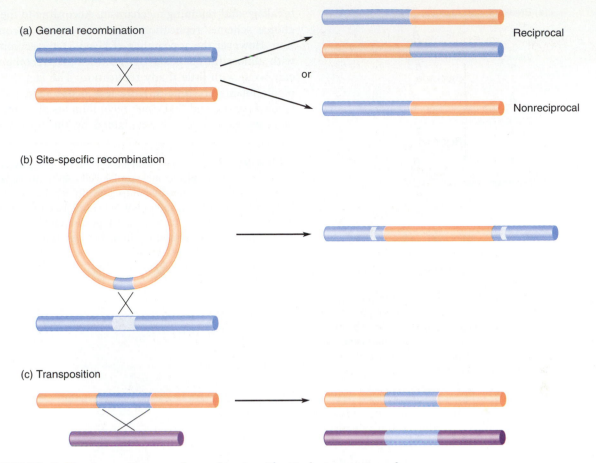

(a) General recombination

Reciprocal

or

Nonreciprocal

(b) Site-specific recombination

(c) Transposition

▶ **FIGURE 23–1** The three main types of recombination. The X's depict genetic exchanges.

sequences may make the event more probable at some points than at others. General recombination is usually a bimolecular reaction between independent DNA molecules; it is the type of recombination that occurs between homologous chromosomes during meiosis. General recombination is accompanied by specific pairing (synapsis) of homologous sequences and is usually *reciprocal,* producing both of the products of a genetic exchange, although nonreciprocal exchanges can and often do occur in prokaryotes and viruses.

Site-specific recombination is another, more specialized type of recombination in which genetic exchanges are limited to highly select sites within the genome (Figure 23–1b). Site-specific recombination essentially involves a single reciprocal exchange between two very short regions of sequence homology. The best-characterized example of this kind of recombination is the insertion of λ phage DNA into the *E. coli* chromosome (Chapter 10).

Transpositional recombination (or transposition) is the third major type of recombination. It involves nonreciprocal exchanges between sites on a single DNA molecule or between two different DNA molecules, with no homology between the recombination sites (Figure 23–1c). This type of recombination promotes the movement of transposable genetic elements from one genetic location to another and is responsible for a variety of genetic rearrangements.

In this section, we will discuss the molecular basis of general recombination. Examples of site-specific and transpositional recombination can be found in Chapter 10, along with a discussion of their possible mechanisms.

Recombination by Breakage and Rejoining

Two mechanisms for the recombination of DNA molecules were debated during the 1950s. One mechanism was **breakage and rejoining** (▶ Figure 23–2a), in which recombining DNAs were assumed to break at the same point and exchange nucleotide sequences. The second mechanism, copy choice (Figure 23–2b), hypothesized that information contained in a DNA sequence could be transferred from a chromosome to its homolog without an exchange of structural parts; a sudden switching of the growing strands during DNA replication was thought to allow each of the paired DNA molecules to copy part of the sequence of the other.

The first conclusive test of these models was carried out by Matthew Meselson and Jean Weigle in 1961. To distinguish between the two proposed mechanisms, Meselson and Weigle took advantage of the fact that no DNA would be exchanged under the copy choice scheme (only information would be exchanged). In contrast, recombinants in the breakage-and-rejoining mechanism would receive some of their atoms from each parental molecule entering the cross.

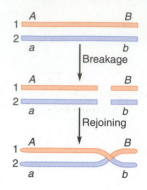

(a) Breakage and rejoining

(b) Copy choice

▶ **FIGURE 23–2** Two possible mechanisms for genetic recombination. (a) Breakage and rejoining involves a physical exchange of parts between homologous chromosomes. (b) In copy choice, replication (indicated by the broken line) generates a new daughter chromosome using first one of the two homologous chromosomes and then the other as a template.

Meselson and Weigle tested these models by infecting *E. coli* cells with a mix of two genetically dissimilar λ phages containing different isotopes of carbon and nitrogen. They used an unlabeled double mutant *c mi* (where *c* and *mi* are the genes for clear and minute plaques, respectively) and a $^{13}C^{15}N$-labeled wild-type strain, c^+mi^+, which was produced from cells growing in a $^{13}C^{15}N$-labeled medium. ^{13}C and ^{15}N are both heavy isotopes, so their presence in the DNA would increase the density of the labeled phage.

The progeny phages resulting from the mixed infection were centrifuged in a CsCl density gradient and gave the density distribution shown in ▶ Figure 23–3. Although the majority of the progeny had the parental genotypes (Figure 23–3a), some (about 1.5%) were recombinant (Figure 23–3b). Most of the parental-type progeny were completely light (unlabeled), since the parental phages infected a host growing in an unlabeled medium. However, a few heavy and some partially heavy phages were produced. The heavy and partially heavy modes in the distribution of the originally heavy c^+mi^+ parental type consist of phage particles whose protein coats are completely light and whose DNA components are probably either fully labeled, having failed to replicate prior to maturation, or half-labeled, having replicated only once in the unlabeled medium. Similar density modes appear in the distribution of the c^+mi recombinant type, but in this case the heavy and partially heavy bands are shifted to a slightly lower density. In contrast, the distribution of the $c mi^+$ recombinant type resembles that of the originally light *c mi* parental type—it has only a single mode in the lowest density region of the centrifuge tube.

These results can be explained only in terms of the breakage-and-rejoining mechanism. According to the copy-choice scheme, replication must accompany recombination. However the appearance of heavy c^+mi recombinants with almost fully labeled DNA shows that recombination can occur with little if any replication. The fact that these recombinants are only slightly less dense than the fully-labeled parent, and that $c mi^+$ recombinants are almost free of heavy isotope, can be explained by the breakage-and-

▶ **FIGURE 23–3** Density distributions of the progeny of the cross $^{13}C^{15}N$-labeled $c^+mi^+ \times$ unlabeled *c mi* in phage lambda. The graphs show the density modes for fully-labeled (HH) DNA, half-labeled (HL) DNA, and unlabeled (LL) DNA. (a) Parental-type distributions. (b) Recombinant-type distributions. (c) Result of a reciprocal breakage and rejoining between fully-labeled and unlabeled parental DNAs. *Adapted from:* M. Meselson and J. J. Weigle, *Proc. Natl. Acad. Sci. USA* 47 (1961): 857. Used with permission of M. Meselson.

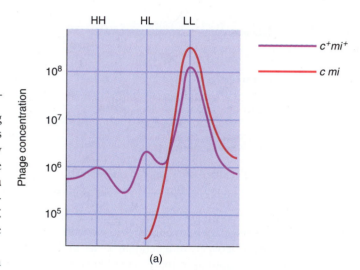

(a)

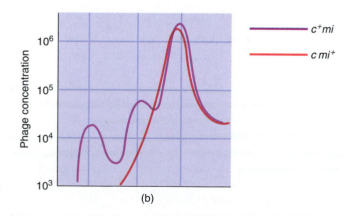

(b)

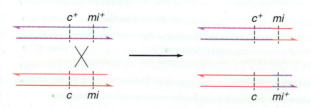

rejoining mechanism and the relative locations of the *c* and *mi* genes (Figure 23–3c). Both genes must be located toward one end of λ DNA, with *mi* having a more terminal position. A reciprocal exchange between fully-labeled and unlabeled parental DNA in the region between these genes could then produce an almost fully-labeled *c⁺mi* product and an almost completely unlabeled *c mi⁺* molecule containing only a small amount of heavy isotope at its end.

Models of General Recombination

Several models have been developed to explain the mechanism of general recombination at the DNA level. One of the most influential was the model proposed by Robin Holliday in 1964 (▶ Figure 23–4). The Holliday model has as its centerpiece a hybrid DNA intermediate called the chi (χ) structure (or Holliday intermediate), which consists of

▶ **FIGURE 23–4** The Holliday model of recombination. The homologous DNA molecules consist of strands of opposite polarity (+ and −). *Adapted from:* H. Potter and D. Dressler, *Proc. Nat. Acad. Sci. USA* 73 (1976): 3000–04.

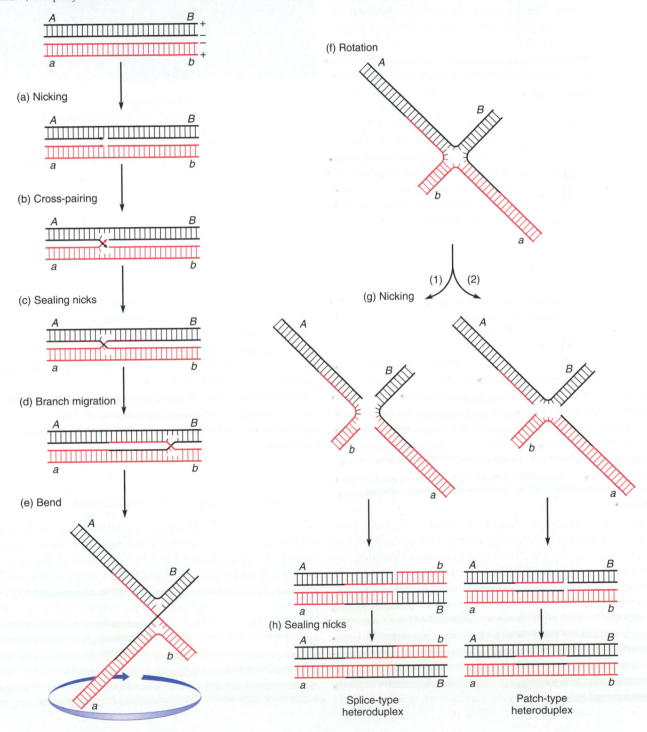

two DNA duplexes held together at a region of homology by two of the four strands of the recombining molecules. Recombination is depicted as a breakage-and-rejoining process that occurs through a sequence of three main steps: (1) the cleavage of one of the homologous strands of each of the paired DNA molecules followed by unwinding, (2) the exchange and cross-pairing of the unwound broken strands to form the cross-bridged Holliday intermediate (Figure 23–4b and c), and (3) the cleavage of the previously unbroken, uncrossed strands to convert the chi structure into separate recombinant molecules (Figure 23–4g, alternative 1).

The second of these steps is the process most closely resembling synapsis. The formation of the cross-bridged intermediate allows the precise alignment of the recombining DNA molecules and thus avoids a gain or loss of genetic material. The recombining molecules are held together in exact register by complementary base pairing in a heteroduplex region of overlap. This heteroduplex region can extend over many base pairs and may include minor differences in base sequence between the parental DNA strands.

Once the crossed-strand connection between the recombining molecules has been established, the resulting cross-bridge can move along each duplex by the process of **branch migration** (Figure 23–4d). During branch migration, the strands in the cross-bridge exchange complementary chains, each displacing the other from its original duplex. The effect of this zipperlike movement is to extend the region of hybrid overlap over distances of several thousand base pairs, much longer (in molecular terms) than the primary pairing event.

To resolve the chi intermediate into its recombinant products, a second pair of single-strand breaks must be introduced (step 3). The outcome of this cleavage depends on which strands are cut (see Figure 23–4g, which shows two planar views of the chi structure). If the cleavage involves the previously unbroken strands (alternative 1), the result is a pair of **splice-type** heteroduplex molecules—these joint molecules are recombinant for the genes that flank the region of overlap. However, if the cleavage involves the previously broken strands (alternative 2), the result is a pair of **patch-type** heteroduplex molecules that are parental for the flanking genes—strand exchange can thus occur between DNA duplexes without a recombination of the flanking regions.

Electron microscopic studies on recombining DNA have found χ-shaped structures resembling the postulated chi intermediate (▶ Figure 23–5) and have thus provided experimental support for the idea that crossing-over proceeds through strand exchange between DNA molecules.

In the Holliday model, the initial cleavage is assumed to occur at precisely the same point in the homologous strands of the two recombining molecules. Even though single-strand breaks do occur occasionally in DNA, it is difficult to imagine how two such breaks could occur simultaneously, conveniently located across from each other on separate DNAs. To avoid the requirement for two identi-

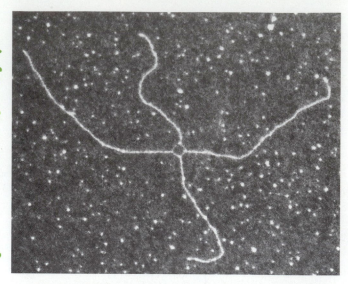

▶ **FIGURE 23–5** Electron micrograph of *E. coli* plasmid DNA, showing a striking resemblance to the Holliday chi intermediate. Note the single-stranded connections at the intersection of the four "arms." (Plasmid DNA is normally circular; this DNA has been cleaved by a restriction endonuclease at a specific site.)

cal breaks in homologous strands, and still provide a mechanism for homologous pairing, more recent models have departed from the Holliday model in explaining the origin of the chi structure. One of these is the Meselson-Radding model, which is shown in ▶ Figure 23–6. In this model, recombination is initiated by a single-strand break in only one of the interacting duplexes. After unwinding, one of the loose ends invades a homologous region in the second duplex, where it base-pairs with the complementary strand and displaces the other strand. This single-strand invasion creates a loop called a **D-loop** (D for DNA displacement) and a heteroduplex region of DNA. Displacement continues until a second break at some point in the D-loop forms an intermediate that can produce the cross-bridged chi structure.

Enzymology of General Recombination

The original model proposed by Holliday has undergone several modifications over the years as a result of experimental as well as theoretical advances. Most of the experimental advances have resulted from the identification of enzymes that catalyze different steps in the recombination pathway. Several enzymes that participate in the process have been identified in *E. coli*. One is the **RecBCD enzyme**, which is the product of the *recB*, *recC*, and *recD* genes of *E. coli*. RecBCD is a multifunctional enzyme that unwinds the DNA duplex during recombination and also acts as a nuclease by cutting single-stranded DNA near the sequence 5′ GCTGGTGG 3′, which is called a **Chi site**. There are

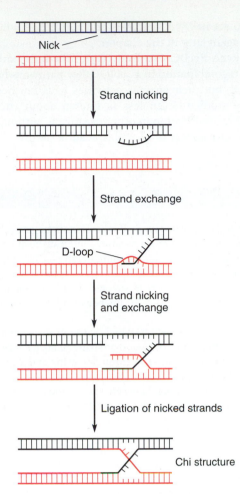

► FIGURE 23–6 The Meselson-Radding model of recombination. Recombination is initiated by a single-strand break, which is followed by unwinding and invasion of the homologous duplex by the free strand. A chi structure is ultimately formed and is resolved as in the Holliday model.

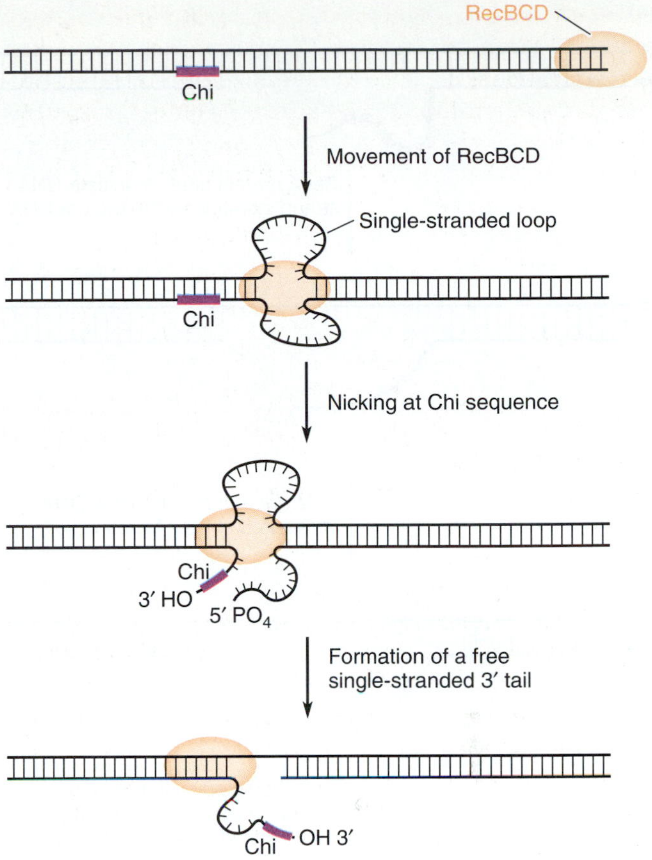

► FIGURE 23–7 Model of the role of the RecBCD protein in genetic recombination in *E. coli*. The RecBCD protein attaches to an end of duplex DNA and unwinds the duplex, forming a twin loop. When the enzyme encounters a properly oriented Chi sequence, it cuts the strand containing Chi to break one of the loops. Further unwinding produces a DNA molecule with a gap and a free single-stranded 3′ tail.

about 1200 Chi sites in *E. coli* DNA; each can serve as the source of a single-stranded end for initiating recombination.

► Figure 23–7 shows a model of the dual role of RecBCD during recombination. The enzyme binds to the free end of double-stranded DNA and then travels along the duplex, using the energy from ATP hydrolysis to unwind the double helix as it goes. Rewinding occurs after the enzyme passes, but it proceeds at a slower rate than unwinding, so that two single-stranded loops appear on either side of the reaction center. Once the enzyme encounters a Chi site, it cuts the single-stranded loop containing the Chi sequence into two single-stranded tails. The tail with the Chi sequence near its 3′ end then lengthens and the other tail rewinds as the enzyme progresses along the DNA, leaving behind a gap in the duplex structure. The DNA molecule is thus left with a gap and a free single-stranded 3′ tail, which can function as sites for strand exchange during recombination.

The **RecA protein** in *E. coli* is also involved in recombination. This enzyme, which is specified by the *recA* gene,

can catalyze strand exchange between DNAs when there is a single-stranded region and a free 3′ end in a region where the two molecules are homologous. When the recombining DNAs are both linear duplex molecules, the strand-transfer activities of this enzyme lead to the formation of a χ-shaped intermediate identical to that predicted in the Holliday model.

The role of the RecA protein during recombination and its possible dependence on RecBCD are shown in ► Figure 23–8. Recombination is initiated when RecA binds to the exposed single-stranded tail produced by RecBCD and mediates pairing with a homologous region in another molecule. When RecA protein monomers bind to single-stranded DNA, they polymerize along the length of the chain to form a DNA-protein filament (Figure 23–8a); a single-strand-binding protein assists in the formation of a stable nucleoprotein filament. RecA in the filament then unwinds a homologous region in the paired DNA duplex, using the energy provided by ATP hydrolysis (Figure

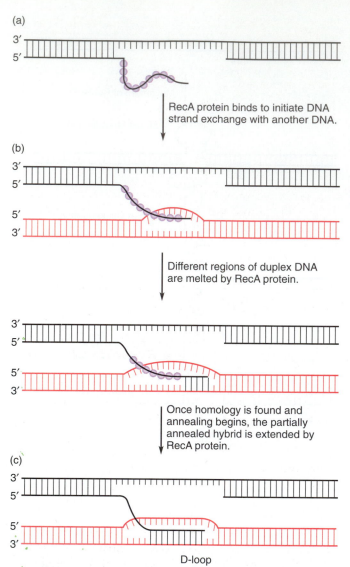

(a)

3′
5′

RecA protein binds to initiate DNA
strand exchange with another DNA.

(b)

3′
5′

5′
3′

Different regions of duplex DNA
are melted by RecA protein.

3′
5′

5′
3′

Once homology is found and
annealing begins, the partially
annealed hybrid is extended by
RecA protein.

(c)

3′
5′

5′
3′

D-loop

▶ **FIGURE 23–8** Promotion of recombination by the RecA
protein. (a) RecA protein first binds to single-stranded DNA
forming a DNA-protein filament. (b) RecA in the filament
unwinds a segment of duplex DNA. (c) Annealing between the
single-stranded tail and the homologous duplex forms a D-loop,
displacing RecA protein from the filament.

23–8b). During unwinding, the exposed complementary
sequence pairs with the transferred strand, forming a D-
loop and displacing RecA protein from the filament (Figure
23–8c). The result is a base-paired heteroduplex region
that holds the recombining molecules together.

Other reactions involved in the recombination process,
such as those shown in the Holliday and Meselson-Radding
models, require the action of additional enzymes. Nucleases
are needed to nick the DNA to cleave the D-loop and to
convert the chi structure into separate recombinant molecules. These nicks are repaired by forming a new phosphodiester bond between the adjoining broken ends at each
nicked site. Some of the enzymes involved in recombina-

tion also act in DNA synthesis and repair and will be considered again later in the chapter.

The RecA-RecBCD pathway is not the only pathway for
general recombination in *E. coli*. Other pathways have been
identified, but their enzyme activities have not been as intensively studied. Even less is known about the mechanisms of recombination in other organisms, although RecA-
like activities appear to be widespread in nature.

To Sum Up

1. There are three major types of recombination: general recombination, site-specific recombination, and transpositional recombination. General recombination involves DNA regions that
   share extensive homology, while DNA regions with very limited homology take part in site-specific recombination. Transpositional recombination occurs between DNA regions that are
   not homologous.
2. In general recombination, which is the most common type of
   recombination, any regions that share extensive sequence homology can be recombined through pairing and a (usually) reciprocal exchange.
3. General recombination between homologous chromosomes involves the breakage of the paired chromosomes, the physical
   exchange of chromosome segments, and the rejoining of the
   broken segment.
4. The precise molecular mechanism of general recombination is
   not known. The Holliday model proposes that crossing-over
   proceeds in a stepwise manner, beginning with the cleavage
   and unwinding of one of the strands in each of the paired DNA
   molecules. In the intermediate χ-shaped structure, the two recombining DNA molecules are held together in exact register
   by complementary pairing between two of their four polynucleotide chains. In the final step, the χ-shaped intermediate is
   processed in one of two related ways to produce either of two
   kinds of heteroduplex product DNA molecules, one with a
   parental gene arrangement flanking the heteroduplex region
   and the other with recombinant flanking genes.
5. Other proposed models for general recombination differ in the
   first step. In the Meselson-Radding model, only one homolog
   undergoes single-strand breakage. The unwound single strand
   then invades the complementary region in the homologous
   chromosome, forming a D-loop as it displaces one strand of
   that homolog and base pairs with the other. Breakage in the D-
   loop and the subsequent sealing of the homologous ends results in the formation of the χ-shaped structure.
6. Various enzymes that participate in general recombination
   have been identified in *E. coli*. Rec proteins catalyze the steps of
   the various recombination models. Some of these recombination enzymes are also important in DNA synthesis and repair.

ENZYMATIC REPAIR SYSTEMS:
AN OVERVIEW

All DNA, whether in a bacterial cell or in a cell of a higher
organism, is vulnerable to damage by chemical and physical

agents in the environment. To reverse this damage and to maintain the functional stability of their DNA, organisms have evolved a variety of repair mechanisms of differing complexity. Some (direct repair) are catalyzed by specific enzymes that recognize and reverse the damage directly without altering the DNA backbone. Others (excision repair) restore the DNA duplex by more radical corrective surgery, relying on the information contained in the complementary strand to replace a damaged base or backbone segment.

Even in the absence of mutagens, spontaneous errors in base pairing continually occur through deamination and tautomerism, and they are also rectified by various error-correcting mechanisms. During and following DNA replication, enzymes "edit" the nucleotide sequence that forms the new strand. These enzymes usually recognize the mispaired bases, excise them from the strand, and insert the correct bases.

The continual repair of induced and spontaneous damage to DNA preserves its information content. Because extensive repair systems operate in both prokaryotic and eukaryotic cells, the expressed incidence of mutation is much less than the actual incidence of damage to the DNA. Mutations can thus be viewed as an escape from the normal cellular repair processes (for example, if extensive damage from a mutagen swamps the repair systems) or as a mistake made during the repair process (mutation would then be considered equivalent to misrepair).

The following sections will discuss the molecular basis of DNA repair and misrepair. Our discussion will be illustrated by the repair system of *E. coli*, the organism in which the majority of experimental studies of repair have been carried out. A variety of repair systems are known to operate in *E. coli*, including processes that act on deleted bases and nucleotides, mispaired bases, chemically altered bases, cross-linkages, and single- and double-strand breaks. However, the best-studied example by far is the repair of damage caused by UV light.

To Sum Up

Both spontaneous and induced damage to DNA can usually be repaired by cellular enzymatic repair systems. Mutation occurs when the repair processes fail to recognize a damaged area or when they repair it incorrectly.

◆

▨ DIRECT REPAIR

In the direct repair of damaged DNA, a covalent modification is simply reversed without replacing the affected base or nucleotide. A well-studied example is the removal of a methyl group from O^6-methylguanine. Recall that when guanine is methylated (or ethylated) at its O^6 position, the modified base tends to pair like adenine (see Figure 22–21a). To restore the site to its nonmutant condition, the lesion is reversed by the action of the enzyme O^6-methylguanine-DNA methyltransferase. During repair, a sulfhydryl group of the enzyme combines with and removes the methyl group from the O^6 position, restoring the normal GC base pair at this location (▶ Figure 23–9). The transfer of the methyl group is a rapid, error-free process, but it leads to the suicidal inactivation of the enzyme. Paradoxically, cells do not have a way to rid the enzyme of the methyl group after it has been transferred, so the active transferase (unlike most other enzymes) is not regenerated upon completion of its catalytic cycle.

Photoreactivation

The most extensively characterized example of direct repair is a repair process known as **photoreactivation**. This process is catalyzed by a light-activated enzyme called **photolyase**, which functions in the repair of DNA damage caused by ultraviolet (UV) radiation.

UV radiation causes genetic damage by inducing photo-

▶ **FIGURE 23–9** Direct-repair reaction catalyzed by O^6-methylguanine methyl transferase. A sulfhydryl group of the enzyme accepts a methyl group from O^6-methylguanine, resulting in the suicidal inactivation of the enzyme.

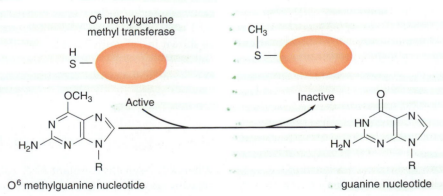

The Double-Strand Break Repair Model of General Recombination

In the Meselson-Radding model in Figure 23–6, the initial steps in general recombination are assumed to proceed by a breakage-and-rejoining process in which only one of the two strands in a DNA molecule is initially nicked. There is now a growing body of evidence indicating that more than one mechanism is responsible for general recombination and that one or possibly several of these mechanisms can involve initial breaks in *both* strands of a recombining DNA as well as a limited amount of DNA synthesis and repair. One of these recombination mechanisms is shown in the first figure, which describes recombination in terms of a **double-strand break repair** process. This model was first proposed by J. W. Szostak and his colleagues to explain various examples of nonreciprocal recombination (see the section on mismatch repair and gene conversion later in this chapter); however, specific features of the model also provide an explanation for certain forms of DNA repair. In this model, recombination is initiated after an endonuclease makes a double-strand cut in one duplex and digestion by exonu-

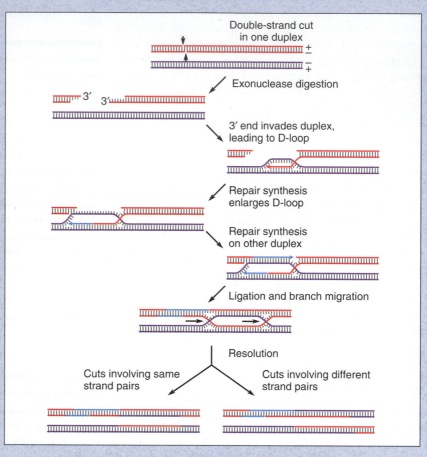

Double-strand cut in one duplex

Exonuclease digestion

3′ end invades duplex, leading to D-loop

Repair synthesis enlarges D-loop

Repair synthesis on other duplex

Ligation and branch migration

Resolution

Cuts involving same strand pairs

Cuts involving different strand pairs

The double-strand break repair model.

chemical reactions in DNA. Recall that the purine and pyrimidine bases in DNA absorb UV radiation, with maximum absorption at 260 nm. One consequence of this absorption is the photochemical fusion of pairs of adjacent pyrimidines in a DNA strand to form **pyrimidine dimers**. The most frequently occurring dimer is the **thymine dimer**, in which the thymine residues are cross-linked by covalent bonds at carbons 5 and 6, forming a cyclobutane ring (▶ Figure 23–10). The thymine dimer distorts the double-helical DNA structure, interfering with the ability of the thymines to form hydrogen bonds with the adenines on the opposite strand. This failure to form hydrogen bonds interferes with DNA replication and transcription and, unless quickly repaired, can cause the cell to die.

In the presence of visible light, cells can readily reverse the dimer-forming process by photoreactivation. The photoreactivating enzyme uses the energy of light (300–500 nm) to break the cyclobutane ring joining the pyrimidines, as shown in ▶ Figure 23–11. The photolyase first binds to the DNA region that contains the dimer. Absorption of visible light activates the bound enzyme, which splits the bonds holding the thymines together and restores each to its monomer form.

Example 23.1

When a UV-sensitive mutant of *E. coli* was exposed to increasing doses of UV light, the following data were obtained:

cleases creates a gap flanked by 3′ single-stranded ends. One 3′ end invades a homologous duplex, generating a D-loop. The D-loop is extended by DNA synthesis until the other 3′ end can pair with a complementary sequence. DNA synthesis from the second 3′ end, followed by ligation and possible branch migration, completes the repair process and converts the structure into a cross-branched intermediate with two Holliday junctions. If both junctions are resolved in the same way (for example, by cutting the inner (−) strands of both), nonrecombinant products are produced for the outside markers; but if the junctions are resolved in opposite ways, that is, by cutting the inner strands at one junction and the outer (+) strands at the other, recombinant products are formed (see art at left).

Evidence supporting double-strand breaks in recombination has been observed in genetic transformation studies on the yeast *Saccharomyces cerevisiae*. When *E. coli* plasmids carrying an inserted yeast gene are introduced into a yeast cell, they cannot replicate since they lack a yeast replication origin. However, these plasmids can transform the cell by recombining with a site on a yeast chromosome that is homologous to the gene that they

carry. Although the frequency of transformation is fairly low, it can be greatly increased by first creating a gap in the transforming DNA by using restriction enzymes to introduce two double-strand breaks in the inserted yeast

gene, as shown in the second figure. Despite the loss of the gene segment between the pair of two-strand breaks, all the recombinants contain full-sized genes, indicating that a repair process has corrected the gap (see below).

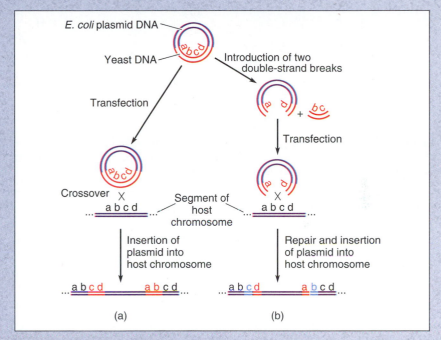

Double-strand break repair during integration of recombinant bacterial plasmids into the yeast genome. (a) The recombinant plasmid, which contains yeast sequences, typically inserts itself at a homologous site in the yeast genome. (b) If a double-strand break is made in the plasmid yeast sequences, the recombinant plasmid can still integrate as long as some yeast sequences remain. The gap is repaired (shown in blue) from chromosomal information during integration.

UV dose (Joules/m²)	Fraction of cells surviving
0	1
0.1	0.1
0.2	0.01
0.3	0.001

According to the Poisson distribution, a survival of 1/e (37%) occurs when there is, on average, one lethal "hit" (inactivating lesion) per cell genome. (a) What fraction of the cells have incurred exactly three lethal hits after a dose of 0.2 J/m²? (b) Given that under the conditions employed a UV dose of 400 J/m² will convert 1% of all adjacent pyrimidine pairs to pyrimidine dimers, calculate how many pyrimidine dimers constitute a lethal hit in this mutant strain.

(Assume that an *E. coli* cell has 4×10^6 base pairs in its genome, which consists of 50% AT.)

Solution: (a) Employing the Poisson distribution, the probability that a cell will receive no lethal hit at a dose of 0.2 J/m² is $P(0) = 0.01 = e^{-m}$, where m = the average number of lethal hits per cell genome. Solving for m, we get $m = 4.6$. Thus the probability of exactly 3 lethal hits is

$$P(3) = \frac{(4.6)^3}{3!} e^{-4.6} = 0.16$$

(b) Since the average number of lethal hits should be directly proportional to the dose, 1 lethal hit should correspond to a dose of approximately 0.2/4.6 = 0.04 J/m².

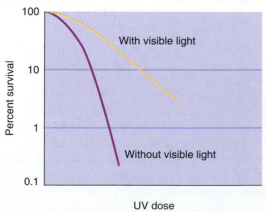

FIGURE 23–10 Formation of a thymine dimer by ultraviolet light. A photochemical reaction causes adjacent thymine bases along a DNA strand to become cross-linked by covalent bonds, forming a cyclobutane ring (red) and distorting the structure of DNA in the process. For simplicity, only a single strand is shown.

▶ **FIGURE 23–11** Photoreactivation is catalyzed by the enzyme photolyase and results in the direct repair of pyrimidine dimers.

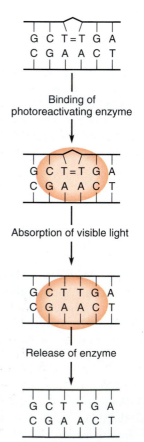

Moreover, out of 4×10^6 pyrimidines in *E. coli* DNA, one-fourth or 10^6 should be located next to another pyrimidine on the same strand, and of these, 1% or 10^4 will undergo dimerization. Thus the number of pyrimidine dimers per lethal hit will be $(10^4)(0.04)/400 = 1$.

Follow-Up Problem 23.1

The survival data in Example 23.1 are expected for single-hit kinetics in which the formation of one pyrimidine dimer is sufficient to kill a cell. The survival curve for single-hit kinetics consists of a straight line when the surviving fraction is plotted on an exponential coordinate as a function of UV dose. In contrast, UV irradiation of wild-type *E. coli* results in a survival curve that has an initial plateau region followed by continued downward curvature, as seen in the following figure. (Note that the y axis is exponential.)

(a) Why does the curve have an initial plateau? (b) Why does the survivorship curve change in the presence of visible light?

To Sum Up

In direct repair of damaged DNA, a single specific enzyme reverses a mutagen-induced covalent modification. Examples include the removal of a methyl group from a base or cleavage of a pyrimidine dimer (photoreactivation).

EXCISION REPAIR

In contrast to direct repair, excision repair is a multistep enzymatic process in which a damaged base or nucleotide is first removed from DNA by hydrolysis. This reaction produces a gap that must be repaired in one strand of the DNA. Usually DNA polymerase I fills this gap, using the opposite strand as a template, and then DNA ligase seals the remaining nick (▶ Figure 23–12).

▶ **FIGURE 23–12** Repair synthesis of a gap in one of the DNA strands. DNA polymerase I fills the gap, using the opposite strand as a template. DNA ligase then seals the remaining nick.

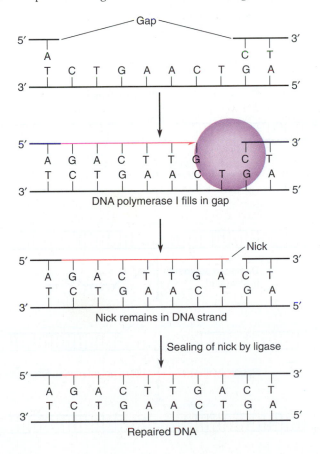

Base-Excision Repair

In this type of excision repair, an enzyme called a DNA glycosylase cleaves the glycosidic bond between an unusual or modified base and its sugar. This creates an apurinic or apyrimidinic site that needs further enzymatic processing to complete the repair.

Base-excision repair is exemplified by the sequence of steps involved in the removal of uracil from DNA (▶ Figure 23–13). This process begins with the cleavage of uracil from its sugar by uracil-DNA glycosylase. Mutants lacking this enzyme cannot remove the uracil residues produced by the deamination of cytosine and are characterized by a high rate of GC → AT transitions. Once the uracil is removed, the resulting AP site is repaired by the combined actions of nuclease, polymerase, and ligase enzymes. An AP endonuclease first cleaves the phosphodiester bond next to the AP site. The abasic sugar is then removed from the nick, either by another AP endonuclease or by an exonuclease, and replaced by the correct nucleotide through repair synthesis, utilizing DNA polymerase and ligase.

The removal of uracil by base-excision repair can occur only in DNA. RNA normally contains uracil as one of its bases, so in RNA a repair enzyme would be unable to distinguish between a nonmutant uracil residue and a uracil derived from the deamination of cytosine. Since it is clearly advantageous for an organism to recognize and repair the defects caused by mutation, why is thymine a normal constituent of DNA but not of RNA? One possible explanation involves a balance between the energy cost of adding a methyl group to the pyrimidine ring (the sole difference between U and T) and the general need for repair. RNA lacks the elaborate repair systems typical of DNA, indicating that a highly stable sequence is not as essential to RNA function. When the DNA sequence of a gene is changed, all the mRNA copies of the gene are also changed. However, an error in transcription affects only a small percentage of the mRNA molecules, leaving most of the mRNA copies functional. Cells can therefore get along with the less repair-efficient structure of RNA and don't need to incur the extra metabolic expense of inserting a methyl group onto uracil to form thymine.

Nucleotide-Excision Repair

In this type of excision repair, the damaged nucleotide is excised enzymatically as an oligonucleotide containing a few neighboring residues. The resulting gap is patched by repair synthesis. In *E. coli*, the excision of damaged nucleotides is catalyzed by the enzyme **ABC excinuclease**. This enzyme consists of three subunits coded for by the *uvrA*, *uvrB*, and *uvrC* genes (so designated because of their involvement in UV repair). Many forms of damage are excised by this enzyme, including pyrimidine dimers, which are recognized by their distortion of the DNA helix. The role of the excinuclease in the removal of a thymine dimer is shown in ▶ Figure 23–14. The ABC excinuclease first

▶ **FIGURE 23–13** Repair of deaminated cytosine by base-excision repair. A DNA glycosylase removes the uracil that is produced by deamination, creating an AP site. An AP endonuclease cuts the DNA adjacent to the AP site. The abasic sugar is then removed, either by another AP endonuclease or by an exonuclease, and the resulting gap is filled in by repair synthesis.

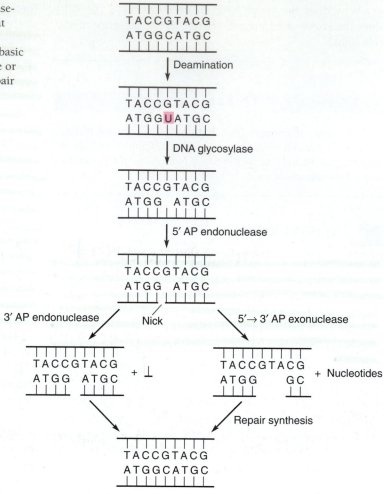

▶ **FIGURE 23–14** Removal of a thymine dimer by nucleotide-excision repair using ABC excinuclease. A dimerized region is recognized by the excinuclease, which excises a 12-13 base segment containing the dimer. The remaining gap is then repaired by the joint actions of DNA polymerase and DNA ligase.

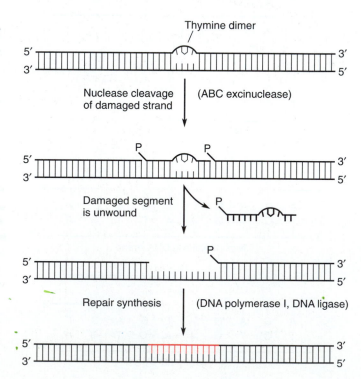

Diseases Caused by Faulty DNA Repair

Several human genetic disorders have been linked to faulty DNA repair. Xeroderma pigmentosum is a severe condition that can cause neurological damage and usually leads to early death from cancer. Mutation in any one of at least nine different genes may lead to this disorder, which results from a deficiency in the ability to carry out excision repair on thymine dimers. Other genetic disorders known to result from faulty DNA repair include Bloom syndrome, Fanconi anemia, ataxia telangiectasia, and Cockayne syndrome.

Bloom syndrome (BS) and Fanconi anemia (FA) are rare autosomal recessive DNA repair disorders that share some common phenotypic symptoms. Cultured cells from affected individuals grow slowly due to a slower-than-normal rate of DNA replication, and karyotype analysis reveals many abnormal chromosomes that contain chromatid breaks, acentric fragments, and dicentric chromosomes. Patients are unusually sensitive to ionizing radiation, such as X-rays, and to chemicals that cause cross-linking of bases in DNA. Skin lesions, various immunological deficiencies, severe respiratory and digestive tract infections, and malignant tumors are common in affected individuals. The exact molecular defect that causes each of these diseases is still unresolved, but evidence obtained so far indicates that BS is the result of a defective DNA ligase.

Ataxia telangiectasia (AT) is a rare autosomal recessive condition that, like BS and FA, is associated with an increased frequency of chromosome breakage, hypersensitivity to ionizing radiation, immunological deficiencies, respiratory infections, and an increased risk of cancer. Muscle control is progressively lost with age (ataxia) due to damage in the cerebellum, the portion of the brain concerned with balance, coordination, and posture. Skin redness (telangiectasia) results from the dilation of capillaries (and also occurs in persons affected with BS). Additional symptoms include premature aging and endocrine disorders. The exact nature of the DNA repair defect responsible for this disorder remains unknown.

Cockayne syndrome is another rare autosomal recessive condition caused by an as yet unidentified defect in DNA repair. The disorder is associated with photosensitivity, mental retardation, dwarfism, neurological degeneration, and early death. Unlike the other DNA repair disorders described here, Cockayne syndrome does not seem to be associated with an increased incidence of cancer in affected individuals, although this may be more the result of a shortened lifespan than a failure of cancer to begin to develop.

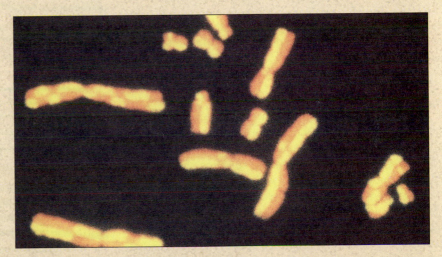

Sister chromatid exchange in differentially stained mitotic chromosomes viewed under fluorescence microscopy. A patient with Bloom syndrome would show a highly-increased rate of sister chromatid exchange.

binds to the damaged site through an ATP-dependent reaction. The enzyme then makes two incisions by hydrolyzing phosphodiester bonds on either side of the modified nucleotides. These cuts free a 12–13 base segment containing the lesion, leaving a gap that is subsequently filled by DNA polymerase and sealed by DNA ligase.

Nucleotide-excision repair also occurs in eukaryotes; however, the repair mechanisms in eukaryotes are not as well understood. A connection between excision repair and UV-induced damage in DNA has been established in the human hereditary disorder **xeroderma pigmentosum** (▶ Figure 23–15). The severe form of this disease is inherited as

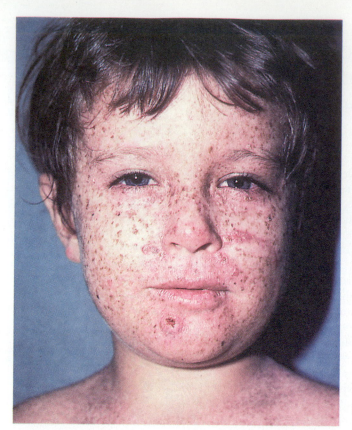

▶ **FIGURE 23–15** Xeroderma pigmentosum is a disease caused by a defect in DNA repair. Affected individuals exhibit extensive tumors in areas of skin that are exposed to ultraviolet light.

an autosomal recessive condition and is characterized by very heavy freckling with open sores in areas of the skin that are exposed to the ultraviolet rays of sunlight. Persons with this disorder are much more sensitive than normal people to the damaging effects of UV radiation and nearly always express some type of skin cancer by adulthood. While cultured cells derived from normal individuals can repair UV damage to DNA, cells from individuals with xeroderma pigmentosum have a greatly reduced repair capacity. Originally, it was thought that xeroderma pigmentosum cells lacked the enzyme that catalyzes the first step (the incision step) in nucleotide-excision repair. However, studies now indicate that mutations in any of several genes involved in UV repair can lead to this disorder.

To Sum Up

DNA that has been damaged by a mutagen can be repaired by processes that involve the excision of a damaged base or nucleotide. In base-excision repair, a DNA glycosylase removes an unusual or modified base, creating an apurinic or apyrimidinic site. The abasic sugar is then removed and replaced by the correct nucleotide. In nucleotide-excision repair, a short DNA region containing a damaged nucleotide is excised and the resulting gap is filled in by DNA polymerase.

POSTREPLICATION (RECOMBINATIONAL) REPAIR

When the exposure to a mutagenic agent, such as UV, is moderate, the direct-repair and excision repair processes are probably able to repair most of the damage efficiently prior to the next round of DNA replication. The subsequent duplication of the DNA molecule then proceeds unimpeded, and the damage is not passed on to the daughter helices. However, if a mutagenic lesion (for example, a dimer) for some reason is not repaired before that area of DNA is replicated, a different repair system is called into play. This system, called **postreplication repair**, is most easily studied in bacteria that carry mutations for the genes that code for direct- and excision-repair enzymes, since such cells are forced to utilize the postreplicative repair process.

▶ Figure 23–16 shows a model for postreplication repair. In the first step shown in the diagram, a replicating double helix is about to encounter a thymine dimer on one of the template strands. The dimer interferes with the reading of the template by DNA polymerase III and thus blocks elongation of the daughter strand. It is thought that in many instances of this type, the polymerase simply skips over the damaged site and resumes reading the template on the other side, leaving a single-stranded gap opposite the modified nucleotides.

Since the genetic information at the lesion is now lost from both strands of the damaged helix, the gap is filled by the correct sequence from the homologous duplex. This repair is accomplished through a recombination process in which multiple copies of the RecA protein bind to the single-stranded gap and mediate pairing with the homologous region in the sister helix. The RecA protein then promotes strand exchange by inserting an undamaged segment from the sister helix into the postreplicative gap. Of course, a new gap is now left on the homologous duplex, but unlike the first gap, this gap is opposite an undamaged region that can be used as a template by DNA polymerase I. The polymerase fills in this second gap, and a ligase seals it into place, completing the repair. The dimer is still present on one strand of the daughter helices, so in a sense, this system is not really a repair. Rather, it can be considered a means of stalling for time until a photoreactivating enzyme or an excinuclease can correct the damage.

Postreplication repair is referred to by some investigators as **recombination repair**, because it involves transferring part of a strand using the enzymes involved in recombination. For example, mutants for the gene coding for the RecA protein in *E. coli* (the *recA* gene) exhibit a deficiency in recombination and an increased sensitivity to UV radiation; irradiated DNA isolated from these mutants contains

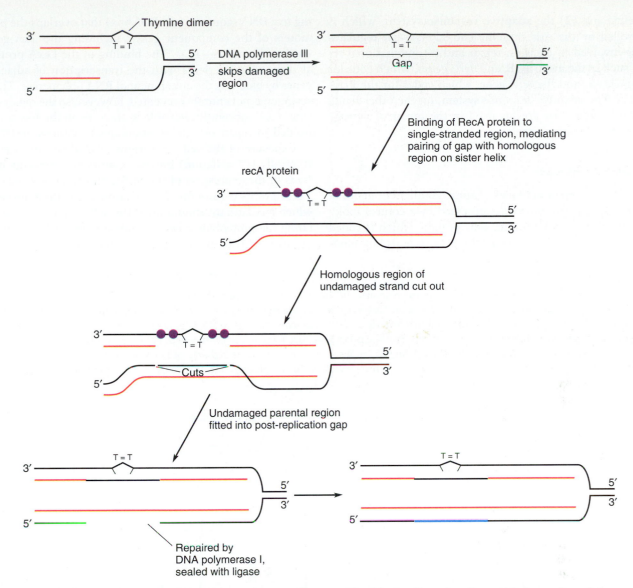

▶ **FIGURE 23–16** Model of postreplication (recombinational) repair. The thymine dimer interferes with the reading of the template by DNA polymerase III during replication. The polymerase skips over the damaged region, leaving a postreplication gap (top right). The RecA protein binds to the single-stranded region and mediates homologous pairing with the undamaged parental strand of the sister helix. After the undamaged strand is transferred into the postreplication gap, DNA polymerase I fills in the gap left in the sister helix.

---

an unusually large number of gaps. These findings implicate the *recA* gene in both repair and recombination.

To Sum Up

Post-replication repair acts only if DNA damage is not repaired prior to DNA replication. The gap left opposite the lesion during replication is filled in with the corresponding section from the homologous molecule. The gap left in the homologous molecule is repaired by DNA polymerase I, using the complementary (undamaged) strand as a template.

◆

◼ INDUCIBLE REPAIR SYSTEMS

Bacterial cells typically contain small amounts (20–100 molecules per cell) of the enzymes involved in DNA repair, so they are immediately available to repair moderate levels of DNA damage. The ability to synthesize these enzymes is enhanced, however, when the cells are exposed to UV radiation or to certain chemical agents. The damaging effects of these agents act as inducing signals to activate genes that code for repair proteins.

Most of the inducible genes that function in DNA repair in *E. coli* belong to two regulatory systems: (1) the **SOS response** system, which is controlled by the LexA and RecA proteins, and (2) the **adaptive response** system, which is controlled by the Ada protein. The SOS system regulates genes involved in excision repair and recombinational repair, such as the *uvrA*, *uvrB*, and *uvrC* genes, which code for excinuclease, and the *recA* gene, which codes for the RecA protein. The adaptive response system, on the other hand, regulates genes that are involved in the repair of damage caused by alkylating agents.

The SOS Response

In *E. coli*, the repair activities of some 20 different genes of the SOS system are normally repressed by a control molecule called the **LexA protein**. This protein, which is coded for by the *lexA* gene, represses the SOS response by bind-ing to a DNA sequence (the SOS box) that overlaps the promoters of the component genes, including the *lexA* gene itself (▶ Figure 23–17a). The binding of the LexA protein at these sites interferes with the transcription of adjacent genes by blocking the attachment of RNA polymerase. Transcription is not entirely prevented, however, so the repair enzymes are constantly available in the cells in the low levels needed to repair occasional spontaneous damage to DNA.

Exposure of the cells to various DNA-damaging agents (typically UV radiation) initiates a sequence of events that leads to the expression of the SOS genes (Figure 23–17b). These events require the participation of the RecA protein, which mediates the induction of the SOS response by facilitating the proteolytic cleavage and subsequent inactivation of LexA. This cleavage reaction is autocatalytic (the LexA

▶ **FIGURE 23–17** Model of the inducible repair response. (a) In the uninduced state, the LexA protein blocks transcription of several repair genes, including *recA* and the *uvr* genes (involved in excision repair). Blockage by LexA is not complete, however, so transcription still occurs in amounts sufficient to allow repair of spontaneous damage to DNA. Since LexA also represses its own synthesis, its levels are kept low enough to ensure that the other genes are not completely blocked. (b) Replication following exposure to a mutagen leaves postreplication gaps. The binding of the RecA protein to the damaged area of DNA on the template strand opposite a gap activates the enzyme so that it facilitates the cleavage of LexA. Cleavage inactivates the LexA repressor, allowing the *recA* and *uvr* genes to be fully expressed.

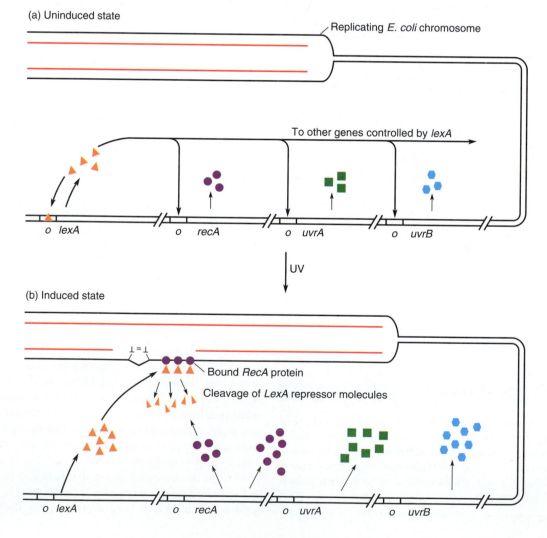

protein cleaves itself), but cleavage is greatly accelerated in the presence of RecA. To effect the cleavage of LexA, the RecA protein must first be activated by binding to single-stranded DNA produced in response to DNA damage. Thus by activating the RecA protein, mutagenic damage leads to the destruction of the LexA repressor and hence to the induction of the SOS response.

The induction of the SOS genes enables *E. coli* to respond quickly and in a proportional manner to varying degrees of damage caused by UV radiation and other mutagenic agents. The cells are thus able to regulate the production of repair enzymes according to their need. Besides controlling the level of DNA repair, the SOS response is also responsible for a high rate of misrepair due to error-prone replication past DNA lesions. This so-called **error-prone repair** functions as a last resort for rescuing cells from DNA damage when there is no intact DNA template available to direct accurate synthesis, and it involves two additional SOS genes: *umuC* and *umuD* (for *UV* non*mutable*). The protein products of these genes somehow allow DNA chain growth across damaged segments, even though an accurate reading of the defective template is impossible. This repair process avoids leaving gaps in the DNA, but it risks inserting the wrong base opposite a lesion such as a UV-induced dimer or an apurinic or apyrimidinic site. Although the role of the products of the *umuC* and *umuD* genes is unknown, their involvement in error-prone repair is thought to be responsible for most base substitutions occurring in *E. coli* as a result of UV-induced mutagenesis.

The Adaptive Response

In *E. coli*, the so-called adaptive response is directed primarily toward correcting the primary lesions of alkylating agents, that is, the O^6-methylguanine and 3-methyladenine that are the major products of alkylation in DNA. Thus two of the known genes in the adaptive response system, the *ada* and *alka* genes, code for O^6-methylguanine-DNA methyltransferase (the Ada protein) and 3-methyladenine glycosylase, respectively.

The dual role of the Ada protein is a particularly interesting feature of the adaptive response system. As we have seen, the Ada protein functions as a methyltransferase by catalyzing the repair of O^6-methylguanine (see Figure 23–9) and becomes methylated during this process. Its second role is as a positive regulator of the adaptive response system. In its unmethylated form, the active enzyme is a weak activator of transcription of the *ada* and *alkA* genes, and thus it helps to maintain the low-level response needed for repair of spontaneous DNA damage. However, the self-methylation that accompanies repair converts the protein from a weak to a strong transcriptional activator, further increasing the cellular levels of these enzymes. Exposure of a cell to alkylating agents thus enhances the activating function of the Ada protein and serves as an inducing signal for the adaptive response.

To Sum Up

1. Extensive damage to DNA by a mutagen enhances the activity of genes that code for repair enzymes. These inducible DNA repair systems include the SOS response, which regulates activity by genes involved in excision-repair and postreplication repair, and the adaptive response system, which stimulates activity by genes that repair damage caused by alkylating agents. Genes that function in direct repair and in excision of methylated bases are among those whose activity is enhanced by the adaptive response.

2. For the most part, the SOS system regulates the activity of genes involved in repair to coincide with the level of damage to the DNA. It also calls into play a last-resort system—the error-prone repair process—for cells that are unable to repair lesions through normal means. This process occurs when certain gene products allow DNA polymerase to replicate across a lesion even though the defective region cannot be read accurately. Most UV-induced mutations in *E. coli* are the result of error-prone repair.

PROOFREADING AND MISMATCH REPAIR

The processes that we have considered so far evolved primarily to repair the damage caused by environmental agents such as UV and alkylating agents. Organisms have also evolved two types of enzymatic mechanisms to guard against replication errors: (1) **error-avoidance mechanisms** and (2) **error-correction mechanisms.**

Error avoidance is largely the responsibility of the DNA polymerases, which avoid replication errors first by their exceptional accuracy in the selection of the proper nucleotide substrates and second by their **proofreading** capability, through which they recognize and excise their own mistakes. If the polymerase inserts the wrong nucleotide during replication, the enzyme stalls and allows the mismatch to be repaired (▶ Figure 23–18). The enzyme first excises the incorrect nucleotide by a $3' \rightarrow 5'$ exonuclease action and then guides the incorporation of the correct nucleotide by its usual $5' \rightarrow 3'$ polymerase activity. The replication accuracy gained by proofreading is substantial. Replication errors prior to proofreading occur at a rate of about one mismatch in 10^5 nucleotides polymerized, but proofreading reduces the error rate to about one in 10^7.

Error correction is the responsibility of a mismatch-repair system that acts on the misincorporated nucleotides that have managed to slip by proofreading. **Mismatch repair** is basically a backup mechanism that corrects errors in base pairing after a DNA strand has been synthesized. Correction occurs through an excision-repair process in which a mismatched base pair is recognized and the incorrect member of the pair is replaced with the proper nucleotide, using the old (parental) strand as a template. Identification of the misincorporated nucleotide in a mismatched pair is based on which DNA strand is newly synthesized. In *E. coli*, mismatch recognition and excision is accomplished by a

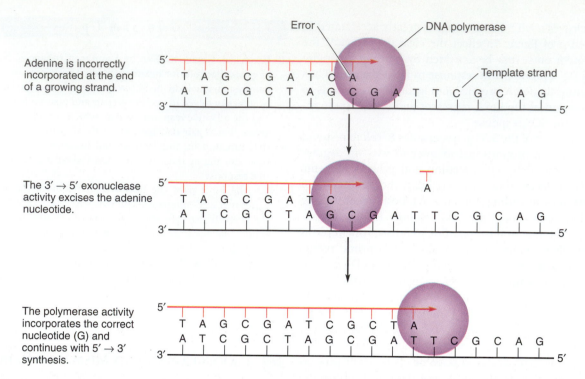

Adenine is incorrectly incorporated at the end of a growing strand.

The 3′ → 5′ exonuclease activity excises the adenine nucleotide.

The polymerase activity incorporates the correct nucleotide (G) and continues with 5′ → 3′ synthesis.

▶ **FIGURE 23–18** Proofreading by DNA polymerase. The 3′ → 5′ exonuclease activity of the enzyme excises the incorrect nucleotide addition (such as the A nucleotide in the top diagram). The polymerase activity of the enzyme then guides the addition of the correct nucleotide in the 5′ to 3′ direction, using the complementary strand as a template.

methyl-directed mismatch repair system that requires the protein products of four genes, *mutH, mutL, mutS,* and *mutU.* A model depicting the role of these products is shown in ▶ Figure 23–19. The process is initiated by the binding of the MutS and MutH proteins: MutS recognizes and binds to base-pair mismatches, while MutH binds to hemimethylated GATC sites located on either side of a MutS-bound mismatch. When activated by MutS and MutL, the MutH protein cleaves the DNA strand containing the misincorporated nucleotide at one of the neighboring GATC sites. The MutU protein then unwinds the strand so that the segment containing the mismatch can be removed by an exonuclease. The resulting gap is then filled in by DNA polymerase I and sealed by DNA ligase.

The *E. coli* system identifies the strand that is to be cut by the methylation pattern of the DNA. In certain strains of *E. coli,* specific enzymes attach methyl groups to the adenine in the sequence GATC wherever it appears in the newly synthesized DNA strand. However, there is a short time lag that occurs between replication and methylation, and during this delay, the old (parental) strands are methylated but the newly synthesized strands are not. The mismatch repair system in *E. coli* operates during this time interval, using the transient difference in methylation to ensure that only the newly synthesized strand, which carries the mistake, is repaired.

It is not clear how the mismatch repair systems in eukaryotes identify the proper strand to repair. Some eukaryotes, such as mammals, do not methylate A residues in their DNA, so mechanisms other than the *E. coli* system must be involved in distinguishing new strands from old.

Mismatch Repair and Gene Conversion

In addition to correcting base-pairing mistakes that occur during DNA replication, mismatch repair systems appear to recognize and process mispaired bases that occur in the heteroduplex intermediates formed during general recombination. The discovery of the apparent conversion of one allele into another during the recombination process gave one of the first clues that some form of mismatch repair operates in recombination. This phenomenon, called **gene conversion,** is best studied in *Neurospora* and certain other fungi, where the products of a single meiosis are retained together within an ascus, allowing them to be analyzed genetically (see Chapter 13). In these organisms, gene conversion is detected by a non-Mendelian segregation ratio of alleles among the meiotic products, as illustrated in ▶ Figure 23–20 for the different ascospore ratios of the *Neurospora m* gene and its wild-type allele. We would ordinarily expect to see a 4:4 segregation pattern for the alleles (Figure 23–20a). However, in about 1% (or less) of the asci, non-Mendelian ratios of the form 6:2 or 5:3 are observed. The 6:2 ratio is the result of **chromatid conversion,** in which an allele on one chromatid is converted into the alternative allele present on the homologous chromosome (Figure 23–20b). Both members of a pair of spores will then carry copies of the converted allele. The 5:3 ratio is

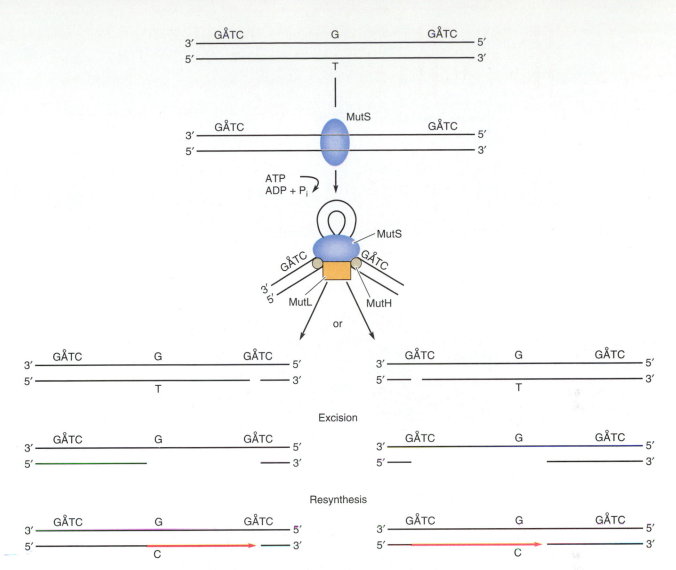

▶ **FIGURE 23–19** A model of mismatch repair. Å is a methylated adenine that targets a palindromic GATC site for cleavage; these sites can be a kilobase or more away from the mismatch. The MutS protein interacts with the mismatch site, and MutL forms a protein-protein interface between the MutS and MutH proteins. MutH nicks opposite a GATC site, and the mismatch is corrected on the unmethylated strand. Correction involves exonuclease action to excise a stretch of nucleotides and resynthesis of DNA to fill in the gap. Subsequent action by ligase seals the remaining nick. (The reaction scheme on the left results from the MutH protein nicking the right-hand target site; the scheme on the right would result from nicking the left-hand target.) *Adapted from:* P. Modrich, "Mechanisms and biological effects of mismatch repair." Reproduced, with permission, from the *Annual Review of Genetics*, Volume 25, © 1992 by Annual Reviews Inc.

the result of **half-chromatid conversion,** in which only half of the chromatid seems to be affected (Figure 23–20c); only one member of a pair of spores then carries the converted allele. Both chromatid and half-chromatid conversion occur at too high a frequency to be accounted for by mutation. The discovery that about 50% of the chromatid and half-chromatid conversions are associated with the recombination of gene loci on either side of the converted allele indicates that they are somehow part of a recombination process.

Several different models of gene conversion have been proposed. Most account for the phenomenon by some form of repair during recombination. (For example, the double-strand break repair model explains many cases of gene con-

version through the removal of DNA containing possible mismatches and the subsequent repair synthesis that are part of the recombination process.) A model of gene conversion that involves a form of mismatch repair is shown in ▶ Figure 23–21; here the locus of the converted allele is in a heteroduplex region created during recombination. After the heteroduplex intermediate is formed, repair enzymes recognize the base mismatches and excise the mismatched region on one of the two DNA strands of each involved chromatid. Each resulting gap is then filled in by DNA polymerase and sealed with DNA ligase. In the example in Figure 23–21, mismatch repair "converts" the m allele to m^+ on both recombining chromatids, so resolution of the recombination intermediate leads to a $6:2$ ratio of $m^+:m$ al-

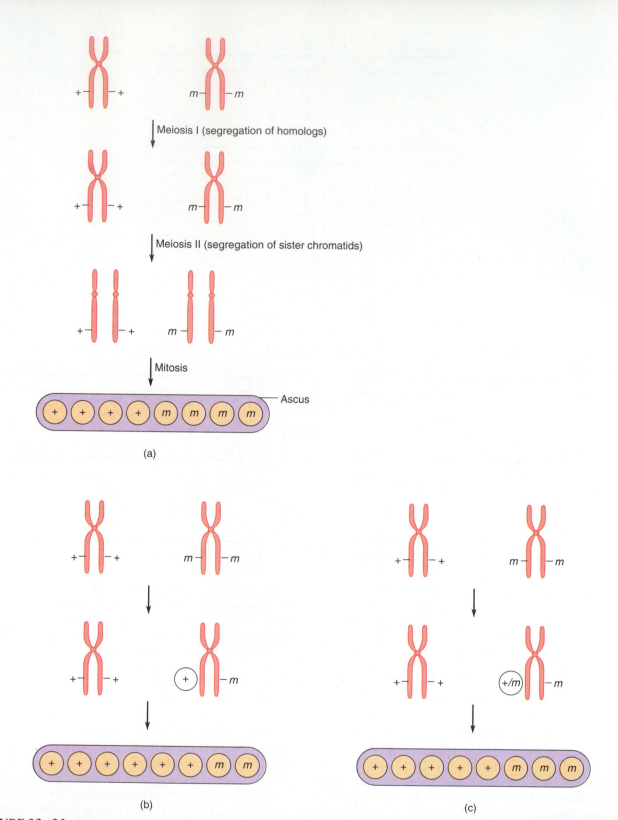

▶ **FIGURE 23–20** Gene conversion. (a) The 4:4 spore ratio in the tetrad shows Mendelian segregation of the m^+, m allele pair during meiosis. (b) Chromatid conversion (conversion of an m allele to m^+) and the resulting 6:2 ratio of ascospores. (c) Half-chromatid conversion and the resulting 5:3 spore ratio.

leles, with half of the converted chromatids being recombinant for genes that flank the affected region. Half-chromatid conversion can also occur by the mechanism shown in Figure 23–21 if mismatch repair results in the conversion of only one allele. Only one spore pair would then be affected, leading to a 5:3 segregation ratio.

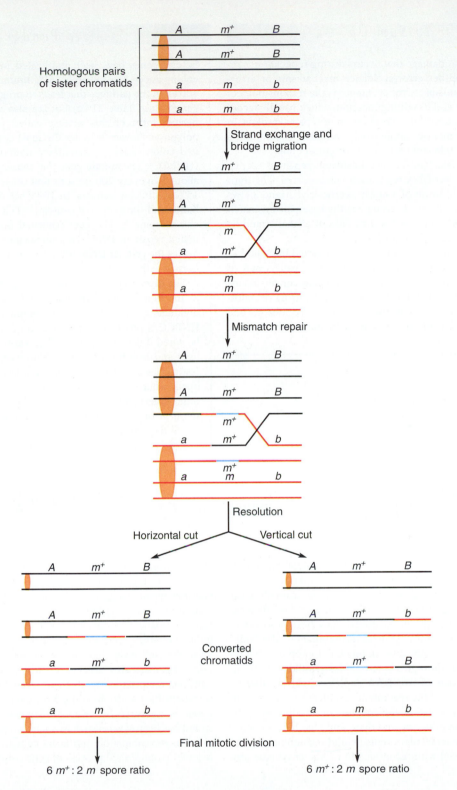

▶ **FIGURE 23–21** Explanation of gene conversion in terms of a mismatch repair mechanism. After strand exchange and bridge migration, the m^+ and m alleles lie within heteroduplex regions. The mismatches in these regions are then repaired on one of the DNA strands of each involved chromatid, giving a 6:2 ratio of $m^+:m$ alleles. Half the converted chromosomes have a parental arrangement of flanking genes (left) and half have a recombinant arrangement (right).

1. Spontaneous DNA damage that occurs during replication is repaired through either error-avoidance mechanisms or error-correction mechanisms. Error avoidance is the responsibility of DNA polymerase and its editing function, which usually recognizes errors that occur during replication or repair synthesis. In *E. coli,* DNA polymerases immediately recognize and cut out a misplaced nucleotide and then insert the correct one.

2. Error-correction mechanisms use mismatch repair to correct mispaired bases after DNA replication has occurred. The transient lack of methylation of a newly synthesized strand allows repair proteins to selectively recognize the base-pair mismatch on the new strand. The error is then cut out of that strand by an excision-and-repair-type process.

3. Gene conversion, in which one allele is converted into another during recombination, shows that mismatch repair is part of general recombination. One explanation for gene conversion involves the formation of a heteroduplex region during recombination, followed by repair of the mismatched regions on one DNA strand of each of the homologous recombining chromatids.

Several genes have been implicated in hereditary forms of colon cancer. For example, *APC* mutations are not only involved in the pathway that leads to progressive colon tumorigenesis (see Table 21–6) but are also responsible for an inherited form of this disease called familial adenomatous polyposis. However, by far the most common form of inherited colon cancer is hereditary nonpolyposis colon cancer (HNPCC). The mutant gene that causes HNPCC is carried by about 1 in every 200 persons and causes perhaps as many as 15% of all colon cancers. In 1993, scientists announced they had found the gene that causes HNPCC on the short arm of chromosome 2. The gene's normal function is in the mismatch repair of DNA. This impairment of mismatch repair makes the cellular DNA 100 times more mutable than that of normal cells.

The discovery of the gene that causes HNPCC could lead to the first genetic screening test for cancer-susceptible individuals in a family or in the population at large. Because HNPCC is relatively common and because colon cancer can be cured if detected early, such a test would quite likely save lives and health-care dollars. However, any form of genetic testing has both pros and cons. If a test for HNPCC is developed, what uses and abuses of it do you foresee? Tests for other inherited forms of cancer will probably be developed in the near future. Do you think it effective or desirable to test for all the various inherited cancers?

Chapter Summary

Recombination produces genetic variation by rearranging existing DNA sequences. There are three main types of recombination: general recombination, site-specific recombination, and transposition. General recombination is the most common type; it involves the breakage and rejoining of paired (synapsed) DNAs that share extensive sequence homology. Several models have been proposed to account for the mechanisms of general recombination. The Holliday model has been one of the most influential; it proposes that recombination proceeds through a heteroduplex (Holliday) DNA intermediate that consists of two DNA duplexes held together by two of the four strands of the recombining molecules. Various enzymes have been implicated in the creation and resolution of the Holliday intermediate in *E. coli,* and their study has led to a better understanding of the molecular events behind recombination.

Intimately connected with recombination is the enzymatic process of DNA repair. Various enzymes are involved in the repair of DNA damage. Some of these enzymes repair DNA damage directly, without altering the DNA backbone, while others participate in excision repair, using the information contained in the complementary strand to replace a damaged base or backbone segment. Certain repair processes, such as the SOS system in bacteria, are inducible—they occur in response to the effects of environmental mutagens and utilize enzymes that are required for recombination. Organisms also have proofreading and mismatch-repair systems that remove incorrectly incorporated bases; DNA polymerase is responsible for proofreading during replication, while mismatch repair is catalyzed by various enzymes shortly after the DNA strand has been synthesized. Mismatch repair has also been implicated in recombination and is thought to be responsible for the recombination-linked process of gene conversion.

Questions and Problems

Recombination at the Molecular Level

1. Define the following terms and distinguish between members of paired terms:
   breakage and rejoining and copy choice
   general, site-specific, and transpositional recombination
   Holliday intermediate
   patch-type and splice-type heteroduplexes

2. Sketch the density profiles that Meselson and Weigle would have observed had they used genetic markers that were located in the center of the λ chromosome.

3. In contrast with the chi structure formed during recombination between linear DNA duplexes (see Figure 23–4), the recombination of circular duplex DNAs results in an intermediate figure-eight structure. Moreover, resolution of this structure can result in the production of either two patch-type heteroduplex circles of the original size or a single, larger composite circle. Using the Holliday model as a guide, diagram the process of recombination between two circular DNA duplexes, showing the development of the figure-eight intermediate and the resolution of the structure to form both types of products.

4. While examining the DNA of a plasmid under the electron microscope, you observe several pairs of plasmid DNA circles that are joined together in the form of figure-eights.

You believe that the figure-eights represent intermediates in some form of general (homologous), site-specific, or transpositional (nonhomologous) recombination. You decide to determine which type of recombination is occurring by treating a sample of DNA containing the figure-eights with a restriction endonuclease that you know has a single cleavage site in the plasmid. Exposure of the DNA to this enzyme should convert the figure-eights to chi structures that can be viewed under the electron microscope. Which of the following groups of chi structures would best represent what you would observe if the recombination process were (a) homologous? (b) site-specific? (c) nonhomologous? Explain your answers.

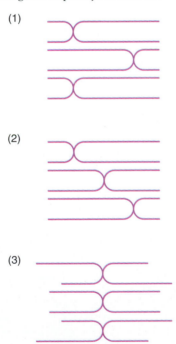

Direct Repair

5. What is the role of visible light in the direct repair of UV-induced DNA damage?

6. Samples of a UV-sensitive mutant of phage T4 were irradiated with various doses of UV, and the surviving fraction was determined by plating the phage samples on bacteria incubated ei-

ther in the dark or in the presence of visible light. Samples of wild-type T4 were treated in the same manner to serve as a control. The results are:

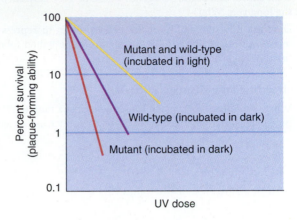

Provide an explanation for the difference in UV sensitivity between the mutant and wild-type.

7. When the UV survival curves of two different phages are compared, the phage with the larger genome size usually exhibits the greater UV sensitivity. (a) Explain this relationship. (b) Suppose that a UV dose of 1 Joule/m^2 gives 5% survival of phage X and 37% survival of phage Y. All else being equal, how much larger is the genome size of phage X than of phage Y?

Excision Repair

8. Three mechanisms are known that can repair the damage caused by thymine dimerization: photoreactivation, base-excision repair (involving a DNA glycosylase), and nucleotide-excision repair. (a) Order these mechanisms according to their metabolic energy demands. Which of these mechanisms is the most energy efficient? (b) Which of these mechanisms do you believe has the greatest chance of introducing mutations through misrepair? Why?

9. Methylation of DNA, especially of cytosine, is common. What would be the consequence of deamination of 5-methylcytosine? Would the resulting nucleotide be repaired? If not, what kind(s) of mutation(s) would result?

10. As a rule, RNA viruses accumulate mutations more rapidly than DNA viruses. Why?

11. In addition to the excision repair process initiated by ABC excinuclease, it is also conceivable that thymine dimers in *E. coli* are removed and replaced in a process that involves an endonuclease nick followed by a single pass of DNA polymerase I catalyzing the process of nick translation. What DNA structural feature would be present during the excinuclease-catalyzed process but not during nick translation? How might this feature be used to experimentally test the relative importance of the two mechanisms in UV repair?

Postreplication (Recombinational) Repair

12. *E. coli uvr* and *recA* mutants are very sensitive to UV light. It has been observed experimentally that *uvr recA* double mutants are much more sensitive to UV than are the single mutants, while *uvr* double mutants (e.g., *uvrA uvrB*) are no more sensitive to UV than are the single mutants. Postulate an explanation for this observation.

Inducible Repair Systems

13. Although it is indispensable during an emergency, the constant expression of the SOS genes would be very harmful over the long run. Why?

14. Describe how the ability of the LexA protein to repress its own synthesis enables the SOS response to be halted following the repair of damaged DNA.

Proofreading and Mismatch Repair

15. What would be the expected consequences of a mutated DNA polymerase that has a decreased $3' \rightarrow 5'$ exonuclease activity?

16. In *Neurospora,* crosses between a mutant and wild-type can produce a 6:2 or 5:3 tetrad ratio as a consequence of gene conversion. Assuming that the mutant allele is the result of a simple base substitution, which ratio would you expect to occur most frequently? Explain your answer.

17. In oligonucleotide-directed mutagenesis (Figure 22–24), the proportion of progeny DNA molecules that carry the mutation often falls well below the expected 50%. Provide an explanation.

THE GENETIC BASIS OF EVOLUTION

Biological evolution is the dual process of genetic change and diversification of organisms through time. By this process related populations can diverge from one another in their genetic characteristics and give rise to new species. The idea that populations can change over time and produce different species and that all present-day species were derived in this manner from a common ancestor (or from a small number of independently arising ancestors) provides a rational framework for organizing the vast array of biological knowledge. Evolution is thus one of the main unifying concepts of biology.

Charles Robert Darwin, 1809–1882.

The first modern ideas on the mechanisms of evolution emerged in the 1850s with the works of Charles Darwin and Alfred Russel Wallace. Darwin and Wallace envisioned evolution as occurring through a selection process based on reproductive ability, which Darwin called **natural selection.** Darwin presented the concept of evolution by natural selection in his book, *On the Origin of Species by Means of Natural Selection* published in 1859. His theory was based on three major premises:

1. *Organisms tend to over-reproduce.* Under suitable conditions, all organisms are capable of producing many more of their kind by asexual and/or sexual reproduction. This potential for reproduction is tremendous. An oyster, for example, can produce as many as 100 million eggs in a single spawning. Most plants are also prolific—some can produce over a million seeds in one season. Even very slow breeders, like elephants, can produce several offspring during their lifespan and can fill an environmental space in a surprisingly short period of time when conditions permit.

2. *There is a struggle for survival.* Despite the tremendous potential for population growth, adult populations in nature remain relatively constant over long periods of time, being held in check by environmental limits that include predation, disease, and competition for food and space. As a result, most offspring produced by a population die before reaching adulthood and thus fail to reproduce themselves.

3. *Organisms vary in their ability to survive and reproduce.* At the heart of the struggle for survival is the presence of genetic variation; different phenotypes have different survival and reproductive abilities and give rise to the condition that Darwin called the "survival of the fittest." Individuals whose characteristics are better suited to their needs in their particular environment are better able to survive and reproduce (are more fit in a reproductive sense) and will therefore be represented by more descendants in succeeding generations. There is thus a natural selection favoring the survival and reproduction of the better-adapted variants in the population.

Although natural selection clearly acts on individual organisms, its genetic consequences can be realized only at the population level. An individual organism cannot evolve, since it has a finite lifespan and a fixed genetic makeup at birth. A population, in contrast, can maintain itself over long periods of time and can change in genetic composition from one generation to the next. The long-term sequence of changes in the hereditary characteristics of a population is the genetic process of evolution.

The Darwin-Wallace theory of evolution had one major defect: It could not explain the mechanisms of heredity and was therefore unable to account for the origin of variation and the role of genetics in the evolutionary process. Neither Darwin nor Wallace was familiar with Mendel's work. It was therefore left to other scientists after the turn of the century to incorporate Mendel's discoveries into the Darwinian model of natural selection. The result of this effort was the **synthetic theory** of evolution (or **neo-Darwinism**, as it is sometimes called), which is now the basis of many of the current concepts of evolution. The synthetic theory brings together contributions from many different areas of biology and treats the population, rather than the individual, as the fundamental unit of evolutionary change. In the synthetic theory, evolution is explained in terms of changes in allele frequencies in populations, and the mechanisms of evolution are described in terms of the forces that alter these frequencies, such as mutation, migration, and natural selection.

In the following chapters, we will consider many of the topics in genetics that are essential to the synthetic theory of evolution. We will begin by describing the genetics of populations, including the concept of allele frequencies. We will then consider the forces that are responsible for genetic change in populations. Finally, we will look at evolution from the perspective of a molecular biologist and consider the process of evolution at the molecular level.

CHAPTER 24

Genes in Populations

Our discussion of genetics so far has focused on the individual—a cell, organism, or strain—and our study of gene transmission has been limited to the results of specific matings, such as those encountered in the analysis of a pedigree or breeding experiment. We will now view genetics from a somewhat different perspective and consider the fate of genes in populations under the less restrictive conditions that prevail in nature.

Population genetics is the study of the genetic mechanisms operating in natural populations. Population geneticists seek to understand the means by which genetic variation is derived and maintained and to describe the nature of the processes of adaptive change. The study of population genetics is complicated by the large amount of genetic variation that exists in natural populations. Because of random mutation, no two members of a population are exactly alike at all gene loci. Even monozygotic twins tend to differ as a result

Color variants of the blood star, *Henricia levinscula*.

of somatic mutations that occur during development. The variation pattern is further complicated by recombination, which breaks up existing gene combinations and produces new genotypes in each succeeding generation. Thus from a genetic point of view, each member of a population is probably unique, having a gene combination that will never again be formed in its entirety during the species' history.

Because of the potential variety and transitory existence of genotypes, any attempt to describe the transmission of genes and their distribution among the progeny becomes almost hopelessly complex unless an effort is made to reduce the problem to its simplest terms. One common simplification is to view a population as a potentially large assemblage of genes, or **gene pool**, consisting of the entire set of alleles present in the population at a given time. Reproduction occurring in each generation can then be regarded as a sampling process in which alleles in the form of gametes are drawn from the parental gene pool and combined in different ways to form the genotypes of the progeny. This sampling process is repeated over time, so that the reconstituted gene pool of the progeny will, in turn, serve as the source of alleles for the gene pool of the next generation, and so on. The genetic composition of a population at any moment thus depends on the kinds and frequencies of alleles in the sample derived from the gene pool of the previous generation.

In this chapter, we examine the structure of the gene pool and see how it can be analyzed and described. We first introduce the relationship between allele and genotype frequencies and the techniques used to estimate allele frequencies in populations. This discussion is followed by a description of the factors that are important in determining how individuals select their mates and an analysis of how the mating system affects the way in which alleles are organized into genotypes among the progeny.

ALLELE FREQUENCY

Despite the simplifications inherent in the gene pool model, a complete description of the gene pool of any population (considering the large number of gene loci involved) is still a monumental task. Another more obvious simplification is to restrict our attention to events at one or just a few gene loci. When only a single locus is involved, we define the allele (or gene) frequency as

$$\text{allele (or gene) frequency} = \frac{\substack{\text{number of copies} \\ \text{of a given allele}}}{\substack{\text{total number of copies} \\ \text{of all alleles at the locus}}}$$

This fraction gives the number of copies of an allele (relative to the total at that locus) present at any moment among the diploid genotypes in the population. As long as the various genotypes are equally viable and fertile, this fraction also measures the proportion of all the gametes of the population that carries the allele in question.

Calculating Allele Frequencies from Genotype Frequencies

Allele frequencies can be calculated directly from genotype frequencies by expressing the relative contribution of each diploid genotype to the allele in question. For example, consider the case of a single autosomal locus with two alleles A and a, which are distributed in a diploid population as follows:

	Alleles		Genotypes		
	A	a	AA	Aa	aa
Frequencies	p	q	D	H	R

Since only two alleles are being considered, $p + q = 1$ and $D + H + R = 1$. If the alleles segregate normally during meiosis, their frequencies among the gametes of this population can be related to the frequencies of the existing parental genotypes as follows:

$$p = D + \tfrac{1}{2}H$$
$$q = R + \tfrac{1}{2}H$$

The frequency of heterozygotes, H, is divided by 2 in both equations since half the gametes of the heterozygotes carry the A allele and the other half carry the a allele. These equations thus permit us to calculate p and q if we know the frequencies of the different genotypes. For example, if a population consists of 10% AA, 20% Aa, and 70% aa, $p = 0.1 + \tfrac{1}{2}(0.2) = 0.2$ and $q = 0.7 + \tfrac{1}{2}(0.2) = 1 - p = 0.8$. The reverse is not necessarily true—we cannot calculate D, H, and R from values of p and q unless more is known about the mating system of the population. We will consider the problem of mating systems and the calculation of genotype frequencies from allele frequencies later in this chapter.

We can extend this procedure to a locus with more than two alleles. To illustrate, let us assume that N total alleles $A_1, A_2, \ldots, A_N$ occur in a population and have frequencies of $f(A_1), f(A_2), \ldots, f(A_N)$, respectively. The genotypes in the population are then $A_1A_1, A_1A_2, A_1A_3, \ldots, A_NA_N$. Each allele is carried in duplicate in one homozygous genotype and occurs in a single dose in $N - 1$ heterozygous genotypes. The frequency of the ith allele can thus be expressed as the sum of the contributions of the different genotypes:

$$f(A_i) = f(A_iA_i) + \tfrac{1}{2}f(A_iA_1) + \tfrac{1}{2}f(A_iA_2) + \cdots + \tfrac{1}{2}f(A_iA_N)$$

This general expression can be applied to any number of alleles—in summary, the frequency of a given allele is equal to the frequency of the homozygous genotype for this allele plus half the sum of all the frequencies of the heterozygous carriers of that allele.

Example 24.1

In a certain population, the genotypes for the alleles A_1, A_2, and A_3 occur in the following ratio: $1\ A_1A_1 : 2\ A_2A_2 : 1\ A_3A_3 : 2\ A_1A_2 : 4\ A_2A_3 : 2\ A_1A_3$. What are the frequencies of the three alleles in this population?

Solution: We observe that the sum of the terms in the ratio gives $1 + 2 + 1 + 2 + 4 + 2 = 12$. The frequencies of A_1A_1, A_2A_2, A_3A_3, A_1A_2, A_2A_3, and A_1A_3 are therefore $\frac{1}{12}$, $\frac{2}{12}$, $\frac{1}{12}$, $\frac{2}{12}$, $\frac{4}{12}$, and $\frac{2}{12}$, respectively. We can then calculate the frequencies of the alleles as follows:

$$f(A_1) = f(A_1A_1) + \tfrac{1}{2}f(A_1A_2) + \tfrac{1}{2}f(A_1A_3) = \tfrac{1}{12} + \tfrac{2}{24} + \tfrac{2}{24} = \tfrac{3}{12}$$

$$f(A_2) = f(A_2A_2) + \tfrac{1}{2}f(A_1A_2) + \tfrac{1}{2}f(A_2A_3) = \tfrac{2}{12} + \tfrac{2}{24} + \tfrac{4}{24} = \tfrac{5}{12}$$

$$f(A_3) = f(A_3A_3) + \tfrac{1}{2}f(A_1A_3) + \tfrac{1}{2}f(A_2A_3) = \tfrac{1}{12} + \tfrac{2}{24} + \tfrac{4}{24} = \tfrac{4}{12}$$

Follow-Up Problem 24.1

Suppose that in a different population, the six genotypes in Example 24.1 all have the same frequency. What are the frequencies of the three alleles in this population?

◆

Measuring Allelic Variability

Researchers ordinarily determine the extent of allelic variability in a population by estimating the number of different alleles that are present and their respective frequencies; to do this, the researcher must first be able to identify the different genotypes. The procedure is relatively straightforward for traits that show incomplete dominance or codominance—in these cases, each genotype is expressed as a distinguishable phenotype, so that the allele frequencies can be determined directly from the phenotype frequencies.

The MN blood group is one example of a codominant trait that has been studied extensively in humans. This blood group system consists of two very weak antigens, type M and type N, that can be detected by purified immune serum that contains anti-M and anti-N antibodies. These antigens are specified by the codominant alleles L^M and L^N, yielding three blood types: M ($L^M L^M$), MN ($L^M L^N$), and N ($L^N L^N$). Most individuals fail to react (or react very weakly) to the M and N antigens, so they are of no importance in blood transfusions, but they have been extremely useful for genetic studies.

The results of a few studies on the MN blood group are given in ■ Table 24–1. Two basic patterns emerge from these results. First, it is apparent that there are differences among the individuals in each population. Most popula-

tions are genetically variable for this trait, that is, both blood group alleles are present at significant frequencies. Traits, such as the MN blood group, for which two or more phenotypes commonly occur within a population, are said to exhibit **polymorphism**. Second, as we see from the MN data in Table 24–1, the allele frequencies tend to differ among the populations. These differences are particularly evident in populations that are separated by wide geographic barriers. Thus, more than 60% of the Eskimos tested on Baffin Island were type M, and less than 3% were type N, while type N is the most common among the Aborigines of Australia.

Unlike the alleles of the MN blood group system, alleles at many other loci are dominant in their expression. When dominance is involved, heterozygotes have the same phenotype as dominant homozygotes, so it is impossible to calculate allele frequencies directly from phenotype frequencies. Various techniques have been used to detect and measure this concealed genetic variation. In some cases, **crossing techniques** have been useful for detecting carriers of recessive alleles. Heterozygote frequencies can then be measured to provide a means for calculating allele frequencies. **Cytological techniques** have also helped to detect concealed variability in instances where an allele is known to be associated with an identifiable chromosome rearrangement (for example, an inversion). In recent years, **biochemical techniques** have been used to detect allelic variability at the molecular level (the methods and results of this approach will be discussed at length in Chapter 26). Another method for determining allele frequencies when dominance is involved is based on a mathematical model that is derived for a particular pattern of mating. We will consider such a model in the following section.

To Sum Up

1. The study of the genetic mechanisms operating in natural populations is called population genetics. Population geneticists attempt to explain the origin and maintenance of genetic variation in populations and describe the genetic processes involved in evolution.
2. Allele (or gene) frequencies, expressed as proportions of the total alleles at a given locus, are used to characterize the genetic

■ **TABLE 24–1** Frequencies of genotypes and alleles for the MN blood group locus among various human populations.

Population	Location	Genotype Frequency			Allele Frequency	
		$L^M L^M$	$L^M L^N$	$L^N L^N$	L^M	L^N
Aborigine	S. Australia	0.024	0.304	0.672	0.176	0.824
Bengali	India	0.354	0.508	0.138	0.608	0.392
Eskimo	Baffin Island	0.662	0.310	0.028	0.817	0.183
German	Berlin	0.284	0.499	0.217	0.533	0.467
Japanese	Tokyo	0.285	0.510	0.205	0.540	0.460
Polynesian	Hawaii	0.125	0.417	0.458	0.333	0.667

structure of a population. An allele frequency can be computed from the pertinent genotype frequencies by adding the frequency of the homozygote for the allele in question to half the sum of the frequencies of the heterozygous carriers for the allele.

3. Determining an allele frequency when dominance is involved requires a method for detecting and measuring the frequency of carriers of the recessive allele. Various methods are employed for detecting this concealed genetic variation, including genetic crosses and cytological and biochemical techniques. In certain cases, the allele frequency can be based on a mathematical model derived for a particular system of mating.

GENE TRANSMISSION WITH RANDOM MATING

We have seen that allele frequencies can be computed directly from genotype frequencies. It is also possible to calculate genotype frequencies from allele frequencies; but first we need precise information about the mating system of the population. The **mating system** of a population is the pattern with which individuals select their mates. The simplest mating system to describe, and the one that serves as the standard for comparison is **random mating**. Mating is said to be random when mates are selected without regard to genotype. In other words, random mating occurs when the genotypes of prospective mates are statistically uncorrelated, so that there is no tendency for individuals to select mates that are similar to or different from themselves in genotype. Since the genotypes of each pair of mates are statistically independent, the frequency of a particular mating can then be expressed as the product of the frequencies of the combined genotypes. For example, if the frequency of the AA genotype is $\frac{1}{2}$ and mating is random, any individual in the population has a 50% chance of selecting an AA mate, so the frequency of $AA \times AA$ matings will be $(\frac{1}{2})(\frac{1}{2}) = \frac{1}{4}$.

Example 24.2

In a certain population, the genotypes AA, Aa, and aa occur at frequencies 0.6, 0.4, and 0, respectively. If this population mates at random, what proportion of all matings should be between mates of different genotypes?

Solution: Three types of matings are possible in this population: $AA \times AA$, $AA \times Aa$, and $Aa \times Aa$. Of these matings, only $AA \times Aa$ involves different genotypes. Since this mating type includes two possibilities, AA male $\times Aa$ female and AA female $\times Aa$ male, each occurring at an expected frequency of $(0.6)(0.4) = 0.24$, the total frequency of this mating in this population is $2(0.24) = 0.48$.

Follow-Up Problem 24.2

In a different population, the genotypes AA, Aa, and aa occur at frequencies 0.1, 0.2, and 0.7, respectively. If this

population mates at random, what proportion of all matings should occur between mates with identical genotypes?

To extend our example further, suppose that all other individuals in the population are aa in genotype, so that the population consists of the two genotypes, AA and aa, in equal proportions. In this case, three kinds of matings are possible, in a ratio of $1\, AA \times AA : 2\, AA \times aa : 1\, aa \times aa$ (► Figure 24–1). If all matings, on the average, produce the same number of offspring (i.e., there is no selection), the genotype frequencies in the next generation can be calculated from the average contributions of these different mating types. Thus, in our example, $AA \times AA$ produces one-fourth of all progeny, all AA in genotype; $AA \times aa$ produces one-half of all progeny, all Aa in genotype; and $aa \times aa$ produces one-fourth of all progeny, all aa in genotype. The collective progeny formed by these matings therefore occur in a ratio of $1\, AA : 2\, Aa : 1\, aa$. As we shall see later, the same progeny ratio is produced by numerous other mixtures of parental genotypes under random mating, such as $1\, AA : 2\, Aa : 1\, aa$, $2\, AA : 1\, Aa : 2\, aa$, and any other combination in which the ratio of alleles among the parents is $1\, A : 1\, a$.

To avoid giving the impression that random mating must always result in a $1 : 2 : 1$ ratio, ► Figure 24–2 shows the same example, but with a starting ratio of $p\, AA : q\, aa$. In this case, the progeny genotypes appear in frequencies of $p^2\, AA$, $2pq\, Aa$, and $q^2\, aa$. As we shall see, this general result is expected for any combination of parental genotypes in which the ratio of alleles is $p\, A : q\, a$.

The preceding examples illustrate two important characteristics of the mating process. First, the genotypic composition of the progeny generation depends on the allele frequencies of the parents. The allele frequencies of the parents are the critical factors in determining the genetic composition of a population, not the parental genotype frequencies. For example, a population with an allele ratio of $1\, A : 1\, a$ will produce a progeny population of $1\, AA : 2\, Aa :$

► **FIGURE 24–1** Results of random mating in a population that consists of a mixture of AA and aa genotypes in a ratio of $1 : 1$.

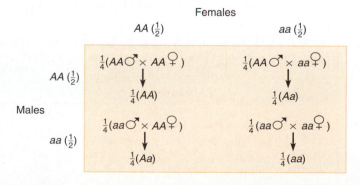

Total offspring $= \frac{1}{4} AA + \frac{1}{2} Aa + \frac{1}{4} aa$

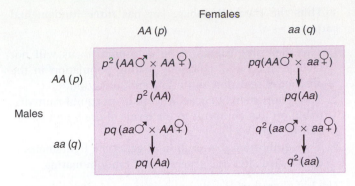

Females

	AA (p)	aa (q)
AA (p)	p^2 (AA♂ × AA♀) ↓ p^2 (AA)	pq(AA♂ × aa♀) ↓ pq(Aa)
aa (q)	pq (aa♂ × AA♀) ↓ pq (Aa)	q^2 (aa♂ × aa♀) ↓ q^2 (aa)

Males

Total offspring = $p^2AA + 2pqAa + q^2aa$

▶ **FIGURE 24–2** Results of random mating in a population that consists of a mixture of *AA* and *aa* genotypes in a ratio of $p:q$.

1 *aa* after one generation of random mating, regardless of whether the genotypes among the parents are 1 *AA*:1 *aa*, 1 *AA*:2 *Aa*:1 *aa*, or 2 *AA*:1 *Aa*:2 *aa*.

Second, in contrast to specific genetic crosses, we cannot predict the genotypic composition of the parental generation by knowing only the genotype frequencies of their offspring. For example, if we know that the progeny occur in a ratio of 1 *AA*:2 *Aa*:1 *aa*, we cannot determine the genotypic ratio among the parents; we can only deduce that the allele ratio must have been 1 *A*:1 *a*. It is therefore impossible to reconstruct the genetic history of a population using only the genotypic proportions of a given generation.

Random-Mating Equilibrium: The Hardy-Weinberg Law

In 1908, G. H. Hardy, an English mathematician, and W. Weinberg, a German physician, demonstrated that genotype frequencies in a population tend to stabilize after one generation of random mating and remain constant in all subsequent generations as long as allele frequencies do not change. Since the genotype frequencies stay constant from generation to generation, we say that an equilibrium is established in the population. The equilibrium is determined solely by the pattern of mating frequencies, and is expected to occur in a large population in which factors that change allele frequencies (such as selection, mutation, and migration) are absent or can be ignored.

The concept that a randomly mating population attains an equilibrium after one generation is known as the **Hardy-Weinberg law** in honor of the men who originally formulated the idea. A formal proof of the Hardy-Weinberg law is given in ■ Table 24–2 for a large randomly mating population with genotype frequencies of *D*, *H*, and *R*. We see that after one generation of random mating, a condition is established in which genotype frequencies are given by the terms of the binomial expansion $(p + q)^2 = p^2 + 2pq + q^2$, and are thus determined solely by existing allele frequencies. For example, if the adult frequencies of *AA*, *Aa*, and *aa* are 0.10, 0.20, and 0.70, respectively, and mating is random, the corresponding genotype frequencies in the next and all subsequent generations will be 0.04, 0.32, and 0.64, as long as allele frequencies remain constant at $p = 0.2$ and $q = 0.8$.

The general relationships between allele and genotype frequencies at equilibrium are shown in ▶ Figure 24–3, where genotype frequency is plotted versus allele frequency, *q*. The actual relationships that are plotted are $D = (1 - q)^2$, $H = 2q(1 - q)$, and $R = q^2$, since $p = 1 - q$. As ▶ Figure 24–4 illustrates, these relationships describe the distribution of genotype frequencies resulting from the random combination of gametes during random mating. Since mates are selected without regard to genotype, an individual gamete has a probability *p* of combining with an *A*-carrying gamete from a randomly selected mate and a probability *q* of combining with an *a*-carrying gamete. The relationship $(p + q)^2$ therefore describes the chance distribution that is expected whenever alleles are combined in a random fashion through random mating.

The equilibrium that is established through random mating not only depends on the existing distribution of allele frequencies but also requires that the allele frequencies remain constant. Changes in the allele frequencies can be

■ **TABLE 24–2** General proof of the Hardy-Weinberg equilibrium law. The genotype frequencies of the offspring produced after one generation of random mating correspond to p^2, $2pq$, and q^2, regardless of the genotype frequencies in the parental population.

Type of Mating	Frequency of Mating	Offspring		
		AA	*Aa*	*aa*
AA × *AA*	D^2	D^2	0	0
AA × *Aa*	$2DH$	DH	DH	0
Aa × *Aa*	H^2	$\frac{1}{4}H^2$	$\frac{1}{2}H^2$	$\frac{1}{4}H^2$
AA × *aa*	$2DR$	0	$2DR$	0
Aa × *aa*	$2HR$	0	HR	HR
aa × *aa*	R^2	0	0	R^2
Total	1.00	$(D + \frac{1}{2}H)^2 = p^2$	$2(D + \frac{1}{2}H)(\frac{1}{2}H + R) = 2pq$	$(\frac{1}{2}H + R)^2 = q^2$

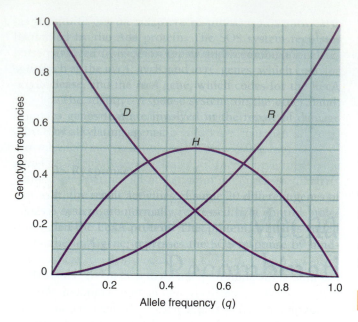

FIGURE 24–3 Genotype frequencies expected for random mating in populations with different allele frequencies.

caused by factors that we have excluded from our model, such as small population size, reproductive inequalities among the genotypes, and the introduction of alleles into the population through mutation and migration. Allele frequencies are not altered by the mating scheme. Although random mating may produce an initial change in genotype frequencies, as they adjust to fit the values of p^2, $2pq$, and q^2, allele frequencies are not expected to change. Contrary to popular misconception, the frequency of an allele does not increase simply because the allele is dominant. Dominance refers only to the ability of an allele to express itself in the heterozygote, not to its ability to increase numerically in the population.

▶ **FIGURE 24–4** Random association of alleles in the gametes of a population undergoing random mating. At equilibrium, the gamete types occur at frequencies corresponding to the terms of the expansion $(p + q)^2$.

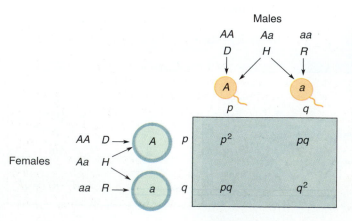

Thus the Hardy-Weinberg law has three fundamental parts:

1. The allele frequencies at an autosomal locus will not change in a large, randomly mating population in the absence of selection, mutation, and migration.
2. The genotype frequencies will attain an equilibrium distribution that conforms to the terms of the binomial expansion $(p + q)^2$.
3. The equilibrium distribution of genotype frequencies is established after one generation of random mating.

The last characteristic is important to a population because it provides resilience to change. If for any reason a population should depart from its binomial array of genotype frequencies, the equilibrium distribution will be reestablished in only a single generation of random mating.

Example 24.3

One assumption that is embodied in the Hardy-Weinberg law is that the allele frequencies at an autosomal locus are the same for both males and females. If they are not the same, then equilibrium genotype frequencies are not established in a population until after two generations of random mating. Show that this one-generation delay in attaining equilibrium occurs for a population that consists of an equal mixture of AA males and aa females.

Solution: A population of AA males and aa females will produce offspring that consist of Aa heterozygotes of both sexes. The offspring of this first generation do not conform to the Hardy-Weinberg frequencies, but after another generation of random mating, they will produce an equilibrium population of $\frac{1}{4}$ AA, $\frac{1}{2}$ Aa, and $\frac{1}{4}$ aa. Thus it takes one generation of random mating just to achieve an equality of allele frequencies between the sexes; the next and all subsequent generations of random mating will then correspond to the Hardy-Weinberg principle.

Follow-Up Problem 24.3

In general, a genetic equilibrium is formed whenever the effects of matings that act to increase the frequency of each genotypic class are exactly counterbalanced by the matings that act to reduce these same frequencies. In the case of two alleles, only two types of matings can act to change the genotype frequencies: AA × aa and Aa × Aa. The first type of mating increases the frequency of heterozygotes in relation to homozygotes by producing only Aa offspring. The second type of mating has the opposite effect, producing homozygotes ($\frac{1}{4}$ AA + $\frac{1}{4}$ aa) as well as heterozygotes ($\frac{1}{2}$ Aa). Therefore, the frequency of Aa × Aa matings must be twice the frequency of AA × aa matings for an equilibrium to be established. Show that this relationship holds for the Hardy-Weinberg equilibrium.

Extensions of the Hardy-Weinberg Equilibrium

The Hardy-Weinberg law can be extended to include multiple alleles. The genotype frequencies that result from random mating for a locus with three alleles are shown in ▶ Figure 24–5. In general, multiple alleles with frequencies $p, q, r, \ldots$ produce equilibrium genotype frequencies that conform to the terms of the expansion $(p + q + r + \cdots)^2$, in which $p^2, q^2, r^2, \ldots$ are the frequencies of the homozygotes and $2pq, 2pr, 2qr, \ldots$ are the frequencies of the heterozygotes. As in the case of a single locus with two alleles, equilibrium is established after a single generation of random mating.

With a few modifications, the Hardy-Weinberg law also applies to the transmission of X-linked genes. Since the homogametic sex (which we will assume to be female) has two X chromosomes, its X-linked genes will follow the same general pattern as is observed for genes on the autosomes (see ▶ Figure 24–6). In other words, with two alleles A and a, females should attain a random-mating equilibrium in which the genotypes AA, Aa, and aa have frequencies p^2, $2pq$, and q^2. The primary difference between the equilibrium established at an X-linked locus and the equilibrium for an autosomal gene involves the males. Since males carry only a single dose of each X-linked gene, their genotype frequencies are the same as the allele frequencies, p and q. Thus, when the male sex is heterogametic, the incidence of an X-linked recessive phenotype is q among males. The ratio of females exhibiting a recessive trait to males exhibiting the trait is

$$\frac{q^2}{q} = q$$

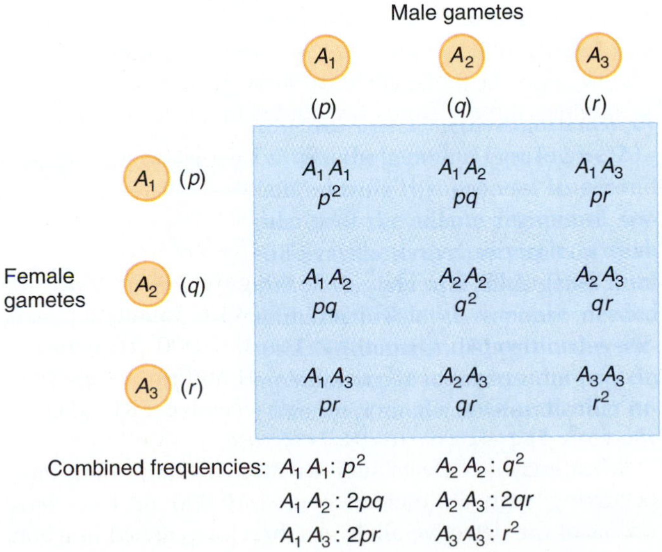

Male gametes
A_1 A_2 A_3
(p) (q) (r)
Female gametes
A_1 (p)
A_2 (q)
A_3 (r)
A_1A_1 p^2
A_1A_2 pq
A_1A_3 pr
A_1A_2 pq
A_2A_2 q^2
A_2A_3 qr
A_1A_3 pr
A_2A_3 qr
A_3A_3 r^2
Combined frequencies: A_1A_1: p^2
A_1A_2: $2pq$
A_1A_3: $2pr$
A_2A_2: q^2
A_2A_3: $2qr$
A_3A_3: r^2

▶ **FIGURE 24–5** Genotype frequencies produced by three alleles after one generation of random mating. Allele frequencies in the parental generation are p, q, and r.

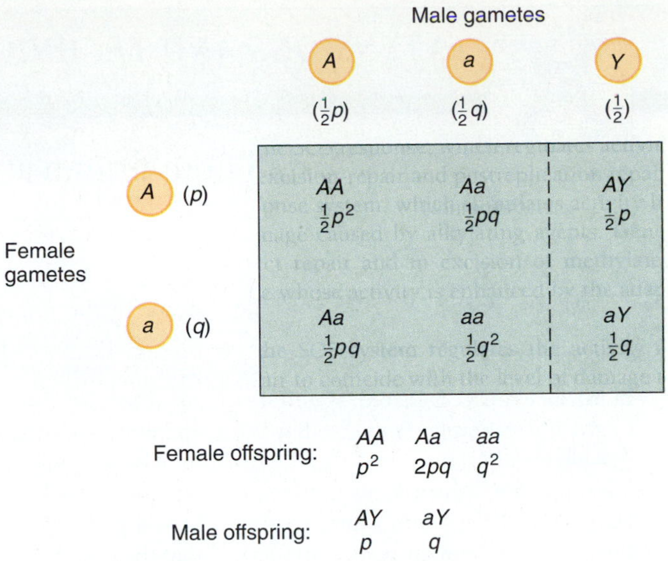

Male gametes
A a Y
$(\frac{1}{2}p)$ $(\frac{1}{2}q)$ $(\frac{1}{2})$
Female gametes
A (p)
a (q)
AA $\frac{1}{2}p^2$
Aa $\frac{1}{2}pq$
AY $\frac{1}{2}p$
Aa $\frac{1}{2}pq$
aa $\frac{1}{2}q^2$
aY $\frac{1}{2}q$
Female offspring: AA p^2 Aa $2pq$ aa q^2
Male offspring: AY p aY q

▶ **FIGURE 24–6** Genotype frequencies for a sex-linked pair of alleles in male and female offspring after one generation of random mating. The genotype frequencies in females correspond to the frequencies expected at an autosomal locus. The genotype frequencies in males are equal to the allele frequencies.

This ratio illustrates why rare X-linked recessive characteristics appear so much more frequently among males than among females. For example, hemophilia occurs in one male in 10,000. The frequency of the recessive allele for hemophilia is thus $q = 1/10{,}000 = 0.0001$. We would therefore expect to observe only one hemophilic female for every 10,000 similarly afflicted males.

The reverse situation holds for rare X-linked dominant traits. In this case, the ratio of females exhibiting the trait in question to males with the trait will be

$$\frac{p^2 + 2pq}{p} = p + 2q = 1 + q$$

If the dominant trait is rare in the population, q approaches 1 and $1 + q$ approaches 2. About twice as many females as males are therefore expected to have the trait.

Applications of the Hardy-Weinberg Equilibrium

The relationships expressed by the Hardy-Weinberg law greatly simplify the calculations of allele and genotype frequencies. When there are only two alleles (for example, A and a), it is theoretically possible to evaluate the allelic and genotypic proportions in a population from knowledge of the frequency of only one homozygote. For example, suppose that the frequency of aa homozygotes in a randomly mating population is $q^2 = 0.36$. The frequency of the a allele must then be the square root of 0.36, so $q = 0.6$. Since $p = 1 - q$, the frequency of the A allele is $p = 0.4$. Thus, the fre-

Attainment of Equilibrium at Two or More Loci

The attainment of an equilibrium through random mating is considerably more complex when two or more loci are considered simultaneously. For example, suppose that the A,a alleles at locus 1 and the B,b alleles at locus 2 have both attained a Hardy-Weinberg equilibrium, so that the genotype frequencies are $p_1^2 (AA)$, $2p_1q_1 (Aa)$, and $q_1^2 (aa)$ at locus 1 and $p_2^2 (BB)$, $2p_2q_2 (Bb)$, and $q_2^2 (bb)$ at locus 2. The fact that each locus when considered separately has attained an equilibrium does not necessarily mean that these loci when considered jointly are in a state of equilibrium. A joint equilibrium for two gene loci is established in a randomly mating population when the genotypes of the zygotes occur in proportions that are the products of the frequencies of their respective alleles. That is, the equilibrium distribution among the zygotes equals the product of the equilibrium distributions of each separate locus:

$$(p_1^2 + 2p_1q_1 + q_1^2)(p_2^2 + 2p_2q_2 + q_2^2)$$

or

$$p_1^2 p_2^2 (AABB), \ 2p_1q_1p_2^2 (AaBB)$$
$$4p_1q_1p_2q_2 (AaBb), \ldots, q_1^2 q_2^2 (aabb)$$

Because the nonallelic genes are associated randomly in a population at equilibrium, the gamete types will also occur in the proportions determined by the products of their respective gene frequencies, that is, $p_1p_2 (AB)$, $p_1q_2 (Ab)$, $q_1p_2 (aB)$, and $q_1q_2 (ab)$. The apparent statistical independence of nonallelic genes at equilibrium applies to linked genes as well as unlinked genes. For this reason, we say that the population is in a state of **linkage equilibrium** when these relationships hold for all alleles at the two loci.

A major difference between the effects of random mating on a single locus and on two separate loci is the time required to attain the equilibrium state. Unlike a single locus, which achieves an equilibrium in a single generation of random mating, two loci considered jointly approach their combined equilibrium state in a gradual manner over several generations. As an example, consider a population that consists of equal numbers of AABB and aabb individuals of both sexes. For simplicity, we'll assume that the two gene pairs are unlinked. All of the allelic frequencies in this starting population are equal to $\frac{1}{2}$, since $f(A) = f(a)$ and $f(B) = f(b)$. Thus, the gamete frequencies at equilibrium are expected to be $\frac{1}{4} AB$, $\frac{1}{4} Ab$, $\frac{1}{4} aB$, and $\frac{1}{4} ab$. Initially, only two types of gametes are formed ($\frac{1}{2} AB$ and $\frac{1}{2} ab$); these types unite in the next generation (generation 1) to produce three diploid genotypes (AABB, AaBb, and aabb) in a ratio of 1:2:1. Of these genotypes, $\frac{1}{4} AABB$ yields $\frac{1}{4} AB$ gametes, $\frac{1}{4} aabb$ yields $\frac{1}{4} ab$ gametes, and $\frac{1}{2} AaBb$ yields $(\frac{1}{2})(\frac{1}{4} AB + \frac{1}{4} Ab + \frac{1}{4} aB + \frac{1}{4} ab)$ gamete types, so the genotypes collectively form gametes in the proportions $\frac{3}{8} AB$, $\frac{1}{8} Ab$, $\frac{1}{8} aB$, and $\frac{3}{8} ab$. If we extend our calculations over several generations, we get the results shown in ■ Table 24–3. Observe the gradual increase in the frequencies of the Ab and aB gametes and the corresponding reduction in the frequencies of the AB and ab types. Eventually, an equilibrium is approached in which all the gamete frequencies are equal.

Note in the accompanying graph that much the same pattern would result if the genes were linked, except that with linked genes, the time it takes to reach an equilibrium is longer. The graph demonstrates that linkage prolongs the time needed to achieve the equilibrium state. In general, the length of time it takes two linked loci to reach a linkage equilibrium is inversely proportional to the distance between the loci—the closer the two loci are on a chromosome, the smaller the chance of recombination and, consequently, the greater the number of generations required for the alleles at these loci to become randomly distributed in the gametes.

---

quency of AA homozygotes must be $p^2 = 0.16$ and that of heterozygotes is $2pq = 0.48$.

This procedure has been especially helpful in medical genetics for estimating the frequency of carriers of certain rare, recessive alleles that produce a genetic disorder when they are in the homozygous state. Many of these disorders arise from enzymatic defects. One example is the recessive disorder phenylketonuria (PKU), which results from the deficiency of an enzyme needed to convert the amino acid phenylalanine to tyrosine (see Chapter 11). About one child in 10,000 births has PKU (see ▶ Figure 24–7). Since the pairing of the unaffected parents of PKU infants is essentially random, we can equate the 1/10,000 frequency of PKU offspring to q^2, and thus $q = 0.01$ and $p = 1 - q = 0.99$. The frequency of carriers of the recessive PKU allele is therefore $2pq = 2(0.99)(0.01) = 0.0198$.

Most enzyme defects such as PKU are quite rare, often occurring with an incidence of 1/10,000 or less. Most copies of the defective allele are therefore carried in a concealed state by heterozygotes. The relative importance of

■ **TABLE 24–3** Gradual approach to random-mating equilibrium for two unlinked genes in a population started with equal numbers of *AABB* and *aabb* genotypes.

	Gamete Frequencies			
Generation	*AB*	*Ab*	*aB*	*ab*
0	$\frac{1}{2}$	0	0	$\frac{1}{2}$
1	$\frac{3}{8}$	$\frac{1}{8}$	$\frac{1}{8}$	$\frac{3}{8}$
2	$\frac{5}{16}$	$\frac{3}{16}$	$\frac{3}{16}$	$\frac{5}{16}$
3	$\frac{9}{32}$	$\frac{7}{32}$	$\frac{7}{32}$	$\frac{9}{32}$
4	$\frac{17}{64}$	$\frac{15}{64}$	$\frac{15}{64}$	$\frac{17}{64}$
5	$\frac{33}{128}$	$\frac{31}{128}$	$\frac{31}{128}$	$\frac{33}{128}$
⋮	⋮	⋮	⋮	⋮
⋮	⋮	⋮	⋮	⋮
Limit	$\frac{1}{4}$	$\frac{1}{4}$	$\frac{1}{4}$	$\frac{1}{4}$

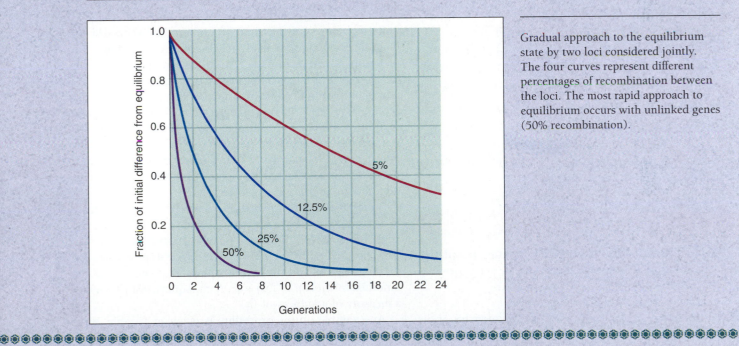

Gradual approach to the equilibrium state by two loci considered jointly. The four curves represent different percentages of recombination between the loci. The most rapid approach to equilibrium occurs with unlinked genes (50% recombination).

heterozygotes as a source of detrimental alleles can be seen more clearly by expressing the ratio of heterozygotes to recessive homozygotes as

$$\frac{H}{R} = \frac{2pq}{q^2} = \frac{2(1 - q)}{q}$$

Observe that when *q* is small (as in the case of most recessive enzyme defects), heterozygotes outnumber recessive homozygotes by a factor of $2/q$. When $q = 0.01$, for example,

there are almost 200 times more heterozygotes than recessive homozygotes in the population.

Equilibrium relationships are also useful when working with multiple alleles, especially when complete dominance is shown by one or more of the alleles. Consider the case of three alleles with a simple hierarchy of complete dominance such that *A* is dominant over *A'* and *A''*, and *A'* is dominant over *A''*. Three phenotypes are then possible, with the following dominance relationships: type A (*AA*, *AA'*, and *AA''*) > type A' (*A'A'* and *A'A''*) > type A'' (*A''A''*).

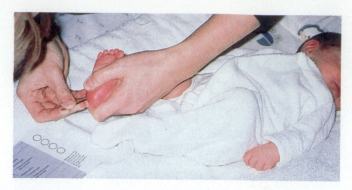

▶ **FIGURE 24–7** Infants with PKU have an excellent chance for a normal healthy life, providing the disease is detected and treated early. Here a nurse is taking a blood sample from an infant's foot to be used in a simple chemical test for PKU.

Suppose that 51% of individuals in a population are determined to be type A in phenotype, 40% are type A′, and 9% are type A″. What are the allele frequencies in this population? Since mating is random, we can arrive at the following conclusions:

1. The frequency of $A''A''$ is $r^2 = 0.09$.
2. The frequency of the A'' allele is $r = \sqrt{0.09} = 0.3$.
3. The sum of the frequencies of type A′ and type A″ phenotypes is $q^2 + 2qr + r^2 = (q + r)^2 = 0.40 + 0.09 = 0.49$.
4. The frequency of the A′ allele is $q = \sqrt{0.49} - r = 0.4$.
5. The frequency of the A allele is $p = 1 - q - r = 0.3$.

Having calculated the allele frequencies, we can substitute these values into the terms of the expansion $(p + q + r)^2$ to estimate the frequencies of the various genotypes in the population.

Example 24.4

If 36% of all persons within a randomly mating population are of blood type O and 45% are of type A, what percentages are expected to be of type B and type AB?

Solution: Let p, q, and r represent the frequencies of the three blood group alleles I^A, I^B, and I^O, respectively. Since the frequency of type O persons is $r^2 = 0.36$, the frequency of the I^O allele must be $r = 0.6$. We can also write that $p + r = \sqrt{f(A) + f(O)}$. Therefore, $p = \sqrt{0.45 + 0.36} - 0.6 = \sqrt{0.81} - 0.6 = 0.3$. From these results, we can calculate that $q = 1 - p - r = 1 - 0.3 - 0.6 = 0.1$. The frequencies of blood types B and AB then become

$$f(B) = f(I^B I^B) + f(I^B I^O) = q^2 + 2qr$$
$$= (0.1)^2 + 2(0.1)(0.6) = 0.13$$

and

$$f(AB) = f(I^A I^B) = 2pq = 2(0.3)(0.1) = 0.06$$

Therefore, we would expect 13% type B and 6% type AB persons in this population.

Follow-Up Problem 24.4

Calculate the frequencies of the A, B, AB, and O blood types in a randomly mating population in which the three blood group alleles are equal in frequency.

◆

Testing Goodness-of-Fit

One of the uses of the Hardy-Weinberg law is to test for conditions of random mating in sample populations. For codominant and incompletely dominant traits, experiments of this type are relatively straightforward. They usually involve comparing the observed numbers of the various genotypic classes with those expected on the basis of random mating. If the observed numbers conform to the expected numbers calculated from the Hardy-Weinberg formula, the population is considered to be in genetic equilibrium, since the data indicate that the genotypes were derived from random mating in preceding generations.

To test for agreement with values predicted by the Hardy-Weinberg equilibrium, we usually begin by calculating the allele frequencies from the observed genotype frequencies. The expected numbers in the different genotypic classes are then derived by substituting the allele frequencies into the binomial formula $p^2 + 2pq + q^2$. A chi-square test then gives a measure of goodness-of-fit.

An example of this procedure is given in ■ Table 24–4 for the results of a study on the MN blood group alleles, L^M

■ **TABLE 24–4** Chi-square test of goodness-of-fit to the Hardy-Weinberg law for blood group frequencies at the MN locus in a sample of 140 Pueblo tribal members.

	M ($L^M L^M$)	MN ($L^M L^N$)	N ($L^N L^N$)	Total
Observed numbers	83	46	11	140
Expected proportions	p^2 (0.573)	$2pq$ (0.368)	q^2 (0.059)	1.0 1.0
Expected numbers	80.2	51.5	8.3	140
$(O - E)^2/E$	0.098	0.587	0.878	$\chi^2 = 1.56$

Note: The expected proportions were derived using $p = 0.757$ and $q = 0.243$.

and L^N. To determine whether these data are consistent with the hypothesis of random mating, we first obtain estimates of the L^M and L^N allele frequencies from the data:

$$f(L^M) = p = \frac{83}{140} + \frac{1}{2}\left(\frac{46}{140}\right) = 0.593 + \frac{1}{2}(0.328) = 0.757$$

$$f(L^N) = q = 1 - 0.757 = 0.243$$

We can then use these allele frequencies to calculate the expected frequencies for the M, MN, and N blood types, assuming random mating:

$$f(M) = p^2 = (0.757)^2 = 0.573$$

$$f(MN) = 2pq = 2(0.757)(0.243) = 0.368$$

$$f(N) = q^2 = (0.243)^2 = 0.059$$

Since the expected and observed results are in close agreement, these data obviously conform to the Hardy-Weinberg law. This conclusion is verified by the chi-square value calculated in Table 24–4. The value of $\chi^2 = 1.56$ corresponds to a probability between 0.2 and 0.3, which indicates that the observed and expected values are not significantly different. Note that there is only one degree of freedom in this case. The allele frequencies in natural populations, unlike those in controlled genetic crosses, do not have values that are theoretically predictable, so the values must be calculated from the observations themselves. Even though the example has three phenotypic classes, we lose an additional degree of freedom by estimating p from the experimental data. The general rule in these situations is that the number of degrees of freedom equals the number of phenotypic classes minus the number of alleles.

Natural populations do not always agree so closely with the Hardy-Weinberg expectations. A lack of agreement between the observed and expected genotypic frequencies can be caused by small population size, nonrandom mating, unequal viability and fertility of genotypes, and other departures from the ideal conditions that are assumed in the model. The Hardy-Weinberg law is not very sensitive to some of these departures, however, and the mere fact that a population conforms to the equilibrium law does not mean that all of the assumed conditions are met.

To Sum Up

1. Genotype and mating frequencies depend on the mating system in the population, that is, the way in which mates are selected for breeding. If mating is random, so that mates are selected without regard to genotype, the frequency of each type of mating can be expressed as the product of the genotype frequencies of the mates involved.
2. Allele frequencies tend to be unaffected by the mating system of the population and by whether the allele is dominant or recessive. However, changes in allele frequencies can result from factors such as selection, mutation, migration, and chance (or sampling) variation in small populations.
3. The Hardy-Weinberg law states that in the absence of the factors that can change allele frequency, a large population will attain an equilibrium in genotype frequencies after one generation of random mating. This equilibrium is characterized by a constant distribution of genotype frequencies that corresponds to the terms of a binomial expansion of the allele frequencies.
4. The Hardy-Weinberg principle can be readily extended to include multiple alleles as well as alleles at a sex-linked locus. As in the case of an autosomal locus with two alleles, both of these extensions result in a genetic equilibrium after a single generation of random mating; this equilibrium is characterized by a random distribution of alleles among existing genotypes. For three alleles, the genotype frequencies at an autosomal locus correspond to the terms of a trinomial distribution of allele frequencies. For an X-linked gene, the heterogametic sex will have genotype frequencies that correspond to the allele frequencies, while the homogametic sex will have genotype frequencies that show the same distribution expected for an autosomal locus.

DEPARTURES FROM THE HARDY-WEINBERG EQUILIBRIUM: SMALL POPULATIONS AND NONRANDOM MATING

The Hardy-Weinberg law is based on four main assumptions: (1) a very large (ideally infinite) population, (2) random mating, (3) equally viable and fertile genotypes, and (4) no changes in allele frequency due to mutation or migration. Strictly speaking, these conditions can never be met in any real population over an extended period of time. Nevertheless, enough populations meet these conditions sufficiently well over a few generations to make the Hardy-Weinberg law a useful model in practice. All that is needed to approximate these conditions is a fairly large population (several hundred or more) in which mating is approximately random and the short-term effects of mutation, migration, and selection are negligible. We will now consider two major exceptions to the Hardy-Weinberg law—small population size and nonrandom mating—to see what happens to the genetic structure of a population when the assumed conditions of a large population and random mating are not even approximately met. These discussions will focus on the exceptions that disrupt the equilibrium established through random mating by reducing the genetic variation within the population.

Random Genetic Drift

The Hardy-Weinberg law describes a situation in which genetic variation in a population is maintained from one generation to the next. One mechanism that can disrupt this equilibrium and lead to a reduction in genetic variation is by way of chance alterations in allele frequencies. Chance changes in allele frequencies occur to some extent in all finite populations, but they are especially important in small breeding groups (consisting of less than 100 individuals, as a general rule). These alterations in allele frequencies that occur by chance in small populations have been termed **random genetic drift** (or simply **genetic drift**), since the measured values of p and q tend to drift about in a random fashion during successive generations.

The random process of genetic drift is essentially a matter of sampling. Unlike the ideal population assumed in the Hardy-Weinberg law, real populations are limited in size, rather than infinite, so that allele frequencies are subject to sampling variation in each generation. Allele frequency values will therefore tend to vary according to some frequency (or sampling) distribution. The sampling distribution can be derived if we consider a randomly mating population in the absence of any factor other than sampling variation that can alter allele frequencies. Now assume that the parents produce a progeny generation consisting of N diploid individuals. Since each offspring must receive an allele from each parent, the progeny generation is formed from a sample of $2N$ gametes. The frequency of an allele, p', among the progeny can then have any of $2N + 1$ possible values:

$$0, \frac{1}{2N}, \frac{2}{2N}, \frac{3}{2N}, \ldots, \frac{x}{2N}, \ldots, \frac{2N-1}{2N}, 1$$

As long as the $2N$ gametes are sampled at random (that is, mating is random), the probability that the gamete sample contains exactly x A alleles (that is, the probability that the frequency of the A allele is $p' = x/2N$) will be given by the binomial probability formula:

$$P(x) = \frac{(2N)!}{x!(2N-x)!} p^x q^{2N-x} \qquad (24.1)$$

where p and q are the frequencies of the A and a alleles in the parental gene pool. This formula describes a distribution of allele frequencies with a mean value of p and a variance of $\sigma_p^2 = pq/2N$ (▶ Figure 24–8). There is thus an inverse relationship between the sample size, $2N$, and the expected variation in allele frequencies.

■ Table 24–5 gives a numerical example of the chance

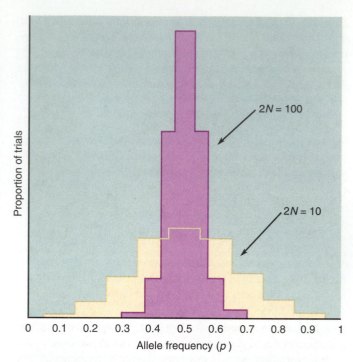

▶ **FIGURE 24–8** Sampling distributions of allele frequencies for two populations of different sizes. In both cases, the gamete samples are assumed to be selected from parental gene pools in which $p = q = 0.5$. Both distributions are approximately normal, having a mean allele frequency of 0.5 and a variance of $pq/2N$, so the variance of the small population is 10 times greater than the variance of the large population.

distribution of allele frequencies for $N = 2$. In this example, it is assumed that the parental generation also consists of two individuals, both of whom are Aa in genotype, so that allele frequencies in the parental gene pool are $p = q = \frac{1}{2}$. The six sibling pairs that are possible among the progeny

■ **TABLE 24–5** Chance distribution of allele frequencies in two offspring from heterozygous parents ($Aa \times Aa$).

Distribution of Offspring			Number of Alleles in Sample	Frequencies of Alleles in Sample	Probability of Offspring
AA	*Aa*	*aa*			
2	0	0	4 A, 0 a	$p = 1, q = 0$	$\frac{4!}{4!0!}(\frac{1}{2})^4(\frac{1}{2})^0 = \frac{1}{16}$
1	1	0	3 A, 1 a	$p = \frac{3}{4}, q = \frac{1}{4}$	$\frac{4!}{3!1!}(\frac{1}{2})^3(\frac{1}{2})^1 = \frac{4}{16}$
0 / 1	2 or 0	0 / 1	2 A, 2 a	$p = \frac{1}{2}, q = \frac{1}{2}$	$\frac{4!}{2!2!}(\frac{1}{2})^2(\frac{1}{2})^2 = \frac{6}{16}$
0	1	1	1 A, 3 a	$p = \frac{1}{4}, q = \frac{3}{4}$	$\frac{4!}{1!3!}(\frac{1}{2})^1(\frac{1}{2})^3 = \frac{4}{16}$
0	0	2	0 A, 4 a	$p = 0, q = 1$	$\frac{4!}{0!4!}(\frac{1}{2})^0(\frac{1}{2})^4 = \frac{1}{16}$

correspond to the six randomly paired combinations of *AA*, *Aa*, and *aa* genotypes in the table. When these combinations are arranged according to the frequency of the *A* allele, the number of cross results reduces to five, corresponding to the five possible values of p' (0, $\frac{1}{4}$, $\frac{1}{2}$, $\frac{3}{4}$, and 1), whose probabilities can be computed from the terms of the binomial expansion $(\frac{1}{2} + \frac{1}{2})^4$.

Example 24.5

Suppose that a population of 50 progeny is formed from a parental population in which the frequency of the *A* allele is $p = 0.50$. What is the probability of a random change in allele frequency among the progeny exceeding ±0.10?

Solution: As a general rule, when *N* is large (about 15 or greater) and the value of *p* is close to $\frac{1}{2}$, the distribution of allele frequencies (given by equation 24.1) can be approximated by a normal distribution with a mean of *p* and a standard deviation of $\sqrt{pq/2N}$. Since $p = 0.5$ in our example,

$$\sigma_P = \sqrt{(0.5)(0.5)/2(50)} = 0.05$$

and the probability that p' will fall within the limits of 0.5 ± 0.10 (which equals $p \pm 2\sigma_P$) is 0.95, assuming a normal curve. The probability that p' will undergo a change greater than ±0.10 to fall outside this range is thus 0.05.

Follow-Up Problem 24.5

Suppose that a population begins with two heterozygotes that produce four offspring in the next generation. What is the probability that the frequency of the *A* allele among the progeny is increased to $p = \frac{3}{4}$ simply as a result of random drift?

We can better appreciate the operation of chance in determining allele frequencies if we understand that the Hardy-Weinberg equilibrium (at least as it pertains to allele frequencies) is neither inherently stable nor unstable but conforms to a **neutral** equilibrium. Like a ball that comes to rest on a level table top, the alleles take on whatever frequencies they possess at the moment; there is no tendency to return to any particular set of values once a disturbance has ceased. Thus, if a small population mates at random, genotype frequencies will retain specific values of p^2, $2pq$, and q^2 only as long as *p* and *q* remain the same. If the allele frequencies are accidentally altered, say from *p* and *q* to p' and q', the genotypes will automatically attain a new equilibrium state with frequencies p'^2, $2p'q'$, and q'^2 in the next generation. This new distribution will endure until there is another disturbance, but there is no tendency to restore original conditions.

Since no stabilizing forces act on *p* and *q* to maintain some specific set of values (other than 0 and 1), initially identical populations tend to diverge in allele frequency as a result of chance. For example, three experimental populations with equal size and with the same initial allele frequency might diverge as shown in ▶ Figure 24–9 over the course of several generations. Although all three groups begin with the same allele frequency, their genetic compositions diverge rapidly. Population (a) eventually reaches **fixation** of the *A* allele ($p = 1$) and thereafter consists of all *AA* individuals, while population (b) experiences **loss** of this allele ($p = 0$) and becomes strictly *aa* in genotype. Population (c) retains both alleles, at least for the first eight generations, and continues to undergo erratic fluctuations in allele frequency. Once a population has experienced fixation or loss of an allele, its members are all homozygous and the allele frequency will no longer change. A useful analogy is the path of a ball in a pinball machine ▶ Figure 24–10). Once the ball enters the trough on either side, the situation is comparable to the fixation or loss of an allele, since the path of the ball can no longer vary.

These examples illustrate two important consequences of genetic drift: (1) a decrease in variation within a population as its members become homozygous for an allele and (2) an increase in variation among populations as different

---

▶ **FIGURE 24–9** Random drift in the frequency of an allele in three small populations. All three populations started with $p = 0.5$. In populations (a) and (b), the ultimate outcome is complete homozygosity resulting from the random fixation (a) or loss (b) of the *p* allele. Population (c) still retains both alleles.

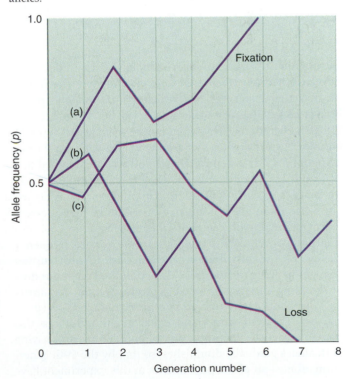

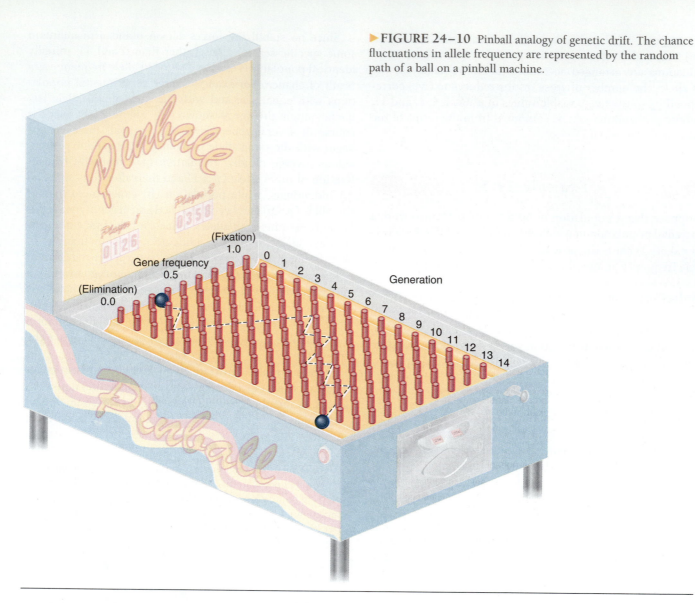

▶ FIGURE 24–10 Pinball analogy of genetic drift. The chance fluctuations in allele frequency are represented by the random path of a ball on a pinball machine.

populations diverge in allele frequency. These effects are demonstrated in ▶ Figure 24–11, which shows the results of an actual experiment in which 107 *Drosophila* populations consisting of 8 males and 8 females were followed for 19 generations. Each population was started with 16 *bw/bw*[75] heterozygotes (where *bw* is the gene for brown eyes and *bw*[75] is a neutral allele) and was maintained at a constant size by randomly choosing 8 males and 8 females for parents in each generation. Sampling variation at first gives rise to an allele frequency distribution that is essentially binomial in character. As this dispersive process continues, however, each population in the various frequency classes of the distribution is subjected to still further changes in allele frequencies. This continued compounding of sampling variation eventually produces a flat (or nearly flat) distribution in which all but the terminal classes ($p = 0$ and $p = 1$) occur with about equal likelihood, Once this phase is reached, the proportion of populations still having both alleles gradually diminishes, so that by the 19th generation, almost all of the populations in this experiment have reached fixation for one allele or the other.

Bottleneck and Founder Effects

In addition to the pattern of drift that occurs over time whenever matings are restricted to small groups, two other situations can lead to random changes in allele frequencies. One of these, known as the **bottleneck effect**, occurs in populations (even moderately large ones) that undergo periodic fluctuations in size. When population sizes are at their lowest values (due, for example, to a natural catastrophe or some infectious disease), the populations are so depleted in numbers that sampling error becomes a distinct possibility. Random changes can then alter the genetic composition of the population. These periodic reductions in number thus serve as bottlenecks that can lead to a much higher degree of sampling variation than would be expected in a stable population of the same average size.

Sampling error is also believed to be important in a phenomenon known as the **founder effect**. This situation occurs whenever a population is established by a small group of colonizers. The founders represent only a small fraction of the genetic variability in the species and may differ sub-

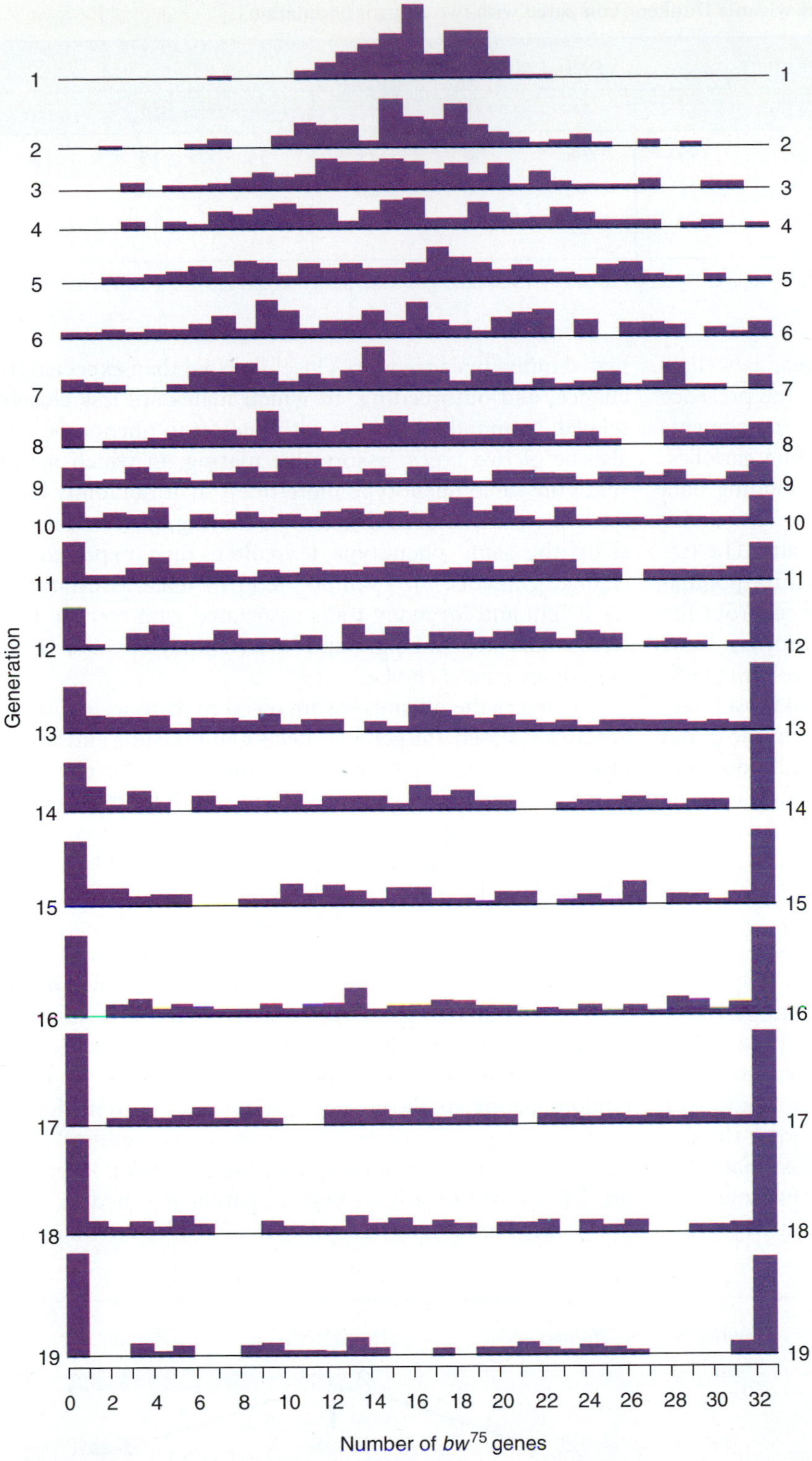

► **FIGURE 24–11** Distribution of allele frequencies over 19 consecutive generations in different lines of *D. melanogaster*. Each line consists of 16 individuals, 8 males and 8 females. *Source:* P. Buri, "Gene frequency in small populations of mutant *Drosophila*," *Evolution* 10 (1956): 386. Used with permission.

Generation

Number of *bw*^75 genes

stantially in allele frequencies from the parent population from which they were derived. For example, an allele may be absent in the founding group or may occur at such a low frequency that it is easily lost before the new population grows appreciably in size.

Founder effects in humans have been studied in various small religious isolates in which the members have tended to marry within the group and have remained more or less

separated genetically from the surrounding population. The religious sect known as the Dunkers, for example, was established from 27 families that came to the United States from western Germany over 200 years ago. Its members have since lived in small farming communities in eastern Pennsylvania, and one small community with about 300 members was studied by a team of researchers headed by Bentley Glass. The researchers analyzed the relative blood group frequencies for the ABO, MN, and Rh systems, as well as some other easily identifiable traits, including the presence or absence of mid-digital hair (hair on the middle segments of the fingers) and attached versus free-hanging earlobes. They then compared the results with corresponding data obtained from the surrounding American population and from the general population in western Germany. The frequencies for the different blood groups for all three populations are shown in ■ Table 24–6. It is apparent from the ABO and MN blood group data that the Dunkers show significant differences in blood group frequencies from both of the control populations. Such differences can most likely be attributed to sampling variation that existed in the initial founding group and, to a certain extent, to the random genetic drift that probably occurred after the Dunkers had migrated to the United States.

Nonrandom Mating

In addition to the effects of a small population size, departures from a Hardy-Weinberg equilibrium can also result from bias in the selection of mates, due to some form of nonrandom mating with regard to genotype. From a genetic perspective, bias in mate selection occurs whenever mating is influenced by the degree of common ancestry (that is, the degree of genetic relationship) and/or the degree of phenotypic resemblance of mates. When mating is influenced by common ancestry, the departures from random mating can be of two types: **inbreeding**, in which the mated individuals are more closely related than expected by chance, and **outbreeding**, in which mates are less closely related. Nonrandom mating with respect to phenotype can also be of two types: **assortative mating**, in which mates share the same phenotype more often than randomly chosen individuals, and **disassortative mating**, in which mates share the same phenotype less often than expected by chance. Humans, for example, tend to mate assortatively for height and for many traits associated with race. In this case, similarities in height and in certain racial characteristics enhance mate choice.

As long as the phenotypes involved in mate selection are genetically based, the genetic effects of inbreeding and assortative mating are qualitatively the same, as are the effects of outbreeding and disassortative mating. Both inbreeding and assortative mating are biased toward partners having similar genotypes (for example, $AA \times AA$, $Aa \times Aa$, and $aa \times aa$) and increase the chance that the offspring will acquire identical alleles. Inbreeding and assortative mating therefore result in increased homozygosity. On the other hand, outbreeding and disassortative mating are biased toward the pairing of dissimilar genotypes (for example, $AA \times Aa$, $AA \times aa$, and $Aa \times aa$) and thus lead to increased heterozygosity.

Among the most striking genotypic effects of nonrandom mating are the changes produced by extreme forms of inbreeding, such as **self-fertilization** (or **selfing**). The results of several generations of selfing are shown in ▶ Figure 24–12. When a homozygous parent is selfed, it produces only homozygous offspring that are identical in geno-

▶ **FIGURE 24–12** Reduction in the frequency of heterozygotes during four generations of self-fertilization. The population is started with all heterozygotes. During each generation of selfing, homozygotes produce only homozygous offspring of the same genotype, while heterozygotes produce all three genotypes in a 1:2:1 ratio. As a result, the frequency of heterozygotes is reduced by a factor of $\frac{1}{2}$ in each generation.

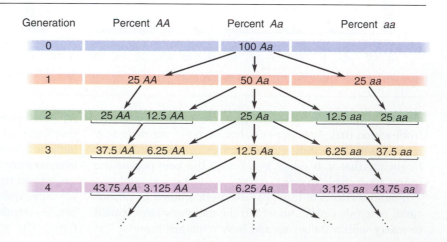

type to the parent. In contrast, heterozygotes form a progeny of $\frac{1}{4}$ *AA*, $\frac{1}{2}$ *Aa*, and $\frac{1}{4}$ *aa*. The proportion of heterozygotes is therefore reduced by half in each generation. If this mode of reproduction were to continue, the frequency of heterozygotes, H_t, after t generations of selfing would become

$$H_t = H_0(\tfrac{1}{2})^t \qquad (24.2)$$

where H_0 is the frequency of heterozygotes when $t = 0$. Thus, when $t = 1$, $H_1 = \frac{1}{2}H_0$; when $t = 2$, $H_2 = \frac{1}{4}H_0$, and so on. Ultimately, the frequency of heterozygotes will approach zero, and all of the remaining individuals will be homozygous for one allele or the other. This increase in homozygosity also occurs with other, less extreme forms of inbreeding. ▶ Figure 24–13 compares the approach toward homozygosity that is produced by continued selfing with certain other forms of inbreeding that occur in organisms with separate sexes. Observe that sib (brother-sister) mating, which is the most intense form of inbreeding in higher animals, produces homozygosity at a significantly slower rate than self-fertilization. It takes approximately nine generations of repeated sib mating to achieve a 90% increase in the frequency of homozygotes, while only about three generations of selfing are needed to achieve the same result. As we would expect, the less intense forms of inbreeding produce an even slower approach to homozygosity.

Despite the qualitative similarities between inbreeding and assortative mating and between outbreeding and disassortative mating, nonrandom mating based on phenotype differs from that based on genotype in one fundamental respect: Assortative and disassortative mating directly affect only the genes that determine the traits involved in mate selection. Even though the *A* and *a* alleles might segregate out only homozygous types as a result of assortative mating, it is possible for other gene pairs (*B* and *b*, *C* and *c*, and so on) to be maintained in equilibrium through random mating. With inbreeding and outbreeding, on the other hand, the genetic similarities extend over the entire genome, so that all gene loci are affected. Thus, with close inbreeding, alleles at all gene loci will undergo fixation, giving rise to different pure-breeding lines. Many pure-breeding lines are thus possible, each differing from the others in the alleles that are present in the homozygous state at the various loci. For example, one gene pair, *A* and *a*, can produce two pure-breeding lines: *AA* and *aa*. With two gene pairs, *A,a* and *B,b*,

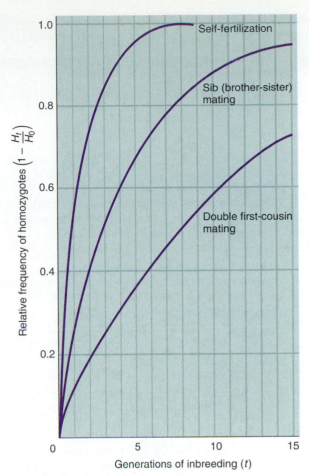

▶ **FIGURE 24–13** Proportional increase in homozygosity under different systems of inbreeding. Continued selfing subdivides a population into distinct lines, each perpetuated by a single parent. In sibling mating, two parents (a brother and sister) perpetuate each line in every generation. Two pairs of parents are used to perpetuate each line in first-cousin matings.

four pure-breeding lines are possible: *AABB*, *AAbb*, *aaBB*, and *aabb* (see ▶ Figure 24–14). In general, if there are *n* gene pairs, close inbreeding can theoretically produce 2^n different pure-breeding lines.

With the exception of disassortative mating, which will be discussed in greater detail in Chapter 25, none of these mating systems change the allelic composition of a popula-

▶ **FIGURE 24–14** Formation of pure-breeding lines upon continued selfing of a tetrahybrid plant. Four inbred lines will be produced once all the pertinent loci are homozygous.

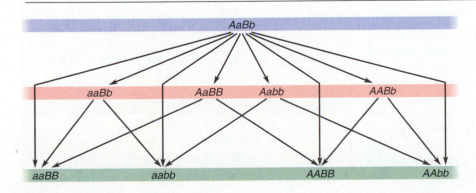

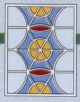

EXTENSIONS AND TECHNIQUES

Measuring the Inbreeding Coefficient from Pedigrees

When exact pedigrees are known, the inbreeding coefficient of an individual can be calculated from the pedigree data. Various rules have been developed to help simplify these calculations; the following two rules are the most basic.

1. *One common ancestor.* When the parents of an individual have only one ancestor in common, the inbreeding coefficient of the individual is given by

$$F = (\tfrac{1}{2})^n$$

where n is the number of ancestors on the path from one of the individual's parents to the common ancestor and back to the other parent. For example, consider the pedigree for the half-sib mating in Figure 24–15. The path diagram includes only the common ancestor and the individuals who are directly involved in the transmission of genes from that ancestor. Note that only three individuals need to be included in the path in this example: the common ancestor and the parents of the indi-

vidual in question. Consequently, $n = 3$ for a half-sib mating, and as we determined previously,

$$F = (\tfrac{1}{2})^3 = \tfrac{1}{8}$$

2. *More than one common ancestor.* When the parents of an individual have more than one ancestor in common, the inbreeding coefficient for that individual is the sum of the F values derived separately for each of the common (but unrelated) ancestors. In symbolic form,

$$F = \sum (\tfrac{1}{2})^n$$

where $\sum$ indicates the sum. For example, let us consider the pedigree for a mating between full sibs (see next pedigree). Note that the path diagram for this pedigree can be divided into two distinct paths representing the transmission of genes from each of the grandparents of the individual. In this case, $n = 3$ for each path, so the individual in question has a $(\tfrac{1}{2})^3$ chance of receiving alleles that are identical by descent from the grandmother plus a $(\tfrac{1}{2})^3$ chance of receiving such alleles from the grandfather. Consequently, for a full-sib mating, $F = (\tfrac{1}{2})^3 + (\tfrac{1}{2})^3 = \tfrac{1}{4}$.

---

tion; they only determine how gametes combine into zygotes and thus affect genotype frequencies. The fact that allele frequencies are unchanged by the mating process is shown in ■ Table 24–7, which illustrates two successive generations of self-fertilization. After one generation of selfing, genotype frequencies have changed from *D, H,* and *R* in the parental generation to $D' = D + \tfrac{1}{4}H$, $H' = \tfrac{1}{2}H$, and $R' = R + \tfrac{1}{4}H$ in the progeny generation. However, the overall allele frequencies have not changed—they remain at $p' = D' + \tfrac{1}{2}H' = D + \tfrac{1}{4}H + \tfrac{1}{4}H = p$ and $q' = 1 - p = q$.

Inbreeding Effects in Small Populations

Inbreeding is one of the most pervasive forms of nonrandom mating. It results not only from a systematic choice of relatives as mates but also from random mating when there is a reduction in the effective breeding size of a population. In organisms with separate sexes, each individual has two parents, four grandparents, eight great-grandparents, and so on for a total of 2^t ancestors t generations back. It should be apparent from the exponential nature of the progression that t does not have to be very large before the theoretical number of separate ancestors in some generation exceeds

■ TABLE 24–7 Frequencies of (a) genotypes and (b) alleles of parents and offspring under conditions of systematic self-fertilization. It is assumed that the genotypes are equally viable and produce, on the average, the same number of offspring when selfed. Note that inbreeding by itself does not alter allele frequencies—only genotype frequencies are changed.

(a) Changes in genotype frequencies:

Parent		Offspring (Genotypes and Frequencies)		
Genotype	Frequency	AA	Aa	aa
AA	D	D	—	—
Aa	H	$\tfrac{1}{4}H$	$\tfrac{1}{2}H$	$\tfrac{1}{4}H$
aa	R	—	—	R
	$(D+H+R)$	$D+\tfrac{1}{4}H$	$\tfrac{1}{2}H$	$R+\tfrac{1}{4}H$
	$= 1$	$= D'$	$= H'$	$= R'$

(b) Constancy of allele frequencies:

$$f(A) = p' = D' + \tfrac{1}{2}H' = D + \tfrac{1}{4}H + \tfrac{1}{2}(\tfrac{1}{2}H)$$
$$= D + \tfrac{1}{2}H = p$$
$$f(a) = q' = R' + \tfrac{1}{2}H' = R + \tfrac{1}{4}H + \tfrac{1}{2}(\tfrac{1}{2}H)$$
$$= R + \tfrac{1}{2}H = q$$

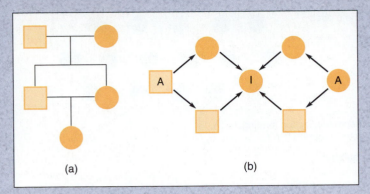

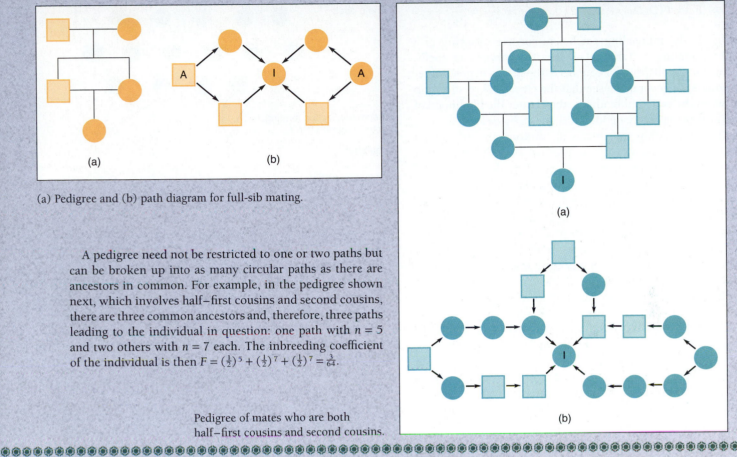

(a) Pedigree and (b) path diagram for full-sib mating.

A pedigree need not be restricted to one or two paths but can be broken up into as many circular paths as there are ancestors in common. For example, in the pedigree shown next, which involves half–first cousins and second cousins, there are three common ancestors and, therefore, three paths leading to the individual in question: one path with $n = 5$ and two others with $n = 7$ each. The inbreeding coefficient of the individual is then $F = (\frac{1}{2})^5 + (\frac{1}{2})^7 + (\frac{1}{2})^7 = \frac{3}{64}$.

Pedigree of mates who are both half–first cousins and second cousins.

the size of any real population. Therefore, all potential mates in a finite population must share at least one common ancestor at some point in the more or less distant past; moreover, the smaller the size of the population, the closer their probable relationship will be. Thus, in small populations, close inbreeding is largely unavoidable and may be quite common, even when mates are selected at random with regard to genotype.

The effect of population size on the severity of inbreeding can be described in terms of a commonly used measure of inbreeding intensity, called the **inbreeding coefficient** (F), which was first defined by Sewall Wright in 1922. The inbreeding coefficient is the probability that two alleles in an individual are identical by descent, that is, that they are copies of the same DNA of a common ancestor. We can further clarify the meaning of F by examining the pedigree for a half-sib mating (see ▶ Figure 24–15). In this pedigree, the DNA molecules carrying the alleles in the common ancestor are designated as D_1 and D_2. The state of the alleles, A or a, is not given, nor is the phenotype of any individual. The paths in the pedigree show that there are two ways for individual I to acquire alleles that are copies of the same DNA in the grandfather (the common ancestor in this case): one

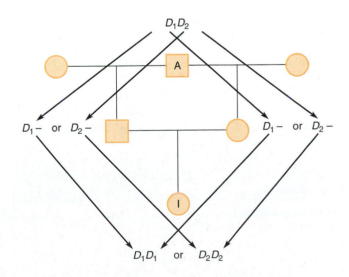

▶ **FIGURE 24–15** Pedigree of a mating between half-sibs. The path diagram shows two circular paths by which individual I can acquire alleles that are copies of the same DNA (D_1 or D_2) in the grandfather (A), the common ancestor.

gives I two copies of D_1 and the other gives I two copies of D_2. The inbreeding coefficient is equal to the probability of individual I receiving two copies of D_1 or two copies of D_2 from the common ancestor, or

$$F_1 = P(D_1D_1) + P(D_2D_2) = \frac{1}{16} + \frac{1}{16} = \frac{1}{8}$$

which is the inbreeding coefficient for the offspring of a half-sib mating. ■ Table 24–8 lists the inbreeding coefficients for the offspring of couples with various degrees of relatedness. In general, we see that the closer the relationship between the mated individuals, the greater the likelihood of their offspring receiving identical alleles by descent.

In a population that mates at random, the inbreeding coefficient represents the proportion of individuals that have acquired alleles that are identical by descent. Even if the common ancestry of the matings is not known, F can nevertheless be estimated in such a population on the basis of the population size. For example, consider a population of some fixed size N that is formed in each generation by the random union of $2N$ gametes. It might be helpful to picture a population of hermaphroditic marine invertebrates that release their gametes into the water at one time during each generation to unite at random during fertilization (► Figure 24–16). Each diploid individual in this population has two DNA molecules that carry the gene of interest. The gametes of this population will therefore carry the copies of $2N$ DNA molecules ($D_1, D_2, D_3, \ldots, D_{2N}$). When these gametes unite during fertilization, there are two ways for a zygote to receive alleles that are identical by descent: (1) the random combination of two alleles that are copies of the same DNA molecule from one of the parents (for example, D_1D_1, D_2D_2, etc.) and (2) the random combination of two alleles that are not derived by replication of the same DNA in this gene pool but are identical by descent from a previous generation. Since each kind of DNA makes up $1/2N$ of the entire gene pool, each allele has a $1/2N$ chance of combining with another allele that is a copy of the same parental DNA molecule. At any given generation (generation t), $1/2N$ of zygotes formed will have copies of the same allele from the parental generation (generation $t - 1$). Of the remaining proportion, $1 - 1/2N$, a fraction F_{t-1} will have

■ **TABLE 24–8** Inbreeding coefficients for offspring of couples of varied degrees of relatedness.

Degree of Relatedness	Inbreeding Coefficient of Offspring
First-degree relatives (parent-child, full sibling)	$\frac{1}{4}$
Second-degree relatives (aunt-nephew, uncle-niece, half-sibling)	$\frac{1}{8}$
Third-degree relatives (first cousins)	$\frac{1}{16}$
Unrelated individuals	0

Generation $t - 1$				
Diploid parents:	1	2	3 …	N
Genomic DNA:	$D_1 D_2$	$D_3 D_4$	$D_5 D_6$ …	$D_{2N-1} D_{2N}$

Gamete pool:

D_2 D_4 D_5 D_6 … D_{2N}

D_1 D_2 D_3 D_4 D_6 … D_{2N}

D_1 D_2 D_4 D_5 D_6 …

Random fertilization forming zygotes in generation t:	$D_2 D_2$	$D_5 D_6$
Probability:	$\frac{1}{2N}$	$1 - \frac{1}{2N}$
Contribution to F_t:	1	F_{t-1}
Total F_t:	$\frac{1}{2N}$ +	$(1 - \frac{1}{2N}) F_{t-1}$

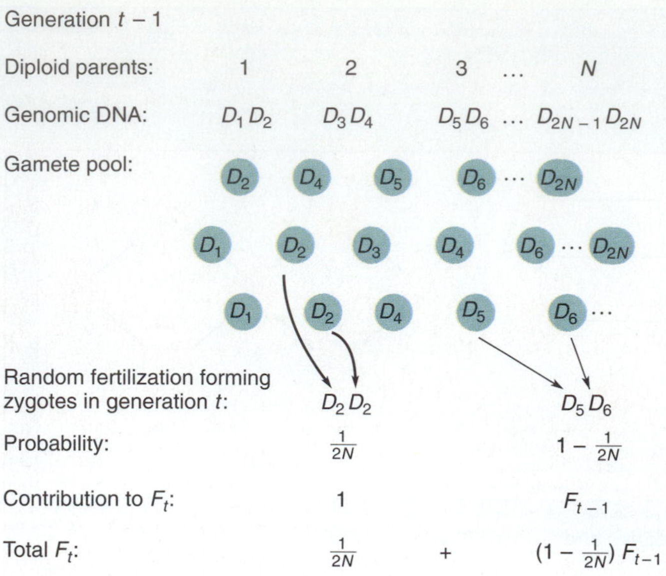

► **FIGURE 24–16** Random fertilization of gametes in a population of finite size. The parental population is assumed to consist of N hermaphroditic marine invertebrates that release their gametes into the water at one time to undergo random fertilization. A fraction $1/2N$ of the fertilization acts will produce zygotes having alleles that are copies of the same DNA in the parents. Of the remaining $1 - 1/2N$ fertilizations, a fraction F_{t-1} will produce zygotes having alleles that are identical by descent from a previous generation.

alleles that are identical by descent from an earlier generation, where F_{t-1} is the average inbreeding coefficient of the parents. Combining these two sources of allelic identity, the total inbreeding coefficient at generation t, F_t, becomes

$$F_t = \frac{1}{2N} + \left(1 - \frac{1}{2N}\right) F_{t-1}$$

Multiplying both sides by -1 and adding 1 leads to

$$1 - F_t = 1 - \frac{1}{2N} - \left(1 - \frac{1}{2N}\right) F_{t-1}$$

or

$$1 - F_t = \left(1 - \frac{1}{2N}\right)(1 - F_{t-1})$$

Since only those individuals without identical alleles can be heterozygous, $1 - F_t$ must be proportional to the frequency of heterozygotes at generation t, H_t. Similarly, $1 - F_{t-1}$ must be proportional to H_{t-1}. Making this substitution (and noting that proportionality constants cancel), we get

$$H_t = \left(1 - \frac{1}{2N}\right) H_{t-1}$$

Thus the heterozygotes in the population will decrease in frequency by a factor of $1 - 1/2N$ in each generation. Thus

$$H_1 = \left(1 - \frac{1}{2N}\right)H_0$$

$$H_2 = \left(1 - \frac{1}{2N}\right)H_1 = \left(1 - \frac{1}{2N}\right)^2 H_0$$

and so on, so that after t generations, H_t will be related to H_0 by

$$H_t = H_0\left(1 - \frac{1}{2N}\right)^t \qquad (24.3)$$

Note that as t increases, $(1 - 1/2N)^t$ will approach zero. Consequently, heterozygotes will eventually be eliminated from a small population, even though mating is random. The loss of heterozygosity due to inbreeding effects can also be predicted when the individuals in the population are not hermaphroditic, as we assumed in this model. For bisexual forms, in which self-fertilization is impossible, $1/2N$ in equation 24.3 reduces approximately to $1/(2N + 1)$, which approaches $1/2N$ when the population is fairly large. The loss of heterozygosity in randomly mating populations is shown in ▶ Figure 24−17 for different population sizes. Observe that when N is sufficiently small, we cannot assume a genetic equilibrium through random mating, even for relatively short periods of time.

The overall effects of inbreeding on the genotypic composition of a population are shown in ■ Table 24−9 in terms of departures from the Hardy-Weinberg equilibrium. With a given amount of inbreeding (as measured by the inbreeding coefficient, F), the frequencies of the genotypes AA, Aa, and aa become $p^2 + Fpq$, $2pq − 2Fpq$, and $q^2 + Fpq$, respectively. The first term in each of these expressions gives the expected Hardy-Weinberg frequency and the second term gives the deviation from that frequency due to inbreeding effects. Inbreeding reduces the frequency of heterozygotes by $2Fpq$, with half of this value being added to each homozygous class. Therefore, once the heterozygotes are eliminated from the population ($F = 1$), the AA and aa homozygotes will occur at frequencies of $p^2 + pq = p$ and $q^2 + pq = q$, respectively.

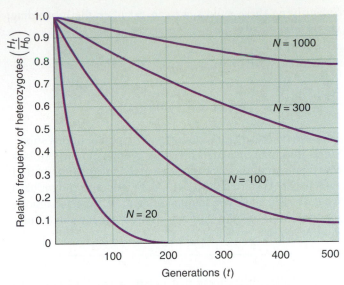

▶ **FIGURE 24−17** Reduction in the relative frequency of heterozygotes in populations of different sizes undergoing random mating. The curves are derived from equation 24.3.

Example 24.6

If inbreeding is the only factor contributing to changes in the genetic structure of a population, what will the inbreeding coefficient of this population be, once the genotypes AA, Aa, and aa attain frequencies of 0.752, 0.096, and 0.152, respectively?

Solution: First note that the frequency of heterozygotes under the conditions of inbreeding can be expressed as $H = 2pq(1 − F)$. Solving for F, we get

$$1 - F = \frac{H}{2pq} \qquad \text{or} \qquad F = 1 - \frac{H}{2pq}$$

H in this population is 0.096, and the allele frequencies can be calculated as

$$p = D + \frac{1}{2}H = 0.752 + \frac{1}{2}(0.096) = 0.8$$

$$q = 1 - 0.8 = 0.2$$

■ **TABLE 24−9** Genotype frequencies in randomly mating populations with varying degrees of inbreeding.

Generations of Inbreeding	Inbreeding Coefficient	Genotype Frequencies		
		AA	Aa	aa
None (Hardy-Weinberg equilibrium)	$F = 0$	p^2	$2pq$	q^2
One or more	$0 < F < 1$	$p^2 + Fpq$	$2pq(1 − F)$	$q^2 + Fpq$
Infinite number (complete homozygosity)	$F = 1$	$p^2 + pq = p$	0	$q^2 + pq = q$

Substituting these values, we can calculate the inbreeding coefficient:

$$F = 1 - \frac{0.096}{2(0.8)(0.2)} = 0.7$$

Follow-Up Problem 24.6

Suppose that a large, randomly mating population is subdivided into many isolated breeding groups of constant size that conform to the conditions assumed in equation 24.3. If each isolate is started with 2 *AA*, 4 *Aa*, and 2 *aa* individuals, what average genotype frequencies will be expected among these isolated breeding groups after 10 generations of random mating?

The inbreeding effects that occur in small populations are particularly evident in religious isolates. The Amish sect, which originated as an offshoot of the Mennonite Church, is one group that has been intensively studied. Because of the closed nature of the group, marriages have been restricted largely to members of the Amish sect. This restriction has led to a high degree of inbreeding and homozygosis, even though the marriages do not always involve close relatives. A medical survey revealed that certain recessive disorders occur within the group at levels that are considerably higher than in the noninbred neighboring populations. One example is the Ellis–van Creveld syndrome, a rare form of dwarfism that is characterized by extra fingers and disproportionately shortened limbs. The incidence of this disorder is extraordinarily high among the Amish in Lancaster County, Pennsylvania, where it exists at a frequency of approximately 0.005. Elsewhere, the frequency is extremely low.

The effect of population size on the incidence of a deleterious recessive trait can be seen by dividing the expected frequency of recessive homozygotes under conditions of inbreeding ($F > 0$ in Table 24–8) by the Hardy-Weinberg frequency for this class. We get

$$\frac{q^2 + Fpq}{q^2} = 1 + \frac{F(1 - q)}{q}$$

where $p = 1 - q$. This formula represents the factor by which inbreeding increases the frequency of recessive homozygotes above that expected in a large, randomly mating population. The increase is directly proportional to F, which, as we have seen, varies inversely with population size. In large populations, F will approach zero, so that little change in the frequency of homozygotes is expected. The effect is therefore observed mainly in very small populations, where inbreeding is common and F can have values significantly greater than zero. The increase is also directly proportional to the factor $(1 - q)/q$ so that inbreeding will have its greatest effect when the trait in question is rare, that is, when q has a value much less than 1.

We should bear in mind that inbreeding in itself is not necessarily harmful. If, by chance, the initial population were substantially free of detrimental recessive genes, close inbreeding could occur without substantial risk of expressing a recessive disorder. This was apparently the case among the pharaohs of ancient Egypt, among whom brother-sister matings were the rule (▶ Figure 24–18). The opposite effect is also possible. If rare, beneficial recessive genes occur in a population, close inbreeding would act to improve their chances of being expressed in homozygous combinations.

To Sum Up

1. Departures from the Hardy-Weinberg equilibrium can occur when one or more of the assumptions in the Hardy-Weinberg law are not met. The causes of these departures include small population size, nonrandom mating, unequal viability and fertility of genotypes, and the introduction of alleles by mutation and migration.

2. Genetic drift involves random fluctuations in allele frequencies that occur as a result of sampling variation. Since the gametes carry only a sample of the parental gene pool, allele frequencies are subject to sampling error when the population is small. When the sampling process is repeated over many generations, the compounding of sampling variation can lead to an increase in homozygosity through the random fixation or loss of alleles in small breeding groups. Different populations therefore diverge in genetic character as a result of genetic drift, with some breeding groups undergoing fixation and others undergoing loss of a given allele.

3. Nonrandom mating occurs whenever there is a correlation between the ancestry or phenotypes of mates. The correlation is positive for inbreeding and assortative mating, in which mates are more similar in ancestry (inbreeding) or phenotype (assortative mating) than randomly chosen individuals. The correlation is negative for outbreeding and disassortative mating, in which mates are less similar in ancestry (outbreeding) or phenotype (disassortative mating) than expected by chance. Inbreeding and assortative mating result in increased homozygosity, compared to random mating, while outbreeding and disassortative mating result in increased heterozygosity.

4. The inbreeding coefficient, F, measures the probability that two alleles are identical by descent, that is, that they are copies of the same DNA in a common ancestor. The value of F varies with the degree of relatedness between mates and with the number of generations of inbreeding. The value of F is 0 for a noninbred individual and is equal to 1 when inbreeding is complete and all individuals in a population are homozygous for alleles that are identical by descent.

5. The smaller the population, the greater the chance that two randomly selected individuals have an ancestor in common. Inbreeding can therefore occur in small populations, even if mating is random. Inbreeding effects result in an overall decrease in heterozygosity and a corresponding increase in homozygosity in small populations. They are one of the factors responsible for the abnormally high frequencies of rare recessive disorders in religious isolates.

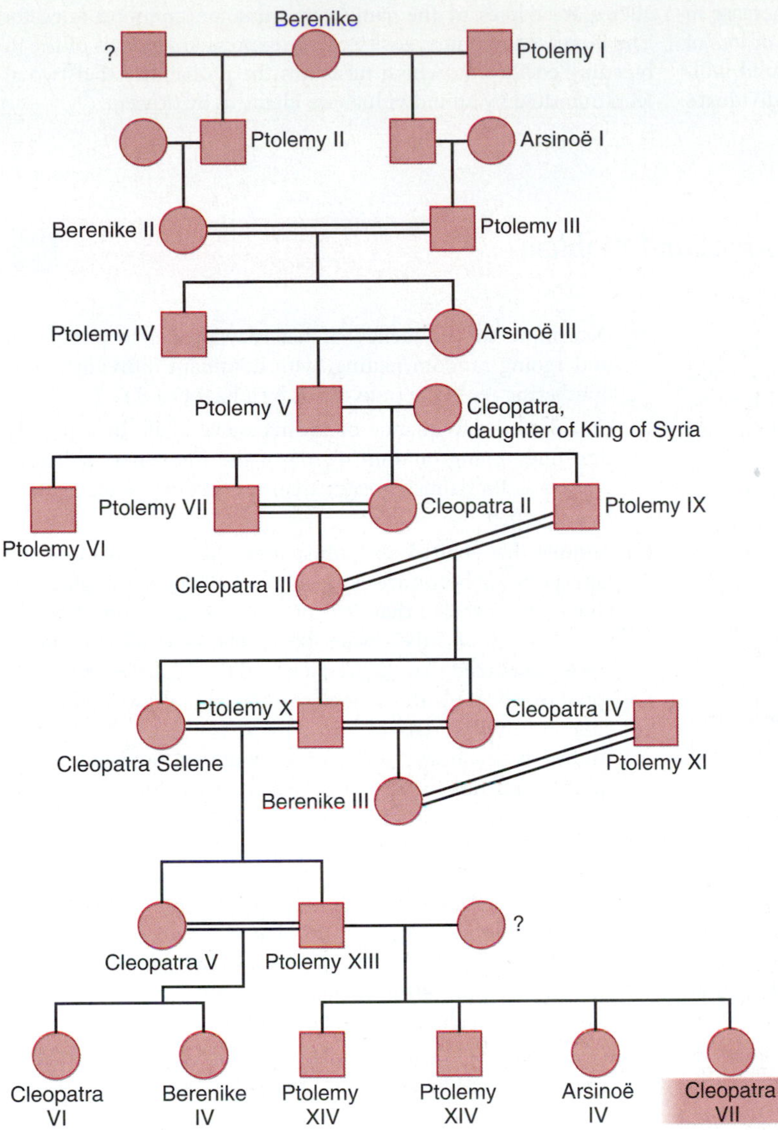

FIGURE 24–18 Pedigree of the Ptolemaic dynasty of Egypt. Incest and inbreeding (indicated by double lines) were common. Cleopatra VII, the great Cleopatra, is most famous for her affairs with Julius Caesar and Marc Antony. She had one son by Caesar and three children by Antony. *Adapted from: Michael R. Cummings, Human Heredity: Principles and Issues,* 3d ed. (West Publishing Company, 1994), p. 491.

Pedigree labels: Berenike; ?; Ptolemy I; Ptolemy II; Arsinoë I; Berenike II; Ptolemy III; Ptolemy IV; Arsinoë III; Ptolemy V; Cleopatra, daughter of King of Syria; Ptolemy VI; Ptolemy VII; Cleopatra II; Ptolemy IX; Cleopatra III; Ptolemy X; Cleopatra IV; Cleopatra Selene; Ptolemy XI; Berenike III; Cleopatra V; Ptolemy XIII; ?; Cleopatra VI; Berenike IV; Ptolemy XIV; Ptolemy XIV; Arsinoë IV; Cleopatra VII.

Chapter Summary

Genetic variability is a common feature of natural populations. Population geneticists attempt to determine the extent of this variability by identifying the alleles at each locus and measuring their respective frequencies. An allele (or gene) frequency can be computed directly from the pertinent genotype frequencies by adding the frequency of the homozygote for the allele to half the sum of the frequencies of the heterozygotes for the allele. Both the allele and the genotype frequencies depend on the particular system of mating in the population, which is determined by the pattern and relative frequencies with which the genotypes mate. If mating is random, that is, if mates are selected without regard to genotype, each mating will occur at a frequency equal to the product of the genotype frequencies of the mates involved.

The Hardy-Weinberg law states that in the absence of the factors that can change allele frequencies, a large population will attain an equilibrium in genotype frequencies after one generation of random mating. For an autosomal locus with two alleles, the Hardy-Weinberg equilibrium is characterized by a constant distribution of genotype frequencies that corresponds to the terms of the binomial expansion $(p + q)^2$, where p and q are the frequencies of the alleles. Extensions of the Hardy-Weinberg law can be applied to multiple alleles and sex-linked loci, and the genotype frequencies predicted by this principle can be used to test populations for conditions of random mating.

In small populations, departures from the Hardy-Weinberg equilibrium often occur as a result of random genetic drift and inbreeding. These departures are characterized by an increase in the overall frequency of homozygotes due to the fixation or loss of an allele and the eventual production of completely inbred individuals. Not only is homozygosity attained, but every individual's alleles are copies of the same gene in some common ancestor. The increase in homozygosity can be expressed in terms of the inbreeding coefficient, which measures the probability that two alleles inherited by an individual are identical by descent.

Questions and Problems

Allele Frequency

1. A sample of 1400 persons living in New York City found 408 persons of blood type M, 694 of type MN, and 298 of type N. Determine the frequencies of the M, MN, and N blood types in this sample and the frequencies of the L^M and L^N alleles.

2. In a certain population, the genotypes for the alleles A_1, A_2, and A_3 occur in the following ratio: 1 A_1A_1:3 A_2A_2:1 A_3A_3: 1 A_1A_2:1 A_1A_3:1 A_2A_3. What is the ratio of the three alleles in this population?

Gene Transmission with Random Mating

3. Distinguish among the terms allele frequency, genotype frequency, and mating frequency. Show how the three are related in a population undergoing random mating.

4. Criticize the following statements: (a) Since fertilization is random, the frequency of the recessive genotype in a population will always equal the square of the frequency of the recessive allele. (b) Since only the dominant allele is expressed in heterozygotes, the dominant trait will always be greater in frequency than the recessive trait. (c) Since alleles always segregate in a 1:1 ratio, a population will tend to reach an equilibrium in which half of the individuals will be heterozygous.

5. A large population consists of the genotypes AA, Aa, and aa at frequencies of 0.1, 0.6, and 0.3, respectively. (a) What are the allele frequencies in this population? (b) Calculate the expected allele and genotype frequencies after one generation of random mating.

6. If a randomly mating population contains 1% recessive homozygotes, what percentage is heterozygous?

7. In a certain population, 80% of all persons are Rh+, and 20% are Rh−. Assume that mating is random for the Rh factor. (a) Of all marriages in this population, what percentage is expected to involve an Rh+ husband and an Rh− wife? (b) In what percentage of all marriages is one person Rh+ and the other Rh−?

8. A farmer planted a large field of corn from seed that consisted of a mixture of purple (dominant) and yellow (recessive) kernels. The mature plants that developed were allowed to pollinate at random. Of the seeds produced in the next generation, 91% were purple and 9% were yellow. (a) What are the allele frequencies in this population? (b) From these results, can the farmer deduce the phenotypic ratio among the seeds that were initially planted? Why or why not? (c) Suppose that the seeds planted by the farmer were also the result of random pollination. Given this additional information, compute the expected ratio of purple to yellow kernels among the seeds that the farmer initially planted.

9. Determine the frequency of heterozygotes in a population undergoing random mating, with dominant individuals outnumbering recessive individuals by a factor of 8 to 1.

10. Calculate the frequency of the recessive allele in a population undergoing random mating if the frequency of heterozygotes is four times greater than the frequency of recessive homozygotes.

11. Assume that spotted and nonspotted (black) wings in a certain species of beetle are expressions of alleles at a single locus. One study revealed that 36% of a large population of beetles were spotted and 64% were black. The geneticist doing the study concluded that the frequency of the allele for spotting is equal to 0.6. (a) Is the geneticist justified in drawing this conclusion from the data given? Why or why not? (b) Previous studies have shown that beetles mate at random for the spotted and nonspotted phenotypes. Does this added information change your answer to part (a)? Explain why or why not. (c) The results of another study revealed that among matings between spotted beetles, about 20% of the offspring were black, while black × black matings produced only black progeny. Based on this information, what are your estimates of the allele frequencies in the beetle population? Justify any difference between the allele frequency for spotting that you obtain and the value calculated by the geneticist.

12. Given $p + q = 1$, verify the following equalities:
    (a) $p + 2q = 1 + q$ (b) $p(1 + q) = 1 - q^2$
    (c) $q^3 + pq^2 = q^2$ (d) $p^2q + pq^2 = pq$

13. The following are the frequencies of genotypes AA, Aa, and aa in four different populations: population 1: 0.64, 0.20, and 0.16; population 2: 0.40, 0, 0.60; population 3: 0.01, 0.18, 0.81; and population 4: $\frac{1}{3}$, $\frac{1}{3}$, and $\frac{1}{3}$. (a) Calculate the allele frequencies for each of the four populations. (b) Determine which of the four populations are not in random-mating equilibrium, and calculate the genotype frequencies that these populations would have after one generation of random mating.

14. Suppose that studies on coat colors in two large herds of shorthorn cattle give the following data:

	Number of Cattle			
	Red	Roan	White	Total
Herd 1:	112	56	32	200
Herd 2:	98	84	18	200

Which herd conforms and which herd does not conform to a population in random-mating equilibrium with respect to coat color frequencies?

15. Show that for a rare autosomal recessive disorder, the frequency of heterozygous carriers is approximately equal to twice the frequency of the recessive allele.

16. Cystic fibrosis is an autosomal recessive disorder in which the mucus-secreting tissues of the affected individual are abnormal. Intestinal obstruction is an early symptom of the disease, and clogged respiratory passages often occur. The incidence of affected persons in the United States white population has been estimated to be 1 in 2500. (a) Determine the expected frequency of the allele for cystic fibrosis and the frequency of carriers of the disorder in this population. (b) What will be the frequency of marriages that involve two carriers?

17. Color blindness is the result of a recessive X-linked gene. Forty males in a sample of 1000 males are found to be color-blind. (a) What percentage of females is expected to be color-blind in this population? (b) What percentage of females is expected to be heterozygous? (c) What percentage of marriages in this population can produce only normal offspring?

18. Two disorders, A and B, are known to be determined by different X-linked genes. Disorder A occurs 100 times more frequently among males than among females. Disorder B, in contrast, is expressed in about 1.98 times as many females as males. (a) Which trait is dominant and which is recessive? (b) For both disorders, calculate the allele frequencies in the population.

19. The following are the observed ABO blood group phenotypes among a group of 600 American Indian students: 200 A, 196 B, 104 AB, and 100 O. (a) Calculate the frequencies of the I^A, I^B, and I^O alleles in this group, assuming random mating. (b) Determine the expected numbers of the blood group phenotypes if the students were selected from a random-mating population.

20. Suppose that a person of unknown blood type is involved in an accident and is given a transfusion of type A blood. If the frequencies of the I^A, I^B, and I^O alleles are 0.6, 0.1, and 0.3, respectively, and if mating in the population is random for this trait, what is the chance that the individual will possess antibodies against the donated blood cells? (Consider only the ABO system.)

21. The results of ABO blood tests reveal that the frequencies of the I^A, I^B, and I^O alleles in a particular randomly mating population are 0.2, 0.1, and 0.7, respectively. (a) What proportion of the matings in this population is expected to be between two persons with type A blood? (b) What proportion of type O children in this population will have a type A mother and a type B father?

*22. A population that mates at random is in linkage equilibrium for the A,a and B,b loci. The allele frequencies in this population have been determined to be $f(A) = 0.6$, $f(a) = 0.4$, $f(B) = 0.3$, and $f(b) = 0.7$. (a) Calculate the frequencies of the gamete types produced by this population. (b) If an individual is selected at random from this population, what is the probability that the genotype of the selected individual is AAbb? (c) Does your answer in part (b) indicate anything about the chromosomal location of the two loci relative to each other? Explain why or why not.

*23. Assume that a population that mates at random consists of equal numbers of the three genotypes AABB, AaBb, and aabb

and that the A,a and B,b loci are carried on different chromosomes. (a) Calculate the frequencies of the gamete types produced by this population. (b) What are the allele frequencies in this population? (c) Calculate the genotype frequencies among the offspring of this population. (d) Are the progeny genotypes in equilibrium for the A,a and B,b loci when they are considered jointly? when each locus is considered separately? (e) Predict the genotype frequencies for this population at equilibrium.

Departures from the Hardy-Weinberg Equilibrium: Small Populations and Nonrandom Mating

24. How is it possible for the members of a small population to select mates at random and still undergo inbreeding?

25. A population has allele frequencies of $p = 0.4$ and $q = 0.6$. What are the expected frequencies of the genotypes AA, Aa, and aa in the population when the inbreeding coefficient equals 0.5?

26. A strictly self-fertilizing plant population is started with genotypes in the following ratio: 1 AA : 2 Aa : 1 aa. What will be the genotype frequencies after (a) 1, (b) 2, (c) 3, and (d) t generations of selfing?

27. How many generations of selfing are required to reduce the frequency of heterozygotes in a plant population to just under 0.1% of its initial value?

28. An autosomal recessive disorder occurs at a frequency of 1/10,000 in a large, randomly mating population. The incidence of this disorder is 50 times higher in a neighboring religious isolate. If the allele frequencies are the same in both populations, what is the inbreeding coefficient for the isolate?

29. A sample that is selected at random from a population consists of 52 AA, 28 Aa, and 20 aa individuals. If inbreeding is the only process that contributes to changes in genotype frequency in this population, what is the value of the inbreeding coefficient?

30. Consider a small population of 50 hermaphroditic marine invertebrates, all diploid and unrelated to each other. (a) What would be the inbreeding coefficient of the next generation if gametes are released into the water at the same time and undergo random fertilization? (b) If the population size remained constant and there were no outside forces acting to change the frequencies of alleles or genotypes, what would be the inbreeding coefficient after ten generations? (c) How many generations are required to attain an inbreeding coefficient of 0.25?

31. Verify that only about 10 generations of mating are required to reduce the frequency of heterozygotes by 10% in the population in problem 30.

32. What is the founder effect? Suppose that a group is founded by four individuals derived at random from a population that has genotypes in a ratio of 1 AA : 2 Aa : 1 aa. What is the probability that the starting group differs in allele frequency from the population from which it was derived?

33. Suppose that a population is founded by a group that consists of four individuals: two males of genotype AA and one female each of genotype AA and Aa. Each female produces two offspring that survive to adulthood. (a) What is the probability

*An asterisk denotes that the question refers to material in the Extensions and Techniques text.

that the progeny will have the same allele frequencies as their parents? (b) If each mated couple in every generation of this population continues to produce only two offspring, what is the probability that the allele frequencies will remain the same over a period of two generations? Over a period of t generations?

34. If the frequency of an allele is 0.05, determine the probability that it is lost due to random drift in just one generation from a population of size (a) 10 and (b) 50.

35. Consider a population with genotype frequencies of 35% *AA*, 50% *Aa*, and 15% *aa* among both males and females. The parental population produces 100 progeny. What is the probability that the frequency of the *A* allele in the progeny will be between 0.67 and 0.53?

36. Suppose that a plant that is initially heterozygous for three independently assorting gene loci (*A,a*; *B,b*; and *C,c*) undergoes repeated self-fertilization. (a) What is the probability that the *A,a* locus is still heterozygous after two generations of selfing? (b) What is the probability that any two of the loci are still heterozygous after two generations of selfing? (c) What is the probability that none of the loci will be heterozygous after two generations of selfing?

*37. Calculate the inbreeding coefficient for individual I in each of the following pedigrees:

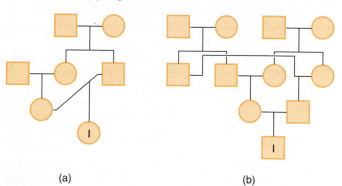

(a)   (b)

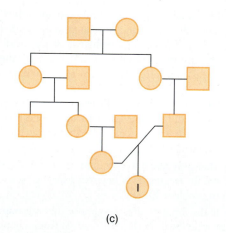

(c)

*38. The following is a pedigree of Roan Gauntlet, a famous shorthorn bull. Calculate the inbreeding coefficient of Roan Gauntlet. (*Note:* the pedigree consists of four different circular paths, two containing Lord Raglan and two containing Champion of England.)

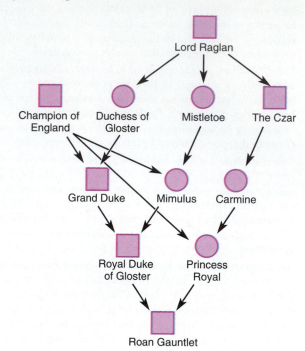

CHAPTER 25

Genetic Processes of Evolution

The Hardy-Weinberg law describes a static situation in which the types and frequencies of alleles and their distribution among the offspring remain constant from generation to generation. It clearly describes a state in which evolution is not occurring. In order to evolve, a population must undergo genetic change, which can come about only through alterations in allele frequencies. Thus the factors that are responsible for evolutionary change are the very factors that were assumed not to operate in a population at a Hardy-Weinberg equilibrium.

Four basic evolutionary forces can modify the frequencies of alleles: mutation, selection, migration, and genetic drift. The first three forces are **deterministic processes** that produce changes in allele frequency that are predictable in both magnitude and direction. Genetic drift, on the other hand, is a **random process** characterized by chance fluctuations in allele frequency that occur mainly in small populations. Of these four forces, selection is

Some finches as drawn by Charles Darwin.

thought to be the dominant evolutionary force, since it is the only elemental process of evolution that has a direct and causal bearing on adaptation.

In this chapter, we will consider these evolutionary forces and describe the mechanisms by which they act on the genetic structure of a population. Since mutation and genetic drift have already been discussed, the main focus of this chapter will be the role of natural selection in adaptive evolution and how it interacts with the other three forces.

▦ REPRODUCTIVE FITNESS AND SELECTIVE CHANGE

Selection is the differential reproduction of genotypes. In other words, selection occurs when individuals of a specific genotype produce more (or fewer) surviving offspring, on the average, than individuals of other genotypes. In nature, these genotypic differences in reproductive success stem from inequalities in fertility and survival ability. Inequalities in fertility can arise through variability in the onset and duration of reproductive periods or through differences in mating success and in the number of functional gametes produced. Unequal survival ability, on the other hand, may occur through a host of different environmental effects, including differences in the ability of certain genotypes to compete for essential resources, to escape predation, or to withstand the rigors of their physical environment.

In addition to the selective process that occurs in nature, humans have practiced **artificial selection** on their crop plants and domesticated animals throughout the greater part of their history. Careful breeding programs in which individuals with desirable qualities are saved for reproductive purposes have modified wild species of plants and animals and developed numerous breeds and varieties to suit our needs and wishes. Since many domesticated stocks are unable to compete effectively with their wild relatives, they are permitted to flourish only through human intervention in the natural selective process. The sections that follow, however, will mainly be limited to the forms of selection that occur in nature.

Measurement of Fitness

The **fitness** of a genotype is a measure of the individual's ability to survive and reproduce. Ideally, fitness should reflect the adaptability of the organism to environmental change and thus its ability to survive over many generations. Unfortunately, no such measure has been devised that can be deduced without recourse to a known evolutionary record. For this reason, fitness is usually expressed in terms of the short-term reproductive success of a population over one or just a few generations. One useful measure of the fitness of a given genotype is by means of a reproductive rate, which incorporates both the survivorship probability and the number of offspring produced by an average individual of the genotype in question. This approach to measuring fitness can be illustrated most clearly by a popula-

tion that reproduces in discrete, nonoverlapping generations, such as annual plants. Suppose, for example, that each parent produces offspring during a single reproductive period and dies before its offspring reach the reproductive age. Let

l_i = the probability that a zygote of genotype i survives to the reproductive age

m_i = the average number of offspring produced per surviving adult of genotype i at the reproductive age

The projected average number of offspring produced per zygote of genotype i is then $m_i l_i$. For example, if each zygote has a 50% chance of reaching sexual maturity and produces an average of 2 offspring during its reproductive period, an average zygote of this genotype is then expected to contribute a total of $(2)(0.5) = 1$ offspring to the next generation. Note that by multiplying m_i by l_i, we obtain a measure of the future genetic contribution of a zygote of this genotype, since by forming offspring, the individual is contributing genes to the next generation.

We can now extend our definition of fitness to a population consisting of more than a single genotype. To illustrate, let us assume a randomly mating population in which there are three genotypes (AA, Aa, aa) with frequencies p^2, $2pq$, and q^2 and in which fertility rates for these genotypes (m_{AA}, m_{Aa}, m_{aa}) and their probabilities of survival (l_{AA}, l_{Aa}, l_{aa}) have values of (3, 4, 10) and (0.9, 0.7, 0.1), respectively. The net reproductive rates ($m_{AA}l_{AA}$, $m_{Aa}l_{Aa}$, $m_{aa}l_{aa}$) are then (2.7, 2.8, 1.0). If we multiply each of these values by the relative frequency of the genotype in the population, the proportionate genetic contributions of the three genotypes become $(2.7)p^2 : (2.8)2pq : (1.0)q^2$. Observe that the aa individuals in this example have the lowest net reproductive rate, even though they are the most prolific as adults. This example illustrates the importance of taking both the survival ability and fertility into account before making any judgment about the genetic contributions of the genotypes involved.

The product $m_i l_i$ is a measure of absolute fitness expressed in the form of a reproductive rate. Usually, however, we are not immediately concerned with the actual values of these rates but only with how the reproductive ability of one genotype compares to that of another. We therefore use a measure of **relative fitness** (designated by w), which is the reproductive rate of a specified genotype divided by the reproductive rate of the genotype with the greatest absolute fitness. Looking again at our numerical example, we see that the value of $m_{Aa}l_{Aa} = 2.8$ is the largest rate. The relative fitness values for the three genotypes then become

$$w_{AA} = \frac{2.7}{2.8} = 0.96 \qquad w_{Aa} = \frac{2.8}{2.8} = 1.0 \qquad w_{aa} = \frac{1.0}{2.8} = 0.36$$

Thus, if mating is random, so that the ratio of genotypes among the zygotes is $p^2 : 2pq : q^2$, the relative contributions at reproduction can be expressed as $(0.96)p^2 : 2pq : (0.36)q^2$.

Example 25.1

Studies have shown that the survival of *Drosophila* larvae in food that contains 0.1% octanoate depends, among other things, on the fly's genotype at the glucose 6-phosphate dehydrogenase (G6PD) locus. At this locus are two alleles that affect survival: *F*, which codes for an enzyme form that migrates in a fast-moving band during gel electrophoresis, and *S*, which codes for a slow-moving form. In one experiment, larvae from *F/S* × *F/S* crosses were grown in the selective medium. The survivors were collected and analyzed to determine the presence of the *F* and *S* alleles. Among the 2400 survivors examined, 520 were *F/F* in genotype, 1250 were *F/S*, and 630 were *S/S*. Calculate the relative fitness values for each of these genotypes in the selective medium.

Solution: Since the study was restricted to a single environment and to the offspring of a single genotype, the survival component of fitness can be evaluated only in this particular selective medium. We can ignore other possible effects of G6PD on total fitness or assume that these factors are constant and independent of genotype under the conditions employed in the experiment. We determine the relative fitness values by first noting that the ratio of genotypes among the survivors (520:1250:630) corresponds to the ratio ($w_{FF}p^2N : w_{FS}2pqN : w_{SS}q^2N$), where $N = 2400$ (the total number) and p^2, $2pq$, and q^2 are the expected genotype frequencies, which are $\frac{1}{4}$, $\frac{1}{2}$, and $\frac{1}{4}$ for zygotes produced by a monohybrid cross. If we divide each observed number in the ratio by its expected value before selection (i.e., p^2N, $2pqN$, or q^2N), we should then obtain the ratio of relative fitness values:

$$\frac{520}{600} : \frac{1250}{1200} : \frac{630}{600} = 0.867 : 1.042 : 1.05$$

which reduces to approximately $(0.83 : 1.0 : 1.0)$. The relative fitness values for the three genotypes are therefore approximately $w_{SS} = w_{FS} = 1.0$ and $w_{FF} = 0.83$.

Another important component of fitness is the generation time—the average age of parents at reproduction. Suppose that a recessive allele in *Drosophila* prolongs the developmental time of homozygotes, resulting in a longer generation time. (a) How would this allele affect the fitness of the homozygotes? (Will the fitness increase or decrease? Why?) (b) Suggest a way to incorporate the concept of generation time into the relative fitness assigned to these flies.

Relative fitness, as we have defined it, can have values only between 0 and 1. A fitness of 0 implies that the geno-type fails to reproduce, as would be the case if it were lethal. In this situation, we say that **complete selection** occurs. A relative fitness between 0 and 1 indicates a form of **partial selection**, in which the genotype is able to reproduce but at a less-than-optimal rate; most forms of selection that operate in nature fall into this category. At the other extreme is a relative fitness of 1—this is the maximum relative fitness that any genotype can have, and it is assigned to the genotype with the optimal reproductive value.

Although selection merely implies a difference in reproductive success, it is customary to speak of selection as though it were operating against one or more genotypes (or genes). The degree of disadvantage is measured by the **selection coefficient, s**, which is the proportional reduction in the relative fitness of the genotype in question. We can therefore express the relative fitness of an adaptively inferior genotype as $w = 1 - s$, where s can vary from 0 (no selection) to 1 (complete selection). Thus s represents a measure of selection intensity.

Selection operates on a genotype through its phenotype. The effectiveness of selection therefore depends on the degree of dominance exhibited by an allele. In the sections that follow, we will consider three types of selection: (1) selection against recessive genotypes, (2) selection against dominant genotypes, and (3) selection against homozygous genotypes (selection favoring heterozygotes). The first two types of selection lead, in theory, to the eventual elimination of the disfavored allele. In contrast, the third type gives rise to a state of equilibrium characterized by constant allele frequencies. For now, we will assume a diploid, randomly mating population in which the fitness of each genotype remains constant over time, independent of the population size and genetic makeup. Later in this chapter, we will relax our restrictions somewhat and consider what happens when the fitness values themselves are functions of the genotype frequencies.

1. Selection occurs when some genotypes produce more (or fewer) surviving offspring, on the average, than others. These differences in reproductive ability result from inequalities in fertility and survival ability.

2. The fitness of a genotype is a measure of its reproductive success in a given environment. Fitness can be expressed in the form of an absolute reproductive rate, which incorporates the fertility and survival ability of the genotype in question, or as a relative rate compared with the reproductive rate of the genotype with the largest reproductive potential.

3. Selection against a given genotype can vary in intensity from complete selection, in which the genotype fails to reproduce, through various degrees of partial selection, in which the genotype reproduces but at a less-than-optimal rate. The selection process operates on a genotype through its phenotype, so the effectiveness of selection on a diploid genotype will depend on the type of dominance shown by the disfavored allele.

DYNAMICS OF SELECTION

Although Darwin and Wallace first proposed that natural selection is the fundamental process of evolution in the 1850s, the basic genetic mechanisms of selection were not understood until much later, after the discovery of Mendel's work. Much of the classical theory on the genetic mechanisms of selection was formulated in the 1930s by R. A. Fisher, J. B. S. Haldane, and Sewall Wright. We will consider some of this theory in the following sections, where we will describe the effects of selection on the frequency of an allele at a single gene locus. These discussions will focus on the mechanisms by which detrimental alleles are eliminated from a population and how rapidly these changes occur over time.

Complete Selection

Many alleles cause premature lethality (prereproductive death) or infertility when present in the homozygous state but have little or no effect on the fitness of heterozygotes. For these alleles, complete selection occurs against the recessive homozygotes. Though obviously an extreme form of selection, complete selection is a useful model for many recessive metabolic diseases in humans. It also applies to the continued elimination of a recessive phenotype in various cases of artificial selection. For example, complete selection occurs when a breeder culls all red-and-white calves from a herd of Holstein cattle; red-and-white spotting is caused by a recessive allele (see Chapter 2), so only recessive homozygotes are affected by the selection process.

The basic selection model for complete selection against recessives is developed in ■ Table 25–1. This model assumes three genotypes (AA, Aa, aa) with relative fitness values (1, 1, 0). Moreover, selection is assumed to operate after zygote formation but prior to the age of reproduction, so that no aa individual reaches sexual maturity. If we start with a zygote population at generation 0 in which the ratio of ($AA:Aa:aa$) is ($p_0^2:2p_0q_0:q_0^2$) and the ratio of ($A:a$) is ($p_0:q_0$), the genotypic ratio of the adult population must then change to ($p_0^2:2p_0q_0:0$) due to the loss of the recessive homozygotes. Although the recessive homozygotes are all eliminated in the first generation, the recessive allele is not, since it is still present in the heterozygous carriers among the breeding adults.

The consequences of a single generation of complete selection can be calculated in terms of the reduction in the frequency of the recessive allele. Since the recessive homozygotes fail to reach sexual maturity, all matings are restricted to crosses involving dominant genotypes (A– × A–). The frequency of the recessive allele among the zygotes in generation 1, q_1, is thus equal to the probability that an offspring will receive a copy of the a allele from a dominant parent:

$$q_1 = \frac{\text{frequency of } a \text{ among } A\text{– parents}}{\text{total frequency of } A \text{ and } a \text{ among } A\text{– parents}} = \frac{p_0q_0}{p_0^2 + 2p_0q_0} = \frac{q_0}{1 + q_0}$$

Furthermore, since the heterozygotes are indistinguishable in phenotype from the dominant homozygotes, matings should be random among the surviving adults. The fre-

■ **TABLE 25–1** Selection model for one generation of complete selection against a recessive allele.

	Algebraic Model			Arithmetic Example		
Generation 0						
Allele frequency:	$f(a) = q_0$			$f(a) = \frac{1}{4}$		
Genotype frequencies:	AA	Aa	aa	AA	Aa	aa
Zygotes:	p_0^2	$2p_0q_0$	q_0^2	$\frac{9}{16}$	$\frac{6}{16}$	$\frac{1}{16}$
Fitness:	1	1	0	1	1	0
Adults:	$\dfrac{p_0^2}{\text{total}}$	$\dfrac{2p_0q_0}{\text{total}}$	0	$\frac{9}{15}$	$\frac{6}{15}$	0
	Adult total $= p_0^2 + 2p_0q_0$			Adult total $= \frac{15}{16}$		
Generation 1						
Allele frequency:	$f(a) = q_1 = \dfrac{p_0q_0}{p_0^2 + 2p_0q_0}$			$f(a) = \dfrac{\frac{1}{4}}{1 + \frac{1}{4}} = \frac{1}{5}$		
	$= \dfrac{q_0}{1 + q_0}$			or		
				$f(a) = 0 + \frac{1}{2}(\frac{6}{15})$		
				$= \frac{1}{5}$		

Assumptions:
1. Recessive homozygotes are formed normally among zygotes but fail to reproduce.
2. The surviving dominant AA and Aa genotypes mate at random in the adult stage, so zygotes are produced according to random-mating frequencies.

quency of *aa* zygotes in generation 1 can therefore be computed from the square of the allele frequency as

$$q_1^2 = \left(\frac{q_0}{1 + q_0}\right)^2$$

The pattern is expected to be the same in the next generation, so the frequency of the recessive allele among zygotes after two generations of complete selection becomes

$$q_2 = \frac{q_1}{1 + q_1} = \frac{q_0/(1 + q_0)}{1 + [q_0/(1 + q_0)]} = \frac{q_0}{1 + 2q_0}$$

Continuing this process over *t* generations yields

$$q_t = \frac{q_0}{1 + tq_0} \qquad (25.1)$$

This equation describes a simple numerical progression. For example, suppose that we start with a population in which the genotypes (*AA*, *Aa*, *aa*) have frequencies of $(\frac{1}{4}, \frac{1}{2}, \frac{1}{4})$, so that $q_0 = \frac{1}{2}$. Solving equation 25.1 successively for $t = 0, 1, 2, 3, 4, \ldots$, we get $q_t = \frac{1}{2}, \frac{1}{3}, \frac{1}{4}, \frac{1}{5}, \frac{1}{6}, \ldots$. There is thus only a gradual decline in allele frequency, despite the fact that recessive homozygotes fail to reproduce in each generation.

The number of generations of complete selection that are required to reduce the frequency of the recessive allele from its initial value q_0 to some designated value q_t can also be derived by solving equation 25.1 for *t*:

$$t = \frac{(q_0 - q_t)}{q_0 q_t} = \frac{1}{q_t} - \frac{1}{q_0} \qquad (25.2)$$

Observe that the number of generations required to reduce the recessive allele by a specified amount is inversely related to the allele frequency. Therefore, a reduction in *q* produces an increase in *t* of the same order of magnitude. For example, the time required to reduce the allele frequency from, say, 0.1 to 0.05 is

$$t = \frac{1}{0.05} - \frac{1}{0.1} = 10 \text{ generations}$$

whereas the time needed to reduce the allele frequency from 0.01 to 0.005 is

$$t = \frac{1}{0.005} - \frac{1}{0.01} = 100 \text{ generations}$$

As this example shows, a reduction in the frequency of the detrimental allele is accompanied by a corresponding decrease in the efficiency of selection against it. When *q* is small, most recessive alleles are in heterozygous combinations and are not exposed to the effects of selection. The reduction in the frequency of the recessive allele is due solely to the selective elimination of *aa* offspring of *Aa* × *Aa* matings and occurs at a very low rate.

The decline in the efficiency of selection at low allele frequencies is important to breeders, since it limits the effectiveness of any program designed to eliminate harmful recessive alleles from breeding stocks of crop plants and do-

mesticated animals. Even when complete selection is practiced and the recessive homozygotes are not allowed to mate, *aa* individuals will continue to be produced for a very long time as a result of occasional *Aa* × *Aa* matings. This decline in selection efficiency also points out the futility of programs designed to eliminate recessive disorders in humans by sterilization or therapeutic abortions or by persuading persons with such a disorder not to have children (or prohibiting them). Such disorders are usually so rare to begin with that even the practice of complete selection will have little effect.

Example 25.2

As we learned in Chapter 11, Tay-Sachs disease is a fatal recessive disorder, in which affected individuals are unable to produce the enzyme hexosaminidase A. The absence of this enzyme results in the accumulation of a specific fatty substance (ganglioside GM_2) within the nerve cells and the subsequent deterioration of the central nervous system. The allele for this disorder is most common in certain Jewish populations, in which it has been estimated to occur at a frequency as high as 1/80. Ignoring any factors that might increase the incidence of this disease, how many generations would be required for selection against homozygous recessives to reduce the frequency of the Tay-Sachs allele in these populations to 1/800?

Solution: We are asked to compute a value for *t* (time in generations) for the change from $q_0 = 1/80$ to $q_t = 1/800$. With complete selection against this recessive trait, we can calculate the value as

$$t = \frac{1}{1/800} - \frac{1}{1/80} = 800 - 80 = 720 \text{ generations}$$

Follow-Up Problem 25.2

(a) Continuing to ignore any factor other than selection that might alter the frequency of the Tay-Sachs allele, predict the incidence of Tay-Sachs disease in the Jewish populations in Example 25.2 after 10 generations of selection against recessive homozygotes. (b) By what percent will the incidence of Tay-Sachs disease be reduced as a result of this selection process?

Complete selection is significantly more effective when acting on an allele that shows complete or partial dominance. When a lethal allele shows complete dominance, both homozygous and heterozygous carriers of the allele fail to survive or are effectively sterile. The genotypes (*AA*, *Aa*, *aa*) will then have fitness values of (0, 0, 1). Since only recessive homozygotes are able to reproduce in this case, any dominant lethal gene that might arise in a population will be lost in a single generation.

Partial Selection: The General Selection Model

Complete selection is the most severe form of selection. The selective processes that occur in nature are usually much less severe, often with genotypes that differ only slightly from one another in terms of relative fitness. In cases of partial selection, the relative intensity of selection will then depend on the exact fitness distribution of existing genotypes.

The effect of one generation of partial selection is usually expressed in terms of the change in allele frequency. When only two alleles are considered, this change can be expressed either as $\Delta q = q' - q$ or $\Delta p = p' - p$, where q, q', p, and p' are the frequencies of the two alleles in successive generations. Since $p + q = p' + q' = 1$, $\Delta q = -\Delta p$; thus the changes in the two allele frequencies differ from one another only in direction.

The basic relationship between the change in allele frequency and the distribution of fitness values can be derived using the general selection model in ■ Table 25–2 and the following identity:

$$\begin{aligned} \Delta q &= q' - q = q' - q'q - q + q'q \\ &= q'(1 - q) - q(1 - q') = q'p - qp' \\ &= pq\left(\frac{q'}{q} - \frac{p'}{p}\right) \end{aligned}$$

The difference in the relative increases in q and p, $q'/q - p'/p$, depends directly on the fitness values of the contributing genotypes. The exact dependence is derived in Table 25–2, from which we obtain

$$\frac{q'}{q} = \frac{qw_{aa} + pw_{Aa}}{\overline{w}} \quad \text{and} \quad \frac{p'}{p} = \frac{pw_{AA} + qw_{Aa}}{\overline{w}}$$

where $\overline{w}$ is the mean fitness, expressed as

$$\overline{w} = p^2 w_{AA} + 2pq w_{Aa} + q^2 w_{aa}$$

Making this substitution for $q'/q - p'/p$ in the identity for Δq, the change in allele frequency becomes

$$\Delta q = pq\left[\frac{(w_{aa} - w_{Aa})q - (w_{AA} - w_{Aa})p}{\overline{w}}\right] \quad (25.3)$$

We can make three important observations from equation 25.3. First, Δq is directly proportional to the product pq. Selection is therefore most effective at intermediate allele frequencies, that is, when neither p nor q is close to 0 or to 1. Second, Δq is inversely proportional to the mean fitness, $\overline{w}$. The response to selection is therefore greatest in highly selective environments, in which $\overline{w}$ is small in comparison to its maximum value of 1. Third, Δq is directly proportional to the fitness difference $(w_{aa} - w_{Aa})q - (w_{AA} - w_{Aa})p$. Since $pq/\overline{w}$ is always positive, the fitness difference determines the direction of selective change. If $(w_{aa} - w_{Aa})q - (w_{AA} - w_{Aa})p$ is positive, q will increase in value; if it is negative, q will decrease in value.

■ Table 25–3 gives a numerical example of the basic selection model in terms of the genotype frequencies before and after one generation of selection. It is assumed in this example that selection occurs through the differential survival of genotypes during development to adults and that zygotes are produced according to Hardy-Weinberg frequencies.

One common example that is covered by the general selection model is selection against recessives. In this special case, only recessive homozygotes are at a selective disadvantage, so the fitness values are as follows:

Genotype:	AA	Aa	aa
Relative fitness:	1	1	$1 - s$

The values of $\overline{w}$ and $(w_{aa} - w_{Aa})q - (w_{AA} - w_{Aa})p$ are then

$$\overline{w} = p^2 + 2pq + q^2(1 - s) = 1 - sq^2$$
$$(w_{aa} - w_{Aa})q - (w_{AA} - w_{Aa})p = -sq$$

By making these substitutions in equation 25.3, we get the following solution for selection against recessive homozygotes:

$$\Delta q = -\frac{s(1 - q)q^2}{1 - sq^2} \quad (25.4)$$

We see that a decrease in allele frequency depends directly on q^2 in this case. Thus when q is small, q^2 is very small (for example, when $q = 0.1$, $q^2 = 0.01$), and thus selection

■ **TABLE 25–2** The general selection model.

	AA	Aa	aa	Total
Frequency before selection:	p^2	$2pq$	q^2	1
Fitness:	w_{AA}	w_{Aa}	w_{aa}	
Proportionate contribution:	$p^2 w_{AA}$	$2pq w_{Aa}$	$q^2 w_{aa}$	$\overline{w}$
Frequency after selection:	$(p^2)\dfrac{w_{AA}}{\overline{w}}$	$(2pq)\dfrac{w_{Aa}}{\overline{w}}$	$(q^2)\dfrac{w_{aa}}{\overline{w}}$	1
Allele frequencies in the next generation:	$p' = f(AA) + \frac{1}{2}f(Aa)$		$q' = f(aa) + \frac{1}{2}f(Aa)$	
	$= (p^2)\dfrac{w_{AA}}{\overline{w}} + (pq)\dfrac{w_{Aa}}{\overline{w}}$		$= (q^2)\dfrac{w_{aa}}{\overline{w}} + (pq)\dfrac{w_{Aa}}{\overline{w}}$	
	$= p\left[\dfrac{pw_{AA} + qw_{Aa}}{\overline{w}}\right]$		$= q\left[\dfrac{qw_{aa} + pw_{Aa}}{\overline{w}}\right]$	

■ TABLE 25–3 Application of the selection model to starting allele frequencies of $p = \frac{1}{4}$ and $q = \frac{3}{4}$.

	AA	Aa	aa	Total
Frequency before selection:	$\frac{1}{16}$	$\frac{6}{16}$	$\frac{9}{16}$	1
Fitness:	1	1	$\frac{1}{3}$	
Proportionate contribution:	$\frac{1}{16}(1)$	$\frac{6}{16}(1)$	$\frac{9}{16}(\frac{1}{3})$	$\frac{10}{16}$
Frequency after selection:	$\frac{1}{10} = 0.1$	$\frac{6}{10} = 0.6$	$\frac{9}{10}(\frac{1}{3}) = 0.3$	1
Allele frequency in the next generation:	$q' = 0.3 + \frac{1}{2}(0.6) = 0.6$			
Change in allele frequency:	$\Delta q = q' - q = 0.60 - 0.75 = -0.15$			
or, directly obtained by using equation 25.3,	$\Delta q = (\frac{1}{4})(\frac{3}{4})\dfrac{(\frac{1}{3} - 1)(\frac{3}{4}) - (1 - 1)(\frac{1}{4})}{\frac{10}{16}} = (3)\dfrac{-(\frac{2}{3})(\frac{3}{4})}{10} = -0.15$			

against recessives becomes a very inefficient process. This inefficiency is illustrated in ▶ Figure 25–1, which shows graphs of q versus number of generations, t, for different selection intensities. These plots result from iterations of equation 25.4. Although this equation does not have a simple analytical solution, it can be solved numerically using the process of iteration. Given a selection intensity, s, and an initial allele frequency, q_0, we can first solve for the allele frequency in generation 1, q_1. We can then use our value of q_1 to solve for q_2, and so on. The plots that are produced by this method show that selection against recessives is characterized by a rapid selection rate in the beginning, when q is large, and by a slow rate at the end, when q is small.

Selection against a dominant allele is another important form of selection that is included in the general selection model. In this case, both heterozygotes and dominant homozygotes are selected against, so the fitness values are as follows:

Genotype:	AA	Aa	aa
Relative fitness:	$1 - s$	$1 - s$	1

The values of $\bar{w}$ and $(w_{aa} - w_{Aa})q - (w_{AA} - w_{Aa})p$ are then

$$\bar{w} = p^2(1 - s) + 2pq(1 - s) + q^2 = 1 - s(1 - q^2)$$
$$(w_{aa} - w_{Aa})q - (w_{AA} - w_{Aa})p = sq$$

If we make these substitutions in equation 25.3 and replace q with $1 - p$ and Δq with $-\Delta p$, we get the following solution for selection against dominants:

$$\Delta p = -\frac{sp(1 - p)^2}{1 - s[1 - (1 - p)^2]} \qquad (25.5)$$

The decrease in allele frequency now depends directly on $(1 - p)^2 = q^2$, which is the square of the frequency of the favored recessive allele. Thus when p is large, with a value close to 1, the value of $(1 - p)^2$ will approach 0—very few individuals have the favored aa genotype, so selection is a very inefficient process. When p is small, $(1 - p)^2$ approaches 1, and the dominant allele is reduced approximately as $\Delta p = -sp$ in each generation. These characteristics are illustrated in ▶ Figure 25–2, which shows graphs of p versus t for

▶ **FIGURE 25–1** Selection against recessives. Different intensities of selection against a recessive detrimental allele that occurs initially (in generation 0) at a frequency of 0.9.

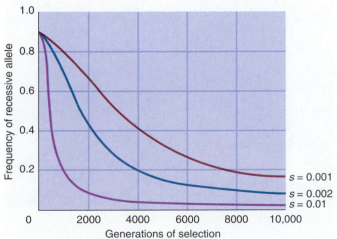

▶ **FIGURE 25–2** Selection against dominants. Different intensities of selection against a dominant detrimental allele that occurs initially (in generation 0) at a frequency of 0.9.

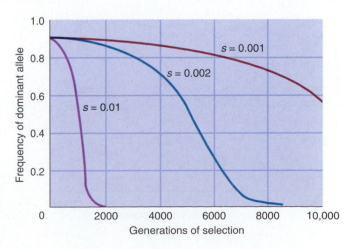

different selection intensities. Again, the selection equation does not have a simple analytical solution, so the plots are produced by iteration. We see from the graphs that selection against a dominant allele gets off to a slow start when p is large. The selective process accelerates, however, as the dominant allele declines in frequency and more *aa* homozygotes appear in the population.

Experimental Support for Selection Theory

Numerous attempts have been made to obtain support for the general theory of selection from the results of laboratory experiments. Such studies have generally shown that agreement between theory and observation is most satisfactory in the case of complete selection. One such example is shown in ▶ Figure 25–3. In this experiment, the starting population of *Drosophila* was heterozygous for an autosomal lethal gene allelic to the recessive eye-color mutation light (*lt*). Since the value of the selection coefficient cannot exceed 1, the increased rate of reduction in allele frequency over that expected for a recessive lethal gene can be explained only on the basis of a slight reduction in the fitness of the heterozygotes. The fact that experimental results on nonlethal genes tend to be even less satisfactory in their agreement with the basic selection models implies that selection in real populations is generally more complex than predicted by classical theory.

▶ **FIGURE 25–3** Selection against an autosomal lethal gene (*lt*) in an experimental population of *D. melanogaster*. Circles represent observed frequencies. The top line gives the expected results for selection against a recessive lethal; the lower line designates the results expected for a lethal gene that reduces heterozygote fitness by 10%. *Source:* B. Wallace, "The elimination of an autosomal lethal from an experimental population of *Drosophila melanogaster*," *Am. Nat.* 97 (1963): 66. Copyright © 1963 by The University of Chicago. All rights reserved.

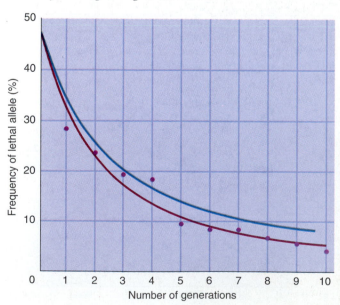

Several examples of selection have also been documented in nature. Many are the result of human-induced changes in the environment, such as the selection for antibiotic resistance in bacteria and for insecticide resistance in insects during prolonged exposure to these chemical agents. As a result of insecticide resistance, progressively greater doses of insecticides are needed to overcome the increase in resistance in certain insect pests, thus increasing the cost of pest control as well as posing a serious threat to human health and the survival of beneficial wildlife.

The rapid increase in the melanic form of the peppered moth *Biston betularia* is probably the most spectacular example of selection that has been recorded in nature. This change occurred in several urban areas of England when the surrounding terrain became darkened with the soot of industrial pollution. During the first part of the nineteenth century, the peppered moth was typified by a light-colored form that blended quite nicely with the lichen-covered tree trunks dotting the countryside (▶ Figure 25–4). In 1848,

▶ **FIGURE 25–4** Typical and melanic forms of *Biston betularia* on (a) an unpolluted, lichen-covered tree trunk and (b) a soot-covered tree trunk.

(a)

(b)

the first melanic form was collected in Manchester. At that time, the melanic form probably comprised less than 1% of the population. Fifty years later, the melanic form had increased in frequency to about 95%. Since the species produces one generation per year, this spectacular change actually occurred in 50 generations.

Much of our knowledge on the nature of the selective agent for industrial melanism has come from the extensive studies of H. B. D. Kettlewell. For example, in 1958 Kettlewell released both light- and dark-colored forms of the peppered moth into industrialized and nonindustrialized areas and then recaptured samples using light traps. The results of his studies clearly demonstrate that the melanic form has the selective advantage in sooty environments. Moreover, photographs taken by Kettlewell and Niko Tinbergen revealed that a major cause of the selective action was differential predation by birds. In these well-documented cases, the melanic form possessed the selective advantage of protective coloration on the soot-darkened trees in industrialized areas, while the opposite was true of the light-colored form.

Breeding experiments have indicated that these color differences are largely determined by a single pair of alleles in which the melanic condition is the dominant trait. Dominance of the melanic form accounts, in large part, for the rapid rise of the dark-colored moths as a result of selection.

Example 25.3

Let us assume for the purposes of calculation that melanism is determined by a dominant allele at a single locus and that the light-colored form of the peppered moth (the recessive form) makes up 49% of a randomly mating population when it emerges as an adult and has only half the reproductive potential of the melanic form. (a) Calculate $\bar{w}$, the average fitness of the moth population. (b) Determine the frequencies of both forms of the moth in the next generation.

Solution: (a) Since the frequency of the recessive form is $q^2 = 0.49$, the frequencies of the recessive and dominant alleles are $q = \sqrt{0.49} = 0.7$ and $p = 0.3$. Also, because the recessive form has only half the reproductive potential of the melanic form, the relative fitness values are $w_{AA} = w_{Aa} = 1$ and $w_{aa} = 0.5$. The average fitness becomes

$$\bar{w} = (1)(0.3)^2 + (1)(2)(0.3)(0.7) + (0.5)(0.7)^2 = 0.755$$

(b) The frequency of the recessive allele in the next generation is

$$q' = \frac{w_{aa}q^2 + w_{Aa}pq}{\bar{w}} = \frac{(0.5)(0.49) + (1)(0.3)(0.7)}{0.755} = 0.6$$

Thus, the frequency of the light-colored moth in the next generation is $q'^2 = 0.36$ and the frequency of the melanic form is $1 - 0.36 = 0.64$. Note that the light-colored form has decreased in frequency by $0.49 - 0.36 = 0.13$ in one generation.

Follow-Up Problem 25.3

Suppose that pollution control reverses the situation in Example 25.3, so that the light-colored form of the peppered moth has the selective advantage. If we assume, as before, that the light-colored form makes up 49% of a randomly mating population when it emerges as an adult, but now assume that it has twice the reproductive potential of the melanic form, (a) calculate the average fitness of the moth population and (b) determine the frequencies of both forms of the moth in the next generation.

To Sum Up

1. When there is complete selection against a recessive trait, the recessive homozygotes fail to reproduce. Despite the severity of this selection process, recessive individuals will continue to be produced for a very long time by matings between heterozygotes. The result is a slow, gradual decline in the frequency of the recessive allele in the population. In contrast, complete selection against a dominant trait will, in theory, eliminate the dominant allele in one generation.

2. The effects of selection at any arbitrary intensity can be described in terms of a general selection model that gives the change in allele frequency per generation as a function of the frequencies and fitness values of the genotypes involved. The general selection model indicates that the effectiveness of selection will be greatest when differences in fitness between the genotypes are large, the mean fitness of the population is small, and allele frequencies are close to neither 0 nor 1.

3. Selection against the recessive homozygote at any arbitrary intensity is characterized by a rapid reduction in the frequency of the recessive phenotype when the frequency of the recessive allele is large and by a very slow decrease when q is small and most recessive alleles are "hidden" in heterozygotes. In contrast, partial selection against dominants gets off to a slow start when p is large but accelerates as the dominant allele declines in frequency.

4. The process of selection has been studied extensively in experimental populations. A number of examples of selection have also been documented in nature: several, such as industrial melanism, are the result of human-induced changes in the environment. Many of the examples observed in nature involve genetic mechanisms that are more complex than those assumed in the general selection model.

EQUILIBRIA AND POLYMORPHISM

In addition to producing systematic changes in allele frequency, which we generally associate with evolution, selection alone or in combination with other evolutionary forces can also give rise to conditions of stable equilibria. The maintenance of equilibria is a very important role of natural selection. In an environment that has remained comparatively stable for some time, the most frequently occurring genotypes are the ones that are best suited for the conditions at hand. The majority of selective changes are then directed toward eliminating the adaptively inferior genotypes that arise by mutation and through the segregation and recombination of genes during the sexual reproductive process. In this role, selection helps to create and preserve a balanced or equilibrium state. In this section, we will consider a few of the selective mechanisms that can act on a single locus and lead to a state of equilibrium in a homogeneous environment.

Balance between Mutation and Selection

As was pointed out in Part IV, many, if not most, mutations that occur in an established population are detrimental in character and will be selected against. Consequently, mutation and selection often act as opposing processes, with mutation lowering fitness through the continued introduction of detrimental alleles to the gene pool and selection counteracting this effect by eliminating the less fit genotypes. The opposing effects of mutation and selection eventually establish an equilibrium condition in which the number of detrimental alleles arising by mutation in each generation equals the number lost through selection.

As an example, we will consider the equilibrium between mutation and selection in the case of selection against recessives. From equation 25.4, we see that approximately $s(1-q)q^2$ of the a alleles will be lost in each generation as a result of selection against the recessive homozygote. (The term sq^2 in the denominator of equation 25.4 is negligible at the low levels of q expected in the vicinity of the equilibrium state, as is back mutation.) However, if the A allele mutates to its recessive form at a rate of μ per generation, $\mu p = \mu(1-q)$ a alleles will be restored to the population each generation through mutation. A balance will exist between these opposing processes when $\mu(1-q) = s(1-q)q^2$. The approximate condition for equilibrium is therefore

$$\hat{q}^2 = \frac{\mu}{s} \qquad \text{or} \qquad \hat{q} = \sqrt{\frac{\mu}{s}}$$

where $\hat{q}$ is the frequency of the a allele at equilibrium. If the recessive trait were lethal ($s = 1$), these expressions would reduce to $\hat{q}^2 = \mu$ and $\hat{q} = \sqrt{\mu}$. For example, if $\mu = 10^{-6}$ and the trait were lethal, $q = 0.001$, that is, about 1 in every 1,000 gametes would carry a mutant allele of this type. Most of these mutant alleles are in heterozygous combinations and are not directly exposed to the effects of selection.

The opposing effects of selection and mutation are shown in the graph of allele frequency versus generation number in ▶ Figure 25–5. A relaxation of selection against a recessive trait results in the establishment of a new equilibrium—although the selection pressure has been reduced, the mutation rate has not, so the frequency of the disfavored allele increases. Such changes in equilibrium occur in human populations when the treatment of a recessive disorder, such as PKU, enables affected individuals to reproduce. The recessive homozygotes can then marry and have children, thereby adding recessive mutant alleles to the genetic load of future generations. Observe that if selection were eliminated entirely, the allele frequency would eventually attain an equilibrium determined by the opposing pressures of forward and back mutation ($A \overset{\mu}{\underset{v}{\rightleftharpoons}} a$). Once the equilibrium between opposing mutation pressures is established, the ratio of the mutant and wild-type alleles will then equal the ratio of the forward and back mutation rates ($\hat{q}/\hat{p} = \mu/v$).

▶ FIGURE 25–5 Effects of relaxing the selection intensity against a detrimental recessive allele. The graph shows idealized curves for the reduction in allele frequency due to intense and relaxed selection and for the increase in frequency resulting from mutation alone. An equilibrium between mutation and intense selection is established at point A. The relaxation of selection results in an increase in the equilibrium frequency of the allele to point B. The complete elimination of selection will eventually result in the mutational equilibrium between forward mutation (at rate μ) and reverse mutation (at rate v) at point C.

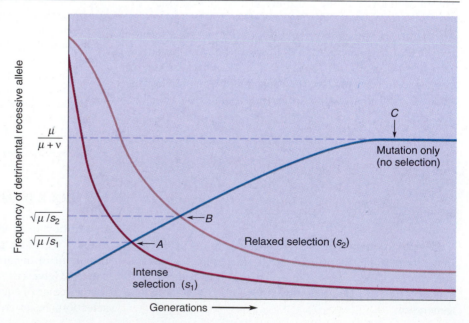

Example 25.4

Example 25.2 gives the frequency of the Tay-Sachs gene in certain Jewish populations as 1/80. If this detrimental gene is maintained through a balance between mutation and complete selection against the recessive trait, what is the mutation rate from the dominant normal allele to the recessive Tay-Sachs allele?

Solution: With complete selection against recessives, $s = 1$, so the mutation rate (μ) from the dominant to the recessive allele equals q^2. Thus, for Tay-Sachs disease

$$\mu = \left(\frac{1}{80}\right)^2 = \frac{1}{6400}$$

or 1.5×10^{-4} per generation. Because this value is unusually large for a mutation rate, geneticists have proposed alternative mechanisms, such as heterozygote advantage, to account for the high incidence of the allele in these Jewish populations.

When an equilibrium exists between mutation and selection against a dominant allele, the equilibrium frequency of the disadvantaged allele is approximately $\hat{p} = \nu/s$, where s is the selection coefficient and ν is the mutation rate ($a \xrightarrow{\nu} A$). Suppose that a disease has two possible genetic causes: an autosomal dominant allele and an autosomal recessive allele, both arising with a mutation rate of 10^{-5} per generation. (a) If individuals with either the dominant or recessive form of this disease have only half the reproductive potential of normal individuals, what will be the frequency of each form of this disease once an equilibrium is established between opposing selection and mutation pressures? (b) Which of the two forms makes a greater contribution to the overall frequency of the disease? Why?

Selection Favoring Heterozygotes

Selection alone can also be important in the maintenance of genetic variability. One example is where heterozygotes show superior fitness, a condition called **overdominance** in fitness. We can represent an overdominant locus by the following fitness values:

Genotype:	AA	Aa	aa
Relative fitness:	$1 - s_A$	1	$1 - s_a$

If we assume random mating, $\bar{w}$ and $(w_{aa} - w_{Aa})q - (w_{AA} - w_{Aa})p$ are

$$\bar{w} = (1 - s_A)p^2 + 2pq + (1 - s_a)q^2 = 1 - s_Ap^2 - s_aq^2$$

$$(w_{aa} - w_{Aa})q - (w_{AA} - w_{Aa})p = s_Ap - s_aq$$

When these values are substituted into the general selection model (equation 25.3), replacing p with $1 - q$ and rearranging terms gives

$$\Delta q = \frac{q(1 - q)[s_A - (s_A + s_a)q]}{1 - s_A(1 - q)^2 - s_aq^2} \qquad (25.6)$$

We see that when $s_A = (s_A + s_a)q$ in this relationship, $\Delta q = 0$. An equilibrium will thus develop when the allele frequencies have the following values:

$$\hat{q} = \frac{s_A}{s_A + s_a} \qquad \text{and} \qquad \hat{p} = \frac{s_a}{s_A + s_a}$$

This equilibrium is shown in ▶ Figure 25–6 in the form of a Δq versus q plot of equation 25.6; we see that Δq is positive for values of q less than $\hat{q}$ and negative for values of q greater than $\hat{q}$. The point $q = \hat{q}$ is therefore a **stable equilibrium**, since any perturbation of q away from $\hat{q}$ will result in a change in the allele frequency back to its equilibrium value.

The effects of this equilibrium on genotype frequencies should also be noted. A numerical example is shown in ■ Table 25–4 in which random mating is assumed and selection occurs through the differential survival of genotypes prior to the age of reproduction. Although genotype frequencies change from the zygotes to the breeding adults in the same generation, allele frequencies do not change, so the equilibrium genotype frequencies are restored in the zygotes in the next generation. This seesaw process is expected to continue as long as the equilibrium conditions hold. In

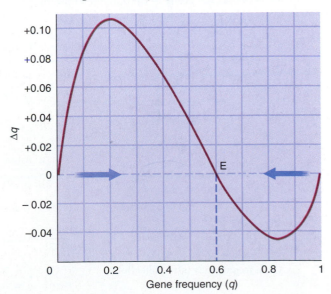

▶ **FIGURE 25–6** Change in allele frequency with overdominance when $s_A = 0.6$ and $s_a = 0.4$. An equilibrium will be established at $q = s_A/(s_A + s_a) = 0.6$. Point E represents a stable equilibrium point since Δq is positive when q is less than 0.6 and negative when q is greater than 0.6.

Stability of an Equilibrium: Heterozygote Disadvantage

When developing a selection model, it is often necessary to test for the stability of equilibrium points. An equilibrium is said to be stable if a system, when perturbed, returns to its equilibrium value. An example of a stable system, as we have seen, is the case of heterozygote superiority (equation 25.6). A physical model depicting stable and unstable equilibria is shown in the following figure:

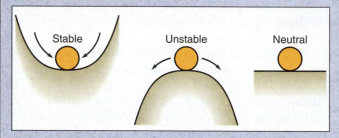

In this model, a ball at rest at the lowest point of a concave surface would be at a stable equilibrium point, since the

ball would return to its original state after a disturbance, whereas a ball at rest at the highest point of a convex surface would be at an unstable equilibrium point. The model also depicts a system at neutral equilibrium, which like a ball on a flat surface will come to rest without further reaction once a disturbance has ceased. (We have already considered the concept of a neutral equilibrium in Chapter 24 when we discussed the equilibrium of allele frequencies in a large population.)

The presence of an unstable equilibrium is exemplified by the case of heterozygote disadvantage, represented by the following:

Genotype:	AA	Aa	aa
Relative fitness:	$w_{AA} = 1$	$w_{Aa} = 1 - s$	$w_{aa} = 1$

If we assume a randomly mating population

$$\bar{w} = p^2 + (1 - s)2pq + q^2 = 1 - 2sq(1 - q)$$

$$(w_{aa} - w_{Aa})q - (w_{AA} - w_{Aa})p = 2s(q - \tfrac{1}{2})$$

an extreme example of this process, both homozygous genotypes, AA and aa, die before reaching the reproductive age and only Aa heterozygotes survive to the adult stage. Nevertheless, $Aa \times Aa$ matings restore genotype frequencies among the zygotes at $\frac{1}{4}\, AA$, $\frac{1}{2}\, Aa$, and $\frac{1}{4}\, aa$, thus maintaining an equilibrium condition with $\hat{p} = \hat{q} = \frac{1}{2}$ by losing half of the population through selection in each generation.

Heterozygote advantage is an important mechanism by which selection can maintain two or more alleles at relatively high frequencies in a population. The result is a stable variation pattern known as a **balanced polymorphism**, in which genotypes with different fitness values can coexist indefinitely in the same population. A well-known example of a balanced polymorphism in humans is the sickle-cell

■ **TABLE 25–4** Genotype frequencies before and after selection at the equilibrium established with $s_A = 0.1$ and $s_a = 0.2$.

Allele frequencies (before selection):	$f(A) = p = 0.2/(0.1 + 0.2) = \frac{2}{3}$			
	$f(a) = q = 0.1/(0.1 + 0.2) = \frac{1}{3}$			
Genotype frequencies:	AA	Aa	aa	Total
Before selection:	$(\frac{2}{3})^2 = \frac{4}{9}$ $= 0.444$	$2(\frac{2}{3})(\frac{1}{3}) = \frac{4}{9}$ $= 0.444$	$(\frac{1}{3})^2 = \frac{1}{9}$ $= 0.112$	1
Fitness:	$1 - 0.1 = 0.9$	1	$1 - 0.2 = 0.8$	
Contribution:	$\frac{4}{9}(0.9) = 0.400$	$\frac{4}{9} = 0.444$	$\frac{1}{9}(0.8) = 0.089$	0.933
After selection:	$\frac{0.400}{0.933} = 0.429$	$\frac{0.444}{0.933} = 0.476$	$\frac{0.089}{0.933} = 0.095$	1
Allele frequencies (after selection):	$f(A) = p = 0.429 + \frac{1}{2}(0.476) = \frac{2}{3}$			
	$f(a) = q = 0.095 + \frac{1}{2}(0.476) = \frac{1}{3}$			

Substituting these values into the basic selection equation, we get

$$\Delta q = \frac{2sq(1 - q)(q - \frac{1}{2})}{1 - 2sq(1 - q)}$$

A plot of this equation is:

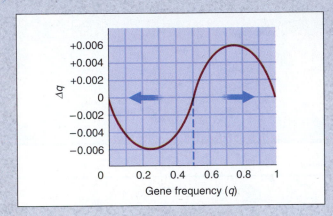

We see that an equilibrium will develop at $q = \frac{1}{2}$. The equilibrium is unstable, however, since Δq is negative for all values of q less than $\frac{1}{2}$ and positive for all values of q greater than $\frac{1}{2}$. Thus, any perturbation from $q = \frac{1}{2}$ will result in a continued movement away from the equilibrium point.

Heterozygote inferiority is a characteristic feature of many chromosome rearrangements, particularly translocations (see Chapter 12). As we have seen, translocation heterozygotes are semisterile due to the production of aneuploid gametes. Therefore, any translocation chromosome that appears in a population at less than the equilibrium level, such as one that enters the gamete pool through mutation, would tend to be selected against. The reverse is also true: If a population with a normal chromosome arrangement were swamped with translocation homozygotes (for example, through sudden mixing with a neighboring population), the translocation chromosome would become fixed. The latter effect is the theoretical basis of a proposed program for insect control, which suggests that genes conferring sensitivity to an insecticide be introduced into a population when they are linked to a translocation chromosome. If the number of translocated types released is sufficiently large to raise the resulting frequency of the translocation above the equilibrium, the genes for insecticide sensitivity would be expected to go to fixation.

polymorphism prevalent in parts of Africa and southern Asia. Recall from Chapter 11 that homozygotes for the hemoglobin S allele ($Hb^S Hb^S$) have sickle-shaped red blood cells and experience severe erythrocyte destruction, leading to sickle-cell anemia. Without proper medical care, these homozygotes tend to die in early childhood. They are clearly less fit than the heterozygotes ($Hb^A Hb^S$), who have a milder form of sickling without anemia. However, in regions of the world where falciparum malaria is prevalent, homozygotes for the normal hemoglobin allele ($Hb^A Hb^A$) are also less fit than heterozygotes, since they are less resistant to the malarial parasite and often die from this disease. Thus, the heterozygotes have a selective advantage and are more likely to reproduce than either homozygous type.

The sickle-cell polymorphism has been linked to a differential resistance to malaria by several forms of evidence, including the correlation between the geographic distributions of the sickle-cell allele and falciparium malaria (see ▶ Figure 25–7), as well as hospital records showing a higher death rate from this form of malaria among homozygotes with normal hemoglobin than among heterozygotes. In areas where malaria is not a problem or does not exist, heterozygotes do not have a selective advantage, and selection then occurs exclusively against the sickle-cell allele. This is the case among African-Americans, in whom the frequency of Hb^S has fallen to a comparatively low level.

Example 25.5

About 1 child in 2500 Caucasians is born with the recessive disorder cystic fibrosis. This disease rarely affects other races. There is some evidence to suggest that the abnormally high incidence of cystic fibrosis in white populations is a result of a reproductive advantage for the heterozygotes. By what value must the reproductive potential of the heterozygotes exceed that of the dominant homozygotes in order to maintain the incidence of cystic fibrosis at its present level through heterozygote advantage? (Assume that the fitness of affected individuals is zero.)

Solution: The frequency of the allele for cystic fibrosis must be

$$q = \sqrt{\frac{1}{2500}} = \frac{1}{50}$$

At equilibrium,

$$q = \frac{s_A}{s_A + s_a} = \frac{s_A}{s_A + 1}$$

(assuming that $s_a = 1$ and that the proposed mechanism of heterozygote advantage is correct). Moreover, since $w_{Aa} = 1$, s_A in this expression (which equals $1 - w_{AA}$) must represent

▶ **FIGURE 25−7** The distribution patterns of (a) the sickle-cell allele and (b) falciparum malaria. There is a good correlation in the occurrence of the sickle-cell allele and falciparum malaria over much of their areas of distribution.

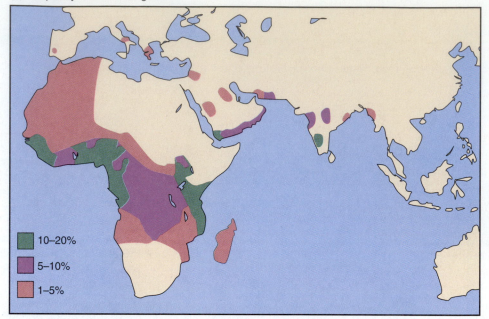

(a) Frequency of sickle-cell gene

■ 10–20%
■ 5–10%
■ 1–5%

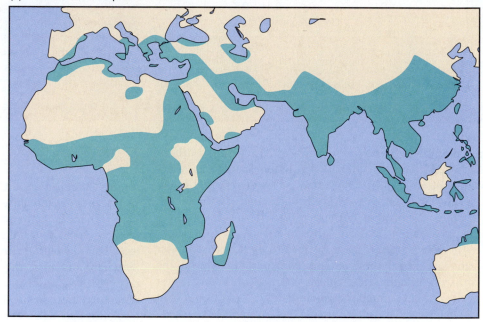

(b) Distribution of falciparum malaria

the fractional excess in fitness of the heterozygotes over the dominant homozygotes. Solving for s_A, we obtain

$$\frac{s_A}{s_A + 1} = \frac{1}{50}$$

which gives $50s_A = s_A + 1$ or

$$s_A = \frac{1}{49}$$

The reproductive potential of the heterozygotes must therefore exceed that of the dominant homozygotes by

$$\frac{1}{49} = 0.02 = 2\%$$

Follow-Up Problem 25.5

Populations that maintain genetic disorders by heterozygote advantage can incur a considerable genetic load (reduction in average fitness) through lethality, sterility, and other means that reduce the reproductive rate. For example, suppose that the relative fitness values of the genotypes *AA*, *Aa*, and *aa* are 0.6, 1.0, and 0, respectively, and that the decrease in fitness of homozygotes is due to a reduction in survival prior to reaching sexual maturity. What fraction of the zygotes produced by this population would fail to reproduce in each generation as a result of selection favoring heterozygotes?

Disassortative Mating and Frequency-Dependent Selection

Recall from Chapter 24 that disassortative mating occurs when there is a tendency for mates to be less similar in phenotype than randomly chosen individuals. Unlike the other patterns of nonrandom mating that we have considered, disassortative mating is typically associated with selection and is therefore often accompanied by allele frequency changes. The selective changes that occur under this pattern of mating are illustrated by an extreme form of disassortative mating in which only dominant × recessive matings occur. In this system, only $AA \times aa$ and $Aa \times aa$. matings are possible. The first type produces only Aa offspring, while the second type produces Aa and aa offspring in equal proportions. Thus only heterozygotes and recessive homozygotes are present among the progeny. In the next generation and in all subsequent generations, matings are therefore restricted to $Aa \times aa$ and result in an equilibrium ratio of $\frac{1}{2} Aa$: $\frac{1}{2} aa$. Observe that unless the initial frequency of recessives equals the initial frequency of dominants, selection will occur against one or more genotypes in the first generation, since not all the individuals of the genotype in excess will be able to mate. Allele frequencies are therefore expected to change and will become $p = \frac{1}{4}$ and $q = \frac{3}{4}$ at equilibrium.

The maintenance of dimorphism for style length in many flowering plants illustrates this situation. Plants such as the primrose, *Primula vulgaris*, have two flower types: *pin*, with long pistils and short stamens, and *thrum*, with short pistils and long stamens (▶ Figure 25–8). Style dimorphism in these plants is controlled by a multigene complex, or **supergene**, that consists of several closely linked loci concerned with flower morphology and pollen tube growth. Despite their dependence on the multigene complex, the pin and thrum phenotypes behave, as a rule, as though they are determined by a single pair of alleles, *S* and *s*, with pin types recessive (*ss*) and thrum types dominant (*S–*). This difference in flower morphology—a condition called **heterostyly** (meaning "different styles")—promotes outcrossing. Insects that pollinate these flowers generally transfer pollen only from thrum (*S–*) plants to pin (*ss*) plants and vice versa. These disassortative matings thus perpetuate a dimorphism between the pin and thrum types at approximately a 1:1 ratio.

The genetic equilibrium that accompanies disassortative mating results from a selective process called **frequency-dependent selection**. This form of selection occurs when the fitness values are inversely related to the genotype frequencies and leads to a situation in which a genotype gains a selective advantage when it is rare (it has a minority advantage) and is selected against when it is common. The result is a state of balanced polymorphism in which the affected genotypes remain in the population at stable equilibrium frequencies.

Several situations other than disassortative mating result in frequency-dependent selection. One is **competition for essential resources**, such as food or suitable living space. Since the resources required for continued survival and reproduction are present in limited amounts, the fitness values of the genotypes that are directly involved in the competition for them will be adversely affected. If the genotypes differ sufficiently in their resource requirements (or preferences), competition will be greater among members of the same genotype than among members of different genotypes. The fitness of each genotype will therefore decline as its members increase in abundance and the effects of competition become more severe.

Another mechanism for frequency-dependent selection

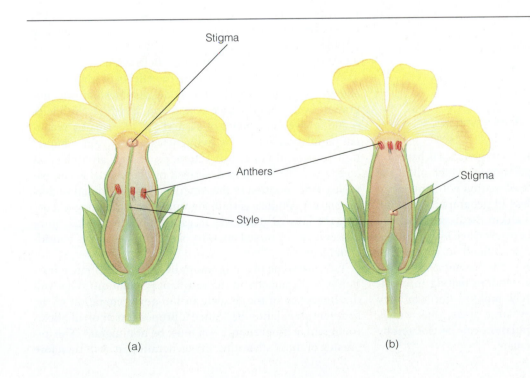

(a) (b)

▶ **FIGURE 25–8** Two types of flower morphology in the primrose (*Primula*). (a) The pin type, which has a highly placed stigma and low-placed anthers. (b) The thrum type, with the reverse morphology. This difference in style morphology, known as heterostyly, helps to prevent selfing.

is **selective predation**, in which the most frequently occurring phenotypes in a prey population are preferentially attacked by predators. This mechanism is thought to operate primarily in populations that serve as major food sources for vertebrate predators. Predators of this kind often develop a search image and learn to attack prey of the phenotype in greatest abundance preferentially. By selectively feeding on the more abundant types of prey, predators waste little time in the search and pursuit of less abundant phenotypes. Selective predation also gives the less frequently occurring forms of prey a chance to reproduce and increase their numbers until they too reach levels high enough to be favored by the predator. Thus, each distinct type of prey within a population will be at an advantage when it is rare and at a disadvantage when it is common.

To Sum Up

1. Most (but not all) mutant alleles arising in a population that is already attuned to its environment are detrimental in character and are selected against. Thus selection and mutation are often opposing forces, and their joint effects can result in an equilibrium that maintains the detrimental allele in the population.
2. Selection acting alone can lead to an equilibrium condition called a balanced polymorphism, in which both the selectively advantageous and the detrimental phenotypes are maintained at relatively high frequencies in a population. A balanced polymorphism can arise from selection favoring the heterozygote, as in the case of the sickle-cell polymorphism prevalent in parts of Africa and southern Asia. Heterozygotes for the sickle-cell allele have a selective advantage in these areas, since they do not experience the severe anemic condition characteristic of homozygotes for this allele and are more resistant to falciparum malaria than are homozygotes for the normal allele.
3. A balanced polymorphism can also arise from some form of frequency-dependent selection, in which the genotypes are at an advantage when they are rare and are selected against when they are common. Causes of frequency-dependent selection include disassortative mating, competition for essential resources, and selective predation.

✵ DIFFERENTIATION OF POPULATIONS

Much of the basic theory of population genetics is based on the idea of a large, isolated population in which all of the individuals have an equal chance of mating with any member of the opposite sex, regardless of geographic location. While such idealized conditions aid in the development of comparatively simple models, they nevertheless give rise to inconsistencies. For instance, it is difficult to conceive of a very large population, especially one having a wide geographic distribution, that is also random-mating throughout. Organisms are not infinitely mobile, nor can they transmit their gametes over very large areas; thus widely separated individuals in the population can be isolated by distance and not get a chance to mate.

Furthermore, individual members of the population often aggregate into groups, if for no other reason than to share a favorable habitat area. For example, many fish aggregate into schools, some birds form flocks, trees are often clustered in groves, and grasses are often clumped into meadows. While members of the same local group can mate at random, matings between members of different groups will occur less frequently and will depend on the migration rate. Any species can thus be thought of as being ultimately subdivided into partially isolated, local breeding populations referred to in general as **demes**.

This breeding structure—a large population subdivided into many smaller, local breeding populations—permits all the evolutionary forces (including migration) to act concurrently throughout the entire assemblage. In response to these pressures, local populations tend to diverge in allele frequencies, giving rise to interdemic differences. These differences result from the combined effects of processes that promote local differentiation, such as random genetic drift and selection in different environments, and therefore depend on the size of the breeding groups, the nature of their local environments, and the amount of migration between them. Subdivision is thus an important contributing factor to genetic variation, since it permits genetic differences to develop not only among individuals of the same deme but also among members of the different demes comprising the larger population.

Effects of Migration

Up until now, we have considered the effects of various evolutionary processes on isolated populations that do not exchange genes with neighboring groups. However, most breeding populations are not completely isolated from others of the same species, so some exchange of genes normally takes place. If populations differ in genetic composition because of evolutionary divergence, the flow of genes from one population to another through migration can then serve as an evolutionary force by changing allele frequencies.

A useful model for describing the effects of migration is shown in ▶ Figure 25–9. This model assumes one-way migration from a large source population, in which the alleles A and a have frequencies of p_m and q_m, to a smaller recipient population with allele frequencies p and q. Since migration occurs only in one direction, p_m and q_m are treated as constants. The situation described by this model is comparable to the situation in the American black population during slavery. While a certain amount of racial mixing did occur, the flow of genes was in essence unidirectional, since descendants of mixed ancestry were almost always considered black.

Let us assume that migrant individuals contribute a fraction m of all gametes in the recipient population. What will the frequency of the A allele in the next generation of the recipient population be? Since a proportion m of all alleles is migrant, a proportion $1 - m$ must be nonmigrant. The frequency of the A allele after one generation, p', will therefore

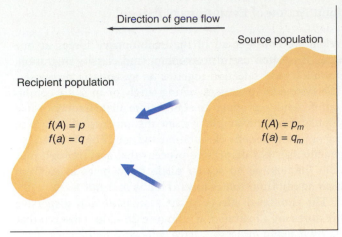

Source population

Recipient population

$f(A) = p$
$f(a) = q$

$f(A) = p_m$
$f(a) = q_m$

Direction of gene flow

▶ **FIGURE 25–9** Unidirectional recurrent migration from a large source population to a smaller recipient group. Recurrent migration of this type gradually alters the genetic composition of the recipient population in the direction of the source population.

be the average contribution of the two groups to the allele frequency: $p' = mp_m + (1 - m)p$. The change in the allele frequency in the recipient population after one generation of migration ($\Delta p = p' - p$) then becomes

$$\Delta p = -m(p - p_m) \qquad (25.7)$$

The change in allele frequency thus depends on the rate of migration, measured by m, and on the difference in allele frequencies between natives and migrants. As long as migrant genes continue to enter the recipient population, $p - p_m$ will continue to decline until $\Delta p = 0$. The allele frequency will then be the same in the two populations.

Continued migration thus counteracts genetic divergence by causing populations to become more alike in allele frequencies. If migration occurred in a subdivided population and was unopposed by any other evolutionary process, the eventual result would be a homogeneous collection of demes, all having the same allele frequencies expected if the demes had been consolidated into one large, randomly mating population.

The homogenizing effects of migration are rarely unopposed by other evolutionary processes, however. One process that promotes divergence (in opposition to migration) is natural selection. Since local populations are geographically separated, they tend to experience different environmental conditions. Selection pressures will therefore differ from group to group in promoting adaptation to local conditions. Such differential selection tends to cause divergence in the genetic structure of different groups. Migration opposes such local differentiation, but if there is some restriction to gene exchange, the different populations can retain their distinctive characteristics even with continued migration between them.

In human ethnic groups, for example, unrestricted gene exchange has been prevented by social, religious, and cultural differences as well as by geographic separation. The effect of these restrictions can be seen in the distribution pattern of the I^B allele (see ▶ Figure 25–10). This allele of the ABO system either is absent or is present in very low frequencies (5 to 10%) in the native populations of Australia and North and South America. In Africa, it is present in rather high frequencies (15 to 20%) in the equatorial regions and decreases to 10 to 15% (and even lower in certain isolated localities) in both the northerly and southerly directions. Of particular interest are the gradual changes in the frequency of the I^B allele along the east-west axis in

▶ **FIGURE 25–10** Distribution of the B allele (I^B) of the ABO blood group in the human populations of the world.

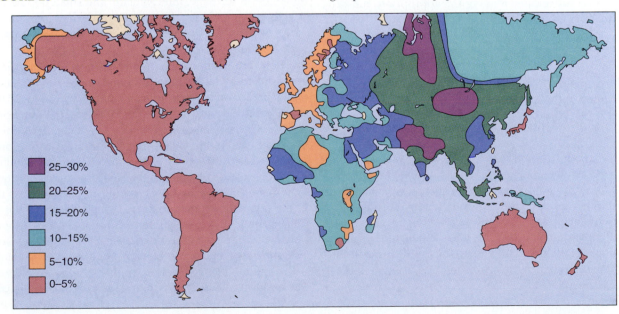

25–30%

20–25%

15–20%

10–15%

5–10%

0–5%

Europe and Asia. This gradient in allele frequency is apparently a result of invasions of the European region by the Huns and other Asiatic tribes between the fifth and sixteenth centuries. The highest I^B frequencies occur in central Asia, where the peak frequency is about 30%. The frequency gradually declines to a low of about 5% in western Europe and even less among the Basques living in the Pyrenees between France and Spain. The gradual change in allele frequency, rather than sharp discontinuities, indicates a partial breakdown of the cultural barriers to gene exchange between natives and migrants. Such a breakdown often occurs during times of war and depends to some extent on the length of the contact between the native populations and the invading armies.

Example 25.6

Because of a certain amount of racial admixture, United States blacks have European as well as African ancestry. Blood tests reveal that the frequency of the Fy^a allele at the Duffy blood group locus is 0.081 in blacks in New York City. The allele has a frequency of 0.43 in Europeans but is essentially absent from African populations. Estimate the average fraction of genes of European descent among New York City blacks.

Solution: We can treat this situation as a case of one-way migration in which the European genes have been incorporated into a base population of basically African ancestry. Let m in equation 25.7 represent the proportion of European (migrant) genes in the black population. Solving for m, we get

$$m = -\frac{\Delta p}{p - p_m} = \frac{p' - p}{p_m - p}$$

In our example, p', p, and p_m represent the respective frequencies of the Fy^a allele among New York City blacks, Africans, and Europeans. Making the appropriate substitutions, we obtain

$$m = \frac{0.081 - 0}{0.43 - 0} = 0.188$$

Therefore, about 19% of the ancestry of the black population sampled in this study is European.

Follow-Up Problem 25.6

Consider a population in which there is complete selection against recessive homozygotes. Suppose that the recessive allele enters this population at the rate of $m = 0.005$ per generation because of migration from a neighboring group. If the frequency of the recessive allele among migrants is essentially constant at $q_m = 0.01$, what will be the frequency of this allele in the population once an equilibrium is established between opposing selection and migration pressures?

Joint Action of Evolutionary Forces

When we consider all of the evolutionary forces at one time, it is often useful to combine and classify migration, mutation, and selection together as **systematic processes**. The systematic processes, acting singly or in some combination, tend to produce an equilibrium allele frequency in a population. For example, we saw this process take place in the opposing actions of mutation and selection. In contrast, genetic drift is a **dispersive process** that acts through random fluctuations to scatter allele frequencies away from their equilibrium values toward limits of 0 and 1.

The combined effects of the systematic and dispersive pressures are illustrated in ▶ Figure 25-11a. Observe that the two major processes have opposing tendencies. Ultimately, a balance is reached when the dispersion of allele frequencies owing to genetic drift is held in check by the systematic forces. The equilibrium condition that then exists can be described by a probability curve, which gives the probability that a local population has some particular allele frequency. We can think of such a curve as showing the distribution of allele frequencies that we might observe at a single point in time in a large assemblage of local populations of the same size existing under the same conditions of mutation, migration, and selection. Another way to interpret the curve is to think of the distribution as showing the variation in p (or q) that might occur under equilibrium conditions in the same population over a large number of generations. We can therefore use the model to describe both spatial and temporal interdemic variation occurring through the joint action of the evolutionary forces.

The relative importance of the dispersive and systematic forces at equilibrium depends on the effective breeding size of the populations (N) and on the magnitudes of the migration rate (m), the intensity of selection (s), and the mutation rate (μ). The effects of these forces are shown in Figure 25-11b, which displays three basic types of probability curves for large, intermediate, and small populations. Whether a population is large, intermediate, or small is based on how the products $4Nm$, $4Ns$, and $4N\mu$ compare with 1. When $4Nm$, $4Ns$, and $4N\mu$, singly or in any combination, are significantly greater than 1, the population is judged to be large. In large populations, the systematic forces predominate, and all breeding groups tend to have allele frequencies at or near the equilibrium value. Since there is little tendency for an allele to change in frequency from one generation to the next, continued evolutionary "progress" depends on the rare occurrence of new beneficial mutations and on changes in the environmental conditions.

When $4Nm$, $4Ns$, and $4N\mu$ are all significantly less than 1, the populations are considered small. In small populations, genetic drift predominates, so most of these populations will undergo fixation or loss of the allele in question. Therefore, the curve for a small population is U-shaped, with 0 and 1 being the most probable allele frequencies. The resulting homozygous populations are so depleted of

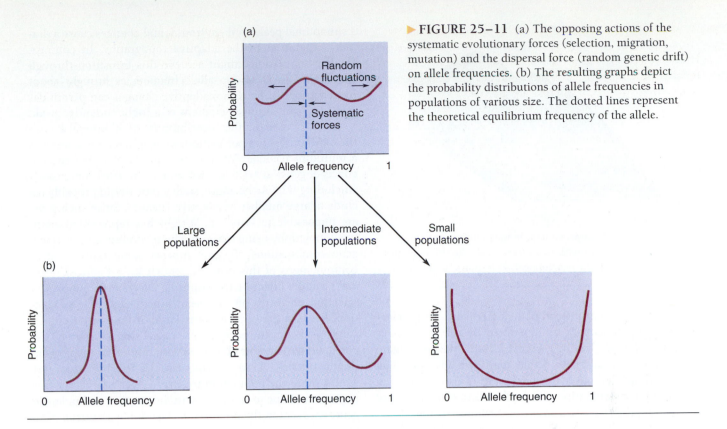

(a)

Random
fluctuations

Systematic
forces

Probability

0   Allele frequency   1

▶ **FIGURE 25–11** (a) The opposing actions of the systematic evolutionary forces (selection, migration, mutation) and the dispersal force (random genetic drift) on allele frequencies. (b) The resulting graphs depict the probability distributions of allele frequencies in populations of various size. The dotted lines represent the theoretical equilibrium frequency of the allele.

Large
populations

Intermediate
populations

Small
populations

(b)

Probability

0   Allele frequency   1

Probability

0   Allele frequency   1

Probability

0   Allele frequency   1

genetic variation that they would be unable to respond to any change in selective pressures, and the end result would be extinction.

When $4Nm$, $4Ns$, and $4N\mu$ are intermediate in value, both dispersive and systematic forces are important in defining the equilibrium condition. Thus populations of intermediate size show an appreciable scatter of allele frequencies, although a maximum probability occurs in the vicinity of the equilibrium value. In such populations, the allele frequencies tend to vary but are still influenced by systematic pressures. This situation provides for considerable genetic divergence without depleting the local populations of the allelic variation needed to form adaptive combinations of genes.

Adaptive Landscapes: The Shifting Balance Theory

The preceding analysis considered evolutionary change in terms of the joint action of evolutionary forces on a single gene locus. However, it should be apparent that evolution involves changes at many loci, not just one. The most comprehensive analysis of such changes was conducted by the geneticist Sewall Wright, using a model that he called the **shifting balance theory**. In this model, the average fitness of a population is represented by a series of peaks and valleys in a two-dimensional field of possible gene combinations. The stepwise construction of the model can be shown by first considering a single locus with two alleles. Assume that heterozygotes are selectively superior so that the mean

fitness can be given as $\overline{w} = 1 - s_A(1-q)^2 - s_a q^2$ (see equation 25.6). This equation is graphed in ▶ Figure 25–12, where the arrows show how selection would change allele frequency. Note that selection acts to maximize the average fitness of the population. Thus, q will change in such a way that the population moves up the "hill" to the point where $\overline{w}$ is maximum.

▶ **FIGURE 25–12** A two-dimensional plot showing improvement in mean fitness for the case where heterozygotes are favored. The arrows indicate the movements of a large population in response to selection to attain an equilibrium allele frequency $\hat{q} = s_A/(s_A + s_a)$, which corresponds to the maximum value of $\overline{w}$.

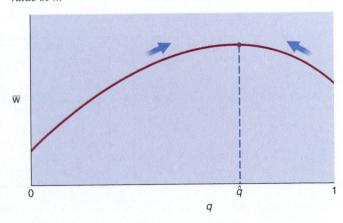

$\overline{w}$

0                    $\hat{q}$ 1

q

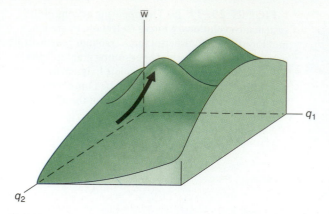

▶ **FIGURE 25–13** Improvement in mean fitness in a three-dimensional fitness plot with two fitness peaks. In this case, the population moves up the nearest peak in response to selection.

Now suppose that instead of a single locus, we consider two loci simultaneously, as illustrated by the $\overline{w}$ versus q_1 and q_2 plot for two loci in ▶ Figure 25–13. We now obtain a three-dimensional surface that may contain more than a single peak. Although the peaks are not necessarily of equal height, a population will find itself in the gradient of one of the peaks and will move accordingly in that direction. If there is another peak with a higher value of $\overline{w}$, selection alone cannot achieve the transition to the higher fitness value. A large population can thus be stranded on

a suboptimal peak until environmental changes cause a dramatic alteration in the adaptive topography. In contrast, smaller populations might achieve this transition through the chance variations in allele frequencies brought about by genetic drift. Thus, nonadaptive changes can permit descent into the stronger gradient of a higher adaptive peak.

In order to show the contributions of additional loci to the mean fitness of a population, we would have to construct a multidimensional plot. For example, a plot showing the effects of genes at 1000 loci would require 1001 dimensions (including $\overline{w}$). Furthermore, such a plot would probably include a large number of adaptive "peaks." Since such plots are impossible to visualize, Wright has represented adaptive topography using maps in which contour lines connect gene combinations of equal fitness. Some representative contour maps of this type are shown in ▶ Figure 25–14. Parts a and b illustrate the opposing effects of mutation and selection pressures on the mean fitness of a large population. Because the population is large, it tends to occupy a single peak, but the portion of the peak that is occupied depends on the amount of genetic variability that the population maintains. Thus, an increase in the mutation rate would cause the population to spread over a larger surface (as in part a) due to an increase in the number of ill-adapted genotypes within the population. The opposite is true of an increase in selection intensity (part b).

Figure 25–14c illustrates the effects of environmental change. What was once a fitness peak may now be a valley.

▶ **FIGURE 25–14** Sewall Wright's shifting balance theory, in which multidimensional fitness plots are represented by contour maps. The dashed lines connect gene combinations of equal fitness. Populations are shown in green.

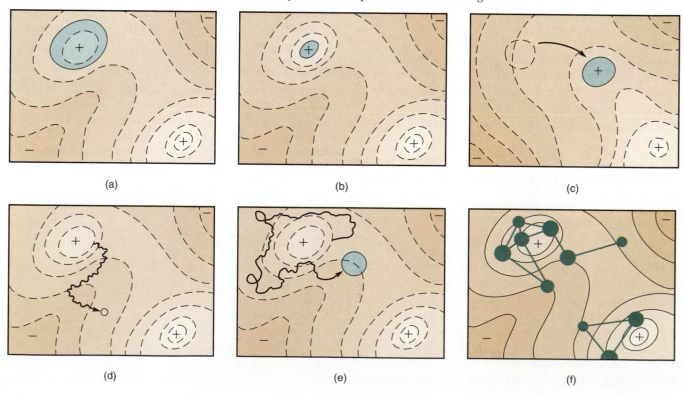

If the population contains sufficient genetic variability, it will respond to these selective changes and move up a neighboring peak. An adaptive response of this type is the primary mode of evolutionary change in large populations.

In Figure 25–14d we see the consequences of small population size. In this case, nonselective changes in allele frequencies move the population down the slope into a neighboring valley. Although the population may enter the gradient of a higher adaptive peak, the random fixation and loss of genes has so depleted its genetic variability that it is incapable of responding to new selective pressures. The result is extinction.

Finally, Figures 25–14e and f show an effective means of moving from peak to peak without the necessity of environmental change. The population in part e is intermediate in size, so both random and systematic forces are assumed to act. In this case, the population might wander down the slope through random changes in allele frequencies, but because of relatively strong selection pressures, it still remains in the vicinity of an adaptive ridge, where it might encounter the gradient of a higher peak. This type of genetic change is very slow. A more favorable condition is shown in part f, where the population is assumed to be divided into numerous demes of intermediate size with migration between them. All evolutionary forces, including migration, act on this subdivided population, so progress is considerably more rapid, with different demes coming under the influence of different peaks. The increase in heterogeneity afforded by subdivision into local groups thus enhances the overall rate of evolutionary change. When the environment itself is changing in time and space, the variability present in the gene pools of the semi-isolated groups offers greater survival opportunity and provides a mechanism for the rapid occupancy of all available adaptive peaks.

Example 25.7

One important genetic effect associated with the formation and subsequent divergence of isolated subpopulations is an overall increase in the average frequency of homozygous genotypes compared with the frequency expected for a single, freely intermating population. The increase in homozygosity that accompanies the formation of isolates—a phenomenon called **Wahlund's principle**—can be reversed in certain cases by the fusion of the formerly isolated groups. The problem we will consider here illustrates the genetic effects associated with the formation and fusion of isolates.

Suppose that two equal-sized isolates are formed from a large randomly mating population in which the allele frequencies are $p = q = \frac{1}{2}$. Assume that the respective frequencies of genotypes *AA, Aa,* and *aa* are 0.64, 0.32, and 0.04 in isolate 1 and 0.04, 0.32, and 0.64 in isolate 2. Compare the average genotype frequencies in the two isolates with the frequencies that the genotypes would have if the two

isolates were consolidated into a single, randomly mating population.

Solution: The average genotype frequencies for the two isolates are $f(AA) = (0.64 + 0.04)/2 = 0.34$, $f(Aa) = (0.32 + 0.32)/2 = 0.32$, and $f(aa) = (0.04 + 0.64)/2 = 0.34$. The average allele frequencies are then $f(A) = f(a) = 0.34 + 0.32/2 = 0.5$, the value expected for isolates that are random samples of the parent population. If the isolates were to be consolidated into one randomly mating group, the genotype frequencies would become $f(AA) = (0.5)^2 = 0.25$, $f(Aa) = 2(0.5)(0.5) = 0.50$, and $f(aa) = (0.5)^2 = 0.25$. Note that the overall frequency of homozygotes is greater among the isolated groups than in a freely intermating population with the same allele frequencies. As this example shows, the breakdown of isolating barriers can lead to a reduction in the overall frequency of harmful recessive traits.

Follow-Up Problem 25.7

Hybrids that are produced by crossing different inbred (homozygous) lines often show an increase in characteristics relating to size, growth rate, and fertility. This increase in performance by hybrids is called **heterosis** or **hybrid vigor**. While heterosis is best known among cultivated plants, such as corn, in which the increases in yield are of particular economic benefit, it has also been reported for a variety of other, less familiar plant and animal species. The widespread occurrence of heterosis has led some geneticists to postulate that it could be, in part, responsible for the increase in human height that has been observed over the past several decades. (a) Assuming that heterosis is an important factor in promoting human growth, use Wahlund's principle to explain how past changes in human population structure and mobility could have led to an increase in height. (b) What other genetic and nongenetic factors could have contributed to an overall increase in human height?

Formation of Species

The evolutionary processes that promote differentiation of populations of the same species are also responsible in large part for the diversification of different species. To a population geneticist, the species is the most inclusive population. Each species, as we have seen, is typically subdivided into numerous local populations, or demes, that are linked by their potential for gene exchange. Populations of the same species are capable of interbreeding, but they do not ordinarily exchange genes with populations of other species. Different species are therefore said to be **reproductively isolated**.

Several mechanisms prevent interbreeding between species. Usually, more than one of these mechanisms operates in any single case. As shown in ■ Table 25–5, these isolat-

■ **TABLE 25–5** Important reproductive isolating mechanisms.

Prezygotic mechanisms: Fertilization and zygote formation are prevented.
1. Ecological: The populations live in the same region but occupy different habitats or become sexually mature at different times.
2. Behavioral (in animals): Incompatible mating behavior prevents mating.
3. Anatomical: Incomplete structure of reproductive organs (genitalia in animals, flowers in plants) prevents mating.
4. Physiological: Gametes fail to survive in alien reproductive tracts.

Postzygotic mechanisms: Zygotes are formed but fail to survive or fail to produce viable and fertile offspring.
1. Hybrid inviability.
2. Hybrid sterility (partial or complete): Hybrids fail to reproduce because gonads fail to develop properly or chromosomes (or chromosome segments) segregate abnormally during meiosis.
3. Inviability and/or sterility of $F_2:F_1$ hybrids are normal, but F_2 hybrids include many weak or sterile individuals.

ing mechanisms act at two levels: **prezygotic** and **postzygotic**. The prezygotic isolating mechanisms include various anatomical, physiological, ecological, and behavioral barriers to mating or cross-fertilization between species; these barriers prevent the formation of hybrid zygotes. The postzygotic isolating mechanisms act after fertilization has taken place and result in the formation of inviable, aberrant, or sterile hybrids. The postzygotic mechanisms are accidental byproducts of the genetic differentiation that occurs when populations adapt to different conditions. As the populations diverge in genetic character, their genes are less likely to act harmoniously in a hybrid. Chromosome mutations can also arise and can lead to problems during meiosis in hybrid individuals (see Chapter 12). In any case, the hybrids of such matings would be selected against. Although the production of unfit hybrids serves the same general purpose of preventing gene exchange as does the inability to form hybrids, it is nevertheless more wasteful of reproductive effort. In regions where different species are in contact, we would therefore expect natural selection to promote the development of prezygotic barriers to gene exchange in populations that are already isolated from one another by postzygotic mechanisms.

Because of these intrinsic barriers to gene exchange, species behave as separate evolutionary units and take different evolutionary paths. Two major types of changes in species are generally recognized—**phyletic evolution** (or **anagenesis**) and **speciation** (or **cladogenesis**). Phyletic evolution occurs within a lineage or single line of descent. The phyletic component of species modification represents the gradual change in the overall genetic makeup of an already established species due to systematic and dispersive evolutionary forces, and it is indicated by each separate line of the

tree diagram in ▶ Figure 25–15. Speciation, on the other hand, is the splitting of a lineage into two or more separate lines of descent and is represented by the branch points in Figure 25–15. Speciation thus results in a multiplication of the number of existing species. It is the single most pervasive pattern of evolution.

How does speciation occur? The prevailing view is that speciation can take place during periods when gene flow between the segments of a population is cut off or at least substantially reduced. According to this view, extrinsic barriers to gene flow enable population segments to diverge genetically to a point where they acquire one or more reproductive isolating mechanisms (▶ Figure 25–16). The extrinsic barriers to gene flow can include geographic barriers, such as deserts, rivers, mountains, and so on (Figure 25–16a), or possibly a simple switching of a parasite from one type of host to another (Figure 25–16b). Once gene flow is restricted, genetic differentiation can occur freely until normal interbreeding is impossible. At this point, the two groups can be regarded as separate species.

The development of complex reproductive isolating mechanisms simply as a byproduct of genetic differentiation can take thousands or even millions of years. Speciation by this route is therefore likely to be gradual. In contrast, mechanisms of species formation involving chromosome rearrangements can occur over relatively short periods of time. For example, polyploidy, as we observed in Chapter 12, can give rise to a new species in just one or a few generations. These accelerated modes of speciation (often

---

▶ **FIGURE 25–15** Hypothetical phylogeny showing the origin and diversification of species. Each branch point represents a speciation event, while the changes that occur within each line of descent form the phyletic component of evolution.

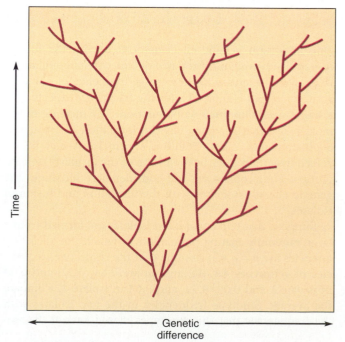

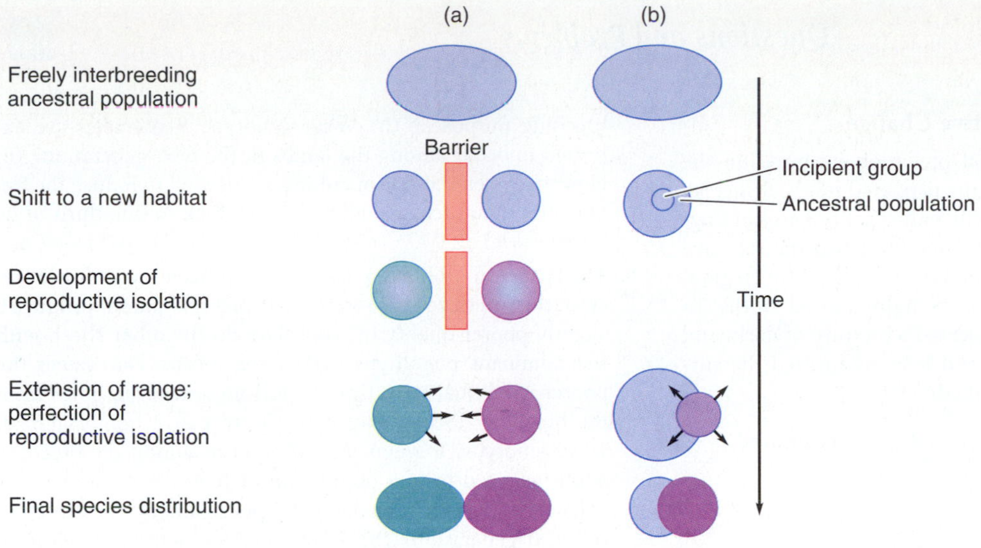

FIGURE 25–16

FIGURE 25–16 Two general mechanisms of speciation. (a) Reproductive isolating mechanisms evolve after the population has been divided by a geographic barrier. (b) Reproductive isolation evolves while the incipient group is still in contact with the ancestral population, for example, during the switching of a parasite to a new host.

(a)   (b)

Freely interbreeding ancestral population

Barrier

Shift to a new habitat

— Incipient group
— Ancestral population

Development of reproductive isolation

Time

Extension of range; perfection of reproductive isolation

Final species distribution

---

called **quantum speciation** to distinguish them from more gradual mechanisms) typically occur through chromosome mutations that result in postzygotic isolating mechanisms and thus allow new species to form at a very rapid rate.

To Sum Up

1. Species are often subdivided into smaller local breeding populations called demes. This breeding structure allows all of the evolutionary forces to act concurrently and is an important contributing factor to genetic variation; it permits genetic differences to develop between the individuals of different demes, as they adapt differentially to their various local environments, as well as between individuals of the same deme.

2. Migration can lead to a flow of genes from one local population to another and can thus serve as an evolutionary force by changing allele frequencies. If migration were to proceed in the absence of any other evolutionary force, it would eventually eliminate any genetic differences between populations.

3. Migration, mutation, and selection are systematic processes that, acting alone or jointly, tend to produce an equilibrium frequency for each affected allele. The effects of these systematic processes tend to be opposed by the dispersive process of genetic drift, which scatters allele frequencies away from their potential equilibrium values toward values of 0 or 1. The combined effects of systematic and dispersive processes depend, among other things, on the size of the breeding groups, the nature of their local environments, and the amount of migration between them.

4. A species consists of one or more populations that are actually or potentially interbreeding but are reproductively isolated from other such groups. New species arise when physically separated populations of an existing species accumulate sufficient genetic differences to prevent interbreeding if they should come into contact at some later time.

Chapter Summary

The process of evolution requires genetic change. Mutation, migration, genetic drift, and natural selection modify allele frequencies and are thus the basic forces responsible for evolution. Natural selection—the differential reproduction of genotypes—is the principal mechanism of adaptive change; it occurs through genotype-specific differences in survival and fertility in a particular environment and is measured in terms of the fitness (the relative reproductive contribution) of the various genotypes.

The effect of natural selection on allele frequency depends on the fitness and frequencies of the genotypes and thus on the degree of dominance and frequency of the allele. In general, selection is most effective as an evolutionary force when there are large differences in fitness between the genotypes, the mean fitness of the population is small, and the allele frequency is close to neither 0 nor 1. If selection occurs against a detrimental recessive or dominant allele that is introduced into the population by recurrent mutation, the frequency of the allele will decline until an equilibrium is produced between the opposing selection and mutation pressures. An equilibrium can also be established when selection favors heterozygotes, as in the case of the sickle-cell trait, or when a form of frequency-dependent selection gives a genotype a minority advantage, giving rise to a balanced polymorphism in which both alleles are maintained in the population.

Species are often subdivided into local breeding populations called demes. Partitioning the gene pool into semi-isolated breeding populations promotes evolutionary divergence, since it permits all evolutionary forces to act concurrently and in a differential manner on each local group. New species arise when migration between populations is restricted. Genetic divergence can then occur freely until sufficient differences accumulate in the isolated populations to prevent interbreeding.

Reproductive Fitness and Selective Change

1. Kettlewell studied the differential predation by birds on the light- and dark-colored forms of the peppered moth, *Biston betularia,* in a woodland near a heavily industrialized area of England. It was proposed that the melanic form had the selective advantage against the soot-darkened tree trunks of this region, since it was less conspicuous than the light-colored moths. To test this hypothesis, Kettlewell released a mixture of dark- and light-colored moths in the area and later recaptured the survivors. The following data were obtained:

	Released	Recaptured
Dark-colored moths	154	82
Light-colored moths	73	16

Use these results to calculate the relative fitness values of the light- and dark-colored moths in terms of their differential survival in this industrialized area.

2. Dobzhansky studied changes in the frequencies of two gene arrangements, Standard (*ST*) and Chiricahua (*CH*), on chromosome III of *Drosophila pseudoobscura.* He determined the number of each genotype in both the egg samples and the freshly emerging adults and obtained the following data:

	CH/CH	CH/ST	ST/ST	Total
Egg sample	42	88	20	150
Young adults	16	83	31	130

Assuming that the egg sample represents the frequencies before selection, determine the relative fitness values of each genotype for young adults.

3. It is fairly easy to measure fitness for asexually reproducing haploid populations such as bacteria. If bacterial strains 1 and 2 grow exponentially, their numbers at time *t* will be

$$N_1 = N_{1,0}e^{w_1 t} \quad \text{and} \quad N_2 = N_{2,0}e^{w_2 t}$$

where w_1 and w_2 are the corresponding fitness values. (a) Show how w_1 and w_2 relate to the generation (doubling) times of the two strains. (b) Suppose that we inoculate 100 ml of nutrient medium with equal numbers of bacterial strains 1 and 2. Two hours later, we observe that strain 1 cells outnumber strain 2 cells by a factor of 4. Compute the difference between the fitness values of the two strains.

Dynamics of Selection

4. In a population of 100,000,000 people, 10,000 have a serious genetic disease caused by a recessive allele. If these individuals are kept from reproducing, how many generations would it take for the frequency of this disorder to be reduced to 1 in 1,000,000 persons? (Ignore the effects of mutation.)

5. Suppose that a farmer decides to cull recessive homozygotes from his flock of sheep because of the poor quality of their wool, leaving only dominant homozygotes and heterozygotes free to mate. The farmer starts with a large flock of lambs consisting of $\frac{9}{16}$ *AA,* $\frac{6}{16}$ *Aa,* and $\frac{1}{16}$ *aa.* (a) What genotype frequencies are expected to occur among the adult sheep that are left for breeding purposes? (b) What genotype frequencies are expected to occur among the lambs in the next generation? (c) How many generations of culling are needed to reduce the frequency of the recessive allele in this flock to one-third of its initial value?

6. The farmer in problem 5 discovers that heterozygotes who receive the recessive allele from their paternal parent produce a slightly poorer quality of wool than do the other sheep with the dominant phenotype. (a) If the farmer can detect the poorer wool quality in time to cull these heterozygotes from the breeding flock (along with the recessive homozygotes), what genotypic frequencies will appear among the breeding adults selected from a population of $\frac{9}{16}$ *AA,* $\frac{6}{16}$ *Aa,* and $\frac{1}{16}$ *aa?* What genotype frequencies will appear among the lambs of the next generation? (b) Show that the method of selection described in this problem will reduce the frequency of the recessive allele by one-half in each generation. (c) Propose a suitable mechanism for the apparent paternal effect on wool quality. Explain how your hypothesis might be tested.

7. Calculate (a) Δq in equation 25.4 and (b) Δp in equation 25.5 for allele frequencies of 0, 0.2, 0.3, 0.4, 0.6, 0.7, 0.8, and 1 and $s = 0.01$. Plot your results in graphical form as Δq versus q and Δp versus p. Describe any differences in the selective efficiencies of the two models and indicate the range of allele frequencies at which each type of selection is most effective.

8. Show that with complete selection against a recessive homozygote, the average fitness of the population becomes $p(1 + q)$.

9. The models described in this chapter for the selective elimination of an allele were restricted to cases of complete dominance. Suppose that selection is acting on an allele that is incompletely dominant, so that the heterozygotes have a fitness value that is exactly intermediate between the homozygous genotypes (i.e., $w_{AA} = 1$, $w_{Aa} = 1 - s$, and $w_{aa} = 1 - 2s$). Using the general selection model (equation 25.3), derive the equation for selection against an incompletely dominant gene.

Equilibria and Polymorphism

10. Distinguish between the equilibrium associated with a balanced polymorphism (such as the one involving the sickle-cell gene) and the random-mating equilibrium described by the Hardy-Weinberg law. In what ways are they similar?

11. Assume that the rate of forward mutation ($A \xrightarrow{\mu} a$) is three times faster than the rate of back mutation ($a \xrightarrow{v} A$). What are the expected frequencies of the two alleles once an equilibrium is established between forward and back mutation?

12. Ichthyosis congenita is a recessive lethal condition in humans that is characterized by abnormal, leathery skin with deep, bleeding fissures. This abnormality arises with a mutation rate of approximately 10^{-5} per gamete per generation in human populations. What is the most likely frequency of the heterozygous carriers of the recessive lethal gene for this disorder?

13. Would you expect a recessive lethal gene that is carried on the X chromosome to be maintained by the opposing mutation and selection pressures at a higher or lower frequency than a recessive lethal gene that is carried on an autosome? Explain your answer.

14. Neurofibromatosis, the syndrome that is the subject of the film and play *Elephant Man*, is a dominant genetic disorder that is characterized by tumorlike formations on the skin and in the nervous tissue. The autosomal gene for this disorder has one of the highest known mutation rates in humans, approximately 10^{-4} per gamete per generation. (a) Show that among 10^6 births, 200 new cases of neurofibromatosis are expected to occur as a result of mutation. (The trait and hence the gene for this abnormality are rare in the population.) (b) It has been estimated that individuals with neurofibromatosis have only half the reproductive potential of homozygotes for the normal allele. What is the expected frequency of the disorder, assuming a balance between selection and mutation? (Hint: Since the frequency of the allele is rare, you can assume that essentially all individuals born with this disorder are heterozygous.)

15. Suppose that a program to control malaria in certain parts of Africa eliminates the selective advantage of heterozygotes carrying the sickle-cell gene. Assume that heterozygotes comprised 20% of the adults in the affected regions prior to the institution of malaria control. If the reproductive capacity of recessive homozygotes (individuals with sickle-cell anemia) is essentially zero, how many generations of malaria control are required to reduce the frequency of the heterozygotes to one-fifth of the initial value?

16. Studies of the allele for sickle-cell hemoglobin (Hb^S) revealed that in a certain region in Africa, the genotypes $Hb^A Hb^A$, $Hb^A Hb^S$, and $Hb^S Hb^S$ had fitness values of 0.84, 1.0, and 0, respectively. (a) What are the equilibrium frequencies of the Hb^A and Hb^S alleles? (b) Assuming that selection in this case involves differential survival between birth and adulthood, what are the expected frequencies of the three genotypes among infants at birth? What are they for adults?

17. In natural populations, random changes in gene frequencies occur not only as a result of chance variation in the segregation ratios of alleles but also because of the random nature of survival. For example, suppose that the zygotes produced by a population consist of an equal mixture of *AA* and *Aa* genotypes and that individuals of both genotypes have a 10% chance of surviving to the reproductive age. Among 10 surviving adults, 6 are *AA* in genotype and 4 are *Aa*. (a) Calculate the change in allele frequency that has occurred between the zygote and adult stages. (b) Calculate the probability of this change occurring. (c) Does your answer in part (a) represent a selective or nonselective change in allele frequency? Explain.

*18. Consider a case of frequency-dependent selection in which the genotypes *AA*, *Aa*, and *aa* have the following relative fitness values: 1, 1, and $1 + s - tq$, where s and t are constants and q is the frequency of the *a* allele. (a) Using the general selection model (equation 25.3), derive the equation for the one-generation change in q in terms of s, t, and q. (b) Show that an equilibrium will be established in this case and show graphically whether the equilibrium is stable or unstable. (c) Determine the s/t range suitable for equilibrium.

*19. This question deals with selection against heterozygotes (see the Extensions and Techniques feature in this chapter). One example of selection against heterozygotes in the human popu-

lation involves a severe hemolytic anemia (called erythroblastosis fetalis) that can develop when an Rh-positive offspring is born to an Rh-negative mother whose immune system has been sensitized to the Rh factor during an earlier pregnancy. Each of these Rh-positive offspring must have received a recessive allele from his or her Rh-negative mother, so the effect is restricted to heterozygotes. (a) Given that selection against heterozygotes can lead to a stable state only when the value of q reaches either 0 or 1, suggest reasons why both the Rh alleles are still present in Caucasian populations at relatively high frequencies. (b) In 1942, Haldane suggested that Rh-negative mothers might compensate for the loss of Rh-positive children through hemolytic disease by producing more offspring than they normally would. Describe how such compensation could serve to balance the losses from hemolytic anemia and also retain both alleles in the population.

Differentiation of Populations

20. Distinguish between a race, a species, and a deme. Rank them in order of increasing complexity.

21. Some species, called sibling species, are so similar in appearance that they are difficult to distinguish by most morphological criteria. Why would a biologist classify such obviously similar organisms as different species and yet group others as diverse as the Chihuahua and the Saint Bernard breeds of dogs into one species?

22. Distinguish between the dispersive and the systematic evolutionary forces. How does a combination of systematic and dispersive pressures allow the development and maintenance of variation both within and between local breeding groups?

23. A selectively neutral allele does not impart a selective advantage or disadvantage on its bearer. Explain how a neutral allele could increase in frequency and eventually become fixed in a population through (a) migration, (b) mutation, and (c) genetic drift. Could selection ever be involved in changing the frequency of a neutral allele? Explain why or why not.

24. Studies of a local frog population revealed that 9% of the frogs exhibit the recessive spotted condition. In each generation, 5% of this population is derived by migration from a large neighboring population that is homozygous for the nonspotted phenotype. (a) If frogs mate at random for spotting, what is the expected frequency of spotted frogs in the local population after one generation of migration subsequent to the study? after two generations? (b) What proportion of spotted frogs is expected to appear eventually in the local population if migration continues for a very long time with no other forces acting to change the gene frequency?

25. Suppose that two communities, each with different frequencies of M, MN, and N blood types, combine to form one freely intermarrying population. Blood tests reveal that in 4% of all marriages in this amalgamated group, both husband and wife are type M, in 16% both are type MN, and in 16% one is type M and the other is type MN. (a) From these results, calculate the expected frequency of the M blood type among the progeny of this population. (b) Given that mating is random for blood type, calculate the frequencies of the M, MN, and N blood types among the adults, just after amalgamation of the two communities. (c) What allele and genotype frequencies are expected among the first generation of progeny of this population?

*An asterisk denotes that the question or problem uses material in the Extensions and Techniques text.

26. Height in a particular plant is influenced strongly by the presence of the two alleles *A* and *a*. When the plants are fully grown, the average heights of the three genotypes are 28 cm (*AA*), 48 cm (*Aa*), and 36 cm (*aa*). A population of this plant species is subdivided into two isolates of equal size. In isolate I, the frequency of allele *A* is 0.1; in isolate II, it is 0.9. (a) Assuming random pollination within each isolate, calculate the average plant height in each isolate. (b) Calculate the average height for the entire population (including both isolates). (c) Suppose that there is a breakdown of the barriers separating isolates I and II, so that they combine into one large, randomly mating population. Calculate the average plant height for the combined population after a generation of random mating. (d) Explain the genetic basis for the difference in average plant height in the subdivided and freely intermating populations.

27. The fundamental theorem of natural selection, first described by R. A. Fisher in 1930, states that in a large population, the change in the mean fitness ($\Delta\bar{w}$) is equal to the population's existing genetic variance in fitness (σ_w^2). Since the variance cannot be negative, the fundamental theorem implies that the mean fitness of a population will never decline but will always increase to a maximum under natural selection. Using Wright's shifting balance theory, explain how the fitness of a population can continue to change, even after a maximum is reached, in apparent contradiction to the fundamental theorem.

Molecular Evolution

Evolutionary relationships among organisms have long been depicted by phylogenetic trees based mainly on fossil records and comparative morphology. Because of the incomplete nature of the fossil record and the lack of knowledge of the genetic and evolutionary significance of morphological differences, many phylogenies based on these traditional methods are open to controversy. Once Watson and Crick provided a molecular concept of the gene, it was only natural for researchers to investigate evolutionary relationships at the molecular level. In the last three decades, a plethora of molecular biological studies of evolution have focused on evolutionary history, the genetic bases of major evolutionary events such as speciation, and the nature and kinds of genetic variation present in natural populations. Most of these studies fall into two main types: one in which the goal has been to identify and quantify genetic differences at the molecular level, using molecular biology as

A guinea pig and a rat. Molecular studies are attempting to determine whether the guinea pig is a rodent.

a tool to augment classical evolutionary investigations, and a second in which the object has been to investigate the molecular mechanisms responsible for these differences.

In this chapter, we will discuss the results of both types of studies. We will first consider the use of molecular techniques to measure the extent of genetic variation in different groups of organisms and follow this with a discussion of the possible adaptive importance of this variation to natural populations. We will then look at the mechanisms of molecular evolution and the use of molecular data in inferring phylogenetic relationships among different taxonomic groups and in reconstructing the evolutionary history of these organisms.

▓ MOLECULAR POLYMORPHISMS

The experimental measurement of genetic variability in natural populations has long been central to the study of evolution. The major questions in this area deal with how much genetic variation exists in natural populations, how much of it is relevant to natural selection, and how measurements of variation can be used to discover and describe evolutionary trends and patterns.

Prior to 1960, the study of the variation concealed in populations was based mainly on the genetic analysis of lethal and morphological traits. The detection of such variation relies on traditional genetic analysis using genetic crosses that promote homozygosity to expose the presence of recessive mutant genes. Then, in the mid-1960s, experimental developments in molecular genetics made it possible to search for genetic variability in natural populations by analyzing the protein products of structural genes. With the appropriate molecular techniques, amino acid changes can be detected in heterozygotes regardless of whether they produce an observable difference in external appearance. By evaluating the phenotype closer to the level of the gene itself, it is possible to bypass many of the problems associated with dominance and measure the proportion of heterozygous gene loci directly. Moreover, if the protein samples are randomly chosen with respect to function, then the genes that encode these proteins can be taken to represent a random sample of all structural gene loci. Thus, unlike traditional procedures, protein analyses are not biased specifically toward any particular kind of gene or phenotypic effect (although they are biased toward structural genes—a point that we will return to later).

The ideal approach to identifying protein differences is to determine the complete amino acid sequences of the polypeptide chains being compared. Although it has been accomplished in certain cases, complete amino acid sequencing is normally too expensive and too time-consuming to be performed on a wide enough scale to study intrapopulational variability (current developments in adapting mass spectrometry to protein sequencing may change this situation). Therefore, we usually have to be satisfied with swifter

and less costly methods that give only partial information on amino acid changes.

One common alternative procedure is gel electrophoresis, which is limited in the amount of protein variability that it can detect. Recall that gel electrophoresis can distinguish only between protein variants that differ sufficiently in charge (and, to a certain extent, in size and shape) to become separated in an electric field. For instance, the replacement of a negatively charged amino acid with one that is positive or neutral (such as the replacement of glutamic acid with lysine or glycine) would probably alter the rate of migration of the protein sufficiently to be detected; however, the replacement of an amino acid with one of the same charge (such as the replacement of glycine with alanine) might not. Since only about one-third of all amino acid replacements result in a change in charge, gel electrophoresis tends to underestimate the amount of protein variability. It is also limited by the availability of suitable stains for differentiating various proteins. However, since a staining technique has nothing to do with the extent of a protein's variability or its affect on phenotype, this limitation is usually not a problem as long as the loci chosen for analysis represent a random sample of all structural genes.

▶ Figure 26–1 shows the idealized electrophoretic patterns expected for molecular variants of monomeric and dimeric enzymes. These molecular variants, called **allozymes**, are alternative enzyme forms that are coded for by different alleles at the same locus. ▶ Figure 26–2 shows the results of an actual study in which a wild population of *D. pseudoobscura* was analyzed for electrophoretic variation at the *esterase-5* locus. In this study, samples 1, 2, 3, 5, and 6 represent homozygous genotypes: sample 1 contains the standard allele (*est-5* $^{1.00}$), samples 2, 3, and 6 contain a variant allele denoted *est-5* $^{0.95}$, and sample 5 has another variant allele, *est-5* $^{1.12}$. Sample 4 contains protein from a genotype heterozygous for the alleles *0.95* and *1.12*. The three bands in sample 4 indicate that esterase-5 is a dimer, so three subunit associations are possible in a heterozygote: *0.95–0.95*, *1.12–1.12*, and *0.95–1.12*. The latter has an electrophoretic mobility midway between the two homozygous forms. Notice that alleles at this level of analysis show only codominance—each allele expresses its protein product. Samples 10 and 11 may also represent heterozygous genotypes, since the bands from a heterozygote appear as a diffuse thick band, rather than as two distinct bands, when the mobilities of the allozymes are not too different.

Population geneticists have recently turned their attention to nucleic acids, since all genetic variation is ultimately due to differences in nucleotide sequences. Unlike electrophoretic analysis of polypeptides, the detection of variation by nucleotide sequencing is not restricted to structural genes. Determination of the nucleotide sequences in a random sample of DNA regions therefore gives an unbiased measure of the degree of genetic variation in a population, regardless of the function (if any) of the regions studied. Although nucleotide sequencing is fast becoming a popular

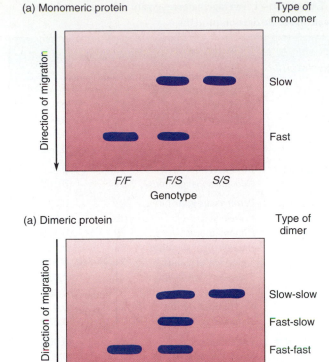

(a) Monomeric protein

Direction of migration

Type of monomer

Slow

Fast

F/F   F/S   S/S
Genotype

(a) Dimeric protein

Direction of migration

Type of dimer

Slow-slow

Fast-slow

Fast-fast

F/F   F/S   S/S
Genotype

▶ **FIGURE 26–1** Electrophoretic patterns of single proteins coded by loci having two alleles. (a) A monomeric protein, which consists of a single polypeptide, has only two forms in heterozygotes: a fast-migrating form associated with allele *F* and a slow-migrating form associated with allele *S*. (b) A dimeric protein, which consists of two polypeptide chains, can have three forms in heterozygotes: a slowly migrating homodimer (slow-slow), a rapidly migrating homodimer (fast-fast), and a heterodimer (fast-slow) with an intermediate electrophoretic mobility.

technology, particularly for the development of molecular phylogenies (to be discussed later), the large-scale sequencing that would be necessary to survey at least 15 to 20 loci in each of several populations is not yet practical in terms of time and cost.

Restriction enzyme techniques provide a simpler way to estimate the extent of DNA variability within a population. Although the analysis is not as complete as would be obtained from sequencing, measurement of restriction fragment length polymorphism (RFLP, see Chapter 15) is much less costly and time-consuming. Variation in organellar or nuclear genomes can be studied, although the large quantity of DNA in a eukaryotic nucleus precludes analysis of anything but a very small fraction of it.

Measurement of Molecular Variation

There are two common ways to measure the extent of genetic variation in a population. One way is to determine the

proportion of polymorphic loci. A locus is **polymorphic** if two or more allelic forms commonly occur in a population. A locus is **monomorphic** if essentially all genes at that locus are of one type. The usual practice is to measure the frequencies of the variants at each locus and regard all loci at which the frequency of the most common variant is less than 0.99 (that is, less common variants have a combined frequency of greater than 0.01) as being polymorphic; all other loci are then considered monomorphic. For example, if 12 out of a total of 30 loci examined had at least two variant forms that each segregated at frequencies greater than 0.01, the proportion of polymorphic loci would then be $12/30 = 0.4$.

The **average heterozygosity**, $\bar{H}$, which is the average frequency of heterozygous genotypes over all loci, is generally a more useful (and less arbitrary) measure of genetic diversity. $\bar{H}$ is the average of H_k, the heterozygosity at the kth locus, which can be calculated for a randomly mating population once allele frequencies are known. In general, if a locus has n alleles (protein variants) at frequencies of q_1, q_2, q_3, . . . , q_n, then the expected (Hardy-Weinberg) frequency of homozygotes at this locus is $\sum_{i=1}^{n} q_i^2$. The equations for H_k and $\bar{H}$ then become

$$H_k = 1 - \sum_{i=1}^{n} q_i^2$$

and

$$\bar{H} = \sum_{k=1}^{m} H_k / m \qquad (26.1)$$

where m is the total number of loci studied. For example, suppose that out of three loci sampled from a randomly mating population, two are polymorphic with three alleles each, having frequencies of 0.5, 0.4, and 0.1 at locus 1 and 0.7, 0.2, and 0.1 at locus 2, and one is monomorphic (locus

▶ **FIGURE 26–2** Adult esterases from *D. pseudoobscura*. Sample 1: *est-5*$^{1.00}$/*est-5*$^{1.00}$ (standard strain); samples 2 and 3: *est-5*$^{0.95}$/*est-5*$^{0.95}$; sample 4: *est-5*$^{0.95}$/*est-5*$^{1.12}$; sample 5: *est-5*$^{1.12}$/*est-5*$^{1.12}$; sample 6: *est-5*$^{0.95}$/*est-5*$^{0.95}$. *Source:* J. L. Hubby and R. C. Lewontin, *Genetics* 54 (1966): 577. Used with permission.

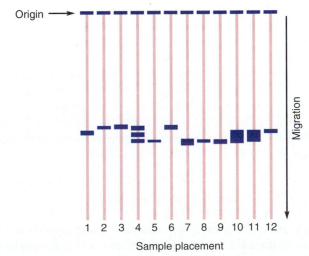

Origin →

Migration

1 2 3 4 5 6 7 8 9 10 11 12
Sample placement

3). The heterozygosities at these three loci would then be $H_1 = 1 - (0.25 + 0.16 + 0.01) = 0.58$, $H_2 = 1 - (0.49 + 0.04 + 0.01) = 0.46$, and $H_3 = 0.00$. The average heterozygosity for this set of loci is then

$$\bar{H} = \frac{0.58 + 0.46 + 0.00}{3} = 0.35$$

Electrophoretic variants of human and *Drosophila* proteins were the first to be analyzed, with reports by Harry Harris, Richard Lewontin, and other researchers in 1966. Harris, for example, found that of the enzyme products of 10 randomly chosen gene loci in humans, 3 were polymorphic. The polymorphic loci included red-cell phosphatase, with three alleles at frequencies of 0.36, 0.60, and 0.04, phosphoglucomutase, with two alleles at frequencies of 0.74 and 0.26, and adenylate kinase, with two alleles at frequencies of 0.95 and 0.05. Using equation 26.1, these data give gene heterozygosity values of 0.509, 0.385, and 0.095, respectively, and the average heterozygosity for all ten loci is 0.099.

Since electrophoretic variants constitute only a fraction of the total protein polymorphisms possible at a locus, these pioneering studies suggested that rather high levels of variability exist in natural populations. This conclusion has been consistently reinforced as additional species have been studied and electrophoretic procedures have been refined.

Example 26.1

Blood proteins coded for by 22 gene loci were electrophoretically studied in ten gorillas. All the gorillas were homozygous at 19 loci; polymorphism was detected at the other three loci. The data for these three loci are shown below; the numbers 85, 95, 97, 98, 100, and 105 identify the various genes by their electrophoretic positions, and the numbers in parentheses represent their allele frequencies:

Ak locus:	98 (0.20)	100 (0.80)
Dia locus:	85 (0.65)	95 (0.35)
6-Pgd locus:	97 (0.15)	105 (0.85)

Determine the (a) proportion of polymorphic loci, and (b) the average heterozygosity for this set of gorilla loci.

Solution: (a) The proportion of polymorphic loci is $3/22 = 0.14$. (b) Using equation 26.1, the heterozygosity per locus values are as follows: $H_{Ak} = 1 - (0.04 + 0.64) = 0.32$; $H_{Dia} = 1 - (0.42 + 0.12) = 0.46$; and $H_{6\text{-}Pgd} = 1 - (0.02 + 0.72) = 0.26$. The average heterozygosity for all loci is then $\bar{H} = (0.32 + 0.46 + 0.26)/22 = 0.047$.

Follow-Up Problem 26.1

On the basis of equation 26.1, what is the maximum heterozygosity for a locus with (a) two alleles; (b) three alleles?

■ **TABLE 26–1** Average heterozygosity values derived from electrophoretic data on proteins.*

Group	Species	Number of Loci	$\bar{H}$
Primates	*Homo sapiens*	121	0.143
	Gorilla gorilla	22	0.046
	Pan troglodytes	43	0.013
	Macaca cyclopis	29	0.041
	M. fascicularis	29	0.096
Ungulates	*Alces alces*	23	0.020
	Odocoileus virginianus	28	0.100
Rodents	*Peromyscus guardia*	25	0.014
	P. dickeyi	25	0.000
	P. maniculatus	29	0.128
	Eutamias panaminitus	36	0.055
	Sigmodon arizonae	24	0.033
	Thomomys umbrinus	27	0.031
	T. bottae	27	0.091
	Geomys personatus	24	0.027
	G. bursarius	24	0.063
Carnivores	*Vulpes vulpes*	21	0.000
	Mustela erminea	21	0.000
	Martes foina	21	0.000
	Meles meles	21	0.000
Lizards	*Anolis* (16 species)	22–25	0.010–0.120
Newts	*Taricha rivularis*	40	0.068
Bony fish	*Etheostoma spectabile*	26	0.069
	Oncorhynchus nerka	23	0.018
	Salmo salar	37	0.035
Fruitflies	*Drosophila nebulosa*	30	0.218
	D. willistoni	30	0.183
	D. pseudoobscura	46	0.136
	D. tropicalis	30	0.155
Landsnails	*Sphincterochilia aharoni*	29	0.067
Plants	*Pinus torreyana*	59	0.000
Bacteria	*Escherichia coli*	20	0.472

*At least 20 loci are needed because of the large interlocus variation in heterozygosity values.

Source: Data from numerous sources, given in M. Nei and D. Graur, "Extent of protein polymorphism and the neutral mutation theory," *Evolutionary Biology* 17 (1984): 83–85. Used with permission of Plenum Publishing Corporation.

■ Table 26–1 gives heterozygosity values derived from protein polymorphisms for a wide variety of species for which available gel staining methods can detect at least 20 different loci. Most of the values for $\bar{H}$ are between about 0.05 to 0.10, although higher values are found in the *Drosophila* species listed. Two-dimensional electrophoretic procedures (such as isoelectric focusing and SDS PAGE methods; see the Extensions and Techniques feature in Chapter 6) allow higher-resolution protein separations and thereby reveal more hidden genetic variability than is apparent from routine gel electrophoresis, but so far these

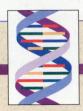

FOCUS ON HUMAN GENETICS

Are There Human Races?

Genetically, races are subdivisions of a species that differ in the frequencies of some of their alleles. This definition is obviously subjective, and its application depends on the judgment of the individual making the determination. Races are often defined in terms of observed morphological differences—in size, shape, and color in the case of humans, for example—and cultural differences, especially language and behavior. Such definitions make the important assumption that these differences between groups reflect significant genetic differences. Just how valid is this assumption—how much genetic variation underlies populations that have been classified as different races, and how significant is that variation compared with the amount of variation within each population?

Techniques such as protein electrophoresis and DNA fingerprinting, which measure genetic variation, have been used to measure the amount of molecular variability in different human populations. The results may surprise you. For example, the protein products of over 200 different gene loci have been studied electrophoretically; about three-quarters of these loci are monomorphic, meaning that no differences have been detected in the gene product they encode. The remaining 25% of the proteins studied are polymorphic, that is, at least two different alleles exist for the gene locus, each encoding an electrophoretically different form of the protein. If these results are representative of all human gene loci, we can conclude that variation in about one-quarter of human genes accounts for all the genetic differences observed between individuals and populations.

The question of whether there are human races can be approached by determining what fraction of molecular polymorphisms are attributable to differences between populations rather than differences among individuals within the same population. Estimates are that over 90% of the variation in human populations occurs within individual populations; only about 7% can be explained by differences between what are classified (by other criteria) as different races. Perhaps conclusions based on these findings are premature—it is possible, for example, that the genes analyzed to date do not include those that are responsible for the observed differences in skin color and facial features that we commonly use to define races. However, it is also possible that those genes are included in the 7% that are polymorphic at the molecular level and that there is not as much genetic difference between what we define as racial groups as one might think.

---

newer techniques have been applied to only a limited number of populations and proteins.

Some recent estimates of genetic variation based on DNA polymorphisms are given in ■ Table 26–2 for a variety of DNA regions and organisms. The variation determined by RFLP analysis is measured in terms of the average haplotypic diversity ($\bar{h}$), which is calculated in the same manner as the average heterozygosity ($\bar{H}$). The variation determined by nucleotide sequencing is expressed in terms of the average nucleotide diversity, which is defined as the average number of nucleotide differences per nucleotide site between two sequences. Notice that the nucleotide sequences of different individuals are seldom, if ever, identical. Also note that the diversity estimated by restriction analysis is about the same in value as the diversity measured directly by sequencing.

The Adaptive Importance of Molecular Evolution

An increasing amount of data suggests that some of the polymorphisms detected by molecular techniques are preserved through some form of balancing selection that main-

■ **TABLE 26–2** Estimates of DNA polymorphisms for selected organisms from RFLP analysis (R) or DNA sequencing (S).

Organism	DNA or Gene Region	Method	Base Pairs	Diversity ($\bar{h}$ or $\bar{H}$)
Human	mtDNA	R	16,500	0.004
	β-globin	R	35,000	0.002
	growth hormone	R	50,000	0.002
Chimpanzee	mtDNA	R	16,500	0.013
Gorilla	mtDNA	R	16,500	0.006
Peromyscus	mtDNA	R	16,500	0.004
Fruitfly	mtDNA	R	11,000	0.008
	Adh gene	R	12,000	0.006
	Adh exons	S	765	0.006
Sea urchin	H4 gene	S	~ 1,300	0.019
Influenza virus	hemagglutinin	S	320	0.510

Source: M. Nei, *Molecular Evolutionary Genetics* (New York: Columbia University Press, 1987), p. 627. Used with permission.

tains two or more genetic variants at stable equilibrium frequencies in a population. For example, selection favoring heterozygotes (overdominance) and frequency-dependent selection (see Chapter 25) are forms of balancing selection.

Selection in heterogeneous environments can also maintain protein polymorphism; studies of the adaptive significance of different alleles at the *lap* locus of the marine mussel *Mytilus edulis* provide an excellent example. The *lap* locus consists of three major alleles that code for aminopeptidase-I, an enzyme involved in regulating intracellular amino acid levels as part of an osmotic response to the salinity of the environment. During the larval stage, these mussels move to the open ocean (high salinity), and lysosomal enzymes degrade cellular protein to small peptides, whose terminal amino acids are cleaved by aminopeptidase-I. The pool of free amino acids therefore builds up as the larvae osmotically counter the increasing salinity of the water. Genotypes containing the *lap*^{94} allele exhibit a significantly higher specific activity of the aminopeptidase-I enzyme than do other genotypes, and thus they have a higher rate of increase in the concentration of free amino acids when they move from low- to high-salt environments. The frequency of this allele in oceanic populations is therefore quite high (▶ Figure 26–3). Conversely, the decrease in environmental salinity accompanying the mussels' move back to estuarian regions requires a rapid decrease in the intracellular pool of amino acids, which are ultimately excreted as amine waste products. As expected, *lap*^{94} genotypes have been found to have higher amine excretion rates than other genotypes.

Figure 26–3 shows that the frequency of the *lap*^{94} genotype is much lower in estuarian regions than in the open ocean. The fact that *Mytilus* populations are so genotypically differentiated over relatively short distances implies intense pressure from natural selection. Larvae returning to estuarian regions have the high frequency of *lap*^{94} characteristic of oceanic populations, but the *lap*^{94} frequency in resident estuarian adult populations is significantly lower, indicating that selection against this allele occurs after the organisms return to regions of low salinity. In autumn, nutrient stress and the high amine excretion rate of *lap*^{94} immigrant genotypes exhaust their nitrogen reserves and lead to a high mortality rate. Both *lap*^{94} homozygotes and heterozygotes are affected, suggesting that *lap*^{94} behaves as a dominant gene in the selection process.

To Sum Up

1. Genetic variability among individuals in natural populations can be measured at the molecular level in terms of differences either in the amino acid composition of a particular protein or in the DNA itself. One advantage of such molecular measures is that they are not biased toward any particular kinds of genes or phenotypic effects.

2. Gel electrophoresis has been extensively used to measure protein variability. This technique detects polymorphism in a protein regardless of its adaptive significance (or lack thereof) or its observable phenotypic effects. Electrophoresis tends to underestimate the amount of variability, however, since only variants with a difference in charge are distinguishable.

3. Nucleic acid sequencing is becoming popular as a method of detecting variability at the molecular level. Any nucleic acid region, regardless of its coding function, can be analyzed. Restriction enzyme polymorphisms are also used to detect variability in DNA when sequencing is not practical.

▶ **FIGURE 26–3** The frequency of the *lap*^{94} allele among populations of the mussel *Mytilus edulis* in estuarian regions of Long Island Sound and in the Atlantic Ocean. *Source:* R. K. Keohn and T. J. Hilbish, *Am. Sci.* 75 (1987): 135. Used with permission.

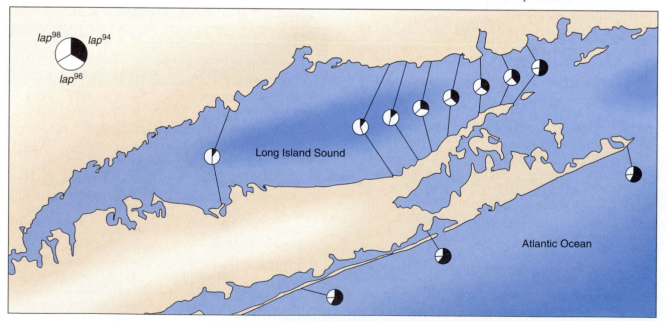

4. Data on protein polymorphism can be used to measure the extent of genetic variation in two ways: as the proportion of loci having two or more allelic forms (the proportion of polymorphic loci) or as the average frequency of heterozygous genotypes over all loci studied (the average heterozygosity). In nucleic acid sequencing, variation is expressed as the average number of nucleotide differences between two sequences (the average nucleotide diversity). DNA polymorphism determined from RFLP analysis is expressed as the average haplotypic diversity.

5. Some of the polymorphisms detected by molecular techniques have an adaptive significance and are maintained in populations by some form of balancing selection. Selection in heterogeneous environments is an important adaptive factor for maintaining molecularly detected polymorphisms.

❖ EVOLUTION BY NEUTRAL DRIFT

Although some molecular polymorphisms have been shown to have adaptive importance, it now appears that many molecular variants are neutral or nearly neutral in their selective effects. This observation led the Japanese geneticist Motoo Kimura and his colleagues to propose the **neutral theory** of molecular evolution. According to this theory, most molecular polymorphisms are selectively neutral in character and occur through the joint effects of mutation and genetic drift. Most molecular variation thus represents **transient polymorphisms** that originate from recurrent mutation pressure and exist temporarily as alleles undergo random fixation or loss. The effects of random drift on neutral mutations are shown in ▶ Figure 26–4. Over a long period of time, many mutations are expected to occur, but most of them are lost within the first few generations sim-

ply by chance. Only a few of the mutations drift to high frequencies, and fewer still reach fixation. Nevertheless, the frequency of each neutral allele changes very slowly, so that at any one time, a significant amount of genetic variation is present in the form of transient polymorphisms.

The overall rate of fixation of neutral genes is limited solely by the mutation rate. We can see why this is so by considering a diploid population of size N (with 2N total alleles at each locus) that produces neutral mutations at a rate μ per gene per generation. Therefore, the total rate at which *new* mutations are introduced per generation at any given locus would be $2N\mu$. If each mutant gene has a probability P of eventual fixation, then the overall rate of fixation, K, becomes

$$K = 2N\mu P$$

Once a mutant allele comes into existence, it makes up $1/2N$ of the genes at its locus. If the alleles are neutral, each gene should have the same probability of becoming fixed as any other gene at this locus, so this probability is simply

$$P = \frac{1}{2N}$$

Thus the rate of fixation becomes

$$K = 2N\mu\left(\frac{1}{2N}\right) = \mu \qquad (26.2)$$

It may be surprising that the rate of fixation is the same as the mutation rate per gene and is independent of the population size, but this can be explained by the inverse relationship between the fixation probability and the population size, the effects of which are shown in ▶ Figure 26–5. Note that when the population size is reduced, fewer mutations occur, since there are then fewer total genes in the

---

▶ **FIGURE 26–4** The effects of random genetic drift on neutral mutations. Allele frequency changes slowly, so a considerable amount of transient polymorphism exists at any one time.

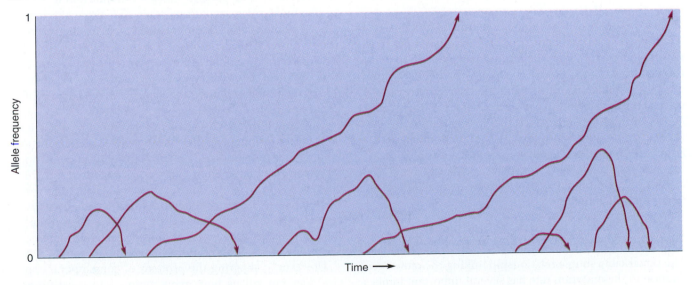

Large population

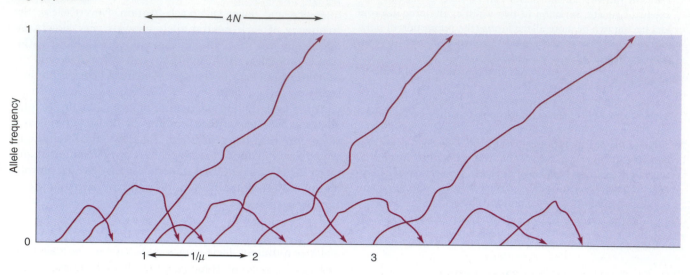

Small population

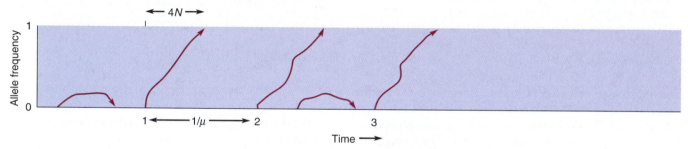

Time ⟶

▶ **FIGURE 26–5** The effect of population size on the chance of fixation of a neutral allele. Although fewer neutral mutations occur in a small population, they reach fixation (or elimination) faster than neutral mutations that arise in a large population.

population. However, a greater fraction of these mutations reach fixation, and they do so at a faster rate. Kimura has shown that the average time (in generations) for fixation of a neutral mutation is approximately $4N$. We would therefore expect neutral mutant genes, once they are formed, to undergo fixation more rapidly in small populations than in large ones.

Strictly speaking, a mutation is neutral if it does not affect the fitness of an organism (for example, $s = 0$). In the neutral theory, however, the meaning of a neutral mutation is broadened to include any allele for which $4Ns \ll 1$; thus both the population size (N) and the selection pressure (s) are taken into account. Recall from Chapter 25 that if either N or s (or both) is small, such that $4Ns$ is significantly less than unity, random evolutionary forces predominate and an allele will change in frequency largely as a result of chance. Thus, even a moderately beneficial (or detrimental) allele can behave as though it were selectively neutral if the population is small enough.

The Rate of Neutral Evolution: Molecular Clocks

The remarkably simple relationship linking the rate of gene fixation to the mutation rate has several important implica-

tions for molecular evolution. According to the neutral theory, most changes that occur over evolutionary time are the result of fixation of selectively neutral mutations by random drift. Furthermore, since the rate of neutral gene fixation depends only on the mutation rate, the rate of molecular evolution should be roughly constant in time and it should be possible to date evolutionary events. Once the actual geological time of one of the events in a phylogenetic tree is known (from the paleontological record, for example), the time of occurrence of all other events in the phylogeny can be determined. Thus we have a **molecular clock**—a constant rate of molecular evolution that can be used to time evolutionary events.

One way to estimate the rate of evolution of a particular gene is to compare the amino acid sequences in the **homologous** (evolutionarily related) **proteins** of contemporary species. Because amino acid replacements result from alterations in the nucleotide sequence of a structural gene, the number of sequence differences in the proteins of two species should provide an estimate of the number of mutational changes that have occurred along each line of descent since the species diverged from a common ancestor.

However, comparing the proteins of contemporary species does not tell us how many amino acid substitutions may have occurred at any given residue site. For example, two substitutions may have occurred during the period of divergence along one line of descent (for example, leu → ser → pro), but only one along the other line of descent (for example, leu → phe). Therefore, it is generally assumed that substitutions, being random, follow a Poisson distribution. If k is the rate of substitution per amino acid site per year, the probability that amino acid substitution does not occur at a given residue site during a period of t years is e^{-kt}. Moreover, since species behave as independent evolutionary units, the probability that substitution does not take place at the same site in the proteins of two different species during time t must be $(e^{-kt})(e^{-kt}) = e^{-2kt}$. Letting n be the total number of amino acid residue sites that can be compared in the two species and n_d the number of residue sites that differ in amino acids, we can write

$$\frac{n_d}{n} = 1 - e^{-2kt} \quad \text{or} \quad k = -\left(\frac{1}{2t}\right)\ln\left(1 - \frac{n_d}{n}\right)$$

Thus, if n and n_d are measured in homologous proteins in two different species and their time of evolutionary divergence, t, can be estimated from the fossil record, k can be evaluated directly. For example, the β-hemoglobin chain in humans differs from homologous chains in a number of other mammalian species in 12–14% of its amino acids. Since mammals probably diverged from a common ancestor some 80 million years ago, we can calculate an approximate substitution rate:

$$k = -\frac{1}{160 \times 10^6}\ln(1 - 0.13) = 8.7 \times 10^{-10} \text{ per year}$$

Once k has been determined, it can be related to another measure of evolutionary change known as the **unit evolutionary period** (UEP). The UEP is the time required for two proteins to diverge by 1% of their amino acid sequence. It is calculated as

$$\text{UEP} = \frac{0.01}{2k} = \frac{0.005}{k}$$

Example 26.2

Comparison of the α-hemoglobin peptide chains of humans and carp shows that the two have 51.4% of their amino acids in common. The divergence of the two species is thought to have occurred 375 million years ago. Estimate the rate of amino acid substitution per year as the α-peptide chains of these two species diverged, assuming that amino acid substitution proceeds as a Poisson process.

Solution: If amino acid substitutions have been random, then the proportion of amino acid sites that differ between the two species can be related to the rate of amino acid substitution, k, by

$$\frac{n_d}{n} = 1 - e^{-2kt}$$

Thus,

$$k = -\frac{1}{750 \times 10^6}\ln(1 - 0.486) = 8.9 \times 10^{-10} \text{ per year}$$

Follow-Up Problem 26.2

Calculate the unit evolutionary period for the α-hemoglobin chain from the information in Example 26.2.

———————————————————◆———————————————————

▶ Figure 26–6 shows the rates of evolution of four different proteins: fibrinopeptides, hemoglobin, cytochrome c, and histones. Each of these proteins has evolved at a relatively constant rate when viewed over the entire period of evolution. The difference between homologous proteins of any two organisms depends almost entirely on their time of divergence. Note that the four protein families exhibit dramatic differences in their rates of change over time, however. According to the neutral theory, these differences are explained by the functional constraints peculiar to each

▶ **FIGURE 26–6** Relationship between the accumulated number of amino acid substitutions and the divergence time in fibrinopeptides, hemoglobin, cytochrome c, and histone IV. The number of amino acid substitutions has been corrected for the occurrence of multiple mutations at any given site. Times since divergence of two lines of organisms were obtained from the geological record. *Source:* After R. E. Dickerson, "The structural history of an ancient protein," *Scientific American* 226 (1972): 58–72. Copyright © 1972 by Scientific American. All rights reserved. Used with permission.

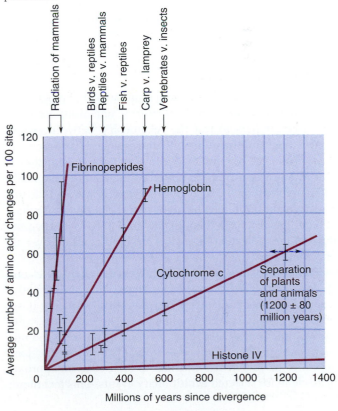

kind of protein. Histones have evolved at about $\frac{1}{100}$ of the rate of hemoglobin and about $\frac{1}{40}$ of the rate of cytochrome c. According to neutralists, the very slow rate of amino acid substitution in histones results from all parts of these molecules having critical roles in binding with DNA—they have very few nonessential neutral sites. Fibrinopeptides, on the other hand, have no critical functions, so almost any amino acid change is selectively neutral. Thus the fewer essential sequences a protein has, the faster its rate of evolution. Adaptive evolution still occurs, but the main role of natural selection at the molecular level is to conserve the established function of the molecule by protecting it from deleterious mutations.

Because of the implications of the molecular clock for the study of evolutionary biology, it is of critical importance to firmly establish the correctness of the neutral theory, which is the theoretical basis of the clock. Although Figure 26–6 demonstrates that the rate of evolution of certain proteins is roughly constant over long periods of time, the rate-constancy feature is not as striking when a greater variety of proteins is studied over shorter evolutionary spans. The per-codon substitution rate and the rate of change for the entire molecule are often not constant for any particular protein. These results could simply mean that the amino acid positions that are variable change from time to time. However, it could also mean that molecular evolution does not proceed by fixation of neutral alleles. By assuming that the number of substitutions along any arm of a phylogenetic tree is a Poisson random variable and using codon substitution data for several proteins simultaneously, C. H. Langley and W. M. Fitch constructed a phylogenetic tree and statistically tested the hypothesis of constant, uniform change over all branches. They obtained a highly significant test statistic that contradicted evolutionary rate constancy. This result does not favor the neutral theory, although the proportion of the phylogeny contributing to the statistical significance is not known.

Extensive data on amino acid sequences of hemoglobins and myoglobins for a wide variety of organisms have allowed researchers to test the relationship between evolutionary time and the number of amino acid substitutions in these proteins. ▶ Figure 26–7 shows that the relationship is indeed linear, with the exception of the data for mammals versus reptiles or birds. It does appear, however, that the rate of substitution may be higher for the period more than 300 million years ago than for the time since then, suggesting that the molecular clock has not been constant over all of evolutionary time.

From these kinds of analyses, most researchers have concluded that the rate of evolution differs, sometimes greatly, among lineages of organisms and that not all sequence positions are equally subject to change. Nevertheless, the average amount of molecular change for many proteins and nucleic acids that have evolved over long periods of time does appear to be sufficiently constant to be used as an approximate clock for selected evolutionary events. This technique is especially useful for obtaining a rough idea of divergence time when fossil records are absent or unreliable.

To Sum Up

1. Many molecular variants are selectively neutral; that is, they have no adaptive significance. The neutral theory of molecular evolution states that most molecular polymorphisms are maintained in populations not by balancing selection but rather through the joint effects of mutation and genetic drift. According to this theory, most variants are transient, existing only while alleles are undergoing random fixation or loss. Even an allele with some adaptive value can behave as though it is selectively neutral if the population size is small.

2. The overall rate of fixation of selectively neutral alleles is equal to the mutation rate per allele per generation. According to the neutral theory, the rate of molecular evolution should thus be constant over time, and evolutionary events can be dated by reference to the actual geological time of one of the events. The neutral theory thus provides the idea of an evolutionary clock.

3. The rates of evolution of certain proteins have been found to be constant over long periods of evolutionary time, indicating that the differences in a given protein among various organisms depend almost entirely on the accumulation of selectively neutral mutations. According to the neutral theory, the role of natural selection is to conserve the established function of the molecule, rather than to promote any kind of change between organisms. Thus, proteins that contain many amino acids that are not critical to function will evolve faster than proteins containing fewer nonessential residues.

MOLECULAR PHYLOGENIES

Recent studies at the molecular level have yielded considerable information on the phylogenetic relationships of various species. Among the most useful results of these studies are phylogenetic trees constructed from sequence differences in the DNA, RNA, and proteins of distantly related organisms. These trees use branching lines to suggest evolutionary relationships, and the length of each branch is directly proportional to the number of nucleotide (or amino acid) changes per sequence position separating the various taxonomic units.

There are several methods for inferring phylogenetic trees, each based on its own set of assumptions. Once the sequence information has been obtained, the analysis always begins with alignment of the sequences, matching regions of homology and omitting regions that lack homology (for example, due to insertion or deletion) or are of uncertain homology. One can then use either a **distance matrix** or a **maximum parsimony** method to construct the tree. In the distance matrix methods, the phylogeny is based on **evolutionary distances**, which are calculated from sequence differences between the taxonomic units. The distance values are usually arranged in a matrix. For example, if we con-

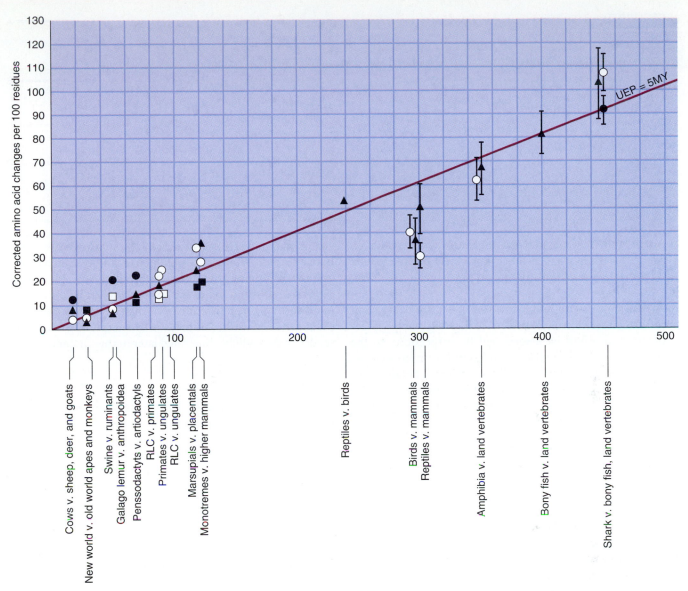

► FIGURE 26–7 Amino acid substitutions between various pairs of species versus paleontological time. The total number of amino acid substitutions in hemoglobin (triangles for alpha chains, circles for beta chains) and myoglobin (squares) is plotted as a function of the time since the ancestors of the species diverged. The straight line is the best linear regression fit to all data points. The vertical error bars for the older divergence points extend over ± two standard deviations. *Adapted from:* R. E. Dickerson and I. Geis, *Hemoglobin* (Menlo Park, CA: Benjamin Cummings, 1983). Used with permission.

sider four species, A, B, C, and D, a distance matrix might appear as follows:

	A	B	C
B	d_{AB}		
C	d_{AC}	d_{BC}	
D	d_{AD}	d_{BD}	d_{CD}

The *d* values represent the evolutionary distances separating the paired taxa. If we assume that the rate of evolution at the molecular level is constant over all lineages, so that a linear relationship exists between evolutionary distance and divergence time, a phylogenetic tree can be constructed directly from these distance values. At the simplest level, this

assumption means that the two most similar species, which are separated by the smallest evolutionary distance, are connected first; then the species most similar to the previous two is added, and so on, with each species separated from related taxa by the average evolutionary distance. This approach is used in the example in ► Figure 26–8, where it is assumed that A and B are the most closely related species (that is, d_{AB} has the smallest value), followed by C. Since A and B are assumed to have the smallest evolutionary distance, they are the first to be connected; each is separated from their point of divergence (branch point) by half the distance between them. C is then connected to a branch point by one-half the average distance between C and the

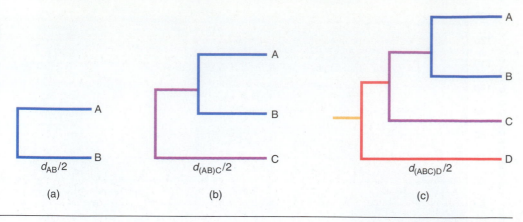

► **FIGURE 26–8** Stepwise construction of a phylogeny for four hypothetical species A, B, C, and D whose distance matrix is given in the text. The tree is constructed by adding one species at a time, beginning in diagram (a).

A-B pair. Finally, D is added; D is connected to a branch point by half the average distance between D and the others. A numerical example of this procedure is given in ► Figure 26–9.

When only a small number of sequences are being compared, the order in which species are added to the phylogenetic tree can often be determined by inspection. As the number of sequences increases, however, the order becomes less clear. Adding species in different orders generates trees with distinct topologies, so the problem becomes how to pick the tree that requires the smallest number of evolutionary changes to explain the sequence differences. Mathematical methods of maximum parsimony use computerized algorithms to accomplish this minimization. Maximum parsimony methods infer ancestral molecular sequences from data on extant species and select the topology that minimizes the number of mutations (nucleotide or amino acid substitutions) over the tree as a whole. As long as the number of changes per site is small, this approach does not require the assumption of a constant rate of evolution.

The question of rate constancy has plagued analysis of evolutionary divergence, especially the study of early evolutionary events. A recently developed technique called **evolutionary parsimony** attempts to incorporate different rates of sequence substitutions in the various evolutionary lineages into the tree-generating method, instead of assuming rate constancy. The use of this method in interpreting nucleotide sequence data will be illustrated later in this chapter.

Amino Acid Sequences

The amino acid sequences of proteins were among the first sequences to be compared in evolutionary studies. The use of amino acid sequencing techniques for constructing phylogenies is exemplified by studies of cytochrome c.

Cytochrome c is a protein that plays a vital role in cellular respiration in aerobic prokaryotes and eukaryotes. Its relatively simple structure consists of a single polypeptide chain that averages slightly longer than 100 amino acid sites. The amino acid sequences of cytochrome c from a number of different organisms are shown in ► Figure 26–10. A comparison of these sequences shows that a high degree of simi-

► **FIGURE 26–9** Stepwise construction of a hypothetical phylogenetic tree from matrix data shown in (a). The two most similar species, A and B, are connected first; the branching point is at $d_{AB}/2 = 2$. Next, the distance between C and the AB pair and between D and the AB pair are computed:

$$d_{(AB)C} = \frac{d_{AC} + d_{BC}}{2} = \frac{8 + 6}{2} = 7 \qquad \text{and} \qquad d_{(AB)D} = \frac{d_{AD} + d_{BD}}{2} = \frac{15 + 17}{2} = 16$$

Of these values $d_{(AB)C}$ is smallest, so C is added to the tree next, connected to a branch point by a distance of $\frac{7}{2} = 3.5$. Finally, $d_{(ABC)D}$ is calculated as

$$\frac{d_{AD} + d_{BD} + d_{CD}}{3} = \frac{15 + 17 + 13}{3} = 15$$

and D is added to the tree at a distance of 7.5. The resulting phylogeny is shown in (b).

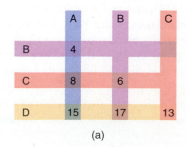

(a)

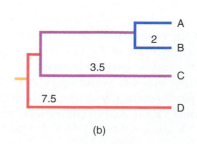

(b)

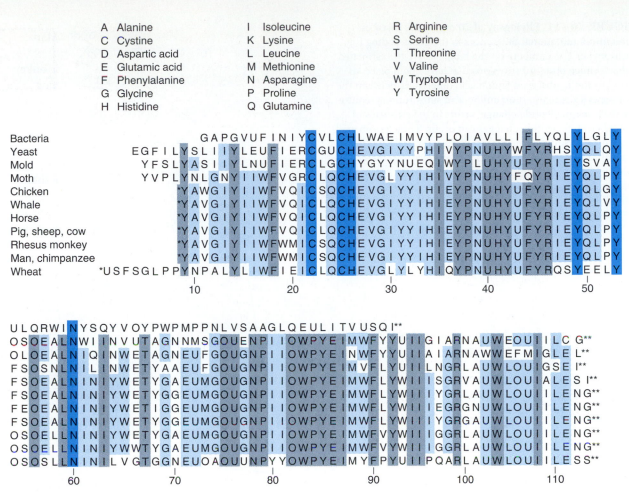

▶ **FIGURE 26–10** Amino acid sequences of cytochrome c from different organisms. Invariable sites are indicated by the darkest shading. Increased variability is indicated by a progressive reduction in the intensity of shading. Note that a high degree of similarity exists at many sites even among very diverse groups of organisms. *Source:* L. S. Dillon, *Evolution: Concepts and Consequences* (St. Louis: C. V. Mosby Co., 1978). Used with permission.

larity exists even among groups as diverse as vertebrates, molds, and higher plants (such as wheat). The similarities in sequence are most evident for the least divergent organisms shown and appear to decline with distance on an evolutionary scale. Thus, disparities in sequence tend to be greater for more distantly related species in most classification schemes. For instance, there is no difference between the cytochrome c of humans and that of chimpanzees and only one amino acid difference (at position 66) exists between the cytochrome c molecules of humans and rhesus monkeys. Greater differences are found in comparisons with other organisms; for example, the cytochrome c of humans and horses differ at 11 amino acid sites.

A phylogenetic tree based on sequence differences in cytochrome c is shown in ▶ Figure 26–11. In this tree, the length of each branch is directly proportional to the **minimal mutational distance**, which is the minimum number of nucleotide changes in the DNA needed to account for the observed amino acid differences in the protein. For example, a change from methionine (AUG) to isoleucine (AUU, AUC, or AUA) at the same residue location could occur with a minimum of one base change, while a change from methionine (AUG) to glutamine (CAA or CAG) would require at least two base substitutions. A comparison of the number of amino acid differences and the minimal mutational distance in cytochrome c is shown in ■ Table 26–3. Note that the minimal mutational distance tends to be greater than the corresponding number of amino acid differences, with the greatest discrepancy involving the most distantly related species.

Nucleotide Sequences

Measures of the divergence of populations can also be based on direct comparisons of their nucleic acids. Until recently, differences in the DNAs of related species were determined by the results of **DNA hybridization.**

In a typical DNA hybridization experiment, single-copy (unique-sequence) DNA is used. The denatured, fragmented DNA from one kind of organism, say species A, is trapped on an agar or membrane filter. These filter-bound fragments are incubated at 60°C in a solution containing a constant

FIGURE 26–11 Phylogeny of 20 organisms based on the minimum mutational differences in the genes coding for cytochrome c. The numbers on the branches are the estimated number of nucleotide substitutions that have taken place, as inferred from the observed amino acid differences between the various species. Since certain amino acid substitutions require more than one nucleotide change, errors in the estimates of the number of nucleotide changes can occur. These errors are corrected by dividing them among all lineages, yielding the fractional mutations shown. A different correctional method would yield a tree that differs in some details but is generally similar to this one. *Source:* W. F. Fitch and E. Margoliash, "Construction of phylogenetic trees," *Science* 155 (1967): 279–284. Copyright © 1967 by the AAAS. Used with permission.

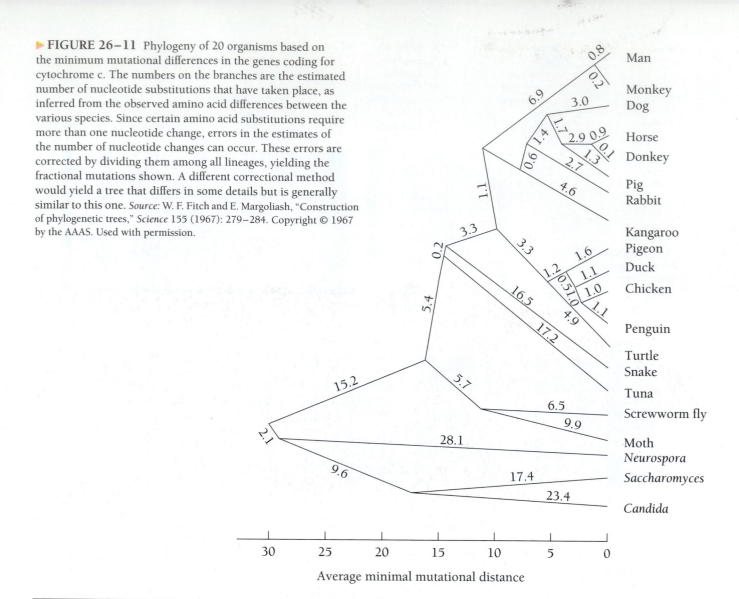

Average minimal mutational distance

TABLE 26–3 Comparison of the number of amino acid differences to the minimal mutational distance in cytochrome c.

Organism	Number of Amino Acid Differences (relative to humans)	Minimal Mutational Distance
Human	—	—
Chimpanzee	0	0
Rhesus monkey	1	1
Rabbit	9	12
Dog	10	13
Pig	10	13
Penguin	11	18
Horse	12	17
Moth	24	36
Yeast	38	56

Source: W. M. Fitch and E. Margoliash, "Construction of phylogenetic trees," *Science* 155 (1967): 279–284. Copyright © 1967 by the AAAS. Used with permission.

concentration of labeled (for example, with ^{3}H or ^{32}P) complementary single-stranded fragments from species A and varying amounts of single-stranded fragments from a second species, species B. The degree to which the DNA of species B is complementary to that of species A is measured by the extent to which species B DNA competes with the labeled type A DNA for the filter-bound fragments. The results of a hybridization experiment using labeled human DNA and the DNA of 11 other species are shown in ■ Table 26–4.

The DNA hybridization technique does not have sufficient resolution to detect short regions of nucleotide pair mismatch. However, the thermostability of the hybrid duplexes decreases as the proportion of mismatched bases increases, so the hybrid molecules are heated to determine their melting temperatures (see Chapter 7). An imperfectly formed hybrid helix, having fewer hydrogen bonds than native DNA, will have a lower melting temperature than the native helices. The change in melting temperature has been found to be directly proportional to the frequency of unpaired nucleotides, with a 1° change in T_m being equivalent

■ **TABLE 26–4** Degree of similarity between human DNA and DNA from 11 other species, based on the degree of inhibition of human-human DNA binding.

Organism	Degree of Taxonomic Differentiation	Percent Inhibition of Human-Human DNA Binding
Human	—	100
Chimpanzee	family	100
Gibbon	family	94
Rhesus monkey	superfamily	88
Capuchin monkey	superfamily	83
Tarsier	suborder	65
Slow loris	suborder	58
Lemur	suborder	47
Tree shrew	suborder	28
Mouse	order	21
Hedgehog	order	19
Chicken	class	10

Source: Data from B. H. Hoyer and R. B. Roberts, *Molecular Genetics* (New York: Academic Press, 1967), pp. 425–479.

to about 1.0–1.5% mismatched base pairs. ■ Table 26–5 shows estimates of T_D, the difference in melting temperature between hybrid and native DNAs, for comparisons of several primates.

In the last several years, **restriction site analysis** has also been used to estimate the difference between the DNAs of related species. The extent to which different DNAs share the same ancestral restriction sequence at a number of given sites is taken as a measure of their relatedness. To use this method, all the locations of the particular restriction site must have been mapped on the DNA so that the proportion shared between the two species can be determined. It is therefore easier to use this method with organellar DNAs, where the length of the DNA is fixed and a restriction enzyme cuts it into a set of defined fragments.

The technique with the greatest potential for measuring the extent of DNA divergence is **nucleic acid sequencing.** Accurate estimates of the average number of nucleotide substitutions per site in DNA sequence comparison studies are difficult to obtain, however, because of the possibility of repeated substitutions at the same site over long evolutionary times and different rates of substitution for the different base pairs and for the three codon positions. The mathematical treatment is therefore more complex than that used to study change in amino acid sequences. However, nucleotide sequencing is easier than sequencing peptides, and since all genetic change ultimately originates in DNA, researchers have gradually shifted their attention from proteins to nucleic acids. Both nuclear and organellar nucleic acids have been used in studies of evolutionary divergence. Mitochondrial DNA has changed much faster than nuclear DNA, so it is especially useful for study of the genetic relationship among closely related species. Chloroplast DNA, on the other hand, has changed slowly, making it useful for the study of more distant taxa.

▶ Figure 26–12 shows an evolutionary tree constructed by Carl Woese and his colleagues; it is based on the average number of accumulated nucleotide changes per nucleotide site in the 16S rRNAs of 21 different species. This RNA has evolved very slowly, making it particularly suitable for clarifying early evolutionary events. It is universal, and its highly conserved structure makes it possible to identify homologous positions in compared sequences relatively easily. In addition, the molecule is large enough to represent a significant sample of genome evolution. The tree constructed by Woese proposes that there are three kingdoms (Archaebacteria, Eubacteria, and Eukaryotes) that descended from a common early ancestor (the progenote). Notice that, according to this scheme, humans are more closely related to plants than *E. coli* is to *B. subtilis,* another eubacterium. Eukaryotic diversity is dominated by the protists (*Prorocentrum, Oxytricha, Dictyostelium, Trypanosoma,* and *Euglena*), with little sequence divergence between the animals and plants. The Archaebacteria seem to be quite different from the Eubacteria in their origins. Other work has shown that in many respects (for example, membrane lipids, the presence of transposable elements, eukaryotic-like RNA polymerases, rRNA, tRNA and ribosomes, and the occurrence of introns) the Archaebacteria are more similar to eukaryotes than they are to eubacteria.

Using 16S rRNA and an evolutionary parsimony method, Jim Lake has concluded that there are five, rather than three, kingdoms of extant organisms (but not the five you probably learned in general biology!), as shown in ▶ Figure 26–13.

■ **TABLE 26–5** Measurements of the difference in DNA melting temperature (T_D) for different species of primates.

	Human	Chimp	Pigmy Chimp	Gorilla	Orangutan	Gibbon
Chimpanzee	1.8					
Pigmy chimp	1.9	0.7				
Gorilla	2.4	2.1	2.3			
Orangutan	3.6	3.7	3.7	3.8		
Gibbon	5.2	5.1	5.6	5.4	5.1	
Baboon	7.7	7.7	8.0	7.5	7.6	7.4

Source: C. G. Sibley and J. E. Ahlquist, *J. Mol. Evol.* 20 (1984): 2–15. Used with permission.

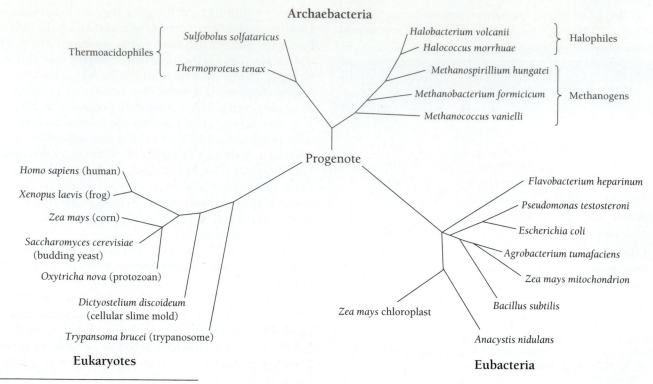

Archaebacteria

Thermoacidophiles {
 Sulfobolus solfataricus
 Thermoproteus tenax
}

Halobacterium volcanii
Halococcus morrhuae
} Halophiles

Methanospirillium hungatei
Methanobacterium formicicum
Methanococcus vanielli
} Methanogens

Progenote

Homo sapiens (human)
Xenopus laevis (frog)
Zea mays (corn)
Saccharomyces cerevisiae
(budding yeast)
Oxytricha nova (protozoan)
Dictyostelium discoideum
(cellular slime mold)
Trypansoma brucei (trypanosome)

Eukaryotes

Flavobacterium heparinum
Pseudomonas testosteroni
Escherichia coli
Agrobacterium tumafaciens
Zea mays mitochondrion
Bacillus subtilis
Zea mays chloroplast
Anacystis nidulans

Eubacteria

0.1 mutations per sequence position

▶ **FIGURE 26–12** Evolutionary tree based on comparisons of 16S rRNA from different species, constructed using a distance matrix method. *Source:* N. R. Pace, G. J. Olsen, and C. R. Woese, "Ribosomal RNA phylogeny and the primary lines of evolutionary descent," *Cell* 45 (1986): 325. Copyright © 1986 by Cell Press. Used with permission.

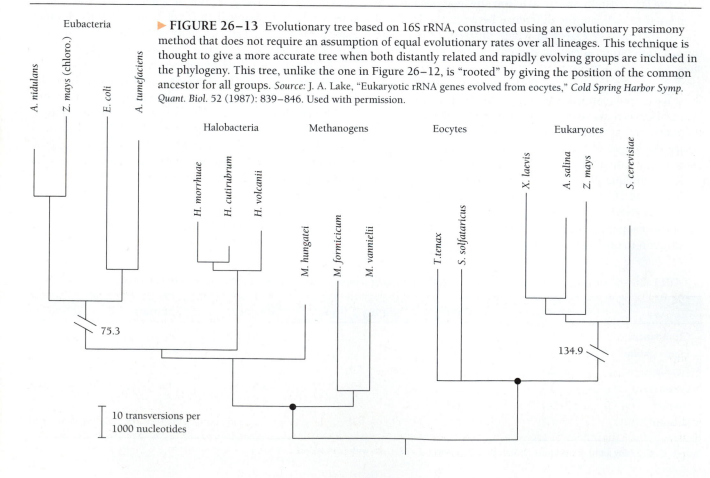

▶ **FIGURE 26–13** Evolutionary tree based on 16S rRNA, constructed using an evolutionary parsimony method that does not require an assumption of equal evolutionary rates over all lineages. This technique is thought to give a more accurate tree when both distantly related and rapidly evolving groups are included in the phylogeny. This tree, unlike the one in Figure 26–12, is "rooted" by giving the position of the common ancestor for all groups. *Source:* J. A. Lake, "Eukaryotic rRNA genes evolved from eocytes," *Cold Spring Harbor Symp. Quant. Biol.* 52 (1987): 839–846. Used with permission.

Eubacteria

A. nidulans
Z. mays (chloro.)
E. coli
A. tumefaciens

Halobacteria
H. morrhuae
H. cutirubrum
H. volcanii

Methanogens
M. hungatei
M. formicicum
M. vannielii

Eocytes
T. tenax
S. solfataricus

Eukaryotes
X. laevis
A. salina
Z. mays
S. cerevisiae

75.3

134.9

10 transversions per 1000 nucleotides

Human Phylogeny

By using mitochondrial instead of nuclear DNA, researchers can more easily deduce phylogenetic relationships of organisms, such as humans, that have evolved relatively recently. The phylogenetic tree below is based on comparisons of mitochondrial DNA and was constructed using a maximum parsimony method. An application of this method was reported in 1987 by R. L. Cann, M. Stoneking, and A. C. Wilson and suggests that humans evolved in Africa. The team worked in Wilson's laboratory. The tree shows two primary lines of descent, one leading to Africans only and the other to some African and to all other populations. You may have heard of this idea of a common African root for all humans—the idea that we all share a common mother, an ancestral "mitochondrial Eve," who lived in Africa about 200,000 years ago.

The report of the Wilson team raised a storm of controversy among researchers and the public. In early 1992, Stoneking and others showed that the conclusions reported in 1987 are statistically flawed. Wilson's group had looked at about 100 computer-generated trees and had chosen the one that minimized the mutational distances. However, looking at only 100 possible trees is not nearly enough to ensure an accurate conclusion. Tens of thousands of additional computer runs done since 1987 have yielded many trees just as good, if not better

(more parsimonious), than the one shown below. Some have an African origin, others a non-African origin. Therefore, the conclusion at this time is not that the origin of humans is non-African, but rather that one cannot tell from the available data and methods of computer analysis.

Other pieces of evidence do suggest an African origin. For example, Africans have the greatest diversity in their mitochondrial DNA of any human population on any continent. This indicates that humans must have lived in Africa longer than anywhere

else, to accumulate the largest number of mutations. Preliminary studies of nuclear DNA also indicate an African origin, but more data on nuclear DNA will probably be needed before the question of human origin is answered. Mitochondrial DNA sequences are not strongly correlated with geography (especially in non-African groups), so it is possible that multiple mtDNA types have been maintained in widely separated populations since those populations diverged. This factor may make evolutionary interpretation of mitochondrial data impossible.

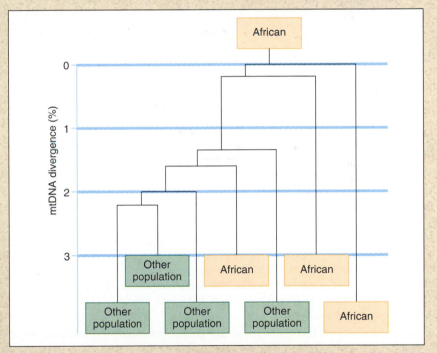

The phylogenetic tree developed by R. L. Cann, M. Stoneking, and A. C. Wilson (1987).

Since the old term prokaryote includes four of Lake's five groups, it is not a useful evolutionary classification under his scheme. Similarly, since his method separates archaebacteria into three separate groups, it also becomes an improper taxonomic grouping. The dispute between Woese's and Lake's interpretation of 16S rRNA sequence data demonstrates that alignment of sequences is extremely complicated, somewhat subjective, and easily biased by inclusion

of very distantly related sequences that may not adhere to the rate constancy assumption.

To Sum Up

1. Phylogenetic trees depicting evolutionary relationships between organisms can be constructed using data on the number of residue differences separating the various taxonomic units

per nucleotide or amino acid position. Various mathematical methods can be applied to such data to determine the topological (and evolutionary) relationships that minimize the number of mutations needed to explain the sequence differences among species.

2. Different assumptions underlie the various methods used to deduce the order in which species are added to a phylogenetic tree. Most require assuming a constant rate of molecular evolution, which may not be valid for evolutionary lineages that are very distantly related. Some techniques, such as evolutionary parsimony, attempt to incorporate different rates instead of assuming a constant rate of substitution.

3. The data used in the construction of phylogenetic trees can be obtained from analysis of the amino acid sequence of a given protein or from measures of DNA differences. DNA comparisons can be based on DNA hybridization, restriction site analysis, or nucleic acid sequencing, which is fast becoming the preferred approach. Both nuclear and mitochondrial DNAs can be sequenced. Sequencing mitochondrial DNA is a good method for measuring the relationship among closely related species, since mitochondrial DNA has evolved much more rapidly than nuclear DNA.

Chapter Summary

Genetic variability among individuals in natural populations can be measured at the molecular level by detecting sequence differences in specific proteins and nucleic acids. The proportion of polymorphic loci and the average heterozygosity are two measures of genetic variation that are often used to quantify this molecular variation. Such molecular studies have shown that the amount of variation in protein-coding genes is substantial. Some of the polymorphisms detected by molecular techniques have adaptive significance and are maintained in natural populations by some form of balancing selection.

Many molecular polymorphisms are selectively neutral and are maintained by the combined effects of mutation and genetic drift in a manner consistent with the neutral theory of molecular evolution. According to this theory, most molecular variants are transient and exist only while alleles are undergoing random fixation or loss. The overall rate of fixation of these selectively neutral alleles is constant and is equal to the mutation rate, thus providing the basis for a molecular clock that can be used to date evolutionary events.

Comparisons of homologous sequences in proteins and DNA have been used to determine the degree of relatedness of different species. The construction of phylogenetic trees depicting evolutionary relationships can be based on these comparisons by using the number of residue differences as a measure of the evolutionary distance between species.

Questions and Problems

Molecular Polymorphisms

1. What kinds of codon substitutions are not distinguishable by electrophoretic studies on proteins?

2. Three loci in a population of deer mice are studied electrophoretically. The *Got-1* locus has three alleles with frequencies of 0.031, 0.094, and 0.875; the *Mdh-1* locus has two alleles with frequencies of 0.031 and 0.969; and the *Est-1* locus has four alleles with frequencies of 0.062, 0.063, 0.375, and 0.500. Calculate the average heterozygosity for this set of loci.

3. Genetic variability in *Drosophila melanogaster* was measured using different restriction enzymes on three different DNA regions. Several haplotypes were identified, as follows:

DNA Region 1		DNA Region 2		DNA Region 3	
Haplotype	Frequency	Haplotype	Frequency	Haplotype	Frequency
m	0.1	N1	0.55	1	0.22
m_a	0.3	N2	0.11	2	0.46
m_b	0.5	N3	0.06	3	0.32
m_c	0.1	N4	0.06		
		N5	0.11		
		N6	0.11		

Determine the average haplotypic diversity for this set of DNAs.

Evolution by Neutral Drift

4. What is the unit evolutionary period (UEP)? How is the UEP related to the rate of amino acid substitution (k)?

5. Would you expect the rate of nucleotide substitution to be faster or slower than the rate of amino acid substitution? Explain your answer.

6. Of the 141 amino acids in the human hemoglobin α peptide, the data below give the mean number that are different in the horse, rabbit, and mouse. For each of the three comparisons, calculate the yearly rate of evolutionary change. (Assume that the divergence time of mammals from a common ancestor is approximately 80 million years.)

	Human	Horse	Rabbit	Mouse
Human	—	18	25	17

7. If the rate of amino acid substitution in a certain protein is 2×10^{-9} per year and the proportion of the amino acids that differ between two species is 0.2, how long has it been since the two species diverged?

8. Of the approximately 1000 amino acids in the proteins examined so far, humans and chimps differ in 23. Assuming a divergence time of about 6 million years, determine the yearly rate of evolutionary change.

Molecular Phylogenies

9. The table below gives the minimum numbers of nucleotide differences in the gene that encodes cytochrome c in four species. Construct a phylogenetic tree from this distance matrix.

	Human	Monkey	Dog
Monkey	1		
Dog	13	12	
Tuna	31	32	29

10. Use the following distance matrix to construct a phylogenetic tree relating the four species A, B, C, and D.

	A	B	C
B	0.20		
C	0.12	0.30	
D	0.55	0.70	0.65

11. Two homologous sections of the α- and β-chains of human hemoglobin are shown below. Using the genetic code dictionary, determine the minimum number of nucleotide changes that have occurred in the evolution of the corresponding sections of the hemoglobin α and β genes since their divergence from a common ancestral gene.

α Lys-Phe-Leu-Ala-Ser-Val-Ser-Thr-Val-Leu-Thr-Ser-Lys-Tyr-Arg

β Lys-Val-Val-Ala-Gly-Val-Ala-Asn-Ala-Leu-Ala-His-Lys-Tyr-His

12. DNA hybridization reveals nucleotide differences of about 9.6% between humans and green monkeys, 15.8% between humans and capuchins, and 16.5% between green monkeys and capuchins. Construct a phylogenetic tree based on these data.

13. When DNA sequencing is used to gather information for determining phylogenetic trees, the DNA is sometimes obtained as cDNA made from mRNA isolated from cells of the various organisms. Do you think this method provides an accurate measure of evolutionary divergence? Explain your answer.

APPENDIX A

SOLUTIONS TO FOLLOW-UP PROBLEMS

CHAPTER 2

2.1 (a) Let *A* be the allele for wild-type coat color, and *a* be the allele for platinum. Wild-type color is dominant to platinum, since only wild-type showed up in the F_1 and approximately $\frac{3}{4}$ of the F_2 expressed this color. The crosses are summarized as follows:

P: wild type × platinum
       (*AA*)   (*aa*)

F_1:   wild type
               (*Aa*)

F_2:   33 wild type + 10 platinum
          (*A*−)   (*aa*)

(b) The genotypes in the F_2 are expected to consist of $\frac{1}{4}$ *AA*, $\frac{2}{4}$ *Aa*, and $\frac{1}{4}$ *aa*. Thus, two out of every three dominant (*A*−) offspring, on the average, should be heterozygous. The number of heterozygotes expected among the F_2 is therefore $(\frac{2}{3})(33) = 22$.

2.2 (a) The trait is dominant. The mating I-1 × I-2 produces both types of offspring, so both I-1 and I-2, which show the trait, must be heterozygous *Aa*.

(b)

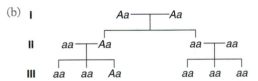

2.3 (a) Rose comb and black feathers are dominant traits. Both of the parents have rose combs and black feathers and produce offspring with single combs and red feathers. Therefore, both parents must be heterozygous and must carry recessive alleles for the single-comb and red-feather traits.

(b) Letting *A* = the rose comb allele, *a* = the single comb allele, *B* = the black color allele, and *b* = the red color allele, the offspring of the cross *A*−*B*− × *A*−*B*− are produced in a ratio of approximately 9 *A*−*B*− : 3 *A*−*bb* : 3 *aaB*− : 1 *aabb*. Since the *aabb* offspring must receive both an *a* and a *b* allele from each parent, the genotypes of the parents are *AaBb* × *AaBb*.

2.4 As in Example 2.4, the failure to produce a 1:1:1:1 ratio shows that the genes are linked. In this case, the *OP* and *op* gamete types produced by the heterozygote in the test cross are the most frequent, since round, smooth (*OoPp*) and elongate, fuzzy (*oopp*) offspring are in the majority. This result suggests that the parental types in this cross have *O* and *P* genes on one homolog and their alleles, *o* and *p*, on the other homolog. The *Op* and *oP* gene combinations present in the round, fuzzy

(*Oopp*) and elongate, smooth (*ooPp*) offspring are recombinant types produced by crossing over.

CHAPTER 3

3.1 The man has a $\frac{2}{3}$ chance of carrying the allele for galactosemia, since he is a normal offspring of heterozygous parents. His wife has a $\frac{1}{3}$ chance of carrying the allele (her mother has a $\frac{2}{3}$ chance of being heterozygous, since she has a sister who is affected and has a $\frac{1}{2}$ chance of transmitting this allele to her daughter). The first child has a $\frac{1}{4}$ chance of receiving an allele for galactosemia from both parents (if both parents are carriers). Therefore, the chance that the first child has galactosemia is $(\frac{2}{3})(\frac{1}{3})(\frac{1}{4}) = \frac{1}{18}$.

3.2 This situation conforms to the binomial formula, even though the probabilities of the two mutually exclusive possibilities in question (a normal boy and an affected girl) do not add to 1. In this case, *p* (the chance of a child's being a normal boy) = *P*(*A*− from *Aa* × *Aa*) × *P*(boy) = (3/4)(1/2) = 3/8 and *q* (the chance of a child's being an affected girl) = (1/4)(1/2) = 1/8. The overall probability of the event having 2 normal boys and 2 affected girls) is then $(4!/2!2!)(3/8)^2(1/8)^2 = 27/2048$.

3.3 The expected 3:1 ratio would result in 75 purple-flowered and 25 white-flowered offspring. This gives a chi-square value of $[(60 − 75)^2/75] + [(40 − 25)^2/25] = 3 + 9 = 12$. In this case, df = 1. The calculated chi-square value is greater than 6.64 in the chi-square table, which corresponds to a *P* value of less than 0.01. Thus, the deviations between the observed and expected values are significant, and the hypothesis proposed to account for the data (a simple 3:1 ratio from a monohybrid cross) must be rejected.

CHAPTER 4

4.1 (a) There are now four alleles in the series. Letting $N = 4$, we obtain $(4)(5)/2 = 10$ different genotypes.

(b) There are six distinguishable blood types distributed among the genotypes as follows:

Blood types:	A_1	A_2	B	A_1B	A_2B	O
Genotypes:	$I^{A_1}I^{A_1}$	$I^{A_2}I^{A_2}$	$I^B I^B$	$I^{A_1}I^B$	$I^{A_2}I^B$	$I^O I^O$
	$I^{A_1}I^{A_2}$	$I^{A_2}I^O$	$I^B I^O$			
	$I^{A_1}I^O$					

4.2 Flower color in sweet peas is determined by two gene pairs with complementary effects. Only plants with the *A*−*B*− genotype have purple flowers. The flowers of plants with other genotypes (*A*−*bb*, *aaB*−, and *aabb*) are white. Because a 9:7 ratio is produced in the F_2, the purple-flowered F_1s must be *AaBb* in genotype. Two possible matings between homozygous parental varieties can produce *AaBb* offspring: *AABB* × *aabb* and *AAbb* × *aaBB*. The first of these matings could not have occurred in this situation, since one of the parents (*AABB*) would have to possess purple flowers. However, the second possibility can account for the results observed in the cross, since both *AAbb* and *aaBB* plants are white. The crossing sequence is as follows:

P: *AAbb* × *aaBB*
          (white)   (white)

F$_1$:   *AaBb*
             (Purple)

F$_2$: 9 *A–B–* : (3 *A–bb* + 3 *aaB–* + 1 *aabb*) = 9 purple : 7 white

4.3 None of the patterns is consistent with the pedigree. The cross involving I-1 and I-2 rules out an X-linked recessive gene and a sex-influenced autosomal gene dominant in males. The cross III-5 and III-6 rules out an X-linked dominant gene and a sex-influenced autosomal gene dominant in females.

CHAPTER 5

5.1 The phenotypic distribution of the offspring follows a binomial distribution with $2n + 1 = 9$ phenotypic classes. The number of segregating gene pairs is thus $n = 4$. To produce a symmetrical phenotypic distribution that follows the observed pattern, both parents must be heterozygous for all four gene pairs: *AaBbCcDd* × *AaBbCcDd*.

5.2 The phenotypic distribution of the F$_2$ has a mean of $(3600 + 1875)/2 = 2737.5$ g and a standard deviation of $\sqrt{52,900} = 230$ g. For a normal distribution, we would expect 95% of the body weights to fall within ±2 standard deviation units of the mean. This range would correspond to $2737.5 ± 460$ g, or 2277.5 g to 3197.5 g.

CHAPTER 6

6.1 Chymotrypsin cleaves only on the carboxyl side of Phe, Trp, and Tyr. Therefore, the free Met released upon hydrolysis by chymotrypsin must be from the original carboxyl end of the polypeptide. Trypsin cleaves only on the carboxyl side of Arg and Lys, and thus the free Arg released upon hydrolysis by trypsin must have originated at the original amino end of the polypeptide. From the overlapping peptide fragments, the primary sequence of the chain can be deduced to be:

Chymotrypsin
treatment: Arg-Ala-Lys-Phe
                       Glu-Arg-Tyr
                                Gly-Trp
                                        Met
Trypsin
treatment: Arg
                   Ala-Lys
                        Phe-Glu-Arg
                              Tyr-Gly-Trp-Met
Primary
structure: Arg-Ala-Lys-Phe-Glu-Arg-Tyr-Gly-Trp-Met

6.2 The α-helix has a pitch (vertical rise per turn) of 0.54 nm. Thus an α-helix 30 nm in length contains (30 nm)/(0.54 nm/turn) = 55.5 turns.

CHAPTER 7

7.1 According to this hypothesis, each DNA molecule would consist of a particular sequence of A, T, G, and C at four adjacent residue sites, repeated along the length of the chain. If any such sequence is possible, the first residue site could be filled by any of the four bases, the next by any of the three remaining bases, the third by any of the two remaining bases, and the fourth by the last base, giving $(4)(3)(2)(1) = 4! = 24$ different primary sequences.

7.2 An (A + G) : (C + T) ratio of 1 means that the concentration of purines equals the concentration of pyrimidines. Although this relationship is true for the double helix, single-stranded DNA molecules can also possess equal amounts of these two types of bases. The fact that this ratio is 1 does not imply that the DNA is double- or single-stranded.

CHAPTER 8

8.1 During exponential growth, the number of population doublings can be calculated directly from t/t_d, where t is the total time and t_d is the doubling time. We can therefore express the exponential growth of a bacterial population as $N = N_0 2^{t/t_d}$. Solving for time t, we get $t = t_d (\log N/N_0)/(\log 2)$. If we assume that the population in Example 8.1 started exponential growth 1 hour after inoculation, then the total time required (from inoculation) to achieve a factor increase in population size of $N/N_0 = 10^9/1.5 \times 10^4 = 6.67 \times 10^4$ will be $t = 1 + (0.91)(\log 6.67 \times 10^4)/(\log 2) = 15.6$ hours. Note that one hour is added to account for the lag phase, when the cells did not grow.

8.2 In conservative replication, each DNA duplex serves as the template for the synthesis of a complementary molecule, so no DNA molecules of intermediate density will be formed. Thus after two generations in the unlabeled medium, $\frac{1}{4}$ of the DNA molecules are expected to be heavy (fully labeled) and $\frac{3}{4}$ will be light.

CHAPTER 9

9.1 (a) The $C_0t_{1/2}$ values for the pure components should be $(500)(0.6) = 300$, $(30)(0.3) = 9$, and $(0.002)(0.1) = 0.0002$ for the slow-, intermediate-, and fast-renaturing forms, respectively. Assuming that the slow-renaturing DNA is non-repetitive, it has a sequence complexity of $(100)(300)/(0.0002) = 1.5 \times 10^8$ bp. The haploid genome size is therefore $(1.5 \times 10^8)/0.6 = 2.5 \times 10^8$ bp.

(b) Using slow-renaturing DNA as the nonrepetitive standard, the sequence complexity of the intermediate form will be $(1.5 \times 10^8)(9)/(300) = 4.5 \times 10^6$ bp.

(c) The repetition frequency for the fast-renaturing component is $(0.1)(2.5 \times 10^8)/(100) = 2.5 \times 10^5$.

9.2 The number of base pairs of DNA in the chromosome of this organism is $(3.4 \times 10^6 \text{ nm})/(0.34 \text{ nm/bp}) = 10^7$ bp. Since replication is bidirectional, the minimum time required for S would then be $(10^7 \text{ bp})/(2)(2000 \text{ bp/min}) = 2,500$ minutes, or approximately 42 hours.

CHAPTER 10

10.1 If the plasmids are selected at random for replication, two possible outcomes occur with equal likelihood: (1) only one of the plasmids replicates and (2) both plasmids replicate. When only one of the plasmids replicates, half of the progeny cells will receive only one plasmid type. When both plasmids replicate, 1/3 of the progeny cells will receive only one plasmid type (see Example 10.1). The total probability that a cell will receive only one plasmid type is therefore $(1/2)(1/2) + (1/2)(1/3) = 5/12$.

10.2 (a) To lyse and produce P2 progeny, a cell must be infected by 1 (or more) P2 and no P4 phages. Therefore the probability that a cell lyses and produces P2 is $1 - e^{-2}$ (for 1 or more P2) $\times e^{-2}$ (for no P4) $= (1 - e^{-2})(e^{-2}) = (0.865)(0.135) = 0.117$, or 11.7%.

(b) To lyse and produce P4 progeny, a cell must be infected by 1 (or more) of both P2 and P4 phages. Therefore the probability that a cell lyses and produces P4 is $1 - e^{-2}$ (for 1 or more P2) $\times 1 - e^{-2}$ (for 1 or more P4) $= (1 - e^{-2})(1 - e^{-2}) = (0.865)(0.865) = 0.748$, or 74.8%.

CHAPTER 11

11.1 (a) The litter size will be reduced. Studies reveal that litter sizes from crosses of yellow mice are, in fact, only three-fourths as large as typical mouse litters.

(b) From the cross $AA^Y \times AA^Y$, the progeny ratio at fertilization will be $1\ AA : 2\ AA^Y : 1\ A^Y A^Y$, while the ratio at birth will be $1\ AA : 2\ AA^Y$, since the $A^Y A^Y$ condition is lethal. This monohybrid cross thus gives a phenotypic ratio of $2 : 1$, rather than the usual $3 : 1$.

11.2

Growth on minimal medium plus:

Mutant strain	C	A	T	S
1	–	–	+	+
2	–	–	–	+
3	–	+	+	+

11.3 (a) Mutations show a Poisson distribution. Thus, the proportion of colonies without any mutations would be $P(0) = e^{-m} = 1 - 87/100 = 0.13$. Solving for m, we get $m = -\ln(0.13) = 2$.

(b) Of the 100 total colonies, $(100)(2e^{-2}) = 26$ are expected to have undergone exactly 1 mutation during growth. Example 11.3 showed that one-half of all clones with one mutation will contain exactly one mutant cell, so we would expect $26/2 = 13$ of the colonies to contain exactly one a^- cell.

11.4 Since three generations are required to attain a mutant frequency of 1.5×10^{-5}, the number of generations required to reach a mutant frequency of 4×10^{-5} will be $t = (3)(4 \times 10^{-5})/(1.5 \times 10^{-5}) = 8$ generations.

CHAPTER 12

12.1 Since two of the three homologs in these trisomic plants carry the p allele, the gamete ratio among the pollen is $2\ p : 1\ P$. The gamete ratio among the eggs is $1\ pp : 2\ Pp : 2\ p : 1\ P$. We can then predict the results of the cross as follows:

	1 (P)	2 (Pp)	2 (p)	1 (pp)
1 (P)	1 PP	2 PPp	2 Pp	1 Ppp
2 (p)	2 Pp	4 Ppp	4 pp	2 ppp

These results yield $\frac{12}{18}$ ($P--$ and $P-$) : $\frac{6}{18}$ (ppp and pp), which reduces to 2 purple : 1 white.

12.2 The $AAA'A'$ parent in this cross can form three gamete genotypes in the ratio of $1\ AA : 4\ AA' : 1\ A'A'$, while the $AAAA'$ parent can form two gamete genotypes in the ratio of $1\ AA : 1\ AA'$. Random combination of the gametes from each of the two parents yields offspring in the following genotypic (and thus phenotypic) ratio: $1\ AAAA : 5\ AAAA' : 5\ AAA'A' : 1\ AA'A'A'$.

12.3 (a) With $n = 7$, $2n = 14$, $3n = 21$, $4n = 28$, and $7n = 49$.

(b) By means of polyploidy.

12.4 One possible sequence of inversion events is the following (parentheses are placed about the sequence affected in each case):

(ABCD)EFGH $\rightarrow$ D(CBAEFG)H $\rightarrow$ DG(FEAB)CH $\rightarrow$ DGBAEFCH

CHAPTER 13

13.1 Each Mp/mP parent in the cross will produce the following gamete frequencies:

$$f(Mp) = f(mP) = (1 - R)/2 = (1 - 0.08)/2 = 0.46$$

and

$$f(MP) = f(mp) = R/2 = 0.08/2 = 0.04$$

The remaining phenotype frequencies will then be

$$f(M-pp) = f(Mp/Mp) + 2f(Mp/mp) = (0.46)^2 + (2)(0.46)(0.04) = 0.2484$$
$$f(mmP-) = f(mP/mP) + 2f(mP/mp) = (0.46)^2 + (2)(0.46)(0.04) = 0.2484$$
$$f(M-P-) = 1 - (2)(0.2484) - 0.0016 = 0.5016$$

13.2 Based on the marker that differs in the corresponding parental and double-recombinant classes, the middle gene locus must be

(a) b
(b) f
(c) h
(d) j

13.3 Applying the pattern shown in Table 13–4 to this cross, the expected frequencies of the recombinant classes will be:

$$f(RAT/rat \text{ and } rat/rat) = CR_{r-a}R_{a-t} = (0.8)(0.05)(0.08) = 0.0032$$
$$f(RAt/rat \text{ and } raT/rat) = R_{r-a} - \gamma = 0.05 - 0.0032 = 0.0468$$
$$f(rAT/rat \text{ and } Rat/rat) = R_{a-t} - \gamma = 0.08 - 0.0032 = 0.0768$$

The expected frequency of the parental class then becomes

$$f(RaT/rat \text{ and } rAt/rat) = 1 - 0.0768 - 0.0468 - 0.0032 = 0.8732$$

CHAPTER 14

14.1 The direction and order of gene transfer in Example 14.1 will occur if each Hfr strain carries an F factor inserted at the corresponding numbered location in the map below (the direction of transfer is indicated by the arrow):

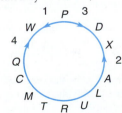

14.2 Of the two possible arrangements (*abc* and *bac*), *bac* is more consistent with the data. If *abc* were the correct arrangement, a minimum of two crossovers could give rise to wild-type recombinants in either of the two reciprocal crosses, as seen in the merozygotes below:

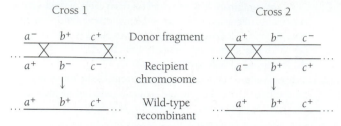

On the other hand, if *bac* is the correct arrangement, a minimum of four crossovers is needed to produce wild-type recombinants in cross 1, while two crossovers are needed in cross 2:

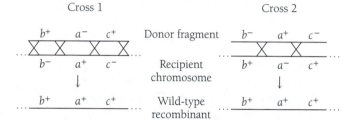

The production of many fewer wild-type recombinants in cross 1 than in cross 2 is consistent with the arrangement *bac*.

14.3 Since strain *a* cannot produce wild-type recombinants with deletion mutant 2 or 3, the point mutation in strain *a* must be located in region ii of the deletion map. Similarly, strain *b* fails to produce wild-type recombinants with deletion mutant 3 or 4, and strain *c* fails to give wild-type recombinants with either 1 or 2, so mutation *b* must be in region iii and mutation *c* must be in region i of the map.

CHAPTER 15

15.1 The data can be explained in terms of a linear 4.1-kb DNA molecule in which three *Alu*I restriction sites alternate with two *Eco*RI sites. Since the 0.8-kb and 0.5-kb *Alu*I fragments are also represented in the double digest, they must be located at the ends of the molecule, bounded internally by an *Alu*I restriction site. By adding the lengths of the various fragments in the double digest, we obtain:

0.5 + 0.4		= 0.9 kb
0.4 + 0.3		= 0.7 kb
0.3 + 1.5		= 1.8 kb
1.5 + 0.6		= 2.1 kb
0.6 + 0.8		= 1.4 kb

These fragments, along with the terminal 0.5-kb and 0.8-kb fragments, account for all of the fragments observed in the single enzyme digests. The overlapping character of the fragments can be used to construct the map:

*Alu*I	*Eco*RI	*Alu*I	*Eco*RI	*Alu*I
0.5	0.4	0.3	1.5	0.6	0.8

15.2 Haplotype C seems to be the best marker for the gene causing the disorder. Every individual carrying the C haplotype has the disease. This haplotype does not show complete linkage to the gene, however, since individual II-3 is affected by the disorder but does not carry the C haplotype (indicating that recombination has occurred between the gene and the haplotype marker).

CHAPTER 16

16.1 The base composition of the RNA would reflect that of the DNA strand complementary to the template strand, which may or may not correspond directly with the combined base compositions of both strands. For example, if the template strand is guanine-rich (say, 50% G, 10% C, 25% A, and 15% T), both complementary DNA and RNA strands would then be cytosine-rich (50% C, 10% G, 15% A, and 25% T or U). The total base composition of the corresponding double-stranded DNA region would then be 30% G, 30% C, 20% A, and 20% T, reflecting only the average GC (50 + 10)/2 and AU (25 + 15)/2 contents of the RNA transcripts.

16.2 Equal amounts of ADP and UDP in the reaction mixture would produce a synthetic RNA containing eight different triplet sequences, AAA, AAU, AUA, UAA, AUU, UAU, UUA, and UUU, all at the same frequency: $(\frac{1}{2})^3 = \frac{1}{8}$ or 12.5%. Referring to Table 16–3, we see that 25% of these codons (AUU and AUA) would code for isoleucine, 12.5% (UUA) would code for leucine, 12.5% (UUU) would code for phenylalanine, and 12.5% (UAU) would code for tyrosine. One of the possible codons (UAA) is a terminator or stop codon, and the remaining two (AAU and AAA) code for asparagine and lysine, respectively.

CHAPTER 17

17.1 GCA and GCG could be recognized by 3′ CGU; GCC and GCU could be recognized by 3′ CGG; and GCA, GCC, and GCU could be recognized by 3′ CGI. The minimum number of tRNAs required to recognize this set of codons is two: one containing 3′ CGU and another containing 3′ CGG, or one containing 3′ CGI and another containing 3′ CGC.

CHAPTER 18

18.1 Using the formula $N = \ln(1 - P)/\ln(1 - f)$, we obtain:

(a) $$N = \frac{\ln (0.10)}{\ln \left(1 - \dfrac{15 \times 10^3}{1.5 \times 10^8} \right)} \simeq 2.3 \times 10^4$$

(b) $$N = \frac{\ln (0.01)}{\ln \left(1 - \dfrac{15 \times 10^3}{1.5 \times 10^8} \right)} \simeq 4.6 \times 10^4$$

18.2 Starting with the 6-amino acid section beginning with Thr gives $4 \times 1 \times 2 \times 2 \times 1 \times 2 = 32$ probes.

CHAPTER 19

19.1 (a) Transcription occurs inductively, because of the wild-type condition.

(b) Transcription occurs constitutively, because of the i^c mutation.

(c) Transcription occurs constitutively, because of the i^c mutation.

(d) Transcription cannot occur, because of the p^- mutation.

(e) Transcription cannot occur, because of the i^s mutation.

(f) Transcription occurs constitutively, because of the o^c mutation.

19.2 Transcription from the promoter p_{RE} occurs during the establishment of lysogeny and is activated by gpcII. The mRNA initiated at p_{RE} is translated very efficiently, resulting in a buildup of gpcI and the suppression of the lytic response. Transcription from the promoter p_{RM} occurs once lysogeny is established and is activated by gpcI. The mRNA initiated at p_{RM} is used to form the cI repressor needed to maintain a stable lysogenic state.

CHAPTER 20

20.1 The calico (tortoiseshell) coloration occurs in females that are heterozygous for an X-linked pair of alleles (see Chapter 4, end-of-chapter Problem 5). In such a heterozygote, dosage compensation inactivates one of the X chromosomes and thus one of the coat color alleles in each cell. As a result, roughly half the cells express only the yellow allele, while the other half express only the black allele, giving the calico pattern.

CHAPTER 21

21.1 Yes, these early cellular divisions and protein synthesis should still occur, because they depend on the products of maternal genes, such as *bcd* and *nos,* that have been expressed by the mother during oogenesis and transported into the egg.

21.2 Assuming the normal cell is homozygous for the *rb-1*+ allele, the hybrid cell should be normal. The dominant *rb-1*+ allele will mask the expression of the recessive mutant oncogene.

CHAPTER 22

22.1 The mRNA codon for Met is AUG, corresponding to the DNA triplet base pair sequence ATG/TAC. An A/T → C/G transversion at the first position would give CTG/GAC on the DNA, CUG on the mRNA, and Leu as the amino acid: an A/T → T/A transversion at this position would give TTG/AAC on the DNA, UUG on the mRNA, and, again, Leu as the amino acid. A T/A → G/C transversion at the second position would give AGG/UCC on the DNA, AGG on the mRNA, and Arg as the amino acid, while a T/A → A/T transversion would give AAG/TTC on the DNA, AAG on the mRNA, and Lys as the amino acid. A G/C → T/A transversion at the third position would give ATT/TAA on the DNA, AUU on the mRNA, and Ile as the amino acid; a G/C → C/G transversion would give ATC/TAG on the DNA, AUC on the mRNA, and, again, Ile as the amino acid.

22.2 Since hydroxylamine can act only on GC base pairs, the original transition in this experiment is most likely GC → AT. This transition could be reversed by base analogues but not by hydroxylamine.

CHAPTER 23

23.1 (a) The initial plateau results from the inability of a single "hit" (pyrimidine dimer) to kill a cell. Because of an efficient repair system, small doses are not effective in killing wild-type cells. However, as the dose increases, the repair system becomes saturated and leaves some potentially lethal lesions unrepaired, thus accounting for the eventual reduction in survivorship at the higher doses.

(b) In the absence of visible light, only those repair processes that do not need to be activated by visible light (e.g., excision repair) function within the cells. However, when the cells are irradiated in the presence of visible light, photoreactivation also occurs, increasing the overall efficiency of repair.

CHAPTER 24

24.1 Since the six genotypes in this example are equal in frequency, the three alleles must also have the same frequency ($\frac{1}{3}$). This conclusion can also be derived mathematically. Consider the A_1 allele as an example. Since each genotype has a frequency of $\frac{1}{6}$, the frequency of the A_1 allele becomes:

$$f(A_1) = f(A_1A_1) + \tfrac{1}{2}f(A_1A_2) + \tfrac{1}{2}f(A_1A_3)$$
$$= \tfrac{1}{6} + \tfrac{1}{12} + \tfrac{1}{12}$$
$$= \tfrac{1}{3}$$

24.2 Three types of matings between individuals of identical genotype are possible: $AA \times AA$, $Aa \times Aa$, and $aa \times aa$. The expected frequencies of these matings in a randomly mating population are: $f(AA \times AA) = (0.1)^2 = 0.01$, $f(Aa \times Aa) = (0.2)^2 = 0.04$, and $f(aa \times aa) = (0.7)^2 = 0.49$. The proportion of all matings between mates of identical genotypes is then $0.01 + 0.04 + 0.49 = 0.54$.

24.3 At equilibrium, the frequencies of matings that can alter genotype frequencies are: $f(AA \times aa) = 2p^2q^2$ and $f(Aa \times Aa) = (2pq)^2 = 4p^2q^2$. Note that $f(Aa \times Aa) = 2f(AA \times aa)$—precisely the relationship needed to maintain a genetic equilibrium.

24.4 When the three blood group alleles are equal in frequency, $p = q = r = \frac{1}{3}$. The blood group frequencies will then be:

$$f(A) = p^2 + 2pr = (\tfrac{1}{3})^2 + 2(\tfrac{1}{3})(\tfrac{1}{3}) = \tfrac{1}{3}$$
$$f(B) = q^2 + 2qr = (\tfrac{1}{3})^2 + 2(\tfrac{1}{3})(\tfrac{1}{3}) = \tfrac{1}{3}$$
$$f(AB) = 2pq = 2(\tfrac{1}{3})(\tfrac{1}{3}) = \tfrac{2}{9} \text{ and}$$
$$f(O) = r^2 = (\tfrac{1}{3})^2 = \tfrac{1}{9}$$

24.5 Two possible outcomes give a frequency of $p' = 3/4$: one in which two of the offspring are AA and two are Aa and another in which three of the offspring are AA and one is aa. The probability of the first outcome is $(4!/2!2!)(1/4)^2(1/2)^2 = 3/32$, and the probability of the second outcome is $(4!/3!1!)(1/4)^2(1/4)^2 = 1/64$. Thus the total probability that $p' = 3/4$ is $3/32 + 1/64 = 7/64$. Note that the same result can be derived using equation 24.1, with $N = 4$, $x = 2$, and $p = q = 1/2$.

24.6 Each isolated breeding group in this example starts with $p = q = 1/2$ and $N = 8$. The average frequency of heterozygotes will decrease to $H_{10} = (1/2)(1 - 1/16)^{10} = 0.2624$. Since the two alleles have the same average frequency, the frequencies of the homozygotes will increase proportionally to $D_{10} = 0.25 + 0.2376/2 = 0.3688$ and $R_{10} = 0.3688$.

CHAPTER 25

25.1 (a) An increase in the generation time will tend to decrease fitness, since the reproductive rate is inversely related to the generation time.

(b) One approach is to express the fitness of a genotype as $W^{1/T}$, where W is the average number of offspring produced per parent during its lifespan and T is the generation time—the average age of reproduction. The fitness will thus increase with W and decrease with T.

25.2 (a) After 10 generations of complete selection, the frequency of the Tay-Sachs allele would become $q = (1/80)/[1 + (10)(1/80)] = 1/90 = 0.011$. The expected frequency of

Tay-Sachs disease would then be $(1/90)^2 = 1/8100 = 0.000123$.

(b) $[(1/80)^2 - (1/90)^2]/(1/80)^2 = 0.21$, or 21%.

25.3 (a) $w_{AA} = w_{Aa} = 0.5$ and $w_{aa} = 1$. The average fitness will then be $\bar{w} = (0.5)(0.3)^2 + (0.5)(2)(0.3)(0.7) + (1)(0.7)^2 = 0.745$.

(b) The frequency of the recessive allele in the next generation is $q' = [(1)(0.49) + (0.5)(0.3)(0.7)]/0.745 = 0.799$. Thus the frequency of the light-colored moth in the next generation is $(q')^2 = 0.638$ and the frequency of the melanic form is $1 - 0.638 = 0.252$.

25.4 (a) For the recessive disease, $\hat{q}^2 = 2 \times 10^{-5}$, and for the dominant disease, $\hat{p}^2 + 2\hat{p}\hat{q} = (2 \times 10^{-5})^2 + 2(2 \times 10^{-5})(1 - 2 \times 10^{-5}) \simeq 4 \times 10^{-5}$.

(b) The dominant form makes a greater contribution because the disease is expressed in heterozygotes.

25.5 The selection coefficients are $s_A = 0.4$ and $s_a = 1.0$. Thus, at equilibrium, the allele frequencies are $\hat{p} = 1.0/1.4 = 0.714$ and $\hat{q} = 0.4/1.4 = 0.286$. In this population, $\bar{w}$ is the fraction of zygotes that survive to reproductive age. Therefore, the fraction that fail to reproduce is $1 - \bar{w} = 1 - [(0.6)(0.714)^2 + (1)(2)(0.714)(0.286)] = 1 - 0.714 = 0.286$.

25.6 The change in the allele frequency per generation due to migration is $\Delta q_m = -m(q - q_m)$, and the change in the allele frequency per generation due to selection is $\Delta q_s = -q^2/(1 + q) \simeq -q^2$ (for small values of q). Thus, the net change is $\Delta q = -m(q - q_m) - q^2$. At equilibrium, $\Delta q = 0$, so $q^2 + mq - mq_m = 0$. Solving for $\hat{q}$ yields

$$\hat{q} = \frac{-m + \sqrt{m^2 + 4mq_m}}{2} = 0.005$$

25.7 (a) For much of recorded history, local human populations tended to be isolated by social, religious, and cultural differences as well as by geographic separation. There is now much greater mobility and freedom in the selection of mates, especially in the developed countries, leading to the breakdown of many of the traditional barriers to gene exchange and thus allowing for greater heterozygosity.

(b) Increases in the quality and quantity of food as well as various medical advances, such as the use of antibiotics, have undoubtedly contributed to an overall increase in human height.

CHAPTER 26

26.1 The maximum heterozygosity for a locus occurs when all allele frequencies are equal.

(a) In the case of two alleles, the maximum heterozygosity would be $H = 1 - [(1/2)^2 + (1/2)^2] = 1/2$.

(b) In the case of three alleles, the maximum heterozygosity would be $H = 1 - [(1/3)^2 + (1/3)^2 + 1/3)^2] = 2/3$.

26.2 For the α-hemoglobin chain, $k = 8.9 \times 10^{-10}$ per year. Thus the unit evolutionary period for the α chain is UEP $= 0.005/(8.9 \times 10^{-10}) = 5.6 \times 10^6$ years.

ANSWERS TO SELECTED END-OF-CHAPTER PROBLEMS

CHAPTER 1

3. (a) Telophase: chromosomes are clustered at the ends of the spindle, and a cell plate is developing.
   (b) Mid-to-late prophase; chromosomes are relatively short and distinct, and the nucleolus and nuclear envelope are starting to disappear.
   (c) Late anaphase; chromosomes are migrating on the spindle and nearing opposite poles.

4. (a) Interphase.
   (b) Prophase.
   (c) Anaphase.

7. Assuming that cytokinesis has occurred by the close of telophase, the numbers of chromosomes, centromeres, and chromatids will be as follows:

	(a)	(b)	(c)	(d)
Chromosomes:	20	20	40	20
Centromeres:	20	20	40	20
Chromatids:	40	40	0* (40)	0* (20)

   *Technically, chromatids do not exist once centromere division has occurred (the structures are now called daughter chromosomes).

9. (a) Occurs before mitosis and meiosis I.
   (b) Occurs during prophase of meiosis I.
   (c) Occurs during metaphase of mitosis and meiosis II.
   (d) Occurs during anaphase of meiosis I.
   (e) Limited to mitosis.

12. (a) $2^4 = 16$
    (b) $2^8 = 256$

CHAPTER 2

2. White. Two dominant parents can produce a recessive offspring if they are both heterozygous, but two recessive parents cannot produce a dominant offspring.

4. Black coat color is dominant to white. Using gene symbols $A- =$ black and $aa =$ white, the parental genotypes are, in order,
   $Aa \times Aa \rightarrow 3 A-:1 aa$;
   $AA \times AA$ (or $AA \times Aa$) $\rightarrow$ all $A-$;
   $AA \times aa \rightarrow$ all Aa;
   $Aa \times aa \rightarrow 1 Aa:1 aa$; and
   $aa \times aa \rightarrow$ all aa.

7. $\frac{3}{4}$. A normally pigmented individual who has an albino parent must be heterozygous, so both the husband and wife are Aa.

11. (a) All F_1 progeny have purple flowers and long stems.
    (b) $\frac{9}{16}$ purple, long : $\frac{3}{16}$ purple, short : $\frac{3}{16}$ white, long : $\frac{1}{16}$ white, short.
    (c) $\frac{1}{4}$ purple, long : $\frac{1}{4}$ purple, short : $\frac{1}{4}$ white, long : $\frac{1}{4}$ white, short.

13. $Aagg \times aaGg$

17. (a) If the C,c and Wx,wx nonallelic genes had assorted independently, the testcross would be expected to produce a $1:1:1:1$ phenotypic ratio in the offspring. The failure to produce such a ratio indicates that the genes are linked. The colored, starchy and colorless, smooth offspring, which occur in the highest frequency, are produced by the combination of a parental-type gamete (either $C Wx$ or $c wx$) from the heterozygote in the cross with a $c wx$ gamete from the recessive (testcross) parent. The colored, smooth and colorless, starchy offspring, which occur in the lowest frequency, are produced by the combination of a recombinant-type gamete (either $C wx$ or $c Wx$) produced by crossing-over with a $c wx$ gamete.
    (b) The phenotypic ratio of the offspring is approximately $2:2:1:1$. Hence, the gamete ratio of the heterozygous parent in the testcross is approximately $2 C Wx : 2 c wx : 1 C wx : 1 c Wx$.

19. (a) Normal Vv offspring.
    (b) Normal L_1Vv offspring.
    (c) Normal Vv offspring.
    (d) Virescent L_2Vv offspring.
    (e) $\frac{3}{4}$ normal $(V-):\frac{1}{4}$ virescent (vv) offspring.
    (f) Virescent L_2 offspring.

CHAPTER 3

1. (a) In an ordinary deck of 52 cards, there are 4 possible aces and 4 possible kings to choose from. Therefore, the probability of getting an ace or a king is $4/52 + 4/52 = 8/52 = 2/13$.
   (b) $(1/13)(1/13) = 1/169$
   (c) $(4/52)(3/51)(2/50)(1/49) = 1/270,725$
   (d) $3/6 = 1/2$
   (e) There are $6 \times 6 = 36$ possible outcomes; 5 of these outcomes give numbers adding to 8. Therefore, the probability is $5/36$.

3. (a) $2^3 = 8$
   (b) $2^4 = 16$
   (c) $2^5 = 32$
   (d) 2^n

7. Letting $A- =$ normal pigmentation, $aa =$ albinism, $B- =$ free lobes, and $bb =$ attached lobes, the parents in this cross have the genotypes $AaBb \times AaBb$. Assuming that the genes for these traits segregate and assort independently, the probability of a normally pigmented, free-lobed son is therefore $P(A-B-$ from $AaBb \times AaBb) \times P(\text{son}) = (9/16)(1/2) = 9/32$.

9. The son is a normal $(A-)$ child of heterozygous $(Aa \times Aa)$ parents. Therefore, the probability that he is a carrier is $P(Aa \mid A-) = 2/3$.

12. (a) $(6!/4!2!)(1/2)^4(1/2)^2 = 15/64$

(b) $(1/2)^4(1/2)^2 = 1/64$

(c) $P(\geq 4 \text{ girls}) = P(4 \text{ girls and } 2 \text{ boys } or \ 5 \text{ girls and } 1 \text{ boy } or \ 6 \text{ girls}) = (6!/4!2!)(1/2)^6 + (6!/5!1!)(1/2)^6 + (6!/6!0!)(1/2)^6 = 11/32$

(d) $P(6 \text{ girls } or \ 6 \text{ boys}) = 1/64 + 1/64 = 1/32$

14. (a) For any single pup, $P(\text{black, short}) = (3/4)(3/4) = 9/16$. Thus, for a litter of 5 pups, $P(2 \text{ black, short}) = (5!/2!3!)(9/16)^2(7/16)^3 = 0.265$.

(b) $(5!/2!1!1!1!)(9/16)^2(3/16)(3/16)(1/16) = 0.0417$

16. (a) The mean number is $(10{,}000)(1/10{,}000) = 1$ per year

(b) $P(1) = e^{-1} = 0.368$

18. (a) 1

(b) 2

(c) 3

(d) 5

(e) 8

19. $\chi^2 = 0.199$

CHAPTER 4

1. (a) Wild-type > steelblu > silverblu

(b) 3

(c) 2, 1

(d) Wild-type × steelblu

3. (a) 1 wild-type : 1 chinchilla

(b) 3 chinchilla : 1 Himalayan

(c) 2 wild-type : 1 chinchilla : 1 albino

7. Since the woman has a color-blind father, she must be heterozygous for the color-blind allele. Therefore, half of her sons will inherit this allele and be color-blind. (The genotype of the father is unimportant in this case, since he contributes only the Y chromosome to his sons.

9. (a) The cross is $BW \times bb$, which yields a ratio of 1 Bb (barred male) : 1 bW (nonbarred female).

(b) The parental genotypes are $BW \times Bb$. The cross $BW \times B-$ can produce a nonbarred female offspring, bW, only if the male in the cross is Bb.

10. X-linked recessive inheritance is a possibility for pedigrees (1) and (2). X-linked dominant inheritance is not a possibility for any of the pedigrees.

14. (a) 3 round : 1 wrinkled

(b) 1 large, numerous : 2 medium, several : 1 small, few

15. (a) Blue Andalusian is the heterozygous expression of a pair of alleles showing incomplete dominance.

(b) No, because Blue Andalusians are heterozygous.

20. First father cannot be type AB, but he could be type O, or type A or B if heterozygous; second cannot be AB; third cannot be B or O; fourth cannot be A or O; fifth cannot be A or O; and sixth cannot be B or O.

23. No; Mrs. X, who is type AB, cannot be the parent of an O child.

25. (a) $AaBb$ female × $Aabb$ male

(b) $aaBb$ male × $AAbb$ female

28. Nonallelic: the results are consistent with the results of a dihybrid cross in which two gene pairs interact to produce different eye colors.

32. (a) $\frac{12}{16}$ gray : $\frac{3}{16}$ chestnut : $\frac{1}{16}$ sorrel

(b) $EeGg$ mare × $eegg$ stallion

35. The cross is B_2B_2 man × B_1B_2 woman.
$P(B_1- \text{ male}) = 1/2 \ (B_1-) \times 1/2 \ (\text{male}) = 1/4$.

37. Let H = the allele for hen-feathering and h = the allele for cock-feathering (limited to males).

(a) hh male × Hh female.

(b) Mating is $HhBW \times Hhbb$, which yields $\frac{1}{2}$ hen-feathered, non-barred females $(--bW) : \frac{3}{8}$ hen-feathered, barred males $(H-Bb) : \frac{1}{8}$ cock-feathered, barred males $(hhBb)$.

40. Sex-linked trait: (c) and (e) are true for an X-linked trait; (d) and (e) are true for a Y-linked trait. Sex-influenced trait: (a), (b), (c), and (e) are true for an autosomal gene. Sex-limited trait: (a), (b), (c), and (e) are true for an autosomal gene.

CHAPTER 5

2. (a)

Eye color:	light blue	blue	blue-green	hazel	light brown	brown	dark brown
Contributing genes:	0	1	2	3	4	5	6

(b) 1 light blue : 6 blue : 15 blue-green : 20 hazel : 15 light brown : 6 brown : 1 dark brown, applying a binomial distribution.

4. If a dominant gene at a single locus is the genetic basis for the trait, the F_1 value would be the same as that of the dominant parent, 68.2%, and the F_2 value would be $(3/4)(68.2) + (1/4)(19.6) = 56.05\%$.

8. (a) $\frac{1}{16}$

(b) $\frac{1}{64}$

(c) $\frac{1}{256}$

(d) $\left(\frac{1}{2}\right)^{2n}$, or $\left(\frac{1}{4}\right)^n$

9. (a) $\frac{9}{16}$

(b) $\frac{27}{64}$

(c) $\frac{81}{256}$

(d) $\left(\frac{3}{4}\right)^n$

10. Parents: $aabbcc = 10$ cm and $AABBCC = 16$ cm. $F_1 : AaBbCc = 16$ cm. $F_2 : 27/64 \ A-B-C- = 16$ cm : $27/64 \ (A-B-cc + A-bbC- + aaB-C-) = 14$ cm : $9/64 \ (A-bbcc + aaB-cc + aabbC-) = 12$ cm : $1/64 \ aabbcc = 10$ cm.

13. (a) Decrease heritability, since it would reduce the overall genetic variance.

(b) Increase heritability, since it would reduce the overall phenotypic variance.

17. (a) $AaBbCc = 22$ inches

(b) Solving for the terms of the binomial formula $[(n!/x!(n-x)!] \ (3/4)^x(1/4)^{n-x}$ for $n = 3$ and $x = 3, 2, 1,$ and 0, we get $27/64$ (22 inches), $27/64$ (18 inches), $9/64$ (14 inches), and $1/64$ (10 inches).

(c) The mean of the $F_2 = (22)(27/64) + (18)(27/64) + (14)(9/64) + (10)(1/64) = 19$ inches. The variance of the $F_2 = (22 - 19)^2(27/64) + (18 - 19)^2(27/64) + (14 - 19)^2(9/64) + (10 - 19)^2(1/64) = 9$ inches^2.

CHAPTER 6

1. (a) Peptide

(b) Cysteine

(c) Primary

2. (a) 20^2

(b) 20^3

(c) 20^4

(d) 20^n

4. L-T-W-I-D-R-V-A-S

8. (a) Secondary

(b) Hydrogen

(c) α-helix
β-sheet

(d) Tertiary

(e) Polar (charged), nonpolar

(f) Oligomeric

(g) Quaternary

11. (200 residue pairs) (0.35 nm repeat per residue) = 70 nm
(120 average residue weight)(400 total residues) = 48,000

12. (Hydrogen bonds between N—H and C=O groups of different peptide bonds, (2) hydrogen bonds between hydrogen donor groups and hydrogen acceptor groups of polar amino acid side chains, (3) disulfide bonds, (4) ionic bonds between charged groups of amino acid side chains, and (5) hydrophobic interactions between nonpolar side chains.

16. Like alanine, glycine is relatively small and uncharged, so its substitution would have only minimal effects on the structure of the active site. Glutamic acid, however, is large and negatively charged and could greatly alter the structure and activity of the active site if it were incorporated in place of alanine.

20. Such a protein would most likely function in a nonpolar cellular environment, such as in the plasma membrane.

CHAPTER 7

1. (a) False; transformation involves naked DNA.

(b) False; transforming activity is destroyed by DNase.

(c) False; it indicates that the protein part of the phage does not enter the cell.

4. (a) Adenine, guanine, and cytosine.

(b) Thymine, uracil

(c) Complementary

(d) Hydrogen, phosphodiester

(e) $3'$, $5'$

(f) B, A, and Z

6. 15% A, 15% T, 35% G, and 35% C.

11. If we assume that the DNA is an ideal B form (see Table 7–2) with 10 base pairs per turn, there would be 2,000,000/10 = 200,000 complete turns.

16. (a) %GC = (86.5 − 69.3)/0.41 = 42%

(b) Buoyant density = 1.660 + (0.00098)(42) = 1.701

19. Since the DNAs of lambda and T2 phages are not complementary, their single strands cannot renature to form hybrid molecules. Slow cooling regenerates only lambda duplex molecules and T2 duplex molecules.

21. The DNA of this virus is probably single-stranded.

CHAPTER 8

3. During logarithmic growth (the straight-line portion of the graph), the increase in log N per unit time (the slope of the line) should equal (log 2)/t_d, where t_d is the doubling (or generation) time. The graph indicates that log N increases by approximately 0.6 during each hour of growth. This gives a doubling time of t_d = (log 2)/(slope) = 0.301/0.6 = 0.50 hour (30 minutes).

8. Base pairs are separated by a distance of 0.34 nm = 0.00034 μm. There are thus 1360/0.00034 = 4 × 10⁶ total base pairs in an E. coli chromosome.

10. Although hydrogen bonds are relatively weak, energy is still needed to break these bonds during strand separation. ATP is the most common source of energy in living cells, so it is the most likely provider of energy for helicase activity.

12. (a) All intermediate (no fully heavy or fully light DNA).

(b) $\frac{1}{4}$ intermediate and $\frac{3}{4}$ light.

(c) If conservative, $\frac{1}{2}$ heavy and $\frac{1}{2}$ light after one round of replication; $\frac{1}{8}$ heavy and $\frac{7}{8}$ light after three rounds.

(d) If dispersive, all within an intermediate band after one round of replication; all within a broad band, located between intermediate and light DNA, after three rounds.

15. (a) For bidirectional replication, (3) is correct, since replication forks would sweep around the circle in both directions from a, past b and d, and finally to c.

(b) For unidirectional replication, (1) is correct, since a single replication fork would sweep around the circle in one direction only.

CHAPTER 9

1. Chromosomes (c) and (f) are acrocentric, (a) is metacentric, (b) is submetacentric, and (e) is telocentric. Chromosome (c) has satellites.

5. (a) 11.2 pg, 11.2 pg, 5.6 pg.

(b) 46 chromosomes yield 92 chromatids, giving 11.2/92 = 0.12 pg per chromatid equivalent at G_2 and at anaphase. At telophase, each cell contains 46 chromatid equivalents, giving 5.6/46 = 0.12 pg per chromatid equivalent.

(c) The average DNA content (in pg/chromosome) will be 11.2/46 = 0.24 at G_2, 11.2/92 = 0.12 at anaphase, and 5.6/46 = 0.12 at the close of telophase.

7. (a) 11/24 = 46%, 8/24 = 33%, 4/24 = 17%, and 1/24 = 4%.

(b) (60)(0.4) = 24 min., (60)(0.3) = 18 min., (60)(0.2) = 12 min., and (60)(0.1) = 6 min., respectively.

9. Chloroplasts usually contain many copies of their relatively small genomes. They therefore contain more total DNA but fewer different genes than bacteria.

CHAPTER 10

2. (a) Autonomously, integrated, episomes

(b) Supercoiled

(c) Uni-, rolling-circle

(d) Origin of replication (ori)

(e) Plasmid incompatibility

4. (a) (4 × 10⁶ bp)/(100 minutes) = 4 × 10⁴ bp/minute. The rate of transfer is about the same as the rate of chromosome replication at a single replication fork.

(b) The transfer of the entire bacterial chromosome takes 100 minutes, so kt = (0.075)(100) = 7.5. The probability that an Hfr donor cell will transfer the entire chromosome therefore equals $e^{-7.5}$ = 5.5 × 10⁻⁴.

8. (a) Transposition

(b) Inverted, direct

(c) Excision, nonmutant

(d) Insertion sequences

11. A multiply drug-resistant plasmid can be formed by the acquisition of several transposons, each carrying one or more drug-resistant genes. The plasmid could spread throughout a bacterial population by conjugation if the plasmid carries a resistance-transfer segment (see Figure 10.15).

13. (a) Icosahedral, filamentous (helical), and binal (having a head and tail)

(b) Protein capsomers, nucleic acid (DNA or RNA) tail proteins

(c) Multiplicity of infection (MOI)

(d) Adsorption, penetration (injection), synthesis, assembly, and release (lysis)

(e) Lag, log (or exponential), eclipse

(f) Concatemers

(g) Exponentially, linearly

(h) Temperate, lysogenic, prophage

(i) Induction

14. DNA volume $= \pi r^2 h = \pi (1 \text{ nm})^2 (2 \times 10^5 \text{ bp})(0.34 \text{ nm/bp}) = 2.1 \times 10^5 \text{ nm}^3$, which gives $(2.1 \times 10^5 \text{ nm}^3)(10^{-7} \text{ cm/nm})^3 = 2.1 \times 10^{-16} \text{ cm}^3$. Note that the DNA has approximately the same volume as the space within the phage head.

16. (a) (300 plaque-forming units)/(0.2 ml) = 1500 per ml

(b) $(1500)(10^8) = 1.5 \times 10^{11}$ per ml

19. Virulent phages, like T2, lyse their host cells and produce a clear plaque. Temperate phages, like λ, occasionally undergo lysogenic growth, resulting in the formation of a few lysogenic cells that are immune to further infection by the phage; these lysogenic cells grow and give a cloudy (turbid) appearance to the plaque.

22. Phage λ: $(49,000)(0.34) = 1.67 \times 10^4$ nm.

E. coli: $(4 \times 10^6)(0.34) = 1.36 \times 10^6$ nm.

Human genomic DNA: $(2.9 \times 10^9)(0.34) = 9.86 \times 10^8$ nm, or approximately 1 m.

CHAPTER 11

3. (a) $\frac{1}{9}$. Both the husband and wife, who are normal, have a $\frac{2}{3}$ chance of carrying the recessive allele for this disease. Therefore, the probability that their first child would have the disease is $\frac{2}{3}$ (that the husband is a carrier) $\times \frac{2}{3}$ (that the wife is a carrier) $\times \frac{1}{4}$ (that both the husband and wife transmit the gene, given that they are both carriers).

(b) $\frac{1}{4}$.

5. The products of somatic mutation in plants can often be propagated vegetatively (i.e., by asexual means).

9. The pathway contains the steps: $\rightarrow \alpha$-ketoisocaproic acid $\rightarrow$ leucine. The pathway is blocked at the first step in strain 1 and at the second step in strain 2.

10. $B \rightarrow D \rightarrow A \rightarrow E \rightarrow C$. Mutant strain 1 is blocked in step $D \rightarrow E$, strain 2 in step $E \rightarrow C$, strain 3 in step $B \rightarrow D$, and strain 4 in step $A \rightarrow E$.

13. (a) Enzyme 2, enzyme 1

(b) *SSbb* (brown) $\times$ *ssBB* (scarlet) $\rightarrow$ *SsBb* (red)

(c) $\frac{9}{16}$ *S–B–* (red) : $\frac{3}{16}$ *S–bb* (brown) : $\frac{3}{16}$ *ssB–* (scarlet) : $\frac{1}{16}$ *ssbb* (colorless)

16. More than one metabolic block can lead to the production of melanin. If *AAbb* and *aaBB* individuals are albino because of blocks at two different enzymatic steps, they would produce *AaBb* offspring who have normal alleles for both enzymatic steps and thus normal pigmentation.

18. Since the mutation for HbS affects the β chain, the mutation for HbX must involve the α chain. Heterozygotes could then inherit a normal copy of each of the genes for the α and β chains and experience neither form of hemoglobin disease.

21. Many changes in the coding sequence of a gene can produce a mutation; however, for a true reversion, there is only one way to restore the wild-type sequence—the reverse of the original mutation.

23. (a) 200

(b) 100

(c) 100

25. One approach would be to use a fluctuation test in which the various samples are exposed to high concentrations of ethidium bromide. If the experiment results in large plate-to-plate variations in the number of resistant clones, the origin of resistance is most probably spontaneous mutation.

27. We would expect 48 mutant cells. The mutation rate is 5×10^{-5} per cell generation.

CHAPTER 12

2. $2n - 1 = 23$

4. (a) 2 colorless (*rr*) : 1 red (*Rr*).

(b) 3 colorless (2 *rr* + 1 *rrr*) : 3 red (2 *Rrr* + 1 *Rr*).

6. The cell with 69 chromosomes is triploid (3*n*). The cell with 52 chromosomes is aneuploid (2*n* + 6).

8. The loss of a Y chromosome during the initial cleavage stages of an XY zygote, followed by the separation of the cells to form an XY and XO twin pair.

12. 14; 7 nullisomics of genome A and 7 nullisomics of genome B are possible, one for each pair of chromosomes.

14. (1/36)(1/36) = 1/1296

16. One approach would be to obtain monoploid cells of this plant using anther culture, expose these cells to a mutagen, select for a virus-resistant strain, and regenerate the monoploid plant. A fertile, diploid virus-resistant strain of this plant could then be formed by applying colchicine.

21. Evolution in such cases may have occurred through a combination of mechanisms that alter chromosome number, such as polyploidy, aneuploidy, and centric fusions.

23. Cross the plants to produce a deficiency heterozygote. If the deficiency is large enough, it would be possible to identify the site of the deficiency cytologically from the pairing arrangement of the chromosomes in the heterozygote during prophase I of meiosis.

26. It is located on the short arm of the Y chromosome.

CHAPTER 13

3. *Rs/rs* = 46%, *rS/rs* = 46%, *RS/rs* = 4%, and *rs/rs* = 4%.

5. Not linked; about 25% of the testcross offspring are colored (*CcDd*), as expected for independent assortment.

7. The gametes produced by the dihybrid *AaBb* plant will occur in a ratio of 0.2 *AB* : 0.3 *Ab* : 0.3 *aB* : 0.2 *ab*. Random fertilization will thus produce a phenotypic ratio in the offspring of 0.54 *A–B–* : 0.21 *A–bb* : 0.21 *aaB–* : 0.04 *aabb*.

9.

a	8	*b*	6	*c*	2	*d*	4	*e*

11. (a) Coupling

(b) 35 map units

(c) *BC/BC* × *bc/bc*

13. The phenotypic classes would be the same among males and females and occur in the following proportions: 43% bar eyes ($s^+B/sB^+ + s^+B/Y$), 43% sable body ($sB^+/sB^+ + sB^+/Y$), 7% wild-type ($s^+B^+/sB^+ + s^+B^+/Y$), and 7% sable, bar ($sB/sB^+ + sB/Y$).

15. The gamete frequencies produced by the s^+t/st^+ parent in the cross are $f(s^+t) = f(st^+) = 0.455$ and $f(s^+t^+) = f(st) = 0.045$, while the gamete frequencies produced by the s^+t^+/st parent are $f(s^+t^+) = f(st) = 0.455$ and $f(s^+t) = f(st^+) = 0.045$. Upon fertilization, offspring phenotypes will be produced in the following ratio: 0.520475 $s^+–t^+–$: 0.229525 $s^+–tt$: 0.229525 $sst^+–$: 0.020475 $sstt$.

17. The *a* and *b* genes are linked in repulsion with a genetic map distance of 20 map units. The *b* and *c* genes are unlinked and assort independently. The *c* and *d* genes are linked in coupling with a genetic map distance of 40 map units.

21. (a) Distance gl-$v = 100(28 + 179 + 183 + 23)/1000 = 41.3$, distance v-$lg = 100(179 + 69 + 70 + 183)/1000 = 50.1$, and distance gl-$lg = 100(28 + 69 + 70 + 23)/1000 = 19.0$. The linkage map is

$$\overset{lg}{\vdash}\quad 19.0 \quad\overset{gl}{\vdash}\quad 41.3 \quad\overset{v}{\dashv}$$

(b) $lg\ gl^+v/lg^+gl\ v^+$

23. (a) $9\ c^+d^+/cd\ h^+/h : 9\ c^+d^+/cd\ h/h : 9\ cd/cd\ h^+/h : 9\ cd/cd\ h/h :$
$1\ c^+d/cd\ h^+/h : 1\ c^+d/cd\ h/h : 1\ cd/cd\ h^+/h : 1\ cd/cd\ h/h$

(b) $52.6875\ c^+$-d^+-h^+- : $17.5625\ c^+$-d^+-hh : $15.1875\ ccddh^+$- :
$5.0625\ ccddhh : 3.5625\ c^+$-$ddh^+$- : $3.5625\ ccd^+$-h^+- : 1.1875
c^+-$ddhh : 1.1875\ ccd^+$-hh

27. (a) 7

(b) We can detect single recombinants for a-b, b-c, and c-d; double recombinants for a-b and b-c, a-b and c-d, and b-c and c-d; and triple recombinants involving all three regions.

(c) 10^6

29. (a) Class 1 is PD, class 2 is NPD, classes 3 and 4 are T.

(b) Independent assortment is indicated by an equality in the frequency of PD and NPD tetrads.

(c) %SDS for $A,a = (100)(120)/660 = 18.2\%$, and %SDS for B,b $= (100)(55)/660 = 8.3\%$.

31. The genes are linked, since PD > NPD. The genetic distance between the genes is

$$\frac{100\left(4 + 2 + \dfrac{11 + 50 + 6}{2}\right)}{500} = 7.9 \text{ map units}$$

33. Distance d-$e =$

$$\frac{100\left(5 + 1 + \dfrac{42 + 10 + 11 + 10 + 12}{2}\right)}{500} = 9.7 \text{ map units}$$

Distance e-$f =$

$$\frac{100\left(12 + 5 + 20 + \dfrac{42 + 90 + 11 + 10 + 1}{2}\right)}{500} = 22.8 \text{ map units}$$

Distance d-$f =$

$$\frac{100\left(20 + \dfrac{90 + 10 + 11 + 10 + 12 + 1}{2}\right)}{500} = 17.4 \text{ map units}$$

The linkage map is

$$\overset{e}{\vdash}\quad 9.7 \quad\overset{d}{\vdash}\quad 17.4 \quad\overset{f}{\dashv}$$

36. (a) The maternal inheritance pattern suggests cytoplasmic inheritance.

(b) Yes; if the arg gene is on a nuclear chromosome, the heterokaryon test will give spores that are arg^+poky^+ only if $poky$ is inherited as an extranuclear trait.

CHAPTER 14

2. The map is circular:

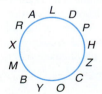

4. (a) Zygotic induction results from the lack of λ repressor protein in the F^- nonlysogens (see Chapter 19). The λ prophage in the donor chromosome is no longer repressed as it enters the F^- nonlysogen, resulting in induction and subsequent cell lysis.

(b) To map the prophage, conduct an interrupted mating experiment using the time of induction and lysis of the recipient cell as a genetic marker.

6. BARNEY

9. (a) The correct order is acb. The much lower frequency of wild-type recombinants in the first cross indicates that a minimum of four crossovers, rather than two, is required to form a wild-type genotype in the recipient chromosome when $++c$ is used as the donor strain. At least four crossovers would be required to form a fully wild-type cell if c is the middle marker.

(b) The number of wild-type recombinants produced when $++d$ is used as the donor strain would be much lower than in the reciprocal cross.

12. The map is circular, and all the markers are equidistant from one another:

15. (a) $4!/2!2! = 6$

(b) $5!/2!3! = 10$

(c) $N!/2!(N - 2)! = N(N - 1)/2$

16. Three regions are defined by the data: region I contains deletions 1 and 4, region II contains deletions 3 and 4, and region III contains deletions 2 and 3. The topological map is:

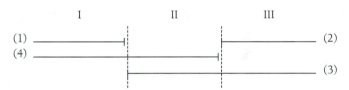

17. bac

20. Mutant loci 1 and 5 are in one cistron, mutant loci 3 and 4 are in a second cistron, and mutant locus 2 is in a third cistron.

CHAPTER 15

2. (a) Chromosome 1 — E_3, chromosome 11 — E_1, chromosome 17 — E_4, chromosome X — E_2

(b) E_4—All cell lines will have this enzyme in order to grow on HAT medium.

4. CATG, ACGT, TTAA, GATC, TCGA, TGCA, AACGTT, GTCGAC, AND CTGCAG.

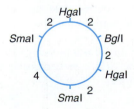

11. The map is circular:

16. No. The sequence would continue to serve as an *Mst*II restriction site, since the position altered in hemoglobin C disease can contain any base and still be recognized by the enzyme.

17. All the bands from child X match bands of either man Y or woman Y, while several bands of the child cannot be assigned to either man X or woman X. Thus child X is actually the biological child of couple Y.

19. 5′ GAACATCAGGATCGTACCAGGGTAA 3′

21. (a) Repulsion.
    (b) III-5 is recombinant; the frequency of recombination is 1/9 = 11.1%, or 11.1 map units.
    (c) 0.053285
    (d) The odds ratio is 22.17, giving a lod score of 1.346.

CHAPTER 16

1. (a) Transcription.
   (b) Promoters.
   (c) Consensus.
   (d) Sigma, core enzyme.
   (e) RNA polymerase II, RNA polymerase I, RNA polymerase III.

2. 5′ GUCUUACGCUAG 3′

5. TATCTGTAT

8. (a) $5 \times 3 = 15$
   (b) 432 (6 codons (Leu) × 2 codons (Tyr) × 6 codons (Arg) × 1 codon (Trp) × 6 codons (Ser))

11. Val: UUG, UGU, GUU
    Gly: GGU, GUG, UGG

12. UUU = $(0.6)^3$ = 21.6%; UUC, UCU, and CUU = $(0.6)^2(0.4)$ = 14.4% each; UCC, CUC, and CCU = $(0.6)(0.4)^2$ = 9.6% each; and CCC = $(0.4)^3$ = 6.4%.

13. The actual percentage of Phe would be 36% (UUU = 14.4%, UUC = 21.6%). Setting the frequency of Phe = 1, the relative amounts of the other amino acids would be:
    Leu = [14.4 (CUU) + 9.6 (CUC)]/36 = 2/3
    Ser = [14.4 (UCU) + 9.6 (UCC)]/36 = 2/3 and
    Pro = [9.6 (CUU) + 6.4 (CCC)]/36 = 4/9

17. (a) Exons, introns.
    (b) Heterogeneous nuclear RNA (hnRNA).
    (c) Spliceosome.
    (d) Transesterification.
    (e) 7-Methylguanosine cap, polyadenylate (poly A) tail.
    (f) Posttranscriptional modification or processing.

20. The three modifications are capping, polyadenylation, and splicing (see text for details).

22. If all the nucleotides covered make up part of the intron, the minimum size of an intron would be 71 (16 + 40 + 15) nucleotides in length.

26. (a) The genome of this virus is probably RNA.
    (b) In this case, the virus is most probably a DNA-containing virus or a retrovirus.

CHAPTER 17

1. Both eukaryotic and prokaryotic ribosomes are small ribonucleoprotein particles consisting of a small subunit with one rRNA component and a large subunit with more than one rRNA component. They differ in size: prokaryotic ribosomes are 70S, with subunits of 50S and 30S, while eukaryotic ribosomes are 80S, with subunits of 60S and 40S. They also differ in subunit composition: in prokaryotic ribosomes, the 30S component consists of one 16S rRNA and 21 proteins and the 50S component contains two rRNAs—one 23S and one 5S—and 31 proteins; in eukaryotic ribosomes, the 40S component consists of one 18S rRNA and 33 proteins and the 60S component contains three rRNAs—one 28S, one 5.8S, and one 5S—and 49 proteins.

3. In prokaryotes, all three rRNA components are processed from a single 30S pre-rRNA transcript, thus ensuring an equal number of each component within the cell.

4. The small ribosomal subunit contains regions involved in binding mRNA, tRNA, and various initiation factors. Regions in the large ribosomal submit are involved in translocation, catalyzing peptide bond formation, and binding tRNA and various protein factors.

7. Some polypeptide residue sites that normally contain His would now contain Arg.

8. (a) CAG, AAU, and GCG
   (b) Val—Leu—Arg

10. Three. The tRNA anticodon 3′AAU would be required to recognize UUA and UUG. Two anticodons are needed to recognize the CU− codons, one (3′GAU) to recognize CUA and CUG and another (3′GAG) to recognize CUU and CUC (or one (3′GAC) to recognize CUG and another (3′GAI) to recognize CUA, CUC, and CUU).

14. (1) In prokaryotes, binding of the mRNA to the ribosome is facilitated by the Shine-Delgarno sequence; in eukaryotes, it is facilitated by the 5′ cap.
    (2) In prokaryotes, 3 protein factors (IFs) are required for initiation; in eukaryotes, many IFs (possibly 10 or more) are required.
    (3) In prokaryotes, GTP is required as an energy source in initiation; in eukaryotes, both ATP and GTP are required.
    (4) In prokaryotes, the initiator tRNA is charged with formylmethionine; in eukaryotes, it is charged with methionine.
    (5) In prokaryotes, protein synthesis occurs on a 70S ribosome; in eukaryotes, it occurs on an 80S ribosome.
    (6) In prokaryotes, the message is usually translated from a polycistronic mRNA; in eukaryotes, it is translated from a monocistronic mRNA.
    (7) In prokaryotes, two release factors are required to recognize stop codons during termination (a third stimulates the activity of the others); in eukaryotes, only one is required.

18. Recent evidence indicates that the catalytic activity for this reaction is provided by 23S rRNA—another example of a critical biological function catalyzed by a ribozyme.

20. The major modifications include the attachment of different chemical groups to amino acid residues (such as acetylation, glycosylation, hydroxylation, methylation, and phosphorylation) and the proteolytic cleavage of the polypeptide chain (such as the cleavage of the signal sequence during the passage of the protein through the ER of eukaryotes or the plasma membrane of prokaryotes).

CHAPTER 18

2. (a) All the *BamHI* sites can be cut with *Sau3A*, since the *Sau3A* enzyme cuts to the 5′ side of the GATC sequence regardless of the adjacent base. However, only 1/4 × 1/4 = 1/16 of the *Sau3A* sites can be cut with *BamHI* since the latter enzyme requires that a guanine be present on each 5′ side of the GATC sequence.

   (b) The hybrid sequence reads 5′ $\frac{GATCC}{CTAGG}$. The hybrid's ends are to the 5′ side of G on the top strand and between the G residues on the bottom strand; these sites can be cut by *Sau3A* but not by *BamHI* (the *BamHI* palindrome has been lost.)

5. The general map is:

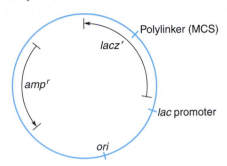

The polylinker region is a multiple cloning site (MCS; see Figure 18–3b) containing several restriction sites into which donor DNA can be inserted. The *amp^r* gene confers ampicillin resistance to cells transformed by a plasmid. Insertion of donor DNA inactivates the *lacz'* gene (controlled by the *lac* promoter), providing a metabolic detection system for detecting cells that have incorporated a recombinant plasmid (blue/white assay—white colonies contain a recombinant plasmid, blue colonies contain a nonrecombinant plasmid). The *ori* permits replication of the recombinant plasmid in the cloning experiment. The donor sequence can be retrieved after cloning by cutting it from the MCS region.

10. The presence of two *EcoRI* sites on the cloning vector means that insertion of the donor fragments would be a replacement (of the section between the *EcoRI* sites). Insertion inactivates the *amp^r* gene, allowing transformed *E. coli* cells to be selected through replica plating by their sensitivity to ampicillin but resistance to tetracycline. cDNA probes made of the different *EcoRI*-cut restriction fragments could then be used in a Southern blot of reisolated plasmid DNAs to determine which colonies contain the desired fragment; alternatively, the electrophoretic mobilities of the reisolated recombinant plasmids could be compared with those of standards created at the beginning of the experiment after ligation to produce the recombinant plasmids.

13. (a) $$N = \frac{\ln (0.10)}{\ln \left(1 - \dfrac{5 \times 10^3}{13.5 \times 10^6}\right)} \approx 6{,}220$$

   (b) $$N = \frac{\ln (0.01)}{\ln \left(1 - \dfrac{5 \times 10^3}{13.5 \times 10^6}\right)} \approx 12{,}430$$

17. Use the primary sequence of the viral protein to design a cDNA probe 18–20 nucleotides long (see Example 18.2); this size provides adequate specificity to determine whether the gene is homologous to the probe. Probe a Southern blot of the *EcoRI* re-

striction digest of the viral genome to locate the desired viral gene, which can then be isolated and amplified for sequencing.

19. Expression in *E. coli* requires the construction of a bacterial expression vector that contains bacterial transcriptional control signals. A pBR322 expression vector like the one shown in Figure 18–16 could be used; it contains the promoter and first seven codons of the *E. coli lacz* gene, including the Shine-Dalgarno sequence required for binding *E. coli* ribosomes to mRNA.

21. Infectious recombinant virions can be produced if the recombinant retroviral vector is used to transfect cells carrying a "helper" retrovirus—a provirus that carries the normal genes analogous to those defective in the vector. It is critical that the vector be defective so that no infectious nonrecombinant RNA tumor progeny viruses will be produced. (Only the recombinant genomes are packaged into virions in this system.)

CHAPTER 19

3. (a) Constitutive enzyme synthesis—the repressor is unable to bind to the altered operator region.

   (b) Constitutive enzyme synthesis—the altered repressor is unable to bind to the operator region.

   (c) Absence (repression) of enzyme synthesis—the altered repressor is unable to bind to the inducer.

   (d) Absence of enzyme synthesis—RNA polymerase is unable to initiate transcription.

5. (a) $i^c p^+ o^+ z^+ y^+ a^+$ and $i^+ p^+ o^c z^+ y^+ a^+$

   (b) $i^s p^+ o^+ z^+ y^+ a^+$ and $i^+ p^- o^+ z^+ y^+ a^+$

6. Three of the genotypes: $i^c p^+ o^+ z^+ y^+ a^+$, $i^+ p^+ o^c z^+ y^+ a^+$, and $i^+ p^- o^+ z^+ y^+ a^+$ will produce mRNA when present together with a wild-type operon; genotype $i^s p^+ o^+ z^+ y^+ a^+$ will not, since i^s is a dominant mutation.

10. cAMP acts as an allosteric activator of the CAP protein, which is required in its active form for the initiation of transcription of a number of inducible operons in *E. coli*, including the *lac*, *ara*, and *gal* operons. The intracellular concentration of cAMP tends to vary inversely with the glucose concentration, making these operons subject to catabolite repression.

12. A *cap^-* mutation would eliminate transcription of the *lac* operon. This mutation would be recessive, since the wild-type gene in a partial diploid *cap^-/cap^+* would provide the missing CAP protein and allow transcription of the *lac* operon to occur.

16. Transcription and translation cannot occur together in eukaryotes. Eukaryotic transcription and translation occur in different cell compartments—transcription in the nucleus and translation in the cytoplasm.

18. Such an arrangement helps to coordinate transcription and protein synthesis and provides a close coupling of these processes during cell growth.

21. The genes controlled by gpN and gpQ are normally turned off unless these proteins act positively to turn them on.

CHAPTER 20

1. The coordinate transcriptional regulation of genes with related functions is a fundamental principle of gene regulation in prokaryotes, where genes with related functions are organized into an operon, under the control of a single promoter. In contrast,

related eukaryotic genes have not been demonstrated to be clustered together into functional units of transcription; each gene seems to be transcribed separately from a different promoter. Transcriptional regulation in eukaryotes is also more complex than in bacteria, in that different RNA polymerases synthesize the different classes of RNAs (in contrast to prokaryotes, where one RNA polymerase synthesizes all RNAs). However, structural and functional similarities between the major subunits of eukaryotic and bacterial RNA polymerases suggest a common ancestral origin for the eukaryotic and prokaryotic enzymes.

Another significant difference between transcriptional regulation in prokaryotes and eukaryotes involves the requirement of eukaryotic (but not prokaryotic) RNA polymerases for specific protein (transcription) factors to activate RNA synthesis. In both cases, specific DNA sequences upstream of the transcriptional startpoint are involved in initiating transcription at the correct startpoint. Transcription in prokaryotes is regulated by a promoter that extends approximately 35 bases upstream of the startpoint; a eukaryotic gene has the same basic promoter organization, although the promoter region is larger. In addition, enhancers up to several thousand nucleotides upstream or downstream of (or within) a gene can influence the rate of transcription in eukaryotes.

4. Given the complexity of transcriptional initiation in eukaryotes, an introduced gene is not likely to be functional if it is inserted into a random chromosomal position. The gene would need to be targeted to a chromosomal position containing a promoter region and the other upstream DNA sequences that are recognized in forming a transcriptional initiation complex.

8. The most important evidence that the DNA-binding and transcriptional activation domains of transcription factors function independently of each other is the observation that the N-terminal transcriptional activation domain of factors such as Sp1 can be deleted with no effect on DNA binding.

12. The wild-type GAL4 protein is a transcription factor that binds to UAS_G to activate transcription of the GAL1 and GAL10 structural genes, which encode enzymes involved in the metabolism of galactose. This is an inducible system, however, since in the absence of galactose, the regulatory GAL80 protein binds to DNA-bound GAL4, inactivating the latter's transcriptional activation domain. The most likely nature of a $GAL4^c$ mutation is therefore a defect in the GAL4 protein domain that recognizes GAL80; since GAL80 then would not be able to complex with GAL4, transcription would be constitutive.

15. The steroid hormone receptors that have been studied are similar to other transcription factors in that (1) independent functional domains determine DNA-binding and transcriptional activation and (2) the recognition specificity of a gene to transcriptional activation by a receptor lies entirely within its DNA-binding domain. These receptors also contain a third domain responsible for binding the ligand (hormone). The DNA-binding domain is highly conserved, in most cases consisting of two CC zinc fingers. Less is known about the structural motifs that might be common to the ligand and transcriptional activation domains of steroid hormone receptors.

19. The primary target of DNA methylation seems to be CpG sequences, in which the cytosine is methylated. Methylation is correlated with a transcriptionally inactive state, although the correlation is not complete.

CHAPTER 21

3. Since actinomycin D blocks transcription, these results clearly indicate that transcription of genes that are expressed between the 6th and 11th hours of development is absolutely required for gastrulation to occur at the 15th hour of development.

6. The occurrence of identical quintuplets in humans proves that human embryonic cells remain totipotent for at least three division cycles because three divisions are needed to obtain eight blastomeres. If the eight blastomeres then accidentally separate, instead of remaining together in one embryo, they will give more than enough cells for identical quintuplets to develop.

9. (a) Mating type a: The absence of a functional α1 protein means that the cell would be unable to activate its α-specific genes.
   (b) Mating type a: The absence of the α2 protein means that transcription of the a-specific genes could not be repressed.
   (c) Mating type α.
   (d) Mating type α: Deletion of HMLE means that the HML locus cannot be silenced and expression of the α1 and α2 proteins occurs.

10. Each B lymphocyte produces just one type of antibody. Prior to its differentiation, however, it can theoretically produce $(m_L \times n_L)(m_H \times n_H \times p)$ different antibodies, where m is the number of variable light (L) or heavy (H) copies, n is the number of copies of the joining region in light or heavy encoding regions, and p is the number of copies of the D segment.

17. Since DNA methylation is associated with an inactive genetic state, inhibition of methylation by L-ethionine could cause liver cancer by allowing excessive expression (a quantitative response) of one or more proto-oncogenes in the liver cells.

20. Yes, the amplification of the expression of a proto-oncogene can cause cancer if the gene comes under the influence of a promoter or enhancer that leads to its hyperexpression; mutation in the proto-oncogene is not required.

26. Integration of the viral genome in the vicinity of a proto-oncogene could activate that gene, causing its overexpression. Alternatively, integration of the viral genome into a proto-oncogene could mutate that gene, leading to oncogenesis.

CHAPTER 22

2. (a) Cytosine.
   (b) Guanine.
   (c) Silent.
   (d) Missense, neutral.
   (e) Nonsense.
   (f) Suppression.
   (g) Cytosine, uracil.

6. Only arginine in the case of transition. Either arginine, glycine, serine, leucine, or cysteine in the case of a transversion.

9. UUA, UUG, UCA, UCG, UAU, UAC, UGU, UGC, UGG, CAA, AAA, GAA, CAG, AAG, GAG, CGA, AGA, GGA.

12. AGA

14. Leaky mutants are a result of a partial metabolic block, due to a mutational alteration in a gene that reduces the activity of the encoded enzyme.

16. GC → TA (or CG → AT) will tend to be the most common mutation.

23. Introducing a new restriction site during site-directed mutagenesis enables the investigator to quickly identify the desired clones.

CHAPTER 23

3.

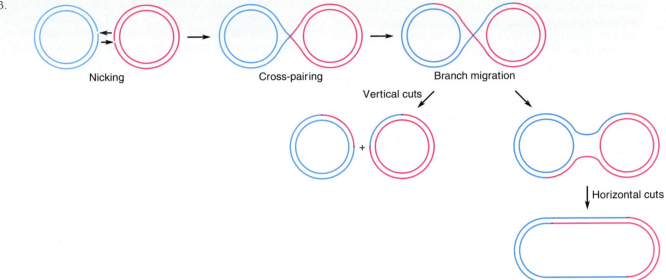

Nicking   Cross-pairing   Branch migration

Vertical cuts

Horizontal cuts

5. Visible light is needed for the activation of the photoreactivating enzyme, photolyase.

9. Deamination of 5-methylcytosine produces thymine, resulting in a GC → AT transition. The transition would not be repaired since thymine is a normal constituent of DNA.

10. Mutations accumulate at a faster rate because RNA lacks the elaborate repair systems typical of DNA.

13. Continued functioning of the SOS system could lead to an excessive incorporation of mismatched bases due to the expression of the $umuC$ and $umuD$ genes.

15. Any decrease in $3' \rightarrow 5'$ exonuclease activity will tend to decrease the proofreading capability of the DNA polymerase and thus increase the frequency of misincorporated bases.

CHAPTER 24

1. $f(M) = 408/1400 = 0.291$, $f(MN) = 694/1400 = 0.496$, $f(N) = 298/1400 = 0.213$; $f(L^M) = 0.291 + 0.496/2 = 0.539$, $f(L^N) = 0.213 + 0.496/2 = 0.461$.

5. (a) $p = f(A) = 0.1 + 0.6/2 = 0.4$, $q = f(a) = 1 - p = 0.6$
   (b) $p = 0.4$, $q = 0.6$, $f(AA) = p^2 = 0.16$, $f(Aa) = 2pq = 0.48$, and $f(aa) = q^2 = 0.36$

7. (a) 0.8 (husband) × 0.2 (wife) = 16%
   (b) 0.8 (male) × 0.2 (female) + 0.2 (male) × 0.8 (female) = 32%

9. $(p^2 + 2pq)/q^2 = (1 - q^2)/q^2 = 8$, so that $9q^2 = 1$. Thus, $q^2 = 1/9$, $q = 1/3$, $p = 2/3$, and $2pq = 4/9$.

10. $2pq = 4q^2$. Thus, $2(1 - q) = 4q$ and $q = 1/3$.

14. In both herds, $p = 0.7$ and $q = 0.3$, giving an expected ratio of $p^2 : 2pq : q^2 = 0.49 : 0.42 : 0.09$ according to the Hardy-Weinberg law. Thus, only herd 2 (which has a ratio of 98:84:18) conforms to a population in random-mating equilibrium.

17. (a) $q = 40/1000 = 0.04$, so $q^2 = 16\%$.
   (b) $2pq = 2(0.96)(0.04) = 0.076$, or about 8%.
   (c) Homozygous normal female × any male = $p^2 = 92\%$.

19. (a) $r = f(I^o) = \sqrt{100/600} = 0.4$; $(p + r)^2 = f(A) + f(O) = 0.50$, so $p + r = \sqrt{0.50} = 0.7$ and $p = 0.7 - 0.4 = 0.3$; $q = 1 - q - r = 0.3$.

   (b) $f(A) = p^2 + 2pr = 0.33$, giving 198 type A; $f(B) = q^2 + 2qr$, giving 198 type B; $f(AB) = 2pq = 0.18$, giving 108 type AB and 96 type O.

25. $f(AA) = p^2(1 - F) + pF = (0.16)(0.5) + (0.4)(0.5) = 0.28$, $f(Aa) = 2pq(1 - F) = (0.48)(0.4) = 0.24$, and $f(aa) = q^2(1 - F) + qF = (0.36)(0.5) + (0.6)(0.5) = 0.48$.

27. Ten generations.

29. $p = 0.52 + 0.28/2 = 0.66$ and $q = 0.34$, so $2pq = 0.449$. Since $f(Aa) = 2pq(1 - F)$, then $F = 1 - f(Aa)/2pq = 1 - 0.28/0.449 = 0.376$.

32. The founder effect is a phenomenon that occurs when a population is established by a small group of colonizers. Due to random sampling error, the allele frequencies in the small group may differ substantially from those of the parent population from which they were derived. Allele frequencies in the base population are $p = q = 0.5$. The probability that $p = q = 0.5$ in the founder population is P{(2 AA and 2 aa) or (1 AA, 2 Aa, and 1 aa) or (4 Aa)} $= 6(0.25)^2(0.25)^2 + 12(0.25)(0.50)^2(0.25) + (0.50)^4 = 0.273$. Therefore, the probability that the founder population differs from the base population in allele frequency is $1 - 0.273 = 0.727$.

34. (a) $(0.95)^{20} = 0.3585$
   (b) $(0.95)^{100} = 0.00592$

CHAPTER 25

1. The fitnesses are $82/154 = 0.532$ and $16/73 = 0.219$. Therefore, the relative fitness values are w (dark-colored moths) = 1 and w (light-colored moths) = $0.219/0.532 = 0.41$.

4. $q_o^2 = 10^{-4}$ and $q_t^2 = 10^{-6}$, so $q_o = 10^{-2}$ and $q_t = 10^{-3}$. Thus, employing equation 25.2, $t = 1/10^{-3} - 1/10^{-2} = 900$ generations.

8. With complete selection against a recessive homozygote, $\bar{w} = p^2 + 2pq + q^2(1 - s) = p^2 + 2pq$, since $s = 1$. This gives $\bar{w} = p(p + 2q) = p(p + q + q) = p(1 + q)$, since $p + q = 1$.

10. The equilibrium associated with a balanced polymorphism is attained through forces of natural selection and retains less-favored phenotypes along with more adapted ones. It is only

indirectly affected by the mating system. The Hardy-Weinberg equilibrium is attained through random mating and involves no selection or other evolutionary forces. The two equilibria are similar in that both maintain variability in a population.

12. Using the equation $\hat{q}^2 = u/s$, we get $\hat{q}^2 = 10^{-5}$, with $s = 1$. Thus, $\hat{q} = 0.0032$ and $\hat{p} = 0.9968$, so that $f(\text{heterozygotes}) = 2\hat{p}\hat{q} = 0.0063$.

13. At a lower frequency, since an X-linked recessive gene is expressed in all males who carry the gene and is, therefore, more directly exposed to selection than an autosomal recessive gene.

17. (a) q in zygotes = 0.25, q in adults = 0.20, giving $\Delta q = 0.05$.
    (b) $P(6\ AA\ \text{and}\ 4\ Aa) = (10!/6!4!)(1/2)^4(1/2)^4 = 105/512$.
    (c) Nonselective. No selection is involved, since both genotypes have the same survival rate. Random sampling error has occurred in forming the small adult population.

20. A deme is composed of freely intermating individuals that can also mate with individuals of another local population but do so less frequently because of isolation by distance. A race is a group of genetically distinct populations that share similar allele frequencies. Both demes and races are linked by their potential for gene exchange. A species is group of populations that are reproductively isolated from populations of other species.

22. Systematic evolutionary forces include migration, mutation, and selection. These forces, alone or in combination, act in an orderly, directional manner to produce equilibrium allele frequencies in a population. The dispersive force of genetic drift causes random fluctuations that scatter allele frequencies away from their equilibrium values, resulting in the fixation or loss of an allele.

25. (a) $f(\text{M offspring}) = f(\text{M} \times \text{M}) + (1/2) f(\text{M} \times \text{MN}) + (1/4) f(\text{MN} \times \text{MN}) = 0.04 + (1/2)(0.16) + (1/4)(0.16) = 0.16$.
    (b) For random mating, $f(\text{M}) = \sqrt{f(\text{M} \times \text{M})} = 0.2$ and $f(\text{MN}) = \sqrt{f(\text{MN} \times \text{MN})} = 0.4$. Thus, $f(\text{N}) = 1 - 0.2 - 0.4 = 0.4$.
    (c) $p = f(L^M) = f(\text{M}) + (1/2) f(\text{MN}) = 0.4$ and $q = f(L^N) = 0.6$. Therefore, the genotype frequencies expected among the offspring are $f(\text{M}) = (0.4)^2 = 0.16$, $f(\text{MN}) = 2(0.4)(0.6) = 0.48$, and $f(\text{N}) = (0.6)^2 = 0.36$.

CHAPTER 26

2. The heterozygosity per locus values are as follows: $H_{Got\text{-}1} = 1 - (0.000961 + 0.008836 + 0.765625) = 0.224578$
   $H_{Mdh\text{-}1} = 1 - (0.000961 + 0.938961) = 0.060078$ and
   $H_{Est\text{-}1} = 1 - (0.003844 + 0.003969 + 0.140625 + 0.250000) = 0.601562$
   The average heterozygosity for the three loci is then $\bar{H} = (0.224578 + 0.060078 + 0.601562)/3 = 0.295406$.

4. The UEP is the time required for two proteins to diverge by 1% of their amino acid sequence. The UEP can be related to the rate of amino acid substitution (k) by the formula UEP $= 0.01/2k = 0.005/k$.

7. Applying the formula $t = -(1/2k)\ln(1 - n_d/n)$, the time of divergence becomes $t = -(1/4 \times 10^{-9})\ln(1 - 0.2) = (2.5 \times 10^8)(0.223) = 5.58 \times 10^7$ years.

9.

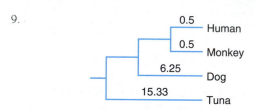

12.

APPENDIX C

FURTHER READINGS

CHAPTER 1 Genetics: Early History and Cytological Foundations

Alberts, B., D. Bray, J. Lewis, M. Raff, K. Roberts, and J. D. Watson. 1989. *Molecular biology of the cell.* New York: Garland.

Carlson, E. A. 1973. *The gene: A critical history.* Philadelphia: Saunders.

Darnell, J., H. Lodish, and D. Baltimore. 1986. *Molecular cell biology.* New York: Scientific American Books.

Dunn, L. C. 1965. *A short history of genetics.* New York: McGraw-Hill.

Gardner, E. J. 1972. *History of biology.* Minneapolis: Burgess.

Hyams, J. S., and B. R. Brinkley, eds. 1989. *Mitosis: molecules and mechanisms.* San Diego: Academic Press.

McIntosh, J. R., and K. L. McDonald. 1989. The mitotic spindle. *Sci. Am.* 261: 48–56.

Moens, P. B., ed. 1987. *Meiosis.* New York: Academic Press.

Rhoades, M. M. 1950. Meiosis in maize. *J. Hered.* 41: 59–67.

Sturtevant, A. H. 1965. *A history of genetics.* New York: Harper & Row.

Sutton, W. S. 1903. The chromosomes in heredity. *Biol. Bull.* 4: 231–51.

Swanson, C. P., T. Merz, and W. J. Young. 1981. *Cytogenetics.* Englewood Cliffs, N.J.: Prentice-Hall.

CHAPTER 2 Mendelian Principles

Bateson, W. 1909. *Mendel's principles of heredity.* Cambridge: Cambridge University Press.

Carlson, E. A. 1973. *The gene: A critical history.* Philadelphia: Saunders.

Douglas, L., and E. Novitski. 1977. What chance did Mendel's experiments give him of noticing linkage? *Heredity* 38: 253–57.

Dunn, L. C. 1965. *A short history of genetics.* New York: McGraw-Hill.

Gardner, E. J. 1972. *History of biology.* Minneapolis: Burgess.

Grant, V. 1975. *Genetics of flowering plants.* New York: Columbia University Press.

Hutt, F. B. 1964. *Animal genetics.* New York: Ronald Press.

McKusick, V. A. 1994. *Mendelian inheritance in man.* 11th ed. Baltimore: Johns Hopkins Press.

Mendel, G. [1865] 1959. Experiments in plant hybridization. Reprinted in *Classical papers in genetics,* ed. J. A. Peters. Englewood Cliffs. N.J.: Prentice-Hall.

Olby, R. C. 1966. *Origins of Mendelism.* London: Constable.

Rennie, J. 1994. Grading the gene tests. *Sci. Am.* 270: 88–97.

Sandler, I., and L. Sandler. 1985. A conceptual ambiguity that contributed to the neglect of Mendel's paper. *Hist. Phil. Life Sci.* 7: 3–70.

Stern, C., and E. R. Sherwood. 1966. *The origins of genetics, A Mendel source book.* San Francisco: W. H. Freeman.

Sturtevant, A. H. 1965. *A history of genetics.* New York: Harper & Row.

CHAPTER 3 Chance and Mendelian Inheritance

Batschelet, E. 1975. *Introduction to mathematics for life scientists.* Heidelberg: Springer-Verlag Berlin.

Elandt-Johnson, R. C. 1971. *Probability methods and statistical methods in genetics.* New York: Wiley.

Grossman, S. I., and J. E. Turner. 1974. *Mathematics for the biological sciences.* New York: Macmillan.

Jeffreys, W. H., and J. O. Berger. 1992. Ockham's razor and Bayesian analysis. *Am. Sci.* 80: 64–72.

Mosimann, J. E. 1968. *Elementary probability for the biological sciences.* New York: Appleton-Century-Crofts.

CHAPTER 4 Extensions of Mendelian Analysis

Atwood, S. S., and J. T. Sullivan. 1943. Inheritance of a cyanogenetic glucoside and its hydrolyzing enzyme in *Trifolium repens. J. Hered.* 34: 311–20.

Bridges, C. B. 1925. Sex in relation to chromosomes and genes. *Am. Nat.* 59: 127–37.

Charlesworth, B. 1991. The evolution of sex chromosomes. *Science* 251: 1030–33.

Corcos, A. F. 1983. Pattern baldness: Its genetics revisited. *Amer. Biol. Teacher* 45: 371–75.

Ginsberg, V. 1972. Enzymatic basis for blood groups. *Methods Enzymol.* 36: 131–49.

Lyon, M. F. 1990. Evolution of the X chromosome. *Nature* 348: 585–86.

McKusick, V. A. 1965. The royal hemophilia. *Sci. Am.* 213: 88–95.

Mollon, J. D. 1986. Understanding color vision. *Nature* 321: 12–13.

Morgan, T. H. 1910. Sex-limited inheritance in *Drosophila*. *Science* 32: 120–22.

Nathans, J. 1989. The genes for color vision. *Sci. Am.* 260: 42–49.

Searle, A. G. 1968. *Comparative genetics of coat colour in mammals.* London: Logos Press.

Sutton, W. S. 1903. The chromosomes in heredity. *Biol. Bull.* 4: 231–37.

CHAPTER 5   Quantitative Inheritance

Bodmer, W. F., and L. L. Cavalli-Sforza. 1976. *Genetics, evolution and man.* San Francisco: W. H. Freeman.

Bouchard, T. J., Jr., and M. McGue. 1981. Familial studies of intelligence: A review. *Science* 212: 1055–59.

Bouchard, T. J., Jr., D. T. Lykken, M. McGue, N. L. Segal, and A. Tellegen. 1990. Sources of human psychological differences: The Minnesota study of twins reared apart. *Science* 250: 223–28.

Brues, A. M. 1946. A genetic analysis of human eye color. *Am. J. Physical Anthropology* (new series) 4: 1–36.

Cavalli-Sforza, L. L., and W. F. Bodmer. 1971. *The genetics of human populations.* San Francisco: W. H. Freeman.

East, E. M. 1910. A Mendelian interpretation of variation that is apparently continuous. *Am. Nat.* 44: 65–82.

———. 1916. Studies on size inheritance in *Nicotiana. Genetics* 1: 164–76.

Elseth, G. D., and K. D. Baumgardner. 1981. *Population biology.* New York: D. Van Nostrand.

Falconer, D. S. 1989. *Introduction to quantitative genetics.* London: Longman.

Feldman, M. W., and R. C. Lewontin. 1975. The heritability hangup. *Science* 190: 1163–68.

Genes & behavior issue, 1994. *Science* 264: 1686–1739.

Holt, S. B. 1961. Inheritance of dermal ridge patterns. In *Recent Advances in Human Genetics,* ed. L. S. Penrose 101–19. London: J. and A. Churchill.

LeVay, S., and D. H. Hamer. 1994. Evidence for a biological influence in male homosexuality. *Sci. Am.* 270: 43–55.

Marx, Jean. 1990. Dissecting the complex diseases. *Science* 247: 1540–42.

Mather, K. and J. L. Jinks. 1977. *Introduction to Biometrical Genetics.* Ithaca, New York: Cornell University Press.

McGue, M., T. J. Bouchard, Jr., W. G. Iacono, and D. T. Lykken. 1993. Behavioral genetics of cognitive ability: A life-span perspective. In *Nature, nurture and psychology,* eds. R. Plomin and G. E. McClearn. Washington, D.C.: American Psychological Association.

Weir, B. S., E. J. Eisen, M. M. Goodman, and G. Namkoong, eds. 1988. *Proceedings of the Second International Conference on Quantitative Genetics.* Sunderland, Mass.: Sinauer.

CHAPTER 6   Proteins

Alberts, B., D. Bray, J. Lewis, M. Raff, K. Roberts, and J. D. Watson. 1994. *Molecular biology of the cell.* New York: Garland.

Anfinsen, C. G. 1973. Principles that govern the folding of protein molecules. *Science* 181: 223–30.

Branden, C., and J. Tooze. 1991. *Introduction to protein structure.* New York: Garland.

Chothia, C., and A. V. Finkelstein. 1990. The classification and origins of protein folding patterns. *Ann. Rev. Biochem.* 59: 1007–39.

Doolittle, R. F. 1985. Proteins. *Sci. Am.* 253: 88–99.

———. 1993. Evolutionarily mobile modules in proteins. *Sci. Am.* 269: 50–56.

Doty, P. 1957. Proteins. *Sci. Am.* 197: 173–84.

Ellis, R. J., and S. M. van der Vies. 1991. Molecular chaperones. *Ann. Rev. Biochem.* 60: 321–47.

Kendrew, J. C. 1961. The three-dimensional structure of a protein molecule. *Sci. Am.* 205: 96–110.

Lehninger, A. L., D. L. Nelson, and M. M. Cox. 1993. *Principles of biochemistry.* New York: Worth.

Matthews, C. R. 1993. Pathways of protein folding. *Ann. Rev. Biochem.* 62: 653–83.

Richards, F. M. 1991. The protein folding problem. *Sci. Am.* 264: 54–63.

Voet, D., and J. G. Voet. 1990. *Biochemistry.* New York: Wiley.

Zubay, G. 1993. *Biochemistry.* Dubuque, Iowa: Wm. C. Brown.

CHAPTER 7   Nucleic Acids

Avery, O. T., C. M. MacLeod, and M. McCarty. 1944. Studies on the chemical nature of the substance inducing transformation of *Pneumococcal* types. *J. Exp. Medicine.* 79: 137–58.

Baver, W. R., F. H. C. Crick, and J. H. White. 1980. Supercoiled DNA. *Sci. Am.* 243: 118–33.

Chargaff, E., E. Vischer, R. Doniger, C. Green, and F. Misani. 1949. The composition of the deoxypentose nucleic acids of thymus and spleen. *J. Biol. Chem.* 177: 405–16.

Dickerson, R. E. 1983. The DNA helix and how it is read. *Sci. Am.* 249: 94–111.

Dickerson, R. E., H. R. Drew, B. N. Conner, R. M. Wing, A. V. Fratini, and M. L. Kopka. 1982. The anatomy of A-, B-, and Z-DNA. *Science* 216: 475–85.

Driscoll, R. J., M. G. Youngquist, and D. D. Baldeschwieler. 1990. Atomic-scale imaging of DNA using scanning tunneling microscopy. *Nature* 346: 294–96.

Felsenfeld, G. 1985. DNA. *Sci. Am.* 253: 58–78.

Franklin, R. E., and R. G. Gosling. 1953. Molecular configuration in sodium thymonucleate. *Nature* 171: 740–41.

Griffith, F. 1928. Significance of pneumococcal types. *J. Hyg.* 27: 113–59.

Hall, S. S. 1993. Old school ties: Watson, Crick, and 40 years of DNA. *Science* 259: 1532–33.

d'Herelle, F. 1921. Le bacteriophage. Son role dans l'immunite. Paris: Masson. English translation (1922). The bacteriophage; its role in immunity. Baltimore: Williams & Wilkens.

Hershey, A. D., and M. Chase. 1952. Independent functions of viral protein and nucleic acid in growth of bacteriophage. *J. Gen. Physiol.* 36: 39–56.

Lehninger, A. L., D. L. Nelson, and M. M. Cox. 1993. *Principles of biochemistry.* New York: Worth.

McCarty, M. 1985. *The transforming principle: Discovering that genes are made of DNA.* New York: Norton.

Meselson, M., and F. W. Stahl. 1958. The replication of DNA in *Escherichia coli. Proc. Natl. Acad. Sci., U.S.* 44: 671–82.

Voet, D., and J. G. Voet. 1990. *Biochemistry.* New York: Wiley.

Wang, J. C. 1982. DNA topoisomerases. *Sci. Am.* 247: 94–109.

Watson, J. D. 1968. *The Double Helix.* New York: Atheneum.

Watson, J. D., and F. H. C. Crick. 1953. Molecular structure of nucleic acids: A structure for deoxyribose nucleic acid. *Nature* 171: 737–38.

Watson, J. D., and F. H. C. Crick. 1953. Genetical implications of the structure of deoxyribonucleic acid. *Nature* 171: 964–67.

Wilkens, M. H. F., A. R. Stokes, and H. R. Wilson. 1953. Molecular structure of desoxypentose nucleic acid. *Nature* 171: 738–40.

Zubay, G. 1993. *Biochemistry.* Dubuque, Iowa: Wm. C. Brown.

CHAPTER 8   Bacteria and Bacterial DNA

Baker, T. A., and S. H. Wickner. 1992. Genetics and enzymology of DNA replication in *Escherichia coli. Ann. Rev. Genet.* 26: 447–77.

Cairns, J. 1966. The bacterial chromosome. *Sci. Am.* 214: 36–44.

Hayes, W. 1968. *The genetics of bacteria and their viruses.* New York: Wiley.

Ingraham, J. L., O. Maaloe, and F. C. Neidhardt. 1983. *The growth of the bacterial cell.* Sunderland, Mass: Sinauer.

Joyce, C. M., and T. A. Steitz. 1994. Function and structure relationships in DNA polymerases. *Ann. Rev. Biochem.* 63: 777–822.

Kornberg, A., and T. A. Baker. 1992. *DNA replication.* New York: W. H. Freeman.

Marians, K. J. 1992. Prokaryotic DNA replication. *Ann Rev. Biochem.* 61: 673–719.

Matson, S. W., and K. A. Kaiser-Rogers. 1990. DNA helicases. *Ann. Rev. Biochem.* 59: 289–329.

McHenry, C. S. 1988. DNA polymerase II holoenzymes of *Escherichia coli. Ann. Rev. Biochem.* 57: 519–50.

Meselson, M., and F. W. Stahl. 1958. The replication of DNA in *Escherichia coli. Proc. Nat. Acad. Sci., U.S.* 44: 671–92.

Ogawa, T., and T. Okazaki. 1980. Discontinuous DNA replication. *Ann. Rev. Biochem.* 49: 421–57.

CHAPTER 9   Eukaryotes and Eukaryotic Chromosomes

Blackburn, E. H. 1990. Telomeres and their synthesis. *Science* 249: 489–90.

_____. 1992. Telomerases. *Ann. Rev. Biochem.* 61: 113–29.

Britten, R. J., and D. E. Kohne. 1970. Repeated segments of DNA. *Sci. Am.* 222: 24–31.

Coverley, D., and R. A. Laskey. 1994. Regulation of eukaryotic DNA replication. *Ann. Rev. Biochem.* 63: 745–76.

Joyce, C. M., and T. A. Steitz. 1994. Function and structure relationships in DNA polymerases. *Ann. Rev. Biochem.* 63: 777–822.

Kellog, D. R., M. Moritz, and B. M. Alberts. 1994. The centrosome and cellular organization. *Ann. Rev. Biochem.* 63: 639–74.

Kornberg, A., and T. A. Baker. 1992. *DNA replication.* San Francisco: W. H. Freeman.

Kornberg, A., and A. Klug. 1981. The nucleosome. *Sci. Am.* 244: 52–64.

Manuelidis, L. 1990. A view of interphase chromosomes. *Science* 250: 1533–40.

Margulis, L. 1981. *Symbiosis in cell evolution.* San Francisco: W. H. Freeman.

Moyzis, R. K. 1991. The human telomere. *Sci. Am.* 265: 48–55.

Murray, A. W., and M. W. Kirschner. 1991. What controls the cell cycle. *Sci. Am.* 264: 56–63.

Norbury, C., and P. Nurse. 1992. Animal cell cycles and their control. *Ann. Rev. Biochem.* 61: 441–70.

Reed, S. I. 1992. The role of p34 kinases in the GI to S-phase transition. *Ann. Rev. Biol.* 8: 529–61.

Richards, R. I., and G. R. Sutherland. 1992. Fragile X Syndrome: The molecular picture comes into focus. *Trends in Genetics* 8: 249–55.

Schulman, I., and K. S. Bloom. 1991. Centromeres: An integrated protein/DNA complex required for chromosome movement. *Ann. Rev. Cell Biol.* 7: 311–36.

Wang, T. S.-F. 1991. Eukaryotic DNA polymerases. *Ann. Rev. Biochem.* 60: 513–52

Zakian, V. A. 1989. Structure and function of telomeres. *Ann. Rev. Genet.* 23: 579–604.

CHAPTER 10   Plasmids, Transposable Elements, and Viruses

Berg, D. E., and M. M. Howe. 1989. *Mobile DNA.* Washington, D.C.: American Society for Microbiology.

Bukhari, A. I., J. A. Shapiro, and S. L. Adhya, eds. 1977. *DNA insertion elements, plasmids and episomes.* Cold Spring Harbor, N.Y.: Cold Spring Harbor Laboratory Press.

Cairns, J., G. Stent, and J. Watson, eds. 1966. *Phage and the origins of molecular biology.* Cold Spring Harbor, N.Y.: Cold Spring Harbor Laboratory Press.

Cohen, S. N., and J. A. Shapiro. 1980. Transposable genetic elements. *Sci. Am.* 242: 40–49.

Doering, H. P., and P. Starlinger. 1984. Barbara McClintock's controlling elements: Now at the DNA level. *Cell* 39: 253–59.

Federoff, N. V. 1984. Transposable genetic elements in maize. *Sci. Am.* 250: 84–99.

Hardy, K., ed. 1986. *Bacterial plasmids.* Washington, D.C.: American Society for Microbiology.

Ippen-Ihler, K. A., and E. G. Minkley. 1986. The conjugation system of F, the fertility factor of *Escherichia coli. Ann. Rev. Genet.* 20: 593–624.

Jacob, F., and E. Wollman. 1961. *Sexuality and the genetics of bacteria.* New York: Academic Press.

Klecker, N. 1990. Regulation of transposition in bacteria. *Ann. Rev. Cell Biol.* 6: 297–327.

Mizuuchi, K., and R. Craigie. 1986. Mechanism of bacteriophage Mu transposition. *Ann. Rev. Genet.* 20: 385–429.

Murialdo, H. 1991. Bacteriophage lambda DNA maturation and packaging. *Ann. Rev. Biochem.* 60: 125–53.

Shapiro, J. A. 1983. *Mobile genetic elements.* New York: Academic Press.

Stent, G. 1963. *Molecular biology of bacterial viruses.* San Francisco: W. H. Freeman.

CHAPTER 11   Gene Mutations

Beadle, G., and E. L. Tatum. 1941. Genetic control of biochemical reactions in *Neurospora. Proc. Natl. Acad. Sci., U.S.* 27: 499–506.

Cairns, J., J. Overbaugh, and S. Miller. 1988. The origin of mutants. *Nature* 335: 142–45.

Charlesworth, D., B. Charlesworth, J. J. Bull, A. Grafen, R. Holliday, R. F. Rosenberg, L. M. Van Valen, A. Danchin, I. Tessman, and J. Cairns. 1988. Origin of mutants disputed (correspondence). *Nature* 336: 525–28.

Culotta, E. 1994. A boost for "adaptive" mutation. *Science* 265: 318–319.

Foster, P. 1992. Directed mutation: Between unicorns and goats. *J. Bacteriol.* 147: 1711–16.

Friedmann, T. 1971. Prenatal diagnosis of genetic disease. *Sci. Am.* 225: 34–42.

Fuchs, F. 1980. Genetic amniocentesis. *Sci. Am.* 242: 47–53.

Green, J. E., A. Dorfmann, S. Jones, S. Bender, L. Patton, and J. D. Schulman. 1988. Chorionic villus sampling: Experience with an initial 940 cases. *Obstet. Gynecol.* 71: 208–12.

Keller, E. F. 1991. Between language and science: The question of directed mutation in molecular genetics. *Perspectives in Biology and Medicine* 35: 292–306.

Luria, S. E., and M. Delbruck. 1943. Mutations of bacteria from virus sensitivity to virus resistance. *Genetics* 28: 491–511.

MacPhee, D. 1993. Directed evolution reconsidered. *American Scientist* 81: 554–61.

————. 1993. Directed mutation: Paradigm postponed. *Mutation Research* 285: 109–16.

Novick, A., and L. Szilard. 1951. Experiments on spontaneous and chemically induced mutations of bacteria growing in the chemostat. *Cold Spring Harbor Symp. Quant. Biol.* 16: 337–43.

Rowley, P. 1984. Genetic screening: Marvel or menace? *Science* 225: 138–44.

Shapiro, A. 1946. The kinetics of growth and mutation in bacteria. *Cold Spring Harbor Symp. Quant. Biol.* 11: 228–35.

CHAPTER 12   Chromosome Mutations

Blakeslee, A. F. 1934. New Jimson weeds from old chromosomes. *J. Hered.* 25: 80–108.

Borgaonkar, D. S. 1989. *Chromosome variation in man: A catalogue of chromosomal variants and anomalies.* New York: Liss.

deGrouchy, J., and C. Turleau. 1984. *Clinical atlas of human chromosomes.* New York: Wiley.

Dellarco, V., P. Voytek, and A. Hollander. 1985. *Aneuploidy: Etiology and mechanisms.* New York: Plenum.

Epstein, C. J. 1988. Mechanisms of the effects of aneuploidy in mammals. *Ann. Rev. Genet.* 22: 51–75.

Feldman, M., and E. R. Sears. 1981. The wild gene resources of wheat. *Sci. Am.* 244: 102–13.

Hassold, T. J., and P. A. Jacobs. 1984. Trisomy in man. *Ann. Rev. Genet.* 18: 69–98.

Hulse, J. H., and D. Spurgeon. 1974. Triticale. *Sci. Am.* 231: 72–80.

Lewis, W. H., ed. 1980. *Polyploidy: Biological relevance.* New York: Plenum.

Lovell-Badge, R. 1992. Testis determination: Soft talk and kinky sex. *Current opinion in genetics and development* 2: 596–601.

Orr, H. A. 1990. Why polyploidy is rarer in animals than in plants (revisited). *Am. Nat.* 136: 759–70.

Patterson, D. 1987. The causes of Down syndrome. *Sci. Am.* 257: 52–61.

Shepherd, J. F. 1982. The regeneration of potato plants from protoplasts. *Sci. Am.* 246: 154–66.

Simmons, N. W. 1976. *Evolution of crop plants.* New York: Longman.

CHAPTER 13   Mapping Genes in Eukaryotes

Ashburner, M. 1991. *DIS69: The genetic maps of* Drosophila. Cambridge, Mass.: Drosophila Information Service.

Bateson, W., E. R. Saunders, and R. Punnett. 1905. Experimental studies in the physiology of heredity. *Rep. Evol. Comm. R. Soc.* II: 1–55 and 80–99.

Crow, J. F. 1988. A. diamond anniversary: The first chromosome map. *Genetics* 118: 1–3.

————. 1990. Mapping functions. *Genetics* 125: 669–71.

Dujon, B. 1981. Mitochondrial genetics and functions. In *The molecular biology of the yeast* Saccharomyces, ed. J. N. Strathern, E. W. Woods, and J. R. Broach. Cold Spring Harbor, N.Y.: Cold Spring Harbor Laboratory Press.

Fincham, J. R. S., P. R. Day, and A. Radford., 1979. *Fungal genetics.* Oxford: Blackwell Scientific.

Gilham, N. W. 1978. *Organelle heredity.* New York: Raven.

Grivell, L. A. 1983. Mitochondrial DNA. *Sci. Am.* 248: 78–89.

Kemp, R. 1970. *Cell division and heredity.* London: Edward Arnold.

Lyon, M. F. 1990. L. C. Dunn and mouse genetic mapping. *Genetics* 125: 231–36.

O'Brien, S. J., ed. 1984. *Genetic maps.* Cold Spring Harbor, N.Y.: Cold Spring Harbor Laboratory Press.

Peters, J. A., ed. 1959. *Classic papers in genetics.* Englewood Cliffs, N.J.: Prentice-Hall.

Sorsa, V. V. 1988. *Chromosome maps of* Drosophila. Boca Raton, Fla.: CRC Press.

Stahl, F. W. 1969. *The mechanics of inheritance.* Englewood Cliffs, N.J.: Prentice-Hall.

Wallace, D. C. 1992. Diseases of the mitochondrial DNA. *Ann. Rev. Biochem.* 61: 1175–1212.

———. 1992. Mitochondrial genetics: A paradigm for aging and degenerative diseases. *Science* 256: 628–32.

CHAPTER 14 Mapping Genes in Prokaryotic Systems

Adelberg, E. A. 1966. *Papers on bacterial genetics.* Boston: Little, Brown.

Archer, L. J. 1973. *Bacterial transformation.* New York: Academic Press.

Bachmann, B. J. 1990. Linkage map of *Escherichia coli* K-12, ed. 8. *Microbiological Reviews* 54: 130–97.

Benzer, S. 1962. The fine structure of the gene. *Sci. Am.* 206: 70–87.

Birge, E. A. 1988. *Bacterial and bacteriophage genetics.* New York: Springer-Verlag.

Brock, T. D. 1990. *The emergence of bacterial genetics.* Cold Spring Harbor, N.Y.: Cold Spring Harbor Laboratory Press.

Curtiss, R. 1969. Bacterial conjugation. *Ann. Rev. Microbiol.* 23: 69–136.

Drlica, K., and M. Riley, eds. 1990. *The bacterial chromosome.* Washington, D.C.: American Society for Microbiology.

Fincham, J. 1966. *Genetic complementation.* Menlo Park, Calif.: Benjamin.

Freifelder, D. 1987. *Microbial genetics.* Boston: Science Books International.

Hayes, W. 1968. *The genetics of bacteria and their viruses.* New York: Wiley.

Hotchkiss, R. D., and M. Gabor. 1970. Bacterial transformation with special reference to recombination processes. *Ann. Rev. Genet.* 4: 193–224.

Ippen-Ihler, K. A., and E. G. Minkley, Jr. 1986. The conjugation system of F, the fertility factor of *Escherichia coli. Ann. Rev. Genet.* 20: 593–624.

Jacob, F., and E. L. Wollman. 1951. *Sexuality and the genetics of bacteria.* New York: Academic Press.

Lederberg, J. 1987. Genetic recombination in bacteria: A discovery account. *Ann. Rev. Genet.* 21: 23–46.

Low, K. B., and D. Porter. 1978. Modes of gene transfer and recombination in bacteria. *Ann. Rev. Genet.* 12: 249–87.

Miller, J. H. 1992. *A short course in bacterial genetics.* Cold Spring Harbor, N.Y.: Cold Spring Harbor Laboratory Press.

O'Brien, S. J., ed. 1984. *Genetic maps.* Cold Spring Harbor, N.Y.: Cold Spring Harbor Laboratory Press.

Stahl, F. W. 1989. The linkage map of phage T4. *Genetics* 123: 245–48.

Stent, G. S., and R. Calendar. 1978. *Molecular genetics: An introductory narrative.* San Francisco: W. H. Freeman.

CHAPTER 15 Cellular and Molecular Approaches to Mapping

Amos, B., and J. Pemberton. 1993. DNA fingerprinting in non-human populations. *Curr. Opin. Genet. Dev.* 2: 857–60.

Arnheim, N., and H. Erlich. 1992. Polymerase chain reaction strategy. *Ann. Rev. Biochem.* 61: 650–60.

Ayala, F. J., and B. Black. 1993. Science and the courts. *Am. Sci.* 81: 230–39.

Botstein, D., R. L. White, M. Skolnick, and R. W. Davis. 1992. Construction of a genetic linkage map in man using restriction fragment length polymorphisms. *J. NIH Res.* 4: 66–74.

Cantor, C. R. 1990. Orchestrating the human genome project. *Science* 248: 49–51.

Chaleff, R. S., and P. S. Carlson. 1974. Somatic cell genetics of higher plants. *Ann. Rev. Genet.* 8: 267–78.

Cooper, N. G., ed. 1994. *The human genome project: deciphering the blueprint of heredity.* Mill Valley, Calif.: University Science Books.

Donis-Keller, H. 1990. *Human genome mapping techniques.* New York: Stockton Press.

Ephrussi. B., and M. C. Weiss. 1969. Hybrid somatic cells. *Sci. Am.* 220: 26–35.

Erlich, H. A., ed. 1989. *PCR technology.* New York: Stockton Press.

Erlich, H. A., and N. Arnheim. 1992. Genetic analysis using the polymerase chain reaction. *Ann. Rev. Genet.* 26: 479–506.

Genome Issue, 1994. *Science* 265: 2031–90.

Hagelberg, E., I. C. Gray, and A. J. Jeffreys. 1991. The identification of the skeletal remains of a murder victim by DNA analysis. *Nature* 352: 427–29.

Hartl. D., and E. Lozovskaya. 1992. The *Drosophila* genome project: Current status of the physical map. *Comp. Biochem. Physiol.* (B) 103: 1–8.

Kirby, L. T. 1990. *DNA fingerprinting: An introduction.* New York: Stockton Press.

Lewin, B. 1994. *Genes V.* New York: Oxford University Press.

McKusick, V. A. 1971. The mapping of human chromosomes. *Sci. Am.* 224: 104–13.

———. 1992. *Mendelian inheritance in man: Catalogs of autosomal dominant, autosomal recessive, and X-linked phenotypes.* 10th ed., Baltimore: Johns Hopkins University Press.

Morell, V. 1993. Huntington's gene finally found. *Science* 260: 28–30.

Mullis, K. B. 1990. The unusual origin of the polymerase chain reaction. *Sci. Am.* 262: 56–65.

Paabo, Svante. 1993. Ancient DNA. *Sci. Am.* 269: 86–92.

Puck, T. T., and F.-T. Kao. 1982. Somatic cell genetics and its application to medicine. *Ann. Rev. Genet.* 16: 225–72.

Reiss, J., and D. N. Cooper. 1990. Application of the polymerase chain reaction to the diagnosis of human disease. *Human Genetics* 85: 1–8.

Rennie, J. 1994. Grading the gene tests. *Sci. Am.* 270: 88–97.

Risch, N. 1992. Genetic linkage: Interpreting lod scores. *Science* 255: 803–04.

Ruddle, F. H., and R. S. Kucherlapati. 1974. Hybrid cells and human genes. *Sci. Am.* 231: 36–44.

Verma, R. S., and A. Babu. 1989. *Human chromosomes: Manual of basic techniques.* Elmsford, N.Y.: Pergamon Press.

White, R., and J.-M. Lalouel. 1988. Chromosome mapping with DNA markers. *Sci. Am.* 258: 40–48.

Wilson, G. G., and N. E. Murray. 1991. Restriction and modification systems. *Ann. Rev. Genet.* 25: 585–627.

CHAPTER 16   RNA Synthesis and the Genetic Code

Baltimore, D. 1985. Retroviruses and retrotransposons: The role of reverse transcriptase in shaping the eukaryotic genome. *Cell* 40: 481–82.

Belfort, M. 1990. Phage T4 introns: self-splicing and mobility. *Ann. Rev. Genet.* 24: 363–86.

Chambon, P. 1981. Split genes. *Sci. Am.* 244: 60–71.

Cech, T. R. 1986. RNA as an enzyme. *Sci. Am.* 255: 64–75.

———. 1990. Self-splicing of group I introns. *Ann. Rev. Biochem.* 59: 543–68.

Crick, F.H.C. 1962. The genetic code. *Sci. Am.* 207: 66–74.

———. 1966. The genetic code III. *Sci. Am.* 215: 55–62.

Darnell, J. E., Jr. 1985. RNA. *Sci. Am.* 253: 68–78.

Doolittle, W. F. 1987. The origins and functions of intervening sequences in DNA: A review. *Am. Nat.* 130: 915–28.

Dreyfuss, G., M. J. Matunis, S. Pinol-Roma, and C. G. Burd. 1993. hnRNP proteins and the biogenesis of mRNA. *Ann. Rev. Biochem.* 62: 289–321.

The Genetic Code. 1966. *Cold Spring Harbor Sympos. Quant. Biol.* 31.

Guthrie, C. 1991. Messenger RNA splicing in yeast: Clues to why the spliceosome is a ribonucleoprotein. *Science* 253: 157–163.

Lambowitz, A. M., and M. Belfort. 1993. Introns as mobile genetic elements. *Ann. Rev. Biochem.* 62: 587–622.

McKeown, M. 1992. Alternative mRNA splicing. *Ann Rev. Cell Biol.* 8: 133–55.

Miller, O. L. 1973. The visualization of genes in action. *Sci. Am.* 228: 34–42.

Nirenberg, M. W. 1963. The genetic code II. *Sci. Am.* 208: 80–94.

Simpson, L. 1990. RNA editing—a novel genetic phenomenon. *Science* 235: 766–71.

Steitz, J. A. 1988. "Snurps." *Sci. Am.* 258: 56–63.

Symons, R. H. 1992. Small catalytic RNAs. *Ann. Rev. Biochem.* 61: 641–71.

Wahle, E., and W. Keller. 1993. The biochemistry of 3′-end cleavage and polyadenylation of messenger RNA precursors. *Ann. Rev. Biochem.* 61: 419–40.

Young, R. 1991. RNA polymerase II. *Ann. Rev. Biochem.* 60: 689–715.

CHAPTER 17   Protein Synthesis and the RNA Decoding System

Bonitz, S. G., R. Berlani, G. Coruzzi, M. Li, G. Macino, F. G. Nobrega, M. P. Nobrega, B. E. Thalenfeld, and A. Tzagoloff. 1980. Codon recognition rules in yeast mitochondria. *Proc. Natl. Acad. Sci., U.S.* 77: 3167–70.

Caron, F., and E. Meyer. 1985. Does *Paramecium primaurelia* use a different genetic code in its macronucleus? *Nature* 314: 185–88.

Carter, C. W., Jr. 1993. Cognition, mechanism, and evolutionary relationships in aminoacyl-tRNA synthetases. *Ann. Rev. Biochem.* 62: 715–48.

Crick, F. H. C. 1966. Codon-anticodon pairing: The wobble hypothesis. *J. Mol. Biol.* 19: 548–55.

Ellis, R. J., and S. M. van der Vies. 1991. Molecular chaperones. *Ann. Rev. Biochem.* 60: 321–47.

Frank, J., A. Verschoor, M. Radermacher, and T. Wagenknecht. 1990. Morphologies of eubacterial and eucaryotic ribosomes as determined by three-dimensional electron microscopy. In *The Ribosome: Structure, Function, and Evolution,* ed. W. H. Hill et al., 107–13. Washington, D.C.: American Society for Microbiology.

Hedrick, J. P., and F. U. Hartl. 1993. Molecular chaperone functions of heat-shock proteins. *Ann. Rev. Biochem.* 62: 349–84.

Hill, W. H. et al., eds. 1990. *The Ribosome: Structure, Function, and Evolution.* Washington, D.C.: American Society for Microbiology.

Kozak, M. 1992. Regulation of translation in eukaryotic systems. *Ann. Rev. Cell Biol.* 8: 197–225.

Lake, J. A. 1981. The ribosome. *Sci. Am.* 245: 84–97.

Landry, S. J., and L. M. Gierasch. 1991. Recognition of nascent polypeptides for targeting and folding. *Trends in Biochem. Sci.* 16: 159–63.

Miller, O. L. 1973. The visualization of genes in action. *Sci. Am.* 228: 34–42.

Nilsson, B., and S. Anderson. 1991. Proper and improper folding of proteins in the cellular environment. *Ann. Rev. Microbiol.* 45: 607–35.

Noller, H. F. 1991. Ribosomal RNA and translation. *Ann. Rev. Biochem.* 60: 191–227.

Pryer, N. K., L. J. Wuestehube, and R. Schekman. 1992. Vesicle-mediated protein sorting. *Ann. Rev. Biochem.* 61: 471–516.

Rich, A., and S. H. Kim. 1978. The three-dimensional structure of transfer RNA. *Sci. Am.* 238: 52–73.

Richards, F. M. 1991. The protein folding problem. *Sci. Am.* 264: 54–63.

Saks, M. E., J. R. Sampson, and J. N. Abelson. 1994. The transfer RNA identity problem: a search for rules. *Science* 263: 191–97.

Van Knippenberg, P. H. 1990. Aspects of translation initiation in *Escherichia coli.* In *The Ribosome: Structure, Function, and*

Evolution, ed. W. H. Hill et al., 265–74. Washington, D.C.: American Society for Microbiology.

CHAPTER 18   Gene Cloning and the Analysis of Gene Function

Biotechnology and Ecology, 1993. *The Amicus Journal.* New York: The National Resources Defense Council.

Capecchi, M. R. 1994. Targeted gene replacement. *Sci. Am.* 270: 52–59.

Cooper, N. G., ed. 1994. *The human genome project: Deciphering the blueprint of heredity.* Mill Valley, Calif.: University Science Books.

Friedfelder, D., ed. 1978. *Readings from Scientific American: Recombinant DNA.* San Francisco: W. H. Freeman.

Gasser, C. S., and R. T. Fraley. 1992. Transgenic crops. *Sci. Am.* 266: 62–69.

Lewin, B. 1994. *GENES V.* New York: Oxford University Press.

Micklos, D. A., and G. A. Freyer. 1990. *DNA science: A first course in recombinant DNA technology.* Cold Spring Harbor, N.Y.: Cold Spring Harbor Laboratory Press.

Morgan, R. A., and W. French. 1993. Human gene therapy. *Ann. Rev. Biochem.* 62: 191–217.

Murray, A. W., and J. W. Szostak. 1987. Artificial chromosomes. *Sci. Am.* 257: 62–68.

Old, R. W., and S. B. Primrose. 1989. *Principles of gene manipulation.* Oxford: Blackwell Scientific Publications.

Pasten, I., V. Chaudhary, and D. J. FitzGerald. 1992. Recombinant toxins as novel therapeutic agents. *Ann. Rev. Biochem.* 61: 331–54.

Sedivy, J. M., and A. L. Joyner. 1992. *Gene targeting.* New York: W. H. Freeman.

Singer, M. and P. Berg. 1991. *Genes and genomes.* Mill Valley, Calif.: University Science Books.

Sinsheimer, R. L. 1977. Recombinant DNA. *Ann. Rev. Biochem.* 46: 415–38.

Watson, J. D., M. Gilman, J. Witkowski, and M. Zoller. 1992. *Recombinant DNA.* New York: W. H. Freeman.

Zilinskas, R. A., and B. K. Zimmerman, eds. 1986. *The gene-splicing wars: Reflections on the recombinant DNA controversy.* New York: Macmillan.

CHAPTER 19   Regulation in Prokaryotes

Das, A. 1993. Control of transcription termination by RNA-binding proteins. *Ann. Rev. Biochem.* 62: 893–930.

Eguchi, Y., T. Itoh, and J.-I. Tomazawa. 1991. Antisense RNA. *Ann. Rev. Biochem.* 60: 631–52.

Geiduschek, E. P. 1991. Regulation of the late genes of bacteriophage T4. *Ann. Rev. Genet.* 25: 437–60.

Harrison, S. C., and A. K. Aggarwal. 1990. DNA recognition by proteins with the helix-turn-helix motif. *Ann. Rev. Biochem.* 59: 933–69.

Hendrix, R. W., J. W. Roberts, F. W. Stahl, and R. A. Weisman. 1983. *Lambda II.* Cold Spring Harbor, N.Y.: Cold Spring Harbor Laboratory Press.

Jacob, F., and J. Monod. 1961. Genetic regulatory mechanisms in the synthesis of proteins. *J. Mol. Biol.* 3: 318–56.

Kolb, A., S. Busby, H. Buc, S. Garges, and S. Adhya. 1993. Transcriptional regulation by cAMP and its receptors. *Ann. Rev. Biochem.* 62: 749–95.

Kolter, R., and C. Yanofsky. 1982. Attenuation in amino acid biosynthetic operons. *Ann. Rev. Genet.* 16: 113–34.

Lewin, B. 1994. *GENES V.* New York: Oxford University Press.

Miller, J. H., and W. S. Reznikoff, eds. 1978. *The operon.* Cold Spring Harbor, N.Y.: Cold Spring Harbor Laboratory Press.

Murialdo, H. 1991. Bacteriophage lambda DNA maturation and packaging. *Ann. Rev. Biochem.* 60: 125–153.

Platt, T. 1981. Termination of transcription and its regulation in the tryptophan operon of *E. coli. Cell* 24: 10–23.

Ptashne, M. 1992. *A genetic switch: Phage λ and higher organisms.* Cambridge, Mass.: Cell Press and Blackwell Scientific Publications.

Raibaud, O., and M. Schwartz. 1984. Positive control of transcription initiation in bacteria. *Ann. Rev. Genet.* 18: 173–206.

Simons, R. W., and N. Kleckner. 1988. Biological regulation by antisense RNA in prokaryotes. *Ann. Rev. Genet.* 22: 567–600.

Studier, F. W. 1972. Bacteriophage T7. *Science* 176: 367–76.

Weintraub, H. M. 1990. Antisense RNA and DNA. *Sci. Am.* 262: 40–46.

CHAPTER 20   Regulation in Eukaryotes I: Transcriptional Activation

Alberts, B., D. Bray, J. Lewis, M. Raff, K. Roberts, and J. D. Watson. 1994. *Molecular biology of the cell.* New York: Garland Publishing.

Atchison, M. L. 1988. Enhancers: Mechanisms of action and cell specificity. *Ann. Rev. Cell Biol.* 4: 127–53.

Bulmer, K. J., and J. Thorner. 1991. Receptor G protein signaling in yeast. *Ann. Rev. Physiol.* 53: 37–57.

Coleman, J. E. Zinc proteins: Enzymes, storage proteins, transcription factors, and replication proteins. *Ann. Rev. Biochem.* 61: 897–946.

Conaway, R. C. 1993. General initiation factors for RNA polymerase II. *Ann. Rev. Biochem.* 62: 161–90.

Fantl, W. J., D. E. Johnson, and L. T. Williams. 1993. Signaling by receptor tyrosine kinases. *Ann. Rev. Biochem.* 62: 453–81.

Harrison, S. C., and A. K. Aggarwal. 1990. DNA recognition by proteins with the helix-turn-helix motif. *Ann. Rev. Biochem.* 59: 933–69.

Herschbach, B. M. and A. D. Johnson. 1993. Transcriptional repression in eukaryotes. *Ann. Rev. Cell Biol.* 9: 479–509.

Kaziro, Y., H. Itoh, T. Kozasa, M. Nakafuko, and T. Satoh. 1991. Structure and function of signal-transducing GTP-binding proteins. *Ann. Rev. Biochem.* 60: 349–400.

Kornberg, R. A. 1992. Chromatin structure and transcription. *Ann. Rev. Cell Biol.* 8: 563–87.

Kozak, M. 1992. Regulation of transcription in eukaryotic systems. *Ann. Rev. Cell Biol.* 8: 197–225.

Lewin, B. 1994. *GENES V.* Oxford University Press.

Linder, M. E., and A. G. Gilman. 1992. G proteins. *Sci. Am.* 267: 56–65.

Lucas, P. C., and D. K. Granner. 1992. Hormone response domains in gene transcription. *Ann. Rev. Biochem.* 61: 1131–73.

Lyon, M. 1992. Some milestones in the history of X-chromosome inactivation. *Ann. Rev. Genet.* 26: 17–28.

McKnight, S. L. 1991. Molecular zippers in gene regulation. *Sci. Am.* 264: 54–64.

Pabo, C. O., and R. T. Sauer. 1992. Transcription factors: Structural families and principles of DNA recognition. *Ann. Rev. Biochem.* 61: 1053–95.

Paranjape, S. M., R. T. Kamakaka, and J. T. Kadonaga. 1994. Role of chromatin structure in the regulation of transcription by RNA polymerase II. *Ann. Rev. Biochem.* 63: 265–97.

Ptashne, M. 1992. *A genetic switch: Phage λ and higher organisms.* Cambridge, Mass.: Cell Press and Blackwell Scientific Publications.

Rhodes, D., and A. Klug. 1993. Zinc fingers. *Sci. Am.* 268: 56–65.

Sapienza, C., and K. Peterson. 1993. Imprinting the genome: Imprinted genes, imprinting genes, and a hypothesis for their interaction. *Ann. Rev. Genet.* 27: 7–32.

Yamamoto, K., and S. McKnight, eds. 1992. *Transcriptional regulation.* Cold Spring Harbor, N.Y.: Cold Spring Harbor Laboratory Press.

Watson, J. D., M. Gilman, J. Witkowski, and M. Zoller. 1992. *Recombinant DNA.* New York: W. H. Freeman.

CHAPTER 21   Regulation in Eukaryotes II: Development and Cancer

Alberts, B., D. Bray, J. Lewis, M. Raff, K. Roberts, and J. D. Watson. 1994. *Molecular biology of the cell.* New York: Garland Publishing.

Beardsley, T. 1991. Smart genes. *Sci. Am.* 265: 86–95.

Blau, H. M. 1992. Differentiation requires continuous active control. *Ann. Rev. Biochem.* 61: 1213–30.

Davies, D. R., E. A. Padlan, and S. Sheriff. 1993. Antibody-antigen complexes. *Ann. Rev. Biochem.* 62: 439–74.

DePomerai, D. 1985. *From gene to animal.* Cambridge: Cambridge University Press.

DeRobertis, E. M., G. Oliver, and C. V. E. Wright. 1990. Homeobox genes and the vertebrate body plan. *Sci. Am.* 263: 46–52.

Gehring, W. J., M. Muller, M. Affolter, A. Percival-Smith, M. Billeter, Y. Q. Qian, G. Otting, and K. Wüthrich. 1990. The structure of the homeodomain and its functional implications. *Trends in Genetics* 6: 323–29.

Gehring, W. J., M. Affolter, and T. Bürglin. 1994. Homeodomain proteins. *Ann. Rev. Biochem.* 63: 487–526.

Gellert, M. 1992. Molecular analysis of V(D)J recombination. *Ann. Rev. Genet.* 26: 425–46.

Gilbert, S. F. 1988. *Developmental biology.* Sunderland, Mass.: Sinauer Association.

Grunstein, M. 1992. Histones as regulators of genes. *Sci. Am.* 267: 68–74B.

Kurjan, J. 1992. Pheromone response in yeast. *Ann. Rev. Biochem.* 61: 1097–1129.

Lawrence, P. 1992. *The making of a fly.* Oxford: Blackwell Scientific.

Levine, A. J. 1993. The tumor suppressor genes. *Ann. Rev. Biochem.* 62: 623–52.

Lewin, B. 1994. *Genes V.* Oxford: Oxford University Press.

Linder, M. E., and A. G. Gilman. 1992. G proteins. *Sci. Am.* 267: 56–65.

Lowry, D. R., and B. M. Willumsen. 1993. Functions and regulation of *RAS. Ann. Rev. Biochem.* 62: 851–92.

Marcu, K. B., S. A. Bossone, and A. J. Patel. 1992. *Myc* function and regulation. *Ann. Rev. Biochem.* 61: 809–60.

Molecular biology of signal transduction. 1988. *Cold Spring Harbor Sympos. Quant. Biol.* 53.

Nusslein-Volhard, C. 1991. Determination of the embryonic axes of *Drosophila. Development Supplement* 1: 1–10.

Pankratz, M. J., and H. Jackle. 1990. Making stripes in the *Drosophila* embryo. *Trends Genet.* 6: 287–92.

Pelech, S. L. 1993. Networking with protein kinases. *Curr. Biol.* 3: 513–15.

Ptashne, M. 1992. *A genetic switch: Phage λ and higher organisms.* Cambridge, Mass.: Cell Press and Blackwell Scientific Publications.

Rennie, J. 1991. Homeobox harvest. *Sci. Am.* 264: 24.

Rustgi, A. K., and D. K. Podolsky. 1992. The molecular basis of colon cancer. *Ann. Rev. Medicine* 43: 61–68.

Schatz, D. G., M. A. Oettinger, and M. S. Schlissel. 1992. V(D)J recombination: Molecular biology and regulation. *Ann. Rev. Immun.* 10: 359–83.

Watson, J. D., M. Gilman, J. Witkowski, and M. Zoller. 1992. *Recombinant DNA.* New York: W. H. Freeman.

CHAPTER 22   Molecular Basis of Mutation

Drake, J. W. 1991. Spontaneous mutation. *Ann. Rev. Genet.* 25: 125–46.

Loeb, L. A., and B. D. Preston. 1986. Mutagenesis by apurinic/apyrimidinic sites. *Ann. Rev. Genet.* 20: 201–30.

Marx, J. 1990. Animal carcinogen testing challenged. *Science* 250: 743–45.

Singer, B., and J. T. Kusmierak. 1982. Chemical mutagenesis. *Ann. Rev. Biochem.* 52: 655–93.

Smith, M. 1985. In vitro mutagenesis. *Ann. Rev. Genet.* 19: 423–62.

Sutherland, G. R., and R. I. Richards. 1994. Dynamic mutations. *Am. Sci.* 82: 157–63.

CHAPTER 23   Mechanisms of Recombination and Repair

Cox, M. M. 1987. Enzymes of general recombination. *Ann. Rev. Biochem.* 56: 229–62.

Demple, B., and L. Harrison. 1994. Repair of oxidative damage to DNA: Enzymology and biology. *Ann. Rev. Biochem.* 63: 915–48.

Holliday, R. 1974. Molecular aspects of genetic exchange and gene conversion. *Genetics* 78: 273–87.

Kowalczykowski, S. C., and A. K. Eggleston. 1994. Homologous pairing and DNA strand-exchange proteins. *Ann. Rev. Biochem.* 63: 991–1043.

Kucherlapati, R. S., and G. R. Smith, eds. 1988. *Genetic Recombination.* Washington, D.C.: American Society for Microbiology.

Meselson, M., and J. J. Weigle. 1961. Chromosome breakage accompanying genetic recombination in bacteriophage. *Proc. Natl. Acad. Sci., U.S.* 47: 857–68.

Modrich, P. 1991. Mechanisms and biological effects of mismatch repair. *Ann. Rev. Genet.* 25: 229–53.

Sancar, A., and G. B. Sancar. 1988. DNA repair enzymes. *Ann. Rev. Biochem.* 57: 29–67.

Smith, G. R. 1987. Mechanisms and control of homologous recombination in *Escherichia coli. Ann. Rev. Genet.* 21: 179–201.

Stahl, F. W. 1987. Genetic recombination. *Sci. Am.* 256: 91–101.

West, S. C. 1992. Enzymes and molecular mechanisms of genetic recombination. *Ann. Rev. Biochem.* 61: 603–40.

CHAPTER 24   Genes in Populations

Bodmer, W. F., and L. L. Cavalli-Sforza. 1971. *Genetics, evolution and man.* San Francisco: W. H. Freeman.

Cavalli-Sforza, L. L., and W. F. Bodmer. 1971. *The genetics of human populations.* San Francisco: W. H. Freeman.

Crow, J. F., and M. Kimura. 1970. *An introduction to population genetics theory.* New York: Harper & Row.

Elseth, G. D., and K. D. Baumgardner. 1981. *Population biology.* New York: D. Van Nostrand.

Hardy, G. H. 1908. Mendelian proportions in a mixed population. *Science* 28: 49–50.

Hartl, D. L. 1980. *Principles of population genetics.* Sunderland, Mass.: Sinauer.

Wallace, B. 1981. *Basic population genetics.* New York: Columbia University Press.

CHAPTER 25   Genetic Processes of Evolution

Allison, A. C. 1954. Protection afforded by sickle-cell trait against subtertian malarial infection. *British Med. J.* 1: 290–94.

Bodmer, W. F., and L. L. Cavalli-Sforza. 1971. *Genetics, evolution and man.* San Francisco: W. H. Freeman.

Crow, J. F., and M. Kimura. 1970. *An introduction to population genetics theory.* New York: Harper & Row.

Dobzhansky, T. 1951. *Genetics and the origin of species.* New York: Columbia University Press.

Dobzhansky, T., F. J. Ayala, G. L. Stebbins, and J. W. Valentine. 1977. *Evolution.* San Francisco: W. H. Freeman.

Elseth, G. D., and K. D. Baumgardner. 1981. *Population biology.* New York: D. Van Nostrand.

Futuyma, D. J. 1979. *Evolutionary biology.* Sunderland, Mass.: Sinauer.

Mettler, L. E., T. G. Gregg, and H. E. Schaffer. 1988. *Population genetics and evolution.* Englewood Cliffs, N.J.: Prentice-Hall.

Roughharden, J. 1979. *Theory of population genetics and evolutionary ecology: An introduction.* New York: Macmillan.

Wallace, B. 1981. *Basic population genetics.* New York: Columbia University Press.

White, M. J. D. 1978. *Modes of speciation.* San Francisco: W. H. Freeman.

CHAPTER 26   Molecular Evolution

Dobzhansky, T., F. J. Ayala, G. L. Stebbins, and J. W. Valentine. 1977. *Evolution.* San Francisco: W. H. Freeman.

Doolittle, R. F., and P. Bork. 1993. Evolutionarily mobile modules in proteins. *Sci. Am.* 269: 50–56.

Elseth, G. D., and K. D. Baumgardner. 1981. *Population biology.* New York: D. Van Nostrand.

Gillespie, J. H. 1991. *The causes of molecular evolution.* New York: Oxford University Press.

Hillis, D. M., J. P. Huelsenbeck, and C. W. Cunningham. 1994. Application and accuracy of molecular phylogenies. *Science* 264: 671–77.

Hillis, D. M., and C. Moritz. 1990. *Molecular systematics.* Sunderland, Mass.: Sinauer.

Klein, J., N. Takahata, and F. J. Ayala. 1993. MHC polymorphism and human origins. *Sci. Am.* 269: 78–83.

Li, W.-H., and D. Grauer. 1991. *Fundamentals of molecular evolution.* Sunderland, Mass.: Sinauer.

Nei, M. 1987. *Molecular evolutionary genetics.* New York: Oxford University Press.

Selander, R. K., A. G. Clark, and T. S. Whittam. 1991. *Evolution at the molecular level.* Sunderland, Mass.: Sinauer.

Terzaghi, E. A., A. S. Wilkins, and D. Penny. 1984. *Molecular evolution.* Boston: Jones and Bartlett.

GLOSSARY

A

A-DNA A right-handed double helical DNA conformation that is thicker and more compact than the predominant B form, found only under relatively dehydrated conditions.

Abortive transduction Transduction in which the donor fragment is not incorporated into the recipient chromosome and is thus unable to replicate.

Acentric fragment A chromosome fragment that lacks a centromere.

Acrocentric chromosome A chromosome that has its centromere very near one end.

Active site The region of a protein that has a specific shape that is responsible for the function of the protein (e.g., the substrate binding site of an enzyme).

Adenine A purine base found in DNA and RNA.

Allele One of two or more alternative forms of a gene, any one of which can occur at a particular chromosomal locus.

Allele frequency The number of copies of a given allele in a population divided by the total number of copies of all alleles at that locus; also called the gene frequency.

Allelic genes See Allele.

Allopolyploids Polyploids whose multiple chromosome sets are initially derived from different species.

Allosteric proteins Proteins that can take on different conformational states with different functions.

Allosteric site A binding site on a protein that is separate from the active site and is involved in the regulation of the protein's activity.

Allozymes Alternative enzyme forms that are coded for by different alleles at the same gene locus.

α-helix A polypeptide secondary structure in which the amino acid chain is wound in a helical configuration.

Ames test A test to detect mutagenicity of chemical substances, based on the ability of a chemical to induce mutation in the bacterium *Salmonella*.

Amino acid A carboxylic acid containing an amino group; the monomer unit of a polypeptide chain.

Amino terminus The end of a polypeptide chain having a free α-amino group.

Aminoacyl-tRNA A tRNA molecule to which an amino acid is bound; also called a charged tRNA.

Aminoacyl-tRNA synthetase A member of a class of enzymes that catalyze the bonding of an amino acid to the 3′ acceptor arm of a tRNA; each is specific for a particular amino acid and the appropriate tRNA.

Amniocentesis A prenatal screening method in which a sample of amniotic fluid is removed with a syringe and fetal cells that were shed into the amniotic fluid are cultured and studied.

Amphidiploid A fertile allotetraploid resulting from chromosome doubling in a hybrid; it has the normal diploid chromosome complement of both ancestral species.

Anagenesis Evolution within a lineage or single line of descent; also called phyletic evolution.

Anaphase The third stage of mitosis, in which each centromere divides and the identical daughter chromosomes are drawn to opposite poles by the spindle fibers.

Anaphase I The third phase of meiosis I, in which the homologous centromeres of each bivalent move toward opposite poles of the spindle.

Anaphase II The third phase of meiosis II, in which the centromeres divide and the daughter chromosomes move to opposite poles.

Anaphase lag The failure of a chromosome to migrate properly during anaphase; normal cells and cells lacking a chromosome are the result.

Aneuploidy The loss or gain of less than a complete set of chromosomes.

Annealing Renaturation of DNA and/or RNA; a process whereby two single-stranded regions of DNA or DNA and RNA come together to form a double helix.

Antibody A complex protein molecule that recognizes, combines with, and neutralizes the effects of antigens; one of the body's major forms of defense against infection.

Anticodon A triplet nucleotide sequence on a tRNA molecule that recognizes a particular mRNA codon at the ribosome.

Antigens Substances (often foreign proteins or protein derivatives) that induce certain cells to produce antibodies against them.

Antiparallel The opposite orientations of the two strands of a double helix, so that the 5′ end of one strand aligns with the 3′ end of the other strand.

Antisense mRNA mRNA in the reverse orientation, complementary in sequence to the normal mRNA.

Antitermination A mechanism in which a positive control system regulates transcriptional termination and allows RNA polymerase to bypass the terminator region.

Antiterminators Regulatory proteins that bind to specific DNA sequences and modify RNA polymerase as it passes these sites so that it no longer recognizes the early transcription terminators that follow.

Aporepressor An inactive repressor molecule that must be activated by combining with the corepressor before it can bind to the operator and regulate transcription of an operon.

Arcocentric chromosome A chromosome in which the centromere is very near one end.

ARS elements See Autonomous replicating sequences.

Ascospores The haploid cells produced by a cross between fungi of opposite mating type, found in certain fungi in which spores are produced in a sac called an ascus.

Ascus (plural, **asci**) An elongated sac that contains the products (ascospores) resulting from a cross between fungi of opposite mating types.

Assortative mating Mating in which the mated individuals share the same phenotype more often than randomly chosen individuals.

Autonomous replicating sequences (ARS elements) AT-rich sequences of DNA that are associated with multiple replication origins in eukaryotes.

Autopolyploids Polyploids whose multiple chromosome sets are derived from the same species.

Autoradiography A method for detecting radiolabeled materials such as polynucleotides. A photographic film that is sensitive to radioactivity is laid over a sample containing a radiolabeled cellular structure; the decay of the radioisotope forms a pattern on the film that corresponds to the distribution of the radioactive material over the structure.

Autosomes Chromosomes other than the sex chromosomes.

Auxotroph A mutant microorganism that is unable to use inorganic salts and a carbon source (minimal medium) to synthesize an organic molecule required for metabolism and growth; thus, a nutritionally dependent organism that will grow only if the minimal medium is supplemented with one or more specific organic substances.

B

B-DNA The predominant conformation of DNA in solution, very similar to the form deduced by Watson and Crick.

Back mutation See Reverse mutation.

Bacteriophages Viruses of bacteria; also called phages.

Balanced polymorphism A stable variation pattern in which genotypes with different fitness values are maintained in the same population by natural selection.

Barr body An inactive, condensed, heterochromatic X chromosome found in the somatic cells of female mammals; also called the sex chromatin body.

Basal general transcription factors (TFs) A series of transcription factors that together with an RNA polymerase constitute the basal transcription apparatus; they assemble at the startpoint in a prescribed order to form a complex that is required for the transcription of a eukaryotic gene.

Base analogue A compound that is so similar in molecular structure to one of the DNA bases that it can be mistakenly incorporated into the DNA, where it acts as a mutagen.

Base stacking The stacking of the planar purine and pyrimidine rings of a polynucleotide chain to reduce their contact with water; this stacking adds rigidity to the chain.

β-bend A tight loop often formed by a polypeptide when it changes direction; it is stabilized by a hydrogen bond and often involves a proline or a glycine residue.

β-sheet A polypeptide secondary structure consisting of two or more polypeptide segments lying side by side in a sheetlike arrangement.

Binominal distribution A probability distribution for an event that has only two possible outcomes, e.g., success and failure.

Bivalent A synapsed pair of homologous chromosomes containing four chromatids; also called a tetrad.

Blending inheritance The (now discredited) idea that the characteristics of offspring are intermediate between those of their parents because of a physical blending of hereditary parental factors in the offspring.

Bottleneck effect Alteration of the genetic composition of a population due to random changes that occur during a periodic reduction in the population size.

Buoyant density The mass-to-volume ratio of a substance in a standard fluid, for example, a CsCl solution; the buoyant density of DNA can be measured by equilibrium density-gradient centrifugation and depends on its GC content.

C

Callus An undifferentiated clump of plant cells derived through cell culture.

Capsid The protein coat that surrounds the genome of a virus.

Capsomers The protein subunits of a virus capsid.

Carboxyl terminus The end of a polypeptide chain having a free α-carboxyl group.

Carcinogens Cancer-inducing agents.

Catabolite repression Inhibition of transcription of an inducible operon in the presence of glucose.

cDNA See Complementary DNA.

cDNA library A collection of all the genes expressed by a certain type of cell, stored in a cloning vector.

CentiMorgan (cM) A unit of map length corresponding to a 1% recombination frequency; also called a map unit.

Centric fusion The fusion of the long arms of two nonhomologous acrocentric chromosomes to a common centromere; also called Robertsonian fusion.

Centrioles Cylindrical organelles located at the poles of the spindle in most animal and many lower plant cells; they function in the development of the spindle.

Centromere The region of a chromosome where identical chromatids are joined and where the spindle fibers attach during cell division.

Charged tRNA A tRNA molecule bound to an amino acid; also called an aminoacyl-tRNA.

Charon phages Replacement vectors that are phage λ derivatives in which the middle region of the phage DNA is replaced by a *lacz'* marker containing unique cloning sites.

Chiasma (plural, **chiasmata**) A microscopically visible point of contact between paired chromatids, observed during the diplotene stage of prophase I; regarded as cytological evidence of crossing over.

Chorionic villus biopsy A prenatal screening method in which a sample of cells from the outer fetal membrane (chorion) is cultured and studied.

Chromatid One of two identical longitudinal halves of a duplicated chromosome; the two sister chromatids are held together at the region of the centromere.

Chromatin The complex of proteins and nucleic acids that makes up nuclear chromosomes.

Chromatosome A structure consisting of a nucleosome core particle and a molecule of histone H1, combined with two full turns (166 base pairs) of chromosomal DNA.

Chromosome The carrier of genetic information in cells, made up of DNA and protein. A eukaryotic chromosome is visible as an entity under the light microscope only during mitosis or meiosis, when the chromatin exists in a highly condensed form.

Chromosome aberration A structural chromosome alteration resulting from disruption and rearrangement of genes on a more extensive scale than point mutations.

Chromosome arm The segment between the centromere and one of the ends of a chromosome.

Chromosome banding Staining techniques that allow chromosome regions to be stained differentially so that each individual chromosome in the normal genome can be identified.

Chromosome jumping A variation of the chromosome walking technique. A long fragment resulting from a partial diges-

tion is circularized by joining with a vector; treatment with another restriction enzyme forms fragments that contain sequences from the ends of the long fragment now in close proximity. These fragments form a "jumping" library to which the chromosome walking technique can be applied more efficiently.

Chromosome mosaic An individual having a mixture of aneuploid and normal cell lines as a result of aneuploidy in somatic cells after fertilization.

Chromosome puff A swelled region originating from a site of active transcription on a polytene chromosome.

Chromosome walking Using a set of overlapping cloned fragments that span a long stretch of DNA to locate a gene of interest.

Cis dominant The dominance of a chromosomal locus over genes adjacent to it on the same chromosome but not over genes located on the homologous chromosome.

Cistron The genetic region that functions as an independent unit in a complementation test; now synonymous with gene.

Cladogenesis The splitting of a lineage into two or more separate lines of descent; also called speciation (see Speciation).

Codominance A situation in which two contrasting alleles are fully functional and express themselves individually when they occur in the heterozygous state.

Codon Sequence of the three bases in DNA or mRNA that serves as a code word for a particular amino acid.

Coefficient of coincidence The ratio of the observed frequency of double recombination to the frequency expected if the crossovers were truly independent; a measure of the degree of interference in multiple crossing over.

Cohesive ends Complementary single-stranded regions at the ends of some DNA molecules.

Col plasmids Toxin-producing plasmids carried by Col+ strains of *E. coli*; they prevent the growth of bacteria that lack a Col plasmid of the same type.

Colchicine An alkaloid extracted from the autumn crocus; it prevents the polymerization of spindle microtubules and thus blocks cell division at metaphase.

Colicinogens See Col plasmids.

Colinearity Linear correspondence between the amino acid sequence of a polypeptide and the coding sequence of its gene.

Competence The ability of bacteria to take up DNA from the surrounding medium during transformation.

Complementary Having a nucleotide sequence that permits pairing with the nucleotides of another chain, that is, so that adenine (A) pairs with thymine (T) and guanine (G) pairs with cytosine (C).

Complementary DNA (cDNA) Synthetic DNA made from a specific RNA template through the action of the enzyme reverse transcriptase.

Complementation test (1) An experiment for determining whether two different mutations affecting the same trait lie in the same or in different genes—a test for allelism. (2) A test for determining whether two mutations within a gene locus are in the same cistron.

Complete dominance The expression of one allele or character in a heterozygote to the exclusion of a contrasting (recessive) allele or characteristic.

Composite transposons Transposons that carry drug resistance or other genes flanked on each side by insertion sequences.

Concatemer A long DNA molecule that is an intermediate in the replication of some viral genomes; contains repeated copies of the viral genome.

Condensation The coiling of the duplicated chromatin threads into the highly coiled fibers of chromosomes during the prophase of mitosis and the prophase I of meiosis.

Condensed Term used to describe the tightly coiled state of chromosomes.

Conditional mutant A mutant gene that has the wild-type phenotype under certain (permissive) environmental conditions and a mutant phenotype under other (restrictive) conditions.

Conditional probability The probability that an event will occur given that another event has already occurred.

Conidium (plural, **conidia**) An asexual reproductive cell of a fungus.

Conjugation A bacterial mating process that involves a one-way, plasmid-directed transfer of genetic material from one bacterium to another during cell-to-cell contact.

Consanguineous mating A mating between relatives.

Consensus sequence An idealized sequence in which each position is represented by the base that is most often found when several actual sequences are compared.

Conservative replication A proposed DNA replication mechanism in which the two newly synthesized strands would form a separate helix, leaving the parental molecule intact.

Conservative transposition Transposition in which both parental strands of the transposon are excised from the donor DNA and inserted into the target site.

Constitutive heterochromatin Heterochromatin that is commonly found in chromosome centromeres and telomeres and remains in a permanently condensed, dark-staining condition throughout the life cycle of the organism.

Contact inhibition The density-dependent inhibition of the growth of normal (non-cancerous) cells, mediated by cells touching one another. Loss of contact inhibition is a distinguishing feature of cancerous cells.

Copy number The number of copies of each kind of plasmid normally maintained in a cell.

Corepressor A molecule (often the end product of a metabolic pathway) that combines allosterically with an aporepressor to activate it.

Cosmids Plasmid replacement vectors that contain the λ *cos* sites.

Cot curve A graph describing the renaturation of single-stranded DNA fragments with time; used to determine the degree of repetitiveness in the DNA base sequences.

Coupling phase (Cis arrangement) The arrangement of linked nonallelic genes in a heterozygote in which the wild-type alleles of both genes are present on one chromosome and the mutant alleles are located on the homologous chromosome.

Crossing-over A process in which paired homologous segments of chromosomes exchange structural parts through breakage and reunion.

Crossover suppressor An inversion that has an apparent inhibitory effect on the recombination of nonallelic genes; actually, such inversions result in nonfunctional recombination products.

Cytogenetics The area of genetics that is concerned with the study of chromosomes.

Cytokinesis The period in the life cycle of a eukaryotic cell in which the cytoplasm divides to separate the two daughter cells.

Cytoplasmic inheritance The inheritance pattern associated with organellar chromosomes (see Maternal inheritance).

Cytosine A pyrimidine base found in DNA and RNA.

D

Degeneracy A property of the genetic code; specifically, the coding of one amino acid by more than one codon of the genetic code.

Deletion (1) A chromosome structural mutation involving the loss of a segment of the chromosome; (2) a gene (point) mutation involving the loss of a single base pair.

Deletion map A topological representation of genes, showing regions where DNA deletions cause mutations.

Deletion mapping The localization of the points of mutation along a chromosome by a series of matings of point mutants with known deletion mutants.

Deme A partially isolated, local breeding population.

Denaturation Loss of the folded structure of a nucleic acid or protein molecule.

Deoxyribonucleic acid (DNA) A molecule consisting of (usually) two polynucleotide chains in which the sugar is deoxyribose and each base is adenine, guanine, cytosine, or thymine; the fundamental informational molecule that provides the cell with a set of instructions in the form of a genetic code.

Deoxyribose The pentose sugar found in deoxyribonucleotides.

Determination The process by which cells in a developing zygote commit themselves irreversibly to a particular fate.

Diakinesis The last stage of prophase I, in which homologous chromosomes finish coiling and the chiasmata move to the ends of the chromosome arms.

Dicentric chromosome A chromosome with two centromeres.

Diploid Containing two complete sets of chromosomes.

Diplonema The mid-to-late stage of prophase I, in which the pairing forces between homologous chromosomes relax and sister chromatids become clearly visible.

Diplotene stage See Diplonema.

Disassortative mating Mating in which the mated individuals share the same phenotype less often than expected by chance.

Disjunction The process in which homologous duplicated chromosomes (meiosis I) or identical daughter chromosomes (meiosis II and mitosis) move apart during nuclear division.

Dispersed Term used to describe the loosely coiled state of chromosomes.

Dispersive replication A proposed DNA replication mechanism in which old and new material would be interspersed along each strand of a newly formed double helix.

Disulfide bond A bond that is made up of two sulfur atoms (—S—S—); it can be formed between the sulfhydryl (—SH) groups of two cysteine residues in the same or in different polypeptide chains.

DNA See Deoxyribonucleic acid.

DNA cloning The production of many copies of a particular DNA fragment through its incorporation into a vector and replication of the recombinant vector molecule in recipient cells.

DNA complexity The total length of different sequences in a DNA preparation, calculated by summing the lengths of all unique sequences plus the unit lengths of all repeated sequences.

DNA fingerprinting A technique for making individual DNA identifications based on RFLP analysis and the large number of VNTR alleles that can occur in a population.

DNA gyrase A particular kind of topoisomerase II produced only in bacteria. It introduces negative supercoils into DNA using energy supplied by the breakdown of ATP; it also prevents unwound DNA strands from twisting during replication.

DNA hybridization Annealing of complementary DNA sections or single strands derived from different sources; used to determine the degree of similarity of the base sequences of the two DNAs.

DNA ligase An enzyme that catalyzes the formation of a bond between a 3′ hydroxyl group and a 5′ phosphate group.

DNA methylation Addition of a methyl group (or groups) to DNA.

DNA polymerase One of a group of enzymes that catalyzes DNA strand elongation in the 5′ to 3′ direction, using the complementary strand as a template.

Docking protein See SRP receptor.

Domain A polypeptide subregion that has folded into a compact globular structure that appears separate from the rest of the chain.

Dominance interaction An interaction between alleles (see Codominance, Complete dominance, Incomplete dominance).

Dominant The character or allele that is expressed in a heterozygote to the exclusion of a contrasting (recessive) character or allele.

Dosage compensation The mechanism by which males and females produce equal quantities of a product coded by an X-linked gene; in mammals this is accomplished by inactivation of one of the X chromosomes in the cells of the female.

Double helix The structure of DNA proposed by James Watson and Francis Crick: two helical, antiparallel polynucleotide chains connected by hydrogen bonds between opposing bases on the two chains.

Drift See Genetic drift.

Duplication Repetition of a chromosome segment, either in the same chromosome or in a different chromosome.

Dyad A single, duplicated chromosome that appears as a doubled rod during prophase of mitosis or meiosis.

E

Eclipse phase The first phase of intracellular phage lytic growth; it includes an initial growth lag and a period of DNA replication and is defined by the absence of infective phage particles.

Effector A molecule that binds to the allosteric site of a protein and either increases or inhibits its activity.

Electrophoresis A method for separating charged molecules (proteins and nucleic acids) according to their migration in an externally applied electric field; migration rates depend on the charge, size, and shape of the molecules.

Endogenote The homologous segment of a recipient chromosome in a merozygote.

Endolysin See Lysozyme.

Endonuclease An enzyme that catalyzes the breakage of a 5′-3′ phosphodiester bond in the interior of a polynucleotide chain, producing a nick.

Enhancers Sequence elements that are located outside of promoters and can greatly increase the rate of transcription.

Environmental variance The component of phenotypic variance that is due to environmental differences.

Episomes Genetic elements that can replicate autonomously in the cytoplasm of the cell that harbors the element or can be inserted into the chromosome of the host cell.

Epistasis A form of interaction between nonallelic genes in which there is masking of the expression of one or both members of a pair of alleles by a nonallelic gene.

Epistatic interaction Modification of the expected phenotypic ratio as a result of interaction between two or more non-allelic gene pairs that affect the same character.

Equational division A cellular division, such as the second meiotic division, in which the daughter cells have the same number of chromosomes as their progenitors.

Euchromatin Loosely packed, lighter-staining areas of interphase chromatin; thought to contain transcriptionally active genes, as opposed to heterochromatin, in which genes are inactive.

Eugenics Selective human breeding for the purpose of improving the human species.

Eukaryotic cell Any cell other than a bacterium; a cell that contains a distinct nucleus, separated from the cytoplasm by a nuclear envelope, and various organelles.

Excinuclease An enzyme that excises a damaged section of DNA by hydrolyzing phosphodiester bonds on either side of the modified region.

Excision repair A multistep repair process in which a damaged base or nucleotide is removed by DNA hydrolysis, the resulting gap is filled in by a DNA polymerase, and the final nick is sealed by DNA ligase.

Exogenote A segment of donor chromosome in a recipient cell that has a DNA segment (endogenote) homologous to the donor fragment; the endogenote and exogenote form the merozygote.

Exon Any segment within a split gene that is represented by a corresponding sequence in the mature RNA.

Exonuclease An enzyme that sequentially excises nucleotides from an exposed end of a polynucleotide chain through the hydrolysis of phosphodiester bonds.

Expression library See cDNA library.

Expression vectors Derivatives of cloning vectors that also contain bacterial transcriptional control signals.

Expressivity The degree of expression of genotype as a particular phenotype.

Extranuclear inheritance See Cytoplasmic inheritance.

F

F factor A conjugative plasmid found in the donor cells of *E. coli* where it can replicate autonomously in the cell or be integrated into the chromosome of its host.

F' factor A modified F factor that carries bacterial genes in addition to its own.

F pilus (plural, **pili**) A hairlike appendage that protrudes from the cell wall of bacterial donor cells; it connects the donor and recipient cells to initiate conjugation.

Facultative heterochromatin Chromatin regions that appear as heterochromatin at some stages of the life cycle and as euchromatin at others.

Feedback inhibition A form of effector control of enzyme activity in which the end product of a metabolic pathway allosterically inhibits the enzyme that (often) catalyzes the first step of the pathway.

Fibrous proteins Filamentous proteins that tend to be insoluble in aqueous solutions and often have structural or contractile roles.

Fine-structure linkage map A map that shows the relative positions of mutation sites within a gene.

First-division segregation (FDS) pattern A linear arrangement of ascospores in an ordered tetrad (*AAaa* or *aaAA*) that indicates that alleles separated at the first meiotic division, meaning that no recombination has occurred between that allele pair and the centromere.

Fitness A measure of the relative reproductive success of a genotype.

Fixation Attainment of an allele frequency value of one, resulting in a population that consists solely of one homozygous genotype.

Foldback DNA A type of extremely rapid renaturing DNA that consists of palindromic sequences that can fold back on themselves.

Folded genome structure A chromosome structure in which a molecule of DNA is folded into a number of independent supercoiled loops stabilized by attachment to a protein scaffold.

Forward mutation A mutation that changes a wild-type gene into a mutant allele.

Founder effect A form of genetic drift (sampling error) occurring when a population is established by a small group of colonizers whose alleles differ substantially in frequency from the parent population from which the colonizers were derived.

Frameshift mutation A mutation that causes a nucleotide addition or deletion and thus shifts the reading frame by one nucleotide.

G

Gene The fundamental unit of inheritance; a segment of DNA that typically codes for a particular protein product.

Gene cloning Amplification of a gene by linking it to a self-replicating vector such as a plasmid or viral genome and propagating the resulting recombinant DNA molecule in a host such as bacteria.

Gene conversion A meiotic DNA repair process in which one allele directs the change of another allele to its own form.

Gene families Groups of distinct but related genes that encode similar protein products and have similar nucleotide sequences; the members of a gene family are thought to have evolved from a common ancestral gene.

Gene frequency See Allele frequency.

Gene locus The specific location on a chromosome of a particular gene.

Gene pool The entire set of alleles present in a population at a given time.

Gene targeting Introducing a mutation into a cloned gene in vitro and studying its effects by returning the altered gene to the original organism and directing its integration into a particular region of the genome.

General recombination The exchange of genetic information between DNA regions that share extensive sequence homology.

General transcription factors (TFs) Protein factors that function in the initiation of transcription and respond to specific DNA sequences that are found in promoters recognized by different eukaryotic RNA polymerases.

Generalized transduction Transfer of fragments of bacterial chromosomes produced during the lytic cycle of infection by a temperate phage. A phage-encoded nuclease fragments the bacterial chromosome, and small pieces are sometimes packaged into mature phage; if this phage infects another cell, it injects the donor fragment into the new host.

Genetic code The set of correspondences between nucleotide triplets in mRNA and amino acids incorporated into a polypeptide chain.

Genetic drift A change in allele frequency that occurs in small populations as a result of random sampling error during reproduction.

Genetic linkage maps Diagrams that reflect the arrangement of identifiable genetic markers along a chromosome, and some measure of the distances between them.

Genetic markers Traditionally, mutations that identify loci on the basis of mutant phenotype; more recently, differences in DNA that can be detected by enzyme digestion or other biochemical techniques, even with no difference in phenotype.

Genetic transformation A process in which genetic material of a donor cell is taken up directly by a recipient cell without direct contact by the cells and without mediation by a vector.

Genetic variance The component of phenotypic variance that is due to the presence of different genotypes in the population.

Genome The entire complement of genes carried by a virus, cell, or organism.

Genomic imprinting A phenomenon in which genes function differently depending on whether they are inherited from the father or the mother.

Genomic library A collection of all randomly generated fragments from a digest of DNA, inserted and stored in a cloning vector.

Genotype The genetic constitution of an organism.

Genotype frequency The proportion of individuals of a particular genotype in a population.

Genotype-environment interaction variance The component of phenotypic variance that results from the varying effects of environmental changes on different genotypes.

Globular proteins Compactly folded proteins that are usually soluble in aqueous solutions, including most enzymes, antibodies, protein hormones, and transport proteins.

Guanine A purine base found in DNA and RNA.

H

Haploid Containing only one complete set of chromosomes.

Haplotype A particular combination of restriction site polymorphisms at two or more closely linked sites identified by a single probe.

HAT method A technique that selects for hybrid cells based on their ability to grow in the absence of specific nutrients and their resistance to various drugs.

Headful mechanism A process in which an endonuclease cuts branches of concatemeric phage DNA into lengths that just fit into the head of the developing progeny.

Helicases Enzymes that unwind and separate parental DNA strands during replication.

Helix-loop-helix motif A structural motif found in many DNA-binding proteins; it consists of two α-helices separated by a loop of variable length, with interactions between the hydrophobic faces of the helices mediating dimerization.

Helix-turn-helix motif A structural motif found in many DNA-binding proteins. It consists of two α-helical protein regions separated by a β-turn, with one helix lying in the major groove of the DNA and the other at an angle across the DNA.

Hemizygote A diploid genotype or individual that has only a single copy of a particular gene, for example an X-linked gene in the heterogametic sex.

Heritability (narrow sense) A measure of the proportion of the total variance that is caused by additive gene affects.

Heterocatalytic transfer Transfer of information between different molecules (DNA or RNA) of the same generation.

Heterochromatin Tightly coiled, dark-staining regions of interphase chromatin; heterochromatin is thought to be genetically inactive.

Heteroduplex A DNA duplex molecule that contains one or more incorrectly matched base pairs.

Heterogametic sex The sex that has unmatched sex chromosomes and therefore produces two different kinds of gametes with respect to the sex chromosomes.

Heterogeneous nuclear RNA (hnRNA) The long RNA molecules that are the primary products of transcription of eukaryotic structural genes containing all the alternating exon and intron segments of the original template DNA.

Heterogenote A partially diploid bacterial cell in which the two copies of the duplicated chromosome region carry different alleles.

Heterokaryon A cell with two genetically distinct nuclei, formed as a result of cell fusion without accompanying nuclear fusion.

Heterokaryon test A test for distinguishing between nuclear and cytoplasmic inheritance in *Neurospora* and other filamentous fungi. A cell with a known genetic marker is fused with a cell carrying a mutation that is believed to be organellar; since the two cytoplasms mix but the nuclei remain separate, the presence of both markers in a resulting conidiospore rules out nuclear inheritance of the gene in question.

Heterosis The increase in characteristics such as size, growth rate, and fertility displayed by hybrid offspring derived from inbred (homozygous) parental lines; also called hybrid vigor.

Heterozygous Carrying contrasting alleles for a particular pair of genes.

Highly repetitive DNA DNA composed of short (5 bp to 300 bp) sequences that are repeated 10^5 to 10^7 times in the haploid genome.

Histones Small, positively charged (basic) proteins that contain relatively large amounts of lysine and arginine and are the most prominent protein component of chromatin.

hnRNA See Heterogeneous nuclear RNA.

Holandric traits Traits determined by Y-linked genes.

Homeo box A DNA sequence of approximately 180 base pairs located near the 3′ end of certain genes; it encodes proteins with a sequence of 60 highly basic amino acids (the homeo domain) that binds to DNA through a helix-turn-helix motif.

Homeo domain A sequence of 60 highly basic amino acids that binds to DNA through a helix-turn-helix motif; proteins containing this sequence are thought to control major developmental events by acting as transcription factors.

Homeologous chromosomes Chromosomes that are derived from different species but still retain partial homology, indicative of some original ancestral homology.

Homeotic genes "Master genes" that control the activities of many genes that dictate the development of major body structures.

Homocatalytic transfer Transfer of information between successive generations of the same molecule (DNA or RNA); also called replicative transfer.

Homogametic sex The sex that has matched sex chromosomes and therefore produces only one type of gamete with respect to the sex chromosomes.

Homologous chromosomes The matched members of a pair of diploid chromosomes; one is maternally derived and one is paternally derived.

Homologs Homologous chromosomes.

Homopolymer tailing technique Use of an enzyme to synthesize single-stranded tails at the 3′ ends of blunt-ended DNA fragments.

Homozygous Having two identical genes for a particular character.

Host-controlled restriction and modification A system in which a host bacterium modifies its DNA by means of methylation in a particular pattern; foreign DNA lacking this pattern is broken down by the restriction endonuclease of the bacterial cell.

Human genetics The study of heredity and variation in human populations through standard scientific techniques.

Hybrid (1) A heterozygote, (2) a progeny individual from a cross between two species or varieties, (3) a duplex nucleic acid molecule made up of strands derived from different sources.

Hybrid cell The result of somatic cell fusion followed by nuclear fusion.

Hybrid vigor See Heterosis.

Hydrogen bond A weak electrostatic bond between an electronegative atom such as nitrogen or oxygen and a hydrogen atom that is covalently bonded to another electronegative atom.

Hydrophobic interactions Interactions between the nonpolar regions of two molecules or of a single molecule when in a polar solvent.

Hyperchromic shift The increase in the absorption of ultraviolet light that accompanies denaturation of DNA.

Hypersensitive site DNA regions, often found upstream of transcriptionally active genes, that are many times more sensitive to nuclease action than are the transcriptionally active genes themselves.

Hypha (plural, **hyphae**) The long, threadlike structure (composed of cells attached end-to-end) produced by a germinating conidium.

I

Imaginal disks Groups of undifferentiated cells that are set aside during embryogenesis in *Drosophila;* the adult structures develop from these cells.

In situ "In place"; in the natural or original position.

In situ hybridization A cytological procedure that combines nucleic acid hybridization and a hybrid detection procedure such as autoradiography or fluorescence microscopy to determine the chromosomal location of a labeled sequence.

In vitro site-directed mutagenesis Experimental introduction of a mutation of a base pair at a particular location within a DNA sequence; see also oligonucleotide-directed site-specific mutagenesis.

Inborn errors of metabolism Gene-controlled enzymatic failures that lead to biochemical disorders.

Inbreeding Mating in which the mated individuals are more closely related than would be expected by chance.

Inbreeding coefficient The probability that two alleles in an individual are identical by descent, that is, that they are copies of the same DNA of a common ancestor; a commonly used measure of inbreeding intensity.

Incompatibility group A collection of plasmids that cannot stably replicate together in a single cell.

Incomplete dominance The situation where contrasting alleles in a heterozygote show a phenotype that is intermediate between the two corresponding homozygous genotypes.

Independent assortment The segregation of nonallelic genes during meiosis so as to give all possible gametic gene combinations in equal frequency; typical of genes located on different chromosome pairs, in which case the segregation of one pair of chromosomes has no influence on the segregation of any other chromosome pair.

Independent events In a group of events, the occurrence of one event does not influence the chance of occurrence of any other event.

Inducer An environmental substance (effector molecule) that indirectly triggers the transcription of an operon by allosteric inactivation of the repressor molecule.

Inducible operon An operon whose transcription is "turned on" by the action of the effector molecule on the regulatory protein.

Induction of prophage The process by which a prophage leaves the bacterial chromosome, thereby initiating the lytic cycle of phage growth.

Informational molecules Polymeric molecules whose sequential structures carry information essential for living processes; for example, nucleic acids and proteins.

Insertion sequences The simplest transposons, consisting of fewer than 2000 base pairs and encoding only transposition functions.

Interference The lack of independence of multiple crossovers. See Negative interference and Positive interference.

Intergenic suppression Suppression in which the suppressor mutation occurs in a different gene than the first mutation.

Interlocus interaction An interaction between nonallelic genes.

Interphase The period in the life cycle of a eukaryotic cell in which chromosome duplication occurs; it is the interval between two successive mitoses.

Intragenic suppression Suppression in which the original and suppressor mutation occur in the same gene.

Intrasome A nucleoprotein complex on which strand exchange takes place during λ integration; it consists of integrase, integration host factor, and *att* site DNAs.

Introns Noncoding DNA segments that are interspersed in coding sequences.

Inversion The reversal of a segment within a chromosome.

Inverted repeat (IR) sequences A DNA segment in which a nucleotide sequence is followed by its complementary sequence in reverse order.

Ionic interactions Interactions between charged groups in the same or in different molecules.

Isoenzymes Different forms of a protein that catalyze the same reaction in an organism; also called isozymes.

Isoschizomers Restriction enzymes that are produced by different bacteria but recognize the same DNA cleavage site.

Isozymes See Isoenzymes.

K

Karyokinesis See Mitosis.

Karyotype The chromosomal complement of a eukaryotic organism or cell, usually viewed during mitotic metaphase.

Kinetochore A protein structure that attaches the centromere to the mitotic spindle during cell division.

L

Latent period The intracellular phase of viral development; the time between infection and the release of mature phage particles.

Law of large numbers As the sample size becomes very large, the observed relative frequency of an event approaches its theoretical probability.

Leptonema The early stage of prophase I, in which chromosomes begin to condense.

Leptotene stage See Leptonema.

Leucine zipper A structural motif found in many DNA-binding proteins. It consists of two complementary helices containing regions in which leucine occurs about every seventh position; the leucines all align on one side of each helix and interlock with each other to form a "zipper."

Ligase See DNA ligase.

LINES **(long interspersed elements)** Repeated DNA sequences of 5000 to 7000 bp that alternate with even greater lengths of unique sequences in the genomes of higher organisms.

Linkage The tendency of nonallelic genes located on the same chromosome to assort into gametes in the same combinations as they were inherited, rather than independently.

Linkage group A group of genes that show linkage with one another and correspond to a particular chromosome.

Linkage maps See Genetic linkage maps.

Linked genes Nonallelic genes that show linkage, measured as a departure from independent assortment.

Locus See Gene locus.

Lod score method A probability technique used in determining whether genes are linked.

Long terminal repeat (LTR) sequences Long repeated sequences found at the ends of certain duplex DNA molecules, such as those produced by retroviruses.

Lyonization The inactivation of an X chromosome in mammals through condensation into a Barr (sex chromatin) body.

Lysogenic cells The bacterial host of a prophage; the cell is capable of spontaneous lysis due to the uncoupling of the prophage from the bacterial chromosome.

Lysogenic cycle The method of phage reproduction through existence as a prophage with replication of the prophage occurring in synchrony with replication of the host chromosome.

Lysozyme An enzyme that degrades the peptidoglycan component of the bacterial cell wall.

Lytic cycle The method of viral reproduction consisting of the production of progeny viruses and lysis of the infected cell.

M

Major groove The larger of the two grooves located on the exterior of the DNA molecule and extending from one end of the molecule to the other in a helical fashion.

Map unit A unit of distance in a genetic map; the classical map unit is equivalent to a 1% recombination frequency (also called a centiMorgan (cM)); in *E. coli* one map unit is equivalent to one minute of transfer during conjugation.

Marker retention A technique for mapping mitochondrial genes in yeast. A mutagen is used to induce random deletions in a multiply drug-resistant strain, and each resulting strain is tested to determine which markers are retained; the marker retention data from different strains can be combined to form a map of the DNA molecule.

Maternal inheritance The extranuclear inheritance of a trait exclusively through the female parent; see Cytoplasmic inheritance.

MCS. See Multiple cloning sites.

Meiocyte A primordial germ cell that undergoes meiosis to produce gametes.

Meiosis The cellular division process undergone by meiocytes to produce sex cells (gametes); it consists of two successive nuclear divisions and produces gametes (or spores) that contain half the number of chromosomes in the original cell.

Meiosis I A reduction division process that produces daughter cells each containing a haploid number of duplicated chromosomes (dyads).

Meiosis II An equational division process similar to mitosis; the haploid products of meiosis I are converted into four haploid cells containing unduplicated chromosomes.

Meiotic nondisjunction See Nondisjunction.

Melting The thermal denaturation of DNA.

Merozygote A partially diploid bacterial cell, containing a donor chromosome segment (exogenote) in addition to the homologous segment of a recipient chromosome (endogenote).

Messenger RNA (mRNA) The end product of transcription of a structural gene; the form of RNA whose codon sequence is translated into an amino acid sequence in a protein.

Metabolic block A nonfunctioning reaction in a metabolic pathway, which is the result of a defect in the enzyme that normally catalyzes the reaction.

Metacentric chromosome A chromosome in which the centromere is in a median position.

Metaphase The second stage of mitosis, in which the individual duplicated chromosomes align themselves on the equatorial plane of the spindle.

Metaphase I The second phase of meiosis I, in which bivalents align at the center of the spindle.

Metaphase II The second stage of meiosis II, in which the individual duplicated chromosomes align on the spindle equator.

Minimal medium A growth medium for microorganisms that contains only inorganic salts and a carbon source.

Minimal nucleosome See Chromatosome.

Minimum mutational distance The minimum number of nucleotide pair changes in the DNA needed to account for the observed amino acid differences in a homologous protein in a different organism.

Minor groove The smaller of the grooves located on the exterior of the DNA molecule and extending from one end of the molecule to the other in a helical fashion.

Missense mutation A mutant in which a single base change in a DNA triplet leads to a single amino acid substitution in the encoded polypeptide.

Mitosis The period in the life cycle of a eukaryotic cell in which the nucleus divides into two nuclei that are genetically and cytologically identical to each other and to the original nucleus.

Mitotic nondisjunction See Nondisjunction.

Moderately repetitive DNA DNA consisting of sequence families that are repeated 10 to 10^5 times in the haploid genome.

Molecular clock A constant rate of molecular evolution that can be used to time evolutionary events.

Molecular genetics The area of genetics that is concerned with the molecular structure and expression of genes.

Monocistronic mRNA mRNA molecules that contain the information of a single gene and code for a single polypeptide.

Monohybrid Hybrid for a single pair of alleles.

Monomorphic locus A locus for which all alleles in a population are essentially of the same type.

Monoploid A cell having a single set of chromosomes, or an organism composed of such cells.

Monosomy An aneuploid condition involving the loss of a single chromosome; in a diploid organism this gives a chromosome number of $2n - 1$.

Mosaic An organism composed of two or more genetically different types of cells.

mRNA See Messenger RNA.

Motif A regular pattern of substructure in a protein, resulting from interaction between adjacent segments of secondary structure.

Multimeric protein A protein that consists of two or more polypeptide subunits.

Multinomial distribution A probability distribution for an event that has more than two possible outcomes.

Multiple allelism The existence of three or more allelic states of a gene.

Multiple cloning sites Several different restriction sites located in the same DNA segment.

Mutagen Any agent that is capable of increasing the rate of mutation.

Mutant A cell or organism that carries a mutant gene.

Mutant frequency The ratio of mutant to total cells in a specific cell population.

Mutation (1) A process that produces a gene or a chromosome complement that is different from wild type, (2) the gene or chromosome complement that results from such a process.

Mutation rate A measure of the probability that a given gene will undergo mutation, defined in reference to some unit of time.

Mutually exclusive events A series of alternative events only one of which can occur at any one time.

Mycelium A multinucleate mass of cytoplasm that is formed by hyphae and releases masses of conidia.

N

N terminus See Amino terminus.

Negative control Genetic regulation by switching off the expression of a gene or group of genes.

Negative eugenics Discouraging people with "undesirable" traits from reproducing.

Negative interference Interference in which one crossover increases the chance of another crossover nearby, giving more double recombinants than would be expected for independent events.

Negative supercoil A supercoil formed when DNA becomes underwound; the form of most cellular DNA.

Neutral mutation A missense mutation that does not change the function of the encoded protein.

Nick A broken 5′–3′ phosphodiester bond in the interior of a polynucleotide chain.

Nick translation A process in which DNA polymerase I excises nucleotides from a nicked polynucleotide chain in the 5′ to 3′ direction and simultaneously adds new nucleotides behind it, using the exposed 3′ end of the nicked strand as a primer.

Nonalleles (nonallelic genes) Genes that occupy different chromosomal positions.

Nondisjunction The failure of homologous chromosomes (meiosis I) or sister chromatids (meiosis II, mitosis) to separate at anaphase.

Nonparental ditype (NPD) tetrad A tetrad (ascus) that contains ascospores with only recombinant genotypes.

Nonrandom mating See Assortative mating, Disassortative mating, and Inbreeding.

Nonsense mutation A mutation that converts a codon for an amino acid to a termination codon, leading to premature termination of polypeptide synthesis.

Nonsense suppressor A suppressor mutation that alters a tRNA anticodon so that it recognizes a stop codon in a strand of mRNA; the mutant tRNA thus reads a nonsense mRNA codon as an amino acid and avoids premature polypeptide termination.

Northern blotting A procedure similar to Southern blotting, but used for analyzing RNA molecules that have been separated on an electrophoretic gel and identifying specific RNA sequences.

Nucleocapsid The capsid and nucleic acid of a virus.

Nucleoid The region of a prokaryotic cell or eukaryotic cellular organelle that contains the genetic information of the cell or organelle.

Nucleolar organizer regions Chromosomal regions in a eukaryotic cell that contain the genes that code for ribosomal RNA.

Nucleolus (plural, **nucleoli**) An organelle in a eukaryotic nucleus that contains ribosomal RNA and multiple copies of the genes coding for rRNA.

Nucleoside A nitrogenous base bound to a sugar molecule, but lacking a phosphate group.

Nucleosome A subunit of chromatin, containing DNA and histone proteins.

Nucleosome core particle The fundamental structural component of a nucleosome; it consists of a histone octamer protein core with a length of DNA wrapped around it.

Nucleotide The monomer unit of a nucleic acid, consisting of a nitrogenous base, a sugar molecule, and a phosphate group.

Nucleus The portion of a eukaryotic cell that contains the genetic information of the cell.

O

Okazaki fragments Short segments of DNA formed by the discontinuous nature of replication of one of the parental DNA strands.

Oligomeric protein A protein that consists of two or more polypeptide subunits.

Oligonucleotide-directed site-specific mutagenesis A method for selectively introducing a mutation at a particular location in a gene. A chemically synthesized oligonucleotide containing a single mutant nucleotide is used as a primer to initiate replication of the gene; half the progeny DNA molecules then carry the mutation.

Oncogene A potential cancer-producing gene, of host or viral origin; it overly stimulates cell division.

Oncogenic Cancer-causing.

Oncogenic virus A cancer-causing virus.

One gene—one enzyme hypothesis The hypothesis that the function of a gene is to control the synthesis of one enzyme (or one protein).

One gene—one polypeptide hypothesis The more general hypothesis that the function of most genes is to code for a polypeptide that, either alone or folded together with other polypeptides, is a functional protein.

Open reading frame A long sequence of nucleotides that begins with a start codon and is free of stop codons until its terminus; it indicates a protein-coding gene.

Operator A DNA region of an operon that serves as the binding site for an active repressor protein.

Operon Two or more adjacent structural genes that are transcribed into a single molecule of mRNA, plus the adjacent transcriptional control sites.

Ordered tetrad An ascus in which the meiotic products are arranged in a linear order reflecting the linear order of meiotic segregation events.

Overlapping genes A situation in which the same DNA sequence is used by more than one gene.

P

Pachynema The mid-prophase I stage in which paired homologous chromosomes exchange structural parts through crossing-over.

Pachytene stage See Pachynema.

Palindrome A nucleotide sequence that reads the same when both strands are followed in the 5′ to 3′ direction.

Pangenesis The (now discredited) theory that each part of the body of an organism contains its own minute hereditary elements called gemmules, which travel via the bloodstream to the gonads, where they are passed on to the next generation.

Paracentric inversion An inversion that does not include the centromere.

Parental ditype (PD) tetrad A tetrad (ascus) that contains ascospores with only the parental genotypes.

Parental type gamete A gamete that contains the parental arrangement of nonallelic genes.

Particulate inheritance The theory that discrete hereditary units ("particles") do not blend by mixing with units from another individual in forming offspring, but instead retain their identities from one generation to the next.

Pedigree A diagrammatic representation of two or more generations of related individuals, showing the pattern of transmission of a particular genetic trait.

Penetrance The percentage of individuals of a certain genotype that show the expected phenotype.

Peptide bond An amide linkage joining the α-amino group of one amino acid to the α-carboxyl group of another amino acid.

Peptidyl transferase A component of the large ribosomal subunit that catalyzes the formation of a peptide bond between amino acids during chain elongation.

Pericentric inversion An inversion that includes the centromere.

Permissive conditions Conditions under which conditional mutations have no effect (i.e., produce a wild-type phenotype).

Phasmids A class of replacement vectors that consists of λ phage genomes containing one or more plasmid molecules.

Phenocopy An environmentally induced phenotype that resembles one known to be genetically determined.

Phenotype The outward appearance of an organism; the characteristic expressed by a genotype.

Phosphatase An enzyme that catalyzes the removal of phosphate.

Phosphodiester bond The chemical linkage between adjacent nucleosides in a polynucleotide such as DNA or RNA, formed by a phosphate group that connects the 3′ carbon of one nucleoside to the 5′ carbon of another.

Photolyase See Photoreactivation.

Photoreactivation The direct repair of DNA damage by a light-activated enzyme (photolyase).

Phyletic evolution Evolution within a lineage or single line of descent; also called anagenesis.

Physical linkage maps Chromosome maps that directly reflect the DNA nucleotide arrangement; they are based on DNA sequencing and similar procedures, not on the results of genetic crosses.

Pitch The vertical rise per turn of a helix.

Plaque A clear area on a bacterial lawn, left by lysis of the bacteria through successive infections by a single original phage particle and its descendants.

Plaque assay A method for measuring phage concentration. A highly diluted phage suspension is added to a concentrated solution of indicator bacteria and spread on an agar plate; the number of plaques that develop from lysed, infected bacteria can be used to calculate the original phage concentration.

Plasmids Small genetic elements that replicate autonomously in the cytoplasm of a prokaryotic or eukaryotic cell.

Pleated sheet See β-sheet.

Pleiotropy Phenomenon whereby a single mutant gene is observed to influence several different traits.

Point mutation A mutation that is caused by a change of a single base pair.

Poisson distribution A probability distribution that is applicable to situations in which the probability of an event is very small but the number of trials is large.

Polycistronic mRNA mRNA molecules that carry the information of two or more genes.

Polygenic trait See Quantitative trait.

Polylinker DNA A DNA segment containing multiple cloning sites.

Polymerase chain reaction An in vitro amplification procedure in which a reaction catalyzed by a DNA polymerase replicates portions of a DNA molecule; used to increase a small sample of DNA for procedures such as DNA fingerprinting and sequencing.

Polymorphic locus A locus for which two or more allelic forms commonly occur in a population.

Polynucleotide phosphorylase An enzyme that catalyzes the formation of single-stranded RNA from nucleoside diphosphates without using a template to specify their sequence.

Polypeptide An unbranched polymer of amino acids linked to one another through peptide bonds.

Polyploidy The gain of one or more entire sets of chromosomes.

Polyribosome A complex of ribosomes attached to the same mRNA and producing multiple copies of the coded polypeptide; also called a polysome.

Polysome See Polyribosome.

Polytene chromosomes Large (multistranded) chromosomes found in the secretory tissues of diptera; formed by successive chromosome replication without accompanying chromatid separation or cell division.

Population genetics The study of genetic variation in populations and the responses of gene and genotype frequencies to mating patterns and evolutionary forces.

Position effect A change in the expression of a gene due to a change in chromosome location.

Positive control Genetic regulation by switching on the expression of a gene or group of genes.

Positive eugenics Encouraging people with "desirable" traits to have children.

Positive interference Interference in which one crossover reduces the chance of another crossover nearby, giving fewer double crossovers than expected on the basis of independence of crossover events.

Positive supercoil A supercoil formed when DNA becomes overwound.

Postreplication repair Repair systems that act on DNA that has replicated before a mutagenic lesion is repaired.

Primary structure The particular sequence of monomer units (amino acids or nucleotides) in a polypeptide or polynucleotide chain.

Primase An enzyme that initiates DNA replication by catalyzing the DNA-dependent synthesis of a short RNA primer (see also Primer).

Primer A segment of a DNA or an RNA chain that is elongated with deoxyribonucleotides by DNA polymerase during DNA replication.

Primosome A group of proteins including primase that catalyzes primer synthesis.

Probability distribution A graphical, tabular, or mathematical representation of the possible values of a random variable and the corresponding probability of each value.

Probe A radiolabeled or chemically labeled DNA or RNA strand that is complementary to the DNA sequence of the region one wishes to detect.

Prokaryotic cell A bacterial cell; a cell that has no nuclear envelope and hence no distinct nucleus.

Promoter A regulatory site that serves as the binding site for RNA polymerase in the initiation of transcription.

Prophage An intracellular phage chromosome that is integrated into the chromosome of the host bacterial cell.

Prophage induction The release of the prophage from the bacterial chromosome to initiate a lytic cycle.

Prophase The first stage of mitosis, characterized by the coiling of the duplicated chromatin threads into shorter, thicker, cytologically visible chromosomes and the disappearance of the nuclear envelope and nucleolus.

Prophase I The first stage of meiosis I; in this stage, the chromosomes condense, and homologous chromosomes pair and undergo crossing-over.

Prosthetic group A nonpolypeptide group conjugated with a protein.

Protein A macromolecule composed of one or more polypeptide chains folded together in a specific fashion.

Protein chaperones Proteins that recognize and bind to folding polypeptides and thus mediate their correct assembly.

Protein kinase An enzyme that catalyzes phosphorylation of proteins and uses ATP or GTP as a phosphate source.

Proto-oncogene A cellular gene that is the normally functioning allele of an oncogene.

Protoplast A cell whose wall has been removed.

Prototroph A wild-type microorganism that will grow in minimal medium, being able to synthesize for itself all organic compounds required for growth and thus requiring no growth supplements in the medium.

Provirus The genome of a virus that is integrated into a chromosome of its host cell.

Pseudogene A DNA sequence that is homologous with a known gene, but is never transcribed.

Pulse-labeling A technique in which a replicating chromosome is labeled with short bursts of a radioactive precursor.

Pulsed-field gel electrophoresis A type of gel electrophoresis that uses pulses of current and alternates the direction of the electric field.

Pure-breeding Producing offspring identical to the parent after self-fertilization; a group of genetically identical individuals that, when intercrossed, always produce offspring that are identical to their parents.

Purine A nitrogenous base that consists of a nine-member heterocyclic ring structure from which adenine and guanine are derived.

Pyrimidine The six-member ring compound from which cytosine, thymine, and uracil are derived.

Pyrimidine dimers Two adjacent pyrimidines in a DNA strand that are covalently bonded together.

Q

Quantitative trait A trait that is determined by the cumulative effects of many genes (called polygenes) whose action can be very sensitive to environmental factors; usually characterized by a continuous distribution of phenotypes in a population rather than discrete phenotypic classes.

Quaternary structure The spatial relationship between the folded subunit chains of an oligomeric protein.

R

R plasmids Plasmids whose genes give the host cell resistance to a variety of antibiotics and heavy metals.

Random genetic drift See Genetic drift.

Random mating Selection of mates without regard to genotype or phenotype.

Random sampling error The chance deviation of observed results from those expected.

Recessive The character or allele that is not expressed in a heterozygote because it is masked by a dominant character or allele.

Reciprocal crosses A pair of crosses of the type female genotype A × male genotype B and female genotype B × male genotype A; for example, *AA* female × *aa* male and *aa* female × *AA* male are reciprocal crosses.

Reciprocal translocation The interchange of parts between nonhomologous chromosomes.

Recombinant DNA A novel DNA molecule constructed from segments of DNA derived from two or more different sources.

Recombinant type gamete A gamete that contains a gene combination that does not reflect the parental arrangement of nonallelic genes.

Recombination Any process that generates new nonallelic gene combinations.

Recombinational repair See Postreplication repair.

Reduction division A cellular division in which the daughter cells have half the number of chromosomes of the original cell, such as the first meiotic division.

Regulatory transcription factors (TFs) Transcription factors that react to specific sequences in certain promoters and/or are active in all types of cells.

Renaturation See Annealing.

Repetitive DNA See Highly repetitive DNA and Moderately repetitive DNA.

Replica plating A technique in which cell colonies from a master plate are transferred to one or more replica plates for various studies; the colonies are located in the same positions on the master plate and all the replica plates.

Replicative transposition Transposition in which the transposon duplicates itself, so that one copy remains at the old site while another copy inserts itself at the new site.

Replicon Unit of replication of a eukaryotic chromosome, consisting of a length of DNA under the influence of a single origin of replication.

Reporter gene A gene that encodes an easily measured protein and is coupled to a regulatory DNA region to analyze the function of the region in different tissues and at different developmental stages.

Repressible operon An operon whose transcription is "turned off" by the action of the effector molecule on the regulatory protein.

Repressor protein A regulatory molecule that binds to an operator site and prevents transcription of an operon.

Repulsion phase (trans arrangement) The arrangement of linked nonallelic genes in a heterozygote in which one wild-type gene and one mutant nonallele are located on each member of the chromosome pair.

Response elements (RE) Enhancer sequences that activate transcription in response to particular extracellular signals.

Restriction endonucleases A group of bacterial enzymes that cleave DNA strands within a recognition sequence specific to each enzyme.

Restriction fragment length polymorphism (RFLP) Variations in the lengths of DNA fragments produced by restriction analysis due to variations in restriction sites within a population.

Restriction map A map of a DNA molecule that shows the locations of the recognition sites for particular restriction endonucleases.

Restrictive conditions Conditions under which conditional mutations produce mutant phenotypes.

Retroposons Eukaryotic transposons that share many of the basic features of retroviruses; also, inserted DNA copies of nonviral cellular RNAs.

Retrovirus An RNA virus that employs the enzyme reverse transcriptase to synthesize a DNA copy of the viral genome within the host cell.

Reverse (back) mutation A mutation that changes a mutant allele back to the wild-type through reversal of the original mutational change.

Reverse transcriptase An enzyme used by retroviruses to form DNA copies of their RNA in infected cells.

Revertant The individual formed by a reverse mutation.

Ribonucleic acid (RNA) A molecule consisting of (usually) one polynucleotide chain in which the sugar is ribose and each base is adenine, guanine, cytosine, or uracil.

Ribose The pentose sugar found in ribonucleotides (the precursors of ribonucleic acid).

Ribosomal RNA (rRNA) The product of transcription of ribosomal RNA genes; complexes with ribosomal proteins to form ribosomes.

RNA polymerase An enzyme that catalyzes the synthesis of RNA using a DNA strand as the template.

RNA replicase An RNA-dependent RNA polymerase encoded by a + strand RNA viral genome; it is produced by translation shortly after infection by an RNA virus occurs, and it copies the viral RNA to make − strands.

Robertsonian fusion See Centric fusion.

Rolling-circle replication A unidirectional form of DNA replication in which one strand of a circular DNA molecule becomes detached from the origin and forms a "tail" on the circle as it synthesizes its complementary strand; also called sigma replication.

rRNA See Ribosomal RNA.

S

Satellite Chromosomal material that is attached to the tips of some chromosomes by a narrow constriction.

Satellite DNA A portion of DNA that separates from the main band during centrifugation; this DNA contains highly repetitive sequences, having a GC content that gives it a different density from main band DNA.

Second-division segregation (SDS) pattern A linear arrangement of ascospores in an ordered tetrad (*AaAa*, *aAaA*, *aAAa*, or *AaaA*) that indicates that the *A* and *a* alleles were segregated during the second meiotic division, showing that crossing over has occurred between that allele pair and the centromere.

Secondary structure The three-dimensional conformation formed by the regular folding of a polynucleotide or polypeptide chain, often in the form of a helix.

Sedimentation coefficient A measure of the rate at which a particle will sediment in a unit centrifugal field of force, generally proportional to particle size.

Self-fertilization (selfing) The union of gametes derived from a single individual.

Semiconservative replication A DNA replication mechanism in which each newly synthesized strand forms a double helix with its complementary parental (template) strand.

Sequence family DNA regions that are similar enough in base sequence to hybridize with one another under standard renaturation conditions.

Sex chromatin body See Barr body.

Sex chromosome A chromosome whose presence or absence is correlated with the sex of an individual.

Sex pilus See F pilus.

Sexduction Transfer of genes to a recipient cell by the F′ factor during conjugation.

Sex-influenced trait A trait determined by alleles (usually autosomal) whose dominance relationship is affected by the sex hormones.

Sex-limited trait A trait expressed exclusively by one sex, usually determined by an autosomal gene.

Sex-linked genes Genes that are carried on either the X chromosome or the Y chromosome, but not on both; usually taken to mean an X-linked gene.

Shifting balance theory A population model in which the average fitness of a population is represented by a series of peaks and valleys in a two-dimensional field of possible gene combinations.

Shine–Dalgarno sequence A purine-rich sequence located upstream of the initiation complex in prokaryotic mRNA; it pairs with a complementary sequence in the 16S rRNA portion of the ribosome 30S subunit during the initiation of translation.

Shotgun approach Cloning all randomly generated fragments from a digest of genomic DNA.

Shuttle vectors Yeast expression vectors that can replicate in both *E. coli* and yeast.

Signal-recognition particle (SRP) A ribonucleoprotein component that binds to the signal sequence of a secretory polypeptide and halts translation; the SRP-ribosome complex then diffuses to a receptor on the endoplasmic reticulum, where translation of the secretory protein resumes, threading the protein across the ER membrane.

Signal sequence A hydrophobic N-terminal sequence of 15 to 35 amino acids found in membrane and secretory proteins; during translation, it directs the ribosome to the membrane surface and initiates the passage of protein through the membrane.

Silent mutation A mutation that produces a DNA base substitution but, because of the degeneracy of the genetic code, does not change the amino acid specified by the altered codon.

SINES (short interspersed elements) Repeated DNA sequences of about 300 bp that alternate with longer unique sequences in the genomes of higher organisms.

Single-strand binding proteins Proteins that temporarily bind to single-stranded DNA regions during replication to prevent the creation of hairpin loops and keep the parental helix from reforming.

Sister chromatids The two chromatids that are derived from the duplication of one chromosome during interphase.

Site-specific recombination The reciprocal exchange of genetic information between DNA regions that share very short sections of sequence homology.

snRNA Small nuclear RNA molecules that recognize certain regions of intron-containing transcripts during the splicing reactions that convert modified pre-mRNA into mature mRNA; see also snRNPs.

snRNPs ("snurps") Ribonucleoprotein particles in which snRNA is usually found.

Solenoid structure The cylindrically coiled condensed chromosome fiber formed by the coiling of the nucleosome fiber.

Somatic cell genetics An area of cell research in which somatic cells are cultured and used in genetic investigation, such as methods for localizing genes by using tissue culture techniques to assign genes to individual chromosomes or chromosome segments.

Somatic cell hybridization The fusion of somatic cells from different species to form a hybrid cell whose nucleus contains the chromosomes of each species.

Somatic doubling A mitotic failure in which chromosome replication without cell division leads to a doubling of the chromosome number in somatic tissue.

Somatic mutation A mutation that occurs in a somatic cell.

Southern blotting A method for transferring DNA segments from an agarose gel (where they have been separated electrophoretically) to the surface of a membrane filter where nucleic acid hybridization can occur.

Southwestern blotting A technique for detecting clones that produce DNA-binding proteins that bind specifically to a particular DNA sequence. Lysed colonies are exposed to radioactively labeled DNA containing the sequence for which the DNA-binding protein is specific; clones expressing the binding protein are then detected by autoradiography.

Specialized transduction Transduction that is restricted to the transfer of bacterial genes located adjacent to the site of prophage insertion.

Speciation The splitting of a lineage into two or more genetically distinct lines of descent as a consequence of the development of reproductive isolation; also called cladogenesis.

Spheroplasts Osmotically sensitive cells produced by damaging the outer membrane of Gram-negative bacteria.

Spindle An organized system of microtubules that attaches to the centromere regions of duplicated chromosomes and draws them to opposite poles; responsible for chromosome movement during meiosis and mitosis.

Spliceosomes Ribonucleoprotein complexes that contain the various factors involved in splicing.

Splicing The excision of introns from primary RNA transcripts and the joining together of the exon regions into mature mRNA.

Split gene A gene locus that is subdivided into coding (exon) and noncoding (intron) nucleotide sequences.

SRP receptor A protein that is located on the surface of the endoplasmic reticulum and combines with the SRP (signal recognition particle)-ribosome complex when it reaches the membrane surface.

Stringent response The reversible inhibition of rRNA and tRNA synthesis under conditions of amino acid starvation.

Structural gene A nonregulatory gene that codes for the amino acid sequence of a polypeptide.

Submetacentric chromosome A chromosome in which the centromere is located in a submedian position (neither at the center nor at one end).

Supercoiling Higher level coiling of a DNA double helix; a common form of tertiary structure in DNAs whose ends are not free to rotate.

Supergene A multigene complex that consists of several closely linked loci.

Superhelicity See Supercoiling.

Suppression A process in which a second mutation at a different site from the first restores the wild-type phenotype but leaves the genotype in a mutant condition; see also Intragenic suppression and Intergenic suppression.

Suppressor mutation A second mutation at a different site that suppresses the effects of a first mutation (see Suppression).

Synapsis A process in which homologous chromosomes pair and align themselves lengthwise during prophase I.

Synaptonemal complex The nucleoprotein structure that holds paired homologous chromosomes in close contact during synapsis.

Synkaryon A hybrid cell with a single nucleus containing the chromosomes of both parental cell lines.

Syntenic genes Nonallelic genes that are located on the same chromosome pair; syntenic genes may or may not show genetic linkage, depending on the distance separating them.

T

Targeted gene replacement A form of gene targeting in which a wild-type gene is replaced by a mutant allele.

Tautomerism The spontaneous and reversible change in the distribution of protons in a nitrogenous base, changing the hydrogen-bonding properties of the base.

Telocentric chromosome A chromosome in which the centromere is at a terminal location.

Telomerase An enzyme that prevents the shortening of the ends of chromosomes during replication by catalyzing synthesis of the repeated telomeric sequences.

Telomere A special DNA region at the end of a chromosome that helps to preserve its integrity during replication.

Telophase The fourth stage of mitosis, in which the daughter cells reach the opposite poles of the spindle, the spindle fibers disappear, and the nuclear envelope and nucleoli reappear.

Telophase I The final phase of meiosis I; the haploid sets of duplicated chromosomes reach opposite poles so that each daughter cell receives only one member of each original pair of chromosomes in the meiocyte.

Telophase II The final stage of meiosis II; the haploid sets of daughter chromosomes reach opposite poles so that each meiotic product contains a complete haploid set of nonhomologous chromosomes.

Temperate viruses Viruses that establish a long-term, usually harmless relationship with their host; the viral DNA replicates in synchrony with the host genome, so that the viral genome is passed on to each daughter cell.

Temperature-sensitive mutation A mutation that results in increased heat or cold sensitivity and thus limits the temperature range in which the organism can grow.

Template strand A DNA sequence that serves as a "mold" or "pattern" for producing a daughter strand of DNA or a strand of RNA.

Terminal redundancy Repetition of the same DNA sequence at the beginning and end of a chromosome.

Terminalization The process in which the chiasmata move toward the ends of the chromosome arms during the diplotene stage of prophase I.

Tertiary hydrogen bonds Hydrogen bonds, some involving unpaired, invariant bases, that help stabilize the tertiary structure of a nucleotide chain; found in the cloverleaf configuration of tRNA.

Tertiary structure The three-dimensional conformation formed by the folding of the secondary structure of a nucleic acid or protein.

Testcross A dominant × recessive mating often used to determine whether the dominant individual is homozygous or heterozygous.

Tetrad A synapsed pair of homologous chromosomes containing four chromatids; also called a bivalent. Also, the four haploid products produced by meiosis in certain fungi and contained within a single ascus.

Tetraploid A cell containing four sets of chromosomes, or an organism composed of such cells.

Tetratype (T) tetrad A tetrad (ascus) that contains four types of spores, two of the parental genotypes and two of the recombinant genotypes.

Three-point testcross A cross in which one parent is trihybrid and the other is a recessive homozygote.

Threshold trait A trait that has a polygenic basis but exhibits discrete phenotypic classes as a result of differences in expression of particular combinations of polygenes that either exceed or fall below certain threshold values.

Thymine A pyrimidine base found in DNA.

Thymine dimer A pyrimidine dimer involving two thymine bases.

Topoisomerases Enzymes that convert DNA into the relaxed state by catalyzing the removal of supercoils through the breakage and rejoining of phosphodiester bonds.

Transcriptase An RNA-dependent RNA polymerase encoded by certain − strand RNA viruses; it catalyzes synthesis of + strands on the − strand template.

Transcription The process in which DNA is used as a template for production of complementary RNA molecules.

Transcription factor (TF) A stimulatory protein that is part of a transcription initiation complex that binds to a regulatory eukaryotic DNA region near the startpoint for RNA synthesis and thus enables the RNA polymerase to begin transcription of the DNA molecule.

Transducing phage A phage particle that transfers bacterial DNA during transduction.

Transduction The transfer of DNA from one cell to another by way of a viral vector, with no physical contact between donor and recipient cells.

Transfer RNA (tRNA) A form of RNA that mediates the transfer of information from mRNA to the polypeptide; each tRNA molecule carries a specific amino acid to the ribosome and has an anticodon that recognizes the specific mRNA codon for that amino acid.

Transformation (1) See Genetic transformation. (2) The process by which a cell changes into a cancerous state.

Transgenic organism An organism in which foreign DNA was integrated into the genome early enough in development to transform all cell lineages; the foreign gene would then be passed on to progeny.

Transition A base substitution in which a purine in one strand of DNA is replaced by another purine and the pyrimidine in the complementary strand is replaced by another pyrimidine.

Translation The production at a ribosome of a polypeptide whose amino acid sequence is determined by the sequence of codons in an mRNA molecule; the transfer of genetic information from mRNA to protein.

Translocation (1) The transfer of a chromosome segment to a nonhomologous chromosome; (2) The movement of a peptidyl-tRNA from the A site to the P site of a ribosome during protein synthesis, accompanied by a shift of the mRNA by one codon.

Transmission genetics The study of the patterns of inheritance of genes between generations.

Transposable element See Transposon.

Transposase A transposon-encoded enzyme that cuts the target DNA sequence prior to insertion of a transposon.

Transposition See Transpositional recombination.

Transpositional recombination Nonreciprocal genetic exchange between DNA sequences with no homology between recombination sites.

Transposon Any DNA segment that is capable of moving from one chromosomal location to another.

Transversion A base substitution in which a purine replaces a pyrimidine and a pyrimidine replaces a purine.

Triploid A cell containing three sets of chromosomes, or an organism composed of such cells.

Trisomy An aneuploid condition that involves the gain of a single chromosome; in a diploid organism this gives a chromosome number of $2n + 1$.

tRNA See Transfer RNA.

True-breeding See Pure-breeding.

Tumor suppressor genes Genes that normally inhibit cell proliferation.

Two-point testcross A testcross in which a dihybrid is mated with a recessive homozygote.

Unequal crossing-over Crossing-over between asymmetrically paired homologs during prophase I, resulting in one homolog with a tandem repeat of a chromosome segment and the other with a deletion of the same region.

Unique sequence DNA DNA that contains a nucleotide sequence that lacks repetition.

Unit evolutionary period (UEP) The time required for two proteins to diverge by 1% of their amino acid sequence.

Unordered tetrad An ascus in which the meiotic products are randomly placed.

Upstream activator sequences Yeast DNA sequences that are analogous to enhancers; they can function in either orientation and at various distances upstream of promoters.

Uracil A pyrimidine base found in RNA.

Variable expressivity See Expressivity.

Variation in the number of tandem repeats (VNTR) A variation in the number of short, repeated segments found between two restriction sites of a DNA molecule; cuts at these sites thus produce DNA fragments whose lengths vary within a population.

Variegation The occurrence within a tissue of sectors with different phenotypes.

Virion The mature nonreplicating form of a virus.

Virulent phage A bacterial virus that is limited to the lytic cycle of growth, being incapable of forming a prophage.

Western blotting An immunochemical procedure for the detection of gene function. Lysed colonies are exposed to a radioactively labeled antibody probe specific for the protein product of the cloned gene; autoradiography then reveals the locations of colonies producing the protein.

Wild-type The allele or genotype or phenotype that is found in nature or in the standard laboratory strain for a given organism.

Wobble hypothesis The idea that the first and second bases of a codon bond to the anticodon according to strict base-pairing rules, but there is some flexibility (or wobble) in pairing at the third position.

X-linked genes Genes located exclusively on the X chromosome.

Yeast artificial chromosomes Yeast chromosomes artificially constructed through recombinant DNA techniques; they contain an origin of replication, telomeric regions, and a centromere, enabling them to replicate and segregate their DNA normally during cell division.

Y-lined genes Genes located exclusively on the Y chromosome.

Z–DNA A form of DNA characterized by a left-handed helix, thinner and more extended than the predominant B form.

Zinc finger A structural motif found in many DNA-binding proteins. Coordinate binding of a zinc atom to cysteine and histidine residues gives a tetrahedral complex from which amino acid chains loop out in fingerlike projections.

Zygonema Part of the early stage of prophase I; in this stage, homologous chromosomes begin to align lengthwise through synapsis.

Zygote A fertilized egg.

Zygotene stage See Zygonema.

Zygotic combinations The various gene combinations produced by fertilization; entities within a Punnett square.

INDEX

Inversion(s), 259–260, 264–265
Inversion loop, 264–265
Inverted repeat (IR) sequences, 131–132, 194–195
In vitro site-directed mutagenesis, 417, 423, 436–437,556–557
Ionic interactions in proteins, 104
IQ, 86
IS (insertion sequences), 189, 194, 196–201
Isoaccepting tRNA(s), 397
Isoelectric focusing, 100
Isoenzymes (Isozymes), 106–107
Isoschizomers, 338

J

Jacob, Francois, 3, 307, 308, 458–459
Jenkins, J. B., 87
Jimsonweed (*Datura stramonium*), 247–248
Johannsen, W. L., 24

K

Karyokinesis, 10, See also Mitosis
Karyotype analysis, 265–267
Keto bases, 541
Kettlewell, H. B. D., 621
Khorana, H. G., 3, 375
Kimura, Motoo, 645
Kinases, 181–182, 412, 497, 525–527
Kinetochore, 164, 166
Klinefelter syndrome, 249–250
Knockout experiments, 437–438, 443–445
Koehn, R. K., 644
Koller, Th., 168–169
Kornberg, A., 3, 155
Krumlauf, R., 513

L

L1 family of retroposons, 389–390
lac operon, 455–468
    catabolite repression of, 467–468
    effect of cAMP on, 467–468
    mutants of, 462–464
Lactose metabolism, in *E. coli*, 456–457
    gene control and, See *lac* operon
    as marker for selection of transformed cells, 418–424, 448
Lake, J. A., 408, 653, 654
Lalouel, J. M., 358
Lamarck, Jean Baptiste, 3, 5–6
λ*dgal* phage particle, 313–315

Lambda phage, See Phage lambda
λZAP vector, 424–425
Langley, C. H., 648
Lariat intermediate, 381–382, 384
Lateral elements, 12, 16
Law of large numbers, 43
Leader sequence, 374, 469–471
Leading strand, 157–159, 181
Leaf disk method, 446–447
Lederberg, Esther, 236
Lederberg, Joshua, 3, 236
Leptotene stage (leptonema), 12, 14
Lesch-Nyhan syndrome, 57, 59, 224, 347, 430, 546
Lethal genes, 216–217, 554, 616–617, 620
Leucine zipper (Zip) motif, 488–492
Leukemia, 531
Levan, A., 3
Lewin, B., 400, 495, 529
Lewontin, R. C., 641, 642
LexA repressor, 576–577
Library
    cDNA (expression), 428–429
    gene, 428
    genomic, 428
    jumping, 432
    size of, 428
Lichter, P., 343
Ligase, DNA, 154, 157–159, 178, 418–419
LINEs (long interspersed elements), 174
Linkage, 35–37, 274–277, See also Genetic linkage maps; Mapping; Physical linkage maps
    χ^2 test for, 276–277
    coupling in, 275–276
    degree of, 273
    detection of, 275–278, 294–297, 312–313
        by lod score method, 355–360
    discovery of, 3, 36
    as exception to Mendelism, 35–37, 276–277
    and random mating equilibrium, 594–595
    repulsion in, 275–276
    and syntenic genes, 334–335, 360
    in testcrosses, 36–37
    transduction and, 312–313
    transformation and, 317–318
Linkage analysis
    by cotransduction, 312–313
    by cotransformation, 318–319
    from dihybrid cross results, 36–37, 277–278
    in phage, 319–323
    from testcross results, 36–37, 275–277

Linkage group, 273
Linkage maps, See Mapping; Genetic linkage maps; Physical linkage maps
Linked genes, 35; See also Linkage
Linkers, in recombinant DNA work, 419–421, 424–425
Lipofection, 437
Locus, 34
Lod score method, 355–360
Log phase, of bacterial growth, 147–148
Lollipop-shaped structures, 194–195
Long interspersed elements (LINEs), 174
Long terminal repeats (LTRs), 387–388, 441
Luria, Salvadore, 234–235
Lymphoma, Burkitt's, 530–531
Lyonization, 502
Lyon, Mary, 502
Lysis, 205
Lysogeny, 208–210
    establishment by phage lambda, 208–210, 475–479
    establishment by other temperate phages, 209
    genetic basis of, 475–480
Lysozyme, 205
Lytic cycle, 208–210
    genetic basis of, in lambda, 476–479
    and transduction, 310–313

M

MacLeod, C. M., 3, 116–117
Maize, See Corn
Malaria, 625–626
Maple syrup urine disease, 224
Mapping, See also Genetic linkage maps, Physical linkage maps
    of bacterial chromosomes, 307–310
    cellular methods of, 331–335
    of centromere, 21–22
    and correcting for multiple crossovers, 280–282
    crossover-distance concept in, 274–275, 278–282
    deletion, 324–327
    of extranuclear genes, 298–301
    of a gene, 3, 323–359
    and human chromosomes, 331–335, 344–348, 355–359
    by in situ hybridization, 343
    molecular methods of, 332, 335–360
    by petite analysis, 289–300
    of phage chromosomes, 319–323
    recombination-distance concept of, 3, 277–282, 284–285
    of restriction sites, 339–343

CREDITS

ART

All art rendered by Precision Graphics unless otherwise noted. **7** Carlyn Iverson. **8** Elizabeth Morales-Denney. From Daniel P. Chiras, *Biology: The Web of Life* (West Publishing Company, 1992). Used with permission. **23** Carlyn Iverson. **38** Carlyn Iverson. **54** Carlyn Iverson. **66** Carlyn Iverson. **95** G&S. **97** G&S. **98 (top)** and **(bottom)** G&S. **110** G&S. **121 (top)** and **(bottom)** G&S. **122** G&S. **123** G&S. **124** G&S. **129 (top)** G&S. **130** G&S. **132** G&S. **146 (b)** Carlyn Iverson. **183 (top)** Carlyn Iverson. **183 (b)** Elizabeth Morales-Denny from *Chemistry for Today: General, Organic, and Biochemistry*, Second Edition, by Spencer L. Seager and Michael R. Slabaugh (West Publishing, 1993), p. 625. Used with permission. **184 (b)** Elizabeth Morales-Denny from *Biology: The Web of Life* by Daniel D. Chiras (West Publishing, 1993), p. 78. Used with permission. **204** Carlyn Iverson. **216 (top)** and **(bottom)** G&S. **219** G&S. **221** G&S. **223** G&S. **225 (b)** Carlyn Iverson in Daniel D. Chiras, *Human Biology: Health, Homeostasis, and the Environment*, Second Edition (West Publishing Co., 1995), p. 94. Used with permission. **226 (top)** Carlyn Iverson in Daniel D. Chiras, *Human Biology: Health, Homeostasis, and the Environment*, Second Edition (West Publishing Co., 1995), p. 129. Used with permission. **226 (bottom)** G&S. **228** G&S. **248** Carlyn Iverson. **254** Carlyn Iverson. **255 (art)** Carlyn Iverson. **261 (bottom art)** Carlyn Iverson. **280** G&S. **282 (comp set portion)** G&S. **289** Carlyn Iverson. **298** Carlyn Iverson. **313** G&S. **321 (top)** and **(bottom)** G&S. **325** G&S. **327** G&S. **336** G&S. **371** G&S. **372** G&S. **379** G&S. **381** G&S. **398** G&S. **403** G&S. **413** Carlyn Iverson. **455** G&S. **457** G&S. **460 (top)** and **(bottom)** G&S. **461** G&S. **465** G&S. **473** G&S. **500 (left)** G&S. **510 (art)** Carlyn Iverson. **511 (top)** and **(bottom)** Carlyn Iverson. **541** G&S. **542** G&S. **543** G&S. **550** G&S. **551** G&S. **552** G&S. **553** G&S. **570** G&S. **613** Carlyn Iverson. **627** Carlyn Iverson. **651** G&S. **652** G&S. **654 (top)** and **(bottom)** G&S.

PHOTOS

v (left) Carol Elseth. **v (right)** Baumgardner nieces and nephews. **ix** VU/© M. Schliwa. **x** © Will & Deni McIntyre, Photo Researchers. **xi** © Biophoto Associates/Science Source. **xii** This photo courtesy of The Little Tikes Company (A Rubbermaid Company). Used with permission. **xiii** Photo Researchers/© Science Source. **xiv** VU/William F. DeGrado/DuPont Merck. **xv** © Ken Eward/Science Source/Photo Researchers. **xvi** VU/© Stan Elems. **1** VU/© M. Schliwa. **2 (a)** VU/© Arthur M. Siegelman. **2 (b)** VU/© Jack M. Bostrack. **2 (c)** VU/© R. Calentine. **2 (d)** Science VU/© Jackson. **2 (e)** VU/© James W. Richardson. **2 (f)** VU/© D. Cavagnaro. **4** Kersting, "Der Geiger Nicolo Paganini," Staatlichen Kunstsammlungen Dresden. **11 (top photo)** VU/© Jack M. Bostrack. **11 (second photo)** VU/© R. Calentine. **11 (third and fourth photos)** VU/© Jack Bostrack. **11 (bottom photo)** VU/© R. Calentine. **Photos on 14 and 15** VU/From *An Introduction to Genetic Analysis*, 5th ed., by Griffiths, Miller, Suzuki, Lewontin, and Gelbart. Copyright © W. H. Freeman and Co. Used with permission. **16 (left)** VU/M. Westergaard and D. von Wettstein, *Ann. Rev. Genetics* 6 (1972): 74–110. **16 (right)** VU/© B. John, Cabisco. **20** Science Photo Library/Photo Researchers. **21** VU/© John D. Cunningham. **22** Photo by K. Libal. From Moravske Museum, Bruno. **24 (a)** and **(b)** VU/© Cabisco. **42** © Jack Wilburn/Animals Animals. **47** VU/© Kirtley-Perkins. **53** VU/© John D. Cunningham. **55 (a)** and **(b)** VU/© Cabisco. **56** © Mary Evans Picture Library/Photo Researchers. **58** VU/© Jay M. Pasachoff. **63 (a)–(d)** Science VU/© T. Somes. **74** VU/© D. Cavagnaro. **87** John B. Jenkins, *Human Genetics* (Harper & Row, 1990), p. 412. **93** VU/Ken Eward/Science Source. **94** Science VU/© R. Feldman, NIH. **109 (a)** and **(b)** Reproduced from Bennett and Stietz, *Journal of Molecular Biology* 140 (1980): 211, with permission from Academic Press Ltd. **114** © Will & Deni McIntyre, Photo Researchers. **118** Lee Simon/Photo Researchers. **125 (a)** VU/© George B. Chapman and Priscilla Devadoss. **127 (a)** VU/© Peter Arnold, Inc./CSHL. **127 (b)** Science VU/Cold Spring Harbor Laboratory. **128 (c)** VU/© Ken Eward/Science Source. **129 (a)** and **(b)** VU/© Ken Eward/Science Source. **131** David Parker, Science Photo Library, Photo Researchers. **133** VU/© Ken Eward/Science Source. **137 (a)** VU/© K. G. Murti. **137 (b)** CNRI/Science Photo Library/Photo Researchers. **144** Dr. Gopal Murti/Science Photo Library/Photo Researchers. **145** VU/© David M. Phillips. **146 (a)** VU/© George Musil. **148** VU/© Michael G. Gabridge. **150** Reprinted with permission from *Endeavour* 22:144, J. Cairns. Copyright 1963, Pergamon Press, Ltd. Photo courtesy of J. Cairn. **153** Meselson and Stahl, *Proc. National Academy of Science, U.S.* 44 (1958): 671. **162** © Biophoto Associates/Science Source. **165 (Generation 1 and 2 Autoradiographs)** J. H. Taylor, *Molecular Genetics, Part I*. Academic Press, 1963. **169 (a)–(d)** Reprinted with permission from F. Thoma and Th. Koller, *Journal of Molecular Biology* 149 (1981): 709–733. Copyright: Academic Press, Inc. (London) Ltd. **170 (top)** Science VU/VU. **170 (bottom)** VU/© J. R. Paulsen, U. K. Laemmli, D. W. Fawcett. **171 (Human chromosome photo)** VU/© K. G. Murti. **173** VU/Grant R. Sutherland. **183 (a)** VU/© T. Kanariki and D. W. Fawcett. **187** VU/© Alfred Pasieka/Photo Researchers. **189** © Omikron/Science Source/Photo Researchers. **198 (a)** © Stanley Cohen/Science Photo Library/Photo Researchers. **202 (upper left)** VU/CMSP/© 1992 NIH. **202 (upper right)** VU/© Scott Camazine/S. S. Billota Best/Photo Researchers. **202 (middle left)** Science VU-CDC/VU. **202 (middle right)** VU/© K. G. Murti. **202 (bottom)** VU/T. F. Anderson, E. L. Wollman, and F. Jacob/Photo Researchers. **203 (a)** VU/© Cabisco. **205** © Lee D. Simon/Science Source/Photo Researchers. **207** VU/A. B. Dowsett/Science Photo Library. **214** VU/© John Sohlden, (Utica, Michigan, USA). **215** VU/© William S. Ormerod, Jr. **225 (a)** VU/© SIU. **227** VU/© SIU (MR) 1992. **229** VU/© Stanley Flegler. **246** © Dept. of Energy/Photo Researchers. **249 (a)** Dr. Ira Rosenthal, Department of Pediatrics, University of Chicago. **249 (b)** This photo courtesy of The Little Tikes Company (A Rubbermaid Company). Used with permission. **252 (a) and (b)** Courtesy of W. Atlee Burpee & Co. **255 (top left photo)** VU/© D. Cavagnaro. **255 (top right and middle left photos)** VU/Dr. J. G. Waines. **255 (middle right photo)** Science VU/© T. S. Cox, USDA. **255 (bottom photo)** VU/© John D. Cunningham. **261 (top photo)** VU/© David M. Phillips. **266** © James King-Holmes/ICRF/Science Photo Library/Photo Researchers. **268** Science VU/VU. **273** Applied Biosystems Division of Perkin-Elmer. **274** © CNRI/SPL/Science Source/Photo Researchers. **285 (left)** VU/© Bernd Wittich. **285 (right)** © Michael P. Gadomski, Photo Researchers. **286 (left) and (right)** VU/Carolina Biological Supply. **290** VU/© James Richardson. **299** VU/© D. Long. **306** Photo Researchers/© Lee Simon. **310** © Anderson/Omikron/SS/Photo Researchers. **320** From G. S. Stent, *Molecular Biology of Bacterial Viruses.* Copyright 1963 by W. H. Freeman & Company. Used with permission. **326** Dr. Seymour Benzer. **331** VU/David Parker/Science Photo Library. **332** VU/© David M. Phillips/Visuals Unlimited. **343** Peter Lichter and David Ward, *Science* 247 (1990): 65. Copyright © 1990 by the AAAS. Used with permission. **350 (left)** VU/© SIU. **350 (right)** David Parker/Science Photo Library/Photo Researchers. **352** VU/© Kjell Sandved. **359** James Holmes/Cellmark Diagnostics/Science Photo Library/Photo Researchers.

360 PhotoEdit/© David Young-Wolff. **365** Oscar Miller/Science Photo Library/Photo Researchers. **367** © Ken Eward/Science Source/Photo Researchers. **385 (top) and (middle)** VU/© K. G. Murti. **385 (bottom)** VU/© George Musil. **389** Science VU/Visuals Unlimited. **394** Photo Researchers/© Science Source. **404 (left)** VU/© E. Kiseleva, D. Fawcett. **404 (right)** VU/Custom Medical Stock Photo/© 1991 Science Photo Library. **417** VU/Custom Medical Stock/© 1993 SPL. **421** VU/© K. G. Murti. **443 (top)** © Jon Gordon/Phototake NYC. **443 (bottom)** Science VU/Jackson Lab. **445** © Jon Gordon/Phototake NYC. **446** Philippe Plailly/Science Photo Library/Photo Researchers. **449 (top left)** VU/© John D. Cunningham. **449 (top right)** VU/© C. L. Case. **449 (bottom)** Courtesy of Calgene Fresh, Inc. **453** Photo Researchers/© Nancy Kedersha/Immunogen/Science Photo Library. **454** Science VU—NIH, R. Feldman/VU. **478** Proceedings of the National Academy of Sciences 39 (1953): 628. **483** VU/William F. DeGrado/DuPont Merck. **490** © Ken Eward/Science Source/Photo Researchers. **491 (top), (bottom left), and (bottom right)** VU/William F. DeGrado/DuPont Merc. **498** VU/© J. M. Bostrack. **499** Marty Snyderman. **500 (right)** Cold Spring Harbor. **503** R. Olson. **505** © Ken Eward/Photo Researchers. **506** VU/© F. R. Turner. **507 (top), (middle), and (bottom)** VU/© F. R. Turner. **509 (a) and (b)** M. Hulskamp and D. Tautz, *BioEssays* 13 (1991): 261, ICSU Press. **510 (photo)** VU/© Cabisco. **518 (a)–(c)** From A. Admit, R. Mariuzza, S. E. Phillips, and R. J. Poljak, *Science* 233 (1986): 747–753. Copyright 1986 by the AAAS. **524** Science VU/© G. Steven Martin. **534 (top)** VU/© Bernd Wittich. **534 (bottom)** VU/© SIU. **538** Nancy Kedersha/Immunogen/Science Photo Library/Photo Researchers. **539** VU/© Jack M. Bostrack. **546** Geoff Tomkinson/Science Library/Photo Researchers. **560** © Ken Eward/Science Source/Photo Researchers. **564** VU/Huntington Potter and David Dressler, Harvard Medical School. **573** Dr. Sheldon Wolff and Judy Bodycote. **574** VU/© Kenneth E. Greer. **585** VU/National Library of Medicine/Science Photo Library/Custom Medical Stock. **587** VU/© Stan Elems. **596** 1990 Science Photo Library/Custom Medical Stock Photo. All rights reserved. **609** The Bettmann Archive. **620 (a)** Science VU/Visuals Unlimited. **620 (b)** VU/© John D. Cunningham. **639** Courtesy of Dr. Song-Kun Shyue and Dr. Wen-Hsiung Li. **right back endsheet** VU/© K. G. Murti.